# Applied Calculus

## FIFTH EDITION

**Geoffrey C. Berresford**
Long Island University

**Andrew M. Rockett**
Long Island University

BROOKS/COLE
CENGAGE Learning·

Australia • Brazil • Japan • Korea • Mexico • Singapore • Spain • United Kingdom • United States

## BROOKS/COLE
### CENGAGE Learning™

*Applied Calculus,* **Fifth Edition**
Geoffrey C. Berresford
Andrew M. Rockett

Publisher: Richard Stratton

Sponsoring Editors: Molly Taylor, Cathy Cantin

Development Editor: Maria Morelli

Associate Editor: Jeannine Lawless

Media Editor: Peter Galuardi

Senior Content Manager: Maren Kunert

Marketing Specialist: Ashley Pickering

Marketing Coordinator: Angela Kim

Marketing Communications Manager:
 Mary Anne Payumo

Project Managers, Editorial Production:
 Kerry Falvey, Nancy Blodget

Editorial Assistant: Laura Collins

Creative Director: Rob Hugel

Art & Design Manager: Jill Haber

Senior Manufacturing Coordinator: Diane Gibbons

Senior Rights Acquisition Account Manager:
 Katie Huha

Production Service: Matrix Productions

Text Designer: Jerilyn Bockorick

Photo Researcher: Naomi Kornhauser

Copy Editor: Betty Duncan

Illustrator: Nesbitt Graphics, ICC Macmillan Inc.

Cover Design Director: Tony Saizon

Cover Designer: Faith Brosnan

Cover Credit: © Glen Allison/Getty Images

Cover Description: USA, California, Altamont, rows
 of wind turbines

Compositor: ICC Macmillan Inc.

**Photo credits:** Chapter opener graphic: © Chad
Baker/Getty Images; p. 2: © AP Photo/Anja
Niedringhaus; p. 76: © Gabe Palmer/Corbis;
p. 176: © Thinkstock Images/Jupiterimages;
p. 256: © Kenneth Garrett/Getty Images;
p. 257: © Bettmann/Corbis; p. 322: © Lee Frost/
Robert Harding/Jupiterimages; p. 408: © Alan
Schein Photography/Corbis; p. 460: © Mitchell
Funk/Getty Images; p. 556: © Bill Bachmann/The
Image Works; p. 626: © Digital Vision/Alamy;
p. 684: © AP Photo/Tom Strattman; p. 742: © Tim
Boyle/Getty Images.

For product information and technology assistance, contact us at
**Cengage Learning Customer & Sales Support, 1-800-354-9706.**

For permission to use material from this text or product,
submit all requests online at **www.cengage.com/permissions.**
Further permissions questions can be e-mailed to
**permissionrequest@cengage.com.**

Library of Congress Control Number: 2008926130

ISBN-13: 978-0-547-16978-1

ISBN-10: 0-547-16978-7

**Brooks/Cole**
10 Davis Drive
Belmont, CA 94002-3098
USA

Cengage Learning is a leading provider of customized learning solutions
with office locations around the globe, including Singapore, the United
Kingdom, Australia, Mexico, Brazil, and Japan. Locate your local office at
**www.cengage.com/global.**

Cengage Learning products are represented in Canada by Nelson
Education, Ltd.

For your course and learning solutions, visit **www.cengage.com.**

Purchase any of our products at your local college store or at our preferred
online store **www.ichapters.com.**

Printed in China
2 3 4 5 6 7 12 11 10

# Contents

## 1 FUNCTIONS

## 2 DERIVATIVES AND THEIR USES

## 3 FURTHER APPLICATIONS OF DERIVATIVES

## 4    EXPONENTIAL AND LOGARITHMIC FUNCTIONS

## 5    INTEGRATION AND ITS APPLICATIONS

## 6    INTEGRATION TECHNIQUES

# 7 CALCULUS OF SEVERAL VARIABLES

# 8 TRIGONOMETRIC FUNCTIONS

# 9 DIFFERENTIAL EQUATIONS

CONTENTS

## 10    SEQUENCES AND SERIES

## 11    PROBABILITY

# Preface

A scientific study of yawning found that more yawns occurred in calculus class than anywhere else.* This book hopes to remedy that situation. Rather than being another dry recitation of standard results, our presentation exhibits some of the many fascinating and useful applications of mathematics in business, the sciences, and everyday life. Even beyond its utility, however, there is a beauty to calculus, and we hope to convey some of its elegance and simplicity.

This book is an introduction to calculus and its applications to the management, social, behavioral, and biomedical sciences, and other fields. The seven-chapter *Brief Applied Calculus* contains more than enough material for a one-semester course, and the eleven-chapter *Applied Calculus* contains additional chapters on trignometry, differential equations, sequences and series, and probability for a two-semester course. The only prerequisites are some knowledge of algebra, functions, and graphing, which are reviewed in Chapter 1.

## CHANGES IN THE FIFTH EDITION

First, what has *not* changed is the essential character of the book: simple, clear, and mathematically correct explanations of calculus, alternating with relevant and engaging examples.

**Exercises**  We have added many new exercises, including new *Applied Exercises* and *Conceptual Exercises*, and have updated others with new data. Many exercises now have *sources* (book or journal names or website addresses) to establish their factual basis and enable further research. In Chapter 1 we have added *regression (modeling) exercises*, in which students use calculators to fit equations to actual data (see, for example, pages 19 and 32). Throughout the book we have added what may be termed *Wall Street exercises* (pages 205 and 315), applications based on financial data from sources that are provided. The regression exercises in Chapter 1 illustrate the methods used to develop the models in the Applied Exercises throughout the book.

**New or Modified Topics**  We have expanded our treatment of the following topics: *limits involving infinity* (pages 83–85), *graphing rational functions* (pages 184–187), and *elasticity of demand* (pages 309–315). To show how to solve the regression (modeling) exercises in Chapter 1 we have added (optional) examples on regression (*linear* on page 13, *power* on page 27, *quadratic* on page 43, and *exponential* on page 62). In addition to these expanded applications, we have included some more difficult exercises (see,

*Ronald Baenninger, "Some Comparative Aspects of Yawning in Betta splendens, Homo sapiens, Panthera leo, and Papoi spinx," *Journal of Comparative Psychology* **101** (4).

for example, pages 136 and 161), and provided a complete proof of the Chain Rule based on Carathédory's definition of the derivative (page 163). To accommodate these additions without substantially lengthening the book we have tightened the exposition in every chapter.

**Pedagogy** We have redrawn many graphs for improved accuracy and clarity. We have relocated some examples immediately to the right of the boxes that summarize results, calling them *Brief Examples*, thereby providing immediate reinforcement of the concepts (see, for example, pages 21 and 23).

## FEATURES

**Realistic Applications** The basic nature of courses using this book is very "applied" and therefore this book contains an unusually large number of applications, many appearing in no other textbook. We explore learning curves in airplane production (pages 26–27 and 31), corporate operating revenues (page 49), the age of the Dead Sea Scrolls (pages 282–283), the distance traveled by sports cars (pages 334–335), lives saved by seat belts (page 378), as well as the cost of a congressional victory (page 511). These and many other applications convincingly show that mathematics is more than just the manipulation of abstract symbols and is deeply connected to everyday life.

**Graphing Calculators (Optional)** Using this book does *not* require a graphing calculator, but having one will enable you to do many problems more easily and at the same time deepen your understanding by allowing you to concentrate on concepts. Throughout the book are **Graphing Calculator Explorations** and **Graphing Calculator Exercises** (marked by the symbol ), which

> *explore interesting applications,* such as when men and women will achieve equal pay (page 18),

> *carry out otherwise "messy" calculations,* such as the population growth comparisons on pages 268 and 272, and

> *show the advantages and limitations of technology,* such as the differences between $\ln x^2$ and $2 \ln x$ on page 279.

While any graphing calculator (or a computer) may be used, the displays shown in the text are from the Texas Instruments *TI-84*, except for a few from the *TI-89*. A discussion of the essentials of graphing calculators follows this preface. For those not using a graphing calculator, the Graphing Calculator Explorations have been carefully planned so that *most can also be read simply for enrichment* (as with the concavity and maximization problems on pages 195 and 216). Students, however, *will* need a calculator with keys like $\boxed{y^x}$ and $\boxed{\ln}$ for powers and natural logarithms.

**Graphing Calculator Programs (Optional)** Some topics require extensive calculation, and for them we have created (optional) graphing calculator programs for use with this book. We provide these programs for free to all students and faculty (see "How to Obtain Graphing Calculator Programs" later in this preface). The topics covered are: Riemann sums (page 350), trapezoidal approximation (page 447), Simpson's rule (page 451), slope fields (page 633), Euler's method (pages 675 and 677), and Newton's method (page 731). These programs allow the student to concentrate on the results rather than the computation.

**Spreadsheets (Optional)** While access to a computer is *not* necessary for this book, the **Spreadsheet Explorations** allow deeper exploration of some topics. We have included spreadsheet explorations of: nondifferentiable functions (pages 167–168), maximizing an enclosed area (pages 213–214), elasticity of demand (page 313), consumption of natural resources (page 343), improper integrals (page 436), and graphing a function of two variables (page 468). Ancillary materials for Microsoft Excel are also available (see "Resources for the Student" later in this preface).

**Enhanced Readability** We have added space around all in-line mathematics to make them stand out from the narrative. An elegant four-color design increases the visual appeal and readability. For the sake of continuity, references to earlier material are minimized by restating results whenever they are used. Where references are necessary, explicit page numbers are given.

**Application Previews** Each chapter begins with an **Application Preview** that presents an interesting application of the mathematics developed in that chapter. Each is self-contained (although some exercises may later refer to it) and serves to motivate interest in the coming material. Topics include: world records in the mile run (pages 3–4), Stevens' law of psychophysics (page 177), cigarette smoking (pages 323–324), and predicting personal wealth (page 627).

**Practice Problems** Learning mathematics requires your active participation—"mathematics is not a spectator sport." Throughout the readings are short pencil-and-paper **Practice Problems** designed to consolidate your understanding of one topic before moving ahead to another, such as using negative exponents (page 22) or finding and checking an indefinite integral (page 325).

**Annotations** Notes to the right of many mathematical formulas and manipulations state the results *in words,* assisting the important skill of *reading mathematics,* as well as providing explanations and justifications for the steps in calculations (see page 100) and interpretations of the results (see page 198).

**Extensive Exercises** Anyone who ever learned any mathematics did so by solving many many problems, and the exercises are the most essential part of the learning process. The exercises (see, for instance, pages 286–289) are graded from routine drills to significant applications, and some conclude with *Explorations and Excursions* that extend and augment the material presented in the text. The Conceptual Exercises were described earlier in this preface. Exercises marked with the symbol require a graphing calculator.

Answers to odd-numbered exercises and answers to all Chapter Review exercises are given at the end of the book (full solutions are given in the *Student Solutions Manual*).

**Explorations and Excursions**  At the end of some exercise sets are optional problems of a more advanced nature that carry the development of certain topics beyond the level of the text, such as: the Beverton-Holt recruitment curve (page 20), average and marginal cost (page 192), elasticity of supply (page 317), competitive and complementary commodities (page 487), and the calculus of the normal distribution, including a direct and complete verification that the area under the standard normal curve is 1 (pages 789–790).

**Conceptual Exercises**  These short problems are true/false, yes/no, or fill-in-the-blank quick-answer questions to reinforce understanding of a subject without calculations (see, for example, page 93). We have found that students actually enjoy these simple and intuitive questions at the end of a long challenging assignment.

 This "Be Careful" icon warns students of possible misunderstandings (see page 52) or particular difficulties (see page 127).

**Just-in-Time Review**  We understand that many students have weak algebra skills. Therefore, rather than just "reviewing" material that they never mastered in the first place, we keep the review chapter brief and then reinforce algebraic skills throughout the exposition with blue annotations immediately to the right of the mathematics in every example. We also review exponential and logarithmic functions again just before they are differentiated in Section 4.3, and the sine and cosine functions just before they are differentiated in Section 8.3. This puts the material where it is relevant and more likely to be remembered.

**Levels of Reinforcement**  Because there are many new ideas and techniques in this book, learning checks are provided at several different levels. As noted above, **Practice Problems** encourage mastery of new skills directly after they are introduced. **Section Summaries** briefly state both essential formulas and key concepts (see page 202). **Chapter Summaries** review the major developments of the chapter and are keyed to particular chapter review exercises (see pages 250–251). **Hints and Suggestions** at the end of each chapter summary unify the chapter, give specific reminders of essential facts or "tricks" that might be otherwise overlooked or forgotten, and list a selection of the review exercises for a **Practice Test** of the chapter material (see page 251). **Cumulative Reviews** at the end of groups of chapters unify the materials developed up to that point (see page 255).

**Accuracy and Proofs**  All of the answers and other mathematics have been carefully checked by several mathematicians. The statements of definitions and theorems are mathematically accurate. Because the treatment is applied rather than theoretical, intuitive and geometric justifications have often been preferred to formal proofs. Such a justification or proof accompanies every important mathematical idea; we never resort to phrases like "it can be shown that ...". When proofs are given, they are correct and honest.

**Philosophy** We wrote this book with several principles in mind. One is that to learn something, it is best to begin doing it as soon as possible. Therefore, the preliminary material is brief, so that students begin calculus without delay. An early start allows more time during the course for interesting applications and necessary review. Another principle is that the mathematics should be done with the applications. Consequently, every section contains applications (there are no "pure math" sections).

**Prerequisites** The only prerequisite for most of this book is some knowledge of algebra, graphing, and functions, and these are reviewed in Chapter 1. Other review material has been placed in relevant locations in later chapters.

**Resources on the Web** Additional materials available on the Internet at www.cengage.com/math/berresford include:

Suggestions for **Projects and Essays,** open-ended topics that ask students (individually or in groups) to research a relevant person or idea, to compare several different mathematical ideas, or to relate a concept to their lives (such as marginal and average cost, why two successive 10% increases don't add up to a 20% increase, elasticity of supply of drugs and alcohol, and arithmetic versus geometric means).

An expanded collection of **Application Previews,** short essays that were used in an earlier edition to introduce each section. Topics include *Exponential Functions and the World's Worst Currency; Size, Shape, and Exponents;* and *The Confused Creation of Calculus.*

## HOW TO OBTAIN GRAPHING CALCULATOR PROGRAMS AND EXCEL SPREADSHEETS

The optional graphing calculator programs used in the text have been written for a variety of Texas Instruments Graphing Calculators (including the *TI-83, TI-84, TI-85, TI-86, TI-89,* and *TI-92*), and may be obtained for free, in any of the following ways:

- If you know someone who already has the programs on a Texas Instruments graphing calculator like yours, you can easily transfer the programs from their calculator to yours using the black cable that came with the calculator and the LINK button.
- You may download the programs and instructions from the Cengage website at www.cengage.com/math/berresford onto a computer and then to your calculator using a USB cable.

The Microsoft Excel spreadsheets used in the Spreadsheet Explorations may be obtained for free by downloading the spreadsheet files from the Cengage website at www.cengage.com/math/berresford.

## RESOURCES FOR THE INSTRUCTOR

**Instructor's Solutions Manual** The Instructor's Solutions Manual contains worked-out solutions for all exercises in the text. It is available on the Instructor's book companion website.

**Computerized Test Bank** Create, deliver and customize tests and study guides in minutes with this easy-to-use assessment software on CD. The thousands of algorithmic questions in the test bank are derived from the textbook exercises, ensuring consistency between exams and the book.

**WebAssign** Instant feedback, grading precision, and ease of use are just three reasons why WebAssign is the most widely used homework system in higher education. WebAssign's homework delivery system lets instructors deliver, collect, grade and record assignments via the web. And now, this proven system has been enhanced to include additional resources for instructors and students.

## RESOURCES FOR THE STUDENT

**Student Solutions Manual** Need help with your homework or to prepare for an exam? The Student Solutions Manual contains worked-out solutions for all odd-numbered exercises in the text. It is a great resource to help you work through those tough problems.

**DVD Lecture Series** These comprehensive, instructional lecture presentations serve a number of uses. They are great if you need to catch up after missing a class, need to supplement online or hybrid instruction, or need material for self-study or review.

**Microsoft Excel Guide by Revathi Narasimhan** This guide provides list of exercises from the text that can be completed after each step-by-step Excel example. No prior knowledge of Excel is necessary.

**WebAssign** WebAssign, the most widely used homework system in higher education, offers instant feedback and repeatable problems—everything you could ask for in an online homework system. WebAssign's homework system lets you practice and submit homework via the web. It is easy to use and loaded with extra resources.

## ACKNOWLEDGMENTS

We are indebted to many people for their useful suggestions, conversations, and correspondence during the writing and revising of this book. We thank Chris and Lee Berresford, Anne Burns, Richard Cavaliere, Ruth Enoch, Theodore Faticoni, Jeff Goodman, Susan Halter, Brita and Ed Immergut, Ethel Matin, Gary Patric, Shelly Rothman, Charlene Russert, Stuart Saal, Bob Sickles, Michael Simon, John Stevenson, and all of our "Math 6" students at C.W. Post for serving as proofreaders and critics over the past years.

We had the good fortune to have had supportive and expert editors at Cengage Learning: Molly Taylor (senior sponsoring editor), Maria Morelli (development editor), Kerry Falvey (production editor), Roger Lipsett (accuracy reviewer), and Holly McLean-Aldis (proofreader). They made the difficult tasks seem easy, and helped beyond words. We also express our gratitude to the many others at Cengage Learning who made important contributions too numerous to mention.

The following reviewers have contributed greatly to the development of the fifth edition of this text:

| | |
|---|---|
| Frederick Adkins | *Indiana University of Pennsylvania* |
| David Allen | *Iona College, NY* |
| Joel M. Berman | *Valencia Community College, FL* |
| Julane Crabtree | *Johnson Community College, KS* |
| Biswa Datta | *Northern Illinois University* |
| Allan Donsig | *University of Nebraska—Lincoln* |
| Sally Edwards | *Johnson Community College, KS* |
| Frank Farris | *Santa Clara University, CA* |
| Brad Feldser | *Kennesaw State University, GA* |
| Abhay Gaur | *Duquesne University, PA* |
| Jerome Goldstein | *University of Memphis, TN* |
| John B. Hawkins | *Georgia Southern University* |
| John Karloff | *University of North Carolina* |
| Todd King | *Michigan Technical University* |
| Richard Leedy | *Polk Community College, FL* |
| Sanjay Mundkur | *Kennesaw State University, GA* |
| David Parker | *Salisbury University, MD* |
| Shahla Peterman | *University of Missouri—Rolla* |
| Susan Pfiefer | *Butler Community College, KS* |
| Daniel Plante | *Stetson University, FL* |
| Xingping Sun | *Missouri State University* |
| Jill Van Valkenburg | *Bowling Green State University* |
| Erica Voges | *New Mexico State University* |

We would also like to thank the reviewers of the previous edition:

John A. Blake, *Oakwood College*; Dave Bregenzer, *Utah State University*; Kelly Brooks, *Pierce College*; Donald O. Clayton, *Madisonville Community College*; Charles C. Clever, *South Dakota State University*; Dale L. Craft, *South Florida Community College*; Kent Craghead, *Colby Community College*; Lloyd David, *Montreat College*; John Haverhals, *Bradley University*; Randall Helmstutler, *University of Virginia*; Heather Hulett, *University of Wisconsin—La Crosse*; David Hutchison, *Indiana State University*; Dan Jelsovsky, *Florida Southern College;* Alan S. Jian, *Solano Community College*; Dr. Hilbert Johs, *Wayne State College*; Hideaki Kaneko, *Old Dominion University*; Michael Longfritz, *Rensselear Polytechnic Institute*; Dr. Hank Martel, *Broward Community College*; Kimberly McGinley Vincent, *Washington State University*; Donna Mills, *Frederick Community College*; Pat Moreland, *Cowley College*; Sue Neal, *Wichita State University*; Cornelius Nelan, *Quinnipiac University*; Catherine A. Roberts, *University of Rhode Island*; George W. Schultz, *St. Petersburg College*; Paul H. Stanford, *University of Texas—Dallas*; Jaak Vilms, *Colorado State University*; Jane West, *Trident Technical College*; Elizabeth White, *Trident Technical College*; Kenneth J. Word, *Central Texas College*.

Finally, and most importantly, we thank our wives, Barbara and Kathryn, for their encouragement and support.

## COMMENTS WELCOMED

With the knowledge that any book can always be improved, we welcome corrections, constructive criticisms, and suggestions from every reader.

geoffrey.berresford@liu.edu
andrew.rockett@liu.edu

# A User's Guide to Features

## ▼ Application Preview

Found on every chapter opener page, Application Previews motivate the chapter. They offer a unique "mathematics in your world" application or an interesting historical note. A page with further information on the topic, and often a related exercise number, is referenced.

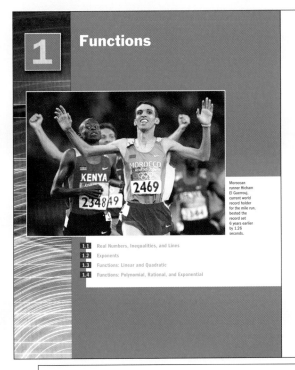

Moroccan runner Hicham El Guerrouj, current world record holder for the mile run, bested the record set 6 years earlier by 1.26 seconds.

1.1 Real Numbers, Inequalities, and Lines
1.2 Exponents
1.3 Functions: Linear and Quadratic
1.4 Functions: Polynomial, Rational, and Exponential

**Application Preview**

**World Record Mile Runs**

The dots on the graph below show the world record times for the mile run from 1865 to the 1999 world record of 3 minutes 43.13 seconds, set by the Moroccan runner Hicham El Guerrouj. These points fall roughly along a line, called the **regression line**. In this section we will see how to use a graphing calculator to find a regression line (see Example 8 and Exercises 69–74), based on a method called **least squares**, whose mathematical basis will be explained in Chapter 7.

Notice that the times do not level off as you might expect, but continue to decrease.

**History of the Record for the Mile Run**

| Time | Year | Athlete | Time | Year | Athlete | Time | Year | Athlete |
|------|------|---------|------|------|---------|------|------|---------|
| 4:36.5 | 1865 | Richard Webster | 4:09.2 | 1931 | Jules Ladoumegue | 3:54.1 | 1964 | Peter Snell |
| 4:29.0 | 1868 | William Chinnery | 4:07.6 | 1933 | Jack Lovelock | 3:53.6 | 1965 | Michel Jazy |
| 4:28.8 | 1868 | Walter Gibbs | 4:06.8 | 1934 | Glenn Cunningham | 3:51.3 | 1966 | Jim Ryun |
| 4:26.0 | 1874 | Walter Slade | 4:06.4 | 1937 | Sydney Wooderson | 3:51.1 | 1967 | Jim Ryun |
| 4:24.5 | 1875 | Walter Slade | 4:06.2 | 1942 | Gunder Hägg | 3:51.0 | 1975 | Filbert Bayi |
| 4:23.2 | 1880 | Walter George | 4:06.2 | 1942 | Arne Andersson | 3:49.4 | 1975 | John Walker |
| 4:21.4 | 1882 | Walter George | 4:04.6 | 1942 | Gunder Hägg | 3:49.0 | 1979 | Sebastian Coe |
| 4:18.4 | 1884 | Walter George | 4:02.6 | 1943 | Arne Andersson | 3:48.8 | 1980 | Steve Ovett |
| 4:18.2 | 1894 | Fred Bacon | 4:01.6 | 1944 | Arne Andersson | 3:48.53 | 1981 | Sebastian Coe |
| 4:17.0 | 1895 | Fred Bacon | 4:01.4 | 1945 | Gunder Hägg | 3:48.40 | 1981 | Steve Ovett |
| 4:15.6 | 1895 | Thomas Conneff | 3:59.4 | 1954 | Roger Bannister | 3:47.33 | 1981 | Sebastian Coe |
| 4:15.4 | 1911 | John Paul Jones | 3:58.0 | 1954 | John Landy | 3:46.31 | 1985 | Steve Cram |
| 4:14.4 | 1913 | John Paul Jones | 3:57.2 | 1957 | Derek Ibbotson | 3:44.39 | 1993 | Noureddine Morceli |
| 4:12.6 | 1915 | Norman Taber | 3:54.5 | 1958 | Herb Elliott | 3:43.13 | 1999 | Hicham El Guerrouj |
| 4:10.4 | 1923 | Paavo Nurmi | 3:54.4 | 1962 | Peter Snell | | | |

Source: USA Track & Field

The equation of the regression line is $y = -0.356x + 257.44$, where $x$ represents years after 1900 and $y$ is the time in seconds. The regression line can be used to predict the world mile record in future years. Notice that the most recent world record would have been predicted quite accurately by this line, since the rightmost dot falls almost exactly on the line.

---

**EXAMPLE 4** FINDING A COMPANY'S COST FUNCTION

An electronics company manufactures pocket calculators at a cost of $9 each, and the company's fixed costs (such as rent) amount to $400 per day. Find a function $C(x)$ that gives the total cost of producing $x$ pocket calculators in a day.

**Solution**

Each calculator costs $9 to produce, so $x$ calculators will cost $9x$ dollars, to which we must add the fixed costs of $400.

$$C(x) = 9x + 400$$
Total cost   Unit cost   Number of units   Fixed cost

## ◄ Real World Icon

This globe icon marks examples in which calculus is connected to every-day life.

---

**Graphing Calculator Exploration**

a. Graph the parabolas $y_1 = x^2$, $y_2 = 2x^2$, and $y_3 = 4x^2$ on the window $[-5, 5]$ by $[-10, 10]$. How does the shape of the parabola change when the coefficient of $x^2$ increases?

b. Graph $y_4 = -x^2$. What did the negative sign do to the parabola?

c. Predict the shape of the parabolas $y_5 = -2x^2$ and $y_6 = \frac{1}{2}x^2$. Then check your predictions by graphing the functions.

## ◄ Graphing Calculator Explorations

To allow for optional use of the graphing calculator, the Explorations are boxed. Most can also be read simply for enrichment. Exercises and examples that are designed to be done with a graphing calculator are marked with an icon.

## ▼ Spreadsheet Explorations

Boxed for optional use, these spreadsheets will enhance students' understanding of the material using Excel, an alternative for those who prefer spreadsheet technology. See "Integrating Excel" on page xx for a list of exercises that can be done with Excel.

### Practice Problems ⊼

Students can check their understanding of a topic as they read the text or do homework by working out a Practice Problem. *Complete solutions* are found at the end of each section, just before the Section Summary.

### Be Careful ▼

The "Be Careful" icon marks places where the authors help students avoid common errors.

## ▼ Section Summary

Found at the end of every section, these summaries briefly state the main ideas of the section, providing a study tool or reminder for students.

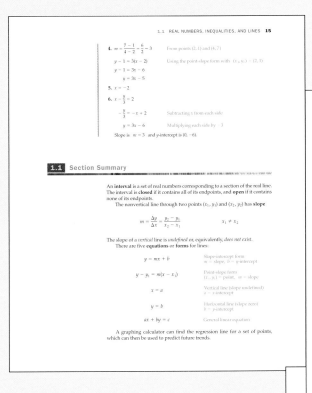

**4.** $m = \dfrac{7-1}{4-2} = \dfrac{6}{2} = 3$    From points (2, 1) and (4, 7)

$y - 1 = 3(x - 2)$    Using the point-slope form with $(x_1, y_1) = (2, 1)$

$y - 1 = 3x - 6$

$y = 3x - 5$

**5.** $x = -2$

**6.** $x - \dfrac{y}{3} = 2$

$-\dfrac{y}{3} = -x + 2$    Subtracting $x$ from each side

$y = 3x - 6$    Multiplying each side by $-3$

Slope is $m = 3$ and $y$-intercept is (0, −6).

### 1.1 Section Summary

An **interval** is a set of real numbers corresponding to a section of the real line. The interval is **closed** if it contains all of its endpoints, and **open** if it contains none of its endpoints.

The nonvertical line through two points $(x_1, y_1)$ and $(x_2, y_2)$ has **slope**

$$m = \frac{\Delta y}{\Delta x} = \frac{y_2 - y_1}{x_2 - x_1} \qquad x_1 \neq x_2$$

The slope of a *vertical* line is *undefined* or, equivalently, *does not exist*.
There are five **equations** or **forms** for lines:

| | |
|---|---|
| $y = mx + b$ | Slope-intercept form $m$ = slope, $b$ = $y$-intercept |
| $y - y_1 = m(x - x_1)$ | Point-slope form $(x_1, y_1)$ = point, $m$ = slope |
| $x = a$ | Vertical line (slope undefined) $a$ = $x$-intercept |
| $y = b$ | Horizontal line (slope zero) $b$ = $y$-intercept |
| $ax + by = c$ | General linear equation |

A graphing calculator can find the regression line for a set of points, which can then be used to predict future trends.

---

### Exercises ◄

The *Applied Exercises* are labeled with general and specific titles so instructors can assign problems appropriate for the class. *Conceptual Exercises* encourage students to "think outside the box," and *Explorations and Excursions* push students further.

$2^{1/6}C = 1.516C$ dollars—that is, about 1.5 times as much. Therefore, to increase capacity by 100% costs only about 50% more.*

**81.** Use the rule of .6 to find how costs change if a company wants to quadruple $(x = 4)$ its capacity.

**82.** Use the rule of .6 to find how costs change if a company wants to triple $(x = 3)$ its capacity.

**83–84. ALLOMETRY: Heart Rate** It is well known that the hearts of smaller animals beat faster than the hearts of larger animals. The actual relationship is approximately

(Heart rate) = 250(Weight)$^{-1/4}$

where the heart rate is in beats per minute and the weight is in pounds. Use this relationship to estimate the heart rate of:

**83.** A 16-pound dog.

**84.** A 625-pound grizzly bear.
*Source: Biology Review 41*

**85–86. BUSINESS: Learning Curves in Airplane Production** Recall (pages 26–27) that the learning curve for the production of Boeing 707 airplanes is $150n^{-0.322}$ (thousand work-hours). Find how many work-hours it took to build:

**85.** The 50th Boeing 707.

**86.** The 250th Boeing 707.

**87. GENERAL: Richter Scale** The Richter scale (developed by Charles Richter in 1935) is widely used to measure the strength of earthquakes. Every increase of 1 on the Richter scale corresponds to a 10-fold increase in ground motion. Therefore, an increase on the Richter scale from $A$ to $B$ means that ground motion increases by a factor of $10^{B-A}$ (for $B > A$). Find the increase in ground motion between the following earthquakes:

**a.** The 1994 Northridge, California, earthquake, measuring 6.8 on the Richter scale, and the 1906 San Francisco earthquake, measuring 8.3. (The San Francisco earthquake resulted in 500 deaths and a 3-day fire that destroyed 4 square miles of San Francisco.)

*(continues)*

*Although the rule of .6 is only a rough "rule of thumb," it can be somewhat justified on the basis that the equipment of such industries consists mainly of containers, and the cost of a container depends on its surface area (square units), which increases more slowly than its capacity (cubic units).*

**b.** The 2004 earthquake near Sumatra (Indonesia), measuring 9.0 on the Richter scale, and the 2008 Sichuan (China) earthquake, measuring 7.9. (The Sumatra earthquake caused a 50-foot-high tsunami, or "tidal wave," that killed 170,000 people in 11 countries. The death toll from the Sichuan earthquake was more than 70,000.)

**88. GENERAL: Richter Scale** (*continuation*) Every increase of 1 on the Richter scale corresponds to an approximately 30-fold increase in *energy released*. Therefore, an increase on the Richter scale from $A$ to $B$ means that the energy released increases by a factor of $30^{B-A}$ (for $B > A$).

**a.** Find the increase in *energy released* between the earthquakes in Exercise 87a.

**b.** Find the increase in *energy released* between the earthquakes in Exercise 87b.

**89–90. GENERAL: Waterfalls** Water falling from a waterfall that is $x$ feet high will hit the ground with speed $\frac{60}{11}x^{1/2}$ miles per hour (neglecting air resistance).

**89.** Find the speed of the water at the bottom of the highest waterfall in the world, Angel Falls in Venezuela (3281 feet high).

**90.** Find the speed of the water at the bottom of the highest waterfall in the United States, Ribbon Falls in Yosemite, California (1650 feet high).

**91–92. ENVIRONMENTAL SCIENCE: Biodiversity** It is well known that larger land areas can support larger numbers of species. According to one study, multiplying the land area by a factor of $x$ multiplies the number of species by a factor of $x^{0.239}$. Use a graphing calculator to graph $y = x^{0.239}$. Use the window [0, 100] by [0, 4].
*Source: Robert H. MacArthur and Edward O. Wilson, The Theory of Island Biogeography*

**91.** Find the multiple $x$ for the land area that leads to *double* the number of species. That is, find the value of $x$ such that $x^{0.239} = 2$. [*Hint:* Either use TRACE or find where $y_1 = x^{0.239}$ INTERSECTs $y_2 = 2$.]

**92.** Find the multiple $x$ for the land area that leads to triple the number of species. That is, find the value of $x$ such that $x^{0.239} = 3$. [*Hint:* Either use TRACE or find where $y_1 = x^{0.239}$ INTERSECTs $y_2 = 3$.]

### Chapter Summary with Hints and Suggestions

Reading the text and doing the exercises in this chapter have helped you to master the following concepts and skills, which are listed by section (in case you need to review them) and are keyed to particular Review Exercises. Answers for all Review Exercises are given at the back of the book, and full solutions can be found in the Student Solutions Manual.

**1.1 Real Numbers, Inequalities, and Lines**

- Translate an interval into set notation and graph it on the real line. (*Review Exercises 1–4.*)

   $[a, b]$   $(a, b)$   $[a, b)$   $(a, b]$

   $(-\infty, b]$   $(-\infty, b)$   $[a, \infty)$   $(a, \infty)$   $(-\infty, \infty)$

- Express given information in interval form. (*Review Exercises 5–6.*)

- Find an equation for a line that satisfies certain conditions. (*Review Exercises 7–12.*)

$$m = \frac{y_2 - y_1}{x_2 - x_1} \qquad y = mx + b$$

$$y - y_1 = m(x - x_1) \qquad x = a \qquad y = b$$

$$ax + by = c$$

- Find an equation of a line from its graph. (*Review Exercises 13–14.*)

- Use straight-line depreciation to find the value of an asset. (*Review Exercises 15–16.*)

- Use real-world data to find a regression line and make a prediction. (*Review Exercise 17.*)

**1.2 Exponents**

- Evaluate negative and fractional exponents without a calculator. (*Review Exercises 18–25.*)

$$x^0 = 1 \qquad x^{-n} = \frac{1}{x^n} \qquad x^{m/n} = \sqrt[n]{x^m} = \left(\sqrt[n]{x}\right)^m$$

- Evaluate an exponential expression using a calculator. (*Review Exercises 26–29.*)

- Use real-world data to find a power regression curve and make a prediction. (*Review Exercise 30.*)

**1.3 Functions: Linear and Quadratic**

- Evaluate and find the domain and range of a function. (*Review Exercises 31–34.*)

   A function $f$ is a rule that assigns to each number $x$ in a set (the domain) a (single) number $f(x)$. The range is the set of all resulting values $f(x)$.

- Use the vertical line test to see if a graph defines a function. (*Review Exercises 35–36.*)

- Graph a linear function: $f(x) = mx + b$ (*Review Exercises 37–38.*)

- Graph a quadratic function: $f(x) = ax^2 + bx + c$ (*Review Exercises 39–40.*)

- Solve a quadratic equation by factoring and by the Quadratic Formula. (*Review Exercises 41–44.*)

| Vertex | $x$-intercepts |
|---|---|
| $x = \dfrac{-b}{2a}$ | $x = \dfrac{-b \pm \sqrt{b^2 - 4ac}}{2a}$ |

- Use a graphing calculator to graph a quadratic function. (*Review Exercises 45–46.*)

---

## End of Chapter Material ►

To help students study, each chapter ends with a **Chapter Summary with Hints and Suggestions** and **Review Exercises.** The last bullet of the Hints and Suggestions lists the Review Exercises that a student could use to self-test. Both even and odd answers are supplied in the back of the book.

### Cumulative Review

There is a Cumulative Review after every 3–4 chapters. Even and odd answers are supplied in the back of the book.

# Integrating Excel

If you would like to use Excel or another spreadsheet software when working the exercises in this text, refer to the chart below. It lists exercises from many sections that you might find instructive to do with spreadsheet technology. Please note that none of these exercises are dependent on Excel. If you would like help using Excel, please consider the *Excel Guide for Finite Mathematics and Applied Calculus,* which is available from Cengage. Additionally, the *Getting Started with Excel* chapter of the guide is available on the website.

| Section | Suggested Exercises | Section | Suggested Exercises |
|---------|--------------------|---------|--------------------|
| 1.1 | 57–74 | 6.1 | 63 and 64 |
| 1.2 | 91–98 | 6.3 | 41 and 42 |
| 1.3 | 71–82 | 6.4 | 9–18, 27–37 |
| 1.4 | 88–92 | | |
| | | 7.1 | 36–40 |
| 2.1 | 77 and 78, 81–84 | 7.2 | 54–56 |
| 2.5 | 45 and 46 | 7.3 | 29–32 |
| 2.7 | 11 and 12 | 7.4 | 13–18, 27–32 |
| | | 7.6 | 31 and 32, 35 and 36 |
| 3.1 | 68–71 and 85 | | |
| 3.2 | 63 and 64 | 8.2 | 36–41 |
| 3.3 | 25–35 | 8.3 | 71–76 |
| 3.4 | 23 and 24 | 8.4 | 49 and 50, 53 and 54 |
| 3.5 | 20 | 8.5 | 20–23 |
| 3.6 | 63 and 64 | | |
| | | 9.3 | 36 |
| 4.1 | 11 and 12, 47–49 | 9.4 | 11–24, 27–30 |
| 4.2 | 31–47 | | |
| 4.3 | 85–87 | 10.1 | 49–58 |
| 4.4 | 36–39 | 10.2 | 9–12, 21–23, 25 and 26 |
| | | 10.4 | 11–24, 33–38 |
| 5.2 | 41 and 42, 45 and 46, 55, 57 and 58 | | |
| | | 11.1 | 29–36 |
| 5.3 | 13–18, 85 and 86 | 11.2 | 37–41 |
| 5.4 | 32, 35 and 36, 69 | 11.3 | 5, 22 and 23 |
| 5.5 | 31 and 32 | 11.4 | 21–26 |
| 5.6 | 77 and 78 | | |

# Graphing Calculator Basics

While the (optional) *Graphing Calculator Explorations* may be carried out on most graphing calculators, the screens shown in this book are from the Texas Instruments *TI-83, TI-84,* and *TI-84 Plus* calculators. Any specific instructions are also for these calculators. (We occasionally show a screen from a *TI-89* calculator, but for illustration purposes only.) To carry out the Graphing Calculator Explorations, you should be familiar with the terms described in *Graphing Calculator Terminology* below. To do the *regression* (or *modeling*) examples in Chapter 1 (again optional), you should be familiar with the techniques in the following section headed *Entering Data.*

## GRAPHING CALCULATOR TERMINOLOGY

The viewing or graphing **WINDOW** is the part of the Cartesian plane shown in the display screen of your graphing calculator. **XMIN** and **XMAX** are the smallest and largest *x*-values shown, and **YMIN** and **YMAX** are the smallest and largest *y*-values shown. These values can be set by using the **WINDOW** or **RANGE** command and are changed automatically by using any of the **ZOOM** operations. **XSCALE** and **YSCALE** define the distance between tick marks on the *x*- and *y*-axes.

Viewing Window [−10, 10] by [−10, 10]

The viewing window is always [XMIN, XMAX] by [YMIN, YMAX]. We will set XSCALE and YSCALE so that there are a reasonable number of tick marks (generally 2 to 20) on each axis. The *x*- and *y*-axes will not be visible if the viewing window does not include the origin.

**Pixel,** an abbreviation for *picture element*, refers to a tiny rectangle on the screen that can be darkened to represent a dot on a graph. Pixels are arranged in a rectangular array on the screen. In the above window, the axes and tick marks are formed by darkened pixels. The size of the screen and number of pixels varies with different calculators.

**TRACE** allows you to move a flashing pixel, or *cursor,* along a curve in the viewing window with the $x$- and $y$-coordinates shown at the bottom of the screen.

> *Useful Hint:* To make the $x$-values in TRACE take simple values like .1, .2, and .3, choose XMIN and XMAX to be multiples of one less than the number of pixels across the screen. For example, on the *TI-84,* which has 95 pixels across the screen, using an $x$-window like [–9.4, 9.4] or [–4.7, 4.7] or [940, –940] will TRACE with simpler $x$-values than the standard windows stated in this book.

**ZOOM IN** allows you to magnify any part of the viewing window to see finer detail around a chosen point. **ZOOM OUT** does the opposite, like stepping back to see a larger portion of the plane but with less detail. These and other **ZOOM** commands change the viewing window.

**VALUE** or **EVALUATE** finds the value of a previously entered expression at a specified $x$-value.

**SOLVE** or **ROOT** finds the $x$-value that solves $f(x) = 0,$ equivalently, the $x$-intercepts of a curve. When applied to a difference $f(x) - g(x),$ it finds the $x$-value where the two curves meet (also done by the **INTERSECT** command).

**MAX** and **MIN** find the maximum and minimum values of a previously entered curve between specified $x$-values.

**NDERIV** or **DERIV** or $dy/dx$ approximates the *derivative* of a function at a point. **FnInt** or $\int f(x)dx$ approximates the definite integral of a function on an interval.

In **CONNECTED MODE** your calculator will darken pixels to connect calculated points on a graph to show it as a continuous or "unbroken" curve. However, this may lead to "false lines" in a graph that should have breaks or "jumps." False lines can be eliminated by using **DOT MODE.**

The **TABLE** command lists in table form the values of a function, just as you have probably done when graphing a curve. The $x$-values may be chosen by you or by the calculator.

The **Order of Operations** used by most calculators evaluates operations in the following order: first powers and roots, then operations like **LN** and **LOG,** then multiplication and division, then addition and subtraction—left to right within each level. For example, $5 \wedge 2x$ means $(5 \wedge 2)x,$ *not* $5 \wedge (2x).$ Also, $1/x + 1$ means $(1/x) + 1,$ *not* $1/(x + 1).$ See your calculator's instruction manual for further information.

 **BE CAREFUL:** Some calculators evaluate $1/2x$ as $(1/2)x$ and some as $1/(2x).$ When in doubt, use parentheses to clarify the expression.

Much more information can be found in the manual for your graphing calculator. Other features will be discussed later as needed.

## ENTERING DATA

To ensure that your calculator is properly prepared to accept data, turn off all statistical plots by pressing 2ND Y= 4 ENTER

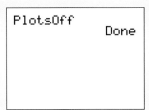

clear all data lists by pressing 2ND + 4 ENTER

set up the statistical data editor by pressing STAT 5 ENTER

and finally clear all functions by pressing Y= and repeatedly using CLEAR and the down-arrow key ▽ until all functions are cleared. You will need to carry out the preceding steps only once if you are using your calculator only with this book.

To enter the data shown in the following table (which is taken from Example 8 on pages 13–14), first put the $x$-values into list L1 by pressing STAT ENTER and entering each value followed by an ENTER. Then put the $y$-values into list L2 by pressing the right-arrow key ▷ and entering each $y$-value followed by an ENTER.

| $x$ | 1 | 2 | 3 | 4 | 5 | 6 |
|---|---|---|---|---|---|---|
| $y$ | 47 | 50.7 | 54 | 58.5 | 61.5 | 64.9 |

Turn on Plot1 by pressing $\boxed{\text{2ND}}$ $\boxed{\text{Y=}}$ $\boxed{\text{ENTER}}$ $\boxed{\text{ENTER}}$ and then use the direction-arrow keys and $\boxed{\text{ENTER}}$ (along with $\boxed{\text{2ND}}$ $\boxed{1}$ for list L1 and $\boxed{\text{2ND}}$ $\boxed{2}$ for list L2) to set Plot1 as shown below. Press $\boxed{\text{ZOOM}}$ $\boxed{9}$ (which gives ZoomStat) to plot this data on an appropriate viewing window. You may then use $\boxed{\text{TRACE}}$ and the arrow keys to check the values of your data points.

When you are finished studying the data you have entered, press $\boxed{\text{Y=}}$ $\boxed{\triangle}$ $\boxed{\text{ENTER}}$ to turn Plot1 off and then $\boxed{\text{2ND}}$ $\boxed{+}$ $\boxed{4}$ $\boxed{\text{ENTER}}$ to clear all lists.

# Applied
# Calculus

# 1

# Functions

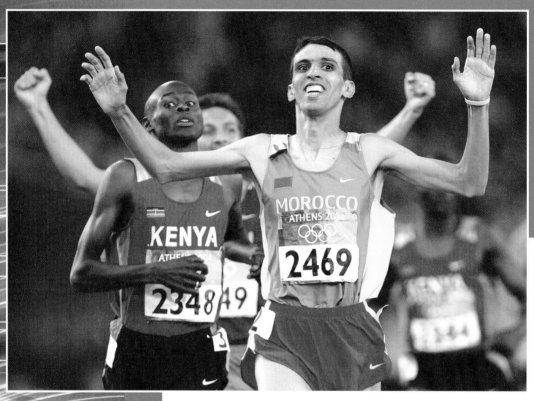

Moroccan
runner Hicham
El Guerrouj,
current world
record holder
for the mile run,
bested the
record set
6 years earlier
by 1.26
seconds.

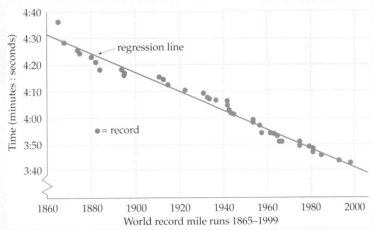

## World Record Mile Runs

The dots on the graph below show the world record times for the mile run from 1865 to the 1999 world record of 3 minutes 43.13 seconds, set by the Moroccan runner Hicham El Guerrouj. These points fall roughly along a line, called the **regression line.** In this section we will see how to use a graphing calculator to find a regression line (see Example 8 and Exercises 69–74), based on a method called **least squares,** whose mathematical basis will be explained in Chapter 7.

Notice that the times do not level off as you might expect, but continue to decrease.

### History of the Record for the Mile Run

| Time | Year | Athlete | Time | Year | Athlete | Time | Year | Athlete |
|------|------|---------|------|------|---------|------|------|---------|
| 4:36.5 | 1865 | Richard Webster | 4:09.2 | 1931 | Jules Ladoumegue | 3:54.1 | 1964 | Peter Snell |
| 4:29.0 | 1868 | William Chinnery | 4:07.6 | 1933 | Jack Lovelock | 3:53.6 | 1965 | Michel Jazy |
| 4:28.8 | 1868 | Walter Gibbs | 4:06.8 | 1934 | Glenn Cunningham | 3:51.3 | 1966 | Jim Ryun |
| 4:26.0 | 1874 | Walter Slade | 4:06.4 | 1937 | Sydney Wooderson | 3:51.1 | 1967 | Jim Ryun |
| 4:24.5 | 1875 | Walter Slade | 4:06.2 | 1942 | Gunder Hägg | 3:51.0 | 1975 | Filbert Bayi |
| 4:23.2 | 1880 | Walter George | 4:06.2 | 1942 | Arne Andersson | 3:49.4 | 1975 | John Walker |
| 4:21.4 | 1882 | Walter George | 4:04.6 | 1942 | Gunder Hägg | 3:49.0 | 1979 | Sebastian Coe |
| 4:18.4 | 1884 | Walter George | 4:02.6 | 1943 | Arne Andersson | 3:48.8 | 1980 | Steve Ovett |
| 4:18.2 | 1894 | Fred Bacon | 4:01.6 | 1944 | Arne Andersson | 3:48.53 | 1981 | Sebastian Coe |
| 4:17.0 | 1895 | Fred Bacon | 4:01.4 | 1945 | Gunder Hägg | 3:48.40 | 1981 | Steve Ovett |
| 4:15.6 | 1895 | Thomas Conneff | 3:59.4 | 1954 | Roger Bannister | 3:47.33 | 1981 | Sebastian Coe |
| 4:15.4 | 1911 | John Paul Jones | 3:58.0 | 1954 | John Landy | 3:46.31 | 1985 | Steve Cram |
| 4:14.4 | 1913 | John Paul Jones | 3:57.2 | 1957 | Derek Ibbotson | 3:44.39 | 1993 | Noureddine Morceli |
| 4:12.6 | 1915 | Norman Taber | 3:54.5 | 1958 | Herb Elliott | 3:43.13 | 1999 | Hicham El Guerrouj |
| 4:10.4 | 1923 | Paavo Nurmi | 3:54.4 | 1962 | Peter Snell | | | |

*Source: USA Track & Field*

The equation of the regression line is $y = -0.356x + 257.44$, where $x$ represents years after 1900 and $y$ is the time in seconds. The regression line can be used to predict the world mile record in future years. Notice that the most recent world record would have been predicted quite accurately by this line, since the rightmost dot falls almost exactly on the line.

Linear trends, however, must not be extended too far. The downward slope of this line means that it will eventually "predict" mile runs in a fraction of a second, or even in *negative* time (see Exercises 57 and 58 on page 17). *Moral:* In the real world, linear trends do not continue indefinitely. This and other topics in "linear" mathematics will be developed in Section 1.1.

## 1.1 REAL NUMBERS, INEQUALITIES, AND LINES

### Introduction

Quite simply, *calculus is the study of rates of change.* We will use calculus to analyze rates of inflation, rates of learning, rates of population growth, and rates of natural resource consumption.

In this first section we will study **linear** relationships between two variable quantities—that is, relationships that can be represented by **lines.** In later sections we will study **nonlinear** relationships, which can be represented by **curves.**

### Real Numbers and Inequalities

In this book the word "number" means **real number,** a number that can be represented by a point on the number line (also called the **real line**).

The *order* of the real numbers is expressed by **inequalities.** For example, $a < b$ means "$a$ is to the *left* of $b$" or, equivalently, "$b$ is to the *right* of $a$."

| Inequalities | | Brief Examples |
|---|---|---|
| *Inequality* | *In Words* | |
| $a < b$ | $a$ is less than (smaller than) $b$ | $3 < 5$ |
| $a \le b$ | $a$ is less than or equal to $b$ | $-5 \le -3$ |
| $a > b$ | $a$ is greater than (larger than) $b$ | $\pi > 3$ |
| $a \ge b$ | $a$ is greater than or equal to $b$ | $2 \ge 2$ |

The inequalities $a < b$ and $a > b$ are called **strict inequalities,** and $a \le b$ and $a \ge b$ are called **nonstrict inequalities.**

IMPORTANT NOTE: Throughout this book are many **Practice Problems**— short questions designed to check your understanding of a topic before moving on to new material. Full solutions are given at the end of the section. Solve the following Practice Problem and then check your answer.

| **Practice Problem 1** | Which number is smaller: $\dfrac{1}{100}$ or $-1{,}000{,}000$? | ➤ **Solution on page 14** |

Multiplying or dividing both sides of an inequality by a negative number reverses the direction of the inequality:

$$-3 < 2 \quad \text{but} \quad 3 > -2 \qquad \text{Multiplying by } -1$$

A **double inequality,** such as $a < x < b$, means that *both* the inequalities $a < x$ and $x < b$ hold. The inequality $a < x < b$ can be interpreted graphically as "*x* is between *a* and *b*."

$$a < x < b$$

## Sets and Intervals

**Braces** { } are read "the set of all" and a **vertical bar** | is read "such that."

| **EXAMPLE 1** | **INTERPRETING SETS** |

The set of all ⌐

**a.** $\{x \mid x > 3\}$  means "the set of all *x* such that *x* is greater than 3."

∟ Such that

**b.** $\{x \mid -2 < x < 5\}$  means "the set of all *x* such that *x* is between $-2$ and 5."

| **Practice Problem 2** | **a.** Write in set notation "the set of all *x* such that *x* is greater than or equal to $-7$." |
| | **b.** Express in words: $\{x \mid x < -1\}$.    ➤ **Solution on page 14** |

The set $\{x \mid 2 \le x \le 5\}$ can be expressed in **interval notation** by enclosing the endpoints 2 and 5 in **square brackets,** [2, 5], to indicate that the endpoints are *included.* The set $\{x \mid 2 < x < 5\}$ can be written with **parentheses,** (2, 5), to indicate that the endpoints 2 and 5 are *excluded.* An interval is **closed** if it includes both endpoints, and **open** if it includes neither endpoint. The four types of intervals are shown below: a **solid dot** • on the graph indicates that the point is *included* in the interval; a **hollow dot** ○ indicates that the point is *excluded.*

| Finite Intervals | | | | *Brief Examples* |
|---|---|---|---|---|
| Interval Notation | Set Notation | Graph | Type | |
| $[a, b]$ | $\{\,x \mid a \leq x \leq b\,\}$ | | Closed (includes endpoints) | $[-2, 5]$ |
| $(a, b)$ | $\{\,x \mid a < x < b\,\}$ | | Open (excludes endpoints) | $(-2, 5)$ |
| $[a, b)$ | $\{\,x \mid a \leq x < b\,\}$ | | Half-open or half-closed | $[-2, 5)$ |
| $(a, b]$ | $\{\,x \mid a < x \leq b\,\}$ | | | $(-2, 5]$ |

An interval may extend infinitely far to the *right* (indicated by the symbol $\infty$ for **infinity**) or infinitely far to the *left* (indicated by $-\infty$ for **negative infinity**). Note that $\infty$ and $-\infty$ are not numbers, but are merely symbols to indicate that the interval extends endlessly in that direction. The infinite intervals in the next box are said to be **closed** or **open** depending on whether they *include* or *exclude* their single endpoint.

| Infinite Intervals | | | | *Brief Examples* |
|---|---|---|---|---|
| Interval Notation | Set Notation | Graph | Type | |
| $[a, \infty)$ | $\{\,x \mid x \geq a\,\}$ | | Closed | $[3, \infty)$ |
| $(a, \infty)$ | $\{\,x \mid x > a\,\}$ | | Open | $(3, \infty)$ |
| $(-\infty, a]$ | $\{\,x \mid x \leq a\,\}$ | | Closed | $(-\infty, 5]$ |
| $(-\infty, a)$ | $\{\,x \mid x < a\,\}$ | | Open | $(-\infty, 5)$ |

We use *parentheses* rather than square brackets with $\infty$ and $-\infty$ since they are not actual numbers.

The interval $(-\infty, \infty)$ extends infinitely far in *both* directions (meaning the entire real line) and is also denoted by $\mathbb{R}$ (the set of all real numbers).

$$\mathbb{R} = (-\infty, \infty)$$

### Cartesian Plane

Two real lines or **axes,** one horizontal and one vertical, intersecting at their zero points, define the **Cartesian plane.*** The point where they meet is called the **origin.** The axes divide the plane into four **quadrants,** I through IV, as shown below.

Any point in the Cartesian plane can be specified uniquely by an ordered pair of numbers $(x, y)$; $x$, called the **abscissa** or **x-coordinate,** is the number on the horizontal axis corresponding to the point; $y$, called the **ordinate** or **y-coordinate,** is the number on the vertical axis corresponding to the point.

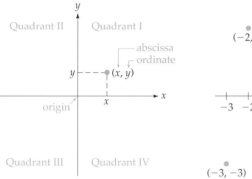

The Cartesian plane

The Cartesian plane with several points.
Order matters: (1, 2) is not the same as (2, 1)

### Lines and Slopes

The symbol **Δ** (read "delta," the Greek letter D) means "the change in." For any two points $(x_1, y_1)$ and $(x_2, y_2)$ we define

$\Delta x = x_2 - x_1$     The change in $x$ is the difference in the $x$-coordinates

$\Delta y = y_2 - y_1$     The change in $y$ is the difference in the $y$-coordinates

Any two distinct points determine a line. A nonvertical line has a **slope** that measures the *steepness* of the line, and is defined as *the change in y divided by the change in x* for any two points on the line.

| Slope of Line Through $(x_1, y_1)$ and $(x_2, y_2)$ | |
| --- | --- |
| $$m = \dfrac{\Delta y}{\Delta x} = \dfrac{y_2 - y_1}{x_2 - x_1}$$ | Slope is the change in $y$ over the change in $x$ ($x_2 \neq x_1$) |

 **BE CAREFUL:** In slope, the $x$-values go in the *denominator*.

*So named because it was originated by the French philosopher and mathematician René Descartes (1596–1650). Following the custom of the day, Descartes signed his scholarly papers with his Latin name Cartesius, hence "Cartesian" plane.

The changes $\Delta y$ and $\Delta x$ are often called, respectively, the "rise" and the "run," with the understanding that a negative "rise" means a "fall." Slope is then "rise over run."

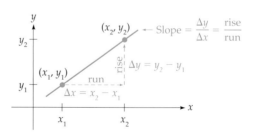

---

EXAMPLE 2

**FINDING SLOPES AND GRAPHING LINES**

Find the slope of the line through each pair of points, and graph the line.

**a.** $(2, 1)$, $(3, 4)$     **b.** $(2, 4)$, $(3, 1)$

**c.** $(-1, 3)$, $(2, 3)$     **d.** $(2, -1)$, $(2, 3)$

*Solution*

We use the slope formula $m = \dfrac{y_2 - y_1}{x_2 - x_1}$ for each pair $(x_1, y_1)$, $(x_2, y_2)$.

**a.** For $(2, 1)$ and $(3, 4)$ the slope is
$$\frac{4 - 1}{3 - 2} = \frac{3}{1} = 3.$$

**b.** For $(2, 4)$ and $(3, 1)$ the slope is $\dfrac{1 - 4}{3 - 2} = \dfrac{-3}{1} = -3.$

**c.** For $(-1, 3)$ and $(2, 3)$ the slope is $\dfrac{3 - 3}{2 - (-1)} = \dfrac{0}{3} = 0.$

**d.** For $(2, -1)$ and $(2, 3)$ the slope is *undefined*:
$$\frac{3 - (-1)}{2 - 2} = \frac{4}{0}.$$

Notice that when the *x*-coordinates are the same [as in part (d) above], the line is *vertical*, and when the *y*-coordinates are the same [as in part (c) above], the line is *horizontal*.

If $\Delta x = 1$, as in Examples 2a and 2b, then the slope is just the "rise," giving an alternative definition for slope:

$$\text{Slope} = \left(\begin{array}{c}\text{Amount that the line rises}\\\text{when } x \text{ increases by 1}\end{array}\right)$$

---

**Practice Problem 3**

A company president is considering four different business strategies, called $S_1$, $S_2$, $S_3$, and $S_4$, each with different projected future profits. The graph on the right shows the annual projected profit for the first few years for each of the strategies. Which strategy yields:

**a.** the highest projected profit in year 1?

**b.** the highest projected profit in the long run?

➤ **Solutions on page 14**

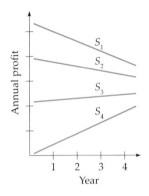

---

## Equations of Lines

The point where a nonvertical line crosses the *y*-axis is called the **y-intercept** of the line. The *y*-intercept can be given either as the *y*-coordinate *b* or as the point (0, *b*). Such a line can be expressed very simply in terms of its slope and *y*-intercept, representing the points by variable coordinates (or "variables") *x* and *y*.

---

**Slope-Intercept Form of a Line**

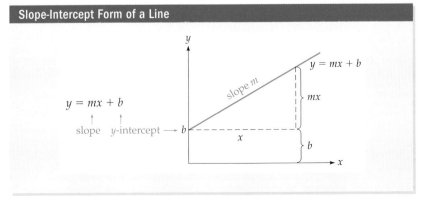

$y = mx + b$
    ↑    ↑
slope  *y*-intercept

*Brief Example*

For the line with slope $-2$ and *y*-intercept 4:

$$y = -2x + 4$$

*y*-intercept (0, 4)

Second point
1 unit "over"
and 2 "down"

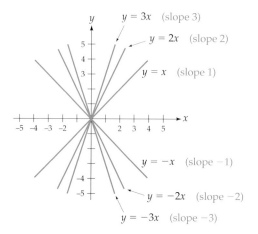

For lines through the origin, the equation takes the particularly simple form, $y = mx$ (since $b = 0$), as illustrated on the left.

The most useful equation for a line is the *point-slope form*.

---

**Point-Slope Form of a Line**

$$y - y_1 = m(x - x_1)$$

$(x_1, y_1)$ = point on the line
$m$ = slope

---

This form comes directly from the slope formula $m = \dfrac{y_2 - y_1}{x_2 - x_1}$ by replacing $x_2$ and $y_2$ by $x$ and $y$, and then multiplying each side by $(x - x_1)$. It is most useful when you know the slope of the line and a point on it.

---

**EXAMPLE 3**

**USING THE POINT-SLOPE FORM**

Find an equation of the line through $(6, -2)$ with slope $-\frac{1}{2}$.

*Solution*

$$y - (-2) = -\frac{1}{2}(x - 6)$$ 
$y - y_1 = m(x - x_1)$ with $y_1 = -2$, $m = -\frac{1}{2}$, and $x_1 = 6$

$$y + 2 = -\frac{1}{2}x + 3$$
Eliminating parentheses

$$y = -\frac{1}{2}x + 1$$
Subtracting 2 from each side

Alternatively, we could have found this equation using $y = mx + b$, replacing $m$ by the given slope $-\frac{1}{2}$, and then substituting the given $x = 6$ and $y = -2$ to evaluate $b$.

**EXAMPLE 4**

**FINDING AN EQUATION FOR A LINE THROUGH TWO POINTS**

Find an equation for the line through the points $(4, 1)$ and $(7, -2)$.

*Solution*

The slope is not given, so we calculate it from the two points.

$$m = \frac{-2 - 1}{7 - 4} = \frac{-3}{3} = -1 \qquad m = \frac{y_2 - y_1}{x_2 - x_1} \quad \text{with } (4, 1) \text{ and } (7, -2)$$

Then we use the point-slope formula with this slope and either of the two points.

$$y - 1 = -1(x - 4) \qquad \begin{array}{l} y - y_1 = m(x - x_1) \text{ with} \\ \text{slope } -1 \text{ and point } (4, 1) \end{array}$$

$$y - 1 = -x + 4 \qquad \text{Eliminating parentheses}$$
$$y = -x + 5 \qquad \text{Adding 1 to each side}$$

**Practice Problem 4**   Find the slope-intercept form of the line through the points $(2, 1)$ and $(4, 7)$.

➤ **Solution on page 15**

Vertical and horizontal lines have particularly simple equations: a variable equaling a constant.

*Vertical Line*
$x = a$

*Horizontal Line*
$y = b$

**EXAMPLE 5**

**GRAPHING VERTICAL AND HORIZONTAL LINES**

Graph the lines $x = 2$ and $y = 6$.

*Solution*

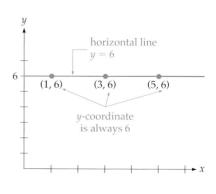

**EXAMPLE 6**

**FINDING EQUATIONS OF VERTICAL AND HORIZONTAL LINES**

**a.** Find an equation for the *vertical* line through (3, 2).

**b.** Find an equation for the *horizontal* line through (3, 2).

*Solution*

**a.** Vertical line       $x = 3$       $x = a$, with $a$ being the
                                          $x$-coordinate from (3, 2)

**b.** Horizontal line     $y = 2$       $y = b$, with $b$ being the
                                          $y$-coordinate from (3, 2)

**Practice Problem 5**     Find an equation for the vertical line through $(-2, 10)$.

➤ Solution on page 15

In a vertical line, the $x$-coordinate does not change, so $\Delta x = 0$, making the slope $m = \Delta y / \Delta x$ *undefined*. Therefore, distinguish carefully between slopes of vertical and horizontal lines:

Vertical line: slope is *undefined*.

Horizontal line: slope *is* defined, and is *zero*.

There is one form that covers *all* lines, vertical and nonvertical.

**General Linear Equation**

$$ax + by = c$$

For constants $a, b, c$, with
$a$ and $b$ not both zero

Any equation that can be written in this form is called a **linear equation,** and the variables are said to **depend linearly** on each other.

**EXAMPLE 7**

**FINDING THE SLOPE AND THE *y*-INTERCEPT FROM A LINEAR EQUATION**

Find the slope and *y*-intercept of the line $2x + 3y = 12$.

*Solution* We write the line in slope-intercept form. Solving for *y*:

$$3y = -2x + 12$$
Subtracting $2x$ from both sides of $2x + 3y = 12$

$$y = -\frac{2}{3}x + 4$$
Dividing each side by 3 gives the slope-intercept form $y = mx + b$

Therefore, the slope is $-\frac{2}{3}$ and the *y*-intercept is $(0, 4)$.

**Practice Problem 6**

Find the slope and *y*-intercept of the line $x - \dfrac{y}{3} = 2$.

➤ **Solution on page 15**

## Linear Regression (Optional)

Given two points, we can find a line through them, as in Example 4. However, some real-world situations involve *many* data points, which may lie approximately but not exactly on a line. How can we find the line that, in some sense, *lies closest* to the points or *best approximates* the points? The most widely used technique is called **linear regression** or **least squares,** and its mathematical basis will be explained in Section 7.4. Even without studying its mathematical basis, however, we can easily find the regression line using a graphing calculator (or spreadsheet or other computer software).

**EXAMPLE 8**

**LINEAR REGRESSION USING A GRAPHING CALCULATOR**

The following graph gives the percentage of U.S. homes with Internet access during recent years.

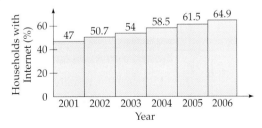

*Source:* Euromonitor International

**a.** Use linear regression to fit a line to the data.

**b.** Interpret the slope of the line.

**c.** Use the regression line to predict Internet access in the year 2012.

*Solution*

**a.** We number the years with $x$-values 1–6, so $x$ stands for *years since 2000* (we could choose other $x$-values instead). We enter the data into lists, as shown in the first screen below (as explained in *Graphing Calculator Basics—Entering Data* on page xxiii), and use *ZoomStat* to graph the data points.

Then (using STAT, CALC, and LinReg), graph the regression along with the data points.

The regression line, which clearly fits the points quite well, is

$$y = 3.6x + 43.5 \qquad \text{Rounded}$$

**b.** Since $x$ is in years, the slope 3.6 means that the percentage of households with Internet access increases by about 3.6 percentage points per year.

**c.** To predict the percentage of U.S. homes with Internet access in the year 2012, we evaluate Y1 at 12 (since $x = 12$ corresponds to 2012). From the screen on the left, if the current trend continues, about 86.8% of homes will have Internet access by 2012.

---

**Solutions to Practice Problems**

**1.** $-1{,}000{,}000$ [the negative sign makes it less than (to the left of) the positive number $\frac{1}{100}$]

**2. a.** $\{x \mid x \geq -7\}$

   **b.** The set of all $x$ such that $x$ is less than $-1$

**3. a.** $S_1$

   **b.** $S_4$

**4.** $m = \dfrac{7 - 1}{4 - 2} = \dfrac{6}{2} = 3$      From points $(2, 1)$ and $(4, 7)$

       $y - 1 = 3(x - 2)$      Using the point-slope form with $(x_1, y_1) = (2, 1)$

       $y - 1 = 3x - 6$

         $y = 3x - 5$

**5.** $x = -2$

**6.** $x - \dfrac{y}{3} = 2$

      $-\dfrac{y}{3} = -x + 2$      Subtracting $x$ from each side

       $y = 3x - 6$      Multiplying each side by $-3$

Slope is $m = 3$ and $y$-intercept is $(0, -6)$.

## 1.1   Section Summary

An **interval** is a set of real numbers corresponding to a section of the real line. The interval is **closed** if it contains all of its endpoints, and **open** if it contains none of its endpoints.

     The nonvertical line through two points $(x_1, y_1)$ and $(x_2, y_2)$ has **slope**

$$m = \frac{\Delta y}{\Delta x} = \frac{y_2 - y_1}{x_2 - x_1} \qquad\qquad x_1 \neq x_2$$

The slope of a *vertical* line is *undefined* or, equivalently, *does not exist*.

     There are five **equations** or **forms** for lines:

$$y = mx + b$$      Slope-intercept form
     $m$ = slope, $b$ = $y$-intercept

$$y - y_1 = m(x - x_1)$$      Point-slope form
     $(x_1, y_1)$ = point, $m$ = slope

$$x = a$$      Vertical line (slope undefined)
     $a$ = $x$-intercept

$$y = b$$      Horizontal line (slope zero)
     $b$ = $y$-intercept

$$ax + by = c$$      General linear equation

     A graphing calculator can find the regression line for a set of points, which can then be used to predict future trends.

## 1.1 Exercises

**1–6.** Write each interval in set notation and graph it on the real line.

**1.** $[0, 6)$    **2.** $(-3, 5]$    **3.** $(-\infty, 2]$    **4.** $[7, \infty)$

**5.** Given the equation $y = 5x - 12$, how will $y$ change if $x$:
   **a.** Increases by 3 units?
   **b.** Decreases by 2 units?

**6.** Given the equation $y = -2x + 7$, how will $y$ change if $x$:
   **a.** Increases by 5 units?
   **b.** Decreases by 4 units?

**7–14.** Find the slope (if it is defined) of the line determined by each pair of points.

**7.** $(2, 3)$ and $(4, -1)$    **8.** $(3, -1)$ and $(5, 7)$

**9.** $(-4, 0)$ and $(2, 2)$    **10.** $(-1, 4)$ and $(5, 1)$

**11.** $(0, -1)$ and $(4, -1)$    **12.** $(-2, \frac{1}{2})$ and $(5, \frac{1}{2})$

**13.** $(2, -1)$ and $(2, 5)$    **14.** $(6, -4)$ and $(6, -3)$

**15–32.** For each equation, find the slope $m$ and $y$-intercept $(0, b)$ (when they exist) and draw the graph.

**15.** $y = 3x - 4$    **16.** $y = 2x$

**17.** $y = -\frac{1}{2}x$    **18.** $y = -\frac{1}{3}x + 2$

**19.** $y = 4$    **20.** $y = -3$

**21.** $x = 4$    **22.** $x = -3$

**23.** $2x - 3y = 12$    **24.** $3x + 2y = 18$

**25.** $x + y = 0$    **26.** $x = 2y + 4$

**27.** $x - y = 0$    **28.** $y = \frac{2}{3}(x - 3)$

**29.** $y = \dfrac{x + 2}{3}$    **30.** $\dfrac{x}{2} + \dfrac{y}{3} = 1$

**31.** $\dfrac{2x}{3} - y = 1$    **32.** $\dfrac{x + 1}{2} + \dfrac{y + 1}{2} = 1$

**33–44.** Write an equation of the line satisfying the following conditions. If possible, write your answer in the form $y = mx + b$.

**33.** Slope $-2.25$ and $y$-intercept 3

**34.** Slope $\frac{2}{3}$ and $y$-intercept $-8$

**35.** Slope 5 and passing through the point $(-1, -2)$

**36.** Slope $-1$ and passing through the point $(4, 3)$

**37.** Horizontal and passing through the point $(1.5, -4)$

**38.** Horizontal and passing through the point $(\frac{1}{2}, \frac{3}{4})$

**39.** Vertical and passing through the point $(1.5, -4)$

**40.** Vertical and passing through the point $(\frac{1}{2}, \frac{3}{4})$

**41.** Passing through the points $(5, 3)$ and $(7, -1)$

**42.** Passing through the points $(3, -1)$ and $(6, 0)$

**43.** Passing through the points $(1, -1)$ and $(5, -1)$

**44.** Passing through the points $(2, 0)$ and $(2, -4)$

**45–48.** Write an equation of the form $y = mx + b$ for each line in the following graphs. [*Hint:* Either find the slope and $y$-intercept or use any two points on the line.]

**45.**

**46.**

**47.**

**48.**
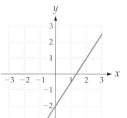

**49–50.** Write equations for the lines determining the four sides of each figure.

**49.**

**50.**

**51.** Show that $y - y_1 = m(x - x_1)$ simplifies to $y = mx + b$ if the point $(x_1, y_1)$ is the $y$-intercept $(0, b)$.

**52.** Show that the linear equation $\dfrac{x}{a} + \dfrac{y}{b} = 1$ has x-intercept $(a, 0)$ and y-intercept $(0, b)$. (The x-intercept is the point where the line crosses the x-axis.)

**53. a.** Graph the lines $y_1 = -x$, $y_2 = -2x$,  and $y_3 = -3x$. on the window $[-5, 5]$ by $[-5, 5]$. Observe how the coefficient of $x$ changes the slope of the line.

*(continues)*

**b.** Predict how the line $y = -9x$ would look, and then check your prediction by graphing it.

**54. a.** Graph the lines $y_1 = x + 2$, $y_2 = x + 1$,  $y_3 = x$, $y_4 = x - 1$, and $y_5 = x - 2$ on the window $[-5, 5]$ by $[-5, 5]$. Observe how the constant changes the position of the line.

**b.** Predict how the lines $y = x + 4$ and $y = x - 4$ would look, and then check your prediction by graphing them.

## Applied Exercises

**55. BUSINESS: Energy Usage** A utility considers demand for electricity "low" if it is below 8 mkW (million kilowatts), "average" if it is at least 8 mkW but below 20 mkW, "high" if it is at least 20 mkW but below 40 mkW, and "critical" if it is 40 mkW or more. Express these demand levels in interval notation. [*Hint:* The interval for "low" is $[0, 8)$.]

**56. GENERAL: Grades** If a grade of 90 through 100 is an A, at least 80 but less than 90 is a B, at least 70 but less than 80 a C, at least 60 but less than 70 a D, and below 60 an F, write these grade levels in interval form (ignoring rounding). [*Hint:* F would be $[0, 60)$.]

**57. ATHLETICS: Mile Run** Read the Application Preview on pages 3–4.

**a.** Use the regression line $y = -0.356x + 257.44$ to predict the world record in the year 2010. [*Hint:* If $x$ represents years after 1900, what value of $x$ corresponds to the year 2010? The resulting $y$ will be in seconds, and should be converted to minutes and seconds.]

**b.** According to this formula, when will the record be 3 minutes 30 seconds? [*Hint:* Set the formula equal to 210 seconds and solve. What year corresponds to this x-value?]

**58. ATHLETICS: Mile Run** Read the Application Preview on pages 3–4. Evaluate the regression line $y = -0.356x + 257.44$ at $x = 720$ and at $x = 722$ (corresponding to the years 2620 and 2622). Does the formula give reasonable times for the mile record in these years? [*Moral:* Linear trends may not continue indefinitely.]

**59. BUSINESS: U.S. Computer Sales** Recently, computer sales in the U.S. have been growing approximately linearly. In 2001 sales were 55.2 million units, and in 2006 sales were 75.7 million units.

**a.** Use the first and last (Year, Sales) data points $(1, 55.2)$ and $(6, 75.7)$ to find the linear relationship $y = mx + b$ between $x = $ Years Since 2000 and $y = $ Sales (in millions).

**b.** Interpret the slope of the line.

**c.** Use the linear relationship to predict sales in the year 2015.

*Source:* Euromonitor International

**60. ECONOMICS: Per Capita Personal Income** In the short run, per capita personal income (PCPI) in the United States grows approximately linearly. In 1990 PCPI was 19.4, and in 2005 it had grown to 34.4 (both in thousands of dollars).

*Source:* Bureau of Economic Analysis

**a.** Use the two given (year, PCPI) data points $(0, 19.4)$ and $(15, 34.4)$ to find the linear relationship $y = mx + b$ between $x = $ years since 1990 and $y = $ PCPI.

**b.** Use your linear relationship to predict PCPI in 2020.

**61. GENERAL: Temperature** On the Fahrenheit temperature scale, water freezes at 32° and boils at 212°. On the Celsius (centigrade) scale, water freezes at 0° and boils at 100°.

**a.** Use the two (Celsius, Fahrenheit) data points $(0, 32)$ and $(100, 212)$ to find the linear relationship $y = mx + b$ between $x = $ Celsius temperature and $y = $ Fahrenheit temperature.

**b.** Find the Fahrenheit temperature that corresponds to 20° Celsius.

**62. BUSINESS: MBA Salaries** Salaries in the United States for new recipients of MBA (master of business administration) degrees have been rising approximately linearly, from $75,200 in 2003 to $92,300 in 2006.

**a.** Use the two (year, salary) data points $(0, 75.2)$ and $(3, 92.3)$ to find the linear relationship $y = mx + b$ between $x = $ years since 2003 and $y = $ salary in thousands of dollars.

*(continues)*

**b.** Use your formula to predict a new MBA's salary in 2015. [*Hint:* Since *x* is years after 2003, what *x*-value corresponds to 2015?]

*Source: Wall Street Journal, College Journal*

**63–64. BUSINESS: Straight-Line Depreciation**
Straight-line depreciation is a method for estimating the value of an asset (such as a piece of machinery) as it loses value ("depreciates") through use. Given the original *price* of an asset, its *useful lifetime*, and its *scrap value* (its value at the end of its useful lifetime), the value of the asset after *t* years is given by the formula:

$$\text{Value} = (\text{Price}) - \left(\frac{(\text{Price}) - (\text{Scrap value})}{(\text{Useful lifetime})}\right) \cdot t$$

$$\text{for} \quad 0 \le t \le (\text{Useful lifetime})$$

**63. a.** A farmer buys a harvester for $50,000 and estimates its useful life to be 20 years, after which its scrap value will be $6000. Use the formula above to find a formula for the value *V* of the harvester after *t* years, for  $0 \le t \le 20$.

**b.** Use your formula to find the value of the harvester after 5 years.

 **c.** Graph the function found in part (a) on a graphing calculator on the window [0, 20] by [0, 50,000]. [*Hint:* Use *x* instead of *t*.]

**64. a.** A newspaper buys a printing press for $800,000 and estimates its useful life to be 20 years, after which its scrap value will be $60,000. Use the formula above Exercise 63 to find a formula for the value *V* of the press after *t* years, for  $0 \le t \le 20$.

**b.** Use your formula to find the value of the press after 10 years.

 **c.** Graph the function found in part (a) on a graphing calculator on the window [0, 20] by [0, 800,000]. [*Hint:* Use *x* instead of *t*.]

 **65. SOCIAL SCIENCE: Age at First Marriage**
Americans are marrying later and later. Based on data for the years 1990 to 2005, the median age at first marriage for men is  $y_1 = 26.3 + 0.058x$, and for women it is  $y_2 = 24 + 0.096x$,  where *x* is the number of years since 1990.

**a.** Graph these lines on the window [0, 40] by [20, 30].

**b.** Use these lines to predict the median marriage ages for men and women in the year 2020. [*Hint:* Which *x*-value corresponds to 2020? Then use TRACE, EVALUATE, or TABLE.]

**c.** Predict the median marriage ages for men and women in the year 2030.

*Source: U.S. Census Bureau*

**66. SOCIAL SCIENCE: Equal Pay for Equal Work**
Women's pay has often lagged behind men's, although Title VII of the Civil Rights Act requires equal pay for equal work. Based on data from 1995–2005, women's annual earnings as a percent of men's can be approximated by the formula  $y = 0.56x + 71.1$,   where *x* is the number of years since 1995. (For example,  $x = 10$  gives  $y = 76.7$,   so in 2005 women's wages were about 76.7% of men's wages.)

**a.** Graph this line on the window [0, 30] by [0, 100].

**b.** Use this line to predict the percentage in the year 2020. [*Hint:* Which *x*-value corresponds to 2020? Then use TRACE, EVALUATE, or TABLE.]

**c.** Predict the percentage in the year 2025.

*Source: U.S. Department of Labor—Women's Bureau*

**67. SOCIAL SCIENCES: Smoking and Income**
Based on a recent study, the probability that someone is a smoker decreases with the person's income. If someone's family income is *x* thousand dollars, then the probability (expressed as a percentage) that the person smokes is approximately  $y = -0.31x + 40$  (for  $10 \le x \le 100$).

**a.** Graph this line on the window [0, 100] by [0, 50].

**b.** What is the probability that a person with a family income of $40,000 is a smoker? [*Hint:* Since *x* is in thousands of dollars, what *x*-value corresponds to $40,000?]

**c.** What is the probability that a person with a family income of $70,000 is a smoker?

Round your answers to the nearest percent.

*Source: Journal of Risk and Uncertainty* **21**(2/3)

**68. ECONOMICS: Does Money Buy Happiness?**
Several surveys in the United States and Europe have asked people to rate their happiness on a scale of  3 = "very happy,"  2 = "fairly happy,"  and  1 = "not too happy,"  and then tried to correlate the answer with the person's income. For those in one income group (making $25,000 to $55,000) it was found that their "happiness" was approximately given by  $y = 0.065x - 0.613$.  Find the reported "happiness" of a person with the following incomes (rounding your answers to one decimal place).

**a.** $25,000  **b.** $35,000  **c.** $45,000

*Source: Review of Economics and Statistics* **85**(4)

**69. BUSINESS: Slot Machine Revenues** The following graph gives the revenues from slot and video poker machines at the Pequot Indians' Mohegan Sun Casino in Connecticut for recent years. (Their accounting is done in fiscal years, running from July to the following June.)

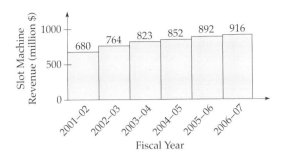

Fiscal Year

a. Number the fiscal years (the bars) with *x*-values 1–6 (so that *x* stands for *years since 2000–01*) and use linear regression to fit a line to the data. State the regression formula. [*Hint:* See Example 8.]

b. Interpret the slope of the line.

c. Use the regression line to predict slot revenues at the casino in the fiscal year 2011–12.

*Source:* Connecticut Division of Special Revenue

 **70. BUSINESS: Satellite Radio Subscribers** The number of satellite radio subscribers in the U.S. for recent quarters is shown in the following table.

| Year | 2005 | 2005 | 2005 | 2005 | 2006 | 2006 | 2006 | 2006 |
|------|------|------|------|------|------|------|------|------|
| Quarter | Q1 | Q2 | Q3 | Q4 | Q1 | Q2 | Q3 | Q4 |
| Subscribers (millions) | 2.2 | 6.3 | 7.2 | 9.1 | 10.4 | 11.5 | 12.1 | 13.8 |

a. Number the data columns with *x*-values 1–8 and use linear regression to fit a line to the data. State the regression formula. [*Hint:* See Example 8.]

b. Interpret the slope of the line. From your answer, what is the *yearly* increase?

c. Use the regression line to predict the number of satellite radio subscribers in the 4th quarter of 2010.

*Source:* Standard & Poor's Industry Surveys

 **71–72. BIOMEDICAL SCIENCES: Life Expectancy** The following tables give the life expectancy for a newborn child born in the indicated year. (Exercise 71 is for males, Exercise 72 for females.)

**71.**

| Birth Year | 1960 | 1970 | 1980 | 1990 | 2000 |
|------------|------|------|------|------|------|
| Life Expectancy (male) | 66.6 | 67.1 | 70.0 | 71.8 | 74.4 |

**72.**

| Birth Year | 1960 | 1970 | 1980 | 1990 | 2000 |
|------------|------|------|------|------|------|
| Life Expectancy (female) | 73.1 | 74.7 | 77.5 | 78.8 | 79.6 |

a. Number the data columns with *x*-values 1–5 and use linear regression to fit a line to the data. State the regression formula. [*Hint:* See Example 8.]

b. Interpret the slope of the line. From your answer, what is the *yearly* change in life expectancy?

c. Use the regression line to predict life expectancy for a child born in 2025 (this might be your child or grandchild).

*Source:* U.S. Census Bureau

 **73. BIOMEDICAL SCIENCES: Future Life Expectancy** Clearly, as people age they should expect to live for fewer additional years. The following graph gives the future life expectancy for Americans of different ages.

Age

a. Using  *x* = Age,  use linear regression to fit a line to the data. State the regression formula. [*Hint:* See Example 8.]

b. Interpret the slope of the line.

c. Use the regression line to estimate future longevity at age 25.

d. Would it make sense to use the regression line to estimate longevity at age 90? What future longevity would the line predict?

*Source:* National Center for Health Statistics

 **74. GENERAL: Seat Belt Use** Because of driver education programs and stricter laws, seat belt use has increased steadily over recent decades. The following table gives the percentage of automobile occupants using seat belts in selected years.

| Year | 1990 | 1995 | 2000 | 2005 |
|------|------|------|------|------|
| Seat Belt Use (%) | 51 | 60 | 71 | 81 |

a. Number the data columns with *x*-values 1–4 and use linear regression to fit a line to the data. State the regression formula. [*Hint:* See Example 8.]

b. Interpret the slope of the line. From your answer, what is the *yearly* increase?

c. Use the regression line to predict seat belt use in 2010.

*(continues)*

**d.** Would it make sense to use the regression line to predict seat belt use in 2015? What percentage would you get?

*Source:* National Highway Traffic Safety Administration

## Conceptual Exercises

**75.** True or False: $\infty$ is the largest number.

**76.** True or False: All negative numbers are smaller than all positive numbers.

**77.** Give two definitions of *slope*.

**78.** Fill in the missing words: If a line slants downward as you go to the right, then its _____ is _____.

**79.** True or False: A vertical line has slope 0.

**80.** True or False: Every line has a slope.

**81.** True or False: Every line can be expressed in the form $ax + by = c$.

**82.** True or False: $x = 3$ is a vertical line.

**83.** A 5-foot-long board is leaning against a wall so that it meets the wall at a point 4 feet above the floor. What is the slope of the board? [*Hint:* Draw a picture.]

**84.** A 5-foot-long ramp is to have a slope of 0.75. How high should the upper end be elevated above the lower end? [*Hint:* Draw a picture.]

## Explorations and Excursions

The following problems extend and augment the material presented in the text.

### More About Linear Equations

**85.** Find the $x$-intercept $(a, 0)$ where the line $y = mx + b$ crosses the $x$-axis. Under what condition on $m$ will a single $x$-intercept exist?

**86.** **i.** Show that the general linear equation $ax + by = c$ with $b \neq 0$ can be written as $y = -\dfrac{a}{b}x + \dfrac{c}{b}$ which is the equation of a line in slope-intercept form.

  **ii.** Show that the general linear equation $ax + by = c$ with $b = 0$ but $a \neq 0$ can be written as $x = \dfrac{c}{a}$, which is the equation of a vertical line.

  [*Note:* Since these steps are reversible, parts (i) and (ii) together show that the general linear equation $ax + by = c$ (for $a$ and $b$ not both zero) includes vertical and nonvertical lines.]

### Beverton-Holt Recruitment Curve

Some organisms exhibit a density-dependent mortality from one generation to the next. Let $R > 1$ be the net reproductive rate (that is, the number of surviving offspring per parent), let $x > 0$ be the density of parents and $y$ be the density of surviving offspring. The *Beverton-Holt recruitment curve* is

$$y = \frac{Rx}{1 + \left(\dfrac{R-1}{K}\right)x}$$

where $K > 0$ is the *carrying capacity* of the environment. Notice that if $x = K$, then $y = K$.

**87.** Show that if $x < K$, then $x < y < K$. Explain what this means about the population size over successive generations if the initial population is smaller than the carrying capacity of the environment.

**88.** Show that if $x > K$, then $K < y < x$. Explain what this means about the population size over successive generations if the initial population is larger than the carrying capacity of the environment.

---

## 1.2  EXPONENTS

### Introduction

Not all variables are related linearly. In this section we will discuss exponents, which will enable us to express many *nonlinear* relationships.

### Positive Integer Exponents

Numbers may be expressed with exponents, as in $2^3 = 2 \cdot 2 \cdot 2 = 8$. More generally, for any positive integer $n$, $x^n$ means the product of $n$ $x$'s.

$$x^n = \overbrace{x \cdot x \cdot \,\cdots\, \cdot x}^{n}$$

The number being raised to the power is called the **base** and the power is the **exponent:**

$$x^n \quad \overset{\text{Exponent or power}}{\underset{\text{Base}}{}}$$

There are several *properties of exponents* for simplifying expressions. The first three are known, respectively, as the addition, subtraction, and multiplication properties of exponents.

| Properties of Exponents | | Brief Examples |
|---|---|---|
| $x^m \cdot x^n = x^{m+n}$ | To *multiply* powers of the same base, *add* the exponents | $x^2 \cdot x^3 = x^5$ |
| $\dfrac{x^m}{x^n} = x^{m-n}$ | To *divide* powers of the same base, *subtract* the exponents (top exponent minus bottom exponent) | $\dfrac{x^5}{x^3} = x^2$ |
| $(x^m)^n = x^{m \cdot n}$ | To raise a power to a power, *multiply* the powers | $(x^2)^3 = x^6$ |
| $(xy)^n = x^n \cdot y^n$ | To raise a product to a power, raise *each factor* to the power | $(2x)^3 = 2^3 \cdot x^3 = 8x^3$ |
| $\left(\dfrac{x}{y}\right)^n = \dfrac{x^n}{y^n}$ | To raise a fraction to a power, raise the numerator *and* denominator to the power | $\left(\dfrac{x}{5}\right)^3 = \dfrac{x^3}{5^3} = \dfrac{x^3}{125}$ |

**Practice Problem 1**   Simplify:   **a.** $\dfrac{x^5 \cdot x}{x^2}$   **b.** $[(x^3)^2]^2$   ➤ Solutions on page 28

Remember:  For exponents in the form $\ x^2 \cdot x^3 = x^5, \ $ *add* exponents.
For exponents in the form $\ (x^2)^3 = x^6, \ $ *multiply* exponents.

### Zero and Negative Exponents

Any number except zero can be raised to a negative integer or zero power.

| Zero and Negative Integer Exponents | | Brief Examples |
|---|---|---|
| For $\ x \neq 0$ | | |
| $x^0 = 1$ | $x$ to the power 0 is 1 | $5^0 = 1$ |
| $x^{-1} = \dfrac{1}{x}$ | $x$ to the power $-1$ is one over $x$ | $7^{-1} = \dfrac{1}{7}$ |
| $x^{-2} = \dfrac{1}{x^2}$ | $x$ to the power $-2$ is one over $x$ squared | $3^{-2} = \dfrac{1}{3^2} = \dfrac{1}{9}$ |
| $x^{-n} = \dfrac{1}{x^n}$ | $x$ to a negative power is one over $x$ to the positive power | $(-2)^{-3} = \dfrac{1}{(-2)^3} = \dfrac{1}{-8} = -\dfrac{1}{8}$ |

Note that $0^0$ and $0^{-3}$ are *undefined.*

The definitions of $x^0$ and $x^{-n}$ are motivated by the following calculations.

$$1 = \frac{x^2}{x^2} = x^{2-2} = x^0$$

The subtraction property of exponents leads to $x^0 = 1$

$$\frac{1}{x^n} = \frac{x^0}{x^n} = x^{0-n} = x^{-n}$$

$x^0 = 1$ and the subtraction property of exponents lead to $x^{-n} = \frac{1}{x^n}$

**Practice Problem 2**   Evaluate:   **a.** $2^0$     **b.** $2^{-4}$                    ➤ Solutions on page 28

A fraction to a negative power means *division* by the fraction, so we "invert and multiply."

$$\left(\frac{x}{y}\right)^{-1} = \frac{1}{\frac{x}{y}} = 1 \cdot \frac{y}{x} = \frac{y}{x}$$

Reciprocal of the original fraction

Therefore, for   $x \neq 0$   and   $y \neq 0$,

$$\left(\frac{x}{y}\right)^{-1} = \frac{y}{x}$$

A fraction to the power $-1$ is the reciprocal of the fraction

$$\left(\frac{x}{y}\right)^{-n} = \left(\frac{y}{x}\right)^n$$

A fraction to the negative power is the reciprocal of the fraction to the positive power

**EXAMPLE 1**        **SIMPLIFYING FRACTIONS TO NEGATIVE EXPONENTS**

**a.** $\left(\dfrac{3}{2}\right)^{-1} = \dfrac{2}{3}$          **b.** $\left(\dfrac{1}{2}\right)^{-3} = \left(\dfrac{2}{1}\right)^3 = \dfrac{2^3}{1^3} = 8$

Reciprocal of the original

**Practice Problem 3**   Simplify: $\left(\dfrac{2}{3}\right)^{-2}$                    ➤ Solution on page 28

## Roots and Fractional Exponents

We may take the square root of any *nonnegative* number, and the cube root of *any* number.

**EXAMPLE 2**

**EVALUATING ROOTS**

**a.** $\sqrt{9} = 3$     **b.** $\sqrt{-9}$ is undefined.     Square roots of negative numbers are not defined

**c.** $\sqrt[3]{8} = 2$     **d.** $\sqrt[3]{-8} = -2$     Cube roots of negative numbers *are* defined

**e.** $\sqrt[3]{\dfrac{27}{8}} = \dfrac{\sqrt[3]{27}}{\sqrt[3]{8}} = \dfrac{3}{2}$

There are *two* square roots of 9, namely 3 and $-3$, but the radical sign $\sqrt{\phantom{x}}$ means just the *positive* one (the "principal" square root).

$\sqrt[n]{a}$ means the principal *n*th root of *a*.     Principal means the positive root if there are two

In general, we may take *odd* roots of *any* number, but *even* roots only if the number is positive or zero.

**EXAMPLE 3**

**EVALUATING ROOTS OF POSITIVE AND NEGATIVE NUMBERS**

Odd roots of negative numbers *are* defined

**a.** $\sqrt[4]{81} = 3$     **b.** $\sqrt[5]{-32} = -2$     Since $(-2)^5 = -32$

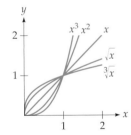

The diagram on the left shows the graphs of some powers and roots of *x* for $x > 0$. Which of these would *not* be defined for $x < 0$? [*Hint:* See Example 2.]

## Fractional Exponents

Fractional exponents are defined as follows:

| Powers of the Form $\frac{1}{n}$ | | *Brief Examples* |
|---|---|---|
| $x^{\frac{1}{2}} = \sqrt{x}$ | Power $\frac{1}{2}$ means the principal square root | $9^{\frac{1}{2}} = \sqrt{9} = 3$ |
| $x^{\frac{1}{3}} = \sqrt[3]{x}$ | Power $\frac{1}{3}$ means the cube root | $125^{\frac{1}{3}} = \sqrt[3]{125} = 5$ |
| $x^{\frac{1}{n}} = \sqrt[n]{x}$ | Power $\frac{1}{n}$ means the principal *n*th root (for a positive integer *n*) | $(-32)^{\frac{1}{5}} = \sqrt[5]{-32} = -2$ |

The definition of $x^{\frac{1}{2}}$ is motivated by the multiplication property of exponents:

$$\left(x^{\frac{1}{2}}\right)^2 = x^{\frac{1}{2} \cdot 2} = x^1 = x$$

Taking square roots of each side of $\left(x^{\frac{1}{2}}\right)^2 = x$ gives

$$x^{\frac{1}{2}} = \sqrt{x}$$

*$x$ to the half power means the square root of $x$*

---

**EXAMPLE 4**

**EVALUATING FRACTIONAL EXPONENTS**

**a.** $\left(\dfrac{4}{25}\right)^{\frac{1}{2}} = \sqrt{\dfrac{4}{25}} = \dfrac{\sqrt{4}}{\sqrt{25}} = \dfrac{2}{5}$

**b.** $\left(-\dfrac{27}{8}\right)^{\frac{1}{3}} = \sqrt[3]{-\dfrac{27}{8}} = -\dfrac{\sqrt[3]{27}}{\sqrt[3]{8}} = -\dfrac{3}{2}$

---

**Practice Problem 4**

Evaluate: **a.** $(-27)^{\frac{1}{3}}$ **b.** $\left(\dfrac{16}{81}\right)^{\frac{1}{4}}$

➤ **Solutions on page 29**

To define $x^{\frac{m}{n}}$ for positive integers $m$ and $n$, the exponent $\frac{m}{n}$ must be fully reduced (for example, $\frac{4}{6}$ must be reduced to $\frac{2}{3}$). Then

$$x^{\frac{m}{n}} = \left(x^{\frac{1}{n}}\right)^m = \left(x^m\right)^{\frac{1}{n}}$$

*Since in both cases the exponents multiply to $\frac{m}{n}$*

Therefore, we define:

---

**Fractional Exponents**

$$x^{\frac{m}{n}} = \left(\sqrt[n]{x}\right)^m = \sqrt[n]{x^m}$$

*$x^{m/n}$ means the $m$th power of the $n$th root, or equivalently, the $n$th root of the $m$th power*

---

Both expressions, $\left(\sqrt[n]{x}\right)^m$ and $\sqrt[n]{x^m}$, will give the same answer. In either case *the numerator determines the power* and *the denominator determines the root*.

$x^{\frac{m}{n}}$  ⟵ Power  ⟵ Root     Power over root

**EXAMPLE 5**

**EVALUATING FRACTIONAL EXPONENTS**

**a.** $8^{2/3} = \sqrt[3]{8^2} = \sqrt[3]{64} = 4$  —— same

First the power, then the root

**b.** $8^{2/3} = \left(\sqrt[3]{8}\right)^2 = (2)^2 = 4$

First the root, then the power

**c.** $25^{3/2} = \left(\sqrt{25}\right)^3 = (5)^3 = 125$

**d.** $\left(\dfrac{-27}{8}\right)^{2/3} = \left(\sqrt[3]{\dfrac{-27}{8}}\right)^2 = \left(\dfrac{-3}{2}\right)^2 = \dfrac{9}{4}$

**Practice Problem 5**

Evaluate: **a.** $16^{3/2}$   **b.** $(-8)^{2/3}$

➤ Solutions on page 29

**IMPORTANT NOTE:** The Graphing Calculator Explorations are optional (unless your instructor says otherwise). At a minimum, looking carefully at the graphs and reading the explanations provide useful information about the mathematics being discussed.

## Graphing Calculator Exploration

**a.** Use a graphing calculator to evaluate $25^{3/2}$. [On some calculators, press 25^(3 ÷ 2).] Your answer should agree with Example 5c above.

**b.** Evaluate $(-8)^{2/3}$. Use the ⎡(−)⎤ key for negation, and parentheses around the exponent. Your answer should be 4. If you get an "error," try evaluating the expression as $[(-8)^{1/3}]^2$ or $[(-8)^2]^{1/3}$. Whichever way works, remember it for evaluating negative numbers to fractional powers in the future.

**EXAMPLE 6**

**EVALUATING NEGATIVE FRACTIONAL EXPONENTS**

**a.** $8^{-2/3} = \dfrac{1}{8^{2/3}} = \dfrac{1}{\left(\sqrt[3]{8}\right)^2} = \dfrac{1}{2^2} = \dfrac{1}{4}$

A negative exponent means the reciprocal of the number to the positive exponent, which is then evaluated as before

**b.** $\left(\dfrac{9}{4}\right)^{-3/2} = \left(\dfrac{4}{9}\right)^{3/2} = \left(\sqrt{\dfrac{4}{9}}\right)^3 = \left(\dfrac{2}{3}\right)^3 = \dfrac{8}{27}$

Interpreting the power 3/2
Reciprocal to the positive exponent
Negative exponent

| Practice Problem 6 | Evaluate: **a.** $25^{-3/2}$   **b.** $\left(\frac{1}{4}\right)^{-1/2}$   **c.** $5^{1.3}$   [*Hint:* Use a calculator.] |

➤ **Solutions on page 29**

 **BE CAREFUL:** While the square root of a product *is* equal to the product of the square roots,

$$\sqrt{a \cdot b} = \sqrt{a} \cdot \sqrt{b}$$

the corresponding statement for *sums* is *not* true:

$$\sqrt{a + b} \quad \text{is } not \text{ equal to} \quad \sqrt{a} + \sqrt{b}$$

For example,

$$\underbrace{\sqrt{9 + 16}}_{\sqrt{25}} \neq \underbrace{\sqrt{9}}_{3} + \underbrace{\sqrt{16}}_{4}$$

The two sides are not equal: one is 5 and the other is 7

Therefore, do not "simplify" $\sqrt{x^2 + 9}$ into $x + 3$. The expression $\sqrt{x^2 + 9}$ *cannot be simplified.* Similarly,

$$(x + y)^2 \quad \text{is } not \text{ equal to} \quad x^2 + y^2$$

The expression $(x + y)^2$ means $(x + y)$ times itself:

$$(x + y)^2 = (x + y)(x + y) = x^2 + xy + yx + y^2 = x^2 + 2xy + y^2$$

This result is worth remembering, since we will use it frequently in Chapter 2.

$$(x + y)^2 = x^2 + 2xy + y^2$$

$(x + y)^2$ is the first number squared plus twice the product of the numbers plus the second number squared

## Learning Curves in Airplane Production

It is a truism that the more you practice a task, the faster you can do it. Successive repetitions generally take less time, following a "learning curve" like that on the left. Learning curves are used in industrial production. For example, it took 150,000 work-hours to build the first Boeing 707 airliner, while later planes $(n = 2, 3, \ldots, 300)$ took less time.*

$$\left(\begin{array}{c}\text{Time to build} \\ \text{plane number } n\end{array}\right) = 150\, n^{-0.322} \text{ thousand work-hours}$$

---

*A work-hour is the amount of work that a person can do in 1 hour. For further information on learning curves in industrial production, see J. M. Dutton *et al.,* "The History of Progress Functions as a Managerial Technology," *Business History Review* **58**.

The time for the 10th Boeing 707 is found by substituting $n = 10$:

$$\begin{pmatrix} \text{Time to build} \\ \text{plane } 10 \end{pmatrix} = 150(10)^{-0.322}$$

$150n^{-0.322}$
with   $n = 10$

$$\approx 71.46 \text{ thousand work-hours}$$   Using a calculator

This shows that building the 10th Boeing 707 took about 71,460 work-hours, which is less than half of the 150,000 work-hours needed for the first. For the 100th 707:

$$\begin{pmatrix} \text{Time to build} \\ \text{plane } 100 \end{pmatrix} = 150(100)^{-0.322}$$

$150n^{-0.322}$
with   $n = 100$

$$\approx 34.05 \text{ thousand work-hours}$$

or about 34,050 work-hours, which is less than the half time needed to build the 10th. Such learning curves are used for determining the cost of a contract to build several planes.

Notice that the learning curve graphed on the previous page decreases less steeply as the number of repetitions increases. This means that while construction time continues to decrease, it does so more slowly for later planes. This behavior, called **diminishing returns,** is typical of learning curves.

## Power Regression (Optional)

Just as we used *linear* regression to fit a *line* to data points, we can use **power regression** to fit a *power curve* like those shown on page 23 to data points. The procedure is easily accomplished using a graphing calculator (or spreadsheet or other computer software), as in the following example.

When do you use power regression instead of linear regression (or some other type)? You should look at a graph of the data and see if it lies more along a *curve* like those shown on page 23 rather than along a line. Furthermore, sometimes there are *theoretical* reasons to prefer a curve. For example, sales of a product may increase linearly for a short time, but then usually grow more slowly because of market saturation or competition, and so are best modeled by a curve.

**EXAMPLE 7**    **POWER REGRESSION USING A GRAPHING CALCULATOR**

The following graph gives the total annual U.S. sales of digital mobile telephones from 2003 through 2007.

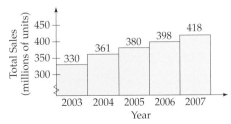

*Source:* Euromonitor International

**a.** Use power regression to fit a power curve to the data and state the regression formula.

**b.** Use the regression formula to predict digital mobile telephone sales in 2012.

*Solution*

**a.** We number the years with $x$-values 1–5, so $x$ stands for *years since 2002* (we could choose other $x$-values instead, but with power regression the $x$-values must all be *positive*). We enter the data into lists, as shown in the first screen below (as explained in *Graphing Calculator Basics—Entering Data* on page xxiii) and use *ZoomStat* to graph the data points.

Then (using STAT, CALC, and PwrReg), we graph the regression curve along with the data points.

The regression curve, which fits the points reasonably well, is

$$y = 328x^{0.143} \qquad \text{Rounded}$$

**b.** To predict sales in 2012, we evaluate Y1 at 10 (since $x = 10$ corresponds to 2012). From the screen on the left, if the current trend continues, digital mobile telephone sales in 2012 will be approximately 456 million units.

---

**Solutions to Practice Problems**

**1. a.** $\dfrac{x^5 \cdot x}{x^2} = \dfrac{x^6}{x^2} = x^4$

   **b.** $[(x^3)^2]^2 = x^{3 \cdot 2 \cdot 2} = x^{12}$

**2. a.** $2^0 = 1$

   **b.** $2^{-4} = \dfrac{1}{2^4} = \dfrac{1}{16}$

**3.** $\left(\dfrac{2}{3}\right)^{-2} = \left(\dfrac{3}{2}\right)^2 = \dfrac{9}{4}$

**4. a.** $(-27)^{1/3} = \sqrt[3]{-27} = -3$

    **b.** $\left(\dfrac{16}{81}\right)^{1/4} = \sqrt[4]{\dfrac{16}{81}} = \dfrac{\sqrt[4]{16}}{\sqrt[4]{81}} = \dfrac{2}{3}$

**5. a.** $16^{3/2} = \left(\sqrt{16}\right)^3 = 4^3 = 64$

    **b.** $(-8)^{2/3} = \left(\sqrt[3]{-8}\right)^2 = (-2)^2 = 4$

**6. a.** $25^{-3/2} = \dfrac{1}{25^{3/2}} = \dfrac{1}{\left(\sqrt{25}\right)^3} = \dfrac{1}{5^3} = \dfrac{1}{125}$

    **b.** $\left(\dfrac{1}{4}\right)^{-1/2} = \left(\dfrac{4}{1}\right)^{1/2} = \sqrt{4} = 2$      **c.** $5^{1.3} \approx 8.103$

## 1.2 Section Summary

We defined zero, negative, and fractional exponents as follows:

$$x^0 = 1 \qquad\qquad \text{for } x \neq 0$$

$$x^{-n} = \frac{1}{x^n} \qquad\qquad \text{for } x \neq 0$$

$$x^{\frac{m}{n}} = \left(\sqrt[n]{x}\right)^m = \sqrt[n]{x^m} \qquad m > 0, \quad n > 0, \quad \frac{m}{n} \text{ fully reduced}$$

With these definitions, the following properties of exponents hold for *all* exponents, whether integral or fractional, positive or negative.

$$x^m \cdot x^n = x^{m+n} \qquad (x^m)^n = x^{m \cdot n} \qquad \left(\frac{x}{y}\right)^n = \frac{x^n}{y^n}$$

$$\frac{x^m}{x^n} = x^{m-n} \qquad (xy)^n = x^n \cdot y^n$$

## 1.2 Exercises

**1–48.** Evaluate each expression *without* using a calculator.

**1.** $(2^2 \cdot 2)^2$

**2.** $(5^2 \cdot 4)^2$

**3.** $2^{-4}$

**4.** $3^{-3}$

**5.** $\left(\dfrac{1}{2}\right)^{-3}$

**6.** $\left(\dfrac{1}{3}\right)^{-2}$

**7.** $\left(\dfrac{5}{8}\right)^{-1}$

**8.** $\left(\dfrac{3}{4}\right)^{-1}$

**9.** $4^{-2} \cdot 2^{-1}$

**10.** $3^{-2} \cdot 9^{-1}$

**11.** $\left(\dfrac{3}{2}\right)^{-3}$

**12.** $\left(\dfrac{2}{3}\right)^{-3}$

**13.** $\left(\dfrac{1}{3}\right)^{-2} - \left(\dfrac{1}{2}\right)^{-3}$

**14.** $\left(\dfrac{1}{3}\right)^{-2} - \left(\dfrac{1}{2}\right)^{-2}$

**15.** $\left[\left(\dfrac{2}{3}\right)^{-2}\right]^{-1}$

**16.** $\left[\left(\dfrac{2}{5}\right)^{-2}\right]^{-1}$

**17.** $25^{1/2}$

**18.** $36^{1/2}$

**19.** $25^{3/2}$

**20.** $16^{3/2}$

**21.** $16^{3/4}$

**22.** $27^{2/3}$

**23.** $(-8)^{2/3}$

**24.** $(-27)^{2/3}$

**25.** $(-8)^{5/3}$

**26.** $(-27)^{5/3}$   **27.** $\left(\dfrac{25}{36}\right)^{3/2}$   **28.** $\left(\dfrac{16}{25}\right)^{3/2}$

**29.** $\left(\dfrac{27}{125}\right)^{2/3}$   **30.** $\left(\dfrac{125}{8}\right)^{2/3}$   **31.** $\left(\dfrac{1}{32}\right)^{2/5}$

**32.** $\left(\dfrac{1}{32}\right)^{3/5}$   **33.** $4^{-1/2}$   **34.** $9^{-1/2}$

**35.** $4^{-3/2}$   **36.** $9^{-3/2}$   **37.** $8^{-2/3}$

**38.** $16^{-3/4}$   **39.** $(-8)^{-1/3}$   **40.** $(-27)^{-1/3}$

**41.** $(-8)^{-2/3}$   **42.** $(-27)^{-2/3}$   **43.** $\left(\dfrac{25}{16}\right)^{-1/2}$

**44.** $\left(\dfrac{16}{9}\right)^{-1/2}$   **45.** $\left(\dfrac{25}{16}\right)^{-3/2}$   **46.** $\left(\dfrac{16}{9}\right)^{-3/2}$

**47.** $\left(-\dfrac{1}{27}\right)^{-5/3}$   **48.** $\left(-\dfrac{1}{8}\right)^{-5/3}$

**49–52.** Use a calculator to evaluate each expression. Round answers to two decimal places.

**49.** $7^{0.39}$   **50.** $5^{0.47}$   **51.** $8^{2.7}$   **52.** $5^{3.9}$

**53–56.** Use a graphing calculator to evaluate each expression.

**53.** $[(0.1)^{0.1}]^{0.1}$   **54.** $\left(1 + \dfrac{1}{1000}\right)^{1000}$

**55.** $\left(1 - \dfrac{1}{1000}\right)^{-1000}$   **56.** $(1 + 10^{-6})^{10^6}$

**57–64.** Write each expression in power form $ax^b$ for numbers $a$ and $b$.

**57.** $\dfrac{4}{x^5}$   **58.** $\dfrac{6}{2x^3}$

**59.** $\dfrac{4}{\sqrt[3]{8x^4}}$   **60.** $\dfrac{6}{\sqrt{4x^3}}$

**61.** $\dfrac{24}{(2\sqrt{x})^3}$   **62.** $\dfrac{18}{(3\sqrt[3]{x})^2}$

**63.** $\sqrt{\dfrac{9}{x^4}}$   **64.** $\sqrt[3]{\dfrac{8}{x^6}}$

Simplify.

**65.** $(x^3 \cdot x^2)^2$   **66.** $(x^4 \cdot x^3)^2$   **67.** $[z^2(z \cdot z^2)^2 z]^3$

**68.** $[z(z^3 \cdot z)^2 z^2]^2$   **69.** $[(x^2)^2]^2$   **70.** $[(x^3)^3]^3$

**71.** $\dfrac{(ww^2)^3}{w^3 w}$   **72.** $\dfrac{(ww^3)^2}{w^3 w^2}$   **73.** $\dfrac{(5xy^4)^2}{25x^3 y^3}$

**74.** $\dfrac{(4x^3 y)^2}{8x^2 y^3}$   **75.** $\dfrac{(9xy^3 z)^2}{3(xyz)^2}$   **76.** $\dfrac{(5x^2 y^3 z)^2}{5(xyz)^2}$

**77.** $\dfrac{(2u^2 vw^3)^2}{4(uw^2)^2}$   **78.** $\dfrac{(u^3 vw^2)^2}{9(u^2 w)^2}$

## Applied Exercises

**79–80.** **ALLOMETRY: Dinosaurs** The study of size and shape is called "allometry," and many allometric relationships involve exponents that are fractions or decimals. For example, the body measurements of most four-legged animals, from mice to elephants, obey (approximately) the following power law:

$$\left(\begin{array}{c}\text{Average body} \\ \text{thickness}\end{array}\right) = 0.4 \,(\text{hip-to-shoulder length})^{3/2}$$

where body thickness is measured vertically and all measurements are in feet. Assuming that this same relationship held for dinosaurs, find the average body thickness of the following dinosaurs, whose hip-to-shoulder length can be measured from their skeletons:

**79.** Diplodocus, whose hip-to-shoulder length was 16 feet.

**80.** Triceratops, whose hip-to-shoulder length was 14 feet.

**81–82.** **BUSINESS: The Rule of .6** Many chemical and refining companies use "the rule of point six" to estimate the cost of new equipment. According to this rule, if a piece of equipment (such as a storage tank) originally cost $C$ dollars, then the cost of similar equipment that is $x$ times as large will be approximately $x^{0.6}C$ dollars. For example, if the original equipment cost $C$ dollars, then new equipment with twice the capacity of the old equipment $(x = 2)$ would cost

$2^{0.6}C = 1.516C$   dollars—that is, about 1.5 times as much. Therefore, to increase capacity by 100% costs only about 50% more.*

**81.** Use the rule of .6 to find how costs change if a company wants to quadruple  $(x = 4)$  its capacity.

**82.** Use the rule of .6 to find how costs change if a company wants to triple  $(x = 3)$  its capacity.

**83–84. ALLOMETRY: Heart Rate**  It is well known that the hearts of smaller animals beat faster than the hearts of larger animals. The actual relationship is approximately

$$(\text{Heart rate}) = 250(\text{Weight})^{-1/4}$$

where the heart rate is in beats per minute and the weight is in pounds. Use this relationship to estimate the heart rate of:

**83.** A 16-pound dog.

**84.** A 625-pound grizzly bear.
*Source: Biology Review 41*

**85–86. BUSINESS: Learning Curves in Airplane Production**  Recall (pages 26–27) that the learning curve for the production of Boeing 707 airplanes is $150n^{-0.322}$ (thousand work-hours). Find how many work-hours it took to build:

**85.** The 50th Boeing 707.

**86.** The 250th Boeing 707.

**87. GENERAL: Richter Scale**  The Richter scale (developed by Charles Richter in 1935) is widely used to measure the strength of earthquakes. Every increase of 1 on the Richter scale corresponds to a 10-fold increase in ground motion. Therefore, an increase on the Richter scale from $A$ to $B$ means that ground motion increases by a factor of $10^{B-A}$ (for $B > A$). Find the increase in ground motion between the following earthquakes:

**a.** The 1994 Northridge, California, earthquake, measuring 6.8 on the Richter scale, and the 1906 San Francisco earthquake, measuring 8.3. (The San Francisco earthquake resulted in 500 deaths and a 3-day fire that destroyed 4 square miles of San Francisco.)

*(continues)*

*Although the rule of .6 is only a rough "rule of thumb," it can be somewhat justified on the basis that the equipment of such industries consists mainly of containers, and the cost of a container depends on its surface area (square units), which increases more slowly than its capacity (cubic units).

**b.** The 2004 earthquake near Sumatra (Indonesia), measuring 9.0 on the Richter scale, and the 2008 Sichuan (China) earthquake, measuring 7.9. (The Sumatra earthquake caused a 50-foot-high tsunami, or "tidal wave," that killed 170,000 people in 11 countries. The death toll from the Sichuan earthquake was more than 70,000.)

**88. GENERAL: Richter Scale** (*continuation*)  Every increase of 1 on the Richter scale corresponds to an approximately *30-fold* increase in *energy released*. Therefore, an increase on the Richter scale from $A$ to $B$ means that the energy released increases by a factor of $30^{B-A}$ (for $B > A$).

**a.** Find the increase in *energy released* between the earthquakes in Exercise 87a.

**b.** Find the increase in *energy released* between the earthquakes in Exercise 87b.

**89–90. GENERAL: Waterfalls**  Water falling from a waterfall that is $x$ feet high will hit the ground with speed  $\frac{60}{11}x^{0.5}$  miles per hour (neglecting air resistance).

**89.** Find the speed of the water at the bottom of the highest waterfall in the world, Angel Falls in Venezuela (3281 feet high).

**90.** Find the speed of the water at the bottom of the highest waterfall in the United States, Ribbon Falls in Yosemite, California (1650 feet high).

**91–92. ENVIRONMENTAL SCIENCE: Biodiversity**  It is well known that larger land areas can support larger numbers of species. According to one study, multiplying the land area by a factor of $x$ multiplies the number of species by a factor of $x^{0.239}$. Use a graphing calculator to graph  $y = x^{0.239}$.  Use the window [0, 100] by [0, 4].
*Source: Robert H. MacArthur and Edward O. Wilson, The Theory of Island Biogeography*

**91.** Find the multiple $x$ for the land area that leads to *double* the number of species. That is, find the value of $x$ such that  $x^{0.239} = 2$. [*Hint:* Either use TRACE or find where $y_1 = x^{0.239}$  INTERSECTs  $y_2 = 2$.]

**92.** Find the multiple $x$ for the land area that leads to *triple* the number of species. That is, find the value of $x$ such that  $x^{0.239} = 3$. [*Hint:* Either use TRACE or find where $y_1 = x^{0.239}$  INTERSECTs  $y_2 = 3$.]

**93–94. GENERAL: Speed and Skidmarks** Police or insurance investigators often want to estimate the speed of a car from the skidmarks it left while stopping. A study found that for standard tires on dry asphalt, the speed (in mph) is given approximately by $y = 9.4x^{0.37}$, where $x$ is the length of the skidmarks in feet. (This formula takes into account the deceleration that occurs even *before* the car begins to skid.) Estimate the speed of a car if it left skidmarks of:

**93.** 150 feet.      **94.** 350 feet.

*Source: Accident Analysis and Prevention* 36

 **95. BUSINESS: MP3 Players** MP3 music players have become increasingly popular in recent years, with the total annual sales of in-home units shown in the following graph.

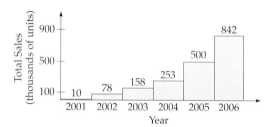

**a.** Number the data bars with $x$-values 1–6 (so that $x$ stands for *years since 2000*), use power regression to fit a power curve to the data, and state the regression formula. [*Hint:* See Example 7.]

**b.** Use the regression formula to predict in-home MP3 sales in the year 2012. [*Hint:* What $x$-value corresponds to 2012?]

*Source: Consumer USA 2008*

 **96. BUSINESS: Mobile Phones** Total mobile phone sales in recent years are given in the following table.

| Year | 2002 | 2003 | 2004 | 2005 | 2006 | 2007 |
|---|---|---|---|---|---|---|
| Annual Sales (millions of units) | 57 | 68.4 | 82 | 101.4 | 113.9 | 123.9 |

**a.** Number the data columns with $x$-values 1–6 (so that $x$ stands for *years since 2001*), use power regression to fit a power curve to the data, and state the regression formula. [*Hint:* See Example 7.]

**b.** Use the regression formula to predict mobile phone sales in the year 2012. [*Hint:* What $x$-value corresponds to 2012?]

*Source: Consumer USA 2008*

 **97. BUSINESS: DVD Sales** Total sales of pre-recorded video DVDs in recent years are given in the following table.

| Year | 2002 | 2003 | 2004 | 2005 | 2006 |
|---|---|---|---|---|---|
| Annual Sales (millions of units) | 619 | 879 | 1212 | 1293 | 1325 |

**a.** Number the data columns with $x$-values 1–5 (so that $x$ stands for *years since 2001*), use power regression to fit a power curve to the data, and state the regression formula. [*Hint:* See Example 7.]

**b.** Use the regression formula to predict DVD sales in the year 2012.

*Source: Standard & Poor's Industry Surveys*

 **98. BUSINESS: Airline Miles** The number of revenue passenger miles flown by JetBlue Airlines in recent years is given in the following table.

| Year | 2003 | 2004 | 2005 | 2006 |
|---|---|---|---|---|
| Passenger Miles Flown (millions) | 11.5 | 15.7 | 20.2 | 23.3 |

**a.** Number the data columns with $x$-values 1–4 (so that $x$ stands for *years since 2002*), use power regression to fit a power curve to the data, and state the regression formula. [*Hint:* See Example 7.]

**b.** Use the regression formula to predict JetBlue's revenue passenger miles in the year 2012.

*Source: Standard & Poor's Industry Surveys*

### Conceptual Exercises

**99.** Should $\sqrt{9}$ be evaluated as 3 or $\pm 3$?

**100–102.** For each statement, either state that it is True (and find a property in the text that shows this) or state that it is False (and give an example to show this).

**100.** $x^m \cdot x^n = x^{m \cdot n}$    **101.** $\dfrac{x^m}{x^n} = x^{m/n}$    **102.** $(x^m)^n = x^{m^n}$

**103–105.** For each statement, state *in words* the values of $x$ for which each exponential expression is defined.

**103.** $x^{1/2}$      **104.** $x^{1/3}$      **105.** $x^{-1}$

**106.** When defining $x^{m/n}$, why did we require that the exponent $\frac{m}{n}$ be fully reduced?

[*Hint:* $(-1)^{2/3} = \left(\sqrt[3]{-1}\right)^2 = 1$, but with an equal but *unreduced* exponent you get $(-1)^{4/6} = \left(\sqrt[6]{-1}\right)^4$. Is this defined?]

## 1.3 FUNCTIONS: LINEAR AND QUADRATIC

### Introduction

In the previous section we saw that the time required to build a Boeing 707 airliner varies, depending on the number that have already been built. Mathematical relationships such as this, in which one number depends on another, are called *functions*, and are central to the study of calculus. In this section we define and give some applications of functions.

### Functions

A **function** is a rule or procedure for finding, from a given number, a new number.[*] If the function is denoted by $f$ and the given number by $x$, then the resulting number is written $f(x)$ (read "$f$ of $x$") and is called *the value of the function $f$ at $x$*. The set of numbers $x$ for which a function $f$ is defined is called the **domain** of $f$, and the set of all resulting function values $f(x)$ is called the **range** of $f$. For any $x$ in the domain, $f(x)$ must be a *single* number.

---

**Function**

A *function* $f$ is a rule that assigns to each number $x$ in a set a number $f(x)$. The set of all allowable values of $x$ is called the *domain*, and the set of all values $f(x)$ for $x$ in the domain is called the *range*.

---

For example, recording the temperature at a given location throughout a particular day would define a *temperature* function:

$$f(x) = \begin{pmatrix} \text{Temperature at} \\ \text{time } x \text{ hours} \end{pmatrix} \qquad \text{Domain would be } [0, 24)$$

Input $x$

Function $f$

Output $f(x)$

A function $f$ may be thought of as a numerical procedure or "machine" that takes an "input" number $x$ and produces an "output" number $f(x)$, as shown on the left. The permissible input numbers form the domain, and the resulting output numbers form the range.

We will be mostly concerned with functions that are defined by *formulas* for calculating $f(x)$ from $x$. If the domain of such a function is not stated, then it is always taken to be the *largest* set of numbers for which the function is defined, called the **natural domain** of the function. To **graph** a function $f$, we plot all points $(x, y)$ such that $x$ is in the domain and $y = f(x)$. We call $x$ the **independent variable** and $y$ the **dependent variable**, since $y$ *depends on* (is calculated from) $x$. The domain and range can be illustrated graphically.

---

[*]In this chapter the word "function" will mean *function of one variable*. In Chapter 7 we will discuss functions of more than one variable.

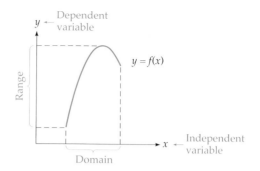

The domain of a function $y = f(x)$ is the
set of all allowable $x$-values, and the range
is the set of all corresponding $y$-values.

**Practice Problem 1**   Find the domain and range of the function graphed below.

➤ Solution on page 44

**EXAMPLE 1**

**FINDING THE DOMAIN AND RANGE**

For the function $f(x) = \dfrac{1}{x-1}$, find:

**a.** $f(5)$     **b.** the domain     **c.** the range

*Solution*

**a.** $f(5) = \dfrac{1}{5-1} = \dfrac{1}{4}$     $f(x) = \dfrac{1}{x-1}$ with $x = 5$

**b.** Domain = $\{x \mid x \neq 1\}$     $f(x) = \dfrac{1}{x-1}$ is defined for
all $x$ except $x = 1$ (since
we can't divide by zero)

 **c.** The graph of the function (from a
graphing calculator) is shown on the
right. From it, and realizing that the
curve continues upward and downward (as may be verified by zooming
out), it is clear that *every* $y$-value is taken
except for $y = 0$ (since the curve does
not touch the $x$-axis). Therefore:

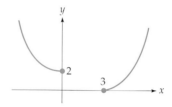

$f(x) = \dfrac{1}{x-1}$ on $[-5, 5]$ by $[-5, 5]$

Range = $\{y \mid y \neq 0\}$     May also be written $\{z \mid z \neq 0\}$
or with any other letter

The range could also be found by solving $y = \dfrac{1}{x-1}$ for $x$, giving

$x = \dfrac{1}{y} + 1$, which again shows that $y$ can take any value except $0$.

**EXAMPLE 2**

**FINDING THE DOMAIN AND RANGE**

For $f(x) = x^2 + 1$, determine:

**a.** $f(-2)$     **b.** the domain     **c.** the range

*Solution*

**a.** $f(-2) = (-2)^2 + 1 = 4 + 1 = 5$     $f(x) = x^2 + 1$ with $x$ replaced by $-2$

**b.** Domain $= \mathbb{R}$     $x^2 + 1$ is defined for *all* real numbers

**c.** The function $f(x) = x^2 + 1$ is a square (which is positive or zero) plus 1 and so will take every $y$-value that is 1 or higher. Therefore:

$$\text{Range} = \{\, y \mid y \geq 1 \,\}$$

The domain and range can be seen from the graph on the left.

Any letters may be used for defining a function or describing the domain and the range. For example, since the range $\{\, y \mid y \geq 1 \,\}$ is a set of numbers, it could be written $\{\, w \mid w \geq 1 \,\}$ or, in interval notation without *any* variables, as $[1, \infty)$.

**Practice Problem 2**

For $g(z) = \sqrt{z-2}$, determine:

**a.** $g(27)$     **b.** the domain     **c.** the range

➤ **Solutions on page 44**

For each $x$ in the domain of a function there must be a *single* number $y = f(x)$, so the graph of a function cannot have two points $(x, y)$ with the same $x$-value but different $y$-values. This leads to the following *graphical* test for functions.

**Vertical Line Test for Functions**

A curve in the Cartesian plane is the graph of a *function* if and only if no vertical line intersects the curve at more than one point.

**EXAMPLE 3**

**USING THE VERTICAL LINE TEST**

This is *not* the graph of a function of $x$ because there is a vertical line (shown dashed) that intersects the curve twice.

This *is* the graph of a function of $x$ because no vertical line intersects the curve more than once.

A graph that has two or more points $(x, y)$ with the same $x$-value but different $y$-values, such as the one on the left above, defines a *relation* rather than a function. We will be concerned exclusively with *functions*, and so we will use the terms "function," "graph," and "curve" interchangeably.

Functions can be classified into several types.

---

**Linear Function**

A *linear function* is a function that can be expressed in the form

$$f(x) = mx + b$$

with constants $m$ and $b$. Its graph is a line with slope $m$ and $y$-intercept $b$.

---

**EXAMPLE 4**

**FINDING A COMPANY'S COST FUNCTION**

An electronics company manufactures pocket calculators at a cost of $9 each, and the company's fixed costs (such as rent) amount to $400 per day. Find a function $C(x)$ that gives the total cost of producing $x$ pocket calculators in a day.

*Solution*

Each calculator costs $9 to produce, so $x$ calculators will cost $9x$ dollars, to which we must add the fixed costs of $400.

$$C(x) = 9x + 400$$

Total cost    Unit cost    Number of units    Fixed cost

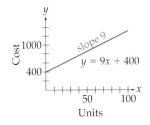

The graph of $C(x) = 9x + 400$ is a line with slope 9 and $y$-intercept 400, as shown on the left. Notice that the **slope** is the same as the **rate of change** of the cost (increasing at the rate of $9 per additional calculator), which is also the company's **marginal cost** (the cost of producing one more calculator is $9). The *slope*, the *rate of change*, and the *marginal* cost are always the same, as we will see in Chapter 2.

The most important property of a linear function is that it always changes by the *same amount* (the slope) whenever the independent variable increases by 1. In the preceding example, the slope has units of *dollars per calculator*.

**Practice Problem 3**

A trucking company will deliver furniture for a charge of $25 plus 5% of the purchase price of the furniture. Find a function $D(x)$ that gives the delivery charge for a piece of furniture that costs $x$ dollars.

➤ **Solution on page 45**

A mathematical description of a real-world situation is called a **mathematical model.** For example, the cost function $C(x) = 9x + 400$ from the previous example is a mathematical model for the cost of manufacturing calculators. In this model, $x$, the number of calculators, should take only whole-number values (0, 1, 2, 3, …), and the graph should consist of discrete dots rather than a continuous curve. Instead, we will find it easier to let $x$ take *continuous* values, and round up or down as necessary at the end.

---

**Quadratic Function**

A *quadratic function* is a function that can be expressed in the form

$$f(x) = ax^2 + bx + c$$

with constants ("coefficients") $a \neq 0$, $b$, and $c$. Its graph is called a *parabola.*

---

The condition $a \neq 0$ keeps the function from becoming $f(x) = bx + c$, which would be linear. Many familiar curves are parabolas.

The center of gravity of a
diver describes a parabola.

A stream of water from a hose
takes the shape of a parabola.

The parabola $f(x) = ax^2 + bx + c$ opens *upward* if the constant $a$ is *positive* and opens *downward* if the constant $a$ is *negative*. The **vertex** of a parabola is its "central" point, the *lowest* point on the parabola if it opens *up* and the *highest* point if it opens *down*.

## Graphing Calculator Exploration

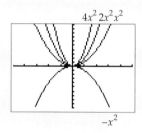

**a.** Graph the parabolas $y_1 = x^2$, $y_2 = 2x^2$, and $y_3 = 4x^2$ on the window $[-5, 5]$ by $[-10, 10]$. How does the shape of the parabola change when the coefficient of $x^2$ increases?

**b.** Graph $y_4 = -x^2$. What did the negative sign do to the parabola?

**c.** Predict the shape of the parabolas $y_5 = -2x^2$ and $y_6 = \frac{1}{3}x^2$. Then check your predictions by graphing the functions.

The $x$-coordinate of the vertex of a parabola may be found by the following formula, which is shown in the diagrams at the top of this page and will be derived in Exercise 63 on page 190.

### Vertex Formula for a Parabola

The vertex of the parabola $f(x) = ax^2 + bx + c$ has $x$-coordinate

$$x = \frac{-b}{2a}$$

For example, the vertex of the parabola $y = 2x^2 + 4x - 5$ has $x$-coordinate

$$x = \frac{-4}{2 \cdot 2} = \frac{-4}{4} = -1 \qquad x = \frac{-b}{2a} \text{ with } a = 2 \text{ and } b = 4$$

as we can see from the graph on the left. The $y$-coordinate of the vertex, $-7$, comes from evaluating $y = 2x^2 + 4x - 5$ at this $x$-coordinate.

**EXAMPLE 5**  **GRAPHING A QUADRATIC FUNCTION**

Graph the quadratic function $f(x) = 2x^2 - 40x + 104$.

*Solution*

Graphing using a graphing calculator is largely a matter of finding an appropriate viewing window, as the following three unsatisfactory windows show.

on $[-10, 10]$ by $[-10, 10]$     on $[0, 20]$ by $[-10, 10]$     on $[0, 20]$ by $[-105, -95]$

To find an appropriate viewing window, we use the vertex formula:

$$x = \frac{-b}{2a} = \frac{-(-40)}{2 \cdot 2} = \frac{40}{4} = 10 \qquad \begin{array}{l} x\text{-coordinate of the vertex, from} \\ x = \frac{-b}{2a} \text{ with } a = 2 \text{ and } b = -40 \end{array}$$

We move a few units, say 5, to either side of $x = 10$, making the $x$-window $[5, 15]$. Using the calculator to EVALUATE the given function at $x = 10$ (or evaluating by hand) gives $y(10) = -96$. Since the parabola opens upward (the coefficient of $x^2$ is positive), the curve rises up from its vertex, so we select a $y$-interval from $-96$ upward, say $[-96, -70]$. Graphing the function on the window $[5, 15]$ by $[-96, -70]$ gives the following result. (Some other graphing windows are just as good.)

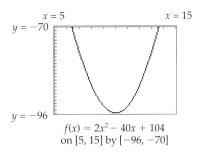

$f(x) = 2x^2 - 40x + 104$
on $[5, 15]$ by $[-96, -70]$

Note that even without a graphing calculator we could sketch the graph of $f(x) = 2x^2 - 40x + 104$ by finding the vertex $(10, -96)$ just as we did above, calculating two more points (say at $x = 5$ and $x = 15$), and drawing an upward-opening parabola from the vertex through these points.

### Solving Quadratic Equations

A value of $x$ that solves an equation $f(x) = 0$ is called a **root of the equation,** or a **zero of the function,** or an **$x$-intercept of the graph** of $y = f(x)$. The roots of a quadratic equation can often be found by factoring.

---

**EXAMPLE 6**  **SOLVING A QUADRATIC EQUATION BY FACTORING**

Solve $2x^2 - 4x = 6$.

*Solution*

$$2x^2 - 4x - 6 = 0$$ Subtracting 6 from each side to get zero on the right

$$2(x^2 - 2x - 3) = 0$$ Factoring out a 2

$$2\underbrace{(x - 3)}_{\substack{\text{Equals } 0 \\ \text{at } x = 3}} \cdot \underbrace{(x + 1)}_{\substack{\text{Equals } 0 \\ \text{at } x = -1}} = 0$$ Factoring $x^2 - 2x - 3$

Finding $x$-values that make each factor zero

$$x = 3, \quad x = -1$$ Solutions

---

## Graphing Calculator Exploration

Find both solutions to the equation in Example 6 by graphing the function $f(x) = 2x^2 - 4x - 6$ and using ZERO or TRACE to find where the curve crosses the $x$-axis. Your answers should agree with those found in Example 6.

**Practice Problem 4**  Solve by factoring or graphing: $9x - 3x^2 = -30$  ➤ **Solution on page 45**

Quadratic equations can often be solved by the *Quadratic Formula,* which is derived on page 45.

The quadratic formula gives these $x$-values

**Quadratic Formula**

The solutions to $ax^2 + bx + c = 0$ are

$$x = \frac{-b \pm \sqrt{b^2 - 4ac}}{2a}$$

The "plus or minus" ($\pm$) sign means calculate *both* ways, first using the $+$ sign and then using the $-$ sign

In a business, it is often important to find a company's **break-even points,** the numbers of units of production where a company's costs are equal to its revenue.

**EXAMPLE 7**    **FINDING BREAK-EVEN POINTS**

A company that installs automobile compact disc (CD) players finds that if it installs $x$ CD players per day, then its costs will be $C(x) = 120x + 4800$ and its revenue will be $R(x) = -2x^2 + 400x$ (both in dollars). Find the company's break-even points. [*Note:* In Section 3.4 we will see how such cost and revenue functions are found.]

*Solution*

$$120x + 4800 = -2x^2 + 400x \qquad \text{Setting } C(x) = R(x)$$

$$2x^2 - 280x + 4800 = 0 \qquad \begin{array}{l}\text{Combining all terms}\\ \text{on one side}\end{array}$$

$$x = \frac{280 \pm \sqrt{(-280)^2 - 4 \cdot 2 \cdot 4800}}{2 \cdot 2} \qquad \begin{array}{l}\text{Quadratic Formula with}\\ a = 2, \ b = -280, \ \text{and}\\ c = 4800\end{array}$$

$$= \frac{280 \pm \sqrt{40,000}}{4} = \frac{280 \pm 200}{4}$$

$$\qquad\qquad\qquad\qquad\qquad\qquad\qquad \begin{array}{l}\text{Working out the}\\ \text{formula on a calculator}\end{array}$$

$$= \frac{480}{4} \text{ or } \frac{80}{4} = 120 \text{ or } 20$$

The company will break even when it installs either 20 or 120 CD players.

Although it is important for a company to know where its break-even points are, most companies want to do better than break even—they want to maximize their profit. **Profit** is defined as *revenue minus cost* (since profit is what is left over after subtracting expenses from income).

| Profit |
| --- |
| Profit = Revenue − Cost |

**EXAMPLE 8**    **MAXIMIZING PROFIT**

For the CD installer whose daily revenue and cost functions were given in Example 7, find the number of units that maximizes profit, and the maximum profit.

*Solution*

The profit function is the revenue function minus the cost function.

$$P(x) = \underbrace{-2x^2 + 400x}_{R(x)} - \underbrace{(120x + 4800)}_{C(x)}$$

$P(x) = R(x) - C(x)$ with
$R(x) = -2x^2 + 400x$ and
$C(x) = 120x + 4800$

$$= -2x^2 + 280x - 4800 \qquad \text{Simplifying}$$

Since this function represents a parabola opening downward (because of the $-2$), it is maximized at its vertex, which is found using the vertex formula.

$$x = \frac{-280}{2(-2)} = \frac{-280}{-4} = 70$$

$x = \dfrac{-b}{2a}$ with

$a = -2$ and $b = 280$

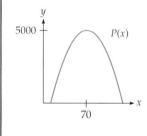

Thus, profit is maximized when 70 CD players are installed. For the maximum profit, we substitute $x = 70$ into the profit function:

$$P(70) = -2(70)^2 + 280 \cdot 70 - 4800$$

$P(x) = -2x^2 + 280x - 4800$
with $x = 70$

$$= 5000 \qquad \text{Multiplying and combining}$$

Therefore, the company will maximize its profit when it installs 70 CD players per day. Its maximum profit will be $5000 per day.

Why doesn't a company make more profit the more it sells? Because to increase its sales it must lower its prices, which eventually leads to lower profits. The relationship among the cost, revenue, and profit functions can be seen graphically as follows.

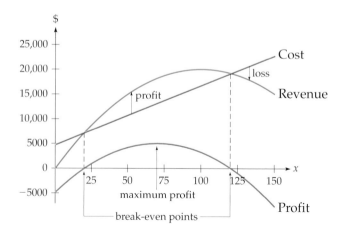

Notice that the break-even points correspond to a profit of zero, and that the maximum profit occurs halfway between the two break-even points.

Not all quadratic equations have (real) solutions.

**EXAMPLE 9**

**USING THE QUADRATIC FORMULA**

Solve $\frac{1}{2}x^2 - 3x + 5 = 0$.

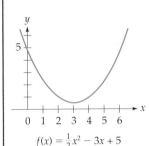

$f(x) = \frac{1}{2}x^2 - 3x + 5$

*Solution*   The Quadratic Formula with $a = \frac{1}{2}$, $b = -3$, and $c = 5$ gives

$$x = \frac{3 \pm \sqrt{9 - 4(\frac{1}{2})(5)}}{2(\frac{1}{2})} = \frac{3 \pm \sqrt{9 - 10}}{1} = 3 \pm \sqrt{-1} \qquad \text{Undefined}$$

Therefore, the equation $\frac{1}{2}x^2 - 3x + 5 = 0$ *has no real solutions* (because of the undefined $\sqrt{-1}$). The geometrical reason that there are no solutions can be seen in the graph on the left: The curve never reaches the $x$-axis, so the function never equals zero.

The quantity $b^2 - 4ac$, whose square root appears in the Quadratic Formula, is called the **discriminant.** If the discriminant is *positive* (as in Example 7), the equation $ax^2 + bx + c = 0$ has *two* solutions (since the square root is added or subtracted). If the discriminant is *zero*, there is only *one* root (since adding or subtracting zero gives the same answer). If the discriminant is *negative* (as in Example 9), then the equation has *no* real roots. Therefore, the discriminant being positive, zero, or negative corresponds to the parabola meeting the $x$-axis at 2, 1, or 0 points, as shown below.

two real roots

one real root

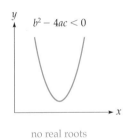

no real roots

## Quadratic Regression (Optional)

For data points that seem to lie along a *parabola,* we use **quadratic regression** to fit a parabola to the points, with the procedure carried out on a graphing calculator.

**EXAMPLE 10**

**QUADRATIC REGRESSION USING A GRAPHING CALCULATOR**

The following graph gives the total annual sales of Macintosh computers from 2001 through 2007.

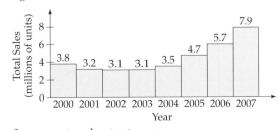

*Source:* systemshootouts.com

**a.** Use quadratic regression to fit a parabola to the data and state the regression function.

**b.** Use the regression function to predict Macintosh sales in 2012.

*Solution*

**a.** We number the years with $x$-values 0–7, so $x$ stands for *years after 2000*. We enter the data into lists, as shown in the first screen below (as explained in *Graphing Calculator Basics—Entering Data* on page xxiii), and use *ZoomStat* to graph the data points.

Then (using STAT, CALC, and QuadReg) we graph the regression along with the data points.

The regression curve, which fits the points reasonably well, is

$$y = 0.20x^2 - 0.88x + 3.9 \qquad \text{Rounded}$$

Y₁(12)
22.78928571

**b.** To predict sales in 2012, we evaluate Y1 at 12 (since $x = 12$ corresponds to 2012). From the screen on the left, if the current trend continues, Macintosh sales in 2012 will be approximately 22.8 million units.

---

### Solutions to Practice Problems

**1.** Domain: $\{\, x \mid x \le 0 \text{ or } x \ge 3 \,\}$; range: $\{\, y \mid y \ge 0 \,\}$

**2. a.** $g(27) = \sqrt{27 - 2} = \sqrt{25} = 5$

  **b.** Domain: $\{\, z \mid z \ge 2 \,\}$

  **c.** Range: $\{\, y \mid y \ge 0 \,\}$

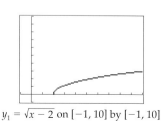

$y_1 = \sqrt{x - 2}$ on $[-1, 10]$ by $[-1, 10]$

**3.** $D(x) = 25 + 0.05x$

**4.**
$$9x - 3x^2 = -30$$
$$-3x^2 + 9x + 30 = 0$$
$$-3(x^2 - 3x - 10) = 0$$
$$-3(x - 5)(x + 2) = 0$$
$$x = 5, x = -2 \qquad \text{or from:}$$

---

## 1.3 Section Summary

In this section we defined and gave examples of *functions*, and saw how to find their domains and ranges. The essential characteristic of a function $f$ is that for any given "input" number $x$ in the domain, there is exactly one "output" number $f(x)$. This requirement is stated geometrically in the *vertical line test*, that no vertical line can intersect the graph of a function at more than one point. We then defined *linear functions* (whose graphs are lines) and *quadratic functions* (whose graphs are parabolas). We solved quadratic equations by factoring, graphing, and using the Quadratic Formula. We maximized and minimized quadratic functions using the vertex formula.

---

## Derivation of the Quadratic Formula

| | |
|---|---|
| $ax^2 + bx + c = 0$ | The quadratic set equal to zero |
| $ax^2 + bx = -c$ | Subtracting $c$ |
| $4a^2x^2 + 4abx = -4ac$ | Multiplying by $4a$ |
| $4a^2x^2 + 4abx + b^2 = b^2 - 4ac$ | Adding $b^2$ |
| $(2ax + b)^2 = b^2 - 4ac$ | Since $4a^2x^2 + 4abx + b^2 = (2ax + b)^2$ |
| $2ax + b = \pm\sqrt{b^2 - 4ac}$ | Taking square roots |
| $2ax = -b \pm \sqrt{b^2 - 4ac}$ | Subtracting $b$ |
| $x = \dfrac{-b \pm \sqrt{b^2 - 4ac}}{2a}$ | Dividing by $2a$ gives the Quadratic Formula |

---

## 1.3 Exercises

**1–8.** Determine whether each graph defines a function of $x$.

**1.**

**2.**

**3.**

**4.**

**5.**

**6.**

**7.**

**8.**

**9–10.** Find the domain and range of each function graphed below.

**9.**

**10.**
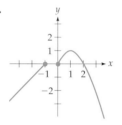

**11–22.** For each function:

**a.** Evaluate the given expression.
**b.** Find the domain of the function.
**c.** Find the range.
   [*Hint:* Use a graphing calculator.]

**11.** $f(x) = \sqrt{x - 1}$;  find $f(10)$

**12.** $f(x) = \sqrt{x - 4}$;  find $f(40)$

**13.** $h(z) = \dfrac{1}{z + 4}$;  find $h(-5)$

**14.** $h(z) = \dfrac{1}{z + 7}$;  find $h(-8)$

**15.** $h(x) = x^{1/4}$;  find $h(81)$

**16.** $h(x) = x^{1/6}$;  find $h(64)$

**17.** $f(x) = x^{2/3}$;  find $f(-8)$

[*Hint for Exercises 17 and 18:* You may need to enter $x^{m/n}$ as $(x^m)^{1/n}$ or as $(x^{1/n})^m$, as discussed on page 25.]

**18.** $f(x) = x^{4/5}$;  find $f(-32)$

**19.** $f(x) = \sqrt{4 - x^2}$;  find $f(0)$

**20.** $f(x) = \dfrac{1}{\sqrt{x}}$;  find $f(4)$

**21.** $f(x) = \sqrt{-x}$;  find $f(-25)$

**22.** $f(x) = -\sqrt{-x}$;  find $f(-100)$

**23–30.** Graph each function "by hand." [*Note:* Even if you have a graphing calculator, it is important to be able to sketch simple curves by finding a few important points.]

**23.** $f(x) = 3x - 2$

**24.** $f(x) = 2x - 3$

**25.** $f(x) = -x + 1$

**26.** $f(x) = -3x + 5$

**27.** $f(x) = 2x^2 + 4x - 16$

**28.** $f(x) = 3x^2 - 6x - 9$

**29.** $f(x) = -3x^2 + 6x + 9$

**30.** $f(x) = -2x^2 + 4x + 16$

**31–34.** For each quadratic function:

**a.** Find the vertex using the vertex formula.

**b.** Graph the function on an appropriate window. (Answers may differ.)

**31.** $f(x) = x^2 - 40x + 500$

**32.** $f(x) = x^2 + 40x + 500$

**33.** $f(x) = -x^2 - 80x - 1800$

**34.** $f(x) = -x^2 + 80x - 1800$

**35–52.** Solve each equation by factoring or the Quadratic Formula, as appropriate.

**35.** $x^2 - 6x - 7 = 0$       **36.** $x^2 - x - 20 = 0$

**37.** $x^2 + 2x = 15$        **38.** $x^2 - 3x = 54$

**39.** $2x^2 + 40 = 18x$      **40.** $3x^2 + 18 = 15x$

**41.** $5x^2 - 50x = 0$       **42.** $3x^2 - 36x = 0$

**43.** $2x^2 - 50 = 0$        **44.** $3x^2 - 27 = 0$

**45.** $4x^2 + 24x + 40 = 4$  **46.** $3x^2 - 6x + 9 = 6$

**47.** $-4x^2 + 12x = 8$      **48.** $-3x^2 + 6x = -24$

**49.** $2x^2 - 12x + 20 = 0$  **50.** $2x^2 - 8x + 10 = 0$

**51.** $3x^2 + 12 = 0$        **52.** $5x^2 + 20 = 0$

**53–62.** Solve each equation using a graphing calculator. [*Hint:* Begin with the window $[-10, 10]$ by $[-10, 10]$ or another of your choice (see Useful Hint in Graphing Calculator Terminology following the Preface) and use ZERO, SOLVE, or TRACE and ZOOM IN.] (In Exercises 61 and 62, round answers to two decimal places.)

**53.** $x^2 - x - 20 = 0$    **54.** $x^2 + 2x - 15 = 0$

**55.** $2x^2 + 40 = 18x$    **56.** $3x^2 + 18 = 15x$

**57.** $4x^2 + 24x + 45 = 9$    **58.** $3x^2 - 6x + 5 = 2$

**59.** $3x^2 + 7x + 12 = 0$    **60.** $5x^2 + 14x + 20 = 0$

**61.** $2x^2 + 3x - 6 = 0$    **62.** $3x^2 + 5x - 7 = 0$

 **63.** Use your graphing calculator to graph the following four equations simultaneously on the window $[-10, 10]$ by $[-10, 10]$:

$$y_1 = 2x + 6$$
$$y_2 = 2x + 2$$
$$y_3 = 2x - 2$$
$$y_4 = 2x - 6$$

(*continues*)

**a.** What do the lines have in common and how do they differ?

**b.** Write the equation of another line with the same slope that lies 2 units below the lowest line. Then check your answer by graphing it with the others.

 **64.** Use your graphing calculator to graph the following four equations simultaneously on the window $[-10, 10]$ by $[-10, 10]$:

$$y_1 = 3x + 4$$
$$y_2 = 1x + 4$$
$$y_3 = -1x + 4 \quad \text{(Use } \boxed{(-)} \text{ to get } -1x.)$$
$$y_4 = -3x + 4$$

**a.** What do the lines have in common and how do they differ?

**b.** Write the equation of a line through this $y$-intercept with slope $\frac{1}{2}$. Then check your answer by graphing it with the others.

## Applied Exercises

**65. BUSINESS: Cost Functions**   A lumberyard will deliver wood for $4 per board foot plus a delivery charge of $20. Find a function $C(x)$ for the cost of having $x$ board feet of lumber delivered.

**66. BUSINESS: Cost Functions**   A company manufactures bicycles at a cost of $55 each. If the company's fixed costs are $900, express the company's costs as a linear function of $x$, the number of bicycles produced.

**67. BUSINESS: Salary**   An employee's weekly salary is $500 plus $15 per hour of overtime. Find a function $P(x)$ giving his pay for a week in which he worked $x$ hours of overtime.

**68. BUSINESS: Salary**   A sales clerk's weekly salary is $300 plus 2% of her total week's sales. Find a function $P(x)$ for her pay for a week in which she sold $x$ dollars of merchandise.

**69. GENERAL: Water Pressure**   At a depth of $d$ feet underwater, the water pressure is $p(d) = 0.45d + 15$ pounds per square inch. Find the pressure at:

**a.** The bottom of a 6-foot-deep swimming pool.
**b.** The maximum ocean depth of 35,000 feet.

**70. GENERAL: Boiling Point**   At higher altitudes, water boils at lower temperatures. This is why at high altitudes foods must be boiled for longer times—the lower boiling point imparts less heat to the food. At an altitude of $h$ thousand feet above sea level, water boils at a temperature of $B(h) = -1.8h + 212$ degrees Fahrenheit. Find the altitude at which water boils at 98.6 degrees Fahrenheit. (Your answer will show that at a high enough altitude, water boils at normal body temperature. This is why airplane cabins must be pressurized—at high enough altitudes one's blood would boil.)

**71–72. GENERAL: Stopping Distance**   A car traveling at speed $v$ miles per hour on a dry road should be able to come to a full stop in a distance of

$$D(v) = 0.055v^2 + 1.1v \quad \text{feet}$$

Find the stopping distance required for a car traveling at:

**71.** 40 mph.

**72.** 60 mph.

*Source: National Transportation Safety Board*

**73. BIOMEDICAL: Cell Growth** The number of cells in a culture after $t$ days is given by $N(t) = 200 + 50t^2$. Find the size of the culture after:

**a.** 2 days.
**b.** 10 days.

**74. ATHLETICS: Juggling** If you toss a ball $h$ feet straight up, it will return to your hand after $T(h) = 0.5\sqrt{h}$ seconds. This leads to the *juggler's dilemma:* Juggling more balls means tossing them higher. However, the square root in the above formula means that tossing them twice as high does not gain twice as much time, but only $\sqrt{2} \approx 1.4$ times as much time. Because of this, there is a limit to the number of balls that a person can juggle, which seems to be about ten. Use this formula to find:

**a.** How long will a ball spend in the air if it is tossed to a height of 4 feet? 8 feet?
**b.** How high must it be tossed to spend 2 seconds in the air? 3 seconds in the air?

**75. GENERAL: Impact Velocity** If a marble is dropped from a height of $x$ feet, it will hit the ground with velocity $v(x) = \frac{60}{11}\sqrt{x}$ miles per hour (neglecting air resistance). Use this formula to find the velocity with which a marble will strike the ground if it is dropped from the top of the tallest building in the United States, the 1454-foot Sears Tower in Chicago.

**76. GENERAL: Tsunamis** The speed of a tsunami (popularly known as a tidal wave, although it has nothing whatever to do with tides) depends on the depth of the water through which it is traveling. At a depth of $d$ feet, the speed of a tsunami will be $s(d) = 3.86\sqrt{d}$ miles per hour. Find the speed of a tsunami in the Pacific basin where the average depth is 15,000 feet.

*Source: William Bascom, Waves and Beaches*

**77−78. GENERAL: Impact Time of a Projectile** If an object is thrown upward so that its height (in feet) above the ground $t$ seconds after it is thrown is given by the function $h(t)$ below, find when the object hits the ground. That is, find the positive value of $t$ such that $h(t) = 0$. Give the answer correct to two decimal places. [*Hint:* Enter the function in terms of $x$ rather than $t$. Use the ZERO operation, or TRACE and ZOOM IN, or similar operations.]

**77.** $h(t) = -16t^2 + 45t + 5$

**78.** $h(t) = -16t^2 + 40t + 4$

**79. BUSINESS: Break-Even Points and Maximum Profit** A company that produces tracking devices for computer disk drives finds that if it produces $x$ devices per week, its costs will be $C(x) = 180x + 16{,}000$ and its revenue will be $R(x) = -2x^2 + 660x$ (both in dollars).

**a.** Find the company's break-even points.
**b.** Find the number of devices that will maximize profit, and the maximum profit.

**80. BUSINESS: Break-Even Points and Maximum Profit** City and Country Cycles finds that if it sells $x$ racing bicycles per month, its costs will be $C(x) = 420x + 72{,}000$ and its revenue will be $R(x) = -3x^2 + 1800x$ (both in dollars).

**a.** Find the store's break-even points.
**b.** Find the number of bicycles that will maximize profit, and the maximum profit.

**81. BUSINESS: Break-Even Points and Maximum Profit** A sporting goods store finds that if it sells $x$ exercise machines per day, its costs will be $C(x) = 100x + 3200$ and its revenue will be $R(x) = -2x^2 + 300x$ (both in dollars).

**a.** Find the store's break-even points.
**b.** Find the number of sales that will maximize profit, and the maximum profit.

**82. SOCIAL SCIENCE: Health Club Attendance** A recent study analyzed how the number of visits a person makes to a health club varies with the monthly membership price. It found that the number of visits per year is given approximately by $v(x) = -0.004x^2 + 0.56x + 42$, where $x$ is the monthly membership price. What monthly price maximizes the number of visits?

*Source: American Economic Review, 2006*

**83. ATHLETICS: Muscle Contraction** The fundamental equation of muscle contraction is of the form $(w + a)(v + b) = c$, where $w$ is the weight placed on the muscle, $v$ is the velocity of contraction of the muscle, and $a$, $b$, and $c$ are constants that depend upon the muscle and the units of measurement. Solve this equation for $v$ as a function of $w$, $a$, $b$, and $c$.

*Source: E. Batschelet, Introduction to Mathematics for Life Scientists*

84. **GENERAL: Longevity**  When a person reaches age 65, the probability of living for another $x$ decades is approximated by the function $f(x) = -0.077x^2 - 0.057x + 1$  (for  $0 \le x \le 3$). Find the probability that such a person will live for another:

   a. One decade.
   b. Two decades.
   c. Three decades.

   *Source: Statistical Abstracts of the United States, 2007*

 85. **BEHAVIORAL SCIENCES: Smoking and Education**  According to a study, the probability that a smoker will quit smoking increases with the smoker's educational level. The probability (expressed as a percent) that a smoker with $x$ years of education will quit is approximately $y = 0.831x^2 - 18.1x + 137.3$  (for  $10 \le x \le 16$).

   a. Graph this curve on the window [10, 16] by [0, 100].
   b. Find the probability that a high school graduate smoker ($x = 12$) will quit.
   c. Find the probability that a college graduate smoker ($x = 16$) will quit.

   *Source: Review of Economics and Statistics LXXVII(1)*

86. **ENVIRONMENTAL SCIENCE: Wind Energy**  The use of wind power is growing rapidly after a slow start, especially in Europe, where it is seen as an efficient and renewable source of energy. Global wind power generating capacity for the years 1990 to 2005 is given approximately by $y = 0.33x^2 - 1.5x + 2.4$  thousand megawatts, where $x$ is the number of years after 1990. (One megawatt would supply the electrical needs of approximately 100 homes).

   a. Graph this curve on the window [0, 20] by [0, 100].
   b. Use this curve to predict the global wind power generating capacity in the year 2010. [*Hint:* Which $x$-value corresponds to 2010? Then use TRACE, EVALUATE, or TABLE.]
   c. Predict the global wind power generating capacity in the year 2015.

   *Source: Worldwatch Institute*

87. **GENERAL: Newsletters**  A newsletter has a maximum audience of 100 subscribers. The publisher estimates that she will lose 1 reader for each dollar she charges. Therefore, if she charges $x$ dollars, her readership will be $(100 - x)$.

   a. Multiply this readership by $x$ (the price) to find her total revenue. Multiply out the resulting quadratic function.

   b. What price should she charge to maximize her revenue? [*Hint:* Find the value of $x$ that maximizes this quadratic function.]

88. **ATHLETICS: Cardiovascular Zone**  Your maximum heart rate (in beats per minute) may be estimated as 220 minus your age. For maximum cardiovascular effect, many trainers recommend raising your heart rate to between 50% and 70% of this maximum rate (called the *cardio zone*).

   a. Write a linear function to represent this upper limit as a function of $x$, your age. Then write a similar linear function to represent the lower limit. Use decimals instead of percents.
   b. Use your functions to find the upper and lower cardio limits for a 20-year-old person. Find the cardio limits for a 60-year-old person.

   *Source: Mayo Clinic*

 89. **BUSINESS: Quest Revenues**  Operating revenues of Quest Communication International for recent years are given in the following graph.

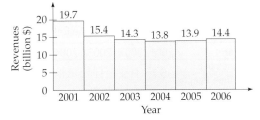

   a. Number the bars with $x$-values 1–6 (so that $x$ stands for *years since 2000*) and use quadratic regression to fit a parabola to the data. State the regression function. [*Hint:* See Example 10.]
   b. Use the regression function to predict Quest operating revenues in the year 2012.

   *Source: Standard & Poor's Industry Surveys*

90. **BUSINESS: Southwest Airlines Revenues**  Operating revenues of Southwest Airlines for recent years are given in the following table.

| Year | 2001 | 2002 | 2003 | 2004 | 2005 | 2006 |
|---|---|---|---|---|---|---|
| Revenues (billion $) | 5.6 | 5.5 | 5.9 | 6.5 | 7.6 | 9.1 |

   a. Number the data columns with $x$-values 1–6 (so that $x$ stands for *years since 2000*) and use quadratic regression to fit a parabola to the data. State the regression function. [*Hint:* See Example 10.]
   b. Use the regression function to predict Southwest's operating revenues in the year 2012.

   *Source: Standard & Poor's Industry Surveys*

**91. SOCIAL SCIENCE: Immigration** The percentage of immigrants in the United States has changed since World War I as shown in the following table.

| Decade | 1930 | 1940 | 1950 | 1960 | 1970 | 1980 | 1990 | 2000 |
|---|---|---|---|---|---|---|---|---|
| Immigrant Percentage | 11.7 | 8.9 | 7 | 5.6 | 4.9 | 6.2 | 7.9 | 11.3 |

a. Number the data columns with $x$-values 0–7 (so that $x$ stands for *decades since 1930*) and use quadratic regression to fit a parabola to the data. State the regression function. [*Hint:* See Example 10.]

b. In 2007, one in eight people living in the United States was an immigrant. Use the regression function to predict the percentage in 2007. How well does this prediction match the actual percentage?

*Sources:* Center for Immigration Studies and U.S. Census Bureau

**92. BIOMEDICAL: Tainted Meat** After falling for many years, the incidence of *E. coli* sickness in the United States from consuming tainted meat began to increase in 2005, as shown in the following table.

| Year | 2001 | 2002 | 2003 | 2004 | 2005 | 2006 |
|---|---|---|---|---|---|---|
| Cases per Million People | 15.7 | 11.9 | 11 | 8.8 | 10.5 | 12.9 |

a. Number the data columns with $x$-values 1–6 (so that $x$ stands for *years since 2000*) and use quadratic regression to fit a parabola to the data. State the regression function. [*Hint:* See Example 10.]

b. Assuming that this trend continues, use the regression function to predict the number of cases per million people in the year 2010.

*Source:* Centers for Disease Control and Prevention

## Conceptual Exercises

**93.** Can the graph of a function have more than one $x$-intercept? Can it have more than one $y$-intercept?

**94.** If a linear function is such that $f(2) = 5$ and $f(3) = 7$, then $f(4) = ?$ [*Hint:* No work necessary.]

**95.** If a linear function is such that $f(4) = 7$ and $f(6) = 11$, then $f(5) = ?$ [*Hint:* No work necessary.]

**96.** The Apocryphal Manufacturing Company makes blivets out of widgets. If a linear function $f(x) = mx + b$ gives the number of widgets that can be made from $x$ blivets, what are the units of the slope $m$ (*widgets per blivet* or *blivets per widget*)?

**97.** In a linear function $f(x) = mx + b$, the slope $m$ has units *blargs per prendle*. What are the units of $x$? What are the units of $y$? [*Hint:* One is in blargs and the other is in prendles, but which is which?]

**98.** For the quadratic function $f(x) = ax^2 + bx + c$, what condition on one of the coefficients will guarantee that the function has a highest value? A lowest value?

**99.** We have discussed quadratic functions that open *up* or open *down*. Can a quadratic function open *sideways*? Explain.

**100.** Explain why, if a quadratic function has two $x$-intercepts, the $x$-coordinate of the vertex will be halfway between them.

---

## 1.4 FUNCTIONS: POLYNOMIAL, RATIONAL, AND EXPONENTIAL

### Introduction

In this section we will define other useful types of functions, including polynomial, rational, exponential, and logarithmic functions, although the latter two types will be discussed more extensively in Sections 4.1 and 4.2. We will also define an important operation, the *composition* of functions.

### Polynomial Functions

A **polynomial function** (or simply a **polynomial**) is a function that can be written in the form

$$f(x) = a_n x^n + a_{n-1} x^{n-1} + \cdots + a_2 x^2 + a_1 x + a_0$$

where $n$ is a nonnegative integer and $a_0, a_1, \ldots, a_n$ are (real) numbers, called **coefficients.** The *domain* of a polynomial is $\mathbb{R}$, the set of all (real) numbers. The **degree** of a polynomial is the highest power of the variable. The following are polynomials.

$$f(x) = 2x^8 - 3x^7 + 4x^5 - 5$$

A polynomial of degree 8 (since the highest power of $x$ is 8)

$$f(x) = -4x^2 - \tfrac{1}{3}x + 19$$

A polynomial of degree 2 (a quadratic function)

$$f(x) = x - 1$$

A polynomial of degree 1 (a linear function)

$$f(x) = 6$$

A polynomial of degree 0 (a constant function)

Cost

Units produced

A cost function may increase at different rates at different production levels.

Polynomials are used to model many situations in which change occurs at different rates. For example, the polynomial graphed on the left might represent the total cost of manufacturing $x$ units of a product. At first, costs rise steeply because of high start-up expenses, then more slowly as the economies of mass production come into play, and finally more steeply as new production facilities need to be built.

Polynomial equations can often be solved by factoring (just as with quadratic equations).

**EXAMPLE 1**

**SOLVING A POLYNOMIAL EQUATION BY FACTORING**

Solve   $3x^4 - 6x^3 = 24x^2$

*Solution*

$$3x^4 - 6x^3 - 24x^2 = 0$$

Rewritten with all the terms on the left side

$$3x^2(x^2 - 2x - 8) = 0$$

Factoring out $3x^2$

$$3x^2 \cdot (x - 4) \cdot (x + 2) = 0$$

Factoring further

Equals   Equals   Equals
zero at   zero at   zero at
$x = 0$   $x = 4$   $x = -2$

Finding the zeros of each factor

$$x = 0, \quad x = 4, \quad x = -2$$

Solutions

As in this example, if a positive power of $x$ can be factored out of a polynomial, then   $x = 0$   is one of the roots.

**Practice Problem 1**   Solve   $2x^3 - 4x^2 = 48x$

➤ Solution on page 64

## Rational Functions

The word "ratio" means fraction or quotient, and the quotient of two polynomials is called a **rational function.** The following are rational functions.

$$f(x) = \frac{3x + 2}{x - 2} \qquad g(x) = \frac{1}{x^2 + 1}$$

A rational function is a polynomial over a polynomial

The domain of a rational function is the set of numbers for which the denominator is not zero. For example, the domain of the function $f(x)$ on the left above is $\{x \mid x \neq 2\}$ (since $x = 2$ makes the denominator zero), and the domain of $g(x)$ on the right is the set of all real numbers $\mathbb{R}$ (since $x^2 + 1$ is never zero). The graphs of these functions are shown below. Notice that these graphs have **asymptotes,** lines that the graphs *approach* but never actually reach.

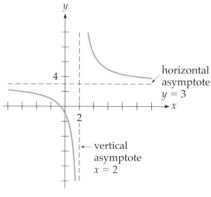

Graph of $f(x) = \dfrac{3x + 2}{x - 2}$

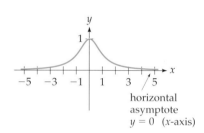

Graph of $g(x) = \dfrac{1}{x^2 + 1}$

 **Practice Problem 2**　What is the domain of $f(x) = \dfrac{18}{(x + 2)(x - 4)}$ ?　➤ **Solution on page 64**

⚠ **BE CAREFUL:** Simplifying a rational function by canceling a common factor from the numerator and the denominator can change the domain of the function, so that the "simplified" and "original" versions may not be equal (since they have different domains). For example, the rational function on the left below is not defined at $x = 1$, while the simplified version on the right *is* defined at $x = 1$, so that the two functions are technically not equal.

$$\underbrace{\frac{x^2 - 1}{x - 1}}_{\substack{\text{Not defined at } x = 1, \\ \text{so the domain is } \{x \mid x \neq 1\}}} = \underbrace{\frac{(x + 1)(x - 1)}{x - 1}}_{} \neq \underbrace{x + 1}_{\substack{\text{Is defined at } x = 1, \\ \text{so the domain is } \mathbb{R}}}$$

However, the functions *are* equal at every *x*-value *except* $x = 1$, and the graphs below are the same except that the rational function omits the point at $x = 1$. We will return to this technical issue when we discuss limits in Section 2.1.

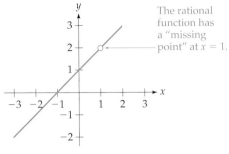

Graph of $y = \dfrac{x^2 - 1}{x - 1}$

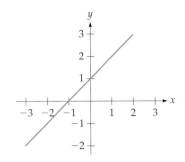

Graph of $y = x + 1$

## Exponential Functions

A function in which the independent variable appears in the exponent, such as $f(x) = 2^x$, is called an **exponential function.**

### GRAPHING AN EXPONENTIAL FUNCTION

Graph the exponential function $f(x) = 2^x$.

*Solution*

This function is defined for *all* real numbers, so its domain is $\mathbb{R}$. Values of the function are shown in the table on the left below, and plotting these points and drawing a smooth curve through them gives the curve on the right below.

| $x$ | $y = 2^x$ |
|---|---|
| 3 | $2^3 = 8$ |
| 2 | $2^2 = 4$ |
| 1 | $2^1 = 2$ |
| 0 | $2^0 = 1$ |
| $-1$ | $2^{-1} = \frac{1}{2}$ |
| $-2$ | $2^{-2} = \frac{1}{4}$ |
| $-3$ | $2^{-3} = \frac{1}{8}$ |

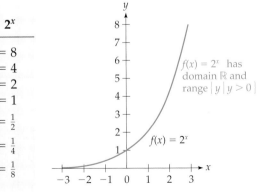

$f(x) = 2^x$ has domain $\mathbb{R}$ and range $\{y \mid y > 0\}$

$f(x) = 2^x$

Exponential functions are often used to model population growth and decline.

World population since the year 1800 can be approximated by an exponential function.

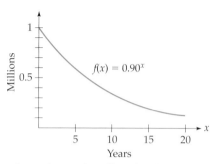

A population of 1 million that declines by 10% each year is modeled by an exponential function.

In mathematics the letter $e$ is used to represent a constant whose value is approximately 2.718. The exponential function $f(x) = e^x$ will be very important beginning in Chapter 4. Another important function is the logarithmic function to the base $e$, written $f(x) = \ln x$. These functions are graphed below.

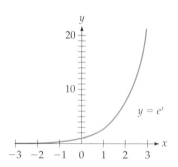

The exponential function $e^x$ has domain $\mathbb{R}$ and range $\{y \mid y > 0\}$.

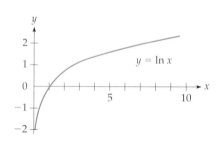

The natural logarithm function has domain $\{x \mid x > 0\}$ and range $\mathbb{R}$.

## Graphing Calculator Exploration

Graph the functions $y_1 = 2^x$, $y_2 = 3^x$, and $y_3 = 4^x$ on the window $[-2, 2]$ by $[0, 5]$.

**a.** Which function rises most steeply?

**b.** Between which two curves would $e^x$ lie? Check your prediction by graphing $y_4 = e^x$. [*Hint:* On some calculators, $e^x$ is obtained by pressing 2nd ln x .]

## Piecewise Linear Functions

The rule for calculating the values of a function may be given in several parts. If each part is linear, the function is called a **piecewise linear function,** and its graph consists of "pieces" of straight lines.

EXAMPLE 3

### GRAPHING A PIECEWISE LINEAR FUNCTION

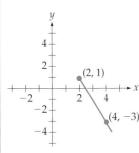

This notation means:

Graph $\quad f(x) = \begin{cases} 5 - 2x & \text{if } x \geq 2 \\ x + 3 & \text{if } x < 2 \end{cases}$ ← Use the top formula for $x \geq 2$
← Use the bottom formula for $x < 2$

*Solution*  We graph one "piece" at a time.

*Step 1:* To graph the first part, $f(x) = 5 - 2x$ if $x \geq 2$, we use the "endpoint" $x = 2$ and also $x = 4$ (or any other $x$-value satisfying $x \geq 2$). The points are $(2, 1)$ and $(4, -3)$, with the $y$-coordinates calculated from $f(x) = 5 - 2x$. Draw the line through these two points, but only for $x \geq 2$ (from $x = 2$ to the *right*), as shown on the left.

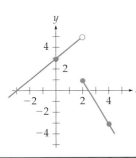

*Step 2:* For the second part, $f(x) = x + 3$ if $x < 2$, the restriction $x < 2$ means that the line ends just *before* $x = 2$. We mark this "missing point" $(2, 5)$ by an "open circle" ○ to indicate that it is *not* included in the graph (the $y$-coordinate comes from $f(x) = x + 3$). For a second point, choose $x = 0$ (or any other $x < 2$), giving $(0, 3)$. Draw the line through these two points, but only for $x < 2$ (to the *left* of $x = 2$), completing the graph of the function.

An important piecewise linear function is the *absolute value* function.

EXAMPLE 4

### THE ABSOLUTE VALUE FUNCTION

The absolute value function is $f(x) = |x|$, and is defined as

$$|x| = \begin{cases} x & \text{if } x \geq 0 \\ -x & \text{if } x < 0 \end{cases}$$

The second line, for *negative x,* attaches a *second* negative sign to make the result *positive*

For example, when applied to either 3 or $-3$, the function gives *positive* 3:

$$|3| = 3$$

Using the top formula (since $3 \geq 0$)

$$|-3| = -(-3) = 3$$

Using the bottom formula (since $-3 < 0$)

To graph the absolute value function, we may proceed as in Example 3, or simply observe that for $x \geq 0$, the function gives $y = x$ (a half-line from the origin with slope 1), and for $x < 0$, it gives $y = -x$ (a half-line on the other side of the origin with slope $-1$), as shown in the following graph.

**Absolute Value Function**

$$|x| = \begin{cases} x & \text{if } x \geq 0 \\ -x & \text{if } x < 0 \end{cases}$$

The absolute value function $f(x) = |x|$ has a "corner" at the origin.

Examples 3 and 4 show that the "pieces" of a piecewise linear function may or may not be connected.

**EXAMPLE 5**

### GRAPHING AN INCOME TAX FUNCTION

Federal income taxes are "progressive," meaning that they take a higher percentage of higher incomes. For example, the 2007 federal income tax for a single taxpayer whose taxable income was less than $77,100 was determined by a three-part rule: 10% of income up to $7825, plus 15% of any amount over $7825 up to $31,850, plus 25% of any amount over $31,850 up to $77,100. For an income of $x$ dollars, the tax $f(x)$ may be expressed as follows:

$$f(x) = \begin{cases} 0.10x & \text{if } 0 \leq x \leq 7825 \\ 782.50 + 0.15(x - 7825) & \text{if } 7825 < x \leq 31,850 \\ 4386.25 + 0.25(x - 31,850) & \text{if } 31,850 < x \leq 77,100 \end{cases}$$

Graphing this by the same technique as before leads to the graph shown below. The slopes 0.10, 0.15, and 0.25 are called the *marginal tax rates.*

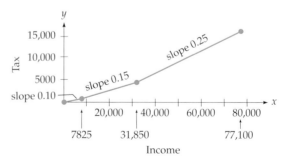

*Source:* Internal Revenue Service

## Graphing Calculator Exploration

In the same way, we can define and graph piecewise *nonlinear* functions. To graph

$$f(x) = \begin{cases} 2x + 3 & \text{if } x < 1 \\ 6 - x & \text{if } 1 \le x \le 4 \\ (x - 4)^2 & \text{if } x > 4 \end{cases}$$

enter

$$y_1 = (2x + 3)(x < 1) + (6 - x)(x \ge 1 \text{ and } x \le 4) + (x - 4)^2(x > 4)$$

or, equivalently,

$$y_1 = (2x + 3)(x < 1) + (6 - x)(x \ge 1)(x \le 4) + (x - 4)^2(x > 4)$$

The inequalities and the word "and" are found in the TEST and LOGIC menus. To avoid "false lines" connecting pieces of the curve that do not touch, change the MODE to DOT graphing. (This function could also have been graphed *by hand* following the procedure explained in Example 3.)

on $[-1, 8]$ by $[-2, 10]$
(4, 2) is included
(4, 0) is excluded

### Composite Functions

Just as we substitute a *number* into a function, we may substitute a *function* into a function. For two functions $f$ and $g$, evaluating $f$ at $g(x)$ gives $f(g(x))$, called the *composition* of $f$ with $g$ evaluated at $x$.*

#### Composite Functions

The *composition* of $f$ with $g$ evaluated at $x$ is $f(g(x))$.

The *domain* of $f(g(x))$ is the set of all numbers $x$ in the domain of $g$ such that $g(x)$ is in the domain of $f$. If we think of the functions $f$ and $g$ as "numerical machines," then the composition $f(g(x))$ may be thought of as a *combined* machine in which the output of $g$ is connected to the input of $f$.

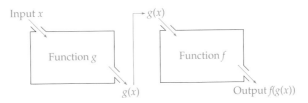

A "machine" for generating the composition of $f$ with $g$.
A number $x$ is fed into the function $g$, and the output
$g(x)$ is then fed into the function $f$, resulting in $f(g(x))$.

*The composition $f(g(x))$ may also be written $(f \circ g)(x)$, although we will not use this notation.

**EXAMPLE 6**

### FINDING COMPOSITE FUNCTIONS

If $f(x) = x^7$ and $g(x) = x^3 - 2x$, find:

**a.** $f(g(x))$          **b.** $f(f(x))$

*Solution*

**a.** $f(g(x))$ $=$ $[g(x)]^7$ $=$ $(x^3 - 2x)^7$

$\underbrace{f(x) = x^7 \text{ with } x}$          $\underbrace{\text{Using}}$
replaced by $g(x)$   $g(x) = x^3 - 2x$

**b.** $f(f(x))$ $=$ $[f(x)]^7$ $=$ $(x^7)^7$ $=$ $x^{49}$

$\underbrace{f(x) = x^7 \text{ with } x}$   $\underbrace{\text{Using}}$
replaced by $f(x)$   $f(x) = x^7$

**EXAMPLE 7**

### FINDING COMPOSITE FUNCTIONS IN DIFFERENT ORDERS

If $f(x) = \dfrac{x + 8}{x - 1}$ and $g(x) = \sqrt{x}$, find $f(g(x))$ and $g(f(x))$.

*Solution*

$$f(g(x)) = \frac{g(x) + 8}{g(x) - 1} = \frac{\sqrt{x} + 8}{\sqrt{x} - 1}$$   $f(x) = \dfrac{x + 8}{x - 1}$ with $x$
replaced by $g(x) = \sqrt{x}$

$$g(f(x)) = \sqrt{f(x)} = \sqrt{\frac{x + 8}{x - 1}}$$   $g(x) = \sqrt{x}$ with $x$
replaced by $f(x) = \dfrac{x + 8}{x - 1}$

 **BE CAREFUL:** The order of composition is important: $f(g(x))$ is not the same as $g(f(x))$. To show this, we evaluate the above $f(g(x))$ and $g(f(x))$ at $x = 4$:

$$f(g(4)) = \frac{\sqrt{4} + 8}{\sqrt{4} - 1} = \frac{2 + 8}{2 - 1} = \frac{10}{1} = 10$$   $f(g(x)) = \dfrac{\sqrt{x} + 8}{\sqrt{x} - 1}$ at $x = 4$

Different answers

$$g(f(4)) = \sqrt{\frac{4 + 8}{4 - 1}} = \sqrt{\frac{12}{3}} = \sqrt{4} = 2$$   $g(f(x)) = \sqrt{\dfrac{x + 8}{x - 1}}$ at $x = 4$

**Practice Problem 3**   If $f(x) = x^2 + 1$ and $g(x) = \sqrt[3]{x}$, find: **a.** $f(g(x))$   **b.** $g(f(x))$

➤ Solutions on page 64

**EXAMPLE 8**    **PREDICTING WATER USAGE**

A planning commission estimates that if a city's population is $p$ thousand people, its daily water usage will be $W(p) = 30p^{1.2}$ thousand gallons. The commission further predicts that the population in $t$ years will be $p(t) = 60 + 2t$ thousand people. Express the water usage $W$ as a function of $t$, the number of years from now, and find the water usage 10 years from now.

*Solution*

Water usage $W$ as a function of $t$ is the *composition* of $W(p)$ with $p(t)$:

$$W(p(t)) = 30[p(t)]^{1.2} = 30(60 + 2t)^{1.2} \qquad \begin{array}{l} W = 30p^{1.2} \text{ with } p \\ \text{replaced by } p(t) = 60 + 2t \end{array}$$

To find water usage in 10 years, we evaluate $W(p(t))$ at $t = 10$:

$$W(p(10)) = 30(60 + 2 \cdot 10)^{1.2} \qquad 30(60 + 2t)^{1.2} \text{ with } t = 10$$
$$= 30(80)^{1.2} \approx 5765 \qquad \text{Using a calculator}$$

Thousand gallons

Therefore, in 10 years the city will need about 5,765,000 gallons of water per day.

## Shifts of Graphs

Sometimes the graph of a composite function is just a horizontal or vertical shift of an original graph. This occurs when one of the functions is simply the addition or subtraction of a constant. The following diagram shows the graph of $y = x^2$ together with various shifts and the functions that generate them.

Vertical shifts

Horizontal shifts

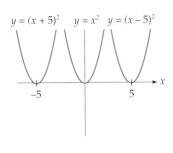

In general, adding to or subtracting from the *x-value* means a *horizontal* shift, while adding to or subtracting from the *function* means a *vertical* shift. These same ideas hold for *any* function: given the graph of $y = f(x)$, adding or subtracting a positive number $a$ to the function $f(x)$ or to the variable $x$ shifts the graph as follows:

| Shifts of Graphs | | |
|---|---|---|
| **Function** | **Shift** | |
| $y = f(x) + a$ | shifted *up* by $a$ units | ⎫ Vertical |
| $y = f(x) - a$ | shifted *down* by $a$ units | ⎭ shifts |
| $y = f(x + a)$ | shifted *left* by $a$ units | ⎫ Horizontal |
| $y = f(x - a)$ | shifted *right* by $a$ units | ⎭ shifts |

Of course, a graph can be shifted both horizontally and vertically, as illustrated by the following shifts of $y = x^2$:

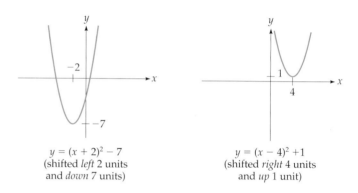

$y = (x + 2)^2 - 7$
(shifted *left* 2 units
and *down* 7 units)

$y = (x - 4)^2 + 1$
(shifted *right* 4 units
and *up* 1 unit)

Such double shifts can be applied to *any* function $y = f(x)$: the graph of $y = f(x + a) + b$ is shifted *left* $a$ units and *up* $b$ units (with the understanding that a *negative* $a$ or $b$ means that the direction is reversed).

 **BE CAREFUL:** Remember that adding a *positive* number to $x$ means a *left shift*.

## Graphing Calculator Exploration

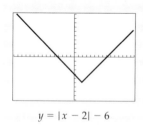

$y = |x - 2| - 6$

The absolute value function $y = |x|$ may be graphed on some graphing calculator as $y_1 = \text{ABS}(x)$.

**a.** Graph $y_1 = \text{ABS}(x - 2) - 6$ and observe that the absolute value function is shifted *right* 2 units and *down* 6 units. (The graph shown is drawn using ZOOM ZSquare.)

**b.** Predict the shift of $y_1 = \text{ABS}(x + 4) + 2$ and then verify your prediction by graphing the function on your calculator.

Given a function $f(x)$, to find an algebraic expression for the "shifted" function $f(x + h)$ we simply replace each occurrence of $x$ by $x + h$.

**EXAMPLE 9**

**FINDING $f(x + h)$ FROM $f(x)$**

If   $f(x) = x^2 - 5x$,   find   $f(x + h)$.

*Solution*

$$f(x + h) = \underbrace{(x + h)^2} - 5\underbrace{(x + h)}$$

$f(x) = x^2 - 5x$ with each $x$ replaced by $x + h$

$$= x^2 + 2xh + h^2 - 5x - 5h$$

Expanding

## Difference Quotients

The quantity   $\dfrac{f(x + h) - f(x)}{h}$   will be very important in Chapter 2 when we begin studying calculus. It is called the **difference quotient,** since it is a quotient whose numerator is a difference. It gives the slope (rise over run) between the points in the curve   $y = f(x)$   at $x$ and at $x + h$.

**EXAMPLE 10**

**FINDING A DIFFERENCE QUOTIENT**

If   $f(x) = x^2 - 4x + 1$,   find and simplify   $\dfrac{f(x + h) - f(x)}{h}$        $(h \neq 0)$

*Solution*

$$\frac{f(x + h) - f(x)}{h} = \frac{\overbrace{(x + h)^2 - 4(x + h) + 1}^{f(x + h)} - \overbrace{(x^2 - 4x + 1)}^{f(x)}}{h}$$

$$= \frac{x^2 + 2xh + h^2 - 4x - 4h + 1 - x^2 + 4x - 1}{h}$$

Expanding

$$= \frac{x^2 + 2xh + h^2 - 4x - 4h + 1 - x^2 + 4x - 1}{h}$$

Canceling

$$= \frac{2xh + h^2 - 4h}{h} = \frac{h(2x + h - 4)}{h}$$

Factoring an $h$ from the top

$$= \frac{h(2x + h - 4)}{h} = 2x + h - 4$$

Canceling $h$ from top and bottom (since $h \neq 0$)

**Practice Problem 4**   If   $f(x) = 3x^2 - 2x + 1$,   find and simplify   $\dfrac{f(x + h) - f(x)}{h}$.

➤ Solution on page 64

EXAMPLE 11

**FINDING A DIFFERENCE QUOTIENT**

If $f(x) = \dfrac{1}{x}$, find and simplify $\dfrac{f(x + h) - f(x)}{h}$ $\qquad (h \neq 0)$

*Solution*

$$\frac{f(x + h) - f(x)}{h} = \frac{\overbrace{\dfrac{1}{x + h}}^{f(x+h)} - \overbrace{\dfrac{1}{x}}^{f(x)}}{h}$$

$$= \frac{1}{h}\left(\frac{1}{x + h} - \frac{1}{x}\right) \qquad \text{Multiplying by } 1/h \\ \text{instead of dividing by } h$$

$$= \frac{1}{h}\left(\frac{x}{(x + h)x} - \frac{x + h}{(x + h)x}\right) \qquad \text{Using the common} \\ \text{denominator } (x + h)x$$

$$= \frac{1}{h}\cdot\frac{\overbrace{x - (x + h)}^{-h}}{(x + h)x} = \frac{1}{h}\cdot\frac{-h}{(x + h)x} \qquad \text{Subtracting fractions,} \\ \text{and simplifying}$$

$$= \frac{1}{\cancel{h}}\cdot\frac{\overset{-1}{\cancel{-h}}}{(x + h)x} = \frac{-1}{(x + h)x} \qquad \text{Canceling } h \ (h \neq 0)$$

## Exponential Regression (Optional)

If data appear to lie along an *exponential* curve, such as those shown on page 54, we may fit an exponential curve to the data using **exponential regression**. The mathematical basis is explained in Section 7.4, with the procedure using a graphing calculator explained below. We will see in Section 4.1 that populations are often modeled by exponential curves, and so quantities that rise and fall with populations, such as goods or services, are appropriately modeled by exponential curves.

EXAMPLE 12

**EXPONENTIAL REGRESSION USING A GRAPHING CALCULATOR**

In 2005 passengers flying on scheduled airlines worldwide flew a total of 2.41 trillion passenger-miles, which is equivalent to 5 million people flying to the moon and back in a year. The following graph shows the number of trillion passenger-miles flown in recent years.

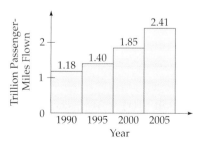

*Source:* Worldwatch

**a.** Use exponential regression to fit a curve to the data and state the regression function.

**b.** Use the regression function to predict world air travel in 2015.

*Solution*

**a.** We number the years with $x$-values 0–3, so $x$ stands for the number of five-year intervals after 1990. We enter the data into lists, as shown in the first screen below (as explained in *Graphing Calculator Basics— Entering Data* on page xxiii), and use *ZoomStat* to graph the data points.

Then (using STAT, CALC, and ExpReg), we graph the regression along with the data points.

The regression curve, which fits the points quite well, is

$$y = 1.15 \cdot 1.27^x \qquad \text{Rounded}$$

**b.** To predict sales in 2015, we evaluate Y1 at 5 (since $x = 5$ means 5 five-year intervals after 1990, giving 2015). From the screen on the right, if the current trend continues, airline flights worldwide will total about 3.8 trillion passenger-miles in 2015.

> **Solutions to Practice Problems**

**1.** $2x^3 - 4x^2 - 48x = 0$

$2x(x^2 - 2x - 24) = 0$

$2x(x + 4)(x - 6) = 0$

$x = 0, \quad x = -4, \quad x = 6$

**2.** $\{\, x \mid x \neq -2, x \neq 4 \,\}$

**3. a.** $f(g(x)) = [g(x)]^2 + 1 = \left(\sqrt[3]{x}\right)^2 + 1 \quad \text{or} \quad x^{2/3} + 1$

　 **b.** $g(f(x)) = \sqrt[3]{f(x)} = \sqrt[3]{x^2 + 1} \quad \text{or} \quad (x^2 + 1)^{1/3}$

**4.** $\dfrac{f(x + h) - f(x)}{h}$

$= \dfrac{3(x + h)^2 - 2(x + h) + 1 - (3x^2 - 2x + 1)}{h}$

$= \dfrac{3x^2 + 6xh + 3h^2 - 2x - 2h + \cancel{1} - 3x^2 + 2x - \cancel{1}}{h}$

$= \dfrac{h(6x + 3h - 2)}{h} = 6x + 3h - 2$

## 1.4 Section Summary

We have introduced a variety of functions: polynomials (which include linear and quadratic functions), rational functions, exponential functions, and piecewise linear functions. Examples of these are shown below and on the following page. You should be able to identify these basic types of functions from their algebraic forms. We also added constants to perform horizontal and vertical *shifts* of graphs of functions, and combined functions by using the "output" of one as the "input" of the other, resulting in *composite* functions.

### A Gallery of Functions

### POLYNOMIALS

Linear function
$f(x) = mx + b$

Quadratic functions
$f(x) = ax^2 + bx + c$

$f(x) = ax^4 + bx^3 + cx^2 + dx + e$

## RATIONAL FUNCTIONS

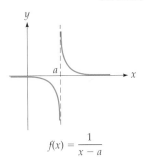

$$f(x) = \frac{1}{x - a}$$

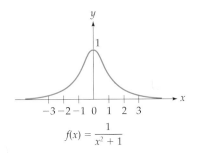

$$f(x) = \frac{1}{x^2 + 1}$$

## EXPONENTIAL FUNCTIONS

$$f(x) = 2^x$$

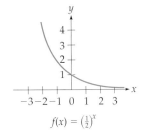

$$f(x) = \left(\tfrac{1}{2}\right)^x$$

## PIECEWISE LINEAR FUNCTIONS

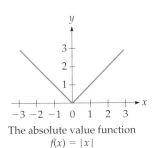

The absolute value function
$$f(x) = |x|$$

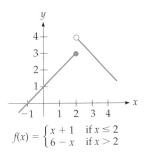

$$f(x) = \begin{cases} x + 1 & \text{if } x \le 2 \\ 6 - x & \text{if } x > 2 \end{cases}$$

## 1.4 Exercises

**1–2.** Find the domain and range of each function graphed below.

**1.**

**2.**

**3–10.** For each function:
**a.** Evaluate the given expression.
**b.** Find the domain of the function.

**c.** Find the range. [*Hint:* Use a graphing calculator. You may have to ignore some false lines on the graph. Graphing in "dot mode" will also eliminate false lines.]

**3.** $f(x) = \dfrac{1}{x + 4}$; find $f(-3)$

**4.** $f(x) = \dfrac{1}{(x - 1)^2}$; find $f(-1)$

**5.** $f(x) = \dfrac{x^2}{x - 1}$; find $f(-1)$

*(See instructions on previous page.)*

**6.** $f(x) = \dfrac{x^2}{x + 2}$; find $f(2)$

**7.** $f(x) = \dfrac{12}{x(x + 4)}$; find $f(2)$

**8.** $f(x) = \dfrac{16}{x(x - 4)}$; find $f(-4)$

**9.** $g(x) = 4^x$; find $g\left(-\frac{1}{2}\right)$

**10.** $g(x) = 8^x$; find $g\left(-\frac{1}{3}\right)$

**11–22.** Solve each equation by factoring.

**11.** $x^5 + 2x^4 - 3x^3 = 0$ **12.** $x^6 - x^5 - 6x^4 = 0$

**13.** $5x^3 - 20x = 0$ **14.** $2x^5 - 50x^3 = 0$

**15.** $2x^3 + 18x = 12x^2$ **16.** $3x^4 + 12x^2 = 12x^3$

**17.** $6x^5 = 30x^4$ **18.** $5x^4 = 20x^3$

**19.** $3x^{5/2} - 6x^{3/2} = 9x^{1/2}$

[*Hint for Exercises 19 and 20:* First factor out a fractional power.]

**20.** $2x^{7/2} + 8x^{5/2} = 24x^{3/2}$

**21.** $2x^{5/2} + 4x^{3/2} = 6x^{1/2}$

**22.** $3x^{7/2} - 12x^{5/2} = 36x^{3/2}$

 **23–24.** Solve each equation using a graphing calculator. Round answers to two decimal places.

**23.** $x^5 - x^4 - 5x^3 = 0$

**24.** $x^6 + 2x^5 - 5x^4 = 0$

**25–30.** Graph each function.

**25.** $f(x) = 3^x$ **26.** $f(x) = \left(\frac{1}{3}\right)^x$

**27.** $f(x) = \begin{cases} 2x - 7 & \text{if } x \geq 4 \\ 2 - x & \text{if } x < 4 \end{cases}$

**28.** $f(x) = \begin{cases} 8 - 2x & \text{if } x \geq 2 \\ x + 2 & \text{if } x < 2 \end{cases}$

**29.** $f(x) = |x - 3| - 3$

**30.** $f(x) = |x + 2| - 2$

 **31–32.** Use a graphing calculator to graph each piecewise *nonlinear* function on the window $[-2, 10]$ by $[-5, 5]$. Where parts of the graph do not touch, state which point is included and which is excluded.

**31.** $f(x) = \begin{cases} x^2 & \text{if } x \leq 2 \\ 6 - x & \text{if } 2 < x < 6 \\ 2x - 17 & \text{if } x \geq 6 \end{cases}$

**32.** $f(x) = \begin{cases} 4 - x^2 & \text{if } x < 3 \\ 2x - 11 & \text{if } 3 \leq x < 7 \\ 8 - x & \text{if } x \geq 7 \end{cases}$

**33–46.** Identify each function as a polynomial, a rational function, an exponential function, a piecewise linear function, or none of these. (Do not graph them; just identify their types.)

**33.** $f(x) = x^5$ **34.** $f(x) = 4^x$

**35.** $f(x) = 5^x$ **36.** $f(x) = x^4$

**37.** $f(x) = x + 2$

**38.** $f(x) = \begin{cases} 3x - 1 & \text{if } x \geq 2 \\ 1 - x & \text{if } x < 2 \end{cases}$

**39.** $f(x) = \dfrac{1}{x + 2}$ **40.** $f(x) = x^2 + 9$

**41.** $f(x) = \begin{cases} x - 2 & \text{if } x < 3 \\ 7 - 4x & \text{if } x \geq 3 \end{cases}$

**42.** $f(x) = \dfrac{x}{x^2 + 9}$ **43.** $f(x) = 3x^2 - 2x$

**44.** $f(x) = x^3 - x^{2/3}$ **45.** $f(x) = x^2 + x^{1/2}$

**46.** $f(x) = 5$

 **47.** For the functions $y_1 = \left(\frac{1}{3}\right)^x$, $y_2 = \left(\frac{1}{2}\right)^x$, $y_3 = 2^x$, and $y_4 = 3^x$:

   **a.** Predict which curve will be the highest for large values of $x$.

   **b.** Predict which curve will be the lowest for large values of $x$.

   **c.** Check your predictions by graphing the functions on the window $[-3, 3]$ by $[0, 5]$.

   **d.** From your graph, what is the common $y$-intercept? Why do all such exponential functions meet at this point?

 **48.** Graph the parabola $y_1 = 1 - x^2$ and the semicircle $y_2 = \sqrt{1 - x^2}$ on the window $[-1, 1]$ by $[0, 1]$. (You may want to adjust the window to make the semicircle look more like a semicircle.) Use TRACE to determine which is the "inside" curve (the parabola or the semicircle) and which is the "outside" curve. These graphs show that when you graph a parabola, you should draw the curve near the vertex to be slightly more "pointed" than a circular curve.

**49–56.** For each pair of functions $f(x)$ and $g(x)$, find **a.** $f(g(x))$ **b.** $g(f(x))$ and **c.** $f(f(x))$

**49.** $f(x) = x^5$; $g(x) = 7x - 1$

**50.** $f(x) = x^8$;  $g(x) = 2x + 5$

**51.** $f(x) = \dfrac{1}{x}$;  $g(x) = x^2 + 1$

**52.** $f(x) = \sqrt{x}$;  $g(x) = x^3 - 1$

**53.** $f(x) = \sqrt{x} - 1$;  $g(x) = x^3 - x^2$

**54.** $f(x) = x^2 + 1$;  $g(x) = x - \sqrt{x}$

**55.** $f(x) = x^2 - x$;  $g(x) = \dfrac{x^3 - 1}{x^3 + 1}$

**56.** $f(x) = x^3 + x$;  $g(x) = \dfrac{x^4 + 1}{x^4 - 1}$

**57–58.** For each pair of functions $f(x)$ and $g(x)$, find and *fully simplify*  **a.** $f(g(x))$  and  **b.** $g(f(x))$

**57.** $f(x) = 2x - 6$;  $g(x) = \dfrac{x}{2} + 3$

**58.** $f(x) = x^3 + 1$;  $g(x) = \sqrt[3]{x - 1}$

**59–62.** For each function, find and simplify  $f(x + h)$.

**59.** $f(x) = 5x^2$

**60.** $f(x) = 3x^2$

**61.** $f(x) = 2x^2 - 5x + 1$

**62.** $f(x) = 3x^2 - 5x + 2$

**63–74.** For each function, find and simplify
$\dfrac{f(x + h) - f(x)}{h}$.  (Assume $h \neq 0$.)

**63.** $f(x) = 5x^2$  **64.** $f(x) = 3x^2$

**65.** $f(x) = 2x^2 - 5x + 1$

**66.** $f(x) = 3x^2 - 5x + 2$

**67.** $f(x) = 7x^2 - 3x + 2$

**68.** $f(x) = 4x^2 - 5x + 3$

**69.** $f(x) = x^3$
[*Hint:* Use  $(x + h)^3 = x^3 + 3x^2h + 3xh^2 + h^3$.]

**70.** $f(x) = x^4$
[*Hint:* Use  $(x + h)^4 = x^4 + 4x^3h + 6x^2h^2 +$
$4xh^3 + h^4$.]

**71.** $f(x) = \dfrac{2}{x}$  **72.** $f(x) = \dfrac{3}{x}$

**73.** $f(x) = \dfrac{1}{x^2}$  **74.** $f(x) = \sqrt{x}$

[*Hint for Exercise 74:* Multiply top and bottom of the fraction by  $\sqrt{x + h} + \sqrt{x}$.]

**75.** Find, rounding to five decimal places:

 **a.** $\left(1 + \dfrac{1}{100}\right)^{100}$

 **b.** $\left(1 + \dfrac{1}{10,000}\right)^{10,000}$

 **c.** $\left(1 + \dfrac{1}{1,000,000}\right)^{1,000,000}$

 **d.** Do the resulting numbers seem to be approaching a limiting value? Estimate the limiting value to five decimal places. The number that you have approximated is denoted $e$, and will be used extensively in Chapter 4.

 **76.** Use the TABLE feature of your graphing calculator to evaluate  $\left(1 + \dfrac{1}{x}\right)^x$  for values of $x$
such as 100, 10,000, 1,000,000, and higher values. Do the resulting numbers seem to be approaching a limiting value? Estimate the limiting value to five decimal places. The number that you have approximated is denoted $e$, and will be used extensively in Chapter 4.

 **77.** How will the graph of  $y = (x + 3)^3 + 6$  differ from the graph of  $y = x^3$?  Check by graphing both functions together.

 **78.** How will the graph of  $y = -(x - 4)^2 + 8$  differ from the graph of  $y = -x^2$?  Check by graphing both functions together.

## Applied Exercises

**79–80.** **SOCIAL SCIENCES: World Population**
The world population (in millions) since the year 1700 is approximated by the exponential function $P(x) = 522(1.0053)^x$,  where $x$ is the number of years since 1700 (for  $0 \leq x \leq 200$).  Using a calculator, estimate the world population in the year:

**79.** 1750  **80.** 1800

*Source: World Almanac 2008*

**81.** **ECONOMICS: Income Tax**  The following function expresses an income tax that is 10% for incomes below $5000, and otherwise is $500 plus 30% of income in excess of $5000.

$$f(x) = \begin{cases} 0.10x & \text{if } 0 \leq x < 5000 \\ 500 + 0.30(x - 5000) & \text{if } x \geq 5000 \end{cases}$$

 **a.** Calculate the tax on an income of $3000.
 **b.** Calculate the tax on an income of $5000.

*(continues)*

**c.** Calculate the tax on an income of $10,000.
**d.** Graph the function.

**82. ECONOMICS: Income Tax** The following function expresses an income tax that is 15% for incomes below $6000, and otherwise is $900 plus 40% of income in excess of $6000.

$$f(x) = \begin{cases} 0.15x & \text{if } 0 \le x < 6000 \\ 900 + 0.40(x - 6000) & \text{if } x \ge 6000 \end{cases}$$

**a.** Calculate the tax on an income of $3000.
**b.** Calculate the tax on an income of $6000.
**c.** Calculate the tax on an income of $10,000.
**d.** Graph the function.

**83–84. GENERAL: Dog-Years** The usual estimate that each human-year corresponds to 7 dog-years is not very accurate for young dogs, since they quickly reach adulthood. Exercises 83 and 84 give more accurate formulas for converting human-years $x$ into dog-years. For each conversion formula:

**a.** Find the number of dog-years corresponding to the following amounts of human time: 8 months, 1 year and 4 months, 4 years, 10 years.
**b.** Graph the function.

*Source: Bull. Acad. Vet. France* **26**

**83.** The following function expresses dog-years as $10\frac{1}{2}$ dog-years per human-year for the first 2 years and then 4 dog-years per human-year for each year thereafter.

$$f(x) = \begin{cases} 10.5x & \text{if } 0 \le x \le 2 \\ 21 + 4(x - 2) & \text{if } x > 2 \end{cases}$$

**84.** The following function expresses dog-years as 15 dog-years per human-year for the first year, 9 dog-years per human-year for the second year, and then 4 dog-years per human-year for each year thereafter.

$$f(x) = \begin{cases} 15x & \text{if } 0 \le x \le 1 \\ 15 + 9(x - 1) & \text{if } 1 < x \le 2 \\ 24 + 4(x - 2) & \text{if } x > 2 \end{cases}$$

**85. BUSINESS: Insurance Reserves** An insurance company keeps reserves (money to pay claims) of $R(v) = 2v^{0.3}$, where $v$ is the value of all of its policies, and the value of its policies is predicted to be $v(t) = 60 + 3t$, where $t$ is the number of years from now. (Both $R$ and $v$ are in millions of dollars.) Express the reserves $R$ as

a function of $t$, and evaluate the function at $t = 10$.

**86. BUSINESS: Research Expenditures** An electronics company's research budget is $R(p) = 3p^{0.25}$, where $p$ is the company's profit, and the profit is predicted to be $p(t) = 55 + 4t$, where $t$ is the number of years from now. (Both $R$ and $p$ are in millions of dollars.) Express the research expenditure $R$ as a function of $t$, and evaluate the function at $t = 5$.

**87. BIOMEDICAL: Cell Growth** One leukemic cell in an otherwise healthy mouse will divide into two cells every 12 hours, so that after $x$ days the number of leukemic cells will be $f(x) = 4^x$.

**a.** Find the approximate number of leukemic cells after 10 days.
**b.** If the mouse will die when its body has a billion leukemic cells, will it survive beyond day 15?

*Source: Annals of NY Academy of Sciences* **54**

**88. GENERAL: Cell Phones** The use of cellular telephones has been growing rapidly, in both developed and developing countries. The number of cell phones worldwide (in millions) is approximated by the exponential function $f(x) = 650(1.45)^x$, where $x$ is the number of years since 2000. Predict the number of cell phones in the year 2010.

*Source: Worldwatch*

**89. BUSINESS: Handheld Computers** Total U.S. sales of handheld computers (also called palmtops or PDAs for *personal digital assistants*) for recent years are shown in the following graph.

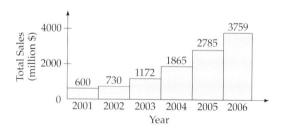

**a.** Number the bars with $x$-values 1–6 (so that $x$ stands for *years since 2000*) and use exponential regression to fit a curve to the data. State the regression formula. [*Hint:* See Example 11.]
**b.** Use the regression function to predict sales in the year 2010.

*Source: Euromonitor*

**90. BUSINESS: Bottled Water** Total U.S. sales of bottled water for recent years are given in the following table.

| Year | 1986 | 1991 | 1996 | 2001 | 2006 |
|------|------|------|------|------|------|
| Total Sales (billion gallons) | 1.4 | 2.2 | 3.3 | 5.1 | 8.2 |

a. Number the data columns with $x$-values 0–4 (so that $x$ stands for *number of five-year intervals since 1986*) and use exponential regression to fit a curve to the data. State the regression formula. [*Hint:* See Example 11.]

b. Use the regression function to predict sales in the year 2016.

*Source:* Beverage Marketing Corporation

**91. ENVIRONMENTAL SCIENCE: Wind Power** Wind power, popular for environmental and energy security reasons, is growing rapidly in the United States and Europe. Total world wind energy–generating capacity, in megawatts, is given in the following table. (One megawatt would power about 1000 average homes.)

| Year | 1990 | 1995 | 2000 | 2005 |
|------|------|------|------|------|
| Wind–Generating Capacity (megawatts) | 1930 | 4780 | 18,450 | 59,091 |

a. Number the data columns with $x$-values 0–3 (so that $x$ stands for the *number of five-year intervals since 1990*) and use exponential regression to fit a curve to the data. State the regression formula. [*Hint:* See Example 11.]

b. Use the regression function to predict world wind-energy capacity in the year 2015.

*Source:* Worldwatch

**92. BUSINESS: U.S. Oil Production** Crude oil production in the United States has been declining for decades because of the depletion of many older fields. Production (in billion barrels per day) for recent years is given in the following table.

| Year | 1985 | 1990 | 1995 | 2000 | 2005 |
|------|------|------|------|------|------|
| Crude Oil Production | 8.97 | 7.36 | 6.56 | 5.82 | 5.18 |

a. Number the data columns with $x$-values 0–4 (so that $x$ stands for the *number of five-year intervals since 1985*) and use exponential regression to fit a curve to the data. State the regression formula. [*Hint:* See Example 11.]

b. Use the regression function to predict crude oil production in the year 2015.

*Source:* Energy Information Administration

## Conceptual Exercises

**93.** How do two graphs differ if their functions are the same except that the domain of one excludes some $x$-values from the domain of the other?

**94.** Which of the following is *not* a polynomial, and why?
$$x^2 + \sqrt{2} \qquad x^{\sqrt{2}} + 1 \qquad \sqrt{2}x^2 + 1$$

**95.** The income tax function graphed on page 56 has segments with various slopes. What would it mean about an income tax if a segment had slope 1?

**96.** If $f(x) = ax$, then $f(f(x)) = ?$

**97.** If $f(x) = x + a$, then $f(f(x)) = ?$

**98.** How do the graphs of $f(x)$ and $f(x) + 10$ differ?

**99.** How do the graphs of $f(x)$ and $f(x + 10)$ differ?

**100.** How do the graphs of $f(x)$ and $f(x + 10) + 10$ differ?

**101.** True or False: If $f(x) = x^2$, then $f(x + h) = x^2 + h^2$.

**102.** True or False: If $f(x) = mx + b$, then $f(x + h) = f(x) + mh$.

## Explorations and Excursions

The following problems extend and augment the material presented in the text.

*Greatest Integer Function*

**103.** For any $x$, the function $\text{INT}(x)$ is defined as the greatest integer less than or equal to $x$. For example, $\text{INT}(3.7) = 3$ and $\text{INT}(-4.2) = -5$.

a. Use a graphing calculator to graph the function $y_1 = \text{INT}(x)$. (You may need to graph it in DOT mode to eliminate false connecting lines.)

b. From your graph, what are the domain and range of this function?

**104.** a. Use a graphing calculator to graph the function $y_1 = 2\,\text{INT}(x)$. [See the previous exercise for a definition of $\text{INT}(x)$.]

b. From your graph, what are the domain and range of this function?

*More About Compositions*

**105. a.** Find the composition $f(g(x))$ of the two linear functions $f(x) = ax + b$ and $g(x) = cx + d$ (for constants $a$, $b$, $c$, and $d$).
  **b.** Is the composition of two linear functions always a linear function?

**106. a.** Is the composition of two quadratic functions always a quadratic function? [*Hint:* Find the composition of $f(x) = x^2$ and $g(x) = x^2$.]
  **b.** Is the composition of two polynomials always a polynomial?

# Chapter Summary with Hints and Suggestions

Reading the text and doing the exercises in this chapter have helped you to master the following concepts and skills, which are listed by section (in case you need to review them) and are keyed to particular Review Exercises. Answers for all Review Exercises are given at the back of the book, and full solutions can be found in the Student Solutions Manual.

## 1.1 Real Numbers, Inequalities, and Lines

■ Translate an interval into set notation and graph it on the real line. (*Review Exercises 1–4.*)

$$[a, b] \quad (a, b) \quad [a, b) \quad (a, b]$$

$$(-\infty, b] \quad (-\infty, b) \quad [a, \infty) \quad (a, \infty) \quad (-\infty, \infty)$$

■ Express given information in interval form. (*Review Exercises 5–6.*)

■ Find an equation for a line that satisfies certain conditions. (*Review Exercises 7–12.*)

$$m = \frac{y_2 - y_1}{x_2 - x_1} \qquad y = mx + b$$

$$y - y_1 = m(x - x_1) \qquad x = a \qquad y = b$$

$$ax + by = c$$

■ Find an equation of a line from its graph. (*Review Exercises 13–14.*)

■ Use straight-line depreciation to find the value of an asset. (*Review Exercises 15–16.*)

■ Use real-world data to find a regression line and make a prediction. (*Review Exercise 17.*)

## 1.2 Exponents

■ Evaluate negative and fractional exponents without a calculator. (*Review Exercises 18–25.*)

$$x^0 = 1 \qquad x^{-n} = \frac{1}{x^n} \qquad x^{m/n} = \sqrt[n]{x^m} = \left(\sqrt[n]{x}\right)^m$$

■ Evaluate an exponential expression using a calculator. (*Review Exercises 26–29.*)

■ Use real-world data to find a power regression curve and make a prediction. (*Review Exercise 30.*)

## 1.3 Functions: Linear and Quadratic

■ Evaluate and find the domain and range of a function. (*Review Exercises 31–34.*)

A function $f$ is a rule that assigns to each number $x$ in a set (the domain) a (single) number $f(x)$. The range is the set of all resulting values $f(x)$.

■ Use the vertical line test to see if a graph defines a function. (*Review Exercises 35–36.*)

■ Graph a linear function: $f(x) = mx + b$ (*Review Exercises 37–38.*)

■ Graph a quadratic function: $f(x) = ax^2 + bx + c$ (*Review Exercises 39–40.*)

■ Solve a quadratic equation by factoring and by the Quadratic Formula. (*Review Exercises 41–44.*)

| Vertex | $x$-intercepts |
|---|---|
| $x = \dfrac{-b}{2a}$ | $x = \dfrac{-b \pm \sqrt{b^2 - 4ac}}{2a}$ |

■ Use a graphing calculator to graph a quadratic function. (*Review Exercises 45–46.*)

■ Construct a linear function from a word problem or from real-life data, and then use the function in an application.
*(Review Exercises 47–50.)*

■ For given cost and revenue functions, find the break-even points and maximum profit.
*(Review Exercises 51–52.)*

■ Use real-world data to find a quadratic regression curve and make a prediction.
*(Review Exercise 53.)*

## 1.4 Functions: Polynomial, Rational, and Exponential

■ Evaluate and find the domain and range of a more complicated function.
*(Review Exercises 54–57.)*

■ Solve a polynomial equation by factoring.
*(Review Exercises 58–61.)*

■ Graph an exponential function.
*(Review Exercise 62.)*

■ Graph a "shifted" function.
*(Review Exercise 63.)*

■ Graph a piecewise linear function.
*(Review Exercises 64–65.)*

■ Given two functions, find their composition.
*(Review Exercises 66–69.)*

$$f(g(x)) \qquad g(f(x))$$

■ For a given function $f(x)$, find and simplify the difference quotient $\dfrac{f(x + h) - f(x)}{h}$.
*(Review Exercises 70–71.)*

■ Solve an applied problem involving the composition of functions. *(Review Exercise 72.)*

■ Solve a polynomial equation.
*(Review Exercises 73–74.)*

■ Use real-world data to find an exponential regression curve and make a prediction.
*(Review Exercise 75.)*

## Hints and Suggestions

■ *(Overview)* In reviewing this chapter, notice the difference between *geometric* objects (points, curves, etc.) and *analytic* objects (numbers, functions, etc.), and the connections between them. You should be able to express geometric objects analytically, and vice versa. For example, given a *graph* of a line, you should be able to find an *equation* for it, and given a quadratic *function*, you should be able to *graph* it.

■ A graphing calculator or a computer with appropriate software can help you to *explore* a concept (for example, seeing how a curve changes as a coefficient or exponent changes), and also to *solve* a problem (for example, eliminating the point-plotting aspect of graphing, or finding a regression line or curve).

■ The Practice Problems help you to check your mastery of the skills presented. Complete solutions are given at the end of each section.

■ The Student Solutions Manual, available separately from your bookstore, provides fully worked-out solutions to selected exercises.

■ **Practice for Test:** Review Exercises 1, 9, 11, 13, 15, 18, 33, 35, 37, 39, 45, 49, 51, 58, 64, 68, 70, and 74.

## Review Exercises for Chapter 1

Practice test exercise numbers are in green.

### 1.1 Real Numbers, Inequalities, and Lines

**1–4.** Write each interval in set notation and graph it on the real line.

1. $(2, 5]$    **2.** $[-2, 0)$    **3.** $[100, \infty)$    **4.** $(-\infty, 6]$

5. **GENERAL: Wind Speed** The United States Coast Guard defines a "hurricane" as winds of at least 74 mph, a "storm" as winds of at least 55 mph but less than 74 mph, a "gale" as winds of at least 38 mph but less than 55 mph, and a "small craft warning" as winds of at least 21 mph

but less than 38 mph. Express each of these wind conditions in interval form. [*Hint:* A small craft warning is [21, 38).]

**6.** State in interval form:
  **a.** The set of all positive numbers.
  **b.** The set of all negative numbers.
  **c.** The set of all nonnegative numbers.
  **d.** The set of all nonpositive numbers.

**7–12.** Write an equation of the line satisfying each of the following conditions. If possible, write your answer in the form $y = mx + b$.

**7.** Slope 2 and passing through the point $(1, -3)$

**8.** Slope $-3$ and passing through the point $(-1, 6)$

**9.** Vertical and passing through the point $(2, 3)$

**10.** Horizontal and passing through the point $(2, 3)$

**11.** Passing through the points $(-1, 3)$ and $(2, -3)$

**12.** Passing through the points $(1, -2)$ and $(3, 4)$

**13–14.** Write an equation of the form $y = mx + b$ for each line graphed below.

**13.**

**14.**

**15. BUSINESS: Straight-Line Depreciation**
A contractor buys a backhoe for $25,000 and estimates its useful life to be 8 years, after which its scrap value will be $1000.
  **a.** Use straight-line depreciation to find a formula for the value $V$ of the backhoe after $t$ years, for $0 \le t \le 8$.
  **b.** Use your formula to find the value of the backhoe after 4 years.

**16. BUSINESS: Straight-Line Depreciation** A trucking company buys a satellite communication system for $78,000 and estimates its useful life to be 15 years, after which its scrap value will be $3000.
  **a.** Use straight-line depreciation to find a formula for the value $V$ of the system after $t$ years, for $0 \le t \le 15$.
  **b.** Use your formula to find the value of the system after 8 years.

**17. PERSONAL FINANCE AND MANAGEMENT: Parking Space in Manhattan** Buying a parking space in some parts of New York City has become as expensive as buying an apartment, on a per-square-foot basis. The following table gives the purchase cost per square foot of a parking space in Manhattan in recent years.

| Year | 1998 | 2000 | 2002 | 2004 | 2006 |
|------|------|------|------|------|------|
| Cost ($) | 150 | 410 | 510 | 880 | 1100 |

  **a.** Number the data columns with $x$-values 1–5 (so that $x$ stands for the *number of two-year intervals since 1996*) and use linear regression to fit a line to the data. State the regression formula.
  **b.** Interpret the slope of the line. From your answer, what is the *yearly* increase?
  **c.** Use the regression line to estimate what the cost was in 2001.
  **d.** Use the regression line to predict the cost in 2012.
  **e.** Since an average parking space is 150 square feet, multiply your answers for parts (c) and (d) by 150 to find the actual costs of the parking space in those years.

*Source: New York Times, July 12, 2007*

## 1.2 Exponents

**18–25.** Evaluate each expression without using a calculator.

**18.** $\left(\frac{1}{6}\right)^{-2}$    **19.** $\left(\frac{4}{3}\right)^{-1}$    **20.** $64^{1/2}$    **21.** $1000^{1/3}$

**22.** $81^{-3/4}$    **23.** $100^{-3/2}$    **24.** $\left(-\frac{8}{27}\right)^{-2/3}$    **25.** $\left(\frac{9}{16}\right)^{-3/2}$

**26–27.** Use a calculator to evaluate each expression. Round answers to two decimal places.

**26.** $3^{2.4}$                 **27.** $12^{1.9}$

**28–29. ENVIRONMENTAL SCIENCE: Animal Size**
It is well known that larger islands or continents can support larger animals. One study found the following relationships between $x =$ land size (in square miles) and $y =$ weight (in pounds) of the "top animal" ever to live on that land mass. Use the formula to estimate the size of the top animal for:

  **a.** Hawaii (4000 square miles).
  **b.** North America (9,400,000 square miles).

**28.** $y = 0.86x^{0.47}$ (for cold-blooded meat-eating animals, such as lizards and crocodiles)

**29.** $y = 1.7x^{0.52}$ (for warm-blooded plant-eating animals, which includes many mammals)

*Source: Proceedings of the National Academy of Sciences 98*

**30. BUSINESS: Satellite Radio** The operating revenues for Sirius Satellite Radio for recent years are given in the following table.

| Year | 2003 | 2004 | 2005 | 2006 |
|---|---|---|---|---|
| Operating Revenues (million $) | 12.9 | 66.9 | 242.2 | 637.2 |

**a.** Number the data columns with $x$-values 1–4 (so that $x$ stands for *years since 2002*), use power regression to fit a curve to the data, and state the regression formula.
**b.** Use the regression function to predict Sirius's operating revenue in the year 2012.

*Source:* Standard & Poor's Industry Surveys

## 1.3 Functions: Linear and Quadratic

**31–34.** For each function:

**a.** Evaluate the given expression.
**b.** Find the domain.
**c.** Find the range.

**31.** $f(x) = \sqrt{x - 7}$;  find $f(11)$

**32.** $g(t) = \dfrac{1}{t + 3}$;  find $g(-1)$

**33.** $h(w) = w^{-3/4}$;  find $h(16)$

**34.** $w(z) = z^{-4/3}$;  find $w(8)$

**35–36.** Determine whether each graph defines a function of $x$.

**35.**

**36.**

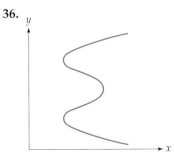

**37–40.** Graph each function.

**37.** $f(x) = 4x - 8$     **38.** $f(x) = 6 - 2x$

**39.** $f(x) = -2x^2 - 4x + 6$

**40.** $f(x) = 3x^2 - 6x$

**41–44.** Solve each equation by **a.** factoring and **b.** the Quadratic Formula.

**41.** $3x^2 + 9x = 0$       **42.** $2x^2 - 8x - 10 = 0$

**43.** $3x^2 + 3x + 5 = 11$     **44.** $4x^2 - 2 = 2$

**45–46.** For each quadratic function:

**a.** Find the vertex using the vertex formula.
**b.** Graph the function on an appropriate window. (Answers may vary.)

**45.** $f(x) = x^2 - 10x - 25$

**46.** $f(x) = x^2 + 14x - 15$

**47. BUSINESS: Car Rentals** A rental company rents cars for $45 per day and $0.25 per mile. Find a function $C(x)$ for the cost of a rented car driven for $x$ miles in a day.

**48. BUSINESS: Simple Interest** If money is borrowed for less than a year, the interest is often calculated as *simple* interest, according to the formula   Interest $= P \cdot r \cdot t$,  where $P$ is the principal, $r$ is the rate (expressed as a decimal), and $t$ is the time (in years). Find a function $I(t)$ for the interest charged on a loan of $10,000 at an interest rate of 8% for $t$ years. Simplify your answer.

**49. ENVIRONMENTAL SCIENCES: Air Temperature** The air temperature decreases by about 1 degree Fahrenheit for each 300 feet of altitude. Find a function $T(x)$ for the temperature at an altitude of $x$ feet if the sea level temperature is 70°.

*Source:* Federal Aviation Administration

**50. ENVIRONMENTAL SCIENCES: Carbon Dioxide Pollution** The burning of fossil fuels (such as oil and coal) added 27 billion tons of carbon dioxide to the atmosphere during 2000, and this annual amount is growing by 0.58 billion tons per year. Find a function $C(t)$ for the amount of carbon dioxide added during the year $t$ years after 2000, and use the formula to find how soon this annual amount will reach 30 billion tons. [*Note:* Carbon dioxide traps solar heat, increasing the earth's temperature, and may lead to flooding of lowland areas by melting the polar ice.]

*Source:* U.S. Environmental Protection Agency

51. **BUSINESS: Break-Even Points and Maximum Profit** A store that installs satellite TV receivers finds that if it installs $x$ receivers per week, its costs will be $C(x) = 80x + 1950$ and its revenue will be $R(x) = -2x^2 + 240x$ (both in dollars).

   **a.** Find the store's break-even points.

   **b.** Find the number of receivers the store should install to maximize profit, and the maximum profit.

52. **BUSINESS: Break-Even Points and Maximum Profit** An air conditioner outlet finds that if it sells $x$ air conditioners per month, its costs will be $C(x) = 220x + 202,500$ and its revenue will be $R(x) = -3x^2 + 2020x$ (both in dollars).

   **a.** Find the outlet's break-even points.

   **b.** Find the number of air conditioners the outlet should sell to maximize profit, and the maximum profit.

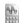 53. **BUSINESS: Boeing Revenues** Operating revenues of Boeing Corporation for recent years are given in the following table.

| Year | 2001 | 2002 | 2003 | 2004 | 2005 | 2006 |
|---|---|---|---|---|---|---|
| Revenues (billion $) | 58.2 | 54.1 | 50.5 | 52.5 | 54.8 | 61.5 |

   **a.** Number the data columns with $x$-values 1–6 (so that $x$ stands for *years since 2000*), use quadratic regression to fit a curve to the data, and state the regression formula. [*Hint:* See Example 10 on pages 43–44.]

   **b.** Use the regression function to predict Boeing's operating revenues in the year 2012.

*Sources:* Standard & Poor's Industry Surveys, Boeing Corporation

## 1.4 Functions: Polynomial, Rational, and Exponential

**54–57.** For each function:
**a.** Evaluate the given expression.
**b.** Find the domain.
**c.** Find the range.

54. $f(x) = \dfrac{3}{x(x-2)}$; find $f(-1)$

55. $f(x) = \dfrac{16}{x(x+4)}$; find $f(-8)$

56. $g(x) = 9^x$; find $g\left(\tfrac{3}{2}\right)$

57. $g(x) = 8^x$; find $g\left(\tfrac{5}{3}\right)$

**58–61.** Solve each equation by factoring.

58. $5x^4 + 10x^3 = 15x^2$     59. $4x^5 + 8x^4 = 32x^3$

60. $2x^{5/2} - 8x^{3/2} = 10x^{1/2}$

61. $3x^{5/2} + 3x^{3/2} = 18x^{1/2}$

**62–65.** Graph each function.

62. $f(x) = 4^x$            63. $f(x) = (x-2)^2 - 4$

64. $f(x) = \begin{cases} 3x - 7 & \text{if } x \geq 2 \\ -x - 1 & \text{if } x < 2 \end{cases}$

(If you use a graphing calculator for Exercises 64 and 65, be sure to indicate any missing points.)

65. $f(x) = \begin{cases} 6 - 2x & \text{if } x > 2 \\ 2x - 1 & \text{if } x \leq 2 \end{cases}$

**66–69.** For each pair of functions $f(x)$ and $g(x)$, find **a.** $f(g(x))$, **b.** $g(f(x))$.

66. $f(x) = x^2 + 1$; $g(x) = \dfrac{1}{x}$

67. $f(x) = \sqrt{x}$; $g(x) = 5x - 4$

68. $f(x) = \dfrac{x+1}{x-1}$; $g(x) = x^3$

69. $f(x) = 2^x$; $g(x) = x^2$

**70–71.** For each function, find and simplify the difference quotient $\dfrac{f(x+h) - f(x)}{h}$.

70. $f(x) = 2x^2 - 3x + 1$     71. $f(x) = \dfrac{5}{x}$

72. **BUSINESS: Advertising Budget** A company's advertising budget is $A(p) = 2p^{0.15}$, where $p$ is the company's profit, and the profit is predicted to be $p(t) = 18 + 2t$, where $t$ is the number of years from now. (Both $A$ and $p$ are in millions of dollars.) Express the advertising budget $A$ as a function of $t$, and evaluate the function at $t = 4$.

73. **a.** Solve the equation $x^4 - 2x^3 - 3x^2 = 0$ by factoring.

    **b.** Use a graphing calculator to graph $y = x^4 - 2x^3 - 3x^2$ and find the $x$-intercepts of the graph. Be sure that you understand why your answers to parts (a) and (b) agree.

74. **a.** Solve the equation $x^3 + 2x^2 - 3x = 0$ by factoring.

    **b.** Use a graphing calculator to graph $y = x^3 + 2x^2 - 3x$ and find the $x$-intercepts of the graph. Be sure that you understand why your answers to parts (a) and (b) agree.

**75.** **SOCIAL SCIENCE: Health Care Expenses**
Health care expenses in the United States have been growing rapidly for several decades. The following table gives per capita annual health care expenses for selected years.

| Year | 1990 | 1995 | 2000 | 2005 |
|---|---|---|---|---|
| Per Capita Expenditure | $2627 | $3546 | $4475 | $6270 |

**a.** Number the data columns with $x$-values 0–3 (so that $x$ stands for the *number of five-year intervals since 1990*) and use exponential regression to fit a curve to the data. State the regression formula. [*Hint:* See Example 12 on pages 62–63.]

**b.** Use the regression function to predict annual health care expenses in the year 2015.

*Source:* U.S. Census Bureau, *Statistical Abstract 2008*

# 2

# Derivatives and Their Uses

### Temperature, Superconductivity, and Limits

It has long been known that there is a coldest possible temperature, called **absolute zero,** the temperature of an object if all heat could be removed from it. On the Fahrenheit temperature scale, absolute zero is about 460 degrees below zero. On the "absolute" or "Kelvin" temperature scale (named after the nineteenth-century scientist Lord Kelvin), absolute zero temperature is assigned the value 0. Absolute zero is a temperature that can be *approached* but never actually *reached*. At temperatures approaching absolute zero, some metals become increasingly able to conduct electricity, with efficiencies approaching 100%, a state called *superconductivity.* The graph below gives the electrical conductivity of aluminum, showing the remarkable fact that as temperature decreases to absolute zero, aluminum becomes *superconducting*—its conductivity approaches a limit of 100% (see Exercise 83 on page 93).

Conductivity of aluminum as a function of temperature

If $f(t)$ stands for the percent conductivity of aluminum at temperature $t$ degrees Kelvin, then the fact that conductivity approaches 100 as temperature decreases to zero may be written

$$\lim_{t \to 0^+} f(t) = 100 \qquad \text{Limit as } t \to 0^+ \text{ of } f(t) \text{ is } 100$$

This is an example of *limits*, which are discussed in Section 2.1. Superconductivity has many commercial applications, from high-speed "mag-lev" trains that "float" on magnetic fields above the tracks to supercomputers and medical imaging devices.

*Source:* Michael Tinkham, *Introduction to Superconductivity*

## 2.1 LIMITS AND CONTINUITY

### Introduction

This chapter introduces the **derivative,** one of the most important concepts in all of calculus. We begin by discussing two preliminary topics, **limits** and **continuity,** both of which will be treated intuitively rather than formally and will be useful when we define the derivative in the next section.

### Limits

The word "limit" is used in everyday conversation to describe the ultimate behavior of something, as in the "limit of one's endurance" or the "limit of one's patience." In mathematics, the word "limit" has a similar but more precise meaning. The notation $x \to 3$ (read: "$x$ approaches 3") means that $x$ takes values *arbitrarily close to 3 without ever equaling 3*. Given a function $f(x)$, if $x$ approaching 3 causes the function to take values approaching (or equaling) some particular number, such as 10, then we will call 10 *the limit of the function* and write

$$\lim_{x \to 3} f(x) = 10 \qquad \text{Limit of } f(x) \text{ as } x \text{ approaches 3 is 10}$$

 **BE CAREFUL:** $x \to 3$ means that $x$ takes values closer and closer to 3 *but never equals 3.*

In practice, the two simplest ways we can approach 3 are *from the left* or *from the right*. For example, the numbers 2.9, 2.99, 2.999, ... approach 3 *from the left*, which we denote by $x \to 3^-$, and the numbers 3.1, 3.01, 3.001, ... approach 3 *from the right*, denoted by $x \to 3^+$. Such limits are called *one-sided limits*.

The following example shows how to find limits from tables of values of the function.

---

**EXAMPLE 1**

### FINDING A LIMIT BY TABLES

Use tables to find $\lim_{x \to 3} (2x + 4)$.     Limit of $2x + 4$ as $x$ approaches 3

*Solution*

We make two tables, as shown below, one with $x$ approaching 3 *from the left*, and the other with $x$ approaching 3 *from the right*.

| $x$ | $2x + 4$ |
|---|---|
| 2.9 | 9.8 |
| 2.99 | 9.98 |
| 2.999 | 9.998 |

Approaching 3 *from the left*

Limit is 10

This table shows $\lim\limits_{x \to 3^-} (2x + 4) = 10$

| $x$ | $2x + 4$ |
|---|---|
| 3.1 | 10.2 |
| 3.01 | 10.02 |
| 3.001 | 10.002 |

Approaching 3 *from the right*

Limit is 10

This table shows $\lim\limits_{x \to 3^+} (2x + 4) = 10$

$f(x) = 2x + 4$

Choosing $x$-values even closer to 3 (such as 2.9999 or 3.0001) would result in values of $2x + 4$ *even closer to 10*, so that both one-sided limits equal 10:

$$\lim_{x \to 3^-} (2x + 4) = 10 \quad \text{and} \quad \lim_{x \to 3^+} (2x + 4) = 10$$

Since approaching 3 from *either* side causes $2x + 4$ to approach the same number, 10, we may state that *the limit* is 10:

$$\lim_{x \to 3} (2x + 4) = 10 \qquad \text{Limit of } 2x + 4 \text{ as } x \text{ approaches 3 is 10}$$

This limit says that as $x$ approaches 3 from *either side*, or even *alternating sides*, as in 2.9, 3.01, 2.999, 3.0001, ... , the values of $2x + 4$ will approach 10. This may be seen in the succession of graphs on the left: as $x \to 3$ (on the $x$-axis), $f(x) \to 10$ (on the $y$-axis).

$x \to 3 \leftarrow x$

Limits can be defined with greater rigor,* but we will use the following intuitive definition to express the idea that $\lim\limits_{x \to c} f(x) = L$ means that $f(x)$ *approaches the number L as x approaches the number c.*

### Limits

The statement

$$\lim_{x \to c} f(x) = L \qquad \text{Limit of } f(x) \text{ as } x \text{ approaches } c \text{ is } L$$

means that the values of $f(x)$ can be made arbitrarily close to $L$ by taking values of $x$ sufficiently close (but not equal) to $c$.

The one-sided limits

$$\lim_{x \to c^-} f(x) = L \qquad \text{Limit of } f(x) \text{ as } x \text{ approaches } c \text{ from the } left \text{ is } L$$

$$\lim_{x \to c^+} f(x) = L \qquad \text{Limit of } f(x) \text{ as } x \text{ approaches } c \text{ from the } right \text{ is } L$$

have similar meanings but with the $x$-values restricted to, respectively, $x < c$ and $x > c$.

*More precisely, the definition of $\lim\limits_{x \to c} f(x) = L$ is that for every number $\varepsilon > 0$ there is a number $\delta > 0$ such that $|f(x) - L| < \varepsilon$ whenever $0 < |x - c| < \delta$.

The limit $\lim_{x \to c} f(x)$ is sometimes called a *two-sided limit* to distinguish it from one-sided limits. A limit may fail to exist, as we will see in Example 5, but if the limit *does* exist, it must be a *single number.* As we saw in the preceding Example if both one-sided limits exist and have the same value, then the (two-sided) limit will exist and *have this same value.* In the Application Preview on page 77, we used the notation $t \to 0^+$ to indicate the one-sided limit as the temperature approached absolute zero through positive values.

The correct limit in Example 1 could have been found simply by *evaluating* the function at $x = 3$:

$$f(3) = 2 \cdot 3 + 4 = 10$$

$f(x) = 2x + 4$ evaluated at $x = 3$ gives the correct limit, 10

However, finding limits by this technique of *direct substitution* is not always possible, as the next Example shows.

---

**EXAMPLE 2**

**FINDING A LIMIT BY TABLES**

Find $\lim_{x \to 0} (1 + x)^{1/x}$ correct to three decimal places.

*Solution*

As before, we make tables with $x$ approaching 0 (the given $x$-value) from the left and from the right, with the values of $(1 + x)^{1/x}$ found from a calculator.

| $x$ | $(1 + x)^{1/x}$ | $x$ | $(1 + x)^{1/x}$ |
|---|---|---|---|
| 0.1 | 2.594 | −0.1 | 2.868 |
| 0.01 | 2.705 | −0.01 | 2.732 |
| 0.001 | 2.717 | −0.001 | 2.720 |
| 0.0001 | 2.718 | −0.0001 | 2.718 |

To use a graphing calculator to find these numbers, enter the function as $(1 + x)^\wedge(1/x)$, and use the TABLE feature.

Since the last row of the tables show that the two one-sided limits have the same value, this common value of the right and left limit is *the limit:*

$$\lim_{x \to 0} (1 + x)^{1/x} \approx 2.718$$

From the agreement in the last columns of the tables

This limit may be seen from the graph of $f(x) = (1 + x)^{1/x}$ shown on the left. Notice that the limiting point on the $y$-axis is *missing* since the function is not defined at $x = 0$ (because the exponent would be 1/0).

---

**Practice Problem 1**

Evaluate $(1 + x)^{1/x}$ at $x = 0.000001$ and $x = -0.000001$. Based on the results, give a better approximation for $\lim_{x \to 0} (1 + x)^{1/x}$.

➤ **Solution on page 89**

The actual value of this particular limit is the number that was discussed on page 54, $e \approx 2.71828$, which will be very important in our later work. Since $(1 + x)^{1/x}$ cannot be evaluated *at* $x = 0$, this limit *requires* the limit process.

Which limits can be evaluated by **direct substitution** (as in Example 1) and which cannot (as in Example 2)? The answer comes from the following "Rules of Limits."

## Rules of Limits

For any constants $a$ and $c$, and any positive integer $n$:

**1.** $\lim\limits_{x \to c} a = a$      The limit of a constant is just the constant

**2.** $\lim\limits_{x \to c} x^n = c^n$      The limit of a power is the power of the limit

**3.** $\lim\limits_{x \to c} \sqrt[n]{x} = \sqrt[n]{c}$    $(c > 0$ if $n$ is even$)$      The limit of a root is the root of the limit

**4.** If $\lim\limits_{x \to c} f(x)$ and $\lim\limits_{x \to c} g(x)$ both exist, then

    **a.** $\lim\limits_{x \to c} [f(x) + g(x)] = \lim\limits_{x \to c} f(x) + \lim\limits_{x \to c} g(x)$      The limit of a sum is the sum of the limits

    **b.** $\lim\limits_{x \to c} [f(x) - g(x)] = \lim\limits_{x \to c} f(x) - \lim\limits_{x \to c} g(x)$      The limit of a difference is the difference of the limits

    **c.** $\lim\limits_{x \to c} [f(x) \cdot g(x)] = [\lim\limits_{x \to c} f(x)] \cdot [\lim\limits_{x \to c} g(x)]$      The limit of a product is the product of the limits

    **d.** $\lim\limits_{x \to c} \dfrac{f(x)}{g(x)} = \dfrac{\lim\limits_{x \to c} f(x)}{\lim\limits_{x \to c} g(x)}$    $(\text{if } \lim\limits_{x \to c} g(x) \neq 0)$      The limit of a quotient is the quotient of the limits

These rules, which may be proved from the definition of limit, can be summarized as follows.

## Summary of Rules of Limits

For functions composed of additions, subtractions, multiplications, divisions, powers, and roots, limits may be evaluated by direct substitution, provided that the resulting expression is defined.

$$\lim_{x \to c} f(x) = f(c)$$      Limit evaluated by direct substitution

The *rules of limits* and also the above *summary* hold for one-sided limits as well as for regular (two-sided) limits.

**EXAMPLE 3**

### FINDING LIMITS BY DIRECT SUBSTITUTION

    **a.** $\lim\limits_{x \to 4} \sqrt{x} = \sqrt{4} = 2$      Direct substitution of $x = 4$ using Rule 3 or the Summary

    **b.** $\lim\limits_{x \to 6} \dfrac{x^2}{x + 3} = \dfrac{6^2}{6 + 3} = \dfrac{36}{9} = 4$      Direct substitution of $x = 6$ (Rules 4, 2, and 1 or the Summary)

**Practice Problem 2**     Find $\lim\limits_{x \to 3} (2x^2 - 4x + 1)$.      ▶ Solution on page 89

If direct substitution into a quotient gives the undefined expression $\frac{0}{0}$, the limit may still exist. Try factoring, simplifying, and *then* using direct substitution.

---

**EXAMPLE 4**

**FINDING A LIMIT BY SIMPLIFYING**

Find $\lim\limits_{x \to 1} \dfrac{x^2 - 1}{x - 1}$.

*Solution*

Direct substitution of $x = 1$ into $\dfrac{x^2 - 1}{x - 1}$ gives $\dfrac{0}{0}$, which is undefined. But factoring and simplifying give

$$\lim_{x \to 1} \frac{x^2 - 1}{x - 1} = \lim_{x \to 1} \underbrace{\frac{(x + 1)(x - 1)}{x - 1}}_{\substack{\text{Factoring the} \\ \text{numerator}}} = \lim_{x \to 1} \underbrace{\frac{(x + 1)(x - 1)}{x - 1}}_{\substack{\text{Canceling the} \\ (x - 1)\text{'s} \\ (\text{since } x \neq 1)}} = \lim_{x \to 1} \underbrace{(x + 1) = 2}_{\substack{\text{Now use} \\ \text{direct} \\ \text{substitution}}}$$

Therefore, the limit *does* exist and equals 2.

---

## Graphing Calculator Exploration

on $[-2, 4]$ by $[-2, 5]$

The graph of $f(x) = \dfrac{x^2 - 1}{x - 1}$ (from Example 4) is shown on the left.

**a.** Can you explain why the graph appears to be a straight line?

**b.** From your knowledge of rational functions, should the graph really be a (complete) line? [*Hint:* See pages 52–53.] Can you see that a point is indeed missing from the graph?*

**c.** Use the graph on the left (where each tick mark is 1 unit) or enter the function and use TRACE on your calculator to verify that the limit as $x$ approaches 1 is indeed 2.

*To have your calculator show the point as missing, choose a window such that the $x$-value in question is midway between XMIN and XMAX, or see the *Useful Hint* in *Graphing Calculator Basics* in the Preface on page xxii.

---

**Practice Problem 3**

Find $\lim\limits_{x \to 5} \dfrac{2x^2 - 10x}{x - 5}$

➤ Solution on page 89

A limit (or the fact that a limit does not exist) may be found from a graph.

**EXAMPLE 5**

**FINDING THAT A LIMIT DOES NOT EXIST**

For the piecewise linear function $f(x) = \begin{cases} x + 1 & \text{if } x \le 3 \\ 8 - 2x & \text{if } x > 3 \end{cases}$ graphed on the left, find the following limits or state that they do not exist.

**a.** $\lim\limits_{x \to 3^-} f(x)$      **b.** $\lim\limits_{x \to 3^+} f(x)$      **c.** $\lim\limits_{x \to 3} f(x)$

We give two solutions, one using the graph and one using the expression for the function. Both methods are important.

*Solution*   (Using the graph)

Use this line for $x \to 3^-$

Use this line for $x \to 3^+$

**a.** $\lim\limits_{x \to 3^-} f(x) = 4$

Approaching 3 *from the left* means using the line on the *left* of $x = 3$, which approaches height 4

**b.** $\lim\limits_{x \to 3^+} f(x) = 2$

Approaching 3 *from the right* means using the line on the *right* of $x = 3$, which approaches height 2

**c.** $\lim\limits_{x \to 3} f(x)$  does not exist

The two one-sided limits both exist, but they have different values (4 and 2), so *the limit does not exist*

*Solution* $\left( \text{Using } f(x) = \begin{cases} x + 1 & \text{if } x \le 3 \\ 8 - 2x & \text{if } x > 3 \end{cases} \right)$

$f(x) = \begin{cases} x + 1 & \text{if } x \le 3 \\ 8 - 2x & \text{if } x > 3 \end{cases}$

**a.** For $x \to 3^-$ we have $x < 3$, so $f(x)$ is given by the *upper* line of the function:
$$\lim_{x \to 3^-} f(x) = \lim_{x \to 3^-} (x + 1) = 3 + 1 = 4$$

$f(x) = \begin{cases} x + 1 & \text{if } x \le 3 \\ 8 - 2x & \text{if } x > 3 \end{cases}$

**b.** For $x \to 3^+$ we have $x > 3$, so $f(x)$ is given by the *lower* line of the function:
$$\lim_{x \to 3^+} f(x) = \lim_{x \to 3^+} (8 - 2x) = 8 - 2 \cdot 3 = 2$$

**c.** $\lim\limits_{x \to 3} f(x)$ does not exist since although the two one-sided limits exist, they are not equal.

Notice that the two methods found the same answers.

**Practice Problem 4**

Explain the difference among $x \to 3^-$, $x \to -3$, and $x \to -3^-$.

➤ **Solution on page 89**

## Limits Involving Infinity

A limit statement such as $\lim\limits_{x \to c} f(x) = \infty$ (the symbol $\infty$ is read "infinity") does *not* mean that the function takes values near "infinity," since there is no number "infinity." It means, instead, that the values of the function *become arbitrarily large* (the graph rises arbitrarily high) near the number $c$. Similarly, the limit statement $\lim\limits_{x \to c} f(x) = -\infty$ means that the values of the function *become arbitrarily small* (the graph falls arbitrarily low) near the number $c$. Such statements, which may also be written with one-sided limits, mean that the graph has a *vertical asymptote* at $x = c$, as in the following Example.

---

**EXAMPLE 6**

**FINDING LIMITS INVOLVING $\pm\infty$**

Describe the asymptotic behavior of $f(x) = \dfrac{3x + 2}{x - 2}$ using limits involving $\pm\infty$.

*Solution*

The graph of $f(x) = \dfrac{3x + 2}{x - 2}$ is shown below. It is undefined at $x = 2$ (the denominator would be zero), as indicated by the vertical dashed line. Notice that as $x$ approaches 2 *from the right*, the curve rises arbitrarily *high*, and as $x$ approaches 2 *from the left*, the curve falls arbitrarily *low*, as expressed in the limit statements below. The tables on the right show these results numerically.

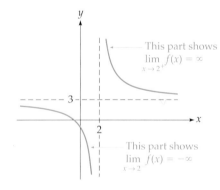

This part shows $\lim\limits_{x \to 2^+} f(x) = \infty$

This part shows $\lim\limits_{x \to 2^-} f(x) = -\infty$

Graph of $f(x) = \dfrac{3x + 2}{x - 2}$

| $x$ | $\dfrac{3x + 2}{x - 2}$ |
|-----|-------------------------|
| 2.1 | 83 |
| 2.01 | 803 |
| 2.001 | 8003 |

This table shows $\lim\limits_{x \to 2^+} f(x) = \infty$

| $x$ | $\dfrac{3x + 2}{x - 2}$ |
|-----|-------------------------|
| 1.9 | $-77$ |
| 1.99 | $-797$ |
| 1.999 | $-7997$ |

This table shows $\lim\limits_{x \to 2^-} f(x) = -\infty$

Since the limit from one side was $\infty$ and from the other was $-\infty$, the (two-sided) limit $\lim\limits_{x \to 2} f(x)$ *does not exist.* (If *both* one-sided limits had yielded $\infty$, we could have stated $\lim\limits_{x \to 2} f(x) = \infty$, and if both had yielded $-\infty$ we could have stated $\lim\limits_{x \to 2} f(x) = -\infty$.)

For $x$-values *arbitrarily far out to the right* (denoted $x \to \infty$), the curve levels off approaching height 3, which we express as $\lim\limits_{x \to \infty} f(x) = 3$, and for $x$-values *arbitrarily far out to the left* (denoted $x \to -\infty$), the curve again levels off at height 3, which we express as $\lim\limits_{x \to -\infty} f(x) = 3$. These results are shown below both graphically and numerically: the graph has a *horizontal asymptote* at $y = 3$ and the tables have (rounded) function values approaching 3.

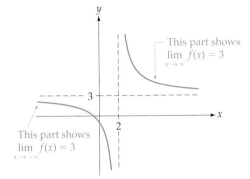

This part shows
$\lim_{x \to \infty} f(x) = 3$

This part shows
$\lim_{x \to -\infty} f(x) = 3$

Graph of $f(x) = \dfrac{3x + 2}{x - 2}$

| $x$ | $\dfrac{3x + 2}{x - 2}$ |
|---|---|
| 100 | 3.082 |
| 1000 | 3.008 |
| 10000 | 3.001 |

This table shows
$\lim_{x \to \infty} f(x) = 3$

| $x$ | $\dfrac{3x + 2}{x - 2}$ |
|---|---|
| $-100$ | 2.922 |
| $-1000$ | 2.992 |
| $-10000$ | 2.999 |

This table shows
$\lim_{x \to -\infty} f(x) = 3$

Summarizing:

### Limits Involving Infinity

| | |
|---|---|
| $\lim_{x \to c^+} f(x) = \infty$ | means that the values of $f(x)$ grow arbitrarily large as $x$ approaches $c$ from the right |
| $\lim_{x \to c^-} f(x) = \infty$ | means that the values of $f(x)$ grow arbitrarily large as $x$ approaches $c$ from the left |
| $\lim_{x \to c} f(x) = \infty$ | means that *both* of the above statements are true |

Similar statements hold if $\infty$ is replaced by $-\infty$ and the words "arbitrarily large" by "arbitrarily small."

| | |
|---|---|
| $\lim_{x \to \infty} f(x) = L$ | means that the values of $f(x)$ become arbitrarily close to the number $L$ as $x$ becomes arbitrarily large |
| $\lim_{x \to -\infty} f(x) = L$ | means that the values of $f(x)$ become arbitrarily close to the number $L$ as $x$ becomes arbitrarily small |

**BE CAREFUL:** To say that a limit *exists* means that the limit is a *number*, and since $\infty$ and $-\infty$ are not numbers, a statement such as $\lim_{x \to c} f(x) = \infty$ means that *the limit does not exist*. The limit statement $\lim_{x \to c} f(x) = \infty$ goes further to explain *why* the limit does not exist: the function values become too large to approach any limit.

**Practice Problem 5**

Use limits involving $\pm\infty$ to describe the asymptotic behavior of the function $f(x) = \dfrac{1}{(x + 2)^2}$ from its graph.

➤ Solution on page 89

### Limits of Functions of Two Variables

Some limits involve two variables, with only one variable approaching a limit.

**EXAMPLE 7**

**FINDING A LIMIT OF A FUNCTION OF TWO VARIABLES**

Find $\lim\limits_{h \to 0} (x^2 + xh + h^2)$.

*Solution*

Only $h$ is approaching zero, so $x$ remains unchanged. Since the function involves only powers of $h$, we may evaluate the limit by direct substitution of $h = 0$:

$$\lim_{h \to 0} (x^2 + xh + h^2) = x^2 + \underbrace{x \cdot 0}_{0} + \underbrace{0^2}_{0} = x^2$$

**Practice Problem 6**

Find $\lim\limits_{h \to 0} (3x^2 + 5xh + 1)$.

➤ Solution on page 89

### Continuity

Intuitively, a function is said to be **continuous** at $c$ if its graph passes through the point at $x = c$ *without a "hole" or a "jump."* For example, the first function below is *continuous* at $c$ (it has no hole or jump at $x = c$), while the second and third are **discontinuous** at $c$.

| Continuous at $c$ | Discontinuous at $c$ | Discontinuous at $c$ |

In other words, a function is *continuous at c* if the curve *approaches the point at* $x = c$, which may be stated in terms of limits:

$$\lim_{x \to c} f(x) = f(c) \qquad \text{Height of the } \textit{curve} \text{ approaches the height of the } \textit{point}$$

This equation means that the quantities on both sides must exist and be *equal*, which we make explicit as follows:

**Continuity**

A function $f$ is continuous at $c$ if the following three conditions hold:

1. $f(c)$ is defined      Function is *defined* at $c$
2. $\lim\limits_{x \to c} f(x)$ exists      Left and right limits exist and agree
3. $\lim\limits_{x \to c} f(x) = f(c)$      Limit and value *at c* agree

$f$ is *discontinuous* at $c$ if one or more of these conditions *fails* to hold.

Condition 3, which is just the statement that the expressions in Conditions 1 and 2 are equal to each other, may by itself be taken as the definition of continuity.

| EXAMPLE 8 | **FINDING DISCONTINUITIES FROM A GRAPH** |

Each function below is *discontinuous at c* for the indicated reason.

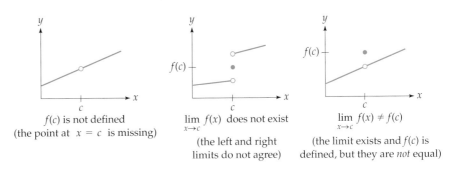

$f(c)$ is not defined
(the point at $x = c$ is missing)

$\lim_{x \to c} f(x)$ does not exist

(the left and right
limits do not agree)

$\lim_{x \to c} f(x) \neq f(c)$

(the limit exists and $f(c)$ is
defined, but they are *not* equal)

| Practice Problem 7 | For each graph below, determine whether the function is continuous at $c$. If it is *not* continuous, indicate the *first* of the three conditions in the definition of continuity (on the previous page) that is violated. |

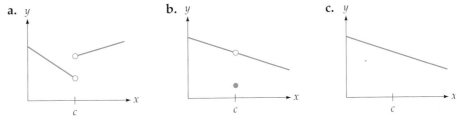

**a.** **b.** **c.**

➤ Solutions on page 89

## Continuity on Intervals

A function is continuous on an *open interval* $(a, b)$ if it is continuous at each point of the interval. A function is continuous on a *closed interval* $[a, b]$ if it is continuous on the open interval $(a, b)$ and has "one-sided continuity" at the endpoints: $\lim_{x \to a^+} f(x) = f(a)$ and $\lim_{x \to b^-} f(x) = f(b)$. A function that is continuous on the entire real line $(-\infty, \infty)$ is said to be *continuous everywhere*, or simply *continuous*.

## Which Functions Are Continuous?

Which functions are continuous? *Linear* and *quadratic* functions are continuous, since their graphs are, respectively, straight lines and parabolas, with no holes or jumps. Similarly, *exponential* functions are continuous and *logarithmic* functions are continuous on their domains, as may be seen from the graphs on pages 53–54. These and other continuous functions can be combined as follows to give other continuous functions.

---

**Continuous Functions**

If functions $f$ and $g$ are continuous at $c$, then the following are also continuous at $c$:

| | |
|---|---|
| **1.** $f \pm g$ | Sums and differences of continuous functions are continuous |
| **2.** $a \cdot f$ [for any constant $a$] | Constant multiples of continuous functions are continuous |
| **3.** $f \cdot g$ | Products of continuous functions are continuous |
| **4.** $f/g$ [if $g(c) \neq 0$] | Quotients of continuous functions are continuous |
| **5.** $f(g(x))$ [for $f$ continuous at $g(c)$] | Compositions of continuous functions are continuous |

---

These statements, which can be proved from the Rules of Limits, show that the following types of functions are continuous:

Every polynomial function is continuous.

Every rational function is continuous except where the denominator is zero.

---

**EXAMPLE 9**

**DETERMINING CONTINUITY**

Determine whether each function is continuous or discontinuous. If discontinuous, state *where* it is discontinuous.

**a.** $f(x) = x^3 - 3x^2 - x + 3$ **b.** $f(x) = \dfrac{1}{(x + 1)^2}$ **c.** $f(x) = e^{x^2 - 1}$

*Solution*

**a.** The first function is continuous since it is a polynomial.

**b.** The second function is discontinuous at $x = -1$ (where the denominator is zero). It is continuous at all other $x$-values.

**c.** The third function is continuous since it is the composition of the exponential function $e^x$ and the polynomial $x^2 - 1$.

---

Calculator-drawn graphs of the functions in Example 9 are shown on the next page. The polynomial (on the left) and the exponential function (on the right) are continuous, although you can't really tell from such graphs. The rational function (in the middle) exhibits the discontinuity at $x = -1$.

$f(x) = x^3 - 3x^2 - x + 3$
on $[-5, 5]$ by $[-5, 10]$

$f(x) = \dfrac{1}{(x + 1)^2}$
on $[-5, 5]$ by $[-5, 10]$

$f(x) = e^{x^2 - 1}$
on $[-2, 2]$ by $[-1, 4]$

## Relationship Between Limits and Continuity

We first defined continuity intuitively, saying that a function is continuous at $c$ if its graph passes through the point at $x = c$ without a "hole" or a "jump." We then used limits to define continuity more precisely, saying that to be continuous at $c$, $f$ must satisfy

$$\lim_{x \to c} f(x) = f(c)$$

However, this is just the equation that we encountered on page 81 in finding which limits may be evaluated by *direct substitution*. We may now use continuity to restate this result more succinctly: The limit $\lim_{x \to c} f(x)$ may be evaluated by direct substitution if and only if $f$ is continuous at $c$.

### Solutions to Practice Problems

1. $\lim_{x \to 0} (1 + x)^{1/x} \approx 2.71828$

2. $\lim_{x \to 3} (2x^2 - 4x + 1) = 2 \cdot 3^2 - 4 \cdot 3 + 1 = 18 - 12 + 1 = 7$

3. $\lim_{x \to 5} \dfrac{2x^2 - 10x}{x - 5} = \lim_{x \to 5} \dfrac{2x(x - 5)}{x - 5} = \lim_{x \to 5} \dfrac{2x(x - 5)}{x - 5} = \lim_{x \to 5} 2x = 10$

4. $x \to 3^-$ means: $x$ approaches (positive) 3 *from the left.*

   $x \to -3$ means: $x$ approaches $-3$ (the ordinary two-sided limit).

   $x \to -3^-$ means: $x$ approaches $-3$ *from the left.*

5. $\lim_{x \to -\infty} f(x) = 0$; $\lim_{x \to 2^-} f(x) = \infty$ and $\lim_{x \to 2^+} f(x) = \infty$, so that $\lim_{x \to 2} f(x) = \infty$; and $\lim_{x \to \infty} f(x) = 0$.

6. $\lim_{h \to 0} (3x^2 + 5xh + 1) = 3x^2 + 5x \cdot 0 + 1$    (using direct substitution)

   $\qquad\qquad\qquad\qquad = 3x^2 + 1$

7. **a.** Discontinuous, $f(c)$ is not defined    **b.** Discontinuous, $\lim_{x \to c} f(x) \neq f(c)$    **c.** Continuous

## 2.1 Section Summary

Intuitively, the limit $\lim_{x \to c} f(x)$ is the number (if it exists) that $f(x)$ approaches as $x$ approaches the number $c$. More precisely, $\lim_{x \to c} f(x) = L$ if the values of $f$ can be made as close as desired to $L$ by taking values of $x$ sufficiently close (but not equal to) $c$. We may find limits by letting $x$ approach $c$ from just one side and then from the other: the limit exists only if these two *one-sided limits* agree (that is, have the same value). Other ways to find limits are by direct substitution (if the function is continuous) or by algebraic simplification. We then used limits with the symbols $\infty$ and $-\infty$ to indicate that a value becomes arbitrarily large or arbitrarily small.

We defined continuity geometrically (no holes or jumps) and analytically ( $\lim\limits_{x \to c} f(x) = f(c)$ ). We discussed continuity for two of the most useful types of functions: polynomials are continuous everywhere, and rational functions are continuous everywhere except where their denominators are zero.

## 2.1 Exercises

**1–4.** Complete the tables and use them to find the given limits. Round calculations to three decimal places. A graphing calculator with a TABLE feature will be very helpful.

**1. a.** $\lim\limits_{x \to 2^-} (5x - 7)$
   **b.** $\lim\limits_{x \to 2^+} (5x - 7)$
   **c.** $\lim\limits_{x \to 2} (5x - 7)$

| x | 5x − 7 | x | 5x − 7 |
|---|---|---|---|
| 1.9 | | 2.1 | |
| 1.99 | | 2.01 | |
| 1.999 | | 2.001 | |

**2. a.** $\lim\limits_{x \to 4^-} (2x + 1)$
   **b.** $\lim\limits_{x \to 4^+} (2x + 1)$
   **c.** $\lim\limits_{x \to 4} (2x + 1)$

| x | 2x + 1 | x | 2x + 1 |
|---|---|---|---|
| 3.9 | | 4.1 | |
| 3.99 | | 4.01 | |
| 3.999 | | 4.001 | |

**3. a.** $\lim\limits_{x \to 1^-} \left( \dfrac{x^3 - 1}{x - 1} \right)$
   **b.** $\lim\limits_{x \to 1^+} \left( \dfrac{x^3 - 1}{x - 1} \right)$
   **c.** $\lim\limits_{x \to 1} \left( \dfrac{x^3 - 1}{x - 1} \right)$

| x | $\dfrac{x^3 - 1}{x - 1}$ | x | $\dfrac{x^3 - 1}{x - 1}$ |
|---|---|---|---|
| 0.9 | | 1.1 | |
| 0.99 | | 1.01 | |
| 0.999 | | 1.001 | |

**4. a.** $\lim\limits_{x \to 1^-} \left( \dfrac{x^4 - 1}{x - 1} \right)$
   **b.** $\lim\limits_{x \to 1^+} \left( \dfrac{x^4 - 1}{x - 1} \right)$
   **c.** $\lim\limits_{x \to 1} \left( \dfrac{x^4 - 1}{x - 1} \right)$

| x | $\dfrac{x^4 - 1}{x - 1}$ | x | $\dfrac{x^4 - 1}{x - 1}$ |
|---|---|---|---|
| 0.9 | | 1.1 | |
| 0.99 | | 1.01 | |
| 0.999 | | 1.001 | |

**5–8.** Find each limit by constructing tables similar to those in Exercises 1–4.

**5.** $\lim\limits_{x \to 0} (1 + 2x)^{1/x}$

**6.** $\lim\limits_{x \to 0} (1 - x)^{1/x}$

**7.** $\lim\limits_{x \to 2} \dfrac{\frac{1}{x} - \frac{1}{2}}{x - 2}$

**8.** $\lim\limits_{x \to 1} \dfrac{\sqrt{x} - 1}{x - 1}$

**9–12.** Find each limit by graphing the function and using TRACE or TABLE to examine the graph near the indicated $x$-value.

**9.** $\lim\limits_{x \to 1} \dfrac{\frac{1}{x} - 1}{1 - x}$

Use window [0, 2] by [0, 5].

**10.** $\lim\limits_{x \to 1.5} \dfrac{2x^2 - 4.5}{x - 1.5}$

Use window [0, 3] by [0, 10].

**11.** $\lim\limits_{x \to 4} \dfrac{x^{1.5} - 4x^{0.5}}{x^{1.5} - 2x}$

**12.** $\lim\limits_{x \to 1} \dfrac{x - 1}{x - \sqrt{x}}$

[*Hint:* Choose a window whose $x$-values are centered at the limiting $x$-value.]

**13–32.** Find the following limits *without* using a graphing calculator or making tables.

**13.** $\lim\limits_{x \to 3} (4x^2 - 10x + 2)$

**14.** $\lim\limits_{x \to 7} \dfrac{x^2 - x}{2x - 7}$

**15.** $\lim\limits_{x \to 5} \dfrac{3x^2 - 5x}{7x - 10}$

**16.** $\lim\limits_{t \to 3} \sqrt[3]{t^2 + t - 4}$

**17.** $\lim\limits_{x \to 3} \sqrt{2}$

**18.** $\lim\limits_{q \to 9} \dfrac{8 + 2\sqrt{q}}{8 - 2\sqrt{q}}$

**19.** $\lim\limits_{t \to 25} [(t + 5)t^{-1/2}]$

**20.** $\lim\limits_{s \to 4} (s^{3/2} - 3s^{1/2})$

**21.** $\lim\limits_{h \to 0} (5x^3 + 2x^2h - xh^2)$

**22.** $\lim_{h \to 0} (2x^2 + 4xh + h^2)$

**23.** $\lim_{x \to 2} \dfrac{x^2 - 4}{x - 2}$

**24.** $\lim_{x \to 1} \dfrac{x - 1}{x^2 + x - 2}$

**25.** $\lim_{x \to -3} \dfrac{x + 3}{x^2 + 8x + 15}$

**26.** $\lim_{x \to -4} \dfrac{x^2 + 9x + 20}{x + 4}$

**27.** $\lim_{x \to -1} \dfrac{3x^3 - 3x^2 - 6x}{x^2 + x}$

**28.** $\lim_{x \to 0} \dfrac{x^2 - x}{x^2 + x}$

**29.** $\lim_{h \to 0} \dfrac{2xh - 3h^2}{h}$

**30.** $\lim_{h \to 0} \dfrac{5x^4h - 9xh^2}{h}$

**31.** $\lim_{h \to 0} \dfrac{4x^2h + xh^2 - h^3}{h}$

**32.** $\lim_{h \to 0} \dfrac{x^2h - xh^2 + h^3}{h}$

**33–36.** For each piecewise linear function $f(x)$ graphed below, find:

**a.** $\lim_{x \to 2^-} f(x)$    **b.** $\lim_{x \to 2^+} f(x)$    **c.** $\lim_{x \to 2} f(x)$

**33.**

**34.**

**35.**

**36.**

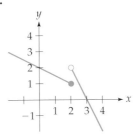

**37–40.** For each piecewise linear function, find:

**a.** $\lim_{x \to 4^-} f(x)$    **b.** $\lim_{x \to 4^+} f(x)$    **c.** $\lim_{x \to 4} f(x)$

**37.** $f(x) = \begin{cases} 3 - x & \text{if } x \leq 4 \\ 10 - 2x & \text{if } x > 4 \end{cases}$

**38.** $f(x) = \begin{cases} 5 - x & \text{if } x < 4 \\ 2x - 5 & \text{if } x \geq 4 \end{cases}$

**39.** $f(x) = \begin{cases} 2 - x & \text{if } x \leq 4 \\ x - 6 & \text{if } x > 4 \end{cases}$

**40.** $f(x) = \begin{cases} 2 - x & \text{if } x < 4 \\ 2x - 10 & \text{if } x \geq 4 \end{cases}$

**41–44.** For each function, find:

**a.** $\lim_{x \to 0^-} f(x)$    **b.** $\lim_{x \to 0^+} f(x)$    **c.** $\lim_{x \to 0} f(x)$

**41.** $f(x) = |x|$

**42.** $f(x) = -|x|$

**43.** $f(x) = \dfrac{|x|}{x}$

**44.** $f(x) = -\dfrac{|x|}{x}$

**45–52.** Use limits involving $\pm\infty$ to describe the asymptotic behavior of each function from its graph.

**45.** $f(x) = \dfrac{1}{3 - x}$

**46.** $f(x) = \dfrac{1}{(x + 2)^2}$

**47.** $f(x) = \dfrac{1}{x^2}$

**48.** $f(x) = \dfrac{1}{x}$

**49.** $f(x) = \dfrac{2x + 1}{x - 1}$

**50.** $f(x) = \dfrac{x - 3}{x + 3}$

**51.** $f(x) = \dfrac{x^2}{(x + 2)^2}$

**52.** $f(x) = \dfrac{2x^2}{(x - 1)^2}$

**53–60.** Determine whether each function is continuous or discontinuous at $c$. If it is discontinuous, indicate the *first* of the three conditions in the definition of continuity (page 86) that is violated.

**53.**

**54.**

**55.**

**56.**

**57.**

**58.**

**59.**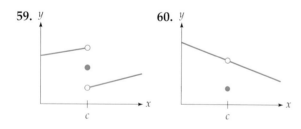

**60.**

**61–64.** For each piecewise linear function:

**a.** Draw its graph (by hand or using a graphing calculator).

**b.** Find the limits as $x$ approaches 3 from the left and from the right.

**c.** Is it continuous at $x = 3$? If not, indicate the first of the three conditions in the definition of continuity (page 86) that is violated.

**61.** $f(x) = \begin{cases} x & \text{if } x \le 3 \\ 6 - x & \text{if } x > 3 \end{cases}$

**62.** $f(x) = \begin{cases} 5 - x & \text{if } x \le 3 \\ x - 2 & \text{if } x > 3 \end{cases}$

**63.** $f(x) = \begin{cases} x & \text{if } x \le 3 \\ 7 - x & \text{if } x > 3 \end{cases}$

**64.** $f(x) = \begin{cases} 5 - x & \text{if } x \le 3 \\ x - 1 & \text{if } x > 3 \end{cases}$

**65–78.** Determine whether each function is continuous or discontinuous. If discontinuous, state where it is discontinuous.

**65.** $f(x) = 7x - 5$

**66.** $f(x) = 5x^3 - 6x^2 + 2x - 4$

**67.** $f(x) = \dfrac{x + 1}{x - 1}$

**68.** $f(x) = \dfrac{x^3}{(x + 7)(x - 2)}$

**69.** $f(x) = \dfrac{12}{5x^3 - 5x}$

**70.** $f(x) = \dfrac{x + 2}{x^4 - 3x^3 - 4x^2}$

**71.** $f(x) = \begin{cases} 3 - x & \text{if } x \le 4 \\ 10 - 2x & \text{if } x > 4 \end{cases}$

[*Hint:* See Exercise 37.]

**72.** $f(x) = \begin{cases} 5 - x & \text{if } x < 4 \\ 2x - 5 & \text{if } x \ge 4 \end{cases}$

[*Hint:* See Exercise 38.]

**73.** $f(x) = \begin{cases} 2 - x & \text{if } x \le 4 \\ x - 6 & \text{if } x > 4 \end{cases}$

[*Hint:* See Exercise 39.]

**74.** $f(x) = \begin{cases} 2 - x & \text{if } x < 4 \\ 2x - 10 & \text{if } x \ge 4 \end{cases}$

[*Hint:* See Exercise 40.]

**75.** $f(x) = |x|$

[*Hint:* See Exercise 41.]

**76.** $f(x) = \dfrac{|x|}{x}$

[*Hint:* See Exercise 43.]

**77.** $f(x) = \begin{cases} x^2 & \text{if } x \le 2 \\ 6 - x & \text{if } 2 < x < 6 \\ 2x - 17 & \text{if } x \ge 6 \end{cases}$

[*Hint for Exercises 77 and 78:* You may have graphed these piecewise nonlinear functions in Exercises 31 and 32 on page 66.]

**78.** $f(x) = \begin{cases} 4 - x^2 & \text{if } x < 3 \\ 2x - 11 & \text{if } 3 \le x < 7 \\ 8 - x & \text{if } x \ge 7 \end{cases}$

**79.** By canceling the common factor,
$\frac{(x-1)(x+2)}{x-1}$ simplifies to $x+2$. At
$x = 1$, however, the function $\frac{(x-1)(x+2)}{x-1}$

is *discontinuous* (since it is undefined where the denominator is zero), whereas $x+2$ is *continuous*. Are these two functions, one obtained from the other by simplification, equal to each other? Explain.

## Applied Exercises

**80. GENERAL: Relativity** According to Einstein's special theory of relativity, under certain conditions a 1-foot-long object moving with velocity $v$ will appear to an observer to have length $\sqrt{1 - (v/c)^2}$, in which $c$ is a constant equal to the speed of light. Find the limiting value of the apparent length as the velocity of the object approaches the speed of light by finding

$$\lim_{v \to c^-} \sqrt{1 - \left(\frac{v}{c}\right)^2}$$

**81. BUSINESS: Interest Compounded Continuously**  If you deposit $1 into a bank account paying 10% interest compounded continuously (see Section 4.1), a year later its value will be

$$\lim_{x \to 0} \left(1 + \frac{x}{10}\right)^{1/x}$$

Find the limit by making a TABLE of values correct to two decimal places, thereby finding the value of the deposit in dollars and cents.

**82. BUSINESS: Interest Compounded Continuously** If you deposit $1 into a bank account paying 5% interest compounded continuously (see Section 4.1), a year later its value will be

$$\lim_{x \to 0} \left(1 + \frac{x}{20}\right)^{1/x}$$

Find the limit by making a TABLE of values correct to two decimal places, thereby finding the value of the deposit in dollars and cents.

**83. GENERAL: Superconductivity** The conductivity of aluminum at temperatures near absolute zero is approximated by the function $f(x) = \frac{100}{1 + .001x^2}$, which expresses the conductivity as a percent. Find the limit of this conductivity percent as the temperature $x$ approaches 0 (absolute zero) from the right. (See the Application Preview on page 77.)

**84. GENERAL: First-Class Mail** In 2008, the U.S. Postal Service would deliver a first-class letter weighing 3 ounces or less for the following prices:

not more than 1 ounce, 42¢; more than 1 ounce but not more than 2 ounces, 59¢; and more than 2 ounces but not more than 3 ounces, 76¢. This information determines the price (in cents) as a function of the weight (in ounces). At which values in open interval (0, 3) is this function discontinuous?
*Source:* U.S. Postal Service

## Conceptual Exercises

**85.** A student once said: "The limit is where a function is *going*, even if it never *gets* there." Explain what the student meant.

**86.** A student once said: "A continuous function *gets* where it's going; a discontinuous function *doesn't*." Explain what the student meant.

**87.** True or False: If $f(2)$ is not defined, then $\lim_{x \to 2} f(x)$ does not exist.

**88.** True or False: If $f(2) = 5$, then $\lim_{x \to 2} f(x) = 5$.

**89.** True or False: If $\lim_{x \to 2^+} f(x) = 7$, then $\lim_{x \to 2} f(x) = 7$.

**90.** True or False: If $\lim_{x \to 2} f(x) = 7$, then $\lim_{x \to 2^+} f(x) = 7$.

**91.** True or False: $\lim_{x \to 1} \frac{x^2 - 1}{x - 1} = \frac{\lim_{x \to 1}(x^2 - 1)}{\lim_{x \to 1}(x - 1)}$

**92.** True or False: If a function is not defined at $x = 5$, then the function is not continuous at $x = 5$.

**93.** If $\lim_{x \to 2} f(x) = 7$ and $f(x)$ is continuous at $x = 2$, then $f(2) = 7$.

**94.** True or False: If a function is defined and continuous at every x-value, then its graph has no jumps or breaks.

**95.** For the function graphed below, find:

 **a.** $\lim_{x \to 0^+} f(x)$  **b.** $\lim_{x \to 0^-} f(x)$  **c.** $\lim_{x \to 0} f(x)$

## 2.2 RATES OF CHANGE, SLOPES, AND DERIVATIVES

### Introduction

In this section we will define the **derivative,** one of the two most important concepts in all of calculus, which measures the **rate of change** of a function or, equivalently, the **slope** of a curve.* We begin by discussing rates of change.

### Average and Instantaneous Rate of Change

We often speak in terms of *rates of change* to express how one quantity changes with another. For example, in the morning the temperature might be "rising at the rate of 3 degrees per hour" and in the evening it might be "falling at the rate of 2 degrees per hour." For simplicity, suppose that in some location the temperature at time $x$ hours is $f(x) = x^2$ degrees. We shall calculate the *average rate of change of temperature* over various time intervals—the change in temperature divided by the change in time. For example, at time 1 the temperature is $1^2 = 1$ degrees, and at time 3 the temperature is $3^2 = 9$ degrees, so the temperature went up by $9 - 1 = 8$ degrees in 2 hours, for an average rate of:

$$\begin{pmatrix} \text{Average rate} \\ \text{of change} \\ \text{from } 1 \text{ to } 3 \end{pmatrix} = \frac{3^2 - 1^2}{3 - 1} = \frac{9 - 1}{2} = \frac{8}{2} = 4 \qquad \begin{array}{l} \text{Average rate} \\ \text{of change} \\ \text{over 2 hours} \end{array}$$

Similarly,

degrees per hour

$$\begin{pmatrix} \text{Average rate} \\ \text{of change} \\ \text{from } 1 \text{ to } 2 \end{pmatrix} = \frac{2^2 - 1^2}{2 - 1} = \frac{4 - 1}{1} = 3 \qquad \begin{array}{l} \text{Average rate} \\ \text{of change} \\ \text{over 1 hour} \end{array}$$

$$\begin{pmatrix} \text{Average rate} \\ \text{of change} \\ \text{from } 1 \text{ to } 1.5 \end{pmatrix} = \frac{1.5^2 - 1^2}{1.5 - 1} = \frac{2.25 - 1}{0.5} = \frac{1.25}{0.5} = 2.5 \qquad \begin{array}{l} \text{Average rate} \\ \text{of change} \\ \text{over 0.5 hour} \end{array}$$

degrees per hour

$$\begin{pmatrix} \text{Average rate} \\ \text{of change} \\ \text{from } 1 \text{ to } 1.1 \end{pmatrix} = \frac{1.1^2 - 1^2}{1.1 - 1} = \frac{1.21 - 1}{0.1} = \frac{0.21}{0.1} = 2.1 \qquad \begin{array}{l} \text{Average rate} \\ \text{of change} \\ \text{over 0.1 hour} \end{array}$$

We see that rate of change of temperature over shorter and shorter time intervals is decreasing from 4 to 3 to 2.5 to 2.1, numbers that seem to be approaching 2 degrees per hour. To verify that this is indeed true, we generalize the process. We have been finding the average over a time interval from 1 to a slightly later time that we now call $1 + h$, where $h$ is a small positive

---

*The second most important concept in calculus is the definite integral, which will be defined in Section 5.3.

number. Notice that in each step we calculated the following expression for successively smaller values of $h$.

$$\begin{pmatrix} \text{Average rate} \\ \text{of change} \\ \text{from } 1 \text{ to } 1 + h \end{pmatrix} = \frac{(1 + h)^2 - 1^2}{h} \qquad \begin{array}{l} \text{For } h \text{ we used} \\ 2, 1, 0.5, \text{ and } 0.1 \end{array}$$

⌐From $1 + h - 1 = h$

If we now use our limit notation to let $h$ approach zero, the amount of time will shrink to an instant, giving what is called the *instantaneous rate of change*:

$$\begin{pmatrix} \text{Instantaneous} \\ \text{rate of change} \\ \text{at time } 1 \end{pmatrix} = \lim_{h \to 0} \frac{(1 + h)^2 - 1^2}{h} \qquad \begin{array}{l} \text{Taking the limit as} \\ h \text{ approaches zero} \end{array}$$

We have been using the function $f(x) = x^2$ and the time $x = 1$, but the same procedure applies to *any* function $f$ and number $x$, leading to the following general definition:

---

### Average and Instantaneous Rate of Change

The *average* rate of change of a function $f$ between $x$ and $x + h$ is

$$\frac{f(x + h) - f(x)}{h} \qquad \begin{array}{l} \text{Difference quotient gives} \\ \text{the } \textit{average} \text{ rate of change} \end{array}$$

The *instantaneous* rate of change of a function $f$ at the number $x$ is

$$\lim_{h \to 0} \frac{f(x + h) - f(x)}{h} \qquad \begin{array}{l} \text{Taking the limit makes} \\ \text{it } \textit{instantaneous} \end{array}$$

---

The fraction is just the difference quotient that we introduced on page 61; the numerator is the change in the *function* between two $x$-values, and the denominator is the change between the two $x$-values: $(x + h) - (x) = h$. We may use the second formula to check our guess that the average rate of change of temperature is indeed approaching 2 degrees per hour.

**EXAMPLE 1**

### FINDING AN INSTANTANEOUS RATE OF CHANGE

Find the instantaneous rate of change of the temperature function $f(x) = x^2$ at time $x = 1$.

*Solution*

$$\lim_{h \to 0} \frac{f(x + h) - f(x)}{h} \qquad \begin{array}{l} \text{Formula for the instantaneous} \\ \text{rate of change of } f \text{ at } x \end{array}$$

$$\lim_{h \to 0} \frac{f(1 + h) - f(1)}{h} \qquad \text{Substituting } x = 1$$

$$= \lim_{h \to 0} \frac{(1 + h)^2 - 1^2}{h}$$

$f(x) = x^2$ gives $f(1 + h) = (1 + h)^2$ and $f(1) = 1^2$

$$= \lim_{h \to 0} \frac{1 + 2h + h^2 - 1}{h}$$

Expanding $(1 + h)^2 = 1 + 2h + h^2$

$$= \lim_{h \to 0} \frac{\cancel{1} + 2h + h^2 - \cancel{1}}{h}$$

Simplifying

$$= \lim_{h \to 0} \frac{h(2 + h)}{h} = \lim_{h \to 0} \frac{\cancel{h}(2 + h)}{\cancel{h}}$$

Factoring out $h$ and canceling (since $h \neq 0$)

$$= \lim_{h \to 0} (2 + h) = 2$$

Evaluating the limit by direct substitution

Since $f(x)$ gives the temperature at time $x$ hours, this means that *the instantaneous rate of change of temperature at time $x = 1$ hour is 2 degrees per hour* (just as we had guessed earlier).

---

 **BE CAREFUL:** Make sure you understand how to find the *units* of the rate of change. In this example, $f$ gave the temperature (degrees) at time $x$ (hours), so the units of the rate of change were *degrees per hour.* In general, for a function $f(x)$, the units of the instantaneous rate of change are *function units per x unit.*

**Practice Problem 1**

If $f(x)$ gives the population of a city in year $x$, what are the units of the instantaneous rate of change? ➤ **Solution on page 104**

### Secant and Tangent Lines

How can we "see" the average and instantaneous rates of change on the graph of a function? First, some terminology. A **secant line** to a curve is a line that passes through two points of the curve. A **tangent line** is a line that passes through a point of the curve and matches exactly the *steepness* of the curve at that point.*

Tangent line to the curve at $P$

Secant line to the curve through $P$ and $Q$

Tangent and secant lines to a curve

*The word "secant" comes from the Latin *secare,* "to cut," suggesting that the secant line "cuts" the curve at two points. The word "tangent" comes from the Latin *tangere,* "to touch," suggesting that the tangent line just "touches" the curve at one point.

If the curve is the graph of a function $f$ and the points $P$ and $Q$ have $x$-coordinates $x$ and $x + h$, respectively (and so $y$-coordinates $f(x)$ and $f(x + h)$, respectively), then the above graph takes the following form:

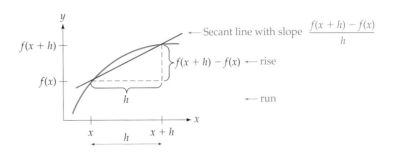

Observe that the *slope* (rise over run) of the secant line is the difference quotient $\frac{f(x + h) - f(x)}{h}$, exactly the same as our earlier definition of the average rate of change of the function between $x$ and $x + h$.

Furthermore, the two points where the secant line meets the curve are separated by a distance $h$ along the $x$-axis. Letting $h \to 0$ forces the second point to approach the first, causing the *secant* line to approach the *tangent* line, the slope of which is then $\lim_{h \to 0} \frac{f(x + h) - f(x)}{h}$. But this is exactly the same as our earlier definition of the *instantaneous rate of change of the function at $x$.*

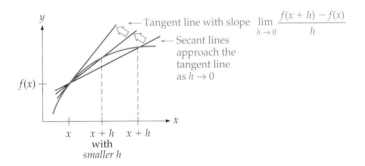

The last two graphs have shown that the slope of the *secant* line is the *average* rate of change of the function, and, more importantly, that the slope of the *tangent* line is the *instantaneous* rate of change of the function.

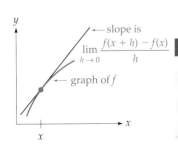

**Slope of the Tangent Line**

The slope of the tangent line to the graph of $f$ at $x$ is

$$\lim_{h \to 0} \frac{f(x + h) - f(x)}{h}$$

The fact that rates of change are related to slopes (and in fact are given by the same formula) should come as no surprise: if a function has a large rate of change, its graph will rise rapidly, giving it a large slope; if a function has only a small rate of change, its graph will rise slowly, giving it a small slope. In fact, *rate of change* and *slope* are just the numerical and graphical versions of exactly the same idea.

In Example 1 we found the instantaneous rate of change of a function. We now use the same formula to find the slope of the tangent line to the graph of a function.

**EXAMPLE 2**

**FINDING THE SLOPE OF A TANGENT LINE**

Find the slope of the tangent line to $f(x) = \dfrac{1}{x}$ at $x = 2$.

*Solution*

$$\lim_{h \to 0} \frac{f(x + h) - f(x)}{h} \qquad \text{Formula for the slope of the tangent line at } x$$

$$\lim_{h \to 0} \frac{f(2 + h) - f(2)}{h} \qquad \text{Substituting } x = 2$$

$$= \lim_{h \to 0} \frac{\dfrac{1}{2 + h} - \dfrac{1}{2}}{h} \qquad \text{Using the function } f(x) = \frac{1}{x}$$

$$= \lim_{h \to 0} \frac{1}{h}\left(\frac{1}{2 + h} - \frac{1}{2}\right) \qquad \text{Since multiplying by } 1/h \text{ is equivalent to dividing by } h$$

$$= \lim_{h \to 0} \frac{1}{h} \cdot \frac{2 - (2 + h)}{(2 + h) \cdot 2} \qquad \text{Combining fractions using the common denominator } (2 + h) \cdot 2$$

$$= \lim_{h \to 0} \frac{1}{h} \cdot \frac{-h}{(2 + h) \cdot 2} \qquad \text{Simplifying the numerator}$$

$$= \lim_{h \to 0} \frac{1}{\cancel{h}} \cdot \frac{\cancel{-h}}{(2 + h) \cdot 2} = \lim_{h \to 0} \frac{-1}{(2 + h) \cdot 2} \qquad \text{Canceling}$$

$$= \frac{-1}{(2 + 0) \cdot 2} = \frac{-1}{4} = -\frac{1}{4} \qquad \text{Evaluating the limit by direct substitution of } h = 0$$

Therefore, the curve $y = \frac{1}{x}$ has slope $-\frac{1}{4}$ at $x = 2$, as shown in the graph on the left.

Tangent line has slope $-\frac{1}{4}$ at $x = 2$

$y = \dfrac{1}{x}$

**EXAMPLE 3**

**FINDING A TANGENT LINE**

Use the result of the preceding example to find the equation of the tangent line to $f(x) = \frac{1}{x}$ at $x = 2$.

*Solution*

From Example 2 we know that the *slope* of the tangent line is $m = -\frac{1}{4}$. The point on the curve at $x = 2$ is $(2, \frac{1}{2})$, the $y$-coordinate coming from $y = f(2) = \frac{1}{2}$. The point-slope formula then gives:

$$y - \tfrac{1}{2} = -\tfrac{1}{4}(x - 2)$$ 
             $y - y_1 = m(x - x_1)$ with $m = -\frac{1}{4}$, $x_1 = 2$, and $y_1 = \frac{1}{2}$

$$y - \tfrac{1}{2} = -\tfrac{1}{4}x + \tfrac{1}{2}$$ 
             Multiplying out

$$y = 1 - \tfrac{1}{4}x$$ 
             Simplifying

       Equation of the tangent line

The first of the following graphs shows the curve $y = \frac{1}{x}$ along with the tangent line $y = 1 - \frac{1}{4}x$ that we found in the preceding Example. The next two graphs show the results of successively "zooming in" near the point of tangency, showing that the curve seems to straighten out and almost *become its own tangent line*. For this reason, the tangent line is called *the best linear approximation to the curve* near the point of tangency.

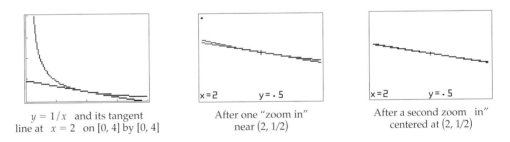

$y = 1/x$ and its tangent line at $x = 2$ on $[0, 4]$ by $[0, 4]$      After one "zoom in" near $(2, 1/2)$      After a second zoom in" centered at $(2, 1/2)$

## Graphing Calculator Exploration

Use a graphing calculator to graph $y = x^3 - 2x^2 - 3x + 4$ (or any function of your choice). Then "zoom in" a few times around a point to see the curve straighten out and almost *become* its own tangent line.

## The Derivative

In Examples 1 and 2 we found an instantaneous rate of change and the slope of a tangent line *at a particular number.* It is much more efficient to carry out the same calculation but keeping $x$ as a *variable,* obtaining a new function that gives the instantaneous rate of change or the slope of the tangent line at *any* value of $x$. This new function is denoted with a *prime*, $f'(x)$ (read: "*f prime of x*"), and is called *the derivative of f at x*.

---

### Derivative

For a function $f$, the *derivative of f at x* is defined as

$$f'(x) = \lim_{h \to 0} \frac{f(x + h) - f(x)}{h}$$  Limit of the difference quotient

(provided that the limit exists). The derivative $f'(x)$ gives the *instantaneous rate of change of f at x* and also the *slope of the graph of f at x*.

---

In general, the units of the derivative are *function units per x unit*.

---

**EXAMPLE 4**  **FINDING A DERIVATIVE FROM THE DEFINITION**

Find the derivative of $f(x) = 2x^2 - 9x + 39$.

*Solution*

$$f'(x) = \lim_{h \to 0} \frac{f(x + h) - f(x)}{h}$$  Definition of the derivative

$$= \lim_{h \to 0} \frac{\overbrace{2(x + h)^2 - 9(x + h) + 39}^{f(x+h)} - \overbrace{(2x^2 - 9x + 39)}^{f(x)}}{h}$$  Using $f(x) = 2x^2 - 9x + 39$

$$= \lim_{h \to 0} \frac{2x^2 + 4xh + 2h^2 - 9x - 9h + 39 - 2x^2 + 9x - 39}{h}$$  Expanding and simplifying

$$= \lim_{h \to 0} \frac{4xh + 2h^2 - 9h}{h} = \lim_{h \to 0} \frac{h(4x + 2h - 9)}{h}$$  Simplifying (since $h \neq 0$)

$$= \lim_{h \to 0} (4x + 2h - 9) = 4x - 9$$  Evaluating the limit by direct substitution

Therefore, the derivative of $f(x) = 2x^2 - 9x + 39$ is $f'(x) = 4x - 9$.

---

The operation of calculating derivatives should be thought of as an operation on *functions*, taking one function [such as $f(x) = 2x^2 - 9x + 39$] and giving another function $[f'(x) = 4x - 9]$. The resulting function is called "the derivative" because it is *derived* from the first, and the process of obtaining it is called "differentiation." If the derivative is defined at $x$, then the original function is said to be **differentiable at $x$.**

---

**EXAMPLE 5**

**USING A DERIVATIVE IN AN APPLICATION**

The following graph shows the annual sales of Apple Macintosh computers for recent years (in hundred thousands of units).

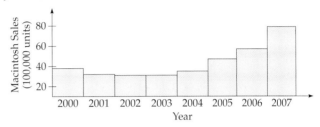

*Source:* Apple

These annual sales are approximated by the function   $f(x) = 2x^2 - 9x + 39$, where $x$ stands for *years since 2000*. Find the instantaneous rate of change of this function at   $x = 6$   and at   $x = 1$   and interpret the results.

*Solution*

The instantaneous rate of change means the *derivative*, so ordinarily we would now take the derivative of the sales function   $f(x) = 2x^2 - 9x + 39$. However, this is just what we did in the previous Example, so we will use the result that the derivative is   $f'(x) = 4x - 9$.   Therefore, we need only evaluate the derivative at the given $x$-values and interpret the results.

$$f'(6) = 4 \cdot 6 - 9 = 15 \qquad \text{Evaluating } f'(x) = 4x - 9 \text{ at } x = 6$$

*Interpretation:* $x = 6$   represents the year 2006 (6 years after 2000). Therefore, in 2006 Macintosh sales were growing at the rate of 15 hundred thousand, or 1.5 million units per year.

$$f'(1) = 4 \cdot 1 - 9 = -5 \qquad \text{Evaluating } f'(x) = 4x - 9 \text{ at } x = 1$$

*Interpretation:* In the year 2001 (corresponding to   $x = 1$),   Macintosh sales were *falling* at the rate of 5 hundred thousand, or 500,000 units per year.

---

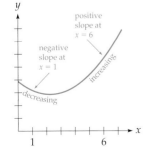

A derivative that is *positive* means that the original quantity is *increasing*, and a derivative that is *negative* means that the original quantity is *decreasing*.

The graph of   $f(x) = 2x^2 - 9x + 39$   on the left shows that the slope of the curve is indeed negative at   $x = 1$   and positive at   $x = 6$.

EXAMPLE 6

**FINDING A DERIVATIVE FROM THE DEFINITION**

Find the derivative of $f(x) = x^3$.

*Solution*

In our solution we will use the expansion $(x + h)^3 = x^3 + 3x^2h + 3xh^2 + h^3$ (found by multiplying together three copies of $(x + h)$).

$$f'(x) = \lim_{h \to 0} \frac{f(x + h) - f(x)}{h} \qquad \text{Definition of } f'(x)$$

$$= \lim_{h \to 0} \frac{(x + h)^3 - x^3}{h} \qquad \text{Using } f(x) = x^3$$

$$= \lim_{h \to 0} \frac{x^3 + 3x^2h + 3xh^2 + h^3 - x^3}{h} \qquad \text{Using the expansion of } (x + h)^3$$

$$= \lim_{h \to 0} \frac{\cancel{x^3} + 3x^2h + 3xh^2 + h^3 - \cancel{x^3}}{h} = \lim_{h \to 0} \frac{h(3x^2 + 3xh + h^2)}{h} \qquad \text{Canceling and then factoring out an } h$$

$$= \lim_{h \to 0} \frac{\cancel{h}(3x^2 + 3xh + h^2)}{\cancel{h}} = \lim_{h \to 0} 3x^2 + 3xh + h^2 \qquad \text{Canceling again}$$

$$= 3x^2 \qquad \text{Evaluating the limit by direct substitution}$$

Therefore, the derivative of $f(x) = x^3$ is $f'(x) = 3x^2$.

## Graphing Calculator Exploration

Some advanced graphing calculators have computer algebra systems that can simplify algebraic expressions. For example, the Texas Instruments *TI-89* calculator will find and simplify the difference quotient for the function $f(x) = x^3$ that we found in the preceding Example as follows:

You enter this ⟶ ... The calculator gives this

From the (simplified) difference quotient $3x^2 + 3xh + h^2$, taking the limit (using direct substitution of $h = 0$) gives the derivative $f'(x) = 3x^2$, just as we found above.

### Leibniz's Notation for the Derivative

Calculus was developed by Isaac Newton (1642–1727) and Gottfried Wilhelm Leibniz (1646–1716) in two different countries, so there naturally developed two different notations for the derivative. Newton denoted derivatives by a dot over the function, $\dot{f}$, a notation that has been largely replaced by our "prime" notation. Leibniz wrote the derivative of $f(x)$ by writing $\dfrac{d}{dx}$ in front of the function: $\dfrac{d}{dx}f(x)$. In Leibniz's notation, the fact that the derivative of $x^3$ is $3x^2$ is written

$$\frac{d}{dx}x^3 = 3x^2 \qquad\qquad \text{The derivative of } x^3 \text{ is } 3x^2$$

The following table shows equivalent expressions in the two notations.

| Prime Notation | | Leibniz's Notation | |
|---|---|---|---|
| $f'(x)$ | $=$ | $\dfrac{d}{dx}f(x)$ | Prime and $\dfrac{d}{dx}$ both mean the derivative |
| $y'$ | $=$ | $\dfrac{dy}{dx}$ | For $y$ a function of $x$ |

Each notation has its own advantages, and we will use both.* Leibniz's notation comes from writing the definition of the derivative as:

$$\frac{dy}{dx} = \lim_{\Delta x \to 0} \frac{f(x + \Delta x) - f(x)}{\Delta x} \qquad \text{Definition of the derivative (page 100) with the change in } x \text{ written as } \Delta x$$

or

$$\frac{dy}{dx} = \lim_{\Delta x \to 0} \frac{\Delta y}{\Delta x} \qquad \begin{array}{l} f(x + \Delta x) - f(x) \text{ is the change in } y, \\ \text{and so can be written } \Delta y \end{array}$$

It is as if the limit turns $\Delta$ (read "Delta," the Greek letter D) into $d$, changing the $\dfrac{\Delta y}{\Delta x}$ into $\dfrac{dy}{dx}$. That is, Leibniz's notation reminds us that the derivative $\dfrac{dy}{dx}$ is the limit of the slope $\dfrac{\Delta y}{\Delta x}$.

 **BE CAREFUL:** Some functions are *not differentiable* (the derivative does not exist) at certain $x$-values. For example, the following diagram shows a function that has a "corner point" at $x = 1$. At this point the slope (and therefore the

---

*Other notations for the derivative are $Df(x)$ and $D_x f(x)$, but we will not use them.

derivative) cannot be defined, so the function is not differentiable at $x = 1$. Other nondifferentiable functions will be discussed in Section 2.7.

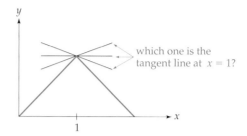

which one is the tangent line at $x = 1$?

Since the tangent line cannot be uniquely defined at $x = 1$, the slope, and therefore the derivative, is undefined at $x = 1$.

The following diagram shows the geometric relationship between a function (upper graph) and its derivative (lower graph). Observe carefully how the *slope of f* is shown by the *sign of f'*.

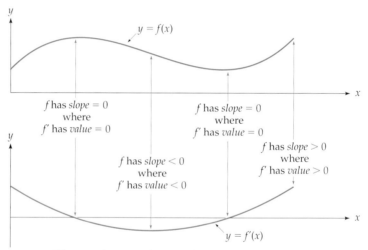

$y = f(x)$

$f$ has *slope* = 0 where $f'$ has *value* = 0

$f$ has *slope* = 0 where $f'$ has *value* = 0

$f$ has *slope* < 0 where $f'$ has *value* < 0

$f$ has *slope* > 0 where $f'$ has *value* > 0

$y = f'(x)$

Wherever the original (upper) function has *slope* zero, the derivative (lower) has *value* zero (it crosses the *x*-axis).

**Practice Problem 2**

The graph shows a function and its derivative. Which is the original function (#1 or #2) and which is its derivative? [*Hint:* Which curve has *slope* zero where the other has *value* zero?]

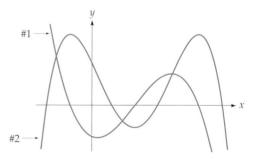

#1

#2

➤ **Solution below**

➤ **Solutions to Practice Problems**

1. The units are *people per year,* measuring the rate of growth of the population.
2. #2 is the original function and #1 is its derivative.

## 2.2 Section Summary

The derivative of the function $f$ at $x$ is

$$f'(x) = \lim_{h \to 0} \frac{f(x + h) - f(x)}{h}$$    Provided that the limit exists

Some students remember the steps as **DESL** (pronounced "diesel"): write the **D**efinition, **E**xpress the numerator in terms of the function, **S**implify, and take the **L**imit.

The derivative $f'(x)$ gives both the *slope of the graph* of the function at $x$ and the *instantaneous rate of change of the function* at $x$. In other words, "derivative," "slope," and "instantaneous rate of change" are merely the mathematical, the geometric, and the analytic versions of the same idea.

The derivative gives the rate of change *at a particular instant,* not an actual change over a period of time. Instantaneous rates of changes are like the speeds on an automobile speedometer—a reading of 50 mph at one moment does not mean that you will travel exactly 50 miles in the next hour, since the actual distance depends upon your speed during the entire hour. The derivative, however, may be interpreted as the *approximate* change resulting from a 1-unit increase in the independent variable. For example, if your speedometer reads 50 mph, then you may say that you will travel *about* 50 miles during the next hour, meaning that this will be true provided that your speed remains steady throughout the hour. (In Chapter 5 we will see how to calculate actual changes from rates of change that do not stay constant.)

## 2.2 Exercises

**1–4.** By imagining tangent lines at points $P_1$, $P_2$, and $P_3$, state whether the slopes are positive, zero, or negative at these points.

**1.**

**2.**

**3.**

**4.**

**5–6.** Use the tangent lines shown at points $P_1$ and $P_2$ to find the slopes of the curve at these points.

**5.**

**6.**

**7–8.** Use the graph of each function $f(x)$ to make a rough sketch of the derivative $f'(x)$ showing where $f'(x)$ is positive, negative, and zero. (Omit scale on $y$-axis.)

**7.**

**8.**

**9–10.** Find the average rate of change of the given function between the following pairs of $x$-values. [*Hint:* See page 94.]

**a.** $x = 1$ and $x = 3$
**b.** $x = 1$ and $x = 2$
**c.** $x = 1$ and $x = 1.5$
**d.** $x = 1$ and $x = 1.1$
**e.** $x = 1$ and $x = 1.01$
**f.** What number do your answers seem to be approaching?

**9.** $f(x) = x^2 + x$        **10.** $f(x) = 2x^2 + 5$

**11–12.** Find the average rate of change of the given function between the following pairs of $x$-values. [*Hint:* See page 94.]

**a.** $x = 2$ and $x = 4$
**b.** $x = 2$ and $x = 3$
**c.** $x = 2$ and $x = 2.5$
**d.** $x = 2$ and $x = 2.1$
**e.** $x = 2$ and $x = 2.01$
**f.** What number do your answers seem to be approaching?

**11.** $f(x) = 2x^2 + x - 2$   **12.** $f(x) = x^2 + 2x - 1$

**13–14.** Find the average rate of change of the given function between the following pairs of $x$-values. [*Hint:* See page 94.]

**a.** $x = 3$ and $x = 5$
**b.** $x = 3$ and $x = 4$
**c.** $x = 3$ and $x = 3.5$
**d.** $x = 3$ and $x = 3.1$
**e.** $x = 3$ and $x = 3.01$
**f.** What number do your answers seem to be approaching?

**13.** $f(x) = 5x + 1$        **14.** $f(x) = 7x - 2$

**15–16.** Find the average rate of change of the given function between the following pairs of $x$-values. [*Hint:* See page 94.]

**a.** $x = 4$ and $x = 6$
**b.** $x = 4$ and $x = 5$
**c.** $x = 4$ and $x = 4.5$
**d.** $x = 4$ and $x = 4.1$
**e.** $x = 4$ and $x = 4.01$
**f.** What number do your answers seem to be approaching?

**15.** $f(x) = \sqrt{x}$        **16.** $f(x) = \dfrac{4}{x}$

**17–20.** Use the formula on page 95 to find the instantaneous rate of change of the function at the given $x$-value. If you did the related problem in Exercises 9–16, compare your answers.

**17.** $f(x) = x^2 + x$ at $x = 1$

**18.** $f(x) = x^2 + 2x - 1$ at $x = 2$

**19.** $f(x) = 5x + 1$ at $x = 3$

**20.** $f(x) = \dfrac{4}{x}$ at $x = 4$

**21–24.** Use the formula on page 97 to find the slope of the tangent line to the curve at the given $x$-value. If you did the related problem in Exercises 9–16, compare your answers. [*Hint:* See Example 2.]

**21.** $f(x) = 2x^2 + x - 2$ at $x = 2$

**22.** $f(x) = 2x^2 + 5$ at $x = 1$

**23.** $f(x) = \sqrt{x}$ at $x = 4$

**24.** $f(x) = 7x - 2$ at $x = 3$

**25–44.** Find $f'(x)$ by using the definition of the derivative.

**25.** $f(x) = x^2 - 3x + 5$   **26.** $f(x) = 2x^2 - 5x + 1$

**27.** $f(x) = 1 - x^2$   **28.** $f(x) = \frac{1}{2}x^2 + 1$

**29.** $f(x) = 9x - 2$   **30.** $f(x) = -3x + 5$

**31.** $f(x) = \dfrac{x}{2}$   **32.** $f(x) = 0.01x + 0.05$

**33.** $f(x) = 4$   **34.** $f(x) = \pi$

**35.** $f(x) = ax^2 + bx + c$
($a$, $b$, and $c$ are constants)

**36.** $f(x) = (x + a)^2$
($a$ is a constant.) [*Hint:* First expand $(x + a)^2$.]

**37.** $f(x) = x^5$
[*Hint:* Use $(x + h)^5 =$
$x^5 + 5x^4h + 10x^3h^2 + 10x^2h^3 + 5xh^4 + h^5$]

**38.** $f(x) = x^4$
[*Hint:* Use $(x + h)^4 =$
$x^4 + 4x^3h + 6x^2h^2 + 4xh^3 + h^4$]

**39.** $f(x) = \dfrac{2}{x}$   **40.** $f(x) = \dfrac{1}{x^2}$

**41.** $f(x) = \sqrt{x}$   **42.** $f(x) = \dfrac{1}{\sqrt{x}}$

[*Hint:* Multiply the numerator and denominator of the difference quotient by $(\sqrt{x + h} + \sqrt{x})$ and then simplify.]

[*Hint:* Multiply the numerator and denominator of the difference quotient by $(\sqrt{x} + \sqrt{x + h})$ and then simplify.]

**43.** $f(x) = x^3 + x^2$   **44.** $f(x) = \dfrac{1}{2x}$

**45.** **a.** Find the equation for the tangent line to the curve $f(x) = x^2 - 3x + 5$ at $x = 2$, writing the equation in slope-intercept form. [*Hint:* Use your answer to Exercise 25.]

 **b.** Use a graphing calculator to graph the curve together with the tangent line to verify your answer.

**46.** **a.** Find the equation for the tangent line to the curve $f(x) = 2x^2 - 5x + 1$ at $x = 2$, writing the equation in slope-intercept form. [*Hint:* Use your answer to Exercise 26.]

**b.** Use a graphing calculator to graph the curve together with the tangent line to verify your answer.

 **47.** **a.** Graph the function $f(x) = x^2 - 3x + 5$ on the window $[-10, 10]$ by $[-10, 10]$. Then use the DRAW menu to graph the TANGENT line at $x = 2$. Your screen should also show the *equation* of the tangent line. (If you did Exercise 45, this equation for the tangent line should agree with the one you found there.)

**b.** Add to your graph the tangent line at $x = 1$, and the tangent lines at any other $x$-values that you choose.

 **48.** **a.** Graph the function $f(x) = 2x^2 - 5x + 1$ on the window $[-10, 10]$ by $[-10, 10]$. Then use the DRAW menu to graph the TANGENT line at $x = 2$. Your screen should also show the *equation* of the tangent line. (If you did Exercise 46, this equation for the tangent line should agree with the one you found there.)

**b.** Add to your graph the tangent line at $x = 0$, and the tangent lines at any other $x$-values that you choose.

**49–54.** For each function:

**a.** Find $f'(x)$ using the definition of the derivative.

**b.** Explain, by considering the original function, why the derivative is a constant.

**49.** $f(x) = 3x - 4$          **50.** $f(x) = 2x - 9$

**51.** $f(x) = 5$          **52.** $f(x) = 12$

**53.** $f(x) = mx + b$
($m$ and $b$ are constants)

**54.** $f(x) = b$
($b$ is a constant)

## Applied Exercises

**55.** **BUSINESS: Temperature** The temperature in an industrial pasteurization tank is $f(x) = x^2 - 8x + 110$ degrees centigrade after $x$ minutes (for $0 \le x \le 12$).

**a.** Find $f'(x)$ by using the definition of the derivative.

**b.** Use your answer to part (a) to find the instantaneous rate of change of the temperature after 2 minutes. Be sure to interpret the sign of your answer.

**c.** Use your answer to part (a) to find the instantaneous rate of change after 5 minutes.

**56.** **GENERAL: Population** The population of a town is $f(x) = 3x^2 - 12x + 200$ people after $x$ weeks (for $0 \le x \le 20$).

**a.** Find $f'(x)$ by using the definition of the derivative.

**b.** Use your answer to part (a) to find the instantaneous rate of change of the population after 1 week. Be sure to interpret the sign of your answer.

**c.** Use your answer to part (a) to find the instantaneous rate of change of the population after 5 weeks.

**57.** **BEHAVIORAL SCIENCE: Learning Theory** In a psychology experiment, a person could memorize $x$ words in $f(x) = 2x^2 - x$ seconds (for $0 \le x \le 10$).

**a.** Find $f'(x)$ by using the definition of the derivative.

**b.** Find $f'(5)$ and interpret it as an instantaneous rate of change in the proper units.

**58.** **BUSINESS: Advertising** An automobile dealership finds that the number of cars that it sells on day $x$ of an advertising campaign is $S(x) = -x^2 + 10x$ (for $0 \le x \le 7$).

**a.** Find $S'(x)$ by using the definition of the derivative.

**b.** Use your answer to part (a) to find the instantaneous rate of change on day $x = 3$.

**c.** Use your answer to part (a) to find the instantaneous rate of change on day $x = 6$.

Be sure to interpret the signs of your answers.

**59.** **SOCIAL SCIENCE: Immigration** The percentage of people in the United States who are immigrants (that is, were born elsewhere) for different decades is shown below.

*Sources:* Center for Immigration Studies and U.S. Census Bureau

These percentages are approximated by the function $f(x) = \frac{1}{2}x^2 - 3.7x + 12$, where $x$ stands for *the number of decades since 1930* (so that, for example, $x = 5$ would stand for 1980).

a. Find $f'(x)$ using the definition of the derivative.
b. Evaluate the derivative at $x = 1$ and interpret the result.
c. Find the rate of change of the immigrant percentage in the year 2000.

60. **BUSINESS: Revenue** Operating revenues for Honeywell International for recent years are shown below.

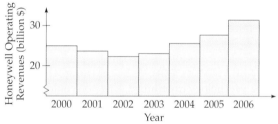

*Source:* Standard & Poor's

These revenues are approximated by the function $f(x) = \frac{1}{2}x^2 - 2x + 25$, where $x$ stands for *years since 2000.*

a. Find $f'(x)$ using the definition of the derivative.
b. Evaluate the derivative at $x = 1$ and interpret the result.
c. Find the rate of change of Honeywell's operating revenues in 2006.

## Conceptual Exercises

61. Describe the difference between the *average* rate of change and the *instantaneous* rate of change of a function. What formula would you use to find the instantaneous rate of change?

62. Describe the difference between a *secant* line and a *tangent* line for the graph of a function. What formula would you use to find the slope of the secant? What formula for the tangent?

63. When we calculate the derivative using the formula $f'(x) = \lim_{h \to 0} \dfrac{f(x + h) - f(x)}{h}$, we eventually evaluate the limit by direct substitution of $h = 0$. Why don't we just substitute $h = 0$ into the formula to begin with?

64. For a function $f(x)$, if $f$ is in widgets and $x$ is in blivets, what are the units of the derivative $f'(x)$, *widgets per blivet* or *blivets per widget*?

65. The derivative $f'(x)$ of a function is in *prendles per blarg*. What are the units of $x$? What are the units of $f$? [*Hint:* One is *prendles* and the other is *blargs*.]

66. The population of a city in year $x$ is given by a function whose derivative is negative. What does this mean about the city?

67. A patient's temperature at time $x$ hours is given by a function whose derivative is positive for $0 < x < 24$ and negative for $24 < x < 48$. Assuming that the patient begins and ends with a normal temperature, is the patient's health improving or deteriorating during the first day? During the second day?

68. Suppose that the temperature outside your house $x$ hours after midnight is given by a function whose derivative is negative for $0 < x < 6$ and positive for $6 < x < 12$. What can you say about the temperature at time 6 a.m. compared to the temperature throughout the first half of the day?

## 2.3 SOME DIFFERENTIATION FORMULAS

### Introduction

In Section 2.2 we defined the *derivative* of a function and used it to calculate instantaneous rates of change and slopes. Even for a function as simple as $f(x) = x^2$, however, calculating the derivative from the definition was rather involved. Calculus would be of limited usefulness if all derivatives had to be found in this way.

In this section we will learn several **rules of differentiation** that will simplify finding the derivatives of many useful functions. The rules are derived from the definition of the derivative, which is why we studied the definition first. We will also learn another important use for differentiation: calculating **marginals** (marginal revenue, marginal cost, and marginal profit), which are used extensively in business and economics.

### Derivative of a Constant

The first rule of differentiation shows how to differentiate a constant function.

| Constant Rule | *Brief Example* |
|---|---|
| For any constant $c$,  $$\frac{d}{dx}c = 0$$ | $$\frac{d}{dx}7 = 0$$ |

$f(x) = c$

derivative
(slope) is zero

In words: *the derivative of a constant is zero.* This rule is obvious geometrically, as shown in the diagram: the graph of a constant function $f(x) = c$ is the horizontal line $y = c$. Since the slope of a horizontal line is zero, the derivative of $f(x) = c$ is zero.

This rule follows immediately from the definition of the derivative. The constant function $f(x) = c$ has the same value $c$ for *any* value of $x$, so, in particular, $f(x + h) = c$ and $f(x) = c$. Substituting these into the definition of the derivative gives

$$f'(x) = \lim_{h \to 0} \frac{\overbrace{f(x + h)}^{c} - \overbrace{f(x)}^{c}}{h} = \lim_{h \to 0} \frac{c - c}{h} = \lim_{h \to 0} \frac{\overbrace{0}}{h} = \lim_{h \to 0} 0 = 0$$

Therefore, the derivative of a constant function $f(x) = c$ is $f'(x) = 0$.

### Power Rule

One of the most useful differentiation formulas in all of calculus is the **Power Rule.** It tells how to differentiate powers such as $x^7$ or $x^{100}$.

| Power Rule | |
|---|---|
| For any constant exponent $n$,  $$\frac{d}{dx}x^n = n \cdot x^{n-1}$$ | To differentiate $x^n$, bring down the exponent as a multiplier and then decrease the exponent by 1 |

A derivation of the Power Rule for positive integer exponents is given at the end of this section. We will use the Power Rule for *all* real numbers *n*, since more general proofs will be given later (see pages 135, 246–247, and 299).

**EXAMPLE 1**

**USING THE POWER RULE**

a.   $\dfrac{d}{dx}x^7 = 7x^{7-1} = 7x^6$

Bring down the exponent ↑   ↑ Decrease the exponent by 1

b.  $\dfrac{d}{dx}x^{100} = 100x^{100-1} = 100x^{99}$

c.  $\dfrac{d}{dx}x^{-2} = -2x^{-2-1} = -2x^{-3}$      The Power Rule holds for negative exponents

d.  $\dfrac{d}{dx}\sqrt{x} = \dfrac{d}{dx}x^{\frac{1}{2}} = \dfrac{1}{2}x^{\frac{1}{2}-1} = \dfrac{1}{2}x^{-\frac{1}{2}}$      and for fractional exponents

e.  $\dfrac{d}{dx}x = \dfrac{d}{dx}x^1 = 1x^{1-1} = x^0 = 1$

This last result is used so frequently that it should be remembered separately.

derivative (slope) is 1

$$\frac{d}{dx}x = 1$$      The derivative of *x* is 1

This result is obvious geometrically, as shown in the diagram.

From now on we will skip the middle step in these examples, differentiating powers in one step:

$$\frac{d}{dx}x^{50} = 50x^{49}$$

$$\frac{d}{dx}x^{2/3} = \frac{2}{3}x^{-1/3} \qquad \tfrac{2}{3}-1$$

**Practice Problem 1**

Find

a.  $\dfrac{d}{dx}x^2$      b.  $\dfrac{d}{dx}x^{-5}$      c.  $\dfrac{d}{dx}\sqrt[4]{x}$      ➤ Solutions on page 117

## Constant Multiple Rule

The Power Rule shows how to differentiate a power such as $x^3$. The *Constant Multiple Rule* extends this result to functions such as $5x^3$, a constant *times* a function. Briefly, to differentiate a constant times a function, we simply "carry along" the constant and differentiate the function.

---

**Constant Multiple Rule**

For any constant $c$,

$$\frac{d}{dx}[c \cdot f(x)] = c \cdot f'(x)$$

The derivative of a constant times a function is the constant times the derivative of the function

---

(provided, of course, that the derivative $f'(x)$ exists). A derivation of this rule is given at the end of this section.

**EXAMPLE 2**

**USING THE CONSTANT MULTIPLE RULE**

**a.** $\dfrac{d}{dx} 5x^3 = 5 \cdot 3x^2 = 15x^2$ 

Carry along the constant | Derivative of $x^3$

**b.** $\dfrac{d}{dx} 3x^{-4} = 3(-4)x^{-5} = -12x^{-5}$

Again we will skip the middle step, bringing down the exponent and immediately multiplying it by the number in front of the $x$.

**EXAMPLE 3**

**CALCULATING DERIVATIVES MORE QUICKLY**

**a.** $\dfrac{d}{dx} 8x^{-1/2} = -4x^{-3/2}$ 

$8(-\frac{1}{2})$

**b.** $\dfrac{d}{dx} 7x = 7 \cdot 1 = 7$

Derivative of $x$

This last example, showing that the derivative of $7x$ is just 7, leads to a very useful general rule.

---

For any constant $c$,

$$\frac{d}{dx}(cx) = c$$

The derivative of a constant times $x$ is just the constant

---

**EXAMPLE 4**    **FINDING DERIVATIVES INVOLVING CONSTANTS**

**a.** $\dfrac{d}{dx}(7x) = 7$    Using $\dfrac{d}{dx}(cx) = c$

**b.** $\dfrac{d}{dx}7 = 0$    For a constant alone, the derivative is zero

**c.** $\dfrac{d}{dx}(7x^2) = 7 \cdot 2x = 14x$    But for a constant *times a function,* the derivative is the constant times the derivative of the function

 **BE CAREFUL:** Be sure to understand the difference between the last two examples: for a constant *alone* [as in part (b) above], the derivative is *zero;* but for a constant *times a function* [as in part (c) above], you *carry along the constant and differentiate the function.*

### Sum Rule

The *Sum Rule* extends differentiation to sums of functions. Briefly, to differentiate a *sum* of two functions, just differentiate the functions separately and add the results.

| Sum Rule | Brief Example |
|---|---|
| $\dfrac{d}{dx}[f(x) + g(x)] = f'(x) + g'(x)$ | $\dfrac{d}{dx}(x^3 + x^5) = 3x^2 + 5x^4$ |

(provided, of course, that both the derivatives $f'(x)$ and $g'(x)$ exist). In words: *the derivative of a sum is the sum of the derivatives.* A derivation of the Sum Rule is given at the end of this section.

A similar rule holds for the *difference* of two functions,

$$\dfrac{d}{dx}[f(x) - g(x)] = f'(x) - g'(x)$$    The derivative of a difference is the difference of the derivatives

(provided that $f'(x)$ and $g'(x)$ exist). These two rules may be combined:

| Sum-Difference Rule | |
|---|---|
| $\dfrac{d}{dx}[f(x) \pm g(x)] = f'(x) \pm g'(x)$ | Use both upper signs or both lower signs |

Similar rules hold for sums and differences of any finite number of terms. Using these rules, we may differentiate any polynomial or, more generally, functions with variables raised to *any* constant powers.

**EXAMPLE 5**

**USING THE SUM-DIFFERENCE RULE**

**a.** $\dfrac{d}{dx}(x^3 - x^5) = 3x^2 - 5x^4$

Derivatives taken separately

**b.** $\dfrac{d}{dx}(5x^{-2} - 6x^{1/3} + 4) = -10x^{-3} - 2x^{-2/3}$

The constant 4 has derivative 0

## Leibniz's Notation and Evaluation of Derivatives

Leibniz's derivative notation $\dfrac{d}{dx}$ is often read *the derivative with respect to x* to emphasize that the independent variable is $x$. To differentiate a function of some *other* variable, the $x$ in $\dfrac{d}{dx}$ is replaced by the other variable. For example:

| *Function* | *Derivative* | |
|---|---|---|
| $f(t)$ | $\dfrac{d}{dt}f(t)$ | Use $\dfrac{d}{dt}$ for the derivative with respect to $t$ |
| $w^3$ | $\dfrac{d}{dw}w^3$ | Use $\dfrac{d}{dw}$ for the derivative with respect to $w$ |

The following two notations both mean the derivative *evaluated* at $x = 2$.

┌─ Derivative

$f'(2)$

└─ Evaluated at $x = 2$

┌─ Derivative

$\dfrac{df}{dx}\Big|_{x=2}$

└─ Evaluated at $x = 2$

Bar | means "evaluated at"

 **BE CAREFUL:** Both notations mean *first* differentiate and *then* evaluate.

**EXAMPLE 6**

**EVALUATING A DERIVATIVE**

If $f(x) = 2x^3 - 5x^2 + 7$, find $f'(2)$.

*Solution*

$f'(x) = 6x^2 - 10x$                    First differentiate

$f'(2) = 6 \cdot 2^2 - 10 \cdot 2 = 24 - 20 = 4$                    Then evaluate

**Practice Problem 2**

If $f(x) = 5x^4 + 1$, find $\left. \dfrac{df}{dx} \right|_{x=-1}$

➤ Solution on page 117

**EXAMPLE 7**

**FINDING A TANGENT LINE**

Find the equation of the tangent line to $f(x) = 2x^3 - 5x^2 + 7$ at $x = 2$.

*Solution*

As we know from pages 96–100, the slope of the tangent line at $x = 2$ is the slope of the curve at $x = 2$, which is the value of the derivative $f'$ at $x = 2$. Ordinarily, we would first differentiate and then evaluate to find $f'(2)$. However, this is exactly what we did in the preceding example, finding $f'(2) = 4$, so the slope of the tangent line is $m = 4$. The point on the curve at $x = 2$ is $(2, 3)$, the $y$-coordinate coming from the original function: $y = f(2) = 3$. The point-slope formula then gives

$$y - 3 = 4(x - 2) \qquad \text{$y - y_1 = m(x - x_1)$ with $m = 4$, $y_1 = 3$, and $x_1 = 2$}$$

$$y - 3 = 4x - 8 \qquad \text{Multiplying out and simplifying}$$

$$y = 4x - 5 \qquad \text{Adding 3 to each side}$$

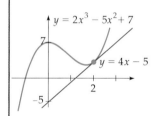

The graph on the left shows that $y = 4x - 5$ is the tangent line to the original function at $x = 2$.

## Derivatives in Business and Economics: Marginals

There is another interpretation for the derivative, one that is particularly important in business and economics. Suppose that a company has calculated its revenue, cost, and profit functions, as defined below.

$$R(x) = \begin{pmatrix} \text{Total revenue (income)} \\ \text{from selling $x$ units} \end{pmatrix} \qquad \text{Revenue function}$$

$$C(x) = \begin{pmatrix} \text{Total cost of} \\ \text{producing $x$ units} \end{pmatrix} \qquad \text{Cost function}$$

$$P(x) = \begin{pmatrix} \text{Total profit from producing} \\ \text{and selling $x$ units} \end{pmatrix} \qquad \text{Profit function}$$

The term **marginal cost** means the additional cost of producing one more unit, $C(x + 1) - C(x)$, which may be written $\dfrac{C(x + 1) - C(x)}{1}$, which is just the difference quotient $\dfrac{C(x + h) - C(x)}{h}$ with $h = 1$. If many units are being produced, then $h = 1$ is a relatively small number compared with $x$, so this difference quotient may be approximated by its limit as $h \to 0$, that is, by the *derivative* of the cost function. In view of this approximation, in calculus the marginal cost is *defined* to be the derivative of the cost function:

$$MC(x) = C'(x)$$   Marginal cost is the derivative of cost

The **marginal revenue** function $MR(x)$ and the **marginal profit** function $MP(x)$ are similarly defined as the derivatives of the revenue and cost functions.

$$MR(x) = R'(x)$$   Marginal revenue is the derivative of revenue

$$MP(x) = P'(x)$$   Marginal profit is the derivative of profit

All of this can be summarized very briefly: "marginal" means "derivative of." We now have three interpretations for the derivative: *slopes, instantaneous rates of change,* and *marginals*.

---

**EXAMPLE 8**   **FINDING AND INTERPRETING MARGINAL COST**

The Pocket EZCie is a miniature key chain flashlight based on LED (light-emitting diode) technology. The cost function (the total cost of producing $x$ Pocket EZCies) is

$$C(x) = 8\sqrt[4]{x^3} + 300$$   Cost function

dollars, where $x$ is the number of Pocket EZCies produced.

**a.** Find the marginal cost function $MC(x)$.

**b.** Find the marginal cost when 81 Pocket EZCies have been produced and interpret your answer.

*Solution*

**a.** The marginal cost function is the derivative of the cost function $C(x) = 8x^{3/4} + 300$, so

$$MC(x) = 6x^{-1/4} = \frac{6}{\sqrt[4]{x}}$$   Derivative of $C(x)$

**b.** To find the marginal cost when 81 Pocket EZCies have been produced, we evaluate the marginal cost function $MC(x)$ at $x = 81$:

$$MC(81) = \frac{6}{\sqrt[4]{81}} = \frac{6}{3} = 2$$   $MC(x) = \frac{6}{\sqrt[4]{x}}$ evaluated at $x = 81$

*Interpretation:* When 81 Pocket EZCies have been produced, the marginal cost is $2, meaning that to produce one more Pocket EZCie costs about $2.

*Source:* EZCie Manufacturing

Cost function showing the marginal cost (slope) at $x = 81$

**EXAMPLE 9**  **FINDING A LEARNING RATE**

Suppose that a psychology researcher finds that the number of names that a person can memorize in $x$ minutes is approximately $f(x) = 6\sqrt[3]{x^2}$. Find the instantaneous rate of change of this function after 8 minutes and interpret your answer.

*Solution*

$$f(x) = 6x^{2/3} \qquad \text{$6\sqrt[3]{x^2}$ in exponential form}$$

$$f'(x) = 6 \cdot \frac{2}{3} x^{-1/3} = 4x^{-1/3} \qquad \begin{array}{l}\text{The instantaneous rate}\\ \text{of change is } f'(x)\end{array}$$

$$f'(8) = 4(8)^{-1/3} = 4\left(\frac{1}{\sqrt[3]{8}}\right) = 4\left(\frac{1}{2}\right) = 2 \qquad \text{Evaluating at } x = 8$$

*Interpretation:* After 8 minutes the person can memorize about two additional names per minute.

## Functions as Single Objects

You may have noticed that calculus requires a more abstract point of view than precalculus mathematics. In earlier courses you looked at functions and graphs as collections of individual points, to be plotted one at a time. Now, however, we are operating on *whole functions* all at once (for example, differentiating the function $x^3$ to obtain the function $3x^2$). In calculus, the basic objects of interest are *functions*, and a function should be thought of as a *single* object.

This is in keeping with a trend toward increasing abstraction as you learn mathematics. You first studied single numbers, then points (pairs of numbers), then functions (collections of points), and now collections of functions (polynomials, differentiable functions, and so on). Each stage has been a generalization of the previous stage as you have reached higher levels of sophistication. This process of generalization or "chunking" of knowledge enables you to express ideas of wider applicability and power.

## Derivatives on a Graphing Calculator

Graphing calculators have an operation called NDERIV (or something similar), standing for **numerical derivative,** which gives an *approximation* of the derivative of a function. Most do so by evaluating the **symmetric difference quotient,** $\dfrac{f(x + h) - f(x - h)}{2h}$ for a small value of $h$, such as $h = 0.001$. The numerator represents the change in the function when $x$ changes by $2h$ (from $x - h$ to $x + h$), and the denominator divides by this change in $x$. Geometrically, the symmetric difference quotient gives the slope of the secant line through two points on the curve $h$ units on either side of the point at $x$. While NDERIV usually approximates the derivative quite closely, it sometimes gives erroneous results, as we will see in later sections. For this reason, using a graphing calculator effectively requires an understanding of both the calculus that underlies it and the technology that limits it.

## Graphing Calculator Exploration

$y_2 \; y_1$

**a.** Use a graphing calculator to graph $y_1 = x^3 - x^2 - 6x + 3$ on $[-5, 5]$ by $[-10, 10]$, as shown on the left.

**b.** Define $y_2$ as the derivative of $y_1$ (using NDERIV) and graph both functions.

**c.** Observe that where $y_1$ is horizontal, the *value* of $y_2$ is zero; where $y_1$ slopes *upward*, $y_2$ is *positive*; and where $y_1$ slopes *downward*, $y_2$ is *negative*. Would you be able to use these observations to identify which curve is the original function and which is the derivative?

**d.** Now check the answer to Example 9 as follows: redefine $y_1$ as $y_1 = 6x^{2/3}$, reset the window to $[0, 10]$ by $[-10, 30]$, GRAPH $y_1$ and $y_2$, and EVALUATE $y_2$ at $x = 8$. Your answer should agree with that of Example 9.

### Solutions to Practice Problems

**1. a.** $\dfrac{d}{dx} x^2 = 2x^{2-1} = 2x$

**b.** $\dfrac{d}{dx} x^{-5} = -5x^{-5-1} = -5x^{-6}$

**c.** $\dfrac{d}{dx} \sqrt[4]{x} = \dfrac{d}{dx} x^{1/4} = \dfrac{1}{4}x^{(1/4)-1} = \dfrac{1}{4}x^{-3/4}$

**2.** $\dfrac{df}{dx} = 20x^3$

$\left.\dfrac{df}{dx}\right|_{x=-1} = 20(-1)^3 = -20$

## 2.3 Section Summary

Our development of calculus has followed two quite different lines—one technical (the *rules* of derivatives) and the other conceptual (the *meaning* of derivatives).

On the conceptual side, derivatives have three meanings:

- Instantaneous rates of change
- Slopes
- Marginals

The fact that the derivative represents all three of these ideas simultaneously is one of the reasons that calculus is so useful.

On the technical side, although we have learned several differentiation rules, we really know how to differentiate only one kind of function, *x to a constant power*:

$$\frac{d}{dx} x^n = nx^{n-1}$$

The other rules,

$$\frac{d}{dx}[c \cdot f(x)] = c \cdot f'(x)$$

and

$$\frac{d}{dx}[f(x) \pm g(x)] = f'(x) \pm g'(x)$$

simply extend the Power Rule to sums, differences, and constant multiples of such powers. Therefore, any function to be differentiated must first be expressed in terms of powers. This is why we reviewed exponential notation so carefully in Chapter 1.

## Verification of the Rules of Differentiation

### Verification of the Power Rule for Positive Integer Exponents

Multiplying $(x + h)$ times itself repeatedly gives

$$(x + h)^2 = x^2 + 2xh + h^2$$

$$(x + h)^3 = x^3 + 3x^2h + 3xh^2 + h^3$$

and in general, for any positive integer $n$,

$$(x + h)^n = x^n + nx^{n-1}h + \tfrac{1}{2}n(n - 1)x^{n-2}h^2 + \cdots + nxh^{n-1} + h^n$$

$$= x^n + nx^{n-1}h + h^2[\tfrac{1}{2}n(n - 1)x^{n-2} + \cdots + nxh^{n-3} + h^{n-2}] \qquad \text{Factoring out } h^2$$

$$= x^n + nx^{n-1}h + h^2 \cdot P \qquad \qquad \begin{array}{l} P \text{ stands for the polynomial} \\ \text{in the square bracket above} \end{array}$$

The resulting formula

$$(x + h)^n = x^n + nx^{n-1}h + h^2 \cdot P$$

will be useful in the following verification. To prove the Power Rule for any positive integer $n$, we use the definition of the derivative to differentiate $f(x) = x^n$.

$$f'(x) = \lim_{h \to 0} \frac{f(x + h) - f(x)}{h} \qquad \text{Definition of the derivative}$$

$$= \lim_{h \to 0} \frac{(x + h)^n - x^n}{h} \qquad \begin{array}{l} \text{Since } f(x + h) = (x + h)^n \\ \text{and } f(x) = x^n \end{array}$$

$$= \lim_{h \to 0} \frac{x^n + nx^{n-1}h + h^2 \cdot P - x^n}{h}$$

Expanding, using the formula derived earlier

$$= \lim_{h \to 0} \frac{nx^{n-1}h + h^2 \cdot P}{h}$$

Canceling the $x^n$ and the $-x^n$

$$= \lim_{h \to 0} \frac{h(nx^{n-1} + h \cdot P)}{h}$$

Factoring out an $h$

$$= \lim_{h \to 0} (nx^{n-1} + h \cdot P)$$

Canceling the $h$ (since $h \neq 0$)

$$= nx^{n-1}$$

Evaluating the limit by direct substitution

This shows that for any positive integer $n$, the derivative of $x^n$ is $nx^{n-1}$.

### Verification of the Constant Multiple Rule

For a constant $c$ and a function $f$, let $g(x) = c \cdot f(x)$. If $f'(x)$ exists, we may calculate the derivative $g'(x)$ as follows:

$$g'(x) = \lim_{h \to 0} \frac{g(x + h) - g(x)}{h}$$

Definition of the derivative

$$= \lim_{h \to 0} \frac{c \cdot f(x + h) - c \cdot f(x)}{h}$$

Since $g(x + h) = c \cdot f(x + h)$ and $g(x) = c \cdot f(x)$

$$= \lim_{h \to 0} \frac{c \cdot [f(x + h) - f(x)]}{h}$$

Factoring out the $c$

$$= c \cdot \lim_{h \to 0} \frac{f(x + h) - f(x)}{h}$$

Taking $c$ outside the limit leaves just the definition of the derivative $f'(x)$.

$$= c \cdot f'(x)$$

Constant Multiple Rule

This shows that the derivative of a constant times a function, $c \cdot f(x)$, is the constant times the derivative of the function, $c \cdot f'(x)$.

### Verification of the Sum Rule

For two functions $f$ and $g$, let their sum be $s = f + g$. If $f'(x)$ and $g'(x)$ exist, we may calculate $s'(x)$ as follows:

$$s'(x) = \lim_{h \to 0} \frac{s(x + h) - s(x)}{h}$$

Definition of the derivative

$$= \lim_{h \to 0} \frac{[f(x + h) + g(x + h)] - [f(x) + g(x)]}{h}$$

Since $s(x + h) = f(x + h) + g(x + h)$ and $s(x) = f(x) + g(x)$

$$= \lim_{h \to 0} \frac{f(x + h) + g(x + h) - f(x) - g(x)}{h}$$

Eliminating the brackets

$$= \lim_{h \to 0} \frac{f(x + h) - f(x) + g(x + h) - g(x)}{h}$$

Rearranging the numerator

$$= \lim_{h \to 0} \left[ \frac{f(x + h) - f(x)}{h} + \frac{g(x + h) - g(x)}{h} \right]$$

Separating the fraction into two parts

$$= \underbrace{\lim_{h \to 0} \frac{f(x + h) - f(x)}{h}}_{f'(x)} + \underbrace{\lim_{h \to 0} \frac{g(x + h) - g(x)}{h}}_{g'(x)}$$

Using Limit Rule 4a on page 81

Recognizing the definition of the derivatives of $f$ and $g$

$$= f'(x) + g'(x)$$

Sum Rule

This shows that the derivative of a sum $f(x) + g(x)$ is the sum of the derivatives $f'(x) + g'(x)$.

## 2.3 Exercises

**1–30.** Find the derivative of each function.

**1.** $f(x) = x^4$   **2.** $f(x) = x^5$   **3.** $f(x) = x^{500}$

**4.** $f(x) = x^{1000}$   **5.** $f(x) = x^{1/2}$   **6.** $f(x) = x^{1/3}$

**7.** $g(x) = \frac{1}{2}x^4$   **8.** $f(x) = \frac{1}{3}x^9$

**9.** $g(w) = 6\sqrt[3]{w}$   **10.** $g(w) = 12\sqrt{w}$

**11.** $h(x) = \frac{3}{x^2}$   **12.** $h(x) = \frac{4}{x^3}$

**13.** $f(x) = 4x^2 - 3x + 2$

**14.** $f(x) = 3x^2 - 5x + 4$

**15.** $f(x) = \frac{1}{x^{1/2}}$   **16.** $f(x) = \frac{1}{x^{2/3}}$

**17.** $f(x) = \frac{6}{\sqrt[3]{x}}$   **18.** $f(x) = \frac{4}{\sqrt{x}}$

**19.** $f(r) = \pi r^2$   **20.** $f(r) = \frac{4}{3}\pi r^3$

**21.** $f(x) = \frac{1}{6}x^3 + \frac{1}{2}x^2 + x + 1$

**22.** $f(x) = \frac{1}{24}x^4 + \frac{1}{6}x^3 + \frac{1}{2}x^2 + x + 1$

**23.** $g(x) = \sqrt{x} - \frac{1}{x}$   **24.** $g(x) = \sqrt[3]{x} - \frac{1}{x}$

**25.** $h(x) = 6\sqrt[3]{x^2} - \frac{12}{\sqrt[3]{x}}$   **26.** $h(x) = 8\sqrt{x^3} - \frac{8}{\sqrt[4]{x}}$

**27.** $f(x) = \frac{10}{\sqrt{x}} - \frac{9\sqrt[3]{x^5}}{5} + 17$

**28.** $f(x) = \frac{9}{2\sqrt[3]{x^2}} - 16\sqrt{x^5} - 14$

**29.** $f(x) = \frac{x^2 + x^3}{x}$   **30.** $f(x) = x^2(x + 1)$

**31–38.** Find the indicated derivatives.

**31.** If $f(x) = x^5$, find $f'(-2)$.

**32.** If $f(x) = x^4$, find $f'(-3)$.

**33.** If $f(x) = 6\sqrt[3]{x^2} - \frac{48}{\sqrt[3]{x}}$, find $f'(8)$.

**34.** If $f(x) = 12\sqrt[3]{x^2} + \frac{48}{\sqrt[3]{x}}$, find $f'(8)$.

**35.** If $f(x) = x^3$, find $\left. \frac{df}{dx} \right|_{x=-3}$

**36.** If $f(x) = x^4$, find $\left. \frac{df}{dx} \right|_{x=-2}$

**37.** If $f(x) = \frac{16}{\sqrt{x}} + 8\sqrt{x}$, find $\left. \frac{df}{dx} \right|_{x=4}$

**38.** If $f(x) = \frac{54}{\sqrt{x}} + 12\sqrt{x}$, find $\left. \frac{df}{dx} \right|_{x=9}$

**39. a.** Find the equation of the tangent line to $f(x) = x^2 - 2x + 2$ at $x = 3$.
   **b.** Graph the function and the tangent line on the window $[-1, 6]$ by $[-10, 20]$.

**40. a.** Find the equation of the tangent line to $f(x) = x^2 - 4x + 6$ at $x = 1$.
   **b.** Graph the function and the tangent line on the window $[-1, 5]$ by $[-2, 10]$.

**41. a.** Find the equation of the tangent line to
$f(x) = x^3 - 3x^2 + 2x - 2$ at $x = 2$.
 **b.** Graph the function and the tangent line on the window $[-1, 4]$ by $[-7, 5]$.

**42. a.** Find the equation of the tangent line to
$f(x) = 3x^2 - x^3$ at $x = 1$.
 **b.** Graph the function and the tangent line on the window $[-1, 3]$ by $[-2, 5]$.

 **43.** Use a graphing calculator to verify that the derivative of a constant is zero, as follows. Define $y_1$ to be a constant (such as $y_1 = 5$) and then use NDERIV to define $y_2$ to be the derivative of $y_1$. Then graph the two functions together on an appropriate window and use TRACE to observe that the derivative $y_2$ is zero (graphed as a line along the $x$-axis), showing that the derivative of a constant is zero.

 **44.** Use a graphing calculator to verify that the derivative of a linear function is a constant, as follows. Define $y_1$ to be a linear function (such as $y_1 = 3x - 4$) and then use NDERIV to define $y_2$ to be the derivative of $y_1$. Then graph the two functions together on an appropriate window and observe that the derivative $y_2$ is a constant (graphed as a horizontal line, such as $y_2 = 3$), verifying that the derivative of $y_1 = mx + b$ is $y_2 = m$.

## Applied Exercises

**45. BUSINESS: Software Costs** Businesses can buy multiple licenses for PowerZip data-compression software at a total cost of approximately $C(x) = 24x^{2/3}$ dollars for $x$ licenses. Find the derivative of this cost function at:

**a.** $x = 8$ and interpret your answer.
**b.** $x = 64$ and interpret your answer.
*Source: Trident Software*

**46. BUSINESS: Software Costs** Media companies can buy multiple licenses for AudioTime audio-recording software at a total cost of approximately $C(x) = 168x^{5/6}$ dollars for $x$ licenses. Find the derivative of this cost function at:

**a.** $x = 1$ and interpret your answer.
**b.** $x = 64$ and interpret your answer.
*Source: NCH Swift Sound*

**47. BUSINESS: Marginal Cost** (*45 continued*) Use a calculator to find the actual cost of the 64th license by evaluating $C(64) - C(63)$ for the cost function in Exercise 45. Is your answer close to the $4 that you found for part (b) of that Exercise?

**48. BUSINESS: Marginal Cost** (*46 continued*) Use a calculator to find the actual cost of the 64th license by evaluating $C(64) - C(63)$ for the cost function in Exercise 46. Is your answer close to the $70 that you found for part (b) of that Exercise?

**49. BUSINESS: Marketing to Young Adults** Companies selling products to young adults often try to predict the size of that population in future years. According to predictions by the Census Bureau, the 18- to 24-year-old population in the United States will follow the curve shown below.

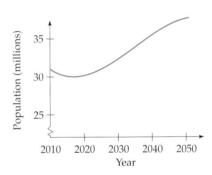

This population is given by $P(x) = -\frac{1}{3}x^3 + 25x^2 - 300x + 31{,}000$ (in thousands), where $x$ is the number of years after 2010. Find the rate of change of this population:

**a.** In the year 2030 and interpret your answer.
**b.** In the year 2010 and interpret your answer.
*Source: 2007 Statistical Abstract of the United States*

**50. BIOMEDICAL: Flu Epidemic** The number of people newly infected on day $t$ of a flu epidemic is $f(t) = 13t^2 - t^3$ (for $0 \le t \le 13$). Find the instantaneous rate of change of this number on:

**a.** Day 5 and interpret your answer.
**b.** Day 10 and interpret your answer.

**51. BUSINESS: Advertising** It has been estimated that the number of people who will see a newspaper advertisement that has run for $x$ consecutive days is of the form   $N(x) = T - \frac{1}{2}T/x$   for $x \geq 1$,   where $T$ is the total readership of the newspaper. If a newspaper has a circulation of 400,000, an ad that runs for $x$ days will be seen by

$$N(x) = 400,000 - \frac{200,000}{x}$$

people. Find how fast this number of potential customers is growing when this ad has run for 5 days.

**52. BIOMEDICAL: Blood Flow** Nitroglycerin is often prescribed to enlarge blood vessels that have become too constricted. If the cross-sectional area of a blood vessel $t$ hours after nitroglycerin is administered is   $A(t) = 0.01t^2$   square centimeters (for   $1 \leq t \leq 5$),   find the instantaneous rate of change of the cross-sectional area 4 hours after the administration of nitroglycerin.

**53. GENERAL: Internet Access** The percentage of U.S. households with broadband Internet access is approximated by   $f(x) = \frac{1}{4}x^2 + 5x + 6$,   where $x$ is the number of years after the year 2000. Find the rate of change of this percentage in the year 2010 and interpret your answer.

*Source: Consumer USA 2008*

**54. GENERAL: Traffic Safety** Traffic fatalities have decreased over recent decades, with the number of fatalities per hundred million vehicle miles traveled given approximately by   $f(x) = \dfrac{2}{\sqrt{x}} + 1$,

where $x$ stands for the number of five-year intervals since 1975 (so, for example,   $x = 2$   would mean 1985). Find the rate of change of this function at:

**a.** $x = 1$   and interpret your answer.
**b.** $x = 4$   and interpret your answer.

*Source: National Highway Traffic Safety Administration*

**55. PSYCHOLOGY: Learning Rates** A language school has found that its students can memorize $p(t) = 24\sqrt{t}$   phrases in $t$ hours of class (for $1 \leq t \leq 10$).   Find the instantaneous rate of change of this quantity after 4 hours of class.

**56. ENVIRONMENTAL SCIENCE: Water Quality** Downstream from a waste treatment plant the amount of dissolved oxygen in the water usually decreases for some distance (due to bacteria consuming the oxygen) and then increases (due

to natural purification). A graph of the dissolved oxygen at various distances downstream looks like the curve below (known as the "oxygen sag"). The amount of dissolved oxygen is usually taken as a measure of the health of the river.

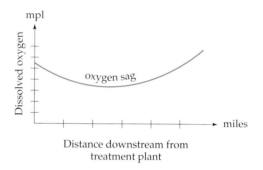

Suppose that the amount of dissolved oxygen $x$ miles downstream is   $D(x) = 0.2x^2 - 2x + 10$   mpl (milligrams per liter) for   $0 \leq x \leq 20$.   Use this formula to find the instantaneous rate of change of the dissolved oxygen:

**a.** 1 mile downstream.
**b.** 10 miles downstream.

Interpret the signs of your answers.

**57–58. ECONOMICS: Marginal Utility** Generally, the more you have of something, the less valuable each additional unit becomes. For example, a dollar is less valuable to a millionaire than to a beggar. Economists define a person's "utility function" $U(x)$ for a product as the "perceived value" of having $x$ units of that product. The *derivative* of $U(x)$ is called the *marginal utility function*,   $MU(x) = U'(x)$.   Suppose that a person's utility function for money is given by the function below. That is, $U(x)$ is the utility (perceived value) of $x$ dollars.

**a.** Find the marginal utility function $MU(x)$.
**b.** Find $MU(1)$, the marginal utility of the first dollar.
**c.** Find $MU(1,000,000)$, the marginal utility of the millionth dollar.

**57.** $U(x) = 100\sqrt{x}$        **58.** $U(x) = 12\sqrt[3]{x}$

**59. GENERAL: Smoking and Education** According to a study, the probability that a smoker will quit smoking increases with the smoker's educational level. The probability (expressed as a percent) that a smoker with $x$ years of education will quit is approximated by the equation $f(x) = 0.831x^2 - 18.1x + 137.3$ (for $10 \leq x \leq 16$).

a. Find $f(12)$ and $f'(12)$ and interpret these numbers. [*Hint:*   $x = 12$   corresponds to a high school graduate.]

b. Find $f(16)$ and $f'(16)$ and interpret these numbers. [*Hint:*   $x = 16$   corresponds to a college graduate.]

*Source: Review of Economics and Statistics LXXVII*

**60. BIOMEDICAL: Lung Cancer**  Asbestos has been found to be a potent cause of lung cancer. According to one study of asbestos workers, the number of lung cancer cases in the group depended on the number $t$ of years of exposure to asbestos according to the function   $N(t) = 0.00437t^{3.2}$.

a. Graph this function on the window [0, 15] by [−10, 30].

b. Find $N(10)$ and $N'(10)$ and interpret these numbers.

*Source: British Journal of Cancer 45*

**61–62. GENERAL: College Tuition**  The following graph shows the average annual college tuition costs (tuition and fees) for a year at a private or public college. The data is given for five-year intervals.

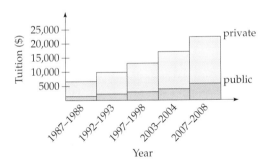

Year

*Source:* The College Board

**61.** The tuition for a *private* college is approximated by the function   $f(x) = 400x^2 + 2500x + 7200$, where $x$ is the number of five-year intervals since the academic year 1987–88 (so the years in the graph are numbered   $x = 0$   through   $x = 4$).

a. Use this function to predict tuition in the academic year 2017–18. [*Hint:* What $x$-value corresponds to that year?]

b. Find the derivative of this function for the $x$-value that you used in part (a) and interpret it as a rate if change in the proper units.

c. From your answer to part (b), estimate how rapidly tuition will be increasing *per year* in 2017–18.

**62.** The tuition shown above for a *public* college is approximated by the function   $f(x) = 200x^2 + 350x + 1600$,   where $x$ is the number of five-year intervals since the academic year 1987–88 (so the years in the graph are numbered   $x = 0$   through   $x = 4$).

a. Use this function to predict tuition in the academic year 2017–18. [*Hint:* What $x$-value corresponds to that year?]

b. Find the derivative of this function for the $x$-value that you used in part (a) and interpret it as a rate if change in the proper units.

c. From your answer to part (b), estimate how rapidly tuition will be increasing *per year* in 2017–18.

## Conceptual Exercises

**63.** Explain, in *two different ways*, without using the rules of differentiation, why the derivative of the constant function   $f(x) = 2$   must be   $f'(x) = 0$. [*Hint:* Think of the slope of the graph of a constant function, and also of the instantaneous rate of change of a function that stays constant.]

**64.** Explain, in *two different ways*, without using the rules of differentiation, why the derivative of the linear function   $f(x) = 3x − 5$   must be   $f'(x) = 3$. [*Hint:* Think of the slope of the line   $y = mx + b$   that represents this function, and also of the instantaneous rate of change of a function that increases linearly.]

**65.** Give an intuitive explanation of the Constant Multiple Rule (page 111) by thinking that if $f(x)$ has a certain rate of change, then what will be the rate of change of the function   $2 \cdot f(x)$   or of the function   $c \cdot f(x)$?

**66.** Give an intuitive explanation of the Sum Rule (page 112) by thinking that if $f(x)$ and $g(x)$ have certain rates of change, then what will be the rate of change of   $f(x) + g(x)$?

**67.** How will the slopes of $f$ and of   $-f$   differ? Explain intuitively and in terms of the rules of differentiation.

**68.** How will the slopes of $f$ and   $f + 10$   differ? Explain intuitively and in terms of the rules of differentiation.

**69.** We have said that the expression $f'(2)$ means *first* differentiate and *then* evaluate. What if we were to first evaluate the function at 2 and *then* differentiate? What would we get? Would the answer depend on the particular function or on the particular number?

**70.** Give an example to show that if a function is positive (at a particular $x$-value) its derivative (at that same $x$-value) need not be positive.

**71.** A recent article studying the relationship between life expectancy and education found that if a function $l(s)$ gives the life expectancy of a person who has had $s$ years of schooling, then $\dfrac{dl}{ds} = 1.7$. Interpret this result.

*Source: Review of Economic Studies, 2005*

**72.** A recent article studying the relationship between the probability of an accident and driving speed found that if a function $p(s)$ gives the probability (as a percent) of an accident for a driver who is exceeding the speed limit by $s\%$, then $\dfrac{dp}{ds} = 13$. Interpret this result.

*Source: Accident Analysis and Prevention, 2006*

## 2.4 THE PRODUCT AND QUOTIENT RULES

### Introduction

In the previous section we learned how to differentiate the sum and difference of two functions—we simply take the sum or difference of the derivatives. In this section we learn how to differentiate the *product* and *quotient* of two functions. Unfortunately, we do not simply take the product or quotient of the derivatives. Matters are a little more complicated.

### Product Rule

To differentiate the product of two functions, $f(x) \cdot g(x)$, we use the *Product Rule*.

| Product Rule |
| --- |

$$\frac{d}{dx}[f(x) \cdot g(x)] = f'(x) \cdot g(x) + f(x) \cdot g'(x)$$

The derivative of a product is the derivative of the first times the second plus the first times the derivative of the second.

(provided, of course, that the derivatives $f'(x)$ and $g'(x)$ both exist). A derivation of the Product Rule is given at the end of this section.

The formula is clearer if we write the functions simply as $f$ and $g$.

$$\frac{d}{dx}(f \cdot g) = f' \cdot g + f \cdot g'$$

Derivative of the first    Second    First    Derivative of the second

| EXAMPLE 1 | **USING THE PRODUCT RULE** |

Use the Product Rule to calculate $\dfrac{d}{dx}(x^3 \cdot x^5)$.

*Solution*

$$\frac{d}{dx}(x^3 \cdot x^5) = 3x^2 \cdot x^5 + x^3 \cdot 5x^4 = 3x^7 + 5x^7 = 8x^7$$

Derivative    Second    First    Derivative
of the first                        of the second

We may check this answer by simplifying the original product, $x^3 \cdot x^5 = x^8$, and then differentiating:

$$\underbrace{\frac{d}{dx}(x^3 \cdot x^5)}_{x^8} = \frac{d}{dx}x^8 = 8x^7 \qquad \text{Agrees with above answer}$$

 **BE CAREFUL:** The derivative of a product is *not* the product of the derivatives: $(f \cdot g)' \neq f' \cdot g'$. For $x^3 \cdot x^5$ the product of the derivatives would be $3x^2 \cdot 5x^4 = 15x^6$, which is *not* the correct answer $8x^7$ that we found above. The Product Rule shows the correct way to differentiate a product.

| EXAMPLE 2 | **USING THE PRODUCT RULE** |

Use the Product Rule to find $\dfrac{d}{dx}[(x^2 - x + 2)(x^3 + 3)]$.

*Solution*

$$\frac{d}{dx}[(x^2 - x + 2)(x^3 + 3)]$$

$$= \underbrace{(2x - 1)}(x^3 + 3) + (x^2 - x + 2)\underbrace{(3x^2)}$$

Derivative of         Derivative
$x^2 - x + 2$          of $x^3 + 3$

$$= 2x^4 + 6x - x^3 - 3 + 3x^4 - 3x^3 + 6x^2 \qquad \text{Multiplying out}$$

$$= 5x^4 - 4x^3 + 6x^2 + 6x - 3 \qquad \text{Simplifying}$$

In the preceding Example we could have *first* multiplied out the function and *then* differentiated. However, our way was slightly easier. Furthermore, we will soon have problems that can *only* be done by the product rule, so the practice will be useful.

| Practice Problem 1 | Use the Product Rule to find $\dfrac{d}{dx}[x^3(x^2 - x)]$. | ➤ Solution on page 132 |

**EXAMPLE 3**

**USING THE PRODUCT RULE**

Use the product rule to find the derivative of $f(x) = \sqrt{x}\,(2x - 4)$.

*Solution*

Writing the function in power form

$$f(x) = x^{1/2}\,(2x - 4)$$

the Product Rule gives

$$f'(x) = \tfrac{1}{2}x^{-1/2}(2x - 4) + x^{1/2}(2) = x^{1/2} - 2x^{-1/2} + 2x^{1/2} \qquad \text{Multiplying out}$$

$\underbrace{\qquad\qquad}$      $\nearrow$
Derivative      Derivative
of $x^{1/2}$      of $2x - 4$

$$= 3x^{1/2} - 2x^{-1/2} = 3\sqrt{x} - \frac{2}{\sqrt{x}} \qquad \begin{array}{l}\text{Combining terms}\\ \text{and writing in}\\ \text{radical form}\end{array}$$

## Quotient Rule

The *Quotient Rule* shows how to differentiate a quotient of two functions.

---

**Quotient Rule**

$$\frac{d}{dx}\left(\frac{f(x)}{g(x)}\right) = \frac{g(x) \cdot f'(x) - g'(x) \cdot f(x)}{[g(x)]^2}$$

← The bottom times the derivative of the top, minus the derivative of the bottom times the top

← The bottom squared

---

(provided that the derivatives $f'(x)$ and $g'(x)$ both exist and that $g(x) \neq 0$). A derivation of the Quotient Rule is given at the end of this section.

    The Quotient Rule looks less formidable if we write the functions simply as $f$ and $g$,

$$\frac{d}{dx}\left(\frac{f}{g}\right) = \frac{g \cdot f' - g' \cdot f}{g^2}$$

or even as*

$$\frac{d}{dx}\left(\frac{\text{top}}{\text{bottom}}\right) = \frac{(\text{bottom}) \cdot \left(\dfrac{d}{dx}\,\text{top}\right) - \left(\dfrac{d}{dx}\,\text{bottom}\right) \cdot (\text{top})}{(\text{bottom})^2}$$

---

*Some students remember this by the mnemonic *Lo D Hi − Hi D Lo over Lo Lo* where *Lo* and *Hi* stand for the denominator and numerator, respectively.

EXAMPLE 4

**USING THE QUOTIENT RULE**

Use the Quotient Rule to find $\dfrac{d}{dx}\left(\dfrac{x^9}{x^3}\right)$.

*Solution*

$$\underset{\text{Bottom}}{} \overset{\substack{\text{Derivative} \\ \text{of the top}}}{} \overset{\substack{\text{Derivative} \\ \text{of the bottom}}}{} \overset{\text{Top}}{}$$

$$\frac{d}{dx}\left(\frac{x^9}{x^3}\right) = \frac{(x^3)(9x^8) - (3x^2)(x^9)}{(x^3)^2} = \frac{9x^{11} - 3x^{11}}{x^6} = \frac{6x^{11}}{x^6} = 6x^5$$

Bottom squared

We may check this answer by simplifying the original quotient and then differentiating:

$$\frac{d}{dx}\left(\frac{x^9}{x^3}\right) = \frac{d}{dx}x^6 = 6x^5 \qquad \text{Agrees with above answer}$$

$x^6$

 **BE CAREFUL:** The derivative of a quotient is *not* the quotient of the derivatives:

$$\left(\frac{f}{g}\right)' \text{ is } not \text{ equal to } \frac{f'}{g'}$$

For the quotient $\dfrac{x^9}{x^3}$ taking the quotient of the derivatives would give $\dfrac{9x^8}{3x^2} = 3x^6$, which is *not* the correct answer $6x^5$ that we found above. The Quotient Rule shows the correct way to differentiate a quotient.

EXAMPLE 5

**USING THE QUOTIENT RULE**

Find $\dfrac{d}{dx}\left(\dfrac{x^2}{x+1}\right)$.

*Solution* The Quotient Rule gives

$$\underset{\text{Bottom}}{} \overset{\substack{\text{Derivative} \\ \text{of the top}}}{} \overset{\substack{\text{Derivative} \\ \text{of the bottom}}}{} \overset{\text{Top}}{}$$

$$\frac{d}{dx}\left(\frac{x^2}{x+1}\right) = \frac{(x+1)(2x) - (1)(x^2)}{(x+1)^2} = \frac{2x^2 + 2x - x^2}{(x+1)^2} = \frac{x^2 + 2x}{(x+1)^2}$$

Bottom squared

**Practice Problem 2**   Find $\dfrac{d}{dx}\left(\dfrac{2x^2}{x^2+1}\right)$.

➤ Solution on page 132

**EXAMPLE 6**    **FINDING THE COST OF CLEANER WATER**

Practically every city must purify its drinking water and treat its wastewater. The cost of the treatment rises steeply for higher degrees of purity. If the cost of purifying a gallon of water to a purity of $x$ percent is

$$C(x) = \frac{2}{100 - x} \qquad \text{for } 80 < x < 100$$

dollars, find the rate of change of the purification costs when the purity is:

**a.** 90%          **b.** 98%

*Solution*

The rate of change of cost is the *derivative* of the cost function:

Derivative of 2

Derivative of $100 - x$

$$C'(x) = \frac{d}{dx}\left(\frac{2}{100-x}\right) = \frac{(100-x)(0) - (-1)(2)}{(100-x)^2} \qquad \text{Differentiating by the Quotient Rule}$$

$$= \frac{0+2}{(100-x)^2} = \frac{2}{(100-x)^2} \qquad \text{Simplifying (the derivative is undefined at } x = 100)$$

**a.** For 90% purity we evaluate at $x = 90$:

$$C'(90) = \frac{2}{(100-90)^2} = \frac{2}{10^2} = \frac{2}{100} = 0.02 \qquad C'(x) = \frac{2}{(100-x)^2} \text{ evaluated at } x = 90$$

*Interpretation:* At 90% purity, the rate of change of the cost is 0.02 dollar, meaning that the costs increase by about *2 cents for each additional percentage of purity.*

**b.** For 98% purity we evaluate $C'(x)$ at $x = 98$:

$$C'(98) = \frac{2}{(100-98)^2} = \frac{2}{2^2} = \frac{2}{4} = \frac{1}{2} = 0.50 \qquad C'(x) = \frac{2}{(100-x)^2} \text{ evaluated at } x = 98$$

*Interpretation:* At 98% purity, the rate of change of the cost is 0.50 dollar, meaning that the costs increase by about *50 cents for each additional percentage of purity.*

Notice that an extra percentage of purity above the 98% level is 25 times as costly as an extra percentage above the 90% purity level.

## Graphing Calculator Exploration

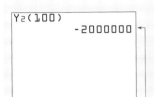

a. On a graphing calculator, enter the cost function from Example 6 as $y_1 = 2/(100 - x)$. Then use NDERIV to define $y_2$ to be the derivative of $y_1$.

b. Graph both $y_1$ and $y_2$ on the window [80, 100] by [−1, 5]. Your graph should resemble the one on the left (but you may have an additional "false" vertical line on the right).

c. Verify the results of Example 6 by evaluating $y_2$ at $x = 90$ and at $x = 98$.

d. Evaluate $y_2$ at $x = 100$, giving (supposedly) the derivative of $y_1$ at $x = 100$. However, in Example 6 we saw that the derivative of $y_1$ is *undefined* at $x = 100$. Your calculator is giving you a "false value" for the derivative, resulting from NDERIV's use of a symmetric difference quotient (see pages 116–117) and a small positive value for $h$. Therefore, to use your calculator effectively, you must also understand calculus.

calculator error!

Not every quotient requires the Quotient Rule. Some are simple enough to be differentiated by the Power Rule.

**EXAMPLE 7**

### DIFFERENTIATING A QUOTIENT BY THE POWER RULE

Find the derivative of $y = \dfrac{5}{x^2}$.

*Solution* $\quad \dfrac{d}{dx}\left(\dfrac{5}{x^2}\right) = \dfrac{d}{dx}(5x^{-2}) = -10x^{-3} = -\dfrac{10}{x^3}$  Differentiated by the power rule

In this example we rewrote the expression before and after the differentiation:

| Begin | Rewrite | Differentiate | Rewrite |
|-------|---------|---------------|---------|
| $y = \dfrac{5}{x^2}$ | $y = 5x^{-2}$ | $\dfrac{dy}{dx} = -10x^{-3}$ | $\dfrac{dy}{dx} = -\dfrac{10}{x^3}$ |

This way is often much easier than using the Quotient Rule if the numerator or denominator is a constant.

## Marginal Average Cost

It is often useful to calculate not just the *total* cost of producing $x$ units of some product, but also the **average cost per unit,** denoted $AC(x)$, which is found by dividing the total cost $C(x)$ by the number of units $x$.

$$AC(x) = \frac{C(x)}{x}$$   Average cost per unit is total cost divided by the number of units

The derivative of the average cost function is called the **marginal average cost,** *MAC*.*

$$MAC(x) = \frac{d}{dx}\left[\frac{C(x)}{x}\right]$$

Marginal average cost is the derivative of average cost

**Marginal average revenue,** *MAR,* and **marginal average profit,** *MAP,* are defined similarly as the derivatives of average revenue per unit, $\frac{R(x)}{x}$, and average profit per unit, $\frac{P(x)}{x}$.

$$MAR(x) = \frac{d}{dx}\left[\frac{R(x)}{x}\right]$$

Marginal average revenue is the derivative of average revenue $\frac{R(x)}{x}$

$$MAP(x) = \frac{d}{dx}\left[\frac{P(x)}{x}\right]$$

Marginal average profit is the derivative of average profit $\frac{P(x)}{x}$

---

**EXAMPLE 8**   **FINDING AND INTERPRETING MARGINAL AVERAGE COST**

POD, or *printing on demand*, is a recent development in publishing that makes it feasible to print small quantities of books (even a single copy), thereby eliminating overstock and storage costs. For example, POD-publishing a typical 200-page book would cost $18 per copy, with fixed costs of $1500. Therefore, the cost function is

$$C(x) = 18x + 1500 \qquad \text{Total cost of producing } x \text{ books}$$

**a.** Find the average cost function.
**b.** Find the marginal average cost function.
**c.** Find the marginal average cost at $x = 100$ and interpret your answer.

*Source*: e-booktime.com

*Solution*

**a.** The average cost function is

$$AC(x) = \underbrace{\frac{18x + 1500}{x}}_{\substack{\text{Total cost divided} \\ \text{by number of units}}} = \underbrace{18 + \frac{1500}{x}}_{\text{Simplifying}} = \underbrace{18 + 1500x^{-1}}_{\text{In power form}}$$

---

*The marginal average cost function is sometimes denoted $\overline{C}'(x)$, with similar notations used for marginal average revenue and marginal average profit.

b. The *marginal* average cost is the derivative of average cost. We could use the Quotient Rule on the first expression above, but it is easier to use the Power Rule on the last expression:

$$MAC(x) = \frac{d}{dx}(18 + 1500x^{-1}) = -1500x^{-2} = -\frac{1500}{x^2}$$

c. Evaluating at $x = 100$:

$$MAC(100) = -\frac{1500}{100^2} = -\frac{1500}{10,000} = -0.15 \qquad -\frac{1500}{x^2} \text{ at } x = 100$$

*Interpretation:* When 100 books have been produced, the average cost per book is decreasing (because of the negative sign) by about *15 cents per additional book produced.* This reflects the fact that while *total* costs rise when you produce more, the *average cost per unit* decreases, because of the economies of mass production.

## Graphing Calculator Exploration

Use a graphing calculator to investigate further the effects of mass production in the preceding Example.

a. Graph the average cost function [any of the expressions for $AC(x)$ from part (a) of the solution] on the window [0, 400] by [0, 50]. Your graph should resemble that shown on the left. TRACE along the average cost curve to see how the average cost drops from the 20s down to the teens as the number of books increases from 100 to 400. Note that although average cost falls, it does so more slowly as the number of units increases (the *law of diminishing returns*).

b. To see exactly how rapidly the average cost declines, graph the *marginal* average cost function on the window [0, 400] by [−1, 1]. Evaluating at $x = 100$ gives −0.15, the answer to part (c) in the above Example. TRACE along this curve to see how the marginal average cost (which is negative since costs are decreasing) approaches zero as the number of units increases.

**EXAMPLE 9**    **FINDING TIME SAVED BY SPEEDING**

A certain mathematics professor drives 25 miles to his office every day, mostly on highways. If he drives at constant speed $v$ miles per hour, his travel time (distance divided by speed) is

$$T(v) = \frac{25}{v}$$

hours. Find $T'(55)$ and interpret this number.

*Solution* Since $T(v) = \dfrac{25}{v}$ is a quotient, we could differentiate it by the Quotient Rule. However, it is easier to write $\dfrac{25}{v}$ as a *power*,

$$T(v) = 25v^{-1}$$

and differentiate using the Power Rule:

$$T'(v) = -25v^{-2}$$

This gives the rate of change of the travel time with respect to driving speed. $T'(v)$ is negative, showing that as speed increases, travel time *decreases*. Evaluating this at speed $v = 55$ gives

$$T'(55) = -25(55)^{-2} = \dfrac{-25}{(55)^2} \approx -0.00826 \qquad \text{Using a calculator}$$

This number, the rate of change of travel time with respect to driving speed, means that when driving at 55 miles per hour, you save only 0.00826 hour for each extra mile per hour of speed. Multiplying by 60 gives the saving in *minutes:*

$$(-0.00826)(60) \approx -0.50 = -\dfrac{1}{2}$$

That is, each extra mile per hour of speed saves only about half a minute, or 30 seconds. For example, speeding by 10 mph would save only about $\frac{1}{2} \cdot 10 = 5$ minutes. One must then decide whether this slight savings in time is worth the risk of an accident or a speeding ticket.

### Solutions to Practice Problems

**1.** $\dfrac{d}{dx}[x^3(x^2 - x)] = 3x^2(x^2 - x) + x^3(2x - 1)$

$$= 3x^4 - 3x^3 + 2x^4 - x^3$$

$$= 5x^4 - 4x^3$$

**2.** $\dfrac{d}{dx}\left(\dfrac{2x^2}{x^2 + 1}\right) = \dfrac{(x^2 + 1)4x - 2x \cdot 2x^2}{(x^2 + 1)^2}$

$$= \dfrac{4x^3 + 4x - 4x^3}{(x^2 + 1)^2} = \dfrac{4x}{(x^2 + 1)^2}$$

## 2.4 Section Summary

The following is a list of the differentiation formulas that we have learned so far. The letters $c$ and $n$ stand for constants, and $f$ and $g$ stand for differentiable functions of $x$.

$$\frac{d}{dx}c = 0$$

$$\frac{d}{dx}x^n = nx^{n-1} \qquad \text{Special case:} \quad \frac{d}{dx}x = 1$$

$$\frac{d}{dx}(c \cdot f) = c \cdot f' \qquad \text{Special case:} \quad \frac{d}{dx}(cx) = c$$

$$\frac{d}{dx}(f \pm g) = f' \pm g'$$

$$\frac{d}{dx}(f \cdot g) = f' \cdot g + f \cdot g'$$

$$\frac{d}{dx}\left(\frac{f}{g}\right) = \frac{g \cdot f' - g' \cdot f}{g^2} \qquad g \neq 0$$

These formulas are used extensively throughout calculus, and you should not proceed to the next section until you have mastered them.

## Verification of the Differentiation Formulas

We conclude this section with derivations of the Product and Quotient Rules, and the Power Rule in the case of *negative* integer exponents. First, however, we need to establish a preliminary result about an arbitrary function $g$:

$$\text{If } g'(x) \text{ exists, then } \lim_{h \to 0} g(x + h) = g(x).$$

We begin with $\lim_{h \to 0} g(x + h)$ and show that it is equal to $g(x)$:

$$\lim_{h \to 0} g(x + h) = \lim_{h \to 0} [g(x + h) - g(x) + g(x)] \qquad \text{Subtracting and adding } g(x)$$

$$= \lim_{h \to 0} \left[ \frac{g(x + h) - g(x)}{h} \cdot h + g(x) \right] \qquad \text{Dividing and multiplying by } h$$

$$= \lim_{h \to 0} \left[ \frac{g(x + h) - g(x)}{h} \cdot h \right] + \lim_{h \to 0} g(x) \qquad \text{The limit of a sum is the sum of the limits}$$

$$= \underbrace{\lim_{h \to 0} \frac{g(x + h) - g(x)}{h}}_{g'(x)} \cdot \underbrace{\lim_{h \to 0} h}_{0} + g(x) \qquad \text{The limit of a product is the product of the limits}$$

$$= g'(x) \cdot 0 + g(x) \qquad \text{Since the first limit above is the definition of } g'(x) \text{ and the second limit is zero}$$

$$= g(x) \qquad \text{Simplifying}$$

This proves the result that if $g'(x)$ exists, then $\lim\limits_{h \to 0} g(x + h) = g(x)$. Replacing $x + h$ by a new variable $y$, this equation becomes

$$\lim_{y \to x} g(y) = g(x) \qquad\qquad y = x + h, \text{ so } h \to 0 \text{ implies } y \to x$$

According to the definition of continuity on page 86 (but with different letters), this equation means that the function $g$ is *continuous* at $x$. Therefore, the result that we have shown can be stated simply:

If a function is *differentiable* at $x$, then it is *continuous* at $x$.

Or, even more briefly:

Differentiability implies continuity.

### Verification of the Product Rule

For two functions $f$ and $g$, let their product be $p(x) = f(x) \cdot g(x)$. If $f'(x)$ and $g'(x)$ exist, we may calculate $p'(x)$ as follows.

$$p'(x) = \lim_{h \to 0} \frac{p(x + h) - p(x)}{h} \qquad\qquad \text{Definition of the derivative}$$

$$= \lim_{h \to 0} \frac{f(x + h)g(x + h) - f(x)g(x)}{h} \qquad\qquad \begin{array}{l} p(x + h) = f(x + h) \cdot g(x + h) \\ \text{and } p(x) = f(x) \cdot g(x) \end{array}$$

$$= \lim_{h \to 0} \frac{f(x + h)g(x + h) - f(x)g(x + h) + f(x)g(x + h) - f(x)g(x)}{h} \qquad \begin{array}{l} \text{Subtracting and adding} \\ f(x)g(x + h) \end{array}$$

$$= \lim_{h \to 0} \left[ \frac{f(x + h)g(x + h) - f(x)g(x + h)}{h} + \frac{f(x)g(x + h) - f(x)g(x)}{h} \right] \qquad \begin{array}{l} \text{Separating the fraction} \\ \text{into two parts} \end{array}$$

$$= \lim_{h \to 0} \frac{[f(x + h) - f(x)]g(x + h)}{h} + \lim_{h \to 0} \frac{f(x)[g(x + h) - g(x)]}{h} \qquad \begin{array}{l} \text{Using Limit Rule 4a on} \\ \text{page 81 and factoring} \end{array}$$

$$= \underbrace{\lim_{h \to 0} \frac{f(x + h) - f(x)}{h}}_{f'(x)} \underbrace{\lim_{h \to 0} g(x + h)}_{g(x)} + \underbrace{f(x)}_{f(x)} \underbrace{\lim_{h \to 0} \frac{g(x + h) - g(x)}{h}}_{g'(x)} \qquad \begin{array}{l} \text{Using Limit Rule 4c} \\ \text{on page 81} \\[4pt] \text{Recognizing the defini-} \\ \text{tions of } f'(x) \text{ and } g'(x) \end{array}$$

$$= f'(x) \cdot g(x) + f(x) \cdot g'(x) \qquad\qquad \text{Product Rule}$$

### Verification of the Quotient Rule

For two functions $f$ and $g$ with $g(x) \neq 0$, let the quotient be $q(x) = \dfrac{f(x)}{g(x)}$. If $f'(x)$ and $g'(x)$ exist, we may calculate $q'(x)$ as follows.

$$q'(x) = \lim_{h \to 0} \frac{q(x + h) - q(x)}{h} \qquad\qquad \begin{array}{l} \text{Definition of} \\ \text{the derivative} \end{array}$$

$$= \lim_{h \to 0} \frac{\dfrac{f(x + h)}{g(x + h)} - \dfrac{f(x)}{g(x)}}{h} \qquad\qquad \begin{array}{l} q(x + h) = \dfrac{f(x + h)}{g(x + h)} \\[8pt] \text{and } q(x) = \dfrac{f(x)}{g(x)} \end{array}$$

$$= \lim_{h \to 0} \frac{1}{h} \left[ \frac{f(x + h)}{g(x + h)} - \frac{f(x)}{g(x)} \right]$$

Since dividing by $h$ is equivalent to multiplying by $1/h$

$$= \lim_{h \to 0} \left[ \frac{1}{h} \cdot \frac{g(x)f(x + h) - g(x + h)f(x)}{g(x + h)g(x)} \right]$$

Subtracting the fractions, using the common denominator $g(x + h)g(x)$

$$= \lim_{h \to 0} \left[ \frac{1}{h} \cdot \frac{g(x)f(x + h) - g(x)f(x) - [g(x + h)f(x) - g(x)f(x)]}{g(x + h)g(x)} \right]$$

Subtracting and adding $g(x)f(x)$

$$= \lim_{h \to 0} \left[ \frac{1}{g(x + h)g(x)} \cdot \frac{g(x)[f(x + h) - f(x)] - [g(x + h) - g(x)]f(x)}{h} \right]$$

Factoring in the numerator; switching the denominators

$$= \lim_{h \to 0} \left[ \underbrace{\frac{1}{g(x + h)g(x)}}_{\text{Approaches } g(x)} \left( g(x) \underbrace{\lim_{h \to 0} \frac{f(x + h) - f(x)}{h}}_{f'(x)} - \underbrace{\lim_{h \to 0} \frac{g(x + h) - g(x)}{h}}_{g'(x)} f(x) \right) \right]$$

Using Limit Rules 4b and 4c on page 81

$$= \frac{1}{[g(x)]^2} [g(x)f'(x) - g'(x)f(x)]$$

Using Limit Rules 1 and 4d on page 81

$$= \frac{g(x)f'(x) - g'(x)f(x)}{[g(x)]^2}$$

Quotient Rule

### Verification of the Power Rule for Negative Integer Exponents

On pages 118–119 we proved the Power Rule for *positive* integer exponents. Using the Quotient Rule, we may now prove the Power Rule for *negative* integer exponents. Any negative integer $n$ may be written as $n = -p$, where $p$ is a *positive* integer. Then

$$\frac{d}{dx} x^n = \frac{d}{dx} \left( \frac{1}{x^p} \right)$$

Since $x^n = x^{-p} = \dfrac{1}{x^p}$

$$= \frac{x^p \cdot 0 - px^{p-1} \cdot 1}{x^{2p}}$$

Using the Quotient Rule, with $\dfrac{d}{dx} 1 = 0$ and $\dfrac{d}{dx} x^p = px^{p-1}$

$$= \frac{-px^{p-1}}{x^{2p}}$$

Simplifying

$$= -px^{\underbrace{p - 1 - 2p}} = -px^{\underbrace{-p - 1}}$$
$$\phantom{= -px^{p-1-2p}} \underset{-p-1}{} \quad \underset{n}{} \quad \underset{n-1}{}$$

Subtracting exponents and simplifying

Since $-p = n$

$$= nx^{n-1}$$

Power Rule

This proves the Power Rule, $\dfrac{d}{dx} x^n = nx^{n-1}$, for negative integer exponents $n$.

## 2.4 Exercises

**1–4.** Find the derivative of each function in two ways:

**a.** Using the *Product* Rule.
**b.** Multiplying out the function and using the
    *Power* Rule.
Your answers to parts (a) and (b) should agree.

**1.** $x^4 \cdot x^6$
**2.** $x^7 \cdot x^2$
**3.** $x^4(x^5 + 1)$
**4.** $x^5(x^4 + 1)$

**5–26.** Find the derivative of each function by using
the Product Rule. Simplify your answers.

**5.** $f(x) = x^2(x^3 + 1)$    **6.** $f(x) = x^3(x^2 + 1)$
**7.** $f(x) = x(5x^2 - 1)$    **8.** $f(x) = 2x(x^4 + 1)$
**9.** $f(x) = \sqrt{x}(6x + 2)$   **10.** $f(x) = 6\sqrt[3]{x}\,(2x + 1)$
**11.** $f(x) = (x^2 + 1)(x^2 - 1)$
**12.** $f(x) = (x^3 - 1)(x^3 + 1)$
**13.** $f(x) = (x^2 + x)(3x + 1)$
**14.** $f(x) = (x^2 + 2x)(2x + 1)$
**15.** $f(x) = x^2(x^2 + 3x - 1)$
**16.** $f(x) = x^3(x^2 - 4x + 3)$
**17.** $f(x) = (2x^2 + 1)(1 - x)$
**18.** $f(x) = (2x - 1)(1 - x^2)$
**19.** $f(x) = \left(\sqrt{x} - 1\right)\left(\sqrt{x} + 1\right)$
**20** $f(x) = \left(\sqrt{x} + 2\right)\left(\sqrt{x} - 2\right)$
**21.** $f(t) = 6t^{4/3}(3t^{2/3} + 1)$
**22.** $f(t) = 4t^{3/2}(2t^{1/2} - 1)$
**23.** $f(z) = (z^4 + z^2 + 1)(z^3 - z)$
**24.** $f(z) = (\sqrt[4]{z} + \sqrt{z})(\sqrt[4]{z} - \sqrt{z})$
**25.** $f(z) = (2z + 4\sqrt{z} - 1)(2\sqrt{z} + 1)$
**26.** $f(z) = (z + 6\sqrt{z})(z - 2\sqrt{z} + 1)$

**27–30.** Find the derivative of each function in two
ways:

**a.** Using the *Quotient* rule.
**b.** Simplifying the original function and using the
    *Power* Rule.
Your answers to parts (a) and (b) should agree.

**27.** $\dfrac{x^8}{x^2}$
**28.** $\dfrac{x^9}{x^3}$
**29.** $\dfrac{1}{x^3}$
**30.** $\dfrac{1}{x^4}$

**31–46.** Find the derivative of each function by using
the Quotient Rule. Simplify your answers.

**31.** $f(x) = \dfrac{x^4 + 1}{x^3}$
**32.** $f(x) = \dfrac{x^5 - 1}{x^2}$

**33.** $f(x) = \dfrac{x + 1}{x - 1}$
**34.** $f(x) = \dfrac{x - 1}{x + 1}$

**35.** $f(x) = \dfrac{3x + 1}{2 + x}$
**36.** $f(x) = \dfrac{x + 1}{2x^2 + 1}$

**37.** $f(t) = \dfrac{t^2 - 1}{t^2 + 1}$
**38.** $f(t) = \dfrac{t^2 + 1}{t^2 - 1}$

**39.** $f(s) = \dfrac{s^3 - 1}{s + 1}$
**40.** $f(s) = \dfrac{s^3 + 1}{s - 1}$

**41.** $f(x) = \dfrac{x^2 - 2x + 3}{x + 1}$
**42.** $f(x) = \dfrac{x^2 + 3x - 1}{x - 1}$

**43.** $f(x) = \dfrac{x^4 + x^2 + 1}{x^2 + 1}$
**44.** $f(x) = \dfrac{x^5 + x^3 + x}{x^3 + x}$

**45.** $f(t) = \dfrac{t^2 + 2t - 1}{t^2 + t - 3}$
**46.** $f(t) = \dfrac{2t^2 + t - 5}{t^2 - t + 2}$

**47–52.** Differentiate each function by rewriting
before and after differentiating, as on page 129.

| Begin | Rewrite | Differentiate | Rewrite |
|-------|---------|---------------|---------|

**47.** $y = \dfrac{3}{x}$

**48.** $y = \dfrac{x^2}{4}$

**49.** $y = \dfrac{3x^4}{8}$

**50.** $y = \dfrac{3}{2x^2}$

**51.** $y = \dfrac{x^2 - 5x}{3}$

**52.** $y = \dfrac{4}{\sqrt{x}}$

**53–58.** Find the derivative of each function.

**53.** $(x^3 + 2)\dfrac{x^2 + 1}{x + 1}$
**54.** $(x^5 + 1)\dfrac{x^3 + 2}{x + 1}$

**55.** $\dfrac{(x^2 + 3)(x^3 + 1)}{x^2 + 2}$
**56.** $\dfrac{(x^3 + 2)(x^2 + 2)}{x^3 + 1}$

**57.** $\dfrac{\sqrt{x} - 1}{\sqrt{x} + 1}$
**58.** $\dfrac{\sqrt{x} + 1}{\sqrt{x} - 1}$

## Applied Exercises

**59. ECONOMICS: Marginal Average Revenue**
Use the Quotient Rule to find a general expression for the marginal average revenue. That is, calculate $\dfrac{d}{dx}\left[\dfrac{R(x)}{x}\right]$ and simplify your answer.

**60. ECONOMICS: Marginal Average Profit**
Use the Quotient Rule to find a general expression for the marginal average profit. That is, calculate $\dfrac{d}{dx}\left[\dfrac{P(x)}{x}\right]$ and simplify your answer.

**61. ENVIRONMENTAL SCIENCE: Water Purification** If the cost of purifying a gallon of water to a purity of $x$ percent is

$$C(x) = \dfrac{100}{100 - x}\text{ cents} \quad \text{for } 50 \le x < 100$$

a. Find the instantaneous rate of change of the cost with respect to purity.
b. Evaluate this rate of change for a purity of 95% and interpret your answer.
c. Evaluate this rate of change for a purity of 98% and interpret your answer.

**62. BUSINESS: Marginal Average Cost** A company can produce computer flash memory devices at a cost of $6 each, while fixed costs are $50 per day. Therefore, the company's cost function is $C(x) = 6x + 50$.
a. Find the average cost function $AC(x) = \dfrac{C(x)}{x}$.
b. Find the marginal average cost function $MAC(x)$.
c. Evaluate $MAC(x)$ at $x = 25$ and interpret your answer.
*Source:* Adco Marketing

**63. ENVIRONMENTAL SCIENCE: Water Purification** *(61 continued)*
a. Use a graphing calculator to graph the cost function $C(x)$ from Exercise 61 on the window [50, 100] by [0, 20]. TRACE along the curve to see how rapidly costs increase for purity ($x$-coordinate) increasing from 50 to near 100.
b. To check your answers to Exercise 61, use the "$dy/dx$" or SLOPE feature of your calculator to find the slope of the cost curve at $x = 95$ and at $x = 98$. The resulting rates of change of

the cost should agree with your answers to Exercise 61(b) and (c). Note that further purification becomes increasingly expensive at higher purity levels.

**64. BUSINESS: Marginal Average Cost** *(62 continued)*
a. Graph the average cost function $AC(x)$ that you found in Exercise 62(a) on the window [0, 50] by [0, 50]. TRACE along the average cost curve to see how the average cost falls as the number of devices increases. Note that although average cost falls, it does so more slowly as the number of units increases.
b. To check your answer to Exercise 62, use the "$dy/dx$" or SLOPE feature of your calculator to find the slope of the average cost curve at $x = 25$. This slope gives the rate of change of the cost, which should agree with your answer to Exercise 62(c). Find the slope (rate of change) for other $x$-values to see that the rate of change of average cost tends toward zero (the law of diminishing returns).

**65. BUSINESS: Marginal Average Cost** A company can produce LCD digital alarm clocks at a cost of $6 each while fixed costs are $45. Therefore, the company's cost function is $C(x) = 6x + 45$.
a. Find the average cost function $AC(x) = \dfrac{C(x)}{x}$.
b. Find the marginal average cost function $MAC(x)$.
c. Evaluate $MAC(x)$ at $x = 30$ and interpret your answer.
*Source:* Casad Company

**66. BUSINESS: Sales** The number of bottles of whiskey that a store will sell in a month at a price of $p$ dollars per bottle is

$$N(p) = \dfrac{2250}{p + 7} \quad (p \ge 5)$$

Find the rate of change of this quantity when the price is $8 and interpret your answer.

**67. GENERAL: Body Temperature** If a person's temperature after $x$ hours of strenuous exercise is $T(x) = x^3(4 - x^2) + 98.6$ degrees Fahrenheit (for $0 \le x \le 2$), find the rate of change of the temperature after 1 hour.

**68. BUSINESS: CD Sales** Suppose that after $x$ months, monthly sales of a compact disc are predicted to be $S(x) = x^2(8 - x^3)$ thousand (for $0 \le x \le 2$). Find the rate of change of the sales after 1 month.

 **69. GENERAL: Body Temperature** *(67 continued)*

a. Graph the temperature function $T(x)$ given in Exercise 67 on the window $[0, 2]$ by $[90, 110]$. TRACE along the temperature curve to see how the temperature rises and then falls as time increases.

b. To check your answer to Exercise 67, use the "$dy/dx$" or SLOPE feature of your calculator to find the slope (rate of change) of the curve at $x = 1$. Your answer should agree with your answer to Exercise 67.

c. TRACE along the temperature curve to estimate the maximum temperature.

 **70. BUSINESS: CD Sales** *(68 continued)*

a. Graph the sales function $S(x)$ given in Exercise 68 on the window $[0, 2]$ by $[0, 12]$. TRACE along the sales curve to see how the sales rise and then fall as $x$, the number of months, increases.

b. To check your answer to Exercise 68, use the "$dy/dx$" or SLOPE feature of your calculator to find the slope (rate of change) of the curve at $x = 1$. Your answer should agree with your answer to Exercise 68.

c. TRACE along the curve to estimate the maximum sales.

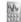 **71. ECONOMICS: National Debt** The national debt (the amount of money that the federal government has borrowed from and therefore owes to the public) is approximately $D(x) = 496x + 5378$ billion dollars, where $x$ is the number of years since 2000. The population of the United States is approximately $P(x) = 2.57x + 279$ million.

a. Enter these functions into your calculator as $y_1$ and $y_2$, respectively, and define $y_3$ to be $y_1 \div y_2$, the national debt divided by the population, so that $y_3$ is the *per capita national debt*, in thousands of dollars (since it is billions divided by millions). Evaluate $y_3$ at 10 and at 20 to find the *per capita national debt in* the years 2010 and 2020. This is the amount that the government would owe each of its citizens if the debt were divided equally among them.

b. Use the numerical derivative operation NDERIV to find the derivative of $y_3$ at 10 and at 20 and interpret your answers.

*Source:* U.S. Census Bureau

**72. GENERAL: Fuel Economy** The gas mileage (in miles per gallon) of a subcompact car is approximately

$$g(x) = \frac{-15x^2 + 1125x}{x^2 - 110x + 3500}$$

where $x$ is the speed in miles per hour (for $35 \le x \le 65$).

a. Find $g'(x)$.

b. Find $g'(40)$, $g'(50)$, and $g'(60)$ and interpret your answers.

c. What does the sign of $g'(40)$ tell you about whether gas mileage increases or decreases with speed when driving at 40 mph? Do the same for $g'(60)$ and 60 mph. Then do the same for $g'(50)$ and 50 mph. From your answers, what do you think is the most economical speed for a subcompact car?

*Source:* U.S. Environmental Protection Agency

**73–74. PITFALLS OF NDERIV ON A GRAPHING CALCULATOR**

a. Find the derivative (by hand) of each function below, and observe that the derivative is undefined at $x = 0$.

b. Find the derivative of each function below by using NDERIV on a graphing calculator and evaluate the derivative at $x = 0$. If your calculator gives you an answer, this is a "false value" for the derivative, since in part (a) you showed that the derivative is undefined at $x = 0$. [For an explanation, see the Graphing Calculator Exploration part (d) on page 129.]

**73.** $y = \dfrac{1}{x^2}$     **74.** $y = \dfrac{1}{x}$

## Conceptual Exercises

**75–80.** Let $f$ and $g$ be differentiable functions of $x$. Assume that denominators are not zero.

**75.** True or False: $\dfrac{d}{dx}(f \cdot g) = f' \cdot g'$

**76.** True or False: $\dfrac{d}{dx}\left(\dfrac{f}{g}\right) = \dfrac{f'}{g'}$

**77.** True or False: $\dfrac{d}{dx}(x \cdot f) = f + x \cdot f'$

**78.** True or False: $\dfrac{d}{dx}\left(\dfrac{f}{x}\right) = \dfrac{x \cdot f' - f}{x^2}$

**79.** Show that the Product Rule may be written in the following form:

$$\frac{d}{dx}(f \cdot g) = (f \cdot g)\left(\frac{f'}{f} + \frac{g'}{g}\right)$$

[*Hint:* Multiply out the right-hand side.]

**80.** Show that the Quotient Rule may be written in the following form:

$$\frac{d}{dx}\left(\frac{f}{g}\right) = \left(\frac{f}{g}\right)\left(\frac{f'}{f} - \frac{g'}{g}\right)$$

[*Hint:* Multiply out the right-hand side and combine it into a single fraction.]

**81–82.** Imagine a country in which everyone is equally wealthy—call the identical amount of money that each person has the *personal wealth*. The *national wealth* of the entire country is then this personal wealth times the population. However, these three quantities, personal wealth, population, and national wealth, may change with time.

**81.** True or False: To find the rate of change of national wealth, you would multiply the rate of change of the population times the rate of change of the personal wealth.

**82.** True or False: To find the rate of change of personal wealth, you would divide the rate of change of the national wealth by the rate of change of the population.

### Explorations and Excursions

The following problems extend and augment the material presented in the text.

#### More About Differentiation Formulas

**83. PRODUCT RULE FOR THREE FUNCTIONS** Show that if *f*, *g*, and *h* are differentiable functions of *x*, then

$$\frac{d}{dx}(f \cdot g \cdot h) = f' \cdot g \cdot h + f \cdot g' \cdot h + f \cdot g \cdot h'$$

[*Hint:* Write the function as $f \cdot (g \cdot h)$ and apply the product rule twice.]

**84.** Derive the Quotient Rule from the Product Rule as follows.

 **a.** Define the quotient to be a single function,

$$Q(x) = \frac{f(x)}{g(x)}.$$

 **b.** Multiply both sides by $g(x)$ to obtain the equation $Q(x) \cdot g(x) = f(x)$.

 **c.** Differentiate each side, using the Product Rule on the left side.

 **d.** Solve the resulting formula for the derivative $Q'(x)$.

 **e.** Replace $Q(x)$ by $\frac{f(x)}{g(x)}$ and show that the resulting formula for $Q'(x)$ is the same as the Quotient Rule.

Note that in this derivation when we differentiated $Q(x)$ we *assumed* that the derivative of the quotient exists, whereas in the derivation on pages 134–135 we *proved* that the derivative exists.

**85.** Find a formula for $\frac{d}{dx}[f(x)]^2$ by writing it as $\frac{d}{dx}[f(x)f(x)]$ and using the Product Rule. Be sure to simplify your answer.

**86.** Find a formula for $\frac{d}{dx}[f(x)]^{-1}$ by writing it as $\frac{d}{dx}\left[\frac{1}{f(x)}\right]$ and using the Quotient Rule. Be sure to simplify your answer.

### Two Biomedical Applications

**87. Beverton-Holt Recruitment Curve** Some organisms exhibit a density-dependent mortality from one generation to the next. Let $R > 1$ be the net reproductive rate (that is, the number of surviving offspring per parent), let $x > 0$ be the density of parents, and *y* be the density of surviving offspring. The *Beverton-Holt recruitment curve* is

$$y = \frac{Rx}{1 + \left(\frac{R-1}{K}\right)x}$$

where $K > 0$ is the *carrying capacity* of the organism's environment. Show that $\frac{dy}{dx} > 0$, and interpret this as a statement about the parents and the offspring.

**88. Murrell's Rest Allowance** Work-rest cycles for workers performing tasks that expend more than 5 kilocalories per minute (kcal/min) are often based on Murrell's formula

$$R(w) = \frac{w-5}{w-1.5} \qquad \text{for } w \geq 5$$

for the number of minutes $R(w)$ of rest for each minute of work expending *w* kcal/min. Show that $R'(w) > 0$ for $w \geq 5$ and interpret this fact as a statement about the additional amount of rest required for more strenuous tasks.

## 2.5 HIGHER-ORDER DERIVATIVES

### Introduction

We have seen that from one function we can calculate a new function, the *derivative* of the original function. This new function, however, can itself be differentiated, giving what is called the **second derivative** of the original function. Differentiating again gives the **third derivative** of the original function, and so on. In this section we will calculate and interpret such **higher-order derivatives.**

### Calculating Higher-Order Derivatives

**EXAMPLE 1**

**FINDING HIGHER DERIVATIVES OF A POLYNOMIAL**

From $f(x) = x^3 - 6x^2 + 2x - 7$ we may calculate

$$f'(x) = 3x^2 - 12x + 2 \qquad \text{"First" derivative of } f$$

Differentiating again gives

$$f''(x) = 6x - 12 \qquad \begin{array}{l}\text{Second derivative of } f, \\ \text{read "} f \text{ double prime"}\end{array}$$

and a third time:

$$f'''(x) = 6 \qquad \begin{array}{l}\text{Third derivative of } f, \\ \text{read "} f \text{ triple prime"}\end{array}$$

and a fourth time:

$$f''''(x) = 0 \qquad \begin{array}{l}\text{Fourth derivative of } f, \\ \text{read "} f \text{ quadruple prime"}\end{array}$$

All further derivatives of this function will, of course, be zero.

We also denote derivatives by replacing the primes by the number of differentiations in *parentheses.* For example, the fourth derivative may be denoted $f^{(4)}(x)$.

**Practice Problem 1**

If $f(x) = x^3 - x^2 + x - 1$, find:

**a.** $f'(x)$     **b.** $f''(x)$     **c.** $f'''(x)$     **d.** $f^{(4)}(x)$     ➤ **Solutions on page 147**

 **BE CAREFUL:** A 4 in *parentheses,* $f^{(4)}(x)$, means the fourth *derivative* of the function, while a 4 *without* parentheses, $f^4(x)$, means the fourth *power* of the function.

Example 1 showed that a polynomial can be differentiated "down to zero," but the same is not true for all functions.

**EXAMPLE 2**

**FINDING HIGHER DERIVATIVES OF A RATIONAL FUNCTION**

Find the first five derivatives of $f(x) = \dfrac{1}{x}$.

*Solution*

$$f(x) = x^{-1} \qquad\qquad f(x) \text{ in power form}$$
$$f'(x) = -x^{-2} \qquad\qquad \text{First derivative}$$
$$f''(x) = 2x^{-3} \qquad\qquad \text{Second derivative}$$
$$f'''(x) = -6x^{-4} \qquad\qquad \text{Third derivative}$$
$$f^{(4)}(x) = 24x^{-5} \qquad\qquad \text{Fourth derivative}$$
$$f^{(5)}(x) = -120x^{-6} \qquad\qquad \text{Fifth derivative}$$

Clearly, we will never get to zero no matter how many times we differentiate.

**Practice Problem 2**

If $f(x) = 16x^{-1/2}$, find: **a.** $f'(x)$ **b.** $f''(x)$ ➤ **Solutions on page 147**

In Leibniz's notation, the second derivative $\dfrac{d}{dx}\dfrac{df}{dx}$ is written $\dfrac{d^2f}{dx^2}$. The superscript goes after the $d$ in the numerator and after the $dx$ in the denominator. The following table shows equivalent statements in the two notations.

| *Prime Notation* | | *Leibniz's Notation* | |
|---|---|---|---|
| $f''(x)$ | $=$ | $\dfrac{d^2}{dx^2} f(x)$ | Second derivative |
| $y''$ | $=$ | $\dfrac{d^2y}{dx^2}$ | |
| $f'''(x)$ | $=$ | $\dfrac{d^3}{dx^3} f(x)$ | Third derivative |
| $y'''$ | $=$ | $\dfrac{d^3y}{dx^3}$ | |
| $f^{(n)}(x)$ | $=$ | $\dfrac{d^n}{dx^n} f(x)$ | $n$th derivative |
| $y^{(n)}$ | $=$ | $\dfrac{d^ny}{dx^n}$ | |

Calculating higher derivatives merely requires repeated use of the same differentiation rules that we have been using.

---

**EXAMPLE 3**

**FINDING A SECOND DERIVATIVE USING THE QUOTIENT RULE**

Find $\dfrac{d^2}{dx^2}\left(\dfrac{x^2 + 1}{x}\right)$.

*Solution*

$$\frac{d}{dx}\left(\frac{x^2 + 1}{x}\right) = \frac{x(2x) - (x^2 + 1)}{x^2} \qquad \text{First derivative, using the Quotient Rule}$$

$$= \frac{2x^2 - x^2 - 1}{x^2} = \frac{x^2 - 1}{x^2} \qquad \text{Simplifying}$$

Differentiating this answer gives the *second* derivative of the original:

$$\frac{d}{dx}\left(\frac{x^2 - 1}{x^2}\right) = \frac{x^2(2x) - 2x(x^2 - 1)}{x^4} \qquad \text{Second derivative (derivative of the derivative)}$$

$$= \frac{2x^3 - 2x^3 + 2x}{x^4} = \frac{2x}{x^4} = \frac{2}{x^3} \qquad \text{Simplifying}$$

*Answer:* $\dfrac{d^2}{dx^2}\left(\dfrac{x^2 + 1}{x}\right) = \dfrac{2}{x^3}$

---

The function in this example was a quotient, so it was perhaps natural to use the Quotient Rule. It is easier, however, to simplify the original function first,

$$\frac{x^2 + 1}{x} = \frac{x^2}{x} + \frac{1}{x} = x + x^{-1}$$

and then differentiate by the Power Rule. So, the first derivative of $x + x^{-1}$ is $1 - x^{-2}$, and differentiating again gives $2x^{-3}$, agreeing with the answer found by the Quotient Rule. *Moral:* **Always simplify before differentiating.**

---

**Practice Problem 3**

Find $f''(x)$ if $f(x) = \dfrac{x + 1}{x}$.

➤ Solution on page 147

---

**EXAMPLE 4**

**EVALUATING A SECOND DERIVATIVE**

If $f(x) = \dfrac{1}{\sqrt{x}}$, find $f''\left(\dfrac{1}{4}\right)$.

First differentiate, then evaluate

*Solution*

$$f(x) = x^{-1/2} \qquad \text{$f(x)$ in power form}$$

$$f'(x) = -\frac{1}{2}x^{-3/2} \qquad \text{Differentiating once}$$

$$f''(x) = \frac{3}{4} x^{-5/2} \qquad \text{Differentiating again}$$

$$f''\left(\frac{1}{4}\right) = \frac{3}{4}\left(\frac{1}{4}\right)^{-5/2} = \frac{3}{4}(4)^{5/2} \qquad \text{Evaluating } f''(x) \text{ at } \frac{1}{4}$$

$$= \frac{3}{4}\left(\sqrt{4}\right)^5 = \frac{3}{4}(2)^5 = \frac{3}{4}(32) = 24$$

---

**Practice Problem 4**     Find $\left. \dfrac{d^2}{dx^2}(x^4 + x^3 + 1) \right|_{x=-1}$

> **Solution on page 147**

### Velocity and Acceleration

There is another important interpretation for the derivative, one that also gives a meaning to the *second* derivative. Imagine that you are driving along a straight road, and let $s(t)$ stand for your distance (in miles) from your starting point after $t$ hours of driving. Then the derivative $s'(t)$ gives the instantaneous rate of change of distance with respect to time (miles per hour). However, "miles per hour" means speed or velocity, so the derivative of the **distance** function $s(t)$ is just the **velocity** function $v(t)$, giving your velocity at any time $t$.

In general, for an object moving along a straight line, with distance measured from some fixed point, measured positively in one direction and negatively in the other (sometimes called "directed distance"),

$$\text{if} \qquad s(t) = \begin{pmatrix} \text{Distance} \\ \text{at time } t \end{pmatrix}$$

$$\text{then} \qquad s'(t) = \begin{pmatrix} \text{Velocity} \\ \text{at time } t \end{pmatrix}$$

Letting $v(t)$ stand for the velocity at time $t$, we may state this simply as:

$$v(t) = s'(t) \qquad \text{The velocity function is the derivative of the distance function}$$

The units of velocity come directly from the distance and time units of $s(t)$. For example, if distance is measured in feet and time in seconds, then the velocity is in *feet per second*, while if distance is in miles and time is in hours, then velocity is in *miles per hour*.

In everyday speech, the word "accelerating" means "speeding up." That is, acceleration means the rate of increase of speed, and since rates of increase are just derivatives, **acceleration** *is the derivative of velocity.* Since velocity is itself the derivative of distance, acceleration is the *second* derivative of distance. Letting $a(t)$ stand for the acceleration at time $t$, we have:

| Distance, Velocity, and Acceleration | |
|---|---|
| $s(t) = \left(\begin{array}{c}\text{Distance}\\\text{at time } t\end{array}\right)$ | |
| $v(t) = s'(t)$ | Velocity is the derivative of distance |
| $a(t) = v'(t) = s''(t)$ | Acceleration is the derivative of velocity, and the second derivative of distance |

Therefore, we now have an interpretation for the *second* derivative: if $s(t)$ represents distance, then the *first* derivative represents velocity, and the *second* derivative represents *acceleration.* (In physics, there is even an interpretation for the third derivative, which gives the rate of change of acceleration: it is called the "jerk," since it is related to motion being "jerky."*)

**EXAMPLE 5**  **FINDING AND INTERPRETING VELOCITY AND ACCELERATION**

A delivery truck is driving along a straight road, and after $t$ hours its distance (in miles) east of its starting point is

$$s(t) = 24t^2 - 4t^3 \qquad \text{for } 0 \le t \le 6$$

**a.** Find the velocity of the truck after 2 hours.
**b.** Find the velocity of the truck after 5 hours.
**c.** Find the acceleration of the truck after 1 hour.

*Solution*

**a.** To find velocity, we differentiate distance:

$$v(t) = 48t - 12t^2 \qquad \text{Differentiating } s(t) = 24t^2 - 4t^3$$

$$v(2) = 48 \cdot 2 - 12 \cdot (2)^2 \qquad \text{Evaluating } v(t) \text{ at } t = 2$$

$$= 96 - 48 = 48 \text{ miles per hour} \qquad \text{Velocity after 2 hours}$$

**b.** At $t = 5$ hours:

$$v(5) = 48 \cdot 5 - 12 \cdot (5)^2 \qquad \text{Evaluating } v(t) \text{ at } t = 5$$

$$= 240 - 300 = -60 \text{ miles per hour} \qquad \text{Velocity after 5 hours}$$

*See T. R. Sandlin, "The Jerk," *Physics Teacher* 28: 36–40, January 1990.

What does the negative sign mean? Since distances are measured *eastward* (according to the original problem), the "positive" direction is east, so a negative velocity means a *westward* velocity. Therefore, at time $t = 5$ the truck is driving *westward at 60 miles per hour* (that is, back toward its starting point).

**c.** The acceleration is

$$a(t) = 48 - 24t \qquad \text{Differentiating } v(t) = 48t - 12t^2$$

$$a(1) = 48 - 24 = 24 \qquad \text{Acceleration after 1 hour}$$

Therefore, after 1 hour the acceleration of the truck is $24 \text{ mi/hr}^2$ (it is "speeding up").

## Graphing Calculator Exploration

Distance, velocity, and acceleration on [0, 6] by [−150, 150]. Which is which?

Use a graphing calculator to graph the distance function $y_1 = 24x^2 - 4x^3$ (use $x$ instead of $t$), the velocity function $y_2 = 48x - 12x^2$, and the acceleration function $y_3 = 48 - 24x$. (Alternatively, you could define $y_2$ and $y_3$ using NDERIV.) Your display should look like the one on the left. By looking at the graph, can you determine which curve represents distance, which represents velocity, and which represents acceleration? [*Hint:* Which curve gives the slope of which other curve?] Check your answer by using TRACE to identify functions 1, 2, and 3.

In Example 5, velocity was in miles per hour (mi/hr), so acceleration, the rate of change of velocity with respect to time, is in miles per hour *per hour*, written $\text{mi/hr}^2$. In general, the units of acceleration are $\text{distance/time}^2$.

**Practice Problem 5**

A helicopter rises vertically, and after $t$ seconds its height above the ground is $s(t) = 6t^2 - t^3$ feet (for $0 \le t \le 6$).

**a.** Find its velocity after 2 seconds.

**b.** Find its velocity after 5 seconds.

**c.** Find its acceleration after 1 second.

[*Hint:* Distances are measured *upward*, so a negative velocity means *downward*.]

➤ Solutions on page 148

### Other Interpretations of Second Derivatives

The second derivative has other meanings besides acceleration. In general, second derivatives measure how the rate of change is itself changing. That is, if the first derivative measures the rate of growth, then the second derivative tells whether the growth is "speeding up" or "slowing down".

EXAMPLE 6  **PREDICTING POPULATION GROWTH**

United Nations demographers predict that $t$ years from the year 2000 the population of the world will be:

$$P(t) = 6250 + 160t^{3/4}$$

million people. Find $P'(16)$ and $P''(16)$ and interpret these numbers.

*Solution* The derivative is

$$P'(t) = 120t^{-1/4} \qquad\qquad \text{Derivative of } P(t)$$

so

$$P'(16) = 120(16)^{-1/4} = 160 \cdot \frac{1}{2} = 80 \qquad \text{Evaluating at } t = 16$$

*Interpretation:* In 2016 ($t = 16$ years from 2000) the world population will be growing at the rate of 80 million people per year.
　　The second derivative is

$$P''(t) = -30t^{-5/4} \qquad\qquad \text{Derivative of } P'(t) = 120t^{-1/4}$$
$$P''(16) = -30(16)^{-5/4} \qquad\qquad \text{Evaluating at } t = 16$$

$$= -30\frac{1}{32} = \frac{-30}{32} \approx -0.94$$

The fact that the first derivative is positive and the second derivative (the rate of change of the derivative) is negative means that the growth is continuing but more slowly.

*Interpretation:* After 16 years, the growth rate is *decreasing* by about 0.94 million (or 940 thousand) people per year each year. In other words, in the following year the population will continue to grow, but at the slower rate of (rounding 0.94 to 1) about $80 - 1 = 79$ million people per year.

*Source:* U.N. Department of Economic and Social Affairs

## Graphing Calculator Exploration

$y_1 = 6250 + 160x^{3/4}$
on [0, 50] by [6000, 10000]

Use a graphing calculator to graph the population function $y_1 = 6250 + 160x^{3/4}$ (using $x$ instead of $t$). Your graph should resemble the one on the left. Can you see from the graph that the first derivative is positive (sloping upward), and that the second derivative is negative (slope decreasing)? You may check these facts numerically using NDERIV.

*Productivity Growth Slows and the Costs of Labor Rise*

WASHINGTON (AP) Worker productivity, a crucial factor in rising living standards, slowed sharply in the last quarter of 2007 while wage

© 2008 *The New York Times*

Statements about first and second derivatives occur frequently in everyday life, and may even make headlines. For example, the newspaper headline on the left (*New York Times*, February 7, 2008) says that productivity grew (the first derivative is positive) but more slowly (the second derivative is negative), following a curve like that shown below.

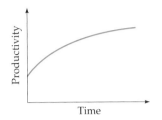

If $y$ represents productivity, then

$$\frac{dy}{dx} > 0 \quad \text{and} \quad \frac{d^2y}{dx^2} < 0$$

**Practice Problem 6**

The following headlines appeared recently in the *New York Times*. For each headline, sketch a curve representing the type of growth described and indicate the correct signs of the first and second derivatives.

**a.** Consumer Prices Rose in October at a Slower Rate

**b.** Households Still Shrinking, but Rate Is Slower

➤ **Solutions on next page**

**Practice Problem 7**

A mathematics journal* included the following statement: "In the fall of 1972 President Nixon announced that the rate of increase of inflation was decreasing. This was the first time a sitting president used the third derivative to advance his case for reelection." Explain why Nixon's announcement involved a third derivative.

➤ **Solution on next page**

**Solutions to Practice Problems**

**1.** **a.** $f'(x) = 3x^2 - 2x + 1$      **b.** $f''(x) = 6x - 2$

     **c.** $f'''(x) = 6$      **d.** $f^{(4)}(x) = 0$

**2.** **a.** $f'(x) = -8x^{-3/2}$      **b.** $f''(x) = 12x^{-5/2}$

**3.** $f(x) = \dfrac{x}{x} + \dfrac{1}{x} = 1 + x^{-1}$      Simplifying first

     $f'(x) = -x^{-2}$

     $f''(x) = 2x^{-3}$

**4.** $\dfrac{d}{dx}(x^4 + x^3 + 1) = 4x^3 + 3x^2$

     $\dfrac{d^2}{dx^2}(x^4 + x^3 + 1) = \dfrac{d}{dx}(4x^3 + 3x^2) = 12x^2 + 6x$

     $(12x^2 + 6x)\big|_{x=-1} = 12 - 6 = 6$

*Notices of the American Mathematical Society*, 43(10): 1108, 1996.

**5. a.** $v(t) = 12t - 3t^2$

$v(2) = 24 - 12 = 12 = 12 \text{ ft/sec}$

**b.** $v(5) = 60 - 75 = -15 \text{ ft/sec}$ or $15 \text{ ft/sec}$ *downward*

**c.** $a(t) = 12 - 6t \qquad a(1) = 12 - 6 = 6 \text{ ft/sec}^2$

**6. a.**

Time
$f' > 0, f'' < 0$

**b.**

Time
$f' < 0, f'' > 0$

**7.** Inflation is itself a derivative since it is the rate of change of the consumer price index. Therefore, its growth rate is a second derivative, and the slowing of this growth would be a third derivative.

---

## 2.5 Section Summary

By simply repeating the process of differentiation we can calculate second, third, and higher derivatives. We also have another interpretation for the derivative, one that gives an interpretation for the second derivative as well. For distance measured along a straight line from some fixed point:

If $\qquad s(t) = $ *distance* at time $t$

then $\qquad s'(t) = $ *velocity* at time $t$

and $\qquad s''(t) = $ *acceleration* at time $t$.

Therefore, whenever you are driving along a straight road, your *speedometer gives the derivative of your odometer reading.*

Velocity is the derivative of distance

We now have *four* interpretations for the derivative: **instantaneous rate of change, slope, marginals,** and **velocity.** It has been said that science is at its best when it unifies, and the derivative, unifying these four different concepts, is one of the most important ideas in all of science. We also saw that the second derivative, which measures the rate of change of the rate of change, can show whether growth is speeding up or slowing down.

**BE CAREFUL:** Remember that derivatives measure just what an automobile speedometer measures: the velocity at a particular *instant*. Although this statement may be obvious for velocities, it is easy to forget when dealing with marginals. For example, suppose that the marginal cost for a product is $15 when 100 units have been produced [which may be written $C'(100) = 15$]. Therefore, costs are increasing at the rate of $15 per additional unit, but only at the instant when $x = 100$. Although this may be used to *estimate* future costs (*about* $15 for each additional unit), it does not mean that one additional unit will increase costs by exactly $15, two more by exactly $30, and so on, since the marginal rate usually changes as production increases. A marginal cost is only an *approximate* predictor of future costs.

## 2.5 Exercises

**1–6.** For each function, find:

**a.** $f'(x)$   **b.** $f''(x)$   **c.** $f'''(x)$   **d.** $f^{(4)}(x)$

**1.** $f(x) = x^4 - 2x^3 - 3x^2 + 5x - 7$

**2.** $f(x) = x^4 - 3x^3 + 2x^2 - 8x + 4$

**3.** $f(x) = 1 + x + \frac{1}{2}x^2 + \frac{1}{6}x^3 + \frac{1}{24}x^4 + \frac{1}{120}x^5$

**4.** $f(x) = 1 + x + \frac{1}{2}x^2 + \frac{1}{6}x^3 + \frac{1}{24}x^4$

**5.** $f(x) = \sqrt{x^5}$      **6.** $f(x) = \sqrt{x^3}$

**7–12.** For each function, find **a.** $f''(x)$ and **b.** $f''(3)$.

**7.** $f(x) = \dfrac{x-1}{x}$      **8.** $f(x) = \dfrac{x+2}{x}$

**9.** $f(x) = \dfrac{x+1}{2x}$      **10.** $f(x) = \dfrac{x-2}{4x}$

**11.** $f(x) = \dfrac{1}{6x^2}$      **12.** $f(x) = \dfrac{1}{12x^3}$

**13–18.** Find the *second* derivative of each function.

**13.** $f(x) = (x^2 - 2)(x^2 + 3)$

**14.** $f(x) = (x^2 - 1)(x^2 + 2)$

**15.** $f(x) = \dfrac{27}{\sqrt[3]{x}}$      **16.** $f(x) = \dfrac{32}{\sqrt[4]{x}}$

**17.** $f(x) = \dfrac{x}{x-1}$      **18.** $f(x) = \dfrac{x}{x-2}$

**19–26.** Evaluate each expression.

**19.** $\dfrac{d^2}{dr^2}(\pi r^2)$      **20.** $\dfrac{d^3}{dr^3}\left(\dfrac{4}{3}\pi r^3\right)$

**21.** $\dfrac{d^2}{dx^2}x^{10}\Big|_{x=-1}$      **22.** $\dfrac{d^2}{dx^2}x^{11}\Big|_{x=-1}$

**23.** $\dfrac{d^3}{dx^3}x^{10}\Big|_{x=-1}$      **24.** $\dfrac{d^3}{dx^3}x^{11}\Big|_{x=-1}$

**25.** $\dfrac{d^2}{dx^2}\sqrt{x^3}\Big|_{x=1/16}$      **26.** $\dfrac{d^2}{dx^2}\sqrt[3]{x^4}\Big|_{x=1/27}$

**27–32.** Find the second derivative of each function.

**27.** $(x^2 - x + 1)(x^3 - 1)$

**28.** $(x^3 + x - 1)(x^3 + 1)$

**29.** $\dfrac{x}{x^2+1}$      **30.** $\dfrac{x}{x^2-1}$

**31.** $\dfrac{2x-1}{2x+1}$      **32.** $\dfrac{3x+1}{3x-1}$

### Applied Exercises

**33. GENERAL: Velocity** After $t$ hours a freight train is $s(t) = 18t^2 - 2t^3$ miles due north of its starting point (for $0 \le t \le 9$).

**a.** Find its velocity at time $t = 3$ hours.
**b.** Find its velocity at time $t = 7$ hours.
**c.** Find its acceleration at time $t = 1$ hour.

**34. GENERAL: Velocity** After $t$ hours a passenger train is $s(t) = 24t^2 - 2t^3$ miles due west of its starting point (for $0 \le t \le 12$).

**a.** Find its velocity at time $t = 4$ hours.
**b.** Find its velocity at time $t = 10$ hours.
**c.** Find its acceleration at time $t = 1$ hour.

**35. GENERAL: Velocity** A rocket can rise to a height of $h(t) = t^3 + 0.5t^2$ feet in $t$ seconds. Find its velocity and acceleration 10 seconds after it is launched.

**36. GENERAL: Velocity** After $t$ hours a car is a distance $s(t) = 60t + \dfrac{100}{t+3}$ miles from its starting point. Find the velocity after 2 hours.

**37. GENERAL: Impact Velocity** If a steel ball is dropped from the top of Taipei 101, the tallest building in the world, its height above the ground $t$ seconds after it is dropped will be $s(t) = 1667 - 16t^2$ feet (neglecting air resistance).

a. How long will it take to reach the ground? [*Hint*: Find when the height equals zero.]
b. Use your answer to part (a) to find the velocity with which it will strike the ground. (This is called the *impact velocity*.)
c. Find the acceleration at any time $t$. (This number is called the *acceleration due to gravity*.)

**38. GENERAL: Impact Velocity** If a marble is dropped from the top of the Sears Tower in Chicago, its height above the ground $t$ seconds after it is dropped will be $s(t) = 1454 - 16t^2$ feet (neglecting air resistance).

a. How long will it take to reach the ground?
b. Use your answer to part (a) to find the velocity with which it will strike the ground.
c. Find the acceleration at any time $t$. (This number is called the *acceleration due to gravity*.)

**39. GENERAL: Maximum Height** If a bullet from a 9-millimeter pistol is fired straight up from the ground, its height $t$ seconds after it is fired will be $s(t) = -16t^2 + 1280t$ feet (neglecting air resistance) for $0 \leq t \leq 80$.

a. Find the velocity function.
b. Find the time $t$ when the bullet will be at its maximum height. [*Hint:* At its maximum height the bullet is moving neither up nor down, and has velocity zero. Therefore, find the time when the velocity $v(t)$ equals zero.]
c. Find the maximum height the bullet will reach. [*Hint:* Use the time found in part (b) together with the height function $s(t)$.]

**40. BIOMEDICAL: Fever** The temperature of a patient $t$ hours after taking a fever reducing medicine is $T(t) = 98 + 8/\sqrt{t}$ degrees Fahrenheit. Find $T(2)$, $T'(2)$, and $T''(2)$, and interpret these numbers.

**41. ECONOMICS: National Debt** The national debt of a South American country $t$ years from now is predicted to be $D(t) = 65 + 9t^{4/3}$ billion dollars. Find $D'(8)$ and $D''(8)$ and interpret your answers.

 **42. ENVIRONMENTAL SCIENCE: Global Temperatures** The burning of oil, coal, and other fossil fuels generates "greenhouse gasses" that trap heat and raise global temperatures. Although predictions depend upon assumptions of countermeasures, one study predicts an increase in global temperature (above the 2000 level) of $T(t) = 0.25t^{1.4}$ degrees Fahrenheit, where $t$ is the number of decades since 2000 (so, for example, $t = 2$ means the year 2020). Find $T(10)$, $T'(10)$, and $T''(10)$, and interpret your answers. [*Note:* Rising temperatures could adversely affect weather patterns and crop yields in many areas.]

*Source:* Intergovernmental Panel on Climate Change, 2007 Synthesis Report

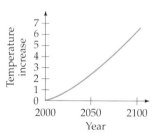

**43. ENVIRONMENTAL SCIENCE: Sea Level** Increasing global temperatures raise sea levels by thermal expansion and the melting of polar ice. Precise predictions are difficult, but a United Nations study predicts a rise in sea level (above the 2000 level) of $L(x) = 0.02x^3 - 0.07x^2 + 8x$ centimeters, where $x$ is the number of decades since 2000 (so, for example, $x = 2$ means the year 2020). Find $L(10)$, $L'(10)$, and $L''(10)$, and interpret your answers. [*Note:* Rising sea levels could flood many islands and coastal regions.]

*Source:* U.N. Environment Programme

 **44. BUSINESS: Profit** The annual profit of the Digitronics company $x$ years from now is predicted to be $P(x) = 5.27x^{0.3} - 0.463x^{1.52}$ million dollars (for $0 \leq x \leq 8$). Evaluate the profit function and its first and second derivatives at $x = 3$ and interpret your answers. [*Hint:* Enter the given function in $y_1$, define $y_2$ to be the derivative of $y_1$ (using NDERIV), and define $y_3$ to be the derivative of $y_2$. Then evaluate each at the stated $x$-value.]

 **45. GENERAL: Windchill Index** The windchill index (revised in 2001) for a temperature of 32 degrees Fahrenheit and wind speed $x$ miles per hour is $W(x) = 55.628 - 22.07x^{0.16}$.

a. Graph the windchill index on a graphing calculator using the window [0, 50] by [0, 40]. Then find the windchill index for wind speeds of $x = 15$ and $x = 30$ mph.

b. Notice from your graph that the windchill index has first derivative negative and second derivative positive. What does this mean about how successive 1-mph increases in wind speed affect the windchill index?

c. Verify your answer to part (b) by defining $y_2$ to be the derivative of $y_1$ (using NDERIV), evaluating it at $x = 15$ and $x = 30$, and interpreting your answers.

*Source:* National Weather Service

 **46. BIOMEDICAL: AIDS** The cumulative number of cases of AIDS (acquired immunodeficiency syndrome) in the United States between 1981 and 2000 is given approximately by the function

$$f(x) = -0.0182x^4 + 0.526x^3 - 1.3x^2 + 1.3x + 5.4$$

in thousands of cases, where $x$ is the number of years since 1980.

a. Graph this function on your graphing calculator on the window [1, 20] by [0, 800]. Notice that at some time in the 1990s the rate of growth began to slow.

b. Find when the rate of growth began to slow. [*Hint:* Find where the second derivative of $f(x)$ is zero, and then convert the $x$-value to a year.]

*Source:* Centers for Disease Control

## Conceptual Exercises

**47–50.** Suppose that the quantity described is represented by a function $f(t)$ where $t$ stands for time. Based on the description:

a. Is the first derivative positive or negative?
b. Is the second derivative positive or negative?

**47.** The temperature is dropping increasingly rapidly.

**48.** The economy is growing, but more slowly.

**49.** The stock market is declining, but less rapidly.

**50.** The population is growing increasingly fast.

**51.** True or False: If $f(x)$ is a polynomial of degree $n$, then $f^{(n+1)}(x) = 0$.

**52.** At time $t = 0$ a helicopter takes off gently and then 60 seconds later it lands gently. Let $f(t)$ be its altitude above the ground at time $t$ seconds.

a. Will $f'(1)$ be positive or negative? Same question for $f''(1)$.
b. Will $f'(59)$ be positive or negative? Same question for $f''(59)$.

**53. GENERAL: Velocity** Each of the following three "stories," labeled **a, b,** and **c,** matches one of the velocity graphs, labeled (i), (ii), and (iii). For each story, choose the most appropriate graph.

a. I left my home and drove to meet a friend, but I got stopped for a speeding ticket. Afterward I drove on more slowly.

b. I started driving but then stopped to look at the map. Realizing that I was going the wrong way, I drove back the other way.

c. After driving for a while I got into some stop-and-go driving. Once past the tie-up I could speed up again.

**54. BUSINESS: Profit** Each of the following three descriptions of a company's profit over time, labeled **a, b,** and **c,** matches one of the graphs, labeled (i), (ii), and (iii). For each description, choose the most appropriate graph.

a. Profits were growing increasingly rapidly.

b. Profits were declining but the rate of decline was slowing.

c. Profits were rising, but more and more slowly.

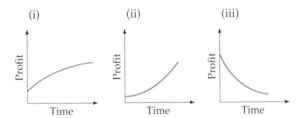

## Explorations and Excursions

The following problems extend and augment the material presented in the text.

*More About Higher-Order Derivatives*

**55.** Find $\dfrac{d^{100}}{dx^{100}}(x^{100} - 4x^{99} + 3x^{50} + 6)$.

[*Hint:* You may use the "factorial" notation: $n! = n(n - 1)\cdots1$. For example, $3! = 3\cdot2\cdot1 = 6$.]

**56.** Find a general formula for $\dfrac{d^n}{dx^n}x^{-1}$.

[*Hint:* Calculate the first few derivatives and look for a pattern. You may use the "factorial" notation: $n! = n(n - 1)\cdots1$. For example, $3! = 3\cdot2\cdot1 = 6$.]

**57.** Verify the following formula for the *second* derivative of a product, where $f$ and $g$ are differentiable functions of $x$:

$$\frac{d^2}{dx^2}(f \cdot g) = f'' \cdot g + 2f' \cdot g' + f \cdot g''$$

[*Hint:* Use the Product Rule repeatedly.]

**58.** Verify the following formula for the *third* derivative of a product, where $f$ and $g$ are differentiable functions of $x$:

$$\frac{d^3}{dx^3}(f \cdot g) = f''' \cdot g + 3f'' \cdot g' + 3f' \cdot g'' + f \cdot g'''$$

[*Hint:* Differentiate the formula in Exercise 57 by the Product Rule.]

## 2.6 THE CHAIN RULE AND THE GENERALIZED POWER RULE

### Introduction

In this section we will learn the last of the general rules of differentiation, the **Chain Rule** for differentiating **composite functions.** We will then prove a very useful special case of it, the **Generalized Power Rule** for differentiating powers of functions. We begin by reviewing composite functions.

### Composite Functions

As we saw on page 57, *composite* functions are simply functions of functions: the composition of $f$ with $g$ evaluated at $x$ is $f(g(x))$.

---

**EXAMPLE 1**   **FINDING A COMPOSITE FUNCTION**

For $f(x) = x^2$ and $g(x) = 4 - x$, find $f(g(x))$.

*Solution*   $f(g(x)) = (4 - x)^2$

$f(x) = x^2$ with $x$
replaced by $g(x) = 4 - x$

---

**Practice Problem 1**   For the same $f(x) = x^2$ and $g(x) = 4 - x$, find $g(f(x))$.

➤ **Solution on page 159**

---

## Graphing Calculator Exploration

$f(g(x)) = (4 - x)^2$

$g(f(x)) = 4 - x^2$

Use a graphing calculator to verify that the compositions $f(g(x))$ and $g(f(x))$ above are different.

**a.** Enter $y_1 = x^2$ and $y_2 = 4 - x$.

**b.** Then define $y_3$ and $y_4$ to be the compositions in the two orders by entering: $y_3 = y_1(y_2)$ and $y_4 = y_2(y_1)$.

**c.** Graph $y_3$ and $y_4$ (but turn "off" $y_1$ and $y_2$) on the window $[-5, 8]$ by $[-7, 10]$ and notice that the graphs are very different.

Besides building compositions out of simpler functions, we can also *decompose* functions into compositions of simpler functions.

**EXAMPLE 2**

**DECOMPOSING A COMPOSITE FUNCTION**

Find functions $f(x)$ and $g(x)$ such that $(x^2 + 1)^5$ is the composition $f(g(x))$.

*Solution*

Think of $(x^2 + 1)^5$ as an inside function $x^2 + 1$ followed by an outside operation $(\quad)^5$. We match the "inside" and "outside" parts of $(x^2 + 1)^5$ and $f(g(x))$.

$$
\underbrace{(x^2 + 1)}_{\text{Inside function}}{}^{\overbrace{5}^{\text{Outside function}}} = f(\underbrace{g(x)}) 
$$

Therefore, $(x^2 + 1)^5$ can be written as $f(g(x))$ with $\begin{cases} f(x) = x^5 \\ g(x) = x^2 + 1 \end{cases}$.

(Other answers are possible.)

Note that expressing a function as a composition involves thinking of the function in terms of "blocks," an inside block that starts a calculation and an outside block that completes it.

**Practice Problem 2**

Find $f(x)$ and $g(x)$ such that $\sqrt{x^5 - 7x + 1}$ is the composition $f(g(x))$.

➤ **Solution on page 159**

## The Chain Rule

If we were asked to differentiate the function $(x^2 - 5x + 1)^{10}$, we could first multiply together ten copies of $x^2 - 5x + 1$ (certainly a long, tedious, and error-prone process), and then differentiate the resulting polynomial. There is, however, a much easier way, using the *Chain Rule*, which shows how to differentiate a composite function of the form $f(g(x))$.

**Chain Rule**

$$\frac{d}{dx} f(g(x)) = f'(g(x)) \cdot g'(x)$$

To differentiate $f(g(x))$, differentiate $f(x)$, then replace each $x$ by $g(x)$, and finally multiply by the derivative of $g(x)$

(provided that the derivatives on the right-hand side of the equation exist). The name comes from thinking of compositions as "chains" of functions. A verification of the Chain Rule is given at the end of this section.

**EXAMPLE 3**

**DIFFERENTIATING USING THE CHAIN RULE**

Use the Chain Rule to find $\dfrac{d}{dx}(x^2 - 5x + 1)^{10}$.

*Solution*

$(x^2 - 5x + 1)^{10}$ is $f(g(x))$ with $\begin{cases} f(x) = x^{10} & \text{Outside function} \\ g(x) = x^2 - 5x + 1 & \text{Inside function} \end{cases}$

Since $f'(x) = 10x^9$, we have

$$f'(g(x)) = 10(g(x))^9 \qquad\qquad \begin{array}{l}f'(x) = 10x^9 \text{ with } x \\ \text{replaced by } g(x)\end{array}$$

$$= 10(x^2 - 5x + 1)^9 \qquad\qquad \text{Using } g(x) = x^2 - 5x + 1$$

Substituting this last expression into the Chain Rule gives:

$$\frac{d}{dx}f(g(x)) = f'(g(x))\,g'(x) \qquad\qquad \text{Chain Rule}$$

$$\frac{d}{dx}(x^2 - 5x + 1)^{10} = 10(x^2 - 5x + 1)^9(2x - 5) \qquad \begin{array}{l}\text{Using } f'(g(x)) = \\ 10(x^2 - 5x + 1)^9 \text{ and} \\ g'(x) = 2x - 5\end{array}$$

This result says that to differentiate $(x^2 - 5x + 1)^{10}$, we bring down the exponent 10, reduce the exponent to 9 (steps familiar from the Power Rule), and finally multiply by the derivative of the inside function.

$$\frac{d}{dx}(x^2 - 5x + 1)^{10} = 10(x^2 - 5x + 1)^9(2x - 5)$$

| Inside function | Bring down the power $n$ | Power $n - 1$ | Derivative of the inside function |

### Generalized Power Rule

Example 3 suggests a general rule for differentiating a function to a power.

**Generalized Power Rule**

$$\frac{d}{dx}[g(x)]^n = n \cdot [g(x)]^{n-1} \cdot g'(x)$$

To differentiate a function to a power, bring down the power as a multiplier, reduce the exponent by 1, and then multiply by the derivative of the inside function

(provided, of course, that the derivative $g'(x)$ exists). The Generalized Power Rule follows from the Chain Rule by reasoning similar to that of Example 3:

the derivative of $f(x) = x^n$ is $f'(x) = nx^{n-1}$, so

$$\frac{d}{dx} f(g(x)) = f'(g(x)) \, g'(x) \qquad \text{Chain Rule}$$

gives

$$\frac{d}{dx} [g(x)]^n = n[g(x)]^{n-1} g'(x) \qquad \text{Generalized Power Rule}$$

---

**EXAMPLE 4**

**DIFFERENTIATING USING THE GENERALIZED POWER RULE**

Find $\dfrac{d}{dx} \sqrt{x^4 - 3x^3 - 4}$.

*Solution*

$$\frac{d}{dx} (x^4 - 3x^3 - 4)^{1/2} = \frac{1}{2} (x^4 - 3x^3 - 4)^{-1/2} (4x^3 - 9x^2)$$

$\underbrace{\phantom{xxxxxxxxx}}$ Inside function    Bring down the $n$      Power $n-1$      Derivative of the inside function

---

Think of the Generalized Power Rule "from the outside in." That is, first bring down the outer exponent and reduce the exponent by 1, and only then multiply by the derivative of the inside function.

 **BE CAREFUL:** It is the *original function* (not the differentiated function) that is raised to the power $n - 1$. Only at the end do you multiply by the derivative of the inside function.

---

**EXAMPLE 5**

**SIMPLIFYING AND DIFFERENTIATING**

Find $\dfrac{d}{dx} \left( \dfrac{1}{x^2 + 1} \right)^3$.

*Solution*

Writing the function as $(x^2 + 1)^{-3}$ gives

$$\frac{d}{dx} (x^2 + 1)^{-3} = -3(x^2 + 1)^{-4}(2x) = -6x(x^2 + 1)^{-4} = -\frac{6x}{(x^2 + 1)^4}$$

Inside function    Bring down the $n$      Power $n-1$      Derivative of the inside function

---

**Practice Problem 3**

Find $\dfrac{d}{dx} (x^3 - x)^{-1/2}$.

➤ **Solution on page 159**

EXAMPLE 6  **FINDING THE GROWTH RATE OF AN OIL SLICK**

An oil tanker hits a reef, and after $t$ days the radius of the oil slick is $r(t) = \sqrt{4t + 1}$ miles. How fast is the radius of the oil slick expanding after 2 days?

*Solution*

To find the rate of change of the radius, we differentiate:

$$\frac{d}{dt}(4t + 1)^{1/2} = \frac{1}{2}(4t + 1)^{-1/2}(4) = 2(4t + 1)^{-1/2}$$

Derivative of $4t + 1$

At $t = 2$ this is

$$2(4 \cdot 2 + 1)^{-1/2} = 2 \cdot 9^{-1/2} = 2 \cdot \frac{1}{3} = \frac{2}{3}$$

*Interpretation:* After 2 days the radius of the oil slick is growing at the rate of $\frac{2}{3}$ of a mile per day.

## Graphing Calculator Exploration

a. On a graphing calculator, enter the radius of the oil slick as $y_1 = \sqrt{4x + 1}$. Then use NDERIV to define $y_2$ to be the derivative of $y_1$. Graph both $y_1$ and $y_2$ on the window [0, 6] by [−1, 5].

b. Verify your answer to Example 6 by evaluating $y_2$ at $x = 2$. Do you get $\frac{2}{3}$?

c. Notice from your graph that the (derivative) function $y_2$ is *decreasing*, meaning that the radius is growing more *slowly*. Estimate when the radius will be growing by only $\frac{1}{2}$ mile per day. [*Hint:* One way is to use TRACE to follow along the curve $y_2$ to the $x$-value (number of days) when the $y$-coordinate is 0.5, zooming in if necessary. Your answer should be between 3 and 4 days. You can also use INTERSECT with $y_3 = 0.5$.]

Some problems require the Generalized Power Rule in combination with another differentiation rule, such as the Product or Quotient Rule.

**EXAMPLE 7** | **DIFFERENTIATING USING TWO RULES**

Find $\dfrac{d}{dx}[(5x - 2)^4(9x + 2)^7]$.

*Solution*

Since this is a product of powers, $(5x - 2)^4$ times $(9x + 2)^7$, we use the Product Rule together with the Generalized Power Rule.

$$\frac{d}{dx}[(5x - 2)^4(9x + 2)^7] = \underbrace{4(5x - 2)^3(5)}_{\text{Derivative of } (5x - 2)^4}(9x + 2)^7 + (5x - 2)^4\underbrace{[7(9x + 2)^6(9)]}_{\text{Derivative of } (9x + 2)^7}$$

$$= 20(5x - 2)^3(9x + 2)^7 + 63(5x - 2)^4(9x + 2)^6 \quad \text{Simplifying}$$
$$\quad\quad \vert_{4 \cdot 5} \quad\quad\quad\quad\quad\quad\quad \vert_{7 \cdot 9}$$

**EXAMPLE 8** | **DIFFERENTIATING USING TWO RULES**

Find $\dfrac{d}{dx}\left(\dfrac{x}{x + 1}\right)^4$.

*Solution*

Since the function is a quotient raised to a power, we use the Quotient Rule together with the Generalized Power Rule. Working from the outside in, we obtain

$$\frac{d}{dx}\left(\frac{x}{x + 1}\right)^4 = 4\left(\frac{x}{x + 1}\right)^3 \underbrace{\frac{(x + 1)(1) - (1)(x)}{(x + 1)^2}}_{\substack{\text{Derivative of the} \\ \text{inside function } \frac{x}{x + 1}}}$$

$$= 4\left(\frac{x}{x + 1}\right)^3 \frac{x + 1 - x}{(x + 1)^2} = 4\left(\frac{x}{x + 1}\right)^3 \frac{1}{(x + 1)^2} \quad \text{Simplifying}$$

$$= 4\frac{x^3}{(x + 1)^3} \frac{1}{(x + 1)^2} = \frac{4x^3}{(x + 1)^5} \quad \text{Simplifying further}$$

**EXAMPLE 9** | **DIFFERENTIATING USING A RULE TWICE**

Find $\dfrac{d}{dz}[z^2 + (z^2 - 1)^3]^5$.

*Solution*

Since this is a function to a power, where the inside function also contains a function to a power, we must use the Generalized Power Rule *twice*.

$$\frac{d}{dz}[z^2 + (z^2 - 1)^3]^5 = 5[z^2 + (z^2 - 1)^3]^4[2z + 3(z^2 - 1)^2(2z)]$$

Derivative of
$z^2 + (z^2 - 1)^3$

$$= 5[z^2 + (z^2 - 1)^3]^4[2z + 6z(z^2 - 1)^2] \qquad \text{Simplifying}$$

$3 \cdot 2$

## Chain Rule in Leibniz's Notation

A composition may be written in two parts:

$$y = f(g(x)) \quad \text{is equivalent to} \quad y = f(u) \quad \text{and} \quad u = g(x)$$

The derivatives of these last two functions are:

$$\frac{dy}{du} = f'(u) \quad \text{and} \quad \frac{du}{dx} = g'(x)$$

The Chain Rule

$$\frac{d}{dx} f(g(x)) = f'(g(x)) \cdot g'(x) \qquad \text{Chain Rule}$$

$y$

$\frac{dy}{dx}$ $\quad$ $\frac{dy}{du}$ $\quad$ $\frac{du}{dx}$

with the indicated substitutions then becomes:

| Chain Rule in Leibniz's Notation |
| --- |
| For $y = f(u)$ with $u = g(x)$, $$\frac{dy}{dx} = \frac{dy}{du} \cdot \frac{du}{dx}$$ |

In this form the Chain Rule is easy to remember, since it looks as if the *du* in the numerator and the denominator cancel:

$$\frac{dy}{dx} = \frac{dy}{du} \cdot \frac{du}{dx}$$

However, since derivatives are not really fractions (they are *limits* of fractions), this is only a convenient device for remembering the Chain Rule.

**BE CAREFUL:** The Product Rule (page 124) showed that the derivative of a product is not the product of the derivatives. We now see where the product of the derivatives *does* appear: it appears in the Chain Rule, when differentiating composite functions. In other words, the product of the derivatives comes not from *products* but from *compositions* of functions.

### A Simple Example of the Chain Rule

The derivation of the Chain Rule is rather technical, but we can show the basic idea in a simple example. Suppose that your company produces steel, and you want to calculate your company's total revenue in dollars per year. You would take the revenue from a ton of steel (dollars per ton) and multiply by your company's output (tons per year). In symbols:

$$\frac{\$}{\text{year}} = \frac{\$}{\text{ton}} \cdot \frac{\text{ton}}{\text{year}}$$

Note that "ton" cancels

If we were to express these rates as derivatives, the equation above would become the Chain Rule.

---

**Solutions to Practice Problems**

1. $g(f(x)) = 4 - f(x) = 4 - x^2$
2. $f(x) = \sqrt{x}, \quad g(x) = x^5 - 7x + 1$
3. $\dfrac{d}{dx}(x^3 - x)^{-1/2} = -\dfrac{1}{2}(x^3 - x)^{-3/2}(3x^2 - 1)$

## 2.6 Section Summary

To differentiate a composite function (a function of a function), we have the Chain Rule:

$$\frac{d}{dx}f(g(x)) = f'(g(x)) \cdot g'(x)$$

or, in Leibniz's notation, writing $y = f(g(x))$ as $y = f(u)$ and $u = g(x)$,

$$\frac{dy}{dx} = \frac{dy}{du} \cdot \frac{du}{dx}$$

The derivative of a composite function is the product of the derivatives

To differentiate a function to a power, $[f(x)]^n$, we have the *Generalized Power Rule* (a special case of the Chain Rule when the "outer" function is a power):

$$\frac{d}{dx}[f(x)]^n = n \cdot [f(x)]^{n-1} \cdot f'(x)$$

For now, the Generalized Power Rule is more useful than the Chain Rule, but in Chapter 5 we will make important use of the Chain Rule.

## Verification of the Chain Rule

Let $f(x)$ and $g(x)$ be differentiable functions. We define $k$ by

$$k = g(x + h) - g(x)$$

or, equivalently,

$$g(x + h) = g(x) + k$$

Then

$$\lim_{h \to 0} \underbrace{[g(x + h) - g(x)]}_{k} = \lim_{h \to 0} k = 0$$

showing that $h \to 0$ implies $k \to 0$ (see page 134). With these relations we may calculate the derivative of the composition $f(g(x))$.

$$\frac{d}{dx} f(g(x)) = \lim_{h \to 0} \frac{f(g(x + h)) - f(g(x))}{h}$$

Definition of the derivative of $f(g(x))$

$$= \lim_{h \to 0} \left[ \frac{f(g(x + h)) - f(g(x))}{g(x + h) - g(x)} \cdot \frac{g(x + h) - g(x)}{h} \right]$$

Dividing and multiplying by $g(x + h) - g(x)$

$$= \lim_{h \to 0} \frac{f(g(x + h)) - f(g(x))}{g(x + h) - g(x)} \lim_{h \to 0} \frac{g(x + h) - g(x)}{h}$$

The limit of a product is the product of the limits (Limit Rule 4c on page 81)

$$= \underbrace{\lim_{k \to 0} \frac{f(g(x) + k) - f(g(x))}{k}}_{f'(g(x))} \underbrace{\lim_{h \to 0} \frac{g(x + h) - g(x)}{h}}_{g'(x)}$$

Using the relations $g(x + h) = g(x) + k$ and $k = g(x + h) - g(x)$ and that $h \to 0$ implies $k \to 0$

$$= f'(g(x)) \cdot g'(x)$$

Chain Rule

The last step comes from recognizing the first limit as the definition of the derivative $f'$ at $g(x)$, and the second limit as the definition of the derivative $g'(x)$. This verifies the Chain Rule,

$$\frac{d}{dx} f(g(x)) = f'(g(x)) \cdot g'(x)$$

Strictly speaking, this verification requires an additional assumption, that the denominator $g(x + h) - g(x)$ is never zero. This assumption can be avoided by using Carathéodory's definition of the derivative, as shown in Exercises 79–80.

## 2.6  Exercises

1–10. Find functions $f$ and $g$ such that the given function is the composition $f(g(x))$.

1. $\sqrt{x^2 - 3x + 1}$    2. $(5x^2 - x + 2)^4$

3. $(x^2 - x)^{-3}$    4. $\dfrac{1}{x^2 + x}$    5. $\dfrac{x^3 + 1}{x^3 - 1}$

6. $\dfrac{\sqrt{x} - 1}{\sqrt{x} + 1}$    7. $\left(\dfrac{x + 1}{x - 1}\right)^4$    8. $\sqrt{\dfrac{x - 1}{x + 1}}$

9. $\sqrt{x^2 - 9} + 5$    10. $\sqrt[3]{x^3 + 8} - 5$

**11–46.** Use the Generalized Power Rule to find the derivative of each function.

**11.** $f(x) = (x^2 + 1)^3$  **12.** $f(x) = (x^3 + 1)^4$

**13.** $g(x) = (2x^2 - 7x + 3)^4$

**14.** $g(x) = (3x^3 - x^2 + 1)^5$

**15.** $h(z) = (3z^2 - 5z + 2)^4$

**16.** $h(z) = (5z^2 + 3z - 1)^3$

**17.** $f(x) = \sqrt{x^4 - 5x + 1}$  **18.** $f(x) = \sqrt{x^6 + 3x - 1}$

**19.** $w(z) = \sqrt[3]{9z - 1}$  **20.** $w(z) = \sqrt[5]{10z - 4}$

**21.** $y = (4 - x^2)^4$  **22.** $y = (1 - x)^{50}$

**23.** $y = \left(\dfrac{1}{w^3 - 1}\right)^4$  **24.** $y = \left(\dfrac{1}{w^4 + 1}\right)^5$

**25.** $y = x^4 + (1 - x)^4$

**26.** $f(x) = (x^2 + 4)^3 - (x^2 + 4)^2$

**27.** $f(x) = \dfrac{1}{\sqrt{3x^2 - 5x + 1}}$  **28.** $f(x) = \dfrac{1}{\sqrt{2x^2 - 7x + 1}}$

**29.** $f(x) = \dfrac{1}{\sqrt[3]{(9x + 1)^2}}$  **30.** $f(x) = \dfrac{1}{\sqrt[3]{(3x - 1)^2}}$

**31.** $f(x) = \dfrac{1}{\sqrt[3]{(2x^2 - 3x + 1)^2}}$

**32.** $f(x) = \dfrac{1}{\sqrt[3]{(x^2 + x - 9)^2}}$

**33.** $f(x) = [(x^2 + 1)^3 + x]^3$

**34.** $f(x) = [(x^3 + 1)^2 - x]^4$

**35.** $f(x) = 3x^2(2x + 1)^5$  **36.** $f(x) = 2x(x^3 - 1)^4$

**37.** $f(x) = (2x + 1)^3(2x - 1)^4$

**38.** $f(x) = (2x - 1)^3(2x + 1)^4$

**39.** $g(z) = 2z(3z^2 - z + 1)^4$

**40.** $g(z) = z^2(2z^3 - z + 5)^4$

**41.** $f(x) = \left(\dfrac{x + 1}{x - 1}\right)^3$  **42.** $f(x) = \left(\dfrac{x - 1}{x + 1}\right)^5$

**43.** $f(x) = x^2\sqrt{1 + x^2}$  **44.** $f(x) = x^2\sqrt{x^2 - 1}$

**45.** $f(x) = \sqrt{1 + \sqrt{x}}$  **46.** $f(x) = \sqrt[3]{1 + \sqrt[3]{x}}$

**47.** Find the derivative of $(x^2 + 1)^2$ in two ways:
  **a.** By the Generalized Power Rule.
  **b.** By "squaring out" the original expression and then differentiating.
  Your answers should agree.

**48.** Find the derivative of $\dfrac{1}{x^2}$ in three ways:
  **a.** By the Quotient Rule.
  **b.** By writing $\dfrac{1}{x^2}$ as $(x^2)^{-1}$ and using the Generalized Power Rule.
  **c.** By writing $\dfrac{1}{x^2}$ as $x^{-2}$ and using the (ordinary) Power Rule.
  Your answers should agree.

**49.** Find the derivative of $\dfrac{1}{3x + 1}$ in two ways:
  **a.** By the Quotient Rule.
  **b.** By writing the function as $(3x + 1)^{-1}$ and using the Generalized Power Rule.
  Your answers should agree. Which way was easier? Remember this for the future.

**50.** Find an expression for the derivative of the composition of three functions, $\dfrac{d}{dx} f(g(h(x)))$.
  [*Hint:* Use the Chain Rule twice.]

**51–52.** Find the *second* derivative of each function.

**51.** $f(x) = (x^2 + 1)^{10}$  **52.** $f(x) = (x^3 - 1)^5$

## *Applied Exercises*

**53. BUSINESS: Cost** A company's cost function is $C(x) = \sqrt{4x^2 + 900}$ dollars, where $x$ is the number of units. Find the marginal cost function and evaluate it at $x = 20$.

 **54. BUSINESS: Cost** (*continuation*) Graph the cost function $y_1 = \sqrt{4x^2 + 900}$ on the window [0, 30] by [−10, 70]. Then use NDERIV to define $y_2$ as the derivative of $y_1$. Verify the answer to Exercise 53 by evaluating the marginal cost function $y_2$ at $x = 20$.

 **55. BUSINESS: Cost** (*continuation*) Find the number $x$ of units at which the marginal cost is 1.75. [*Hint:* TRACE along the marginal cost function $y_2$ to find where the $y$-coordinate is 1.75, giving your answer as the $x$-coordinate rounded to the nearest whole number.]

**56. SOCIOLOGY: Educational Status** A study estimated how a person's social status (rated on a scale where 100 indicates the status of a college graduate) depended on years of education. Based on this study, with $e$ years of education, a person's status is $S(e) = 0.22(e + 4)^{2.1}$. Find $S'(12)$ and interpret your answer.

*Source: Sociometry 34*

57. **SOCIOLOGY: Income Status** A study estimated how a person's social status (rated on a scale where 100 indicates the status of a college graduate) depended upon income. Based on this study, with an income of $i$ thousand dollars, a person's status is $S(i) = 17.5(i - 1)^{0.53}$. Find $S'(25)$ and interpret your answer.

*Source: Sociometry 34*

58. **ECONOMICS: Compound Interest** If $1000 is deposited in a bank paying $r\%$ interest compounded annually, 5 years later its value will be

$$V(r) = 1000(1 + 0.01r)^5 \quad \text{dollars}$$

Find $V'(6)$ and interpret your answer. [*Hint:* $r = 6$ corresponds to 6% interest.]

59. **BIOMEDICAL: Drug Sensitivity** The strength of a patient's reaction to a dose of $x$ milligrams of a certain drug is $R(x) = 4x\sqrt{11 + 0.5x}$ for $0 \le x \le 140$. The derivative $R'(x)$ is called the *sensitivity* to the drug. Find $R'(50)$, the sensitivity to a dose of 50 mg.

 60. **GENERAL: Population** The population of a city $x$ years from now is predicted to be $P(x) = \sqrt[4]{x^2 + 1}$ million people for $1 \le x \le 5$. Find when the population will be growing at the rate of a quarter of a million people per year. [*Hint:* On a graphing calculator, enter the given population function in $y_1$, use NDERIV to define $y_2$ to be the derivative of $y_1$, and graph both on the window [1, 5] by [0, 3]. Then TRACE along $y_2$ to find the $x$-coordinate (rounded to the nearest tenth of a unit) at which the $y$-coordinate is 0.25. You may have to ZOOM IN to find the correct $x$-value.]

61. **BIOMEDICAL: Drug Sensitivity** (*59 continued*) For the reaction function given in Exercise 59, find the dose at which the sensitivity is 25. [*Hint:* On a graphing calculator, enter the reaction function $y_1 = 4x\sqrt{11 + 0.5x}$, and use NDERIV to define $y_2$ to be the derivative of $y_1$. Then graph $y_2$ on the window [0, 140] by [0, 50] and TRACE along $y_2$ to find the $x$-coordinate (rounded to the nearest whole number) at which the $y$-coordinate is 25. You may have to ZOOM IN to find the correct $x$-value.]

62. **BIOMEDICAL: Blood Flow** It follows from Poiseuille's Law that blood flowing through certain arteries will encounter a resistance of

$R(x) = 0.25(1 + x)^4$, where $x$ is the distance (in meters) from the heart. Find the instantaneous rate of change of the resistance at:

**a.** 0 meters. **b.** 1 meter.

63. **ENVIRONMENTAL SCIENCE: Pollution** The carbon monoxide level in a city is predicted to be $0.02x^{3/2} + 1$ ppm (parts per million), where $x$ is the population in thousands. In $t$ years the population of the city is predicted to be $x(t) = 12 + 2t$ thousand people. Therefore, in $t$ years the carbon monoxide level will be

$$P(t) = 0.02(12 + 2t)^{3/2} + 1 \quad \text{ppm}$$

Find $P'(2)$, the rate at which carbon monoxide pollution will be increasing in 2 years.

64. **PSYCHOLOGY: Learning** After $p$ practice sessions, a subject could perform a task in $T(p) = 36(p + 1)^{-1/3}$ minutes for $0 \le p \le 10$. Find $T'(7)$ and interpret your answer.

65. **ENVIRONMENTAL SCIENCE: Greenhouse Gasses and Global Warming** The following graph shows the concentration (in parts per million, or *ppm*) of carbon dioxide ($CO_2$), the most common greenhouse gas, in the atmosphere for recent years, showing that, at least in the short run, $CO_2$ levels rise linearly.

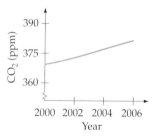

This $CO_2$ concentration is approximated by the function $y_1 = 1.9x + 370$ where $x$ is the number of years since 2000. The function $y_2 = 0.024x + 51.3$ is an estimate of the average global temperature (in degrees Fahrenheit) if the $CO_2$ level is $x$.

**a.** Enter $y_1$ and $y_2$ into your calculator and define $y_3$ to be the composition $y_3 = y_2(y_1)$, so that $y_3$ gives the average global temperature for any $x$ (years since 2000).

**b.** Turn "off" functions $y_1$ and $y_2$ and graph $y_3$ on the window [0, 20] by [60, 62]. Use the operation NDERIV or $dy/dx$ to find the slope, giving the predicted temperature rise per year in the short run.

**c.** Use this slope to find how long it will take global temperatures to rise by 1.8 degrees.

*(continues)*

(It has been estimated that a 1.8-degree temperature rise could raise sea levels by a foot, inundating coastal areas and disrupting world food supplies.*)

*Source:* Worldwatch Institute

**66. GENERAL: Happiness and Temperature** Based on a recent study, the "happiness" of people who live in a country whose average temperature is $t$ degrees Fahrenheit is given by $h(t) = 8.2 - (0.01t - 2.8)^2$, for $35 \le t \le 72$. ("Happiness" was rated from $1 =$ "not at all happy" to $4 =$ "very happy".) Find $h(40)$ and $h'(40)$. Interpret your answers.

*Source: Ecological Economics* **52**

## Conceptual Exercises

**67.** True or False: $\dfrac{d}{dx} f(g(x)) = f'(g'(x)) \cdot g'(x)$

**68.** True or False: $\dfrac{d}{dx} [g(x)]^n = n \cdot [g'(x)]^{n-1} \cdot g'(x)$

**69.** Explain the difference between the Chain Rule and the Generalized Power Rule.

**70.** Explain the difference between the Generalized Power Rule and the Power Rule.

**71.** True or False: $\dfrac{d}{dx} f(5x) = 5 \cdot f'(5x)$

**72.** True or False: $\dfrac{d}{dx} f(x/2) = \dfrac{f'(x/2)}{2}$

**73.** True or False: $\dfrac{d}{dx} \sqrt{g(x)} = \sqrt{g(x)} \cdot g'(x)$

**74.** True or False: $\dfrac{d}{dx} f(x + 5) = f'(x + 5)$

**75.** Imagine a square whose sides are expanding at a given rate (so that the area is also increasing). Since the area is the *square* of the length of a side, is it true that the rate of change of the area is the *square* of the rate of change of the side?

**76.** Imagine a cube whose sides are expanding at a given rate (so that the volume is also increasing). Since the volume is the *cube* of the length of a side, is it true that the rate of change of the volume is the *cube* of the rate of change of the side?

## Explorations and Excursions

The following problems extend and augment the material presented in the text.

*The function $y_1$ giving $CO_2$ levels is widely accepted as accurate. The function $y_2$ giving temperature based on $CO_2$ levels is less well established since factors other than $CO_2$ affect temperature.

*Further Uses of the Chain Rule*

**77.** Suppose that $L(x)$ is a function such that $L'(x) = \dfrac{1}{x}$. Use the Chain Rule to show that the derivative of the composite function $L(g(x))$ is $\dfrac{d}{dx} L(g(x)) = \dfrac{g'(x)}{g(x)}$.

**78.** Suppose that $E(x)$ is a function such that $E'(x) = E(x)$. Use the Chain Rule to show that the derivative of the composite function $E(g(x))$ is $\dfrac{d}{dx} E(g(x)) = E(g(x)) \cdot g'(x)$.

*Carathéodory's Definition of the Derivative and Proof of the Chain Rule*

**79.** The following is an alternate definition of the derivative, due to Constantine Carathéodory.*

> A function $f$ is differentiable at $x$ if there is a function $F$ that is continuous at 0 and such that $f(x + h) - f(x) = F(h) \cdot h$. In this case, $F(0) = f'(x)$.

Show that Carathéodory's definition of the derivative is equivalent to the definition on page 100.

**80.** The Chain Rule (page 153) states that the derivative of $f(g(x))$ is $f'(g(x)) \cdot g'(x)$. Use Carathéodory's definition of the derivative to prove the Chain Rule by giving reasons for the following steps.

a. Since $g$ is differentiable at $x$, there is a function $G$ that is continuous at 0 such that $g(x + h) - g(x) = G(h) \cdot h$, and $G(0) = g'(x)$.

b. Since $f$ is differentiable at $g(x)$, there is a function $F$ that is continuous at 0 such that $f(g(x) + h) - f(g(x)) = F(h) \cdot h$, and $F(0) = f'(g(x))$.

c. For the function $f(g(x))$ we have
$$f(g(x + h)) - f(g(x))$$
$$= f(g(x) + g(x + h) - g(x)) - f(g(x))$$
$$= f(g(x) + (g(x + h) - g(x))) - f(g(x))$$
$$= F(g(x + h) - g(x)) \cdot (g(x + h) - g(x))$$
$$= F(g(x + h) - g(x)) \cdot G(h) \cdot h$$

d. Therefore, the derivative of $f(g(x))$ is
$$F(g(x + 0) - g(x)) \cdot G(0) = F(0) \cdot G(0)$$
$$= f'(g(x)) \cdot g'(x)$$
as was to be proved.

*Greek mathematician, 1873–1950.

## 2.7 NONDIFFERENTIABLE FUNCTIONS

### Introduction

In spite of all of the rules of differentiation, there are **nondifferentiable functions**—functions that cannot be differentiated at certain values. We begin this section by exhibiting such a function (the absolute value function) and showing that it is not differentiable at $x = 0$. We will then discuss general geometric conditions for a function to be nondifferentiable. Knowing where a function is not differentiable is important for understanding graphs and for interpreting answers from a graphing calculator.

### Absolute Value Function

On page 56 we defined the absolute value function,

$$|x| = \begin{cases} x & \text{if } x \geq 0 \\ -x & \text{if } x < 0 \end{cases}$$

Although the absolute value function is *defined* for *all* values of $x$, we will show that it is *not* differentiable at $x = 0$.

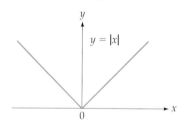

The graph of the absolute value function $f(x) = |x|$ has a "corner" at the origin.

---

**EXAMPLE 1**

**SHOWING NONDIFFERENTIABILITY**

Show that $f(x) = |x|$ is not differentiable at $x = 0$.

*Solution* We have no "rules" for differentiating the absolute value function, so we must use the definition of the derivative:

$$f'(x) = \lim_{h \to 0} \frac{f(x + h) - f(x)}{h}$$

*provided that this limit exists.* It is this provision, which until now we have steadfastly ignored, that will be important in this example. We will show that this limit, and hence the derivative, does not exist at $x = 0$.

For $x = 0$ the definition becomes

$$\lim_{h \to 0} \frac{f(0 + h) - f(0)}{h} = \lim_{h \to 0} \frac{f(h) - f(0)}{h} = \lim_{h \to 0} \underbrace{\frac{|h| - |0|}{h}}_{\substack{\text{Using} \\ f(x) = |x|}} = \lim_{h \to 0} \frac{|h|}{h}$$

For this limit to exist, the two one-sided limits must exist and have the same value. The limit from the right is

$$\lim_{h \to 0^+} \frac{|h|}{h} = \lim_{h \to 0^+} \frac{h}{h} = \lim_{h \to 0^+} 1 = 1$$

Since $|h| = h$        $\frac{h}{h} = 1$
for $h > 0$

The limit from the left is:

$$\lim_{h \to 0^-} \frac{|h|}{h} = \lim_{h \to 0^-} \frac{-h}{h} = \lim_{h \to 0^-} (-1) = -1$$

For $h < 0, |h| = -h$        $\frac{-h}{h} = -1$
(the negative sign
makes the negative
$h$ positive)

Since the one-sided limits do not agree (one is $+1$ and the other is $-1$), the limit $\lim_{h \to 0} \frac{|h|}{h}$ does not exist, so *the derivative does not exist.* This is what we wanted to show—that the absolute value function is not differentiable at $x = 0$.

## Geometric Explanation of Nondifferentiability

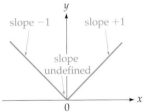

We can give a geometric and intuitive reason why the absolute value function is not differentiable at $x = 0$. Its graph consists of two straight lines with slopes $+1$ and $-1$ that meet in a corner at the origin. To the right of the origin the slope is $+1$ and to the left of the origin the slope is $-1$, but *at* the origin the two conflicting slopes make it impossible to define a *single* slope. Therefore, the slope (and hence the derivative) is undefined at $x = 0$.

## Graphing Calculator Exploration

If your graphing calculator has an operation like ABS or something similar for the absolute value function, graph $y_1 = \text{ABS}(x)$ on the window $[-2, 2]$ by $[-1, 2]$. Use NDERIV to "find" the derivative of $y_1$ at $x = 0$. Your calculator may give a "false value" such as 0, resulting from its *approximating* the derivative by the symmetric difference quotient (see pages 116–117). The correct answer is that the derivative is *undefined* at $x = 0$, as we just showed. This is why it is important to understand nondifferentiability—your calculator may give a misleading answer.

## Other Nondifferentiable Functions

For the same reason, at any **corner point** of a graph, where two different slopes conflict, the function will not be differentiable.

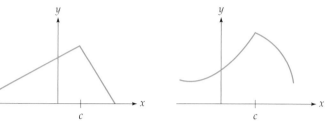

Each of the functions graphed here has a "corner point"
at $x = c$, and so is not differentiable at $x = c$.

There are other reasons, besides a corner point, why a function may not be differentiable. If a curve has a **vertical tangent** line at a point, the slope will not be defined at that $x$-value, since the slope of a vertical line is undefined.

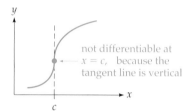

not differentiable at $x = c$, because the tangent line is vertical

We showed on pages 133–134 that if a function is differentiable, then it is continuous. Therefore, if a function is discontinuous (has a "jump") at some point, then it will not be differentiable at that $x$-value.

not differentiable at $x = c$, because it is discontinuous

If a function $f$ satisfies *any* of the following conditions:

**1.** $f$ has a corner point at $x = c$,

**2.** $f$ has a vertical tangent at $x = c$,

**3.** $f$ is discontinuous at $x = c$,

then $f$ will not be differentiable at $c$.

In Chapter 3, when we use calculus for graphing, it will be important to remember these three conditions that make the derivative fail to exist.

**Practice Problem**

For the function graphed below, find the $x$-values at which the derivative is undefined.

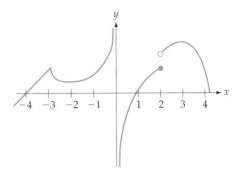

➤ **Solution on next page**

 **BE CAREFUL:** All differentiable functions are continuous (see page 134), but *not* all continuous functions are differentiable—for example, $f(x) = |x|$. These facts are shown in the following diagram.

 ## Spreadsheet Exploration

Another function that is not differentiable is $f(x) = x^{2/3}$. The following spreadsheet* calculates values of the difference quotient $\frac{f(x + h) - f(x)}{h}$ at $x = 0$ for this function. Since $f(0) = 0$, the difference quotient at $x = 0$ simplifies to:

$$\frac{f(x + h) - f(x)}{h} = \frac{f(0 + h) - f(0)}{h} = \frac{f(h)}{h} = \frac{h^{2/3}}{h} = h^{-1/3}$$

For example, cell **B5** evaluates $h^{-1/3}$ at $h = \frac{1}{1000}$ obtaining $\left(\frac{1}{1000}\right)^{-1/3} = 1000^{1/3} = \sqrt[3]{1000} = 10$. Column **B** evaluates this different quotient for the *positive* values of $h$ in column **A**, while column **E** evaluates it for the corresponding negative values of $h$ in column **D**.

*To obtain this and other Spreadsheet Explorations, go to www.cengage.com/math/berresford.

| | B5 | ▼ | | **=** | =A5^(-1/3) | |
|---|---|---|---|---|---|---|
| | A | B | C | D | E |
| 1 | h | (f(0+h)-f(0))/h | | h | (f(0+h)-f(0))/h |
| 2 | 1.0000000 | 1.0000000 | | -1.0000000 | -1.0000000 |
| 3 | 0.1000000 | 2.1544347 | | -0.1000000 | -2.1544347 |
| 4 | 0.0100000 | 4.6415888 | | -0.0100000 | -4.6415888 |
| 5 | 0.0010000 | 10.0000000 | | -0.0010000 | -10.0000000 |
| 6 | 0.0001000 | 21.5443469 | | -0.0001000 | -21.5443469 |
| 7 | 0.0000100 | 46.4158883 | | -0.0000100 | -46.4158883 |
| 8 | 0.0000010 | 100.0000000 | | -0.0000010 | -100.0000000 |
| 9 | 0.0000001 | 215.4434690 | | -0.0000001 | -215.4434690 |

$\underbrace{\qquad\qquad}_{\text{becoming large}}$ $\underbrace{\qquad\qquad}_{\text{becoming small}}$

Notice that the values in column **B** are becoming arbitrarily large, while the values in column **E** are becoming arbitrarily small, so the difference quotient does not approach a limit as $h \to 0$. This shows that the derivative of $f(x) = x^{2/3}$ at 0 *does not exist*, so the function $f(x) = x^{2/3}$ is *not differentiable* at $x = 0$.

**Solution to Practice Problem**

$x = -3, \quad x = 0, \quad \text{and} \quad x = 2$

## 2.7 Exercises

**1–4.** For each function graphed below, find the $x$-values at which the derivative does not exist.

**1.**

**2.**

**3.**

**4.**

**5–8.** Use the definition of the derivative to show that the following functions are not differentiable at $x = 0$.

[*Hint for Exercises 5 and 6:* Modify the calculations on pages 164–165 to apply to these functions.]

**5.** $f(x) = |2x|$          **6.** $f(x) = |3x|$

**7.** $f(x) = x^{2/5}$          **8.** $f(x) = x^{4/5}$

 **9–10.** Use NDERIV to "find" the derivative of each function at $x = 0$.  Is the result correct?

**9.** $f(x) = \dfrac{1}{x}$          **10.** $f(x) = \dfrac{1}{x^2}$

 **11. a.** Show that the definition of the derivative applied to the function $f(x) = \sqrt{x}$ at

$x = 0$  gives  $f'(0) = \lim\limits_{h \to 0} \dfrac{\sqrt{h}}{h}$.

   **b.** Use a calculator to evaluate the difference quotient $\dfrac{\sqrt{h}}{h}$ for the following values of $h$: 0.1, 0.001, and 0.00001. [*Hint:* Enter the calculation into your calculator with $h$ replaced by 0.1, and then change the value of $h$ by inserting zeros.]

   **c.** From your answers to part (b), does the limit exist? Does the derivative of $f(x) = \sqrt{x}$ at $x = 0$  exist?

   **d.** Graph $f(x) = \sqrt{x}$ on the window [0, 1] by [0, 1]. Do you see why the slope at $x = 0$  does not exist?

**12. a.** Show that the definition of the derivative applied to the function $f(x) = \sqrt[3]{x}$ at $x = 0$

gives  $f'(0) = \lim\limits_{h \to 0} \dfrac{\sqrt[3]{h}}{h}$.

   **b.** Use a calculator to evaluate the difference quotient $\dfrac{\sqrt[3]{h}}{h}$ for the following values

of $h$: 0.1, 0.0001, and 0.0000001. [*Hint:* Enter the calculation into your calculator with $h$ replaced by 0.1, and then change the value of $h$ by inserting zeros.]

   **c.** From your answers to part (b), does the limit exist? Does the derivative of $f(x) = \sqrt[3]{x}$ at $x = 0$  exist?

   **d.** Graph $f(x) = \sqrt[3]{x}$ on the window $[-1, 1]$ by $[-1, 1]$. Do you see why the slope at $x = 0$ does not exist?

### *Conceptual Exercises*

**13.** Using your own words, explain geometrically why the derivative is undefined where a curve has a *corner point*.

**14.** Using your own words, explain geometrically why the derivative is undefined where a curve has a *vertical tangent*.

**15.** Using your own words, explain in terms of instantaneous rates of change why the derivative is undefined where a function has a *discontinuity*.

**16.** True or False:  If a function is *continuous* at a number, then it is *differentiable* at that number.

**17.** True or False:  If a function is *differentiable* at a number, then it is *continuous* at that number.

**18.** True or False:  If a function is not differentiable at a point, then its graph cannot have a tangent line at that point.

**19.** Using only straight lines, sketch a function that (a) is continuous everywhere and (b) is differentiable everywhere *except* at $x = 1$  and  $x = 3$.

**20.** Sketch a function that (a) is continuous everywhere, (b) has a tangent line at every point, and (c) is differentiable everywhere *except* at $x = 2$ and  $x = 4$.

## Chapter Summary with Hints and Suggestions

Reading the text and doing the exercises in this chapter have helped you to master the following concepts and skills, which are listed by section (in case you need to review them) and are keyed to particular Review Exercises. Answers for all Review Exercises are given at the back of the book, and full solutions can be found in the Student Solutions Manual.

### 2.1  Limits and Continuity

■  Find the limit of a function from tables. (*Review Exercises 1–2.*)

■  Find left and right limits. (*Review Exercises 3–4.*)

■  Find the limit of a function. (*Review Exercises 5–14.*)

■  Determine whether a function is continuous or discontinuous. (*Review Exercises 15–22.*)

## 2.2  Rates of Change, Slopes, and Derivatives

■ Find the derivative of a function from the *definition* of the derivative. (*Review Exercises 23–26.*)

$$f'(x) = \lim_{h \to 0} \frac{f(x + h) - f(x)}{h}$$

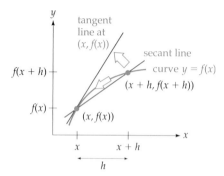

## 2.3  Some Differentiation Formulas

■ Find the derivative of a function using the rules of differentiation. (*Review Exercises 27–32.*)

$$\frac{d}{dx} c = 0 \qquad \frac{d}{dx} x^n = nx^{n-1}$$

$$\frac{d}{dx}(c \cdot f) = c \cdot f' \qquad \frac{d}{dx}(f \pm g) = f' \pm g'$$

■ Calculate and interpret a company's marginal cost.  (*Review Exercise 33.*)

$$MC(x) = C'(x) \qquad MR(x) = R'(x)$$
$$MP(x) = P'(x)$$

■ Find and interpret the derivative of a learning curve.  (*Review Exercise 34.*)

■ Find and interpret the derivative of an area or volume formula.  (*Review Exercises 35–36.*)

## 2.4  The Product and Quotient Rules

■ Find the derivative of a function using the Product Rule or Quotient Rule. (*Review Exercises 37–47.*)

$$\frac{d}{dx}(f \cdot g) = f' \cdot g + f \cdot g'$$

$$\frac{d}{dx}\left(\frac{f}{g}\right) = \frac{g \cdot f' - g' \cdot f}{g^2}$$

■ Use differentiation to solve an applied problem and interpret the answer. (*Review Exercises 48–50.*)

## 2.5  Higher-Order Derivatives

■ Calculate the second derivative of a function. (*Review Exercises 51–60.*)

■ Find and interpret the first and second derivatives in an applied problem. (*Review Exercise 61.*)

■ Find the velocity and acceleration of a rocket. (*Review Exercise 62.*)

$$v(t) = s'(t) \qquad a(t) = v'(t) = s''(t)$$

■ Find the maximum height of a projectile. (*Review Exercise 63.*)

## 2.6  The Chain Rule and the Generalized Power Rule

■ Find the derivative of a function using the Generalized Power Rule. (*Review Exercises 64–73.*)

$$\frac{d}{dx} f^n = n \cdot f^{n-1} \cdot f'$$

$$\frac{d}{dx} f(g(x)) = f'(g(x)) \cdot g'(x)$$

$$\frac{dy}{dx} = \frac{dy}{du} \cdot \frac{du}{dx}$$

■ Find the derivative of a function using *two* differentiation rules.  (*Review Exercises 74–85.*)

■ Find the *second* derivative of a function using the Generalized Power Rule. (*Review Exercises 86–89.*)

■ Find the derivative of a function in several different ways.  (*Review Exercises 90–91.*)

■ Use the Generalized Power Rule to find the derivative in an applied problem and interpret the answer.  (*Review Exercises 92–93.*)

■ Compare the profit from one unit to the marginal profit found by differentiation. (*Review Exercise 94.*)

■ Find where the marginal profit equals a given number.  (*Review Exercise 95.*)

■ Use the Generalized Power Rule to solve an applied problem and interpret the answer. (*Review Exercises 96–97.*)

## 2.7 Nondifferentiable Functions

- See from a graph where the derivative is undefined.  *(Review Exercises 98–101.)*

$$f' \text{ is undefined at } \begin{cases} \text{corner points} \\ \text{vertical tangents} \\ \text{discontinuities} \end{cases}$$

- Prove that a function is not differentiable at a given value.
*(Review Exercises 102–103.)*

## Hints and Suggestions

- *(Overview)* This chapter introduced one of the most important concepts in all of calculus, the *derivative*. First we defined it (using limits), then we developed several "rules of differentiation" to simplify its calculation.

- Remember the four interpretations of the derivative—*slopes, instantaneous rates of change, marginals,* and *velocities.*

- The *second* derivative gives the rate of change of the rate of change, and acceleration.

- Graphing calculators help to find limits, graph curves and their tangent lines, and calculate derivatives (using NDERIV) and second derivatives (using NDERIV twice). NDERIV, however, provides only an *approximation* to the derivative, and therefore sometimes gives a misleading result.

- The units of the derivative are important in applied problems. For example, if $f(x)$ gives the temperature in degrees at time $x$ hours, then the derivative $f'(x)$ is in *degrees per hour.* In general, the units of the derivative $f'(x)$ are "*f*-units" per "*x*-unit."

- **Practice for Test:** Review Exercises 1, 2, 3, 4, 7, 9, 11, 14, 17, 21, 22, 23, 27, 33, 39, 45, 49, 51, 59, 61, 68, 72, 78, 84, 92, 95, 98.

# Review Exercises for Chapter 2

Practice test exercise numbers are in green.

## 2.1 Limits and Continuity

**1–2.** Complete the tables and use them to find each limit (or state that it does not exist). Round calculations to three decimal places.

**1. a.** $\lim\limits_{x \to 2^-} (4x + 2)$     **b.** $\lim\limits_{x \to 2^+} (4x + 2)$

  **c.** $\lim\limits_{x \to 2} (4x + 2)$

| x | 4x + 2 |
|---|--------|
| 1.9 | |
| 1.99 | |
| 1.999 | |

| x | 4x + 2 |
|---|--------|
| 2.1 | |
| 2.01 | |
| 2.001 | |

**2. a.** $\lim\limits_{x \to 0^-} \dfrac{\sqrt{x + 1} - 1}{x}$     **b.** $\lim\limits_{x \to 0^+} \dfrac{\sqrt{x + 1} - 1}{x}$

  **c.** $\lim\limits_{x \to 0} \dfrac{\sqrt{x + 1} - 1}{x}$

| x | $\dfrac{\sqrt{x + 1} - 1}{x}$ |
|---|--------|
| −0.1 | |
| −0.01 | |
| −0.001 | |

| x | $\dfrac{\sqrt{x + 1} - 1}{x}$ |
|---|--------|
| 0.1 | |
| 0.01 | |
| 0.001 | |

**3–4.** For each piecewise linear function, find:

**a.** $\lim\limits_{x \to 5^-} f(x)$   **b.** $\lim\limits_{x \to 5^+} f(x)$   **c.** $\lim\limits_{x \to 5} f(x)$

**3.** $f(x) = \begin{cases} 2x - 7 & \text{if } x \le 5 \\ 3 - x & \text{if } x > 5 \end{cases}$

**4.** $f(x) = \begin{cases} 4 - x & \text{if } x < 5 \\ 2x - 11 & \text{if } x \ge 5 \end{cases}$

**5–12.** Find the following limits (*without* using limit tables).

**5.** $\lim\limits_{x \to 4} \sqrt{x^2 + x + 5}$

**6.** $\lim\limits_{x \to 0} \pi$

**7.** $\lim\limits_{s \to 16} \left( \dfrac{1}{2}s - s^{1/2} \right)$

**8.** $\lim\limits_{r \to 8} \dfrac{r}{r^2 - 30\sqrt[3]{r}}$

**9.** $\lim\limits_{x \to 1} \dfrac{x^2 - x}{x^2 - 1}$

**10.** $\lim\limits_{x \to -1} \dfrac{3x^3 - 3x}{2x^2 + 2x}$

**11.** $\lim\limits_{h \to 0} \dfrac{2x^2 h - xh^2}{h}$

**12.** $\lim\limits_{h \to 0} \dfrac{6xh^2 - x^2 h}{h}$

**13–14.** Use limits involving $\pm\infty$ to describe the asymptotic behavior of each function from its graph.

**13.** $f(x) = \dfrac{1}{(x + 2)^2}$

**14.** $f(x) = \dfrac{3x}{x - 2}$

**15–22.** For each function, state whether it is continuous or discontinuous. If it is discontinuous, state the values of $x$ at which it is discontinuous.

**15.** $f(x) = 2x + 5$

**16.** $f(x) = x^2 - 1$

**17.** $f(x) = \dfrac{1}{x + 1}$

**18.** $f(x) = \dfrac{1}{x^2 + 1}$

**19.** $f(x) = \dfrac{x - 1}{x^2 + x}$

**20.** $f(x) = \dfrac{1}{|x| - 3}$

**21.** $f(x) = \begin{cases} 2x - 7 & \text{if } x \le 5 \\ 3 - x & \text{if } x > 5 \end{cases}$

[*Hint:* See Exercise 3.]

**22.** $f(x) = \begin{cases} 4 - x & \text{if } x < 5 \\ 2x - 11 & \text{if } x \ge 5 \end{cases}$

[*Hint:* See Exercise 4.]

## 2.2 Rates of Change, Slopes, and Derivatives

**23–26.** Find the derivative of each function using the definition of the derivative (that is, as you did in Section 2.2).

**23.** $f(x) = 2x^2 + 3x - 1$

**24.** $f(x) = 3x^2 + 2x - 3$

**25.** $f(x) = \dfrac{3}{x}$

**26.** $f(x) = 4\sqrt{x}$

## 2.3 Some Differentiation Formulas

**27–28.** Find the derivative of each function.

**27.** $f(x) = 6\sqrt[3]{x^5} - \dfrac{4}{\sqrt{x}} + 1$

**28.** $f(x) = 4\sqrt{x^5} - \dfrac{6}{\sqrt[3]{x}} + 1$

**29–32.** Evaluate each expression.

**29.** If $f(x) = \dfrac{1}{x^2}$, find $f'\left( \dfrac{1}{2} \right)$.

**30.** If $f(x) = \dfrac{1}{x}$, find $f'\left( \dfrac{1}{3} \right)$.

**31.** If $f(x) = 12\sqrt[3]{x}$, find $f'(8)$.

**32.** If $f(x) = 6\sqrt[3]{x}$, find $f'(-8)$.

**33. BUSINESS: Educational Software Costs** Educational institutions can buy multiple licenses for Word Search Maker software (for making word puzzles) at a total cost of $C(x) = 24x^{5/6}$ dollars for $x$ software disks. Find the derivative of this cost function at:
   **a.** $x = 1$ and interpret your answer.
   **b.** $x = 64$ and interpret your answer.
   *Source:* Variety Games, Inc.

**34. Learning Curves in Industry** From page 26, the learning curve for building Boeing 707 airplanes is $f(x) = 150x^{-0.322}$, where $f(x)$ is the time (in thousands of hours) that it took to build the $x$th Boeing 707. Find the instantaneous rate of change of this production time for the tenth plane, and interpret your answer.

**35. GENERAL: Geometry** The formula for the area of a circle is $A = \pi r^2$, where $r$ is the radius of the circle and $\pi$ is a constant.

**a.** Show that the derivative of the area formula is $2\pi r$, the formula for the circumference of a circle.

**b.** Give an explanation for this in terms of rates of change.

**36. GENERAL: Geometry** The formula for the volume of a sphere is $V = \frac{4}{3}\pi r^3$, where $r$ is the radius of the sphere and $\pi$ is a constant.

**a.** Show that the derivative of the volume formula is $4\pi r^2$, the formula for the surface area of a sphere.

**b.** Give an explanation for this in terms of rates of change.

## 2.4 The Product and Quotient Rules

**37–46.** Find the derivative of each function.

**37.** $f(x) = 2x(5x^3 + 3)$

**38.** $f(x) = x^2(3x^3 - 1)$

**39.** $f(x) = (x^2 + 5)(x^2 - 5)$

**40.** $f(x) = (x^2 + 3)(x^2 - 3)$

**41.** $y = (x^4 + x^2 + 1)(x^5 - x^3 + x)$

**42.** $y = (x^5 + x^3 + x)(x^4 - x^2 + 1)$

**43.** $y = \dfrac{x - 1}{x + 1}$   **44.** $y = \dfrac{x + 1}{x - 1}$

**45.** $y = \dfrac{x^5 + 1}{x^5 - 1}$   **46.** $y = \dfrac{x^6 - 1}{x^6 + 1}$

**47.** Find the derivative of $f(x) = \dfrac{2x + 1}{x}$ in three different ways, and check that the answers agree:

**a.** By the Quotient Rule

**b.** By writing the function in the form $f(x) = (2x + 1)(x^{-1})$ and using the Product Rule.

**c.** By thinking of another way, which is the easiest of all.

**48. BUSINESS: Sales** The manager of an electronics store estimates that the number of cassette tapes that a store will sell at a price of $x$ dollars is

$$S(x) = \frac{2250}{x + 9}$$

Find the rate of change of this quantity when the price is $6 per tape, and interpret your answer.

**49. BUSINESS: Marginal Average Profit** A company profit function is $P(x) = 6x - 200$ dollars, where $x$ is the number of units.

**a.** Find the average profit function.

**b.** Find the marginal average profit function.

**c.** Evaluate the marginal average profit function at $x = 10$ and interpret your answer.

**50. BUSINESS: Marginal Average Cost** A company can produce a mini optical computer mouse at a cost of $7.50 each while fixed costs are $50. Therefore, the company's cost function is $C(x) = 7.5x + 50$.

**a.** Find the average cost function $AC(x) = \dfrac{C(x)}{x}$.

**b.** Find the marginal average cost function $MAC(x)$.

**c.** Evaluate $MAC(x)$ at $x = 50$ and interpret your answer.

*Source:* Green Pearle International

## 2.5 Higher-Order Derivatives

**51–54.** Find the *second* derivative of each function.

**51.** $f(x) = 12\sqrt{x^3} - 9\sqrt[3]{x}$

**52.** $f(x) = 18\sqrt[3]{x^2} - 4\sqrt{x^3}$

**53.** $f(x) = \dfrac{1}{3x^2}$

**54.** $f(x) = \dfrac{1}{2x^3}$

**55–60.** Evaluate each expression.

**55.** If $f(x) = \dfrac{2}{x^3}$, find $f''(-1)$.

**56.** If $f(x) = \dfrac{3}{x^4}$, find $f''(-1)$.

**57.** $\dfrac{d^2}{dx^2} \left. x^6 \right|_{x = -2}$

**58.** $\left.\dfrac{d^2}{dx^2}\, x^{-2}\right|_{x=-2}$

**59.** $\left.\dfrac{d^2}{dx^2}\, \sqrt{x^5}\right|_{x=16}$

**60.** $\left.\dfrac{d^2}{dx^2}\, \sqrt{x^7}\right|_{x=4}$

**61. GENERAL: Population** The population of a city $t$ years from now is predicted to be $P(t) = 0.25t^3 - 3t^2 + 5t + 200$ thousand people. Find $P(10)$, $P'(10)$, and $P''(10)$ and interpret your answers.

**62. GENERAL: Velocity** A rocket rises $s(t) = 8t^{5/2}$ feet in $t$ seconds. Find its velocity and acceleration after 25 seconds.

**63. GENERAL: Velocity** The fastest baseball pitch on record (thrown by Lynn Nolan Ryan of the California Angels on August 20, 1974) was clocked at 100.9 miles per hour (148 feet per second).

   **a.** If this pitch had been thrown straight up, its height after $t$ seconds would have been $s(t) = -16t^2 + 148t + 5$ feet. Find the maximum height the ball would have reached.

   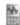 **b.** Verify your answer to part (a) by graphing the height function $y_1 = -16x^2 + 148x + 5$ on the window [0, 10] by [0, 400]. Then TRACE along the curve to find its highest point (or use the MAXIMUM feature of your calculator).

## 2.6 The Chain Rule and the Generalized Power Rule

**64–85.** Find the derivative of each function.

**64.** $h(z) = (4z^2 - 3z + 1)^3$

**65.** $h(z) = (3z^2 - 5z - 1)^4$

**66.** $g(x) = (100 - x)^5$

**67.** $g(x) = (1000 - x)^4$

**68.** $f(x) = \sqrt{x^2 - x + 2}$

**69.** $f(x) = \sqrt{x^2 - 5x - 1}$

**70.** $w(z) = \sqrt[3]{6z - 1}$

**71.** $w(z) = \sqrt[3]{3z + 1}$

**72.** $h(x) = \dfrac{1}{\sqrt[5]{(5x + 1)^2}}$

**73.** $h(x) = \dfrac{1}{\sqrt[5]{(10x + 1)^3}}$

**74.** $g(x) = x^2(2x - 1)^4$

**75.** $g(x) = 5x(x^3 - 2)^4$

**76.** $y = x^3 \sqrt[3]{x^3 + 1}$

**77.** $y = x^4 \sqrt{x^2 + 1}$

**78.** $f(x) = [(2x^2 + 1)^4 + x^4]^3$

**79.** $f(x) = [(3x^2 - 1)^3 + x^3]^2$

**80.** $f(x) = \sqrt{(x^2 + 1)^4 - x^4}$

**81.** $f(x) = \sqrt{(x^3 + 1)^2 + x^2}$

**82.** $f(x) = (3x + 1)^4(4x + 1)^3$

**83.** $f(x) = (x^2 + 1)^3(x^2 - 1)^4$

**84.** $f(x) = \left(\dfrac{x + 5}{x}\right)^4$

**85.** $f(x) = \left(\dfrac{x + 4}{x}\right)^5$

**86–89.** Find the *second* derivative of each function.

**86.** $h(w) = (2w^2 - 4)^5$    **87.** $h(w) = (3w^2 + 1)^4$

**88.** $g(z) = z^3(z + 1)^3$    **89.** $g(z) = z^4(z + 1)^4$

**90.** Find the derivative of $(x^3 - 1)^2$ in two ways:

   **a.** By the Generalized Power Rule.
   **b.** By "squaring out" the original expression and then differentiating.
   Your answers should agree.

**91.** Find the derivative of $g(x) = \dfrac{1}{x^3 + 1}$ in two ways:

   **a.** By the Quotient Rule.
   **b.** By the Generalized Power Rule.
   Your answers should agree.

**92. BUSINESS: Marginal Profit** A company's profit from producing $x$ tons of polyurethane is $P(x) = \sqrt{x^3 - 3x + 34}$ thousand dollars (for $0 \le x \le 10$). Find $P'(5)$ and interpret your answer.

**93. GENERAL: Compound Interest** If $500 is deposited in an account earning interest at $r$ percent annually, after 3 years its value will be $V(r) = 500(1 + 0.01r)^3$ dollars. Find $V'(8)$ and interpret your answer.

**94. BUSINESS: Marginal Profit** (*92 continued*)
Using a graphing calculator, graph the profit function from Exercise 92 on the window [0, 10] by [0, 30].

    **a.** Evaluate the profit function at 4, 5, and 6 and calculate the following actual costs:

        $P(5) - P(4)$ (actual cost of the fifth ton)

        $P(6) - P(5)$ (actual cost of the sixth ton)

    Compare these results with the "instantaneous" marginal profit at $x = 5$ found in Exercise 92.

    **b.** Define another function to be the derivative of the previously entered profit function (using NDERIV) and graph both together. Find (to the nearest tenth of a unit) the $x$-value where the marginal profit reaches 4.

**95.** On a graphing calculator, graph the cost function $C(x) = \sqrt[3]{x^2 + 25x + 8}$ as $y_1$ on the window [0, 20] by [−2, 10]. Use NDERIV to define $y_2$ to be its derivative, graphing both. TRACE along the derivative to find the $x$-value (to the nearest unit) where the marginal cost is 0.25.

**96. BIOMEDICAL: Blood Flow** Blood flowing through an artery encounters a resistance of $R(x) = 0.25(0.01x + 1)^4$, where $x$ is the distance (in centimeters) from the heart. Find the instantaneous rate of change of the resistance 100 centimeters from the heart.

**97. GENERAL: Survival Rate** Suppose that for a group of 10,000 people, the number who survive to age $x$ is $N(x) = 1000\sqrt{100 - x}$. Find $N'(96)$ and interpret your answer.

## 2.7 Nondifferentiable Functions

**98–101.** For each function graphed below, find the values of $x$ at which the derivative does not exist.

**98.**

**99.**

**100.**

**101.**

**102–103.** Use the definition of the derivative to show that the following functions are not differentiable at $x = 0$.

**102.** $f(x) = |5x|$      **103.** $f(x) = x^{3/5}$

# 3 Further Applications of Derivatives

### Stevens' Law of Psychophysics

How accurately can you judge how heavy something is? If you give someone two weights, with one *twice* as heavy as the other, most people will judge the heavier weight as being *less* than twice as heavy. This is one of the oldest problems in experimental psychology—how sensation (perceived weight) varies with stimulus (actual weight). Similar experiments can be performed for perceived brightness of a light compared with actual brightness, perceived effort compared with actual work, and so on. The results will vary somewhat from person to person, but the following diagram shows some typical stimulus-response curves (in arbitrary units).

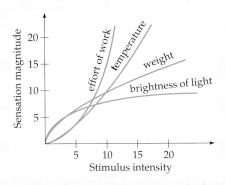

Notice, for example, that perceived effort increases more rapidly than actual work, which suggests that a 10% increase in an employee's work should be rewarded with a *greater* than 10% increase in pay.

Such stimulus-response curves were studied by the psychologist S. S. Stevens* at Harvard, who expressed them as power functions:

$$\text{Response} = a(\text{stimulus})^b$$

or

$$f(x) = ax^b$$

for constants $a$ and $b$

In this chapter we will see that calculus can be very helpful for graphing such functions, including showing whether they "curl upward" like the work and temperature curves or "curl downward" like the weight and brightness curves (see page 205).

---

*See S. S. Stevens, "On the Psychophysical Law," *Psychological Review* **64**.

## 3.1 GRAPHING USING THE FIRST DERIVATIVE

### Introduction

In this chapter we will put derivatives to two major uses: **graphing** and **optimization.** Graphing involves using calculus to find the most important points on a curve, then sketching the curve either by hand or using a graphing calculator. Optimization means finding the largest or smallest values of a function (for example, maximizing profit or minimizing risk). We begin with graphing, since it will form the basis for optimization later in the chapter.

We saw in Chapter 2 that the *derivative* of a function gives the *slope* of its graph. On an interval:

$$f' > 0 \quad \text{means } f \text{ is } \textit{increasing}.$$

$$f' < 0 \quad \text{means } f \text{ is } \textit{decreasing}.$$

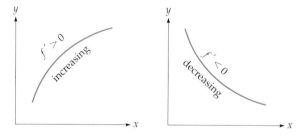

"Increasing" and "decreasing" on a graph mean rising and falling as you move *from left to right*, the same direction in which you are reading this book.

### Relative Extreme Points and Critical Numbers

On a graph, a **relative maximum point** is a point that is at least as *high* as the neighboring points of the curve on either side, and a **relative minimum point** is a point that is at least as *low* as the neighboring points on either side.

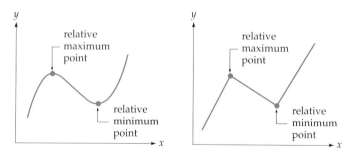

The word "relative" means that although these points may not be the highest and lowest on the *entire* curve, they are the highest and lowest *relative to points nearby*. (Later we will use the terms "absolute maximum" and "absolute minimum" to mean the highest and lowest points on the entire curve.) A curve may have any number of relative maximum and minimum points (collectively, **relative extreme points**), even none. For a function $f$, the relative extreme points may be defined more formally in terms of the values of $f$.

$f$ has a **relative maximum value** at $c$ if $\quad f(c) \geq f(x) \quad$ for all $x$ near $c$.

$f$ has a **relative minimum value** at $c$ if $\quad f(c) \leq f(x) \quad$ for all $x$ near $c$.

By "near $c$" we mean in some open interval containing $c$.

In the first of the two graphs below, the relative extreme points occur where the slope is *zero* (where the tangent line is horizontal), and in the second graph they occur where the slope is *undefined* (at corner points). The $x$-coordinates of such points are called **critical numbers.**

---

**Critical Number**

A *critical number* of a function $f$ is an $x$-value in the domain of $f$ at which either

or
$$f'(x) = 0$$
$$f'(x) \text{ is undefined}$$

Derivative is zero
or undefined

---

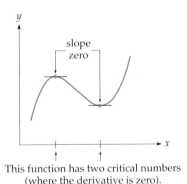

This function has two critical numbers (where the derivative is zero).     This function also has two critical numbers (but where the derivative is undefined).

## Graphing Functions

We graph a function by finding its critical numbers, making a **sign diagram** for the derivative to show the intervals of increase and decrease and the relative extreme points, and then drawing the curve on a graphing calculator or "by hand." Obtaining a reasonable graph even with a graphing calculator requires more than just pushing buttons, as shown in the unsatisfactory graph of the function $f(x) = x^3 - 12x^2 - 60x + 36$ on the left. We will improve on this graph in the following example.

A "useless" graph of $f(x) = x^3 - 12x^2 - 60x + 36$ on $[-10, 10]$ by $[-10, 10]$

**EXAMPLE 1**

### GRAPHING A FUNCTION

Graph the function $f(x) = x^3 - 12x^2 - 60x + 36$.

*Solution*

*Step 1* **Find critical numbers.**

$f'(x) = 3x^2 - 24x - 60$ 

Derivative of $f(x) = x^3 - 12x^2 - 60x + 36$

$= 3(x^2 - 8x - 20) = 3(x - 10)(x + 2) = 0$ 

Factoring and setting equal to zero

Zero at $x = 10$   Zero at $x = -2$

The derivative is zero at $x = 10$ and at $x = -2$, and there are no numbers at which the derivative is undefined (it is a polynomial), so the critical numbers (CNs) are

$$\text{CN} \begin{cases} x = 10 \\ x = -2 \end{cases}$$

Both are in the domain of the orginal function

*Step 2* **Make a sign diagram for the derivative.** A sign diagram for $f'$ begins with a copy of the $x$-axis with the critical numbers written below it and the behavior of $f'$ indicated above it.

Behavior of $f'$

$f' = 0$          $f' = 0$
$x = -2$          $x = 10$

Critical numbers

Since $f'$ is continuous, it can change sign only at critical numbers, so $f'$ must keep the same sign between consecutive critical numbers. We determine the sign of $f'$ in each interval by choosing a **test point** in each interval and substituting it into $f'$. We use the factored form: $f'(x) = 3(x - 10)(x + 2)$. For the first interval, choosing $-3$ for the test point,

$f'(-3) = 3(-3 - 10)(-3 + 2) = 3(\text{negative})(\text{negative}) = (\text{positive}).$

We indicate the sign of $f'$ (the slope of $f$) by arrows: $\nearrow$ for positive slope, $\rightarrow$ for zero slope, and $\searrow$ for negative slope.

Using test point $-3, f'(-3) = 39$, so $f'$ is *positive*

Using test point $0, f'(0) = -60$, so $f'$ is *negative*

Using test point $11, f'(11) = 39$, so $f'$ is *positive*

$f' > 0$      $f' = 0$      $f' < 0$      $f' = 0$      $f' > 0$
$x = -2$                  $x = 10$

The sign diagram shows that to the left of $-2$ the function increases, then between $-2$ and $10$ it decreases, and then to the right of $10$ it increases again. Therefore, the open intervals of increase are $(-\infty, -2)$ and $(10, \infty)$, and the open interval of decrease is $(-2, 10)$.

Arrows ↗ → ↘ indicate a relative *maximum* point, and arrows ↘ → ↗ indicate a relative *minimum* point. We then list these points under the critical numbers.

$$f' > 0 \qquad f' = 0 \qquad f' < 0 \qquad f' = 0 \qquad f' > 0$$
$$\underset{x = -2}{\phantom{+}} \qquad\qquad\qquad \underset{x = 10}{\phantom{+}}$$

↗     →     ↘        ↗

rel max
$(-2, 100)$

→

rel min
$(10, -764)$

The $y$-coordinates of the points were found by evaluating the *original* function at the $x$-coordinate: $f(-2) = 100$ and $f(10) = -764$. [*Hint:* Use 🖩 with TABLE or EVALUATE.]

*Step 3* **Sketch the graph.** Our arrows ↗ → ↘ → ↗ show the general shape of the curve: going up, level, down, level, and up again. The critical numbers show that the graph should include $x$-values from before $x = -2$ to after $x = 10$, suggesting an interval such as $[-10, 20]$. The $y$-coordinates show that we want to go from above $100$ to below $-764$, suggesting an interval of $y$-values such as $[-800, 200]$.

**Using a graphing calculator,** you then would graph on the window $[-10, 20]$ by $[-800, 200]$ (or some other reasonable window), as shown below.

**By hand,** you would plot the relative maximum point (with a "cap" ⌢) and the relative minimum point (with a "cup" ⌣), and then draw an "up-down-up" curve through them.

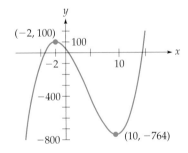

Two important observations:

1. Even with a graphing calculator, you still need calculus to find the relative extreme points that determine an appropriate window.

2. Given a calculator-drawn graph such as that on the left above, you should be able to make a hand-drawn graph such as that on the right, including numbers on the axes and coordinates of important points (using TRACE or TABLE if necessary).

## Graphing Calculator Exploration

Do you see where the "useless" graph shown on page 179 fits into the "useful" graph on the previous page? Check this by graphing the function $y_1 = x^3 - 12x^2 - 60x + 36$ first on the window $[-10, 20]$ by $[-800, 200]$ (obtaining the graph shown on the previous page) and then on the *standard* window $[-10, 10]$ by $[-10, 10]$.

**Practice Problem 1**

Find the critical numbers of $f(x) = x^3 - 12x + 8$. ➤ **Solution on page 187**

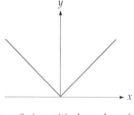

$x = 0$ is a critical number of
$f(x) = |x|$

In Example 1 there were no critical numbers at which the derivative was *undefined* (it was a polynomial). For an example of a function with a critical number where the derivative is *undefined*, think of the absolute value function $f(x) = |x|$. The graph has a "corner point" at the origin (as shown on the left), and so the derivative is undefined at the critical number $x = 0$.

### First-Derivative Test for Relative Extreme Values

The graphical idea from the sign diagram, that ↗ → ↘ (up, level, and down) indicates a relative *maximum* and ↘ → ↗ (down, level, and up) indicates a relative *minimum*, can be stated more formally in terms of the derivative.

**First-Derivative Test**

If a function $f$ has a critical number $c$, then at $x = c$ the function has a

*relative maximum* if $f' > 0$ just before $c$ and $f' < 0$ just after $c$.

*relative minimum* if $f' < 0$ just before $c$ and $f' > 0$ just after $c$.

⚠ **BE CAREFUL:** Remember the order: slopes that are *positive then negative* mean a *max* and slopes that are *negative then positive* mean a *min*.

If the derivative has the *same* sign on both sides of $c$, then the function has *neither* a relative maximum nor a relative minimum at $x = c$.

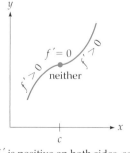

$f'$ is positive on both sides, so
$f$ has *neither* at $x = c$.

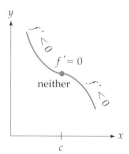

$f'$ is negative on both sides,
$f$ has *neither* at $x = c$.

The diagrams below show that the first-derivative test applies even at critical numbers where the derivative is *undefined* (abbreviated: $f'$ und).

$f'$ is positive then negative, so $f$ has a relative *maximum* at $x = c$.

$f'$ is negative then positive, so $f$ has a relative *minimum* at $x = c$.

$f'$ is positive on both sides, so $f$ has *neither* at $x = c$.

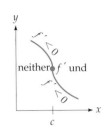

$f'$ is negative on both sides, so $f$ has *neither* at $x = c$.

**EXAMPLE 2**

### GRAPHING A FUNCTION

Graph   $f(x) = -x^4 + 4x^3 - 20$.

*Solution*

$$f'(x) = -4x^3 + 12x^2 \qquad \text{Differentiating}$$
$$= -4x^2(x - 3) = 0 \qquad \text{Factoring and setting equal to zero}$$

Critical numbers:

$$\text{CN} \begin{cases} x = 0 & \text{From} \quad -4x^2 = 0 \\ x = 3 & \text{From} \quad (x - 3) = 0 \end{cases}$$

We make a sign diagram for the derivative:

| $f' = 0$ | | $f' = 0$ | Behavior of $f'$ |
|---|---|---|---|
| $x = 0$ | | $x = 3$ | Critical values |

We determine the sign of   $f'(x) = -4x^2(x - 3)$   using test points in each interval (such as $-1$ in the leftmost interval), and then add arrows.

| $f' > 0$ | $f' = 0$ | $f' > 0$ | $f' = 0$ | $f' < 0$ |
|---|---|---|---|---|
| | $x = 0$ | | $x = 3$ | |

Finally, we interpret the arrows to describe the behavior of the function, which we state under the horizontal arrows.

$f' > 0 \qquad f' = 0 \qquad f' > 0 \qquad f' = 0 \qquad f' < 0$

$x = 0 \qquad\qquad x = 3$

rel max
$(3, 7)$

neither
$(0, -20)$

The open intervals of increase are $(-\infty, 0)$ and $(0, 3)$, and the open interval of decrease is $(3, \infty)$.

**Using a graphing calculator,** we would choose a window such as $[-2, 5]$ by $[-30, 10]$ to include the points $(0, -20)$ and $(3, 7)$, and graph the function.

**By hand,** we would plot $(0, -20)$ and $(3, 7)$, and join them by a curve that goes, according to the arrows, "up-level-up-level-down."

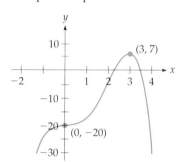

With a graphing calculator, an incomplete sign diagram may be enough to find an appropriate window.

### Graphing Rational Functions

Recall from page 52 that a rational function is a quotient of polynomials $\dfrac{p(x)}{q(x)}$.

On pages 84–85 we graphed a rational function (shown on the left) with its *vertical asymptote* (where the denominator is zero and near which the curve becomes arbitrarily high or low) and *horizontal asymptote* (which the curve approaches as $x \to \infty$ or $x \to -\infty$).

Why do vertical asymptotes occur where the denominator is zero? If the polynomial in the denominator is zero at $x = c$, by continuity it will be *near* zero for $x$-values *near* $c$, and reciprocals of numbers near zero are very large or very small (for example, $\frac{1}{.001} = 1000$ and $\frac{1}{-.001} = -1000$, as you may verify on a calculator). This leads to the following criterion for vertical asymptotes.

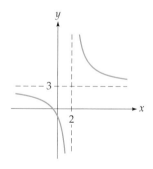

Graph of $f(x) = \dfrac{3x + 2}{x - 2}$

---

**Vertical Asymptotes**

A rational function $\dfrac{p(x)}{q(x)}$ has a **vertical asymptote** $x = c$ if $q(c) = 0$ but $p(c) \ne 0$.

A function has a *horizontal* asymptote if it has a limit as $x \to \infty$ or $x \to -\infty$.

## Horizontal Asymptotes

A function $f(x)$ has a **horizontal asymptote** $y = c$ if $\displaystyle\lim_{x \to \infty} f(x) = c$ or $\displaystyle\lim_{x \to -\infty} f(x) = c$.

Finding horizontal asymptotes by taking limits as $x$ approaches $\pm \infty$ involves using some particular limits, each of which comes from the fact that *the reciprocal of a large number is a small number* (for example, $\frac{1}{1000} = .001$):

$$\lim_{x \to \infty} \frac{1}{x} = 0, \quad \lim_{x \to \infty} \frac{1}{x^2} = 0, \quad \text{or more generally,} \quad \lim_{x \to \infty} \frac{1}{x^n} = 0$$

for any positive integer $n$. These results also hold if the 1 in the numerator is replaced by any other number, and the same results hold for limits as $x$ approaches *negative* infinity.

**EXAMPLE 3**

### GRAPHING A RATIONAL FUNCTION

Graph $f(x) = \dfrac{2x + 3}{x - 1}$.

*Solution*

Since this is a rational function, we first look for vertical asymptotes. The denominator is zero at $x = 1$, and the numerator is not, so the line $x = 1$ is a vertical asymptote, which we show on the graph on the left.

To find *horizontal* asymptotes, take the limit as $x$ approaches $\infty$ or $-\infty$. Dividing the numerator and denominator of the function by $x$ and simplifying makes it easy to find the limit:

$$\lim_{x \to \infty} \frac{2x + 3}{x - 1} = \lim_{x \to \infty} \frac{\frac{2x}{x} + \frac{3}{x}}{\frac{x}{x} - \frac{1}{x}} = \lim_{x \to \infty} \frac{2 + \frac{3}{x}}{1 - \frac{1}{x}} = \frac{2 + 0}{1 - 0} = \frac{2}{1} = 2$$

$\underbrace{\hspace{3cm}}$ Dividing numerator and denominator by $x$ and then simplifying

$\underbrace{\hspace{2cm}}$ Since $\frac{3}{x}$ and $\frac{1}{x}$ both approach 0

$\underbrace{\hspace{2cm}}$ Simplifying: the limit is 2

Since the limit is 2, $y = 2$ is a horizontal asymptote, as drawn on the left. Letting $x \to -\infty$ gives the same limit and so the same horizontal asymptote.

Having found the asymptotes, we take the derivative to find the slope of the function. The quotient rule gives:

$$f'(x) = \frac{(x - 1)(2) - (1)(2x + 3)}{(x - 1)^2} = \frac{2x - 2 - 2x - 3}{(x - 1)^2} = \frac{-5}{(x - 1)^2} \qquad \text{From } f(x) = \frac{2x + 3}{x - 1}$$

This derivative is undefined at $x = 1$, and elsewhere it is always negative since it is a negative number over a square. Therefore, all parts of the curve slope *downward*. The only way that the curve can slope downward everywhere and have the vertical and horizontal asymptotes we found is to have the graph shown on the left. Evaluating at $x = 0$ and plotting the $y$-intercept $(0, -3)$ also helps.

Graph of $f(x) = \dfrac{2x + 3}{x - 1}$

In the preceding example we divided numerator and denominator by $x$ before taking the limit. In other cases we may divide by a *higher* power of $x$, often the highest power that appears in the function. Doing so makes it easy to find the limit.

**EXAMPLE 4**

**GRAPHING A RATIONAL FUNCTION**

Graph $f(x) = \dfrac{3x^2}{x^2 - 4}$.

*Solution*

Factoring the denominator into $(x - 2)(x + 2)$ shows that there are two vertical asymptotes: $x = 2$ and $x = -2$, as shown on the left. To find *horizontal* asymptotes, we take the limit, first dividing numerator and denominator by $x^2$, the highest power of $x$ appearing in the function:

$$\lim_{x \to \infty} \frac{3x^2}{x^2 - 4} = \lim_{x \to \infty} \frac{\frac{3x^2}{x^2}}{\frac{x^2}{x^2} - \frac{4}{x^2}} = \lim_{x \to \infty} \frac{3}{1 - \frac{4}{x^2}} = \frac{3}{1 - 0} = 3 \qquad \text{The limit is 3}$$

$$\underbrace{\qquad\qquad\qquad}_{\substack{\text{Dividing by } x^2 \\ \text{and simplifying}}} \quad \underbrace{\qquad}_{\substack{\text{Since } \frac{4}{x^2} \\ \text{approaches } 0}}$$

Since the limit is 3 (and the limit as $x \to -\infty$ is also 3), $y = 3$ is a horizontal asymptote, as drawn on the left.

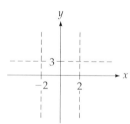

After finding the asymptotes, we take the derivative using the quotient rule:

$$f'(x) = \frac{(x^2 - 4)(6x) - (2x)(3x^2)}{(x^2 - 4)^2} = \frac{6x^3 - 24x - 6x^3}{(x^2 - 4)^2} = \frac{-24x}{(x^2 - 4)^2} \qquad \text{From } f(x) = \frac{3x^2}{x^2 - 4}$$

Since the derivative changes sign, we make a sign diagram, remembering that *the function is undefined at* $x = 2$ *and at* $x = -2$ (indicated by vertical dashed lines) and showing that the derivative is zero at $x = 0$:

Plotting the relative maximum point $(0, 0)$ and drawing the curve to have the asymptotes we found and the slopes shown in the sign diagram gives the curve on the left.

Graph of $f(x) = \dfrac{3x^2}{x^2 - 4}$

The rational functions in the last two examples had both vertical and horizontal asymptotes. However, a rational function may have just one type or no asymptotes at all. For example, the rational function $f(x) = \dfrac{x^2}{1} = x^2$ is in fact a parabola and so has no asymptotes.

**Practice Problem 2**

Below are four rational functions and below them four graphs numbered 1–4. Match the functions with the correct graphs. [*Hint:* Little or no calculation is necessary. Think of asymptotes and intercepts.]

$f(x) = \dfrac{1}{x^2 - 4}$    $g(x) = \dfrac{1}{x^2 + 4}$    $h(x) = \dfrac{x}{x^2 - 4}$    $i(x) = \dfrac{x^2}{x^2 + 4}$

This function has graph ___    This function has graph ___    This function has graph ___    This function has graph ___

1)

2)

3)

4)

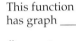 **Solutions below**

**Solutions to Practice Problems**

1. $f'(x) = 3x^2 - 12 = 3(x^2 - 4) = 3(x + 2)(x - 2)$

   $\text{CN} \begin{cases} x = -2 \\ x = 2 \end{cases}$

2. $f(x)$ is 3, $g(x)$ is 4, $h(x)$ is 1, and $i(x)$ is 2.

## 3.1 Section Summary

We have graphed two types of functions: polynomials and rational functions (quotients of polynomials). We used the fact that the derivative gives slope: where the derivative is *positive*, the curve slopes *up*; where the derivative is *negative*, the curve slopes *down*; and where it does one followed by the other, the curve has a relative maximum or minimum point. Such information is most easily displayed on a *sign diagram* for the derivative. We saw that rational functions may have asymptotes (lines that the graph approaches arbitrarily closely), while polynomials do not have asymptotes.

In curve sketching, when do you use $f$ and when do you use $f'$? Use the derivative $f'$ to find useful $x$-values, such as critical numbers, but always return to the *original function f* for the *y-coordinate* of a point to be plotted on the graph.

To sketch the graph of a function $f$:

■ Find the domain (by excluding any $x$-values at which the function is not defined).

■ For a *rational* function, find all asymptotes:

If the denominator is zero at an $x$-value $c$ and the numerator is not, then $x = c$ is a *vertical* asymptote.

If the function approaches a limit $c$ as $x \to \infty$ or as $x \to -\infty$, then $y = c$ is a *horizontal* asymptote.

■ Find the critical numbers (where $f'$ is zero or undefined but where $f$ is defined).

■ List all of these on a sign diagram for the derivative, indicating the behavior of $f'$ on each interval. Add arrows and relative extreme points.

■ Finally, sketch the curve. If you use a graphing calculator, the sign diagram will suggest an appropriate window. If you graph by hand, your sign diagram will show you the shape of the curve.

If there are no relative extreme points, choose a few $x$-values, including any of special interest. On a graphing calculator, use these $x$-values to determine an $x$-interval, and use TABLE or TRACE to find a $y$-interval; then graph the function on the resulting window. By hand, plot the points corresponding to the chosen $x$-values and draw an appropriate curve through the points using your sign diagram.

## 3.1 Exercises

**1–2.** For the functions graphed below:

**a.** Find the intervals on which the derivative is positive.

**b.** Find the intervals on which the derivative is negative.

**3.** Which of the numbers 1, 2, 3, 4, 5, and 6 are critical numbers of the function graphed below?

**1.**

**2.**

**4.** The first column in parts (a) through (d) shows the graphs of four functions, and the second column shows the graphs of their derivatives, but not necessarily in the same order. Write below each derivative the correct function from which it came.

**a.**

function $f_1$

(a) derivative of function: ____

**b.**

function $f_2$

(b) derivative of function: ____

**c.**

function $f_3$

(c) derivative of function: ____

**d.**

function $f_4$

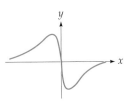

(d) derivative of function: ____

**5–16.** Find the critical numbers of each function.

**5.** $f(x) = x^3 - 48x$

**6.** $f(x) = x^3 - 27x$

**7.** $f(x) = x^3 - 6x^2 - 15x + 30$

**8.** $f(x) = x^3 + 6x^2 - 36x - 60$

**9.** $f(x) = x^4 + 4x^3 - 8x^2 + 1$

**10.** $f(x) = x^4 + 4x^3 - 20x^2 - 12$

**11.** $f(x) = (2x - 6)^4$

**12.** $f(x) = (x^2 + 6x - 7)^2$

**13.** $f(x) = 3x + 5$

**14.** $f(x) = 4x - 12$

**15.** $f(x) = x^3 + x^2 - x + 4$

**16.** $f(x) = x^3 - x^2 + 15$

**17–32.** Sketch the graph of each function "by hand" after making a sign diagram for the derivative and finding all open intervals of increase and decrease.

**17.** $f(x) = x^3 - 3x^2 - 9x + 10$

**18.** $f(x) = x^3 + 3x^2 - 9x - 11$

**19.** $f(x) = x^4 + 4x^3 - 8x^2 + 64$

**20.** $f(x) = x^4 - 4x^3 - 8x^2 + 64$

**21.** $f(x) = -x^4 + 4x^3 - 4x^2 + 1$

**22.** $f(x) = -x^4 - 4x^3 - 4x^2 + 1$

**23.** $f(x) = 3x^4 - 8x^3 + 6x^2$

**24.** $f(x) = 3x^4 + 8x^3 + 6x^2$

**25.** $f(x) = (x - 1)^6$

**26.** $f(x) = (x - 1)^5$

**27.** $f(x) = (x^2 - 4)^2$

**28.** $f(x) = (x^2 - 2x - 8)^2$

**29.** $f(x) = x^2(x - 4)^2$

**30.** $f(x) = x(x - 4)^3$

**31.** $f(x) = x^2(x - 5)^3$

**32.** $f(x) = x^3(x - 5)^2$

**33–62.** Sketch the graph of each rational function after making a sign diagram for the derivative and finding all relative extreme points and asymptotes.

**33.** $f(x) = \dfrac{6}{x + 3}$

**34.** $f(x) = \dfrac{4}{x - 2}$

**35.** $f(x) = \dfrac{3x + 6}{x - 2}$

**36.** $f(x) = \dfrac{2x - 6}{x + 3}$

**37.** $f(x) = \dfrac{12x - 24}{3x + 6}$

**38.** $f(x) = \dfrac{10x + 30}{2x - 6}$

**39.** $f(x) = \dfrac{8}{(x - 2)^2}$

**40.** $f(x) = \dfrac{18}{(x + 3)^2}$

**41.** $f(x) = \dfrac{12}{x^2 - 2x - 3}$

**42.** $f(x) = \dfrac{72}{x^2 - 2x - 8}$

**43.** $f(x) = \dfrac{8}{x^2 + 4}$    This curve is called the Witch of Agnesi

**44.** $f(x) = \dfrac{4x}{x^2 + 4}$    This curve is called Newton's serpentine

**45.** $f(x) = \dfrac{x^2}{x^2 + 1}$

**46.** $f(x) = \dfrac{x^2 - 1}{x^2 + 1}$

**47.** $f(x) = \dfrac{2x}{x^2 + 1}$

**48.** $f(x) = \dfrac{6x}{x^2 + 9}$

**49.** $f(x) = \dfrac{3}{x^2 - 1}$

**50.** $f(x) = \dfrac{9}{x^2 - 9}$

**51.** $f(x) = \dfrac{5x}{x^2 - 9}$

**52.** $f(x) = \dfrac{11x}{x^2 - 25}$

**53.** $f(x) = \dfrac{2x^2 - 4x + 6}{x^2 - 2x - 3}$

**54.** $f(x) = \dfrac{3x^2 + 6x + 3}{x^2 + 2x - 3}$

**55.** $f(x) = \dfrac{2x^2}{x^2 - 1}$

**56.** $f(x) = \dfrac{x^2 + 4}{x^2 - 4}$

**57.** $f(x) = \dfrac{16}{(x + 2)^3}$

**58.** $f(x) = \dfrac{27}{(x - 3)^3}$

**59.** $f(x) = \dfrac{4}{x^2 (x - 3)}$

**60.** $f(x) = \dfrac{4}{x (x - 3)^2}$

**61.** $f(x) = \dfrac{2x^2}{x^4 + 1}$

**62.** $f(x) = \dfrac{x^4 - 2x^2 + 1}{x^4 + 2}$

**63.** Derive the formula $x = \dfrac{-b}{2a}$ for the x-coordinate of the vertex of parabola $y = ax^2 + bx + c$. [*Hint:* The slope is zero at the vertex, so finding the vertex means finding the critical number.]

**64.** Derive the formula $x = -b$ for the x-coordinate of the vertex of parabola $y = a(x + b)^2 + c$. [*Hint:* The slope is zero at the vertex, so finding the vertex means finding the critical number.]

## Applied Exercises

**65. BIOMEDICAL: Bacterial Growth** A population of bacteria grows to size $p(x) = x^3 - 9x^2 + 24x + 10$ after $x$ hours (for $x \geq 0$). Graph this population curve (based on, if you wish, a calculator graph), showing the coordinates of the relative extreme points.

**66. BEHAVIORAL SCIENCE: Learning Curves** A *learning curve* is a function $L(x)$ that gives the amount of time that a person requires to learn $x$ pieces of information. Many learning curves take the form $L(x) = (x - a)^n + b$ (for $x \geq 0$), where $a$, $b$, and $n$ are constants. Graph the learning curve $L(x) = (x - 2)^3 + 8$ (based on, if you wish, a calculator graph), showing the coordinates of all corresponding points to critical numbers.

**67. GENERAL: Airplane Flight Path** A plane is to take off and reach a level cruising altitude of 5 miles after a horizontal distance of 100 miles, as shown in the diagram below. Find a polynomial flight path of the form $f(x) = ax^3 + bx^2 + cx + d$ by following steps i to iv to determine the constants $a$, $b$, $c$, and $d$.

**i.** Use the fact that the plane is on the ground at $x = 0$ [that is, $f(0) = 0$] to determine the value of $d$.

**ii.** Use the fact that the path is horizontal at $x = 0$ [that is, $f'(0) = 0$] to determine the value of $c$.

**iii.** Use the fact that at $x = 100$ the height is 5 and the path is horizontal to determine the values of $a$ and $b$. State the function $f(x)$ that you have determined.

 **iv.** Use a graphing calculator to graph your function on the window [0, 100] by [0, 6] to verify its shape.

 **68. GENERAL: Aspirin** Clinical studies have shown that the analgesic (pain-relieving) effect of aspirin is approximately $f(x) = \dfrac{100x^2}{x^2 + 0.02}$ where $f(x)$ is the percentage of pain relief from $x$ grams of aspirin.

**a.** Graph this "dose-response" curve for doses up to 1 gram, that is, on the window [0, 1] by [0, 100].

**b.** TRACE along the curve to see that the curve is very close to its maximum height of 100% or 1 by the time $x$ reaches 0.65. This means that there is very little added effect in going above 650 milligrams, the amount of aspirin in two regular tablets, notwithstanding the aspirin companies' promotion of "extra strength" tablets.

[*Note:* Aspirin's dose-response curve is extremely unusual in that it levels off quite early, and, even more unusual, aspirin's effect in protecting against heart attacks even *decreases* as the dosage $x$ increases above about 80 milligrams.]

*Source: Consumer Reports*

**69. GENERAL: Drug Interception** Suppose that the cost of a border patrol that intercepts $x$ percent of the illegal drugs crossing a state border is

$$C(x) = \frac{600}{100 - x} \quad \text{million dollars (for } x < 100\text{).}$$

a. Graph this function on [0, 100] by [0, 100].

b. Observe that the curve is at first rather flat, but then rises increasingly steeply as $x$ nears 100. Predict what the graph of the derivative would look like.

c. Check your prediction by defining $y_2$ to be the derivative of $y_1$ (using NDERIV) and graphing both $y_1$ and $y_2$.

**70. BIOMEDICAL: Heart Medication** A cardiac medication is injected into the arm of a patient, and $t$ minutes later the concentration in the heart is $f(t) = \dfrac{4t}{t^2 + 4}$ (milligrams per deciliter of blood). Graph this function on the interval [0, 6], showing the coordinates when the concentration is greatest.

**71. BUSINESS: Long-Run Average Cost** Suppose that a software company produces CDs (computer disks) at a cost of $3 each, and fixed costs are $50. The *cost function*, the total cost of producing $x$ disks, will then be $C(x) = 3x + 50$, and the *average cost per unit* will be the total cost divided by the number of units: $AC(x) = \dfrac{3x + 50}{x}$.

a. Show that $\lim_{x \to \infty} AC(x) = 3$, which is the *unit* or *marginal* cost.

b. Sketch the graph of $AC(x)$, showing the horizontal asymptote. [*Note:* Your graph should be an illustration of the general business principle for linear cost functions: *In the long run, average cost approaches marginal cost.*]

**72. BUSINESS: Long-Run Average Cost** Suppose that a company has a linear cost function (the total cost of producing $x$ units) $C(x) = ax + b$ for constants $a$ and $b$, where $a$ is the *unit* or *marginal* cost and $b$ is the *fixed* cost. Then the *average cost per unit* will be the total cost divided by the

number of units: $AC(x) = \dfrac{ax + b}{x}$. Show that $\lim_{x \to \infty} AC(x) = a$. [*Note:* Since $a$ is the marginal cost, you have proved the general business principle for linear cost functions: *In the long run, average cost approaches marginal cost.*]

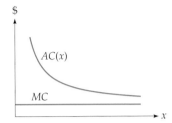

## Conceptual Exercises

**73.** True or False: A *critical number* is where the derivative is zero or undefined.

**74.** True or False: At a critical number the function must be *defined*.

**75.** True or False: A relative maximum point is the highest point on the entire curve.

**76.** True or False: A relative minimum point is the lowest point on the entire curve.

**77.** From the graph of the *derivative* shown below, find the open intervals of increase and of decrease for the *original function*.

**78.** From the graph of the *derivative* shown below, find the open intervals of increase and of decrease for the *original function*.

**79.** Explain why, for a (simplified) rational function, the function and its derivative will be undefined at exactly the same $x$-values.

**80.** We saw on page 182 that the absolute value function is defined at $x = 0$ but its derivative is not. Can a *rational* function have an $x$-value where the function is defined but the derivative is not?

**81.** True or False: If the derivative has the *same sign* immediately on either side of an $x$-value, the function has neither a maximum nor a minimum at that $x$-value.

**82.** True or False: If a function is defined and has a positive derivative for all values of $x$, then $f(a) < f(b)$  whenever  $a < b$.

**83.** Why does the criterion for  $\dfrac{p(x)}{q(x)}$  to have a vertical asymptote at $c$ require that  $q(c) = 0$  *but*  $p(c) \neq 0$ (see page 184)? Why not just  $q(c) = 0$?  Can you think of a rational function that does *not* have a vertical asymptote at a point where its denominator is zero? [*Hint:* Make both $p(c) = 0$  and  $q(c) = 0$.]

**84.** Can a rational function have different horizontal asymptotes as  $x \to \infty$  and as  $x \to -\infty$? [*Hint:* To have a horizontal asymptote other than the $x$-axis, the highest power of $x$ in the numerator and denominator must be the same, such as in  $f(x) = \dfrac{ax^2 + bx + c}{Ax^2 + Bx + C}$. What are the two limits? Can you do the same for higher powers?]

### Explorations and Excursions

The following problems extend and augment the material presented in the text.

*Average and Marginal Cost*

**85.** A company's cost function is
$$C(x) = x^2 + 2x + 4$$
dollars, where $x$ is the number of units.

**a.** Enter the cost function in $y_1$ on a graphing calculator.

**b.** Define $y_2$ to be the *marginal* cost function by defining $y_2$ to be the derivative of $y_1$ (using NDERIV).

**c.** Define $y_3$ to be the company's *average* cost function,
$$AC(x) = \frac{C(x)}{x},$$
by defining  $y_3 = \dfrac{y_1}{x}$.

**d.** Turn off the function $y_1$ so that it will not be graphed, but graph the marginal cost function $y_2$ and the average cost function $y_3$ on the window [0, 10] by [0, 10]. Observe that the marginal cost function pierces the average cost function at its minimum point (use TRACE to see which curve is which function).

**e.** To see that the final sentence of part (d) is true in general, change the coefficients in the cost function $C(x)$, or change the cost function to a cubic or some other function [so that $C(x)/x$ has a minimum]. Again turn off the cost function and graph the other two to see that the marginal cost function pierces the average cost function at its minimum.

**86.** Prove that the marginal cost function intersects the average cost function at the point where the average cost is minimized (as shown in the diagram below) by justifying each numbered step in the following series of equations.

At the $x$-value where the average cost function  $AC(x) = \dfrac{C(x)}{x}$  is minimized, we must have

$$0 \overset{①}{=} \frac{d}{dx} AC(x) \overset{②}{=} \frac{xC'(x) - C(x)}{x^2}$$

$$\overset{③}{=} \frac{1}{x}\left[ \frac{xC'(x) - C(x)}{x} \right]$$

$$\overset{④}{=} \frac{1}{x}\left[ C'(x) - \frac{C(x)}{x} \right] \overset{⑤}{=} \frac{1}{x}[MC(x) - AC(x)].$$

⑥ Finally, explain why the equality of the first and last expressions in the series of equations proves the result.

The marginal cost function pierces the average cost function at its minimum.

## 3.2   GRAPHING USING THE FIRST AND SECOND DERIVATIVES

### Introduction

In the previous section we used the first derivative to find the function's slope and relative extreme points and to draw its graph. In this section we will use the *second* derivative to find the **concavity** or **curl** of the curve, and to define the important concept of **inflection point.** The second derivative also gives us a very useful way to distinguish between maximum and minimum points of a curve.

### Concavity and Inflection Points

A curve that curls upward is said to be **concave up,** and a curve that curls downward is said to be **concave down.** A point where the concavity *changes* (from up to down or down to up) is called an **inflection point.**

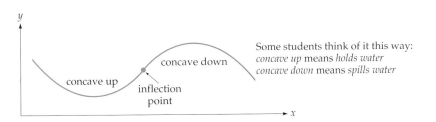

Some students think of it this way:
*concave up* means *holds water*
*concave down* means *spills water*

Concavity shows how a curve *curls* or *bends* away from straightness.

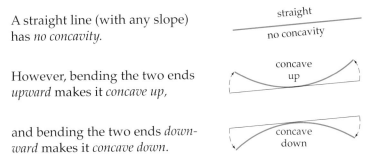

A straight line (with any slope) has *no concavity.*

However, bending the two ends *upward* makes it *concave up,*

and bending the two ends *downward* makes it *concave down.*

As these pictures show, a curve that is concave *up* lies *above* its tangent, while a curve that is concave *down* lies *below* its tangent (except, of course, at the point of tangency).

**Practice Problem 1**

For each of the following curves, label the parts that are concave up and the parts that are concave down. Then find all inflection points.

a.

b.

➤ **Solutions on page 201**

How can we use calculus to determine concavity? The key is the second derivative. The second derivative, being the derivative of the derivative, gives the *rate of change* of the slope, showing whether the slope is *increasing* or *decreasing*. That is, $f'' > 0$ means that the slope is increasing, and so the curve must be *concave up* (as in the diagram on the left below). Similarly, $f'' < 0$ means that the slope is decreasing, and so the curve must be *concave down* (as on the right below).

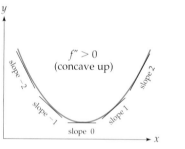

$f'' > 0$ means that the slope is
increasing, so $f$ is *concave up*.

$f'' < 0$ means that the slope is
decreasing, so $f$ is *concave down*.

Since an inflection point is where the concavity changes, the second derivative must be negative on one side and positive on the other. Therefore, *at an inflection point, $f''$ must be either zero or undefined*. All of this may be summarized as follows.

**Concavity and Inflection Points**

On an interval:

$f'' > 0$ means that $f$ is *concave up* (curls upward).

$f'' < 0$ means that $f$ is *concave down* (curls downward).

An *inflection point* is where the concavity *changes*
($f''$ must be zero or undefined).

## Graphing Calculator Exploration

*f* is concave *up*   *f* is concave *down*
*f″* is positive    *f″* is negative

**a.** Use a graphing calculator to graph $y_1 = \sqrt[3]{x}$ on the window $[-2, 2]$ by $[-2, 2]$. Observe where the curve is concave up and where it is concave down.

**b.** Use NDERIV to define $y_2$ to be the derivative of $y_1$, and $y_3$ to be the derivative of $y_2$. Graph $y_1$ and $y_3$ (but turn off $y_2$ so that it will not be graphed).

**c.** Verify that $y_3$ (the second derivative) is positive where $y_1$ is concave up, and negative where $y_1$ is concave down.

**d.** Now change $y_1$ to $y_1 = \dfrac{x^2 + 2}{x^2 + 1}$ and observe that where this curve is concave up or down agrees with where $y_3$ is positive or negative. According to $y_3$, how many inflection points does $y_1$ have? Can you see them on $y_1$?

To find inflection points, we make a sign diagram for the *second* derivative to show where the concavity changes (where $f''$ changes sign). An example will make the method clear.

---

**EXAMPLE 1**  **GRAPHING AND INTERPRETING A COMPANY'S ANNUAL PROFIT FUNCTION**

A company's annual profit after $x$ years is $f(x) = x^3 - 9x^2 + 24x$ million dollars (for $x \geq 0$). Graph this function, showing all relative extreme points and inflection points. Interpret the inflection points.

*Solution*

$$f'(x) = 3x^2 - 18x + 24 \qquad \text{Differentiating}$$
$$= 3(x^2 - 6x + 8) = 3(x - 2)(x - 4) \qquad \text{Factoring}$$

The critical numbers are $x = 2$ and $x = 4$, and the sign diagram for $f'$ (found in the usual way) is

| $f' > 0$ | $f' = 0$ | $f' < 0$ | $f' = 0$ | $f' > 0$ |
|---|---|---|---|---|
|  | $x = 2$ |  | $x = 4$ |  |

↗     rel max (2, 20)     ↘     rel min (4, 16)     ↗

To find the inflection points, we calculate the second derivative:

$$f''(x) = 6x - 18 = 6(x - 3)$$    Differentiating $f'(x) = 3x^2 - 18x + 24$

This is zero at $x = 3$, which we enter on a sign diagram for the *second* derivative.

$$\underline{\qquad\qquad\underset{x = 3}{\overset{f'' = 0}{\mid}}\qquad\qquad}$$

← Behavior of $f''$

← Where $f''$ is zero or undefined

We use test points to determine the sign of $f''(x) = 6(x - 3)$ on either side of 3, just as we did for the first derivative.

$$\underset{f'' < 0}{\qquad}\quad\overset{\llcorner f''(2) = 6(2-3) < 0}{\underset{\underset{x = 3}{\mid}}{f'' = 0}}\quad\overset{\llcorner f''(4) = 6(4-3) > 0}{\underset{f'' > 0}{\qquad}}$$

con dn                    con up

Concave down, concave up
(so concavity *does* change)

IP (3, 18)

IP means *inflection point*. The 18 comes from substituting $x = 3$ into $f(x) = x^3 - 9x^2 + 24x$

**Using a graphing calculator,** we would choose an $x$-interval such as [0, 6] (to include the $x$-values on the sign diagrams) and a $y$-interval such as [0, 30] (to include the origin and the $y$-values on the sign diagrams), and graph the function.

on [0, 6] by [0, 30]

**By hand,** we would plot the relative maximum (⌢), minimum (⌣), and inflection point and sketch the curve according to the sign diagrams, being sure to show the concavity changing at the inflection point.

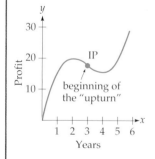

*Interpretation of the inflection point:* Observe what the graph shows—that the company's profit increased (up to year 2), then decreased (up to year 4), and then increased again. The inflection point at $x = 3$ is where the profit *first began to show signs of improvement.* It marks the end of the period of increasingly steep decline and the first sign of an "upturn," where a clever investor might begin to "buy in."

⚠️ **BE CAREFUL:** At an inflection point, the concavity (that is, the sign of $f''$) must *actually change.* For example, a second derivative sign diagram such as

$$\underset{\text{con dn}}{f'' < 0} \quad \underset{\underset{x = 3}{+}}{f'' = 0} \quad \underset{\text{con dn}}{f'' < 0}$$

sign of $f''$ (concavity) does *not* change

would mean that there is *not* an inflection point at $x = 3$, since the concavity is the same on both sides. For there to be an inflection point, the signs of $f''$ on the two sides *must be different.*

**Practice Problem 2**

For each curve, is there an inflection point? [*Hint:* Does the concavity change?]

**a.**

$y$

Is this an inflection point? (yes or no)

$x$

**b.**

$y$

Is this an inflection point? (yes or no)

$x$

➤ **Solutions on page 201**

## Inflection Points in the Real World

Inflection points occur in many everyday situations.*

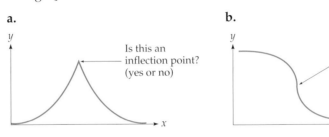

The function in Example 1, while constructed for ease of calculation, is essentially the graph of net income for AT&T over recent years, shown on the left. Notice that the inflection point represents the first sign of the upturn in AT&T's net income, which occurred in 2003.
*Source:* Standard & Poor's

The graph on the left shows the annual sales of portable MP3 players. The inflection point, occurring between 2004 and 2005, marks the end of the period of increasingly rapid sales growth and first sign of approaching market saturation. This is when savvy manufacturers might begin to curtail new investment in MP3 production.
*Source: Consumer USA 2008*

The graph on the left shows the annual sales of analog cameras a few years after the introduction of digital cameras. The inflection point, occurring in 2004, marks the end of the increasingly steep sales decline and the first sign of steadying but lower sales.
*Source:* Euromonitor

*For an interesting application of concavity to economics, see Harry M. Markowitz, "The Utility of Wealth," *Journal of Applied Political Economy* **60**(2). In this article, "concave" means "concave up" and "convex" means "concave down." Markowitz won the Nobel Memorial Prize in Economics in 1990.

Distinguish carefully between slope and concavity: *slope* measures *steepness*, whereas *concavity* measures *curl*. All combinations of slope and concavity are possible. A graph may be

*Increasing* and concave *up*
($f' > 0, f'' > 0$), such as

*Increasing* and concave *down*
($f' > 0, f'' < 0$), such as

*Decreasing* and concave *up*
($f' < 0, f'' > 0$), such as

*Decreasing* and concave *down*
($f' < 0, f'' < 0$), such as

The four quarters of a circle illustrate all four possibilities, as shown at the left.

## EXAMPLE 2

### GRAPHING A FRACTIONAL POWER FUNCTION

Graph $f(x) = 18x^{1/3}$.

*Solution*

The derivative is

$$f'(x) = 6x^{-2/3} = \frac{6}{\sqrt[3]{x^2}}$$

*Undefined* at $x = 0$
(zero denominator)

The sign diagram for $f'$ is

$$\underline{\quad f' > 0 \quad\quad f' \text{ und} \quad\quad f' > 0 \quad}$$
$$x = 0$$
$$\text{neither}$$
$$(0, 0)$$

$f'$ is undefined at $x = 0$ and positive on either side (using test points)

The *second* derivative is

$$f''(x) = -4x^{-5/3} = \frac{-4}{\sqrt[3]{x^5}}$$

*Also undefined* at $x = 0$

The sign diagram for $f''$ is

$$\underline{\quad f'' > 0 \quad\quad f'' \text{ und} \quad\quad f'' < 0 \quad}$$
$$\text{con up} \quad\quad x = 0 \quad\quad \text{con dn}$$
$$\text{IP } (0, 0)$$

Concavity is different on either side of $x = 0$ (using test points), so there *is* an inflection point at $x = 0$

Based on this information, we may graph the function with a graphing calculator or by hand:

**Using a graphing calculator,** we experiment with viewing windows centered at the inflection point $(0, 0)$ until we find one that shows the curve effectively. The graph on $[-2, 2]$ by $[-20, 20]$ is shown below. Many other windows would be just as good.

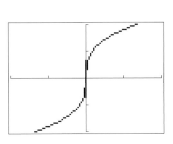

**By hand,** we use the sign diagrams to draw the curve to the *left* of $x = 0$ as *increasing* and concave *up*, and to the right of $x = 0$ as increasing and concave *down*, with the two parts meeting at the origin. The scale comes from calculating the points $(1, 18)$ and $(-1, -18)$.

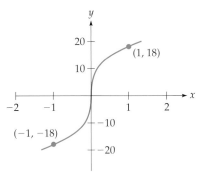

The fact that the derivative is *undefined* at $x = 0$ is shown in the graph by the *vertical tangent* at the origin (since the slope of a vertical line is undefined). This function is the stimulus-response curve for brightness of light (for $x \geq 0$; see page 177), and the vertical tangent indicates the disproportionately large effect of small increases in dim light.

**EXAMPLE 3**

A useless graph of
$f(x) = 36 \sqrt[3]{(x - 1)^2}$
on $[-10, 10]$ by $[-10, 10]$

**GRAPHING USING A GRAPHING CALCULATOR**

Use a graphing calculator to graph $f(x) = 36\sqrt[3]{(x - 1)^2}$.

*Solution*

Using a standard window $[-10, 10]$ by $[-10, 10]$ gives the graph on the left, which is useless.

Instead, we begin by differentiating $f(x) = 36(x - 1)^{2/3}$ and making its sign diagram.

$$f'(x) = 24(x - 1)^{-1/3} = \frac{24}{\sqrt[3]{x - 1}} \qquad \text{Undefined at } x = 1$$

$$\frac{f' < 0 \qquad f' \text{ und} \qquad f' > 0}{\searrow \quad \substack{x = 1 \\ \text{rel min} \\ (1, 0)} \quad \nearrow}$$

Using test points on either side of $x = 1$

The $y$-coordinate in $(1, 0)$ comes from evaluating the original function at $x = 1$

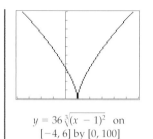

$y = 36\sqrt[3]{(x-1)^2}$ on
$[-4, 6]$ by $[0, 100]$

From the sign diagram we choose an $x$-interval centered at the critical number $x = 1$, such as $[-4, 6]$. After some experimenting, we choose the $y$-interval $[0, 100]$ (beginning at $y = 0$ because the minimum point has $y$-coordinate 0) so that the curve fills the screen. A sharp point on a graph such as the one at $x = 1$ is called a *cusp*, where the function is not differentiable.

## Second-Derivative Test

Determining whether a twice-differentiable function has a relative maximum or minimum at a critical number is merely a question of concavity: concave *up* means a relative *minimum*, and concave *down* means a relative *maximum*.

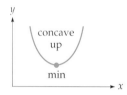

Concave *up* at a critical
number: relative *minimum*.

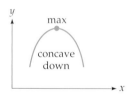

Concave *down* at a critical
number: relative *maximum*.

Since the second derivative determines concavity, we have the following *second-derivative test*, which will be very useful in the next two sections.

**Second-Derivative Test for Relative Extreme Points**

If $x = c$ is a critical number of $f$ at which $f''$ is defined, then

$$f''(c) > 0 \quad \text{means that } f \text{ has a relative } \textit{minimum} \text{ at } x = c.$$

$$f''(c) < 0 \quad \text{means that } f \text{ has a relative } \textit{maximum} \text{ at } x = c.$$

To use the second-derivative test, first find all critical numbers, substitute each into the second derivative (if possible), and determine the sign of the result: a *positive* result means a *minimum* at the critical number, and a *negative* result means a *maximum*. (If the second derivative is zero, then the test is inconclusive, and you should use the first-derivative test (page 182) or make a sign diagram for $f'$.)

---

**EXAMPLE 4**

### USING THE SECOND-DERIVATIVE TEST

Use the second-derivative test to find all relative extreme points of $f(x) = x^3 - 9x^2 + 24x$.

*Solution*

$$f'(x) = 3x^2 - 18x + 24 \qquad \text{The derivative}$$
$$= 3(x^2 - 6x + 8) = 3(x - 2)(x - 4) \qquad \text{Factoring}$$
$$\text{CN} \begin{cases} x = 2 \\ x = 4 \end{cases} \qquad \text{Critical numbers}$$

We substitute each critical number into $f''(x) = 6x - 18$.

At  $x = 2$:   $f''(2) = 6 \cdot 2 - 18 = -6$   (negative)   $\begin{array}{l} f''(x) = 6x - 18 \\ \text{at } x = 2 \end{array}$

Therefore, $f$ has a relative *maximum* at  $x = 2$.

At  $x = 4$:   $f''(4) = 6 \cdot 4 - 18 = 6$   (positive)   $\begin{array}{l} f''(x) = 6x - 18 \\ \text{at } x = 4 \end{array}$

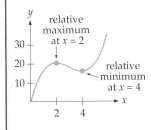

Therefore, $f$ has a relative *minimum* at  $x = 4$.
The relative maximum at  $x = 2$  and the relative minimum at  $x = 4$  are exactly what we found before when we graphed this function on page 196 or as shown on the left.

⚠ **BE CAREFUL:**  The second-derivative test tells us *nothing* if the second derivative is zero at a critical number. This is shown by the three functions below. Each function has a critical number  $x = 0$  at which $f''$ is zero (as you may check), but one has a maximum, one has a minimum, and one has neither.

Three functions showing that a critical number where $f'' = 0$
may be a maximum, a minimum, or neither.

## Solutions to Practice Problems

**1. a.**

**b.**

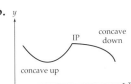

**2. a.**  No (concave up on both sides)

**b.**  Yes (concave down then concave up)

## 3.2 Section Summary

The main developments of this section were the use of the second derivative to determine the *concavity* or *curl* of a function, and finding and interpreting *inflection points*. On an interval:

$$f'' > 0 \quad \text{means that } f \text{ is concave } up.$$

$$f'' < 0 \quad \text{means that } f \text{ is concave } down.$$

To locate inflection points (points at which the concavity changes), we find where the second derivative is zero or undefined, and then make a sign diagram for $f''$ to see whether the concavity *actually changes*.

To graph a function, we use the *first*-derivative sign diagram to find *slope* and *relative extreme points*, and the *second*-derivative sign diagram to find *concavity* and *inflection points*. Then we graph the function, either by hand or on a graphing calculator (using the relative extreme points and inflection points to determine the window).

The following steps may be useful when graphing a function by hand.

### Curve Sketching

1. Find the domain.
2. For a rational function, find and graph any vertical or horizontal asymptotes.
3. Find the first derivative and all $x$-values at which it is zero or undefined.
4. Make a sign diagram for the first derivative, showing "max," "min," and "neither" points (with $y$-coordinates found from $f$).
5. Find the second derivative and all $x$-values at which it is zero or undefined. (This and the following step may not be necessary for rational functions.)
6. Make a sign diagram for the second derivative, showing inflection points (where the concavity *changes*).
7. Based on the sign diagrams, sketch the graph, labeling all maximum, minimum, and inflection points as well as any asymptotes.

The *second-derivative test* shows whether a function has a relative maximum or minimum at a critical value (provided that $f''$ is defined and not zero):

$$f'' > 0 \quad \text{means a relative minimum.}$$

$$f'' < 0 \quad \text{means a relative maximum.}$$

The second-derivative test should be thought of as a simple application of concavity.

## 3.2   Exercises

**1–6.**  For each graph, which of the numbered points are inflection points?

**1.**

**2.**

**3.**

**4.**

**5.**

**6.**

**7–30.**  For each function:
**a.**  Make a sign diagram for the first derivative.
**b.**  Make a sign diagram for the second derivative.
**c.**  Sketch the graph by hand, showing all relative extreme points and inflection points.

**7.**  $f(x) = x^3 + 3x^2 - 9x + 5$

**8.**  $f(x) = x^3 - 3x^2 - 9x + 7$

**9.**  $f(x) = x^3 - 3x^2 + 3x + 4$

**10.**  $f(x) = x^3 + 3x^2 + 3x + 6$

**11.**  $f(x) = x^4 - 8x^3 + 18x^2 + 2$

**12.**  $f(x) = x^4 + 8x^3 + 18x^2 + 8$

**13.**  $f(x) = x^4 + 4x^3 + 15$

**14.**  $f(x) = 2x^4 - 8x^3 + 30$

**15.**  $f(x) = 5x^4 - x^5$       **16.**  $f(x) = (x - 2)^3 + 2$

**17.**  $f(x) = (2x + 4)^5$       **18.**  $f(x) = (3x - 6)^6 + 1$

**19.**  $f(x) = x(x - 3)^2$       **20.**  $f(x) = x^3(x - 4)$

**21.**  $f(x) = x^{3/5}$       **22.**  $f(x) = x^{1/5}$

**23.**  $f(x) = \sqrt[5]{x^4} + 2$       **24.**  $f(x) = \sqrt[5]{x^2} - 1$

**25.**  $f(x) = \sqrt[4]{x^3}$       **26.**  $f(x) = \sqrt{x^5}$

**27.**  $f(x) = \sqrt[3]{(x - 1)^2}$       **28.**  $f(x) = \sqrt[5]{x + 2} + 3$

**29.**  $f(x) = \dfrac{x}{x^2 - 1}$       **30.**  $f(x) = \dfrac{54(x^2 - 1)}{x^2 + 27}$

  **31–40.**  Graph each function using a graphing calculator by first making a sign diagram for just the first derivative. Make a sketch from the screen, showing the coordinates of all relative extreme points and inflection points. Graphs may vary depending on the window chosen.

**31.**  $f(x) = x^{1/2}$       **32.**  $f(x) = x^{3/2}$

**33.**  $f(x) = x^{-1/2}$       **34.**  $f(x) = x^{-3/2}$

**35.**  $f(x) = 9x^{2/3} - 6x$       **36.**  $f(x) = 30x^{1/3} - 10x$

**37.**  $f(x) = 8x - 10x^{4/5}$       **38.**  $f(x) = 6x - 10x^{3/5}$

**39.**  $f(x) = 3x^{2/3} - x^2$       **40.**  $f(x) = 3x^{4/3} - 2x^2$

**41–46.**  For each function, find all critical numbers and then use the second-derivative test to determine whether the function has a relative maximum or minimum at each critical number.

**41.**  $f(x) = x^3 - 6x^2 + 9x - 2$

**42.**  $f(x) = x^3 - 12x + 4$

**43.**  $f(x) = x^4 - 4x^3 + 4x^2 + 1$

**44.**  $f(x) = x^4 - 2x^2 + 1$

**45.**  $f(x) = x + \dfrac{9}{x}$

**46.**  $f(x) = \dfrac{x}{4} + \dfrac{1}{x}$

  **47–48.  FINDING INFLECTION POINTS**  Use a graphing calculator to estimate the $x$-coordinates of the inflection points of each function, rounding your answers to two decimal places. [*Hint:* Graph the second derivative, either calculating it directly or using NDERIV twice, and see where it crosses the $x$-axis.]

**47.**  $f(x) = x^5 - 2x^3 + 3x + 4$

**48.**  $f(x) = x^5 - 3x^3 + 6x + 2$

**49–56.** Sketch the graph of a function $f(x)$ that satisfies the stated conditions. Mark any inflection points by writing *IP* on your graph. [*Note:* There is more than one possible answer.]

**49. a.** $f$ is continuous and differentiable everywhere.
  **b.** $f(0) = 3$
  **c.** $f'(x) > 0$ on $(-\infty, -4)$ and $(0, \infty)$
  **d.** $f'(x) < 0$ on $(-4, 0)$
  **e.** $f''(x) < 0$ on $(-\infty, -2)$
  **f.** $f''(x) > 0$ on $(-2, \infty)$

**50. a.** $f$ is continuous and differentiable everywhere.
  **b.** $f(0) = 4$
  **c.** $f'(x) < 0$ on $(-\infty, -1)$ and $(3, \infty)$
  **d.** $f'(x) > 0$ on $(-1, 3)$
  **e.** $f''(x) > 0$ on $(-\infty, 1)$
  **f.** $f''(x) < 0$ on $(1, \infty)$

**51. a.** $f$ is continuous and differentiable everywhere except at $x = 2$, where it is undefined.
  **b.** $f(0) = -1$
  **c.** Horizontal asymptote $y = 1$ and vertical asymptote $x = 2$
  **d.** $f'(x) < 0$ wherever $f'$ is defined

**52. a.** $f$ is continuous and differentiable everywhere except at $x = -5$ and $x = 5$, where it is undefined.
  **b.** $f(0) = -1$
  **c.** Horizontal asymptote $y = 2$ and vertical asymptotes $x = -5$ and $x = 5$

  **d.** $f''(x) > 0$ on $(-\infty, -5)$ and $(5, \infty)$
  **e.** $f''(x) < 0$ on $(-5, 5)$

**53. a.** $f$ is continuous everywhere and differentiable everywhere except at $x = 0$.
  **b.** $f(0) = 3$
  **c.** $f'(x) > 0$ on $(-\infty, -2)$ and $(0, 2)$
  **d.** $f'(x) < 0$ on $(-2, 0)$ and $(2, \infty)$
  **e.** $f''(x) < 0$ on $(-\infty, 0)$ and $(0, \infty)$

**54. a.** $f$ is continuous everywhere and differentiable everywhere except at $x = 0$.
  **b.** $f(0) = 5$
  **c.** $f'(x) < 0$ on $(-\infty, -3)$ and $(0, 3)$
  **d.** $f'(x) > 0$ on $(-3, 0)$ and $(3, \infty)$
  **e.** $f''(x) > 0$ on $(-\infty, 0)$ and $(0, \infty)$

**55. a.** $f$ is continuous and differentiable everywhere.
  **b.** $f(0) = 6$
  **c.** $f'(x) < 0$ on $(-\infty, -6)$ and $(0, 6)$
  **d.** $f'(x) > 0$ on $(-6, 0)$ and $(6, \infty)$
  **e.** $f''(x) > 0$ on $(-\infty, -3)$ and $(3, \infty)$
  **f.** $f''(x) < 0$ on $(-3, 3)$

**56. a.** $f$ is continuous and differentiable everywhere.
  **b.** $f(0) = 2$
  **c.** $f'(x) > 0$ on $(-\infty, -8)$ and $(0, 8)$
  **d.** $f'(x) < 0$ on $(-8, 0)$ and $(8, \infty)$
  **e.** $f''(x) < 0$ on $(-\infty, -4)$ and $(4, \infty)$
  **f.** $f''(x) > 0$ on $(-4, 4)$

## Applied Exercises

**57.** BUSINESS: Revenue  A company's annual revenue after $x$ years is

$$f(x) = x^3 - 9x^2 + 15x + 25$$

thousand dollars (for $x \geq 0$).
  **a.** Make sign diagrams for the first and second derivatives.
  **b.** Sketch the graph of the revenue function, showing all relative extreme points and inflection points.
  **c.** Give an interpretation of the inflection point.

**58.** BUSINESS: Sales  A company's weekly sales (in thousands) after $x$ weeks are given by
$f(x) = -x^4 + 4x^3 + 70$ (for $0 \leq x \leq 3$).
  **a.** Make sign diagrams for the first and second derivatives.

  **b.** Sketch the graph of the sales function, showing all relative extreme points and inflection points.
  **c.** Give an interpretation of the positive inflection point.

**59.** GENERAL: Temperature  The temperature in a refining tower is $f(x) = x^4 - 4x^3 + 112$ degrees Fahrenheit after $x$ hours (for $0 \leq x \leq 5$).
  **a.** Make sign diagrams for the first and second derivatives.
  **b.** Sketch the graph of the temperature function, showing all relative extreme points and inflection points.
  **c.** Give an interpretation of the positive inflection point.

**60.** BIOMEDICAL: Dosage Curve  The dose-response curve for $x$ grams of a drug is
$f(x) = 8(x - 1)^3 + 8$ (for $x \geq 0$).

a. Make sign diagrams for the first and second derivatives.

b. Sketch the graph of the response function, showing all relative extreme points and inflection points.

**61. PSYCHOLOGY: Stimulus and Response**
Sketch the graph of the brightness response curve $f(x) = x^{2/5}$ for $x \geq 0$, showing all relative extreme points and inflection points.

*Source: S. S. Stevens, On the Psychophysical Law*

**62. PSYCHOLOGY: Stimulus and Response**
Sketch the graph of the loudness response curve $f(x) = x^{4/5}$ for $x \geq 0$, showing all relative extreme points and inflection points.

*Source: S. S. Stevens, On the Psychophysical Law*

**63–64. SOCIOLOGY: Status** Sociologists have estimated how a person's "status" in society (as perceived by others) depends on the person's income and education level. One estimate is that status $S$ depends on income $i$ according to the formula $S(i) = 16\sqrt{i}$ (for $i \geq 0$), and that status depends upon education level $e$ according to the formula $S(e) = \frac{1}{4}e^2$ (for $e \geq 0$).

*Source: Social Forces 50*

**63. a.** Sketch the graph of the function $S(i)$ above.
 **b.** Is the curve concave up or down? What does this signify about the rate at which status increases at higher income levels?

**64. a.** Sketch the graph of the function $S(e)$ above.
 **b.** Is the curve concave up or down? What does this signify about the rate at which status increases at higher education levels?

**65. BUSINESS: Income** AT&T's net income is approximated by the function $I(x) = \frac{1}{3}x^3 - 6x^2 + 27x + 42$, in hundreds of millions of dollars, where $x$ stands for the number of *half years since 2000* (so that, for example, $x = 4$ would correspond to 2002).

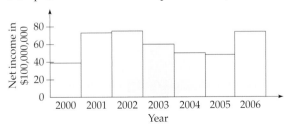

a. Make sign diagrams for the first and second derivatives.

b. Sketch the graph of $I(x)$, showing all relative extreme points and inflection points.

c. Interpret the meaning of the inflection point and determine the year in which it occurred.

*Source: Standard & Poor's*

**66. BUSINESS: Income** Microsoft's net income is approximated by the function $I(x) = -2x^3 + 18x^2 - 30x + 90$, in hundreds of millions of dollars, where $x$ stands for the number of *half-years since 2003* (so that, for example, $x = 4$ would correspond to 2005).

a. Make sign diagrams for the first and second derivatives.

b. Sketch the graph of $I(x)$, showing all relative extreme points and inflection points.

c. Interpret the meaning of the inflection point and determine the year in which it occurred.

*Source: Standard & Poor's*

**67. GENERAL: Airplane Flight Path** In Exercise 67 on page 190 it was found that the flight path that satisfies the conditions in the diagram shown below was $y = -0.00001x^3 + 0.0015x^2$. Find the inflection point of this path, and explain why it represents the point of steepest ascent of the airplane.

**68. BIOMEDICAL: Dose-Response Curves** The relationship between the dosage, $x$, of a drug and the resulting change in body temperature is given by $f(x) = x^2(3 - x)$ for $0 \leq x \leq 3$. Make sign diagrams for the first and second derivatives and sketch this dose-response curve, showing all relative extreme points and inflection points.

## Conceptual Exercises

**69.** For which values of the constant $a$ is the function $f(x) = ax^2$ concave *up*? For which value of $a$ is it concave *down*?

**70.** True or False: If the graph of $f(x)$ is concave up, then the graph of $-f(x)$ will be concave down.

**71.** True or False: Inflection points are simply points where the second derivative equals zero.

**72.** True or False: A polynomial of degree $n$ can have at most $n - 2$ inflection points.

73. From the graph of the *second derivative* shown below, is the point at $x = 1$ an inflection point of the original function?

74. From the graph of the *second derivative* shown below, is the point at $x = 1$ an inflection point of the original function?

75. A company president is looking at a graph of her company's daily sales during the first quarter of the year. On January 15 sales hit an all-time low, and then began to rise; on February 15 there was an inflection point (the only inflection point on the graph); on March 15 the sales hit an all-time high and then began to decline. What was the significance of the inflection point? (Assume that the graph has no straight segments.)

76. An investor is looking at a graph of the value of his portfolio for the first quarter of the year. On January 10 the value was at an all-time high and then began to decline; on February 10 there was an inflection point (the only inflection point on the graph); on March 10 the value was at its lowest point and then began to rise. What was the significance of the inflection point? (Assume that the graph has no straight segments.)

77. True or False: If a graph is concave *up* before an inflection point and concave *down* after it, then the curve has its greatest slope *at* the inflection point.

78. True or False: If a graph is concave *down* before an inflection point and concave *up* after it, then the curve has its smallest slope *at* the inflection point.

### Explorations and Excursions

The following problems extend and augment the material presented in the text.

79. CONCAVITY OF A PARABOLA Show that the quadratic function $f(x) = ax^2 + bx + c$ is concave up if $a > 0$ and is concave down if $a < 0$. Therefore, the rule that a parabola opens up if $a > 0$ and down if $a < 0$ is merely an application of concavity. [*Hint:* Find the second derivative.]

80. INFLECTION POINT OF A CUBIC Show that the general "cubic" (third degree) function $f(x) = ax^3 + bx^2 + cx + d$ (with $a \neq 0$) has an inflection point at $x = \dfrac{-b}{3a}$.

81. INFLECTION POINTS Explain why, at an inflection point, a curve must cross its tangent line (assuming that the tangent line exists).

82. INFLECTION POINTS For a twice-differentiable function, explain why the slope must have a relative maximum or minimum value at an inflection point. [*Hint:* Use the fact that the concavity changes at an inflection point, and then interpret concavity in terms of increasing and decreasing slope.]

## 3.3 OPTIMIZATION

### Introduction

Many problems consist of **optimizing** a function—that is, finding its maximum or minimum value. For example, you might want to maximize your profit, or to minimize the time required to do a task. If you could express your happiness as a function, you would want to maximize it.* One of the

*Expressing happiness in numbers has an honorable past: Plato (*Republic* IX, 587) calculated that a king is exactly 729 (= $3^6$) times as happy as a tyrant.

principal uses of calculus is that it provides a very general technique for optimizing functions.

We will concentrate on *applications* of optimization. Accordingly, we will optimize continuous functions that are defined on closed intervals, or functions that have only one critical number in their domain. Most applications fall into these two categories, and the wide range of examples and exercises in this and the following sections will demonstrate the power of these techniques.

## Absolute Extreme Values

The **absolute maximum value** of a function is the *largest* value of the function on its domain. Similarly, the **absolute minimum value** of a function is the *smallest* value of the function on its domain. An **absolute extreme value** is a value that is either the absolute maximum or the absolute minimum value of the function. (This use of the word "absolute" has nothing to do with its use in the *absolute value function* defined on page 56.) The maximum and minimum values of the function correspond to the highest and lowest points on its graph.

For a given function, both absolute extreme values may exist, or one or both may fail to exist, as the following graphs show.

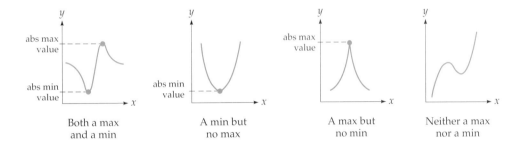

| Both a max and a min | A min but no max | A max but no min | Neither a max nor a min |

When will both extreme values exist? For a *continuous* function (one whose graph is a single unbroken curve) defined on a *closed* interval (that is, including both endpoints), the absolute maximum and minimum values are *guaranteed* to exist. We know from graphing functions that maximum and minimum values can occur only at critical numbers, unless they occur at endpoints, where the curve is "cut off" from rising or falling further. To summarize:

---

**Optimizing Continuous Functions on Closed Intervals**

A continuous function $f$ on a closed interval $[a, b]$ has both an *absolute maximum value* and an *absolute minimum value*. To find them:

**1.** Find all critical numbers of $f$ in $[a, b]$.

**2.** Evaluate $f$ at the critical numbers and at the endpoints $a$ and $b$.

The largest and smallest values found in step 2 will be the absolute maximum and minimum values of $f$ on $[a, b]$.

Simply stated, to find the absolute extreme values, we need to consider only *critical numbers and endpoints.*

---

**EXAMPLE 1**

**OPTIMIZING A CONTINUOUS FUNCTION ON A CLOSED INTERVAL**

Find the absolute extreme values of $f(x) = x^3 - 9x^2 + 15x$ on $[0, 3]$.

*Solution*

The function is continuous (it is a polynomial) and the interval is closed, so both extreme values exist. First we find the critical numbers.

$$f'(x) = 3x^2 - 18x + 15 \qquad \text{The derivative}$$

$$= 3(x^2 - 6x + 5) = 3(x - 1)(x - 5) \qquad \text{Factoring}$$

$$\text{CN} \begin{cases} x = 1 \\ x = 5 \end{cases} \quad \leftarrow \text{Not in the given domain} \\ \phantom{x = 5 \quad \leftarrow} [0, 3], \text{ so we eliminate it}$$

We evaluate the original function $f(x) = x^3 - 9x^2 + 15x$ at the remaining critical numbers and at the endpoints (EP).

$$\text{CN: } x = 1 \quad f(1) = 1 - 9 + 15 \qquad\qquad = 7 \quad \leftarrow \text{Largest function value}$$

$$\text{EP} \begin{cases} x = 0 \quad f(0) = 0 - 0 + 0 \qquad\qquad = 0 \\ x = 3 \quad f(3) = 27 - 9\cdot 9 + 15 \cdot 3 = -9 \quad \leftarrow \text{Smallest function value} \end{cases}$$

The largest (7) and the smallest ($-9$) of the resulting values of the function are the absolute extreme values of $f$ on $[0, 3]$:

Maximum value of $f$ is $\ 7$ (occurring at $\ x = 1$).

Minimum value of $f$ is $-9$ (occurring at $\ x = 3$).

The absolute maximum and minimum values may be seen in the graph on the left, but it is important to realize that we found them without drawing the graph.

---

For this function, one extreme value occurred at a critical number ($x = 1$), and the other at an endpoint ($x = 3$). In other problems, both extreme values might occur at critical numbers or both might occur at endpoints.

Notice how calculus helped in this example. The absolute extreme values could have occurred at *any* x-values in $[0, 3]$. Calculus reduced this infinite list of possibilities (*all* numbers between 0 and 3, not just integers) to a mere three numbers (0, 1, and 3). We then had only to "test" these numbers by substituting them into the function to find which numbers made $f$ largest and smallest.

 **BE CAREFUL:** The derivative helped us to find the x-values where the extremes could occur, but the actual extreme values (7 and $-9$ in the above example) come from evaluating the *original function.*

### Second-Derivative Test for Absolute Extreme Values

Recall from page 200 that the second-derivative test is a way to determine whether a function had a *maximum* or *minimum* at a particular critical number and amounts simply to checking the sign of the second derivative at the critical number.

---

**Second-Derivative Test**

If $x = c$ is a critical number at which $f''$ is defined, then

$$f''(c) > 0 \quad \text{means that } f \text{ has a relative } minimum \text{ at } x = c.$$

$$f''(c) < 0 \quad \text{means that } f \text{ has a relative } maximum \text{ at } x = c.$$

---

Earlier we used the second-derivative test to find *relative* maximum and minimum values. However, if a continuous function has only one critical number in its domain, then the second-derivative test can be used to find *absolute* extreme values (since without a second critical number the function must continue to increase or decrease away from the relative extreme point). Some students refer to this as "the only critical point in town" test.

### Applications of Optimization

If a timber forest is allowed to grow for $t$ years, the value of the timber increases in proportion to the square root of $t$, while maintenance costs are proportional to $t$. Therefore, the value of the forest after $t$ years is of the form

$$V(t) = a\sqrt{t} - bt \qquad \text{\small \textit{a} and \textit{b} are constants}$$

**EXAMPLE 2**

**OPTIMIZING THE VALUE OF A TIMBER FOREST**

The value of a timber forest after $t$ years is $V(t) = 96\sqrt{t} - 6t$ thousand dollars (for $t > 0$). Find when its value is maximized.

*Solution*

$$V(t) = 96t^{1/2} - 6t \qquad \text{\small \textit{V(t) in exponential form}}$$

$$V'(t) = 48t^{-1/2} - 6 = 0 \qquad \text{\small Setting the derivative equal to zero}$$

$$48t^{-1/2} = 6 \qquad \text{\small Adding 6 to each side of } 48t^{-1/2} - 6 = 0$$

$$t^{-1/2} = \frac{6}{48} = \frac{1}{8} \qquad \text{\small Dividing by 48}$$

$$\frac{1}{\sqrt{t}} = \frac{1}{8} \qquad \text{\small Expressing } t^{-1/2} \text{ in radical form}$$

$$\sqrt{t} = 8 \qquad \text{\small Inverting both sides}$$

$$t = 64 \qquad \text{\small Squaring both sides}$$

Since there is a *single* critical number ($t = 0$, which makes the derivative undefined, is not in the domain), we may use the second-derivative test. The second derivative is

$$V''(t) = -24t^{-3/2} = -\frac{24}{\sqrt{t^3}} \qquad \text{Differentiating } V'(t) = 48t^{-1/2} - 6$$

At $t = 64$ this is clearly negative, so by the second-derivative test $V(t)$ is *maximized* at $t = 64$. Finally, we state the answer clearly:

The value of the forest is maximized after 64 years. The maximum value is $V(64) = 96\sqrt{64} - 6 \cdot 64 = 384$ thousand dollars, or $384,000.

This maximum may be seen in the graph on the left.

## Maximizing Profit

The famous economist John Maynard Keynes said, "The engine that drives Enterprise is Profit." Many management problems consist of maximizing profit, and require *constructing* the profit function before maximizing it. Such problems have three economic ingredients. The first is that profit is defined as *revenue minus cost:*

$$\text{Profit} = \text{Revenue} - \text{Cost}$$

The second ingredient is that revenue is *price times quantity.* For example, if a company sells 100 toasters for $25 each, the revenue will obviously be $25 \cdot 100 = $2500$.

$$\text{Revenue} = \binom{\text{Unit}}{\text{price}} \cdot (\text{Quantity})$$

The third economic ingredient reflects the fact that, in general, price and quantity are inversely related: increasing the price decreases sales, while decreasing the price increases sales. To put this another way, "flooding the market" with a product drives the price down, whereas creating a shortage drives the price up. If the relationship between the price $p$ and the quantity $x$ that consumers will buy at that price is expressed as a function $p(x)$, it is called the **price function.**[*]

The price function $p(x)$ shows the inverse relation between price $p$ and quantity $x$.

| **Price Function** |
| --- |
| $p(x)$ gives the price $p$ at which consumers will buy exactly $x$ units of the product. |

The price function, relating price and quantity, may be linear or curved (as shown on the left), but it will always be a *decreasing* function.

*We will use *lowercase p* for price and *capital P* for profit.

In actual practice, price functions are very difficult to determine, requiring extensive (and expensive) market research. In this section we will be given the price function. In the next section we will see how to do without price functions, at least in simple cases.

**EXAMPLE 3**

### MAXIMIZING A COMPANY'S PROFIT

It costs the American Automobile Company $8000 to produce each automobile, and fixed costs (rent and other expenses that do not depend on the amount of production) are $20,000 per week. The company's price function is $p(x) = 22,000 - 70x$, where $p$ is the price at which exactly $x$ cars will be sold.

**a.** How many cars should be produced each week to maximize profit?

**b.** For what price should they be sold?

**c.** What is the company's maximum profit?

*Solution*

**Revenue** is price times quantity, $R = p \cdot x$:

$$R = p \cdot x = \underbrace{(22,000 - 70x)}_{p(x)} x = \underbrace{22,000x - 70x^2}_{\text{Revenue function } R(x)}$$

Replacing $p$ by the price function
$p = 22,000 - 70x$

**Cost** is the cost per car ($8000) times the number of cars ($x$) plus the fixed cost ($20,000):

$$C(x) = 8000x + 20,000 \qquad \text{(Unit cost)} \cdot \text{(Quantity)} + \text{(Fixed costs)}$$

**Profit** is revenue minus cost:

$$P(x) = \underbrace{(22,000x - 70x^2)}_{R(x)} - \underbrace{(8000x + 20,000)}_{C(x)}$$

$$= -70x^2 + 14,000x - 20,000 \qquad \begin{matrix}\text{Profit function} \\ \text{(after simplification)}\end{matrix}$$

**a.** We maximize the profit by setting its derivative equal to zero:

$$P'(x) = -140x + 14,000 = 0 \qquad \begin{matrix}\text{Differentiating} \\ P = -70x^2 + 14,000x - 20,000\end{matrix}$$

$$-140x = -14,000 \qquad \text{Solving}$$

$$x = \frac{-14,000}{-140} = 100 \qquad \text{Only one critical number}$$

$$P''(x) = -140 \qquad \text{From } P'(x) = -140x + 14,000$$

The second derivative is negative, so the profit is maximized at the critical number. (If the second derivative had involved $x$, we would have substituted the critical number $x = 100$.) Since $x$ is the number of cars, the company should produce 100 cars per week (the time period stated in the problem).

**b.** The selling price $p$ is found from the price function:

$$p(100) = 22,000 - 70 \cdot 100 = \$15,000$$

$p(x) = 22,000 - 70x$
evaluated at $x = 100$

**c.** The maximum profit is found from the profit function:

$$P(100) = -70(100)^2 + 14,000(100) - 20,000$$
$$= \$680,000$$

$P(x) = -70x^2 + 14,000x - 20,000$
evaluated at $x = 100$

Finally, state the answer clearly in words.

The company should make 100 cars per week and sell them for \$15,000 each. The maximum profit will be \$680,000.

Actually, automobile dealers seem to prefer prices like \$14,999 as if \$1 makes a difference.

---

## Graphs of the Revenue, Cost, and Profit Functions

The graphs of the revenue and cost functions are shown on the left. At $x$-values where revenue is above cost, there is a profit, and where the cost is above the revenue, there is a loss.

The height of the profit function at any $x$ is the amount by which the revenue is above the cost in the graph. Since profit equals revenue minus cost, we may differentiate each side of $P(x) = R(x) - C(x)$, obtaining

$$P'(x) = R'(x) - C'(x)$$

This shows that setting $P'(x) = 0$ (which we do to maximize profit) is equivalent to setting $R'(x) - C'(x) = 0$, which is equivalent to $R'(x) = C'(x)$. This last equation may be expressed in marginals, $MR = MC$, which is a classic economic criterion for maximum profit.

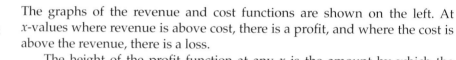

**Classic Economic Criterion for Maximum Profit**

At maximum profit:

$$\left(\begin{array}{c}\text{Marginal}\\\text{revenue}\end{array}\right) = \left(\begin{array}{c}\text{Marginal}\\\text{cost}\end{array}\right)$$

However, there is a hidden flaw in using this criterion to maximize profit: it holds not only at the maximum profit but also at the *minimum* profit [since $P'(x) = 0$ at both]. It is better to calculate the profit function and solve $P'(x) = 0$, as we have done, since we may then calculate $P''(x)$ and use the second derivative test to assure that we are *maximizing* profit, thereby avoiding the costly mistake of minimizing profit.

EXAMPLE 4  **MAXIMIZING THE AREA OF AN ENCLOSURE**

A farmer has 1000 feet of fence and wants to build a rectangular enclosure along a straight wall. If the side along the wall needs no fence, find the dimensions that make the enclosure as large as possible. Also find the maximum area.

*Solution*

The largest enclosure means, of course, the largest area.

We let variables stand for the length and width:

$$x = \text{length (parallel to wall)}$$
$$y = \text{width (perpendicular to wall)}$$

The problem becomes

| | |
|---|---|
| Maximize $A = xy$ | Area is length times width |
| subject to $x + 2y = 1000$ | One $x$ side and two $y$ sides from 1000 feet of fence |

We must express the area $A = xy$ in terms of one variable. We use

| | |
|---|---|
| $x = 1000 - 2y$ | Solving $x + 2y = 1000$ for $x$ |
| $A = xy = (1000 - 2y)y = 1000y - 2y^2$ | Substituting $x = 1000 - 2y$ into $A = xy$ |
| $\underbrace{\qquad\qquad}_{x}$ | |
| $A' = 1000 - 4y = 0$ | Maximizing $A = 1000y - 2y^2$ by setting the derivative equal to zero |
| $y = 250$ | Solving $1000 - 4y = 0$ for $y$ |

Since $A'' = -4$, the second-derivative test shows that the area is indeed *maximized* when $y = 250$. The length $x$ is

$$x = 1000 - 2 \cdot 250 = 500 \qquad \text{Evaluating } x = 1000 - 2y \text{ at } y = 250$$

Length (parallel to the wall) is 500 feet, width (perpendicular to the wall) is 250 feet, and area (length times width) is 125,000 square feet.

 **Spreadsheet Exploration**

In the preceding example you might think that it does not matter how the fence is laid out as long as all 1000 feet are used. To see that the area enclosed really *does* change, and that the maximum occurs at $y = 250$, the following spreadsheet* calculates the area $A(y) = 1000y - 2y^2$ for

*To obtain this and other Spreadsheet Explorations, go to www.cengage.com/math/berresford.

*y*-values from 245 to 255. Notice that the area is largest for $y = 250$, and that for each change of 1 in the *y*-value the change in the area is smallest for widths closest to 250. This verifies *numerically* that the derivative (rate of change) becomes zero as *y* approaches 250.

| B8 | ▼ | = | =1000*A8-2*A8^2 |
|---|---|---|---|

| | A | B |
|---|---|---|
| 1 | Width (perpendicular to wall) | Area Enclosed |
| 2 | | |
| 3 | 245 | 124950 |
| 4 | 246 | 124968 |
| 5 | 247 | 124982 |
| 6 | 248 | 124992 |
| 7 | 249 | 124998 |
| 8 | 250 | 125000 |
| 9 | 251 | 124998 |
| 10 | 252 | 124992 |
| 11 | 253 | 124982 |
| 12 | 254 | 124968 |
| 13 | 255 | 124950 |

largest area

Based on this spreadsheet, how much area will the farmer lose if he mistakenly makes the width 249 or 251 feet instead of 250 (and therefore the length 502 or 498 feet)? Is this loss significant based on an area of 125,000 square feet? This is characteristic of maximization problems where the slope is zero—being *near* the maximizing value is essentially as good as being *at* it.

---

**EXAMPLE 5**    **MAXIMIZING THE VOLUME OF A BOX**

An open-top box is to be made from a square sheet of metal 12 inches on each side by cutting a square from each corner and folding up the sides, as in the diagrams below. Find the volume of the largest box that can be made in this way.

Square sheet                    Corners removed                    Side flaps folded up to make open-top box.

*Solution*

Let $x$ = the length of the side of the square cut from each corner.

The length and width of the base will be $12 - 2x$ (we subtract $2x$ because $x$ is cut from both ends of each edge).

The 12" by 12" square with four $x$ by $x$ corners removed.

The height (or depth) is $x$, the size of the edge folded up.

Therefore, the volume is

$$V(x) = (12 - 2x)(12 - 2x)x \qquad \text{(Length)} \cdot \text{(Width)} \cdot \text{(Height)}$$

Since $x$ is a length, $x > 0$, and since $x$ inches are cut from *both* sides of each 12-inch edge, we must have $2x < 12$, so $x < 6$. The problem becomes

Maximize $\quad V(x) = (12 - 2x)(12 - 2x)x \qquad$ on $0 < x < 6$

$\qquad\qquad V(x) = (144 - 48x + 4x^2)x \qquad$ Multiplying out

$\qquad\qquad\qquad = 4x^3 - 48x^2 + 144x \qquad$ Multiplying out

$\qquad\quad V'(x) = 12x^2 - 96x + 144 \qquad$ Differentiating

$\qquad\qquad\qquad = 12(x^2 - 8x + 12) \qquad$ Factoring

$\qquad\qquad\qquad = 12(x - 2)(x - 6) \qquad$ Factoring

$$\text{CN} \begin{cases} x = 2 \\ x = 6 \end{cases} \leftarrow \text{Not in the domain, so we eliminate it}$$

The second derivative is $V''(x) = 24x - 96$, which at $x = 2$ is

$$V''(2) = 48 - 96 < 0$$

Therefore, the volume is *maximized* at $x = 2$.

Maximum volume is 128 cubic inches. $\qquad$ From $V(x)$ evaluated at $x = 2$

 **BE CAREFUL:** Be sure to use only critical numbers that are *in the domain of the original function*. This was important here and in Example 1 on page 208.

## Graphing Calculator Exploration

Maximum
X=1.9999997 Y=128

Interpret as $x = 2$

Do the previous example (and then modify it) on a graphing calculator as follows:

**a.** Enter the volume as $y_1 = (12 - 2x)(12 - 2x)x$ and graph it on $[0, 6]$ by $[-40, 150]$.

**b.** Use MAXIMUM to maximize $y_1$. Your answer should agree with that found above. (While this may seem faster than doing the problem "by hand," you would still need to have found the volume function and the domain, which was most of the work.)

**c.** What if the beginning size were not a 12-inch by 12-inch square, but a standard 8.5-inch by 11-inch sheet of paper? Go back to $y_1$ and replace the two 12s by 8.5 and 11 and find the $x$ and the maximum volume. (*Answer:* volume about 66 cubic inches)

**d.** What about a 3 by 5 card?

Parts (c) and (d), which would be more difficult to solve by hand, show how useful a graphing calculator is for solving related problems after one has been analyzed using calculus.

## 3.3 Section Summary

There is no single, all-purpose procedure for solving word problems. You must think about the problem, draw a picture if possible, and express the quantity to be maximized or minimized in terms of some appropriate variable. You can become good at it only with practice.

We have two procedures for optimizing continuous functions on intervals:

**1.** If the function has only one critical number in the interval, we find it and use the second-derivative test to show whether the function is maximized or minimized there.

**2.** If the interval is closed, we evaluate the function at all critical numbers and endpoints in the interval; the largest and smallest resulting values will be the maximum and minimum values of the function.

If neither procedure finds the desired optimal value(s), graph the function. The maximum (or minimum) value will be the $y$-coordinate of the highest (or lowest) point on the graph.

# 3.3 Exercises

**1–20.** Find (*without* using a calculator) the absolute extreme values of each function on the given interval.

1. $f(x) = x^3 - 6x^2 + 9x + 8$ on $[-1, 2]$
2. $f(x) = x^3 - 6x^2 + 22$ on $[-2, 2]$
3. $f(x) = x^3 - 12x$ on $[-3, 3]$
4. $f(x) = x^3 - 27x$ on $[-2, 2]$
5. $f(x) = x^4 + 4x^3 + 4x^2$ on $[-2, 1]$
6. $f(x) = x^4 - 4x^3 + 4x^2$ on $[0, 3]$
7. $f(x) = x^4 + 4x^3 + 15$ on $[-4, 1]$
8. $f(x) = 2x^4 - 8x^3 + 30$ on $[-1, 4]$
9. $f(x) = 2x^5 - 5x^4$ on $[-1, 3]$
10. $f(x) = 4x^5 - 5x^4$ on $[0, 2]$
11. $f(x) = 3x^2 - x^3$ on $[0, 5]$
12. $f(x) = 6x^2 - x^3$ on $[0, 5]$
13. $f(x) = 5 - x$ on $[0, 5]$
14. $f(x) = x(100 - x)$ on $[0, 100]$
15. $f(x) = x^2(3 - x)$ on $[-1, 3]$
16. $f(x) = x^3(4 - x)$ on $[-1, 4]$
17. $f(x) = (x^2 - 1)^2$ on $[-1, 1]$
18. $f(x) = \sqrt[3]{x^2}$ on $[-1, 8]$
19. $f(x) = \dfrac{x}{x^2 + 1}$ on $[-3, 3]$
20. $f(x) = \dfrac{1}{x^2 + 1}$ on $[-3, 3]$

21. Find the number in the interval $[0, 3]$ such that the number minus its square is:
    a. As large as possible.
    b. As small as possible.

22. Find the number in the interval $[\frac{1}{3}, 2]$ such that the sum of the number and its reciprocal is:
    a. As large as possible.
    b. As small as possible.

23. **ONE FUNCTION, DIFFERENT DOMAINS**
    a. Graph the function $y_1 = x^3 - 15x^2 + 63x$ on the window $[0, 10]$ by $[0, 130]$. By visual inspection of this function on this domain, where do the absolute maximum and minimum values occur: both at critical numbers, both at endpoints, or one at a critical number and one at an endpoint?
    b. Now change the domain to $[0, 8]$ and answer the same question.
    c. Now change the domain to $[2, 8]$ and answer the same question.
    d. Can you find a domain such that the minimum occurs at a critical number and the maximum at an endpoint?

24. **EXISTENCE OF EXTREME VALUES**
    a. Graph the function $y_1 = 5/x$ on the window $[1, 10]$ by $[0, 10]$. By visual inspection, does the function have an absolute maximum value on this domain? An absolute minimum value?
    b. Now change the $x$ window to $[0, 10]$ and answer the same questions. (We cannot now call $[0, 10]$ the domain, since the function is not defined at 0.)
    c. Based on the screen display, answer the same questions for the domain $(0, \infty)$.
    d. Is there a domain on which this function has an absolute maximum but no absolute minimum?
    e. It was stated in the box on page 207 that a continuous function on a closed interval will always have an absolute maximum and minimum value. Is this claim violated by the function $f(x) = 5/x$ on $[0, 10]$ [part (b)]? Is it violated by $f(x) = 5/x$ on $(0, \infty)$ [part (c)]?

## Applied Exercises

25. **BIOMEDICAL: Pollen Count** The average pollen count in New York City on day $x$ of the pollen season is $P(x) = 8x - 0.2x^2$ (for $0 < x < 40$). On which day is the pollen count highest?

26. **GENERAL: Fuel Economy** The fuel economy (in miles per gallon) of an average American compact car is $E(x) = -0.015x^2 + 1.14x + 8.3$, where $x$ is the driving speed (in miles per hour, $20 \le x \le 60$). At what speed is fuel economy greatest?
    *Source:* U.S. Environmental Protection Agency

27. **GENERAL: Fuel Economy** The fuel economy (in miles per gallon) of an average American midsized car is $E(x) = -0.01x^2 + 0.62x + 10.4$, where $x$ is the driving speed (in miles per hour, $20 \le x \le 60$). At what speed is fuel economy greatest?
    Source: U.S. Environmental Protection Agency

28. **BUSINESS: Copier Repair** A copier company finds that copiers that are $x$ years old require, on average, $f(x) = 1.2x^2 - 4.7x + 10.8$ repairs annually for $0 \le x \le 5$. Find the year that requires the least repairs, rounding your answer to the nearest year.

29. **GENERAL: Driving and Age** Studies have shown that the number of accidents a driver has varies with the age of the driver, and is highest for very young and very old drivers. The number of serious accidents for drivers of age $x$ during 2003 was approximately $f(x) = 0.013x^2 - 1.35x + 48$ for $16 \le x \le 85$. Find the age that has the least accidents, rounding your answer to the nearest year.
    Source: Insurance Information Institute

30. **GENERAL: Water Power** The proportion of a river's energy that can be obtained from an undershot waterwheel is $E(x) = 2x^3 - 4x^2 + 2x$, where $x$ is the speed of the waterwheel relative to the speed of the river. Find the maximum value of this function on the interval [0, 1], thereby showing that only about 30% of a river's energy can be captured. Your answer should agree with the old millwright's rule that the speed of the wheel should be about one-third of the speed of the river.
    Source: U.S. Government Printing Office, *The Energy Book*

31. **ENVIRONMENTAL SCIENCE: Wind Power** Electrical generation from wind turbines (large "windmills") is increasingly popular in Europe and the United States. The fraction of the wind's energy that can be extracted by a turbine is $f(x) = \frac{1}{2}x(2 - x)^2$, where $x$ is the fraction by which the wind is slowed in passing through the turbine. Find the fraction $x$ that maximizes the energy extracted.
    Source: U.S. Government Printing Office, *The Energy Book*

32. **BUSINESS: Cigarette Production** Per capita cigarette production in the United States during recent decades is approximately given

by $f(x) = -0.6x^2 + 12x + 945$, where $x$ is the number of years after 1980. Find the year when per capita cigarette production was at its greatest.
    Source: U.S. Department of Agriculture

33. **GENERAL: Timber Value** The value of a timber forest after $t$ years is $V(t) = 480\sqrt{t} - 40t$ (for $0 \le t \le 50$). Find when its value is maximized.

34. **GENERAL: Longevity and Exercise** A recent study of the exercise habits of 17,000 Harvard alumni found that the death rate (deaths per 10,000 person-years) was approximately $R(x) = 5x^2 - 35x + 104$, where $x$ is the weekly amount of exercise in thousands of calories $(0 \le x \le 4)$. Find the exercise level that minimizes the death rate.
    Source: *Journal of the American Medical Association* 273:15

35. **ENVIRONMENTAL SCIENCE: Pollution** Two chemical factories are discharging toxic waste into a large lake, and the pollution level at a point $x$ miles from factory A toward factory B is $P(x) = 3x^2 - 72x + 576$ parts per million (for $0 \le x \le 50$). Find where the pollution is the least.

36. **BUSINESS: Maximum Profit** City Cycles Incorporated finds that it costs $70 to manufacture each bicycle, and fixed costs are $100 per day. The price function is $p(x) = 270 - 10x$, where $p$ is the price (in dollars) at which exactly $x$ bicycles will be sold. Find the quantity City Cycles should produce and the price it should charge to maximize profit. Also find the maximum profit.

37. **BUSINESS: Maximum Profit** Country Motorbikes Incorporated finds that it costs $200 to produce each motorbike, and that fixed costs are $1500 per day. The price function is $p(x) = 600 - 5x$, where $p$ is the price (in dollars) at which exactly $x$ motorbikes will be sold. Find the quantity Country Motorbikes should produce and the price it should charge to maximize profit. Also find the maximum profit.

38. **BUSINESS: Maximum Profit** A retired potter can produce china pitchers at a cost of $5 each. She estimates her price function to be $p = 17 - 0.5x$, where $p$ is the price at which exactly $x$ pitchers will be sold per week. Find the number of pitchers that she should produce and the price that she should charge in order to maximize profit. Also find the maximum profit.

**39. BUSINESS: Maximum Revenue**
BoxedNGone.com truck rentals calculates that its price function is $p(x) = 240 - 3x$, where $p$ is the price (in dollars) at which exactly $x$ trucks will be rented per day. Find the number of trucks that BoxedNGone should rent and the price it should charge to maximize revenue. Also find the maximum revenue.

**40. BUSINESS: Maximum Revenue** NRG-SUP. com, a supplier of energy supplements for athletes, determines that its price function is $p(x) = 60 - \frac{1}{2}x$, where $p$ is the price (in dollars) at which exactly $x$ boxes of supplements will be sold per day. Find the number of boxes that NRG-SUP will sell per day and the price it should charge to maximize revenue. Also find the maximum revenue.

**41. GENERAL: Parking Lot Design** A company wants to build a parking lot along the side of one of its buildings using 800 feet of fence. If the side along the building needs no fence, what are the dimensions of the largest possible parking lot?

**42. GENERAL: Area** A farmer wants to make two identical rectangular enclosures along a straight river, as in the diagram shown below. If he has 600 yards of fence, and if the sides along the river need no fence, what should be the dimensions of each enclosure if the total area is to be maximized?

**43. GENERAL: Area** A farmer wants to make three identical rectangular enclosures along a straight river, as in the diagram shown below. If he has 1200 yards of fence, and if the sides along the river need no fence, what should be the dimensions of each enclosure if the total area is to be maximized?

**44. GENERAL: Area** What is the area of the largest rectangle whose perimeter is 100 feet?

**45. GENERAL: Package Design** An open-top box is to be made from a square piece of cardboard that measures 18 inches by 18 inches by removing a square from each corner and folding up the sides. What are the dimensions and volume of the largest box that can be made in this way?

**46. GENERAL: Gutter Design** A long gutter is to be made from a 12-inch-wide strip of metal by folding up the two edges. How much of each edge should be folded up in order to maximize the capacity of the gutter? [*Hint:* Maximizing the capacity means maximizing the cross-sectional area, shown below.]

**47. GENERAL: Maximizing a Product** Find the two numbers whose sum is 50 and whose product is a maximum.

**48. GENERAL: Maximizing Area** Show that the largest rectangle with a given perimeter is a square.

**49. GENERAL: Athletic Fields** A running track consists of a rectangle with a semicircle at each end, as shown below. If the perimeter is to be exactly 440 yards, find the dimensions ($x$ and $r$) that maximize the area of the rectangle. [*Hint:* The perimeter is $2x + 2\pi r$.]

**50. GENERAL: Window Design** A Norman window consists of a rectangle topped by a semicircle, as shown below. If the perimeter is to be 18 feet, find the dimensions ($x$ and $r$) that maximize the area of the window. [*Hint:* The perimeter is $2x + 2r + \pi r$.]

**51. BIOMEDICAL: Bacterial Growth** A chemical reagent is introduced into a bacterial population, and $t$ hours later the number of bacteria (in thousands) is $N(t) = 1000 + 15t^2 - t^3$ (for $0 \leq t \leq 15$).

   **a.** When will the population be the largest, and how large will it be?

   **b.** When will the population be growing at the fastest rate, and how fast? (What word applies to such a point?)

**52. GENERAL: Value of a Pulpwood Forest** The value of a pulpwood forest after growing for $x$ years is predicted to be $V(t) = 400x^{0.4} - 40x$ thousand dollars (for $0 \leq x \leq 25$). Use a graphing calculator with MAXIMUM to find when the value will be maximized, and what the maximum value will be.

**53–54. GENERAL: Package Design** Use a graphing calculator (as explained on page 216) to find the side of the square removed and the volume of the box described in Example 5 (pages 214–215) if the square piece of metal is replaced by a:

**53.** 5 by 7 card (5 inches by 7 inches)

**54.** 6 by 8 card (6 inches by 8 inches)

You might try constructing such a box.

### Conceptual Exercises

**55.** What is the difference between an *absolute* maximum value and *relative* maximum value?

**56.** True or False: If a function is defined and continuous on a closed interval, then it has both an absolute maximum value and an absolute minimum value.

**57.** True or False: If a function is continuous on an *open* interval, then it will not have absolute maximum or minimum values.

**58.** Draw the graph of a function defined on $(-\infty, \infty)$ that has no absolute maximum or minimum value. Draw one that has *both* an absolute maximum *and* an absolute minimum value.

**59.** Draw the graph of a function defined on the closed interval [0, 1] that has no absolute maximum or minimum value.

**60.** You are to maximize a function on the interval $[-5, 5]$ and find critical numbers $-1, 0,$ and $6$. Which critical numbers should you use?

**61.** If $f'(5) = 0$ and $f''(5) > 0$, what conclusion can you draw?

**62.** If $f'(5) = 0$ and $f''(5) = 0$, what conclusion can you draw?

**63.** True of False: If $f$ has an absolute maximum value, then $-f$ will have an absolute minimum value.

**64.** True or False: A polynomial of degree $n$ can have at most $n - 1$ critical numbers.

### Explorations and Excursions

The following problems extend and augment the material presented in the text.

**65. BIOMEDICAL: Coughing** When you cough, you are using a high-speed stream of air to clear your trachea (windpipe). During a cough your trachea contracts, forcing the air to move faster, but also increasing the friction. If a trachea contracts from a normal radius of 3 centimeters to a radius of $r$ centimeters, the velocity of the airstream is $V(r) = c(3 - r)r^2$, where $c$ is a positive constant depending on the length and the elasticity of the trachea. Find the radius $r$ that maximizes this velocity. (X-ray pictures verify that the trachea does indeed contract to this radius.)

*Source: Lung* **152**

**66. GENERAL: "Efishency"** At what speed should a fish swim upstream so as to reach its destination with the least expenditure of energy? The energy depends on the friction of the fish through the water and on the duration of the trip. If the fish swims with velocity $v$, the energy has been found experimentally to be proportional to $v^k$ (for constant $k > 2$) times the duration of the trip. A distance of $s$ miles against a current of speed $c$ requires time $\dfrac{s}{v - c}$ (distance divided by speed). The energy required is then proportional to $\dfrac{v^k s}{v - c}$. For $k = 3$, minimizing energy is equivalent to minimizing

$$E(v) = \frac{v^3}{v - c}$$

Find the speed $v$ with which the fish should swim in order to minimize its expenditure $E(v)$. (Your answer will depend on $c$, the speed of the current.)

*Source: E. Batschelet, Introduction to Mathematics for Life Scientists*

# 3.4 FURTHER APPLICATIONS OF OPTIMIZATION

## Introduction

In this section we continue to solve optimization problems. In particular, we will see how to maximize a company's profit if we are not given the price function, provided that we are given information describing how price changes will affect sales. We will also see that sometimes $x$ should be chosen as something *other* than the quantity sold.

**EXAMPLE 1**

### FINDING PRICE AND QUANTITY FUNCTIONS

A store can sell 20 bicycles per week at a price of $400 each. The manager estimates that for each $10 price reduction she can sell two more bicycles per week. The bicycles cost the store $200 each. If $x$ stands for *the number of $10 price reductions*, express the price $p$ and the quantity $q$ as functions of $x$.

*Solution*

Let

$$x = \text{the number of \$10 price reductions}$$

For example, $x = 4$ means that the price is reduced by $40 (four $10 price reductions). Therefore, in general, if there are $x$ $10 price reductions from the original $400 price, then the price $p(x)$ is

$$p(x) = \underbrace{400}_{\substack{\text{Original} \\ \text{price}}} - \underset{\substack{\uparrow \\ \text{Less } x \text{ \$10 price reductions}}}{10x} \qquad\qquad \text{Price}$$

The quantity sold $q(x)$ will be

$$q(x) = \underbrace{20}_{\substack{\text{Original} \\ \text{quantity}}} + \underset{\substack{\uparrow \\ \text{Plus two for each price reduction}}}{2x} \qquad\qquad \text{Quantity}$$

We will return to this example and maximize the store's profit after a practice problem.

**Practice Problem**

A computer manufacturer can sell 1500 personal computers per month at a price of $3000 each. The manager estimates that for each $200 price reduction he will sell 300 more each month. If $x$ stands for *the number of $200 price reductions*, express the price $p$ and the quantity $q$ as functions of $x$.

➤ Solution on page 227

**EXAMPLE 2**  **MAXIMIZING PROFIT (*Continuation of Example 1*)**

Using the information in Example 1, find the price of the bicycles and the quantity that maximize profit. Also find the maximum profit.

*Solution*

In Example 1 we found

$$p(x) = 400 - 10x \qquad \text{Price}$$
$$q(x) = 20 + 2x \qquad \text{Quantity sold at that price}$$

Revenue is price times quantity,   $p(x) \cdot q(x)$:

$$R(x) = (400 - 10x)(20 + 2x) \qquad p(x)q(x)$$
$$= 8000 + 600x - 20x^2 \qquad \text{Multiplying out and simplifying}$$

The cost function is unit cost times quantity:

$$C(x) = \underbrace{200}_{\substack{\text{Unit} \\ \text{cost}}} \underbrace{(20 + 2x)}_{\substack{\text{Quantity} \\ q(x)}} = 4000 + 400x \qquad \begin{array}{l} \text{If there were a fixed cost,} \\ \text{we would add it in} \end{array}$$

Profit is revenue minus cost:

$$P(x) = \underbrace{(8000 + 600x - 20x^2)}_{R(x)} - \underbrace{(4000 + 400x)}_{C(x)}$$

$$= 4000 + 200x - 20x^2 \qquad \text{Simplifying}$$

We maximize profit by setting the derivative equal to zero:

$$200 - 40x = 0 \qquad \text{Differentiating} \quad P = 4000 + 200x - 20x^2$$

The critical number is $x = 5$. The second derivative, $P''(x) = -40$, shows that the profit is *maximized* at $x = 5$. Since $x = 5$ is the number of $10 price reductions, the original price of $400 should be lowered by $50 ($10 five times), from $400 to $350. The quantity sold is found from the quantity function:

$$q(5) = 20 + 2 \cdot 5 = 30 \qquad q(x) = 20 + 2x \quad \text{at} \quad x = 5$$

Finally, we state the answer clearly.

Sell the bicycles for $350 each.

Quantity sold: 30 per week.

Maximum profit: $4500.     From   $P(x) = 4000 + 200x - 20x^2$   at   $x = 5$

Although we did not need to graph the profit function, the diagram on the left does verify that the maximum occurs at $x = 5$.

Exercise 23 will show how a graphing calculator enables you to modify the problem (such as changing the cost per bicycle) and then immediately recalculate the new answer.

### Choosing Variables

Notice that in Example 2 we did not choose $x$ to be the quantity sold, but instead to be *the number of $10 price reductions.* We chose this $x$ because from it we could easily calculate both the new price and the new quantity. Other choices for $x$ are also possible, but in situations where a price change will make one quantity rise and another fall, it is often easiest to choose $x$ to be the *number of such changes.*

 **BE CAREFUL:** If $x$ is the number of price *increases*, then the negative number $x = -2$ means two price *decreases.* Similarly, a negative number of price *decreases* would mean a price *increase.*

---

**EXAMPLE 3**  **MAXIMIZING HARVEST SIZE**

An orange grower finds that if he plants 80 orange trees per acre, each tree will yield 60 bushels of oranges. He estimates that for each additional tree that he plants per acre, the yield of each tree will decrease by 2 bushels. How many trees should he plant per acre to maximize his harvest?

*Solution*

We take $x$ equal to the number of "changes"—that is, let

$$x = \text{the number of added trees per acre}$$

With $x$ extra trees per acre,

Trees per acre:  $80 + x$       Original 80 plus $x$ more
Yield per tree:  $60 - 2x$     Original yield less 2 per extra tree

Therefore, the total yield per acre will be

$$Y(x) = \underbrace{(60 - 2x)}_{\substack{\text{Yield} \\ \text{per tree}}}\underbrace{(80 + x)}_{\substack{\text{Trees} \\ \text{per acre}}} = 4800 - 100x - 2x^2$$

We maximize this by setting the derivative equal to zero:

$$-100 - 4x = 0 \qquad \text{Differentiating } Y = 4800 - 100x - 2x^2$$
$$x = -25 \qquad \text{Negative!}$$

$x = -25$
number of added " trees

The number of *added* trees is negative, meaning that the grower should plant 25 *fewer* trees per acre. The second derivative, $Y''(x) = -4$, shows that the yield is indeed maximized at $x = -25$. Therefore:

Plant 55 trees per acre.       $80 - 25 = 55$

---

 **BE CAREFUL:** Don't forget to use the second-derivative test to be sure you have maximized (or minimized). *Minimizing* your yield is seldom a good idea!

Earlier problems involved maximizing areas and volumes using only a fixed amount of material (such as a fixed length of fence). Instead, we could minimize the amount of materials for a fixed area or volume.

**EXAMPLE 4**

### MINIMIZING PACKAGE MATERIALS

A moving company wishes to design an open-top box with a square base whose volume is exactly 32 cubic feet. Find the dimensions of the box requiring the least amount of materials.

*Solution*

The base is square, so we define

$x$ = length of side of base

$y$ = height

The volume (length · width · height) is $x \cdot x \cdot y$ or $x^2y$, which (according to the problem) must equal 32 cubic feet:

$$x^2y = 32$$

The box consists of a bottom (area $x^2$) and four sides (each of area $xy$). Minimizing the amount of materials means minimizing the surface area of the bottom and four sides:

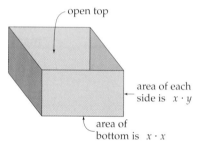

$$A = x^2 + 4xy \qquad \binom{\text{Area of}}{\text{bottom}} + \binom{\text{Area of}}{\text{four sides}}$$

As usual, we must express this area in terms of just *one* variable, so we use the volume requirement to express $y$ in terms of $x$:

$$y = \frac{32}{x^2} \qquad \text{Solving } x^2y = 32 \text{ for } y$$

The area function becomes

$$A = x^2 + 4x\,\frac{32}{x^2} \qquad A = x^2 + 4xy \text{ with } y \text{ replaced by } \frac{32}{x^2}$$

$$= x^2 + \frac{128}{x} \qquad \text{Simplifying}$$

$$= x^2 + 128x^{-1} \qquad \text{Writing } \frac{1}{x} \text{ as } x^{-1}$$

We minimize this by finding the critical number:

$$A'(x) = 2x - 128x^{-2} \qquad \text{Differentiating } A = x^2 + 128x^{-1}$$

$$2x - \frac{128}{x^2} = 0 \qquad \text{Setting the derivative equal to zero}$$

$$2x^3 - 128 = 0 \qquad \text{Multiplying by } x^2 \text{ (since } x > 0)$$

$$x^3 = 64 \qquad \text{Adding 128 and then dividing by 2}$$

$$x = 4 \qquad \text{Taking cube roots}$$

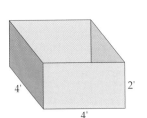

The second derivative

$$A''(x) = 2 + 256x^{-3} = 2 + \frac{256}{x^3}$$   From   $A'(x) = 2x - 128x^{-2}$

is positive at  $x = 4$,  so the area is minimized. Therefore, the dimensions using the least materials are:

Base: 4 feet on each side

Height: 2 feet

Height from  $y = 32/x^2$  at  $x = 4$

## Graphing Calculator Exploration

Minimum
X=3.9999995        Y=48

Interpret as $x = 4$

Use a graphing calculator to solve the previous example as follows:

**a.** Enter the area function to be minimized as   $y_1 = x^2 + 128/x$.

**b.** Graph $y_1$ for $x$-values [0, 10], using TABLE or TRACE to find where $y_1$ seems to "bottom out" to determine an appropriate $y$-interval.

**c.** Graph the function on the window determined in part (b) and then use MINIMUM to find the minimum value. Your answer should agree with that found above.

Notice that either way required first finding the function to be minimized.

### Minimizing the Cost of Materials

How would the preceding problem have changed if the material for the bottom of the box had been more costly than the material for the sides? If, for example, the material for the sides cost $2 per square foot and the material for the base, needing greater strength, cost $4 per square foot, then instead of simply minimizing the surface area, we would minimize *total cost*:

$$\text{Cost} = \left(\begin{array}{c}\text{Area of}\\\text{bottom}\end{array}\right)\left(\begin{array}{c}\text{Cost of bottom}\\\text{per square foot}\end{array}\right) + \left(\begin{array}{c}\text{Area of}\\\text{sides}\end{array}\right)\left(\begin{array}{c}\text{Cost of sides}\\\text{per square foot}\end{array}\right)$$

Since the areas would be just as before, this cost would be

$$\text{Cost} = (x^2)(4) + (4xy)(2) = 4x^2 + 8xy$$

From here on we would proceed just as before, eliminating the $y$ (using the volume relationship $x^2y = 32$) and then setting the derivative equal to zero.

### Maximizing Tax Revenue

Governments raise money by collecting taxes. If a sales tax or an import tax is too high, trade will be discouraged and tax revenues will fall. If, on the other hand, the tax rate is too low, trade may flourish but tax revenues will again fall. Economists often want to determine the tax rate that maximizes revenue for the government. To do this, they must first predict the relationship between the tax on an item and the total sales of the item.

Suppose, for example, that the relationship between the tax rate $t$ on an item and its total sales $S$ is

$$S(t) = 9 - 20\sqrt{t}$$

$t$ = tax rate $(0 \le t \le 0.20)$
$S$ = total sales (millions of dollars)

If the tax rate is $t = 0$ (0%), then the total sales will be

$$S(0) = 9 - 20\sqrt{0} = 9 \qquad \text{\$9 million}$$

If the tax rate is raised to $t = 0.16$ (16%), then sales will be

$$S(0.16) = 9 - 20\sqrt{0.16}$$
$$= 9 - (20)(0.4) = 9 - 8 = 1 \qquad \text{\$1 million}$$

That is, raising the tax rate from 0% to 16% will discourage \$8 million worth of sales. The graph of $S(t)$ on the left shows how total sales decrease as the tax rate increases. With such information (which may be found from historical data), one can find the tax rate that maximizes revenue.

**EXAMPLE 5**

**MAXIMIZING TAX REVENUE**

Economists estimate that the relationship between the tax rate $t$ on an item and the total sales $S$ of that item (in millions of dollars) is

$$S(t) = 9 - 20\sqrt{t} \qquad \text{For } 0 \le t \le 0.20$$

Find the tax rate that maximizes revenue to the government.

*Solution*

The government's revenue $R$ is the tax rate $t$ times the total sales $S(t) = 9 - 20\sqrt{t}$:

$$R(t) = t \cdot \underbrace{(9 - 20t^{1/2})}_{S(t)} = 9t - 20t^{3/2}$$

To maximize this function, we set its derivative equal to zero:

$$9 - 30t^{1/2} = 0 \qquad \text{Derivative of } 9t - 20t^{3/2}$$
$$9 = 30t^{1/2} \qquad \text{Adding } 30t^{1/2} \text{ to each side}$$
$$t^{1/2} = \frac{9}{30} = 0.3 \qquad \text{Switching sides and dividing by 30}$$
$$t = 0.09 \qquad \text{Squaring both sides}$$

This gives a tax rate of $t = 9\%$. The second derivative,

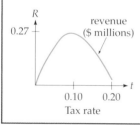

$$R''(t) = -30 \cdot \frac{1}{2} t^{-1/2} = -\frac{15}{\sqrt{t}} \qquad \text{From } R' = 9 - 30t^{1/2}$$

is negative at $t = 0.09$, showing that the revenue is maximized. Therefore:

A tax rate of 9% maximizes revenue for the government.

## Graphing Calculator Exploration

on [−10, 10] by [−10, 10]

The graph of the function from Example 5, $y_1 = 9x - 20x^{3/2}$ (written in $x$ instead of $t$ for ease of entry), is shown on the left on the standard window [−10, 10] by [−10, 10]. This might lead you to believe, erroneously, that the function is maximized at the endpoint (0, 0).

**a.** Why does this graph not look like the graph at the end of the previous example? [*Hint:* Look at the scale.]

**b.** Can you find a window on which your graphing calculator will show a graph like the one at the end of the preceding solution?

This example illustrates one of the pitfalls of graphing calculators—the part of the curve where the "action" takes place may be entirely hidden in one pixel. Calculus, on the other hand, will *always* find the critical value, no matter where it is, and then a graphing calculator can be used to confirm your answer by showing the graph on an appropriate window.

### Solution to Practice Problem

Price: $p(x) = 3000 - 200x$
Quantity: $q(x) = 1500 + 300x$

## 3.4 Exercises

**1. BUSINESS: Maximum Profit** An automobile dealer can sell 12 cars per day at a price of $15,000. He estimates that for each $300 price reduction he can sell two more cars per day. If each car costs him $12,000, and fixed costs are $1000, what price should he charge to maximize his profit? How many cars will he sell at this price? [*Hint:* Let $x$ = the number of $300 price reductions.]

**2. BUSINESS: Maximum Profit** An automobile dealer can sell four cars per day at a price of $12,000. She estimates that for each $200 price reduction she can sell two more cars per day. If each car costs her $10,000, and her fixed costs are $1000, what price should she charge to maximize her profit? How many cars will she sell at this price? [*Hint:* Let $x$ = the number of $200 price reductions.]

**3. BUSINESS: Maximum Revenue** An airline finds that if it prices a cross-country ticket at $200,

it will sell 300 tickets per day. It estimates that each $10 price reduction will result in 30 more tickets sold per day. Find the ticket price (and the number of tickets sold) that will maximize the airline's revenue.

**4. ECONOMICS: Oil Prices** An oil-producing country can sell 1 million barrels of oil a day at a price of $120 per barrel. If each $1 price increase will result in a sales decrease of 10,000 barrels per day, what price will maximize the country's revenue? How many barrels will it sell at that price?

**5. BUSINESS: Maximum Revenue** Rent-A-Reck Incorporated finds that it can rent 60 cars if it charges $80 for a weekend. It estimates that for each $5 price increase it will rent three fewer cars. What price should it charge to maximize its revenue? How many cars will it rent at this price?

6. **ENVIRONMENTAL SCIENCES: Maximum Yield** A peach grower finds that if he plants 40 trees per acre, each tree will yield 60 bushels of peaches. He also estimates that for each additional tree that he plants per acre, the yield of each tree will decrease by 2 bushels. How many trees should he plant per acre to maximize his harvest?

7. **ENVIRONMENTAL SCIENCES: Maximum Yield** An apple grower finds that if she plants 20 trees per acre, each tree will yield 90 bushels of apples. She also estimates that for each additional tree that she plants per acre, the yield of each tree will decrease by 3 bushels. How many trees should she plant per acre to maximize her harvest?

8. **GENERAL: Fencing** A farmer has 1200 feet of fence and wishes to build two identical rectangular enclosures, as in the diagram. What should be the dimensions of each enclosure if the total area is to be a maximum?

9. **GENERAL: Minimum Materials** An open-top box with a square base is to have a volume of 4 cubic feet. Find the dimensions of the box that can be made with the smallest amount of material.

10. **GENERAL: Minimum Materials** An open-top box with a square base is to have a volume of 108 cubic inches. Find the dimensions of the box that can be made with the smallest amount of material.

11. **GENERAL: Largest Postal Package** The U.S. Postal Service will accept a package if its length plus its girth (the distance all the way around) does not exceed 84 inches. Find the dimensions and volume of the largest package with a square base that can be mailed.

*Source*: U.S. Postal Service

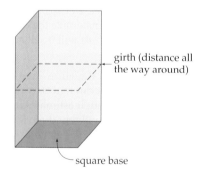

girth (distance all the way around)

square base

12. **GENERAL: Fencing** A homeowner wants to build, along his driveway, a garden surrounded by a fence. If the garden is to be 800 square feet, and the fence along the driveway costs $6 per foot while on the other three sides it costs only $2 per foot, find the dimensions that will minimize the cost. Also find the minimum cost.

$2 per foot

$6 per foot

driveway

13. **GENERAL: Fencing** A homeowner wants to build, along her driveway, a garden surrounded by a fence. If the garden is to be 5000 square feet, and the fence along the driveway costs $6 per foot while on the other three sides it costs only $2 per foot, find the dimensions that will minimize the cost. Also find the minimum cost. (See the diagram above.)

14–15. **ECONOMICS: Tax Revenue** Suppose that the relationship between the tax rate $t$ on imported shoes and the total sales $S$ (in millions of dollars) is given by the function below. Find the tax rate $t$ that maximizes revenue for the government.

14. $S(t) = 4 - 6\sqrt[3]{t}$    15. $S(t) = 8 - 15\sqrt[3]{t}$

16. **BIOMEDICAL: Drug Concentration** If the amount of a drug in a person's blood after $t$ hours is $f(t) = t/(t^2 + 9)$, when will the drug concentration be the greatest?

17. **GENERAL: Wine Appreciation** A case of vintage wine appreciates in value each year, but there is also an annual storage charge. The value of a typical case of investment-grade wine after $t$ years is $V(t) = 2000 + 80\sqrt{t} - 10t$ dollars (for $0 \le t \le 25$). Find the storage time that will maximize the value of the wine.

*Source*: wineeducation.com

18. **GENERAL: Bus Shelter Design** A bus stop shelter, consisting of two square sides, a back, and a roof, as shown in the following diagram, is to have volume 1024 cubic feet. What are the dimensions that require the least amount of materials?

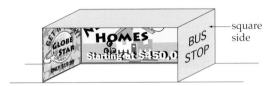

19. **GENERAL: Area** Show that the rectangle of fixed area whose perimeter is a minimum is a square.

20. **POLITICAL SCIENCE: Campaign Expenses** A politician estimates that by campaigning in a county for $x$ days, she will gain $2x$ (thousand) votes, but her campaign expenses will be $5x^2 + 500$ dollars. She wants to campaign for the number of days that maximizes the number of votes per dollar, $f(x) = \dfrac{2x}{5x^2 + 500}$. For how many days should she campaign?

21. **GENERAL: Page Design** A page of 96 square inches is to have margins of 1 inch on either side and $1\frac{1}{2}$ inches at the top and bottom, as in the diagram. Find the dimensions of the page that maximize the print area.

total area: 96 in.²

22. **GENERAL: Page Design** A page of 54 square inches is to have margins of 1 inch on either side and $1\frac{1}{2}$ inches at the top and bottom, as in the diagram above. Find the dimensions of the page that maximize the print area.

23. **BUSINESS: Exploring a Profit Maximization Problem** Use a graphing calculator to further explore Example 2 (page 222) as follows:

   a. Enter the price function $y_1 = 400 - 10x$ and the quantity function $y_2 = 20 + 2x$ into your graphing calculator.
   b. Make $y_3$ the revenue function by defining $y_3 = y_1y_2$ (price times quantity).
   c. Make $y_4$ the cost function by defining $y_4 = 200y_2$ (unit cost times quantity).

   d. Make $y_5$ the profit function by defining $y_5 = y_3 - y_4$ (revenue minus cost).
   e. Turn off $y_1$, $y_2$, $y_3$, and $y_4$ and graph the profit function $y_5$ on the window $[0, 10]$ by $[0, 10{,}000]$ and then use MAXIMUM to maximize it. Your answer should agree with that found in Example 2.

   Now change the problem!

   f. What if the store finds that it can buy the bicycles from another wholesaler for $150 instead of $200? In $y_4$, change the 200 to 150. Then graph the profit $y_5$ (you may have to turn off $y_4$ again) and maximize it. Find the new price and quantity by evaluating $y_1$ and $y_2$ (using EVALUATE) at the new $x$-value.
   g. What if cycling becomes more popular and the manager estimates that she can sell 30 instead of 20 bicycles per week at the original $400 price? Go back to $y_2$ and change 20 to 30 (keeping the change made earlier) and graph and find the price and quantity that maximize profit now.

   Notice how flexible this setup is for changing any of the numbers.

24. **BUSINESS: Maximizing Profit** An electronics store can sell 35 cellular telephones per week at a price of $200. The manager estimates that for each $20 price reduction she can sell 9 more per week. The telephones cost the store $100 each, and fixed costs are $700 per week.

   a. If $x$ is the number of $20 price reductions, find the price $p(x)$ and enter it in $y_1$. Then enter the quantity function $q(x)$ in $y_2$.
   b. Make $y_3$ the revenue function by defining $y_3 = y_1y_2$ (price times quantity).
   c. Make $y_4$ the cost function by defining $y_4$ as unit cost times $y_2$ plus fixed costs.
   d. Make $y_5$ the profit function by defining $y_5 = y_3 - y_4$ (revenue minus cost).
   e. Turn off $y_1$, $y_2$, $y_3$, and $y_4$ and graph the profit function $y_5$ for $x$-values $[-10, 10]$, using TABLE or TRACE to find an appropriate $y$-interval. Then use MAXIMUM to maximize it.
   f. Use EVALUATE to find the price and the quantity for this maximum profit.

### Conceptual Exercises

25. What is the relationship between *price, quantity,* and *revenue*?

26. What is the relationship between *revenue, cost,* and *profit*?

**27.** If the current price is $50 and you find that profit is maximized when there are −3 price reductions of $5, what should the price be?

**28.** If the current price is $75 and you find that profit is maximized when there are −5 price increases of $3, what should the price be?

**29.** What will be the result of minimizing cost if the cost function is linear?

**30.** Suppose that, as a variation on Example 4 on pages 224–225, you wanted to find the open-top box with volume 32 cubic feet and a square base that requires the *greatest* amount of materials. Would such a problem have a solution? [*Hint:* No calculation necessary.]

**31.** If a company wants to increase its profit, why not just produce more of whatever it makes? Explain.

**32.** If a government wants to increase revenue, why not just increase the tax rate? Explain.

**33.** In a revenue, cost, and profit problem, is maximizing the revenue the same as maximizing the profit? Explain.

**34.** In a revenue, cost, and profit problem, is minimizing the cost the same as maximizing the profit? Explain.

### Explorations and Excursions

The following problems extend and augment the material presented in the text.

*Sources: F. Happensteadt and C. Peskin, Mathematics in Medicine and the Life Sciences and R. Anderson and R. May, Infectious Diseases of Humans: Dynamics and Control*

**35. BIOMEDICAL: Contagion**  If an epidemic spreads through a town at a rate that is proportional to the number of uninfected people and to the square of the number of infected people, then the rate is $R(x) = cx^2(p - x)$, where $x$ is the number of infected people and $c$ and $p$ (the population) are positive constants. Show that the rate $R(x)$ is greatest when two-thirds of the population is infected.

**36. BIOMEDICAL: Contagion**  If an epidemic spreads through a town at a rate that is proportional to the number of infected people and to the number of uninfected people, then the rate is $R(x) = cx(p - x)$, where $x$ is the number of infected people and $c$ and $p$ (the population) are positive constants. Show that the rate $R(x)$ is greatest when half of the population is infected.

## 3.5   OPTIMIZING LOT SIZE AND HARVEST SIZE

### Introduction

In this section we discuss two important applications of optimization, one economic and one ecological. The first concerns the most efficient way for a business to order merchandise (or for a manufacturer to produce merchandise), and the second concerns the preservation of animal populations that are harvested by people. Either of these applications may be read independently of the other.

### Minimizing Inventory Costs

A business encounters two kinds of costs in maintaining inventory: **storage costs** (warehouse and insurance costs for merchandise not yet sold) and **reorder costs** (delivery and bookkeeping costs for each order). For example, if a furniture store expects to sell 250 sofas in a year, it could order all 250 at once (incurring high storage costs), or it could order them in many small lots, say 50 orders of five each, spaced throughout the year (incurring high reorder costs). Obviously, the best order size (or **lot size**) is the one that minimizes the total of storage plus reorder costs.

**EXAMPLE 1**

**MINIMIZING INVENTORY COSTS**

A furniture showroom expects to sell 250 sofas a year. Each sofa costs the store $300, and there is a fixed charge of $500 per order. If it costs $100 to store a sofa for a year, how large should each order be and how often should orders be placed to minimize inventory costs?

*Solution*

Let

$$x = \text{lot size} \qquad \text{Number of sofas in each order}$$

**Storage Costs** If the sofas sell steadily throughout the year, and if the store reorders $x$ more whenever the stock runs out, then its inventory during the year looks like the following graph.

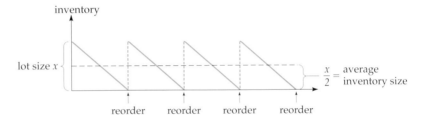

Notice that the inventory level varies from the lot size $x$ down to zero, with an average inventory of $x/2$ sofas throughout the year. Because it costs $100 to store a sofa for a year, the total (annual) storage costs are

$$\begin{pmatrix}\text{Storage} \\ \text{costs}\end{pmatrix} = \begin{pmatrix}\text{Storage} \\ \text{per item}\end{pmatrix} \cdot \begin{pmatrix}\text{Average num-} \\ \text{ber of items}\end{pmatrix}$$

$$= \quad 100 \cdot \frac{x}{2} \quad = 50x$$

**Reorder Costs** Each sofa costs $300, so an order of lot size $x$ costs $300x$, plus the fixed order charge of $500:

$$\begin{pmatrix}\text{Cost} \\ \text{per order}\end{pmatrix} = 300x + 500$$

The yearly supply of 250 sofas, with $x$ sofas in each order, requires $\frac{250}{x}$ orders. (For example, 250 sofas at 5 per order require $\frac{250}{5} = 50$ orders.) Therefore, the yearly reorder costs are

$$\begin{pmatrix}\text{Reorder} \\ \text{costs}\end{pmatrix} = \begin{pmatrix}\text{Cost} \\ \text{per order}\end{pmatrix} \cdot \begin{pmatrix}\text{Number} \\ \text{of orders}\end{pmatrix} = (300x + 500) \cdot \left(\frac{250}{x}\right)$$

**Total Cost** $C(x)$ is storage costs plus reorder costs:

$$C(x) = \begin{pmatrix}\text{Storage} \\ \text{costs}\end{pmatrix} + \begin{pmatrix}\text{Reorder} \\ \text{costs}\end{pmatrix}$$

$$= 100 \frac{x}{2} + (300x + 500)\left(\frac{250}{x}\right)$$    Using the storage and reorder costs found earlier

$$= 50x + 75{,}000 + 125{,}000x^{-1}$$    Simplifying

To minimize $C(x)$, we differentiate:

$$C'(x) = 50 - 125{,}000x^{-2} = 50 - \frac{125{,}000}{x^2}$$    Differentiating $C = 50x + 75{,}000 + 125{,}000x^{-1}$

$$50 - \frac{125{,}000}{x^2} = 0$$    Setting the derivative equal to zero

$$50x^2 = 125{,}000$$    Multiplying by $x^2$ and adding 125,000 to each side

$$x^2 = \frac{125{,}000}{50} = 2500$$    Dividing each side by 50

$$x = 50$$    Taking square roots $(x > 0)$ gives lot size 50

$$C''(x) = 250{,}000x^{-3} = 250{,}000 \frac{1}{x^3}$$    $C''$ is positive, so $C$ is minimized at $x = 50$

At 50 sofas per order, the yearly 250 will require $\frac{250}{50} = 5$ orders. Therefore:

Lot size is 50 sofas, with orders placed five times a year.

## Graphing Calculator Exploration

Minimum
X=50.000024   Y=80000

Interpret as $x = 50$

a. Verify the answer to the previous example by graphing the total cost function $y_1 = 50x + 75{,}000 + 125{,}000/x$ on the window $[0, 200]$ by $[-40{,}000, 150{,}000]$ and using MINIMUM.

b. Notice that the curve is rather flat to the right of $x = 50$. From this observation, if you cannot order exactly 50 at a time, would it be better to order somewhat more than 50 or somewhat less than 50?

### Modifications and Assumptions

If the number of orders per year is not a whole number, say 7.5 orders per year, we just interpret it as 15 orders in 2 years, and handle it accordingly.

We made two major assumptions in Example 1. We assumed that there was a steady demand, and that orders were equally spaced throughout the year. These are reasonable assumptions for many products, while for seasonal products such as bathing suits or winter coats, separate calculations can be done for the "on" and "off" seasons.

### Production Runs

Similar analysis applies to manufacturing. For example, if a book publisher can estimate the yearly demand for a book, she may print the yearly total all at once, incurring high storage costs, or she may print them in several smaller runs throughout the year, incurring setup costs for each run. Here the setup costs for each printing run play the role of the reorder costs for a store.

**EXAMPLE 2**

**MINIMIZING INVENTORY COSTS FOR A PUBLISHER**

A publisher estimates the annual demand for a book to be 4000 copies. Each book costs $8 to print, and setup costs are $1000 for each printing. If storage costs are $2 per book per year, find how many books should be printed per run and how many printings will be needed if costs are to be minimized.

*Solution*

Let
$$x = \text{the number of books in each run}$$

*Storage Costs*   As in Example 1, an average of $\dfrac{x}{2}$ books are stored throughout the year, at a cost of $2 each, so annual storage costs are

$$\begin{pmatrix}\text{Storage}\\\text{costs}\end{pmatrix} = \left(\frac{x}{2}\right) \cdot 2 = x$$

*Production Costs*   The cost per run is

$$\begin{pmatrix}\text{Costs}\\\text{per run}\end{pmatrix} = 8x + 1000 \qquad \text{\scriptsize $x$ books at \$8 each, plus \$1000 setup costs}$$

The 4000 books at $x$ books per run will require $\dfrac{4000}{x}$ runs. Therefore, production costs are

$$\begin{pmatrix}\text{Production}\\\text{costs}\end{pmatrix} = (8x + 1000)\left(\frac{4000}{x}\right) \qquad \text{\scriptsize Cost per run times number of runs}$$

*Total Cost*   The total cost is storage costs plus production costs:

$$C(x) = x + (8x + 1000)\left(\frac{4000}{x}\right) \qquad \text{\scriptsize Storage + production}$$

$$= x + 32{,}000 + 4{,}000{,}000x^{-1} \qquad \text{\scriptsize Multiplying out}$$

We differentiate, set the derivative equal to zero, and solve, just as before (omitting the details), obtaining $x = 2000$. The second-derivative test will show that costs are minimized at 2000 books per run. The 4000 books require $\frac{4000}{2000} = 2$ printings. Therefore, the publisher should:

Print 2000 books per run, with two printings.

## Maximum Sustainable Yield

The next application involves industries such as fishing, in which a naturally occurring animal population is "harvested." Harvesting some of the population means more food and other resources for the remaining population so that it will expand and replace the harvested portion. But taking too large a harvest will kill off the animal population (like the bowhead whale, hunted almost to extinction in the nineteenth century). We want to find the **maximum sustainable yield,** the largest amount that may be harvested year after year and still have the population return to its previous level the following year.

For some animals one can determine a **reproduction function** $f(p)$ which gives the expected population a year from now if the present population is $p$.

---

**Reproduction Function**

A reproduction function $f(p)$ gives the population
a year from now if the current population is $p$.

---

For example, the reproduction function $f(p) = -\frac{1}{4}p^2 + 3p$ (where $p$ and $f(p)$ are measured in thousands) means that if the population is now $p = 6$ (thousand), then a year from now the population will be

$$f(6) = -\frac{1}{4}6^2 + 3 \cdot 6 = -9 + 18 = 9 \qquad \text{(thousand)}$$

Therefore, during the year the population will increase from 6000 to 9000.

If, on the other hand, the present population is $p = 10$ (thousand), a year later the population will be

$$f(10) = -\frac{1}{4} \cdot 10^2 + 3 \cdot 10 = -25 + 30 = 5 \qquad \text{(thousand)}$$

That is, during the year the population will decline from 10,000 to 5000 (perhaps because of inadequate food to support such a large population). In actual practice, reproduction functions are very difficult to calculate, but can sometimes be estimated by analyzing previous population and harvest data.[*]

Suppose that we have a reproduction function $f$ and a current population of size $p$, which will therefore grow to size $f(p)$ next year. The *amount of growth* in the population during that year is

$$\begin{pmatrix} \text{Amount} \\ \text{of growth} \end{pmatrix} = f(p) - p$$

Current population
Next year's population

Harvesting this amount removes only the *growth*, returning the population to its former size $p$. The population will then repeat this growth, and taking the

---

[*]For more information, see J. Blower, L. Cook, and J. Bishop, *Estimating the Size of Animal Populations* (London: George Allen and Unwin Ltd., 1981).

same harvest $f(p) - p$ will cause this situation to repeat itself year after year. The quantity $f(p) - p$ is called the *sustainable yield*.

---

**Sustainable Yield**

For reproduction function $f(p)$, the sustainable yield is

$$Y(p) = f(p) - p$$

---

We want the population size $p$ that maximizes the sustainable yield $Y(p)$. To maximize $Y(p)$, we set its derivative equal to zero:

$$Y'(p) = f'(p) - 1 = 0 \qquad \text{Derivative of } Y = f(p) - p$$
$$f'(p) = 1 \qquad \text{Solving for } f'(p)$$

For a given reproduction function $f(p)$, we find the maximum sustainable yield by solving this equation (provided that the second-derivative test gives $Y''(p) = f''(p) < 0$).

---

**Maximum Sustainable Yield**

For reproduction function $f(p)$, the population $p$ that results in the maximum sustainable yield is the solution to

$$f'(p) = 1$$

(provided that $f''(p) < 0$). The maximum sustainable yield is then

$$Y(p) = f(p) - p$$

---

Once we calculate the population $p$ that gives the maximum sustainable yield, we wait until the population reaches this size and then harvest, year after year, an amount $Y(p)$.

 **BE CAREFUL:** Note that to find the maximum sustainable yield we set the derivative $f'(p)$ equal to 1, not 0. This is because we are maximizing not the reproductive function $f(p)$ but rather the yield function $Y(p) = f(p) - p$.

**EXAMPLE 3**  **FINDING MAXIMUM SUSTAINABLE YIELD**

The reproduction function for the American lobster in an East Coast fishing area is $f(p) = -0.02p^2 + 2p$ (where $p$ and $f(p)$ are in thousands). Find the population $p$ that gives the maximum sustainable yield and find the size of the yield.

*Solution*

We set the derivative of the reproduction function equal to 1:

$$f'(p) = -0.04p + 2 = 1 \qquad \text{Differentiating } f(p) = -0.02p^2 + 2p$$
$$-0.04p = -1 \qquad \text{Subtracting 2 from each side}$$
$$p = \frac{-1}{-0.04} = 25 \qquad \text{Dividing by } -0.04$$

The second derivative is $f''(p) = -0.04$, which is negative, showing that $p = 25$ (thousand) is the population that gives the maximum sustainable yield. The actual yield is found from the yield function $Y(p) = f(p) - p$:

$$Y(p) = \underbrace{-0.02p^2 + 2p}_{f(p)} - p = -0.02p^2 + p$$

$$Y(25) = -0.02(25)^2 + 25 \qquad \text{Evaluating at } p = 25$$
$$= -12.5 + 25 = 12.5 \qquad \text{(thousand)}$$

The population size for the maximum sustainable yield is 25,000, and the yield is 12,500 lobsters. _25,000 from $p = 25$_

## Graphing Calculator Exploration

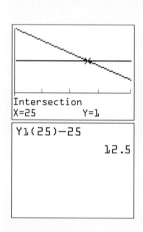

Intersection
X=25        Y=1

Y1(25)−25
            12.5

Solve the previous example on a graphing calculator as follows:

**a.** Enter the reproduction function as $y_1 = -0.02x^2 + 2x$.

**b.** Define $y_2$ to be the derivative of $y_1$ (using NDERIV).

**c.** Define $y_3 = 1$.

**d.** Turn off $y_1$ and graph $y_2$ and $y_3$ on the window [0, 40] by [−0.5, 2].

**e.** Use INTERSECT to find where $y_2$ and $y_3$ meet, thereby solving $f' = 1$. (You should find $x = 25$, as above.)

**f.** Find the yield by evaluating $y_1 - x$ at the $x$-value found in part (e).

In this problem, solving "by hand" was probably easier, but this graphing calculator method may be preferable if the reproduction function is more complicated.

## 3.5 Exercises

### Lot Size

**1.** A supermarket expects to sell 4000 boxes of sugar in a year. Each box costs $2, and there is a fixed delivery charge of $20 per order. If it costs $1 to store a box for a year, what is the order size and how many times a year should the orders be placed to minimize inventory costs?

**2.** A supermarket expects to sell 5000 boxes of rice in a year. Each box costs $2, and there is a fixed delivery charge of $50 per order. If it costs $2 to store a box for a year, what is the order size and how many times a year should the orders be placed to minimize inventory costs?

**3.** A liquor warehouse expects to sell 10,000 bottles of scotch whiskey in a year. Each bottle costs $12, plus a fixed charge of $125 per order. If it costs $10 to store a bottle for a year, how many bottles should be ordered at a time and how many orders should the warehouse place in a year to minimize inventory costs?

**4.** A wine warehouse expects to sell 30,000 bottles of wine in a year. Each bottle costs $9, plus a fixed charge of $200 per order. If it costs $3 to store a bottle for a year, how many bottles should be ordered at a time and how many orders should the warehouse place in a year to minimize inventory costs?

5. An automobile dealer expects to sell 800 cars a year. The cars cost $9000 each plus a fixed charge of $1000 per delivery. If it costs $1000 to store a car for a year, find the order size and the number of orders that minimize inventory costs.

6. An automobile dealer expects to sell 400 cars a year. The cars cost $11,000 each plus a fixed charge of $500 per delivery. If it costs $1000 to store a car for a year, find the order size and the number of orders that minimize inventory costs.

## Production Runs

7. A toy manufacturer estimates the demand for a game to be 2000 per year. Each game costs $3 to manufacture, plus setup costs of $500 for each production run. If a game can be stored for a year for a cost of $2, how many should be manufactured at a time and how many production runs should there be to minimize costs?

8. A toy manufacturer estimates the demand for a doll to be 10,000 per year. Each doll costs $5 to manufacture, plus setup costs of $800 for each production run. If it costs $4 to store a doll for a year, how many should be manufactured at a time and how many production runs should there be to minimize costs?

9. A producer of audio tapes estimates the yearly demand for a tape to be 1,000,000. It costs $800 to set up the machinery for the tape, plus $10 for each tape produced. If it costs the company $1 to store a tape for a year, how many should be produced at a time and how many production runs will be needed to minimize costs?

10. A compact disc manufacturer estimates the yearly demand for a CD to be 10,000. It costs $400 to set the machinery for the CD, plus $3 for each CD produced. If it costs the company $2 to store a CD for a year, how many should be burned at a time and how many production runs will be needed to minimize costs?

## Maximum Sustainable Yield

11. Marine ecologists estimate the reproduction curve for swordfish in the Georges Bank fishing grounds to be $f(p) = -0.01p^2 + 5p$, where $p$ and $f(p)$ are in hundreds. Find the population that gives the maximum sustainable yield, and the size of the yield.

12. The reproduction function for the Hudson Bay lynx is estimated to be $f(p) = -0.02p^2 + 5p$, where $p$ and $f(p)$ are in thousands. Find the population that gives the maximum sustainable yield, and the size of the yield.

13. The reproduction function for the Antarctic blue whale is estimated to be $f(p) = -0.0004p^2 + 1.06p$, where $p$ and $f(p)$ are in thousands. Find the population that gives the maximum sustainable yield, and the size of the yield.

14. The reproduction function for the Canadian snowshoe hare is estimated to be $f(p) = -0.025p^2 + 4p$, where $p$ and $f(p)$ are in thousands. Find the population that gives the maximum sustainable yield, and the size of the yield.

15. The reproduction function for king salmon in the Bering Sea is $f(p) = -\frac{1}{8}p^2 + 7.5p$, where $p$ and $f(p)$ are in thousand metric tons. Find the population that gives the maximum sustainable yield, and the size of the yield.

*Source: Ecological Modeling, 2006*

16. The reproduction function for the anchovy in the Peruvian fishery is $f(p) = -0.011p^2 + 1.33p$, where $p$ and $f(p)$ are in million metric tons. Find the population that gives the maximum sustainable yield, and the size of the yield. [*Note:* This catch was exceeded in the early 1970s, and this, along with other factors, caused a collapse of the Peruvian fishing economy.]

*Source: Journal of the Fisheries Research Board of Canada* **30**

17. A conservation commission estimates the reproduction function for rainbow trout in a large lake to be $f(p) = 50\sqrt{p}$, where $p$ and $f(p)$ are in thousands and $p \leq 1000$. Find the population that gives the maximum sustainable yield, and the size of the yield.

18. The reproduction function for oysters in a large bay is $f(p) = 30\sqrt[3]{p^2}$, where $p$ and $f(p)$ are in pounds and $p \leq 10,000$. Find the size of the population that gives the maximum sustainable yield, and the size of the yield.

19. The reproduction function for the Pacific sardine (from Baja California to British Columbia) is $f(p) = -0.0005p^2 + 2p$, where $p$ and $f(p)$ are in hundred metric tons. Find the population that gives the maximum sustainable yield, and the size of the yield.

*Source: Ecology* **48**

**20. BUSINESS: Exploring a Lot Size Problem**
Use a graphing calculator to explore Example 1 (pages 231–232) as follows:

a. Enter the total cost function, unsimplified, as $y_1 = 100(x/2) + (300x + 500)(250/x)$.

b. Graph $y_1$ on the window [0, 200] by [0, 150,000] and use MINIMUM to minimize it. Your answer should agree with the $x = 50$ found in Example 1.

c. Suppose that business improves, and the showroom expects to sell 350 per year instead of 250. In $y_1$, change the 250 to 350 and minimize it.

d. Suppose that a modest recession decreases sales (so change the 350 back to 250), and that inflation has driven the cost of storage up to $125 (so change the 100 to 125), and minimize $y_1$.

### Conceptual Exercises

**21.** If lot size becomes very large, will storage costs increase or decrease? Will reorder costs increase or decrease?

**22.** If lot size becomes very small, will storage costs increase or decrease? Will reorder costs increase or decrease?

**23.** If computer printing technology reduces setup costs (as it already has), would you expect print-run size to increase or decrease?

**24.** If lot size is $x$, we say that the average inventory size is $x/2$. What assumptions are we making?

**25.** For a reproduction function $f(p)$, explain the serious consequences that will result if $f(p) < p$ for all values of $p$.

**26.** Explain why a reproduction function $f(p)$ must satisfy $f(0) = 0$.

**27.** Although reproduction functions can sometimes be estimated by analyzing past population and harvest data, they are, in practice, very difficult to find. Give one reason for this difficulty.

**28.** Is it realistic to have a reproduction function that is increasing for all values of $p > 0$, such as $f(p) = 5p$ or $f(p) = p^2$?

## 3.6 IMPLICIT DIFFERENTIATION AND RELATED RATES

### Introduction

A function written in the form $y = f(x)$ is said to be defined *explicitly*, meaning that $y$ is defined by a rule or *formula $f(x)$ in $x$ alone*. A function may instead be defined **implicitly,** meaning that $y$ is defined by an *equation in $x$ and $y$*, such as $x^2 + y^2 = 25$. In this section we will see how to differentiate such **implicit functions** when ordinary "explicit" differentiation is difficult or impossible. We will then use implicit differentiation to find rates of change.

### Implicit Differentiation

The equation $x^2 + y^2 = 25$ defines a circle. While a circle is not the graph of a function (it violates the vertical line test, see page 35), the top half by itself defines a function, as does the bottom half by itself. To find these two functions, we solve $x^2 + y^2 = 25$ for $y$:

$$y^2 = 25 - x^2$$

Subtracting $x^2$ from each side of $x^2 + y^2 = 25$

$$y = \pm\sqrt{25 - x^2}$$

Plus or minus since when squared either one gives $25 - x^2$

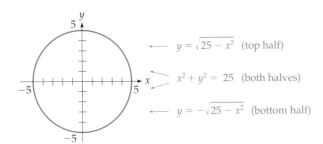

The *positive* square root defines the top half of the circle (where $y$ is positive), and the *negative* square root defines the bottom half (where $y$ is negative). The equation $x^2 + y^2 = 25$ defines *both* functions at the same time.

To find the slope anywhere on the circle, we could differentiate the "top" and "bottom" functions separately. However, it is easier to find both answers at once by differentiating *implicitly*, that is by differentiating both sides of the equation $x^2 + y^2 = 25$ with respect to $x$. Remember, however, that $y$ is a *function* of $x$, so differentiating $y^2$ means differentiating a *function* squared, which requires the Generalized Power Rule:

$$\frac{d}{dx}\, y^n = n \cdot y^{n-1}\, \frac{dy}{dx}$$

**EXAMPLE 1**

**DIFFERENTIATING IMPLICITLY**

Use implicit differentiation to find $\dfrac{dy}{dx}$ when $x^2 + y^2 = 25$.

*Solution*

We differentiate both sides of the equation with respect to $x$:

$$\frac{d}{dx}\, x^2 + \frac{d}{dx}\, y^2 = \frac{d}{dx}\, 25 \qquad \text{Differentiating } x^2 + y^2 = 25$$

$$2x \;+\; 2y\,\frac{dy}{dx} = 0 \qquad \text{Using the Generalized Power Rule on } y^2$$

Solving for $\dfrac{dy}{dx}$:

$$2y\,\frac{dy}{dx} = -2x \qquad \text{Subtracting } 2x$$

$$\frac{dy}{dx} = -\frac{x}{y} \qquad \begin{array}{l}\text{Canceling the 2's}\\ \text{and dividing by } y\end{array}$$

Therefore, $\dfrac{dy}{dx} = -\dfrac{x}{y}$ when $x$ and $y$ are related by $x^2 + y^2 = 25$.

Notice that the formula for $\dfrac{dy}{dx}$ involves both $x$ and $y$. Implicit differentiation enables us to find derivatives that would otherwise be difficult or impossible to calculate, but at a "cost"—the result may depend on both $x$ and $y$.

**BE CAREFUL:** Remember that $x$ and $y$ play different roles: $x$ is the *independent* variable, and $y$ is a *function*. Therefore, we must include a $\dfrac{dy}{dx}$ (from the Generalized Power Rule) when differentiating $y^n$, but not when differentiating $x^n$ since $\dfrac{dx}{dx} = 1$.

---

**EXAMPLE 2**

**EVALUATING AN IMPLICIT DERIVATIVE (*Continuation of Example 1*)**

Find the slope of the circle $x^2 + y^2 = 25$ at the points (3, 4) and (3, −4).

*Solution*

We simply evaluate the derivative $\dfrac{dy}{dx} = -\dfrac{x}{y}$ (found in Example 1) at the given points.

At (3, 4): $\dfrac{dy}{dx} = -\dfrac{3}{4}$ ← $x$
← $y$

At (3, −4): $\dfrac{dy}{dx} = -\dfrac{3}{-4} = \dfrac{3}{4}$

Note that the negative sign in $\dfrac{dy}{dx} = -\dfrac{x}{y}$ gives the slope the correct sign: negative at (3, 4) and positive at (3, −4).

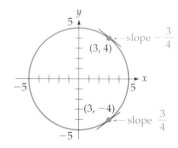

---

# Graphing Calculator Exploration

a. Graph the entire circle of the previous Example by graphing $y_1 = \sqrt{25 - x^2}$ and $y_2 = -\sqrt{25 - x^2}$. You may have to adjust the window (or use Zoom ZSquare) to make the circle look "circular."

b. Verify the answer to Example 2 by finding the derivatives of $y_1$ and $y_2$ at $x = 3$.

c. Can you find the derivative at $x = 5$? Why not?

---

 **BE CAREFUL:** Derivatives should be evaluated only at points on the curve, so we evaluate $\dfrac{dy}{dx}$ only at $x$- and $y$-values *satisfying the original equation.* (It is easy to check that $x = 3$ and $y = \pm 4$ *do* satisfy $x^2 + y^2 = 25$.) Evaluating at a point not on the curve, such as (2, 3), would give a meaningless result.

The following are typical "pieces" that might appear in implicit differentiation problems. Studying them carefully will help you to do longer problems.

**EXAMPLE 3**

**FINDING DERIVATIVES—IMPLICIT AND EXPLICIT**

**a.** $\dfrac{d}{dx} y^3 = 3y^2 \dfrac{dy}{dx}$ 

Differentiating $y^3$, so *include* $\dfrac{dy}{dx}$

**b.** $\dfrac{d}{dx} x^3 = 3x^2$ 

Differentiating $x^3$, so *no* $\dfrac{dx}{dx}$

**c.** $\dfrac{d}{dx}(x^3 y^5) = 3x^2 \cdot y^5 + x^3 \cdot 5y^4 \dfrac{dy}{dx}$ 

Using the Product Rule

$$\underbrace{\quad}_{\frac{d}{dx} x^3} \qquad \underbrace{\quad}_{\frac{d}{dx} y^5}$$

$$= 3x^2 y^5 + 5x^3 y^4 \dfrac{dy}{dx}$$

Try to do problems such as this in one step, putting the constants in front from the start

**Practice Problem**

Find: **a.** $\dfrac{d}{dx} x^4$ **b.** $\dfrac{d}{dx} y^2$ **c.** $\dfrac{d}{dx}(x^2 y^3)$

➤ **Solutions on page 246**

Implicit differentiation involves three steps.

> **Finding $\dfrac{dy}{dx}$ by Implicit Differentiation**
>
> **1.** Differentiate both sides of the equation *with respect to x*. When differentiating a $y$, include $\dfrac{dy}{dx}$.
>
> **2.** Collect all terms involving $\dfrac{dy}{dx}$ on one side, and all others on the other side.
>
> **3.** Factor out the $\dfrac{dy}{dx}$ and solve for it by dividing.

**EXAMPLE 4**

**FINDING AND EVALUATING AN IMPLICIT DERIVATIVE**

For $y^4 + x^4 - 2x^2 y^2 = 9$ **a.** find $\dfrac{dy}{dx}$ **b.** evaluate it at $x = 2, y = 1$

*Solution*

$$4y^3 \dfrac{dy}{dx} + 4x^3 - 4xy^2 - 4x^2 y \dfrac{dy}{dx} = 0$$

Differentiating with respect to $x$, putting constants first

$$4y^3 \frac{dy}{dx} - 4x^2 y \frac{dy}{dx} = -4x^3 + 4xy^2 \qquad \text{Collecting } dy/dx \text{ terms on the left, others on the right}$$

$$(4y^3 - 4x^2 y) \frac{dy}{dx} = -4x^3 + 4xy^2 \qquad \text{Factoring out } \frac{dy}{dx}$$

$$\frac{dy}{dx} = \frac{-4x^3 + 4xy^2}{4y^3 - 4x^2 y} \qquad \leftarrow \begin{array}{l} \text{Answer for} \\ \text{part (a)} \end{array} \qquad \text{Dividing by } 4y^3 - 4x^2 y \text{ to solve for } dy/dx$$

$$= \frac{-x^3 + xy^2}{y^3 - x^2 y} \qquad \text{Dividing by 4}$$

$$\frac{dy}{dx} = \frac{-(2)^3 + (2)(1)^2}{(1)^3 - (2)^2(1)} = \frac{-6}{-3} = 2 \qquad \text{Evaluating at } x = 2, y = 1 \text{ gives the answer for part (b)}$$

Note that in the example above the given point *is* on the curve, since $x = 2$ and $y = 1$ satisfy the original equation:

$$1^4 + 2^4 - 2 \cdot 2^2 \cdot 1 = 1 + 16 - 8 = 9$$

In economics, a **demand equation** is the relationship between the price $p$ of an item and the quantity $x$ that consumers will demand at that price. (All prices are in dollars unless otherwise stated.)

**EXAMPLE 5**  **FINDING AND INTERPRETING AN IMPLICIT DERIVATIVE**

For the demand equation $x = \sqrt{1900 - p^3}$, use implicit differentiation to find $dp/dx$. Then evaluate it at $p = 10$ and interpret your answer.

*Solution*

$$x^2 = 1900 - p^3 \qquad \text{Simplifying by squaring both sides of } x = \sqrt{1900 - p^3}$$

$$2x = -3p^2 \frac{dp}{dx} \qquad \text{Differentiating both sides with respect to } x$$

$$\frac{dp}{dx} = -\frac{2x}{3p^2} \qquad \text{Solving for } \frac{dp}{dx}$$

To find the value of $x$ we substitute $p = 10$ into the original equation:

$$x = \sqrt{1900 - 10^3} = \sqrt{1900 - 1000} = \sqrt{900} = 30 \qquad \begin{array}{l} x = \sqrt{1900 - p^3} \\ \text{with } p = 10 \end{array}$$

Then

$$\frac{dp}{dx} = -\frac{60}{300} = -0.2 \qquad \qquad \frac{dp}{dx} = -\frac{2x}{3p^2} \text{ with } p = 10, x = 30$$

*Interpretation:* $dp/dx = -0.2$ says that the rate of change of price with respect to quantity is $-0.2$, so that increasing the quantity by 1 means decreasing the price by 0.20 (or 20 cents). Therefore, each 20-cent price decrease brings approximately one more sale (at the given value of $p$).

Our first step of squaring both sides of the original equation was not necessary, but it made the differentiation easier by avoiding the Generalized Power Rule.

Notice that this particular demand function $x = \sqrt{1900 - p^3}$ can be solved *explicitly* for $p$:

$$x^2 = 1900 - p^3 \qquad \text{Squaring}$$

$$p^3 = 1900 - x^2 \qquad \text{Adding } p^3 \text{ and subtracting } x^2$$

$$p = (1900 - x^2)^{1/3} \qquad \text{Taking cube roots}$$

We can differentiate this *explicitly* with respect to $x$:

$$\frac{dp}{dx} = \frac{1}{3}(1900 - x^2)^{-2/3}(-2x) \qquad \text{Using the Generalized Power Rule}$$

$$= -\frac{2}{3}x(1900 - x^2)^{-2/3} \qquad \text{Simplifying}$$

Evaluating at the value of $x$ found in Example 5 gives

$$\frac{dp}{dx} = -\frac{2}{3} \cdot 30(1900 - 30^2)^{-2/3} \qquad \text{Substituting } x = 30$$

$$= -20(1000)^{-2/3} = -\frac{20}{100} = -0.2 \qquad \text{Simplifying}$$

This agrees with the answer by implicit differentiation. Which way was easier?

## Related Rates

Sometimes *both* variables in an equation will be functions of a *third* variable, usually $t$ for time. For example, for a seasonal product such as winter coats, the price $p$ and weekly sales $x$ will be related by a demand equation, and both price $p$ and quantity $x$ will depend on the time of year. Differentiating both sides of the demand equation with respect to time $t$ will give an equation relating the derivatives $dp/dt$ and $dx/dt$. Such "related rates" equations show how fast one quantity is changing relative to another. First, an "everyday" example.

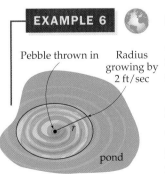

**EXAMPLE 6**

Pebble thrown in   Radius growing by 2 ft/sec

*r*

pond

**FINDING RELATED RATES**

A pebble thrown into a pond causes circular ripples to radiate outward. If the radius of the outer ripple is growing by 2 feet per second, how fast is the area of its circle growing at the moment when the radius is 10 feet?

*Solution*

The formula for the area of a circle is $A = \pi r^2$. Both the area $A$ and the radius $r$ of the circle increase with time, so both are functions of $t$. We are told that the radius is increasing by 2 feet per second $(dr/dt = 2)$, and we want to know how fast the area is changing ($dA/dt$). To find the relationship between $dA/dt$ and $dr/dt$, we differentiate both sides of $A = \pi r^2$ with respect to $t$.

$$\frac{dA}{dt} = 2\pi r \cdot \frac{dr}{dt}$$

From $A = \pi r^2$, writing the 2 before the $\pi$

$$\frac{dA}{dt} = 2\pi \cdot \underbrace{10}_{r} \cdot \underbrace{2}_{\frac{dr}{dt}} = \underbrace{40\pi \approx 125.6}_{\substack{\text{Using} \\ \pi \approx 3.14}}$$

Substituting $r = 10$ and $\frac{dr}{dt} = 2$

Therefore, at the moment when the radius is 10 feet, the area of the circle is growing at the rate of about 126 square feet per second.

We should be ready to interpret any *rate* as a derivative, just as we interpreted the radius growing by 2 feet per second as $dr/dt = 2$ and the growth rate of the area as $dA/dt$. The general procedure is as follows.

---

**To Solve a Related Rate Problem**

1. Determine the quantities that are changing with time.
2. Find an equation that relates these quantities (a diagram may be helpful).
3. Differentiate both sides of this equation implicitly with respect to $t$.
4. Substitute into the new equation any given values for the variables and for the derivatives (interpreted as rates of change).
5. Solve for the remaining derivative and interpret the answer as a rate of change.

---

**EXAMPLE 7**    **USING RELATED RATES TO FIND PROFIT GROWTH**

A boat yard's total profit from selling $x$ outboard motors is $P = -x^2 + 1000x - 2000$. If the outboards are selling at the rate of 20 per week, how fast is the profit changing when 400 motors have been sold?

*Solution*

Profit $P$ and quantity $x$ both change with time, so both are functions of $t$. We differentiate both sides of $P = -x^2 + 1000x - 2000$ with respect to $t$ and then substitute the given data.

$$\frac{dP}{dt} = -2x\frac{dx}{dt} + 1000\frac{dx}{dt}$$

Differentiating with respect to $t$

$$= -2 \cdot \underbrace{400}_{x} \cdot \underbrace{20}_{dx/dt} + 1000 \cdot \underbrace{20}_{dx/dt}$$

Substituting $x = 400$ (number sold) and $dx/dt = 20$ (sales per week)

$$= -16{,}000 + 20{,}000 = 4000$$

Therefore, the company's profits are growing at the rate of $4000 per week.

**EXAMPLE 8**  **USING RELATED RATES TO PREDICT POLLUTION**

A study of urban pollution predicts that sulfur oxide emissions in a city will be $S = 2 + 20x + 0.1x^2$ tons, where $x$ is the population (in thousands). The population of the city $t$ years from now is expected to be $x = 800 + 20\sqrt{t}$ thousand people. Find how rapidly the sulfur oxide pollution will be increasing 4 years from now.

*Solution* Finding the rate of increase of pollution means finding $\dfrac{dS}{dt}$.

$$\frac{dS}{dt} = 20\frac{dx}{dt} + 0.2x\frac{dx}{dt} \qquad \begin{array}{l} S = 2 + 20x + 0.1x^2 \\ \text{differentiated with respect to } t \\ (x \text{ is also a function of } t) \end{array}$$

We then find $dx/dt$ from the other given equation:

$$\frac{dx}{dt} = 10t^{-1/2} \qquad \begin{array}{l} x = 80 + 20t^{1/2} \text{ differentiated} \\ \text{with respect to } t \end{array}$$

$$= 10 \cdot 4^{-1/2} = 10 \cdot \frac{1}{2} = 5 \qquad \begin{array}{l} \text{Substituting the given } t = 4 \\ \text{gives } dx/dt = 5 \end{array}$$

$\dfrac{dS}{dt}$ then becomes

$$\frac{dS}{dt} = \underbrace{20 \cdot 5}_{dx/dt} + 0.2\underbrace{(800 + 20\sqrt{4})}_{x}\underbrace{5}_{dx/dt} \qquad \begin{array}{l} \frac{dS}{dt} = 20\frac{dx}{dt} + 0.2x\frac{dx}{dt} \\ \text{with } dx/dt = 5 \text{ and} \\ x = 800 + 20t^{1/2} \text{ at } t = 4 \end{array}$$

$$= 100 + 0.2(840)5 = 100 + 840 = 940$$

Therefore, in 4 years the sulfur oxide emissions will be increasing at the rate of 940 tons per year.

## Graphing Calculator Exploration

Y4=nDeriv(Y3,X,X)

$y_2$
$y_4$

$y_1$

X=4    Y=940.00001

Interpret as 940

Verify the answer to the preceding example on a graphing calculator as follows:

a. Define $y_1 = 2 + 20x + 0.1x^2$ (the sulfur oxide function).

b. Define $y_2 = 800 + 20\sqrt{x}$ (the population function, using $x$ for ease of entry).

c. Define $y_3 = y_1(y_2)$ (the composition of $y_1$ and $y_2$, giving pollution in year $x$).

d. Define $y_4$ to be the derivative of $y_3$ (using NDERIV, giving rate of change of pollution).

e. Graph these on the window [0, 10] by [−200, 1500]. (Which function does *not* appear on the screen?)

f. Evaluate $y_4$ at $x = 4$ to verify the answer to Example 8.

**Solution to Practice Problem**

a. $\dfrac{d}{dx}\, x^4 = 4x^3$    b. $\dfrac{d}{dx}\, y^2 = 2y\,\dfrac{dy}{dx}$    c. $\dfrac{d}{dx}(x^2 y^3) = 2xy^3 + 3x^2 y^2\,\dfrac{dy}{dx}$

## 3.6   Section Summary

An equation in $x$ and $y$ may define one or more functions $y = f(x)$, which we may need to differentiate. Instead of solving the equation for $y$, which may be difficult or impossible, we can differentiate *implicitly*, differentiating both sides of the original equation with respect to $x$ (writing a $dy/dx$ or $y'$ whenever we differentiate $y$) and solving for the derivative $dy/dx$. The derivative at any point of the curve may then be found by substituting the coordinates of that point.

Implicit differentiation is especially useful when several variables in an equation depend on an underlying variable, usually $t$ for time. Differentiating the equation implicitly with respect to this underlying variable gives an equation involving the rates of change of the original variables. Numbers may then be substituted into this "related rate equation" to find a particular rate of change.

## Verification of the Power Rule for Rational Powers

On page 109 we stated the Power Rule for differentiation:

$$\frac{d}{dx}\, x^n = nx^{n-1}$$

Although we have proved it only for *integer* powers, we have been using the Power Rule for *all* constant powers $n$. Using implicit differentiation, we may now prove the Power Rule for *rational* powers. (Recall that a rational number is of the form $p/q$, where $p$ and $q$ are integers with $q \neq 0$.) Let $y = x^n$ for a rational exponent $n = p/q$, and let $x$ be a number at which $x^{p/q}$ is differentiable. Then

| | |
|---|---|
| $y = x^n = x^{p/q}$ | Since $n = p/q$ |
| $y^q = x^p$ | Raising each side to the power $q$ |
| $qy^{q-1}\dfrac{dy}{dx} = px^{p-1}$ | Differentiating each side implicitly with respect to $x$ |
| $\dfrac{dy}{dx} = \dfrac{px^{p-1}}{qy^{q-1}}$ | Dividing each side by $qy^{q-1}$ |

$$= \frac{px^{p-1}}{q\left(x^{\frac{p}{q}}\right)^{q-1}} = \frac{p}{q}\frac{x^{p-1}}{x^{p-\frac{p}{q}}}$$

Using $y = x^{p/q}$ and multiplying out the exponents in the denominator

$$= \frac{p}{q}x^{\underbrace{p-1-\left(p-\frac{p}{q}\right)}_{-1+\frac{p}{q}}} = \frac{p}{q}x^{\frac{p}{q}-1} = nx^{n-1}$$

Subtracting powers, simplifying, and replacing $p/q$ by $n$ (twice)

This is what we wanted to show, that the derivative of $y = x^n$ is $dy/dx = nx^{n-1}$ for any rational exponent $n = p/q$. This proves the Power Rule for rational exponents.

## 3.6 Exercises

**1–20.** For each equation, use implicit differentiation to find $dy/dx$.

**1.** $y^3 - x^2 = 4$
**2.** $y^2 = x^4$
**3.** $x^3 = y^2 - 2$
**4.** $x^2 + y^2 = 1$
**5.** $y^4 - x^3 = 2x$
**6.** $y^2 = 4x + 1$
**7.** $(x + 1)^2 + (y + 1)^2 = 18$
**8.** $xy = 12$
**9.** $x^2y = 8$
**10.** $x^2y + xy^2 = 4$
**11.** $xy - x = 9$
**12.** $x^3 + 2xy^2 + y^3 = 1$
**13.** $x(y - 1)^2 = 6$
**14.** $(x - 1)(y - 1) = 25$
**15.** $y^3 - y^2 + y - 1 = x$
**16.** $x^2 + y^2 = xy + 4$
**17.** $\frac{1}{x} + \frac{1}{y} = 2$
**18.** $\sqrt[3]{x} + \sqrt[3]{y} = 2$
**19.** $x^3 = (y - 2)^2 + 1$
**20.** $\sqrt{xy} = x + 1$

**21–28.** For each equation, find $dy/dx$ evaluated at the given values.

**21.** $y^2 - x^3 = 1$ at $x = 2, y = 3$
**22.** $x^2 + y^2 = 25$ at $x = -3, y = 4$
**23.** $y^2 = 6x - 5$ at $x = 1, y = -1$
**24.** $xy = 12$ at $x = 6, y = 2$
**25.** $x^2y + y^2x = 0$ at $x = -2, y = 2$
**26.** $y^2 + y + 1 = x$ at $x = 1, y = -1$
**27.** $x^2 + y^2 = xy + 7$ at $x = 3, y = 2$
**28.** $\sqrt[3]{x} + \sqrt[3]{y} = 3$ at $x = 1, y = 8$

**29–36.** For each demand equation, use implicit differentiation to find $dp/dx$.

**29.** $p^2 + p + 2x = 100$
**30.** $p^3 + p + 6x = 50$
**31.** $12p^2 + 4p + 1 = x$
**32.** $8p^2 + 2p + 100 = x$
**33.** $xp^3 = 36$
**34.** $xp^2 = 96$
**35.** $(p + 5)(x + 2) = 120$
**36.** $(p - 1)(x + 5) = 24$

## Applied Exercises on Implicit Differentiation

**37. BUSINESS: Demand Equation** A company's demand equation is $x = \sqrt{2000 - p^2}$, where $p$ is the price in dollars. Find $dp/dx$ when $p = 40$ and interpret your answer.

**38. BUSINESS: Demand Equation** A company's demand equation is $x = \sqrt{2900 - p^2}$, where $p$ is the price in dollars. Find $dp/dx$ when $p = 50$ and interpret your answer.

**39. BUSINESS: Sales** The number $x$ of MP3 music players that a store will sell and their price $p$ (in dollars) are related by the equation $2x^2 = 15,000 - p^2$. Find $dx/dp$ at $p = 100$ and interpret your answer. [*Hint:* You will have to find the value of $x$ by substituting the given value of $p$ into the original equation.]

40. **BUSINESS: Supply** The number $x$ of automobile tires that a factory will supply and their price $p$ (in dollars) are related by the equation $x^2 = 8000 + 5p^2$. Find $dx/dp$ at $p = 80$ and interpret your answer. [*Hint:* You will have to find the value of $x$ by substituting the given value of $p$ into the original equation.]

41. **BUSINESS: Work Hours** A management consultant estimates that the number $h$ of hours per day that employees will work and their daily pay of $p$ dollars are related by the equation $60h^5 + 2{,}000{,}000 = p^3$. Find $dh/dp$ at $p = 200$ and interpret your answer.

42. **BIOMEDICAL: Bacteria** The number $x$ of bacteria of type X and the number $y$ of type Y that can coexist in a cubic centimeter of nutrient are related by the equation $2xy^2 = 4000$. Find $dy/dx$ at $x = 5$ and interpret your answer.

43. **BUSINESS: Sales** If a company spends $r$ million dollars on research, its sales will be $s$ million dollars, where $r$ and $s$ are related by $s^2 = r^3 - 55$.

    a. Find $ds/dr$ by implicit differentiation and evaluate it at $r = 4$, $s = 3$. [*Hint:* Differentiate the equation with respect to $r$.]

    b. Find $dr/ds$ by implicit differentiation and evaluate it at $r = 4$, $s = 3$. [*Hint:* Differentiate the original equation with respect to $s$.]

    c. Interpret your answers to parts (a) and (b) as rates of change.

44. **BIOMEDICAL: Muscle Contraction** When a muscle lifts a load, it does so according to the "fundamental equation of muscle contraction," also known as Hill's equation, $(L + m)(V + n) = k$, where $L$ is the load that the muscle is lifting, $V$ is the velocity of contraction of the muscle, and $m$, $n$, and $k$ are constants. Use implicit differentiation to find $dV/dL$.
    *Source:* E. Batschelet, *Introduction to Mathematics for Life Scientists*

## Exercises on Related Rates

**45–52.** In each equation, $x$ and $y$ are functions of $t$. Differentiate with respect to $t$ to find a relation between $dx/dt$ and $dy/dt$.

45. $x^3 + y^2 = 1$

46. $x^5 - y^3 = 1$

47. $x^2 y = 80$

48. $xy^2 = 96$

49. $3x^2 - 7xy = 12$

50. $2x^3 - 5xy = 14$

51. $x^2 + xy = y^2$

52. $x^3 - xy = y^3$

## Applied Exercises on Related Rates

53. **GENERAL: Snowballs** A large snowball is melting so that its radius is decreasing at the rate of 2 inches per hour. How fast is the volume decreasing at the moment when the radius is 3 inches? [*Hint:* The volume of a sphere of radius $r$ is $V = \frac{4}{3}\pi r^3$.]

54. **GENERAL: Hailstones** A hailstone (a small sphere of ice) is forming in the clouds so that its radius is growing at the rate of 1 millimeter per minute. How fast is its volume growing at the moment when the radius is 2 millimeters? [*Hint:* The volume of a sphere of radius $r$ is $V = \frac{4}{3}\pi r^3$.]

55. **BIOMEDICAL: Tumors** The radius of a spherical tumor is growing by $\frac{1}{2}$ centimeter per week.
    Find how rapidly the volume is increasing at the moment when the radius is 4 centimeters. [*Hint:* The volume of a sphere of radius $r$ is $V = \frac{4}{3}\pi r^3$.]

56. **BUSINESS: Profit** A company's profit from selling $x$ units of an item is $P = 1000x - \frac{1}{2}x^2$ dollars. If sales are growing at the rate of 20 per day, find how rapidly profit is growing (in dollars per day) when 600 units have been sold.

57. **BUSINESS: Revenue** A company's revenue from selling $x$ units of an item is given as $R = 1000x - x^2$ dollars. If sales are increasing at the rate of 80 per day, find how rapidly revenue is growing (in dollars per day) when 400 units have been sold.

**58. SOCIAL SCIENCE: Accidents** The number of traffic accidents per year in a city of population $p$ is predicted to be $T = 0.002p^{3/2}$. If the population is growing by 500 people a year, find the rate at which traffic accidents will be rising when the population is $p = 40,000$.

**59. SOCIAL SCIENCE: Welfare** The number of welfare cases in a city of population $p$ is expected to be $W = 0.003p^{4/3}$. If the population is growing by 1000 people per year, find the rate at which the number of welfare cases will be increasing when the population is $p = 1,000,000$.

**60. GENERAL: Rockets** A rocket fired straight up is being tracked by a radar station 3 miles from the launching pad. If the rocket is traveling at 2 miles per second, how fast is the distance between the rocket and the tracking station changing at the moment when the rocket is 4 miles up? [*Hint:* The distance $D$ in the illustration satisfies $D^2 = 9 + y^2$. To find the value of $D$, solve $D^2 = 9 + 4^2$.]

**61. BIOMEDICAL: Poiseuille's Law** Blood flowing through an artery flows faster in the center of the artery and more slowly near the sides (because of friction). The speed of the blood is $V = c(R^2 - r^2)$ millimeters (mm) per second, where $R$ is the radius of the artery, $r$ is the distance of the blood from the center of the artery, and $c$ is a constant. Suppose that arteriosclerosis is narrowing the artery at the rate of $dR/dt = -0.01$ mm per year.

Artery

Find the rate at which blood flow is being reduced in an artery whose radius is $R = 0.05$ mm with $c = 500$. [*Hint:* Find $dV/dt$, considering $r$ to be a constant. The units of $dV/dt$ will be mm per second per year.]

*Source:* E. Batschelet, *Introduction to Mathematics for Life Scientists*

**62. IMPLICIT AND EXPLICIT DIFFERENTIATION** The equation $x^2 + 4y^2 = 100$ describes an ellipse.

a. Use implicit differentiation to find its slope at the points $(8, 3)$ and $(8, -3)$.

b. Solve the equation for $y$, obtaining *two* functions, and differentiate both to find the slopes at $x = 8$. [Answers should agree with part (a).]

c. Use a graphing calculator to graph the two functions found in part (b) on an appropriate window. Then use NDERIV to find the derivatives at (or near) $x = 8$. [Answers should agree with parts (a) and (b).]

Notice that differentiating implicitly was easier than solving for $y$ and then differentiating.

**63. RELATED RATES: Speeding** A traffic patrol helicopter is stationary a quarter of a mile directly above a highway, as shown in the diagram below. Its radar detects a car whose line-of-sight distance from the helicopter is half a mile and is increasing at the rate of 57 mph. Is the car exceeding the highway's speed limit of 60 mph?

**64. RELATED RATES: Speeding** (*63 continued*)

In Exercise 63 you found that the car's speed was $\dfrac{(0.5)(57)}{\sqrt{(0.5)^2 - (0.25)^2}}$ mph. Enter this expression into a graphing calculator and then replace both occurrences of the line-of-sight distance 0.5 by 0.4 and calculate the new speed of the car. What if the line-of-sight distance were 0.3 mile?

65. **BUSINESS: Sales** The number $x$ of printer cartridges that a store will sell per week and their price $p$ (in dollars) are related by the equation $x^2 = 4500 - 5p^2$. If the price is falling at the rate of \$1 per week, find how the sales will change if the current price is \$20.

66. **BUSINESS: Supply** The number $x$ of handbags that a manufacturer will supply per week and their price $p$ (in dollars) are related by the equation $5x^3 = 20{,}000 + 2p^2$. If the price is rising at the rate of \$2 per week, find how the supply will change if the current price is \$100.

### Conceptual Exercises

67. For the function defined explicitly by $y = \sqrt[3]{x^2 - 1}$, define it *implicitly* by an equation without square roots and with zero on the right-hand side.

68. For the function defined implicitly by $2x + 3y = 1$, define $y$ *explicitly* as a function of $x$.

69. Suppose that you have a formula that relates the amount of gas used (denoted by $x$) to the distance driven (denoted by $y$) in your car. State, in everyday language, what $\dfrac{dy}{dx}$ and $\dfrac{dx}{dy}$ would mean.

70. Suppose that you have a formula that relates the value of an investment (denoted by $x$) to the day of the year (denoted by $y$). State, in everyday language, what $\dfrac{dy}{dx}$ and $\dfrac{dx}{dy}$ would mean.

71. In Example 1 on page 239 we found that for $x^2 + y^2 = 25$ the slope was given by $\dfrac{dy}{dx} = -\dfrac{x}{y}$. Does this mean that at the point (10, 10) the slope is $-1$? Explain.

72. In general, implicit differentiation gives an expression for the derivative that involves both $x$ and $y$. Under what conditions will the expression involve only $x$?

73. Write in calculus notation: The population of a city is shrinking at the rate of 1500 per year. (Be sure to define your variables.)

74. Write in calculus notation: The value of my car is falling at the rate of \$2000 per year. (Be sure to define your variables.)

75. Write in calculus notation: The rate of change of revenue is twice as great as the rate of change in profit. (Be sure to define your variables.)

76. If two quantities, $x$ and $y$, are related by a linear equation $y = mx + b$, how are the rates of change $\dfrac{dx}{dt}$ and $\dfrac{dy}{dt}$ related?

## Chapter Summary with Hints and Suggestions

Reading the text and doing the exercises in this chapter have helped you to master the following concepts and skills, which are listed by section (in case you need to review them) and are keyed to particular Review Exercises. Answers for all Review Exercises are given at the back of the book, and full solutions can be found in the Student Solutions Manual.

### 3.1 Graphing Using the First Derivative

### 3.2 Graphing Using the First and Second Derivatives

- Graph a polynomial, showing all relative extreme points and inflection points.
  (*Review Exercises 1–8.*)

  $f'$ gives slope, $f''$ gives concavity

- Graph a fractional power function, showing all relative extreme points and inflection points.
  (*Review Exercises 9–12.*)

- Graph a rational function showing all asymptotes and relative extreme points.
  (*Review Exercises 13–18.*)

- Graph a rational function showing all asymptotes, relative extreme points, and inflection points.
  (*Review Exercises 19–20.*)

## 3.3  Optimization

- Find the absolute extreme values of a given function on a given interval.
  *(Review Exercises 21–30.)*

- Maximize the efficiency of a tugboat or a flying bird.   *(Review Exercises 31–32.)*

## 3.4  Further Applications of Optimization

- Solve a geometric optimization problem.
  *(Review Exercises 33–38.)*

- Maximize profit for a company.
  *(Review Exercise 39.)*

- Maximize revenue from an orchard.
  *(Review Exercise 40.)*

- Minimize the cost of a power cable.
  *(Review Exercise 41.)*

- Maximize tax revenue to the government.
  *(Review Exercise 42.)*

- Minimize the materials used in a container.
  *(Review Exercises 43–45.)*

## 3.5  Optimizing Lot Size and Harvest Size

- Find the lot size that minimizes production costs or inventory costs.
  *(Review Exercises 46–47.)*

- Find the population size that allows the maximum sustainable yield.
  *(Review Exercises 48–49.)*

## 3.6  Implicit Differentiation and Related Rates

- Find a derivative by implicit differentiation.
  *(Review Exercises 50–53.)*

- Find a derivative by implicit differentiation and evaluate it.   *(Review Exercises 54–57.)*

- Solve a geometric related rates problem.
  *(Review Exercise 58.)*

- Use related rates to find the growth in profit or revenue.   *(Review Exercises 59–60.)*

## Hints and Suggestions

- Do not confuse *relative* and *absolute* extremes: a function may have several *relative* maximum

points (high points compared to their neighbors), but it can have at most one *absolute* maximum value (the largest value of the function on its entire domain). *Relative* extremes are used in graphing, and *absolute* extremes are used in optimization.

- Graphing calculators can be very helpful for graphing functions. However, you must first find a window that shows the interesting parts of the curve (relative extreme points and inflection points), and that is where calculus is essential.

- We have two procedures for optimizing continuous functions. Both begin by finding all critical numbers of the function in the domain. Then:

  **1.** If the function has only one critical number, find the sign of the second derivative there: a *positive* sign means an absolute *minimum*, and a *negative* sign means an absolute *maximum*.

  **2.** If the interval is closed, the maximum and minimum values may be found by evaluating the function at all critical numbers in the interval and endpoints—the largest and smallest resulting values are the maximum and minimum values of the function.

  A good strategy is this: Find all critical numbers in the domain. If there is only one, use procedure 1 above. Otherwise, try to define the function on a closed interval and use procedure 2. If all else fails, make a sketch of the graph.

- Don't forget to use the second-derivative test in applied problems. Your employer will not be happy if you accidentally *minimize* your company's profits.

- In implicit differentiation problems, remember which is the *function* (usually $y$) and which is the independent variable (usually $x$).

- In related rate problems, begin by looking for an equation that relates the variables, and then differentiate it with respect to the underlying variable (usually $t$).

- **Practice for Test:** Review Exercises 3, 11, 17, 19, 23, 27, 34, 36, 39, 43, 45, 47, 53, 55.

# Review Exercises for Chapter 3

Practice test exercise numbers are in green.

## 3.1 and 3.2 Graphing

**1–12.** Graph each function showing all relative extreme points and inflection points.

**1.** $f(x) = x^3 - 3x^2 - 9x + 12$

**2.** $f(x) = x^3 + 3x^2 - 9x - 7$

3. $f(x) = x^4 - 4x^3 + 15$

**4.** $f(x) = x^4 + 4x^3 + 17$

**5.** $f(x) = x(x + 3)^2$

**6.** $f(x) = x(x - 6)^2$

**7.** $f(x) = x(x - 4)^3$

**8.** $f(x) = x(x + 4)^3$

**9.** $f(x) = \sqrt[7]{x^5} + 1$

**10.** $f(x) = \sqrt[7]{x^6} + 1$

11. $f(x) = \sqrt[7]{x^4} + 1$

**12.** $f(x) = \sqrt[7]{x^3} + 1$

**13–18.** Graph each function showing all asymptotes and relative extreme points.

**13.** $f(x) = \dfrac{3 - x}{x - 1}$   **14.** $f(x) = \dfrac{2x - 4}{x + 1}$

**15.** $f(x) = \dfrac{1}{(x + 1)^3}$   **16.** $f(x) = \dfrac{x(x - 2)}{(x - 1)^2}$

17. $f(x) = \dfrac{x^2}{x^2 - 4}$   **18.** $f(x) = \dfrac{8}{(x + 1)(x - 2)^2}$

**19–20.** Graph each function showing all asymptotes, relative extreme points, and inflection points.

19. $f(x) = \dfrac{4x^2}{x^2 + 3}$   **20.** $f(x) = \dfrac{36x}{(x - 1)^2}$

## 3.3 and 3.4 Optimization

**21–30.** Find the absolute extreme values of each function on the given interval.

**21.** $f(x) = 2x^3 - 6x$ on $[0, 5]$

**22.** $f(x) = 2x^3 - 24x$ on $[0, 5]$

23. $f(x) = x^4 - 4x^3 - 8x^2 + 64$ on $[-1, 5]$

**24.** $f(x) = x^4 - 4x^3 + 4x^2 + 1$ on $[0, 10]$

**25.** $h(x) = (x - 1)^{2/3}$ on $[0, 9]$

**26.** $f(x) = \sqrt{100 - x^2}$ on $[-10, 10]$

27. $g(w) = (w^2 - 4)^2$ on $[-3, 3]$

**28.** $g(x) = x(8 - x)$ on $[0, 8]$

**29.** $f(x) = \dfrac{x}{x^2 + 1}$ on $[-3, 3]$

**30.** $f(x) = \dfrac{x}{x^2 + 4}$ on $[-4, 4]$

**31. GENERAL: Fuel Efficiency** At what speed should a tugboat travel upstream so as to use the least amount of fuel to reach its destination? If the tugboat's speed through the water is $v$ and the speed of the current (relative to the land) is $c$, then the energy used is proportional to

$E(v) = \dfrac{v^2}{v - c}$.   Find the velocity $v$ that minimizes the energy $E(v)$. Your answer will depend upon $c$, the speed of the current.

*Source:* UMAP Modules, COMAP

**32. GENERAL: Bird Flight** Let $v$ be the flying speed of a bird, and let $w$ be its weight. The power $P$ that the bird must maintain during flight is

$P = \dfrac{aw^2}{v} + bv^3$,   where $a$ and $b$ are positive constants depending on the shape of the bird and the density of the air. Find the speed $v$ that minimizes the power $P$.

*Source:* E. Batschelet, *Introduction to Mathematics for Life Scientists*

**33. GENERAL: Fencing** A homeowner wants to enclose three adjacent rectangular pens of equal size along a straight wall, as in the following diagram. If the side along the wall needs no fence, what is the largest total area that can be enclosed using only 240 feet of fence?

**34. GENERAL: Maximum Area** A homeowner wants to enclose three adjacent rectangular pens of equal size, as in the following diagram. What is

the largest total area that can be enclosed using only 240 feet of fence?

**35. GENERAL: Unicorns** To celebrate the acquisition of Styria in 1261, Ottokar II sent hunters into the Bohemian woods to capture a unicorn. To display the unicorn at court, the king built a rectangular cage. The material for three sides of the cage cost 3 ducats per running cubit, while the fourth was to be gilded and cost 51 ducats per running cubit. In 1261 it was well known that a happy unicorn requires an area of 2025 square cubits. Find the dimensions that would keep the unicorn happy at the lowest cost.

36. **GENERAL: Box Design** An open-top box with a square base is to have a volume of exactly 500 cubic inches. Find the dimensions of the box that can be made with the smallest amount of materials.

**37. GENERAL: Packaging** Find the dimensions of the cylindrical tin can with volume $16\pi$ cubic inches that can be made from the least amount of tin. (*Note:* $16\pi \approx 50$ cubic inches.)

$V = \pi r^2 h$
$A = 2\pi r^2 + 2\pi rh$

**38. GENERAL: Packaging** Find the dimensions of the open-top cylindrical tin can with volume $8\pi$ cubic inches that can be made from the least amount of tin. (*Note:* $8\pi \approx 25$ cubic inches.)

open top

$V = \pi r^2 h$
$A = \pi r^2 + 2\pi rh$

39. **BUSINESS: Maximum Profit** A computer dealer can sell 12 personal computers per week at a price of $2000 each. He estimates that each $400 price decrease will result in three more sales per week. If the computers cost him $1200 each, what price should he charge to maximize his profit? How many will he sell at that price?

**40. ENVIRONMENTAL SCIENCES: Farming** A peach tree will yield 100 pounds of peaches now, which will sell for 40 cents a pound. Each week that the farmer waits will increase the yield by 10 pounds, but the selling price will decrease by 2 cents per pound. How long should the farmer wait to pick the fruit in order to maximize her revenue?

**41. GENERAL: Minimum Cost** A cable is to connect a power plant to an island that is 1 mile offshore and 3 miles downshore from the power plant. It costs $5000 per mile to lay a cable underwater and $3000 per mile to lay it underground. If the cost of laying the cable is to be minimized, find the distance $x$ downshore from the island where the cable should meet the land.

**42. ECONOMICS: Tax Revenue** Economists[*] have found that if cigarettes are taxed at rate $t$, then cigarette sales will be $S(t) = (64 - 51.26t)/(1 + t)$ (billion dollars annually, before taxes).
   **a.** Use a graphing calculator to find the tax rate $t$ that maximizes revenue to the government.
   **b.** Multiply this tax rate times $4 (a typical pretax price of a pack of cigarettes) to find the actual tax per pack.

   *Source:* Chaloupka (ed.), *The Economic Analysis of Substance Use and Abuse*

43. **GENERAL: Packaging** A 12-ounce soft drink can has volume 21.66 cubic inches. If the top and bottom are twice as thick as the sides, find the dimensions (radius and height) that minimize the amount of metal used in the can.

**44. GENERAL: Box Design** A standard 8.5- by 11-inch piece of paper can be made into a box with a lid by cutting $x$- by $x$-inch squares from two corners and $x$- by 5.5-inch rectangles from the

[*]Michael Grossman, Gary Becker (winner of the 1992 Nobel Memorial Prize in Economics), Frank Chaloupka, and Kevin Murphy.

other corners and then folding, as shown below. What value of $x$ maximizes the volume of the box, and what is the maximum volume?

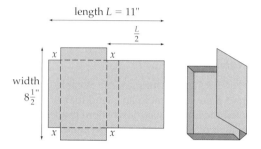

length $L = 11"$

$\frac{L}{2}$

width $8\frac{1}{2}"$

45. **GENERAL: Box Design** A standard 6- by 8-inch card can be made into a box with a lid by cutting $x$- by $x$-inch squares from two corners and $x$- by 4-inch rectangles from the other corners and then folding. (See the diagram above, but use length 8 and width 6.) What value of $x$ maximizes the volume of the box, and what is the maximum volume?

## 3.5 Optimizing Lot Size and Harvest Size

46. **BUSINESS: Production Runs** A wallpaper company estimates the demand for a certain pattern to be 900 rolls per year. It costs $800 to set up the presses to print the pattern, plus $200 to print each roll. If the company can store a roll of wallpaper for a year at a cost of $4, how many rolls should it print at a time and how many printing runs will it need in a year to minimize production costs?

47. **BUSINESS: Lot Size** A motorcycle shop estimates that it will sell 500 motorbikes in a year. Each bike costs $300, plus a fixed charge of $500 per order. If it costs $200 to store a motorbike for a year, what is the order size and how many orders will be needed in a year to minimize inventory costs?

48. **MAXIMUM SUSTAINABLE YIELD** The reproduction function for the North American duck is estimated to be $f(p) = -0.02p^2 + 7p$, where $p$ and $f(p)$ are measured in thousands. Find the size of the population that allows the maximum sustainable yield, and also find the size of the yield.

49. **MAXIMUM SUSTAINABLE YIELD** Ecologists estimate the reproduction function for striped bass in an East Coast fishing ground to be $f(p) = 60\sqrt{p}$, where $p$ and $f(p)$ are measured in thousands and $p \leq 1000$. Find the size of the population that allows the maximum sustainable yield, and also the size of the yield.

## 3.6 Implicit Differentiation and Related Rates

**50–53.** For each equation, use implicit differentiation to find $dy/dx$.

50. $6x^2 + 8xy + y^2 = 100$

51. $8xy^2 - 8y = 1$

52. $2xy^2 - 3x^2y = 0$     53. $\sqrt{x} - \sqrt{y} = 10$

**54–57.** For each equation, find $dy/dx$ evaluated at the given values.

54. $x + y = xy$   at   $x = 2, y = 2$

55. $y^3 - y^2 - y = x$   at   $x = 2, y = 2$

56. $xy^2 = 81$   at   $x = 9, y = 3$

57. $x^2y^2 - xy = 2$   at   $x = -1, y = 1$

58. **GENERAL: Melting Ice** A cube of ice is melting so that each edge is decreasing at the rate of 2 inches per hour. Find how fast the volume of the ice is decreasing at the moment when each edge is 10 inches long.

59. **BUSINESS: Profit** A company's profit from selling $x$ units of a product is $P = 2x^2 - 20x$ dollars. If sales are growing at the rate of 30 per day, find the rate of change of profit when 40 units have been sold.

60. **BUSINESS: Revenue** A company finds that its revenue from selling $x$ units of a product is $R = x^2 + 500x$ dollars. If sales are increasing at the rate of 50 per month, find the rate of change of revenue when 200 units have been sold.

61. **BIOMEDICAL: Medication** You swallow a spherical pill whose radius is 0.5 centimeter (cm), and it dissolves in your stomach so that its radius decreases at the rate of 0.1 cm per minute. Find the rate at which the volume is decreasing (the rate at which the medication is being made available to your system) when the radius is

   **a.** 0.5 cm

   **b.** 0.2 cm

## Cumulative Review for Chapters 1–3

The following exercises review some of the basic techniques that you learned in Chapters 1–3. Answers to all of these cumulative review exercises are given in the answer section at the back of the book.

1. Find an equation for the line through the points $(-4, 3)$ and $(6, -2)$. Write your answer in the form $y = mx + b$.

2. Simplify $\left(\frac{4}{25}\right)^{-1/2}$.

3. Find, correct to three decimal places:
$$\lim_{x \to 0} (1 + 3x)^{1/x}$$

4. For the function $f(x) = \begin{cases} 4x - 8 & \text{if } x < 3 \\ 7 - 2x & \text{if } x \geq 3 \end{cases}$

   a. Draw its graph.

   b. Find $\lim_{x \to 3^-} f(x)$.

   c. Find $\lim_{x \to 3^+} f(x)$.

   d. Find $\lim_{x \to 3} f(x)$.

   e. Is $f(x)$ continuous or discontinuous, and if it is discontinuous, where?

5. Use the definition of the derivative, $f'(x) =$
$$\lim_{h \to 0} \frac{f(x + h) - f(x)}{h}, \quad \text{to find the derivative of}$$
$f(x) = 2x^2 - 5x + 7$.

6. Find the derivative of $f(x) = 8\sqrt{x^3} - \dfrac{3}{x^2} + 5$.

7. Find the derivative of $f(x) = (x^5 - 2)(x^4 + 2)$.

8. Find the derivative of $f(x) = \dfrac{2x - 5}{3x - 2}$.

9. The population of a city $x$ years from now is predicted to be $P(x) = 3600x^{2/3} + 250{,}000$ people. Find $P'(8)$ and $P''(8)$ and interpret your answers.

10. Find $\dfrac{d}{dx}\sqrt{2x^2 - 5}$ and write your answer in radical form.

11. Find $\dfrac{d}{dx}[(3x + 1)^4(4x + 1)^3]$.

12. Find $\dfrac{d}{dx}\left(\dfrac{x - 2}{x + 2}\right)^3$ and simplify your answer.

13. Make sign diagrams for the first and second derivatives and draw the graph of the function $f(x) = x^3 - 12x^2 - 60x + 400$. Show on your graph all relative extreme points and inflection points.

14. Make sign diagrams for the first and second derivatives and draw the graph of the function $f(x) = \sqrt[3]{x^2} - 1$. Show on your graph all relative extreme points and inflection points.

15. A homeowner wishes to use 600 feet of fence to enclose two identical adjacent pens, as in the diagram below. Find the largest total area that can be enclosed.

16. A store can sell 12 telephone answering machines per day at a price of $200 each. The manager estimates that for each $10 price reduction she can sell 2 more per day. The answering machines cost the store $80 each. Find the price and the quantity sold per day to maximize the company's profit.

17. For $y$ defined implicitly by
$$x^3 + 9xy^2 + 3y = 43$$
find $\dfrac{dy}{dx}$ and evaluate it at the point $(1, 2)$.

18. A large spherical balloon is being inflated at the rate of 32 cubic feet per minute. Find how fast the radius is increasing at the moment when the radius is 2 feet.

# 4

# Exponential and Logarithmic Functions

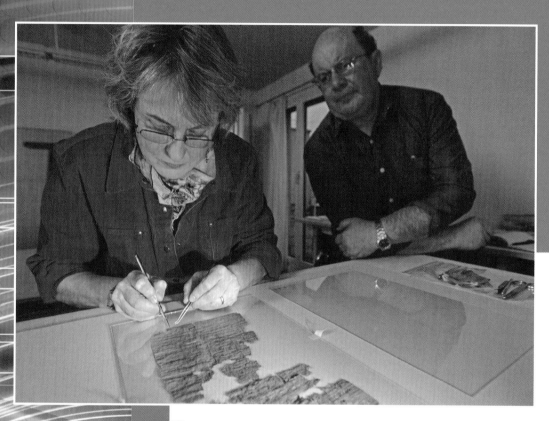

### Carbon 14 Dating and the Shroud of Turin

Carbon 14 dating is a method for estimating the age of ancient plants and animals. While a plant is alive, it absorbs carbon dioxide, and with it both "ordinary" carbon and carbon 14, a radioactive form of carbon produced by cosmic rays in the upper atmosphere. When the plant dies, it stops absorbing both types of carbon, and the carbon 14 decays exponentially at a known rate, as shown in the following graph.

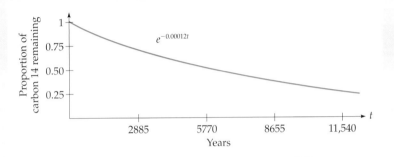

The time since the plant died can then be estimated by comparing the amount of carbon 14 that remains with the amount of ordinary carbon remaining, since they were originally in a known proportion. The comparison technique involves exponential and logarithmic functions, and is explained on pages 282–283. Animal remains can also be dated in this way, since their diets contain plant matter.

In 1988, the Roman Catholic Church used carbon 14 dating to determine the authenticity of the Shroud of Turin, believed by many to be the burial cloth of Jesus. This was done by estimating the age of the flax plants from which the linen shroud was woven. In Exercise 32 on page 287 you will find whether the shroud is old enough to be the burial cloth of Christ.

The usual method to find the amount of carbon 14 remaining in a sample is to burn the sample near a Geiger counter to measure the radioactivity. For the shroud, this would have required a handkerchief-sized sample, and so instead a more advanced method was employed, using a tandem accelerator mass spectrometer and requiring only a postage-stamp-sized sample.

Carbon 14 dating assumes, however, that the ratio of carbon 14 to ordinary carbon in the atmosphere has remained steady over the centuries. This assumption seemed difficult to verify until the discovery in 1955 of a bristlecone pine tree in the White Mountains of California that was over 2000 years old (according to its growth rings). Each annual ring had absorbed carbon and carbon 14 from the air, and so provided a year-by-year record of their ratio. The analysis showed that the ratio has remained relatively steady over the centuries, and where variations did occur, it enabled scientists to construct a table for making corrections in the original carbon 14 dates. (Incidentally, bristlecone pine trees that are 4900 years old have been found, making them the oldest living things on earth.) For extremely old remains, such as dinosaur fossils, archeologists use longer lasting radioactive elements, such as potassium 40.

Carbon 14 dating, for which its inventor Willard Libby received a Nobel prize in 1960, has become an invaluable tool in the social and biological sciences.

## 4.1  EXPONENTIAL FUNCTIONS

### Introduction

In Chapter 1 we introduced **exponential** and **logarithmic functions,** two of the most useful functions in all of mathematics. In this chapter we develop their properties and apply them to a wide variety of problems. We begin with exponential functions, showing how they are used to model the processes of growth and decay.

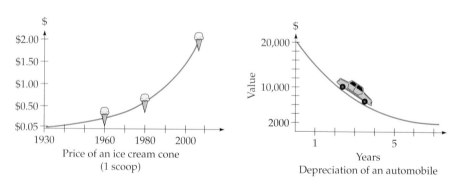

We will also define the very important mathematical constant *e.*

### Exponential Functions

A function that has a variable in an exponent, such as $f(x) = 2^x$, is called an *exponential function.* The number being raised to the power is called the **base.**

$$f(x) = 2^x$$

Exponent

Base

More formally:

| Exponential Functions | *Brief Examples* |
|---|---|
| For any number $a > 0$, the function $$f(x) = a^x$$ is called an *exponential function* with base $a$ and exponent (or power) $x$. | $f(x) = 2^x$ has base 2 <br> $f(x) = \left(\frac{1}{2}\right)^x$ has base $\frac{1}{2}$ |

The table below shows some values of the exponential function $f(x) = 2^x$, and its graph (based on these points) is shown on the right.

| $x$ | $y = 2^x$ |
|---|---|
| $-3$ | $2^{-3} = \frac{1}{8}$ |
| $-2$ | $2^{-2} = \frac{1}{4}$ |
| $-1$ | $2^{-1} = \frac{1}{2}$ |
| $0$ | $2^0 = 1$ |
| $1$ | $2^1 = 2$ |
| $2$ | $2^2 = 4$ |
| $3$ | $2^3 = 8$ |

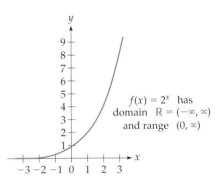

$f(x) = 2^x$ has domain $\mathbb{R} = (-\infty, \infty)$ and range $(0, \infty)$

Graph of $y = 2^x$

Clearly, the exponential function $2^x$ is quite different from the parabola $x^2$. In particular, it is not symmetric about the $y$-axis and has the $x$-axis as a horizontal asymptote $\left(\lim\limits_{x \to -\infty} 2^x = 0\right)$.

The following table shows some values of the exponential function $f(x) = \left(\frac{1}{2}\right)^x$ and its graph is shown to the right of the table. Notice that it is the mirror image of the curve $y = 2^x$, with both passing through the point $(0, 1)$.

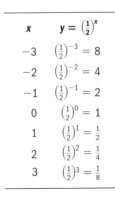

| $x$ | $y = \left(\frac{1}{2}\right)^x$ |
|---|---|
| $-3$ | $\left(\frac{1}{2}\right)^{-3} = 8$ |
| $-2$ | $\left(\frac{1}{2}\right)^{-2} = 4$ |
| $-1$ | $\left(\frac{1}{2}\right)^{-1} = 2$ |
| $0$ | $\left(\frac{1}{2}\right)^0 = 1$ |
| $1$ | $\left(\frac{1}{2}\right)^1 = \frac{1}{2}$ |
| $2$ | $\left(\frac{1}{2}\right)^2 = \frac{1}{4}$ |
| $3$ | $\left(\frac{1}{2}\right)^3 = \frac{1}{8}$ |

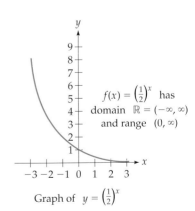

$f(x) = \left(\frac{1}{2}\right)^x$ has domain $\mathbb{R} = (-\infty, \infty)$ and range $(0, \infty)$

Graph of $y = \left(\frac{1}{2}\right)^x$

We can define an exponential function $f(x) = a^x$ for any positive base $a$. We always take the base to be positive, so for the rest of this section *the letter $a$ will stand for a positive constant*.

Exponential functions with bases $a > 1$ are used to model *growth*, as in populations or savings accounts, and exponential functions with bases $a < 1$ are used to model *decay*, as in depreciation. (For base $a = 1$, the graph is a horizontal line, since $1^x = 1$ for all $x$.)

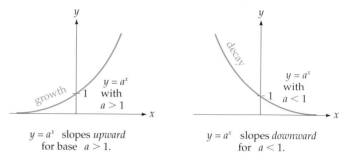

$y = a^x$ slopes *upward* for base $a > 1$.

$y = a^x$ slopes *downward* for $a < 1$.

## Compound Interest

Money invested at compound interest grows exponentially. (The word "compound" means that the interest is added to the account, earning more interest.) Banks always state *annual* interest rates, but the compounding may be done more frequently. For example, if a bank offers 8% compounded quarterly, then each quarter you get 2% (one quarter of the annual 8%), so that 2% of your money is added to the account each quarter. If you begin with $P$ dollars (the **principal**), at the end of the first quarter you would have $P$ dollars plus 2% of $P$ dollars:

$$\begin{pmatrix} \text{Value after} \\ \text{1 quarter} \end{pmatrix} = P + 0.02P = P \cdot (1 + 0.02) \qquad \text{Factoring out the } P$$

Notice that increasing a quantity by 2% is the same as multiplying it by $(1 + 0.02)$. Since a year has 4 quarters, $t$ years will have $4t$ quarters. Therefore, to find the value of your account after $t$ years, we simply multiply the principal by $(1 + 0.02)$ a total of $4t$ times, obtaining:

$$\begin{pmatrix} \text{Value after} \\ t \text{ years} \end{pmatrix} = P \cdot \overbrace{(1 + 0.02) \cdot (1 + 0.02) \cdots (1 + 0.02)}^{4t \text{ times}}$$

$$= P \cdot (1 + 0.02)^{4t}$$

The 8%, which gave the $\frac{0.08}{4} = 0.02$ quarterly rate, can be replaced by any interest rate $r$ (written in decimal form), and the 4 can be replaced by any number $m$ of compounding periods per year, leading to the following general formula.

### Compound Interest

For $P$ dollars invested at annual interest rate $r$ compounded $m$ times a year for $t$ years,

$$\begin{pmatrix} \text{Value after} \\ t \text{ years} \end{pmatrix} = P \cdot \left(1 + \frac{r}{m}\right)^{mt}$$

$r$ = annual rate
$m$ = periods per year
$t$ = number of years

For example, for monthly compounding we would use $m = 12$ and for daily compounding $m = 365$ (the number of days in the year).

**EXAMPLE 1**  **FINDING A VALUE UNDER COMPOUND INTEREST**

Find the value of $4000 invested for 2 years at 12% compounded quarterly.

*Solution*

$$4000 \cdot \left(1 + \underbrace{\frac{0.12}{4}}_{0.03}\right)^{4 \cdot 2} = 4000(1 + 0.03)^8$$

$P \cdot \left(1 + \dfrac{r}{m}\right)^{mt}$
with  $P = 4000$,
$r = 0.12$,  $m = 4$,
and  $t = 2$

$$= 4000 \cdot 1.03^8 \approx 5067.08 \qquad \text{Using a calculator}$$

The value after 2 years will be $5067.08.

!  **BE CAREFUL:** Always enter the interest rate into your calculator as a *decimal*.

We may interpret the formula $4000(1 + 0.03)^8$ intuitively as follows: multiplying the $4000 principal by $(1 + 0.03)$ means that you keep the original amount (the "1") plus some interest (the 0.03), and the exponent 8 means that this is done a total of 8 times.

**Practice Problem 1**  Find the value of $2000 invested for 3 years at 24% compounded monthly.

➤ **Solution on page 268**

## Graphing Calculator Exploration

Intersection
X=34.596676    Y=5408.5309

In the long run, which is more important—principal or interest rate? The graph shows the value of $1000 at 5% interest, together with the value of a mere $200 at the higher rate of 10% (both compounded annually). The fact that the (initially) lower curve eventually surpasses the higher one illustrates a general fact: the higher interest rate will eventually prevail, regardless of the size of the initial investment.

## Present Value

The value to which a sum will grow under compound interest is often called its *future value*. That is, Example 1 showed that the future value of $4000 (at 12% compounded quarterly for 2 years) is $5067.08.

Reversing the order, we can speak of the *present value* of a future payment. Example 1 shows that if a payment of $5067.08 is to be made in 2 years, its *present value* is $4000 (at this interest rate). That is, a promise to pay you $5067.08 in 2 years is worth exactly $4000 now, since $4000 deposited in a

Future value

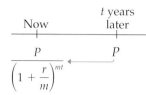

Present value

bank now would be worth that much in 2 years (at the stated interest rate). To find the *future* value we *multiply* by $(1 + r/m)^{mt}$, and so to find the *present* value, we *divide* by $(1 + r/m)^{mt}$.

**Present Value**

For a future payment of $P$ dollars at annual interest rate $r$ compounded $m$ times a year to be paid in $t$ years,

$$\begin{pmatrix}\text{Present} \\ \text{value}\end{pmatrix} = \frac{P}{\left(1 + \dfrac{r}{m}\right)^{mt}}$$

$r$ = annual rate
$m$ = periods per year
$t$ = number of years

**EXAMPLE 2**  **FINDING PRESENT VALUE**

Find the present value of $5000 to be paid 8 years from now at 10% interest compounded semiannually.

*Solution*

For semiannual compounding $(m = 2)$, the formula gives

$$\frac{P}{\left(1 + \dfrac{r}{m}\right)^{mt}} = \frac{5000}{\left(1 + \dfrac{0.10}{2}\right)^{2 \cdot 8}}$$

$P = 5000$, $r = 0.10$,
$m = 2$, and $t = 8$

$$= \frac{5000}{(1 + 0.05)^{16}} \approx 2290.56$$

Using a calculator

Therefore, the present value of the $5000 is just $2290.56.

## Depreciation by a Fixed Percentage

Depreciation by a fixed percentage means that a piece of equipment loses the same percentage of its current value each year. Losing a percentage of value is like compound interest but with a *negative* interest rate. Therefore, we use the compound interest formula $P(1 + r/m)^{mt}$ with $m = 1$ (since depreciation is annual) and with $r$ being negative.

**EXAMPLE 3**  **DEPRECIATING AN ASSET**

A car worth $15,000 depreciates in value by 40% each year. How much is it worth after 3 years?

*Solution*

The car loses 40% of its value each year, which is equivalent to an interest rate of *negative* 40%. The compound interest formula gives

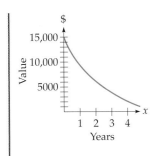

$$15,000(1 - 0.40)^3 = 15,000(0.60)^3 = \underbrace{\$3240}_{\text{Using a calculator}}$$

$P(1 + r/m)^{mt}$ with $P = 15,000$, $r = -0.40$, $m = 1$, and $t = 3$

The exponential function $f(x) = 15,000(0.60)^x$, giving the value of the car after $x$ years of depreciation, is graphed on the left. Notice that a yearly loss of 40% means that 60% of the value is retained each year.

**Practice Problem 2**

A printing press, originally worth $50,000, loses 20% of its value each year. What is its value after 4 years?                    ▶ **Solution on page 268**

The above graph shows that depreciation by a fixed percentage is quite different from "straight-line" depreciation. Under straight-line depreciation the same *dollar* value is lost each year, while under fixed-percentage depreciation the same *percentage* of value is lost each year, resulting in larger dollar losses in the early years and smaller dollar losses in later years. Depreciation by a fixed percentage (also called the **declining balance method**) is one type of **accelerated depreciation.** The method of depreciation that one uses depends on how one chooses to estimate value, and in practice is often determined by the tax laws.

## The Number *e*

Imagine that a bank offers 100% interest, and that you deposit $1 for 1 year. Let us see how the value changes under different types of compounding.

For *annual* compounding, your $1 would in a year grow to $2 (the original dollar plus a dollar interest).

For *quarterly* compounding, we use the compound interest formula with $P = 1$, $r = 1$ (for 100%), $m = 4$, and $t = 1$:

$$1\left(1 + \frac{1}{4}\right)^{4 \cdot 1} = 1(1 + 0.25)^4 = (1.25)^4 \approx 2.44 \qquad P\left(1 + \frac{r}{m}\right)^{mt}$$

or $2.44, an improvement of 44 cents over annual compounding.

For *daily* compounding, the value after a year would be

$$\left(1 + \frac{1}{365}\right)^{365} \approx 2.71$$

$m = 365$ periods
$\dfrac{r}{m} = \dfrac{100\%}{365} = \dfrac{1}{365}$

an increase of 27 cents over quarterly compounding. Clearly, if the interest rate, the principal, and the amount of time stay the same, the value increases as the compounding is done more frequently.

In general, if the compounding is done $m$ times a year, the value of the dollar after a year will be

$$\left(\begin{array}{c} \text{Value of \$1 after 1 year at 100\%} \\ \text{interest compounded } m \text{ times a year} \end{array}\right) = \left(1 + \frac{1}{m}\right)^m$$

The following table shows the value of $\left(1 + \frac{1}{m}\right)^m$ for various values of $m$.

| Value of $1 at 100% Interest Compounded $m$ Times a Year for 1 Year | | |
|---|---|---|
| $m$ | $\left(1 + \dfrac{1}{m}\right)^m$ | *Answer (rounded)* | |
| 1 | $\left(1 + \dfrac{1}{1}\right)^1 = 2.00000$ | Annual compounding |
| 4 | $\left(1 + \dfrac{1}{4}\right)^4 \approx 2.44141$ | Quarterly compounding |
| 365 | $\left(1 + \dfrac{1}{365}\right)^{365} \approx 2.71457$ | Daily compounding |
| 10,000 | $\left(1 + \dfrac{1}{10,000}\right)^{10,000} \approx 2.71815$ | |
| 100,000 | $\left(1 + \dfrac{1}{100,000}\right)^{100,000} \approx 2.71827$ | |
| 1,000,000 | $\left(1 + \dfrac{1}{1,000,000}\right)^{1,000,000} \approx 2.71828$ | Answers agree to five decimal places |
| 10,000,000 | $\left(1 + \dfrac{1}{10,000,000}\right)^{10,000,000} \approx 2.71828$ | |

Notice that as $m$ increases, the values in the right-hand column seem to be settling down to a definite value, approximately 2.71828. That is, as $m$ approaches infinity, the limit of $\left(1 + \frac{1}{m}\right)^m$ is approximately 2.71828. This particular number is very important in mathematics, and is given the name $e$ (just as 3.14159... is given the name $\pi$). In the following definition we use the letter $n$ to state the definition in its traditional form.

**The Constant $e$**

$$e = \lim_{n \to \infty} \left(1 + \frac{1}{n}\right)^n = 2.71828...$$

The dots mean that the decimal expansion goes on forever

The same $e$ appears in probability and statistics in the formula for the "bell-shaped" or "normal" curve. Its value has been calculated to several million decimal places, and its value to 15 decimal places is $e \approx 2.718281828459045$.

## Continuous Compounding of Interest

This kind of compound interest, the limit as the compounding frequency approaches infinity, is called **continuous compounding.** We have shown that $1 at 100% interest compounded continuously for 1 year would be worth precisely $e$ dollars (about $2.72). The formula for continuous compound interest at other rates is as follows (a justification for it is given at the end of this section).

### Continuous Compounding

For $P$ dollars invested at annual interest rate $r$ compounded continuously for $t$ years,

$$\begin{pmatrix} \text{Value after} \\ t \text{ years} \end{pmatrix} = Pe^{rt}$$

**EXAMPLE 4**  **FINDING VALUE WITH CONTINUOUS COMPOUNDING**

Find the value of $1000 at 8% interest compounded continuously for 20 years.

*Solution*

We use the formula $Pe^{rt}$ with $P = 1000$, $r = 0.08$, and $t = 20$.

$$Pe^{rt} = \underbrace{1000}_{P} \cdot e^{\overset{\uparrow\;\uparrow}{(0.08)(20)}}_{r\;\;n} = 1000 \cdot \underbrace{e^{1.6}}_{4.95303} \approx \$4953.03 \qquad \begin{array}{l} e^{1.6} \text{ is usually found using} \\ \text{the } \boxed{\text{2nd}} \text{ and } \boxed{\text{LN}} \text{ keys} \end{array}$$

## Present Value with Continuous Compounding

As before, the value that a sum will attain in $t$ years is often called its *future* value, and the current value of a future sum is its *present* value. Under continuous compounding, to find future value we multiply $P$ by $e^{rt}$, and so to find *present* value we *divide* by $e^{rt}$.

### Present Value with Continuous Compounding

For a future payment of $P$ dollars at annual interest rate $r$ compounded continuously to be paid in $t$ years,

$$\begin{pmatrix} \text{Present} \\ \text{value} \end{pmatrix} = \frac{P}{e^{rt}} = Pe^{-rt}$$

EXAMPLE 5  **FINDING PRESENT VALUE WITH CONTINUOUS COMPOUNDING**

The present value of $5000 to be paid in 10 years, at 7% interest compounded continuously, is

$$\frac{5000}{e^{0.07 \cdot 10}} = \frac{5000}{e^{0.7}} \approx \$2482.93 \qquad \text{Using a calculator}$$

## Intuitive Meaning of Continuous Compounding

Under quarterly compounding, your money is, in a sense, earning interest throughout the quarter, but the interest is not added to your account until the end of the quarter. Under continuous compounding, the interest is added to your account *as it is earned*, with no delay. The extra earnings in continuous compounding come from this "instant crediting" of interest, since then your interest starts earning more interest immediately.

## How to Compare Interest Rates

How do you compare different interest rates, such as 16% compounded quarterly and 15.8% compounded continuously? You simply see what each will do for a deposit of $1 for 1 year.

EXAMPLE 6  **COMPARING INTEREST RATES**

Which gives a better return: 16% compounded quarterly or 15.8% compounded continuously?

*Solution*

For 16% compounded quarterly (on $1 for 1 year),

$$1 \cdot \left(1 + \frac{0.16}{4}\right)^4 = (1 + 0.04)^4 \approx 1.170 \qquad \text{Using } P\left(1 + \frac{r}{m}\right)^{mt}$$

For 15.8% compounded continuously,

$$1 \cdot e^{0.158 \cdot 1} = e^{0.158} \approx 1.171 \qquad \overset{\text{Better}}{\nearrow} \qquad \text{Using } Pe^{rt}$$

Therefore, 15.8% compounded continuously is better. (The difference is only a tenth of a cent, but it would be more impressive for principals larger than $1 or periods longer than 1 year.)

The actual percentage increase during 1 year is called the **effective rate of interest,** the **annual percentage rate (APR),** or the **annual percentage yield (APY).** For example, the "continuous" rate of 15.8% used above gives a return of $1.171, and after subtracting the dollar invested, leaves a gain of 0.171 on the original $1, which means an APR of 17.1%. The *stated* rate (here, 15.8%) is called the **nominal** rate of interest. That is, a nominal rate of 15.8% compounded continuously is equivalent to an APR of 17.1%. The 1993 Truth in Savings Act requires that banks always state the annual percentage rate.

**Practice Problem 3**    From the previous example, what is the annual percentage rate for 16% compounded quarterly? [*Hint:* No calculation is needed.]

➤ **Solution on next page**

### The Function  $y = e^x$

The number $e$ gives us a new exponential function $f(x) = e^x$. This function is used extensively in business, economics, and all areas of science. The table below shows the values of $e^x$ for various values of $x$. These values lead to the graph of $f(x) = e^x$ shown at the right.

| $x$ | $y = e^x$ |
|---|---|
| $-3$ | $e^{-3} \approx 0.05$ |
| $-2$ | $e^{-2} \approx 0.14$ |
| $-1$ | $e^{-1} \approx 0.37$ |
| $0$ | $e^{0} = 1$ |
| $1$ | $e^{1} \approx 2.72$ |
| $2$ | $e^{2} \approx 7.39$ |
| $3$ | $e^{3} \approx 20.09$ |

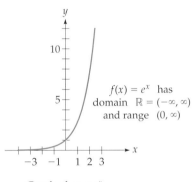

$f(x) = e^x$ has
domain $\mathbb{R} = (-\infty, \infty)$
and range $(0, \infty)$

Graph of $y = e^x$.

Notice that $e^x$ is never zero, and is positive for all values of $x$, even when $x$ is negative. We restate this important observation as follows:

$e$ to any power is positive.

The following graph shows the function $f(x) = e^{kx}$ for various values of the constant $k$. For positive values of $k$ the curve rises, and for negative values of $k$ the curve falls (as you move to the right). For higher values of $k$ the curve rises more steeply. Each curve has the $x$-axis as a horizontal asymptote and crosses the $y$-axis at 1.

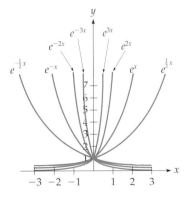

$f(x) = e^{kx}$ for various values of $k$.

## Graphing Calculator Exploration

Intersection
X=4.2582441,Y=19.548134

The most populous states are California and Texas, with New York third and Florida fourth but gaining. According to data from the Census Bureau, $x$ years after 2005 the population of New York will be $19.3e^{0.003x}$ and the population of Florida will be $17.8e^{0.022x}$ (both in millions).

**a.** Graph these two functions on the window [0, 10] by [0, 25]. [Use the 2nd and LN keys for entering $e$ to powers.]

**b.** Use INTERSECT to find the $x$-value where the curves intersect.

**c.** From your answer to part (b), predict the year in which Florida will (or did) overtake New York as the third largest state. [*Hint:* $x$ is years after 2005.]

*Source:* U.S. Census Bureau

### Exponential Growth

All exponential growth, whether continuous or discrete, has one common characteristic: the amount of growth is proportional to the size. For example, the interest that a bank account earns is proportional to the size of the account, and the growth of a population is proportional to the size of the population. This is in contrast, for example, to a person's height, which does not increase exponentially. That is, exponential growth occurs in those situations where a quantity grows *in proportion to its size.* This is why exponential functions are so important.

### Solutions to Practice Problems

**1.** $2000(1 + 0.24/12)^{12 \cdot 3} = 2000(1 + 0.02)^{36} = 2000(1.02)^{36} \approx 2000(2.039887) \approx$ \$4079.77

**2.** $50{,}000(1 - 0.20)^4 = 50{,}000(0.8)^4 = 50{,}000(0.4096) = \$20{,}480$

**3.** 17% (from 1.170)

## 4.1   Section Summary

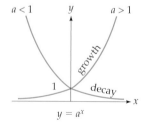

$a < 1$      $y$      $a > 1$

growth

1

decay

$y = a^x$

Exponential functions have exponents that involve variables. The exponential functions $f(x) = a^x$ slope *upward* or *downward* depending upon whether the base $a$ (which must be positive) satisfies $a > 1$ or $a < 1$.

The formula $P(1 + r/m)^{mt}$ gives the value after $t$ years for an investment of $P$ dollars that increases at annual rate $r$ compounded $m$ times a year. This same formula with $m = 1$ and a *negative* growth rate $r$ governs depreciation by a fixed percentage.

By considering a 100% increase not given all at once but divided up into smaller and smaller successive increases, we defined a new constant $e$:

$$e = \lim_{n \to \infty} \left(1 + \frac{1}{n}\right)^n \approx 2.71828$$

Exponential functions with base $e$ are used extensively in modeling many types of growth, such as the growth of populations and interest that is compounded continuously. For interest compounded *continuously*, the formula is $Pe^{rt}$.

## Justification of the Formula for Continuous Compounding

The compound interest formula $Pe^{rt}$ is derived as follows: $P$ dollars invested for $t$ years at interest rate $r$ compounded $m$ times a year yields

$$P\left(1 + \frac{r}{m}\right)^{mt} \qquad \text{From the formula on page 260}$$

Define $n = \dfrac{m}{r}$, so that $m = rn$. Replacing $m$ by $rn$ and letting $m$ (and therefore $n$) approach $\infty$, this becomes

$$P\left(1 + \frac{r}{rn}\right)^{rnt} = P\left[\left(1 + \frac{1}{n}\right)^n\right]^{rt} \to Pe^{rt} \qquad \text{Letting } n \to \infty$$

$r$ cancels

Approaches $e$ as $n \to \infty$

The limit shown on the right is the continuous compounding formula $Pe^{rt}$.

## 4.1  Exercises

**1–2.** Use a calculator to evaluate, rounding to three decimal places.

**1. a.** $e^2$    **b.** $e^{-2}$    **c.** $e^{1/2}$

**2. a.** $e^3$    **b.** $e^{-3}$    **c.** $e^{1/3}$

**3–4.** Express as a power of $e$.

**3. a.** $e^5 e^{-2}$    **b.** $\dfrac{e^5}{e^3}$    **c.** $\dfrac{e^5 e^{-1}}{e^{-2} e}$

**4. a.** $e^{-3} e^5$    **b.** $\dfrac{e^4}{e}$    **c.** $\dfrac{e^2 e}{e^{-2} e^{-1}}$

**5–8.** Graph each function. If you are using a graphing calculator, make a hand-drawn sketch from the screen.

**5.** $y = 3^x$    **6.** $y = 5^x$    **7.** $y = \left(\frac{1}{3}\right)^x$    **8.** $y = \left(\frac{1}{5}\right)^x$

**9–10.** Evaluate each expression using a calculator.

**9.** $e^{1.74}$    **10.** $e^{-0.09}$

**11.** $e^x$ **Versus** $x^n$  Which curve is eventually higher, $x$ to a power or $e^x$?

**a.** Graph $x^2$ and $e^x$ on the window [0, 5] by [0, 20]. Which curve is higher?

**b.** Graph $x^3$ and $e^x$ on the window [0, 6] by [0, 200]. Which curve is higher for large values of $x$?

**c.** Graph $x^4$ and $e^x$ on the window [0, 10] by [0, 10,000]. Which curve is higher for large values of $x$?

**d.** Graph $x^5$ and $e^x$ on the window [0, 15] by [0, 1,000,000]. Which curve is higher for large values of $x$?

**e.** Do you think that $e^x$ will exceed $x^6$ for large values of $x$? Based on these observations, can you make a conjecture about $e^x$ and *any* power of $x$?

12. **Linear Versus Exponential Growth**
    a. Graph $y_1 = x$ and $y_2 = e^{0.01x}$ in the window [0, 10] by [0, 10]. Which curve is higher for $x$ near 10?
    b. Then graph the same curves on the window [0, 1000] by [0, 1000]. Which curve is higher for $x$ near 1000?

A function such as $y_1$ represents *linear* growth, and $y_2$ represents *exponential* growth, and the result here is true in general: exponential growth always beats linear growth (eventually, no matter what the constants).*

## Applied Exercises

13. **BUSINESS: Interest** Find the value of $1000 deposited in a bank at 10% interest for 8 years compounded:
    a. annually.    b. quarterly.
    c. continuously.

14. **BUSINESS: Interest** Find the value of $1000 deposited in a bank at 12% interest for 8 years compounded:
    a. annually.    b. quarterly.
    c. continuously.

15. **PERSONAL FINANCE: Interest** A loan shark lends you $100 at 2% compound interest per week (this is a *weekly*, not an annual rate).
    a. How much will you owe after 3 years?
    b. In "street" language, the profit on such a loan is known as the "vigorish" or the "vig." Find the shark's vig.

16. **GENERAL: Compound Interest** In 1626, Peter Minuit purchased Manhattan Island from the native Americans for $24 worth of trinkets and beads. Find what the $24 would be worth in the year 2010 if it had been deposited in a bank paying 5% interest compounded quarterly.

17. **PERSONAL FINANCE: Annual Percentage Rate (APR)** Find the error in the ad shown below, which appeared in a New York newspaper. [*Hint:* Check that the nominal rate is equivalent to the effective rate. For daily compounding, some banks use 365 days and some use 360 days in the year. Try it both ways.]

18. **PERSONAL FINANCE: Annual Percentage Rate (APR)** Find the error in the following ad, which appeared in a Washington, D.C., newspaper. Assume that the compounding is done daily. [*Hint:* Check that the nominal rate is equivalent to the effective rate. For daily compounding, some banks use 365 days and some use 360 days in the year. Try it both ways.]

19. **PERSONAL FINANCE: Present Value** A rich uncle wants to make you a millionaire. How much money must he deposit in a trust fund paying 8% compounded quarterly at the time of your birth to yield $1,000,000 when you retire at age 60?

20. **PERSONAL FINANCE: Present Value** The cost of a four-year private college education (after financial aid) has been estimated to be $65,000. How large a trust fund, paying 6% compounded quarterly, must be established at a child's birth to ensure sufficient funds at age 18?

*Source: U.S. News & World Report*

21. **GENERAL: Consumer Fraud** Buying a "vacation timeshare" means buying the right to use a vacation property for a fixed period each year. Suppose that you pay $500 for a vacation timeshare and receive a "money-back guarantee" that at any time the company will buy back your timeshare, or if not, give you a $1000 bond. The deception, however, is that the bond is not redeemable for 45 years. Find the real value of the "guarantee"—that is, find the present value of $1000 in 45 years (assume a 6% interest rate compounded annually).

22. **PERSONAL FINANCE: Zero-Coupon Bonds** Santa Rosa County recently sold $1000 zero-coupon bonds** maturing in 4 years with an

*The realization that populations grow exponentially while food supplies grow only linearly caused the great nineteenth-century essayist Thomas Carlyle to dub economics the "dismal science." He was commenting not on how interesting economics is, but on the grim conclusions that follow from populations outstripping their food supplies.

**A zero-coupon bond makes no payments (coupons) until it matures, at which time it pays its face value of $1000, having sold originally at a lower price.

annual yield of 6%. Find the price. [*Hint:* The price is the present value of $1000 4 years from now at the stated interest rate.]

*Source:* fmsbonds.com

23. **PERSONAL FINANCE: Zero-Coupon Bonds** FUJI Holdings recently sold zero-coupon $1000 bonds* maturing in 3 years with an annual yield of 10%. Find the price. [*Hint:* The price is the present value of $1000 3 years from now at the stated interest rate.]

*Source:* Reuters

24. **PERSONAL FINANCE: Baseball Cards** A mint-condition 1910 Honus Wagner** baseball card sold at auction in 2007 for $2.35 million, its value having grown by 11% annually. Predict its value in the year 2017.

*Source: Los Angeles Times*, February 28, 2007

25. **GENERAL: Compound Interest** Which is better: 10% interest compounded quarterly or 9.8% compounded continuously?

26. **GENERAL: Compound Interest** Which is better: 8% interest compounded quarterly or 7.8% compounded continuously?

27. **PERSONAL FINANCE: Depreciation** A Toyota Corolla automobile lists for $15,450 and depreciates by 35% per year. Find its value after:

a. 4 years.     b. 6 months.

*Source:* Toyota

28. **BUSINESS: Art Appreciation** In 2006 Paul Cézanne's painting *Still Life with Fruits and Ginger Jar* sold at auction for $37 million, its value having grown by 12.6% annually. Predict its value in the year 2016.

*Source: New York Times*, November 8, 2006

29. **SOCIAL SCIENCE: Population** The population of the world $x$ years after the year 2005 is predicted to be $6.44e^{0.0123x}$ billion people (for $0 \leq x \leq 20$). Use this formula to predict the world population in the year 2015.

*Source:* U.N. Fund for Population Activities

30. **SOCIAL SCIENCE: Population** The most populous country is China, with a (2005) population of 1.3 billion, growing at the rate of 0.6% a year. The 2005 population of India was 1.13 billion, growing at 1.5% a year. Assuming that these growth rates continue, which population will be larger in the year 2025?

*Source:* U.N.—Population Division

31. **GENERAL: Nuclear Meltdown** The probability of a "severe core meltdown accident" at a nuclear reactor in the United States within the next $n$ years is $1 - (0.9997)^{100n}$. Find the probability of a meltdown:

a. within 25 years.     b. within 40 years.

*Source:* Nuclear Regulatory Commission

(The 1986 core meltdown in the Chernobyl reactor in the Soviet Union spread radiation over much of Eastern Europe, leading to an undetermined number of fatalities.)

32. **BIOMEDICAL SCIENCE: Mosquitoes** Female mosquitoes *(Culex pipiens)* feed on blood (only the females drink blood) and then lay several hundred eggs. In this way each mosquito can, on the average, breed another 300 mosquitoes in about 9 days. Find the number of great-grandchildren mosquitoes that will be descended from one female mosquito, assuming that all eggs hatch and mature.

33. **ENVIRONMENTAL SCIENCE: Light** According to the Bouguer–Lambert Law, the proportion of light that penetrates ordinary seawater to a depth of $x$ feet is $e^{-0.44x}$. Find the proportion of light that penetrates to a depth of:

a. 3 feet.     b. 10 feet.

34–35. **BIOMEDICAL: Drug Dosage** If a dosage $d$ of a drug is administered to a patient, the amount of the drug remaining in the tissues $t$ hours later will be $f(t) = de^{-kt}$, where $k$ (the "absorption constant") depends on the drug.

*Source:* T. R. Harrison, ed., *Principles of Internal Medicine*

34. For the immunosuppressant cyclosporine, the absorption constant is $k = 0.012$. For a dose of $d = 400$ milligrams, use the previous formula to find the amount of cyclosporine remaining in the tissues after:

a. 24 hours.     b. 48 hours.

35. For the cardioregulator digoxin, the absorption constant is $k = 0.018$. For a dose of $d = 2$ milligrams, use the previous formula to find the amount remaining in the tissues after:

a. 24 hours.     b. 48 hours.

---

*A zero-coupon bond makes no payments (coupons) until it matures, at which time it pays its face value of $1000, having sold originally at a lower price.

**Honus Wagner, regarded as the greatest player of his time, played for the Pittsburgh Pirates. Baseball cards were then sold with cigarettes, and Wagner became even more famous when he refused to accept payment for the use of his picture as a protest against the use of tobacco.

**36. BIOMEDICAL: Bacterial Growth** A colony of bacteria in a petri dish doubles in size every hour. At noon the petri dish is just covered with bacteria. At what time was the petri dish:

**a.** 50% covered? [*Hint:* No calculation needed.]
**b.** 25% covered?

**37. BUSINESS: Advertising** A company finds that $x$ days after the conclusion of an advertising campaign the daily sales of a new product are $S(x) = 100 + 800e^{-0.2x}$. Find the daily sales 10 days after the end of the advertising campaign.

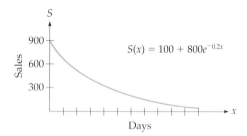

**38. BUSINESS: Quality Control** A company finds that the proportion of its light bulbs that will burn continuously for longer than $t$ weeks is $e^{-0.01t}$. Find the proportion of bulbs that burn for longer than 10 weeks.

**39. GENERAL: Temperature** A covered mug of coffee originally at 200 degrees Fahrenheit, if left for $t$ hours in a room whose temperature is 70 degrees, will cool to a temperature of $70 + 130e^{-1.8t}$ degrees. Find the temperature of the coffee after:

**a.** 15 minutes.     **b.** half an hour.

**40. BEHAVIORAL SCIENCE: Learning** In many psychology experiments the percentage of items that are remembered after $t$ time units is

$$p(t) = 100\,\frac{1 + e}{1 + e^{t+1}}$$

Such curves are called "forgetting" curves. Find the percentage remembered after:

**a.** 0 time units.     **b.** 2 time units.

**41. BIOMEDICAL: Epidemics** The Reed–Frost model for the spread of an epidemic predicts that the number $I$ of newly infected people is $I = S(1 - e^{-rx})$, where $S$ is the number of susceptible people, $r$ is the effective contact rate, and $x$ is the number of infectious people. Suppose that a school reports an outbreak of measles with $x = 10$ cases, and that the effective contact rate is $r = 0.01$. If the number of susceptibles is $S = 400$, use the Reed–Frost model to estimate

how many students will be newly infected during this stage of the epidemic.

*Source: Human Biology* 24

**42. SOCIAL SCIENCE: Election Cost** The cost of winning a seat in the House of Representatives in recent years has been approximately $805e^{0.0625x}$ thousand dollars, where $x$ is the number of years since 2000. Estimate the cost of winning a House seat in the year 2016.

*Source: Center for Responsive Politics*

**43. PERSONAL FINANCE: Rate of Return** An investment of $8000 grows to $10,291.73 in 4 years. Find the annual rate of return for annual compounding. [*Hint:* Use $P(1 + r/m)^{mt}$ with $m = 1$ and solve for $r$ (rounded).]

**44. PERSONAL FINANCE: Rate of Return** An investment of $9000 grows to $10,380.65 in 2 years. Find the annual rate of return for quarterly compounding. [*Hint:* Use $P(1 + r/m)^{mt}$ with $m = 4$ and solve for $r$ (rounded).]

**45. ATHLETICS: Olympic Games** When the Olympic Games were held near Mexico City in the summer of 1968, many athletes were concerned that the high elevation would affect their performance. If air pressure decreases exponentially by 0.4% for each 100 feet of altitude, by what percentage did the air pressure decrease in moving from Tokyo (the site of the 1964 Summer Olympics, at altitude 30 feet) to Mexico City (altitude 7347 feet)?

*Source: CRC Handbook of Chemistry and Physics*

**46. BIOMEDICAL: Radioactive Contamination** The core meltdown and explosions at the nuclear reactor in Chernobyl in 1986 released large amounts of strontium 90, which decays exponentially at the rate of 2.5% per year. Areas downwind of the reactor will be uninhabitable for 100 years. What percent of the original strontium 90 contamination will still be present after:

**a.** 50 years?     **b.** 100 years?

*Source: Environmental Protection Agency*

**47. GENERAL: Population** As stated earlier, the most populous state is California, with Texas second but gaining. According to the Census Bureau, $x$ years after 2005 the population of California will be $36e^{0.013x}$ and the population of Texas will be $22e^{0.019x}$ (all in millions).

**a.** Graph these two functions on a calculator on the window [0, 100] by [0, 100].

**b.** In which year is Texas projected to overtake California as the most populous state? [*Hint:* Use INTERSECT.]

*Source: U.S. Census Bureau*

**48. GENERAL: St. Louis Arch** The Gateway Arch in St. Louis is built around a mathematical curve called a "catenary." The height of this catenary above the ground at a point $x$ feet from the center line is

$$y = 688 - 31.5(e^{0.01033x} + e^{-0.01033x})$$

a.  Graph this curve on a calculator on the window $[-400, 400]$ by $[0, 700]$.
b.  Find the height of the Gateway Arch at its highest point, using the fact that the top of the arch is 5 feet higher than the top of the central catenary.

*Source:* National Park Service

**49. PERSONAL FINANCE: Earnings and Calculus** A recent study found that one's earnings are affected by the mathematics courses one has taken. In particular, compared to someone making \$40,000 who had taken *no* calculus, a comparable person who had taken $x$ years of calculus would be earning  $\$40,000e^{0.195x}$. Find the salary of a person who has taken $x = 2$  years of calculus. [*Note:* Other mathematics courses were included in the study, but calculus courses brought the greatest increase in salary.]

*Source: Review of Economics and Statistics 86*

## Conceptual Exercises

**50.** In the box on page 259, why do we require that the base $a$ be *positive*? [*Hint:* How would we define $f(x) = a^x$ for  $x = 0.5$  if $a$ were negative?]

**51.** Does the graph of $a^x$ for   $a > 1$   have a horizontal asymptote? [*Hint:* Look at the graph on page 260.]

**52.** Does the graph of $a^x$ for   $a < 1$   have a horizontal asymptote? [*Hint:* Look at the graph on page 260.]

**53.** Which will be largest for very large values of $x$: $x^2$, $e^x$,   or   $x^{1000}$?

**54.** True or False: If   $x < y$,   then   $e^x < e^y$.

**55.** What is the most important characteristic of exponential growth?

**56.** Assuming the same fixed nominal interest rate, put the following types of compounding in order of increasing benefit to the depositor: daily, continuously, quarterly, semiannually.

**57.** Explain why 5% compounded *continuously* is better than 5% compounded *monthly.*

**58.** A banker once told one of the authors that continuous compounding means the same thing as daily compounding. Was he right? Explain.

**59.** A company is considering two ways to depreciate a truck: *straight line* or *by a fixed percentage*. If each way begins with a value of \$25,000 and ends 5 years later with a value of \$1000, which way will result in the bigger change in value during the *first* year? During the *last* year?

**60.** Suppose that you have a bank account with interest compounded continuously, but you can't remember the continuously compounded interest rate. If at the end of the year you had 10% more than you began with, was the continuously compounded rate *more* than 10% or *less* than 10%?

---

## 4.2   LOGARITHMIC FUNCTIONS

### Introduction

In this section we introduce logarithmic functions, emphasizing the *natural* logarithm function. We then apply natural logarithms to a wide variety of problems, from doubling money under compound interest to drug dosage.

### Logarithms

The word **logarithm** (abbreviated **log**) means *power* or *exponent*. The number being raised to the power is called the **base** and is written as a subscript. For example, the expression

$$\log_{10} 1000 \qquad \text{Read "log (base 10) of 1000"}$$

Base

means the *exponent* to which we raise 10 to get 1000. Since $10^3 = 1000$, the exponent is 3, so the *logarithm* is 3:

$$\log_{10} 1000 = 3 \qquad \text{Since } 10^3 = 1000$$

We find logarithms by writing them as exponents and then finding the exponent.

---

**EXAMPLE 1**

**FINDING A LOGARITHM**

Evaluate $\log_{10} 100$.

*Solution*

$$\log_{10} 100 = y \qquad \text{is equivalent to} \qquad 10^y = 100 \qquad \text{Exponent 2 works, so 2 is the logarithm}$$

The logarithm $y$ is the exponent that solves

Therefore

$$\log_{10} 100 = 2 \qquad \text{Since } 10^2 = 100$$

---

**EXAMPLE 2**

**FINDING A LOGARITHM**

Evaluate $\log_{10} \dfrac{1}{10}$.

*Solution*

$$\log_{10} \frac{1}{10} = y \qquad \text{is equivalent to} \qquad 10^y = \frac{1}{10} \qquad \text{Exponent } -1 \text{ works, so } -1 \text{ is the logarithm}$$

The logarithm $y$ is the exponent that solves

Therefore

$$\log_{10} \frac{1}{10} = -1 \qquad \text{Since } 10^{-1} = \frac{1}{10}$$

---

**Practice Problem 1**     Evaluate $\log_{10} 10,000$.     ➤ Solution on page 284

Logarithms to the base 10 are called **common logarithms** and are often written without the base, so that $\log 100$ means $\log_{10} 100$. The LOG key on many calculators evaluates common logarithms.

We may find logarithms to other bases as well. Any positive number other than 1 may be used as a base.

**EXAMPLE 3**

**FINDING A LOGARITHM**

Find $\log_9 3$.

*Solution*

$$\log_9 3 = y \quad \text{is equivalent to} \quad 9^y = 3 \qquad \text{Exponent } \tfrac{1}{2} \text{ works,} \\ \text{so } \tfrac{1}{2} \text{ is the logarithm}$$

Therefore

$$\log_9 3 = \frac{1}{2} \qquad \text{Since} \quad 9^{1/2} = \sqrt{9} = 3$$

In general, for any positive base $a$ other than 1,

| Logarithms |
|---|
| $\log_a x = y \quad \text{is equivalent to} \quad a^y = x$ $\qquad$ $\log_a x$ means the exponent to which we raise $a$ to get $x$ |

**Practice Problem 2**

Find:

**a.** $\log_5 125$ $\qquad$ **b.** $\log_8 2$ $\qquad\qquad$ ➤ **Solutions on page 284**

## Natural Logarithms

The most widely used of all bases is $e$, the number (approximately 2.718) that we defined on page 264. Logarithms to the base $e$ are called **natural logarithms** or **Napierian logarithms.**\* The natural logarithm of $x$ is written $\ln x$ ("$n$" for "natural") instead of $\log_e x$, and may be found using the ⬚ LN key on a calculator.

| Natural Logarithms |
|---|
| $\ln x = \text{logarithm of } x \text{ to the base } e$ $\qquad$ $\ln x$ means $\log_e x$ |

In words: $\ln x$ is the power to which we raise $e$ to get $x$.

 **BE CAREFUL:** Since $e$ to any power is positive, $\ln x$ is defined only for *positive values of x.*

**Practice Problem 3**

Use a calculator to find $\ln 8.34$. $\qquad\qquad$ ➤ **Solution on page 284**

---

\*After John Napier (1550–1617), a Scottish mathematician and, incidentally, the inventor of the decimal point.

The table below shows some values of the natural logarithm function $f(x) = \ln x$, and its graph (based on these points) is shown on the right.

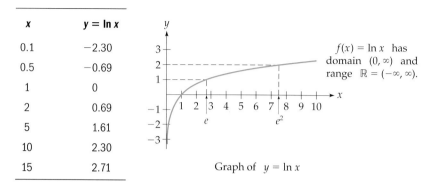

| x | y = ln x |
|---|---|
| 0.1 | −2.30 |
| 0.5 | −0.69 |
| 1 | 0 |
| 2 | 0.69 |
| 5 | 1.61 |
| 10 | 2.30 |
| 15 | 2.71 |

$f(x) = \ln x$ has domain $(0, \infty)$ and range $\mathbb{R} = (-\infty, \infty)$.

Graph of $y = \ln x$

Notice that the graph of $\ln x$ is always increasing and has the $y$-axis as a vertical asymptote ( $\lim\limits_{x \to 0^+} \ln x = -\infty$ ). A definition of natural logarithms without recourse to exponents is given in Exercise 74 on page 363. The natural logarithm function may be used for modeling growth that continually slows.

The following graph shows logarithm functions for several different bases. Notice that each passes through the point $(1, 0)$, since $a^0 = 1$. We will concentrate on the *natural* logarithm function, since it is the one most used in applications.

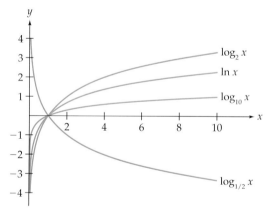

Since logs are exponents, each property of exponents (see page 21) can be restated as a property of logarithms. The first three properties show that some natural logarithms can be found without using a calculator.

| Properties of Natural Logarithms | |
|---|---|
| **1.** $\ln 1 = 0$ | The natural log of 1 is 0 (since $e^0 = 1$) |
| **2.** $\ln e = 1$ | The natural log of $e$ is 1 (since $e^1 = e$) |
| **3.** $\ln e^x = x$ | The natural log of $e$ to a power is just the power (since $e^x = e^x$) |
| **4.** $e^{\ln x} = x$ $\qquad x > 0$ | $e$ raised to the natural log of a number is just the number |

The first two properties are special cases of the third, with $x = 0$ and $x = 1$. Since logarithms are exponents, the third property simply says that the exponent of $e$ that gives $e^x$ is $x$, which is obvious when you think about it. Since $\ln x$ is the power of $e$ that gives $x$, raising $e$ to that power must give $x$, which is the fourth property.

**EXAMPLE 4**

**USING THE PROPERTIES OF NATURAL LOGARITHMS**

a. $\ln e^7 = 7$

b. $\ln e^{3x} = 3x$

c. $\ln \sqrt[3]{e^2} = \ln e^{2/3} = \frac{2}{3}$

The natural log of $e$ to a power is just the power

d. $e^{\ln 5} = 5$

$e$ raised to the natural log of 5 is just 5

---

## Graphing Calculator Exploration

```
ln(e^(17))
              17
e^(ln(29))
              29
```

a. Evaluate $\ln e^{17}$ to verify that the answer is 17. Change 17 to other numbers (positive, negative, or zero) in order to verify that $\ln e^x = x$ for any $x$.

b. Evaluate $e^{\ln 29}$ to verify that the answer is 29. Change the 29 to another positive number to verify that $e^{\ln x} = x$. What about negative numbers, or zero?

The next four properties enable us to simplify logs of products, quotients, and powers. For positive numbers $M$ and $N$ and any number $P$:

| Properties of Natural Logarithms |
| --- |

5. $\ln (M \cdot N) = \ln M + \ln N$ — The log of a *product* is the *sum* of the logs

6. $\ln \left(\dfrac{1}{N}\right) = -\ln N$ — The log of 1 *over* a number is the *negative* of the log of the number

7. $\ln \left(\dfrac{M}{N}\right) = \ln M - \ln N$ — The log of a *quotient* is the *difference* of the logs

8. $\ln (M^P) = P \cdot \ln M$ — The log of a number to a power is the power *times* the log

Property 8 can be stated simply: *logarithms bring down exponents*. A justification for these four properties is given at the end of this section.

**EXAMPLE 5**

**USING THE PROPERTIES OF NATURAL LOGARITHMS**

a. $\ln(2 \cdot 3) = \ln 2 + \ln 3$      Property 5

b. $\ln \frac{1}{7} = -\ln 7$      Property 6

c. $\ln \frac{2}{3} = \ln 2 - \ln 3$      Property 7

d. $\ln(2^3) = 3 \ln 2$      $\ln(2^3) = 3 \ln 2$

Properties 1 through 8 are very useful for simplifying functions.

**EXAMPLE 6**

**SIMPLIFYING A FUNCTION**

$$f(x) = \ln(2x) - \ln 2$$

$$= \ln 2 + \ln x - \ln 2 \qquad \text{Since } \ln(2x) = \ln 2 + \ln x \text{ by Property 5}$$

$$= \ln x \qquad \text{Canceling}$$

**EXAMPLE 7**

**SIMPLIFYING A FUNCTION**

$$f(x) = \ln\left(\frac{x}{e}\right) + 1$$

$$= \ln x - \ln e + 1 \qquad \text{Since } \ln(x/e) = \ln x - \ln e \text{ by Property 7}$$

$$= \ln x - 1 + 1 \qquad \text{Since } \ln e = 1 \text{ by Property 2}$$

$$= \ln x \qquad \text{Canceling}$$

**EXAMPLE 8**

**SIMPLIFYING A FUNCTION**

$$f(x) = \ln(x^5) - \ln(x^3)$$

$$= 5 \ln x - 3 \ln x \qquad \text{Bringing down exponents by Property 8}$$

$$= 2 \ln x \qquad \text{Combining}$$

## Graphing Calculator Exploration

**a.** Graph $y = \ln x^2$ on the window $[-5, 5]$ by $[-5, 5]$.

**b.** Change the function to $y = 2 \ln x$.

$\ln x^2 \rightarrow$

$\leftarrow 2 \ln x$

Explain why the two graphs are different. (Doesn't Property 8 on page 277 say that $\ln x^2 = 2 \ln x$?)

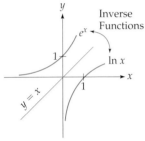

$y$

Inverse
Functions

$e^x$

$\ln x$

$x$

$y = x$

$y = e^x$ and $y = \ln x$
are inverse functions.

On page 276 we saw the natural logarithm properties $\ln e^x = x$ and $e^{\ln x} = x$. These show that the natural logarithm function $\ln x$ and the exponential function $e^x$ are *inverse functions* in that either one "undoes" or "inverts" the other. (Notationally, $e^x$ raises $x$ up to the exponent, and logarithms bring down exponents, so it is reasonable that they "undo" each other.) This inverse relationship between $e^x$ and $\ln x$ is equivalent to the geometric fact that their graphs are reflections of each other in the diagonal line $y = x$, as shown on the left.

## Doubling Under Compound Interest

How soon will money invested at compound interest double in value? The solution to this question makes important use of the property $\ln (M^P) = P \cdot \ln M$ ("logs bring down exponents").

---

**EXAMPLE 9**  **FINDING DOUBLING TIME**

A sum is invested at 12% interest compounded quarterly. How soon will it double in value?

*Solution*

We use the formula $P(1 + r/m)^{mt}$ with $r = 0.12$ and $m = 4$. Since double $P$ dollars is $2P$ dollars, we want to solve

$$P\left(1 + \frac{0.12}{4}\right)^{4t} = 2P \qquad \text{$t$ is the number of years}$$

$\underbrace{\qquad}_{0.03}$  $P$ dollars doubled

$$1.03^{4t} = 2 \qquad \begin{array}{l}\text{Simplifying } 1 + 0.03 \text{ to } 1.03 \\ \text{and canceling the } P\text{'s}\end{array}$$

The variable is in the *exponent*, so we take logarithms to bring it down.

$$\ln{(1.03^{4t})} = \ln 2 \qquad \text{Taking natural logs of both sides}$$

$$4t \cdot \ln 1.03 = \ln 2 \qquad \begin{array}{l}\text{Bringing down the exponent}\\ \text{(Property 8 of logarithms)}\end{array}$$

$$t = \frac{\ln 2}{4 \cdot \ln 1.03} \qquad \begin{array}{l}\text{Dividing each side by 4 and}\\ \text{by} \quad \ln 1.03 \quad \text{to solve for } t\end{array}$$

$$t = \frac{0.6931}{4 \cdot 0.0296} \approx 5.9 \qquad \begin{array}{l}\text{Using a calculator to evaluate the}\\ \text{logs and the rest of the expression}\end{array}$$

A sum at 12% compounded quarterly doubles in about 5.9 years.

Of course, for quarterly compounding you will not get the $2P$ dollars until the end of the last quarter, by which time it will be slightly more.

Notice that the principal $P$ canceled out after the first step, showing that *any* sum will double in the same amount of time: 1 dollar will double into 2 dollars in exactly the same time that a million dollars will double into 2 million dollars.

To find how soon the value *triples*, we would solve $P(1 + r/m)^{mt} = 3P$, and similarly for any other multiple. To find how soon the value would *increase by 50%* (that is, become 1.5 times its original value), we would solve

$$P\left(1 + \frac{r}{m}\right)^{mt} = 1.5P \qquad \text{Multiplying by 1.5 increases } P \text{ by 50\%}$$

**Practice Problem 4**

A sum is invested at 10% interest compounded semiannually (twice a year). How soon will it increase by 60%? ➤ **Solution on page 284**

**EXAMPLE 10**  **FINDING TRIPLING TIME WITH CONTINUOUS COMPOUNDING**

A sum is invested at 15% compounded continuously. How soon will it triple?

*Solution*

The idea is the same as before, but for continuous compounding we use the formula $Pe^{rt}$:

$$Pe^{0.15t} = 3P \qquad 3P \text{ since the value triples}$$

$$e^{0.15t} = 3 \qquad \text{Canceling the } P\text{'s}$$

$$\ln e^{0.15t} = \ln 3 \qquad \text{Taking the natural log of both sides}$$

$$\underbrace{\quad}_{0.15t}$$

$$0.15t = \ln 3$$   Since the natural log of $e$ to a power is just the power

$$t = \frac{\ln 3}{0.15}$$   Dividing by 0.15 to solve for $t$

$$\approx \frac{1.0986}{0.15} \approx 7.3$$   Using a calculator

A sum at 15% compounded continuously triples in about 7.3 years.

## Drug Dosage

Drug concentration in the blood

The amount of a drug that remains in a person's bloodstream decreases exponentially with time. If the initial concentration is $c$ (milligrams per milliliter of blood), the concentration $t$ hours later will be

$$C(t) = ce^{-kt}$$

where the "absorption constant" $k$ measures how rapidly the drug is absorbed.

Every medicine has a minimum concentration below which it is not effective. When the concentration falls to this level, another dose should be administered. If doses are administered regularly every $T$ hours over a period of time, the concentration will look like the following graph.

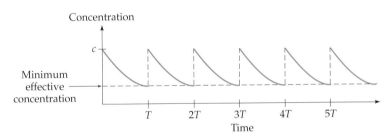

Drug concentration with repeated doses

The problem is to determine the time $T$ at which the dose should be repeated so as to maintain an effective concentration.

**EXAMPLE 11**    **CALCULATING DRUG DOSAGE**

The absorption constant for penicillin is $k = 0.11$, and the minimum effective concentration is 2 milligrams. If the original concentration is $c = 5$ milligrams, find when another dose should be administered in order to maintain an effective concentration.

*Solution*

The concentration formula $C(t) = ce^{-kt}$ with $c = 5$ and $k = 0.11$ is

$$C(t) = 5e^{-0.11t}$$

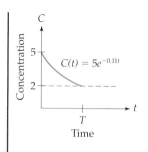

To find the time when this concentration reaches the minimum effective level of 2, we solve

$$5e^{-0.11t} = 2$$

$$e^{-0.11t} = 0.4 \qquad \text{Dividing by 5}$$

$$\underbrace{\ln e^{-0.11t}}_{-0.11t} = \ln 0.4 \qquad \begin{array}{l}\text{Taking natural logs to}\\ \text{bring down the exponent}\end{array}$$

$$-0.11t = \ln 0.4$$

$$t = \frac{\ln 0.4}{-0.11} \approx \frac{-0.9163}{-0.11} \approx 8.3 \qquad \begin{array}{l}\text{Solving for } t \text{ and}\\ \text{using a calculator}\end{array}$$

The concentration will reach the minimum effective level in 8.3 hours, so another dose should be taken approximately every 8 hours.

The last two examples have led to equations of the form $e^{at} = b$. Such equations occur frequently in applications, and are solved by taking the natural log of each side to bring down the exponent. Such equations may also be solved on a graphing calculator by graphing the function as $y_1$ and the constant as $y_2$ and using INTERSECT to find where they meet.

## Carbon 14 Dating

As discussed at the beginning of this chapter, all living things absorb small amounts of radioactive carbon 14 from the atmosphere. When they die, the carbon 14 stops being absorbed and decays exponentially into ordinary carbon. Therefore, the proportion of carbon 14 still present in a fossil or other ancient remain can be used to estimate how old it is. The proportion of the original carbon 14 that will be present after $t$ years is

$$\left(\begin{array}{l}\text{Proportion of carbon 14}\\ \text{remaining after } t \text{ years}\end{array}\right) = e^{-0.00012t}$$

EXAMPLE 12  **DATING BY CARBON 14**

The Dead Sea Scrolls, discovered in a cave near the Dead Sea in what was then Jordan, are among the earliest documents of Western civilization. Estimate the age of the Dead Sea Scrolls if the animal skins on which some were written contain 78% of their original carbon 14.

*Solution*

The proportion of carbon 14 remaining after $t$ years is $e^{-0.00012t}$. We equate this formula to the actual proportion (expressed as a decimal):

$$e^{-0.00012t} = 0.78 \qquad \text{Equating the proportions}$$

$$\ln e^{-0.00012t} = \ln 0.78 \qquad \text{Taking natural logs}$$

$$-0.00012t = \ln 0.78 \qquad \begin{array}{l}\ln e^{-0.00012t} = -0.00012t \\ \text{by Property 3}\end{array}$$

$$t = \frac{\ln 0.78}{-0.00012} \approx \frac{-0.24846}{-0.00012} \approx 2071 \qquad \begin{array}{l}\text{Solving for } t \text{ and} \\ \text{using a calculator}\end{array}$$

Therefore, the Dead Sea Scrolls are approximately 2070 years old.

## Graphing Calculator Exploration

Intersection
X=2070.5113  Y=.78

Solve Example 12 by graphing $y_1 = e^{-0.00012x}$ and $y_2 = 0.78$ on the window $[0, 4000]$ by $[-0.3, 1]$ and using INTERSECT to find where they meet. (Solving with a graphing calculator requires choosing an appropriate window, a difficulty that does not occur when solving by logarithms.)

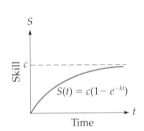

$S(t) = c(1 - e^{-kt})$

## Behavioral Science: Learning Theory

Your ability to do a task generally improves with practice. Frequently, your skill after $t$ units of practice is given by a function of the form

$$S(t) = c(1 - e^{-kt})$$

where $c$ and $k$ are positive constants.

**EXAMPLE 13**  **ESTIMATING LEARNING TIME**

After $t$ weeks of training, your secretary can type

$$S(t) = 100(1 - e^{-0.25t})$$

words per minute. How many weeks will he take to reach 80 words per minute?

*Solution* We solve for $t$ in the following equation:

$$100(1 - e^{-0.25t}) = 80 \qquad \text{Setting } S(t) \text{ equal to 80}$$

$$1 - e^{-0.25t} = 0.80 \qquad \text{Dividing by 100}$$

$$-e^{-0.25t} = -0.20 \qquad \text{Subtracting 1}$$

$$e^{-0.25t} = 0.20 \qquad \text{Multiplying by } -1$$

$$-0.25t = \ln 0.20 \qquad \text{Taking natural logs}$$

$$t = \frac{\ln 0.20}{-0.25} \qquad \text{Solving for } t$$

$$\approx \frac{-1.6094}{-0.25} \approx 6.4 \qquad \text{Using a calculator}$$

He will reach 80 words per minute in about $6\frac{1}{2}$ weeks.

Proportion who heard the news

$p(t) = 1 - e^{-kt}$

Time

## Social Science: Diffusion of Information by Mass Media

When a news bulletin is repeatedly broadcast over radio and television, the proportion of people who hear the bulletin within $t$ hours is

$$p(t) = 1 - e^{-kt}$$

for some constant $k$.

---

**EXAMPLE 14**  **PREDICTING THE SPREAD OF INFORMATION**

A storm warning is broadcast, and the proportion of people who hear the bulletin within $t$ hours of its first broadcast is $p(t) = 1 - e^{-0.30t}$. When will 75% of the people have heard the bulletin?

*Solution*

Equating the proportions gives $1 - e^{-0.30t} = 0.75$. Solving this equation as in the previous examples (omitting the details) gives $t \approx 4.6$. Therefore, it takes about $4\frac{1}{2}$ hours for 75% of the people to hear the news.

---

### Solutions to Practice Problems

**1.** $\log_{10} 10{,}000 = 4$   (Since $10^4 = 10{,}000$)

**2. a.** $\log_5 125 = 3$   (Since $5^3 = 125$)

   **b.** $\log_8 2 = \frac{1}{3}$   (Since $8^{1/3} = \sqrt[3]{8} = 2$)

**3.** $\ln 8.34 \approx 2.121$ (rounded)

**4.** $P\left(1 + \dfrac{0.10}{2}\right)^{2t} = 1.60 \cdot P$

$$1.05^{2t} = 1.6$$

$$\ln(1.05^{2t}) = \ln 1.6$$

$$2t \cdot \ln 1.05 = \ln 1.6$$

$$t = \frac{\ln 1.6}{2 \cdot \ln 1.05} \approx 4.8$$

In about 4.8 years.

## 4.2 Section Summary

We define logarithms as follows:

$$\log_a x = y \quad \text{is equivalent to} \quad a^y = x \qquad \text{Logarithm to the base } a$$
$$\ln x = y \quad \text{is equivalent to} \quad e^y = x \qquad \text{Natural logarithm (base } e)$$

Natural logarithms have the following properties, which are also listed inside the back cover.

**1.** $\ln 1 = 0$      **2.** $\ln e = 1$

**3.** $\ln e^x = x$      **4.** $e^{\ln x} = x$

**5.** $\ln (M \cdot N) = \ln M + \ln N$      **6.** $\ln \left(\dfrac{1}{N}\right) = -\ln N$

**7.** $\ln \left(\dfrac{M}{N}\right) = \ln M - \ln N$      **8.** $\ln (M^P) = P \cdot \ln M$

## Justification of the Properties of Logarithms

Properties 1–4 were justified on page 277.

Property 5 follows from the addition property of exponents, $e^x \cdot e^y = e^{x+y}$ ("the exponent of a product is the sum of the exponents"). Since logs are exponents, this can be restated "the log of a product is the sum of the logs," which is Property 5.

Property 7 follows from the subtraction property of exponents, $e^x/e^y = e^{x-y}$ ("the exponent of a quotient is the difference of the exponents"). This translates into "the log of a quotient is the difference of the logs," which is Property 7.

Property 6 is just a special case of Property 7 (with $M = 1$), along with Property 1.

Property 8 comes from the property of exponents, $(e^x)^y = e^{x \cdot y}$, which says that the exponent $y$ can be "brought down" and multiplied by the $x$. Since logs are exponents, this says that in $\ln (M^P)$ the exponent $P$ can be brought down and multiplied by the logarithm, which gives $P \cdot \ln M$, and this is just Property 8.

Each property of logs is equivalent to a property of exponents:

| *Logarithmic Property* | *Exponential Property* |
| --- | --- |
| $\ln 1 = 0$ | $e^0 = 1$ |
| $\ln (M \cdot N) = \ln M + \ln N$ | $e^x \cdot e^y = e^{x+y}$ |
| $\ln \left(\dfrac{1}{N}\right) = -\ln N$ | $\dfrac{1}{e^y} = e^{-y}$ |
| $\ln \left(\dfrac{M}{N}\right) = \ln M - \ln N$ | $\dfrac{e^x}{e^y} = e^{x-y}$ |
| $\ln M^P = P \cdot \ln M$ | $(e^x)^y = e^{xy}$ |

## 4.2 Exercises

**1–4.** Find each logarithm *without* using a calculator or tables.

**1. a.** $\log_5 25$  **b.** $\log_3 81$  **c.** $\log_3 \frac{1}{3}$

   **d.** $\log_3 \frac{1}{9}$  **e.** $\log_4 2$  **f.** $\log_4 \frac{1}{2}$

**2. a.** $\log_3 27$  **b.** $\log_2 16$  **c.** $\log_{16} 4$

   **d.** $\log_4 \frac{1}{4}$  **e.** $\log_2 \frac{1}{8}$  **f.** $\log_9 \frac{1}{3}$

**3. a.** $\ln (e^{10})$  **b.** $\ln \sqrt{e}$  **c.** $\ln \sqrt[3]{e^4}$

   **d.** $\ln 1$  **e.** $\ln (\ln (e^e))$  **f.** $\ln \left(\frac{1}{e^3}\right)$

**4. a.** $\ln (e^{-5})$  **b.** $\ln e$  **c.** $\ln \sqrt[3]{e}$

   **d.** $\ln \sqrt{e^5}$  **e.** $\ln \left(\frac{1}{e}\right)$  **f.** $\ln (\ln e)$

**5–14.** Use the properties of natural logarithms to simplify each function.

**5.** $f(x) = \ln (9x) - \ln 9$

**6.** $f(x) = \ln \left(\frac{x}{2}\right) + \ln 2$

**7.** $f(x) = \ln (x^3) - \ln x$

**8.** $f(x) = \ln (4x) - \ln 4$

**9.** $f(x) = \ln \left(\frac{x}{4}\right) + \ln 4$

**10.** $f(x) = \ln (x^5) - 3 \ln x$

**11.** $f(x) = \ln (e^{5x}) - 2x - \ln 1$

**12.** $f(x) = \ln (e^{-2x}) + 3x + \ln 1$

**13.** $f(x) = 8x - e^{\ln x}$

**14.** $f(x) = e^{\ln x} + \ln (e^{-x})$

 **15–16.** Find the domain and range of each function:

**15.** $f(x) = \ln (x^2 - 1)$  **16.** $f(x) = \ln (1 - x^2)$

### Applied Exercises

**17. PERSONAL FINANCE: Interest** An investment grows at 24% compounded monthly. How many years will it take to:

   **a.** double?  **b.** increase by 50%?

**18. PERSONAL FINANCE: Interest** An investment grows at 36% compounded monthly. How many years will it take to:

   **a.** double?  **b.** increase by 50%?

**19. PERSONAL FINANCE: Interest** A bank offers 7% compounded continuously. How soon will a deposit:

   **a.** triple?  **b.** increase by 25%?

**20. PERSONAL FINANCE: Interest** A bank offers 6% compounded continuously. How soon will a deposit:

   **a.** quadruple?  **b.** increase by 75%?

**21. PERSONAL FINANCE: Depreciation** An automobile depreciates by 30% per year. How soon will it be worth only half its original value? [*Hint:* Depreciation is like interest but at a negative rate *r*.]

**22. GENERAL: Population Growth** During 2006 and 2007, the population of Georgia increased by 2.9% annually. Assuming that this trend continues, in how many years will the population double?

*Source:* U.S. Census Bureau

**23. ECONOMICS: Energy Output** The world's output of primary energy (petroleum, natural gas, coal, hydroelectricity, and nuclear electricity) is increasing at the rate of 2.4% annually. How long will it take to increase by 50%?

*Source:* Worldwatch

**24. PERSONAL FINANCE: Art Appreciation** In 2004, Pablo Picasso's painting *Boy with a Pipe* sold for $104 million, shattering the record for an auctioned painting, having increased in value by 16% annually. At this rate, how long will it take for the painting to double in value?

*Source: New York Times*

**25–26. BUSINESS: Advertising** After *t* days of advertisements for a new laundry detergent, the proportion of shoppers in a town who have seen the ads is $1 - e^{-0.03t}$. How long must the ads run to reach:

**25.** 90% of the shoppers?

**26.** 99% of the shoppers?

**27. BEHAVIORAL SCIENCE: Forgetting** The proportion of students in a psychology experiment who could remember an eight-digit number correctly for $t$ minutes was $0.9 - 0.2 \ln t$ (for $t > 1$). Find the proportion that remembered the number for 5 minutes.

**28. SOCIAL SCIENCE: Diffusion of Information by Mass Media** Election returns are broadcast in a town of 1 million people, and the number of people who have heard the news within $t$ hours is $1{,}000{,}000(1 - e^{-0.4t})$. How long will it take for 900,000 people to hear the news?

**29. BEHAVIORAL SCIENCE: Learning** After $t$ weeks of practice, a typing student can type $100(1 - e^{-0.4t})$ words per minute (wpm). How soon will the student type 80 wpm?

**30. BIOMEDICAL: Drug Dose** If the original concentration of a drug in a patient's bloodstream is $c$ (milligrams per liter), $t$ hours later the concentration will be $C(t) = ce^{-kt}$, where $k$ is the absorption constant. If the original concentration of the asthma medication theophylline is $c = 20$ and the absorption constant is $k = 0.23$, when should the drug be readministered so that the concentration does not fall below the minimum effective concentration of 5?
*Source: T. R. Harrison, ed., Principles of Internal Medicine*

**31–32. GENERAL: Carbon 14 Dating** The proportion of carbon 14 still present in a sample after $t$ years is $e^{-0.00012t}$.
*Source: A. Dickin, Radiogenic Isotope Geology*

**31.** Estimate the age of the cave paintings discovered in the Ardéche region of France if the carbon with which they were drawn contains only 2.3% of its original carbon 14. They are the oldest known paintings in the world.

**32.** Estimate the age of the Shroud of Turin, believed by many to be the burial cloth of Christ (see the Application Preview on pages 257–258), from the fact that its linen fibers contained 92.3% of their original carbon 14.

**33–34. GENERAL: Potassium 40 Dating** The radioactive isotope potassium 40 is used to date very old remains. The proportion of potassium 40 that remains after $t$ million years is $e^{-0.00054t}$. Use

this function to estimate the age of the following fossils.
*Source: A. Dickin, Radiogenic Isotope Geology*

**33.** The most complete skeleton of an early human ancestor ever found was discovered in Kenya. Use the formula above to estimate the age of the remains if they contained 99.91% of their original potassium 40.

**34. DATING OLDER WOMEN** Use the formula above to estimate the age of the partial skeleton of *Australopithecus afarensis* (known as "Lucy") that was found in Ethiopia if it had 99.82% of its original potassium 40.

**35. ENVIRONMENTAL SCIENCE: Radioactive Waste** Hospitals use radioactive tracers in many medical tests. After the tracer is used, it must be stored as radioactive waste until its radioactivity has decreased enough for it to be disposed of as ordinary chemical waste. For the radioactive isotope iodine 131, the proportion of radioactivity remaining after $t$ days is $e^{-0.087t}$. How soon will the proportion of radioactivity decrease to 0.001 so that it can be disposed of as ordinary chemical waste?
*Source: New York Times*

**36. ENVIRONMENTAL SCIENCE: Rain Forests** It has been estimated that the world's tropical rain forests are disappearing at the rate of 1.8% per year. If this rate continues, how soon will the rain forests be reduced to 50% of their present size? (Rain forests not only generate much of the oxygen that we breathe, but also contain plants with unique medical properties, such as the rosy periwinkle, which has revolutionized the treatment of leukemia.)
*Source: Science 253*

**37–38. SOCIAL SCIENCE: Education and Income** According to a study, each additional year of education increases one's income by 16%. Therefore, with $x$ extra years of education, your income will be multiplied by a factor of $1.16^x$.
*Source: American Economic Review 84*

**37.** How many additional years of education are required to *double* your income? That is, find the $x$ that satisfies $1.16^x = 2$.

**38.** How many additional years of education are required to increase your income by 50%? That is, find the $x$ that satisfies $1.16^x = 1.5$.

**39–47.** Solve the following exercises on a graphing calculator by graphing an appropriate exponential function (using $x$ for ease of entry) together with a constant function and using INTERSECT to find where they meet. You will have to choose an appropriate window.

39. **GENERAL: Inflation** At 2% inflation, prices increase by 2% compounded annually. How soon will prices:
    **a.** double?     **b.** triple?

40. **GENERAL: Inflation** At 5% inflation, prices increase by 5% compounded annually. How soon will prices:
    **a.** double?     **b.** triple?

41. **PERSONAL FINANCE: Interest** A bank account grows at 6% compounded quarterly. How many years will it take to:
    **a.** double?     **b.** increase by 50%?

42. **PERSONAL FINANCE: Interest** A bank account grows at 7% compounded continuously. How many years will it take to:
    **a.** double?     **b.** increase by 25%?

43. **BUSINESS: Advertising** After a sale has been advertised for $t$ days, the proportion of shoppers in a city who have seen the ad is $1 - e^{-0.08t}$. How long must the ad run to reach:
    **a.** 50% of the shoppers?
    **b.** 60% of the shoppers?

44. **BIOMEDICAL: Drug Dosage** If the original concentration of a drug in a patient's bloodstream is 5 (milligrams per milliliter), and if the absorption constant is 0.15, then $t$ hours later the concentration will be $5e^{-0.15t}$. When should the drug be readministered so that the concentration does not fall below the minimum effective concentration of 2.7?

45. **GENERAL: Carbon 14 Dating** In 1991 two hikers in the Italian Alps found the frozen but well-preserved body of the most ancient human ever found, dubbed "Iceman." Estimate the age of Iceman if his grass cape contained 53% of its original carbon 14. (Use the carbon 14 decay function stated in Exercises 31–32.)

46. **GENERAL: Potassium 40 Dating** Estimate the age of the oldest known dinosaur, a dog-sized creature called *Herrerasaurus* found in Argentina in 1988, if volcanic material found with it contained 88.4% of its original potassium 40. (Use the potassium 40 decay function given in Exercises 33–34.)

47. **ENVIRONMENTAL SCIENCE: Nuclear Waste** More than 60 years after the beginning of the nuclear age, we do not have a safe or permanent way to dispose of long-lived radioactive waste. Among the most hazardous radioactive waste is irradiated fuel from nuclear power plants, totaling 245,000 tons in 2000 and growing by 11.3% annually. At this rate, how long will it take for this amount to double?
    *Source:* Worldwatch

48. **SOCIAL SCIENCE: Cell Phone Usage** In 2006 the number of cell phone subscribers worldwide increased by 24%. At this rate, what is the doubling time for cell phone subscribers? Express your answer in years and months.

49. **PERSONAL FINANCE: Bonds** The European Bank for Reconstruction and Development recently sold fifteen-year $1000 bonds with an annual yield of 9.25%. After how much time could you sell them for twice what you paid for them? Give your answer in years and months.
    *Source:* ebrd.com

50. **PERSONAL FINANCE: Bonds** The Central Bank of Brazil recently sold $1000 bonds maturing in twelve years with an annual yield of 9.75%. After how much time could you sell them for twice what you paid for them? Give your answer in years and months.
    *Source:* shop4bonds.com

## Conceptual Exercises

**51–54.** In each pair of equations, one is true and one is false. Choose the correct one.

51. $\ln 1 = 0$   or   $\ln 0 = 1$.

52. $\ln (x + y) = \ln x \cdot \ln y$
    or   $\ln (x \cdot y) = \ln x + \ln y$.

53. $\dfrac{\ln x}{\ln y} = \ln (x - y)$   or   $\ln \dfrac{x}{y} = \ln x - \ln y$.

54. $(\ln x)^n = n \cdot \ln x$   or   $\ln (x^n) = n \cdot \ln x$.

55. If $y = 10^x$, then $x = $ ?

56. If $y = e^x$, then $x = $ ?

**57.** Why can't we define logs of negative numbers, such as $\ln(-2)$? [*Hint:* If $\ln(-2) = x$, what is the equivalent exponential statement? What is the sign of $e^x$?]

**58.** Why can't we define the logarithm of zero? [*Hint:* If $\ln 0 = x$, what is the equivalent exponential statement? What is the sign of $e^x$?]

**59.** Which type of compounding would give the shortest doubling time for a fixed interest rate: *daily, continuous,* or *annual*? Which would give the longest?

**60.** Recall that the concentration of a drug in the bloodstream after $t$ hours is $ce^{-kt}$, where $k$ is called the "absorbtion constant." If one drug has a larger absorbtion constant than another, will it require *more* or *less* time between doses? (Assume that both drugs have the same value of $c$.)

## Explorations and Excursions

The following problems extend and augment the material presented in the text.

**61. BIOMEDICAL: Half-Life of a Drug** The time required for the amount of a drug in one's body to decrease by half is called the "half-life" of the drug.

 **a.** For a drug with absorption constant $k$, derive the following formula for its half-life:
 $$\left(\begin{matrix}\text{Half-}\\\text{life}\end{matrix}\right) = \frac{\ln 2}{k}$$
 [*Hint:* Solve the equation $e^{-kt} = \frac{1}{2}$ for $t$ and use the properties of logarithms.]

 **b.** Find the half-life of the cardioregulator digoxin if its absorption constant is $k = 0.018$ and time is measured in hours.

 *Source:* T. R. Harrison, ed., *Principles of Internal Medicine*

**62. BIOMEDICAL: Population Growth** The *Gompertz growth curve* models the size $N(t)$ of a population at time $t \geq 0$ as $N(t) = Ke^{-ae^{-bt}}$ where $K$ and $b$ are positive constants. Show that if $N_0 = N(0)$ is the *initial population* at time $t = 0$, then $a = \ln(K/N_0)$.

 *Source:* F. Brauer and C. Castillo-Chávez, *Mathematical Models in Population Biology and Epidemiology*

**63. BIOMEDICAL: Heterozygosity** The frequency $f(x)$ of heterozygotes (averaged across many populations) after $x$ generations is given by $f(x) = H_0\left(1 - \frac{1}{2N}\right)^x$ where $H_0$ is the initial heterozygosity and $N$ is the number of individuals in the population. For a population of size $N = 500$, how many generations are needed to reduce the frequency by 6%?

 *Source:* R. Frankham, J. Ballou, and D. Briscoe, *Introduction to Conservation Genetics*

**64. BIOMEDICAL: Ricker Recruitment** The population dynamics of many fish (such as salmon) can be described by the *Ricker curve* $y = axe^{-bx}$ for $x \geq 0$ where $a > 1$ and $b > 0$ are constants, $x$ is the size of the parental stock, and $y$ is the number of recruits (offspring). Determine the size of the equilibrium population for which $y = x$.

 *Source:* J. Fish. Res. Board Can. **11**

**65–66. BUSINESS: Rule of 72** If a sum is invested at interest rate $r$ compounded continuously, the doubling time (the time in which it will double in value) is found by solving the equation $Pe^{rt} = 2P$. The solution (by the usual method of canceling the $P$ and taking logs) is $t = \dfrac{\ln 2}{r} \approx \dfrac{0.69}{r}$. For *annual* compounding, the doubling time should be somewhat longer, and may be estimated by replacing 69 by 72.

---

**Rule of 72**

For $r\%$ interest compounded annually, the doubling time is approximately $\dfrac{72}{r}$ years.

---

For example, to estimate the doubling time for an investment at 8% compounded annually we would divide 72 by 8, giving $\frac{72}{8} = 9$ years. The 72, however, is only a rough "upward adjustment" of 69, and the rule is most accurate for interest rates around 9%. For each interest rate:

 **a.** Use the rule of 72 to estimate the doubling time for annual compounding.

 **b.** Use the compound interest formula $P(1 + r)^t$ to find the actual doubling time for annual compounding.

**65.** 6%   (*See instructions above.*)

**66.** 1% (This shows that for interest rates very different from 9% the rule of 72 is less accurate.)

**67. ECONOMICS: Interest-Growth Times** Find a formula for the time required for an investment to grow to $k$ times its original size if it grows at interest rate $r$ compounded annually.

**68. ECONOMICS: Interest-Growth Times** Find a formula for the time required for an investment to grow to $k$ times its original size if it grows at interest rate $r$ compounded continuously.

## 4.3 DIFFERENTIATION OF LOGARITHMIC AND EXPONENTIAL FUNCTIONS

### Introduction

You may have noticed that Sections 4.1 and 4.2 contained no calculus. Their purpose was to develop the properties of logarithmic and exponential functions. In this section we differentiate these new functions and use their derivatives for graphing, optimization, and finding rates of change. We emphasize *natural* (base *e*) logs and exponentials, since most applications use these exclusively. Verifications of the differentiation rules are given at the end of the section.

### Derivatives of Logarithmic Functions

The rule for differentiating the natural logarithm function is as follows:

**Derivative of  ln *x***

$$\frac{d}{dx}\ln x = \frac{1}{x}$$

The derivative of  ln *x*  is 1 over *x*

---

**EXAMPLE 1**

**DIFFERENTIATING A LOGARITHMIC FUNCTION**

Differentiate  $f(x) = x^3 \ln x$.

*Solution*

The function is a *product*,  $x^3$  times  ln *x*,  so we use the Product Rule.

$$\frac{d}{dx}(x^3 \ln x) = 3x^2 \ln x + x^3 \frac{1}{x} = 3x^2 \ln x + x^2$$

Derivative of the first

Second left alone

First left alone

Derivative of ln *x*

From  $x^3 \frac{1}{x} = x^2$

---

**Practice Problem 1**

Differentiate  $f(x) = \dfrac{\ln x}{x}$.

➤ Solution on page 298

The preceding rule, together with the Chain Rule, shows how to differentiate the natural logarithm of a *function*. For any differentiable function $f(x)$ that is positive:

**Derivative of  ln *f(x)***

$$\frac{d}{dx}\ln f(x) = \frac{f'(x)}{f(x)}$$

The derivative of the natural log of a function is the derivative of the function over the function

Notice that the right-hand side does not involve logarithms at all.

| EXAMPLE 2 | **DIFFERENTIATING A LOGARITHMIC FUNCTION** |

$$\frac{d}{dx}\ln(x^2 + 1) = \frac{2x}{x^2 + 1}$$

← Derivative of $x^2 + 1$

← Original function (without the ln)

As we observed, the answer does not involve logarithms.

| Practice Problem 2 | Find $\dfrac{d}{dx}\ln(x^3 - 5x + 1)$. | ➤ **Solution on page 298** |

| EXAMPLE 3 | **DIFFERENTIATING A LOGARITHMIC FUNCTION** |

Find the derivative of $f(x) = \ln(x^4 - 1)^3$.

*Solution*  We need the rule for differentiating the natural logarithm of a function, together with the Generalized Power Rule [for differentiating $(x^4 - 1)^3$].

$$\frac{d}{dx}\ln(x^4 - 1)^3 = \frac{\dfrac{d}{dx}(x^4 - 1)^3}{(x^4 - 1)^3}$$  Using $\dfrac{d}{dx}\ln f = \dfrac{f'}{f}$

$$= \frac{3(x^4 - 1)^2\, 4x^3}{(x^4 - 1)^3}$$  Using the Generalized Power Rule

$$= \frac{12x^3}{x^4 - 1}$$  Dividing top and bottom by $(x^4 - 1)^2$

*Alternative Solution*  It is easier if we simplify first, using Property 8 of logarithms (see the inside back cover) to bring down the exponent 3:

$$\ln(x^4 - 1)^3 = 3\ln(x^4 - 1)$$  Using $\ln(M^P) = P \cdot \ln M$

Now we differentiate the simplified expression:

$$\frac{d}{dx}\, 3\ln(x^4 - 1) = 3\,\frac{4x^3}{x^4 - 1} = \frac{12x^3}{x^4 - 1}$$  Same answer as before

*Moral:* Changing $\ln(\cdots)^n$ to $n\ln(\cdots)$ simplifies differentiation.

## Derivatives of Exponential Functions

The rule for differentiating the exponential function $e^x$ is as follows:

**Derivative of $e^x$**

$$\frac{d}{dx}e^x = e^x$$  The derivative of $e^x$ is simply $e^x$

For $y = e^x$, $y' = y$.

This shows the rather surprising fact that $e^x$ is its own derivative. Stated another way, the function $e^x$ is unchanged by the operation of differentiation.

This rule can be interpreted graphically: if $y = e^x$, then $y' = e^x$, so that $y = y'$. This means that on the graph of $y = e^x$, the slope $y'$ always equals the $y$-coordinate, as shown in the graph on the left. Since $y'$ and $y''$ both equal $e^x$, they are always positive and the graph is always increasing and concave upwards.

## Graphing Calculator Exploration

$e^x$ or $\dfrac{d}{dx} e^x$ ?

**a.** Define $y_1 = e^x$ and $y_2$ as the derivative of $y_1$ (using NDERIV) and graph them together on the window $[-3, 3]$ by $[-1, 10]$.

**b.** Why does the screen show only one curve?

**c.** Use TRACE to compare the values of $y_1$ and $y_2$ at some chosen $x$-value. Do the $y$-values agree *exactly*? If not, explain the slight discrepancy. [*Hint:* Is NDERIV really the derivative?]

---

**EXAMPLE 4**

**FINDING A DERIVATIVE INVOLVING $e^x$**

Find $\dfrac{d}{dx}\left(\dfrac{e^x}{x}\right)$.

*Solution*

Since the function is a quotient, we use the Quotient Rule:

Derivative of $e^x$
Derivative of $x$

$$\frac{d}{dx}\left(\frac{e^x}{x}\right) = \frac{x \cdot e^x - 1 \cdot e^x}{x^2} = \frac{xe^x - e^x}{x^2}$$

---

**EXAMPLE 5**

**EVALUATING A DERIVATIVE INVOLVING $e^x$**

If $f(x) = x^2 e^x$, find $f'(1)$.

*Solution*

$$f'(x) = 2xe^x + x^2 e^x \qquad \text{Using the Product Rule on } x^2 \cdot e^x$$
$$f'(1) = 2(1)e^1 + (1)^2 e^1 \qquad \text{Substituting } x = 1$$
$$= 2e + e = 3e \qquad \text{Simplifying}$$

In these problems we leave our answers in their "exact" forms, leaving $e$ as $e$. Later, in applied problems, we will approximate our answers using $e \approx 2.718$ or a calculator.

**Practice Problem 3**      If $f(x) = xe^x$, find $f'(1)$.                              ➤ **Solution on page 298**

The rule for differentiating $e^x$, together with the Chain Rule, shows how to differentiate $e^{f(x)}$. For any differentiable function $f(x)$:

---

**Derivative of** $e^{f(x)}$

$$\frac{d}{dx} e^{f(x)} = e^{f(x)} \cdot f'(x)$$        The derivative of $e$ to a function is $e$ to the function times the derivative of the function

---

That is, to differentiate $e^{f(x)}$ we simply "copy" the original $e^{f(x)}$ and then multiply by the derivative of the exponent.

**EXAMPLE 6**      **DIFFERENTIATING AN EXPONENTIAL FUNCTION**

$$\frac{d}{dx} e^{x^4+1} = e^{x^4+1}(4x^3) = 4x^3 e^{x^4+1}$$        Reversing the order

Copied         Derivative of the exponent

**EXAMPLE 7**      **DIFFERENTIATING AN EXPONENTIAL FUNCTION**

Find $\dfrac{d}{dx} e^{x^2/2}$.

*Solution*

The exponent $x^2/2$ should first be rewritten as $\frac{1}{2}x^2$, a constant times $x$ to a power, since then its derivative is easily seen to be $x$.

$$\frac{d}{dx} e^{x^2/2} = \frac{d}{dx} e^{\frac{1}{2}x^2} = e^{\frac{1}{2}x^2}(x) = xe^{\frac{1}{2}x^2} = xe^{x^2/2}$$        Rewriting the exponent in its original form

Derivative of the exponent

**Practice Problem 4** Find $\dfrac{d}{dx} e^{1+x^3/3}$     ➤ **Solution on page 298**

The formulas for differentiating natural logarithmic and exponential functions are summarized as follows, with $f(x)$ written simply as $f$.

| *Logarithmic Formulas* | *Exponential Formulas* | |
|---|---|---|
| $\dfrac{d}{dx} \ln x = \dfrac{1}{x}$ | $\dfrac{d}{dx} e^x = e^x$ | Top formulas apply only to $\ln x$ and $e^x$ |
| $\dfrac{d}{dx} \ln f = \dfrac{f'}{f}$ | $\dfrac{d}{dx} e^f = e^f \cdot f'$ | Bottom formulas apply to $\ln$ and $e$ of a *function* |

⚠ **BE CAREFUL:** Do *not* take the derivative of $e^x$ by the Power Rule,

$$\frac{d}{dx} x^n = nx^{n-1}$$

The Power Rule applies to $x^n$, *a variable to a constant power*, whereas $e^x$ is *a constant to a variable power*. The two types of functions are quite different, as their graphs show.

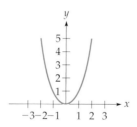

The graph of $x^2$
(a variable to a
constant power)

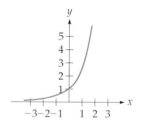

The graph of $e^x$
(a constant to a
variable power)

Each type of function has its own differentiation formula.

$$\frac{d}{dx} x^n = nx^{n-1}$$

For a variable
$x$ to a constant
power $n$

$$\frac{d}{dx} e^x = e^x$$

For the constant
$e$ to a variable
power $x$

**EXAMPLE 8**

**DIFFERENTIATING A LOGARITHMIC AND EXPONENTIAL FUNCTION**

Find the derivative of   $\ln(1 + e^x)$.

*Solution*

$$\frac{d}{dx}\ln(1 + e^x) \;=\; \underbrace{\frac{\dfrac{d}{dx}(1 + e^x)}{1 + e^x}}_{\text{Using }\frac{d}{dx}\ln f = \frac{f'}{f}} \;=\; \underbrace{\frac{e^x}{1 + e^x}}_{\substack{\text{Working out} \\ \text{the numerator}}}$$

Functions of the form $e^{kx}$ (for constant $k$) arise in many applications. The derivative of $e^{kx}$ is as follows.

$$\frac{d}{dx}e^{kx} = e^{kx} \cdot \underbrace{k}_{\text{Derivative of the exponent}} = ke^{kx} \qquad\qquad \text{Using } \frac{d}{dx}e^f = e^f \cdot f'$$

This result is so useful that we record it as a separate formula.

---

**Derivative of   $e^{kx}$**

$$\frac{d}{dx}e^{kx} = ke^{kx} \qquad\qquad\qquad \text{For any constant } k$$

---

This formula says that the rate of change (the derivative) of $e^{kx}$ is proportional to itself. That is, the function  $y = e^{kx}$  satisifies the *differential equation*

$$y' = ky$$

We noted this earlier when we observed that in exponential growth a quantity *grows in proportion to itself* (as in populations and savings accounts).

These differentiation formulas enable us to find instantaneous rates of change of logarithmic and exponential functions. In many applications the variable stands for time, so we use $t$ instead of $x$.

**EXAMPLE 9**

**FINDING A RATE OF IMPROVEMENT OF A SKILL**

After $t$ weeks of practice a pole vaulter can vault

$$H(t) = 15 - 11e^{-0.1t}$$

feet. Find the rate of change of the athlete's jumps after

**a.**  0 weeks (at the beginning of training)

**b.**  12 weeks

*Solution*

We differentiate to find the rate of change

$$H'(t) = -11(-0.1)e^{-0.1t} \quad = \quad 1.1e^{-0.1t}$$

Differentiating
$15 - 11e^{-0.1t}$

Using $\dfrac{d}{dt}\, e^{kt} = ke^{kt}$   Simplifying

**a.** For the rate of change after 0 weeks:

$$H'(0) = 1.1e^{-0.1(0)} = 1.1e^0 = 1.1$$

$H'(t) = 1.1e^{-0.1t}$
with $t = 0$

**b.** After 12 weeks:

$$H'(12) = 1.1e^{-0.1(12)}$$
$$= 1.1e^{-1.2} \approx 1.1(0.30) = 0.33$$

$H'(t) = 1.1e^{-0.1t}$
with $t = 12$

Using a calculator

At first, the vaults increased by 1.1 feet per week. After 12 weeks, the gain was only 0.33 foot (about 4 inches) per week.

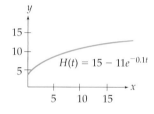

This result is typical of learning a new skill: early improvement is rapid, later improvement is slower. This trend is called *diminishing returns,* and may be seen in the leveling off of the polevault heights in the graph on the left.

## Maximizing Consumer Expenditure

The amount of a commodity that consumers will buy depends on the price of the commodity. For a commodity whose price is $p$, let the consumer demand be given by a function $D(p)$. Multiplying the number of units $D(p)$ by the price $p$ gives the total *consumer expenditure* for the commodity.

---

**Consumer Demand and Expenditure**

Let $D(p)$ be the consumer demand at price $p$. Then the consumer expenditure is

$$E(p) = p \cdot D(p)$$

---

**EXAMPLE 10**  **MAXIMIZING CONSUMER EXPENDITURE**

If consumer demand for a commodity is $D(p) = 10{,}000e^{-0.02p}$ units per week, where $p$ is the selling price, find the price that maximizes consumer expenditure.

*Solution*

Using the preceding formula for consumer expenditure,

$$E(p) = p \cdot 10{,}000e^{-0.02p} = 10{,}000pe^{-0.02p}$$

$E(p) = p \cdot D(p)$

To maximize $E(p)$ we differentiate:

$$E'(p) = \underbrace{10{,}000e^{-0.02p}}_{\substack{\text{Derivative} \\ \text{of } 10{,}000p}} + \underbrace{10{,}000p(-0.02)e^{-0.02p}}_{\substack{\text{Derivative} \\ \text{of } e^{-0.02p}}}$$

Using the Product Rule to differentiate $E(p) = 10{,}000p \cdot e^{-0.02p}$

$$= 10{,}000e^{-0.02p} - 200pe^{-0.02p}$$

Simplifying

$$= 200e^{-0.02p}(50 - p)$$

Factoring

$$\text{CN: } p = 50$$

Critical number from $(50 - p)$ (since $e$ to a power is never zero)

We calculate $E''$ for the second-derivative test:

$$E''(p) = 200(-0.02)e^{-0.02p}(50 - p) + 200e^{-0.02p}(-1)$$

From $E'(p) = 200e^{-0.02p} \cdot (50 - p)$ using the Product Rule

$$= -4e^{-0.02p}(50 - p) - 200e^{-0.02p}$$

Simplifying

At the critical number $p = 50$,

$$E''(50) = -4e^{-0.02(50)}(50 - 50) - 200e^{-0.02(50)}$$

Substituting $p = 50$

$$= -200e^{-1} = \frac{-200}{e}$$

Simplifying

$E''$ is negative, so the expenditure $E(p)$ is maximized at $p = 50$:

Consumer expenditure is maximized at price $50.

## Graphing Calculator Exploration

Maximum
X=49.999987  Y=183939.72

Interpret as $x = 50$

Use a graphing calculator to verify the answer to the previous example by graphing $E(x) = 10{,}000xe^{-0.02x}$ (using $x$ instead of $p$) on the window [0, 150] by [−50,000, 200,000] and using MAXIMUM.

How did we choose the graphing window? We first found the critical number "by hand" (setting the derivative equal to zero and finding, in this example, 50) and chose $x$-values around it. For the $y$-values, we evaluated the function at the critical number (using EVALUATE, finding in this example $y \approx 184{,}000$) and chose $y$-values including it. Notice how graphing calculators and calculus are most effective when used together.

### Graphing Logarithmic and Exponential Functions

To graph logarithmic and exponential functions using a graphing calculator, we first find critical points and possible inflection points, and then graph the function on a window including these points. (If graphing "by hand," we would make sign diagrams for the first and second derivatives and then sketch the graph, as in Chapter 3.)

**EXAMPLE 11**    **GRAPHING AN EXPONENTIAL FUNCTION**

Graph $f(x) = e^{-x^2/2}$.

*Solution*

As before, we write the function as $f(x) = e^{-\frac{1}{2}x^2}$. The derivative is

$$f'(x) = e^{-\frac{1}{2}x^2}(-x) = -xe^{-\frac{1}{2}x^2} \qquad \text{Using } \frac{d}{dx}e^f = e^f \cdot f'$$

Derivative of the exponent

CN:   $x = 0$                Critical number is 0

$y = 1$                From $y = e^{-\frac{1}{2}x^2}$ evaluated at $x = 0$

The second derivative is

$$f''(x) = (-1)e^{-\frac{1}{2}x^2} - xe^{-\frac{1}{2}x^2}(-x) \qquad \text{From } f'(x) = -x \cdot e^{-\frac{1}{2}x^2}$$
$$\text{using the Product Rule}$$
$$= -e^{-\frac{1}{2}x^2} + x^2e^{-\frac{1}{2}x^2} \qquad \text{Simplifying}$$
$$= e^{-\frac{1}{2}x^2}(-1 + x^2) \qquad \text{Factoring}$$
$$= (x^2 - 1)e^{-\frac{1}{2}x^2} \qquad \text{Rearranging}$$
$$= (x + 1)(x - 1)e^{-\frac{1}{2}x^2} \qquad \text{Factoring}$$
$$x = \pm 1 \qquad \text{Where } f'' = 0$$
$$y = e^{-\frac{1}{2}} \approx 0.6 \qquad \text{From } y = e^{-\frac{1}{2}x^2}$$
$$\text{evaluated at } x = \pm 1$$

$f(x) = e^{-x^2/2}$ on $[-3, 3]$ by $[-1, 2]$

Based on these values, we choose the graphing window as follows. For the $x$-values we choose $[-3, 3]$ (to include 0 and $\pm 1$ and beyond), and for the $y$-values we choose $[-1, 2]$ (to include 1 and 0.6 and above and below). This window gives the graph on the left. (Many other windows would be just as good, and after seeing the graph you might want to adjust the window.)

This function, multiplied by the constant $1/\sqrt{2\pi}$, is the famous "bell-shaped curve" of statistics, used for predictions of many things from the IQs of newborn babies to stock prices.

**Solutions to Practice Problems**

Derivative of $\ln x$

Derivative of $x$

**1.** $f'(x) = \dfrac{d}{dx}\left(\dfrac{\ln x}{x}\right) = \dfrac{x\left(\frac{1}{x}\right) - 1 \ln x}{x^2}$

$= \dfrac{1 - \ln x}{x^2}$

**2.** $\dfrac{d}{dx}\ln(x^3 - 5x + 1) = \dfrac{3x^2 - 5}{x^3 - 5x + 1}$

**3.** $f'(x) = e^x + xe^x$

$f'(1) = e^1 + 1e^1 = 2e$     (which is approximately $2 \cdot 2.718 = 5.436$)

**4.** $\dfrac{d}{dx}e^{1+x^3/3} = \dfrac{d}{dx}e^{1+\frac{1}{3}x^3} = e^{1+\frac{1}{3}x^3}(x^2) = x^2e^{1+x^3/3}$

## 4.3 Section Summary

In this section we developed the formulas for differentiating logarithmic and exponential functions:

$$\frac{d}{dx}\ln x = \frac{1}{x} \qquad\qquad \frac{d}{dx}e^x = e^x$$

$$\frac{d}{dx}\ln f(x) = \frac{f'(x)}{f(x)} \qquad\qquad \frac{d}{dx}e^{f(x)} = e^{f(x)}\cdot f'(x)$$

The last of these has a useful special case when the exponent is of the form $f(x) = kx$ for a constant $k$:

$$\frac{d}{dx}e^{kx} = ke^{kx}$$

We used these formulas together with our other differentiation formulas (the product, quotient, and chain rules) to use derivatives to find rates of change and to optimize functions.

## Verification of the Differentiation Formulas

### Verification of the Power Rule for Arbitrary Powers

On pages 118–119 we proved the Power Rule $\dfrac{d}{dx}x^n = nx^{n-1}$ for positive integer exponents, and on pages 246–247 we extended it to rational exponents. We can now show that the Power Rule holds for *all* exponents. We begin by using one of the "inverse" properties of logarithms, $x = e^{\ln x}$ (for $x > 0$), but with $x$ replaced by $x^n$:

$$f(x) = x^n = e^{\ln x^n} = e^{n\ln x} \qquad\text{Using the property that logs bring down exponents}$$

Differentiating:

$$f'(x) = \underbrace{e^{n\ln x}}_{x^n}\underbrace{\left(n\frac{1}{x}\right)}_{\text{Derivative of the exponent}} \qquad\text{Differentiating } f(x) = e^{n\ln x} \text{ by the } e^{f(x)} \text{ formula (above)}$$

$$= \underbrace{x^n\cdot\frac{1}{x}}_{x^{n-1}}\cdot n = nx^{n-1} \qquad\text{Replacing } e^{n\ln x} \text{ by } x^n, \text{ reordering, and combining } x\text{'s}$$

This is what we wanted to show—that the derivative of $f(x) = x^n$ is $f'(x) = nx^{n-1}$ for *all* exponents $n$. (The equation $x^n = e^{n\ln x}$ can be taken as a *definition* of $x$ to a power.)

### Verification of the Differentiation Formula for ln *x* and ln *f*(*x*)

The formula for the derivative of the natural logarithm function comes from applying the definition of the derivative to $f(x) = \ln x$.

$$f'(x) = \lim_{h \to 0} \frac{f(x+h) - f(x)}{h} = \lim_{h \to 0} \frac{\ln(x+h) - \ln x}{h}$$
Definition of $f'(x)$

$$= \lim_{h \to 0} \frac{1}{h}[\ln(x+h) - \ln x]$$
Dividing by $h$ is equivalent to multiplying by $1/h$

$$= \lim_{h \to 0} \frac{1}{h}\ln\left(\frac{x+h}{x}\right) = \lim_{h \to 0} \frac{1}{h}\ln\left(1 + \frac{h}{x}\right)$$
Using Property 7 of logarithms (inside back cover), and simplifying

$$= \lim_{h \to 0} \frac{1}{x}\frac{x}{h}\ln\left(1 + \frac{h}{x}\right) = \lim_{h \to 0} \frac{1}{x}\ln\left(1 + \frac{h}{x}\right)^{x/h}$$
Dividing and multiplying by $x$, and then using Property 8 of logarithms

$$= \lim_{n \to \infty} \frac{1}{x}\ln\underbrace{\left(1 + \frac{1}{n}\right)^n}_{\substack{\text{Approaches} \\ e \text{ as } n \to \infty}}$$
Defining $n = x/h$, so that $h \to 0$ implies $n \to \infty$ (for $h > 0$)

$$= \frac{1}{x}\ln e = \frac{1}{x}$$
Since $\ln e = 1$ (the same conclusion follows if $h < 0$)

This is the result that we wanted to show—that the derivative of the natural logarithm function $f(x) = \ln x$ is $f'(x) = \frac{1}{x}$.

For the rule to differentiate the natural logarithm of a (positive) *function*, we begin with

$$\frac{d}{dx}f(g(x)) = f'(g(x))g'(x)$$
Chain Rule (from page 153) for differentiable functions $f$ and $g$

$$\frac{d}{dx}\ln(g(x)) = \frac{1}{g(x)}g'(x) = \frac{g'(x)}{g(x)}$$
Taking $f(x) = \ln x$, so $f'(x) = \frac{1}{x}$

Replacing $g$ by $f$, this is exactly the formula we wanted to show,

$$\frac{d}{dx}\ln f(x) = \frac{f'(x)}{f(x)}$$

### Verification of the Differentiation Formula for *e*ˣ and *e*ᶠ⁽ˣ⁾

To derive the rule for differentiating $e^x$ we begin with

$$\ln e^x = x$$
Property 3 of natural logarithms

and differentiate both sides:

$$\frac{\frac{d}{dx}e^x}{e^x} = 1$$
Using $\frac{d}{dx}\ln f = \frac{f'}{f}$ on the left side

Multiplying each side by $e^x$ gives

$$\frac{d}{dx} e^x = e^x$$

This is the rule for differentiating $e^x$. This rule together with the Chain Rule gives the rule for differentiating $e^{f(x)}$, just as before.

## 4.3 Exercises

**1–44.** Find the derivative of each function.

**1.** $f(x) = x^2 \ln x$

**2.** $f(x) = \dfrac{\ln x}{x^3}$

**3.** $f(x) = \ln x^2$

**4.** $f(x) = \ln(x^3 + 1)$

**5.** $f(x) = \ln \sqrt{x}$

**6.** $f(x) = \sqrt{\ln x}$

**7.** $f(x) = \ln(x^2 + 1)^3$

**8.** $f(x) = \ln(x^4 + 1)^2$

**9.** $f(x) = \ln(-x)$

**10.** $f(x) = \ln(5x)$

**11.** $f(x) = \dfrac{e^x}{x^2}$

**12.** $f(x) = x^3 e^x$

**13.** $f(x) = e^{x^3 + 2x}$

**14.** $f(x) = 2e^{7x}$

**15.** $f(x) = e^{x^3/3}$

**16.** $f(x) = \ln(e^x - 2x)$

**17.** $f(x) = x - e^{-x}$

**18.** $f(x) = x \ln x - x$

**19.** $f(x) = \ln e^{2x}$

**20.** $f(x) = \ln e^x$

**21.** $f(x) = e^{1 + e^x}$

**22.** $f(x) = \ln(e^x + e^{-x})$

**23.** $f(x) = x^e$

**24.** $f(x) = ex$

**25.** $f(x) = e^3$

**26.** $f(x) = \sqrt{e}$

**27.** $f(x) = \ln(x^4 + 1) - 4e^{x/2} - x$

**28.** $f(x) = x^2 e^x - 2 \ln x + (x^2 + 1)^3$

**29.** $f(x) = x^2 \ln x - \frac{1}{2}x^2 + e^{x^2} + 5$

**30.** $f(x) = e^{-2x} - x \ln x + x - 7$

**31.** $f(x) = e^x \ln x^2$

**32.** $f(x) = e^x \ln(x + 1)$

**33.** $f(x) = \ln \dfrac{1}{x^2}$

**34.** $f(x) = \ln \dfrac{1}{e^{x^2}}$

**35.** $f(t) = (e^{2t} + 1)^3$

**36.** $f(t) = \sqrt{e^{2t} + 4}$

**37.** $f(t) = \sqrt{t^2 + 2 \ln t}$

**38.** $f(t) = (2t + \ln t)^3$

**39.** $f(z) = \dfrac{e^z}{z^2 - 1}$

**40.** $f(z) = \dfrac{e^z}{1 + e^z}$

**41.** $f(z) = \dfrac{10}{1 + e^{-2z}}$

**42.** $f(z) = \dfrac{12}{1 + 2e^{-z}}$

**43.** $f(x) = \dfrac{e^x + e^{-x}}{e^x - e^{-x}}$

**44.** $f(x) = \dfrac{e^x - e^{-x}}{e^x + e^{-x}}$

**45–50.** For each function, find the indicated expressions.

**45.** $f(x) = \dfrac{\ln x}{x^5}$, find **a.** $f'(x)$ **b.** $f'(1)$

**46.** $f(x) = x^4 \ln x$, find **a.** $f'(x)$ **b.** $f'(1)$

**47.** $f(x) = \ln(x^4 + 48)$, find **a.** $f'(x)$ **b.** $f'(2)$

**48.** $f(x) = x^2 \ln x - x^2$, find **a.** $f'(x)$ **b.** $f'(e)$

**49.** $f(x) = \ln(e^x - 3x)$, find **a.** $f'(x)$ **b.** $f'(0)$

**50.** $f(x) = \ln(e^x + e^{-x})$, find **a.** $f'(x)$ **b.** $f'(0)$

**51–54.** For each function:

**a.** Find $f'(x)$.

**b.** Evaluate the given expression and approximate it to three decimal places.

**51.** $f(x) = 5x \ln x$, find and approximate $f'(2)$.

**52.** $f(x) = e^{x^2/2}$, find and approximate $f'(2)$.

**53.** $f(x) = \dfrac{e^x}{x}$, find and approximate $f'(3)$.

**54.** $f(x) = \ln(e^x - 1)$, find and approximate $f'(3)$.

**55–56.** Find the *second* derivative of each function.

**55.** $f(x) = e^{-x^5/5}$

**56.** $f(x) = e^{-x^6/6}$

**57–58.** By calculating the first few derivatives, find a formula for the $n$th derivative of each function ($k$ is a constant).

**57.** $f(x) = e^{kx}$

**58.** $f(x) = e^{-kx}$

**59–62.** Use your graphing calculator to graph each function on a window that includes all relative extreme points and inflection points, and give the coordinates of these points (rounded to two decimal places). [*Hint:* Use NDERIV once or twice with ZERO.] (Answers may vary depending on the graphing window chosen.)

**59.** $f(x) = e^{-2x^2}$

**60.** $f(x) = 1 - e^{-x^2/2}$

**61.** $f(x) = \ln(1 + x^2)$

**62.** $f(x) = e^x + e^{-x}$

**63–66.** Use your graphing calculator to graph each function on the indicated interval, and give the coordinates of all relative extreme points and inflection points (rounded to two decimal places). [*Hint:* Use NDERIV once or twice together with ZERO.] (Answers may vary depending on the graphing window chosen.)

**63.** $f(x) = \dfrac{x^2}{e^x}$ for $-1 \le x \le 8$

**64.** $f(x) = \dfrac{x}{e^x}$ for $-1 \le x \le 5$

**65.** $f(x) = x \ln|x|$ for $-2 \le x \le 2$

[*Hint for Exercises 65–66:* $|x|$ is sometimes entered as ABS $(x)$.]

**66.** $f(x) = x^2 \ln|x|$ for $-2 \le x \le 2$

**67–68.** Use implicit differentiation to find $dy/dx$.

**67.** $y^2 - ye^x = 12$  **68.** $y^2 - x \ln y = 10$

## Applied Exercises

**69. PERSONAL FINANCE: Earnings and Calculus** A recent study found that one's earnings are affected by the mathematics courses one has taken. In particular, compared to someone making $40,000 who had taken *no* calculus, a comparable person who had taken $x$ years of calculus would be earning $\$40,000e^{0.195x}$. Find the rate of change of this function at $x = 1$ and interpret your answer.

*Source: Review of Economics and Statistics* 86

**70. ENVIRONMENTAL SCIENCE: Solar Cells** Cumulative world production of solar cells for generating electricity is predicted to be $5e^{0.24t}$ thousand megawatts, where $t$ is the number of years since 2005. Find the rate of change of this quantity at the value of $t$ corresponding to 2015 and interpret your answer. [*Note:* Three thousand megawatts is enough to power a million homes.]

*Source:* Earth Policy Institute

**71. PERSONAL FINANCE: Compound Interest** A sum of $1000 at 5% interest compounded continuously will grow to $V(t) = 1000e^{0.05t}$ dollars in $t$ years. Find the rate of growth after:

**a.** 0 years (the time of the original deposit).
**b.** 10 years.

**72. PERSONAL FINANCE: Depreciation** A $10,000 automobile depreciates so that its value after $t$ years is $V(t) = 10,000e^{-0.35t}$ dollars. Find the instantaneous rate of change of its value:

**a.** when it is new ($t = 0$).
**b.** after 2 years.

**73. GENERAL: Population** The world population (in billions) is predicted to be $P(t) = 6.45e^{0.0175t}$, where $t$ is the number of years after 2005. Find the instantaneous rate of change of the population in the year 2015.

*Source:* U.N.—Population Division

**74. BEHAVIORAL SCIENCE: Ebbinghaus Memory Model** According to the Ebbinghaus model of memory, if one is shown a list of items, the percentage of items that one will remember $t$ time units later is $P(t) = (100 - a)e^{-bt} + a$, where $a$ and $b$ are constants. For $a = 25$ and $b = 0.2$, this function becomes $P(t) = 75e^{-0.2t} + 25$. Find the instantaneous rate of change of this percentage:

**a.** at the beginning of the test ($t = 0$).
**b.** after 3 time units.

*Source:* Gorfein and Hoffman, *Memory and Learning*

**75. BIOMEDICAL: Drug Dosage** A patient receives an injection of 1.2 milligrams of a drug, and the amount remaining in the bloodstream $t$ hours later is $A(t) = 1.2e^{-0.05t}$. Find the instantaneous rate of change of this amount:

**a.** just after the injection (at time $t = 0$).
**b.** after 2 hours.

**76. GENERAL: Temperature** A covered cup of coffee at 200 degrees, if left in a 70-degree room, will cool to $T(t) = 70 + 130e^{-2.5t}$ degrees in $t$ hours. Find the rate of change of the temperature:

**a.** at time $t = 0$.  **b.** after 1 hour.

**77. BUSINESS: Sales** The weekly sales (in thousands) of a new product are predicted to be $S(x) = 1000 - 900e^{-0.1x}$ after $x$ weeks. Find the rate of change of sales after:

**a.** 1 week.  **b.** 10 weeks.

**78. SOCIAL SCIENCE: Diffusion of Information by Mass Media** The number of people in a town of 50,000 who have heard an important news bulletin within $t$ hours of its first broadcast is $N(t) = 50,000(1 - e^{-0.4t})$. Find the rate of change of the number of informed people:

**a.** at time $t = 0$.  **b.** after 8 hours.

**79–80. ECONOMICS: Consumer Expenditure** If consumer demand for a commodity is given by the function below (where $p$ is the selling price in dollars), find the price that maximizes consumer expenditure.

**79.** $D(p) = 5000e^{-0.01p}$     **80.** $D(p) = 8000e^{-0.05p}$

**81–82. BUSINESS: Maximizing Revenue** Each of the following functions is a company's price function, where $p$ is the price (in dollars) at which quantity $x$ (in thousands) will be sold.

**a.** Find the revenue function $R(x)$.
   [*Hint:* Revenue is price times quantity, $p \cdot x$.]
**b.** Find the quantity and price that will maximize revenue.

**81.** $p = 400e^{-0.20x}$     **82.** $p = 4 - \ln x$

**83. GENERAL: Temperature** A mug of beer chilled to 40 degrees, if left in a 70-degree room, will warm to a temperature of $T(t) = 70 - 30e^{-3.5t}$ degrees in $t$ hours.

**a.** Find $T(0.25)$ and $T'(0.25)$ and interpret your answers.
**b.** Find $T(1)$ and $T'(1)$ and interpret your answers.

**84. SOCIAL SCIENCE: Diffusion of Information by Mass Media** The number of people in a city of 200,000 who have heard a weather bulletin within $t$ hours of its first broadcast is $N(t) = 200{,}000(1 - e^{-0.5t})$.

**a.** Find $N(0.5)$ and $N'(0.5)$ and interpret your answers.
**b.** Find $N(3)$ and $N'(3)$ and interpret your answers.

 **85–86. ATHLETICS: World's Record 100-Meter Run** In 1987 Carl Lewis set a new world's record of 9.93 seconds for the 100-meter run. The distance that he ran in the first $x$ seconds was

$$11.274[x - 1.06(1 - e^{-x/1.06})] \text{ meters}$$

for $0 \leq x \leq 9.93$. Enter this function as $y_1$, and define $y_2$ as its derivative (using NDERIV), so that $y_2$ gives his velocity after $x$ seconds. Graph them on the window $[0, 9.93]$ by $[0, 100]$.

*Source: SIAM Review 35*

**85.** Trace along the velocity curve to verify that Lewis's maximum speed was about 11.27 meters per second. Find how quickly he reached a speed of 10 meters per second, which is 95% of his maximum speed.

**86.** Define $y_3$ as the derivative of $y_2$ (using NDERIV) so that $y_3$ gives his acceleration after $x$ seconds, and graph $y_2$ and $y_3$ on the window $[0, 9.93]$ by

$[0, 20]$. Evaluate both $y_2$ and $y_3$ at $x = 0.1$ and also at $x = 9.93$ (using EVALUATE). Interpret your answers.

 **87. ATHLETICS: How Fast Do Old Men Slow Down?** The fastest times for the marathon (26.2 miles) for male runners aged 35 to 80 are approximated by the function

$$f(x) = \begin{cases} 106.2e^{0.0063x} & \text{if } x \leq 58.2 \\ 850.4e^{0.000614x^2 - 0.0652x} & \text{if } x > 58.2 \end{cases}$$

in minutes, where $x$ is the age of the runner.

*Source: Review of Economics and Statistics LXXVI*

**a.** Graph this function on the window $[35, 80]$ by $[0, 240]$.
   [*Hint:* On some graphing calculators, enter $y_1 = (106.2e^{0.0063x}) (x \leq 58.2) + (850.4e^{0.000614x^2 - 0.0652x}) (x > 58.2)$.]
**b.** Find $f(35)$ and $f'(35)$ and interpret these numbers. [*Hint:* Use NDERIV or $dy/dx$.]
**c.** Find $f(80)$ and $f'(80)$ and interpret these numbers.

## Conceptual Exercises

**88.** Why can't we differentiate $e^x$ by the *Power Rule?*

**89.** Choose the correct answer: $\dfrac{d}{dx} e^x =$

   **a.** $xe^{x-1}$     **b.** $e^x$     **c.** $0$

**90.** Choose the correct answer: $\dfrac{d}{dx} e^{f(x)} =$

   **a.** $e^{f(x)}f'(x)$     **b.** $e^{f'(x)}f(x)$     **c.** $e^{f'(x)}f'(x)$

**91.** Choose the correct answer: $\dfrac{d}{dx} e^5 =$

   **a.** $5e^4$     **b.** $e^5$     **c.** $0$

**92.** Choose the correct answer: $\dfrac{d}{dx} x^e =$

   **a.** $ex^{e-1}$     **b.** $x^e$     **c.** $0$

**93.** Choose the correct answer: $\dfrac{d}{dx} \ln x =$

   **a.** $\dfrac{x}{1}$     **b.** $\dfrac{1}{x}$     **c.** $0$

**94.** Choose the correct answer: $\dfrac{d}{dx} \ln f(x) =$

   **a.** $\dfrac{f(x)}{f'(x)}$     **b.** $\dfrac{f'(x)}{f(x)}$     **c.** $\dfrac{f'(x)}{f'(x)}$

**95.** Choose the correct answer: $\dfrac{d}{dx} \ln 5 =$

   **a.** $\dfrac{5}{1}$     **b.** $\dfrac{1}{5}$     **c.** $0$

**96.** True or False: If a function involves natural logarithms, then its derivative will involve natural logarithms. Explain.

**97.** Explain why it is obvious, without any calculation, that $\dfrac{d}{dx} e^{\ln x} = 1$.

**98.** Explain why it is obvious, without any calculation, that $\dfrac{d}{dx} \ln e^x = 1$.

### *Explorations and Excursions*

The following problems extend and augment the material presented in the text.

**99.** **BIOMEDICAL: Population Growth** The *Gompertz growth curve* models the size $N(t)$ of a population at time $t \ge 0$ as $N(t) = Ke^{-ae^{-bt}}$ where $K$ and $b$ are positive constants. Show that $\dfrac{dN}{dt} = bN \ln\left(\dfrac{K}{N}\right)$ and interpret this derivative to make statements about the population growth when $N < K$ and when $N > K$.

*Source:* S. Brauer and C. Castillo-Chávez, *Mathematical Models in Population Biology and Epidemiology*

**100.** **BIOMEDICAL: Ricker Recruitment** The population dynamics of many fish (such as salmon) can be described by the *Ricker curve* $y = axe^{-bx}$ for $x \ge 0$ where $a > 1$ and $b > 0$ are constants, $x$ is the size of the parental stock, and $y$ is the number of recruits (offspring). Determine the size of the parental stock that maximizes the number of recruits.

*Source: J. Fish. Res. Board Can.* **11**

**101.** **BIOMEDICAL: Reynolds Number** An important characteristic of blood flow is the "Reynolds number." As the Reynolds number increases, blood flows less smoothly. For blood flowing through certain arteries, the Reynolds number is

$$R(r) = a \ln r - br$$

where $a$ and $b$ are positive constants and $r$ is the radius of the artery. Find the radius $r$ that maximizes the Reynolds number $R$. (Your answer will involve the constants $a$ and $b$.)

**102.** **BIOMEDICAL: Drug Concentration** If a drug is injected intramuscularly, the concentration

of the drug in the bloodstream after $t$ hours will be

$$A(t) = \frac{c}{b - a}\left(e^{-at} - e^{-bt}\right)$$

If the constants are $a = 0.4$, $b = 0.6$, and $c = 0.1$, find the time of maximum concentration.

**103.** **CHANGE OF BASE FORMULA FOR LOGARITHMS** Derive the formula

$$\log_a x = \frac{\ln x}{\ln a} \quad \text{(for } a > 0 \text{ and } x > 0)$$

which expresses logarithms to *any* base $a$ in terms of *natural* logarithms, as follows:

**a.** Define $y = \log_a x$, so that $x = a^y$, and take the natural logarithms of both sides of the last equation and obtain $\ln x = y \ln a$.

**b.** Solve the last equation for $y$ to obtain $y = \dfrac{\ln x}{\ln a}$ and then use the original definition of $y$ to obtain the stated change of base formula.

**104.** **CHANGE OF BASE FORMULA FOR EXPONENTS** Derive the formula

$$a^x = e^{(\ln a)x} \quad \text{(for } a > 0\text{)}$$

by identifying the property of natural logarithms (from the inside back cover) that justifies the first equality in the following sequence:

$$a^x = \left(e^{\ln a}\right)^x = e^{(\ln a)x}$$

**105–106.** Use the formula derived in Exercise 104 to write each expression in the form $e^{bx}$ for some number $b$.

**105.** $2.23^x$        **106.** $0.63^x$

[*Note:* The graphing calculator operation EXPREG gives expressions like those above (see page 63), and Exercises 104–106 show how to express such results using $e$.]

## Exponential and Logarithmic Functions to Other Bases

The rules for differentiating exponential functions with (positive) base $a$ are shown below:

**Derivatives of $a^x$ and $a^{f(x)}$**

$$\frac{d}{dx}a^x = (\ln a)\,a^x$$

$$\frac{d}{dx}a^{f(x)} = (\ln a)a^{f(x)}f'(x) \qquad \text{For a differentiable function } f$$

For example,

$$\frac{d}{dx}2^x = (\ln 2)2^x$$

$$\frac{d}{dx}5^{3x^2+1} = (\ln 5)5^{3x^2+1}(6x) = 6(\ln 5)x\,5^{3x^2+1}$$

These formulas are more complicated than the corresponding base $e$ formulas (page 294), which is why $e$ is called the "natural" base: it makes the derivative formulas simplest. These formulas reduce to the natural (base $e$) formulas if $a = e$.

---

**107–108.** Use the formulas above to find the derivative of each function.

**107. a.** $f(x) = 10^x$   **b.** $f(x) = 3^{x^2+1}$   **c.** $f(x) = 2^{3x}$   **108. a.** $f(x) = 5^x$   **b.** $f(x) = 2^{x^2-1}$   **c.** $f(x) = 3^{4x}$
      **d.** $f(x) = 5^{3x^2}$   **e.** $f(x) = 2^{4-x}$   **d.** $f(x) = 9^{5x^2}$   **e.** $f(x) = 10^{1-x}$

---

The rules for differentiating logarithmic functions with (positive) base $a$ are as follows:

**Derivatives of $\log_a x$ and $\log_a f(x)$**

$$\frac{d}{dx}\log_a x = \frac{1}{(\ln a)x}$$

$$\frac{d}{dx}\log_a f(x) = \frac{f'(x)}{(\ln a)f(x)} \qquad \text{For a differentiable function } f > 0$$

For example,

$$\frac{d}{dx}\log_5 x = \frac{1}{(\ln 5)x}$$

$$\frac{d}{dx}\log_2 (x^3 + 1) = \frac{3x^2}{(\ln 2)(x^3 + 1)}$$

These formulas are more complicated than the corresponding base $e$ formulas (page 294), and again the simplicity of the base $e$ formulas is why $e$ is called the "natural" base. As before, these formulas reduce to the natural (base $e$) formulas if $a = e$.

---

**109–110.** Use the formulas on the previous page to find the derivative of each function.

**109. a.** $\log_2 x$    **b.** $\log_{10}(x^2 - 1)$    **c.** $\log_3(x^4 - 2x)$     **110. a.** $\log_3 x$    **b.** $\log_2(x^2 + 1)$    **c.** $\log_{10}(x^3 - 4x)$

---

## 4.4   TWO APPLICATIONS TO ECONOMICS: RELATIVE RATES AND ELASTICITY OF DEMAND

### Introduction

In this section we define **relative rates of change** and see how they are used in economics. (*Relative* rates are not the same as the *related* rates discussed in Section 3.6.) We then define the very important economic concept of **elasticity of demand.**

### Relative Versus Absolute Rates

The derivative of a function gives its rate of change. For example, if $f(t)$ is the cost of a pair of shoes at time $t$ years, then $f'(t)$ is the rate of change of cost (in dollars per year). That is, $f' = 3$ would mean that the price of shoes is increasing at the rate of \$3 per year. Similarly, if $g(t)$ is the price of a new automobile at time $t$ years, then $g' = 300$ would mean that automobile prices are increasing at the rate of \$300 per year.

Does this mean that car prices are rising 100 times as fast as shoe prices? In absolute terms, yes. However, this does not take into account the enormous price difference between automobiles and shoes.

### Relative Rates of Change

If shoe prices are increasing at the rate of \$3 per year, and if the current price of a pair of shoes is \$60, the *relative* rate of increase is $\frac{3}{60} = \frac{1}{20} = 0.05$, which means that shoe prices are increasing at the relative rate of 5% per year. Similarly, if the price of an average automobile is \$15,000, then an increase of \$300 relative to this price is $\frac{300}{15,000} = \frac{1}{50} = 0.02$, for a *relative* rate of 2% per year. Therefore, in a *relative* sense (that is, as a fraction of the current price), car prices are increasing *less* rapidly than shoe prices.

In general, if $f(t)$ is the price of an item at time $t$, then the rate of change is $f'(t)$, and the *relative* rate of change is $f'(t)/f(t)$, the derivative divided by the function. We will sometimes call the derivative $f'(x)$ the "absolute" rate of change to distinguish it from the relative rate of change $f'(x)/f(x)$.

Relative rates are often more meaningful than absolute rates. For example, it is easier to grasp the fact that the gross domestic product is growing at the relative rate of 3% a year than that it is growing at the absolute rate of \$400,000,000,000 per year.

The expression $f'(x)/f(x)$ is the derivative of the natural logarithm of $f(x)$:

$$\frac{d}{dx} \ln f(x) = \frac{f'(x)}{f(x)}$$

This provides an alternative expression for the relative rate of change, in terms of logarithms.

---

**Relative Rate of Change**

$$\begin{pmatrix} \text{Relative rate of} \\ \text{change of } f(t) \end{pmatrix} = \frac{d}{dt} \ln f(t) = \frac{f'(t)}{f(t)} \qquad \text{For a differentiable function } f > 0$$

---

We use the variable $t$ since it often stands for time. Both formulas in the box give the same result, and the middle expression is sometimes called the *logarithmic derivative*, since it is found by first taking the logarithm and then the derivative.

The relative rate of change, being a ratio or a percent, does not depend on the units of the function (whether dollars or euros, pounds or kilos). Therefore, relative rates can be compared between different products, and even between different nations. This is in contrast to absolute rates of change (that is, derivatives), which *do* depend on the units (for example, dollars per year).

**EXAMPLE 1**    **FINDING A RELATIVE RATE OF CHANGE**

If the gross domestic product $t$ years from now is predicted to be $G(t) = 8.2e^{\sqrt{t}}$ trillion dollars, find the relative rate of change 25 years from now.

We give two solutions, showing the use of both formulas.

***Solution***   (using the $\dfrac{d}{dt} \ln f(t)$ formula)

First we simplify:

$$\ln G(t) = \ln 8.2e^{\sqrt{t}} \qquad \text{Taking natural logs}$$

$$= \ln 8.2 + \ln e^{\sqrt{t}} \qquad \begin{array}{l}\text{Log of a product is} \\ \text{the sum of the logs}\end{array}$$

$$= \ln 8.2 + \sqrt{t} \qquad \begin{array}{l}\ln e^{\sqrt{t}} = \sqrt{t} \quad \text{by Property 3} \\ \text{of logs (see inside back cover)}\end{array}$$

$$= \ln 8.2 + t^{1/2} \qquad \text{In exponent form}$$

Then we differentiate:

$$\frac{d}{dt}(\ln 8.2 + t^{1/2}) = 0 + \frac{1}{2}t^{-1/2} = \frac{1}{2}t^{-1/2} \qquad \begin{array}{l}\ln 8.2 \text{ is a constant, so} \\ \text{its derivative is zero}\end{array}$$

Finally, we evaluate at the given time $t = 25$:

$$\frac{1}{2}(25)^{-1/2} = \frac{1}{2}\frac{1}{\sqrt{25}} = \frac{1}{2}\frac{1}{5} = \frac{1}{10} = 0.10 \qquad \tfrac{1}{2}t^{-1/2} \text{ evaluated at } t = 25$$

Therefore, in 25 years the gross domestic product will be increasing at the relative rate of 0.10, or 10%, per year.

*Alternative Solution* (using the $\dfrac{f'(t)}{f(t)}$ formula)

$$G(t) = 8.2e^{\sqrt{t}} = 8.2e^{t^{1/2}} \qquad \text{Writing } G(t) \text{ with fractional exponents}$$

$$G'(t) = 8.2e^{t^{1/2}}\left(\frac{1}{2}t^{-1/2}\right) \qquad \text{Differentiating}$$

Derivative of the exponent

Therefore, the relative rate of change $\dfrac{G'(t)}{G(t)}$ is

$$\frac{G'(t)}{G(t)} = \frac{8.2e^{t^{1/2}}\left(\dfrac{1}{2}t^{-1/2}\right)}{8.2e^{t^{1/2}}} = \frac{1}{2}t^{-1/2} \qquad \begin{array}{l}\text{Same result as with}\\\text{the first formula}\end{array}$$

At $t = 25$,

$$\frac{1}{2}(25)^{-1/2} = \frac{1}{2}\frac{1}{\sqrt{25}} = \frac{1}{2}\frac{1}{5} = \frac{1}{10} = 0.10 \qquad \text{Again the same}$$

Therefore, the relative growth rate is 10%, just as we found before.

Both formulas give the same answer, so you should use the one that is easier to apply in your particular problem. The $\dfrac{d}{dt}\ln f(t)$ formula sometimes allows simplification before the differentiation, while the $\dfrac{f'(t)}{f(t)}$ formula often involves simplification afterward.

**Practice Problem 1**

An investor estimates that if a piece of land is held for $t$ years, it will be worth $f(t) = 300 + t^2$ thousand dollars. Find the relative rate of change at time $t = 10$ years. [*Hint:* Use the $\dfrac{f'(t)}{f(t)}$ formula.]

➤ Solution on page 314

## Graphing Calculator Exploration

Enter the function from Practice Problem 1 into a graphing calculator as $y_1 = 300 + x^2$, and then turn *off* the function so that it will not graph. Then graph $y_2 = \dfrac{d}{dx}\ln y_1$ and $y_3 = \dfrac{y_1'}{y_1}$ (using NDERIV) together on the window [0, 20] by [0, 0.1]. Why do you get only one curve for the two functions?

### Elasticity of Demand

Farmers are aware of the paradox that an abundant harvest usually brings *lower* total revenue than a poor harvest. The reason is simply that the larger quantities in an abundant harvest result in lower prices, which in turn cause increased demand, but the demand does not increase enough to compensate for the lower prices.

Revenue is price times quantity, $R = p \cdot q$, and when one of these quantities rises, the other falls. The question is whether the rise in one is enough to compensate for the fall in the other. For example, if a 1% price decrease brings a 2% quantity increase, revenue will rise, but if the 1% price decrease brings only a $\frac{1}{2}$% quantity increase, revenue will fall. The concept of *elasticity of demand* was invented to analyze such problems. (Economists use the word *demand* instead of *quantity*.)

Roughly speaking, we may think of elasticity as *the percentage change in demand divided by the percentage change in price*:

| Understanding Elasticity of Demand | *Brief Examples* |
|---|---|
| $$E = \frac{\text{Percent change in demand}}{\text{Percent change in price}}$$ | If a 1% change in price brings a 2% change in demand: $\quad E = \dfrac{2\%}{1\%} = 2$ <br><br> If a 1% change in price brings only a $\frac{1}{2}$% change in demand: $\quad E = \dfrac{\frac{1}{2}\%}{1\%} = \dfrac{1}{2}$ |

We classify demand as elastic or inelastic depending on whether elasticity is greater than 1 or less than 1:

> Demand is *elastic* if $\quad E > 1$
>
> Demand is *inelastic* if $\quad E < 1$
>
> Demand is *unit-elastic* if $\quad E = 1$

In the first Brief Example above, $E = 2$, so demand was *elastic*, while in the second example, $E = \frac{1}{2}$, so demand was *inelastic*.

**Practice Problem 2**

Suppose that Starbooks, purveyors of lattes and literature, raise prices by 5% and demand falls by 15%. Find the elasticity and identify the demand as elastic or inelastic.

> Solution on page 314

Intuitively, we may think of elasticity of demand as measuring how *responsive* demand is to price changes: *elastic* means *responsive* and *inelastic* means *unresponsive*. That is, for elastic demand, a price cut will bring a large

| Good or Service | Elasticity |
| --- | --- |
| Clothing | 0.20 |
| Housing | 0.30 |
| Gasoline | 0.43 |
| Movies | 0.87 |
| Automobiles | 1.87 |
| Restaurant meals | 2.27 |
| Fresh fruit | 3.02 |

*Source:* Houthaker and Taylor, *Consumer Demand in the United States, Review of Economics and Statistics* **62**

increase in demand, so total revenue will rise. On the other hand, for inelastic demand, a price cut will bring only a slight increase in demand, so total revenue will *fall*.

Economists calculate elasticity of demand for many products, and some typical elasticities are shown in the table on the left. Notice that for *necessities* (clothing, food), demand is inelastic since consumers need them even if prices rise, while for luxuries (restaurant meals) demand is elastic since consumers can cut back or find substitutes in response to price increases.

The preceding formula for elasticity was useful for understanding the concept but is not entirely satisfactory since it does not reflect how elasticity may change as price changes. In fact, the elasticities in the table are only approximately correct since prices are always changing. To see how elasticity changes as prices change (which will be useful in applications), we need a more precise definition. First, we must define the *demand function*, $x = D(p)$, which gives the *demand* for an item if the price of the item is $p$.

---

**Demand Function**

The demand function

$$x = D(p)$$

gives the quantity $x$ of an item that will be demanded by consumers if the price is $p$.

---

Law of downward-sloping demand

Since, in general, demand falls as prices rise, the slope of the demand function is negative, as shown on the left. This is known as the *law of downward-sloping demand*.

### Calculating Elasticity of Demand

To analyze changes in demand compared to changes in price, we calculate the relative rate of change of demand divided by the relative rate of change of price. Using the derivative-of-the-logarithm formula,

$$\frac{\left(\begin{array}{c}\text{Relative rate of}\\\text{change of demand}\end{array}\right)}{\left(\begin{array}{c}\text{Relative rate of}\\\text{change of price}\end{array}\right)} = \frac{\dfrac{d}{dp}\ln D(p)}{\dfrac{d}{dp}\ln p} = \frac{\dfrac{D'(p)}{D(p)}}{\dfrac{1}{p}} = \underbrace{\frac{pD'(p)}{D(p)}}_{\text{Simplified}}$$

Because most demand functions are downward-sloping, the derivative $D'(p)$ is generally negative. Economists prefer to work with positive numbers, so the *elasticity of demand* is taken to be the negative of this quantity (in order to make it positive).*

*Some economists omit the negative sign.

> ### Elasticity of Demand
>
> For a demand function $D(p)$, the elasticity of demand is
>
> $$E(p) = \frac{-p \cdot D'(p)}{D(p)}$$
>
> Demand is *elastic* if $E(p) > 1$ and *inelastic* if $E(p) < 1$.

Elasticity, being composed of *relative* rates of change, does not depend on the units of the demand function. Therefore, elasticities can be compared between different products, and even between different countries.

**EXAMPLE 2**

D

$D(p) = 81 - p^2$

Demand

81

Price

$p$

**FINDING ELASTICITY OF DEMAND FOR COMMUTER BUS SERVICE**

A bus line estimates the demand function for its daily commuter tickets to be $D(p) = 81 - p^2$ (in thousands of tickets), where $p$ is the price in dollars $(0 \le p \le 9)$. Find the elasticity of demand when the price is:

**a.** $3     **b.** $6

*Solution*

$$E(p) = \frac{-pD'(p)}{D(p)} \qquad \text{Definition of elasticity}$$

$$= \frac{-p(-2p)}{81 - p^2} \qquad \begin{array}{l}\text{Substituting } D(p) = 81 - p^2 \\ \text{so } D'(p) = -2p\end{array}$$

$$= \frac{2p^2}{81 - p^2} \qquad \text{Simplifying}$$

**a.** Evaluating at $p = 3$ gives

$$E(3) = \frac{2(3)^2}{81 - (3)^2} = \frac{18}{81 - 9} = \frac{18}{72} = \frac{1}{4} \qquad E(p) = \frac{2p^2}{81 - p^2} \text{ with } p = 3$$

*Interpretation:* The elasticity is less than 1, so demand for tickets is *inelastic* at a price of $3. This means that a small price change (up or down from this level) will cause only a *slight* change in demand. More precisely, elasticity of $\frac{1}{4}$ means that a 1% price change will cause only about a $\frac{1}{4}$% change in demand.

**b.** At the price of $6, the elasticity of demand is

$$E(6) = \frac{2(6)^2}{81 - (6)^2} = \frac{72}{81 - 36} = \frac{8}{5} = 1.6 \qquad E(p) = \frac{2p^2}{81 - p^2} \text{ with } p = 6$$

*Interpretation:* The elasticity is greater than 1, so demand is *elastic* at a price of $6. This means that a small change in price (up or down from this level) will cause a relatively *large* change in demand. In particular, an elasticity of 1.6 means that a price change of 1% will cause about a 1.6% change in demand.

The changes in demand are, of course, in the opposite direction from the changes in price. That is, if prices are *raised* by 1% (from the $6 level), demand will *fall* by 1.6%, whereas if prices are *lowered* by 1%, demand will *rise* by 1.6%. In the future we will assume that the *direction* of the change is clear, and say simply that a 1% change in price will cause about a 1.6% change in demand.

**Practice Problem 3**

For the demand function $D(p) = 90 - p$, find the elasticity of demand $E(p)$ and evaluate it at $p = 30$ and $p = 75$. (Be sure to complete this Practice Problem, because the results will be used shortly.)

> **Solution on page 314**

## Graphing Calculator Exploration

```
Y2(3)
            .25
Y2(6)
            1.6
```

For any demand function $y_1$ (written in terms of $x$), define $y_2$ as the elasticity function for $y_1$ by defining $y_2 = -x \cdot y_1' / y_1$ (using NDERIV). Try entering the demand function from Example 2 (as shown on the left) or Practice Problem 3 in $y_1$ and then evaluating $y_2$ at appropriate numbers to check the answers found there.

### Using Elasticity to Increase Revenue

In Example 2 we found that at a price of $3, demand is inelastic $(E = \frac{1}{4} < 1)$, and so demand responds only *weakly* to price changes. Therefore, to increase revenue the company should *raise* prices, since the higher prices will drive away only a relatively small number of customers. On the other hand, at a price of $6, demand is elastic $(E = 1.6 > 1)$, and so demand is very responsive to price changes. In this case, to increase revenue the company should *lower* prices, since this will attract more than enough new customers to compensate for the price decrease. In general:

**Elasticity and Revenue**

To increase revenue:

If demand is elastic $(E > 1)$, you should *lower* prices.

If demand is inelastic $(E < 1)$, you should *raise* prices.

This should be intuitively clear: In simplest terms, you should *lower* prices if demand *will* change and *raise* prices if demand *won't* change. This statement shows why elasticity of demand is important to any company that cuts prices in an attempt to boost revenue, or to any utility that raises prices in order to increase revenue. Elasticity shows whether the strategy will succeed or fail.

## Spreadsheet Exploration

The following spreadsheet* is based on the demand function $D(p) = 90 - p$ from Practice Problem 3 on the previous page, and shows the elasticity of demand along with the demand and revenue for various prices.

| D6 | | ▼ | = | =$A6/(90-$A6) |
|---|---|---|---|---|
| | **A** | **B** | **C** | **D** |
| 1 | Price | Demand | Revenue | Elasticity |
| 2 | | | | |
| 3 | 30 | 60 | 1800 | 0.500 |
| 4 | 35 | 55 | 1925 | 0.636 |
| 5 | 40 | 50 | 2000 | 0.800 |
| 6 | 45 | 45 | 2025 | 1.000 |
| 7 | 50 | 40 | 2000 | 1.250 |
| 8 | 55 | 35 | 1925 | 1.571 |
| 9 | 60 | 30 | 1800 | 2.000 |
| 10 | 65 | 25 | 1625 | 2.600 |
| 11 | 70 | 20 | 1400 | 3.500 |
| 12 | 75 | 15 | 1125 | 5.000 |

*Inelastic:* Revenue increases as price increases {rows 3–5}

*Elastic:* Revenue decreases as price increases {rows 7–12}

Notice that where elasticity is less than 1 (the first three rows), revenue rises as the price increases, but where elasticity is greater than 1 (the last six rows), revenue falls as the price increases.

*To obtain this and other Spreadsheet Explorations, go to www.cengage.com/math/berresford.

The borderline case, $E = 1$ (called *unit-elasticity*), is where revenue cannot be raised, which will be the case if revenue is at its maximum. Therefore, elasticity must be unitary when revenue is maximized.

> At maximum revenue, elasticity of demand must equal 1.

We could use this fact as a basis for a new method for maximizing revenue, but instead, we will stick with our earlier (and easier) method of maximizing functions by finding critical numbers.

**Practice Problem 4**   According to the preceding spreadsheet, at what price is revenue maximized, and what is the elasticity of demand at that price?

➤ **Solution below**

Demand is inelastic    Demand is elastic

Here    $E = \frac{1}{2}$

Here    $E = 5$

90

Demand

30    75 90    Price

Revenue

30    75 90    Price

Demand is inelastic    Demand is elastic

**BE CAREFUL:** Do not confuse elasticity of demand with the slope of the demand curve; the two ideas are quite different. For example, in Practice Problem 3 on page 312, we found that the linear demand function $D(p) = 90 - p$ (which has slope $-1$ all along it) has elasticity $\frac{1}{2}$ at one point and elasticity 5 at another, as shown on the left. In fact, the elasticities vary from 0 (at the top) to approaching infinity (at the bottom).

The second graph shows the revenue function for this demand function

$$R = p(90 - p) = 90p - p^2$$

From $R = p \cdot D(p)$ with $D(p) = 90 - p$

Notice that for values of $p$ to the left of center, where demand is inelastic, increasing $p$ will increase revenue, while for values of $p$ to the right of center, where demand is elastic, you should *lower* $p$ to increase revenue, just stated in the box on page 312.

**Solutions to Practice Problems**

1.  $\dfrac{f'(t)}{f(t)} = \dfrac{2t}{300 + t^2}$

    At $t = 10$,   $\dfrac{2 \cdot 10}{300 + 10^2} = \dfrac{20}{400} = \dfrac{1}{20} = 0.05$ or 5%

2.  $E = \dfrac{15\%}{5\%} = 3$,   so demand is elastic.

3.  $E(p) = \dfrac{-pD'(p)}{D(p)} = \dfrac{-p(-1)}{90 - p} = \dfrac{p}{90 - p}$

    At $p = 30$,   $E(30) = \dfrac{30}{90 - 30} = \dfrac{30}{60} = \dfrac{1}{2} = 0.5$   (demand is inelastic)

    At $p = 75$,   $E(75) = \dfrac{75}{90 - 75} = \dfrac{75}{15} = 5$   (demand is elastic)

4.  At $p = 45$.  Elasticity is 1.

## Verification of the Relationship Between Elasticity and Revenue

We may verify the relationship between elasticity of demand and revenue as follows: revenue is price $p$ times quantity $x$.

$$R = px = p \cdot D(p)$$

Using $x = D(p)$

Differentiating will show how revenue responds to price changes.

$$R'(p) = D(p) + pD'(p) \qquad \text{Using the Product Rule on } p \cdot D(p)$$

$$= D(p)\left[1 + \frac{pD'(p)}{D(p)}\right] \qquad \text{Factoring out } D(p)$$

$$= D(p)\left[1 - \frac{-pD'(p)}{D(p)}\right] \qquad \begin{array}{l}\text{Replacing the plus sign}\\\text{by two minus signs}\end{array}$$

This is the definition of elasticity $E(p)$

$$= D(p)[1 - E(p)] \qquad \text{Replacing } \frac{-pD'(p)}{D(p)} \text{ by } E(p)$$

If demand is *elastic*, $E > 1$, then the quantity in brackets is negative, and so the derivative $R'(p)$ is negative, showing that revenue *decreases* as price increases. Therefore, to increase revenue, one should *lower* prices. On the other hand, if the demand is *inelastic*, $E < 1$, then the quantity in brackets is positive, and so the derivative $R'(p)$ is positive, showing that revenue *increases* as price increases. In this case, to increase revenue, one should *raise* prices. This proves the statements on page 312 in the box headed "Elasticity and Revenue."

## 4.4 Exercises

### Exercises on Relative Rates

**1–10.** For each function:

**a.** Find the relative rate of change.

**b.** Evaluate the relative rate of change at the given value(s) of $t$.

**1.** $f(t) = t^2$, $t = 1$ and $t = 10$

**2.** $f(t) = t^3$, $t = 1$ and $t = 10$

**3.** $f(t) = 100e^{0.2t}$, $t = 5$

**4.** $f(t) = 100e^{-0.5t}$, $t = 4$

**5.** $f(t) = e^{t^2}$, $t = 10$

**6.** $f(t) = e^{t^3}$, $t = 5$

**7.** $f(t) = e^{-t^2}$, $t = 10$

**8.** $f(t) = e^{-t^3}$, $t = 5$

**9.** $f(t) = 25\sqrt{t - 1}$, $t = 6$

**10.** $f(t) = 100\sqrt[3]{t + 2}$, $t = 8$

### Applied Exercises on Relative Rates

**11. APPLE STOCK PRICE** Since the development of the iPod, the stock price of Apple has been growing rapidly and has been approximately $3.2e^{0.5x}$, where $x$ is the number of years since 2000 (for $2 \le x \le 7$). Find the relative growth rate of Apple's stock price at any time during that period.

*Source: BigCharts.com*

**12. RESEARCH IN MOTION STOCK PRICE** During the first 11 months of 2007, the stock price of Research in Motion, the maker of the BlackBerry communications device, was growing at a record pace, and was given approximately by $32e^{0.11x}$, where $x$ is the number of the month (so $x = 1$ stands for January). Find the relative growth rate of the Research in Motion stock price during that period.

*Source: BigCharts.com*

**13–14. ECONOMICS: National Debt** If the national debt of a country (in trillions of dollars) $t$ years from now is given by the indicated function, find the relative rate of change of the debt 10 years from now.

**13.** $N(t) = 0.5 + 1.1e^{0.01t}$    **14.** $N(t) = 0.4 + 1.2e^{0.01t}$

 **15–16. GENERAL: Population** The population (in millions) of a city $t$ years from now is given by the indicated function.

**a.** Find the relative rate of change of the population 8 years from now.
**b.** Will the relative rate of change ever reach 1.5%?

**15.** $P(t) = 4 + 1.3e^{0.04t}$    **16.** $P(t) = 6 + 1.7e^{0.05t}$

## Exercises on Elasticity of Demand

**17–28.** For each demand function $D(p)$:

**a.** Find the elasticity of demand $E(p)$.
**b.** Determine whether the demand is elastic, inelastic, or unit-elastic at the given price $p$.

**17.** $D(p) = 200 - 5p$, $p = 10$

**18.** $D(p) = 60 - 8p$, $p = 5$

**19.** $D(p) = 300 - p^2$, $p = 10$

**20.** $D(p) = 100 - p^2$, $p = 5$

**21.** $D(p) = \dfrac{300}{p}$, $p = 4$

**22.** $D(p) = \dfrac{500}{p}$, $p = 2$

**23.** $D(p) = \sqrt{175 - 3p}$, $p = 50$

**24.** $D(p) = \sqrt{100 - 2p}$, $p = 20$

**25.** $D(p) = \dfrac{100}{p^2}$, $p = 40$

**26.** $D(p) = \dfrac{600}{p^3}$, $p = 25$

**27.** $D(p) = 4000e^{-0.01p}$, $p = 200$

**28.** $D(p) = 6000e^{-0.05p}$, $p = 100$

## Applied Exercises on Elasticity of Demand

**29. AUTOMOBILE SALES** An automobile dealer is selling cars at a price of $12,000. The demand function is $D(p) = 2(15 - 0.001p)^2$, where $p$ is the price of a car. Should the dealer raise or lower the price to increase revenue?

**30. LIQUOR SALES** A liquor distributor wants to increase its revenues by discounting its bestselling liquor. If the demand function for this liquor is $D(p) = 60 - 3p$, where $p$ is the price per bottle, and if the current price is $15, will the discount succeed?

**31. CITY BUS REVENUES** The manager of a city bus line estimates the demand function to be $D(p) = 150,000\sqrt{1.75 - p}$, where $p$ is the fare in dollars. The bus line currently charges a fare of $1.25, and it plans to raise the fare to increase its revenues. Will this strategy succeed?

**32. NEWSPAPER SALES** The demand function for a newspaper is $D(p) = 80,000\sqrt{75 - p}$, where $p$ is the price in cents. The publisher currently charges 50 cents, and it plans to raise the price to increase revenues. Will this strategy succeed?

**33. ELECTRICITY RATES** An electrical utility asks the Federal Regulatory Commission for permission to raise rates to increase revenues. The utility's demand function is

$$D(p) = \frac{120}{10 + p}$$

where $p$ is the price (in cents) of a kilowatt-hour of electricity. If the utility currently charges 6 cents per kilowatt-hour, should the commission grant the request?

**34. OIL PRICES** A Middle Eastern oil-producing country estimates that the demand for oil (in millions of barrels per day) is $D(p) = 9.5e^{-0.04p}$, where $p$ is the price of a barrel of oil. To raise its revenues, should it raise or lower its price from its current level of $120 per barrel?

**35. OIL PRICES** A European oil-producing country estimates that the demand for its oil (in millions of barrels per day) is $D(p) = 3.5e^{-0.06p}$, where $p$ is the price of a barrel of oil. To raise its revenues, should it raise or lower its price from its current level of $120 per barrel?

**36–37. LIQUOR AND BEER** The demand functions for distilled spirits and for beer are given below, where $p$ is the retail price and $D(p)$ is the demand in gallons per capita. For each demand function, find the elasticity of demand for any price $p$. [*Note:* You will find, in each case, that demand is inelastic. This means that taxation, which acts like a price increase, is an ineffective way of discouraging liquor consumption, but is an effective way of raising revenue.]

*Source: Journal of Consumer Research* **12**

**36.** $D(p) = 3.509p^{-0.859}$ (for distilled spirits)

**37.** $D(p) = 7.881p^{-0.112}$ (for beer)

 **38–39. AUTOMOBILE SALES** The demand function for automobiles in a dealership is given below, where $p$ is the selling price.

**a.** Use the method described in the Graphing Calculator Exploration on page 312 to find the elasticity of demand at a price of $12,000.

**b.** Should the dealer raise or lower the price from this level to increase revenue?

**c.** Find the price at which elasticity equals 1. [*Hint:* Use INTERSECT.]

**38.** $D(p) = 3\sqrt{20 - 0.001p}$ **39.** $D(p) = \dfrac{200}{8 + e^{0.0001p}}$

**40. ELASTICITIES FOR LINEAR DEMAND** For the linear demand function $D(p) = 90 - p$ used in Practice Problem 3 and graphed on page 314:

**a.** Find a formula for the elasticity of demand at *any* value of $p$. [*Hint:* Use the formula on page 311 with the given demand function.]

**b.** Show that the elasticity is zero at the *top* of the line (that is, for $p = 0$).

**c.** Show that the elasticity approaches infinity as $p$ approaches the *bottom* of the line.

**d.** Show that the elasticity is $\frac{1}{2}$ at the midpoint of the line.

**e.** Show that statements (b), (c), and (d) are true for *any* linear demand function $D(p) = a - bp$ for any positive constants $a$ and $b$? [*Hint:* Find an expression for the elasticity at any value of $p$, as you did in part (a).]

**41. CIGARETTE SALES** In 1998 the State of New Jersey raised cigarette taxes, with the result that the price of a pack of cigarettes rose from $2.40 to $2.80. Demand then fell from 52 to 47.5 million packs. Use these data to estimate the elasticity of demand for cigarettes. [*Hint:* Find the percentage change in price and the percentage change in demand and use the formula in the "Understanding Elasticity of Demand" box on page 309.]

*Source:* P. Samuelson and W. Nordhaus, *Economics*

## Conceptual Exercises

**42–45.** For each function, calculate "in your head" the relative rate of change.

**42.** $f(x) = e^x$      **43.** $f(x) = e^{1.2x}$

**44.** $f(x) = x$      **45.** $f(x) = x^n$

**46–47.** For each demand function, calculate "in your head" the elasticity.

**46.** $D(p) = e^{-p}$      **47.** $D(p) = e^{-2p}$

**48.** If demand is inelastic and you raise prices, would you expect demand to increase or decrease, and strongly or weakly?

**49.** If demand is elastic and you lower prices, would you expect demand to increase or decrease, and strongly or weakly?

**50.** For which of the following two items would you expect demand to be elastic and for which inelastic: *heating oil, olive oil.*

**51.** For which of the following two items would you expect demand to be elastic and for which inelastic: *cigarettes, jewelry.*

## Explorations and Excursions

The following problems extend and augment the material presented in the text.

**52. CONSTANT ELASTICITY**

**a.** Show that for a demand function of the form $D(p) = c/p^n$, where $c$ and $n$ are positive constants, the elasticity is constant.

**b.** What type of demand function has elasticity equal to 1 for every value of $p$?

**53. LINEAR ELASTICITY** Show that for a demand function of the form $D(p) = ae^{-cp}$, where $a$ and $c$ are positive constants, the elasticity of demand is $E(p) = cp$.

**54–55. ELASTICITY OF SUPPLY** A supply function $S(p)$ gives the total amount of a product that producers are willing to supply at a given price $p$. The *elasticity of supply* is defined as

$$E_s(p) = \frac{p \cdot S'(p)}{S(p)}$$

Elasticity of supply measures the relative increase in supply resulting from a small relative increase in price. It is less useful than elasticity of demand, however, since it is not related to total revenue.

**54.** Use the preceding formula to find the elasticity of supply for a supply function of the form $S(p) = ae^{cp}$, where $a$ and $c$ are positive constants.

**55.** Use the preceding formula to find the elasticity of supply for a supply function of the form $S(p) = ap^n$, where $a$ and $n$ are positive constants.

## Chapter Summary with Hints and Suggestions

Reading the text and doing the exercises in this chapter have helped you to master the following concepts and skills, which are listed by section (in case you need to review them) and are keyed to particular Review Exercises. Answers for all Review Exercises are given at the back of the book, and full solutions can be found in the Student Solutions Manual.

### 4.1 Exponential Functions

- Find the value of money invested at compound interest. *(Review Exercise 1.)*

$$P\left(1 + \frac{r}{m}\right)^{mt} \qquad Pe^{rt}$$

- Determine which of two banks gives better interest. *(Review Exercise 2.)*
- Depreciate an asset. *(Review Exercise 3.)*

$$P(1 + r)^t \qquad \text{(for depreciation, } r \text{ is negative)}$$

- Determine which of two drugs provides more medication. *(Review Exercise 4.)*
- Predict the world's largest city. *(Review Exercise 5.)*
- Predict computer memory capacity. *(Review Exercise 6.)*

### 4.2 Logarithmic Functions

- Find doubling (and other) times for compound interest. *(Review Exercises 7–8.)*

$$\ln M^P = P \cdot \ln M \qquad \ln e^x = x$$

- Date a fossil by potassium 40. *(Review Exercises 9–10.)*
- Predict the spread of information, or oil demand. *(Review Exercises 11–12.)*
- Find the time required for a given percentage change in a bank deposit or in advertising results. *(Review Exercises 13–14.)*

### 4.3 Differentiation of Logarithmic and Exponential Functions

- Find the derivative of a logarithmic or exponential function. *(Review Exercises 15–30.)*

$$\frac{d}{dx}\ln x = \frac{1}{x} \qquad \frac{d}{dx}e^x = e^x \qquad \frac{d}{dx}\ln f = \frac{f'}{f}$$

$$\frac{d}{dx}e^f = e^f \cdot f' \qquad \frac{d}{dx}e^{kx} = ke^{kx}$$

- Graph an exponential or a logarithmic function. *(Review Exercises 31–32.)*

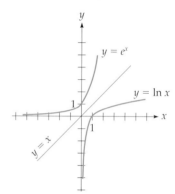

- Find the rate of change of sales, amount of medication, learning, temperature, or diffusion of information. *(Review Exercises 33–37.)*
- Find when a company maximizes its present value. *(Review Exercise 38.)*
- Maximize a company's revenue. *(Review Exercises 39–40.)*
- Maximize consumer expenditure for a product. *(Review Exercise 41.)*
- Graph an exponential or logarithmic function. *(Review Exercises 42–43.)*
- Maximize consumer expenditure or maximize revenue. *(Review Exercises 44–45.)*

### 4.4 Two Applications to Economics: Relative Rates and Elasticity of Demand

- Find the relative rate of change of a country's gross domestic product. *(Review Exercises 46–47.)*

$$\left(\begin{array}{c}\text{Relative}\\\text{rate}\end{array}\right) = \frac{d}{dt}\ln f = \frac{f'}{f}$$

■ Find the elasticity of demand for a product, and its consequences.    *(Review Exercises 48–50.)*

$$\left(\begin{array}{c}\text{Elasticity}\\\text{of demand}\end{array}\right) = \frac{-p \cdot D'(p)}{D(p)}$$

■ Find the relative rate of change of population. *(Review Exercise 51.)*

■ Find the elasticity of demand for a product, how it affects revenue, and what price gives unit-elasticity. *(Review Exercise 52.)*

## Hints and Suggestions

■ *(Overview)* Exponential and logarithmic functions should be thought of as just other types of functions, like polynomials, but having their own differentiation rules. In fact, in a sense they are more "natural" than polynomials because they give natural growth rates.

■ A graphing calculator helps by drawing graphs of exponential and logarithmic functions, and finding intersection points (for example, where one population overtakes another). It is also useful for checking derivatives and graphically verifying maximum and minimum values.

■ Interest rates are always *annual* rates unless stated otherwise. To use $P(1 + r/m)^{mt}$ for finding value under compound interest, $r$ is the *annual interest rate*, $m$ is the number of *compounding periods*, and $t$ is the *number of years*.

■ When do you use $Pe^{rt}$ and when do you use $P(1 + r/m)^{mt}$? Use $Pe^{rt}$ if the word "continuous" occurs, and use $P(1 + r/m)^{mt}$ if it does not. $P(1 + r)^t$ with negative $r$ gives depreciation. Dividing $P$ by $e^{rt}$ or $(1 + r/m)^{mt}$ gives present value.

■ The Power Rule $\frac{d}{dx}x^n = nx^{n-1}$ is for differentiating a *variable to a constant power*, and $\frac{d}{dx}e^x = e^x$ is for differentiating the *constant e to a variable power*.

■ **Practice for Test:** Review Exercises 1, 3, 5, 7, 8, 11, 13, 15, 19, 21, 23, 27, 31, 33, 37, 41, 45, 47, 49, 51.

# Review Exercises for Chapter 4

Practice test exercise numbers are in green.

## 4.1 Exponential Functions

1. **PERSONAL FINANCE: Interest** Find the value of $10,000 invested for 8 years at 8% interest if the compounding is done:
   **a.** quarterly.    **b.** continuously.

2. **PERSONAL FINANCE: Interest** One bank offers 6% compounded quarterly and a second offers 5.98% compounded continuously. Where should you take your money?

3. **BUSINESS: Depreciation** An $800,000 computer loses 20% of its value each year.
   **a.** Give a formula for its value after $t$ years.
   **b.** Find its value after 4 years.

4. **BIOMEDICAL: Drug Concentration** If the concentration of a drug in a patient's bloodstream is $c$ (milligrams per milliliter), then $t$ hours later the concentration will be $C(t) = ce^{-kt}$, where $k$ is a constant (the "elimination constant"). Two drugs are being compared: drug A with initial concentration $c = 2$ and elimination constant $k = 0.2$, and drug B with initial concentration $c = 3$ and elimination constant $k = 0.25$. Which drug will have the greater concentration 4 hours later?

5. **GENERAL: Population** The largest city in the world is Tokyo, with São Paulo (Brazil) smaller but growing faster. According to the Census Bureau, $x$ years after 2000 the population of Tokyo will be $27.5e^{0.0034x}$ and the population of São Paulo will be $17.4e^{0.011x}$ (both in millions). Graph both functions on a calculator with the window [0, 100] by [0, 50]. When will São Paulo overtake Tokyo as the world's largest city (assuming that these rates continue to hold)?

6. **GENERAL: Moore's Law of Computer Memory** The amount of information that can be stored on a computer chip can be measured in megabits (a "bit" is a binary digit, 0 or 1, and a "megabit" is a million bits). The first 1-megabit chips became available in 1987, and 4-megabit chips became available in 1990. This quadrupling

of capacity every 3 years is expected to continue, so that chip capacity will be $C(t) = 4^{t/3}$ megabits where $t$ is the number of years after 1987. Use this formula (known as Moore's Law, after Gordon Moore, a founder of the Intel Corporation) to predict chip capacity in the year 2011. [*Hint:* What value of $t$ corresponds to 2011?]

## 4.2 Logarithmic Functions

7. **PERSONAL FINANCE: Interest** Find how soon an investment at 10% interest compounded semiannually will:
   **a.** double in value.     **b.** increase by 50%.

8. **PERSONAL FINANCE: Interest** Find how soon an investment at 7% interest compounded continuously will:
   **a.** double in value.     **b.** increase by 50%.

**9–10. GENERAL: Fossils** In the following exercises, use the fact that the proportion of potassium 40 remaining after $t$ million years is $e^{-0.00054t}$.

9. In 1984 in the Wind River Basin of Wyoming, scientists discovered a fossil of a small, three-toed horse, an ancestor of the modern horse. Estimate the age of this fossil if it contained 97.3% of its original potassium 40.

10. Estimate the age of a skull found in 1959 in Tanzania (dubbed "Nutcracker Man" because of its huge jawbone) that contained 99.9% of its original potassium 40.

11. **SOCIAL SCIENCE: Diffusion of Information by Mass Media** In a city of a million people, election results broadcast over radio and television will reach $N(t) = 1{,}000{,}000(1 - e^{-0.3t})$ people within $t$ hours. Find when the news will have reached 500,000 people.

12. **ECONOMICS: Oil Demand** The demand for oil in the United States is increasing by 3% per year. Assuming that this rate continues, how soon will demand increase by 50%?

13. **PERSONAL FINANCE: Interest** If a bank offers 6.5% interest, in how many years will a deposit increase by 50% if the compounding is done:
   **a.** quarterly?     **b.** continuously?

14. **BUSINESS: Advertising** After the opening of a new store has been advertised for $t$ days, the proportion of people in a city who have seen the ad is $p(t) = 1 - e^{-0.032t}$. How long must the ad run to reach:
   **a.** 30% of the people?
   **b.** 40% of the people?

## 4.3 Differentiation of Logarithmic and Exponential Functions

**15–30.** Find the derivative of each function

15. $f(x) = \ln 2x$
16. $f(x) = \ln(x^2 - 1)^2$
17. $f(x) = \ln(1 - x)$
18. $f(x) = \ln\sqrt{x^2 + 1}$
19. $f(x) = \ln\sqrt[3]{x}$
20. $f(x) = \ln e^x$
21. $f(x) = \ln x^2$
22. $f(x) = x \ln x - x$
23. $f(x) = e^{-x^2}$
24. $f(x) = e^{1-x}$
25. $f(x) = \ln e^{x^2}$
26. $f(x) = e^{x^2 \ln x - x^2/2}$
27. $f(x) = 5x^2 + 2x \ln x + 1$
28. $f(x) = 2x^3 + 3x \ln x - 1$
29. $f(x) = 2x^3 - 3xe^{2x}$
30. $f(x) = 4x - 2x^2 e^{2x}$

**31–32.** Graph each function, showing all relative extreme points and inflection points.

31. $f(x) = \ln(x^2 + 4)$
32. $f(x) = 16e^{-x^2/8}$

33. **BUSINESS: Sales** The weekly sales (in thousands) of a new product after $x$ weeks of advertising is $S(x) = 2000 - 1500e^{-0.1x}$. Find the rate of change of sales after:
   **a.** 1 week.     **b.** 10 weeks.

34. **BIOMEDICAL: Drug Dosage** A patient receives an injection of 1.5 milligrams of a drug, and the amount remaining in the bloodstream $t$ hours later is $A(t) = 1.5e^{-0.08t}$. Find the instantaneous rate of change of this amount:
   **a.** immediately after the injection (time $t = 0$).
   **b.** after 5 hours.

35. **BEHAVIORAL SCIENCE: Learning** In a test of short-term memory, the percent of subjects who remember an eight-digit number for at least $t$ seconds is $P(t) = 100 - 200\ln(t + 1)$. Find the rate of change of this percent after 5 seconds.

36. **GENERAL: Temperature** A thermos bottle that is chilled to 35 degrees Fahrenheit and then left in a 70-degree room will warm to a temperature of $T(t) = 70 - 35e^{-0.1t}$ degrees after $t$ hours. Find the rate of change of the temperature:
   **a.** at time $t = 0$.     **b.** after 5 hours.

37. **SOCIAL SCIENCE: Diffusion of Information by Mass Media** The number of people in a town of 30,000 who have heard an important news

bulletin within $t$ hours of its first broadcast is $N(t) = 30{,}000(1 - e^{-0.3t})$. Find the instantaneous rate of change of the number of informed people after:

**a.** 1 hour.  **b.** 8 hours.

**38. BUSINESS: Maximizing Present Value**
A new company is growing so that its value $t$ years from now will be $50t^2$ dollars. Therefore, its present value (at the rate of 8% compounded continuously) is $V(t) = 50t^2e^{-0.08t}$ dollars (for $t > 0$). Find the number of years that maximizes the present value.

**39–40. BUSINESS: Maximizing Revenue**
The given function is a company's price function, where $x$ is the quantity (in thousands) that will be sold at price $p$ dollars.

**a.** Find the revenue function $R(x)$.
[*Hint:* Revenue is price times quantity, $p \cdot x$.]
**b.** Find the quantity and price that will maximize revenue.

**39.** $p = 200e^{-0.25x}$    **40.** $p = 5 - \ln x$

**41. ECONOMICS: Maximizing Consumer Expenditure** Consumer demand for a commodity is estimated to be $D(p) = 25{,}000e^{-0.02p}$ units per month, where $p$ is the selling price in dollars. Find the selling price that maximizes consumer expenditure.

**42–43.** Use your graphing calculator to graph each function on a window that includes all relative extreme points and inflection points, and give the coordinates of these points (rounded to two decimal places). [*Hint:* Use NDERIV once or twice with ZERO.] (Answers may vary depending on the graphing window chosen.)

**42.** $f(x) = \dfrac{x^4}{e^x}$    **43.** $f(x) = x^3 \ln|x|$

**44. ECONOMICS: Consumer Expenditure**
If consumer demand for a commodity is $D(p) = 200e^{-0.0013p}$ (where $p$ is the selling price in dollars), find the price that maximizes consumer expenditure.

**45. BUSINESS: Maximizing Revenue** A manufacturer finds that the price function for autofocus cameras is $p(x) = 20e^{-0.0077x}$, where $p$ is the price in dollars at which quantity $x$ (in thousands) will be sold. Find the quantity $x$ that maximizes revenue.

## 4.4 Two Applications to Economics: Relative Rates and Elasticity of Demand

**46–47. ECONOMICS: Relative Rate of Change**
The gross domestic product of a developing country is forecast to be $G(t) = 5 + 2e^{0.01t}$ million dollars $t$ years from now. Find the relative rate of change:

**46.** 20 years from now.    **47.** 10 years from now.

**48. ECONOMICS: Elasticity of Demand** A South American country exports coffee and estimates the demand function to be $D(p) = 63 - 2p^2$. If the country wants to raise revenues to improve its balance of payments, should it raise or lower the price from the present level of $3 per pound?

**49. ECONOMICS: Elasticity of Demand** A South African country exports gold and estimates the demand function to be $D(p) = 400\sqrt{600 - p}$. If the country wants to raise revenues to improve its balance of payments, should it raise or lower the price from the present level of $350 per ounce?

**50. ECONOMICS: Elasticity of Demand** The demand function for cigarettes is of the form $D(p) = 1.2p^{-0.44}$, where $p$ is the price of a pack of cigarettes and $D(p)$ is the demand measured in packs per day per capita. Find the elasticity of demand. [*Note:* You will find that demand is inelastic. This means that taxation, which acts like a price increase, is an ineffective way of discouraging smoking, but is an effective way of raising revenue.]
*Source: American Economic Review 70*

**51. SOCIAL SCIENCES: Relative Rate of Change**
The population of a city $x$ years from now is projected to be $P(x) = 3.25 + 0.04x + 0.002x^3$ million people (for $0 \le x \le 10$). Find the relative rate of change 9 years from now.

**52. ECONOMICS: Elasticity of Demand** A boat dealer finds that the demand function for outboard motor boats near a large lake is $D(p) = 200 - 20p + p^2 - 0.03p^3$ (for $0 \le p \le 15$), where $p$ is the selling price in thousands of dollars.

**a.** Use a graphing calculator to find the elasticity of demand at a price of $10,000. [*Hint:* What value of $p$ corresponds to $10,000?]
**b.** Should the dealer raise or lower the price from this level to increase revenue?
**c.** Find the price at which elasticity equals 1.

# 5

# Integration and Its Applications

### Cigarette Smoking

Most cigarettes today have filters to absorb some of the toxic material or "tar" in the smoke before it is inhaled. The tobacco near the filter acts like an additional filter, absorbing tar until it is itself smoked, at which time it releases all of its accumulated toxins. A typical cigarette consists of 8 centimeters of tobacco followed by a filter.

As the cigarette is smoked, tar is typically inhaled at the rate of $r(x) = 300e^{0.025x} - 240e^{0.02x}$ milligrams (mg) of tar per centimeter (cm) of tobacco, where $x$ is the distance along the cigarette.* The amount of tar inhaled from any particular segment of the cigarette is the area under the graph of this function over that interval, which can be calculated by a process called *definite integration*, as explained in this chapter.

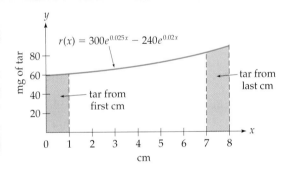

The results of integrating over the first and last centimeter of the cigarette are as follows (omitting the details—see Exercise 94 on page 364). The numbers represent milligrams of tar from the beginning and end of the cigarette.

$$\begin{pmatrix} \text{Tar from} \\ \text{first cm} \end{pmatrix} = \int_0^1 (300e^{0.025x} - 240e^{0.02x})\, dx \approx 61$$

Integrate from 0 to 1 for the first centimeter

$$\begin{pmatrix} \text{Tar from} \\ \text{last cm} \end{pmatrix} = \int_7^8 (300e^{0.025x} - 240e^{0.02x})\, dx \approx 83$$

Integrate from 7 to 8 for the last centimeter

*The actual rate depends upon the type of cigarette and the proportion of time that it is smoked rather than left to burn in the air. For further information, see Helen Marcus-Roberts, and Maynard Thompson (Eds.), *Modules in Applied Mathematics*, Vol. 4, *Life Science Models*, Springer-Verlag, pp. 238–249.

Notice that the last centimeter releases significantly more tar (about 36% more) than the first centimeter.

*Moral:* The Surgeon General has determined that smoking is hazardous to your health, and the last puffs are 36% more hazardous than the first.

This is just one of the many applications of integration discussed in this chapter.

## 5.1 ANTIDERIVATIVES AND INDEFINITE INTEGRALS

### Introduction

We have been studying differentiation and its uses. We now consider the reverse process, ***antidifferentiation,*** which, for a given derivative, essentially recovers the original function. Antidifferentiation has many uses. For example, differentiation turns a cost function into a marginal cost function, and so antidifferentiation turns marginal cost back into cost. Later we will use antidifferentiation for other purposes, such as finding areas.

### Antiderivatives and Indefinite Integrals

We begin with a simple example of antidifferentiation. Since the derivative of $x^2$ is $2x$, an *anti*derivative of $2x$ is $x^2$:

An antiderivative of $2x$ is $x^2$     Since the derivative of $x^2$ is $2x$

There are, however, other antiderivatives of $2x$. Each of the following is an antiderivative of $2x$:

$x^2 + 1$     $x^2 - 17$     $x^2 + e$     Since the derivative of each is $2x$

Clearly, we may add *any* constant to $x^2$ and the derivative will still be $2x$. Therefore, $x^2 + C$ is an antiderivative of $2x$ for *any* constant C. Furthermore, it can be shown that there are no other antiderivatives of $2x$, and so the *most general* antiderivative of $2x$ is $x^2 + C$. The most general antiderivative of $2x$ is called the **indefinite integral** of $2x$, and is written with the $2x$ between an **integral sign** $\int$ and a **dx:**

$$\int 2x\, dx = x^2 + C$$
The indefinite integral of $2x$ is $x^2 + C$ (because the derivative of $x^2 + C$ is $2x$)

Integral sign ⌐   ⌐ Integrand   ⌐ Arbitrary constant

The function to be integrated (here $2x$) is called the **integrand.** The $dx$ reminds us that the variable of integration is $x$. The constant C is called an **arbitrary constant** because it may take any value, positive, negative, or zero.

An indefinite integral (sometimes called simply an **integral**) can be checked by differentiation: the derivative of the answer must equal the integrand (as is the case with $x^2 + C$ and $2x$).

---

**Indefinite Integral**

$$\int f(x)\, dx = g(x) + C \qquad \text{The integral of } f(x) \text{ is } g(x) + C$$

if and only if

$$g'(x) = f(x) \qquad \text{The derivative of } g(x) \text{ is } f(x)$$

---

## Integration Rules

There are several "rules" that simplify integration. The first, which is one of the most useful rules in all of calculus, shows how to integrate $x$ to a constant power. It comes from "reversing" the Power Rule for differentiation (see page 109).

---

**Power Rule for Integration**  *Brief Example*

$$\int x^n\, dx = \frac{1}{n+1} x^{n+1} + C \qquad (n \neq -1) \qquad \int x^3\, dx = \frac{1}{4}x^4 + C$$

---

In words: To integrate $x$ to a power, add 1 to the power and multiply by 1 over the new power (for any power other than $-1$).

 **BE CAREFUL:** To avoid mistakes, you should always check your answer by differentiating it to be sure you get back the original integrand. In this case, differentiating the answer $\frac{1}{4}x^4 + C$ *does* give the original integrand $x^3$, so the answer is correct.

**Practice Problem 1**   Find $\int x^2\, dx$ and check your answer by differentiation.

➤ **Solution on page 332**

The proof of the Power Rule for Integration consists simply of differentiating the right-hand side.

$$\frac{d}{dx}\left( \frac{1}{n+1} x^{n+1} + C \right) = \frac{1}{n+1}(n+1)x^n = x^n \qquad (n+1)\text{'s cancel}$$

The power $n + 1$ brought down ⟶     ⌞The power decreased by 1     ⌝Simplified

Since the derivative is the integrand $x^n$, the Power Rule for Integration is correct.

To integrate functions like $\sqrt{x}$ and $\dfrac{1}{x^2}$ we first express them as powers.

---

**EXAMPLE 1**

**EXPRESSING AS A POWER BEFORE INTEGRATING**

$$\int \sqrt{x}\, dx = \int x^{1/2}\, dx = \frac{2}{3} x^{3/2} + C \qquad \text{Using the Power Rule for Integration with } n = 1/2$$

$$n = \frac{1}{2}$$

$$n + 1 = \frac{1}{2} + 1 = \frac{3}{2}$$

$$\frac{1}{n+1} = \frac{1}{3/2} = \frac{2}{3}$$

---

Note that in the answer, the multiple $\frac{2}{3}$ and the exponent $\frac{3}{2}$ are reciprocals of each other. Can you see, from the Power Rule, why this will always be so?

---

**EXAMPLE 2**

**EXPRESSING AS A POWER BEFORE INTEGRATING**

$$\int \frac{1}{x^2}\, dx = \int x^{-2}\, dx = \frac{1}{-1} x^{-1} + C = -x^{-1} + C$$

$$n = -2$$

$$n + 1 = -1$$

Simplified answer

---

**Practice Problem 2**

Find $\displaystyle\int \frac{dx}{x^3}$. $\left[ \textit{Hint: } \text{Equivalent to } \displaystyle\int \frac{1}{x^3}\, dx. \right]$    ▶ **Solution on page 332**

---

**EXAMPLE 3**

**INTEGRATING 1**

$$\int 1\, dx = \int x^0\, dx = \frac{1}{1} x^1 + C = x + C \qquad \text{Using } x^0 = 1$$

$$n = 0 \qquad n + 1 = 1$$

---

This result is so useful that it should be memorized.

$$\int 1\, dx = x + C \qquad \text{The integral of 1 is } x + C$$

**EXAMPLE 4**

**INTEGRATING WITH OTHER VARIABLES**

**a.** $\displaystyle\int t^3 \, dt = \frac{1}{4} t^4 + C$

Using the Power Rule ($n = 3$) with $dt$ since the variable is $t$ ("integrating with respect to $t$")

**b.** $\displaystyle\int u^{-4} \, du = \frac{1}{-3} u^{-3} + C = -\frac{1}{3} u^{-3} + C$

Using the Power Rule ($n = -4$) with $du$ since the variable is $u$ ("integrating with respect to $u$")

**Practice Problem 3**

Find $\displaystyle\int z^{-1/2} \, dz$.

▶ Solution on page 332

 **BE CAREFUL:** Notice what happens if we try to integrate $x^{-1}$:

$$\int x^{-1} \, dx = \frac{1}{0} x^0 + C$$

Undefined because of the $\frac{1}{0}$

$n = -1$ ↑            $n + 1 = 0$

The Power Rule for Integration fails for the exponent $-1$ because it leads to the undefined expression $\frac{1}{0}$. For this reason, the Power Rule for Integration includes the restriction "$n \neq -1$." It can integrate any power of $x$ *except* $x^{-1}$. (In the next section we will see how to integrate $x^{-1}$.)

The Constant Multiple and Sum Rules for differentiation (pages 111 and 112) lead immediately to analogous rules for simplifying integrals. The first rule says that the sum of two functions may be integrated one at a time.

**Sum Rule for Integration**

$$\int [f(x) + g(x)] \, dx = \int f(x) \, dx + \int g(x) \, dx$$

The integral of a sum is the sum of the integrals

**EXAMPLE 5**

**USING THE SUM RULE**

$$\int (x^2 + x^3) \, dx = \int x^2 \, dx + \int x^3 \, dx$$

Using the Sum Rule to break the integral into two integrals

$$= \frac{1}{3} x^3 + \frac{1}{4} x^4 + C$$

Using the Power Rule on each (one "$+C$" is enough)

The second rule says that a constant may be moved across the integral sign.

### Constant Multiple Rule for Integration

For any constant $k$,

$$\int k \cdot f(x)\, dx = k \int f(x)\, dx$$

The integral of a constant times a function is the constant times the integral of the function

**EXAMPLE 6**

### USING THE CONSTANT MULTIPLE RULE

$$\int 6x^2\, dx = 6 \int x^2\, dx$$

Using the Constant Multiple Rule to move the 6 across the integral sign

$$= 6 \cdot \frac{1}{3} x^3 + C$$

Using the Power Rule with $n = 2$

$$= 2x^3 + C$$

Simplifying

**EXAMPLE 7**

### INTEGRATING A CONSTANT

$$\int 7\, dx = 7 \int 1\, dx = 7x + C$$

Moving the constant outside

The integral of 1 is $x$ (plus $C$)

This leads to a very useful general rule. For any constant $k$,

### Integral of a Constant

$$\int k\, dx = kx + C$$

The integral of a constant is the constant times $x$ (plus $C$)

The Sum Rule can be extended to integrate the sum or difference of *any* number of terms, writing only one $+C$ at the end (since any number of arbitrary constants can be added together to give just one).

**EXAMPLE 8**

### INTEGRATING A SUM OF POWERS

$$\int (6x^2 - 3x^{-2} + 5)\, dx$$

$$= 6 \int x^2\, dx - 3 \int x^{-2}\, dx + 5 \int 1\, dx$$

Breaking up the integral and moving constants outside

$$= 6 \cdot \frac{1}{3} x^3 - 3 \cdot \frac{1}{-1} x^{-1} + 5x + C$$

Integrating each separately

From integrating the 1

$$= 2x^3 + 3x^{-1} + 5x + C$$

Simplifying

**EXAMPLE 9**

**REWRITING BEFORE INTEGRATING**

$$\int \left( \frac{3\sqrt{x}}{2} - \frac{2}{\sqrt{x}} \right) dx = \int \left( \frac{3}{2} x^{1/2} - 2x^{-1/2} \right) dx \qquad \text{Writing as powers of } x$$

$$= \frac{3}{2} \cdot \frac{2}{3} x^{3/2} - 2 \cdot \frac{2}{1} x^{1/2} + C \qquad \text{Integrating each term separately}$$

$$= x^{3/2} - 4x^{1/2} + C \qquad \text{Simplifying}$$

**Practice Problem 4**

Find $\int \left( \sqrt[3]{w} - \frac{4}{w^3} \right) dw.$

▶ Solution on page 332

Some integrals are so simple that they can be integrated "at sight."

**EXAMPLE 10**

**INTEGRATING "AT SIGHT"**

**a.** $\int 4x^3 \, dx = x^4 + C$      By remembering that $4x^3$ is the derivative of $x^4$

**b.** $\int 7x^6 \, dx = x^7 + C$      By remembering that $7x^6$ is the derivative of $x^7$

**Practice Problem 5**

Integrate "at sight" by noticing that each integrand is of the form $nx^{n-1}$ and integrating to $x^n$ without working through the Power Rule.

**a.** $\int 5x^4 \, dx$      **b.** $\int 3x^2 \, dx$

▶ Solutions on page 332

## Algebraic Simplification of Integrals

Sometimes an integrand needs to be multiplied out or otherwise rewritten before it can be integrated.

**EXAMPLE 11**

**EXPANDING BEFORE INTEGRATING**

Find $\int x^2(x + 6)^2 \, dx.$

*Solution*

$$\int x^2(x + 6)^2 \, dx = \int x^2 \underbrace{(x^2 + 12x + 36)}_{(x + 6)^2} \, dx \qquad \text{"Squaring out"} \\ \text{the } (x + 6)^2$$

$$= \int (x^4 + 12x^3 + 36x^2) \, dx \qquad \text{Multiplying out}$$

$$= \frac{1}{5}x^5 + 12 \cdot \frac{1}{4}x^4 + 36 \cdot \frac{1}{3}x^3 + C \qquad \begin{array}{l}\text{Integrating each} \\ \text{term separately}\end{array}$$

$$= \frac{1}{5}x^5 + 3x^4 + 12x^3 + C \qquad \text{Simplifying}$$

---

**Practice Problem 6**    Find $\displaystyle\int \frac{6t^2 - t}{t} \, dt.$      [*Hint:* First simplify the integrand.]

➤ **Solution on page 332**

---

Since differentiation turns a cost function into a marginal cost function, integration turns a marginal cost function back into a cost function. To evaluate the constant, however, we need the fixed costs.

**EXAMPLE 12**     **RECOVERING COST FROM MARGINAL COST**

A company's marginal cost function is $MC(x) = 6\sqrt{x}$ and the fixed cost is $1000. Find the cost function.

*Solution*

We integrate the marginal cost to find the cost function.

$$C(x) = \int MC(x) \, dx = \int 6\sqrt{x} \, dx = 6 \int x^{1/2} \, dx \qquad \text{Integrating}$$

$$= 6 \cdot \frac{2}{3}x^{3/2} + K = 4x^{3/2} + K \qquad \begin{array}{l}\text{From the Power Rule} \\ \text{(using } K \text{ to avoid} \\ \text{confusion with } C \\ \text{for cost)}\end{array}$$

It remains to find the value of the constant $K$. The cost function evaluated at $x = 0$ always gives the fixed cost (because when nothing is produced, only the fixed cost remains):

$$\underbrace{C(0) = K}_{\text{Fixed cost}} \qquad \begin{array}{l}\text{Evaluating } C(x) = 4x^{3/2} + K \\ \text{at } x = 0 \text{ gives } K\end{array}$$

Therefore, $K$ equals the fixed cost, which is given as 1000, so $K = 1000$. The completed cost function is obtained by replacing $K$ by 1000.

$$C(x) = 4x^{3/2} + 1000 \qquad \begin{array}{l}C(x) = 4x^{3/2} + K \\ \text{with } K = 1000\end{array}$$

**BE CAREFUL:** In the preceding example, the arbitrary constant turned out to be the fixed cost. This will not always be so, as we will see in the next section.

We found the cost function in two steps: integrating the marginal cost and then evaluating the arbitrary constant. Other problems that involve recovering a quantity from its *marginal* (or rate of change) together with a particular value are solved in the same way:

1. Integrate the *marginal* or *rate of change*.
2. Use the given fixed value to evaluate the arbitrary constant.

For example, given the rate of growth of an economy and its present size, we can find the size of the economy at any time in the future.

**EXAMPLE 13**  **RECOVERING A QUANTITY FROM ITS RATE OF CHANGE**

The gross domestic product (GDP) of a country is $78 billion and growing at the rate of $4.4t^{-1/3}$ billion dollars per year after $t$ years. Find a formula for the GDP after $t$ years. Then use your formula to find the GDP after 8 years.

*Solution*

$$G(t) = \int 4.4t^{-1/3}\, dt = 4.4 \int t^{-1/3}\, dt \qquad \text{Integrating the rate of change}$$

$$= 4.4 \left(\frac{3}{2}\right) t^{2/3} + C = 6.6t^{2/3} + C \qquad \begin{array}{l}\text{Using the Power Rule and}\\ \text{then simplifying}\end{array}$$

As before, evaluating $G(t) = 6.6t^{2/3} + C$ at $t = 0$ gives $G(0) = C$, which shows that the constant $C$ is the GDP at time $t = 0$, which is given as $78 billion. Therefore

$$G(t) = 6.6t^{2/3} + 78 \qquad G(t) = 6.6t^{2/3} + C \text{ with } C = 78$$

For the GDP after 8 years, we evaluate $G(t)$ at $t = 8$:

$$G(8) = 6.6 \cdot \underbrace{8^{2/3}}_{4} + 78 = 6.6(4) + 78 = 104.4 \qquad \begin{array}{l}G(t) = 6.6t^{2/3} + 78\\ \text{at } t = 8\end{array}$$

Therefore, the GDP after 8 years is $104.4 billion.

## Geometrical Interpretation of the Arbitrary Constant

Notice that the $+C$, which at first may have seemed like a pointless formality, enabled us to give the cost function its correct fixed cost in Example 12 and to give the GDP its correct current value in Example 13.

The arbitrary constant can be interpreted *geometrically*. In $\int 2x\, dx = x^2 + C$, the solution $x^2 + C$ is actually a whole *collection* of curves, one for

$y = x^2 + C$ with
tangent lines at $x = 1$

each value of C. The graph on the left shows three of these curves, $x^2 + 2$, $x^2$, and $x^2 - 2$, corresponding to C being 2, 0, and $-2$. Notice that each curve has the same slope at $x = 1$ (their tangent lines are parallel). In fact, all three curves have the same slope at *any* given x-value (we call these *parallel* curves). This is because each comes from integrating 2x, so each function has *derivative* 2x, giving each the same slope. This is the geometric meaning of integrating a function: the result is a collection of parallel curves that all have the same slope (given by the original function) at any particular x-value.

## Integrals as Continuous Sums

We have seen that differentiation is a kind of "breaking down," changing total cost into marginal (per unit) cost. Integration, the reverse process, is therefore a kind of "adding up," recovering the total from its parts. For example, integration recovers the total cost from marginal (per unit) cost, and integration adds up the growth of an economy over several years to give its total size.

To put this another way, ordinary addition is for summing "chunks," such as $2 plus $3 equals $5, whereas integration is for summing *continuous* change, such as the slow but steady growth of an economy over time. In other words:

> Integration is continuous summation.

## Solutions to Practice Problems

**1.** $\int x^2 \, dx = \dfrac{1}{3}x^3 + C$ 　　　　　 Check: $\dfrac{d}{dx}\left(\dfrac{1}{3}x^3 + C\right) = x^2$

**2.** $\int \dfrac{1}{x^3} \, dx = \int x^{-3} \, dx = -\tfrac{1}{2}x^{-2} + C$

**3.** $\int z^{-1/2} \, dz = \tfrac{1}{1/2}z^{1/2} + C = 2z^{1/2} + C$

**4.** $\int (w^{1/3} - 4w^{-3}) \, dw = \int w^{1/3} \, dw - 4\int w^{-3} \, dw$

$$= \tfrac{3}{4}w^{4/3} - 4\left(-\tfrac{1}{2}\right)w^{-2} + C$$

$$= \tfrac{3}{4}w^{4/3} + 2w^{-2} + C$$

**5. a.** $\int 5x^4 \, dx = x^5 + C$ 　　　　　 **b.** $\int 3x^2 \, dx = x^3 + C$

**6.** $\int \dfrac{6t^2 - t}{t} \, dt = \int \dfrac{t(6t - 1)}{t} \, dt$

$$= \int (6t - 1) \, dt$$

$$= 6\int t \, dt - \int 1 \, dt = 6 \cdot \tfrac{1}{2}t^2 - t + C$$

$$= 3t^2 - t + C$$

## 5.1   Section Summary

In this section we have discussed both the *techniques* of integration and the *meanings* or *uses* of integration.

On the *technical* side, we defined indefinite integration as the reverse process of differentiation. The Power Rule enables us to integrate powers.

$$\int x^n \, dx = \frac{1}{n+1} x^{n+1} + C \qquad n \neq -1$$

(Increasing the exponent by 1 should seem quite reasonable: differentiation *lowers* the exponent by 1, so integration, the reverse process, should *raise* it by 1.) The Sum and Constant Multiple Rules extended integration to more complicated functions such as polynomials. Sometimes algebra is necessary to express a function in power form.

 **BE CAREFUL:** Don't forget the C.

As for the *uses* of integration, it recovers cost, revenue, and profit from their marginals, and in general it recovers any quantity from its rate of change (together with a particular value). The inverse relationship between integration and differentiation is shown in the following diagram.

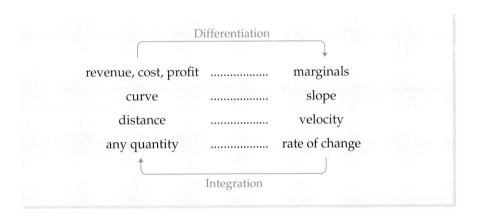

## 5.1   Exercises

**1–40.** Find each indefinite integral.

**1.** $\int x^4 \, dx$    **2.** $\int x^7 \, dx$    **3.** $\int x^{2/3} \, dx$    **10.** $\int \frac{dz}{\sqrt[3]{z}}$    **11.** $\int 6x^5 \, dx$    **12.** $\int 9x^8 \, dx$

**4.** $\int x^{3/2} \, dx$    **5.** $\int \sqrt{u} \, du$    **6.** $\int \sqrt[3]{u} \, du$    **13.** $\int (8x - 5) \, dx$    **14.** $\int (2 - 4x) \, dx$

**7.** $\int \frac{dw}{w^4}$    **8.** $\int \frac{dw}{w^2}$    **9.** $\int \frac{dz}{\sqrt{z}}$    **15.** $\int (8x^3 - 3x^2 + 2) \, dx$

**16.** $\int (12x^3 + 3x^2 - 5)\, dx$

**17.** $\int \left(6\sqrt{x} + \dfrac{1}{\sqrt[3]{x}}\right) dx$   **18.** $\int \left(3\sqrt{x} + \dfrac{1}{\sqrt{x}}\right) dx$

**19.** $\int \left(16\sqrt[3]{x^5} - \dfrac{16}{\sqrt[3]{x^5}}\right) dx$

**20.** $\int \left(14\sqrt[4]{x^3} - \dfrac{3}{\sqrt[4]{x^3}}\right) dx$

**21.** $\int \left(10\sqrt[3]{t^2} + \dfrac{1}{\sqrt[3]{t^2}}\right) dt$   **22.** $\int \left(21\sqrt{t^5} + \dfrac{6}{\sqrt{t^5}}\right) dt$

**23.** $\int \left(\dfrac{4}{z^3} + \dfrac{1}{\sqrt{z}}\right) dz$   **24.** $\int \left(\dfrac{1}{z^2} + \dfrac{1}{\sqrt[3]{z}}\right) dz$

**25.** $\int (x - 1)^2\, dx$   **26.** $\int (x + 2)^2\, dx$

**27.** $\int 12x^2\,(x - 1)\, dx$   **28.** $\int x^2\,(8x + 3)\, dx$

**29.** $\int (1 + 10w)\,\sqrt{w}\, dw$

**30.** $\int (1 - 7w)\,\sqrt[3]{w}\, dw$

**31.** $\int \dfrac{6x^3 - 6x^2 + x}{x}\, dx$

**32.** $\int \dfrac{4x^4 + 4x^2 - x}{x}\, dx$

**33.** $\int (x - 2)(x + 4)\, dx$

**34.** $\int (x + 5)(x - 3)\, dx$

**35.** $\int (r - 1)(r + 1)\, dr$

**36.** $\int (3s + 1)(3s - 1)\, ds$

**37.** $\int \dfrac{x^2 - 1}{x + 1}\, dx$   **38.** $\int \dfrac{x^2 - 1}{x - 1}\, dx$

**39.** $\int (t + 1)^3\, dt$   **40.** $\int (t - 1)^3\, dt$

**41. a.** Verify that $\int x^2\, dx = \frac{1}{3}x^3 + C$.
  **b.** Graph the five functions $\frac{1}{3}x^3 - 2$, $\frac{1}{3}x^3 - 1$, $\frac{1}{3}x^3$, $\frac{1}{3}x^3 + 1$, and $\frac{1}{3}x^3 + 2$ (the solutions for five different values of $C$) on the window $[-3, 3]$ by $[-5, 5]$. Use TRACE to see how the constant shifts the curve vertically.
  **c.** Find the slopes (using NDERIV or $dy/dx$) of several of the curves at a particular $x$-value and check that in each case the slope is the square of the $x$-value. This verifies that the derivative of each curve is $x^2$, and so each is an integral of $x^2$.

 **42. a.** Graph the five functions $\ln x - 2$, $\ln x - 1$, $\ln x$, $\ln x + 1$, and $\ln x + 2$ on the window $[0, 4]$ by $[-3, 3]$.
  **b.** Find the slope (using NDERIV or $dy/dx$) of several of the curves at a particular $x$-value and check that in each case the slope is the reciprocal of the $x$-value. This suggests that the derivative of each function is $1/x$.
  **c.** Based on part (b), conjecture what is the indefinite integral of the function $1/x$ (for $x > 0$).

---

### Applied Exercises

**43. BUSINESS: Cost** A company's marginal cost function is $MC = 20x^{3/2} - 15x^{2/3} + 1$, where $x$ is the number of units, and fixed costs are $4000. Find the cost function.

**44. BUSINESS: Cost** A company's marginal cost function is $MC = 21x^{4/3} - 6x^{1/2} + 50$, where $x$ is the number of units, and fixed costs are $3000. Find the cost function.

**45. BUSINESS: Revenue** A company's marginal revenue function is $MR = 12\sqrt[3]{x} + 3\sqrt{x}$, where $x$ is the number of units. Find the revenue function. (Evaluate $C$ so that revenue is zero when nothing is produced.)

**46. BUSINESS: Revenue** A company's marginal revenue function is $MR = 15\sqrt{x} + 4\sqrt[3]{x}$, where $x$ is the number of units. Find the revenue function. (Evaluate $C$ so that revenue is zero when nothing is produced.)

**47. GENERAL: Velocity** A Porsche 997 Turbo Cabriolet can accelerate from a standing start to a speed of $v(t) = -0.24t^2 + 18t$ feet per second after $t$ seconds (for $0 \le t < 40$).
*Source: Porsche*
  **a.** Find a formula for the distance that it will travel from its starting point in the first $t$ seconds. [*Hint:* Integrate velocity to find distance,

and then use the fact that distance is 0 at time $t = 0$.]

**b.** Use the formula that you found in part (a) to find the distance that the car will travel in the first 10 seconds.

**48. GENERAL: Velocity** A Mercedes-Benz Brabus Rocket can accelerate from a standing start to a speed of $v(t) = -0.15t^2 + 14t$ feet per second after $t$ seconds (for $0 \leq t < 40$).

*Source: Mercedes-Benz*

**a.** Find a formula for the distance that it will travel from its starting point in the first $t$ seconds. [*Hint:* Integrate velocity to find distance, and then use the fact that distance is 0 at time $t = 0$.]

**b.** Use the formula that you found in part (a) to find the distance that the car will travel in the first 10 seconds.

**49. GENERAL: Learning** A person can memorize words at the rate of $3/\sqrt{t}$ words per minute.

**a.** Find a formula for the total number of words that can be memorized in $t$ minutes. [*Hint:* Evaluate $C$ so that 0 words have been memorized at time $t = 0$.]

**b.** Use the formula that you found in part (a) to find the total number of words that can be memorized in 25 minutes.

**50. BIOMEDICAL: Temperature** A patient's temperature is 104 degrees Fahrenheit and is changing at the rate of $t^2 - 3t$ degrees per hour, where $t$ is the number of hours since taking a fever-reducing medication $(0 \leq t \leq 3)$.

**a.** Find a formula for the patient's temperature after $t$ hours. [*Hint:* Evaluate the constant $C$ so that the temperature is 104 at time $t = 0$.]

**b.** Use the formula that you found in part (a) to find the patient's temperature after 3 hours.

**51. ENVIRONMENTAL SCIENCE: Pollution** A chemical plant is adding pollution to a lake at the rate of $40\sqrt{t^3}$ tons per year, where $t$ is the number of years that the plant has been in operation.

**a.** Find a formula for the total amount of pollution that will enter the lake in the first $t$ years of the plant's operation. [*Hint:* Evaluate $C$ so that no pollution has been added at time $t = 0$.]

**b.** Use the formula that you found in part (a) to find how much pollution will enter the lake in the first 4 years of the plant's operation.

**c.** If all life in the lake will cease when 400 tons of pollution have entered the lake, will the lake "live" beyond 4 years?

**52. BUSINESS: Appreciation** A $20,000 art collection is increasing in value at the rate of $300\sqrt{t}$ dollars per year after $t$ years.

**a.** Find a formula for its value after $t$ years. [*Hint:* Evaluate $C$ so that its value at time $t = 0$ is $20,000.]

**b.** Use the formula that you found in part (a) to find its value after 25 years.

**53. GENERAL: Young-Adult Population** The U.S. Census Bureau estimates that the number of young adults (ages 18 to 24) in the United States to be 31,000 (in thousands) in 2010 and changing at the rate of $-x^2 + 50x - 300$ thousand per year, where $x$ is the number of years since 2010. Find a formula for the size of this population at any time $x$ and use your formula to predict the size of the young-adult population in 2040. [*Hint:* Keep all calculations in units of *thousands.*]

*Source: 2007 Statistical Abstract of the United States*

**54. ENVIRONMENTAL SCIENCE: Global Temperatures** In 2000 the average surface temperature of the earth was 57.8 (degrees Fahrenheit) and, according to one study, increasing at the rate of $0.014t^{0.4}$ degrees per year, where $x$ is the number of years since 2000. Find a formula for the temperature at any time $t$ and use your formula to predict the temperature in 2100. [*Note:* Such global warming is expected to raise sea levels and cause widespread coastal flooding.]

*Source: Intergovernmental Panel on Climate Change, 2007 Synthesis Report*

**55. BUSINESS: Income** In 2000 AT&T's net income was 4200 million dollars and changing at the rate of $800x^2 - 4800x + 5400$ million dollars per year, where $x$ is the number of years since 2000 (for $x \leq 6$). Find a formula for AT&T's net income at any time $x$ and use your formula to find their income in 2006.

*Source: Standard & Poor's*

**56. BUSINESS: Income** In 2003 Microsoft's net income was 9 billion dollars and changing at the rate of $-4.8x^2 + 14.4x - 6$ billion dollars per year, where $x$ stands for the number of years since 2003 (for $x \leq 3$). Find a formula for Microsoft's net income at any time $x$ and use your formula to find their income in 2006.

*Source: Standard & Poor's*

### Review Exercise

This exercise will be important in the next section.

**57.** Find $\dfrac{d}{dx}\ln(-x)$.

### Conceptual Exercises

**58.** What power of $x$ *cannot* be integrated by the Power Rule?

**59.** What is the integral of *zero*? $\int 0\,dx = ?$

**60.** One student tells another: "Once you've got one antiderivative, you've got them all." What did the student mean?

**61.** If $f(x)$ and $g(x)$ have the same derivative, how are $f(x)$ and $g(x)$ related?

**62.** Show that the integral of a product is *not* the product of the integrals by carrying out the following steps:

   **a.** Find the integral of the product $\int x \cdot x\,dx$ by evaluating $\int x^2\,dx$.

   **b.** Find the corresponding product of the integrals $\int x\,dx \cdot \int x\,dx$.

   **c.** Do the answers for parts (a) and (b) agree?

**63.** You can move a *constant* across the integral sign, but can you do the same with a *variable*? Find out by carrying out the following steps:

   **a.** Find $\int x \cdot x\,dx$ by evaluating $\int x^2\,dx$.

   **b.** Find $x \cdot \int x\,dx$.

   **c.** Do the answers for parts (a) and (b) agree?

**64.** Integrating $x^2$, you get $\frac{1}{3}x^3 + C$, and then differentiating $\frac{1}{3}x^3 + C$ gives back exactly the $x^2$ you began with. What if you begin with $x^2$ but *first* differentiate and *then* integrate—do you get back exactly what you began with?

**65.** Show that the integral of a quotient is *not* the quotient of the integrals by carrying out the following steps:

   **a.** Find the integral of the quotient $\displaystyle\int \frac{x}{x}\,dx$ by evaluating $\int 1\,dx$.

   **b.** Find the corresponding quotient of the integrals $\dfrac{\int x\,dx}{\int x\,dx}$.

   **c.** Do the answers for parts (a) and (b) agree?

**66.** Supply the most appropriate single word: Integrating the *rate of change* of a population gives the _____.

**67.** Supply the most appropriate single word: Integrating the *marginal cost* gives the _____.

**68.** For each of the following casual, everyday sentences, fill in the blank with either the word *differentiate* or the word *integrate*, whichever is more appropriate.

   **a.** To find the value of an investment, we _____ how fast it is growing.

   **b.** To find how quickly it is getting warmer, we _____ temperature.

   **c.** To find distance, we _____ speed.

   **d.** To find acceleration, we _____ speed.

**69.** For each of the following casual, everyday sentences, fill in the blank with one of the words *differentiate* or *integrate*, whichever is more appropriate.

   **a.** To find marginal cost, we _____ cost.

   **b.** To find profit, we _____ marginal profit.

   **c.** To find how fast a stock price is rising, we _____ price.

   **d.** To find a population, we _____ how fast it is growing.

---

## 5.2   INTEGRATION USING LOGARITHMIC AND EXPONENTIAL FUNCTIONS

### Introduction

In the previous section we defined integration as the reverse of differentiation, and we introduced several integration formulas. In this section we develop integration formulas involving logarithmic and exponential functions. One of these formulas will answer a question that we could not answer earlier—namely, how to integrate $x^{-1}$, the only power not covered by the Power Rule.

## The Integral $\int e^{ax}\, dx$

On page 295 we saw that to *differentiate* $e^{ax}$, we *multiply* by $a$ to get $ae^{ax}$. Therefore, to *integrate* $e^{ax}$, the reverse process, we must *divide* by $a$. That is, for any $a \neq 0$:

| Integrating an Exponential Function | *Brief Example* |
|---|---|
| $$\int e^{ax}\, dx = \frac{1}{a} e^{ax} + C$$ | $$\int e^{2x}\, dx = \frac{1}{2} e^{2x} + C$$ |

In words: The integral of $e$ to a constant times $x$ is 1 over the constant times the original function (plus $C$).

As always, we may check the answer by differentiation.

$$\frac{d}{dx}\left(\frac{1}{2} e^{2x} + C\right) = \frac{1}{2} \cdot 2 \cdot e^{2x} = e^{2x}$$

Using $\frac{d}{dx} e^{ax} = ae^{ax}$ and canceling the 2's

The result is the integrand $e^{2x}$, so the integration is correct.

The proof of this rule consists simply of differentiating the right-hand side.

$$\frac{d}{dx}\left(\frac{1}{a} e^{ax} + C\right) = \frac{1}{a} ae^{ax} = e^{ax}$$

Using $\frac{d}{dx} e^{ax} = ae^{ax}$ and canceling the $a$'s

The result is the integrand, so the integration formula is correct.

⚠ **BE CAREFUL:** To use the formula in the preceding box, the exponent of $e$ cannot be more complicated than a constant times $x$. In Section 5.6 we will integrate more general exponential expressions.

**EXAMPLE 1** **INTEGRATING EXPONENTIAL FUNCTIONS**

**a.** $\displaystyle\int e^{\frac{1}{2}x}\, dx = 2e^{\frac{1}{2}x} + C$   $a = \frac{1}{2}$   so   $\frac{1}{a} = \frac{1}{\frac{1}{2}} = 2$

$\quad\quad 1/a$

**b.** $\displaystyle\int 6e^{-3x}\, dx = 6\int e^{-3x}\, dx = 6\left(-\frac{1}{3}\right)\cdot e^{-3x} + C = -2e^{-3x} + C$

$\quad\quad\quad\quad\quad a = -3 \quad\quad \frac{1}{a} = -\frac{1}{3}$

**c.** $\displaystyle\int e^{x}\, dx = 1e^{x} + C = e^{x} + C$   $a = 1,$   so   $\frac{1}{a} = 1$

Each of these answers may be checked by differentiation. The last one says that $e^x$ is the integral of itself, just as $e^x$ is the derivative of itself.

$$\int e^x \, dx = e^x + C \qquad \text{The integral of } e^x \text{ is } e^x \text{ (plus } C\text{)}$$

**Practice Problem 1**

Find: **a.** $\displaystyle\int 12e^{4x}\, dx$ **b.** $\displaystyle\int e^{x/3}\, dx$

[*Hint:* For part (b), write $e^{x/3}$ as $e^{\frac{1}{3}x}$, then integrate, and rewrite again.]

➤ **Solutions on page 344**

 **BE CAREFUL:** When using these new integration formulas to solve applied problems, be careful to evaluate the constant $C$ correctly. It will not always be equal to the initial value of the function.

**EXAMPLE 2**  **FINDING TOTAL FLU CASES FROM THE RATE OF CHANGE**

An influenza epidemic hits a large city and spreads at the rate of $12e^{0.2t}$ new cases per day, where $t$ is the number of days since the epidemic began. The epidemic began with 4 cases.

**a.** Find a formula for the total number of flu cases in the first $t$ days of the epidemic.

**b.** Use your formula to find the number of cases during the first 30 days.

*Solution*

**a.** To find the total number of cases, we integrate the growth rate $12e^{0.2t}$.

$$\underbrace{f(t)}_{\substack{\text{Total cases}\\ \text{in first } t \text{ days}}} = \int 12e^{0.2t}\, dt = 12 \int e^{0.2t}\, dt \qquad \text{Taking out the constant}$$

$$= 12\,\underset{\underset{5}{\uparrow}}{\frac{1}{0.2}}\, e^{0.2t} + C = 60e^{0.2t} + C \qquad \text{Using the } \int e^{ax}\, dx \text{ formula}$$

Evaluating $f(t)$ at $t = 0$ must give the initial number of cases:

$$f(0) = 60\underbrace{e^{0.2(0)}}_{\substack{e^0 = 1}} + C = \underbrace{60 + C}_{}$$

$$\underbrace{\phantom{f(0)}}_{\substack{\text{Initial}\\ \text{number}\\ \text{of cases}}}$$

$f(t) = 60e^{0.2t} + C$ evaluated at $t = 0$ (the beginning of the epidemic)

This initial number must equal the given initial number, 4:

$$60 + C = 4$$  Initial number from the formula set equal to the given initial number

$$C = -56$$  Solving for $C$

Replacing $C$ by $-56$ gives the total number of flu cases within $t$ days:

$$\underbrace{f(t)}_{\substack{\text{Number of cases} \\ \text{in first } t \text{ days}}} = \underbrace{60e^{0.2t} - 56}_{\substack{\text{Answer to} \\ \text{part (a)}}}$$  $f(t) = 60e^{0.2t} + C$ with $C = -56$

**b.** To find the number within 30 days, we evaluate at $t = 30$:

$$f(30) = 60e^{0.2(30)} - 56$$  $f(t) = 60e^{0.2t} + C$ at $t = 30$

$$= 60e^6 - 56 \approx 24{,}150$$  Using a calculator

Therefore, within 30 days the epidemic will have grown to more than 24,000 cases.

## Evaluating the Constant *C*

Notice that we did *not* simply replace the constant $C$ by the initial number of cases, 4. (This would have given the wrong initial number of cases.) Instead, we evaluated the function at the initial time $t = 0$ and set it equal to the initial number of cases:

$$\underbrace{60e^0 + C}_{\substack{f(t) \text{ evaluated} \\ \text{at } t = 0}} = \underbrace{4}_{\substack{\text{Given} \\ \text{initial value}}}$$

We then solved to find the correct value of $C$, $C = -56$, which we then substituted into the formula. In general:

To evaluate the constant $C$:

1. Evaluate the integral at the given number (usually $t = 0$) and set the result equal to the stated initial value.
2. Solve for $C$.
3. Write the answer with $C$ replaced by its correct value.

## The Integral $\displaystyle\int \frac{1}{x}\, dx$

The differentiation formula $\dfrac{d}{dx} \ln x = \dfrac{1}{x}$ can be read "backward" as an integration formula:

$$\int \frac{1}{x}\, dx = \ln x + C$$  The integral of 1 over $x$ is the natural log of $x$ (plus $C$)

This formula, however, is restricted to $x > 0$, for only then is $\ln x$ defined. For $x < 0$ we can differentiate $\ln(-x)$, giving

$$\frac{d}{dx}\ln(-x) = \frac{-1}{-x} = \frac{1}{x} \qquad \text{Using } \frac{d}{dx}\ln f = \frac{f'}{f} \text{ and simplifying}$$

This result says that for $x < 0$, the integral of $1/x$ is $\ln(-x)$. The negative sign in $\ln(-x)$ serves only to make the already negative $x$ positive, and this could be accomplished just as well with absolute value bars.

$$\int \frac{1}{x}\,dx = \ln|x| + C \qquad \text{The integral of 1 over } x \text{ is the natural logarithm of the absolute value of } x$$

This formula holds for negative *and* positive values of $x$, since in both cases $\ln|x|$ is defined. The integral can be written in three different ways, all of which have the same answer.

$$\int \frac{1}{x}\,dx \;=\; \int \frac{dx}{x} \;=\; \int x^{-1}\,dx \;=\; \ln|x| + C$$

---

**EXAMPLE 3**

**INTEGRATING USING THE LN RULE**

$$\int \frac{5}{2x}\,dx \;=\; \int \frac{5}{2}\frac{1}{x}\,dx \;=\; \frac{5}{2}\int \frac{1}{x}\,dx \;=\; \frac{5}{2}\ln|x| + C$$

$\uparrow$
Taking out
the constant

Using the
ln formula

---

**EXAMPLE 4**

**INTEGRATING NEGATIVE POWERS**

$$\int (x^{-1} + x^{-2})\,dx = \int x^{-1}\,dx + \int x^{-2}\,dx = \ln|x| - x^{-1} + C$$

From the
natural log
formula

From the
Power Rule
with $n = -2$

---

**Practice Problem 2**    Find $\displaystyle\int \frac{3}{4x}\,dx.$    ➤ Solution on page 344

---

**EXAMPLE 5**

**FINDING TOTAL SALES FROM THE SALES RATE**

An electronics dealer estimates that during month $t$ of a sale, a discontinued computer will sell at a rate of approximately $25/t$ per month, where $t = 1$ corresponds to the beginning of the sale, at which time none have been sold.

Find a formula for the total number of computers that will be sold up to month $t$. Will the store's inventory of 64 computers be sold by month $t = 12$?

*Solution*

To find the total sales, we integrate the sales rate $\dfrac{25}{t}$:

$$\underbrace{S(t)}_{\substack{\text{Total sales in} \\ \text{first } t \text{ months}}} = \int \frac{25}{t}\, dt = 25 \int \frac{1}{t}\, dt = 25 \ln t + C \qquad \begin{array}{l}\text{Omitting absolute} \\ \text{values, since } t > 0\end{array}$$

To evaluate $C$, we evaluate at the given starting time $t = 1$:

$$\underbrace{S(1)}_{\substack{\text{Initial number} \\ \text{of sales}}} = 25 \underbrace{\ln 1}_{0} + C = C$$

The initial number of sales must be zero, giving $C = 0$, and substituting this into the sales function gives

$$S(t) = 25 \ln t \qquad\qquad\qquad S(t) = 25 \ln t + C \text{ with } C = 0$$

$$\text{Total number sold up to month } t$$

To find the number sold up to month 12, we evaluate at $t = 12$:

$$S(12) = 25 \ln 12 \approx 62 \qquad\qquad \text{Using a calculator}$$

Therefore, all but two of the 64 computers will be sold by month 12.

In this example the initial time (the beginning of the sale) was given as $t = 1$ rather than the more usual $t = 0$. The initial time will be clear from the problem.

## Consumption of Natural Resources

Just as the world population grows exponentially, so does the world's annual consumption of natural resources. We can estimate the total consumption at any time in the future by integrating the rate of consumption, and from this predict when the known reserves will be exhausted.

EXAMPLE 6      **FINDING A FORMULA FOR TOTAL CONSUMPTION FROM THE RATE**

The annual world consumption of tin is predicted to be $0.26e^{0.01t}$ million metric tons per year, where $t$ is the number of years since 2008. Find a formula for the total tin consumption within $t$ years of 2008 and estimate when the known world reserves of 6.1 million metric tons will be exhausted.

*Source:* U.S. Geological Survey

*Solution*

To find the total consumption, we integrate the rate $0.26e^{0.01t}$.

$$\underbrace{C(t)}_{\substack{\text{Total tin consumed} \\ \text{in first } t \text{ years} \\ \text{after 2008}}} = \int 0.26e^{0.01t}\, dt = 0.26 \int e^{0.01t}\, dt \qquad \text{Taking out the constant}$$

$$= 0.26\, \underbrace{\frac{1}{0.01}}_{100}\, e^{0.01t} + C = 26e^{0.01t} + C \qquad \text{Using the } \int e^{ax}\, dx \text{ formula}$$

The total consumed in the first zero years must be zero, so $C(0) = 0$.

$$C(0) = 26\underbrace{e^0}_{0} + \underbrace{C = 26 + C}_{\substack{\uparrow \\ \text{Must equal zero}}} \qquad C(t) = 26e^{0.01t} + C \text{ with } t = 0$$

From $26 + C = 0$ we find $C = -26$. We substitute this into the formula for $C(t)$:

$$C(t) = \underbrace{26e^{0.01t} - 26}_{\substack{\uparrow \\ \text{Formula for total consumption} \\ \text{within first } t \text{ years since 2008}}} \qquad C(t) = 26e^{0.01t} + C \text{ with } C = -26$$

To predict when the total world reserves of 6.1 million metric tons will be exhausted, we set this function equal to 6.1 and solve for $t$:

$$26e^{0.01t} - 26 = 6.1 \qquad \text{Setting } C(t) \text{ equal to 6.1}$$

$$26e^{0.01t} = 32.1 \qquad \text{Adding 26}$$

$$e^{0.01t} = \frac{32.1}{26} \approx 1.235 \qquad \text{Dividing by 26}$$

$$\ln \underbrace{e^{0.01t}}_{0.01t} = \ln \underbrace{1.235}_{0.211} \qquad \begin{array}{l}\text{Taking natural logs} \\ \text{and using } \ln e^x = x\end{array}$$

$$t \approx \frac{0.211}{0.01} = 21.1 \qquad \text{Solving } 0.01t = 0.211 \text{ for } t$$

Therefore, the known world supply of tin will be exhausted in about 21 years after 2008, which means in about the year 2029 (assuming that consumption continues at the predicted rate).

Realistically, tin will not suddenly disappear at this date, but its price will rise significantly as this date approaches because of scarcity. The theoretical depletion date is an indicator as to when such price rises will occur, unless other resources are found.

## Spreadsheet Exploration

Since integration is continuous summation, instead of integrating we could add up the annual tin consumption for each year to see when the total will reach the known reserves of 6.1 million metric tons. The following spreadsheet* shows this annual consumption for each year (using the formula $0.26e^{0.01t}$ from the preceding example) in column **B** continuing into column **F**, with the cumulative totals shown in column **C** continuing into column **G**.

| | B2 | | = | =0.26*EXP(0.01*A2) | | |
|---|---|---|---|---|---|---|
| | A | B | C | D E | F | G |
| 1 | Year | Annual Consumption | Cumulative Consumption | Year | Annual Consumption | Cumulative Consumption |
| 2 | 1 | 0.26261304 | 0.26261304 | 13 | 0.29609538 | 3.62761574 |
| 3 | 2 | 0.26525235 | 0.52786539 | 14 | 0.29907119 | 3.92668692 |
| 4 | 3 | 0.26791818 | 0.79578357 | 15 | 0.30207690 | 4.22876383 |
| 5 | 4 | 0.27061080 | 1.06639437 | 16 | 0.30511283 | 4.53387665 |
| 6 | 5 | 0.27333049 | 1.33972486 | 17 | 0.30817926 | 4.84205591 |
| 7 | 6 | 0.27607750 | 1.61580236 | 18 | 0.31127651 | 5.15333243 |
| 8 | 7 | 0.27885213 | 1.89465449 | 19 | 0.31440490 | 5.46773732 |
| 9 | 8 | 0.28165464 | 2.17630912 | 20 | 0.31756472 | 5.78530204 |
| 10 | 9 | 0.28448531 | 2.46079444 | 21 | 0.32075630 | 6.10605834 |
| 11 | 10 | 0.28734444 | 2.74813888 | 22 | 0.32397995 | 6.43003829 |
| 12 | 11 | 0.29023230 | 3.03837117 | 23 | 0.32723600 | 6.75727429 |
| 13 | 12 | 0.29314918 | 3.33152036 | 24 | 0.33052478 | 7.08779907 |
| 14 | | | | | | |

Known reserves
exhausted

Since the cumulative consumption reaches 6.1 million metric tons in year 21, we conclude that the known world reserves of tin will be exhausted sometime in 2029, in agreement with Example 6.

The integration method of Example 6 has two advantages over addition: it is often easier (especially for a large number of years) and it includes changes in the amount *during* each year rather than just at the end of each year.

*To obtain this and other Spreadsheet Explorations, go to www.cengage.com/math/berresford.

Incidentally, tin is used mainly for coating steel (a "tin" can is actually a steel can with a thin protective coating of tin to prevent rust). The predicted unavailability of tin is already causing major changes in the food-packaging industry.

## Power Rule for Integration, Revisited

Our new integration formula shows how to integrate $x^{-1}$, the only power not covered by the Power Rule. Therefore, we can now write one "combined" formula for the integral $\int x^n \, dx$ for *any power n*.

**Integrals of Powers of $x$**

$$\int x^n \, dx = \begin{cases} \dfrac{1}{n+1}x^{n+1} + C & \text{if } n \neq -1 \\[2mm] \ln|x| + C & \text{if } n = -1 \end{cases}$$

Use the Power Rule if $n$ is *other* than $-1$

Use the ln formula if $n$ equals $-1$

It is a curious fact that every power of $x$ integrates to another power of $x$, with the single exception of $x^{-1}$, which integrates to an entirely different kind of function, the natural logarithm.

**Solutions to Practice Problems**

**1. a.** $\displaystyle\int 12e^{4x} \, dx = 12\int e^{4x} \, dx = 12 \cdot \tfrac{1}{4}e^{4x} + C = 3e^{4x} + C$

**b.** $\displaystyle\int e^{x/3} \, dx = \int e^{\frac{1}{3}x} \, dx = 3e^{\frac{1}{3}x} + C = 3e^{x/3} + C$

**2.** $\displaystyle\int \frac{3}{4x} \, dx = \frac{3}{4}\int \frac{1}{x} \, dx = \frac{3}{4}\ln|x| + C$

## 5.2 Section Summary

We have three integration formulas:

$$\int x^n \, dx = \frac{1}{n+1}x^{n+1} + C \qquad\qquad n \neq -1$$

$$\int e^{ax} \, dx = \frac{1}{a}e^{ax} + C \qquad\qquad a \neq 0$$

$$\int \frac{1}{x} \, dx = \int \frac{dx}{x} = \int x^{-1} \, dx = \ln|x| + C$$

What is the difference between the bottom formula *with* and *without* the absolute value bars?

$$\int \frac{1}{x} \, dx = \ln|x| + C \qquad \text{versus} \qquad \int \frac{1}{x} \, dx = \ln x + C$$

Both are correct, but the second holds only for *positive* $x$-values (so that $\ln x$ is defined; see page 275), while the first holds for *negative* as well as positive $x$-values. We will sometimes use the second when we know that $x$ is positive.

To evaluate $C$ in an application, we set the function (evaluated at the given number) equal to the stated initial value and solve for $C$.

## 5.2  Exercises

**1–34.** Find each indefinite integral.

**1.** $\displaystyle\int e^{3x}\,dx$

**2.** $\displaystyle\int e^{4x}\,dx$

**3.** $\displaystyle\int e^{x/4}\,dx$

**4.** $\displaystyle\int e^{x/3}\,dx$

**5.** $\displaystyle\int e^{0.05x}\,dx$

**6.** $\displaystyle\int e^{0.02x}\,dx$

**7.** $\displaystyle\int e^{-2y}\,dy$

**8.** $\displaystyle\int e^{-3y}\,dy$

**9.** $\displaystyle\int e^{-0.5x}\,dx$

**10.** $\displaystyle\int e^{-0.4x}\,dx$

**11.** $\displaystyle\int 6e^{2x/3}\,dx$

**12.** $\displaystyle\int 24e^{-2u/3}\,du$

**13.** $\displaystyle\int -5x^{-1}\,dx$

**14.** $\displaystyle\int -\frac{1}{2}x^{-1}\,dx$

**15.** $\displaystyle\int \frac{3\,dx}{x}$

**16.** $\displaystyle\int \frac{dx}{2x}$

**17.** $\displaystyle\int \frac{3}{2v}\,dv$

**18.** $\displaystyle\int \frac{2}{3v}\,dv$

**19.** $\displaystyle\int (4e^{2x}-6x)\,dx$

**20.** $\displaystyle\int (6e^{3x}+4x)\,dx$

**21.** $\displaystyle\int \left(e^{3x}-\frac{3}{x}\right)dx$

**22.** $\displaystyle\int \left(e^{2x}-\frac{2}{x}\right)dx$

**23.** $\displaystyle\int (3e^{0.5t}-2t^{-1})\,dt$

**24.** $\displaystyle\int (5e^{0.5t}-4t^{-1})\,dt$

**25.** $\displaystyle\int (x^2+x+1+x^{-1}+x^{-2})\,dx$

**26.** $\displaystyle\int (x^{-2}-x^{-1}+1-x+x^2)\,dx$

**27.** $\displaystyle\int (5e^{0.02t}-2e^{0.01t})\,dt$

**28.** $\displaystyle\int (3e^{0.05t}-2e^{0.04t})\,dt$

**29.** $\displaystyle\int \frac{e^{w}-w}{2}\,dw$

**30.** $\displaystyle\int \frac{e^{w}+w^2}{3}\,dw$

**31.** $\displaystyle\int \frac{z^3+z}{z^2}\,dz$

**32.** $\displaystyle\int \frac{z^2+1}{z}\,dz$

**33.** $\displaystyle\int \frac{xe^{x}+1}{x}\,dx$

**34.** $\displaystyle\int \frac{1}{x}(1-xe^{x})\,dx$

**35–40.** Find each indefinite integral. [*Hint:* Use some algebra first.]

**35.** $\displaystyle\int \frac{(x+1)^2}{x}\,dx$

**36.** $\displaystyle\int \frac{(x-1)^2}{x}\,dx$

**37.** $\displaystyle\int \frac{(t-1)(t+3)}{t^2}\,dt$

**38.** $\displaystyle\int \frac{(t+2)(t-4)}{t^2}\,dt$

**39.** $\displaystyle\int \frac{(x-2)^3}{x}\,dx$

**40.** $\displaystyle\int \frac{(x+2)^3}{x}\,dx$

## Applied Exercises

**41–42. BIOMEDICAL: Epidemics**  A flu epidemic hits a college community, beginning with five cases on day $t = 0$.  The rate of growth of the epidemic (new cases per day) is given by the following function $r(t)$, where $t$ is the number of days since the epidemic began.

**a.** Find a formula for the total number of cases of flu in the first $t$ days.

**b.** Use your answer to part (a) to find the total number of cases in the first 20 days.

**41.** $r(t) = 18e^{0.05t}$

**42.** $r(t) = 20e^{0.04t}$

**43. BUSINESS: Sales**  In an effort to reduce its inventory, a music store runs a sale on its least popular compact discs (CDs). The sales rate (CDs sold per day) on day $t$ of the sale is predicted to be $50/t$ (for $t \geq 1$), where $t = 1$ corresponds to the beginning of the sale, at which time none of the inventory of 200 CDs had been sold.

**a.** Find a formula for the total number of CDs sold up to day $t$.

**b.** Will the store have sold its inventory of 200 CDs by day $t = 30$?

**44. BUSINESS: Sales**  In an effort to reduce its inventory, a bookstore runs a sale on its least popular mathematics books. The sales rate (books sold per day) on day $t$ of the sale is predicted to be $60/t$ (for $t \geq 1$), where $t = 1$ corresponds to the beginning of the sale, at which time none of the inventory of 350 books had been sold.

    **a.** Find a formula for the number of books sold up to day $t$.

    **b.** Will the store have sold its inventory of 350 books by day $t = 30$?

**45. ENVIRONMENTAL SCIENCE: Consumption of Natural Resources** World consumption of silver is running at the rate of $20e^{0.02t}$ thousand metric tons per year, where $t$ is measured in years and $t = 0$ corresponds to 2008.

    **a.** Find a formula for the total amount of silver that will be consumed within $t$ years of 2008.

    **b.** When will the known world resources of 570 thousand metric tons of silver be exhausted? (Silver is used extensively in photography.)

    *Source:* U.S. Geological Survey

**46. ENVIRONMENTAL SCIENCE: Consumption of Natural Resources** World consumption of copper is running at the rate of $15e^{0.04t}$ million metric tons per year, where $t$ is measured in years and $t = 0$ corresponds to 2008.

    **a.** Find a formula for the total amount of copper that will be used within $t$ years of 2008.

    **b.** When will the known world resources of 940 million metric tons of copper be exhausted?

    *Source:* U.S. Geological Survey

**47. GENERAL: Cost of Maintaining a Home** The cost of maintaining a home generally increases as the home becomes older. Suppose that the rate of cost (dollars per year) for a home that is $x$ years old is $200e^{0.4x}$.

    **a.** Find a formula for the total maintenance cost during the first $x$ years. (Total maintenance should be zero at $x = 0$.)

    **b.** Use your answer to part (a) to find the total maintenance cost during the first 5 years.

**48. BIOMEDICAL: Cell Growth** A culture of bacteria is growing at the rate of $20e^{0.8t}$ cells per day, where $t$ is the number of days since the culture was started. Suppose that the culture began with 50 cells.

    **a.** Find a formula for the total number of cells in the culture after $t$ days.

    **b.** If the culture is to be stopped when the population reaches 500, when will this occur?

**49. GENERAL: Freezing of Ice** An ice cube tray filled with tap water is placed in the freezer, and the temperature of the water is changing at the rate of $-12e^{-0.2t}$ degrees Fahrenheit per hour after $t$ hours. The original temperature of the tap water was 70 degrees.

    **a.** Find a formula for the temperature of water that has been in the freezer for $t$ hours.

    **b.** When will the ice be ready? (Water freezes at 32 degrees.)

**50. SOCIAL SCIENCE: Divorces** The divorce rate in the United States has been declining in recent years. The number of divorces per year is predicted to be $1.07e^{-0.025t}$ million, where $t$ is the number of years since 2005.

    **a.** Find a formula for the total number of divorces within $t$ years of 2005.

    **b.** Use your formula to find the total number of divorces from 2005 to 2015.

    *Source:* National Center for Health Statistics

**51. BUSINESS: Total Savings** A factory installs new equipment that is expected to generate savings at the rate of $800e^{-0.2t}$ dollars per year, where $t$ is the number of years that the equipment has been in operation.

    **a.** Find a formula for the total savings that the equipment will generate during its first $t$ years.

    **b.** If the equipment originally cost $2000, when will it "pay for itself"?

**52. BUSINESS: Total Savings** A company installs a new computer that is expected to generate savings at the rate of $20{,}000e^{-0.02t}$ dollars per year, where $t$ is the number of years that the computer has been in operation.

    **a.** Find a formula for the total savings that the computer will generate during its first $t$ years.

    **b.** If the computer originally cost $250,000, when will it "pay for itself"?

**53. BUSINESS: Value of an Investment** A real estate investment, originally worth $5000, grows continuously at the rate of $400e^{0.05t}$ dollars per year, where $t$ is the number of years since the investment was made.

    **a.** Find a formula for the value of the investment after $t$ years.

    **b.** Use your formula to find the value of the investment after 10 years.

**54. BUSINESS: Value of an Investment** A biotechnology investment, originally worth $20,000, grows continuously at the rate of $1000e^{0.10t}$ dollars per year, where $t$ is the number of years since the investment was made.

a. Find a formula for the value of the investment after $t$ years.
b. Use your formula to find the value of the investment after 7 years.

 55. **GENERAL: Consumption of Natural Resources** World consumption of lead is running at the rate of $6.1e^{0.01t}$ million metric tons per year, where $t$ is measured in years and $t = 0$ corresponds to 2008.

a. Find a formula for the total amount of lead that will be consumed within $t$ years of 2008.
b. Use a graphing calculator to find when the world's known resources of 140 million metric tons of lead will be exhausted. [*Hint:* Use INTERSECT.] Lead has many uses, from batteries to shields against radioactivity.

*Source:* U.S. Geological Survey

56. **GENERAL: Total Savings** A homeowner installs a solar water heater that is expected to generate savings at the rate of $70e^{0.03t}$ dollars per year, where $t$ is the number of years since it was installed.

a. Find a formula for the total savings within the first $t$ years of operation.
b. Use a graphing calculator to find when the heater will "pay for itself" if it cost $800. [*Hint:* Use INTERSECT.]

57. **BUSINESS: Stock Price** The success of the iPod has raised the price of Apple stock significantly. The rate of increase has been approximately $1.6e^{0.5t}$ dollars per year, where $t$ is the number of years since 2001 (for $t \leq 6$).

a. Find a formula for the total increase in Apple stock within $t$ years of 2001.
b. Use your formula to find the total increase from 2001 to 2007. [*Note:* In 2001 the stock price was $9.]

*Source:* BigCharts.com

58. **BUSINESS: Stock Price** The stock price of Research In Motion (makers of the BlackBerry communications device) increased through most of 2007 at the rate of $3.52e^{0.11t}$ dollars per month, where $t$ is in months and $t = 1$ represents January (for $t \leq 11$).

a. Find a formula for the total increase in the value of the stock within the first $t$ months of 2007.

b. Use your formula to find the total increase through November. [*Note:* In January the stock price was $43.]

*Source:* BigCharts.com

## Conceptual Exercises

**59–63.** Choose the correct answer.

59. $\int e^x \, dx = ?$

a. $\frac{1}{x+1} e^{x+1} + C$     b. $e^x + C$     c. $e^x x + C$

60. $\int x^{-1} \, dx = ?$

a. $\ln |x| + C$     b. $\frac{1}{0} x^0 + C$     c. $x^{-1} x + C$

61. $\int x^e \, dx = ?$

a. $\frac{1}{e+1} x^{e+1} + C$     b. $x^e + C$     c. $x^e x + C$

62. $\int e^3 \, dx = ?$

a. $\frac{1}{4} e^4 + C$     b. $e^3 + C$     c. $e^3 x + C$

63. $\int e^{-1} \, dx = ?$

a. $\ln |e| + C$     b. $\frac{1}{0} e^0 + C$     c. $e^{-1} x + C$

64. Are there any powers of $x$ that we still cannot integrate?

65. Find a formula for $\int e^{ax+b} \, dx$ where $a$ and $b$ are constants.

66. Does it follow from the box on page 338 that
$$\int e^{x^5} \, dx = e^{x^5} + C?$$

67. Does it follow from the box on page 340 that
$$\int \frac{1}{x^5} \, dx = \ln |x^5| + C?$$

68. Which one of these formulas is correct?

a. $\int \ln x \, dx = \dfrac{1}{|x|} + C$

b. $\int \ln |x| \, dx = \dfrac{1}{x} + C$

c. $\int \dfrac{1}{x} \, dx = \ln |x| + C$

d. $\int \dfrac{1}{\ln x} \, dx = |x| + C$

## 5.3 DEFINITE INTEGRALS AND AREAS

### Introduction

We begin this section by calculating areas under curves, leading to a definition of the **definite integral** of a function. The **Fundamental Theorem of Integral Calculus** then provides an easier way to calculate definite integrals using *indefinite* integrals. Finally, we will illustrate the wide variety of applications of definite integrals.

### Area Under a Curve

The diagram below shows a continuous nonnegative function. We want to calculate the *area* under the curve and above the $x$-axis between the vertical lines $x = a$ and $x = b$.

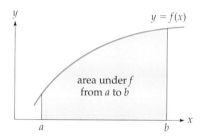

We begin by *approximating* the area by rectangles. In the graph on the left below, the area under the curve is approximated by *five* rectangles with equal bases and with heights equal to the height of the curve at the left-hand edge of the rectangle. (These are called *left* rectangles.) Five rectangles, however, do not give a very accurate approximation for the area under the curve: they underestimate the actual area by the small white spaces just above the rectangles.

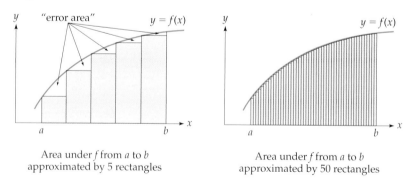

Area under $f$ from $a$ to $b$
approximated by 5 rectangles

Area under $f$ from $a$ to $b$
approximated by 50 rectangles

In the second graph, this same area is approximated by *fifty* rectangles, giving a much better approximation: the white *"error area"* between the curve and the rectangles is so small as to be almost invisible.

These diagrams suggest that more rectangles give a better approximation. In fact, the *exact* area under the curve is defined as the *limit* of the approximations as the number of rectangles *approaches infinity*. The following example shows how to carry out such an approximation for the area under a given curve, and afterward we will find the *exact* area by letting the number of rectangles approach infinity.*

**EXAMPLE 1**

**APPROXIMATING AREA BY RECTANGLES**

Approximate the area under the curve $f(x) = x^2$ from 1 to 2 by five rectangles. Use rectangles with equal bases and with heights equal to the height of the curve at the left-hand edge of the rectangles.

*Solution*

For five rectangles, we divide the distance from $a = 1$ to $b = 2$ into five equal parts, so each rectangle has width (or change in $x$)

$$\Delta x = \frac{2 - 1}{5} = \frac{1}{5} = 0.2 \qquad \text{For } n \text{ rectangles,} \quad \Delta x = \frac{b - a}{n}$$

Along the $x$-axis beginning at $x = 1$ we mark successive points with spacing $\Delta x = 0.2$, giving points 1, 1.2, 1.4, 1.6, 1.8, and 2, as shown below. Above each of the resulting subintervals we draw a rectangle whose height is the height of the curve at the left-hand edge of that rectangle. The curve is $y = x^2$, so these heights are the squares of the corresponding $x$-values.

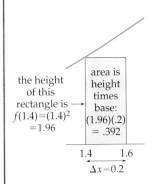

the height of this rectangle is → $f(1.4)=(1.4)^2$ $=1.96$

area is height times base: $(1.96)(.2)$ $= .392$

Middle rectangle

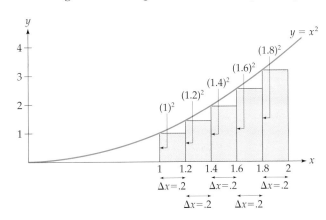

An enlarged view of the middle rectangle is shown on the left. The sum of the areas of the five rectangles, height times base $\Delta x = 0.2$ for each rectangle, is

| 1st rectangle | 2nd rectangle | 3rd rectangle | 4th rectangle | 5th rectangle | Height times base for each rectangle |

$$(1)^2 \cdot (0.2) + (1.2)^2 \cdot (0.2) + (1.4)^2 \cdot (0.2) + (1.6)^2 \cdot (0.2) + (1.8)^2 \cdot (0.2)$$

Multiplying out and summing

$$= 0.2 + 0.288 + 0.392 + 0.512 + 0.648 = 2.04$$

Therefore, the area under the curve is approximately 2.04 square units.

---

*This idea of dividing an area into many small parts and adding up their contributions to find a total was recommended by Tolstoy as the proper way to study history: "Only by taking infinitesimally small units for observation (the differentia of history, that is, the individual tendencies of men) and attaining to the art of integrating them (that is, finding the sum of these infinitesimals) can we hope to arrive at the laws of history." (From *War and Peace*, book 11, chapter 1)

As we saw earlier, using only five rectangles does not give a very accurate approximation for the true area under the curve. For greater accuracy we use more rectangles, calculating the area in the same way. The following table gives the "rectangular approximation" for the area under the curve in Example 1 for larger numbers of rectangles. The calculations were carried out on a graphing calculator using the program on page 361, rounding answers to three decimal places.

| Number of Rectangles | Sum of Areas of Rectangles | |
|---|---|---|
| 5 | 2.04 | ← Found in Example 1 |
| 10 | 2.185 | The sum of the areas is approaching $2\frac{1}{3}$ |
| 100 | 2.318 | |
| 1000 | 2.332 | |
| 10,000 | 2.333 | |

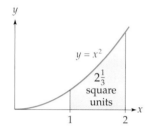

The areas in the right-hand column are approaching $2.333\ldots = 2\frac{1}{3}$, which is the *exact* area under the curve (as we will verify later). Therefore, the area under the curve $f(x) = x^2$ from 1 to 2 is $2\frac{1}{3}$ square units, as shown on the left.

Areas are given in "square units," meaning that if the units on the graph are inches, feet, or some other units, then the area is in *square* inches, *square* feet, or, in general, some other *square* units.

## Graphing Calculator Exploration

In the previous example the rectangles touched the curve at the *left-hand* edge of the rectangle (and are called *left* rectangles). Similarly, we may use *midpoint* rectangles (which touch the curve at the *midpoint* of the rectangle) or *right* rectangles (which touch the curve at their right-hand edge). The graphing calculator program* RIEMANN draws these (and other) types of rectangles for any function and calculates the sum of the areas, as shown below. The first screen shows five *midpoint* rectangles for $f(x) = x^2$ on the interval [1, 2]. The second screen shows the sum of the areas for various numbers of midpoint rectangles.

5 *midpoint* rectangles for $f(x) = x^2$ on [1, 2]

number of rectangles    sum of the areas

Notice from the second screen that the areas approach $2\frac{1}{3}$ very quickly, especially when compared with the previous table for *left* rectangles. Do you see why midpoint rectangles are more accurate?

*To obtain this and other programs, go to www.cengage.com/math/berresford.

### Definite Integral

Approximating the area under a nonnegative function $f$ by $n$ rectangles means multiplying heights $f(x)$ by widths $\Delta x$ and adding, obtaining

$$f(x_1) \cdot \Delta x + f(x_2) \cdot \Delta x + f(x_3) \cdot \Delta x + \cdots + f(x_n) \cdot \Delta x$$

The general procedure is as follows:

---

**Area Under f from a to b Approximated by n Left Rectangles**

1. Calculate the rectangle width   $\Delta x = \dfrac{b - a}{n}$.

2. Find $x$-values $x_1$, $x_2$, $\ldots$ , $x_n$ by successive additions of $\Delta x$ beginning with $x_1 = a$.

3. Calculate the sum:

$$f(x_1) \cdot \Delta x + f(x_2) \cdot \Delta x + f(x_3) \cdot \Delta x + \cdots + f(x_n) \cdot \Delta x$$

---

The sum in step 3 is called a **Riemann sum,** after the great German mathematician Georg Bernhard Riemann (1826–1866).* The *limit* of the Riemann sum as the number $n$ of rectangles approaches infinity gives the **area under the curve,** and is called the **definite integral of the function $f$ from $a$ to $b$,** written $\displaystyle\int_a^b f(x)\,dx.$   Formally:

---

**Definite Integral**

Let $f$ be a continuous function on an interval $[a, b]$. The definite integral of $f$ from $a$ to $b$ is defined as

$$\int_a^b f(x)\,dx = \lim_{n \to \infty}\left[ f(x_1) \cdot \Delta x + f(x_2) \cdot \Delta x + \cdots + f(x_n) \cdot \Delta x \right]$$

where $\Delta x = \dfrac{b - a}{n}$,   and   $x_1, x_2, \ldots, x_n$ are $x$-values beginning with $x_1 = a$ and obtained by successive additions of $\Delta x$. If $f$ is nonnegative on $[a, b]$, then the definite integral gives the *area under the curve from $a$ to $b$.*

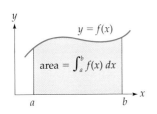

---

The numbers $a$ and $b$ are called the **lower** and **upper limits of integration.**

---

*Actually, Riemann sums are slightly more general, allowing the subintervals to have different widths, with each width multiplied by the function evaluated at *any* $x$-value within that subinterval. For a continuous function, any Riemann sum will approach the same limiting value as the rectangles become arbitrarily narrow, so we may restrict our attention to the particular Riemann sums defined above (sometimes called "left Riemann sums").

### Fundamental Theorem of Integral Calculus

A function followed by a vertical bar $\left.\phantom{x}\right|_a^b$ with numbers $a$ and $b$ means evaluate the function at the *upper* number $b$ and then subtract the evaluation at the *lower* number $a$.

$$F(x)\Big|_a^b = \underbrace{F(b)}_{\substack{\text{Evaluation at}\\\text{upper number}}} - \underbrace{F(a)}_{\substack{\text{Evaluation at}\\\text{lower number}}}$$

---

**EXAMPLE 2**     **USING THE EVALUATION NOTATION**

$$x^2\Big|_3^5 = \underbrace{(5)^2}_{\substack{x^2\\\text{at } x = 5}} - \underbrace{(3)^2}_{\substack{x^2\\\text{at } x = 3}} = \underbrace{25 - 9}_{\text{Simplifying}} = 16$$

Evaluation notation

---

**Practice Problem 1**     Evaluate $\sqrt{x}\,\Big|_4^{25}$          ▶ Solution on page 358

The following **Fundamental Theorem of Integral Calculus** shows how to evaluate definite integrals by using *indefinite* integrals. A geometric and intuitive justification of the theorem is given at the end of this section.

---

**Fundamental Theorem of Integral Calculus**

For a continuous function $f$ on an interval $[a, b]$,

$$\int_a^b f(x)\,dx = F(b) - F(a)$$

The right-hand side may be written $F(x)\Big|_a^b$

where $F$ is any antiderivative of $f$.

---

The theorem is "fundamental" in that it establishes a deep and unexpected connection between definite integrals (limits of Riemann sums) and antiderivatives. It says that definite integrals can be evaluated in two simple steps:

1. Find an *indefinite* integral of the function (omitting the $+C$).
2. *Evaluate* the result at $b$ and *subtract* the evaluation at $a$.

**EXAMPLE 3**

**FINDING A DEFINITE INTEGRAL BY THE FUNDAMENTAL THEOREM**

Find $\displaystyle\int_1^2 x^2\,dx$.

*Solution*

Because $x^2$ is continuous on [1, 2], we can use the Fundamental Theorem.

$$\int_1^2 x^2\,dx \;=\; \underbrace{\frac{1}{3}x^3\Big|_1^2}_{\substack{\text{Integrating}\\ x^2}} \;=\; \underbrace{\frac{1}{3}2^3}_{\substack{\text{Evaluating}\\ \frac{1}{3}x^3 \text{ at } x=2}} \;-\; \underbrace{\frac{1}{3}1^3}_{\substack{\text{and at}\\ x=1}} \;=\; \underbrace{\frac{8}{3}-\frac{1}{3}}_{\text{Subtracting}} \;=\; \frac{7}{3}$$

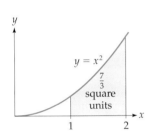

Since definite integrals of nonnegative functions give areas, this result means that the area under $f(x) = x^2$ from $x = 1$ to $x = 2$ is $\frac{7}{3}$ square units.

On pages 349–350 we found this same area by the much more laborious process of calculating Riemann sums and taking the limit as the number of rectangles approached infinity, and our answer here, $\frac{7}{3}$, agrees with the answer there, $2\frac{1}{3}$ square units. Whenever possible, we will calculate definite integrals and find areas in this much simpler way, using the Fundamental Theorem of Integral Calculus.

## Graphing Calculator Exploration

∫f(x)dx=2.3333333

a. Graph $y_1 = x^2$ on the window [0, 2] by [−1, 4].

b. Verify the result of Example 3 by having your graphing calculator find the definite integral of $x^2$ from 1 to 2. [*Hint:* Use a command like FnInt or $\int f(x)\,dx$.]

**Practice Problem 2**

Find the area under $y = x^3$ from 0 to 2 by evaluating $\displaystyle\int_0^2 x^3\,dx$.

➤ Solution on page 358

**EXAMPLE 4**

**FINDING THE AREA UNDER A CURVE**

Find the area under $y = e^{2x}$ from $x = 0$ to $x = 1$.

*Solution*

Because $e^{2x}$ is nonnegative and continuous on [0, 1], the area is given by the definite integral:

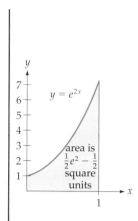

$$\int_0^1 e^{2x}\,dx = \underbrace{\frac{1}{2}e^{2x}\Big|_0^1}_{\substack{\text{Indefinite integral}\\ \text{(using the formula at }x=1\text{ and}\\ \text{for }\int e^{ax}\,dx)}} = \underbrace{\frac{1}{2}e^2 - \frac{1}{2}e^0}_{\substack{\text{Evaluating}\\ x=1\text{ and}\\ \text{at }x=0}} = \underbrace{\frac{1}{2}e^2 - \frac{1}{2}}_{\substack{\text{Simplifying}\\ \text{(using }e^0=1)}}$$

Therefore, the area is $\frac{1}{2}e^2 - \frac{1}{2}$ square units.

We leave the answer in this "exact" form (in terms of the number $e$). In an application we would use a calculator to approximate this answer as 3.19 square units.

 **BE CAREFUL:** Notice the difference between *definite* and *indefinite* integrals: *indefinite* integrals (the subject of the previous two sections) are *functions plus arbitrary constants*, while *definite* integrals are particular *numbers* (hence the word *definite*). The Fundamental Theorem of Integral Calculus shows how to use *indefinite* integrals to evaluate *definite* integrals.

**EXAMPLE 5**

**FINDING THE AREA UNDER A CURVE**

Find the area under $f(x) = \dfrac{1}{x}$ from $x = 1$ to $x = e$.

*Solution*

Because $\dfrac{1}{x}$ is nonnegative and continuous on [1, $e$], the area is:

$$\int_1^e \frac{1}{x}\,dx = \ln x\Big|_1^e = \underbrace{\ln e}_{1} - \underbrace{\ln 1}_{0} = 1$$

Therefore, the area is 1 square unit.

From this example we can give an alternate definition of the number $e$: $e$ is the number such that the definite integral of $1/x$ from 1 to $e$ is 1.

## Graphing Calculator Exploration

$\int f(x)dx = .74682413$

Use a graphing calculator to evaluate the definite integral

$$\int_0^1 e^{-x^2}\, dx.$$

(This definite integral, which is important in probability and statistics, cannot be evaluated by the Fundamental Theorem because the function $e^{-x^2}$ has no simple antiderivative. Your calculator may find the answer by approximating the area under the curve by modified Riemann sums.)

Definite integrals have many of the properties of indefinite integrals. These properties follow from interpreting definite integrals as limits of Riemann sums.

### Properties of Definite Integrals

$$\int_a^b c \cdot f(x)\, dx = c \int_a^b f(x)\, dx$$

A constant may be moved across the integral sign

$$\int_a^b [f(x) \pm g(x)]\, dx = \int_a^b f(x)\, dx \pm \int_a^b g(x)\, dx$$

Read both upper signs or both lower signs

The integral of a sum is the sum of the integrals (and similarly for differences)

**EXAMPLE 6**

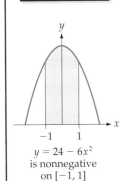

$y = 24 - 6x^2$
is nonnegative
on $[-1, 1]$

### USING THE PROPERTIES OF DEFINITE INTEGRALS

Find the area under $y = 24 - 6x^2$ from $-1$ to $1$.

*Solution*

$$\int_{-1}^{1} (24 - 6x^2)\, dx = \left( 24x - 6 \cdot \frac{1}{3} x^3 \right) \Big|_{-1}^{1}$$

$$= (24 - 2) - (-24 + 2)$$

$$= 44$$

Therefore, the area is 44 square units.

## Total Cost of a Succession of Units

Given a marginal cost function, to find the *total* cost of producing, say, units 100 to 400, we could proceed as follows: *integrate* the marginal cost to find the total cost, *evaluate* at 400 to find the total cost up to unit 400, and *subtract* the evaluation at 100 to leave just the cost of units 100 to 400. However, these steps of integrating, evaluating, and subtracting are just the steps in evaluating a definite integral by the Fundamental Theorem. Therefore, the cost of a succession of units is equal to the definite integral of the marginal cost function.

### Cost of a Succession of Units

For a marginal cost function $MC(x)$:

$$\left(\begin{array}{c}\text{Total cost of}\\ \text{units } a \text{ to } b\end{array}\right) = \int_a^b MC(x)\, dx$$

**EXAMPLE 7**

**FINDING THE COST OF A SUCCESSION OF UNITS**

A company's marginal cost function is $MC(x) = \dfrac{75}{\sqrt{x}}$ where $x$ is the number of units. Find the total cost of producing units 100 to 400.

*Solution*

$$\left(\begin{array}{c}\text{Total cost of}\\ \text{units 100 to 400}\end{array}\right) = \int_{100}^{400}\frac{75}{\sqrt{x}}\,dx = 75\int_{100}^{400}x^{-1/2}\,dx \qquad \text{Integrating marginal cost}$$

$$= (75\cdot 2\cdot x^{1/2})\Big|_{100}^{400}$$

$$= 150\cdot(400)^{1/2} - 150\cdot(100)^{1/2}$$

$$= 150\cdot 20 - 150\cdot 10 = 1500$$

The cost of producing units 100 to 400 is $1500.

Similarly, integrating *any* rate from $a$ to $b$ gives the *total accumulation* at that rate between $a$ and $b$.

### Total Accumulation at a Given Rate

$$\left(\begin{array}{c}\text{Total accumulation at}\\ \text{rate } f \text{ from } a \text{ to } b\end{array}\right) = \int_a^b f(x)\, dx$$

The following diagrams illustrate this idea. In each case, the *curve* represents a *rate*, and the *area under the curve*, given by the definite integral, gives the *total accumulation* at that rate.

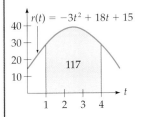

marginal cost

productivity

sales rate

In repetitive tasks, a person's productivity usually increases with practice until it is slowed by monotony.

**EXAMPLE 8**

### FINDING TOTAL PRODUCTIVITY FROM A RATE

A technician can test computer chips at the rate of $-3t^2 + 18t + 15$ chips per hour (for $0 \leq t \leq 6$), where $t$ is the number of hours after 9:00 a.m. How many chips can be tested between 10:00 a.m. and 1:00 p.m.?

*Solution*

The total work accomplished is the integral of this rate from $t = 1$ (10 a.m.) to $t = 4$ (1 p.m.):

$$\int_1^4 (-3t^2 + 18t + 15)\, dt = \left. \left( -t^3 + 18 \cdot \frac{1}{2} t^2 + 15t \right) \right|_1^4$$

$$= (-64 + 144 + 60) - (-1 + 9 + 15) = 117$$

That is, between 10 a.m. and 1 p.m., 117 chips can be tested.

### Integration Notation

The symbol $\Sigma$ (the Greek letter S) is used in mathematics to indicate a *sum*, and so the Riemann sum can be written $\sum_1^n f(x_k)\, \Delta x$. The fact that the Riemann sum approaches the definite integral can be expressed as:

$$\sum_1^n f(x_k)\, \Delta x \longrightarrow \int_a^b f(x)\, dx$$

$\Sigma$ becomes $\int$
$\Delta$ becomes $d$
as $n \to \infty$

That is, the $n$ approaching infinity changes the $\Sigma$ (a Greek S) into an integral sign $\int$ (a "stretched out" S), and the $\Delta$ (a Greek D) into a $d$. In other words, the integral notation reminds us that a definite integral represents a *sum of rectangles* of height $f(x)$ and width $dx$.

### Functions Taking Positive and Negative Values

Riemann sums can be calculated for *any* continuous function, not just non-negative functions. The following diagram illustrates a Riemann sum for a continuous function that takes positive *and* negative values.

The two rightmost rectangles, where $f$ is *negative*, will make a *negative* contribution to the sum. Taking the limit, the definite integral where the function lies *below* the $x$-axis will give the *negative* of the area between the curve and the $x$-axis. Therefore, the definite integral of such a function from $a$ to $b$ will give the *signed area* between the curve and the $x$-axis: the area *above* the axis minus the area *below* the axis, shown as $A_{up}$ *minus* $A_{down}$ in the diagram below.

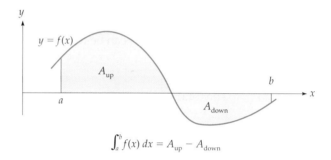

$$\int_a^b f(x)\,dx = A_{up} - A_{down}$$

## Solutions to Practice Problems

**1.** $\sqrt{x}\,\Big|_4^{25} = \sqrt{25} - \sqrt{4} = 5 - 2 = 3$

**2.** $\displaystyle\int_0^2 x^3\,dx = \tfrac{1}{4}x^4\,\Big|_0^2 = \tfrac{1}{4}\cdot 16 - \tfrac{1}{4}\cdot 0 = 4$ square units

## 5.3 Section Summary

We began by approximating the area under a curve from $a$ to $b$ by Riemann sums (sums of rectangles). We then defined the *definite integral* $\displaystyle\int_a^b f(x)\,dx$ as the *limit* of the Riemann sum as the number of rectangles approaches infinity. The Fundamental Theorem of Integral Calculus showed how to evaluate definite integrals much more simply by evaluating *indefinite* integrals:

$$\int_a^b f(x)\,dx = F(x)\,\Big|_a^b \qquad \text{\textit{F} is any antiderivative of \textit{f}}$$

A *definite* integral is a *number*, but an *indefinite* integral is a function plus an arbitrary constant.

As for the *uses* of definite integrals, the definite integral of a nonnegative function gives the *area under the curve:*

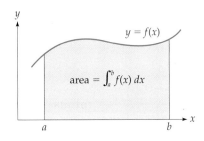

The definite integral of a *rate* gives the *total accumulation* at that rate:

$$\int_a^b f(x)\,dx = \left(\begin{array}{c}\text{Total accumulation at} \\ \text{rate } f \text{ from } a \text{ to } b\end{array}\right)$$

## Verification of the Fundamental Theorem of Integral Calculus

The following is a geometric and intuitive justification of the Fundamental Theorem of Integral Calculus.

For a continuous and nonnegative function $f$ on an interval $[a, b]$, we define a new function $A(x)$ as the *area under f from a to x.*

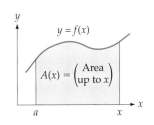

$$A(x) = \left(\begin{array}{c}\text{Area under } f \\ \text{from } a \text{ to } x\end{array}\right)$$

Therefore,

$$A(b) = \left(\begin{array}{c}\text{Area under } f \\ \text{from } a \text{ to } b\end{array}\right)$$

Replacing $x$ by $b$ gives the entire area from $a$ to $b$

and

$$A(a) = 0$$

The area "from $a$ to $a$" must be zero

Subtracting these last two expressions, $A(b) - A(a)$, gives the total area from $a$ to $b$ [since $A(b)$ is the total area, and $A(a)$ is zero]. This same area can be expressed as the definite integral of $f$ from $a$ to $b$, leading to the equation

$$\int_a^b f(x)\,dx = A(b) - A(a)$$

Area under $f$ from $a$ to $b$ expressed in two ways

If we can show that $A(x)$ is an antiderivative of $f(x)$, then the equation above will show that the definite integral can be found by an antiderivative evaluated at the upper and lower limits and subtracted, which will verify the Fundamental Theorem. To show that $A(x)$ is an antiderivative of $f(x)$, we show that $A'(x) = f(x)$, differentiating $A(x)$ by the definition of the derivative:

$$A'(x) = \lim_{h \to 0} \frac{A(x + h) - A(x)}{h}$$

In the numerator, $A(x + h)$ is the area under the curve up to $x + h$ and $A(x)$ is the area up to $x$, so subtracting them, $A(x + h) - A(x)$, leaves just the area from $x$ to $x + h$, as shown below.

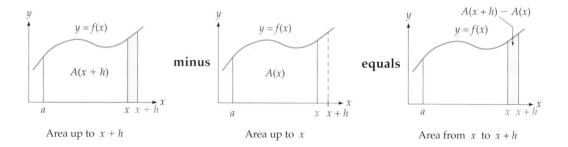

| Area up to $x + h$ | minus | Area up to $x$ | equals | Area from $x$ to $x + h$ |

When $h$ is small, this last area can be approximated by a rectangle of base $h$ and height $f(x)$, where the approximation becomes exact as $h$ approaches zero. Therefore, in the limit we may replace $A(x + h) - A(x)$ by the area of the rectangle, $h \cdot f(x)$:

$$A'(x) = \lim_{h \to 0} \frac{\overbrace{A(x + h) - A(x)}^{\approx\, h \cdot f(x)}}{h} = \lim_{h \to 0} \frac{h \cdot f(x)}{h} = f(x)$$

This equation says that $A'(x) = f(x)$, showing that $A(x)$ *is* an antiderivative of $f(x)$. This completes the verification of the Fundamental Theorem of Integral Calculus.

## 5.3   Exercises

**1–6.** Find the sum of the areas of the shaded rectangles under each graph. Round to two decimal places. [*Hint:* The width of each rectangle is the difference between the $x$-values at its base. The height of each rectangle is the height of the curve at the left edge of the rectangle.]

**2.**

**1.**

**3.**

**4.**

**5.**

**6.**

**7–12.** For each function:
i. *Approximate* the area under the curve from $a$
to $b$ by calculating a Riemann sum with the given
number of rectangles. Use the method described
in Example 1 on page 349, rounding to three
decimal places.
ii. Find the *exact* area under the curve from $a$ to $b$ by
evaluating an appropriate definite integral using
the Fundamental Theorem.

**7.** $f(x) = 2x$ from $a = 1$ to $b = 2$.
For part (i), use 5 rectangles.

**8.** $f(x) = x^2 + 1$ from $a = 0$ to $b = 1$.
For part (i), use 5 rectangles.

**9.** $f(x) = \sqrt{x}$ from $a = 1$ to $b = 4$.
For part (i), use 6 rectangles.

**10.** $f(x) = e^x$ from $a = -1$ to $b = 1$.
For part (i), use 8 rectangles.

**11.** $f(x) = \dfrac{1}{x}$ from $a = 1$ to $b = 2$.
For part (i), use 10 rectangles.

**12.** $f(x) = \dfrac{1}{\sqrt{x}}$ from $a = 1$ to $b = 4$.
For part (i), use 6 rectangles.

**13–18.** Use the graphing calculator program
RIEMANN (see page 350) or the simplified program
given below or a similar program to find the
following Riemann sums.

i. Calculate the Riemann sum for each function for
the following values of $n$: 10, 100, and 1000. Use
left, right, or midpoint rectangles, making a table
of the answers, rounded to three decimal places.
ii. Find the *exact* value of the area under the curve
by evaluating an appropriate definite integral
using the Fundamental Theorem. The values of
the Riemann sums from part (i) should approach
this number.

**13.** $f(x) = 2x$ from $a = 1$ to $b = 2$

**14.** $f(x) = x^2 + 1$ from $a = 0$ to $b = 1$

**15.** $f(x) = \sqrt{x}$ from $a = 1$ to $b = 4$

**16.** $f(x) = e^x$ from $a = -1$ to $b = 1$

**17.** $f(x) = \dfrac{1}{x}$ from $a = 1$ to $b = 2$

**18.** $f(x) = \dfrac{1}{\sqrt{x}}$ from $a = 1$ to $b = 4$

*TI-83/84 Program for (Left) Riemann Sums*

(Enter the function in $y_1$ before executing)

```
0 → S
Disp "USES FUNCTION Y1"
Prompt A, B, N
(B-A)/N → D
A → X
For(I, 1, N)
S + Y1*D → S
X + D → X
End
Disp "RIEMANN SUM IS", S
```

**19–40.** Use a definite integral to find the area under
each curve between the given $x$-values. For Exercises 19–24, also make a sketch of the curve showing
the region.

**19.** $f(x) = x^2$ from $x = 0$ to $x = 3$

**20.** $f(x) = x$ from $x = 0$ to $x = 4$

**21.** $f(x) = 4 - x$ from $x = 0$ to $x = 4$

**22.** $f(x) = 1 - x^2$ from $x = -1$ to $x = 1$

*(See instructions on previous page.)*

**23.** $f(x) = \dfrac{1}{x}$ from $x = 1$ to $x = 2$

**24.** $f(x) = e^x$ from $x = 0$ to $x = 1$

**25.** $f(x) = 8x^3$ from $x = 1$ to $x = 3$

**26.** $f(x) = 6x^2$ from $x = 2$ to $x = 3$

**27.** $f(x) = 6x^2 + 4x - 1$ from $x = 1$ to $x = 2$

**28.** $f(x) = 27 - 3x^2$ from $x = 1$ to $x = 3$

**29.** $f(x) = \dfrac{1}{\sqrt{x}}$ from $x = 4$ to $x = 9$

**30.** $f(x) = \dfrac{1}{x^2}$ from $x = 1$ to $x = 3$

**31.** $f(x) = 8 - 4\sqrt[3]{x}$ from $x = 0$ to $x = 8$

**32.** $f(x) = 9 - 3\sqrt{x}$ from $x = 0$ to $x = 9$

**33.** $f(x) = \dfrac{1}{x}$ from $x = 1$ to $x = 5$

**34.** $f(x) = \dfrac{1}{x}$ from $x = e$ to $x = e^3$

**35.** $f(x) = x^{-1} + x^2$ from $x = 1$ to $x = 2$

**36.** $f(x) = 6e^{2x}$ from $x = 0$ to $x = 2$

**37.** $f(x) = 2e^x$ from $x = 0$ to $x = \ln 3$

**38.** $f(x) = e^{-x}$ from $x = 0$ to $x = 1$

**39.** $f(x) = e^{x/2}$ from $x = 0$ to $x = 2$

**40.** $f(x) = e^{x/3}$ from $x = 0$ to $x = 3$

**41–46.** For each function:

a. Integrate ("by hand") to find the area under the curve between the given $x$-values.

b. Verify your answer to part (a) by having your calculator graph the function and find the area (using a command like FnInt or $\int f(x)\,dx$).

**41.** $f(x) = 12 - 3x^2$ from $x = 1$ to $x = 2$

**42.** $f(x) = 9x^2 - 6x + 1$ from $x = 1$ to $x = 2$

**43.** $f(x) = \dfrac{1}{x^3}$ from $x = 1$ to $x = 4$

**44.** $f(x) = \dfrac{1}{\sqrt[3]{x}}$ from $x = 8$ to $x = 27$

**45.** $f(x) = 2x + 1 + x^{-1}$ from $x = 1$ to $x = 2$

**46.** $f(x) = e^x$ from $x = 0$ to $x = 3$

**47–66.** Evaluate each definite integral.

**47.** $\displaystyle\int_0^1 (x^{99} + x^9 + 1)\,dx$   **48.** $\displaystyle\int_2^4 (1 + x^{-2})\,dx$

**49.** $\displaystyle\int_1^2 (6t^2 - 2t^{-2})\,dt$   **50.** $\displaystyle\int_{-2}^2 (3w^2 - 2w)\,dw$

**51.** $\displaystyle\int_1^4 \dfrac{1}{y^2}\,dy$   **52.** $\displaystyle\int_1^4 \dfrac{1}{\sqrt{z}}\,dz$

**53.** $\displaystyle\int_1^e \dfrac{dx}{x}$   **54.** $\displaystyle\int_1^{e^2} \dfrac{3}{x}\,dx$

**55.** $\displaystyle\int_1^3 (9x^2 + x^{-1})\,dx$   **56.** $\displaystyle\int_1^2 (x^{-1} - 4x^2)\,dx$

**57.** $\displaystyle\int_{-2}^{-1} 3x^{-1}\,dx$   **58.** $\displaystyle\int_{-3}^{-1} (1 + x^{-1})\,dx$

**59.** $\displaystyle\int_0^1 12e^{3x}\,dx$   **60.** $\displaystyle\int_0^2 3e^{x/2}\,dx$

**61.** $\displaystyle\int_{-1}^1 5e^{-x}\,dx$   **62.** $\displaystyle\int_0^1 (6x^2 - 4e^{2x})\,dx$

**63.** $\displaystyle\int_{\ln 2}^{\ln 3} e^x\,dx$   **64.** $\displaystyle\int_0^{\ln 5} e^x\,dx$

**65.** $\displaystyle\int_1^2 \dfrac{(x+1)^2}{x}\,dx$   **66.** $\displaystyle\int_1^2 \dfrac{(x+1)^2}{x^2}\,dx$

 **67–72.** Use a graphing calculator to evaluate each definite integral, rounding answers to three decimal places. [*Hint:* Use a command like FnInt or $\int f(x)\,dx$.]

**67.** $\displaystyle\int_0^2 \dfrac{1}{x^2 + 1}\,dx$   **68.** $\displaystyle\int_{-1}^1 \sqrt{x^4 + 1}\,dx$

**69.** $\displaystyle\int_{-1}^1 e^{x^2}\,dx$   **70.** $\displaystyle\int_{-2}^2 e^{(-1/2)x^2}\,dx$

**71.** $\displaystyle\int_0^4 \sqrt{x}\,e^x\,dx$   **72.** $\displaystyle\int_1^4 x^x\,dx$

**73.** OMITTING THE *C* IN DEFINITE INTEGRALS

a. Evaluate the definite integral $\displaystyle\int_0^3 x^2\,dx$.

b. Evaluate the same definite integral by completing the following calculation, in which the antiderivative includes a constant $C$.

$$\int_0^3 x^2\,dx = \left(\frac{1}{3}x^3 + C\right)\Big|_0^3 = \cdots$$

[The constant $C$ should cancel out, giving the same answer as in part (a).]

c. Explain why the constant will cancel out of *any* definite integral. (We therefore omit the constant in definite integrals. However, be sure to keep the $+C$ in *indefinite* integrals.)

**74.** Show that for any number $a > 0$,

$$\int_1^a \frac{1}{x}\, dx = \ln a$$

This equation is often used as a *definition* of natural logarithms, defining  $\ln a$  as the area under the curve  $y = 1/x$  between 1 and $a$. [*Hint:*  $\ln 1 = 0$.]

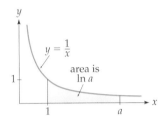

**75–76. GEOMETRIC INTEGRATION** Find

$$\int_0^4 f(x)\, dx$$ for the function $f(x)$ graphed below. [*Hint:* No calculus is necessary—just as we used integrals to find areas, we can use areas to find integrals. The curves shown are quarter circles.]

**75.**

**76.**

## Applied Exercises

**77. GENERAL: Electrical Consumption** On a hot summer afternoon, a city's electricity consumption is  $-3t^2 + 18t + 10$  units per hour, where $t$ is the number of hours after noon $(0 \le t \le 6)$. Find the total consumption of electricity between the hours of 1 and 5 p.m.

**78. BIOMEDICAL: Height** An average child of age $x$ years grows at the rate of  $6x^{-1/2}$  inches per year (for  $2 \le x \le 16$). Find the total height gain from age 4 to age 9.

*Source:* National Center for Health Statistics

**79–80. GENERAL: Repetitive Tasks** After $t$ hours of work, a bank clerk can process checks at the rate of $r(t)$ checks per hour for the function $r(t)$ given below. How many checks will the clerk process during the first three hours (time 0 to time 3)?

**79.** $r(t) = -t^2 + 90t + 5$    **80.** $r(t) = -t^2 + 60t + 9$

**81–82. BUSINESS: Cost** A company's marginal cost function is $MC(x)$ (given below), where $x$ is the number of units. Find the total cost of the first hundred units ($x = 0$  to  $x = 100$).

**81.** $MC(x) = 6e^{-0.02x}$      **82.** $MC(x) = 8e^{-0.01x}$

**83. GENERAL: Price Increase** The price of a double-dip ice cream cone is increasing at the rate of  $18e^{0.08t}$  cents per year, where $t$ is measured in years and  $t = 0$  corresponds to 2000. Find the total change in price between the years 2000 and 2010.

**84. BUSINESS: Sales** An automobile dealer estimates that the newest model car will sell at the rate of  $30/t$  cars per month, where $t$ is measured in months and  $t = 1$  corresponds to the beginning of January. Find the number of cars that will be sold from the beginning of January to the beginning of May.

**85. BUSINESS: Tin Consumption** World consumption of tin is running at the rate of $0.26e^{0.01t}$ million metric tons per year, where $t$ is measured in years and  $t = 0$  corresponds to the beginning of 2008. Find the total consumption of tin from the beginning of 2008 to the beginning of 2018.

*Source:* U.S. Geological Survey

**86. SOCIOLOGY: Marriages** The marriage rate in the United States has been declining recently, with about  $2.1e^{-0.034t}$  million marriages per year, where $t$ is the number of years since 2008. Assuming that this rate continues, find the total number of marriages in the United States from 2008 to 2018.

*Source:* U.S. Census Bureau

**87. BEHAVIORAL SCIENCE: Learning** A student can memorize words at the rate of  $6e^{-t/5}$  words per minute after $t$ minutes. Find the total number of words that the student can memorize in the first 10 minutes.

**88. BIOMEDICAL: Epidemics** An epidemic is spreading at the rate of  $12e^{0.2t}$  new cases per day, where $t$ is the number of days since the epidemic began. Find the total number of new cases in the first 10 days of the epidemic.

**89. GENERAL: Area**

a. Use your graphing calculator to find the area between 0 and 1 under the following curves: $y = x$, $y = x^2$, $y = x^3$, and $y = x^4$.

b. Based on your answers to part (a), conjecture a formula for the area under $y = x^n$ between 0 and 1 for any value of $n > 0$.

c. Prove your conjecture by evaluating an appropriate definite integral "by hand."

**90. GENERAL: Dam Construction** Ever since the Johnstown Dam burst in 1889, killing 2200 people, dam construction has become increasingly scientific.

a. Estimate the amount of concrete needed to build the Snake River Dam by finding the area of the cross section shown in the following diagram and multiplying the result by the 574-foot length of the dam. All dimensions are in feet. [*Hint:* Find the cross-sectional area by integrating and using area formulas.]

b. If a mixing truck carries 300 cubic feet of concrete, about how many truckloads would be needed to build the dam?

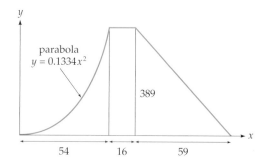

**91. BIOMEDICAL: Drug Absorption** An oral medication is absorbed into the bloodstream at the rate of $5e^{-0.04t}$ milligrams per minute, where $t$ is the number of minutes since the medication was taken. Find the total amount of medication absorbed within the first 30 minutes.

**92. BIOMEDICAL: Aortic Volume** The rate of change of the volume of blood in the aorta $t$ seconds after the beginning of the cardiac cycle is $-kP_0e^{-mt}$ milliliters per second, where $k$, $P_0$, and $m$ are constants (depending, respectively, on the elasticity of the aorta, the initial aortic pressure, and various characteristics of the

cardiac cycle). Find the total change in volume from time 0 to time $T$ (the end of the cardiac cycle). (Your answer will involve the constants $k$, $P_0$, $m$, and $T$.)

*Source: R. E. Klabunde, Cardiovascular Physiology Concepts*

**93. BUSINESS: Sales** A dealer predicts that new cars will sell at the rate of $8xe^{-0.1x}$ sales per week in week $x$. Find the total sales in the first half year (week 0 to week 26).

**94. GENERAL: Cigarette Smoking** Reread, if necessary, the Application Preview on pages 323–324.

a. Evaluate the definite integrals

$$\int_0^1 (300e^{0.025x} - 240e^{0.02x})\, dx$$

and

$$\int_7^8 (300e^{0.025x} - 240e^{0.02x})\, dx$$

to verify the answers given there for the amount of tar inhaled from the first and last centimeters of the cigarette.

b. Evaluate the definite integral

$$\int_0^8 (300e^{0.025x} - 240e^{0.02x})\, dx$$

to find the amount of tar inhaled from smoking the entire cigarette.

*Source: H. Marcus-Roberts and M. Thompson, eds., Modules in Applied Mathematics, Vol. 4, Life Science Models*

**95. GENERAL: Population** A resort community swells at the rate of $100e^{0.4\sqrt{x}}$ new arrivals per day on day $x$ of its "high season." Find the total number of arrivals in the first two weeks (day 0 to day 14).

**96. BEHAVIORAL SCIENCE: Repetitive Tasks** After $t$ hours of work, a medical technician can carry out T-cell counts at the rate of $2x^2e^{-x/4}$ tests per hour. How many tests will the technician process during the first eight hours (time 0 to time 8)?

## Conceptual Exercises

**97.** Suppose that you have a positive, increasing function and you approximate the area under it by a Riemann sum with left rectangles. Will the Riemann sum *overestimate* or *underestimate* the actual area? [*Hint:* Make a sketch.]

**98.** Suppose that you have a positive, decreasing function and you approximate the area under it by a Riemann sum with left rectangles. Will the Riemann sum *over*estimate or *under*estimate the actual area? [*Hint:* Make a sketch.]

**99.** Suppose that you have a positive, increasing, concave up function and you approximate the area under it by a Riemann sum with *midpoint* rectangles. Will the Riemann sum *over*estimate or *under*estimate the actual area? [*Hint:* Make a sketch.]

**100.** Suppose that you have a positive, increasing, concave down function and you approximate the area under it by a Riemann sum with *midpoint* rectangles. Will the Riemann sum *over*estimate or *under*estimate the actual area? [*Hint:* Make a sketch.]

**101.** A friend says that definite and indefinite integrals are exactly the same except that one has numbers plugged in. A second friend disagrees, saying that the essential ideas are entirely different. Who is right?

**102.** Suppose that you have a positive function and you approximate the area under it using Riemann sums with midpoint rectangles. Explain why, if the function is *linear,* you will always get the *exact* area, no matter how many (or few) rectangles you use. [*Hint:* Make a sketch.]

**103.** Evaluate $\displaystyle\int_{1}^{1} \frac{x^{43}\, e^{-17x} + 219\sqrt[3]{x^2}}{\ln \sqrt[29]{6x^3 - x^{-11}} - \pi^3}\, dx$

[*Hint:* No work necessary.]

**104.** Find a formula for $\displaystyle\int_{a}^{b} c\, dx.$ [*Hint:* No calculation necessary—just think of a graph.]

**105.** Finding $\displaystyle\int_{-1}^{1} \frac{1}{x^2}\, dx$ by the Fundamental Theorem of Integral Calculus gives an answer of $-2$, as you should check. However, shouldn't the area under a positive function be positive? Explain.

**106.** How will $\displaystyle\int_{a}^{b} f(x)\, dx$ and $\displaystyle\int_{b}^{a} f(x)\, dx$ differ?

[*Hint:* Assume that they can be evaluated by the Fundamental Theorem of Integral Calculus, and think how they will differ at the "evaluate and subtract" step.]

## Explorations and Excursions

The following problems extend and augment the material presented in the text.

**107. ECONOMICS: Pareto's Law** The economist Vilfredo Pareto estimated that the number of people who have an income between $A$ and $B$ dollars ($A < B$) is given by a definite integral of the form

$$N = \int_{A}^{B} ax^{-b}\, dx \qquad (b \neq 1)$$

where $a$ and $b$ are constants. Evaluate this integral.

*Source:* A. Reynolds, *Income and Wealth*

**108. BIOMEDICAL: Poiseuille's Law** According to Poiseuille's law, the speed of blood in a blood vessel is given by $V = \dfrac{p}{4Lv}(R^2 - r^2)$, where $R$ is the radius of the blood vessel, $r$ is the distance of the blood from the center of the blood vessel, and $p$, $L$, and $v$ are constants determined by the pressure and viscosity of the blood and the length of the vessel. The total blood flow is then given by

$$\begin{pmatrix} \text{Total} \\ \text{blood flow} \end{pmatrix} = \int_{0}^{R} 2\pi \frac{p}{4Lv}(R^2 - r^2)r\, dr$$

Find the total blood flow by finding this integral ($p$, $L$, $v$, and $R$ are constants).

*Source:* E. Batschelet, *Introduction to Mathematics for Life Scientists*

**109–110. BUSINESS: Capital Value of an Asset** The *capital value* of an asset (such as an oil well) that produces a continuous stream of income is the sum of the present value of all future earnings from the asset. Therefore, the capital value of an asset that produces income at the rate of $r(t)$ dollars per year (at a continuous interest rate $i$) is

$$\begin{pmatrix} \text{Capital} \\ \text{value} \end{pmatrix} = \int_{0}^{T} r(t)e^{-it}\, dt$$

where $T$ is the expected life (in years) of the asset.

*Source:* T. Lee, *Income and Value Measurement*

**109.** Use the formula in the preceding instructions to find the capital value (at interest rate $i = 0.06$) of an oil well that produces income at the constant rate of $r(t) = 240{,}000$ dollars per year for 10 years.

**110.** Use the formula in the preceding instructions to find the capital value (at interest rate $i = 0.05$) of a uranium mine that produces income at the rate of $r(t) = 560{,}000t^{1/2}$ dollars per year for 20 years.

## 5.4 FURTHER APPLICATIONS OF DEFINITE INTEGRALS: AVERAGE VALUE AND AREA BETWEEN CURVES

### Introduction

In this section we will use definite integrals for two important purposes: finding average values of functions and finding areas between curves. Average values are used everywhere. Birth weights of babies are compared with average weights, and retirement benefits are determined by average income. Averages eliminate fluctuations, reducing a collection of numbers to a single "representative" number. Areas between curves are used to find quantities from trade deficits to lives saved by seat belts (see pages 377–378).

### Average Value of a Function

The average value of $n$ numbers is found by adding the numbers and dividing by $n$. For example,

$$\begin{pmatrix} \text{Average of} \\ a, b, \text{and } c \end{pmatrix} = \frac{1}{3}(a + b + c)$$

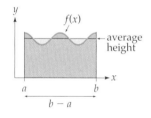

How can we find the average value of a *function* on an interval? For example, if a function gives the temperature over a 24-hour period, how can we calculate the *average temperature?* We could, of course, just take the temperature at every hour and then average these 24 values, but this would ignore the temperature at all of the intermediate times. Intuitively, the average should represent a "leveling off" of the curve to a uniform height, the horizontal line shown on the left.

This leveling should use the "hills" to fill in the "valleys," maintaining the same total area under the curve. Therefore, the area under the horizontal line (a rectangle with base $(b - a)$ and height up to the line) must equal the area under the curve (the definite integral of the function).

$$\underbrace{(b - a)\begin{pmatrix} \text{Average} \\ \text{height} \end{pmatrix}}_{\text{Area under line}} = \underbrace{\int_a^b f(x)\,dx}_{\text{Area under curve}} \qquad \text{Equating the two areas}$$

Therefore:

$$\begin{pmatrix} \text{Average} \\ \text{height} \end{pmatrix} = \frac{1}{b - a}\int_a^b f(x)\,dx \qquad \text{Dividing by } (b - a)$$

This formula gives the average (or "mean") value of a continuous function on an interval.

| Average Value of a Function |
|---|
| $$\begin{pmatrix} \text{Average value} \\ \text{of } f \text{ on } [a, b] \end{pmatrix} = \frac{1}{b - a}\int_a^b f(x)\,dx$$     Average value is the definite integral of the function divided by the length of the interval |

Finding the average value of a function by integrating and dividing by $b - a$ is analogous to averaging $n$ numbers by adding and dividing by $n$ (since integrals are continuous sums). A derivation of this formula by Riemann sums is given at the end of this section.

**EXAMPLE 1**

### FINDING THE AVERAGE VALUE OF A FUNCTION

Find the average value of $f(x) = \sqrt{x}$ from $x = 0$ to $x = 9$.

*Solution*

$$\binom{\text{Average}}{\text{value}} = \frac{1}{9 - 0} \int_0^9 \sqrt{x}\, dx = \frac{1}{9} \int_0^9 x^{1/2}\, dx$$
Integral divided by the length of the interval

$$= \frac{1}{9}\frac{2}{3} x^{3/2}\Big|_0^9 = \frac{2}{27} 9^{3/2} - \frac{2}{27} 0^{3/2}$$
Integrating and evaluating

$$= \frac{2}{27}\left(\sqrt{9}\right)^3 - 0 = \frac{2}{27} 27 = 2$$
Simplifying

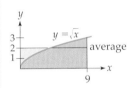

The average value of $f(x) = \sqrt{x}$ over the interval $[0, 9]$ is 2.

**EXAMPLE 2**

### FINDING AVERAGE POPULATION

The population of the United States is predicted to be $P(t) = 309e^{0.0087t}$ million people, where $t$ is the number of years since 2010. Predict the average population between the years 2020 and 2030.

*Source:* U.S. Census Bureau

*Solution*

We integrate from $t = 10$ (year 2020) to $t = 20$ (year 2030).

$$\binom{\text{Average}}{\text{value}} = \frac{1}{20 - 10} \int_{10}^{20} 309e^{0.0087t}\, dt$$
Integral divided by the length of the interval

$$= \frac{309}{10} \int_{10}^{20} e^{0.0087t}\, dt$$

$$= 30.9 \frac{1}{0.0087} e^{0.0087t}\Big|_{10}^{20}$$
Integrating by the $\int e^{ax}\, dx$ formula

$$\approx 3552e^{0.174} - 3552e^{0.087} \approx 352$$
Evaluating, using a calculator

The average population of the United States during the third decade of the twenty-first century will be about 352 million people.

⚠ **BE CAREFUL:** When finding an average value, it is a common mistake to forget to divide by $b - a$. But that would be like finding the average of three numbers by adding them up and forgetting to divide by 3.

**Practice Problem 1**

Find the average value of $f(x) = 3x^2$ from $x = 0$ to $x = 2$.

➤ Solution on page 373

### Area Between Curves: Integrating "Upper Minus Lower"

We know that definite integrals give areas under curves. To calculate the area *between* two curves, we take the area under the *upper* curve and subtract the area under the *lower* curve.

In terms of integrals:

$$\int_a^b f(x)\, dx \qquad - \qquad \int_a^b g(x)\, dx \qquad = \qquad \int_a^b [f(x) - g(x)]\, dx$$

↑ ↑
Upper Lower
curve curve

Therefore, the area between the curves can be written as a single integral:

| Area Between Curves |
| --- |
| The area between two continuous curves $f(x) \geq g(x)$ on $[a, b]$ is $$\left(\begin{array}{c}\text{Area between} \\ f \text{ and } g \text{ on } [a, b]\end{array}\right) = \int_a^b [f(x) - g(x)]\, dx \qquad \text{Integrate "upper minus lower"}$$ |

Finding area by integrating "upper minus lower" works regardless of whether one or both curves dip below the $x$-axis.

**EXAMPLE 3**

**FINDING THE AREA BETWEEN CURVES**

Find the area between $y = 3x^2 + 4$ and $y = 2x - 1$ from $x = -1$ to $x = 2$.

*Solution*

The area is shown in the diagram on the left. We integrate "upper minus lower" between the given *x*-values.

$$\int_{-1}^{2} [\underbrace{(3x^2 + 4)}_{\text{Upper}} - \underbrace{(2x - 1)}_{\text{Lower}}] \, dx$$

$$= \int_{-1}^{2} (3x^2 + 4 - 2x + 1) \, dx$$

$$= \int_{-1}^{2} (3x^2 - 2x + 5) \, dx \qquad \text{Simplifying}$$

$$= (x^3 - x^2 + 5x)\Big|_{-1}^{2} \qquad \text{Integrating}$$

$$= (8 - 4 + 10) - (-1 - 1 - 5) \qquad \text{Evaluating}$$

$$= 21 \text{ square units}$$

 **BE CAREFUL:** When doing other problems, you may need to make a similar rough sketch to see which curve is "upper" and which is "lower."

**Practice Problem 2**

Find the area between
$y = 2x^2 + 1$ and $y = -x^2 - 1$
from $x = -1$ to $x = 1$.

➤ **Solution on page 373**

If the two curves represent *rates* (one unit per another unit), then the area between the curves gives the *total accumulation* at the upper rate minus the lower rate.

**EXAMPLE 4**

**FINDING SALES FROM EXTRA ADVERTISING**

A company marketing high-definition television sets expects to sell them at the rate of $2e^{0.05t}$ thousand sets per month, where *t* is the number of months since they became available. However, with additional advertising using a sports celebrity, they should sell at the rate of $3e^{0.1t}$ thousand sets per month. How many additional sales would result from the celebrity endorsement during the first year?

*Solution*

We integrate the difference of the rates from month $t = 0$ (the beginning of the first year) to month $t = 12$ (the end of the year):

$$\int_0^{12} (3e^{0.1t} - 2e^{0.05t})\, dt = \left( 3\,\underbrace{\frac{1}{0.1}}\, e^{0.1t} - 2\,\underbrace{\frac{1}{0.05}}\, e^{0.05t} \right) \Bigg|_0^{12}$$

$$\underbrace{\phantom{3\frac{1}{0.1}}}_{30} \qquad \underbrace{\phantom{2\frac{1}{0.05}}}_{40}$$

$$= \underbrace{(30e^{1.2} - 40e^{0.6})}_{\text{Evaluating at } t = 12} \quad - \quad \underbrace{(30e^0 - 40e^0)}_{\text{Evaluating at } t = 0}$$

$$\approx \underbrace{26.7 - (-10)}_{\text{Using a calculator}} = 36.7 \qquad \text{In thousands}$$

The celebrity endorsement should result in 36.7 thousand, or 36,700 additional sales during the first year. (The profits from these additional sales must then be compared to the cost of the celebrity endorsement to decide whether it is worthwhile.)

## Area Between Curves That Cross

At a point where two curves cross, the upper curve becomes the lower curve and the lower becomes the upper. The area between them must then be calculated by two (or more) integrals, upper minus lower on *each* interval.

**EXAMPLE 5**

**FINDING THE AREA BETWEEN CURVES THAT CROSS**

Find the area between the curves $y = 12 - 3x^2$ and $y = 4x + 5$ from $x = 0$ to $x = 3$.

*Solution*

A sketch shows that the curves *do* cross. To find the intersection point, we set the functions equal to each other and solve.

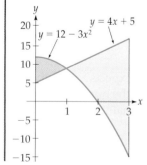

$$4x + 5 = 12 - 3x^2 \qquad \text{Equating the two functions}$$

$$3x^2 + 4x - 7 = 0 \qquad \text{Combining all terms on the left}$$

$$(3x + 7)(x - 1) = 0 \qquad \text{Factoring}$$

$$x = 1 \qquad \begin{array}{l}\text{The other solution, } x = -7/3,\\ \text{is not in the interval } [0, 3]\end{array}$$

The curves cross at $x = 1$, so we must integrate separately over the intervals [0, 1] and [1, 3]. The diagram shows which curve is upper and which is lower on each interval.

$$\int_0^1 \underbrace{[(12 - 3x^2)}_{\text{Upper}} - \underbrace{(4x + 5)]}_{\text{Lower}} \, dx + \int_1^3 \underbrace{[(4x + 5)}_{\text{Upper}} - \underbrace{(12 - 3x^2)]}_{\text{Lower}} \, dx$$

Integrating upper minus lower on each interval

$$= \int_0^1 (-3x^2 - 4x + 7) \, dx + \int_1^3 (3x^2 + 4x - 7) \, dx$$

Simplifying the integrands

$$= (-x^3 - 2x^2 + 7x)\Big|_0^1 + (x^3 + 2x^2 - 7x)\Big|_1^3$$

Integrating and simplifying

$$= -1 - 2 + 7 + 27 + 18 - 21 - (1 + 2 - 7) = 32$$   Evaluating

Therefore, the area between the curves is 32 square units.

## Graphing Calculator Exploration

Signed area

```
fnInt(Y₁-Y₂,X,0,
3)
            -24
fnInt(abs(Y₁-Y₂)
,X,0,3)
    32.00000372
```

Area (with absolute values, interpret as 32)

In Example 5 we used *two* integrals, since "upper" and "lower" switched at $x = 1$.

**a.** To see what happens if you integrate *without* regard to upper and lower, enter $y_1 = 12 - 3x^2$ and $y_2 = 4x + 5$ and graph them on the window $[0, 3]$ by $[-20, 20]$. Have your calculator find the definite integral of $y_1 - y_2$ on the interval $[0, 3]$. (Use a command like FnInt or $\int f(x)\, dx$. You will get a negative answer, which cannot be correct for an area.)

**b.** Explain why the answer was negative. [*Hint:* Look at the graph.]

**c.** Finally, obtain the correct answer for the area by returning to the calculation in part (a) and integrating the *absolute value* of the difference, $|y_1 - y_2|$ [on some calculators, entered as ABS($y_1 - y_2$)].

The *x*-values *a* and *b* are where the curves meet.

## Areas Bounded by Curves

It is sometimes useful to find the *area bounded by two curves*, without being told the starting and ending *x*-values. In such problems the curves completely enclose an area, and the *x*-values for the upper and lower limits of integration are found by setting the functions equal to each other and solving.

**EXAMPLE 6**

### FINDING AN AREA BOUNDED BY CURVES

Find the area bounded by the curves

$$y = 3x^2 - 12 \quad \text{and} \quad y = 12 - 3x^2$$

*Solution*

The $x$-values for the upper and lower limits of integration are not given, so we find them by setting the functions equal to each other and solving.

$$3x^2 - 12 = 12 - 3x^2 \qquad \text{Setting the functions equal to each other}$$

$$6x^2 - 24 = 0 \qquad \text{Combining everything on one side}$$

$$6(x^2 - 4) = 0 \qquad \text{Factoring}$$

$$6(x + 2)(x - 2) = 0 \qquad \text{Factoring further}$$

$$x = -2 \quad \text{and} \quad x = 2 \qquad \text{Solving}$$

The smaller of these, $x = -2$, is the lower limit of integration and the larger, $x = 2$, is the upper limit. To determine which function is "upper" and which is "lower," we choose a "test value" between $x = -2$ and $x = 2$ ($x = 0$ will do), which we substitute into each function to see which is larger. Evaluating each of the original functions at the test point $x = 0$ yields

$$3x^2 - 12 = 3(0)^2 - 12 = -12 \qquad \text{(Smaller)} \qquad \begin{matrix} 3x^2 - 12 \\ \text{at } x = 0 \end{matrix}$$

$$12 - 3x^2 = 12 - 3(0)^2 = 12 \qquad \text{(Larger)} \qquad \begin{matrix} 12 - 3x^2 \\ \text{at } x = 0 \end{matrix}$$

Therefore, $y = 12 - 3x^2$ is the "upper" function (since it gives a higher $y$-value) and $y = 3x^2 - 12$ is the "lower" function. We then integrate upper minus lower between the $x$-values found earlier.

$$\int_{-2}^{2} [\underbrace{12 - 3x^2}_{\text{Upper}} - (\underbrace{3x^2 - 12}_{\text{Lower}})]\, dx = \int_{-2}^{2} (24 - 6x^2)\, dx \qquad \text{Simplifying}$$

$$= (24x - 2x^3) \Big|_{-2}^{2} \qquad \text{Integrating}$$

$$= (48 - 16) - (-48 + 16) = 64 \qquad \text{Evaluating}$$

The area bounded by the two curves is 64 square units.

The two curves $y = 12 - 3x^2$ and $y = 3x^2 - 12$ are shown in the graph on the left. Notice that we were able to calculate the area between them without having to graph them.

**Practice Problem 3**

Find the area bounded by $y = 2x^2 - 1$ and $y = 2 - x^2$.

➤ **Solution on next page**

Integrate *upper minus lower* on three intervals.

For curves that intersect at *more* than two points, several integrals may be needed, integrating "upper minus lower" on each interval, as in Example 5. Test points in each interval will determine the upper and lower curves on that interval.

---

### Solutions to Practice Problems

**1.** $\dfrac{1}{2}\displaystyle\int_0^2 3x^2\,dx = \dfrac{1}{2}\cdot x^3\,\Big|_0^2 = \dfrac{1}{2}\cdot 2^3 - \dfrac{1}{2}\cdot 0^3 = \dfrac{1}{2}\cdot 8 = 4$

**2.** $\displaystyle\int_{-1}^1 [(2x^2+1)-(-x^2-1)]\,dx = \int_{-1}^1 (2x^2+1+x^2+1)\,dx$

$= \displaystyle\int_{-1}^1 (3x^2+2)\,dx = (x^3+2x)\,\Big|_{-1}^1 = (1+2)-(-1-2) = 6\text{ square units}$

**3.** $2x^2 - 1 = 2 - x^2$

$3x^2 - 3 = 0$

$3(x^2 - 1) = 0$

$3(x+1)(x-1) = 0$

$x = 1 \qquad$ and $\qquad x = -1.$

Test value $x = 0$ shows that $2 - x^2$ is "upper" and $2x^2 - 1$ is "lower."

$\displaystyle\int_{-1}^1 [(2-x^2)-(2x^2-1)]\,dx = \int_{-1}^1 (3-3x^2)\,dx = (3x - x^3)\,\Big|_{-1}^1$

$= (3-1)-(-3+1) = 4\text{ square units}$

---

## 5.4 Section Summary

The average value of a continuous function over an interval is defined as the definite integral of the function divided by the length of the interval:

$$\left(\begin{array}{c}\text{Average value}\\ \text{of } f \text{ on } [a, b]\end{array}\right) = \frac{1}{b-a}\int_a^b f(x)\,dx$$

To find the area between two curves:

1. If the $x$-values are not given, set the functions equal to each other and solve for the points of intersection.

2. Use a test point within each interval to determine which curve is "upper" and which is "lower."

3. Integrate "upper minus lower" on each interval.

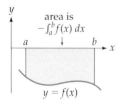

area is
$-\int_a^b f(x)\,dx$

$y = f(x)$

If a curve lies *below* the $x$-axis, as in the diagram on the left, then the "upper" curve is the $x$-axis $(y = 0)$ and the "lower" curve is $y = f(x)$, and so integrating "upper minus lower" results in integrating the *negative* of the function.

$$\int_a^b \underbrace{[0}_{\text{Upper}} - \underbrace{f(x)]}_{\text{Lower}}\,dx = \int_a^b [-f(x)]\,dx = -\int_a^b f(x)\,dx$$

The case of area below the $x$-axis need not be remembered separately if you simply remember one idea:

To find area, always integrate "upper minus lower" over each interval.

Definite integration provides a very powerful method for finding areas, taking us far beyond the few formulas (for rectangles, triangles, and circles) that we knew before studying calculus.

## Derivation of the Average Value Formula from Riemann Sums

The formula for the average value of a function (page 366) can be derived using Riemann sums: For a continuous function $f$ we could define a "sample average" by "sampling" the function at $n$ points and averaging the results. From the interval $[a, b]$, we choose $n$ numbers $x_1, x_2, \ldots, x_n$ from successive subintervals of length $\Delta x = \dfrac{b-a}{n}$, sum the resulting values of the function, and divide by $n$. This gives a "sample average" of the following form:

$$\frac{1}{n}[f(x_1) + f(x_2) + \cdots + f(x_n)] \qquad \begin{array}{l}\text{Sum of } n \text{ values}\\ \text{divided by } n\end{array}$$

$$= \frac{1}{b-a} \cdot \underbrace{\frac{b-a}{n}}_{\Delta x}[f(x_1) + f(x_2) + \cdots + f(x_n)] \qquad \begin{array}{l}\text{Dividing and}\\ \text{multiplying by } b - a\end{array}$$

$$= \frac{1}{b-a}\underbrace{[f(x_1) + f(x_2) + \cdots + f(x_n)] \cdot \Delta x}_{\text{This is a Riemann sum for } \int_a^b f(x)\,dx} \qquad \begin{array}{l}\text{Moving } \Delta x = \dfrac{b-a}{n}\\ \text{to the right}\end{array}$$

To get a more "representative" average, we increase the number $n$ of sample points. Letting $n$ approach infinity makes the above Riemann sum approach the definite integral $\int_a^b f(x)\,dx$, leading to the definition of the average value of a function:

$$\left(\begin{array}{c}\text{Average value}\\ \text{of } f \text{ on } [a, b]\end{array}\right) = \frac{1}{b-a}\int_a^b f(x)\,dx$$

## 5.4 Exercises

### Average Value

**1–26.** Find the average value of each function over the given interval.

**1.** $f(x) = x^2$ on $[0, 3]$  **2.** $f(x) = x^3$ on $[0, 2]$

**3.** $f(x) = 3\sqrt{x}$ on $[0, 4]$  **4.** $f(x) = \sqrt[3]{x}$ on $[0, 8]$

**5.** $f(x) = \dfrac{1}{x^2}$ on $[1, 5]$  **6.** $f(x) = \dfrac{1}{x^2}$ on $[1, 3]$

**7.** $f(x) = 2x + 1$ on $[0, 4]$

**8.** $f(x) = 4x - 1$ on $[0, 10]$

**9.** $f(x) = 36 - x^2$ on $[-2, 2]$

**10.** $f(x) = 9 - x^2$ on $[-3, 3]$

**11.** $f(z) = 3z^2 - 2z$ on $[-1, 2]$

**12.** $f(z) = 4z - 3z^2$ on $[-2, 2]$

**13.** $f(x) = 3$ on $[10, 50]$

**14.** $f(x) = 2$ on $[5, 100]$

**15.** $f(x) = e^{x/2}$ on $[0, 2]$

**16.** $f(x) = e^{-2x}$ on $[0, 1]$

**17.** $f(t) = e^{0.01t}$ on $[0, 10]$

**18.** $f(t) = e^{-0.1t}$ on $[0, 10]$

**19.** $f(x) = \dfrac{1}{x}$ on $[1, 2]$

**20.** $f(x) = \dfrac{1}{x}$ on $[1, 10]$

**21.** $f(x) = x^n$ on $[0, 1]$, where $n$ is a constant $(n > 0)$

**22.** $f(x) = e^{kx}$ on $[0, 1]$, where $k$ is a constant $(k \neq 0)$

**23.** $f(x) = ax + b$ on $[0, 2]$, where $a$ and $b$ are constants

**24.** $f(x) = \dfrac{1}{x}$ on $[1, c]$, where $c$ is a constant $(c > 1)$

 **25.** $f(x) = e^{-x^4}$ on $[-1, 1]$

 **26.** $f(x) = \sqrt{1 + x^4}$ on $[-2, 2]$

### Applied Exercises on Average Value

**27–28. BUSINESS: Sales** A store's sales on day $x$ are given by the function $S(x)$ below. Find the average sales during the first 3 days (day 0 to day 3).

**27.** $S(x) = 200x + 6x^2$  **28.** $S(x) = 400x + 3x^2$

**29. GENERAL: Temperature** The temperature at time $t$ hours is $T(t) = -0.3t^2 + 4t + 60$ (for $0 \leq t \leq 12$). Find the average temperature between time 0 and time 10.

**30. BEHAVIORAL SCIENCE: Practice** After $x$ practice sessions, a person can accomplish a task in $f(x) = 12x^{-1/2}$ minutes. Find the average

time required from the end of session 1 to the end of session 9.

**31. ENVIRONMENTAL SCIENCE: Pollution** The amount of pollution in a lake $x$ years after the closing of a chemical plant is $P(x) = 100/x$ tons (for $x \geq 1$). Find the average amount of pollution between 1 and 10 years after the closing.

**32. GENERAL: Population** The population of the United States is predicted to be $P(t) = 309e^{0.0087t}$ million, where $t$ is the number of years after the year 2010. Find the average population between the years 2010 and 2060.

*Source: U.S. Census Bureau*

33. **BUSINESS: Compound Interest** A deposit of $1000 at 5% interest compounded continuously will grow to $V(t) = 1000e^{0.05t}$ dollars after $t$ years. Find the average value during the first 40 years (that is, from time 0 to time 40).

34. **BIOMEDICAL: Bacteria** A colony of bacteria is of size $S(t) = 300e^{0.1t}$ after $t$ hours. Find the average size during the first 12 hours (that is, from time 0 to time 12).

35. **BUSINESS: Average Income** Microsoft's net income during the years 2003–2007 was approximately $I(x) = -16x^3 + 72x^2 - 60x + 90$ hundred million dollars, where $x$ is the number of years since 2003. Find the average income from 2003 to 2007.

*Source:* Standard & Poor's

36. **BUSINESS: Average Sales** Sales of Apple Macintosh computers in recent years have been approximately $S(x) = 2x^2 - 9x + 39$ hundred thousand units, where $x$ is the number of years since 2000. Find the average sales from 2000 to 2006.

*Source:* Apple

## Area

37. Find the area between the curve $y = x^2 + 1$ and the line $y = 2x - 1$ (shown below) from $x = 0$ to $x = 3$.

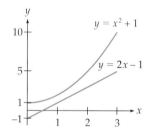

38. Find the area between the curve $y = x^2 + 3$ and the line $y = 2x$ (shown below) from $x = 0$ to $x = 3$.

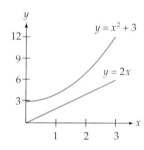

39. Find the area between the curves $y = e^x$ and $y = e^{2x}$ (shown below) from $x = 0$ to $x = 2$. (Leave the answer in its exact form.)

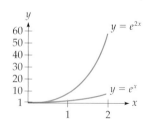

40. Find the area between the curves $y = e^x$ and $y = e^{-x}$ (shown in the following diagram) from $x = 0$ to $x = 1$. (Leave the answer in its exact form.)

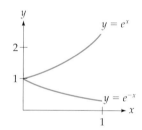

41. **a.** Sketch the parabola $y = x^2 + 4$ and the line $y = 2x + 1$ on the same graph.
    **b.** Find the area between them from $x = 0$ to $x = 3$.

42. **a.** Sketch the parabola $y = x^2 + 5$ and the line $y = 2x + 3$ on the same graph.
    **b.** Find the area between them from $x = 0$ to $x = 3$.

43. **a.** Sketch the parabola $y = 3x^2 - 3$ and the line $y = 2x + 5$ on the same graph.
    **b.** Find the area between them from $x = 0$ to $x = 3$.

44. **a.** Sketch the parabola $y = 3x^2 - 12$ and the line $y = 2x - 11$ on the same graph.
    **b.** Find the area between them from $x = 0$ to $x = 3$.

45–60. Find the area bounded by the given curves.

45. $y = x^2 - 1$ and $y = 2 - 2x^2$

46. $y = x^2 - 4$ and $y = 8 - 2x^2$

47. $y = 6x^2 - 10x - 8$ and $y = 3x^2 + 8x - 23$

**48.** $y = 3x^2 - x - 1$  and  $y = 5x + 8$

**49.** $y = 3x^2 - 12x$  and  $y = 0$

**50.** $y = x^2 - 6x$  and  $y = 0$

**51.** $y = 3x^2$  and  $y = 12$

**52.** $y = x^2$  and  $y = 4$

**53.** $y = x^2$  and  $y = x^3$

**54.** $y = x^3$  and  $y = x^4$

**55.** $y = 4x^3 + 3$  and  $y = 4x + 3$

**56.** $y = x^3$  and  $y = 4x$

**57.** $y = 7x^3 - 36x$  and  $y = 3x^3 + 64x$

**58.** $y = x^n$  and  $y = x^{n-1}$  (for  $n > 1$)

 **59.** $y = e^x$  and  $y = x + 3$

[*Hint for Exercises 59–60:* Use INTERSECT to find the intersection points for the upper and lower limits of integration.]

 **60.** $y = \ln x$  and  $y = x - 2$

## Applied Exercises on Area

**61. GENERAL: Population** The birthrate in Africa has increased from  $17e^{0.02t}$  million births per year to  $22e^{0.02t}$  million births per year, where $t$ is the number of years since 2000. Find the total increase in population that will result from this higher birth rate between 2000 ($t = 0$) and 2020 ($t = 20$).

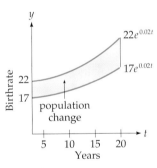

*Source: New York Times*

**62. BUSINESS: Profit from Expansion** A company expects profits of  $60e^{0.02t}$  thousand dollars per month, but predicts that if it builds a new and larger factory, its profits will be  $80e^{0.04t}$  thousand dollars per month, where $t$ is the number of months from now. Find the extra profits resulting from the new factory during the first two years  ($t = 0$  to  $t = 24$).  If the new factory will cost $1,000,000, will this cost be paid off during the first two years?

**63. BUSINESS: Net Savings** A factory installs new machinery that saves  $S(x) = 1200 - 20x$  dollars per year, where $x$ is the number of years since installation. However, the cost of maintaining the new machinery is  $C(x) = 100x$  dollars per year.

**a.** Find the year $x$ at which the maintenance cost $C(x)$ will equal the savings $S(x)$. (At this time, the new machinery should be replaced.)

**b.** Find the accumulated net savings [savings $S(x)$ minus cost $C(x)$] during the period from  $t = 0$ to the replacement time found in part (a).

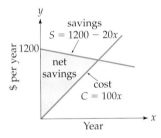

**64. GENERAL: Design** A graphic design consists of a white square containing the blue shape shown below. Find the area of the blue interior.

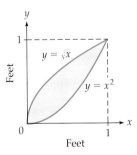

**65. ECONOMICS: Balance of Trade** A country's annual imports are  $I(t) = 30e^{0.2t}$  and its exports are  $E(t) = 25e^{0.1t}$,  both in billions of dollars, where $t$ is measured in years and  $t = 0$  corresponds to the beginning of 2000. Find the country's accumulated trade deficit (imports minus exports) for the 10 years beginning with 2000.

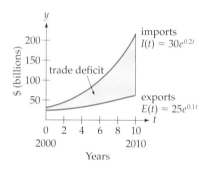

**66. ECONOMICS: Cost of Labor Contracts** An employer offers to pay workers at the rate of $30{,}000e^{0.04t}$ dollars per year, while the union demands payment at the rate of $30{,}000e^{0.08t}$ dollars per year, where $t = 0$ corresponds to the beginning of the contract. Find the accumulated difference in pay between these two rates over the 10-year life of the contract.

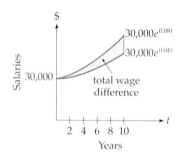

**67–68. BUSINESS: Cumulative Profit** A company's marginal revenue function is $MR(x) = 700x^{-1}$ and its marginal cost function is $MC(x) = 500x^{-1}$

(both in thousands of dollars), where $x$ is the number of units $(x > 1)$. Find the total profit from

**67.** $x = 100$ to $x = 200$

**68.** $x = 200$ to $x = 300$

**69. GENERAL: Lives Saved by Seat Belts** Seat belt use in the United States has risen to 81%, but nonusers still risk needless expense and serious injury. The upper line in the following graph represent an estimate of fatalities per year if seat belts were *not* used, and the lower line is a prediction of actual fatalities with seat belt use ($x$ represents years after 2006). Therefore, the area between the lines represents lives saved by seat belts. Find the area between the curves from 0 to 20, giving an estimate of the number of lives saved by seat belts during the years 2006–2026.

*Source:* National Highway Traffic Safety Commission

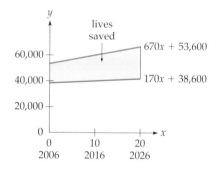

## Review Exercises

These exercises review material that will be helpful in Section 5.6.

**70–74.** Find the derivative of each function.

**70.** $e^{x^3+6x}$

**71.** $e^{x^2+5x}$

**72.** $\ln(x^3 + 6x)$

**73.** $\ln(x^2 + 5x)$

**74.** $\sqrt{x^3 + 1}$

## Conceptual Exercises

**75.** In the graph on page 367, the lowest and highest $y$-values are 0 and 3, the midpoint of which is $1\frac{1}{2}$. Yet the average of the *function* was found to be 2. Looking at the shape of the curve, can you explain why the average of the function is *larger* than the "middle" $y$-value?

**76.** Generalizing your observation in the preceding Exercise, what can you say about the average value of a function compared to its "middle" $y$-value if the curve is concave *down*? What if the curve is concave *up*?

**77.** If a linear function passes through two points $(x_1, y_1)$ and $(x_2, y_2)$, what is the average value of the function on the interval from $x_1$ to $x_2$?

**78.** If the values of a function on an interval are always positive, can the average value of the function over that interval be negative?

**79.** If the values of a function on an interval are always greater than 7, what can you say about the average value of the function on that interval?

**80.** Can the average value of a function on an interval be larger than the *maximum* value of the function on that interval? Can it be smaller than the *minimum* value on that interval?

**81.** If the average value of $f(x)$ on an interval is a number $c$, what will be the average value of the function $-f(x)$ on that interval?

**82.** If two curves cross twice, you can find the area contained by them by evaluating *one* definite integral (integrating "upper minus lower"). What if the curves cross *three* times—how many integrations of "upper minus lower" would you need? What if the curves cross *ten* times?

**83.** Suppose that a company found its sales rate (in sales per day) if it *did* advertise, and also its (lower) sales rate if it did *not* advertise. If you integrated "upper minus lower" over a month, describe the meaning of the number that you would find.

**84.** Suppose that a company found its rate of revenue (dollars per day) and its (lower) rate of costs (also in dollars per day). If you integrated "upper minus lower" over a month, describe the meaning of the number that you would find.

**5.5** **TWO APPLICATIONS TO ECONOMICS: CONSUMERS' SURPLUS AND INCOME DISTRIBUTION**

### Introduction

In this section we discuss several important economic concepts—**consumers' surplus, producers' surplus,** and the **Gini index of income distribution**—each of which is defined as the area between two curves.

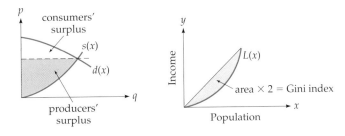

### Consumers' Surplus

Imagine that you really liked pizza and were willing to pay $12 for a pizza pie. If, in fact, a pizza costs only $8, then you have, in some sense, "saved" $4, the $12 that you were willing to pay minus the $8 market price. If one were to add up this difference for all pizzas sold in a given period of time (the price that each consumer was willing to pay minus the price actually paid), the total savings would be called the *consumers' surplus* for that product. The consumers' surplus measures the benefit that consumers derive from an economy in which competition keeps prices low.

## Demand Functions

Price and quantity are inversely related: if the price of an item rises, the quantity sold generally falls, and vice versa. Through market research, economists can determine the relationship between price and quantity for an item. This relationship can often be expressed as a **demand function** (or **demand curve**) $d(x)$, so called because it gives the price at which exactly $x$ units will be demanded.

**Demand Function**

The demand function $d(x)$ for a product gives the price at which exactly $x$ units will be sold.

$$d(x) = \left(\begin{array}{c} \text{Price when} \\ \text{demand is } x \end{array}\right)$$

On page 210 we called $d(x)$ the *price function*.

## Mathematical Definition of Consumers' Surplus

The demand curve gives the price that consumers are *willing* to pay, and the *market price* is what they *do* pay, so the amount by which the demand curve is above the market price measures the benefit or "surplus" to consumers. We add up all of these benefits by integrating, so the area between the demand curve and the market price line gives the *total benefit* that consumers derive from being able to buy at the market price. This total benefit (the shaded area in the diagram on the left) is called the *consumers' surplus*.

**Consumers' Surplus**

For a demand function $d(x)$ and demand level $A$, the market price $B$ is the demand function evaluated at $x = A$, so that $B = d(A)$. The consumers' surplus is the area between the demand curve and the market price.

$$\left(\begin{array}{c} \text{Consumers'} \\ \text{surplus} \end{array}\right) = \int_0^A [d(x) - B]\, dx$$

$\uparrow$ Demand function    $\uparrow$ Market price

**EXAMPLE 1**  **FINDING CONSUMERS' SURPLUS FOR ELECTRICITY**

If the demand function for electricity is $d(x) = 1100 - 10x$ dollars (where $x$ is in millions of kilowatt-hours, $0 \le x \le 100$), find the consumers' surplus at the demand level $x = 80$.

*Solution*

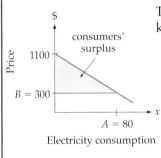

The market price is the demand function $d(x)$ evaluated at $x = 80$.

$$\left(\begin{array}{c}\text{Market} \\ \text{price } B\end{array}\right) = d(80) = 1100 - 10 \cdot 80 = 300 \qquad \begin{array}{l} d(x) = 1100 - 10x \\ \text{at } x = 80 \end{array}$$

The consumers' surplus is the area between the demand curve and the market price line.

$$\left(\begin{array}{c}\text{Consumers'} \\ \text{surplus}\end{array}\right) = \int_0^{80} (\underbrace{1100 - 10x}_{\substack{\text{Demand} \\ \text{curve}}} - \underbrace{300}_{\substack{\text{Market} \\ \text{price}}}) \, dx$$

$$= \int_0^{80} (800 - 10x) \, dx = (800x - 5x^2)\Big|_0^{80}$$

$$= (64{,}000 - 32{,}000) - (0) = 32{,}000$$

Therefore, the consumers' surplus for electricity is $32,000.

## How Consumers' Surplus Is Used

In Example 1, at demand level $x = 80$, the consumers' surplus was $32,000. If electricity usage were to increase to $x = 90$, the market price would then drop to $d(90) = 1100 - 10 \cdot 90 = 200$. We could then calculate the consumers' surplus at this higher demand level (and would find that the answer is $40,500). Therefore, a price decrease from $300 to $200 would mean that consumers would benefit by an additional $40,500 - $32,000 = $8500. This benefit would then be compared to the cost of a new generator to decide whether the expenditure would be worthwhile.

## Graphing Calculator Exploration

```
fnInt(Y₁-Y₂(80),
X,0,80)
              32000
fnInt(Y₁-Y₂(90),
X,0,90)
              40500
```

**a.** Verify the solution to Example 1 on your graphing calculator by entering $y_1 = 1100 - 10x$ and then using FnInt or $\int f(x)\,dx$ to integrate $y_1(x) - y_1(80)$ from 0 to 80.

**b.** Find the consumers' surplus at demand level 90 by integrating $y_1(x) - y_1(90)$ from 0 to 90. [*Hint:* Simply return to the calculation of part (a) and replace 80 by 90. Your answer should agree with that in the preceding paragraph.]

## Producers' Surplus

Just as the consumers' surplus measures the total benefit to consumers, the *producers' surplus* measures the total benefit that producers derive from being able to sell at the market price. Returning to our pizza example, if a pizza producer might be just willing to remain in business if the price of pizzas dropped to $5, the fact that pizzas can be sold for $8 gives the producer a "benefit" of $3. The sum of all such benefits is the *producers' surplus* for a product.

## Supply Functions

Clearly, as the price of an item rises, so does the quantity that producers are willing to supply at that price. The relationship between the price of an item and the quantity that producers are willing to supply at that price can be expressed as a **supply function** (or **supply curve**) $s(x)$.

### Supply Function

The supply function $s(x)$ for a product gives the price at which exactly $x$ units will be supplied.

$$s(x) = \begin{pmatrix} \text{Price when} \\ \text{supply is } x \end{pmatrix}$$

## Mathematical Definition of Producers' Surplus

As before, we integrate to find the total benefit, but now "upper" is the market price and "lower" is the supply curve $s(x)$.

### Producers' Surplus

For a supply function $s(x)$ and demand level $A$, the market price is $B = s(A)$. The producers' surplus is the area between the market price and the supply curve.

$$\begin{pmatrix} \text{Producers'} \\ \text{surplus} \end{pmatrix} = \int_0^A [B - s(x)]\, dx$$

          ↑    ↑

Market  Supply
price  function

**EXAMPLE 2**  **FINDING PRODUCERS' SURPLUS**

For the supply function $s(x) = 0.09x^2$ dollars and the demand level $x = 200$, find the producers' surplus.

*Solution*

The market price is the supply function $s(x)$ evaluated at $x = 200$.

$$\left(\begin{matrix}\text{Market} \\ \text{price } B\end{matrix}\right) = s(200) = 0.09(200)^2 = 3600 \qquad \begin{matrix} s(x) = 0.09x^2 \\ \text{at } x = 200 \end{matrix}$$

The producers' surplus is the area between the market price line and the supply curve.

$$\left(\begin{matrix}\text{Producers'} \\ \text{surplus}\end{matrix}\right) = \int_0^{200} (\underbrace{3600}_{\substack{\text{Market} \\ \text{price}}} - \underbrace{0.09x^2}_{\substack{\text{Supply} \\ \text{function}}})\, dx = (3600x - 0.03x^3)\Big|_0^{200}$$

$$= (720{,}000 - 240{,}000) - (0) = 480{,}000$$

Therefore, the producers' surplus is \$480,000.

## Consumers' Surplus and Producers' Surplus

The demand $x$ at which the supply and demand curves intersect is called the **market demand.** The consumers' surplus and the producers' surplus can be shown together on the same graph. These two areas together give a numerical measure of the total benefit that consumers and producers derive from competition, showing that both consumers and producers benefit from an open market.

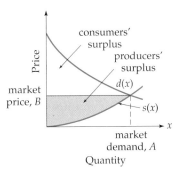

## Gini Index of Income Distribution

In any society, some people make more money than others. To measure the "gap" between the rich and the poor, economists calculate the proportion of the total income that is earned by the lowest 20% of the population, and then the proportion that is earned by the lowest 40% of the population, and so on.

This information (for the United States in the year 2006) is given in the table below (with percentages written as decimals), and is graphed on the right.

| Proportion of Population | Proportion of Income |
|---|---|
| 0.20 | 0.03 |
| 0.40 | 0.12 |
| 0.60 | 0.27 |
| 0.80 | 0.49 |
| 1.00 | 1.00 |

*Source:* U.S. Census Bureau

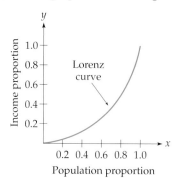

For example, the lowest 20% of the population earns only 3% of the total income, the lowest 40% earns only 12% of the total income, and so on. The curve is known as the *Lorenz curve.**

---

**Lorenz Curve**

The Lorenz curve $L(x)$ gives the proportion of total income earned by the lowest proportion $x$ of the population.

---

## Gini Index

The Lorenz curve may be compared with two extreme cases of income distribution.

1. *Absolute equality of income* means that everyone earns exactly the same income, and so the lowest 10% of the population earns exactly 10% of the total income, the lowest 20% earns exactly 20% of the income, and so on. This gives the Lorenz curve $y = x$ shown below on the left.

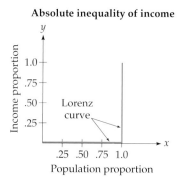

2. *Absolute inequality of income* means that nobody earns any income except one person, who earns all the income. This gives the Lorenz curve shown above on the right.

---

*After Max Otto Lorenz (1880–1962), American statistician and economist.

To measure how the actual distribution differs from absolute equality, we calculate the area between the actual distribution and the line of absolute equality $y = x$. Since this area can be at most $\frac{1}{2}$ (the area of the entire lower triangle), economists multiply the area by 2 to get a number between 0 (absolute equality) and 1 (absolute inequality). This measure is called the *Gini index*.* Note that a higher Gini index means greater *in*equality (greater deviation from the line of absolute equality).

---

**Gini Index**

For a Lorenz curve $L(x)$, the Gini index is

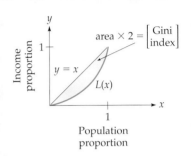

$$\left(\begin{array}{c}\text{Gini} \\ \text{index}\end{array}\right) = 2\int_0^1 [x - L(x)]\, dx$$

The Gini index varies from 0 (absolute equality) to 1 (absolute inequality).

---

**EXAMPLE 3**  **FINDING THE GINI INDEX**

The Lorenz curve for income distribution in the United States in 2006 was approximately $L(x) = x^{2.75}$. Find the Gini index.

*Solution*

First we calculate the area between the curve of absolute equality $y = x$ and the Lorenz curve $y = x^{2.75}$.

$$\int_0^1 (x - x^{2.75})\, dx = \left(\frac{1}{2}x^2 - \frac{1}{3.75}x^{3.75}\right)\Big|_0^1$$

$$= \frac{1}{2}\cdot 1^2 - \frac{1}{3.75}\cdot 1^{3.75} - 0$$

$$\approx 0.5 - 0.267 = 0.233$$

Multiplying by 2 gives the Gini index.

$$\left(\begin{array}{c}\text{Gini} \\ \text{index}\end{array}\right) = 0.47 \qquad\qquad \text{Rounding to 2 decimal places.}$$

**Practice Problem 1**

In 2000, the Gini index for income was 0.44. Since 2000, has income distribution become more equal or less equal? ➤ **Solution on next page**

*After Corrado Gini (1884–1965), Italian statistician and sociologist.

## Graphing Calculator Exploration

Lorenz exponent

In Example 3, how was the Lorenz function of the form $x^n$ found? It was found by a method called "least squares" (discussed in Section 7.4), which minimizes the squared differences between the income proportions (from the table on page 384) and the curve $x^n$ at the x-values 0.2, 0.4, 0.6, and 0.8. This amounts to minimizing the function

$$(0.2^x - 0.03)^2 + (0.4^x - 0.12)^2 + (0.6^x - 0.27)^2 + (0.8^x - 0.49)^2$$

(Notice that here we are using x for the *exponent*.)

**a.** Graph this function on your calculator on the window $[0, 5]$ by $[-0.1, 0.4]$.

**b.** Find the x that minimizes the function. The resulting value is the exponent. Your answer should agree with the exponent in Example 3.

The graph on the left, on the window $[0, 1]$ by $[0, 1]$, shows that the function $x^{2.75}$ fits the points from the table on page 384 very well. (Other types of functions besides $x^n$ could also be used to fit these data.)

The Gini index for *total* wealth can be calculated similarly.

**Practice Problem 2**

The Lorenz curve for total wealth in the United States for 2006 was $L(x) = x^{9.0}$. Calculate the Gini index.  ➤ **Solution below**

*Source:* United Nations University

Practice Problem 2 shows that the Gini index for wealth is greater than the Gini index for income. That is, total wealth in the United States is distributed more unequally than total income. One reason for this is that we have an income tax but no "wealth" tax.

**Solutions to Practice Problems**

**1.** Less equal

**2.** $\int_0^1 (x - x^{9.0})\, dx = \left(\frac{1}{2}x^2 - \frac{1}{10}x^{10}\right)\Big|_0^1 = \frac{1}{2} - \frac{1}{10} - 0 = 0.4$

Gini index for wealth is $2 \times 0.4 = 0.8$.

## 5.5 Exercises

**1–8.** For each demand function $d(x)$ and demand level x, find the consumers' surplus.

**1.** $d(x) = 4000 - 12x, \quad x = 100$

**2.** $d(x) = 500 - x, \quad x = 400$

**3.** $d(x) = 300 - \frac{1}{2}x, \quad x = 200$

**4.** $d(x) = 200 - \frac{1}{2}x, \quad x = 300$

**5.** $d(x) = 350 - 0.09x^2, \quad x = 50$

**6.** $d(x) = 840 - 0.06x^2, \quad x = 100$

**7.** $d(x) = 200e^{-0.01x}, \quad x = 100$

**8.** $d(x) = 400e^{-0.02x}, \quad x = 75$

**9–12.** For each supply function $s(x)$ and demand level $x$, find the producers' surplus.

**9.** $s(x) = 0.02x, \quad x = 100$

**10.** $s(x) = 0.4x, \quad x = 200$

**11.** $s(x) = 0.03x^2, \quad x = 200$

**12.** $s(x) = 0.06x^2, \quad x = 50$

**13–18.** For each demand function $d(x)$ and supply function $s(x)$:

a. Find the market demand (the positive value of $x$ at which the demand function intersects the supply function).

b. Find the consumers' surplus at the market demand found in part (a).

c. Find the producers' surplus at the market demand found in part (a).

**13.** $d(x) = 300 - 0.4x, \quad s(x) = 0.2x$

**14.** $d(x) = 120 - 0.16x, \quad s(x) = 0.08x$

**15.** $d(x) = 300 - 0.03x^2, \quad s(x) = 0.09x^2$

**16.** $d(x) = 360 - 0.03x^2, \quad s(x) = 0.006x^2$

**17.** $d(x) = 300e^{-0.01x}, \quad s(x) = 100 - 100e^{-0.02x}$

**18.** $d(x) = 400e^{-0.01x}, \quad s(x) = 0.01x^{2.1}$

**19–30.** Find the Gini index for the given Lorenz curve.

*Source:* United Nations University-World Income Inequality Database

**19.** $L(x) = x^{3.2}$  (the Lorenz curve for U.S. income in 1929)

**20.** $L(x) = x^3$  (the Lorenz curve for U.S. income in 1935)

**21.** $L(x) = x^{2.1}$  (the Lorenz curve for income in Sweden in 1990)

**22.** $L(x) = x^{15.3}$  (the Lorenz curve for wealth in Great Britain in 1990)

**23.** $L(x) = 0.4x + 0.6x^2$

**24.** $L(x) = 0.2x + 0.8x^3$

**25.** $L(x) = x^n$  (for $n > 1$)

**26.** $L(x) = \frac{1}{2}x + \frac{1}{2}x^n$  (for $n > 1$)

 **27.** $L(x) = \dfrac{e^{x^2} - 1}{e - 1}$

 **28.** $L(x) = 1 - \sqrt{1 - x}$

 **29.** $L(x) = \dfrac{x + x^2 + x^3}{3}$

 **30.** $L(x) = 0.62x^{7.15} + 0.38x^{9.47}$

 **31–32.** The following tables give the distribution of family income in the United States: Exercise 31 is for the year 1977 and Exercise 32 is for 1989. Use the procedure described in the Graphing Calculator Exploration on the previous page to find the Lorenz function of the form $x^n$ for the data. Then find the Gini index. If you do both problems, did family income become more concentrated or less concentrated from 1977 to 1989?

*Source:* Congressional Budget Office

**31.**

| Proportion (Lowest) of Families | Proportion of Income (1977) |
|---|---|
| 0.20 | 0.06 |
| 0.40 | 0.18 |
| 0.60 | 0.34 |
| 0.80 | 0.57 |

**32.**

| Proportion (Lowest) of Families | Proportion of Income (1989) |
|---|---|
| 0.20 | 0.04 |
| 0.40 | 0.14 |
| 0.60 | 0.29 |
| 0.80 | 0.51 |

## Review Exercises

These exercises review material that will be helpful in Section 5.6.

**33–38.** Find the derivative of each function.

**33.** $(x^5 - 3x^3 + x - 1)^4$

**34.** $(x^4 - 2x^2 - x + 1)^5$

**35.** $\ln(x^4 + 1)$

**36.** $\ln(x^3 - 1)$

**37.** $e^{x^3}$

**38.** $e^{x^4}$

*Conceptual Exercises*

**39.** Suppose that for a demand function $d(x)$ we have $d(0) = 1000$. Describe in everyday language what this means about the number 1000.

**40.** Suppose that for a demand function $d(x)$ we have $d(1000) = 0$. Describe in everyday language what this means about the number 1000.

**41.** If $d(x)$ is the demand function for a product, what would it mean about the product if $d(0) = 0$?

**42.** Should demand curves slope *upward* or *downward*? Why?

**43.** Should supply curves slope *upward* or *downward*? Why?

**44.** For a Lorenz curve $L(x)$, what must be the values of $L(0)$ and $L(1)$?

**45.** Can a Lorenz curve look like this? Explain.

**46.** What would it mean about the distribution of income if the Lorenz curve looked like this?

**47.** In the country of Equalia, the Gini index for income is, by law, fixed at 0. State one good thing and one bad thing about living in Equalia.

**48.** Does a higher Gini index mean more *equality* or more *inequality*?

## 5.6   INTEGRATION BY SUBSTITUTION

### Introduction

The Chain Rule (page 153) greatly expanded the range of functions that we could differentiate. In this section we will learn a related technique for integration, called the *substitution method*, which will greatly expand the range of functions that we can integrate. First, however, we must define **differentials.**

### Differentials

One of the notations for the derivative of a function $f(x)$ is $df/dx$. Although written as a fraction, $df/dx$ was not defined as the quotient of two quantities $df$ and $dx$, but as a single object, the *derivative*. We will now define $df$ and $dx$ separately (they are called *differentials*) so that their quotient $df \div dx$ is equal to the derivative $df/dx$. We begin with

$$\frac{df}{dx} = f'$$    Since $df/dx$ and $f'$ are both notations for the derivative

$$df = f' \, dx$$    Multiplying each side by $dx$

This leads to a definition for the differential $df$.

| **Differential** |
| --- |
| For a differentiable function $f(x)$, the differential $df$ is $$df = f'(x) \, dx$$    $df$ is the derivative times $dx$ |

BE CAREFUL: $df$ does not mean $d$ times $f$, but means the derivative of the function $f$ times $dx$. What is $dx$? We have used it as the notation at the end of integrals, but it really represents a small change in $x$, just like the $\Delta x$ that was the base of the rectangles in Riemann sums (see page 349). The reason for finding differentials will be clear shortly.

---

**EXAMPLE 1**

**FINDING DIFFERENTIALS**

| **Function $f(x)$** | **Differential $df$** |
|---|---|
| $f(x) = x^2$ | $df = 2x \, dx$ |
| $f(x) = \ln x$ | $df = \dfrac{1}{x} \, dx$ |
| $f(x) = e^{x^2}$ | $df = e^{x^2}(2x) \, dx$ |
| $f(x) = x^4 - 5x + 2$ | $df = \underbrace{(4x^3 - 5)}_{f'(x)} \, dx$ |

Each differential is the derivative of the function times $dx$

---

**Practice Problem 1**

For $f(x) = x^3 - 4x - 2$, find the differential $df$. ➤ Solution on page 398

The differential formula $df = f' \, dx$ is easy to remember because dividing both sides by $dx$ gives

$$\frac{df}{dx} = f'$$

which simply says "the derivative equals the derivative." We may use other letters besides $f$ and $x$.

---

**EXAMPLE 2**

**CALCULATING DIFFERENTIALS IN OTHER VARIABLES**

| **Function** | **Differential** |
|---|---|
| $u = x^3 + 1$ | $du = 3x^2 \, dx$ |
| $u = e^{2t} + 1$ | $du = 2e^{2t} \, dt$ |

Differentials end with $d$ followed by the variable

---

**Practice Problem 2**

For $u = e^{-5t}$, find the differential $du$. ➤ Solution on page 398

## Substitution Method

Using differential notation, we can state three very useful integration formulas.

### Substitution Formulas

(A) $\displaystyle\int u^n\, du = \frac{1}{n+1}u^{n+1} + C$  $\qquad\qquad\qquad n \neq -1$

(B) $\displaystyle\int e^u\, du = e^u + C$

(C) $\displaystyle\int \frac{du}{u} = \int \frac{1}{u}\,du = \int u^{-1}\,du = \ln|u| + C$

These formulas are easy to remember because they are exactly the formulas that we learned earlier (see pages 338 and 344) except that here we use the letter $u$ to stand for a *function*. The $du$ is the **differential** of the function. Each of these formulas may be justified by differentiating the right-hand side (see Exercises 61–63). A few examples will illustrate their use.

**EXAMPLE 3**

**INTEGRATING BY SUBSTITUTION**

Find $\displaystyle\int (x^2 + 1)^3\, 2x\, dx$.

*Solution*

The integral involves a function to a power:

$$\int (x^2 + 1)^3\, 2x\, dx \qquad\qquad (x^2+1)^3 \text{ is a function to a power}$$

$$\Big\downarrow$$

as does formula (A):  $\displaystyle\int u^n\, du$  $\qquad u^n$ is a function to a power in $\displaystyle\int u^n\, du = \frac{1}{n+1}u^{n+1} + C$

To make the integral "fit" the formula we take  $u = x^2 + 1$  and  $n = 3$.

$$\int \underbrace{(x^2 + 1)^3}_{u^3}\, 2x\, dx \qquad\qquad u = x^2 + 1 \text{ and } n = 3$$

For  $u = x^2 + 1$  the differential is  $du = 2x\, dx$,  which is exactly the remaining part of the integral. We then write the integral with each $x$-expression replaced by its equivalent $u$-expression.

$$\int \underbrace{(x^2 + 1)^3}_{u^3}\, \underbrace{2x\, dx}_{du} = \int \underbrace{u^3\, du}_{\text{Written in terms of } u} \qquad \begin{array}{l}\text{Using } u = x^2 + 1 \\ \text{and } du = 2x\, dx\end{array}$$

The last integral we solve by formula (A):

$$\int u^3 \, du = \frac{1}{4} u^4 + C \qquad\qquad \int u^n \, du = \frac{1}{n+1} u^{n+1} + C$$

$$\text{[formula (A)] with } n = 3$$

Finally, we substitute back to the original variable $x$, using our relationship $u = x^2 + 1$, to get the answer:

$$\frac{1}{4}(x^2 + 1)^4 + C \qquad\qquad \frac{1}{4}u^4 + C \text{ with } u = x^2 + 1$$

The procedure is not as complicated as it might seem. All of these steps may be written together as follows.

$$\underbrace{\int (x^2 + 1)^3}_{u^3} \underbrace{2x \, dx}_{du} \;=\; \int u^3 \, du \;=\; \frac{1}{4}u^4 + C \;=\; \frac{1}{4}(x^2 + 1)^4 + C$$

Choosing $u = x^2 + 1$ therefore $du = 2x \, dx$ 
Substituting $u^3 = (x^2 + 1)^3$ $du = 2x \, dx$ 
Integrating by formula (A) with $n = 3$ 
Substituting back to $x$ using $u = x^2 + 1$

We may check this answer by differentiation (using the Generalized Power Rule or Chain Rule).

$$\frac{d}{dx}\underbrace{\left[\frac{1}{4}(x^2 + 1)^4 + C\right]}_{\text{Answer}} = \frac{1}{4} \cdot 4(x^2 + 1)^3 \underbrace{2x}_{\substack{\text{Derivative} \\ \text{of the inside}}} = \underbrace{(x^2 + 1)^3 \, 2x}_{\text{Integrand}}$$

Since the result of the differentiation agrees with the original integrand, the integration is correct.

## Multiplying Inside and Outside by Constants

If the integral does not exactly match the form $\int u^n \, du$, we may sometimes still solve the integral by multiplying by constants.

**EXAMPLE 4**

### INSERTING CONSTANTS BEFORE SUBSTITUTING

Find $\displaystyle\int (x^2 + 1)^3 x \, dx$.   Same as Example 3 but without the 2

*Solution*

As before, we use formula (A) with $u = x^2 + 1$, which gives $du = 2x \, dx$. But the integral has only an $x \, dx$, not $\underline{2x \, dx}$, which would allow us to substitute $du$.

$$\int \underbrace{(x^2 + 1)^3}_{u^3} \underbrace{x \, dx}_{} \qquad\qquad u = x^2 + 1, \quad \text{so} \quad du = 2x \, dx$$

not $du = 2x \, dx$
because there is no 2

Therefore, the integral is *not* in the form $\int u^3 \, du$. (The integral must fit the formula exactly: *everything* in the integral must be accounted for either by the $u^n$ or by the $du$.) However, we may multiply inside the integral by 2 as long as we compensate by also multiplying by $\frac{1}{2}$, and the $\frac{1}{2}$ may be written *outside* the integral (since constants may be moved across the integral sign), leading to the solution:

Multiplying by $\frac{1}{2}$ and 2

$$\frac{1}{2} \int \underbrace{(x^2 + 1)^3}_{u^3} \underbrace{2x \, dx}_{du} = \frac{1}{2} \int u^3 \, du = \frac{1}{2} \cdot \frac{1}{4} u^4 + C \quad = \frac{1}{8}(x^2 + 1)^4 + C$$

$u = x^2 + 1$     Substituting    Integrating    Substituting back
$du = 2x \, dx$                by formula (A)    to $x$ using
                                                      $u = x^2 + 1$

This method of multiplying inside and outside by a constant is very useful, and may be used with the other substitution formulas as well.

**EXAMPLE 5**

**USING OTHER SUBSTITUTION FORMULAS**

Find $\displaystyle\int e^{x^3} x^2 \, dx$.

*Solution*

The integral involves
$e$ to a function: $\qquad\qquad \displaystyle\int e^{x^3} x^2 \, dx \qquad\qquad\qquad e^{x^3}$ is $e$ to a function

as does formula (B) on
page 390: $\qquad\qquad \displaystyle\int e^u \, du = e^u + C$

Matching exponents of $e$ gives $\;u = x^3$, and the differential is $\;du = 3x^2 \, dx$. The differential requires a 3, which is not in the integral, so we multiply inside by 3 and outside by $\frac{1}{3}$.

$$\int e^{x^3}x^2\,dx = \frac{1}{3}\int \underbrace{e^{x^3}}_{e^u}\,\underbrace{3x^2\,dx}_{du} \quad = \quad \frac{1}{3}\int e^u\,du \quad = \quad \frac{1}{3}e^u + C \quad = \quad \frac{1}{3}e^{x^3} + C$$

|  |  |  |  |  |
|---|---|---|---|---|
| $u = x^3$ | | Substituting | Integrating | Substituting |
| $du = 3x^2\,dx$ | Multiplying | | using | back to $x$ |
| | by $\frac{1}{3}$ and 3 | | formula (B) | using $u = x^3$ |

Why did the 1/3 become part of the answer but the 3 disappeared? The 3 became part of the $du$ ($du = 3x^2\,dx$), which was then "used up" in the integration along with the integral sign.

**EXAMPLE 6**

**RECOVERING COST FROM MARGINAL COST**

A company's marginal cost function is $MC(x) = \dfrac{x^3}{x^4 + 1}$ and fixed costs are $1000. Find the cost function.

*Solution*   Cost is the integral of marginal cost.

$$C(x) = \int \frac{x^3\,dx}{x^4 + 1}$$

The differential of the denominator is $4x^3\,dx$, which except for the 4 is just the numerator. This suggests formula (C), $\displaystyle\int \frac{du}{u} = \ln|u| + C$ with $u = x^4 + 1$. We multiply inside by 4 (to complete the $du = 4x^3\,dx$ in the numerator) and outside by $\frac{1}{4}$.

$$\int \frac{x^3\,dx}{x^4 + 1} = \frac{1}{4}\int \frac{\overset{du}{\overbrace{4x^3\,dx}}}{\underset{u}{x^4 + 1}} = \frac{1}{4}\int \frac{du}{u} = \frac{1}{4}\ln|u| + C = \frac{1}{4}\ln(x^4 + 1) + C$$

|  |  |  |  |  |
|---|---|---|---|---|
| $u = x^4 + 1$ | Multiplying | Substituting | Integrating by | Substituting |
| $du = 4x^3\,dx$ | by 4 and $\frac{1}{4}$ | | formula (C) | back to $x$ |

We dropped the absolute value bars because $x^4 + 1$ is positive. To evaluate the constant $C$, we set the cost function (evaluated at $x = 0$) equal to the given fixed cost.

$$\frac{1}{4}\underbrace{\ln(1)}_{0} + C = 1000$$

$\ln(x^4 + 1) + C$ at $x = 0$
set equal to 1000

This gives $C = 1000$. Therefore, the company's cost function is

$$C(x) = \frac{1}{4}\ln(x^4 + 1) + 1000$$

$C(x) = \frac{1}{4}\ln(x^4 + 1) + C$
with $C = 1000$

## Which Formula to Use

The three formulas apply to three different types of integrals.

(A) $\displaystyle\int u^n \, du = \frac{1}{n+1} u^{n+1} + C \quad (n \neq -1)$
    *Integrates a function to a constant power* (except $-1$) times the differential of the function

(B) $\displaystyle\int e^u \, du = e^u + C$
    *Integrates e to a power* times the differential of the exponent

(C) $\displaystyle\int \frac{du}{u} = \int u^{-1} \, du = \ln |u| + C$
    *Integrates a fraction whose top is the differential of the bottom,* or equivalently, *a function to the power* $-1$ times the differential of the function

To solve an integral by substitution, choose the formula whose left-hand side has the same form as the given integral.

**Practice Problem 3**

For each of the following integrals, choose the most appropriate formula: (A), (B), or (C). (Do not solve the integral.)

**a.** $\displaystyle\int e^{5x^2 - 1} x \, dx$     **b.** $\displaystyle\int \frac{x \, dx}{x^2 + 1}$     **c.** $\displaystyle\int (x^4 - 12)^4 x^3 \, dx$

**d.** $\displaystyle\int (x^4 - 12)^{-1} x^3 \, dx$       ➤ **Solutions on page 398**

 **BE CAREFUL:** Only constants can be adjusted. That is, we may multiply inside and outside only by constants, not variables, since only constants can be moved across the integral sign. Therefore, the *du* in a problem must already be "complete" except for adjusting the constant. Otherwise, the problem cannot be solved by a substitution. For example, the following integral cannot be found by a substitution.

$$\int e^{x^3} x \, dx$$

$e^u$       *not du* (does not have an $x^2$)

$u = x^3$
$du = 3x^2 \, dx$

**Practice Problem 4**

Which of these integrals can be found by a substitution? [*Hint:* See whether only a constant is needed to complete the *du*.]

**a.** $\displaystyle\int (x^3 + 1)^3 x^3 \, dx$     **b.** $\displaystyle\int e^{x^2} \, dx$       ➤ **Solutions on page 398**

| EXAMPLE 7 | **INTEGRATING BY SUBSTITUTION** |

Find $\displaystyle\int \sqrt{x^3 - 3x}\,(x^2 - 1)\,dx$.

*Solution*

Since $\sqrt{x^3 - 3x} = (x^3 - 3x)^{1/2}$ is a *function to a power*, we use the formula for $\int u^n\,du$ with $u = x^3 - 3x$. Comparing the differential $du = (3x^2 - 3)dx = 3(x^2 - 1)\,dx$ with the problem shows that we need to multiply by 3.

$$\int (x^3 - 3x)^{1/2}(x^2 - 1)\,dx = \frac{1}{3}\int \underbrace{(x^3 - 3x)^{1/2}}_{u^{1/2}}\underbrace{3(x^2 - 1)\,dx}_{du} \qquad \text{Multiplying by 3 and } \tfrac{1}{3}$$

$$\begin{aligned} u &= x^3 - 3x \\ du &= (3x^2 - 3)\,dx \\ &= 3(x^2 - 1)\,dx \end{aligned}$$

$$= \underbrace{\frac{1}{3}\int u^{1/2}\,du}_{\text{Substituting}} = \underbrace{\frac{1}{3}\cdot\frac{2}{3}u^{3/2} + C}_{\substack{\text{Integrating by}\\\text{formula (A)}}} = \underbrace{\frac{2}{9}(x^3 - 3x)^{3/2} + C}_{\substack{\text{Substituting}\\\text{back to } x}}$$

| EXAMPLE 8 | **INTEGRATING BY SUBSTITUTION** |

Evaluate $\displaystyle\int e^{\sqrt{x}}\, x^{-1/2}\,dx$.

*Solution*

The integral involves $e^{\sqrt{x}}$, $e$ to a function, so we use the formula $\int e^u\,du = e^u + C$ with $u = x^{1/2}$.

$$\int e^{x^{1/2}}x^{-1/2}\,dx = 2\int \underbrace{e^{x^{1/2}}}_{e^u}\underbrace{\frac{1}{2}x^{-1/2}\,dx}_{du} \qquad \text{Multiplying by } \tfrac{1}{2} \text{ and } 2$$

$$\begin{aligned} u &= x^{1/2} \\ du &= \frac{1}{2}x^{-1/2}\,dx \end{aligned}$$

$$= \underbrace{2\int e^u\,du}_{\text{Substituting}} = \underbrace{2e^u + C}_{\text{Integrating}} = \underbrace{2e^{x^{1/2}} + C}_{\substack{\text{Substituting}\\\text{back to } x}}$$

## Evaluating Definite Integrals by Substitution

Sometimes a *definite* integral requires a substitution. In such cases changing from $x$ to $u$ also requires changing the limits of integration from $x$-values to $u$-values, using the substitution formula for $u$.

**EXAMPLE 9**

**EVALUATING A DEFINITE INTEGRAL BY SUBSTITUTION**

Evaluate $\displaystyle\int_4^5 \frac{dx}{3-x}$.

*Solution*

The differential of the denominator $3 - x$ is $-1 \cdot dx$, which except for the $-1$ is just the numerator. This suggests formula (C), $\displaystyle\int \frac{du}{u} = \ln|u| + C$ with $u = 3 - x$ (from equating the denominators). We multiply inside and outside by $-1$.

$$u = 3 - 5 = -2$$

$$\int_4^5 \frac{dx}{3-x} = -\int_4^5 \frac{-dx}{3-x} = -\int_{-1}^{-2} \frac{du}{u}$$

$$u = 3 - x$$
$$du = -dx$$

$$u = 3 - 4 = -1$$

New upper and lower limits of integration for $u$ are found by evaluating $u = 3 - x$ at the old $x$ limits

$$= -\ln|u|\Big|_{-1}^{-2} = -\ln|-2| - (-\ln|-1|) = -\ln 2 + \ln 1 = -\ln 2$$

---

## Graphing Calculator Exploration

/f(x)dx=-.6931472

Why is the answer to Example 9 *negative*? Graph $f(x) = \dfrac{1}{3-x}$ on the window [3, 6] by [−2, 2] to see why. Have your calculator find the definite integral from 4 to 5. How would you change the integral so that it gives the *area* between the curve and the $x$-axis?

Definite integrals are used to find areas, total accumulations, and average values, and any of these uses may require a substitution.

**EXAMPLE 10**

**FINDING TOTAL POLLUTION FROM A RATE**

Pollution is being discharged into a lake at the rate of $r(t) = 400te^{t^2}$ tons per year, where $t$ is the number of years since measurements began. Find the total amount of pollutant discharged into the lake during the first 2 years.

*Solution*

The total accumulation is the definite integral from $t = 0$ (the beginning) to $t = 2$ (2 years later). Since the integral involves $e$ to a function, we use the formula for $\int e^u\, du$ with $u = t^2$ (by equating exponents).

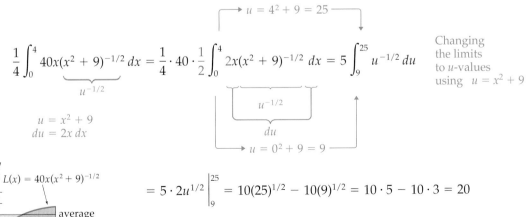

$$\int_0^2 400\, t\underbrace{e^{t^2}}_{e^u}\, dt \;=\; 400 \cdot \frac{1}{2}\int_0^2 2t\, e^{t^2}\, dt = 200 \int_0^4 e^u\, du$$

Taking out the constant

$u = t^2$
$du = 2t\, dt$

Changing the limits to $u$-values using $u = t^2$

$$= 200e^u \Big|_0^4 = 200e^4 - 200e^0 \approx 10{,}720$$

Therefore, during the first 2 years 10,720 tons of pollutant were discharged into the lake.

Notice that the *du* does not need to be all together, but can be in several separate pieces, as long as it is all *there*.

**EXAMPLE 11**  **FINDING AVERAGE WATER DEPTH**

After $x$ months the water level in a newly built reservoir is $L(x) = 40x(x^2 + 9)^{-1/2}$ feet. Find the average depth during the first 4 months.

*Solution*

The average value is the definite integral from $x = 0$ to $x = 4$ (the end of month 4) divided by the length of the interval.

$$\frac{1}{4}\int_0^4 40x\underbrace{(x^2 + 9)^{-1/2}}_{u^{-1/2}}\, dx = \frac{1}{4}\cdot 40 \cdot \frac{1}{2}\int_0^4 2x(x^2 + 9)^{-1/2}\, dx = 5\int_9^{25} u^{-1/2}\, du$$

$u = x^2 + 9$
$du = 2x\, dx$

Changing the limits to $u$-values using $u = x^2 + 9$

$$= 5 \cdot 2u^{1/2} \Big|_9^{25} = 10(25)^{1/2} - 10(9)^{1/2} = 10 \cdot 5 - 10 \cdot 3 = 20$$

That is, the average depth of the reservoir over the last 4 months was 20 feet, as shown on the left.

> ### Solutions to Practice Problems
>
> 1. $df = (3x^2 - 4)\,dx$
> 2. $du = -5e^{-5t}\,dt$
> 3. **a.** (B)          **b.** (C)          **c.** (A)          **d.** (C)
> 4. Neither.
>
> **a.** Try formula (A) with $u = x^3 + 1$, so $du = 3x^2\,dx$. The problem has an $x^3$ for the differential instead of the needed $x^2$.
>
> **b.** Try formula (B) with $u = x^2$, so $du = 2x\,dx$. The problem does not have the $x$ that is needed for the differential.

## 5.6   Section Summary

The substitution method does for integration what the Chain Rule does for differentiation—it allows the use of simpler calculus formulas on complicated mathematical expressions by seeing them in "blocks," here the $u$ and the $du$. In fact, if you use the substitution method to integrate, you should expect to use the Chain Rule to differentiate when you check your answer.

The three substitution formulas are listed on the inside back cover. Most of the work in using these formulas is making a problem "fit" the left-hand side of one of the formulas (choosing the $u$ and adjusting constants to complete the $du$). Once a problem fits a left-hand side, the right-hand side immediately gives the answer (except for substituting back to the original variable).

Note that the $du$ now plays a very important role: the $du$ must be correct if the answer is to be correct. For example, the formula $\int e^u\,du = e^u + C$ should not be thought of as the formula for integrating $e^u$, but as the formula for integrating $e^u\,du$. The $du$ is just as important as the $e^u$.

## 5.6   Exercises

**1–8.** Find each indefinite integral. [Integration formulas (A), (B), and (C) are on the inside back cover, numbered 5–7]

**1.** $\displaystyle\int (x^2 + 1)^9\,2x\,dx$

[*Hint:* Use $u = x^2 + 1$ and formula 5.]

**2.** $\displaystyle\int (x^3 + 1)^4\,3x^2\,dx$

[*Hint:* Use $u = x^3 + 1$ and formula 5.]

**3.** $\displaystyle\int (x^2 + 1)^9 x\,dx$

[*Hint:* Use $u = x^2 + 1$ and formula 5.]

**4.** $\displaystyle\int (x^3 + 1)^4 x^2\,dx$

[*Hint:* Use $u = x^3 + 1$ and formula 5.]

**5.** $\displaystyle\int e^{x^5} x^4\,dx$

[*Hint:* Use $u = x^5$ and formula 7.]

**6.** $\displaystyle\int e^{x^4} x^3\,dx$

[*Hint:* Use $u = x^4$ and formula 7.]

**7.** $\displaystyle\int \frac{x^5\,dx}{x^6 + 1}$

[*Hint:* Use $u = x^6 + 1$ and formula 6.]

**8.** $\displaystyle\int \frac{x^4\,dx}{x^5 + 1}$

[*Hint:* Use  $u = x^5 + 1$  and formula 6.]

**9–12.** Show that each integral *cannot* be found by our substitution formulas.

**9.** $\displaystyle\int \sqrt{x^3 + 1}\; x\,dx$     **10.** $\displaystyle\int \sqrt{x^5 + 9}\; x^2\,dx$

**11.** $\displaystyle\int e^{x^4} x^5\,dx$     **12.** $\displaystyle\int e^{x^3} x^4\,dx$

**13–46.**  Find each indefinite integral by the substitution method or state that it cannot be found by our substitution formulas.

**13.** $\displaystyle\int (x^4 - 16)^5 x^3\,dx$     **14.** $\displaystyle\int (x^5 - 25)^6 x^4\,dx$

**15.** $\displaystyle\int e^{-x^2} x\,dx$     **16.** $\displaystyle\int e^{-x^4} x^3\,dx$

**17.** $\displaystyle\int e^{3x}\,dx$     **18.** $\displaystyle\int e^{5x}\,dx$

**19.** $\displaystyle\int e^{x^2} x^2\,dx$     **20.** $\displaystyle\int e^{x^3} x\,dx$

**21.** $\displaystyle\int \frac{dx}{1 + 5x}$     **22.** $\displaystyle\int \frac{dx}{1 + 3x}$

**23.** $\displaystyle\int (x^2 + 1)^9 5x\,dx$     **24.** $\displaystyle\int (x^2 - 4)^6 3x\,dx$

**25.** $\displaystyle\int \sqrt[4]{z^4 + 16}\; z^3\,dz$     **26.** $\displaystyle\int \sqrt[3]{z^3 - 8}\; z^2\,dz$

**27.** $\displaystyle\int \sqrt[4]{x^4 + 16}\; x^2\,dx$     **28.** $\displaystyle\int \sqrt[3]{x^3 - 8}\; x\,dx$

**29.** $\displaystyle\int (2y^2 + 4y)^5 (y + 1)\,dy$

**30.** $\displaystyle\int (3y^2 - 6y)^3 (y - 1)\,dy$

**31.** $\displaystyle\int e^{x^2 + 2x + 5}(x + 1)\,dx$

**32.** $\displaystyle\int e^{x^3 - 3x + 7}(x^2 - 1)\,dx$

**33.** $\displaystyle\int \frac{x^3 + x^2}{3x^4 + 4x^3}\,dx$

**34.** $\displaystyle\int \frac{x^2 - x}{2x^3 - 3x^2}\,dx$

**35.** $\displaystyle\int \frac{x^3 + x^2}{(3x^4 + 4x^3)^2}\,dx$

**36.** $\displaystyle\int \frac{x^2 - x}{(2x^3 - 3x^2)^3}\,dx$

**37.** $\displaystyle\int \frac{x}{1 - x^2}\,dx$

**38.** $\displaystyle\int \frac{1}{1 - x}\,dx$

**39.** $\displaystyle\int (2x - 3)^7\,dx$

**40.** $\displaystyle\int (5x + 9)^9\,dx$

**41.** $\displaystyle\int \frac{e^{2x}}{e^{2x} + 1}\,dx$

**42.** $\displaystyle\int \frac{e^{3x}}{e^{3x} - 1}\,dx$

**43.** $\displaystyle\int \frac{\ln x}{x}\,dx$    [*Hint:* Let  $u = \ln x.$]

**44.** $\displaystyle\int \frac{(\ln x)^2}{x}\,dx$    [*Hint:* Let  $u = \ln x.$]

**45.** $\displaystyle\int \frac{e^{\sqrt{x}}}{\sqrt{x}}\,dx$    [*Hint:* Let  $u = \sqrt{x}.$]

**46.** $\displaystyle\int \frac{e^{1/x}}{x^2}\,dx$    $\left[\text{*Hint:* Let } u = \dfrac{1}{x}.\right]$

**47–50.**  Find each integral. [*Hint:* Try some algebra.]

**47.** $\displaystyle\int (x + 1)x^2\,dx$

**48.** $\displaystyle\int (x + 4)(x - 2)\,dx$

**49.** $\displaystyle\int (x + 1)^2 x^3\,dx$     **50.** $\displaystyle\int (x - 1)^2 \sqrt{x}\,dx$

**51–60.**  For each definite integral:

**a.** Evaluate it "by hand."
**b.** Check your answer by using a graphing calculator.

**51.** $\displaystyle\int_0^3 e^{x^2} x\,dx$     **52.** $\displaystyle\int_0^2 e^{x^3} x^2\,dx$

*(See instructions on previous page)*

**53.** $\displaystyle\int_0^1 \frac{x}{x^2 + 1}\, dx$

**54.** $\displaystyle\int_2^3 \frac{x^2}{x^3 - 7}\, dx$

**55.** $\displaystyle\int_0^4 \sqrt{x^2 + 9}\; x\, dx$

**56.** $\displaystyle\int_0^3 \sqrt{x^2 + 16}\; x\, dx$

**57.** $\displaystyle\int_2^3 \frac{dx}{1 - x}$

**58.** $\displaystyle\int_3^4 \frac{dx}{2 - x}$

**59.** $\displaystyle\int_1^8 \frac{e^{\sqrt[3]{x}}}{\sqrt[3]{x^2}}\, dx$

**60.** $\displaystyle\int_1^4 \frac{e^{\sqrt{x}}}{\sqrt{x}}\, dx$

**61.** Prove the integration formula

$$\int u^n\, du = \frac{1}{n + 1}\, u^{n+1} + C \qquad (n \neq -1)$$

as follows.

  **a.** Differentiate the right-hand side of the formula with respect to $x$ (remembering that $u$ is a function of $x$).

  **b.** Verify that the result of part (a) agrees with the integrand in the formula (after replacing $du$ in the formula by $u'\, dx$).

**62.** Prove the integration formula $\int e^u\, du = e^u + C$ by following the steps in Exercise 61.

**63.** Prove the integration formula $\displaystyle\int \frac{du}{u} = \ln u + C$

  $(u > 0)$ by following the steps in Exercise 61. (Absolute value bars come from applying the same argument to $-u$ for $u < 0$.)

**64.** Find $\int (x + 1)\, dx$:

  **a.** by using the formula for $\int u^n\, du$ with $n = 1$.

  **b.** by dropping the parentheses and integrating directly.

  **c.** Can you reconcile the two seemingly different answers? [*Hint:* Think of the arbitrary constant.]

## Applied Exercises

**65. BUSINESS: Cost** A company's marginal cost function is $MC(x) = \dfrac{1}{2x + 1}$ and its fixed costs are 50. Find the cost function.

**66. BUSINESS: Cost** A company's marginal cost function is $MC(x) = \dfrac{1}{\sqrt{2x + 25}}$ and its fixed costs are 100. Find the cost function.

**67. GENERAL: Average Value** The population of a city is expected to be $P(x) = x(x^2 + 36)^{-1/2}$ million people after $x$ years. Find the average population between year $x = 0$ and year $x = 8$.

**68. GENERAL: Area** Find the area between the curve $y = xe^{x^2}$ and the $x$-axis from $x = 1$ to $x = 3$. (Leave the answer in its exact form.)

**69. BUSINESS: Average Sales** A company's sales (in millions) during week $x$ are given by

$S(x) = \dfrac{1}{x + 1}$. Find the average sales from week $x = 1$ to week $x = 4$.

**70. BEHAVIORAL SCIENCE: Repeated Tasks** A subject can perform a task at the rate of $\sqrt{2t + 1}$ tasks per minute at time $t$ minutes. Find the total number of tasks performed from time $t = 0$ to time $t = 12$.

**71. BIOMEDICAL: Cholesterol** An experimental drug lowers a patient's blood serum cholesterol at the rate of $t\sqrt{25 - t^2}$ units per day, where $t$ is the number of days since the drug was administered $(0 \leq t \leq 5)$. Find the total change during the first 3 days.

**72. BUSINESS: Total Sales** During an automobile sale, cars are selling at the rate of $\dfrac{12}{x + 1}$ cars per day, where $x$ is the number of days since the sale began. How many cars will be sold during the first 7 days of the sale?

**73. BUSINESS: Total Sales** A real estate office is selling condominiums at the rate of $100e^{-x/4}$ per week after $x$ weeks. How many condominiums will be sold during the first 8 weeks?

**74. BUSINESS: Revenue** An aircraft company estimates its marginal revenue function for helicopters to be $MR(x) = \sqrt{x^2 + 80x}(x + 40)$ thousand dollars, where $x$ is the number of helicopters sold. Find the total revenue from the sale of the first 10 helicopters.

**75–76. ENVIRONMENTAL SCIENCE: Pollution** A factory is discharging pollution into a lake at the rate of $r(t)$ tons per year given below, where $t$ is the number of years that the factory has been in operation. Find the total amount of pollution discharged during the first 3 years of operation.

**75.** $r(t) = \dfrac{t}{t^2 + 1}$

**76.** $r(t) = t\sqrt{t^2 + 16}$

**77. BIOMEDICAL: Weight Gain** An average young female in the United States gains weight at the rate of $14(x - 10)^{-1/2}$ pounds per year, where $x$ is her age $(10 \le x \le 19)$. Find the total weight gain from age 11 to 19.

*Source:* National Center for Health Statistics

**78. BIOMEDICAL: Weight Gain** An average young male in the United States gains weight at the rate of $18(x - 10)^{-1/2}$ pounds per year, where $x$ is his age $(11 \le x \le 20)$. Find the total weight gain from age 11 to 19.

*Source:* National Center for Health Statistics

## Conceptual Exercises

**79.** A friend says that finding differentials is as easy as finding derivatives—you just multiply the derivative by $dx$. Is your friend right?

**80.** A friend says that if you can move *numbers* across the integral sign, you can do the same for *variables* since variables stand for numbers, and in this way you can always "fix" the differential $du$ to be what you want. Is your friend right?

**81.** Explain why the substitution $u = x$ is useless for finding an integral by the substitution method. Give an example.

**82.** What is wrong with the following use of the substitution $u = x^2$?

$$\int e^{x^2}\, dx = \int e^u\, du = e^u + C = e^{x^2} + C$$

**83.** What is wrong with the following use of the substitution $u = x^2 - 1$?

$$\int \frac{1}{x^2 - 1}\, dx = \int \frac{1}{u}\, du = \ln|u| + C = \ln|x^2 - 1| + C$$

**84.** Al and Betty carry out the substitution method for definite integrals slightly differently. Both use the same substitution to change the original "$x$" integral into a "$u$" integral, which they then integrate.

Then Al changes the "$x$" limits of integration into their corresponding "$u$" values, substitutes, and subtracts to get the final answer.

Betty uses the substitution to return the integrand to its "$x$" form and uses the original limits of integration to get the final answer.

Which way is better?

## Explorations and Excursions

The following problems extend and augment the material presented in the text.

*Further Use of the Substitution Method*

**85–94.** The substitution method can be used to find integrals that do not fit our formulas. For example, observe how we find the following integral using the substitution $u = x + 4$ which implies that $x = u - 4$ and so $dx = du$.

$$\int (x - 2)(x + 4)^8\, dx = \int (u - 4 - 2)u^8\, du$$

$$= \int (u - 6)u^8\, du$$

$$= \int (u^9 - 6u^8)\, du$$

$$= \frac{1}{10} u^{10} - \frac{2}{3} u^9 + C$$

$$= \frac{1}{10} (x + 4)^{10} - \frac{2}{3}(x + 4)^9 + C$$

It is often best to choose $u$ to be the quantity that is raised to a power. The following integrals may be found in this way (as well as by the methods of Section 6.1).

**85.** $\displaystyle\int (x + 1)(x - 5)^4\, dx$  **86.** $\displaystyle\int (x - 2)(x + 4)^5\, dx$

**87.** $\displaystyle\int x(x - 2)^6\, dx$  **88.** $\displaystyle\int x(x + 4)^7\, dx$

**89.** $\displaystyle\int \frac{x - 1}{x - 2}\, dx$  **90.** $\displaystyle\int \frac{x - 4}{x - 5}\, dx$

**91.** $\displaystyle\int x\sqrt[3]{x - 4}\, dx$  **92.** $\displaystyle\int (x - 1)\sqrt{x + 2}\, dx$

**93.** $\displaystyle\int \frac{x}{\sqrt{x + 2}}\, dx$  **94.** $\displaystyle\int \frac{x}{\sqrt[3]{x + 1}}\, dx$

## Chapter Summary with Hints and Suggestions

Reading the text and doing the exercises in this chapter have helped you to master the following concepts and skills, which are listed by section (in case you need to review them) and are keyed to particular Review Exercises. Answers for all Review Exercises are given at the back of the book, and full solutions can be found in the Student Solutions Manual.

### 5.1 Antiderivatives and Indefinite Integrals

■ Find an indefinite integral using the Power Rule. *(Review Exercises 1–8.)*

$$\int x^n \, dx = \frac{1}{n+1} x^{n+1} + C \qquad (n \neq -1)$$

■ Solve an applied problem involving integration. *(Review Exercises 9–10.)*

### 5.2 Integration Using Logarithmic and Exponential Functions

■ Find an indefinite integral involving $e^x$ or $\frac{1}{x}$.

*(Review Exercises 11–18.)*

$$\int e^{ax} \, dx = \frac{1}{a} e^{ax} + C$$

$$\int \frac{1}{x} \, dx = \int x^{-1} \, dx = \ln |x| + C$$

■ Solve an applied problem involving integration. *(Review Exercises 19–22.)*

### 5.3 Definite Integrals and Areas

■ Evaluate a definite integral. *(Review Exercises 23–30.)*

$$\int_a^b f(x) \, dx = F(b) - F(a)$$

■ Find the area under a curve. *(Review Exercises 31–36.)*

■ Graph a function and find the area under it. *(Review Exercises 37–38.)*

■ Solve an applied problem using definite integration. *(Review Exercises 39–44.)*

■ Approximate the area under a curve by rectangles (Riemann sum). *(Review Exercises 45–46.)*

$$\int_a^b f(x) \, dx = \lim_{n \to \infty} [f(x_1) \cdot \Delta x + \cdots + f(x_n) \cdot \Delta x]$$

■ Use a Riemann sum program to find Riemann sums. *(Review Exercises 47–48.)*

### 5.4 Further Applications of Definite Integrals: Average Value and Area Between Curves

■ Find the area bounded by two curves. *(Review Exercises 49–56.)*

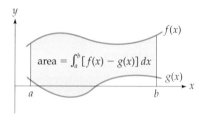

■ Find the average value of a function on an interval. *(Review Exercises 57–60.)*

$$\left( \begin{matrix} \text{Average of } f \\ \text{from } a \text{ to } b \end{matrix} \right) = \frac{1}{b-a} \int_a^b f(x) \, dx$$

■ Solve an applied problem involving average value. *(Review Exercises 61–64.)*

■ Solve an applied problem involving area between curves. *(Review Exercises 65–66.)*

### 5.5 Two Applications to Economics: Consumers' Surplus and Income Distribution

■ Find the consumers' surplus for a product. *(Review Exercises 67–70.)*

$$\left( \begin{matrix} \text{Consumers'} \\ \text{surplus} \end{matrix} \right) = \int_0^A [d(x) - B] \, dx$$

- Find the Gini index of income distribution. (*Review Exercises 71–74.*)

$$\left(\begin{array}{c}\text{Gini} \\ \text{index}\end{array}\right) = 2 \cdot \int_0^1 [x - L(x)]\, dx$$

## 5.6  Integration by Substitution

- Use a substitution to find an integral. (*Review Exercises 75–90.*)

$$\int u^n\, du = \frac{1}{n+1} u^{n+1} + C \qquad (n \neq -1)$$

$$\int e^u\, du = e^u + C \qquad \int \frac{1}{u}\, du = \ln|u| + C$$

- Use a substitution to find a definite integral. (*Review Exercises 91–98.*)

- Use a substitution to find the area under a curve. (*Review Exercises 99–100.*)

- Use a substitution to find the average value of a function.   (*Review Exercises 101–102.*)

- Use a substitution to solve an applied problem. (*Review Exercises 103–104.*)

### Hints and Suggestions

- An *indefinite* integral is a function plus a constant, whereas a *definite* integral is a *number* (the signed area between the curve and the *x*-axis). The Fundamental Theorem of Integral Calculus shows how to evaluate the second in terms of the first.

- To integrate any power of *x* except $x^{-1}$, use the Power Rule; for $x^{-1}$, use the ln rule.

- *Indefinite* integrals have a $+C$. *Definite* integrals do not.

- Differentiation *breaks things down into parts*—for example, turning cost into marginal cost (cost per unit). Integration *combines back into a whole*—for example, combining all of the per-unit costs back into a total cost. In fact, the word "integrate" means "make whole."

- To find the area between two curves, integrate *upper* minus *lower*.

- For average values, don't forget the $\frac{1}{b-a}$.

- The substitution method can only "fix up" the *constant*; the variable part must already be correct (or else the function cannot be integrated by this technique).

- A graphing calculator helps by graphing functions, evaluating definite integrals, and finding areas under curves and total accumulations. With an appropriate program it can calculate Riemann sums with many rectangles.

- **Practice for test:** Review Exercises 1, 5, 9, 15, 21, 23, 27, 31, 37, 39, 43, 49, 55, 57, 61, 67, 71, 79, 81, 91.

# Review Exercises for Chapter 5

Practice test exercise numbers are in green.

## 5.1  Antiderivatives and Indefinite Integrals

**1–8.** Find each indefinite integral.

**1.** $\int (24x^2 - 8x + 1)\, dx$

**2.** $\int (12x^3 + 6x - 3)\, dx$

**3.** $\int (6\sqrt{x} - 5)\, dx$

**4.** $\int (8\sqrt[3]{x} - 2)\, dx$

**5.** $\int (10\sqrt[3]{x^2} - 4x)\, dx$

**6.** $\int (5\sqrt{x^3} - 6x)\, dx$

**7.** $\int (x + 4)(x - 4)\, dx$

**8.** $\int \frac{3x^3 + 2x^2 + 4x}{x}\, dx$

9. **BUSINESS: Cost** A company's marginal cost function is $MC(x) = x^{-1/2} + 4$, where $x$ is the number of units. If fixed costs are 20,000, find the company's cost function.

10. **GENERAL: Population** The population of a town is now 40,000 and $t$ years from now will be growing at the rate of $300\sqrt{t}$ people per year.

    **a.** Find a formula for the population of the town $t$ years from now.
    **b.** Use your formula to find the population of the town 16 years from now.

## 5.2 Integration Using Logarithmic and Exponential Functions

**11–18.** Find each indefinite integral.

11. $\int e^{x/2} \, dx$

12. $\int e^{-2x} \, dx$

13. $\int 4x^{-1} \, dx$

14. $\int \frac{2}{x} \, dx$

15. $\int \left(6e^{3x} - \frac{6}{x}\right) dx$

16. $\int (x - x^{-1}) \, dx$

17. $\int (9x^2 + 2x^{-1} + 6e^{3x}) \, dx$

18. $\int \left(\frac{1}{x^2} + \frac{1}{x} + e^{-x}\right) dx$

19. **ENVIRONMENTAL SCIENCE: Consumption of Natural Resources** World consumption of aluminum is running at the rate of $75e^{0.05t}$ thousand metric tons per year, where $t$ is the number of years since 2008.

    **a.** Find a formula for the total amount of aluminum consumed within $t$ years of 2008.
    **b.** If consumption continues at this rate, when will the known resources of 36,000 thousand metric tons of aluminum be exhausted?

    *Source:* U.S. Geological Survey

20. **PERSONAL FINANCE: Total Savings** A homeowner installs a solar heating system, which is expected to generate savings at the rate of $200e^{0.1t}$ dollars per year, where $t$ is the number of years since the system was installed.

    **a.** Find a formula for the total savings in the first $t$ years.
    **b.** If the system originally cost $1500, when will it "pay for itself"?

21. **ENVIRONMENTAL SCIENCE: Consumption of Natural Resources** World consumption of zinc is running at the rate of $11e^{0.02t}$ million metric tons per year, where $t$ is the number of years since 2008.

    **a.** Find a formula for the total amount of zinc consumed within $t$ years of 2008.
    **b.** If consumption continues at this rate, when will the known resources of 460 million metric tons of zinc be exhausted? (Zinc is used to make protective coatings for iron and steel.)

    *Source:* U.S. Geological Survey

22. **BUSINESS: Profit** A company's profit is growing at the rate of $200x^{-1}$ thousand dollars per month after $x$ months, for $x \geq 1$.

    **a.** Find a formula for the total growth in the profit from month 1 to month $x$.
    **b.** When will the total growth in profit reach 600 thousand dollars?

## 5.3 Definite Integrals and Areas

**23–30.** Evaluate each definite integral.

23. $\int_1^9 \left(x - \frac{1}{\sqrt{x}}\right) dx$

24. $\int_2^5 (3x^2 - 4x + 5) \, dx$

25. $\int_1^{e^4} \frac{dx}{x}$

26. $\int_1^5 \frac{dx}{x}$

27. $\int_0^2 e^{-x} \, dx$

28. $\int_0^2 e^{x/2} \, dx$

29. $\int_0^{100} (e^{0.05x} - e^{0.01x}) \, dx$

30. $\int_0^{10} (e^{0.04x} - e^{0.02x}) \, dx$

**31–36.** For each function:

**a.** Find the area under the curve between the given $x$-values.

**b.** Verify your answer to part (a) by using a graphing calculator to find the area.

31. $f(x) = 6x^2 - 1$, $x = 1$ to $x = 2$

32. $f(x) = 9 - x^2$, $x = -3$ to $x = 3$

33. $f(x) = 12e^{2x}$, $x = 0$ to $x = 3$

**34.** $f(x) = e^{x/2}$, $x = 0$ to $x = 4$

**35.** $f(x) = \dfrac{1}{x}$, $x = 1$ to $x = 100$

**36.** $f(x) = x^{-1}$, $x = 1$ to $x = 1000$

 **37–38.** Use a calculator to graph each function and find the area under it between the given $x$-values.

**37.** $f(x) = \dfrac{10}{x^4 + 1}$ from $x = -2$ to $x = 2$

**38.** $f(x) = e^{x^4}$ from $x = -1$ to $x = 1$

**39. BIOMEDICAL: Weight Gain** An average young person in the United States gains weight at the rate of $16(x - 10)^{-1/2}$ pounds per year, where $x$ is the person's age $(11 \le x \le 19)$. Find the total weight gain from age 11 to 19.

*Source: National Center for Health Statistics*

**40. BEHAVIORAL SCIENCE: Learning** A student can memorize foreign vocabulary words at the rate of $2/\sqrt[3]{t}$ words per minute, where $t$ is the number of minutes since the studying began. Find the number of words that can be memorized from time $t = 1$ to time $t = 8$.

**41. ENVIRONMENTAL SCIENCE: Global Warming** The temperature of the earth is rising, as a result of the "greenhouse effect," in which carbon dioxide prevents the escape of heat from the atmosphere. According to one study, the temperature is rising at the rate of $0.014t^{0.4}$ degrees Fahrenheit per year, find the total rise in temperature over the next 100 years.

*Source: Intergovernmental Panel on Climate Change, 2007 Synthesis Report*

**42. BUSINESS: Cost** A company's marginal cost function is $MC(x) = x^{-1/2} + 4$ dollars, where $x$ is the number of units. Find the total cost of the first 400 units (units $x = 0$ to $x = 400$).

 **43. BUSINESS: Cost** A company's marginal cost function is $MC(x) = 22e^{-\sqrt{x}/5}$ dollars, where $x$ is the number of units. Find the total cost of the first hundred units (units $x = 0$ to $x = 100$).

**44. BEHAVIORAL SCIENCE: Repetitive Tasks** A proofreader can read $15xe^{-0.25x}$ pages per hour, where $x$ is the number of hours worked. Find the total number of pages that can be proofread in 8 hours.

## Exercises on Riemann Sums

**45. a.** Approximate the area under the curve $f(x) = x^2$ from 0 to 2 using ten left rectangles with equal bases.

**b.** Find the *exact* area under the curve between the given $x$-values by evaluating an appropriate definite integral using the Fundamental Theorem.

**46. a.** Approximate the area under the curve $f(x) = \sqrt{x}$ from 0 to 4 using ten left rectangles with equal bases. (Round calculations to three decimal places.)

**b.** Find the *exact* area under the curve between the given $x$-values by evaluating an appropriate definite integral using the Fundamental Theorem.

 **47–48.** Use the graphing calculator program RIEMANN (see page 350), the program on page 361, or a similar program to find the following Riemann sums.

**a.** Calculate the Riemann sum for each function below for the following values of $n$: 10, 100, 1000. Use left, right, or midpoint rectangles, making a table of the answers, keeping three decimal places of accuracy.

**b.** Find the *exact* value of the area under the curve in each exercise below by evaluating an appropriate definite integral using the Fundamental Theorem. Your answers from part (a) should approach the number found from this integral.

**47.** $f(x) = e^x$ from $a = -2$ to $b = 2$

**48.** $f(x) = \dfrac{1}{x}$ from $a = 1$ to $b = 4$

## 5.4 Further Applications of Definite Integrals: Average Value and Area Between Curves

**49–56.** Find the area bounded by each pair of curves.

**49.** $y = x^2 + 3x$ and $y = 3x + 1$

**50.** $y = 12x - 3x^2$ and $y = 6x - 24$

**51.** $y = x^2$ and $y = x$

**52.** $y = x^4$ and $y = x$

**53.** $y = 4x^3$ and $y = 12x^2 - 8x$

**54.** $y = x^3 + x^2$ and $y = x^2 + x$

**55.** $y = e^x$ and $y = x + 5$

**56.** $y = \ln x$ and $y = \dfrac{x^2}{10}$

**57–60.** Find the average value of the function on the given interval.

**57.** $f(x) = \dfrac{1}{x}$ on $[1, 4]$

*(See instructions on previous page.)*

**58.** $f(x) = 6\sqrt{x}$ on $[1, 4]$

 **59.** $f(x) = \sqrt{x^3 + 1}$ on $[0, 5]$

 **60.** $f(x) = \ln(e^{x^2} + 10)$ on $[-2, 2]$

**61. GENERAL: Average Population** The population of the United States is predicted to be $P(t) = 309e^{0.0087t}$ billion, where $t$ is the number of years after the year 2010. Find the average population between the years 2010 and 2110.
*Source:* U.S. Census Bureau

**62. PERSONAL FINANCE: Compound Interest** A deposit of $3000 in a bank paying 6% interest compounded continuously will grow to $V(t) = 3000e^{0.06t}$ dollars after $t$ years. Find the average value during the first 20 years ($t = 0$ to $t = 20$).

 **63. PERSONAL FINANCE: Real Estate** The value of a suburban plot of land being considered for rezoning is assessed at $4.3e^{0.01x^2}$ hundred thousand dollars $x$ years from now. Find the average value over the next 10 years (year 0 to year 10).

**64. PERSONAL FINANCE: Stock Price** The price of a share of stock is expected to be $28e^{0.01x^{1.2}}$ dollars where $x$ is the number of weeks from now. Find the average price over the next year (week 0 to week 52).

**65. GENERAL: Art** An artist wants to paint the interior of the shape shown below on the side of a building. How much area (in square meters) will the artist need to paint?

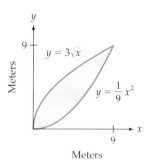

**66. ECONOMICS: Balance of Trade** A country's annual exports will be $E(t) = 40e^{0.2t}$ and its imports will be $I(t) = 20e^{0.1t}$ (both in billions of dollars per year), where $t$ is the number of years from now. Find the accumulated trade surplus (exports minus imports) over the next 10 years.

## 5.5 Two Applications to Economics: Consumers' Surplus and Income Distribution

**67–70. ECONOMICS: Consumers' Surplus** For each demand function $d(x)$ and demand level $x$, find the consumers' surplus.

**67.** $d(x) = 8000 - 24x$, $x = 200$

**68.** $d(x) = 1800 - 0.03x^2$, $x = 200$

 **69.** $d(x) = 300e^{-0.2\sqrt{x}}$, $x = 120$

 **70.** $d(x) = \dfrac{100}{1 + \sqrt{x}}$, $x = 100$

**71–74. ECONOMICS: Gini Index** For each Lorenz curve, find the Gini index.

**71.** $L(x) = x^{3.5}$

**72.** $L(x) = x^{2.5}$

**73.** $L(x) = \dfrac{x}{2 - x}$

**74.** $L(x) = \dfrac{e^{x^4} - 1}{e - 1}$

## 5.6 Integration by Substitution

**75–90.** Find each integral or state that it cannot be evaluated by our substitution formulas.

**75.** $\displaystyle\int x^2 \sqrt[3]{x^3 - 1}\, dx$

**76.** $\displaystyle\int x^3 \sqrt{x^4 - 1}\, dx$

**77.** $\displaystyle\int x \sqrt[3]{x^3 - 1}\, dx$

**78.** $\displaystyle\int x^2 \sqrt{x^4 - 1}\, dx$

**79.** $\displaystyle\int \frac{dx}{9 - 3x}$

**80.** $\displaystyle\int \frac{dx}{1 - 2x}$

**81.** $\displaystyle\int \frac{dx}{(9 - 3x)^2}$

**82.** $\displaystyle\int \frac{dx}{(1 - 2x)^2}$

**83.** $\displaystyle\int \frac{x^2}{\sqrt[3]{8 + x^3}}\, dx$

**84.** $\displaystyle\int \frac{x}{\sqrt{9 + x^2}}\, dx$

**85.** $\displaystyle\int \frac{w + 3}{(w^2 + 6w - 1)^2}\, dw$

**86.** $\displaystyle\int \frac{t - 2}{(t^2 - 4t + 1)^2}\, dt$

**87.** $\displaystyle\int \frac{\left(1 + \sqrt{x}\right)^2}{\sqrt{x}}\, dx$

**88.** $\displaystyle\int \frac{\left(1 + \sqrt[3]{x}\right)^2}{\sqrt[3]{x^2}}\, dx$

**89.** $\displaystyle\int \frac{e^x}{e^x - 1}\, dx$

**90.** $\displaystyle\int \frac{1}{x \ln x}\, dx$

**91–98.** For each definite integral:

**a.** Evaluate it ("by hand") or state that it cannot be evaluated by our substitution formulas.

 **b.** Verify your answer to part (a) by using a graphing calculator.

**91.** $\displaystyle\int_0^3 x\sqrt{x^2 + 16}\, dx$

**92.** $\displaystyle\int_0^4 \frac{dz}{\sqrt{2z + 1}}$

**93.** $\displaystyle\int_0^4 \frac{w}{\sqrt{25 - w^2}}\, dw$

**94.** $\displaystyle\int_1^2 \frac{x + 1}{(x^2 + 2x - 2)^2}\, dx$

**95.** $\displaystyle\int_3^9 \frac{dx}{x - 2}$

**96.** $\displaystyle\int_4^5 \frac{dx}{x - 6}$

**97.** $\displaystyle\int_0^1 x^3\, e^{x^4}\, dx$

**98.** $\displaystyle\int_0^1 x^4\, e^{x^5}\, dx$

**99–100.** Find the area under the given curve between the given $x$-values.

**99.** $y = \dfrac{x^2 + 6x}{\sqrt[3]{x^3 + 9x^2 + 17}}$  from  $x = 1$  to  $x = 3$

**100.** $y = \dfrac{x + 6}{\sqrt{x^2 + 12x + 4}}$  from  $x = 0$  to  $x = 3$

**101–102.** Find the average value of the function on the given interval.

**101.** $f(x) = xe^{-x^2}$  on $[0, 2]$

**102.** $f(x) = \dfrac{x}{x^2 - 3}$  on $[2, 4]$

**103. BUSINESS: Cost** A company's marginal cost function is $MC(x) = \dfrac{1}{\sqrt{2x + 9}}$ and fixed costs are 100. Find the cost function.

**104. BIOMEDICAL: Temperature** An experimental drug changes a patient's temperature at the rate of $\dfrac{3x^2}{x^3 + 1}$ degrees per milligram of the drug, where $x$ is the amount of the drug administered. Find the total change in temperature resulting from the first 3 milligrams of the drug. [*Note:* The rate of change of temperature with respect to dosage is called the "drug sensitivity."]

# 6 Integration Techniques

### Improper Integrals and Eternal Recognition

Suppose that after you become rich and famous, you decide to commission a statue of yourself for your hometown. Your town, however, will accept this selfless gesture only if you pay for the perpetual upkeep of the statue by establishing a fund that will generate $2000 annually for every year in the future. Before deciding whether or not to accept this condition, you of course want to know how much it will cost. (Interestingly, we will see that a fund that will generate income forever does *not* require an infinite amount of money.) On page 265 we found that to realize a yield of $2000 $t$ years from now requires only its *present value* deposited now in a bank. At an interest rate of, say, 5% compounded continuously, $2000 in $t$ years requires a deposit now of

$$\begin{pmatrix} \text{Present value of \$2000 at 5\%} \\ \text{compounded continuously} \end{pmatrix} = 2000e^{-0.05t} \qquad \text{Amount times } e^{-0.05t}$$

Therefore, the size of the fund needed to generate $2000 annually *forever* is found by summing (integrating) this present value over the infinite time interval from zero to infinity ($\infty$).

$$\begin{pmatrix} \text{Size of fund to yield an} \\ \text{annual \$2000 forever} \end{pmatrix} = \int_0^\infty 2000e^{-0.05t}\, dt \qquad \begin{array}{l}\text{Integrating out} \\ \text{to infinity}\end{array}$$

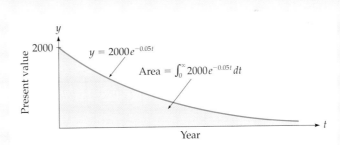

On page 437 we will learn how to evaluate such "improper integrals," finding that the value of this integral is $40,000, which shows that $40,000 deposited in a bank at 5% compounded continuously will generate $2000 every year *forever*. That is, $40,000 would pay for the perpetual upkeep of your statue, buying you (or at least your likeness) a kind of immortality.

Incidentally, to generate $2000 annually for only the first *hundred* years would require $\displaystyle\int_0^{100} 2000e^{-0.05t}\, dt \approx \$39{,}730$ (integrating from 0 to 100). This amount is only $270 less than the amount needed to generate the same sum *forever*. This small additional cost shows that the short term is expensive, but eternity is cheap.

## Introduction

In this chapter we introduce further techniques for finding integrals: **integration by parts, integration by tables,** and **numerical integration.** We also discuss **improper integrals** (integrals over infinite intervals).

You may have felt that integration is "harder" than differentiation. One reason is that integration is an *inverse* process, carried out by *reversing* differentiation, and inverse processes are generally more difficult. Another reason is that, while we have the product and quotient rules for differentiating complicated expressions, there are no product and quotient rules for integration. The method of integration by parts, explained in this section, is in some sense a "Product Rule for integration" in that it comes from interpreting the Product Rule as an integration formula.

## Integration by Parts

For two differentiable functions $u(x)$ and $v(x)$, hereafter denoted simply $u$ and $v$, the Product Rule is

$$(uv)' = u'v + uv'$$

The derivative of a product is the derivative of the first times the second, plus the first times the derivative of the second

If we integrate both sides of this equation, integrating the left side "undoes" the differentiation.

$$uv = \int u'v \, dx + \int uv' \, dx$$

$$\underbrace{\phantom{u'v}}_{du} \qquad \underbrace{\phantom{uv'}}_{dv}$$

Using differential notation,
$du = u' \, dx, \quad dv = v' \, dx$

$$uv = \int v \, du + \int u \, dv$$

Formula above in differential notation

Solving this equation for the second integral $\int u \, dv$ gives

$$\int u \, dv = uv - \int v \, du$$

This formula is the basis for a technique called "integration by parts."

| Integration by Parts |
|---|

For differentiable functions $u$ and $v$,

$$\int u \, dv = uv - \int v \, du$$

We use this formula to solve integrals by a "double substitution," substituting *u* for part of the given integral and *dv* for the rest, and then expressing the integral in the form $uv - \int v\, du$. The point is to choose the *u* and the *dv* so that the resulting integral $\int v\, du$ is *simpler* than the original integral $\int u\, dv$. A few examples will make the method clear.

---

**EXAMPLE 1**

**INTEGRATING BY PARTS**

Use integration by parts to find $\displaystyle\int xe^x\, dx$.

*Solution*

$$\int \underset{u\quad dv}{x\, e^x\, dx}$$

Original integral

We choose $u = x$ and $dv = e^x dx$

$$\begin{bmatrix} u = x & dv = e^x dx \\ \downarrow & \downarrow \\ du = 1\, dx = dx & v = \int e^x dx = e^x \end{bmatrix}$$

Differentiating *u* to find *du* and integrating *dv* to find *v* (omitting the C) (the arrows show which part leads to which other part)

$$= \underset{u\ v}{x\, e^x} - \int \underset{v\ \ du}{e^x\, dx}$$

Replacing the original integral $\int u\, dv$ by the right-hand side of the formula, $u \cdot v - \int v\, du$, with $u = x$, $v = e^x$, and $du = dx$

$$= xe^x - e^x + C$$

$$\text{From } \int e^x\, dx$$

Finding the new integral $\int e^x dx = e^x$ to give the final answer (with +C)

---

The procedure is not as complicated as it might seem. All the steps may be written together as follows:

$$\int \underset{u\ dv}{xe^x\, dx} = \underset{uv}{xe^x} - \int \underset{v\ \ du}{e^x\, dx} = xe^x - e^x + C$$

$$\text{From } \int e^x\, dx$$

$$\begin{bmatrix} u = x & dv = e^x dx \\ du = dx & v = \int e^x dx = e^x \end{bmatrix}$$

We can check this answer by differentiation.

$$\frac{d}{dx}(xe^x - e^x + C) = \overset{\overbrace{\text{Cancel}}}{e^x + xe^x - e^x} = xe^x$$

Differentiating $xe^x$ by the Product Rule

Agrees with the original integrand, so the integration is correct

---

**Remarks on the Integration by Parts Procedure**

i. The differentials $du$ and $dv$ include the $dx$.

ii. We omit the constant $C$ when we integrate $dv$ to get $v$ because one $+C$ at the end is enough.

iii. The integration by parts formula does not give a "final answer," but rather expresses the given integral as $uv - \int v\, du$, a product $u \cdot v$ (already integrated) and a new integral $\int v\, du$. That is, integration by parts "exchanges" the original integral $\int u\, dv$ for another integral $\int v\, du$. The hope is that the second integral will be simpler than the first. In our example we "exchanged" $\int xe^x\, dx$ for the simpler $\int e^x\, dx$, which could be integrated immediately by formula 3 (inside back cover).

iv. Integration by parts should be used only if the simpler formulas 1 through 7 (inside back cover) fail to solve the integral.

---

## How to Choose the *u* and the *dv*

In Example 1, the choice of $\begin{cases} u = x \\ dv = e^x\, dx \end{cases}$ "exchanged" the original integral $\int xe^x\, dx$ for the simpler integral $\int e^x\, dx$. If we had instead chosen $\begin{cases} u = e^x \\ dv = x\, dx \end{cases}$ we would have "exchanged" the original integral for $\int x^2 e^x\, dx$ (as you may check), which is *more* difficult than the original (because of the $x^2$). Therefore, the first choice was the "right" choice in that it led to a solution. Generally, one choice for $u$ and $dv$ will be "best," and finding it may involve some trial and error. While there is no foolproof rule for finding the best $u$ and $dv$, the following guidelines often help.

---

**Guidelines for Choosing *u* and *dv***

1. Choose $dv$ to be the most complicated part of the integral that can be integrated easily.

2. Choose $u$ so that $u'$ is simpler than $u$.

---

**EXAMPLE 2**

**INTEGRATING BY PARTS**

Find $\displaystyle\int x^2 \ln x\, dx$.

*Solution*

None of the easier formulas (1 through 7 on the inside back cover) will solve the integral, as you may easily check. Therefore, we try integration by parts. The integrand is a product, $x^2$ times $\ln x$. The guidelines say to choose $dv$ to be the most complicated part that can be easily integrated. We can integrate $x^2$ but not $\ln x$ (we know how to *differentiate* $\ln x$, but not how to *integrate* it), so we choose $dv = x^2\,dx$, and therefore $u = \ln x$.

$$\int x^2 \ln x\, dx = (\ln x)\frac{1}{3}x^3 - \int \frac{1}{3}x^3\frac{1}{x}\,dx = \frac{1}{3}x^3 \ln x - \frac{1}{3}\int x^2\,dx$$

$$\underbrace{\qquad}_{\substack{u\\dv}} \quad \underbrace{\quad}_{u}\underbrace{\quad}_{v} \quad \underbrace{\quad}_{v}\underbrace{\quad}_{du} \qquad \text{Moving the } \tfrac{1}{3}$$

outside and simplifying

$x^3 \frac{1}{x}$ to $x^2$

$$\left[ \begin{array}{ll} u = \ln x & dv = x^2\,dx \\ du = \dfrac{1}{x}\,dx & v = \int x^2\,dx = \tfrac{1}{3}x^3 \end{array} \right]$$

$$= \frac{1}{3}x^3 \ln x - \frac{1}{3}\frac{1}{3}x^3 + C = \frac{1}{3}x^3 \ln x - \frac{1}{9}x^3 + C$$

We may check this answer by differentiation.

$$\frac{d}{dx}\left(\frac{1}{3}x^3 \ln x - \frac{1}{9}x^3 + C\right) = \underbrace{x^2 \ln x + \frac{1}{3}x^3\frac{1}{x}}_{\substack{\text{Differentiating } \tfrac{1}{3}x^3 \ln x \\ \text{by the Product Rule}}} - \frac{1}{3}x^2$$

$$= x^2 \ln x + \frac{1}{3}x^2 - \frac{1}{3}x^2 = x^2 \ln x$$

Agrees with the original integrand, so the integration is correct

From $x^3\dfrac{1}{x}$    Cancel

**Practice Problem 1**    Use integration by parts to find $\int x^3 \ln x\, dx$.    ➤ **Solution on page 417**

Integration by parts is also useful for integrating products of linear functions to powers.

**EXAMPLE 3**

**INTEGRATING BY PARTS**

Use integration by parts to find $\int (x-2)(x+4)^8\,dx$.

*Solution*

The guidelines recommend that $dv$ be the most complicated part that can be integrated. Both $x - 2$ and $(x + 4)^8$ can be integrated. For example,

$$\int (x + 4)^8\, dx = \frac{1}{9}(x + 4)^9 + C \qquad \text{By the substitution method with } u = x + 4 \text{ (omitting the details)}$$

Since $(x + 4)^8$ is more complicated than $(x - 2)$, we take $dv = (x + 4)^8\, dx$.

$$\int \underbrace{(x - 2)}_{u}\underbrace{(x + 4)^8\, dx}_{dv} = \underbrace{(x - 2)}_{u}\underbrace{\frac{1}{9}(x + 4)^9}_{v} - \int \underbrace{\frac{1}{9}(x + 4)^9}_{v}\underbrace{dx}_{du}$$

$$\begin{bmatrix} u = x - 2 & dv = (x + 4)^8\, dx \\ du = dx & v = \int (x + 4)^8\, dx \\ & = \frac{1}{9}(x + 4)^9 \end{bmatrix}$$

$$= \frac{1}{9}(x - 2)(x + 4)^9 - \frac{1}{9}\int (x + 4)^9\, dx$$

$$\uparrow$$
$$\text{Taking out the } \tfrac{1}{9}$$

$$= \frac{1}{9}(x - 2)(x + 4)^9 - \frac{1}{9}\frac{1}{10}(x + 4)^{10} + C$$

$$\underbrace{\qquad\qquad\qquad}$$
$$\text{Integrating by the substitution method}$$

$$= \frac{1}{9}(x - 2)(x + 4)^9 - \frac{1}{90}(x + 4)^{10} + C$$

Again we could check this answer by differentiation. (Do you see how the Product Rule, applied to the first part of this answer, will give a piece that will cancel with the derivative of the second part?) This integral can also be found by the substitution method (see page 401).

**Practice Problem 2**   Use integration by parts to find $\int (x + 1)(x - 1)^3\, dx$.

➤ **Solution on page 417**

## Present Value of a Continuous Stream of Income

If a business or some other asset generates income continuously at the rate $C(t)$ dollars per year, where $t$ is the number of years from now, then $C(t)$ is called a **continuous stream of income.** * On page 265 we saw that to find the *present value* of a sum (the amount now that will later yield the stated sum) under continuous compounding we multiply by $e^{-rt}$, where $r$ is the interest rate and $t$ is the number of years. (We will refer to a rate with continuous compounding as a "continuous interest rate.") Therefore, the present value of the continuous stream $C(t)$ is found by multiplying by $e^{-rt}$ and summing (integrating) over the time period.

> **Present Value of a Continuous Stream of Income**
>
> The present value of the continuous stream of income $C(t)$ dollars per year, where $t$ is the number of years from now, for $T$ years at continuous interest rate $r$ is
>
> $$\begin{pmatrix} \text{Present} \\ \text{value} \end{pmatrix} = \int_0^T C(t)e^{-rt}\, dt$$

**EXAMPLE 4**

### FINDING THE PRESENT VALUE OF A CONTINUOUS STREAM OF INCOME

A business generates income at the rate of $2t$ million dollars per year, where $t$ is the number of years from now. Find the present value of this continuous stream for the next 5 years at the continuous interest rate of 10%.

*Solution*

$$\begin{pmatrix} \text{Present} \\ \text{value} \end{pmatrix} = \int_0^5 2te^{-0.1t}\, dt \qquad \begin{array}{l} \text{Multiplying } C(t) = 2t \text{ by} \\ e^{-0.1t} \text{ (since } 10\% = 0.1 \text{) and} \\ \text{integrating from 0 to 5 years} \end{array}$$

This is a *definite* integral, but we will ignore the limits of integration until after we have found the *indefinite* integral. None of the formulas 1 through 7 on the inside back cover will find this integral, so we try integration by parts with $u = 2t$ and $dv = e^{-0.1t}\, dt$. (Do you see why the guidelines on page 412 suggest this choice?)

$$\int \underbrace{2t}_{u}\, \underbrace{e^{-0.1t}\, dt}_{dv} = \underbrace{(2t)}_{u}\underbrace{(-10e^{-0.1t})}_{v} - \int \underbrace{(-10e^{-0.1t})}_{v}\, \underbrace{2\, dt}_{du}$$

$$\begin{bmatrix} u = 2t & dv = e^{-0.1t}\, dt \\ du = 2\, dt & v = \int e^{-0.1t}\, dt \\ & \quad = -10e^{-0.1t} \end{bmatrix}$$

$$= -20te^{-0.1t} + 20 \int e^{-0.1t}\, dt$$

$$= -20te^{-0.1t} + 20(-10)e^{-0.1t} + C$$

$$= -20te^{-0.1t} - 200e^{-0.1t} + C$$

---

* $C(t)$ must be continuous, meaning that the income is being paid continuously rather than in "lump-sum" payments. However, even lump-sum payments can be approximated by a continuous stream if the payments are frequent enough or last long enough.

For the *definite* integral, we evaluate this from 0 to 5:

$$(-20te^{-0.1t} - 200e^{-0.1t})\Big|_0^5 = \underbrace{(-20 \cdot 5e^{-0.5} - 200e^{-0.5})}_{\text{Evaluation at } t = 5} - \underbrace{(\overbrace{-20 \cdot 0e^0}^{0} - \overbrace{200e^0}^{200})}_{\text{Evaluation at } t = 0}$$

$$= -300e^{-0.5} + 200 \approx 18 \qquad \begin{array}{l}\text{In millions of}\\ \text{dollars (using}\\ \text{a calculator)}\end{array}$$

The present value of the stream of income over 5 years is approximately $18 million.

This answer means that $18 million at 10% interest compounded continuously would generate the continuous stream $C(t) = 2t$ million dollars for 5 years. This method is often used to determine the fair value of a business or some other asset, since it gives the present value of future income.

## Graphing Calculator Exploration

∫f(x)dx=18.040802

on [0, 5] by [−2, 8]

a. Verify the answer to Example 4 by graphing $2xe^{-0.1x}$ and finding the area under the curve from 0 to 5.

b. Can you explain why the curve increases less steeply farther to the right? [*Hint:* Think of the present value of money to be paid in the more distant future.]

Why is it necessary to learn integration by parts if integrals like the one in Example 4 can be evaluated on graphing calculators? One answer is that you should have a *variety* of ways to approach a problem—sometimes geometrically, sometimes analytically, and sometimes numerically (using a calculator). These various approaches mutually support one another; for example, you can solve a problem one way and check it another way. Furthermore, not all applications involve *definite* integrals. Example 4 might have asked for a *formula* for the present value up to any time $t$, and such formulas may be found by integrating "by hand" but not from most graphing calculators.

 **BE CAREFUL:** You should use integration by parts *only* if the "easier" formulas (1 through 7 on the inside back cover) fail to solve the integral.

**Practice Problem 3**

Which of the following integrals require integration by parts, and which can be found by the substitution formula $\int e^u \, du = e^u + C$? (Do not solve the integrals.)

**a.** $\displaystyle\int xe^x \, dx$

**b.** $\displaystyle\int xe^{x^2} \, dx$

➤ Solution below

### Solutions to Practice Problems

**1.** $\displaystyle\int x^3 \ln x \, dx = (\ln x)\left(\frac{1}{4}x^4\right) - \int \frac{1}{4}x^4 \frac{1}{x} \, dx = (\ln x)\left(\frac{1}{4}x^4\right) - \frac{1}{4}\int x^3 \, dx$

$$\left[\begin{array}{ll} u = \ln x & dv = x^3 \, dx \\ du = \dfrac{1}{x} \, dx & v = \displaystyle\int x^3 \, dx = \dfrac{1}{4}x^4 \end{array}\right]$$

$$= \frac{1}{4}x^4 \ln x - \frac{1}{16}x^4 + C$$

**2.** $\displaystyle\int (x+1)(x-1)^3 \, dx = (x+1)\frac{1}{4}(x-1)^4 - \int \frac{1}{4}(x-1)^4 \, dx$

$$\left[\begin{array}{ll} u = x+1 & dv = (x-1)^3 \, dx \\ du = dx & v = \int(x-1)^3 \, dx \\ & \quad = \frac{1}{4}(x-1)^4 \end{array}\right]$$

$$= \frac{1}{4}(x+1)(x-1)^4 - \frac{1}{4}\cdot\frac{1}{5}(x-1)^5 + C$$

$$= \frac{1}{4}(x+1)(x-1)^4 - \frac{1}{20}(x-1)^5 + C$$

**3. a.** Requires integration by parts

    **b.** Can be solved by the substitution $u = x^2$

## 6.1 Section Summary

The integration by parts formula

$$\int u \, dv = uv - \int v \, du$$

is simply the integration version of the Product Rule. In fact, if you use integration by parts to find an integral, you should expect to use the Product Rule when you check your answer by differentiation. The guidelines on page 412 lead to the following suggestions for choosing $u$ and $dv$ in some commonly occurring integrals.

**For Integrals of the Form:**                    **Choose:**

$$\int x^n e^{ax}\, dx$$                    $u = x^n \qquad dv = e^{ax}\, dx$

$$\int x^n \ln x\, dx$$                    $u = \ln x \qquad dv = x^n\, dx$

$$\int (x + a)(x + b)^n\, dx$$                    $u = x + a \quad dv = (x + b)^n\, dx$

Integration by parts can be useful in any situation involving integrals, such as recovering total cost from marginal cost, calculating areas, average values (the definite integral divided by the length of the interval), continuous accumulations, or present values of continuous income streams.

## 6.1 Exercises

**1–8.** Integration by parts often involves finding integrals like the following when integrating $dv$ to find $v$. Find the following integrals *without* using integration by parts (using formulas 1 through 7 on the inside back cover). Be ready to find similar integrals during the integration by parts procedure.

**1.** $\int e^{2x}\, dx$         **2.** $\int x^5\, dx$

**3.** $\int (x + 2)\, dx$         **4.** $\int (x - 1)\, dx$

**5.** $\int \sqrt{x}\, dx$         **6.** $\int e^{-0.5t}\, dt$

**7.** $\int (x + 3)^4\, dx$         **8.** $\int (x - 5)^6\, dx$

**9–36.** Use integration by parts to find each integral.

**9.** $\int xe^{2x}\, dx$         **10.** $\int xe^{3x}\, dx$

**11.** $\int x^5 \ln x\, dx$         **12.** $\int x^4 \ln x\, dx$

**13.** $\int (x + 2)e^x\, dx$   [*Hint:* Take $u = x + 2$.]

**14.** $\int (x - 1)e^x\, dx$   [*Hint:* Take $u = x - 1$.]

**15.** $\int \sqrt{x} \ln x\, dx$         **16.** $\int \sqrt[3]{x} \ln x\, dx$

**17.** $\int (x - 3)(x + 4)^5\, dx$   **18.** $\int (x + 2)(x - 5)^5\, dx$

**19.** $\int te^{-0.5t}\, dt$         **20.** $\int te^{-0.2t}\, dt$

**21.** $\int \frac{\ln t}{t^2}\, dt$         **22.** $\int \frac{\ln t}{\sqrt{t}}\, dt$

**23.** $\int s(2s + 1)^4\, ds$         **24.** $\int \frac{x + 1}{e^{3x}}\, dx$

**25.** $\int \frac{x}{e^{2x}}\, dx$         **26.** $\int \frac{\ln(x + 1)}{\sqrt{x + 1}}\, dx$

**27.** $\int \frac{x}{\sqrt{x + 1}}\, dx$         **28.** $\int x\sqrt{x + 1}\, dx$

**29.** $\int xe^{ax}\, dx \quad (a \neq 0)$

**30.** $\int (x + b)e^{ax}\, dx \quad (a \neq 0)$

**31.** $\displaystyle\int x^n \ln ax\, dx \qquad (a \neq 0,\ \ n \neq -1)$

**32.** $\displaystyle\int (x + a)^n \ln(x + a)\, dx \qquad (n \neq -1)$

**33.** $\displaystyle\int \ln x\, dx \quad$ [*Hint:* Take $\ \ u = \ln x, \ \ dv = dx$.]

**34.** $\displaystyle\int \ln x^2\, dx \quad$ [*Hint:* Take $\ \ u = \ln x^2, \ \ dv = dx$.]

**35.** $\displaystyle\int x^3 e^{x^2}\, dx \quad$ [*Hint:* Take $\ u = x^2, \ \ dv = xe^{x^2},$
and use a substitution to find $v$ from $dv$.]

**36.** $\displaystyle\int x^3\, (x^2 - 1)^6\, dx$

[*Hint:* Take $\ \ u = x^2, \ \ dv = x(x^2 - 1)^6\, dx,$
and use a substitution to find $v$ from $dv$.]

**37–38.** Find each integral by integration by parts or a substitution, as appropriate.

**37. a.** $\displaystyle\int xe^{x^2}\, dx$      **b.** $\displaystyle\int \frac{(\ln x)^3}{x}\, dx$

    **c.** $\displaystyle\int x^2 \ln 2x\, dx$      **d.** $\displaystyle\int \frac{e^x}{e^x + 4}\, dx$

**38. a.** $\displaystyle\int \sqrt{\ln x}\,\frac{1}{x}\, dx$      **b.** $\displaystyle\int x^2 e^{x^3}\, dx$

    **c.** $\displaystyle\int x^7 \ln 3x\, dx$      **d.** $\displaystyle\int xe^{4x}\, dx$

**39–46.** Evaluate each definite integral using integration by parts. (Leave answers in exact form.)

**39.** $\displaystyle\int_0^2 xe^x\, dx$      **40.** $\displaystyle\int_0^3 xe^x\, dx$

**41.** $\displaystyle\int_1^3 x^2 \ln x\, dx$      **42.** $\displaystyle\int_1^2 x \ln x\, dx$

**43.** $\displaystyle\int_0^2 z(z - 2)^4\, dz$      **44.** $\displaystyle\int_0^4 z(z - 4)^6\, dz$

**45.** $\displaystyle\int_0^{\ln 4} te^t\, dt$      **46.** $\displaystyle\int_1^e \ln x\, dx$

**47–48.** Find in two different ways and check that your answers agree.

**47.** $\displaystyle\int x(x - 2)^5\, dx$

    **a.** Use integration by parts.
    **b.** Use the substitution $\ u = x - 2 \ $ (so $x$ is replaced by $u + 2$) and then multiply out the integrand.

**48.** $\displaystyle\int x(x + 4)^6\, dx$

    **a.** Use integration by parts.
    **b.** Use the substitution $\ u = x + 4 \ $ (so $x$ is replaced by $u - 4$) and then multiply out the integrand.

**49–50.** Derive each formula by using integration by parts on the left-hand side. (Assume $\ n > 0$.)

**49.** $\displaystyle\int x^n e^x\, dx = x^n e^x - n\int x^{n-1} e^x\, dx$

**50.** $\displaystyle\int (\ln x)^n\, dx = x(\ln x)^n - n\int (\ln x)^{n-1}\, dx$

**51.** Use the formula in Exercise 49 to find the integral $\int x^2 e^x\, dx$.
[*Hint:* Apply the formula twice.]

**52.** Use the formula in Exercise 50 to find the integral $\int (\ln x)^2\, dx$.
[*Hint:* Apply the formula twice.]

## Applied Exercises

**53. BUSINESS: Revenue** If a company's marginal revenue function is $\ MR(x) = xe^{x/4},\ $ find the revenue function. [*Hint:* Evaluate the constant $C$ so that revenue is 0 at $\ x = 0$.]

**54. BUSINESS: Cost** A company's marginal cost function is $\ MC(x) = xe^{-x/2}\ $ and fixed costs are 200. Find the cost function. [*Hint:* Evaluate the constant $C$ so that the cost is 200 at $\ x = 0$.]

**55. BUSINESS: Present Value of a Continuous Stream of Income** An electronics company generates a continuous stream of income of $4t$ million dollars per year, where $t$ is the number of years that the company has been in operation. Find the present value of this stream of income over the first 10 years at a continuous interest rate of 10%.

**56–57.** For each exercise:

**a.** Solve *without* using a graphing calculator.

**b.** Verify your answer to part (a) using a graphing calculator.

**56. BUSINESS: Present Value of a Continuous Stream of Income** An oil well generates a continuous stream of income of $60t$ thousand dollars per year, where $t$ is the number of years that the rig has been in operation. Find the present value of this stream of income over the first 20 years at a continuous interest rate of 5%.

**57. BIOMEDICAL: Drug Dosage** A drug taken orally is absorbed into the bloodstream at the rate of $te^{-0.5t}$ milligrams per hour, where $t$ is the number of hours since the drug was taken. Find the total amount of the drug absorbed during the first 5 hours.

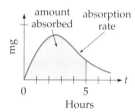

**58. ENVIRONMENTAL SCIENCE: Pollution** Contamination is leaking from an underground waste-disposal tank at the rate of $t \ln t$ thousand gallons per month, where $t$ is the number of months since the leak began. Find the total leakage from the end of month 1 to the end of month 4.

**59. GENERAL: Area** Find the area under the curve $y = x \ln x$ and above the $x$-axis from $x = 1$ to $x = 2$.

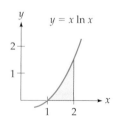

**60. POLITICAL SCIENCE: Fund Raising** A politician can raise campaign funds at the rate of $50te^{-0.1t}$ thousand dollars per week during the first $t$ weeks of a campaign. Find the average amount raised during the first 5 weeks.

**61. BUSINESS: Product Recognition** A company begins advertising a new product and finds that after $t$ weeks the product is gaining customer recognition at the rate of $t^2 \ln t$ thousand customers per week (for $t \geq 1$). Find the total gain in recognition from the end of week 1 to the end of week 6.

**62. GENERAL: Population** The population of a town is increasing at the rate of $400te^{0.02t}$ people per year, where $t$ is the number of years from now. Find the total gain in population during the next 5 years.

**63. ATHLETICS: Fluid Absorption** Runners and other athletes know that their ability to absorb water when exercising varies over time as their electrolyte levels change. Under certain circumstances, the absorption rate of water through the gastrointestinal tract will be $f(x) = 2xe^{-\frac{1}{2}x}$ liters per hour, where $t$ is the number of hours of exercise. Find the number of liters of water absorbed during the first 5 hours of exercise.

*Source:* J. Keener and J. Sneyd, *Mathematical Physiology*

**64. GENERAL: Internet Host Computers** The number of Internet host computers (computers connected directly to the Internet, for networks, bulletin boards, or online services) has been growing at the rate of $f(x) = xe^{0.1x}$ million per year, where $x$ is the number of years since 1990. Find the total number of Internet host computers added during the years 1990–2010.

*Source:* Worldwatch

## Conceptual Exercises

**65.** Choose the correct formula:

**a.** $\int u\, dv = uv + \int v\, du$   or

**b.** $\int u\, dv = uv - \int v\, du$

**66.** Choose the correct formula:

**a.** $\int u\, dv = uv - \int v\, du$   or

**b.** $\int u\, dv = \int v\, du - uv$

**67–69.** Each equation follows from the integration by parts formula by replacing $u$ by $f(x)$ and $v$ by a particular function. What is the function $v$?

**67.** $\int f(x)e^x \, dx = f(x)e^x - \int e^x f'(x) \, dx$

**68.** $\int f(x)\frac{1}{x} \, dx = f(x) \ln x - \int \ln x \, f'(x) \, dx$

**69.** $\int f(x) \, dx = f(x)x - \int x f'(x) \, dx$

**70–72.** According to the guidelines in the text, what is the correct choice for $u$ in each of the following integrals? (Do not carry out the solution, just state what $u$ should be.)

**70.** $\int xe^{3x} \, dx$

**71.** $\int x^7 \ln x \, dx$

**72.** $\int (x + 1)^5 (x + 2) \, dx$

## Explorations and Excursions

The following problems extend and augment the material presented in the text.

### Constants of Integration

**73. a.** Find the integral $\int x^{-1} \, dx$ by integration by parts (using $u = x^{-1}$ and $dv = dx$), obtaining

$$\int x^{-1} \, dx = x^{-1}x - \int (-x^{-2})x \, dx$$

which gives

$$\int x^{-1} \, dx = 1 + \int x^{-1} \, dx$$

**b.** Subtract the integral from both sides of this last equation, obtaining $0 = 1$. Explain this apparent contradiction.

**74.** We omit the constant of integration when we integrate $dv$ to get $v$. Including the constant $C$ in this step simply replaces $v$ by $v + C$, giving the formula

$$\int u \, dv = u(v + C) - \int (v + C) \, du$$

Multiplying out the parentheses and expanding the last integral into two gives

$$\int u \, dv = uv + Cu - \int v \, du - C \int du$$

Show that the second and fourth terms on the right cancel, giving the "old" integration by parts formula $\int u \, dv = uv - \int v \, du$. This shows that including the constant in the $dv$ to $v$ step gives the same formula. *One constant of integration at the end is enough.*

### Repeated Integration by Parts

Sometimes an integral requires two or more integrations by parts. As an example, we apply integration by parts to the integral $\int x^2 e^x \, dx$.

$$\int \underbrace{x^2}_{u} \underbrace{e^x \, dx}_{dv} = \underbrace{x^2}_{u} \underbrace{e^x}_{v} - \int \underbrace{e^x}_{v} \underbrace{2x \, dx}_{du} = x^2 e^x - 2\int xe^x \, dx$$

$$\left[ \begin{array}{ll} u = x^2 & dv = e^x \, dx \\ du = 2x \, dx & v = \int e^x \, dx = e^x \end{array} \right]$$

The new integral $\int xe^x \, dx$ is solved by a second integration by parts. Continuing with the previous solution, we choose new $u$ and $du$:

$$= x^2 e^x - 2\Big( \int \underbrace{xe^x \, dx}_{u \quad dv} \Big) \qquad \left[ \begin{array}{ll} u = x & dv = e^x \, dx \\ du = dx & v = e^x \end{array} \right]$$

$$= x^2 e^x - 2\Big( \underbrace{xe^x}_{uv} - \int \underbrace{e^x \, dx}_{v \quad du} \Big)$$

$$= x^2 e^x - 2(xe^x - e^x) + C$$

$$= x^2 e^x - 2xe^x + 2e^x + C$$

**75–80.** After reading the preceding explanation, find each integral by repeated integration by parts.

**75.** $\int x^2 e^{-x} \, dx$

**76.** $\int x^2 e^{2x} \, dx$

**77.** $\int (x + 1)^2 e^x \, dx$

**78.** $\int (\ln x)^2 \, dx$

**79.** $\int x^2 (\ln x)^2 \, dx$

**80.** $\int x^3 e^x \, dx$

**81–82.** For each definite integral:

**a.** Evaluate it by integration by parts. (Give answer in its *exact* form.)

**b.** Verify your answer to part (a) using a graphing calculator.

**81.** $\int_0^2 x^2 e^x \, dx$

**82.** $\int_1^5 (\ln x)^2 \, dx$

*Repeated Integration by Parts Using a Table*

The solution to a repeated integration by parts problem can be organized in a table. As an example, we solve $\int x^2 e^{3x}\, dx$. We begin by choosing

$$u = x^2 \qquad dv = v'\, dx = e^{3x}\, dx$$

We then make a table consisting of the following three columns:

| Alternating Signs | $u = x^2$ and Its Derivatives | $v' = e^{3x}$ and Its *Antiderivatives* |
|---|---|---|
| + | $x^2$ | $e^{3x}$ |
| − | $2x$ | $\dfrac{1}{3}e^{3x}$ Using |
| + | $2$ | $\dfrac{1}{9}e^{3x}$ the formula |
| − | $0$ | $\dfrac{1}{27}e^{3x}$ for $\int e^{ax}\,dx$ |

Stop when you get to 0

Finally, the solution is found by adding the *signed* products of the diagonals shown in the table:

$$\int x^2 e^{3x}\, dx = \frac{1}{3}x^2 e^{3x} - \frac{2}{9}xe^{3x} + \frac{2}{27}e^{3x} + C$$

**83–88.** After reading the preceding example, find each integral by repeated integration by parts using a table.

**83.** $\displaystyle\int x^2 e^{-x}\, dx$      **84.** $\displaystyle\int x^2 e^{2x}\, dx$

**85.** $\displaystyle\int x^3 e^{2x}\, dx$      **86.** $\displaystyle\int x^3 e^{-x}\, dx$

**87.** $\displaystyle\int (x-1)^3 e^{3x}\, dx$      **88.** $\displaystyle\int (x+1)^2 (x+2)^5\, dx$

## 6.2 INTEGRATION USING TABLES

### Introduction

There are many techniques of integration, and only some of the most useful ones will be discussed in this book. Many of the advanced techniques lead to integration formulas, which can then be collected into a "table of integrals." In this section we will see how to find integrals by choosing an appropriate formula from such a table.*

On the inside back cover is a short table of integrals that we shall use. The formulas are grouped according to the type of integrand (for example, "Forms Involving $x^2 - a^2$ and $a^2 - x^2$"). Look at the table now (formulas 9 through 23) to see how it is organized.

### Using Integral Tables

Given a particular integral, we first look for a formula that fits it exactly.

**EXAMPLE 1**

### INTEGRATING USING AN INTEGRAL TABLE

Find $\displaystyle\int \frac{1}{x^2 - 4}\, dx$.

*Another way is to use a more advanced graphing calculator (see page 428) or a computer software package such as Maple, Mathcad, or Mathematica.

*Solution*

The denominator $x^2 - 4$ is of the form $x^2 - a^2$ (with $a = 2$), so we look in the table of integrals under "Forms Involving $x^2 - a^2$ and $a^2 - x^2$." Formula 15,

$$\int \frac{1}{x^2 - a^2}\, dx = \frac{1}{2a} \ln \left| \frac{x - a}{x + a} \right| + C \qquad \text{Formula 15}$$

with $a = 2$ becomes our answer:

$$\int \frac{1}{x^2 - 4}\, dx = \frac{1}{4} \ln \left| \frac{x - 2}{x + 2} \right| + C \qquad \begin{array}{l}\text{Formula 15 with } a = 2 \\ \text{substituted on both sides}\end{array}$$

Note that the expression $x^2 - a^2$ does not require that the last number be a "perfect square." For example, $x^2 - 3$ can be written $x^2 - a^2$ with $a = \sqrt{3}$.

Integral tables are useful in many applications, such as integrating a rate to find the total accumulation.

EXAMPLE 2    **FINDING TOTAL SALES FROM THE SALES RATE**

A company's sales rate is $\dfrac{x}{\sqrt{x + 9}}$ sales per week after $x$ weeks. Find a formula for the total sales after $x$ weeks.

*Solution*

To find the *total* sales $S(x)$ we integrate the *rate* of sales.

$$S(x) = \int \frac{x}{\sqrt{x + 9}}\, dx$$

In the table on the inside back cover under "Forms Involving $\sqrt{ax + b}$" we find

$$\int \frac{x}{\sqrt{ax + b}}\, dx = \frac{2ax - 4b}{3a^2} \sqrt{ax + b} + C \qquad \text{Formula 13}$$

This formula with $a = 1$ and $b = 9$ gives the integral

$$S(x) = \int \frac{x}{\sqrt{x + 9}}\, dx = \frac{2x - 36}{3} \sqrt{x + 9} + C \qquad \begin{array}{l}\text{Formula 13 with} \\ a = 1 \text{ and } b = 9\end{array}$$

$$= \left( \frac{2}{3}x - 12 \right) \sqrt{x + 9} + C \qquad \text{Simplifying}$$

To evaluate the constant $C$, we use the fact that total sales at time $x = 0$ must be zero: $S(0) = 0$.

$$(-12)\sqrt{9} + C = 0 \qquad \begin{array}{l}(\frac{2}{3}x - 12)\sqrt{x + 9} + C \\ \text{at } x = 0 \text{ set equal to zero}\end{array}$$

$$-36 + C = 0 \qquad \text{Simplifying}$$

Therefore, $C = 36$. Substituting this into $S(x)$ gives the formula for the total sales in the first $x$ weeks.

$$S(x) = \left(\frac{2}{3}x - 12\right)\sqrt{x + 9} + 36$$

$S(x) = (\frac{2}{3}x - 12)\sqrt{x + 9} + C$
with $C = 36$

---

Modern methods of biotechnology are being used to develop many new products, including powerful antibiotics, disease-resistant crops, and bacteria that literally "eat" oil spills. These "gene splicing" techniques require evaluating definite integrals such as the following.

**EXAMPLE 3**  **GENETIC ENGINEERING**

Under certain circumstances, the number of generations of bacteria needed to increase the frequency of a gene from 0.2 to 0.5 is

$$n = 2.5 \int_{0.2}^{0.5} \frac{1}{q^2(1 - q)}\,dq$$

Find $n$ (rounded to the nearest integer).

*Solution* Formula 12 (inside back cover) integrates a similar-looking fraction.

$$\int \frac{1}{x^2(ax + b)}\,dx = -\frac{1}{b}\left(\frac{1}{x} + \frac{a}{b}\ln\left|\frac{x}{ax + b}\right|\right) + C \qquad \text{Formula 12}$$

To make $(ax + b)$ into $(1 - x)$, we take $a = -1$ and $b = 1$, so the left-hand side of the formula becomes

$$\int \frac{1}{x^2(-x + 1)}\,dx \quad \text{or} \quad \int \frac{1}{x^2(1 - x)}\,dx \qquad \begin{array}{l}\text{From formula 12}\\ \text{with } a = -1, \ b = 1\end{array}$$

Except for replacing $x$ by $q$, this is the same as our integral. Therefore, the indefinite integral is found by formula 12 with $a = -1$ and $b = 1$ (which we express in the variable $q$).

$$\int \frac{1}{q^2(1 - q)}\,dq = -\left(\frac{1}{q} - \ln\left|\frac{q}{1 - q}\right|\right) + C \qquad \begin{array}{l}\text{Formula 12 with } a = -1,\\ b = 1, \text{ and } x \text{ replaced by } q\end{array}$$

For the *definite* integral from 0.2 to 0.5, we evaluate and subtract.

$$\underbrace{-\left(\frac{1}{0.5} - \ln\left|\frac{0.5}{1 - 0.5}\right|\right)}_{\text{Evaluation at } q = 0.5} - \underbrace{\left[-\left(\frac{1}{0.2} - \ln\left|\frac{0.2}{1 - 0.2}\right|\right)\right]}_{\text{Evaluation at } q = 0.2}$$

$$= -(2 - \underbrace{\ln 1}_{0}) + \left(5 - \underbrace{\ln\frac{0.2}{0.8}}_{\ln 0.25}\right) = -2 + 5 - \ln 0.25 \approx \underbrace{4.39}_{\substack{\text{Using a}\\ \text{calculator}}}$$

We multiply this by the 2.5 in front of the original integral.

$$(2.5)(4.39) \approx 10.98$$

Therefore, 11 generations are needed to raise the gene frequency from 0.2 to 0.5.

## Graphing Calculator Exploration

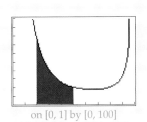

on [0, 1] by [0, 100]
(shaded from 0.2 to 0.5)

**a.** Verify the answer to Example 3 by finding the definite integral of
$$y = \frac{2.5}{x^2(1 - x)}$$ from 0.2 to 0.5.

**b.** From the graph of the function shown on the left, which would require more generations: increasing the gene frequency from 0.1 to 0.2 or from 0.5 to 0.6?

**c.** Can you think of a reason for this? [*Hint:* Genes reproduce from other similar genes.]

Sometimes a substitution is needed to transform a formula to fit a given integral. In such cases both the $x$ and the $dx$ must be transformed. A few examples will make the method clear.

**EXAMPLE 4**

**USING A TABLE WITH A SUBSTITUTION**

Find $\displaystyle\int \frac{x}{\sqrt{x^4 + 1}}\, dx$.

*Solution*

The table of integrals on the inside back cover has no formula involving $x^4$. However, $x^4 = (x^2)^2$, so a formula involving $x^2$, along with a substitution, might work. Formula 18 looks promising:

$$\int \frac{1}{\sqrt{x^2 \pm a^2}}\, dx = \ln\left| x + \sqrt{x^2 \pm a^2} \right| + C$$

The $\pm$ means: use either the upper sign or the lower sign on both sides

$$\int \frac{1}{\sqrt{x^2 + 1}}\, dx = \ln\left| x + \sqrt{x^2 + 1} \right| + C$$

Formula 18 with $a = 1$ and the upper sign

With the substitution

$$x = z^2$$
$$dx = 2z\, dz$$

Differential of $x = z^2$

this becomes

$$\int \frac{1}{\sqrt{z^4 + 1}}\, 2z\, dz = \ln\left| z^2 + \sqrt{z^4 + 1}\right| + C$$

Previous formula with $x = z^2$ and $dx = 2z\, dz$

Dividing by 2 and replacing $z$ by $x$ gives the integral that we wanted:

$$\int \frac{x}{\sqrt{x^4 + 1}}\, dx = \frac{1}{2}\ln\left(x^2 + \sqrt{x^4 + 1}\right) + C$$

Dropping the absolute value bars since $x^2 + \sqrt{x^4 + 1}$ is positive

Given a particular integral, how do we choose a formula?

### How to Choose a Formula

Find a formula that matches the *most complicated part* of the integral, making appropriate substitutions to change the formula into the given integral.

For instance, in Example 4 we matched the $\sqrt{x^4 + 1}$ in the given integral with the $\sqrt{x^2 \pm a^2}$ in the formula, and the rest of the integral followed from the differential.

 **BE CAREFUL:** As indicated in the box, it is usually simpler to change the *formula* into the given *integral* rather than the given integral into the formula. Also, most mistakes come from incorrectly changing the $dx$ into the new differential (using the chosen substitution), so this step should be done with particular care.

**Practice Problem**     Find $\displaystyle\int \frac{t}{9t^4 - 1}\, dt$.     [*Hint:* Use formula 15 with $x = 3t^2$.]

➤ **Solution on page 429**

**EXAMPLE 5**

### USING A TABLE WITH A SUBSTITUTION

Find $\displaystyle\int \frac{e^{-2t}}{e^{-t} + 1}\, dt$.

*Solution*

Looking in the table of integrals under "Forms Involving $e^{ax}$ and $\ln x$," none of the formulas looks anything like this integral. However, replacing

$e^{-t}$ by $x$ would make the denominator of our integral into $x + 1$, so formula 9 might help. This formula with $a = 1$ and $b = 1$ is

$$\int \frac{x}{x + 1}\, dx = x - \ln|x + 1| + C \qquad \text{Formula 9 with} \atop a = 1 \text{ and } b = 1$$

With the substitution

$$x = e^{-t}$$
$$dx = -e^{-t}\, dt \qquad \text{Differential of } x = e^{-t}$$

formula 9 becomes

$$\int \frac{e^{-t}}{e^{-t} + 1}\, (-e^{-t})\, dt = e^{-t} - \ln|e^{-t} + 1| + C$$

or

$$-\int \frac{e^{-2t}}{e^{-t} + 1}\, dt = e^{-t} - \ln(e^{-t} + 1) + C$$

Except for the negative sign, this is the given integral. Multiplying through by $-1$ gives the final answer.

$$\int \frac{e^{-2t}}{e^{-t} + 1}\, dt = -e^{-t} + \ln(e^{-t} + 1) + C$$

## Reduction Formulas

Sometimes we must apply a formula several times to simplify an integral in stages.

**EXAMPLE 6**

**USING A REDUCTION FORMULA**

Find $\displaystyle\int x^3 e^{-x}\, dx$.

*Solution*

In the integral table, we find formula 21.

$$\int x^n e^{ax}\, dx = \frac{1}{a} x^n e^{ax} - \frac{n}{a} \int x^{n-1} e^{ax}\, dx \qquad \text{Formula 21. With } n = 3 \atop \text{and } a = -1, \text{ the left side fits our integral}$$

The right-hand side of this formula involves a new integral, but with a *lower* power of $x$. We will apply formula 21 several times, each time reducing the power of $x$ until we eliminate it completely. Applying formula 21 with $n = 3$ and $a = -1$,

$$\int x^3 e^{-x}\, dx = -x^3 e^{-x} + 3 \int x^2 e^{-x}\, dx \qquad \text{After one application}$$

⌐— The power has been reduced

$$= -x^3 e^{-x} + 3\left(-x^2 e^{-x} + 2 \int x^1 e^{-x}\, dx\right) \qquad \begin{array}{l}\text{Applying formula 21}\\ \text{again (now with } n = 2)\\ \text{to the last integral above}\end{array}$$

$$= -x^3 e^{-x} - 3x^2 e^{-x} + 6 \int x^1 e^{-x}\, dx \qquad \text{Multiplying out}$$

$$= -x^3 e^{-x} - 3x^2 e^{-x} + 6\left(-xe^{-x} + \int \underbrace{x^0 e^{-x}}_{1}\, dx\right) \qquad \begin{array}{l}\text{Using formula 21 a third}\\ \text{time (now with } n = 1)\end{array}$$

$$= -x^3 e^{-x} - 3x^2 e^{-x} - 6xe^{-x} + 6 \int e^{-x}\, dx \qquad \begin{array}{l}\text{Now solve this last integral by the}\\ \text{formula } \int e^{ax}\, dx = \frac{1}{a} e^{ax} + C\end{array}$$

$$= -x^3 e^{-x} - 3x^2 e^{-x} - 6xe^{-x} - 6e^{-x} + C \qquad \begin{array}{l}\text{The solution, after three}\\ \text{applications of formula 21}\end{array}$$

$$= -e^{-x}(x^3 + 3x^2 + 6x + 6) + C \qquad \text{Factoring}$$

We used formula 21 three times, reducing the $x^3$ in steps, first down to $x^2$, then to $x^1$, and finally to $x^0 = 1$, at which point we could solve the integral easily. If the power of $x$ in the integral had been higher, more applications of formula 21 would have been necessary. Formulas such as 21 and 22 are called **reduction formulas,** since they express an integral in terms of a similar integral but with a smaller power of $x$.

## Graphing Calculator Exploration

Some advanced graphing calculators can find *indefinite* integrals. For example, the Texas Instruments *TI-89* graphing calculator finds the integrals in Examples 5 and 6 as follows.

| For Example 5 | For Example 6 |

With some algebra, you can verify that these answers agree with those found in Examples 5 and 6, even though they do not look the same. (Notice that this calculator omits the arbitrary constant of integration.)

> ### Solution to Practice Problem
>
> Formula 15 with the substitution $x = 3t^2$, $dx = 6t\,dt$, and $a = 1$ becomes
>
> $$\int \frac{1}{9t^4 - 1}\,6t\,dt = \frac{1}{2}\ln\left|\frac{3t^2 - 1}{3t^2 + 1}\right| + C$$
>
> Dividing each side by 6 gives the answer:
>
> $$\int \frac{t}{9t^4 - 1}\,dt = \frac{1}{12}\ln\left|\frac{3t^2 - 1}{3t^2 + 1}\right| + C$$

## 6.2 Section Summary

To find a formula that "fits" a given integral, we look for the formula whose left-hand side most closely matches the most complicated part of the integral. Then we choose constants (and possibly a substitution) to make the formula fit exactly. Although we have been using a very brief table, the technique is the same with a more extensive table. Many integral tables have been published, some of which are book-length, containing several thousand formulas.*

## 6.2 Exercises

**1–6.** For each integral, state the number of the integration formula (from the inside back cover) and the values of the constants $a$ and $b$ so that the formula fits the integral. (Do not evaluate the integral.)

**1.** $\displaystyle\int \frac{1}{x^2(5x - 1)}\,dx$    **2.** $\displaystyle\int \frac{x}{2x - 3}\,dx$

**3.** $\displaystyle\int \frac{1}{x\sqrt{-x + 7}}\,dx$    **4.** $\displaystyle\int \frac{x}{\sqrt{-2x + 1}}\,dx$

**5.** $\displaystyle\int \frac{x}{1 - x}\,dx$    **6.** $\displaystyle\int \frac{1}{x\sqrt{1 - 4x}}\,dx$

**7–42.** Find each integral by using the integral table on the inside back cover.

**7.** $\displaystyle\int \frac{1}{9 - x^2}\,dx$
  [*Hint:* Use formula 16 with $a = 3$.]

**8.** $\displaystyle\int \frac{1}{x^2 - 25}\,dx$
  [*Hint:* Use formula 15 with $a = 5$.]

**9.** $\displaystyle\int \frac{1}{x^2(2x + 1)}\,dx$
  [*Hint:* Use formula 12 with $a = 2$, $b = 1$.]

**10.** $\displaystyle\int \frac{x}{x + 2}\,dx$
  [*Hint:* Use formula 9 with $a = 1$, $b = 2$.]

**11.** $\displaystyle\int \frac{x}{1 - x}\,dx$   [*Hint:* Use formula 9.]

**12.** $\displaystyle\int \frac{x}{\sqrt{1 - x}}\,dx$   [*Hint:* Use formula 13.]

**13.** $\displaystyle\int \frac{1}{(2x + 1)(x + 1)}\,dx$

**14.** $\displaystyle\int \frac{x}{(x + 1)(x + 2)}\,dx$

**15.** $\displaystyle\int \sqrt{x^2 - 4}\,dx$    **16.** $\displaystyle\int \frac{1}{\sqrt{x^2 - 1}}\,dx$

**17.** $\displaystyle\int \frac{1}{z\sqrt{1 - z^2}}\,dz$    **18.** $\displaystyle\int \frac{\sqrt{4 + z^2}}{z}\,dz$

*A useful table of integrals containing more than 400 formulas is found in *CRC Standard Mathematical Tables*, CRC Press.

**19.** $\int x^3 e^{2x}\, dx$

**20.** $\int x^{99} \ln x\, dx$

**21.** $\int x^{-101} \ln x\, dx$

**22.** $\int (\ln x)^2\, dx$

**23.** $\int \frac{1}{x(x+3)}\, dx$

**24.** $\int \frac{1}{x(x-3)}\, dx$

**25.** $\int \frac{z}{z^4 - 4}\, dz$

**26.** $\int \frac{z}{9 - z^4}\, dz$

**27.** $\int \sqrt{9x^2 + 16}\, dx$

**28.** $\int \frac{1}{\sqrt{16x^2 - 9}}\, dx$

**29.** $\int \frac{1}{\sqrt{4 - e^{2t}}}\, dt$

**30.** $\int \frac{e^t}{9 - e^{2t}}\, dt$

**31.** $\int \frac{e^t}{e^{2t} - 1}\, dt$

**32.** $\int \frac{e^{2t}}{1 - e^t}\, dt$

**33.** $\int \frac{x^3}{\sqrt{x^8 - 1}}\, dx$

**34.** $\int x^2 \sqrt{x^6 + 1}\, dx$

**35.** $\int \frac{1}{x\sqrt{x^3 + 1}}\, dx$

**36.** $\int \frac{\sqrt{1 - x^6}}{x}\, dx$

**37.** $\int \frac{e^t}{(e^t - 1)(e^t + 1)}\, dt$

**38.** $\int \frac{e^{2t}}{(e^t - 1)(e^t + 1)}\, dt$

**39.** $\int xe^{x/2}\, dx$

**40.** $\int \frac{x}{e^x}\, dx$

**41.** $\int \frac{1}{e^{-x} + 4}\, dx$

**42.** $\int \frac{1}{\sqrt{e^{-x} + 4}}\, dx$

**43–48.** For each definite integral:

**a.** Evaluate it using the table of integrals on the inside back cover. (Leave answers in *exact* form.)

**b.** Use a graphing calculator to verify your answer to part (a).

**43.** $\int_4^5 \sqrt{x^2 - 16}\, dx$

**44.** $\int_0^4 \frac{1}{\sqrt{x^2 + 9}}\, dx$

**45.** $\int_2^3 \frac{1}{x^2 - 1}\, dx$

**46.** $\int_2^4 \frac{1}{1 - x^2}\, dx$

**47.** $\int_3^5 \frac{\sqrt{25 - x^2}}{x}\, dx$

**48.** $\int_3^4 \frac{1}{x\sqrt{25 - x^2}}\, dx$

**49–56.** Find each integral by whatever means are necessary (either substitution or tables).

**49.** $\int \frac{1}{2x + 6}\, dx$

**50.** $\int \frac{x}{x^2 - 4}\, dx$

**51.** $\int \frac{x}{2x + 6}\, dx$

**52.** $\int \frac{1}{4 - x^2}\, dx$

**53.** $\int x\sqrt{1 - x^2}\, dx$

**54.** $\int \frac{x}{\sqrt{1 - x^2}}\, dx$

**55.** $\int \frac{\sqrt{1 - x^2}}{x}\, dx$

**56.** $\int \frac{1}{\sqrt{x^2 - 1}}\, dx$

**57–62.** Find each integral. [*Hint:* Separate each integral into two integrals, using the fact that the numerator is a sum or difference, and find the two integrals by two different formulas.]

**57.** $\int \frac{x - 1}{(3x + 1)(x + 1)}\, dx$

**58.** $\int \frac{x - 1}{x^2(x + 1)}\, dx$

**59.** $\int \frac{x + 1}{x\sqrt{1 + x^2}}\, dx$

**60.** $\int \frac{x - 1}{x\sqrt{x^2 + 4}}\, dx$

**61.** $\int \frac{x + 1}{x - 1}\, dx$ [*Hint:* After separating into two integrals, find one by a formula and the other by a substitution.]

**62.** $\int \frac{x + 1}{\sqrt{x^2 + 1}}\, dx$ [*Hint:* After separating into two integrals, find one by a formula and the other by a substitution.]

### Applied Exercises

**63. BUSINESS: Total Sales** A company's sales rate is $x^2 e^{-x}$ million sales per month after $x$ months. Find a formula for the total sales in the first $x$ months. [*Hint:* Integrate the sales rate to find the total sales and determine the constant $C$ so that total sales are zero at time $x = 0$.]

**64. GENERAL: Population** The population of a city is expected to grow at the rate of $x/\sqrt{x + 9}$ thousand people per year after $x$ years. Find the total change in population from year 0 to year 27.

**65. BIOMEDICAL: Gene Frequency** Under certain circumstances, the number of generations necessary to increase the frequency of a gene from 0.1 to 0.3 is

$$n = 3 \int_{0.1}^{0.3} \frac{1}{q^2(1 - q)}\, dq$$

Find $n$ (rounded to the nearest integer).

**66. BEHAVIORAL SCIENCE: Response Rate**  A subject in a psychology experiment gives responses at the rate of  $t/\sqrt{t+1}$  correct answers per minute after $t$ minutes.

   **a.** Find the total number of correct responses from time  $t=0$  to time  $t=15$.

   **b.** Verify your answer to part (a) using a graphing calculator.

**67. BUSINESS: Cost**  The marginal cost function for a computer chip manufacturer is  $MC(x) = 1/\sqrt{x^2+1}$,  and fixed costs are $2000. Find the cost function.

**68. SOCIAL SCIENCE: Employment**  An urban job placement center estimates that the number of residents seeking employment $t$ years from now will be  $t/(2t+4)$  million people.

   **a.** Find the average number of job seekers during the period  $t=0$  to  $t=10$.

   **b.** Verify your answer to part (a) using a graphing calculator.

## Conceptual Exercises

**69–72.** Which formula (from formulas 9–23 on the inside back cover) and what substitution (of the form $x=f(t)$)  would you use to find the given integral? (Do not carry out the solution, just indicate a formula and a substitution.)

**69.** $\displaystyle\int \frac{e^{100t}}{e^{50t}+7}\,dt$

**70.** $\displaystyle\int \frac{e^{1000t}}{(e^{500t}+5)(e^{500t}+7)}\,dt$

**71.** $\displaystyle\int \frac{t^{49}}{\sqrt{t^{100}+1}}\,dt$

**72.** $\displaystyle\int \frac{\sqrt{1+t^{100}}}{t}\,dt$

**73.** How many uses of formula 21 would you need to find  $\displaystyle\int t^{476}e^{t}\,dt$?

**74.** How many uses of formula 22 would you need to find  $\displaystyle\int (\ln t)^{623}\,dt$?

**75.** State (without carrying them out) two different *methods* to find  $\displaystyle\int \ln t\,dt$.

**76.** State (without carrying them out) two different *methods* to find  $\displaystyle\int te^{t}\,dt$.

**77.** Which *two* formulas can find  $\displaystyle\int \frac{1}{t^2-1}\,dt$?

**78.** Which *two* formulas can find  $\displaystyle\int \frac{1}{1-t^2}\,dt$?

<h2>6.3  IMPROPER INTEGRALS</h2>

### Introduction

In this section we define integrals over intervals that are infinite in length. Such **improper integrals** have many applications, such as the permanent endowments discussed in the Application Preview on page 409 at the beginning of this chapter.

### Limits as $x$ Approaches $\pm\infty$

Recall that the notation $x \to \infty$ ("$x$ approaches infinity") means that $x$ takes on arbitrarily large values.

| $x \to \infty$  and  $x \to -\infty$ | | |
| --- | --- | --- |
| $x \to \infty$ | means: | $x$ takes values arbitrarily far to the *right* on the number line. |
| $x \to -\infty$ | means: | $x$ takes values arbitrarily far to the *left* on the number line. |

Evaluating limits as $x$ approaches positive or negative infinity is simply a matter of thinking about large and small numbers. The reciprocal of a large number is a small number. For example,

$$\frac{1}{1,000,000} = 0.000001$$

One over a million is one one-millionth

Similarly:

$$\lim_{x \to \infty} \frac{1}{x^2} = 0$$

$$\lim_{x \to \infty} e^{-x} = \lim_{x \to \infty} \frac{1}{e^x} = 0$$

As the denominator approaches infinity (with the numerator constant), the value approaches zero

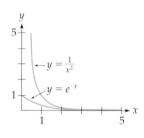

Geometrically, these statements mean that the curves approach the $x$-axis asymptotically as $x \to \infty$, as shown on the left. These examples illustrate the following general rules.

| Limits as $x \to \infty$ | |
|---|---|
| $$\lim_{x \to \infty} \frac{1}{x^n} = 0 \qquad (n > 0)$$ | As $x$ approaches infinity, 1 over $x$ to a *positive* power approaches zero |
| $$\lim_{x \to \infty} e^{-ax} = 0 \qquad (a > 0)$$ | As $x$ approaches infinity, $e$ to a *negative* number times $x$ approaches zero |

**EXAMPLE 1**

**EVALUATING LIMITS**

**a.** $\lim_{b \to \infty} \dfrac{1}{b^2} = 0$    Using the first rule in the box above

**b.** $\lim_{b \to \infty} \left(3 - \dfrac{1}{b}\right) = 3$    Because the $\dfrac{1}{b}$ approaches zero

**c.** $\lim_{b \to \infty} (e^{-2b} - 5) = -5$    Because the $e^{-2b}$ approaches zero

Similar rules hold for $x$ approaching *negative* infinity.

| Limits as $x \to -\infty$ | |
|---|---|
| $$\lim_{x \to -\infty} \frac{1}{x^n} = 0 \quad (\text{for integer } n > 0)$$ | As $x$ approaches *negative* infinity, 1 over $x$ to a positive integer approaches zero |
| $$\lim_{x \to -\infty} e^{ax} = 0 \quad (a > 0)$$ | As $x$ approaches negative infinity, $e$ to a positive number times $x$ approaches zero (because the exponent is approaching $-\infty$) |

## Graphing Calculator Exploration

To see examples of the last two results, graph $y_1 = \dfrac{1}{x^2}$ and $y_2 = e^{0.5x}$ on $[-5, 0]$ by $[0, 5]$. Notice that both curves asymptotically approach the $x$-axis on the left, verifying that both functions approach zero as $x \to -\infty$.

Recall that for a limit to exist, it must be a *number* (that is, a *finite quantitiy*). Therefore, if a function becomes arbitrarily large, we say its limit *does not exist* or is *undefined*. In Chapters 2 and 3 we used notations like $\lim\limits_{x \to \infty} f(x) = \infty$, which does not mean that the limit exists, but rather that the function takes on arbitrarily large values as $x \to \infty$, and so the limit *does not exist*. The following are some examples of limits that are undefined.

### Limits that Do Not Exist

The following limits do not exist:

| | | |
|---|---|---|
| $\lim\limits_{x \to \infty} x^n$ | $(n > 0)$ | As $x$ approaches infinity, $x$ to a positive power has no limit |
| $\lim\limits_{x \to \infty} e^{ax}$ | $(a > 0)$ | As $x$ approaches infinity, $e$ to a positive number times $x$ has no limit |
| $\lim\limits_{x \to \infty} \ln x$ | | As $x$ approaches infinity, the natural logarithm of $x$ has no limit |

**EXAMPLE 2**

### FINDING WHETHER A LIMIT EXISTS

**a.** $\lim\limits_{b \to \infty} b^3$  does not exist.

Because $b^3$ becomes arbitrarily large as $b$ approaches infinity

**b.** $\lim\limits_{b \to \infty} (\sqrt{b} - 1)$  does not exist.

Because $\sqrt{b}$ becomes arbitrarily large as $b$ approaches infinity

**Practice Problem 1**

Evaluate the following limits (if they exist).

**a.** $\lim\limits_{b \to \infty} \left(1 - \dfrac{1}{b}\right)$      **b.** $\lim\limits_{b \to \infty} \left(\sqrt[3]{b} + 3\right)$

➤ Solution on page 439

## Improper Integrals

A definite integral over an interval of infinite length is an *improper integral*. As a first example, we evaluate the improper integral $\displaystyle\int_1^\infty \dfrac{1}{x^2}\,dx$,  which gives the area under the curve  $y = \dfrac{1}{x^2}$  from  $x = 1$  extending arbitrarily far to the right as shown in the graph.

**EXAMPLE 3**

**EVALUATING AN IMPROPER INTEGRAL**

Evaluate $\displaystyle\int_1^\infty \frac{1}{x^2}\,dx$.

*Solution*

To integrate to infinity, we first integrate over a *finite* interval, from 1 to some number $b$ (think of $b$ as some very large number), and then take the limit as $b$ approaches $\infty$. First integrate from 1 to $b$.

$$\int_1^b \frac{1}{x^2}\,dx = \int_1^b x^{-2}\,dx = \left(-x^{-1}\right)\Big|_1^b = \left(-\frac{1}{x}\right)\Big|_1^b \qquad \text{Using the Power Rule}$$

$$= \underbrace{-\frac{1}{b}}_{\text{at } x = b} - \underbrace{\left(-\frac{1}{1}\right)}_{\text{at } x = 1} = -\frac{1}{b} + 1 \qquad \begin{array}{l}\text{Evaluating and}\\ \text{simplifying}\end{array}$$

Then take the limit of this answer as $b \to \infty$.

$$\lim_{b \to \infty}\left(-\frac{1}{b} + 1\right) = 1 \qquad \begin{array}{l}\text{Limit as } b \to \infty \text{ (the } \frac{1}{b}\\ \text{approaches zero)}\end{array}$$

This gives the answer:

$$\int_1^\infty \frac{1}{x^2}\,dx = 1 \qquad \text{Integral from 1 to } \infty \text{ equals 1}$$

$y = \dfrac{1}{x^2}$

area is 1 square unit

Since the limit exists, we say that the improper integral is **convergent.** Geometrically, this procedure amounts to finding the area under the curve from 1 to some number $b$, shown on the left below, and then letting $b \to \infty$ to find the area arbitrarily far to the right.

**Integrating to $b$**

**Integrating to $\infty$**

**Improper Integrals: Integrating to ∞**

If $f$ is continuous and nonnegative for $x \geq a$, we define

$$\int_a^\infty f(x)\,dx = \lim_{b \to \infty}\int_a^b f(x)\,dx$$

provided that the limit exists. The improper integral is said to be *convergent* if the limit exists, and *divergent* if the limit does not exist.

It is possible to define improper integrals for functions that take negative values, and even for discontinuous functions, but we shall not do so in this book, since most applications involve functions that are positive and continuous.

**Practice Problem 2**

Evaluate $\displaystyle\int_{2}^{\infty} \frac{1}{x^2}\,dx.$

➤ **Solution on page 440**

**EXAMPLE 4**

**FINDING WHETHER AN INTEGRAL DIVERGES**

Evaluate $\displaystyle\int_{1}^{\infty} \frac{1}{\sqrt{x}}\,dx.$

*Solution*

Integrating up to $b$:

$$\int_{1}^{b} \frac{1}{\sqrt{x}}\,dx = \int_{1}^{b} x^{-1/2}\,dx \;=\; \underbrace{2\cdot x^{1/2}\Big|_{1}^{b}}_{\substack{\text{Integrating by}\\\text{the Power Rule}}} \;=\; \underbrace{2\sqrt{b}-2\sqrt{1}}_{\text{Evaluating}} = 2\sqrt{b}-2$$

Letting $b$ approach infinity:

$$\lim_{b\to\infty}\left(2\sqrt{b}-2\right) \quad \text{does not exist}$$

Because $\sqrt{b}$ becomes infinite as $b \to \infty$

Therefore,

$$\int_{1}^{\infty} \frac{1}{\sqrt{x}}\,dx \quad \text{is } divergent$$

The integral cannot be evaluated

Notice from Examples 3 and 4 that $\displaystyle\int_{1}^{\infty} \frac{1}{x^2}\,dx$ is *convergent* (its value is 1), but $\displaystyle\int_{1}^{\infty} \frac{1}{\sqrt{x}}\,dx$ is *divergent* (it does not have a finite value), as illustrated in the following diagrams.

**Area under** $\dfrac{1}{x^2}$

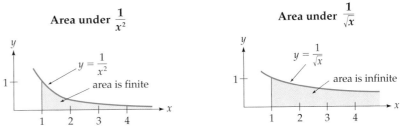

**Area under** $\dfrac{1}{\sqrt{x}}$

Intuitively, the areas differ because the curve $\frac{1}{x^2}$ approaches the $x$-axis more rapidly than does the curve $\frac{1}{\sqrt{x}}$ for large values of $x$, as shown on the previous page, and so has a smaller area under it.

## Spreadsheet Exploration

The following spreadsheet* allows us to numerically "see" that one of these integrals converges while the other diverges. From Example 3 on page 434 we know that $\displaystyle\int_1^\infty \frac{1}{x^2}\,dx = \lim_{b \to \infty}\left(1 - \frac{1}{b}\right)$, while from Example 4 on page 435 we have $\displaystyle\int_1^\infty \frac{1}{\sqrt{x}}\,dx = \lim_{b \to \infty}\left(2\sqrt{b} - 2\right)$. The increasingly large values of $b$ shown in column **A** are used to evaluate $1 - \frac{1}{b}$ in column **B** and $2\sqrt{b} - 2$ in column **C**.

| C7 | | = | =2*$A7^(1/2)-2 |
|---|---|---|---|

| | A | B | C |
|---|---|---|---|
| 1 | $b$ | Integral of $x^{-2}$ | Integral of $x^{-1/2}$ |
| 2 | 1 | 0 | 0.000 |
| 3 | 5 | 0.8 | 2.472 |
| 4 | 10 | 0.9 | 4.325 |
| 5 | 50 | 0.98 | 12.142 |
| 6 | 100 | 0.99 | 18.000 |
| 7 | 500 | 0.998 | 42.721 |
| 8 | 1000 | 0.999 | 61.246 |
| 9 | 5000 | 0.9998 | 139.421 |
| 10 | 10000 | 0.9999 | 198.000 |
| 11 | 50000 | 0.99998 | 445.214 |
| 12 | 100000 | 0.99999 | 630.456 |
| 13 | 500000 | 0.999998 | 1412.214 |

$b$ — areas up to $b$

Is it clear which of these integrals (areas) is converging and which is diverging?

*To obtain this and other Spreadsheet Explorations, go to www.cengage.com/math/berresford.

We may combine the two steps of integrating up to $b$ and letting $b \to \infty$ into a single line. We show how to do this by evaluating the integral from Example 3 again, but more briefly.

$$\int_1^\infty x^{-2}\, dx = \lim_{b \to \infty} \int_1^b x^{-2}\, dx = \lim_{b \to \infty} (-x^{-1}) \Big|_1^b = \lim_{b \to \infty} \left[ -\frac{1}{b} - \left( -\frac{1}{1} \right) \right] = 1$$

Integrating;<br>then use<br>$x^{-1} = \dfrac{1}{x}$  Approaches<br>0  ↑ Same answer<br>as before

## Permanent Endowments

Funds that generate steady income forever are called *permanent endowments.*

---

**EXAMPLE 5**

### FINDING THE SIZE OF A PERMANENT ENDOWMENT

In the Application Preview on page 409 we found that the size of the fund necessary to generate $2000 annually forever (at 5% interest compounded continuously) is $\int_0^\infty 2000e^{-0.05t}\, dt$. Find the size of this permanent endowment by evaluating the integral.

*Solution*

$$\int_0^\infty 2000e^{-0.05t}\, dt = \lim_{b \to \infty} \left( 2000 \int_0^b e^{-0.05t}\, dt \right)$$

Annual<br>income  Continuous<br>interest rate  Moved<br>outside

$$= \lim_{b \to \infty} \left[ 2000(-20)e^{-0.05t} \Big|_0^b \right]$$  Integrating by<br>$\int e^{ax}\, dx = \dfrac{1}{a}e^{ax}$

$$= \lim_{b \to \infty} (-40{,}000e^{-0.05b} + 40{,}000e^0) = 40{,}000$$

Approaches<br>0  1

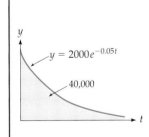

$y = 2000e^{-0.05t}$

40,000

Therefore, the size of the permanent endowment that will pay the $2000 annual maintenance forever is $40,000. (The area represented by this number is shown on the left.)

---

Permanent endowments are used to estimate the ultimate cost of anything that requires continuous long-term funding, from buildings to government agencies to toxic waste sites.

### Finding Improper Integrals Using Substitutions

Solving an improper integral may require a substitution. In such cases we apply the substitution not only to the integrand but also to the differential and the upper and lower limits of integration.

---

**EXAMPLE 6**

**FINDING AN IMPROPER INTEGRAL USING A SUBSTITUTION**

Evaluate $\displaystyle\int_2^\infty \frac{x}{(x^2 + 1)^2}\, dx$.

*Solution*

We use the substitution $u = x^2 + 1$, so $du = 2x\, dx$, requiring multiplication by 2 and by $\frac{1}{2}$. Notice how the substitution changes the limits.

$$\int_2^\infty \frac{x}{(x^2 + 1)^2}\, dx = \frac{1}{2}\int_2^\infty \frac{2x}{(x^2 + 1)^2}\, dx = \frac{1}{2}\int_5^\infty \frac{du}{u^2} = \frac{1}{2}\lim_{b \to \infty}\int_5^b u^{-2}\, du$$

$$\left[\begin{array}{c} u = x^2 + 1 \\ du = 2x\, dx \end{array}\right] \qquad \text{at} \quad x = 2, \quad u^2 + 1 = 5$$

$$= \frac{1}{2}\lim_{b \to \infty}\left[-u^{-1}\right]\Big|_5^b = \frac{1}{2}\lim_{b \to \infty}\left[-\frac{1}{b} - \left(-\frac{1}{5}\right)\right] = \frac{1}{2}\cdot\frac{1}{5} = \frac{1}{10}$$

Integrating      Approaches 0

---

To integrate over an interval that extends arbitrarily far to the *left*, we again integrate over a finite interval and then take the limit.

**Improper Integrals: Integrating to $-\infty$**

If $f$ is continuous and nonnegative for $x \le b$, we define

$$\int_{-\infty}^b f(x)\, dx = \lim_{a \to -\infty}\int_a^b f(x)\, dx$$

provided that the limit exists. The improper integral is *convergent* if the limit exists, and *divergent* if the limit does not exist.

To integrate over the *entire* x-axis, from $-\infty$ to $\infty$, we use two integrals, one from $-\infty$ to 0, and the other from 0 to $\infty$, and then add the results.

---

**Improper Integrals : Integrating from $-\infty$ to $\infty$**

If $f$ is continuous and nonnegative for *all* values of $x$, we define

$$\int_{-\infty}^{\infty} f(x)\,dx = \lim_{a \to -\infty} \int_{a}^{0} f(x)\,dx + \lim_{b \to \infty} \int_{0}^{b} f(x)\,dx$$

The improper integral is *convergent* if both limits exist, and *divergent* if either limit does not exist.

---

**EXAMPLE 7**

**INTEGRATING TO $-\infty$**

Evaluate $\displaystyle\int_{-\infty}^{3} 4e^{2x}\,dx$.

*Solution*

$$\int_{-\infty}^{3} 4e^{2x}\,dx = \lim_{a \to -\infty} \int_{a}^{3} 4e^{2x}\,dx$$

$$= \lim_{a \to -\infty} \left[ 4 \cdot \frac{1}{2} \cdot e^{2x} \Big|_{a}^{3} \right] \qquad \text{Integrating}$$

$$= \lim_{a \to -\infty} \underbrace{(2e^{6} - 2e^{2a})}_{\substack{\text{Approaches } 0 \\ \text{as } a \to -\infty}} = 2e^{6} \qquad \text{Convergent}$$

---

**Practice Problem 3**  Evaluate the improper integral $\displaystyle\int_{-\infty}^{1} 12e^{3x}\,dx$.  ➤ **Solution on next page**

---

**Solutions to Practice Problems**

**1. a.** $\displaystyle\lim_{b \to \infty}\left(1 - \frac{1}{b}\right) = 1$  because $\frac{1}{b}$ approaches 0

**b.** $\displaystyle\lim_{b \to \infty}\left(\sqrt[3]{b} + 3\right)$ does not exist  because $\sqrt[3]{b}$ gets arbitrarily large

**2.** $\int_{2}^{b} x^{-2}\,dx = (-x^{-1})\Big|_{2}^{b} = -\dfrac{1}{b} - \left(-\dfrac{1}{2}\right) = \dfrac{1}{2} - \dfrac{1}{b}$

$\lim_{b \to \infty} \left(\dfrac{1}{2} - \dfrac{1}{b}\right) = \dfrac{1}{2}$

Therefore, $\displaystyle\int_{2}^{\infty} \dfrac{1}{x^2}\,dx = \dfrac{1}{2}$

**3.** $\displaystyle\int_{-\infty}^{1} 12e^{3x}\,dx = \lim_{a \to -\infty}\int_{a}^{1} 12e^{3x}\,dx = \lim_{a \to -\infty}\left[12 \cdot \dfrac{1}{3}e^{3x}\Big|_{a}^{1}\right] = \lim_{a \to -\infty}(4e^3 - 4e^{3a}) = 4e^3$

## 6.3 Section Summary

A definite integral in which one or both limits of integration are infinite is called an "improper" integral. The improper integral of a continuous non-negative function is defined as the *limit* of the integral over a finite interval. The integral is *convergent* if the limit exists, and *divergent* otherwise. This idea of dealing with the infinite by "dropping back to the finite and then taking the limit" is a standard technique in mathematics.

Several particular limits are helpful in evaluating improper integrals.

| *Approaching Infinity* | *Approaching Negative Infinity* |
|---|---|
| $\lim_{x \to \infty} \dfrac{1}{x^n} = 0 \quad (n > 0)$ | $\lim_{x \to -\infty} \dfrac{1}{x^n} = 0 \quad$ (for integer $n > 0$) |
| $\lim_{x \to \infty} e^{-ax} = 0 \quad (a > 0)$ | $\lim_{x \to -\infty} e^{ax} = 0 \quad (a > 0)$ |

Improper integrals give continuous sums over infinite intervals. For example, the total future output of an oil well can be found by integrating the production rate out to infinity, and the value of an asset that lasts indefinitely (such as land) can be found by integrating the present value of the future income out to infinity. Even when infinite duration is unrealistic, $\infty$ is used to represent "long-term behavior."

## 6.3 Exercises

**1–16.** Evaluate each limit (or state that it does not exist).

**1.** $\lim_{x \to \infty} \dfrac{1}{x^2}$

**2.** $\lim_{b \to \infty} \left(\dfrac{1}{\sqrt{b}} - 8\right)$

**3.** $\lim_{b \to \infty} (1 - 2e^{-5b})$

**4.** $\lim_{b \to \infty} (3e^{3b} - 4)$

**5.** $\lim_{x \to \infty} (2 - e^{x/2})$

**6.** $\lim_{x \to \infty} (1 - e^{-x/3})$

**7.** $\lim_{b \to \infty} (3 + \ln b)$

**8.** $\lim_{b \to \infty} (2 - \ln b^2)$

**9.** $\lim_{x \to -\infty} \dfrac{1}{x^3}$

**10.** $\lim_{x \to -\infty} \dfrac{1}{x^2}$

**11.** $\lim_{a \to -\infty} e^{2a}$

**12.** $\lim_{a \to -\infty} e^{-3a}$

**13.** $\lim_{a \to -\infty} e^{-2a}$

**14.** $\lim_{a \to -\infty} e^{\frac{1}{2}a}$

**15.** $\lim_{x \to -\infty} \dfrac{1}{(x^2 + 1)^2}$

**16.** $\lim_{x \to -\infty} \dfrac{1}{e^x + e^{-x}}$

**17–40.** Evaluate each improper integral or state that it is divergent.

**17.** $\displaystyle\int_1^\infty \dfrac{1}{x^3}\,dx$

**18.** $\displaystyle\int_1^\infty \dfrac{1}{\sqrt[3]{x^4}}\,dx$

**19.** $\displaystyle\int_2^\infty 3x^{-4}\,dx$

**20.** $\displaystyle\int_0^\infty e^{-t}\,dt$

**21.** $\displaystyle\int_2^\infty \dfrac{1}{x}\,dx$

**22.** $\displaystyle\int_1^\infty \dfrac{1}{x^{0.99}}\,dx$

**23.** $\displaystyle\int_1^\infty \dfrac{1}{x^{1.01}}\,dx$

**24.** $\displaystyle\int_{10}^\infty e^{-x/5}\,dx$

**25.** $\displaystyle\int_0^\infty e^{-0.05t}\,dt$

**26.** $\displaystyle\int_0^\infty e^{0.01t}\,dt$

**27.** $\displaystyle\int_5^\infty \dfrac{1}{(x - 4)^3}\,dx$

**28.** $\displaystyle\int_0^\infty \dfrac{x}{(x^2 + 1)^2}\,dx$

**29.** $\displaystyle\int_0^\infty \dfrac{x}{x^2 + 1}\,dx$

**30.** $\displaystyle\int_0^\infty \dfrac{x^2}{x^3 + 1}\,dx$

**31.** $\displaystyle\int_0^\infty x^2 e^{-x^3}\,dx$

**32.** $\displaystyle\int_e^\infty (\ln x)^{-2}\dfrac{1}{x}\,dx$

**33.** $\displaystyle\int_{-\infty}^0 e^{3x}\,dx$

**34.** $\displaystyle\int_{-\infty}^0 \dfrac{x^4}{(x^5 - 1)^2}\,dx$

**35.** $\displaystyle\int_{-\infty}^1 \dfrac{1}{2 - x}\,dx$

**36.** $\displaystyle\int_{-\infty}^0 \dfrac{1}{1 - x}\,dx$

**37.** $\displaystyle\int_{-\infty}^\infty \dfrac{e^x}{(1 + e^x)^2}\,dx$

**38.** $\displaystyle\int_{-\infty}^\infty \dfrac{e^{-x}}{(1 + e^{-x})^3}\,dx$

**39.** $\displaystyle\int_{-\infty}^\infty \dfrac{e^x}{1 + e^x}\,dx$

**40.** $\displaystyle\int_{-\infty}^\infty \dfrac{e^{-x}}{1 + e^{-x}}\,dx$

 **41.** Use a graphing calculator to estimate the improper integrals $\displaystyle\int_0^\infty e^{\sqrt{x}}\,dx$ and $\displaystyle\int_0^\infty e^{-x^2}\,dx$ (if they converge) as follows:

**a.** Define $y_1$ to be the definite integral (using FnInt) of $e^{\sqrt{x}}$ from 0 to $x$.

**b.** Define $y_2$ to be the definite integral of $e^{-x^2}$ from 0 to $x$.

**c.** $y_1$ and $y_2$ then give the *areas* under these curves out to any number $x$. Make a TABLE of values of $y_1$ and $y_2$ for $x$-values such as 1, 10, 100, and 500. Which integral converges (and to what number, approximated to five decimal places) and which diverges?

 **42.** Use a graphing calculator to estimate the improper integrals $\displaystyle\int_0^\infty \dfrac{1}{x^2 + 1}\,dx$ and $\displaystyle\int_0^\infty \dfrac{1}{\sqrt{x} + 1}\,dx$ (if they converge) as follows:

**a.** Define $y_1$ to be the definite integral (using FnInt) of $\dfrac{1}{x^2 + 1}$ from 0 to $x$.

**b.** Define $y_2$ to be the definite integral of $\dfrac{1}{\sqrt{x} + 1}$ from 0 to $x$.

**c.** $y_1$ and $y_2$ then give the *areas* under these curves out to any number $x$. Make a TABLE of values of $y_1$ and $y_2$ for $x$-values such as 1, 10, 100, 500, and 10,000. Which integral converges (and to what number, approximated to five decimal places) and which diverges?

## Applied Exercises

**43. GENERAL: Permanent Endowments** Find the size of the permanent endowment needed to generate an annual $12,000 forever at a continuous interest rate of 6%.

**44. GENERAL: Permanent Endowments** Show that the size of the permanent endowment needed to generate an annual $C$ dollars forever at interest rate $r$ compounded continuously is $C/r$ dollars.

**45. GENERAL: Permanent Endowments**

**a.** Find the size of the permanent endowment needed to generate an annual $1000 forever at a continuous interest rate of 10%.

**b.** At this same interest rate, the size of the fund needed to generate an annual $1000 for precisely 100 years is $\int_0^{100} 1000e^{-0.1t}\,dt$. Evaluate this integral (it is not an improper integral), approximating your answer using a calculator.

**c.** Notice that the cost for the first 100 years is almost the same as the cost forever. This illustrates again the principle that in endowments, the short term is expensive, but eternity is cheap.

**46–48. BUSINESS: Capital Value of an Asset** The *capital value of an asset* is defined as the present value of all future earnings. For an asset that may last indefinitely (such as real estate or a corporation), the capital value is

$$\begin{pmatrix} \text{Capital} \\ \text{value} \end{pmatrix} = \int_0^{\infty} C(t)e^{-rt}\,dt$$

where $C(t)$ is the income per year and $r$ is the continuous interest rate. Find the capital value of a piece of property that will generate an annual income of $C(t)$, for the function $C(t)$ given below, at a continuous interest rate of 5%.

**46.** $C(t) = 8000$ dollars

**47.** $C(t) = 50\sqrt{t}$ thousand dollars

**48.** $C(t) = 59t^{0.1}$ thousand dollars

**49. BUSINESS: Oil Well Output** An oil well is expected to produce oil at the rate of $50e^{-0.05t}$ thousand barrels per month indefinitely, where $t$ is the number of months that the well has been in operation. Find the total output over the lifetime of the well by integrating this rate from 0 to $\infty$. [*Note:* The owner will shut down the well when production falls too low, but it is convenient to estimate the total output as if production continued forever.]

**50. GENERAL: Duration of Telephone Calls** Studies have shown that the proportion of telephone calls that last longer than $t$ minutes is approximately $\int_t^{\infty} 0.3e^{-0.3s}\,ds$. Use this formula to find the proportion of telephone calls that last longer than 4 minutes.

**51. AREA** Find the area between the curve $y = 1/x^{3/2}$ and the x-axis from $x = 1$ to $\infty$.

**52. AREA** Find the area between the curve $y = e^{-4x}$ and the x-axis from $x = 0$ to $\infty$.

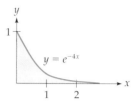

**53. AREA** Find the area between the curve $y = e^{-ax}$ (for $a > 0$) and the x-axis from $x = 0$ to $\infty$.

**54. AREA** Find the area between the curve $y = 1/x^n$ (for $n > 1$) and the x-axis from $x = 1$ to $\infty$.

**55. BEHAVIORAL SCIENCE: Mazes** In a psychology experiment, rats were placed in a T-maze, and the proportion of rats who required more than $t$ seconds to reach the end was $\int_t^{\infty} 0.05e^{-0.05s}\,ds$. Use this formula to find the proportion of rats who required more than 10 seconds.

**56. SOCIOLOGY: Prison Terms** If the proportion of prison terms that are longer than $t$ years is given by the improper integral $\int_t^{\infty} 0.2e^{-0.2s}\,ds$, find the proportion of prison terms that are longer than 5 years.

**57. BUSINESS: Product Reliability** The proportion of light bulbs that last longer than $t$ hours is predicted to be $\int_t^{\infty} 0.001e^{-0.001s}\,ds$. Use this formula to find the proportion of light bulbs that will last longer than 1200 hours.

58. **BUSINESS: Warranties** When a company sells a product with a lifetime guarantee, the number of items returned for repair under the guarantee usually decreases with time. A company estimates that the annual rate of returns after $t$ years will be $800e^{-0.2t}$. Find the total number of returns by summing (integrating) this rate from 0 to $\infty$.

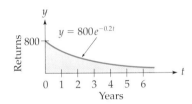

59. **BUSINESS: Sales** A publisher estimates that a book will sell at the rate of $16{,}000e^{-0.8t}$ books per year $t$ years from now. Find the total number of books that will be sold by summing (integrating) this rate from 0 to $\infty$.

60. **BIOMEDICAL: Drug Absorption** To determine how much of a drug is absorbed into the body, researchers measure the difference between the dosage $D$ and the amount of the drug excreted from the body. The total amount excreted is found by integrating the excretion rate $r(t)$ from 0 to $\infty$. Therefore, the amount of the drug absorbed by the body is

$$D - \int_0^\infty r(t)\,dt.$$

If the initial dose is $D = 200$ milligrams (mg), and the excretion rate is $r(t) = 40e^{-0.5t}$ mg per hour, find the amount of the drug absorbed by the body.

## Conceptual Exercises

**61–64.** Which of the two limits exists?

61. **a.** $\lim\limits_{x \to \infty} x^3$  **b.** $\lim\limits_{x \to \infty} \dfrac{1}{x^3}$

62. **a.** $\lim\limits_{x \to -\infty} x^3$  **b.** $\lim\limits_{x \to -\infty} \dfrac{1}{x^3}$

63. **a.** $\lim\limits_{x \to -\infty} e^{3x}$  **b.** $\lim\limits_{x \to -\infty} e^{-3x}$

64. **a.** $\lim\limits_{x \to \infty} e^{3x}$  **b.** $\lim\limits_{x \to \infty} e^{-3x}$

65. True or False: If $f(x)$ is continuous, nonnegative, and $\lim\limits_{x \to \infty} f(x) = 0$, then $\int_1^\infty f(x)\,dx$ converges.

66. True or False: If $f(x)$ is continuous, nonnegative, and $\lim\limits_{x \to \infty} f(x) = 2$, then $\int_1^\infty f(x)\,dx$ diverges.

67. Does $\int_0^\infty 5\,dx$ converge or diverge?

68. Does $\int_0^\infty \dfrac{1}{1000}\,dx$ converge or diverge?

69. True or False: If $f(-x) = f(x)$ for every $x$ and if $\int_0^\infty f(x)\,dx = 7$, then $\int_{-\infty}^0 f(x)\,dx = 7$.

70. True or False: If $\int_{-\infty}^\infty f(x)\,dx$ converges, then $\int_0^\infty f(x)\,dx$ converges.

## Explorations and Excursions

The following problems extend and augment the material presented in the text.

71. **BUSINESS: Preferred Stock** Since preferred stock can remain outstanding indefinitely, the present value per share is the limit of the present value of an annuity* paying that share's dividend $D$ at interest rate $r$:

$$\left(\begin{matrix}\text{Present}\\\text{value}\end{matrix}\right) = \lim_{t \to \infty} D\left(\frac{1 - (1 + r)^{-t}}{r}\right)$$

Find this limit in terms of $D$ and $r$.

72. **BIOMEDICAL: Population Growth** The *Gompertz growth curve* models the size $N(t)$ of a population at time $t \geq 0$ as

$$N(t) = Ke^{-ae^{-bt}}$$

where $K$ and $b$ are positive constants. Find $\lim\limits_{t \to \infty} N(t)$.

*Source:* F. Brauer and C. Castillo-Chávez, *Mathematical Models in Population Biology and Epidemiology*

*See Section 2.4 of *Finite Mathematics and Applied Calculus* (second edition) by the same authors for a derivation of this formula.

**73–74. GENERAL: Permanent Endowments**
The formula for integrating the exponential

function $a^{bx}$ is $\displaystyle\int a^{bx}\,dx = \frac{1}{b\ln a}\,a^{bx} + C$ for

constants $a > 0$ and $b$, as may be verified by using the differentiation formulas on page 305.

73. Use the formula above to find the size of the permanent endowment needed to generate an annual $2000 forever at 5% interest compounded annually.

[*Hint:* Find $\displaystyle\int_0^\infty 2000 \cdot 1.05^{-x}\,dx$.] Compare

your answer with that found in Example 5 (page 437) for the same interest rate but compounded continuously.

74. Use the formula above to find the size of the permanent endowment needed to generate an annual $12,000 forever at 6% interest compounded annually.

[*Hint:* Find $\displaystyle\int_0^\infty 12{,}000 \cdot 1.06^{-x}\,dx$.] Compare

your answer with that found in Exercise 43 (page 441) for the same interest rate but compounded continuously.

## 6.4 NUMERICAL INTEGRATION

### Introduction

In spite of the many techniques of integration, there are still integrals that cannot be found by *any* method (as finite combinations of elementary functions). One example is the integral $\int e^{-x^2}\,dx$, which is closely related to the famous "bell-shaped curve" of probability and statistics. For *definite* integrals, however, it is always possible to *approximate* the actual value by interpreting it as the area under a curve and using a process called **numerical integration.** We will discuss two of the most useful methods of numerical integration, based on approximating areas by **trapezoids** and by **parabolas** (known as **Simpson's Rule**). Both methods can be programmed on a calculator or computer. In fact, when you evaluate a definite integral on a graphing calculator using a command like FnInt, the calculator is using a numerical integration procedure similar to those described here. This section explains the mathematics behind such operations.

Area under $y = f(x)$ from $a$ to $b$ approximated by three rectangles

### Rectangular Approximation and Riemann Sums

In Section 5.3 we approximated the area under a curve by rectangles, and we called the sum of the areas of the rectangles a *Riemann sum.* However, these rectangles underestimate the true area under the curve by the white "error area" just above the rectangles shown on the left. For greater accuracy we could increase the number of rectangles (as we did in Section 5.3), but this would involve more calculation and consequently more roundoff errors. Instead, we will replace the rectangles by shapes that fit the curve more closely.

Area under $y = f(x)$ from $a$ to $b$ approximated by three trapezoids

### Trapezoidal Approximation

We modify the approximating rectangles by allowing their tops to slant with the curve, as shown on the left, giving a much better "fit" to the curve. Such shapes, in which two sides are parallel, are called *trapezoids.* Clearly, approximating the area under the curve by trapezoids is more accurate than approximating it by rectangles: the white "error area" is much smaller.

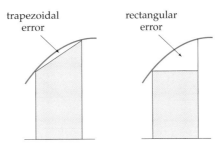

Area under a curve approximated by a
trapezoid (left) and by a rectangle (right).

The area of a trapezoid is the average of the two heights times the width.

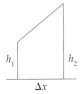

$$\left( \begin{array}{c} \text{Area of a} \\ \text{trapezoid} \end{array} \right) = \underbrace{\frac{h_1 + h_2}{2}}_{\substack{\text{Average} \\ \text{height}}} \cdot \underbrace{\Delta x}_{\text{Width}}$$

If we use trapezoids that all have the same width $\Delta x$, we may add up all of the average heights first and *then* multiply by $\Delta x$. Furthermore, averaging the heights means dividing each height by 2 since $\frac{h_1 + h_2}{2} = \frac{h_1}{2} + \frac{h_2}{2}$. However, a side *between* two trapezoids will be counted twice, once for the trapezoid on either side, thereby canceling the division by 2. Therefore, in adding up the heights, only the two outside heights, at $a$ and $b$, should be divided by 2. This leads to the following procedure for trapezoidal approximation.

---

**Trapezoidal Approximation**

To approximate $\displaystyle\int_a^b f(x)\,dx$ by using $n$ trapezoids:

1. Calculate $\Delta x = \dfrac{b-a}{n}$.             Trapezoid width

2. Find numbers $x_1, x_2, \dots, x_{n+1}$ starting with $x_1 = a$ and successively adding $\Delta x$, ending with $x_{n+1} = b$.

3. The approximation for the integral is

$$\int_a^b f(x)\,dx \approx \left[ \frac{1}{2}f(x_1) + f(x_2) + \cdots + f(x_n) + \frac{1}{2}f(x_{n+1}) \right] \Delta x$$

---

This last formula calculates the total area of the $n$ trapezoids shown on the next page.

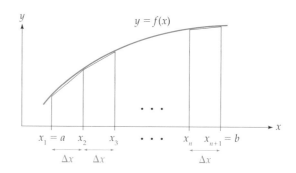

It is easiest to carry out the calculation in a table.

**EXAMPLE 1**

**APPROXIMATING AN INTEGRAL USING TRAPEZOIDS**

Approximate $\displaystyle\int_1^2 x^2\,dx$ using four trapezoids.

*Solution*

The limits of integration are $a = 1$ and $b = 2$, and we are using $n = 4$ trapezoids. The method consists of the following six steps.

1. Calculate the trapezoid width $\Delta x = \dfrac{b-a}{n} = \dfrac{2-1}{4} = 0.25$.

2. List the $x$-values $a$ through $b$ with spacing $\Delta x$.

3. Apply $f(x)$ to each $x$-value.

| $x$ | $f(x) = x^2$ | |
|---|---|---|
| 1 | $(1)^2 = \cancel{1}$ | 0.5 |
| 1.25 | $(1.25)^2 \approx 1.56$ | |
| 1.5 | $(1.5)^2 = 2.25$ | |
| 1.75 | $(1.75)^2 \approx 3.06$ | |
| 2.0 | $(2)^2 = \cancel{4}$ | 2 |

Initial point $a$ — add $\Delta x$, add $\Delta x$, add $\Delta x$, add $\Delta x$ — Final point $b$

4. Take half of first and last entries.

5. Sum last column.

6. Multiply by $\Delta x$.

$$9.37 \cdot (0.25) \approx 2.34$$

Final answer

Therefore, the estimate using four trapezoids is $\displaystyle\int_1^2 x^2\,dx \approx 2.34$.

Earlier (see page 353) we evaluated this integral exactly, obtaining $\displaystyle\int_1^2 x^2\,dx = \frac{7}{3} \approx 2.33$, and we can use this result to assess the accuracy of the trapezoidal method: our approximation of 2.34 is very accurate in spite of the fact that we used only four trapezoids. The relative error (the actual error, 0.01, divided by the actual value, 7/3) is $\dfrac{0.01}{7/3} \approx 0.004$. Expressed as a percentage, this is 0.4% (four tenths of one percent), which is also remarkably small. Notice also that the trapezoidal approximation of 2.34 using four trapezoids is far more accurate than the Riemann sum of 2.04 that we found on page 349 using five (left) rectangles.

### Error in Trapezoidal Approximation

The maximum error in trapezoidal approximation obeys the following formula.

> **Trapezoidal Error**
>
> For the trapezoidal approximation of $\displaystyle\int_a^b f(x)\,dx$ with $n$ trapezoids,
>
> $$(\text{Error}) \le \frac{(b-a)^3}{12n^2} \max_{a \le x \le b} |f''(x)|$$

This formula is very difficult to use because it involves maximizing the absolute value of the second derivative. We will not make further use of it except to observe that the $n^2$ in the denominator means that doubling the number of trapezoids reduces the maximum error by a factor of *four*.

Trapezoidal approximation is most easily carried out on a calculator or a computer, as shown in the following Graphing Calculator Exploration.

## Graphing Calculator Exploration

2.34375

Approximating $\int_1^2 x^2\,dx$ by four trapezoids

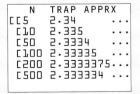

| N | TRAP APPRX | |
|---|---|---|
| 5 | 2.34 | ... |
| 10 | 2.335 | ... |
| 50 | 2.3334 | ... |
| 100 | 2.33335 | ... |
| 200 | 2.3333375 | ... |
| 500 | 2.333334 | ... |

For 500 trapezoids, the approximation is about $2\frac{1}{3}$.

The graphing calculator program* TRAPEZOD calculates the trapezoidal approximation of a definite integral and, for small values of $n$, draws the approximating trapezoids. The first screen on the left shows the approximation of $\displaystyle\int_1^2 x^2\,dx$ using four trapezoids, giving a value 2.34375 that agrees with the 2.34 found in the previous example. Even for only four trapezoids, the curve $y = x^2$ is almost indistinguishable from the tops of the trapezoids, the difference appearing only as a slight thickening of the line where they diverge slightly.

The second screen shows the trapezoidal approximations as the number $n$ of trapezoids increases from 5 to 500. The approximations do approach the *exact* value $2\frac{1}{3}$ found in Example 1 on page 353 by evaluating the definite integral.

Try using a command like FnInt to find the area under $y = x^2$ from $x = 1$ to $x = 2$. How do the answers compare? (FnInt uses a method similar to those described in this section, but is faster since it is built into the calculator.)

*To obtain this and other programs, go to www.cengage.com/math/berresford.

The error formula given on the previous page is of limited usefulness on a computer since maximizing the second derivative is itself subject to error. What is done instead is to calculate the approximations for larger and larger values of $n$ until successive approximations agree to the desired degree of accuracy. For example, the last two results in the above table give answers agreeing to five decimal places, indicating an error of less than 0.00001. While this procedure does not *guarantee* this accuracy, it is often accepted in practice.

### Simpson's Rule (Parabolic Approximation)

For even greater accuracy, we could increase $n$ (resulting in more calculations and more roundoff errors) or we could change the trapezoids, replacing the tops by *curves* chosen to fit the given curve more closely. Replacing the tops of the trapezoids by *parabolas* that pass through three points of the given curve leads to an even more accurate method of approximation, called Simpson's Rule.* The following diagram shows such a curve and its approximation by two parabolas.

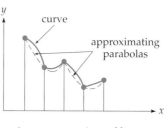

A curve approximated by two
parabolas

Since each parabola spans two intervals, the number of intervals must be even. The area under the approximating parabolas is easily found by integration. The procedure is described as follows and is illustrated in Example 2. A justification of Simpson's Rule is given in Exercise 47.

---

**Simpson's Rule (Parabolic Approximation)**

To approximate $\displaystyle\int_a^b f(x)\,dx$ by Simpson's Rule using $n$ intervals:

1. Calculate $\Delta x = \dfrac{b-a}{n}$.    $n$ must be even

2. Find numbers $x_1, x_2, x_3, \dots, x_{n+1}$ starting with $x_1 = a$ and successively adding $\Delta x$, ending with $x_{n+1} = b$.

3. The approximation for the integral is

$$\int_a^b f(x)\,dx \approx [\,f(x_1) + 4f(x_2) + 2f(x_3) + \cdots + 4f(x_n) + f(x_{n+1})]\frac{\Delta x}{3}$$

$$\underbrace{\phantom{f(x_1) + 4f(x_2) + 2f(x_3) + \cdots + 4f(x_n)}}_{\text{Alternating 4s a nd 2's}}$$

---

*Named after Thomas Simpson (1701–1761), an early user, but not the discoverer, of the formula.

The function values are multiplied by "weights," which, written out by themselves, are

$$1 \quad \underbrace{4 \quad 2 \quad 4 \quad 2 \quad 4 \quad \cdots \quad 4 \quad 2 \quad 4} \quad 1$$

$\uparrow$      Alternating 4's and 2's,     $\uparrow$
Initial 1    beginning and ending with 4    Final 1

**EXAMPLE 2**

**APPROXIMATING AN INTEGRAL USING SIMPSON'S RULE**

Approximate $\displaystyle\int_3^5 \frac{1}{x}\,dx$ by Simpson's Rule with $n = 4$.    $n = 4$ means 2 parabolas

*Solution*

The method consists of the following six steps.

1. Calculate $\Delta x = \dfrac{b - a}{n} = \dfrac{5 - 3}{4} = 0.5$      $n$ must be even

2. List $x$-values $a$ through $b$ with spacing $\Delta x$.

3. Apply $f(x)$ to each $x$-value.

4. Multiply by the weights to get

| $x$ | $f(x) = \dfrac{1}{x}$ | weights | $f(x) \cdot$ weight |
|---|---|---|---|
| *a*   3 | $\dfrac{1}{3} \approx 0.33333$ | 1 ← Initial 1 | 0.33333 |
| add $\Delta x$   3.5 | $\dfrac{1}{3.5} \approx 0.28571$ | 4 | 1.14284 |
| add $\Delta x$   4 | $\dfrac{1}{4} = 0.25$ | 2   Alternating | 0.50000 |
| add $\Delta x$   4.5 | $\dfrac{1}{4.5} \approx 0.22222$ | 4   4's and 2's | 0.88888 |
| add $\Delta x$   *b*   5 | $\dfrac{1}{5} = 0.2$ | 1 ← Final 1 | 0.20000 |

$$3.06505 \cdot \left(\frac{0.5}{3}\right) \approx 0.51084$$

5. Sum last column.

6. Multiply by $\dfrac{\Delta x}{3}$.    Final answer

Therefore, $\displaystyle\int_3^5 \frac{1}{x}\,dx \approx 0.51084$.

This integral, too, can be found exactly. The answer is $\ln 5 - \ln 3 \approx 0.51083$, for an error of only 0.00001. The *relative* error (actual error divided by actual value) is

$$\frac{0.00001}{0.51083} \approx 0.00002 = 0.002\%$$

(two thousandths of one percent) which is also extremely small.

**BE CAREFUL:** Remember that in Simpson's Rule $n$ must be *even*. Also in Simpson's Rule you multiply by $\dfrac{\Delta x}{3}$ while in trapezoidal approximation you multiply by just $\Delta x$.

 ## IQ Distribution

Although it is increasingly clear that human intelligence cannot be measured by a single number, IQ tests are still widely used. (IQ stands for Intelligence Quotient, and is defined as mental age divided by chronological age, multiplied by 100.) The average American IQ is 100, and the distribution of IQs follows the famous "bell-shaped curve" so often used in statistics.*

The proportion of Americans with IQs between two numbers $A$ and $B$ (with $A < B$) is given by the following integral:

$$\begin{pmatrix}\text{Proportion of Americans} \\ \text{with IQs between } A \text{ and } B\end{pmatrix} \approx \frac{1}{\sqrt{2\pi}} \int_{(A-100)/15}^{(B-100)/15} e^{-x^2/2}\, dx$$

For example, the proportion of Americans who have IQs between 115 and 145 is found by substituting $A = 115$ and $B = 145$ into the lower and upper limits in the formula above:

$$\begin{pmatrix}\text{Proportion of IQs} \\ \text{between 115 and 145}\end{pmatrix} = \frac{1}{\sqrt{2\pi}} \int_{1}^{3} e^{-x^2/2}\, dx$$

$\leftarrow 3 \text{ from } \dfrac{B-100}{15} = \dfrac{145-100}{15} = \dfrac{45}{15} = 3$

$\leftarrow 1 \text{ from } \dfrac{A-100}{15} = \dfrac{115-100}{15} = \dfrac{15}{15} = 1$

This integral cannot be found "by hand" (as a finite combination of elementary functions). It can be approximated by Simpson's Rule, just as in Example 2, but it is easier to carry out the calculation on a graphing calculator.

*For those familiar with statistics, IQ scores are normally distributed with mean 100 and standard deviation 15.

## Graphing Calculator Exploration

n=4

.15709468

Approximating $\frac{1}{\sqrt{2\pi}}\int_1^3 e^{-x^2/2}\,dx$
using four intervals

| N | SMPSN APPRX |
|---|---|
| [[4 | .15709468... |
| [10 | .15730028... |
| [20 | .15730504... |
| [50 | .15730534... |
| [100 | .15730535... |
| [200 | .15730535... |

For 100 and 200 intervals
the approximations agree
to 8 decimal places

The graphing calculator program* SIMPSON uses Simpson's Rule to approximate definite integrals and, for small values of $n$, draws the approximating parabolas. The screen on the left shows the approximation of $\frac{1}{\sqrt{2\pi}}\int_1^3 e^{-x^2/2}\,dx$ using $n = 4$ intervals (2 parabolas), giving approximately 0.157. Simpson's Rule is so accurate that the curve is almost indistinguishable from the approximating parabolas, the difference again appearing only as a slight thickening of the line where they diverge slightly.

The second screen shows that Simpson's Rule converges extremely rapidly.

Try using a command like FnInt to find the area under $y = 1/\sqrt{2\pi}\,e^{-x^2/2}$ from $x = 1$ to $x = 3$. How do the answers compare?

*To obtain this and other programs, go to www.cengage.com/math/berresford.

Converting the answer 0.157 to a percentage, about 16% of all Americans have IQs in the range 115 to 145.

16%

115     145

### The Error in Simpson's Rule

The maximum error in Simpson's Rule obeys the following formula.

**Error in Simpson's Rule**

In approximating $\displaystyle\int_a^b f(x)\,dx$ by Simpson's Rule with $n$ intervals,

$$(\text{Error}) \le \frac{(b-a)^5}{180n^4}\,\max_{a\le x\le b}\,|f^{(4)}(x)|$$

This formula is difficult to use because it involves maximizing the absolute value of the fourth derivative. It does, however, show that Simpson's Rule is *exact* for cubics (third-degree polynomials), since cubics have zero fourth derivative, and that doubling the value of $n$ reduces the maximum error by a factor of 16 (because of the $n^4$).

## 6.4 Section Summary

Some definite integrals cannot be evaluated exactly because it is impossible (or very difficult) to find an antiderivative. However, any definite integral can be *approximated* by numerical integration, and two of the most useful methods are trapezoidal approximation and Simpson's Rule (parabolic approximation). Each method involves choosing a number $n$ of intervals and calculating a "weighted average" of function values at the endpoints of these intervals. Higher values of $n$ generally give greater accuracy, but also involve more calculation (and therefore more roundoff errors). In practice, Simpson's Rule is generally the method of choice, since it gives greater accuracy for only slightly more effort. Trapezoidal approximation, however, has the advantage of having a simpler formula for its error. It is particularly appropriate to seek an approximation if the "exact" answer involves logarithms or exponentials, since these functions will probably be approximated anyway in the evaluation step.

It is curious that in Chapter 5 we used integrals to evaluate areas, and here we are using areas to evaluate integrals. This is typical of mathematics, in which any equivalence (such as definite integrals and areas) is exploited in both directions.

### Programs for Trapezoidal Approximation and Simpson's Rule

Below are programs for trapezoidal approximation and for Simpson's Rule for the *TI-83* or *TI-84* calculator. These programs give only single numerical answers, while TRAPEZOD and SIMPSON draw graphs and make tables showing the approximations for several values of $n$.

*TI-83/84 Program for Trapezoidal Approximation*

(Enter the function in $y_1$ before executing.)

Disp "USES FUNCTION $Y_1$"
Prompt A, B, N
$(B - A)/N \rightarrow D$
$(Y_1(A) + Y_1(B))/2 \rightarrow S$
$A + D \rightarrow X$
For(I,1,N − 1)
$S + Y_1 \rightarrow S$
$X + D \rightarrow X$
End
$S*D \rightarrow S$
Disp "TRAP APPROX IS ", S

*TI-83/84 Program for Simpson's Rule*

(Enter the function in $y_1$ before executing.)

Disp "USES FUNCTION $Y_1$"
Disp "N MUST BE EVEN"
Prompt A, B, N
$(B - A)/N \rightarrow D$
$Y_1(A) - Y_1(B) \rightarrow S$
$A \rightarrow X$
For(I,1,N/2)
$S + 4*Y_1(X + D) + 2*Y_1(X + 2D) \rightarrow S$
$X + 2D \rightarrow X$
End
$S*D/3 \rightarrow S$
Disp "SIMP APPROX IS ", S

## 6.4 Exercises

### Exercises on Trapezoidal Approximation

**1–4.** For each definite integral:

**a.** Approximate it "by hand," using trapezoidal approximation with $n = 4$ trapezoids. Round calculations to three decimal places.

**b.** Evaluate the integral exactly using antiderivatives, rounding to three decimal places.

**c.** Find the actual error (the difference between the actual value and the approximation).

**d.** Find the relative error (the actual error divided by the actual value, expressed as a percent).

**1.** $\int_1^3 x^2 \, dx$     **2.** $\int_1^2 x^3 \, dx$

**3.** $\int_2^4 \frac{1}{x} \, dx$     **4.** $\int_1^3 \frac{1}{x} \, dx$

**5–8.** Approximate each integral using trapezoidal approximation "by hand" with the given value of $n$. Round all calculations to three decimal places.

**5.** $\int_0^1 \sqrt{1 + x^2} \, dx, \quad n = 3$

**6.** $\int_0^1 \sqrt{1 + x^3} \, dx, \quad n = 3$

**7.** $\int_0^1 e^{-x^2} \, dx, \quad n = 4$     **8.** $\int_0^1 e^{x^2} \, dx, \quad n = 4$

**9–14.** Approximate each integral using the graphing calculator program TRAPEZOD (see page 447) or another trapezoidal approximation program (see the previous page). Use the following values for the numbers of intervals: 10, 50, 100, 200, 500. Then give an estimate for the value of the definite integral, keeping as many decimal places as the last two approximations agree (when rounded).

**9.** $\int_1^2 \sqrt{\ln x} \, dx$     **10.** $\int_0^1 \ln(x^2 + 1) \, dx$

**11.** $\int_{-1}^1 \sqrt{16 + 9x^2} \, dx$     **12.** $\int_{-1}^1 \sqrt{25 - 9x^2} \, dx$

**13.** $\int_{-1}^1 e^{x^2} \, dx$     **14.** $\int_0^4 \sqrt{1 + x^4} \, dx$

### Applied Exercises on Trapezoidal Approximation

 **15–16. GENERAL: IQs** Use the formula on page 450 and TRAPEZOD or another trapezoidal approximation program (see the previous page) to find the proportion of Americans with IQs between the following two numbers. Use successively higher values of $n$ until answers agree to four decimal places.

**15.** 100 and 130     **16.** 130 and 145

 **17. GENERAL: World Oil Consumption** In recent decades, world consumption of oil has been running at the rate of $f(x) = \sqrt{9 + 0.025x^2}$ billion tons per year, where $x$ is the number of years since 1990. The total amount of oil consumed from 1990 to 2010 is then given by the integral

$\int_0^{20} \sqrt{9 + 0.025x^2} \, dx$. Estimate this amount by approximating the integral using trapezoidal approximation with $n = 20$ trapezoids.

*Source:* Worldwatch

 **18. BUSINESS: Investment Growth** An investment grows at a rate of $3.2 \, e^{\sqrt{t}}$ thousand dollars per year, where $t$ is the number of years since the beginning of the investment. Use TRAPEZOD or another trapezoidal approximation program to estimate the total growth of the investment during the first 3 years (year 0 to year 3). Use successively higher values of $n$ until answers agree to two decimal places.

### Exercises on Simpson's Rule

**19–26.** Estimate each definite integral "by hand," using Simpson's Rule with $n = 4$. Round all calculations to three decimal places. Exercises 19–26 correspond to Exercises 1–8, in which the same integrals were estimated using trapezoids. If you did the corresponding exercise, compare your Simpson's Rule answer with your trapezoidal answer.

**19.** $\int_1^3 x^2 \, dx$     **20.** $\int_1^2 x^3 \, dx$

**21.** $\int_2^4 \frac{1}{x} \, dx$     **22.** $\int_1^3 \frac{1}{x} \, dx$

**23.** $\int_0^1 \sqrt{1 + x^2} \, dx$     **24.** $\int_0^1 \sqrt{1 + x^3} \, dx$

**25.** $\int_0^1 e^{-x^2}\, dx$

**26.** $\int_0^1 e^{x^2}\, dx$

 **27–32.** Approximate each integral using the graphing calculator program SIMPSON (see page 451) or another Simpson's Rule approximation program (see page 452). Use the following values for the numbers of intervals: 10, 20, 50, 100, 200. Then give an estimate for the value of the definite integral, keeping as many decimal places as the last two approximations agree to (when rounded). Exercises 27–32 correspond to Exercises 9–14 in which the same integrals were estimated using trapezoids. If you did the corresponding exercise, compare your Simpson's Rule answer with your trapezoidal answer.

**27.** $\int_1^2 \sqrt{\ln x}\, dx$

**28.** $\int_0^1 \ln(x^2 + 1)\, dx$

**29.** $\int_{-1}^1 \sqrt{16 + 9x^2}\, dx$

**30.** $\int_{-1}^1 \sqrt{25 - 9x^2}\, dx$

**31.** $\int_{-1}^1 e^{x^2}\, dx$

**32.** $\int_0^4 \sqrt{1 + x^4}\, dx$

**33–34. APPROXIMATION OF IMPROPER INTEGRALS** For each improper integral:

a. Make it a "proper" integral by using the substitution $x = \dfrac{1}{t}$ and simplifying.

b. Approximate the proper integral using Simpson's Rule (either "by hand" or using a program) with $n = 4$ intervals, rounding your answer to three decimal places.

**33.** $\int_1^\infty \dfrac{1}{x^3 + 1}\, dx$

**34.** $\int_1^\infty \dfrac{x}{x^3 + 1}\, dx$

## Applied Exercises on Simpson's Rule

 **35. GENERAL: Suspension Bridges** The cable of a suspension bridge hangs in a parabolic curve. The equation of the cable shown below is $y = \dfrac{x^2}{2000}$. Its length in feet is given by the integral

$$\int_{-400}^{400} \sqrt{1 + \left(\dfrac{x}{1000}\right)^2}\, dx$$

Approximate this integral using Simpson's Rule, using successively higher values of $n$ until answers agree to the nearest whole number.

**36. APPROXIMATION OF $\pi$** The number $\pi$ is the ratio of the circumference of a circle to its diameter (since $C = \pi D$). It can be shown that the following definite integral is equal to $\pi$.

$$\int_0^1 \dfrac{4}{x^2 + 1}\, dx = \pi$$

Find $\pi$ by approximating this integral using Simpson's Rule, using successively higher

values of $n$ until answers agree to four decimal places.

 **37. ECONOMICS: Gross World Product** The gross world product (GWP), the total value of all finished goods and services produced worldwide, has been running at the rate of $f(x) = \sqrt{1600 + 10x^2}$ trillion (2006) dollars per year, where $x$ is the number of years since 1990. The total value of all finished goods and services produced during the years 1990 to 2010 is then given by the integral

$$\int_0^{20} \sqrt{1600 + 10x^2}\, dx.$$ Estimate this total GWP by approximating the integral using Simpson's Rule with $n = 20$.

*Source:* Worldwatch

## Conceptual Exercises

**38.** Explain why trapezoidal approximation is better than Riemann sums for approximating definite integrals.

**39.** Explain why the trapezoidal method is *exact* for linear functions. [*Hint:* Either use the error formula on page 447 or think of the graph.]

**40.** True or False: In trapezoidal approximation, the number of trapezoids must be even.

**41.** If you *triple* the number of trapezoids in trapezoidal approximation, by what factor will the maximum error be reduced?

**42.** If you *triple* the size of $n$ in Simpson's Rule, by what factor will the maximum error be reduced?

**43.** Use the *idea* of trapezoidal approximation to approximate the area of the shape shown below.

**44.** Use the *idea* of Simpson's Rule to approximate the area of the shape shown above.

**45.** Since Simpson's Rule is *exact* for cubics (third-degree polynomials), if you use Simpson's Rule to approximate the integral of a cubic, what value of $n$ should you use?

**46.** What are the "weights" for Simpson's Rule when $n = 2$?

## Explorations and Excursions

The following problems extend and augment the material presented in the text.

**47. JUSTIFICATION OF SIMPSON'S RULE** Justify Simpson's Rule by carrying out the following steps, which lead to the formula for Simpson's Rule (page 448) in a simple case.

   **i.** Observe that if the three points shown in the following diagram lie on the parabola $f(x) = ax^2 + bx + c$, then the following three equations hold:

$$a(-d)^2 + b(-d) + c = y_1 \quad \text{Since} \quad f(-d) = y_1$$
$$a(0)^2 + b(0) + c = y_2 \quad \text{Since} \quad f(0) = y_2$$
$$a(d)^2 + b(d) + c = y_3 \quad \text{Since} \quad f(d) = y_3$$

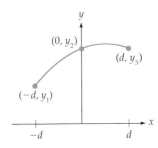

   **ii.** Simplify these three equations to obtain

$$ad^2 - bd + c = y_1$$
$$c = y_2$$
$$ad^2 + bd + c = y_3$$

   **iii.** Add the first and last equation plus four times the middle equation to obtain

$$2ad^2 + 6c = y_1 + 4y_2 + y_3$$

   You will use this equation in step v.

   **iv.** Evaluate the integral $\displaystyle\int_{-d}^{d} (ax^2 + bx + c)\, dx$ and simplify to show that the area under the parabola from $-d$ to $d$ is

$$\text{Area} = \frac{2}{3}ad^3 + 2cd = \left(2ad^2 + 6c\right)\frac{d}{3}$$

   **v.** Use the equation found in step iii to write this area as

$$\text{Area} = (y_1 + 4y_2 + y_3)\frac{d}{3}$$

   **vi.** Use $y_1 = f(-d)$, $y_2 = f(0)$, $y_3 = f(d)$ and the fact that the spacing $\Delta x$ is equal to $d$ to rewrite this last equation as

$$\text{Area} = [f(-d) + 4f(0) + f(d)]\frac{\Delta x}{3}$$

This is exactly the formula for Simpson's Rule using one parabola ($n = 2$). For several parabolas placed next to each other, the function values *between* two neighboring parabolas are added twice (once for each side), and so should have weight 2. Therefore, successive function values are multiplied by the weights given on page 449:

$$1 \quad 4 \quad 2 \quad 4 \quad 2 \quad 4 \quad \ldots \quad 4 \quad 1$$

**48.** While Simpson's Rule is generally more accurate than trapezoidal approximation, show that this is not *always* the case by considering the function $f(x) = 9x^2 - 5x^4$ on the interval $[-1, 1]$ as follows.

   **a.** Find the *exact* area under the curve by integration.

   **b.** Use trapezoidal approximation with two trapezoids to approximate the area.

   **c.** Use Simpson's Rule with $n = 2$ to approximate the area.

   **d.** Which approximation method gave greater accuracy?

# Chapter Summary with Hints and Suggestions

Reading the text and doing the exercises in this chapter have helped you to master the following concepts and skills, which are listed by section (in case you need to review them) and are keyed to particular Review Exercises. Answers for all Review Exercises are given at the back of the book, and full solutions can be found in the Student Solutions Manual.

## 6.1 Integration by Parts

- Find an integral using integration by parts.
  *(Review Exercises 1–14.)*

$$\int u \, dv = uv - \int v \, du$$

- Find an integral by whatever technique is necessary.
  *(Review Exercises 15–22.)*

- Solve an applied problem using integration by parts.
  *(Review Exercises 23–24.)*

$$\left(\begin{array}{c}\text{Present}\\\text{value}\end{array}\right) = \int_0^T C(t)e^{-rt} \, dt$$

$$\left(\begin{array}{c}\text{Total}\\\text{accumulation}\end{array}\right) = \int_0^T r(t) \, dt$$

## 6.2 Integration Using Tables

- Find an integral using a table of integrals.
  *(Review Exercises 25–36.)*

- Solve an applied problem using an integral table.
  *(Review Exercises 37–38.)*

## 6.3 Improper Integrals

- Evaluate an improper integral (if it is convergent).
  *(Review Exercises 39–56.)*

$$\int_a^\infty f(x) \, dx = \lim_{b \to \infty} \int_a^b f(x) \, dx$$

- Solve an applied problem involving an improper integral.
  *(Review Exercises 57–60.)*

- Predict whether an improper integral converges, and then check by evaluating it.
  *(Review Exercises 61–62.)*

## 6.4 Numerical Integration

- Approximate an integral using trapezoidal approximation "by hand."
  *(Review Exercises 63–68.)*

$$\int_a^b f(x) \, dx \approx \left[\frac{1}{2}f(x_1) + f(x_2) + \cdots + f(x_n) + \frac{1}{2}f(x_{n+1})\right] \cdot \Delta x$$

- Use a program to approximate an integral by trapezoidal approximation.
  *(Review Exercises 69–74.)*

- Approximate an integral using Simpson's Rule "by hand."   *(Review Exercises 75–80.)*

$$\int_a^b f(x) \, dx \approx [f(x_1) + 4f(x_2) + 2f(x_3) + \cdots + 4f(x_n) + f(x_{n+1})] \cdot \frac{\Delta x}{3}$$

- Use a program to approximate an integral by Simpson's Rule.   *(Review Exercises 81–86.)*

- Approximate an improper integral using trapezoidal approximation.
  *(Review Exercises 87–88.)*

## Hints and Suggestions

- The unifying idea of this chapter is extensions of the concept of integration.

- Integration by parts takes one integral and gives another integral that, it is hoped, is simpler than the original integral. The formula is simply an integration version of the Product Rule. When using it, try to choose $dv$ (including the $dx$) to be the most complicated part of the integrand that you can integrate, and, if possible, choose the $u$ to be something that simplifies when differentiated.

- There are tables of integrals that are much longer than the one on the inside back cover. Longer tables, however, require much more time to search for the "right" formula. Other techniques (such as a substitution, integration by parts, or use of a formula more than once) may be used with an integral table.

■ To find an integral, try the following methods. First try the "basic" formulas 1 through 4 on the inside back cover. Then try a substitution (formulas 5 through 7). If these methods fail, try integration by parts or an integral table. Remember that some integrals *cannot* be integrated (in terms of elementary functions). A *definite* integral can always be approximated by numerical methods.

■ Before "evaluating" an improper integral, be sure that the integrand is defined over the interval, and that the integral is convergent. If the integral

diverges, then it has no value and we simply state that it is divergent.

■ Numerical integration involves approximating the area under a curve using geometric figures such as trapezoids or parabolas (Simpson's Rule). In practice, the calculations are usually carried out on a calculator or computer, but doing some "by hand" helps to make the method clear.

■ A graphing calculator is very helpful for approximating definite integrals by trapezoidal approximation or Simpson's Rule for large values of $n$. Graphing calculators also have their own built-in numerical procedures for approximating integrals when you use FnInt.

■ **Practice for Test:** Review Exercises 3, 5, 13, 25, 27, 33, 39, 41, 57, 63, 71, 79, 85.

# Review Exercises for Chapter 6

Practice test exercise numbers in green.

## 6.1 Integration by Parts

**1–14.** Find each integral using integration by parts.

**1.** $\displaystyle\int xe^{2x}\, dx$

**2.** $\displaystyle\int xe^{-x}\, dx$

**3.** $\displaystyle\int x^8 \ln x\, dx$

**4.** $\displaystyle\int \sqrt[4]{x}\, \ln x\, dx$

**5.** $\displaystyle\int (x-2)(x+1)^5\, dx$

**6.** $\displaystyle\int (x+3)(x-1)^4\, dx$

**7.** $\displaystyle\int \frac{\ln t}{\sqrt{t}}\, dt$

**8.** $\displaystyle\int x^7 e^{x^4}\, dx$

**9.** $\displaystyle\int x^2 e^x\, dx$

**10.** $\displaystyle\int (\ln x)^2\, dx$

**11.** $\displaystyle\int x(x+a)^n\, dx$   (for constants $a$ and $n > 0$)

**12.** $\displaystyle\int x(1-x)^n\, dx$   (for constant $n > 0$)

**13.** $\displaystyle\int_0^5 xe^x\, dx$

**14.** $\displaystyle\int_1^e x \ln x\, dx$

**15–22.** Find each integral by a substitution or by integration by parts, as appropriate.

**15.** $\displaystyle\int \frac{dx}{1-x}$

**16.** $\displaystyle\int xe^{-x^2}\, dx$

**17.** $\displaystyle\int x^3 \ln 2x\, dx$

**18.** $\displaystyle\int \frac{dx}{(1-x)^2}$

**19.** $\displaystyle\int \frac{\ln x}{x}\, dx$

**20.** $\displaystyle\int \frac{e^{2x}}{e^{2x}+1}\, dx$

**21.** $\displaystyle\int \frac{e^{\sqrt{x}}}{\sqrt{x}}\, dx$

**22.** $\displaystyle\int (e^{2x}+1)^3 e^{2x}\, dx$

**23. BUSINESS: Present Value of a Continuous Stream of Income** A company generates a continuous stream of income of $25t$ million dollars per year, where $t$ is the number of years that the company has been in operation.

   **a.** Find the present value of this stream for the first 10 years at 5% interest compounded continuously. (Do not use a graphing calculator.)

   **b.** Verify your answer to part (a) using FnInt on a graphing calculator.

**24. ENVIRONMENTAL SCIENCE: Pollution** Radioactive waste is leaking out of cement storage vessels at the rate of $te^{0.2t}$ hundred gallons per month, where $t$ is the number of months since the leak began.

   **a.** Find the total leakage during the first 3 months. (Do not use a graphing calculator.)

   **b.** Verify your answer to part (a) using FnInt on a graphing calculator.

## 6.2 Integration Using Tables

**25–36.** Use the integral table on the inside back cover to find each integral.

25. $\int \dfrac{1}{25 - x^2}\, dx$    26. $\int \dfrac{1}{x^2 - 4}\, dx$

27. $\int \dfrac{x}{(x - 1)(x - 2)}\, dx$    28. $\int \dfrac{1}{(x - 1)(x - 2)}\, dx$

29. $\int \dfrac{1}{x\sqrt{x + 1}}\, dx$    30. $\int \dfrac{x}{\sqrt{x + 1}}\, dx$

31. $\int \dfrac{1}{\sqrt{x^2 + 9}}\, dx$    32. $\int \dfrac{1}{\sqrt{x^2 + 16}}\, dx$

33. $\int \dfrac{z^3}{\sqrt{z^2 + 1}}\, dz$    34. $\int \dfrac{e^{2t}}{e^t + 2}\, dt$

35. $\int x^2 e^{2x}\, dx$    36. $\int (\ln x)^4\, dx$

37. **BUSINESS: Cost** A company's marginal cost function is $MC(x) = \dfrac{1}{(2x + 1)(x + 1)}$ and fixed costs are 1000 (all in dollars). Find the company's cost function.

38. **GENERAL: Population** The population of a town is growing at the rate of $\sqrt{t^2 + 1600}$ people per year, where $t$ is the number of years from now.

   **a.** Find the total increase in population during the first 30 years. (Do not use a graphing calculator.)

    **b.** Verify your answer to part (a) using FnInt on a graphing calculator.

## 6.3 Improper Integrals

**39–56.** Find the value of each improper integral or state that it is divergent.

39. $\displaystyle\int_1^\infty \dfrac{1}{x^5}\, dx$    40. $\displaystyle\int_1^\infty \dfrac{1}{x^6}\, dx$

41. $\displaystyle\int_1^\infty \dfrac{1}{\sqrt[5]{x}}\, dx$    42. $\displaystyle\int_1^\infty \dfrac{1}{\sqrt[6]{x}}\, dx$

43. $\displaystyle\int_0^\infty e^{-2x}\, dx$    44. $\displaystyle\int_4^\infty e^{-0.5x}\, dx$

45. $\displaystyle\int_0^\infty e^{2x}\, dx$    46. $\displaystyle\int_4^\infty e^{0.5x}\, dx$

47. $\displaystyle\int_0^\infty e^{-t/5}\, dt$    48. $\displaystyle\int_{100}^\infty e^{-t/10}\, dt$

49. $\displaystyle\int_0^\infty \dfrac{x^3}{(x^4 + 1)^2}\, dx$    50. $\displaystyle\int_0^\infty \dfrac{x^4}{(x^5 + 1)^2}\, dx$

51. $\displaystyle\int_{-\infty}^0 e^{2t}\, dt$    52. $\displaystyle\int_{-\infty}^0 e^{4t}\, dt$

53. $\displaystyle\int_{-\infty}^4 \dfrac{1}{(5 - x)^2}\, dx$    54. $\displaystyle\int_{-\infty}^8 \dfrac{1}{(9 - x)^2}\, dx$

55. $\displaystyle\int_{-\infty}^\infty \dfrac{e^{-x}}{(1 + e^{-x})^4}\, dx$    56. $\displaystyle\int_{-\infty}^\infty \dfrac{e^{-x}}{(1 + e^{-x})^3}\, dx$

57. **GENERAL: Permanent Endowments** Find the size of the permanent endowment needed to generate an annual $6000 forever at an interest rate of 10% compounded continuously.

58. **GENERAL: Automobile Age** Insurance records indicate that the proportion of cars on the road that are more than $x$ years old is approximated by the integral $\int_x^\infty 0.21e^{-0.21t}\, dt$. Find the proportion of cars that are more than 5 years old.

59. **BUSINESS: Book Sales** A publisher estimates that the demand for a certain book will be $12e^{-0.05t}$ thousand copies per year, where $t$ is the number of years since the book's publication. Find the total number of books that will be sold from the publication date onward.

60. **ENVIRONMENTAL SCIENCES: Consumption of Natural Resources** Because of environmental concerns and available substitutes, mercury usage has been declining over recent years. Annual U.S. consumption is estimated to be $280e^{-0.02t}$ metric tons of mercury per year, where $t$ is the number of years since 2008. Assuming that this rate continues, find the total amount of mercury that will be used in the United States from 2008 on.

*Source: U.S. Geological Survey*

**61–62.** For each improper integral, use a graphing calculator to evaluate it (or to show that it diverges) as follows:

   **a.** Define $y_1$ to be the definite integral (using FnInt) of the integrand from 1 to $x$.

   **b.** Make a TABLE of values of $y_1$ for $x$-values such as 1, 10, 100, and 1000. Does the integral converge (and if so, to what number) or does it diverge?

   **c.** Verify your answers to part (b) by evaluating the improper integral "by hand."

61. $\displaystyle\int_1^\infty \dfrac{1}{x^3}\, dx$    62. $\displaystyle\int_1^\infty \dfrac{1}{\sqrt[3]{x}}\, dx$

## 6.4 Numerical Integration

**63–68.** Estimate each integral using trapezoidal approximation with the given value of $n$. (Round all calculations to three decimal places.)

**63.** $\displaystyle\int_0^1 \sqrt{1 + x^4}\, dx, \quad n = 3$

**64.** $\displaystyle\int_0^1 \sqrt{1 + x^5}\, dx, \quad n = 3$

**65.** $\displaystyle\int_0^1 e^{x^2/2}\, dx, \quad n = 4$

**66.** $\displaystyle\int_0^1 e^{-x^2/2}\, dx, \quad n = 4$

**67.** $\displaystyle\int_{-1}^1 \ln(1 + x^2)\, dx, \quad n = 4$

**68.** $\displaystyle\int_{-1}^1 \ln(x^3 + 2)\, dx, \quad n = 4$

**69–74.** Use a trapezoidal approximation program such as TRAPEZOD to approximate each integral. Use successively higher values of $n$ until the results agree to three decimal places (rounded).

**69.** $\displaystyle\int_0^1 \sqrt{1 + x^4}\, dx$     **70.** $\displaystyle\int_0^1 \sqrt{1 + x^5}\, dx$

**71.** $\displaystyle\int_0^1 e^{x^2/2}\, dx$     **72.** $\displaystyle\int_0^1 e^{-x^2/2}\, dx$

**73.** $\displaystyle\int_{-1}^1 \ln(1 + x^2)\, dx$     **74.** $\displaystyle\int_{-1}^1 \ln(x^3 + 2)\, dx$

**75–80.** Estimate each integral using Simpson's Rule with the given value of $n$. (Round all calculations to four decimal places.)

**75.** $\displaystyle\int_0^1 \sqrt{1 + x^4}\, dx, \quad n = 4$

**76.** $\displaystyle\int_0^1 \sqrt{1 + x^5}\, dx, \quad n = 4$

**77.** $\displaystyle\int_0^1 e^{x^2/2}\, dx, \quad n = 4$

**78.** $\displaystyle\int_0^1 e^{-x^2/2}\, dx, \quad n = 4$

**79.** $\displaystyle\int_{-1}^1 \ln(1 + x^2)\, dx, \quad n = 4$

**80.** $\displaystyle\int_{-1}^1 \ln(x^3 + 2)\, dx, \quad n = 4$

**81–86.** Use a Simpson's Rule approximation program such as SIMPSON to approximate each integral. Use successively higher values of $n$ until the rounded results agree to six decimal places.

**81.** $\displaystyle\int_0^1 \sqrt{1 + x^4}\, dx$     **82.** $\displaystyle\int_0^1 \sqrt{1 + x^5}\, dx$

**83.** $\displaystyle\int_0^1 e^{x^2/2}\, dx$     **84.** $\displaystyle\int_0^1 e^{-x^2/2}\, dx$

**85.** $\displaystyle\int_{-1}^1 \ln(1 + x^2)\, dx$     **86.** $\displaystyle\int_{-1}^1 \ln(x^3 + 2)\, dx$

**87–88.** For each improper integral:

**a.** Make it a "proper" integral by using the substitution $x = \dfrac{1}{t}$ and simplifying.

**b.** Approximate the proper integral using trapezoidal approximation with $n = 4$. Keep three decimal places.

**87.** $\displaystyle\int_1^\infty \frac{1}{x^2 + 1}\, dx$

**88.** $\displaystyle\int_1^\infty \frac{x^2}{x^4 + 1}\, dx$

# 7

# Calculus of
# Several Variables

### Gender Pay Gap

Women's wages have always lagged behind men's. The reasons for this are not entirely clear (possibly discrimination or different career choices made by men and women), but what *is* clear is that women's wages have been steadily catching up. In 1980 women working full-time made only 64% of men's salaries, but by 2005 that percentage had risen to 81%. Can we estimate when in the future men's and women's salaries might actually be equal? The mathematical technique of *linear regression* or *least squares*, used extensively in Chapter 1, fits a line to data points. The following graph shows the least squares lines for men's wages and for women's wages.

| Year | Men's Salaries | Women's Salaries |
|------|----------------|------------------|
| 1980 | 785 | 506 |
| 1985 | 782 | 534 |
| 1990 | 763 | 549 |
| 1995 | 732 | 552 |
| 2000 | 772 | 594 |
| 2005 | 767 | 621 |

Median Weekly Earnings of Full-Time and Salary Workers Ages 16 and Older, in 2007 Dollars

*Source:* U.S. Bureau of Labor Statistics

By finding the intersection point of these lines, we may predict that men's and women's salaries will be approximately equal in mid-2032. While in Chapter 1 you may have carried out linear regression by pressing buttons on a graphing calculator, in this chapter we will see how the least squares method *works* and that it is based on the same calculus techniques that we used for optimization in Chapter 3.

## 7.1 FUNCTIONS OF SEVERAL VARIABLES

### Introduction

Many quantities depend on *several* variables. For example, the "windchill" factor announced by the weather bureau during the winter depends on two variables: temperature and wind speed. The cost of a loan depends on *three* variables: amount, duration, and interest rate.

In this chapter we define functions of two or more variables and learn how to differentiate and integrate them. We use derivatives for calculating rates of change and optimizing functions, and integrals for finding volumes, continuous sums, and average values.

Graphing calculators will be less useful in this chapter because of their small screens and limited computing power, but computer-drawn pictures of three-dimensional surfaces will be very useful.

### Functions of Two Variables

A function $f$ that depends on *two* variables, $x$ and $y$, is written $f(x, y)$ (read: $f$ of $x$ and $y$). The **domain** of the function is the set of all ordered pairs $(x, y)$ for which it is defined. The **range** is the set of all resulting values of the function. Formally:

---

**Function of Two Variables**

A function $f$ of two variables is a rule such that to each ordered pair $(x, y)$ in the domain of $f$ there corresponds one and only one number $f(x, y)$.

---

If the domain is not stated, it will always be taken to be the largest set of ordered pairs for which the function is defined (the **natural domain**).

---

**EXAMPLE 1**

**FINDING THE DOMAIN OF A FUNCTION**

For $f(x, y) = \dfrac{\sqrt{x}}{y^2}$ find

**a.** the domain

**b.** $f(9, -1)$

*Solution*

**a.** $\{(x, y) \mid x \geq 0,\ y \neq 0\}$

For $\sqrt{x}/y^2$, $x$ cannot be negative (because of the $\sqrt{\phantom{x}}$), and $y$ cannot be zero

**b.** $f(9, -1) = \dfrac{\sqrt{9}}{(-1)^2} = \dfrac{3}{1} = 3$

$f(x, y) = \sqrt{x}/y^2$ with $x = 9$ and $y = -1$

**EXAMPLE 2**

**FINDING THE DOMAIN OF A FUNCTION INVOLVING LOGARITHMS AND EXPONENTIALS**

For $g(u, v) = e^{uv} - \ln u$, find   **a.** the domain   **b.** $g(1, 2)$

*Solution*

**a.** $\{(u, v) \mid u > 0\}$

*u must be positive so that its logarithm is defined*

**b.** $g(1, 2) = e^{1 \cdot 2} - \underbrace{\ln 1}_{0} = e^2 - 0 = e^2$

$g(u, v) = e^{uv} - \ln u$
with $u = 1$ and $v = 2$

---

**Practice Problem 1**

For $f(x, y) = \dfrac{\ln x}{e^{\sqrt{y}}}$, find   **a.** the domain   **b.** $f(e, 4)$

➤ **Solutions on page 471**

Functions of two variables are used in many applications.

**EXAMPLE 3**

**FINDING A COMPANY'S COST FUNCTION**

A company manufactures three-speed and ten-speed bicycles. It costs $100 to make each three-speed bicycle, it costs $150 to make each ten-speed bicycle, and fixed costs are $2500. Find the cost function, and use it to find the cost of producing 15 three-speed bicycles and 20 ten-speed bicycles.

*Solution*

Let:

$x$ = the number of three-speed bicycles
$y$ = the number of ten-speed bicycles

The cost function is

$$C(x, y) = \underset{\substack{\text{Unit} \quad \text{Quan-}\\ \text{cost} \quad \text{tity}}}{100x} + \underset{\substack{\text{Unit} \quad \text{Quan-}\\ \text{cost} \quad \text{tity}}}{150y} + \underset{\substack{\text{Fixed}\\ \text{costs}}}{2500}$$

The cost of producing 15 three-speed bicycles and 20 ten-speed bicycles is found by evaluating $C(x, y)$ at $x = 15$ and $y = 20$:

$$C(15, 20) = 100 \cdot 15 + 150 \cdot 20 + 2500$$
$$= 1500 + 3000 + 2500 = 7000$$

Producing 15 three-speed and 20 ten-speed bicycles costs $7000.

---

The variables $x$ and $y$ in the preceding example stand for numbers of bicycles, and so should take only integer values. Instead, however, we will allow

$x$ and $y$ to be "continuous" variables, and round to integers at the end if necessary.

Some other "everyday" examples of functions of two variables are as follows:

$$A(l, w) = lw$$   Area of a rectangle of length $l$ and width $w$

$$f(w, v) = kwv^2$$   Length of the skid marks for a car of weight $w$ and velocity $v$ skidding to a stop ($k$ is a constant depending on the road surface)

## Cobb–Douglas Production Functions

A function used to model the output of a company or a nation is called a **production function,** and the most famous is the Cobb–Douglas production function*

$$P(L, K) = aL^bK^{1-b}$$   For constants   $a > 0$   and   $0 < b < 1$

This function expresses the total production $P$ as a function of $L$, the number of units of labor, and $K$, the number of units of capital. (Labor is measured in work-hours, and capital means *invested* capital, including the cost of buildings, equipment, and raw materials.)

---

**EXAMPLE 4**  **EVALUATING A COBB–DOUGLAS PRODUCTION FUNCTION**

Cobb and Douglas modeled the output of the American economy by the function   $P(L, K) = L^{0.75}K^{0.25}$.   Find $P(150, 220)$.

*Solution*

$$P(150, 220) = (150)^{0.75}(220)^{0.25}$$   $P(L, K) = L^{0.75}K^{0.25}$  with $L = 150$  and  $K = 220$

$$\approx (42.9)(3.85) \approx 165$$   Using a calculator

That is, 150 units of labor and 220 units of capital should result in approximately 165 units of production.

---

## Graphing Calculator Exploration

The windchill index announced by the weather bureau during the winter to measure the combined effect of wind and cold is calculated from the formula below, where $x$ is wind speed (miles per hour) and $y$ is temperature (degrees Fahrenheit).

$$W(x, y) = 35.74 + 0.6215y - 35.75x^{0.16} + 0.4275yx^{0.16}$$

*First used by Charles Cobb and Paul Douglas in a landmark study of the American economy published in 1928.

Y1=35.74+.6215*32-35.75X

X=15        Y=21.588989

**a.** Enter this function into your graphing calculator but with $y$ (temperature) replaced by 32 so that it becomes a function of just one variable, $x$ (wind speed). Then graph the function on the window $[0, 45]$ by $[-5, 40]$. The graph shows how the perceived temperature decreases as wind speed increases.

**b.** Evaluate this function at $x = 15$ to find the perceived temperature resulting from a 15-mph wind when the actual temperature is 32 degrees.

**c.** Notice that the graph drops more steeply for low wind speeds than for high wind speeds. What does this mean about the effect of an extra 5 miles per hour of wind on a calm day as opposed to a windy day? (Exercises 36 and 37 continue this analysis.)

*Source:* National Weather Service

## Functions of Three or More Variables

Functions of three (or more) variables are defined analogously. Some examples are:

$$V(l, w, h) = lwh$$

Volume of a rectangular solid of length $l$, width $w$, and height $h$

$$W(P, r, t) = Pe^{rt}$$

Worth of $P$ dollars invested at a continuous interest rate $r$ for $t$ years

$$f(w, x, y, z) = \frac{w + x + y + z}{4}$$

Average of four numbers

**EXAMPLE 5**

**FINDING THE DOMAIN OF A FUNCTION OF THREE VARIABLES**

For $f(x, y, z) = \dfrac{\sqrt{x}}{y} + \ln\dfrac{1}{z}$, find

**a.** the domain

**b.** $f(4, -1, 1)$

*Solution*

**a.** In $f(x, y, z) = \dfrac{\sqrt{x}}{y} + \ln\dfrac{1}{z}$ we must have $x \geq 0$ (because of the square root), $y \neq 0$ (since it is a denominator), and $z > 0$ (so that $1/z$ has a logarithm). Therefore, the domain is

$$\{ (x, y, z) \mid x \geq 0, y \neq 0, z > 0 \}$$

**b.** $f(4, -1, 1) = \dfrac{\sqrt{4}}{-1} + \ln\dfrac{1}{1} = \dfrac{2}{-1} + \underbrace{\ln 1}_{0} = -2$

**EXAMPLE 6**

### FINDING THE VOLUME AND AREA OF A DIVIDED BOX

An open-top box is to have a center divider, as shown in the diagram. Find formulas for the volume $V$ of the box and for the total amount of material $M$ needed to construct the box.

*Solution*

The volume is length times width times height.

$$V = xyz$$

The box consists of a bottom, a front and back, two sides, and a divider, whose areas are shown in the diagram. Therefore, the total amount of material (the area) is

$$M = xy + 2xz + 3yz$$

Bottom — Back and front — Sides and divider

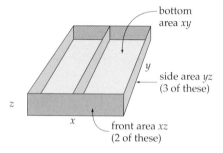

bottom area $xy$

side area $yz$ (3 of these)

front area $xz$ (2 of these)

---

**Practice Problem 2**

Find a formula for the total amount of material $M$ needed to construct an open-top box with three parallel dividers. Use the variables shown in the diagram.

➤ **Solution on page 471**

### Graph of a Function of Two Variables

Graphing a function of two variables requires a **three-dimensional coordinate system.** We draw three perpendicular axes as shown on the right.* We will usually draw only the positive half of each axis, although each axis extends infinitely far in the negative direction as well. The plane at the base is called the *x-y plane.*

A point in a three-dimensional coordinate system is specified by three coordinates, giving its distances from the origin in the $x$, $y$, and $z$ directions. For example, the point

The three-dimensional (r ight-handed") coordinate system

$$(2, 3, 4)$$

$z$-coordinate
$y$-coordinate
$x$-coordinate

The point (2, 3, 4)

*This is called a "right-handed" coordinate system because the $x$, $y$, and $z$ axes correspond to the first two fingers and thumb of the right hand.

is plotted by starting at the origin, moving 2 units in the $x$ direction, 3 units in the $y$ direction, and then 4 units in the (vertical) $z$ direction.

To graph a function of two variables, we choose values for $x$ and $y$, calculate $z$-values from $z = f(x, y)$, and plot the points $(x, y, z)$.

**EXAMPLE 7**

### GRAPHING A FUNCTION OF TWO VARIABLES

To graph $f(x, y) = 18 - x^2 - y^2$, we set $z$ equal to the function.

$$z = 18 - x^2 - y^2 \qquad \text{\small $z$ replaces $f(x, y)$}$$

Then we choose values for $x$ and $y$. Choosing $x = 1$ and $y = 2$ gives

$$z = 18 - 1^2 - 2^2 = 13 \qquad \text{\small $z = 18 - x^2 - y^2$ with $x = 1$ and $y = 2$}$$

for the point

$$(1, 2, 13) \qquad \text{\small The chosen $x = 1$, $y = 2$, and the calculated $z$}$$

Choosing $x = 2$ and $y = 3$ gives

$$z = 18 - 2^2 - 3^2 = 5 \qquad \text{\small $z = 18 - x^2 - y^2$ with $x = 2$ and $y = 3$}$$

for the point

$$(2, 3, 5) \qquad \text{\small The chosen $x = 2$, $y = 3$, and the calculated $z$}$$

These points $(1, 2, 13)$ and $(2, 3, 5)$ are plotted on the graph on the left below. The completed graph of the function is shown on the right.

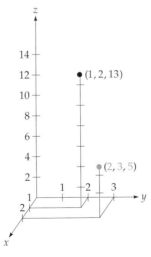

The points $(1, 2, 13)$ and $(2, 3, 5)$
of the function
$f(x, y) = 18 - x^2 - y^2$

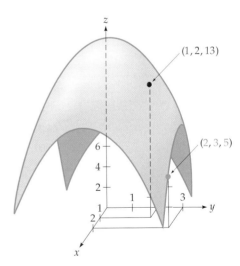

The graph of $f(x, y) = 18 - x^2 - y^2$

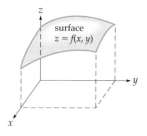

The graph of a function of two variables is a surface whose height above the point $(x, y)$ in the $x$-$y$ plane is $z = f(x, y)$.

In general, the graph of a function of *two* variables is a *surface* above or below the $x$-$y$ plane, just as the graph of a function of *one* variable is a *curve* above or below the $x$-axis.

Graphing functions of two variables involves drawing three-dimensional graphs, which is very difficult. Graphing functions of *more* than two variables requires *more* than three dimensions, and is impossible. For this reason we will not graph functions of several variables. We will, however, often speak of a function of two variables as representing a *surface* in three-dimensional space.

 **Spreadsheet Exploration**

The following spreadsheet* graph of $f(x, y) = 18 - x^2 - y^2$ from the previous example is a chart showing the values of the function that were calculated for values of $x$ and $y$ between $-5$ and $5$.

*To obtain this and other Spreadsheet Explorations, go to www.cengage.com/math/berresford.

Just as with functions of one variable, useful graphs can be constructed based on Re values of a few important points. As we have already seen, these values are easily found using calculus.

## Relative Extreme Points and Saddle Points

Certain points on a surface of such a graph are of special importance.

### Relative Maximum Point

A point $(a, b, c)$ on a surface $z = f(x, y)$ is a *relative maximum point* if $f(a, b) \geq f(x, y)$ for all $(x, y)$ in some region surrounding $(a, b)$.

### Relative Minimum Point

A point $(a, b, c)$ on a surface $z = f(x, y)$ is a *relative minimum point* if $f(a, b) \leq f(x, y)$ for all $(x, y)$ in some region surrounding $(a, b)$.

As before, the term **relative extreme point** means a point that is either a relative maximum or a relative minimum point. A surface can have any number of relative extreme points, even none.

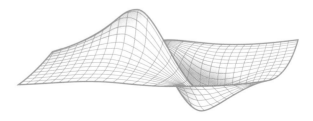

A surface with two relative extreme points: one relative maximum and one relative minimum.

saddle point

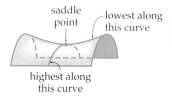

saddle point — lowest along this curve

highest along this curve

The point shown on the left is called a **saddle point** (so named because the diagram resembles a saddle).

A saddle point is a point that is the highest point along one curve of the surface and the lowest point along another curve. A saddle point is *not* a relative extreme point.

If we think of a surface $z = f(x, y)$ as a landscape, then relative maximum and minimum points correspond to "hilltops" and "valley bottoms," and a saddle point corresponds to a "mountain pass" between two peaks.

## Gallery of Surfaces

The following are the graphs of a few functions of two variables.

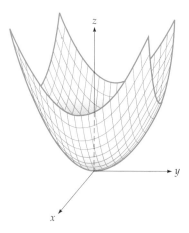

The surface $f(x, y) = x^2 + y^2$ has a relative minimum point at the origin.

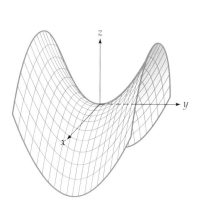

The surface $f(x, y) = y^2 - x^2$ has a saddle point at the origin.

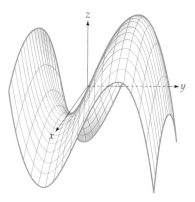

The surface $f(x, y) = 12y + 6x - x^2 - y^3$ has a saddle point and a relative maximum point.

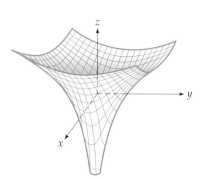

The surface $f(x, y) = \ln(x^2 + y^2)$ has no relative extreme points. It is undefined at $(0, 0)$.

> **Solutions to Practice Problems**
>
> **1. a.** $\{(x, y) \mid x > 0, y \geq 0\}$
>
>   **b.** $f(e, 4) = \dfrac{\ln e}{e^{\sqrt{4}}} = \dfrac{1}{e^2} = e^{-2}$
>
> **2.** $M = xy + 2xz + 5yz$

## 7.1 Section Summary

Just as for a function of one variable, a *function of several variables* gives exactly one value for each point in its domain. For a function of *two* variables, denoted $f(x, y)$, the values $z = f(x, y)$ determine a *surface* above or below the *x-y* plane. The surface may have *relative maximum points* and *relative minimum points* ("hilltops" and "valley bottoms") or even *saddle points* (high points along one curve and low points along another), as shown on the previous pages.

## 7.1 Exercises

**1–8.** For each function, find the domain.

**1.** $f(x, y) = \dfrac{1}{xy}$

**2.** $f(x, y) = \dfrac{\sqrt{x}}{\sqrt{y}}$

**3.** $f(x, y) = \dfrac{1}{x - y}$

**4.** $f(x, y) = \dfrac{\sqrt[3]{x}}{\sqrt[3]{y}}$

**5.** $f(x, y) = \dfrac{\ln x}{y}$

**6.** $f(x, y) = \dfrac{x}{\ln y}$

**7.** $f(x, y, z) = \dfrac{e^{1/y} \ln z}{x}$

**8.** $f(x, y, z) = \dfrac{\sqrt{x} \ln y}{z}$

**9–22.** For each function, evaluate the given expression.

**9.** $f(x, y) = \sqrt{99 - x^2 - y^2}$, find $f(3, -9)$

**10.** $f(x, y) = \sqrt{75 - x^2 - y^2}$, find $f(5, -1)$

**11.** $g(x, y) = \ln(x^2 + y^4)$, find $g(0, e)$

**12.** $g(x, y) = \ln(x^3 - y^2)$, find $g(e, 0)$

**13.** $w(u, v) = \dfrac{1 + 2u + 3v}{uv}$, find $w(-1, 1)$

**14.** $w(u, v) = \dfrac{2u + 4u}{v - u}$, find $w(1, -1)$

**15.** $h(x, y) = e^{xy + y^2 - 2}$, find $h(1, -2)$

**16.** $h(x, y) = e^{x^2 - xy - 4}$, find $h(1, -2)$

**17.** $f(x, y) = xe^y - ye^x$, find $f(1, -1)$

**18.** $f(x, y) = xe^y + ye^x$, find $f(-1, 1)$

**19.** $f(x, y, z) = xe^y + ye^z + ze^x$, find $f(1, -1, 1)$

**20.** $f(x, y, z) = xe^y + ye^z + ze^x$, find $f(-1, 1, -1)$

**21.** $f(x, y, z) = z \ln \sqrt{xy}$, find $f(-1, -1, 5)$

**22.** $f(x, y, z) = z\sqrt{x} \ln y$, find $f(4, e, -1)$

### Applied Exercises

**23. PERSONAL FINANCE: Stock Yield** The *yield* of a stock is defined as $Y(d, p) = \dfrac{d}{p}$ where $d$ is the dividend per share and $p$ is the price of a share of stock. Find the yield of a stock that sells for $140 and offers a dividend of $2.20.

**24. PERSONAL FINANCE: Price-Earnings Ratio** The price-earnings ratio of a stock is defined as $R(P, E) = \dfrac{P}{E}$ where $P$ is the price of a share of stock and $E$ is its earnings. Find the price-earnings ratio of a stock that is selling for $140 with earnings of $1.70.

25. **GENERAL: Scuba Diving** The maximum duration of a scuba dive (in minutes) can be estimated from the formula

$$T(v, d) = \frac{33v}{d + 33}$$

where $v$ is the volume of air (at sea-level pressure) in the tank and $d$ is the depth of the dive. Find $T(90, 33)$.

*Source: U.S. Navy Diving Manual, Vol. 2*

26. **SOCIAL SCIENCE: Cephalic Index** Anthropologists define the *cephalic index* to distinguish the head shapes of different people. For a head of width $W$ and length $L$ (measured from above), the cephalic index is

$$C(W, L) = 100 \frac{W}{L}$$

Calculate the cephalic index for a head of width 8 inches and length 10 inches.

27. **ECONOMICS: Cobb–Douglas Functions** One company's production is estimated to be $P(L, K) = 2L^{0.6}K^{0.4}$. Find $P(320, 150)$.

*Source: Journal of Political Economy 84*

28. **BIOMEDICAL: Body Area** The surface area (in square feet) of a person of weight $w$ pounds and height $h$ feet is approximated by the function $A(w, h) = 0.55 \, w^{0.425}h^{0.725}$. Use this function to estimate the surface area of a person who weighs 160 pounds and who is 6 feet tall. (Such estimates are important in certain medical procedures.)

*Source: Archives of Internal Medicine 17*

29. **ECONOMICS: Cobb–Douglas Functions** Show that the Cobb–Douglas production function $P(L, K) = aL^bK^{1-b}$ satisfies the equation $P(2L, 2K) = 2 \cdot P(L, K)$. This shows that doubling the amounts of labor and capital doubles production, a property called *returns to scale*.

30. **ECONOMICS: Cobb–Douglas Functions** Show that the Cobb–Douglas function $P(L, K) = aL^bK^{1-b}$ with $0 < b < 1$ satisfies

$$P(2L, K) < 2P(L, K) \quad \text{and} \quad P(L, 2K) < 2P(L, K)$$

This shows that doubling the amounts of either labor or capital alone results in *less* than double production, a property called *diminishing returns*.

31. **GENERAL: Telephone Calls** For two cities with populations $x$ and $y$ that are $d$ miles apart, the number of telephone calls per hour between them can be estimated by the function of three variables

$$f(x, y, d) = \frac{3xy}{d^2}$$

(This is called the *gravity model*.) Use the gravity model to estimate the number of calls between two cities of populations 40,000 and 60,000 that are 600 miles apart.

*Source: about.com*

32. **ENVIRONMENTAL SCIENCE: Tag and Recapture Estimates** Ecologists estimate the size of animal populations by capturing and tagging a few animals, and then releasing them. After the first group has mixed with the population, a second group of animals is captured, and the number of tagged animals in this group is counted. If originally $T$ animals were tagged, and the second group is of size $S$ and contains $t$ tagged animals, then the population is estimated by the function of three variables

$$P(T, S, t) = \frac{TS}{t}$$

Estimate the size of a deer population if 100 deer were tagged, and then a second group of 250 contained 20 tagged deer.

*Source: J. Blower, L. Cook, and J. Bishop, Estimating the Size of Animal Populations*

33. **BUSINESS: Cost Function** It costs an appliance company $210 to manufacture each washer and $180 to manufacture each dryer, and fixed costs are $4000. Find the company's cost function $C(x, y)$, using $x$ and $y$ for the numbers of washers and dryers, respectively.

**34–35. GENERAL: Box Design** For each open-top box shown below, find formulas for:

**a.** the volume.

**b.** the total amount of material (the area).

34.

35.

**36. GENERAL: Windchill** Enter the formula for windchill from the Graphing Calculator Exploration on pages 464–465 and again replace $y$ (temperature) by 32 so that it becomes a function of one variable, $x$ (wind speed).

a. Graph the function on the window $[0, 45]$ by $[0, 40]$.
b. Use NDERIV or $dy/dx$ to find the slope of this curve at  $x = 5$.  Interpret the answer.
c. Repeat part (b) but for  $x = 20$,  interpreting the answer.
d. What do the answers indicate about the windchill effect of additional wind on a calm day as opposed to on an already windy day?

*Source: National Weather Service*

**37. GENERAL: Windchill** Enter the formula for windchill index from the Graphing Calculator Exploration on pages 464–465, but with $y$ (temperature) replaced by *several* temperatures: 20, 30, 40, and 50 degrees (using $y_1$, $y_2$, $y_3$, and $y_4$).

a. Graph the functions on the window $[0, 45]$ by $[-10, 60]$.
b. Notice that the lower temperature curves slope downward more steeply than the others. What does this mean about the effect of wind on a colder day?
c. Use NDERIV or $dy/dx$ to find the slope of the lowest and the highest curves at  $x = 10$. Interpret the answers.
d. Do your answers to part (c) support your conclusion in part (b)?

*Source: National Weather Service*

**38. BIOMEDICAL: Oxygen Consumption** The oxygen consumption of a well-insulated non-sweating mammal can be estimated from the formula  $f(t_b, t_a, w) = 2.5(t_b - t_a)w^{-0.67}$,  where $t_b$ is the animal's body temperature, $t_a$ is the air temperature (both in degrees Celsius), and $w$ is the animal's weight (in kilograms). Find the oxygen consumption of a 40-kilogram animal whose body temperature is 35 degrees when the air temperature is 5 degrees.

*Source: D. J. Clow and N. S. Urquhart, Mathematics in Biology*

**39. ATHLETICS: Pythagorean Baseball Standings** Many professional baseball teams (including the Cincinnati Reds and the Boston Red Sox) use Bill James's formula to estimate their probability of winning a league pennant:

$$\left(\begin{array}{c}\text{Probability}\\\text{of winning}\end{array}\right) = \frac{(\text{Runs scored})^2}{(\text{Runs scored})^2 + (\text{Runs allowed})^2}$$

This formula, whose form is reminiscent of the Pythagorean theorem, is considered more accurate

than just the proportion of games won because it takes into consideration the scores of the games. Find this probability for a team that has scored 400 runs and allowed 300 runs.

*Source: B. James, Baseball Abstract 1983*

**40. ATHLETICS: Olympics** A country's population and wealth certainly contribute to its success in the Olympics. The following formula, based on the country's population $p$ and per capita gross domestic product $d$, has proved accurate in predicting the proportion of Olympic medals that a country will win:

$$\left(\begin{array}{c}\text{Proportion}\\\text{of medals}\end{array}\right) = 0.0062 \ln p + 0.0064 \ln d - 0.0652$$

Estimate the proportion of Olympic medals that the United States will win based on a population of 300,000,000 and a per capita gross domestic product of $40,000.

*Source: Review of Economics and Statistics 86*

## Conceptual Exercises

**41.** Which is a *right-handed* coordinate system?

**42.** Fill in the blanks with the correct numbers: To graph a function $f(x)$ of *one* variable we need _____ axes, and to graph a function $f(x, y)$ of *two* variables we need _____ axes.

**43.** Which function has a relative maximum point and which has a relative minimum point?

a. $f(x, y) = x^2 + y^2 - 1$   b. $g(x, y) = 1 - x^2 - y^2$

**44.** True or False: Every function $f(x, y)$ of two variables can be written as the sum of two functions of one variable,  $g(x) + h(y)$.

**45.** True or False: A function $f$ of two variables is a rule such that to each number $f(x, y)$ in the domain there corresponds one and only one ordered pair $(x, y)$.

**46–50.**  For each function, state whether it satisfies:

a. $f(-x, -y) = f(x, y)$  for all $x$ and $y$,
b. $f(-x, -y) = -f(x, y)$  for all $x$ and $y$, or
c. neither of these conditions.

**46.** $f(x, y) = x - y$     **47.** $f(x, y) = x^2 - y^2$
**48.** $f(x, y) = x^2 - y^3$     **49.** $f(x, y) = x^2 y^3$
**50.** $f(x, y) = x^2 y^4$

## 7.2 PARTIAL DERIVATIVES

### Introduction

Functions of several variables have several derivatives, called **partial derivatives,** one for each variable. In this section you will learn how to calculate and interpret these derivatives.

### Partial Derivatives

Before defining partial derivatives, we review the rules governing derivatives and constants.

$$\frac{d}{dx} c = 0$$

For a constant *standing alone*, the derivative is zero.

$$\frac{d}{dx} (cx^3) = c \cdot 3x^2$$

For a constant *multiplying* a function, the constant is carried along.

Carry along the constant    Derivative of $x^3$

These rules will be very useful in this section.

A function $f(x, y)$ has two *partial derivatives*, one with respect to $x$ and the other with respect to $y$.

---

**Partial Derivatives**

$$\frac{\partial}{\partial x} f(x, y) = \lim_{h \to 0} \frac{f(x + h, y) - f(x, y)}{h}$$

Partial derivative of $f$ with respect to $x$ ($x$ is changed by $h$; $y$ is held constant)

$$\frac{\partial}{\partial y} f(x, y) = \lim_{h \to 0} \frac{f(x, y + h) - f(x, y)}{h}$$

Partial derivative of $f$ with respect to $y$ ($y$ is changed by $h$; $x$ is held constant)

(provided that the limits exist).

---

Partial derivatives are written with a "curly" $\partial$, $\partial / \partial x$ instead of $d / dx$, and are often called "partials." (The Greek letter $\partial$ is a lowercase delta, and we have already used the capital delta, $\Delta$, to denote change.)

Rewriting the first formula, but without the $y$'s, gives

$$\lim_{h \to 0} \frac{f(x + h) - f(x)}{h}$$

$$\lim_{h \to 0} \frac{f(x + h, y) - f(x, y)}{h}$$
but omitting the $y$'s

which is just the "ordinary derivative" from page 100. Therefore, the partial derivative with respect to $x$ is just the ordinary derivative with respect to $x$ with $y$ held constant. Similarly, the partial with respect to $y$ is just the ordinary derivative with respect to $y$, but now with $x$ held constant.

$$\frac{\partial}{\partial x}\, f(x, y) = \left(\begin{array}{c}\text{Derivative of } f \text{ with respect} \\ \text{to } x \text{ with } y \text{ held constant}\end{array}\right)$$

$$\frac{\partial}{\partial y}\, f(x, y) = \left(\begin{array}{c}\text{Derivative of } f \text{ with respect} \\ \text{to } y \text{ with } x \text{ held constant}\end{array}\right)$$

**EXAMPLE 1**

### FINDING A PARTIAL DERIVATIVE WITH RESPECT TO *x*

Find $\dfrac{\partial}{\partial x} x^3 y^4$.

*Solution*

The $\partial/\partial x$ means differentiate with respect to $x$, holding $y$ (and therefore $y^4$) constant. We therefore differentiate the $x^3$ and carry along the "constant" $y^4$:

Derivative of $x^3$

$$\frac{\partial}{\partial x} x^3 y^4 = 3x^2 y^4$$

Carry along the "constant" $y^4$

$\partial/\partial x$ means differentiate with respect to $x$, treating $y$ (and therefore $y^4$) as a constant

**EXAMPLE 2**

### FINDING A PARTIAL WITH RESPECT TO *y*

Find $\dfrac{\partial}{\partial y} x^3 y^4$.

*Solution*

$$\frac{\partial}{\partial y} x^3 y^4 = x^3 4y^3 \quad = \quad 4x^3 y^3$$

Carry along the "constant" $x^3$

Derivative of $y^4$

Writing the 4 first

$\partial/\partial y$ means differentiate with respect to $y$, treating $x$ (and therefore $x^3$) as a constant

**Practice Problem 1**   Find   **a.** $\dfrac{\partial}{\partial x} x^4 y^2$    **b.** $\dfrac{\partial}{\partial y} x^4 y^2$

➤ **Solutions on page 484**

**EXAMPLE 3**

### FINDING A PARTIAL DERIVATIVE

Find $\dfrac{\partial}{\partial x} y^4$.

*Solution*

$$\frac{\partial}{\partial x} y^4 = 0$$

The derivative of a constant is zero (since $\partial / \partial x$ means hold $y$ constant)

Partial with respect to $x$ — Function of $y$ alone

---

**Practice Problem 2**   Find  $\dfrac{\partial}{\partial y} x^2.$

➤ Solution on page 484

---

**EXAMPLE 4**

**FINDING A PARTIAL OF A POLYNOMIAL IN TWO VARIABLES**

Find  $\dfrac{\partial}{\partial x} (2x^4 - 3x^3y^3 - y^2 + 4x + 1).$

*Solution*

$$\frac{\partial}{\partial x}(2x^4 - 3x^3y^3 - y^2 + 4x + 1) = 8x^3 - 9x^2y^3 + 4$$

Differentiating with respect to $x$, so each $y$ is held constant

$$\frac{\partial}{\partial x} y^2 = 0$$

---

**Practice Problem 3**   Find  $\dfrac{\partial}{\partial y} (2x^4 - 3x^3y^3 - y^2 + 4x + 1).$

➤ Solution on page 484

---

## Subscript Notation for Partial Derivatives

Partial derivatives are often denoted by subscripts: a subscript $x$ means the partial with respect to $x$, and a subscript $y$ means the partial with respect to $y$.*

| *Subscript Notation* | | $\boldsymbol{\partial}$ *Notation* | |
|:---:|:---:|:---:|:---|
| $f_x(x, y)$ | $=$ | $\dfrac{\partial}{\partial x} f(x, y)$ | $f_x$ means the partial of $f$ with respect to $x$ |
| $f_y(x, y)$ | $=$ | $\dfrac{\partial}{\partial y} f(x, y)$ | $f_y$ means the partial of $f$ with respect to $y$ |

---

*Sometimes subscripts 1 and 2 are used to indicate partial derivatives with respect to the first and second variables: $f_1(x, y)$ means $f_x(x, y)$ and $f_2(x, y)$ means $f_y(x, y)$. We will not use this notation in this book.

**EXAMPLE 5**

### USING SUBSCRIPT NOTATION

Find $f_x(x, y)$ if $f(x, y) = 5x^4 - 2x^2y^3 - 4y^2$.

*Solution*

$$f_x(x, y) = 20x^3 - 4xy^3$$

Differentiating with respect to $x$, holding $y$ constant

**EXAMPLE 6**

### FINDING A PARTIAL INVOLVING LOGS AND EXPONENTIALS

Find both partials of $f = e^x \ln y$.

*Solution*

$$f_x = e^x \ln y$$

The derivative of $e^x$ is $e^x$ (times the "constant" $\ln y$)

$$f_y = e^x \frac{1}{y}$$

The derivative of $\ln y$ is $\frac{1}{y}$ (times the "constant" $e^x$)

**EXAMPLE 7**

### FINDING A PARTIAL OF A FUNCTION TO A POWER

Find $f_y$ if $f = (xy^2 + 1)^4$.

*Solution*

$$f_y = 4(xy^2 + 1)^3(x2y)$$

Partial of the inside with respect to $y$

Using the Generalized Power Rule (the derivative of $f^n$ is $nf^{n-1}f'$, but with $f'$ meaning a *partial*)

$$= 8xy(xy^2 + 1)^3$$

Simplifying

**EXAMPLE 8**

### FINDING A PARTIAL OF A QUOTIENT

Find $\frac{\partial g}{\partial x}$ if $g = \frac{xy}{x^2 + y^2}$.

*Solution*

Partial of the top with respect to $x$
Partial of the bottom with respect to $x$

$$\frac{\partial g}{\partial x} = \frac{(x^2 + y^2)y - 2x \cdot xy}{(x^2 + y^2)^2}$$

Bottom squared

Using the Quotient Rule

$$= \frac{x^2y + y^3 - 2x^2y}{(x^2 + y^2)^2} = \frac{y^3 - x^2y}{(x^2 + y^2)^2}$$

Simplifying

**EXAMPLE 9**

### FINDING A PARTIAL OF THE LOGARITHM OF A FUNCTION

Find $f_x(x, y)$ if $f(x, y) = \ln(x^2 + y^2)$.

*Solution*

$$f_x(x, y) = \frac{2x}{x^2 + y^2}$$

Partial of the bottom with respect to $x$

Derivative of $\ln f$ is $\frac{f'}{f}$

An expression such as $f_x(2, 5)$, which involves both differentiation and evaluation, means *first differentiate and then evaluate.*\*

**EXAMPLE 10**

### EVALUATING A PARTIAL DERIVATIVE

Find $f_y(1, 3)$ if $f(x, y) = e^{x^2 + y^2}$.

*Solution*

$$f_y(x, y) = e^{x^2 + y^2}(2y)$$

Derivative of $e^f$ is $e^f \cdot f'$

Partial of the exponent with respect to $y$

$$f_y(1, 3) = e^{1^2 + 3^2}(2 \cdot 3)$$

Evaluating at $x = 1$ and $y = 3$

$$= 6e^{10}$$

Simplifying

**Practice Problem 4**

Find $f_y(1, 2)$ if $f(x, y) = e^{x^3 + y^3}$.

➤ Solution on page 484

### Partial Derivatives in Three or More Variables

Partial derivatives in three or more variables are defined similarly. That is, the partial derivative of $f(x, y, z)$ with respect to any one variable is the "ordinary" derivative with respect to that variable, holding all other variables constant.

---

\*$f_x(2, 5)$ may be written $\dfrac{\partial f}{\partial x}(2, 5)$ or $\dfrac{\partial f}{\partial x}\bigg|_{(2, 5)}$, again meaning first differentiate, then evaluate.

**EXAMPLE 11**  **FINDING A PARTIAL OF A FUNCTION OF THREE VARIABLES**

$$\frac{\partial}{\partial x}(x^3 y^4 z^5) = 3x^2 y^4 z^5$$

$\partial/\partial x$ means differentiate with respect to $x$, holding $y$ and $z$ constant

Hold constant    Derivative of $x^3$

**Practice Problem 5**  Find $\dfrac{\partial}{\partial y}(x^3 y^4 z^5)$.  ➤ Solution on page 484

**EXAMPLE 12**  **EVALUATING A PARTIAL IN THREE VARIABLES**

Find $f_z(1,1,1)$ if $f(x,y,z) = e^{x^2+y^2+z^2}$.

*Solution*

$$f_z(x,y,z) = e^{x^2+y^2+z^2}(2z)$$  Partial with respect to $z$

$$= 2ze^{x^2+y^2+z^2}$$  Writing the $2z$ first

$$f_z(1,1,1) = 2\cdot 1\cdot e^{1^2+1^2+1^2} = 2e^3$$  Evaluating

### Interpreting Partial Derivatives as Rates of Change

Since partials are just "ordinary" derivatives with the other variable(s) held constant, they give *instantaneous rates of change* with respect to one variable at a time.

**Partials as Rates of Change**

$$f_x(x,y) = \begin{pmatrix} \text{Instantaneous rate of change of } f \text{ with} \\ \text{respect to } x \text{ when } y \text{ is held constant} \end{pmatrix}$$

$$f_y(x,y) = \begin{pmatrix} \text{Instantaneous rate of change of } f \text{ with} \\ \text{respect to } y \text{ when } x \text{ is held constant} \end{pmatrix}$$

This is why they are called *partial* derivatives: not all the variables are changed at once, only a "partial" change is made.

### Cobb–Douglas Production Functions

Recall that a Cobb–Douglas production function $P(L,K) = aL^b K^{1-b}$ expresses production $P$ as a function of $L$ (units of labor) and $K$ (units of capital).

Each partial therefore gives the rate of increase of production with respect to one of these variables while the other is held constant.

---

**EXAMPLE 13**  **INTERPRETING THE PARTIALS OF A COBB–DOUGLAS PRODUCTION FUNCTION**

Find and interpret $P_L(120, 200)$ and $P_K(120, 200)$ for the Cobb–Douglas function $P(L, K) = 20L^{0.6}K^{0.4}$.

*Solution*

$$P_L = 12L^{-0.4}K^{0.4}$$
Partial with respect to $L$ (the 12 is 20 times 0.6)

$$P_L(120, 200) = 12(120)^{-0.4}(200)^{0.4} \approx 14.7$$
Substituting $L = 120$ and $K = 200$, and evaluating using a calculator

*Interpretation:* $P_L = 14.7$ means that production increases by about 14.7 units for each additional unit of labor (when $L = 120$ and $K = 200$). This is called the *marginal productivity of labor.*

$$P_K = 8L^{0.6}K^{-0.6}$$
Partial with respect to $K$ (the 8 is 20 times 0.4)

$$P_K(120, 200) = 8(120)^{0.6}(200)^{-0.6} \approx 5.9$$
Substituting $L = 120$ and $K = 200$, and evaluating using a calculator

*Interpretation:* $P_K = 5.9$ means that production increases by about 5.9 units for each additional unit of capital (when $L = 120$ and $K = 200$). This is called the *marginal productivity of capital.*

---

These numbers show that to increase production, additional units of labor are more than twice as effective as additional units of capital (at the levels $L = 120$ and $K = 200$).

Partial derivatives give the marginals for one product at a time.

---

**Interpreting Partials as Marginals**

Let $C(x, y)$ be the (total) cost function for $x$ units of product A and $y$ units of product B. Then

$$C_x(x, y) = \left(\begin{array}{c} \text{Marginal cost function for product A when} \\ \text{production of product B is held constant} \end{array}\right)$$

$$C_y(x, y) = \left(\begin{array}{c} \text{Marginal cost function for product B when} \\ \text{production of product A is held constant} \end{array}\right)$$

Similar statements hold, of course, for revenue and profit functions: the partials give the marginals for one variable at a time when the other variables are held constant.

EXAMPLE 14    **INTERPRETING PARTIALS OF A PROFIT FUNCTION**

A company's profit from producing $x$ radios and $y$ televisions per day is $P(x, y) = 4x^{3/2} + 6y^{3/2} + xy$. Find the marginal profit functions. Then find and interpret $P_y(25, 36)$.

*Solution*

| | |
|---|---|
| $P_x(x, y) = 6x^{1/2} + y$ | Marginal profit for radios when television production is held constant |
| $P_y(x, y) = 9y^{1/2} + x$ | Marginal profit for televisions when radio production is held constant |
| $P_y(25, 36) = 9\underbrace{(36)^{1/2}}_{6} + 25 = 79$ | Evaluating $P_y$ at $x = 25$ and $y = 36$ |

*Interpretation:* Profit increases by about $79 per additional television (when producing 25 radios and 36 televisions per day).

## Interpreting Partials Geometrically

A function $f(x, y)$ represents a surface in three-dimensional space, and the partial derivatives are the **slopes** along the surface in different directions: $\partial f / \partial x$ gives the slope of the surface "in the $x$-direction," and $\partial f / \partial y$ gives the slope of the surface "in the $y$-direction" at the point $(x, y)$.

To put this colloquially, if you were walking on the surface $z = f(x, y)$, then $\partial f / \partial x$ would be the steepness of the surface *in the x-direction*, and $\partial f / \partial y$ would be the steepness of the surface *in the y-direction* from the point $(x, y)$.

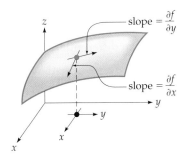

Partial derivatives are slopes.

For example, the following graph shows gridlines in the $x$-direction (roughly up and down the page, with the positive $x$-direction being down) and in the

**482** CHAPTER 7 CALCULUS OF SEVERAL VARIABLES

*y*-direction (roughly across the page, with the positive *y*-direction being to the right).

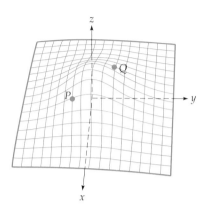

Therefore, at the point *P* on the graph the partials would have the following signs:

$$\frac{\partial f}{\partial x} < 0$$

Walking from *P* in the positive *x*-direction would mean walking *downhill*

$$\frac{\partial f}{\partial y} > 0$$

Walking from *P* in the positive *y*-direction would mean walking *uphill*

**Practice Problem 6**

On the preceding graph at the point *Q,*

**a.** Is $\dfrac{\partial f}{\partial x}$ positive or negative?

**b.** Is $\dfrac{\partial f}{\partial y}$ positive or negative?

[*Hint:* Leaving *Q* in those directions, would you be walking uphill or downhill?]   ➤ **Solutions on page 484**

## Higher-Order Partial Derivatives

We can differentiate a function more than once to obtain *higher-order* partials.

| **Second-Order Partials** | | |
|---|---|---|
| *Subscript Notation* | *∂ Notation* | *In Words* |
| $f_{xx}$   = | $\dfrac{\partial^2}{\partial x^2} f$ | Differentiate twice with respect to *x* |
| $f_{yy}$   = | $\dfrac{\partial^2}{\partial y^2} f$ | Differentiate twice with respect to *y* |

| Subscript Notation | | $\partial$ Notation | In Words |
|---|---|---|---|
| $f_{xy}$ | $=$ | $\dfrac{\partial^2}{\partial y\, \partial x}\, f$ | Differentiate first with respect to $x$, then with respect to $y$ |
| $f_{yx}$ | $=$ | $\dfrac{\partial^2}{\partial x\, \partial y}\, f$ | Differentiate first with respect to $y$, then with respect to $x$ |

Each notation means differentiate first with respect to the variable *closest* to $f$.

Calculating a "second partial" such as $f_{xy}$ is a two-step process: first calculate $f_x$, and then differentiate the *result* with respect to $y$.

**EXAMPLE 15**

**FINDING SECOND-ORDER PARTIALS**

Find all four second-order partials of $f(x, y) = x^4 + 2x^2y^2 + x^3y + y^4$.

*Solution*

First we calculate

$$f_x = 4x^3 + 4xy^2 + 3x^2y \qquad \text{Partial of } f \text{ with respect to } x$$

Then from this we find $f_{xx}$ and $f_{xy}$:

$$f_{xx} = 12x^2 + 4y^2 + 6xy \qquad \text{Differentiating } f_x = 4x^3 + 4xy^2 + 3x^2y \text{ with respect to } x$$

$$f_{xy} = 8xy + 3x^2 \qquad \text{Differentiating } f_x = 4x^3 + 4xy^2 + 3x^2y \text{ with respect to } y$$

Now, returning to the original function $f = x^4 + 2x^2y^2 + x^3y + y^4$, we calculate

$$f_y = 4x^2y + x^3 + 4y^3 \qquad \text{Partial of } f \text{ with respect to } y$$

Then, from this,

$$f_{yy} = 4x^2 + 12y^2 \qquad \text{Differentiating } f_y = 4x^2y + x^3 + 4y^3 \text{ with respect to } y$$

$$f_{yx} = 8xy + 3x^2 \qquad \text{Differentiating } f_y = 4x^2y + x^3 + 4y^3 \text{ with respect to } x$$

Notice that these "mixed partials" are equal:

$$f_{xy} = 8xy + 3x^2$$
$$f_{yx} = 8xy + 3x^2$$

Equal

That is, $f_{xy} = f_{yx}$, so reversing the order of differentiation (first $x$, then $y$, or first $y$, then $x$) makes no difference. This is not true for all functions, but it is true for all the functions that we will encounter in this book, and it is also true for all functions that you are likely to encounter in applications.*

**Practice Problem 7**

For the function $f(x, y) = x^3 - 3x^2y^4 + y^3$, find:

**a.** $f_x$    **b.** $f_{xy}$    **c.** $f_y$    **d.** $f_{yx}$

➤ **Solutions below**

---

**Solutions to Practice Problems**

**1. a.** $\dfrac{\partial}{\partial x} x^4y^2 = 4x^3y^2$    **b.** $\dfrac{\partial}{\partial y} x^4y^2 = x^4 2y = 2x^4y$

**2.** $\dfrac{\partial}{\partial y} x^2 = 0$

**3.** $\dfrac{\partial}{\partial y}(2x^4 - 3x^3y^3 - y^2 + 4x + 1) = 0 - 3x^3 3y^2 - 2y + 0 = -9x^3y^2 - 2y$

**4.** $f_y(x, y) = e^{x^3 + y^3}(3y^2) = 3y^2 e^{x^3 + y^3}$

$f_y(1, 2) = (3 \cdot 4)e^{1^3 + 2^3} = 12e^9$

**5.** $\dfrac{\partial}{\partial y}(x^3y^4z^5) = x^3 4y^3 z^5 = 4x^3y^3z^5$

**6. a.** Positive    **b.** Negative

**7. a.** $f_x = 3x^2 - 6xy^4$    **b.** $f_{xy} = -6x4y^3 = -24xy^3$

**c.** $f_y = -3x^2 4y^3 + 3y^2 = -12x^2y^3 + 3y^2$    **d.** $f_{yx} = -24xy^3$

---

## 7.2 Section Summary

For a function $f(x, y)$ the partial derivatives are:

$$\frac{\partial}{\partial x} f(x, y) = f_x(x, y) = \lim_{h \to 0} \frac{f(x + h, y) - f(x, y)}{h} \qquad \text{Partial derivative of } f \text{ with respect to } x$$

$$\frac{\partial}{\partial y} f(x, y) = f_y(x, y) = \lim_{h \to 0} \frac{f(x, y + h) - f(x, y)}{h} \qquad \text{Partial derivative of } f \text{ with respect to } y$$

provided that the limits exist. The partial derivative with respect to either variable is found by applying the usual differentiation rules to that variable and treating the other variable as a constant. Partial derivatives (or "partials") can be interpreted as instantaneous rates of change:

$$f_x(x, y) = \left( \begin{array}{c} \text{Instantaneous rate of change of } f \\ \text{with respect to } x \text{ when } y \text{ is held constant} \end{array} \right)$$

$$f_y(x, y) = \left( \begin{array}{c} \text{Instantaneous rate of change of } f \\ \text{with respect to } y \text{ when } x \text{ is held constant} \end{array} \right)$$

*$f_{xy} = f_{yx}$ if these partials are continuous. A more detailed statement can be found in a book on advanced calculus.

Partials can also be interpreted as marginals for one product while production of the other is held constant (see page 480). Geometrically, a function $z = f(x, y)$ of two variables represents a *surface* in three-dimensional space, and the partial $\partial f / \partial x$ gives the *slope* or *steepness* along the surface *in the x-direction*, and the partial $\partial f / \partial y$ gives the slope or steepness *in the y-direction*, as shown in the diagram on page 481.

## 7.2 Exercises

**1–16.** For each function, find the partials
**a.** $f_x(x, y)$ and **b.** $f_y(x, y)$.

**1.** $f(x, y) = x^3 + 3x^2y^2 - 2y^3 - x + y$

**2.** $f(x, y) = 2x^4 - 7x^3y^2 - xy + 1$

**3.** $f(x, y) = 12x^{1/2}y^{1/3} + 8$

**4.** $f(x, y) = x^{-1}y + xy^{-2}$

**5.** $f(x, y) = 100x^{0.05}y^{0.02}$    **6.** $f(x, y) = \dfrac{x}{y}$

**7.** $f(x, y) = (x + y)^{-1}$

**8.** $f(x, y) = (x^2 + xy + 1)^4$

**9.** $f(x, y) = \ln(x^3 + y^3)$    **10.** $f(x, y) = x^2e^y$

**11.** $f(x, y) = 2x^3e^{-5y}$    **12.** $f(x, y) = e^{x+y}$

**13.** $f(x, y) = e^{xy}$    **14.** $f(x, y) = \ln(xy^3)$

**15.** $f(x, y) = \ln\sqrt{x^2 + y^2}$   **16.** $f(x, y) = \dfrac{xy}{x + y}$

**17–20.** For each function, find  **a.** $\dfrac{\partial w}{\partial u}$  and **b.** $\dfrac{\partial w}{\partial v}$.

**17.** $w = (uv - 1)^3$    **18.** $w = (u - v)^3$

**19.** $w = e^{(u^2 - v^2)/2}$    **20.** $w = \ln(u^2 + v^2)$

**21–26.** For each function, evaluate the stated partials.

**21.** $f(x, y) = 4x^3 - 3x^2y^2 - 2y^2$,
 find $f_x(-1, 1)$  and  $f_y(-1, 1)$

**22.** $f(x, y) = 2x^4 - 5x^2y^3 - 4y$,
 find $f_x(1, -1)$  and  $f_y(1, -1)$

**23.** $f(x, y) = e^{x^2+y^2}$,  find  $f_x(0, 1)$  and  $f_y(0, 1)$

**24.** $g(x, y) = (xy - 1)^5$,  find  $g_x(1, 0)$  and  $g_y(1, 0)$

**25.** $h(x, y) = x^2y - \ln(x + y)$,  find  $h_x(1, 1)$

**26.** $f(x, y) = \sqrt{x^2 + y^2}$,  find  $f_y(8, -6)$

**27–32.** For each function, find the second-order partials
**a.** $f_{xx}$,  **b.** $f_{xy}$,  **c.** $f_{yx}$,  and  **d.** $f_{yy}$.

**27.** $f(x, y) = 5x^3 - 2x^2y^3 + 3y^4$

**28.** $f(x, y) = 4x^2 - 3x^3y^2 + 5y^5$

**29.** $f(x, y) = 9x^{1/3}y^{2/3} - 4xy^3$

**30.** $f(x, y) = 32x^{1/4}y^{3/4} - 5x^3y$

**31.** $f(x, y) = ye^x - x \ln y$

**32.** $f(x, y) = y \ln x + xe^y$

**33–34.** For each function, calculate the third-order partials
**a.** $f_{xxy}$,  **b.** $f_{xyx}$,  and  **c.** $f_{yxx}$.

**33.** $f(x, y) = x^4y^3 - e^{2x}$   **34.** $f(x, y) = x^3y^4 - e^{2y}$

**35–40.** For each function of three variables, find the partials
**a.** $f_x$,  **b.** $f_y$,  and  **c.** $f_z$.

**35.** $f = xy^2z^3$    **36.** $f = x^2y^3z^4$

**37.** $f = (x^2 + y^2 + z^2)^4$   **38.** $f = (xyz + 1)^3$

**39.** $f = e^{x^2+y^2+z^2}$    **40.** $f = \ln(x^2 - y^3 + z^4)$

**41–44.** For each function, evaluate the stated partial.

**41.** $f = 3x^2y - 2xz^2$,  find  $f_x(2, -1, 1)$

**42.** $f = 2yz^3 - 3x^2z$,  find  $f_z(2, -1, 1)$

**43.** $f = e^{x^2+2y^2+3z^2}$,  find  $f_y(-1, 1, -1)$

**44.** $f = e^{2x^3+3y^3+4z^3}$,  find  $f_y(1, -1, 1)$

## Applied Exercises

**45–46.** BUSINESS: **Marginal Profit** An electronics company's profit $P(x, y)$ from making $x$ DVD players and $y$ CD players per day is given below.

**a.** Find the marginal profit function for DVD players.
**b.** Evaluate your answer to part (a) at  $x = 200$  and  $y = 300$  and interpret the result.

**c.** Find the marginal profit function for CD players.
**d.** Evaluate your answer to part (c) at  $x = 200$  and  $y = 100$  and interpret the result.

**45.** $P(x, y) = 2x^2 - 3xy + 3y^2 + 150x + 75y + 200$

**46.** $P(x, y) = 3x^2 - 4xy + 4y^2 + 80x + 100y + 200$

**47–48. BUSINESS: Cobb–Douglas Production Functions** A company's production is given by the Cobb–Douglas function $P(L, K)$ below, where $L$ is the number of units of labor and $K$ is the number of units of capital.

a. Find $P_L(27, 125)$ and interpret this number.
b. Find $P_K(27, 125)$ and interpret this number.
c. From your answers to parts (a) and (b), which will increase production more: an additional unit of labor or an additional unit of capital?

**47.** $P(L, K) = 270L^{1/3}K^{2/3}$

**48.** $P(L, K) = 225L^{2/3}K^{1/3}$

**49. BUSINESS: Sales** A store's TV sales depend on $x$, the price of the televisions, and $y$, the amount spent on advertising, according to the function $S(x, y) = 200 - 0.1x + 0.2y^2$. Find and interpret the marginals $S_x$ and $S_y$.

**50. ECONOMICS: Value of an MBA** A study found that a businessperson with a master's degree in business administration (MBA) earned an average salary of $S(x, y) = 48{,}340 + 4930x + 3840y$ dollars in 2005, where $x$ is the number of years of work experience before the MBA, and $y$ is the number of years of work experience after the MBA. Find and interpret the marginals $S_x$ and $S_y$.

**51. SOCIOLOGY: Status** A study found that a person's status in a community depends on the person's income and education according to the function $S(x, y) = 7x^{1/3}y^{1/2}$, where $x$ is income (in thousands of dollars) and $y$ is years of education beyond high school.

a. Find $S_x(27, 4)$ and interpret this number.
b. Find $S_y(27, 4)$ and interpret this number.

**52. BIOMEDICAL: Blood Flow** The resistance of blood flowing through an artery of radius $r$ and length $L$ (both in centimeters) is $R(r, L) = 0.08Lr^{-4}$.

a. Find $R_r(0.5, 4)$ and interpret this number.
b. Find $R_L(0.5, 4)$ and interpret this number.

*Source: E. Batschelet, Introduction to Mathematics for Life Scientists*

**53. GENERAL: Highway Safety** The length in feet of the skid marks from a truck of weight $w$ (tons) traveling at velocity $v$ (miles per hour) skidding to a stop on a dry road is $S(w, v) = 0.027wv^2$.

a. Find $S_w(4, 60)$ and interpret this number.
b. Find $S_v(4, 60)$ and interpret this number.

*Source: National Highway Traffic Safety Administration*

**54. GENERAL: Windchill Temperature** The windchill temperature announced by the weather bureau during the cold weather measures how cold it "feels" for a given temperature and wind speed. The formula is $C(t, w) = 35.74 + 0.6215t - 35.75w^{0.16} + 0.4275tw^{0.16}$ where $t$ is the temperature (in degrees Fahrenheit) and $w$ is the wind speed (in miles per hour). Find and interpret $C_w(30, 20)$.

*Source: National Weather Service*

**55. ATHLETICS: Pythagorean Baseball Standings** The Pythagorean standing of a baseball team is given by Bill James's formula $f(x, y) = x^2/(x^2 + y^2)$ where $x$ is the number of runs scored and $y$ is the number of runs allowed.

a. Find $f_x(400, 300)$, giving the rate of change of the standings per additional run scored (for 400 runs scored and 300 allowed).
b. Multiply your answer to part (a) by 20 to find the change that would result from 20 additional runs scored, and then multiply this result by 160 (the approximate number of games per season) to estimate the number of games that would be won by those 20 additional runs.

*Source: B. James, Baseball Abstract 1983*

**56. ATHLETICS: Olympics** A country's share of medals at an Olympic games can be estimated from the formula $f(x, y) = 0.0062 \ln x + 0.0064 \ln y - 0.0652$, where $x$ is the population and $y$ is the per capita gross domestic product (PCGDP) of the country.

a. Find $f_y(x, y)$ and evaluate it at $y = 1000$ to find the rate of change in the proportion of medals per extra dollar when PCGDP is \$1000.
b. Multiply your answer to part (a) by 500 to find the change in the proportion that would result from an additional \$500 in PCGDP, and then multiply this result by 920 (the number of medals at a typical Olympic games) to estimate the number of additional Olympic medals that would be won.

*Source: Review of Economics and Statistics 86*

## Conceptual Exercises

**57.** Find $\dfrac{\partial}{\partial x} e^{\sqrt{1 + 5y^{-23}}} \ln (1 - 17y^{97}/z^{47})$

**58.** Find $\dfrac{\partial}{\partial y} 108^{17x^2}/(x^{23} - 2/z^8)^{13}$

**59.** You are walking on a surface $z = f(x, y)$, and for each unit that you walk in the $x$-direction you rise 3 units and for each unit that you walk in the $y$-direction you fall 2 units. Find the partial derivatives of $f(x, y)$.

**60.** Your company makes two products, called X and Y, in quantities $x$ and $y$. If your profit function satisfies $\dfrac{\partial P}{\partial x} = 12$ and $\dfrac{\partial P}{\partial y} = -8$ (both in dollars), how much money are you making on each of your two products?

**61.** Look at the graph of the surface on page 481. Is $\dfrac{\partial f}{\partial x}$ positive or negative at the point shown? Same question for $\dfrac{\partial f}{\partial y}$.

A *linear function of two variables* is of the form $f(x, y) = ax + by + c$ where $a$, $b$, and $c$ are constants. Find the linear function of two variables satisfying the following conditions.

**62.** $\dfrac{\partial f}{\partial x} = 3$, $\dfrac{\partial f}{\partial y} = -2$, and $f(0, 0) = 5$

**63.** $\dfrac{\partial f}{\partial x} = -1$, $\dfrac{\partial f}{\partial y} = 1$, and $f(0, 0) = 0$

**64.** $\dfrac{\partial f}{\partial x} = 0$, $\dfrac{\partial f}{\partial y} = 0$, and $f(100, 100) = 100$

**65.** $\dfrac{\partial f}{\partial x} = \pi$, $\dfrac{\partial f}{\partial y} = e$, and $f(0, 0) = \ln 2$

**66.** Find all second partial derivatives of a linear function of two variables.

**67–68. SOCIAL SCIENCE: Procrastination** A recent ten-year study of procrastination found that if you have a task to do, your desire to complete the task (denoted $D$) is given by $D = \dfrac{E \cdot V}{T \cdot P}$, where $E$ is the expectation of success, $V$ is the value of completing the task, $T$ is the time needed to complete the task, and $P$ is your tendency to procrastinate, all of which are positive quantities.

*Source: Scientific American, 2007*

**67.** Find the signs of $\dfrac{\partial D}{\partial E}$ and $\dfrac{\partial D}{\partial T}$ and interpret these signs.

**68.** Find the signs of $\dfrac{\partial D}{\partial V}$ and $\dfrac{\partial D}{\partial P}$ and interpret these signs.

## Explorations and Excursions

The following problems extend and augment the material presented in the text.

*Competitive and Complementary Commodities*

**69. ECONOMICS: Competitive Commodities** Certain commodities (such as butter and margarine) are called "competitive" or

"substitute" commodities because one can substitute for the other. If $B(b, m)$ gives the daily sales of butter as a function of $b$, the price of butter, and $m$, the price of margarine:

**a.** Give an interpretation of $B_b(b, m)$.
**b.** Would you expect $B_b(b, m)$ to be positive or negative? Explain.
**c.** Give an interpretation of $B_m(b, m)$.
**d.** Would you expect $B_m(b, m)$ to be positive or negative? Explain.

**70. ECONOMICS: Complementary Commodities** Certain commodities (such as washing machines and clothes dryers) are called "complementary" commodities because they are often used together. If $D(d, w)$ gives the monthly sales of dryers as a function of $d$, the price of dryers, and $w$, the price of washers:

**a.** Give an interpretation of $D_d(d, w)$.
**b.** Would you expect $D_d(d, w)$ to be positive or negative? Explain.
**c.** Give an interpretation of $D_w(d, w)$.
**d.** Would you expect $D_w(d, w)$ to be positive or negative? Explain.

*Harmonic Functions*

A function $f(x, y)$ that has continuous second derivatives and that satisfies $f_{xx} + f_{yy} = 0$ is called a *harmonic function*. Harmonic functions have many interesting properties, including the fact that their value at the center of any circle is the average of their values around the circumference of the circle. So if you stretch the edges of a flexible rubber sheet, the shape will be a harmonic function. Find whether each of the following functions is harmonic.

**71.** $f(x, y) = x^2 - y^2$   **72.** $f(x, y) = xy$
**73.** $f(x, y) = x^2 + y^2$   **74.** $f(x, y) = x^3 - 3xy^2$

*Finding Partial Derivatives from the Definition*

Use the *definition* of partial derivatives on page 474 to find both $\dfrac{\partial}{\partial x} f(x, y)$ and $\dfrac{\partial}{\partial y} f(x, y)$ for each of the following functions. [*Hint:* The method is the same as that on page 100 but applied to just *one* variable at a time.]

**75.** $f(x, y) = 2x^2 + 3y^2 - 3x + 7y + 4$
**76.** $f(x, y) = 3x^2 - 2y^2 + 5x - 4y - 9$
**77.** $f(x, y) = 5x^2 - 3x + 7y + 4$
**78.** $f(x, y) = 4y^2 + 2x - 5y + 12$

## 7.3 OPTIMIZING FUNCTIONS OF SEVERAL VARIABLES

### Introduction

The graph of a function of two variables is a *surface*, with **relative maximum** and **minimum points** ("hilltops" and "valley bottoms") and **saddle points** ("mountain passes"). In this section we will see how to maximize and minimize such functions by finding **critical points** and using a two-variable version of the second-derivative test.

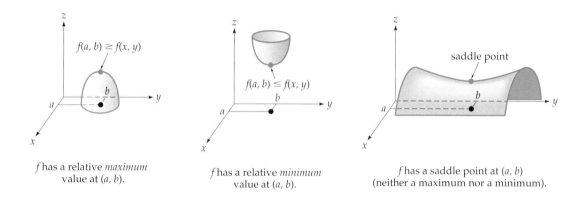

$f$ has a relative *maximum* value at $(a, b)$.

$f$ has a relative *minimum* value at $(a, b)$.

$f$ has a saddle point at $(a, b)$ (neither a maximum nor a minimum).

Such optimization techniques have many applications, such as maximizing profit for a company that makes several products. For simplicity, we will consider only functions whose first and second partials are defined everywhere.

### Critical Points

A critical point (both partials are zero)

At the very top of a smooth hill, the slope or steepness in any direction must be zero. That is, a straight stick would balance horizontally at the top. Since the partials $f_x$ and $f_y$ are the slopes in the $x$- and $y$-directions, these partials must both be zero at a relative maximum point, and similarly for a relative minimum point or a saddle point. A point $(a, b)$ at which both partials are zero is called a *critical point* of the function.*

---

**Critical Point**

$(a, b)$ is a critical point of $f(x, y)$ if

$$f_x(a, b) = 0 \quad \text{and} \quad f_y(a, b) = 0$$

Both first partials are zero

Relative maximum and minimum values can occur only at critical points.

---

*We use the term "critical point" for the *pair* of values $(a, b)$ at which the two partials are zero. The corresponding point on the graph has *three* coordinates $(a, b, f(a,b))$.

**EXAMPLE 1**

**FINDING CRITICAL POINTS**

Find all critical points of

$$f(x, y) = 3x^2 + y^2 + 3xy + 3x + y + 6$$

*Solution*

We want all points at which both partials are zero.

$$f_x = 6x + 3y + 3 \quad \text{and} \quad f_y = 2y + 3x + 1 \qquad \text{Partials}$$

$$\begin{aligned} 6x + 3y + 3 &= 0 \\ 3x + 2y + 1 &= 0 \end{aligned} \left. \begin{aligned} \\ \end{aligned} \right\} \begin{aligned} &\text{Partials} \\ &\text{set equal to zero} \end{aligned}$$

Reordered so the *x*- and *y*-terms line up

To solve these equations simultaneously, we multiply the second by $-2$ so that the *x*-terms drop out when we add.

| | |
|---|---|
| $6x + 3y + 3 = 0$ | First equation |
| $-6x - 4y - 2 = 0$ | Second equation times $-2$ |
| $-y + 1 = 0$ | Adding (*x* drops out) |
| $y = 1$ | From solving $-y + 1 = 0$ |

Substituting $y = 1$ into either equation gives *x*:

| | |
|---|---|
| $6x + 3 + 3 = 0$ | Substituting $y = 1$ into $6x + 3y + 3 = 0$ |
| $6x = -6$ | Simplifying |
| $x = -1$ | Solving |

These *x*- and *y*-values give one critical point.

$$\text{CP: } (-1, 1) \qquad \text{From } x = -1, \ y = 1$$

## Second-Derivative Test for Functions $f(x, y)$: The *D*-Test

To determine whether $f(x, y)$ has a relative maximum, a relative minimum, or a saddle point at a critical point, we use the following **D-test,** which is a generalization of the second-derivative test on page 200.

---

**D-Test**

If $(a, b)$ is a critical point of the function $f$ and if $D$ is defined by

$$D = f_{xx}(a, b) \cdot f_{yy}(a, b) - [f_{xy}(a, b)]^2$$

More briefly,
$D = f_{xx}f_{yy} - (f_{xy})^2$

then $f$ at the point $(a, b)$ has a:

 **i.** relative *maximum* if $D > 0$ and $f_{xx}(a, b) < 0$ 

 **ii.** relative *minimum* if $D > 0$ and $f_{xx}(a, b) > 0$

Different signs for $f_{xx}(a, b)$

 **iii.** *saddle point* if $D < 0$.

⚠ BE CAREFUL: Be sure to apply the *D*-test correctly:

1. The *D*-test is used only *after* finding the critical points. The test is then applied to each critical point, one at a time.

2. $D > 0$ appears only in parts i and ii, and so guarantees that the function has either a relative maximum or a relative minimum. You then use the "old" second-derivative test (checking the sign of $f_{xx}$) to determine which one occurs (maximum or minimum).

3. $D < 0$ means a *saddle point*, regardless of the sign of $f_{xx}$. (A saddle point is neither a maximum nor a minimum.)

4. $D = 0$ means that the *D*-test is *inconclusive*—the function may have a maximum, a minimum, or a saddle point at the critical point.

---

**EXAMPLE 2**

**FINDING RELATIVE EXTREME VALUES OF A POLYNOMIAL**

Find the relative extreme values of

$$f(x, y) = 3x^2 + y^2 + 3xy + 3x + y + 6$$

*Solution*

We find critical points by setting the two partials equal to zero and solving. But we did this for the same function in Example 1, finding one critical point, $(-1, 1)$. For the *D*-test, we calculate the second partials.

$$f_{xx} = 6 \qquad\qquad \text{From } f_x = 6x + 3y + 3$$
$$f_{yy} = 2 \qquad\qquad \text{From } f_y = 2y + 3x + 1$$
$$f_{xy} = 3 \qquad\qquad \text{From } f_x = 6x + 3y + 3$$

Calculating *D*:

$$D = 6 \cdot 2 - (3)^2 = 12 - 9 = 3 \qquad D = f_{xx}f_{yy} - (f_{xy})^2$$

$\quad\quad\quad$ / \ \ $\qquad\qquad\qquad\qquad$ ↖ Positive
$\quad\quad f_{xx}\ f_{yy}\ f_{xy}$

*D* is positive and $f_{xx}$ is positive (since $f_{xx} = 6$), so *f* has a *relative minimum* (part ii of the *D*-test) at the critical point $(-1, 1)$. (If $f_{xx}$ had been negative, there would have been a relative *maximum*.) The relative minimum *value* is found by evaluating $f(x, y)$ at $(-1, 1)$.

$$f(-1, 1) = 3 + 1 - 3 - 3 + 1 + 6 \qquad \begin{array}{l} f = 3x^2 + y^2 + 3xy + 3x + y + 6 \\ \text{evaluated at } x = -1, \ y = 1 \end{array}$$
$$= 5$$

Relative minimum value: $f = 5$ at $x = -1$, $y = 1$.

**EXAMPLE 3**

### FINDING RELATIVE EXTREME VALUES OF AN EXPONENTIAL FUNCTION

Find the relative extreme values of $f(x, y) = e^{x^2 - y^2}$.

*Solution*

$$\left. \begin{array}{l} f_x = e^{x^2 - y^2}(2x) \\ f_y = e^{x^2 - y^2}(-2y) \end{array} \right\} \text{ Partials}$$

$$\left. \begin{array}{l} e^{x^2 - y^2}(2x) = 0 \\ e^{x^2 - y^2}(-2y) = 0 \end{array} \right\} \text{ Partials set equal to zero}$$

$$\text{CP: } (0, 0) \quad \text{Since } \begin{cases} e^{x^2 - y^2}(2x) = 0 \\ e^{x^2 - y^2}(-2y) = 0 \end{cases} \quad \text{only at } x = 0, y = 0$$

For the *D*-test we calculate the second partials:

$$f_{xx} = e^{x^2 - y^2}(2x)(2x) + e^{x^2 - y^2}(2) \qquad \text{From } f_x = e^{x^2 - y^2}(2x) \text{ using the Product Rule}$$

$$= 4x^2 e^{x^2 - y^2} + 2e^{x^2 - y^2} \qquad \text{Simplifying}$$

$$f_{yy} = e^{x^2 - y^2}(-2y)(-2y) + e^{x^2 - y^2}(-2) \qquad \text{From } f_y = e^{x^2 - y^2}(-2y) \text{ using the Product Rule}$$

$$= 4y^2 e^{x^2 - y^2} - 2e^{x^2 - y^2} \qquad \text{Simplifying}$$

$$f_{xy} = e^{x^2 - y^2}(2x)(-2y) = -4xy e^{x^2 - y^2} \qquad \text{From } f_x = e^{x^2 - y^2}(2x) \text{ treating } x \text{ as a constant}$$

Evaluating at the critical point (0, 0):

$$f_{xx}(0, 0) = 0e^0 + 2e^0 = 0 + 2 = 2 \qquad f_{xx} = 4x^2 e^{x^2 - y^2} + 2e^{x^2 - y^2} \text{ at } (0, 0)$$

$$f_{yy}(0, 0) = 0e^0 - 2e^0 = 0 - 2 = -2 \qquad f_{yy} = 4y^2 e^{x^2 - y^2} - 2e^{x^2 - y^2} \text{ at } (0, 0)$$

$$f_{xy}(0, 0) = 0e^0 = 0 \qquad f_{xy} = -4xy e^{x^2 - y^2} \text{ at } (0, 0)$$

Therefore *D* is

$$D = (2)(-2) - (0)^2 = -4 - 0 = -4 \qquad D = f_{xx}f_{yy} - (f_{xy})^2$$

$$\underset{f_{xx}}{\diagup} \quad \underset{f_{yy}}{\diagdown} \quad \underset{f_{xy}}{\diagdown} \qquad \nwarrow \text{Negative}$$

Since *D* is negative, the function has a *saddle point* (part iii of the *D*-test) at the critical point (0, 0).

*f* has no relative extreme values
(it has a saddle point at $x = 0, \ y = 0$).

## Maximizing Profit

If a company makes too few of its products, the resulting lost sales will lower the company's profits. On the other hand, making too many will "flood the market" and depress prices, again resulting in lower profits. Therefore, any realistic profit function must have a maximum at some "intermediate" point $(x, y)$, which therefore must be a *relative* maximum point. Many applied problems are solved in this way: by knowing that the *absolute* extreme values (the highest and lowest values on the entire domain) must exist, and finding them as *relative* extreme points.

Suppose that a company produces two products, A and B, and that the two price functions are

$$p(x) = \left( \begin{array}{c} \text{Price at which exactly } x \text{ units} \\ \text{of product A will be sold} \end{array} \right)$$

and

$$q(y) = \left( \begin{array}{c} \text{Price at which exactly } y \text{ units} \\ \text{of product B will be sold} \end{array} \right)$$

If $C(x, y)$ is the (total) cost function, then the company's profit will be

$$\underbrace{P(x, y)}_{\text{Profit}} = \underbrace{p(x) \cdot x}_{\substack{\text{Price times} \\ \text{quantity for} \\ \text{product A}}} + \underbrace{q(y) \cdot y}_{\substack{\text{Price times} \\ \text{quantity for} \\ \text{product B}}} - \underbrace{C(x, y)}_{\text{Cost}}$$

Revenue for each product (price times quantity) minus the cost function

---

**EXAMPLE 4**

### MAXIMIZING PROFIT FOR A COMPANY

Universal Motors makes compact and midsized cars. The price function for compacts is $p = 17 - 2x$ (for $0 \le x \le 8$), and the price function for midsized cars is $q = 20 - y$ (for $0 \le y \le 20$), both in thousands of dollars, where $x$ and $y$ are, respectively, the numbers of compact and midsized cars produced per hour. If the company's cost function is

$$C(x, y) = 15x + 16y - 2xy + 5$$

thousand dollars, find how many of each car should be produced and the prices that should be charged in order to maximize profit. Also find the maximum profit.

*Solution*

The profit function is

$$P(x, y) = (\underbrace{17 - 2x}_{\text{Price}})\underbrace{x}_{\text{Quantity}} + (\underbrace{20 - y}_{\text{Price}})\underbrace{y}_{\text{Quantity}} - \underbrace{(15x + 16y - 2xy + 5)}_{\text{Cost}}$$

For compacts        For midsized

$$= 17x - 2x^2 + 20y - y^2 - 15x - 16y + 2xy - 5 \qquad \text{Multiplying out}$$
$$= -2x^2 - y^2 + 2xy + 2x + 4y - 5 \qquad \text{Simplifying}$$

We maximize $P(x, y)$ in the usual way:

$$P_x = -4x + 2y + 2$$
$$P_y = -2y + 2x + 4$$

$\left.\begin{array}{c} \\ \\ \end{array}\right\}$ Partials

$$-4x + 2y + 2 = 0 \qquad \text{Partials set equal to zero}$$
$$\underline{2x - 2y + 4 = 0} \qquad \text{Rearranged to line up } x\text{'s and } y\text{'s}$$
$$-2x \qquad + 6 = 0 \qquad \text{Adding (the } y\text{'s cancel)}$$
$$x = 3 \qquad \text{From solving } -2x + 6 = 0$$
$$y = 5 \qquad \begin{array}{l}\text{From substituting } x = 3 \text{ into either} \\ \text{equation (omitting the details)}\end{array}$$

These two values give one critical point.

$$\text{CP: } (3, 5)$$

For the $D$-test we calculate the second partials:

$$P_{xx} = -4 \qquad P_{xy} = 2 \qquad P_{yy} = -2 \qquad \begin{array}{l}\text{From } P_x = -4x + 2y + 2 \\ \text{and } P_y = -2y + 2x + 4\end{array}$$
$$D = (-4)(-2) - (2)^2 = 4 \qquad D = P_{xx}P_{yy} - (P_{xy})^2$$

$D$ is positive and $P_{xx} = -4$ is negative, so profit is indeed *maximized* at $x = 3$ and $y = 5$. To find the prices, we evaluate the price functions:

$$p = 17 - 2 \cdot 3 = 11 \qquad \text{(thousand dollars)} \qquad \begin{array}{l}p = 17 - 2x \\ \text{evaluated at } x = 3\end{array}$$

$$q = 20 - 5 = 15 \qquad \text{(thousand dollars)} \qquad \begin{array}{l}q = 20 - y \\ \text{evaluated at } y = 5\end{array}$$

The profit comes from the profit function:

$$P(3, 5) = \underbrace{-2 \cdot 3^2 - 5^2 + 2 \cdot 3 \cdot 5 + 2 \cdot 3 + 4 \cdot 5 - 5}$$
$$\begin{array}{c}P = -2x^2 - y^2 + 2xy + 2x + 4y - 5 \\ \text{evaluated at } x = 3, \ y = 5\end{array}$$

$$= 8 \text{ (thousand dollars)}$$

Profit is maximized when the company produces 3 compacts per hour, selling them for $11,000 each, and 5 midsized cars per hour, selling them for $15,000 each. The maximum profit will be $8000 per hour.

---

The $D$-test ensures that you have *maximized* profit rather than minimized it.*
Some functions have *more* than one critical point.

*How are the price and cost functions in such problems found? Price functions may be constructed by the methods used on pages 221–223 or by more sophisticated techniques of market research. Cost functions may be found simply as the sum of the unit cost times the number of units for each product, or by regression techniques based on the least squares method described in the following section for constructing functions from data.

**EXAMPLE 5**

**FINDING RELATIVE EXTREME VALUES**

Find the relative extreme values of

$$f(x, y) = x^2 + y^3 - 6x - 12y$$

*Solution*

$$2x - 6 = 0 \qquad\qquad f_x = 0$$
$$3y^2 - 12 = 0 \qquad\qquad f_y = 0$$

The first gives

$$x = 3 \qquad\qquad \text{Solving } 2x - 6 = 0$$

and the second gives

$$3y^2 = 12 \qquad\qquad \begin{array}{l}\text{Adding 12 to each}\\\text{side of } 3y^2 - 12 = 0\end{array}$$

$$y^2 = 4 \qquad\qquad \text{Dividing by 3}$$

$$y = \pm 2 \qquad\qquad \text{Taking square roots}$$

From $x = 3$ and $y = \pm 2$ we get *two* critical points:

$$\text{CP: } (3, 2) \quad \text{and} \quad (3, -2)$$

Calculating the second partials and substituting them into $D$:

$$D = (2)(6y) - (0)^2 = 12y \qquad \begin{array}{l}D = (f_{xx})(f_{yy}) - (f_{xy})^2 \text{ with}\\ f_{xx} = 2, \ f_{yy} = 6y, \ f_{xy} = 0\end{array}$$

We apply the $D$-test to the critical points one at a time.

At $(3, 2)$: $\quad D = 12 \cdot 2 > 0$ $\qquad\qquad \begin{array}{l}D = 12y \text{ evaluated}\\\text{at } (3, 2)\end{array}$

$$f_{xx} = 2 > 0$$

*relative minimum* at $x = 3, \ y = 2$ $\qquad \begin{array}{l}\text{Since } D \text{ and } f_{xx} \text{ are}\\\text{both positive}\end{array}$

At $(3, -2)$: $\ D = 12 \cdot (-2) < 0$ $\qquad \begin{array}{l}D = 12y \text{ evaluated}\\\text{at } (3, -2)\end{array}$

*saddle point* at $x = 3, \ y = -2$ $\qquad\qquad$ Since $D$ is negative

*Answer:*

Relative minimum value: $f = -25$ at $\begin{cases} x = 3 \\ y = 2 \end{cases}$ $\quad \begin{array}{l}f = -25 \text{ from}\\ f = x^2 + y^3 - 6x - 12y\\\text{at } (3, 2)\end{array}$
(saddle point at $x = 3, \ y = -2$)

**Competition and Collusion**

In 1938, the French economist Antoine Cournot published the following comparison of a **monopoly** (a market with only one supplier) and a **duopoly** (a market with two suppliers).

*Monopoly.* Imagine that you are selling spring water from your own spring (or any product whose cost of production is negligible). Since you are the only supplier in town, you have a "monopoly." Suppose that your price function is $p = 6 - 0.01x$, where $p$ is the price in dollars at which you will sell precisely $x$ gallons $(0 \leq x \leq 600)$. Your revenue is then

$$R(x) = (6 - 0.01x)x = 6x - 0.01x^2 \qquad \begin{array}{l}\text{Price } (6 - 0.01x) \\ \text{times quantity } x\end{array}$$

You maximize revenue by setting its derivative equal to zero:

$$R'(x) = 6 - 0.02x = 0$$

$$x = \frac{6}{0.02} = 300 \qquad\qquad \text{Solving } 6 - 0.02x = 0$$

Therefore, you should sell 300 gallons per day. (The second-derivative test will verify that revenue is maximized.) The price will be

$$p = 6 - 0.01 \cdot 300 = 6 - 3 = 3 \qquad \begin{array}{l}p = 6 - 0.01x \\ \text{evaluated at } x = 300\end{array}$$

or $3 dollars per gallon. Your maximum revenue will be

$$R(300) = 6 \cdot 300 - 0.01(300)^2 = \$900 \qquad \begin{array}{l}R(x) = 6x - 0.01x^2 \\ \text{evaluated at } x = 300\end{array}$$

*Duopoly.* Suppose now that your neighbor opens a competing spring water business. (A market such as this with two suppliers is called a "duopoly.") Now both of you must share the same market. If your neighbor sells $y$ gallons per day (and you sell $x$), you must both sell at price

$$p = 6 - 0.01(x + y) = 6 - 0.01x - 0.01y \qquad \begin{array}{l}\text{Price function } p = 6 - 0.01x \\ \text{with } x \text{ replaced by the} \\ \text{combined quantity } x + y\end{array}$$

Each of you calculates revenue as price times quantity:

$$\left(\begin{array}{c}\text{Your} \\ \text{revenue}\end{array}\right) = p \cdot x = (6 - 0.01x - 0.01y)x = 6x - 0.01x^2 - 0.01xy$$

$$\left(\begin{array}{c}\text{Neighbor's} \\ \text{revenue}\end{array}\right) = p \cdot y = (6 - 0.01x - 0.01y)y = 6y - 0.01xy - 0.01y^2$$

You each want to maximize revenue, so you set the partials equal to zero:

$$6 - 0.02x - 0.01y = 0 \qquad \begin{array}{l}\text{Partial of } 6x - 0.01x^2 - 0.01xy \\ \text{with respect to } x, \text{ set equal to zero}\end{array}$$

$$6 - 0.01x - 0.02y = 0 \qquad \begin{array}{l}\text{Partial of } 6y - 0.01xy - 0.01y^2 \\ \text{with respect to } y, \text{ set equal to zero}\end{array}$$

These are easily solved by multiplying one of them by 2 and subtracting the other. The solution (omitting the details) is

$$x = 200 \qquad y = 200$$

so each of you sells 200 gallons per day. The selling price for both will be

$$p = 6 - 0.01(200 + 200) = 6 - 4 = 2 \qquad \begin{array}{l}p = 6 - 0.01 (x + y) \\ \text{evaluated at } x = 200, \; y = 200\end{array}$$

or $2 per gallon. Revenue is price ($2) times quantity (200), resulting in $400 for each of you.

### Comparison of the Monopoly and the Duopoly

The two systems may be compared as follows:

|  | Monopoly | Duopoly |
|---|---|---|
| Quantity: | 300 gallons | 200 gallons each, 400 total |
| Price: | $3 per gallon | $2 per gallon |
| Revenue: | $900 | $400 each, $800 total |

Notice that the duopoly produces *more* than the monopoly (400 gallons versus only 300 in the monopoly), and does so at a *lower price* ($2 versus $3 in the monopoly). Cournot therefore concluded that consumers benefit more from a duopoly than from a monopoly.

However, the smart duopolist will notice that the revenue of $400 is less than half of the monopoly revenue of $900. Therefore, it benefits each duopolist to cooperate and share the market as a single monopoly. This is called *collusion*. It is for this reason that markets with only a few suppliers tend toward collusion rather than competition.

### Optimizing Functions of Three or More Variables

Functions of *more* than two variables are optimized in the same way: by finding critical points (where all of the first partials are zero). For example, to optimize a function $f(x, y, z)$ we would set the three partials equal to zero and solve:

$$f_x = 0$$
$$f_y = 0$$
$$f_z = 0$$

The second-derivative test for functions of three or more variables is very complicated, and we shall not discuss it.

## 7.3  Section Summary

To optimize a function $f(x, y)$ of two variables, first find all critical points, the points $(a, b)$ where both first partials are zero.

$$f_x(a, b) = 0$$
$$f_y(a, b) = 0$$

Then apply the *D*-test (page 489) to each critical point to determine whether the function has a relative maximum, a relative minimum, or a saddle point at that critical point.

## 7.3 Exercises

**1–20.** Find the relative extreme values of each function.

**1.** $f(x, y) = x^2 + 2y^2 + 2xy + 2x + 4y + 7$

**2.** $f(x, y) = 2x^2 + y^2 + 2xy + 4x + 2y + 5$

**3.** $f(x, y) = 2x^2 + 3y^2 + 2xy + 4x - 8y$

**4.** $f(x, y) = 3x^2 + 2y^2 + 2xy + 8x - 4y$

**5.** $f(x, y) = 3xy - 2x^2 - 2y^2 + 14x - 7y - 5$

**6.** $f(x, y) = 2xy - 2x^2 - 3y^2 + 4x - 12y + 5$

**7.** $f(x, y) = xy + 4x - 2y + 1$

**8.** $f(x, y) = 5xy - 2x^2 - 3y^2 + 5x - 7y + 10$

**9.** $f(x, y) = 3x - 2y - 6$

**10.** $f(x, y) = 5x - 4y + 5$

**11.** $f(x, y) = e^{(x^2 + y^2)/2}$    **12.** $f(x, y) = e^{5(x^2 + y^2)}$

**13.** $f(x, y) = \ln(x^2 + y^2 + 1)$

**14.** $f(x, y) = \ln(2x^2 + 3y^2 + 1)$

**15.** $f(x, y) = -x^3 - y^2 + 3x - 2y$

**16.** $f(x, y) = x^3 - y^2 - 3x + 6y$

**17.** $f(x, y) = y^3 - x^2 - 2x - 12y$

**18.** $f(x, y) = -x^2 - y^3 - 6x + 3y + 4$

**19.** $f(x, y) = x^3 - 2xy + 4y$

**20.** $f(x, y) = y^3 - 2xy - 4x$

## Applied Exercises

**21. BUSINESS: Maximum Profit** A company manufactures two products. The price function for product A is $p = 12 - \frac{1}{2}x$ (for $0 \le x \le 24$), and for product B is $q = 20 - y$ (for $0 \le y \le 20$), both in thousands of dollars, where $x$ and $y$ are the amounts of products A and B, respectively. If the cost function is

$$C(x, y) = 9x + 16y - xy + 7$$

thousands of dollars, find the quantities and the prices of the two products that maximize profit. Also find the maximum profit.

**22. BUSINESS: Maximum Profit** A company manufactures two products. The price function for product A is $p = 16 - x$ (for $0 \le x \le 16$), and for product B is $q = 19 - \frac{1}{2}y$ (for $0 \le y \le 38$), both in thousands of dollars, where $x$ and $y$ are the amounts of product A and B, respectively. If the cost function is

$$C(x, y) = 10x + 12y - xy + 6$$

thousand dollars, find the quantities and the prices of the two products that maximize profit. Also find the maximum profit.

**23. BUSINESS: Price Discrimination** An automobile manufacturer sells cars in America and Europe, charging different prices in the two markets. The price function for cars sold in America is $p = 20 - 0.2x$ thousand dollars (for $0 \le x \le 100$), and the price function for cars sold in Europe is $q = 16 - 0.1y$ thousand dollars (for $0 \le y \le 160$), where $x$ and $y$ are the numbers of cars sold per day in America and Europe, respectively. The company's cost function is

$$C = 20 + 4(x + y) \qquad \text{thousand dollars}$$

a. Find the company's profit function. [*Hint:* Profit is revenue from America plus revenue from Europe minus costs, where each revenue is price times quantity.]

b. Find how many cars should be sold in each market to maximize profit. Also find the price for each market.

**24. BIOMEDICAL: Drug Dosage** In a laboratory test the combined antibiotic effect of $x$ milligrams of medicine A and $y$ milligrams of medicine B is given by the function

$$f(x, y) = xy - 2x^2 - y^2 + 110x + 60y$$

(for $0 \le x \le 55, 0 \le y \le 60$). Find the amounts of the two medicines that maximize the antibiotic effect.

**25. PSYCHOLOGY: Practice and Rest** A subject in a psychology experiment who practices a skill for

$x$ hours and then rests for $y$ hours achieves a test score of

$$f(x, y) = xy - x^2 - y^2 + 11x - 4y + 120$$

(for $0 \le x \le 10, 0 \le y \le 4$). Find the numbers of hours of practice and rest that maximize the subject's score.

26. **SOCIOLOGY: Absenteeism** The number of office workers near a beach resort who call in "sick" on a warm summer day is

$$f(x, y) = xy - x^2 - y^2 + 110x + 50y - 5200$$

where $x$ is the air temperature $(70 \le x \le 100)$ and $y$ is the water temperature $(60 \le y \le 80)$. Find the air and water temperatures that maximize the number of absentees.

**27–28. ECONOMICS: Competition and Collusion** Compare the outputs of a monopoly and a duopoly by repeating the analysis on pages 494–496 for the following price function. That is:

a. For a monopoly, calculate the quantity $x$ that maximizes your revenue. Also calculate the price $p$ and the revenue $R$.

b. For the duopoly, calculate the quantities $x$ and $y$ that maximize revenue for each duopolist. Calculate the price $p$ and the two revenues.

c. Are more goods produced under a monopoly or a duopoly?

d. Is the price lower under a monopoly or a duopoly?

27. $p = 12 - 0.005x$ dollars $\quad (0 \le x \le 2400)$

28. $p = a - bx$ dollars (for positive numbers $a$ and $b$ with $0 \le x \le a/b$)

29. **BUSINESS: Maximum Profit** An automobile dealer can sell 8 sedans per day at a price of $20,000 and 4 SUVs (sport utility vehicles) per day at a price of $25,000. She estimates that for each $400 decrease in price of the sedans she can sell two more per day, and for each $600 decrease in price for the SUVs she can sell one more. If each sedan costs her $16,800 and each SUV costs her $19,000, and fixed costs are $1000 per day, what price should she charge for the sedans and the SUVs to maximize profit? How many of each type will she sell at these prices? [*Hint:* Let $x$ be the number of $400 price decreases for sedans and $y$ be the number of $600 price decreases for SUVs, and use the method of Examples 1 and 2 on pages 221–222 for each type of car.]

30. **BUSINESS: Maximum Revenue** An airline flying to a Midwest destination can sell 20 coach-class tickets per day at a price of $250 and six business-class tickets per day at a price of $750. It finds that for each $10 decrease in the price of the coach ticket, it will sell four more per day, and for each $50 decrease in the business-class price, it will sell two more per day. What prices should the airline charge for the coach- and business-class tickets to maximize revenue? How many of each type will be sold at these prices? [*Hint:* Let $x$ be the number of $10 price decreases for coach tickets and $y$ be the number of $50 price decreases for business-class tickets, and use the method of Examples 1 and 2 on pages 221–222 for each type of ticket.]

## Optimizing Functions of Three Variables

31. **BUSINESS: Price Discrimination** An automobile manufacturer sells cars in America, Europe, and Asia, charging a different price in each of the three markets. The price function for cars sold in America is $p = 20 - 0.2x$ (for $0 \le x \le 100$), the price function for cars sold in Europe is $q = 16 - 0.1y$ (for $0 \le y \le 160$), and the price function for cars sold in Asia is $r = 12 - 0.1z$ (for $0 \le z \le 120$), all in thousands of dollars, where $x$, $y$, and $z$ are the numbers of cars sold in America, Europe, and Asia, respectively. The company's cost function is $C = 22 + 4(x + y + z)$ thousand dollars.

a. Find the company's profit function $P(x, y, z)$. [*Hint:* The profit will be revenue from America plus revenue from Europe plus revenue from Asia minus costs, where each revenue is price times quantity.]

b. Find how many cars should be sold in each market to maximize profit. [*Hint:* Set the three partials $P_x$, $P_y$, and $P_z$ equal to zero and solve. Assuming that the maximum exists, it must occur at this point.]

32. **ECONOMICS: Competition and Collusion** Suppose that in the discussion of competition and collusion (pages 494–496), *two* of your neighbors began selling spring water. Use the price function $p = 36 - 0.01x$ (for $0 \le x \le 3600$) and repeat the analysis, but now comparing a monopoly with competition among *three* suppliers (a "triopoly"). That is:

a. For a monopoly, calculate the quantity $x$ that maximizes your revenue. Also calculate the price $p$ and the revenue $R$.

b. For a triopoly, find the quantities $x$, $y$, and $z$ for the three suppliers that maximize revenue for each. Also calculate the price $p$ and the three revenues. [*Hint:* Find the three revenue functions, one for each supplier, and maxmimize each with respect to that supplier's variable.]

c. Are more goods produced under a monopoly or under a triopoly?
d. Is the price lower under a monopoly or under a triopoly?

## Exercises with More Than One Critical Point

**33–38.** Find the relative extreme values of each function.

**33.** $f(x, y) = x^3 + y^3 - 3xy$

**34.** $f(x, y) = x^5 + y^5 - 5xy$

**35.** $f(x, y) = 12xy - x^3 - 6y^2$

**36.** $f(x, y) = 6xy - x^3 - 3y^2$

**37.** $f(x, y) = 2x^4 + y^2 - 12xy$

**38.** $f(x, y) = 16xy - x^4 - 2y^2$

## Conceptual Exercises

**39.** Maximum and minimum points occurred with functions of one variable as well as with functions of two variables. What *new* type of point becomes possible with functions of *two* variables?

**40.** True or False: If $f_x = 0$ and $f_y = 0$ at a point, then that point must be a relative maximum or minimum point.

**41.** True or False: If a function has no critical points, then it has no relative extreme points.

**42.** True or False: If a function has no relative extreme points, then it has no critical points.

**43.** If a function has a critical point at which $D > 0$, what can you conclude?

**44.** If a function has a critical point at which $D < 0$, what can you conclude?

**45.** If a function has a critical point at which $f_{xx} > 0$, what can you conclude?

**46.** In the $D$-test we did not discuss what happens if $D > 0$ and $f_{xx} = 0$. Why not? [*Hint:* Can this case occur?]

**47.** True or False: At a critical point, $D = 0$ means that the point is *not* a relative maximum or minimum point.

**48.** True or False: If $f_{xx}$ and $f_{yy}$ have different signs at a critical point, then that point is a saddle point.

## 7.4   LEAST SQUARES

### Introduction

You may have wondered how the mathematical models in this book were developed. For example, how were the functions that predict populations found from census data, and how were the constants $a$ and $b$ in a Cobb–Douglas production function found from production data? The problem of finding a function that fits a collection of data can be viewed geometrically as the problem of fitting a curve to a collection of points. The simplest case of this problem is fitting a straight line to the points, and the most widely used method for doing this, and the method used by computers and calculators, is called **least squares.** You may have found regression lines and curves in Chapter 1 by pressing buttons on your calculator, and you have certainly used many functions in this book that were developed in this way. Although using a calculator or computer is both reasonable and practical, it is important to understand the method, and that is what this section explains.

## A First Example

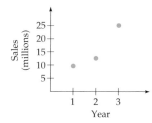

The graph on the left shows a company's annual sales (in millions) over a three-year period. How can we fit a straight line to these three points? Clearly, these points do not lie exactly on a line, and so rather than an "exact" fit, we want the line $y = ax + b$ that fits these three points most closely.

Let $d_1$, $d_2$, and $d_3$ stand for the vertical distances between the three points and the line $y = ax + b$. The line that minimizes the sum of the squares of these vertical deviations is called the *least squares line* or the *regression line*.* Squaring the deviations ensures that none are negative, so that a deviation below the line does not "cancel" one above the line.

 **EXAMPLE 1**  **MINIMIZING THE SQUARED DEVIATIONS**

The table below gives a company's sales (in millions) in year $x$. Find the least squares line for the data. Then use the line to predict sales in year 4.

| $x$ | $y$ |
|-----|-----|
| 1   | 10  |
| 2   | 12  |
| 3   | 25  |

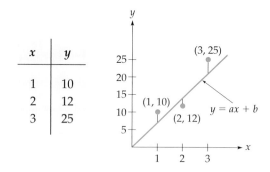

*Solution*

The vertical deviations are found by calculating the heights of the line $y = ax + b$ at each $x$-value minus the $y$-values from the table. These differences are then squared and summed.

$$S = (a \cdot 1 + b - 10)^2 + (a \cdot 2 + b - 12)^2 + (a \cdot 3 + b - 25)^2$$

| Height of the line $y = ax + b$ at $x = 1$ | $y$-value of the point at $x = 1$ | Height of the line $y = ax + b$ at $x = 2$ | $y$-value of the point at $x = 2$ | Height of the line $y = ax + b$ at $x = 3$ | $y$-value of the point at $x = 3$ |
|---|---|---|---|---|---|

*The word "regression" comes from an early use of this technique to determine whether unusually tall parents have unusually tall children. It seems that tall parents do have tall offspring, but not quite as tall, with successive generations exhibiting a "regression" toward the average height of the population.

This sum $S$ depends on $a$ and $b$, the numbers that determine the line $y = ax + b$.  To minimize $S$, we set its partials with respect to $a$ and $b$ equal to zero:

$$\frac{\partial S}{\partial a} = 2(a + b - 10) + 2(2a + b - 12)\cdot 2 + 2(3a + b - 25)\cdot 3$$

Differentiating each part of $S$ by the Generalized Power Rule

$$= 2a + 2b - 20 + 8a + 4b - 48 + 18a + 6b - 150$$

Multiplying out

$$= 28a + 12b - 218$$

Combining terms

$$\frac{\partial S}{\partial b} = 2(a + b - 10) + 2(2a + b - 12) + 2(3a + b - 25)$$

Differentiating each part of $S$ by the Generalized Power Rule

$$= 2a + 2b - 20 + 4a + 2b - 24 + 6a + 2b - 50$$

Multiplying out

$$= 12a + 6b - 94$$

Combining terms

We set the two partials equal to zero and solve:

$$28a + 12b - 218 = 0$$
$$12a + 6b - 94 = 0$$

Partials set equal to zero

$$28a + 12b - 218 = 0$$

First equation

$$\underline{-24a - 12b + 188 = 0}$$

Second multiplied by $-2$

$$4a \qquad - 30 = 0$$

Adding (the $b$'s drop out)

$$a = \frac{30}{4} = 7.5$$

Solving   $4a - 30 = 0$

$$b = \frac{4}{6} \approx 0.67$$

From substituting   $a = 7.5$   into $12a + 6b - 94 = 0$   and solving for $b$

These values for $a$ and $b$ give the least squares line. (The $D$-test would show that $S$ has indeed been minimized.)

$$y = 7.5x + 0.67$$

$\uparrow$
$y = ax + b$  with
$a = 7.5$  and  $b = 0.67$

least squares line
$y = 7.5x + 0.67$

To predict the sales in year 4, we evaluate the least squares line at $x = 4$:

$$y = 7.5 \cdot 4 + 0.67 = 30.67$$

$y = 7.5x + 0.67$
evaluated at  $x = 4$

*Prediction:* 30.67 million sales in year 4.

The slope of the line is 7.5, meaning that the linear trend in the company's sales is a growth of 7.5 million sales per year.

## Graphing Calculator Exploration

on [0, 4] by [0, 30]

Verify that your graphing calculator gives the same least squares line $y = 7.5x + 0.67$ that we found in Example 1 by entering the data from the table on page 500 into your calculator and finding the linear regression line. The $r$ that your calculator might find is a measure of "goodness of fit," with values closer to 1 indicating a better fit.

### Least Squares Line for $n$ Points

Example 1 used only three points, which is too few for most realistic applications. If we carry out the same steps for $n$ points, we would obtain the following formulas (in which $\Sigma$ stands for sum).

---

**Least Squares Line**

For data

| $x$ | $y$ |
|-----|-----|
| $x_1$ | $y_1$ |
| $x_2$ | $y_2$ |
| $\cdot$ | $\cdot$ |
| $\cdot$ | $\cdot$ |
| $\cdot$ | $\cdot$ |
| $x_n$ | $y_n$ |

calculate

$$a = \frac{n\sum xy - \left(\sum x\right)\left(\sum y\right)}{n\sum x^2 - \left(\sum x\right)^2}$$

$$b = \frac{1}{n}\left(\sum y - a\sum x\right)$$

$n$ = number of data points
$\sum x$ = sum of $x$'s
$\sum y$ = sum of $y$'s
$\sum xy$ = sum of products $x \cdot y$
$\sum x^2$ = sum of squares of $x$'s

The least squares line is then $y = ax + b$

---

From now on we will find least squares lines by using these formulas, a derivation of which is given at the end of this section.

⚠ **BE CAREFUL:** $\sum x^2$ is *not* the same as $\left(\sum x\right)^2$.

---

**EXAMPLE 2**  **FINDING THE LEAST SQUARES LINE**

A 1955 study compared cigarette smoking with the mortality rate for lung cancer in several countries. Find the least squares line that fits these data.

Then use the line to predict lung cancer deaths if per capita cigarette consumption is 600 cigarettes per month (a pack a day).

|  | Cigarette Consumption (per capita) | Lung Cancer Deaths (per million) |
|---|---|---|
| Norway | 250 | 90 |
| Sweden | 300 | 120 |
| Denmark | 350 | 170 |
| Canada | 500 | 150 |

*Solution*

The procedure for calculating $a$ and $b$ consists of six steps, beginning with the following table.

**1.** List the $x$- and $y$-values  **2.** Multiply $x \cdot y$  **3.** Square each $x$

| $x$ | $y$ | $xy$ | $x^2$ |
|---|---|---|---|
| 250 | 90 | 22,500 | 62,500 |
| 300 | 120 | 36,000 | 90,000 |
| 350 | 170 | 59,500 | 122,500 |
| 500 | 150 | 75,000 | 250,000 |
| 1400 | 530 | 193,000 | 525,000 ← **4.** Sum each column |
| $\Sigma\, x$ | $\Sigma\, y$ | $\Sigma\, xy$ | $\Sigma\, x^2$ |

**5.** Calculate $a$ and $b$ using the formulas above

$$a = \frac{(4)(193{,}000) - (1400)(530)}{(4)(525{,}000) - 1400^2}$$

$n$ = number of points

$$a = \frac{n\,\Sigma\, xy - (\Sigma\, x)(\Sigma\, y)}{n\,\Sigma\, x^2 - (\Sigma\, x)^2}$$

$$= \frac{30{,}000}{140{,}000} \approx 0.21$$

$$b = \frac{1}{4}[530 - 0.21(1400)] = 59 \qquad b = \frac{1}{n}(\Sigma\, y - a\,\Sigma\, x)$$

**6.** The least squares line is $y = ax + b$ with the $a$ and $b$ values above

$$y = 0.21x + 59 \qquad y = ax + b \text{ with } a = 0.21 \text{ and } b = 59$$

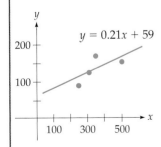

This line is graphed on the left, and shows that lung cancer mortality increases with cigarette smoking.

To predict the lung cancer deaths if per capita cigarette consumption reaches 600, we evaluate the least squares line at $x = 600$.

$$y = 0.21 \cdot 600 + 59 = 185 \qquad y = 0.21x + 59 \text{ with } x = 600$$

*Predicted annual mortality:* 185 deaths per million.

## Criticism of Least Squares

Least squares is the most widely used method for fitting lines to points, but it does have one weakness: the vertical deviations from the line are squared, so one large deviation, when squared, can have an unexpectedly large influence on the line. For example, the graph on the right shows four points and their least squares line. The line fits the points quite closely.

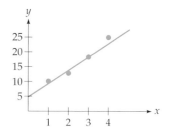

The second graph adds a fifth point, one quite out of line with the others, and shows the least squares line for the five points. The added point has an enormous effect on the line, causing it to slope downward even though all of the other points suggest an upward slope. In actual applications, one should inspect the points for such "outliers" before calculating the least squares line.

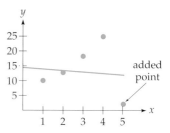

added point

## Fitting Exponential Curves by Least Squares

It is not always appropriate to fit a straight line to a set of points. Sometimes a collection of points will suggest a *curve* rather than a line, such as one of the following exponential curves.

Least squares can be used to fit an exponential curve of the form $y = Be^{Ax}$ to a collection of points as follows. Taking natural logs of both sides of $y = Be^{Ax}$ gives

$$\ln y = \ln(Be^{Ax}) = \ln B + \ln e^{Ax} = \ln B + Ax \qquad \text{Using the properties of natural logarithms}$$

If we introduce a new variable $Y = \ln y$, and also let $b = \ln B$, then $\ln y = \ln B + Ax$ becomes

$$Y = b + Ax \qquad \text{Or, equivalently, } Y = Ax + b$$

Therefore, we fit a straight line to the *logarithms* of the $y$-values. (Now we are minimizing not the squared deviations, but the squared deviations of the logarithms.) The procedure consists of the eight steps shown in the following example.

**EXAMPLE 3**  **FITTING AN EXPONENTIAL CURVE BY LEAST SQUARES**

The world population since 1850 is shown in the table below. Fit an exponential curve to these data and predict the world population in the year 2050.

| Year | Population (billions) |
|------|-----------------------|
| 1850 | 1.26 |
| 1900 | 1.65 |
| 1950 | 2.52 |
| 2000 | 6.08 |

*Source:* U.S. Census Bureau

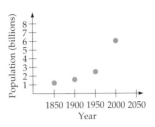

*Solution*   We number the years 1 through 4 (since they are evenly spaced).

**1.** List the $x$- and $y$-values      **2.** Take ln of the $y$-values (call them *capital Y*)      **3.** Multiply $x \cdot Y$ (capital $Y$)      **4.** Square each $x$

| $x$ | $y$ | $Y = \ln y$ | $xY$ | $x^2$ |
|-----|-----|-------------|------|-------|
| 1 | 1.26 | 0.23 | 0.23 | 1 |
| 2 | 1.65 | 0.50 | 1.00 | 4 |
| 3 | 2.52 | 0.92 | 2.76 | 9 |
| 4 | 6.08 | 1.81 | 7.24 | 16 |
| 10 | | 3.46 | 11.23 | 30 |
| ‖ | | ‖ | ‖ | ‖ |
| $\Sigma x$ | | $\Sigma Y$ | $\Sigma xY$ | $\Sigma x^2$ |

**5.** Add each column (except $y$)

**6.** Calculate $A$ and $b$ ($n$ = number of points)

$$A = \frac{4(11.23) - 10(3.46)}{4(30) - 10^2} = \frac{10.32}{20} = 0.516$$

$$A = \frac{n \Sigma xY - (\Sigma x)(\Sigma Y)}{n \Sigma x^2 - (\Sigma x)^2}$$

$$b = \frac{1}{4}(3.46 - 0.516 \cdot 10) = -0.425$$

$$b = \frac{1}{n}(\Sigma Y - A \Sigma x)$$

$$B = e^b = e^{-0.425} \approx 0.65$$

**7.** Calculate $B = e^b$ (since $b = \ln B$)

The exponential curve is

$$y = 0.65e^{0.516x}$$

**8.** The curve is $y = Be^{Ax}$ with the $A$ and $B$ found above

The population in 2050 is this function evaluated at $x = 5$:

$$y = 0.65e^{0.516(5)} \approx 8.6$$

Therefore, the world population in the year 2050 is predicted to be 8.6 billion.

## Graphing Calculator Exploration

Verify that your graphing calculator gives the same curve $y = 0.65e^{0.516x}$ that we found in Example 3 by entering the $x$- and $y$-values from the previous table into your calculator and finding the exponential regression curve. [You may find $y = 0.65 \cdot 1.675^x$, which is the same curve, since $e^{0.516} \approx 1.675$.]

## 7.4 Section Summary

The technique of least squares can be used to determine the equation of a line that "best fits" a collection of points in the sense that it minimizes the sum of the squared vertical deviations. The method is described on pages 502–503. Least squares can also be used to fit a *curve* to a collection of points. The procedure for fitting an exponential curve $y = Be^{Ax}$ is described on pages 504–505. The method is widely used in forecasting and for detecting underlying trends in data.

To determine whether to fit a line or a curve to a collection of points, you should first graph the points. For example, the points in the left-hand graph below suggest a line, while those in the right-hand graph suggest an exponential curve.

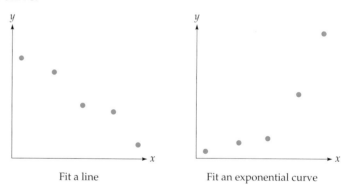

Fit a line          Fit an exponential curve

## Derivation of the Formula for the Least Squares Line

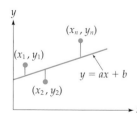

The formulas for $a$ and $b$ in the least squares line come from minimizing the squared vertical deviations, just as in Example 1, but now for the $n$ points $(x_1, y_1), (x_2, y_2), \ldots, (x_n, y_n)$.

$$S = (ax_1 + b - y_1)^2 + (ax_2 + b - y_2)^2 + \cdots + (ax_n + b - y_n)^2$$

The partials are

$$\frac{\partial S}{\partial a} = 2(ax_1 + b - y_1)x_1 + 2(ax_2 + b - y_2)x_2 + \cdots + 2(ax_n + b - y_n)x_n$$

$$= 2ax_1^2 + 2bx_1 - 2x_1y_1 + 2ax_2^2 + 2bx_2 - 2x_2y_2 + \cdots \qquad \text{Multiplying out}$$
$$+ 2ax_n^2 + 2bx_n - 2x_ny_n$$

$$= 2a(x_1^2 + x_2^2 + \cdots + x_n^2) + 2b(x_1 + x_2 + \cdots + x_n) \qquad \text{Regrouping}$$
$$- 2(x_1y_1 + x_2y_2 + \cdots + x_ny_n)$$

$$= 2a \sum x^2 + 2b \sum x - 2 \sum xy \qquad \text{Using } \Sigma \text{ for sum}$$

$$\frac{\partial S}{\partial b} = 2(ax_1 + b - y_1) + 2(ax_2 + b - y_2) + \cdots + 2(ax_n + b - y_n)$$

$$= 2ax_1 + 2b - 2y_1 + 2ax_2 + 2b - 2y_2 + \cdots + 2ax_n + 2b - 2y_n \qquad \text{Multiplying out}$$

$$= 2a(x_1 + x_2 + \cdots + x_n) + 2b(1 + 1 + \cdots + 1) \qquad \text{Regrouping}$$
$$- 2(y_1 + y_2 + \cdots + y_n)$$

$$= 2a \sum x + 2bn - 2 \sum y \qquad \text{Using } \Sigma \text{ for sum}$$

We set the partials equal to zero:

$$a \sum x^2 + b \sum x - \sum xy = 0$$
$$a \sum x + bn - \sum y = 0 \qquad \text{Dividing each by 2}$$

To solve for $a$ we multiply the first of these equations by $n$, the second by $\sum x$, and subtract:

$$an \sum x^2 + bn \sum x - n \sum xy = 0$$

$$a \left( \sum x \right)^2 + bn \sum x - \left( \sum x \right) \left( \sum y \right) = 0 \qquad \begin{array}{l} \textit{Be Careful: } \Sigma \, xy \text{ is not} \\ \text{the same as } (\Sigma\, x)(\Sigma\, y) \end{array}$$

$$a \left( n \sum x^2 - \left( \sum x \right)^2 \right) - \left( n \sum xy - \left( \sum x \right) \left( \sum y \right) \right) = 0 \qquad \begin{array}{l} \text{Subtracting the} \\ \text{last two equations} \end{array}$$

Solving this last equation for $a$ gives:

$$a = \frac{n \sum xy - \left( \sum x \right) \left( \sum y \right)}{n \sum x^2 - \left( \sum x \right)^2} \qquad \text{Formula for } a$$

For $b$ we obtain:

$$b = \frac{1}{n} \left( \sum y - a \sum x \right) \qquad a \sum x + nb - \sum y = 0 \;\; \text{solved for } b$$

These are the formulas in the box on page 502.

## 7.4 Exercises

### Exercises on Least Squares Lines

**1–8.** Find the least squares line for each table of points.

**1.**

| x | y |
|---|---|
| 1 | 2 |
| 2 | 5 |
| 3 | 9 |

**2.**

| x | y |
|---|---|
| 1 | 2 |
| 2 | 4 |
| 3 | 7 |

**3.**

| x | y |
|---|---|
| 1 | 6 |
| 3 | 4 |
| 6 | 2 |

**4.**

| x | y |
|---|---|
| 1 | 9 |
| 4 | 6 |
| 5 | 1 |

**5.**

| x | y |
|---|---|
| 0 | 7 |
| 1 | 10 |
| 2 | 10 |
| 3 | 15 |

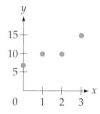

**6.**

| x | y |
|---|---|
| 0 | 5 |
| 1 | 8 |
| 2 | 8 |
| 3 | 12 |

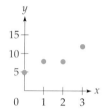

**7.**

| x | y |
|---|---|
| −1 | 10 |
| 0 | 8 |
| 1 | 5 |
| 3 | 0 |
| 5 | −2 |

**8.**

| x | y |
|---|---|
| −2 | 12 |
| 0 | 10 |
| 2 | 6 |
| 4 | 0 |
| 5 | −3 |

### Applied Exercises on Least Squares Lines

**9.** BUSINESS: Sales  A company's annual sales are shown in the following table. Find the least squares line. Use your line to predict the sales in the next year $(x = 5)$.

| Year | 1 | 2 | 3 | 4 |
|---|---|---|---|---|
| Sales (millions) | 7 | 10 | 11 | 14 |

**10.** GENERAL: Automobile Costs  The following table gives the cost per mile of operating a compact car, depending upon the number of miles driven per year. Find the least squares line for these data. Use your answer to predict the cost per mile for a car driven 25,000 miles annually $(x = 5)$.

| Annual Mileage | x | Cost per Mile (cents) |
|---|---|---|
| 5000 | 1 | 50 |
| 10,000 | 2 | 35 |
| 15,000 | 3 | 27 |
| 20,000 | 4 | 25 |

**11. SOCIOLOGY: Crime** A sociologist finds the following data for the number of felony arrests per year in a town. Find the least squares line. Then use it to predict the number of felony arrests in the next year.

| Year | 1 | 2 | 3 | 4 |
|---|---|---|---|---|
| Arrests | 120 | 110 | 90 | 100 |

**12. GENERAL: Farming** A farmer's wheat yield (bushels per acre) depends on the amount of fertilizer (hundreds of pounds per acre) according to the following table. Find the least squares line. Then use the line to predict the yield using 3 hundred pounds of fertilizer per acre.

| Fertilizer | 1.0 | 1.5 | 2.0 | 2.5 |
|---|---|---|---|---|
| Yield | 30 | 35 | 38 | 40 |

**13. BASEBALL: The Disappearance of the .400 Hitter** Between 1901 and 1930, baseball boasted several .400 hitters (Lajoie, Cobb, Jackson, Sisler, Heilmann, Hornsby, and Terry), but only one since then (Ted Williams in 1941). The decline of the "heavy hitter" is evidenced by the following data, showing the highest batting average for each 20-year period (National or American League). Find the least squares line for these data and use it to predict the highest batting average for the period 2001–2020 $(x = 6)$.

*Source:* S. J. Gould, *Triumph and Tragedy in Mudville*

| | x | Highest Average |
|---|---|---|
| 1901–1920 | 1 | .422 |
| 1921–1940 | 2 | .424 |
| 1941–1960 | 3 | .406 |
| 1961–1980 | 4 | .390 |
| 1981–2000 | 5 | .394 |

**14. ECONOMICS: Consumer Price Index** The consumer price index (CPI) is shown in the following table. Fit a least squares line to the data. Then use the line to predict the CPI in the year 2020 $(x = 7)$.

*Source:* U.S. Bureau of Labor Statistics

| | x | CPI |
|---|---|---|
| 1985 | 1 | 107.6 |
| 1990 | 2 | 130.7 |
| 1995 | 3 | 152.4 |
| 2000 | 4 | 172.2 |
| 2005 | 5 | 195.3 |

**15–16. GENERAL: Percentage of Smokers** The following tables show the percentage of smokers among the adult population in the United States for every five years since 1985 (the first table is males, the second females). Find the least squares lines for these data, and use your answer to predict the percentage of that sex who will smoke in the year 2015 $(x = 7)$.

*Source:* U.S. Department of Health and Human Services

**15.**

| | x | Percent Males |
|---|---|---|
| 1985 | 1 | 32 |
| 1990 | 2 | 28 |
| 1995 | 3 | 27 |
| 2000 | 4 | 25 |
| 2005 | 5 | 23 |

**16.**

| | x | Percent Females |
|---|---|---|
| 1985 | 1 | 28 |
| 1990 | 2 | 23 |
| 1995 | 3 | 23 |
| 2000 | 4 | 21 |
| 2005 | 5 | 18 |

**17. BIOMEDICAL SCIENCES: Smoking and Longevity** The following data show the life expectancy of a 25-year-old male based on the number of cigarettes smoked daily. Find the least squares line for these data. The slope of the line estimates the years lost per extra cigarette per day.

| Cigarettes Smoked Daily | Life Expectancy |
|---|---|
| 0 | 73.6 |
| 5 | 69.0 |
| 15 | 68.1 |
| 30 | 67.4 |
| 40 | 65.3 |

**18. ENVIRONMENTAL SCIENCES: Pollution and Absenteeism** The following table shows the relationship between the sulfur dust content of the air (in micrograms per cubic meter) and the number of female absentees in industry. (Only absences of at least seven days were counted.) Find the least squares line for these data. Use your answer to predict absences in a city with a sulfur dust content of 25.

|  | Sulfur | Absences per 1000 Employees |
|---|---|---|
| Cincinnati | 7 | 19 |
| Indianapolis | 13 | 44 |
| Woodbridge | 14 | 53 |
| Camden | 17 | 61 |
| Harrison | 20 | 88 |

## Exercises on Fitting Exponential Curves

**19–26.** Use least squares to find the exponential curve $y = Be^{Ax}$ for the following tables of points.

**19.**

| x | y |
|---|---|
| 1 | 2 |
| 2 | 4 |
| 3 | 7 |

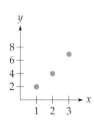

**20.**

| x | y |
|---|---|
| 1 | 3 |
| 2 | 6 |
| 3 | 11 |

**21.**

| x | y |
|---|---|
| 1 | 10 |
| 3 | 5 |
| 6 | 1 |

**22.**

| x | y |
|---|---|
| 1 | 12 |
| 4 | 3 |
| 5 | 2 |

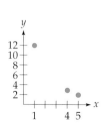

**23.**

| x | y |
|---|---|
| 0 | 1 |
| 1 | 2 |
| 2 | 5 |
| 3 | 10 |

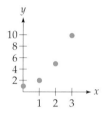

**24.**

| x | y |
|---|---|
| 0 | 2 |
| 1 | 4 |
| 2 | 7 |
| 3 | 15 |

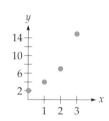

**25.**

| x | y |
|---|---|
| −1 | 20 |
| 0 | 18 |
| 1 | 15 |
| 3 | 4 |
| 5 | 1 |

**26.**

| x | y |
|---|---|
| −2 | 20 |
| 0 | 12 |
| 2 | 9 |
| 4 | 6 |
| 5 | 5 |

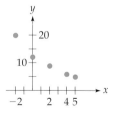

## Applied Exercises on Fitting Exponential Curves

**27. POLITICAL SCIENCE: Cost of a Congressional Victory** The following table shows average amounts spent by winners of seats in the House of Representatives in presidential election years. Fit an exponential curve to these data and use it to predict the cost of a House seat in the year 2012.

*Source:* Center for Responsive Politics

|      | x | Cost ($1000) |
|------|---|--------------|
| 1996 | 1 | 764          |
| 2000 | 2 | 833          |
| 2004 | 3 | 1019         |

**28. GENERAL: Drunk Driving** The following table shows how a driver's blood-alcohol level (% grams per dekaliter) affects the probability of being in a collision. A collision factor of 3 means that the probability of a collision is 3 times as large as normal. Fit an exponential curve to the data. Then use your curve to estimate the collision factor for a blood-alcohol level of 15.

| Blood-Alcohol Level | Collision Factor |
|---------------------|------------------|
| 0                   | 1                |
| 6                   | 1.1              |
| 8                   | 3                |
| 10                  | 6                |

**29. ATHLETICS: Super Bowl Tickets** The following table gives the minimum price of a Super Bowl ticket in 2008 and at preceding five-year intervals. Fit an exponential curve to the data and then use it to predict the minimum cost of a ticket in 2013.

*Source:* National Football League

| Year | x | Price |
|------|---|-------|
| 1993 | 1 | 175   |
| 1998 | 2 | 275   |
| 2003 | 3 | 400   |
| 2008 | 4 | 700   |

**30. GENERAL: Stamp Prices** The following table shows the cost of a first-class postage stamp every 20 years since 1948. Fit an exponential curve to these data and use your answer to predict the cost of a stamp in the year 2028.

*Source:* U.S. Postal Service

|      | x | Postage (cents) |
|------|---|-----------------|
| 1948 | 1 | 3               |
| 1968 | 2 | 6               |
| 1988 | 3 | 25              |
| 2008 | 4 | 42              |

**31. GENERAL: Aging World Population** The number of children in the world (age 14 years and younger) has always exceeded the number of elderly (age 60 years and older), but changes in fertility and mortality are increasing the elderly population much faster than the children's population. The table below gives these populations (in billions) for past years and predicted for 2025. Fit an *exponential curve* to the elderly data and a *line* to the children's data (it is roughly linear). Then evaluate each function at $x = 4$ to see which population will be larger at that time. What year does $x = 4$ correspond to? This aging of the world population has enormous economic consequences.

*Source:* U.N. Department of Economic and Social Affairs

| Year | x | Elderly | Children |
|------|---|---------|----------|
| 1985 | 1 | 0.4     | 1.6      |
| 2005 | 2 | 0.7     | 1.8      |
| 2025 | 3 | 1.2     | 1.9      |

**32. GENERAL: Cost of College Education** The following table shows the cost of one year of tuition at a private four-year college. Fit an exponential curve to the data and then use it to predict the cost for the academic year 2017–2018.

*Source:* The College Board

|         | x | Cost ($1000) |
|---------|---|--------------|
| 1977–78 | 1 | 2.7          |
| 1987–88 | 2 | 7.0          |
| 1997–98 | 3 | 13.8         |
| 2007–08 | 4 | 23.7         |

## Conceptual Exercises

**33.** True or False: The least squares line will always pass through at least *one* of the data points.

**34.** True or False: If the data consists of just *two* points, the least squares line will be the line that passes through the two points. (Assume that the *x*-coordinates of the points are different.)

**35.** True or False: If the data points all lie on a line, then the least squares line for the data will *be* that line. (Assume that the line is not vertical.)

**36.** Suppose that the least squares line for a set of data points is $y = ax + b$. If you *doubled* each *y*-value, what would be the new least squares line? [*Hint:* How has the line been changed?]

**37.** Suppose that the least squares line for a set of data points is $y = ax + b$. If you *added* 5 to each *y*-value, what would be the new least squares line? [*Hint:* How has the line been changed?]

**38.** What do you get if you find the least squares line for just *one* data point? [*Hint:* Try it for the point (1, 2).]

**39.** What goes wrong if you try to find the least squares line for just *two* data points and they have the *same x*-coordinate? [*Hint:* Try it for the points (1, 2) and (1, 5).]

**40.** What goes wrong if you try to fit an exponential curve to data to just *one* data point? [*Hint:* Try it for the point (1, 2).]

**41.** What goes wrong if you try to fit an exponential curve to data and one of the points has a *y*-coordinate that is 0 or a negative number? [*Hint:* Look at the table on page 505.]

**42.** Suppose when you fit an exponential curve to a set of data points you obtain the equation $y = Be^{Ax}$. If you *doubled* each *y*-value, what would be the new exponential curve? [*Hint:* How has the curve been changed?]

## 7.5 LAGRANGE MULTIPLIERS AND CONSTRAINED OPTIMIZATION

### Introduction

In Section 7.3 we optimized functions of several variables. Some problems, however, involve maximizing or minimizing a function subject to a **constraint.** For example, a company might want to maximize production subject to the constraint of staying within its budget, or a soft drink distributor might want to design the least expensive aluminum can subject to the constraint that it hold exactly 12 ounces of soda. In this section we solve such "constrained optimization" problems by the method of **Lagrange multipliers,** invented by the French mathematician Joseph Louis Lagrange (1736–1813).

### Constraints

If a constraint can be written as an equation, such as

$$x^2 + y^2 = 100$$

then by moving all the terms to the left-hand side, it can be written with zero on the right-hand side:

$$\underbrace{x^2 + y^2 - 100}_{g(x,\, y)} = 0$$

In general, any equation can be written with all terms moved to the left-hand side, and we will write all constraints in the form

$$g(x, y) = 0$$

**Practice Problem 1**

Write the constraint $2y^3 = 3x - 1$ in the form $g(x, y) = 0$.

➤ **Solution on page 521**

### First Example of Lagrange Multipliers

We illustrate the method of Lagrange multipliers by an example. The method requires a new variable, and it is customary to use $\lambda$ ("lambda," the Greek letter l, in honor of Lagrange).

**EXAMPLE 1**

### MAXIMIZING THE AREA OF AN ENCLOSURE

A farmer wants to build a rectangular enclosure along an existing stone wall. If the side along the wall needs no fence, find the dimensions of the largest enclosure that can be made using only 400 feet of fence.

*Solution*

We want the enclosure of largest *area*. Let

$$x = \text{width (perpendicular to the wall)}$$

$$y = \text{length (parallel to the wall)}$$

Two widths and one length must be made from the 400 feet of fence, so the constraint is

$$2x + y = 400$$

Therefore, the problem becomes:

maximize  $A = xy$    Area is length times width

subject to  $2x + y - 400 = 0$    Constraint $2x + y = 400$ written with zero on the right

$g(x, y)$

We write a new function $F(x, y, \lambda)$, called the **Lagrange function,** which consists of the function to be maximized plus $\lambda$ times the constraint function:

$$F(x, y, \lambda) = xy + \lambda(2x + y - 400)$$

$$F(x, y, \lambda) = \begin{pmatrix} \text{Function to} \\ \text{be optimized} \end{pmatrix}$$
$$+ \lambda \begin{pmatrix} \text{Constraint} \\ \text{function} \end{pmatrix}$$

$$= xy + \lambda 2x + \lambda y - \lambda 400$$    Multiplied out

The Lagrange function $F(x, y, \lambda)$ is a function of *three* variables, $x$, $y$, and $\lambda$, and we begin as usual by setting its partials with respect to each variable equal to zero.

$$F_x = y + 2\lambda \qquad = 0 \qquad\qquad \text{Partial of } xy + \lambda 2x + \lambda y - \lambda 400 \text{ with respect to } x$$

$$F_y = x + \lambda \qquad = 0 \qquad\qquad \text{Partial of } xy + \lambda 2x + \lambda y - \lambda 400 \text{ with respect to } y$$

$$F_\lambda = \underbrace{2x + y - 400}_{} = 0 \qquad\qquad \text{Partial of } xy + \lambda 2x + \lambda y - \lambda 400 \text{ with respect to } \lambda$$

Constraint $g = 0$

We solve the first two of these equations for $\lambda$:

$$\lambda = -\frac{1}{2}y \qquad\qquad\qquad \text{Solving } y + 2\lambda = 0 \text{ for } \lambda$$

$$\lambda = -x \qquad\qquad\qquad \text{Solving } x + \lambda = 0 \text{ for } \lambda$$

Then we set these two expressions for $\lambda$ equal to each other:

$$-\frac{1}{2}y = -x \qquad\qquad \text{Equating } \lambda = -\frac{1}{2}y \text{ and } \lambda = 2x$$

$$y = 2x \qquad\qquad\qquad \text{Multiplying by } -2$$

We use this relationship $y = 2x$ to eliminate the $y$ in the equation $2x + y - 400 = 0$ (which came from the third partial $F_\lambda = 0$):

$$2x + 2x - 400 = 0 \qquad\qquad 2x + y - 400 = 0 \text{ with } y \text{ replaced by } 2x$$

$$4x = 400 \qquad\qquad\qquad \text{Simplifying}$$

$$x = 100 \qquad\qquad\qquad \text{Dividing by 4}$$

$$y = 200 \qquad\qquad\qquad \text{From } y = 2x \text{ with } x = 100$$

The largest possible enclosure has width 100 feet (perpendicular to the wall) and length 200 feet (parallel to the wall).

## Method of Lagrange Multipliers

In general, the function to be maximized or minimized is called the **objective function,** because the "objective" of the whole procedure is to optimize it. (In Example 1 the objective function was the area function $A = xy$, which was to be maximized.) The variable $\lambda$ is called the **Lagrange multiplier.** The entire method may be summarized as follows. A justification of Lagrange's method is given at the end of this section.

To optimize  $f(x, y)$  subject to  $g(x, y) = 0$:      Objective function and constraint

1. Write  $F(x, y, \lambda) = f(x, y) + \lambda g(x, y)$.      Objective function plus $\lambda$ times the constraint function

2. Set the partials of $F$ equal to zero:
$$F_x = 0 \qquad F_y = 0 \qquad F_\lambda = 0$$
and solve for the critical points.

3. The solution to the original problem (if it exists) will occur at one of these critical points.

It is important to realize that Lagrange's method only finds *critical points*—it does not tell whether the function is maximized, minimized, or neither at the critical point. (The *D*-test, which involves calculating  $D = f_{xx}f_{yy} - (f_{xy})^2$,  is for *un*constrained optimization, and cannot be used with constraints.) In each problem we must *know* that the maximum or minimum (whichever is requested) *does* exist, and it then follows that it must occur at a critical point (found by Lagrange multipliers).

How do we know that the maximum or minimum does exist? Most reasonable applied problems *do* have solutions. In Example 1, for instance, making the width or length too small would make the area (length times width) small, so the area must be largest for some "intermediate" length and width. Therefore, the problem *does* have a solution, which is what Lagrange multipliers find.

We could have solved Example 1 *without* Lagrange multipliers by solving the constraint equation  $2x + y = 400$  for *y* and using it to eliminate the *y* in the objective function, as we did in Sections 3.3 and 3.4. However, Lagrange's method has two advantages: it eliminates the need to solve the constraint equation (which can sometimes be difficult), and it is symmetric in the variables, thereby allowing you to solve for whichever variable is easier to find.

## Hints for Solving the Partial Equations

Solving the partial derivative equations  $F_x = 0$,  $F_y = 0$,  and  $F_\lambda = 0$  can sometimes be difficult. The following strategy (used in each example in this section) may be helpful.

1. Solve each of  $F_x = 0$  and  $F_y = 0$  for $\lambda$.
2. Set the two expressions for $\lambda$ equal to each other.
3. Solve the equation resulting from step 2 together with  $F_\lambda = 0$  to find *x* and *y*.

**EXAMPLE 2**    **DESIGNING THE MOST EFFICIENT CONTAINER**

A container company wants to design an aluminum can that requires the least amount of aluminum but that contains exactly 12 fluid ounces (21.3 cubic inches). Find the radius and height of the can.

*Solution*

Minimizing the amount of aluminum means minimizing the surface area (top plus bottom plus sides) of the cylindrical can. If we let $r$ and $h$ stand for the radius and height (in inches) of the can, then the diagram shows that the area is:

$$A = \underbrace{2\pi r^2}_{\substack{\text{Top and} \\ \text{bottom area}}} + \underbrace{2\pi rh}_{\substack{\text{Side} \\ \text{area}}}$$

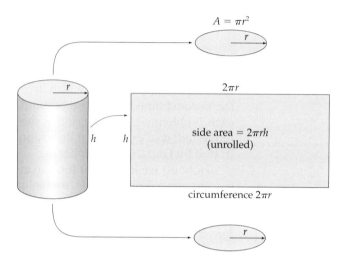

The volume is

$$V = \pi r^2 h$$

Therefore, the problem becomes

minimize $\quad A = 2\pi r^2 + 2\pi rh$ $\qquad$ Area = (top and bottom) + side

subject to $\quad \pi r^2 h = 21.3$ $\qquad$ Volume must equal 21.3 in.$^3$

The Lagrange function is

$$F = 2\pi r^2 + 2\pi rh + \lambda(\pi r^2 h - 21.3)$$

Objective function $A$ plus $\lambda$ times the constraint

$$F_r = 4\pi r + 2\pi h + \lambda 2\pi rh = 0$$
$$F_h = 2\pi r + \lambda \pi r^2 \qquad\quad = 0$$
$$F_\lambda = \pi r^2 h - 21.3 \qquad\quad = 0$$

Partials set equal to zero

Solving, we have

$$\lambda = -\frac{4\pi r + 2\pi h}{2\pi rh} = -\frac{2r + h}{rh}$$

$4\pi r + 2\pi h + \lambda 2\pi rh = 0$
solved for $\lambda$

$$\lambda = -\frac{2\pi r}{\pi r^2} = -\frac{2}{r}$$

$2\pi r + \lambda \pi rh^2 = 0$
solved for $\lambda$

Equating $\lambda$'s yields

$$\frac{2r + h}{rh} = \frac{2}{r}$$    Equating the two expressions for $\lambda$ and multiplying each side by $-1$

$$2r + h = 2h$$    Multiplying each side by $rh$

$$2r = h$$    Subtracting $h$ from each side

Therefore,

$$\pi r^2(2r) = 21.3$$    $\pi r^2 h - 21.3 = 0$ (the third partial equation) with $h = 2r$

$$2\pi r^3 = 21.3$$    Simplifying

$$r^3 = \frac{21.3}{2\pi} \approx 3.39$$    Dividing by $2\pi$ (using a calculator)

$$r \approx \sqrt[3]{3.39} \approx 1.5$$    Taking cube roots

$$h = 3$$    From $h = 2r$ with $r = 1.5$

The most economical 12-fluid-ounce can has radius $r = 1.5$ inches and height $h = 3$ inches.

Notice that the height (3 inches) is twice the radius (1.5 inches), *so the height equals the diameter.* This shows that the most efficient can (least area for given volume) has a "squarish" shape.

It is interesting to consider why so few cans are shaped like this. The most common 12-ounce can for soft drinks is about twice as tall as it is across, requiring about 67% more aluminum than the most efficient can. This results in an enormous waste for the millions of cans manufactured each year. It seems that beverage companies prefer taller cans because they have more area for advertising, and they are easier to hold. Some products, however, are sold in "efficient" cans, with the height equal to the diameter.

**Practice Problem 2**    Which of the cans pictured below is most "efficiently" proportioned?

cola        paint        motor oil        tuna

➤ Solution on page 521

With Lagrange multipliers we can maximize a company's output subject to a budget constraint.

EXAMPLE 3   **MAXIMIZING PRODUCTION**

A company's output is given by the Cobb–Douglas production function $P = 600L^{2/3}K^{1/3}$, where $L$ and $K$ are the numbers of units of labor and capital. Each unit of labor costs the company \$40, and each unit of capital costs \$100. If the company has a total of \$3000 for labor and capital, how much of each should it use to maximize production?

*Solution*

The budget constraint is

$$40L + 100K = 3000$$ 　　Unit cost times the number
　　　　of units for labor and capital

$\underbrace{\phantom{40L}}$ $\underbrace{\phantom{100K}}$ $\underbrace{\phantom{3000}}$
Cost of　Cost of　Total
labor　　capital　budget

Therefore, the problem becomes:

maximize 　$P = 600L^{2/3}K^{1/3}$

subject to 　$40L + 100K - 3000 = 0$ 　　Constraint with zero on the right

$F(L, K, \lambda) = 600L^{2/3}K^{1/3} + \lambda(40L + 100K - 3000)$ 　　Lagrange function

$$\left.\begin{array}{l} F_L = 400L^{-1/3}K^{1/3} + 40\lambda = 0 \\ F_K = 200L^{2/3}K^{-2/3} + 100\lambda = 0 \\ F_\lambda = 40L + 100K - 3000 = 0 \end{array}\right\} \begin{array}{l} \text{Partials set equal} \\ \text{to zero} \end{array}$$

$\lambda = -10L^{-1/3}K^{1/3}$ 　　Solving the first equation for $\lambda$

$\lambda = -2L^{2/3}K^{-2/3}$ 　　Solving the second equation for $\lambda$

$-10L^{-1/3}K^{1/3} = -2L^{2/3}K^{-2/3}$ 　　Equating the two expressions for $\lambda$

$5L^{-1/3}K^{1/3} = L^{2/3}K^{-2/3}$ 　　Dividing each side by $-2$

$5K = L$ 　　Multiplying by $L^{1/3}K^{2/3}$ on each side

$40(5K) + 100K - 3000 = 0$ 　　Substituting $L = 5K$ into $40L + 100K - 300 = 0$

$300K = 3000$ 　　Simplifying

$K = 10$ 　　Solving for $K$

$L = 50$ 　　From $L = 5K$ with $K = 10$

The company should use 50 units of labor and 10 units of capital.

## Meaning of the Lagrange Multiplier

The Lagrange multiplier $\lambda$ has a useful interpretation. If we call the units of the objective function "objective units" and the units of the constraint function "constraint units," then $\lambda$ (or, more precisely, its absolute value) has the following interpretation, which will be justified in the next section (see page 534).

**Interpretation of $\lambda$**

$$|\lambda| = \left( \begin{array}{c} \text{Number of additional objective units} \\ \text{for each additional constraint unit} \end{array} \right)$$

As an illustration, in the preceding example we maximized production (units) subject to a budget constraint (dollars), so $|\lambda|$ gives *the number of additional units produced per additional dollar,* or, in the jargon of economics, *the marginal productivity of money.* We can calculate the value of $\lambda$ from either of the expressions for it.

$$\underbrace{\lambda = -10L^{-1/3}K^{1/3}}_{\text{From Example 3}} = \underbrace{-10(50)^{-1/3}(10)^{1/3}}_{\substack{\text{Substituting} \\ L = 50, \ K = 10}} \underbrace{\approx -5.8}_{\substack{\text{Using a} \\ \text{calculator}}} \qquad \substack{\text{From the} \\ \text{previous page}}$$

Therefore* $|\lambda| \approx 5.8$, meaning that production increases by about 5.8 units for each additional dollar in the budget. For example, an additional \$100 would result in about 580 extra units of production.

**Practice Problem 3**

In Example 1 we maximized the area of an enclosure subject to the constraint of having only 400 feet of fence.

a. Interpret the meaning of $|\lambda|$ in Example 1. [*Hint:* The objective function is area, and the constraint units are feet of fence.]

b. Calculate $|\lambda|$ in Example 1. [*Hint:* Use either expression for $\lambda$ on page 514.]

c. Use this number to approximate the additional area that could be enclosed by an additional 5 feet of fence.

➤ **Solutions on page 521**

## Geometry of Constrained Optimization

A constrained optimization problem can be visualized as a surface (the objective function), with a curve along it determined by the constraint. The *constrained* maximum is the highest point along the *curve,* whereas the *un*constrained maximum is the highest point on the *entire surface.*

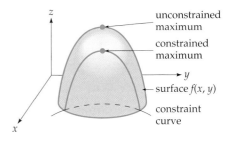

*The negative sign in $\lambda$ comes from our defining the Lagrange function as $F = f + \lambda g$. We could equally well have defined it as $F = f - \lambda g$, in which case $\lambda$ would have been positive.

If there are several critical points, we evaluate the objective function at each of them. The maximum and minimum values (which we assume to exist) are the largest and smallest of the resulting values of the function. In this way we can find both the maximum and minimum values in a constrained optimization problem.

**EXAMPLE 4**

**FINDING BOTH EXTREME VALUES**

Maximize *and* minimize $f(x, y) = 4xy$
subject to the constraint $x^2 + y^2 = 50$

*Solution*

| | |
|---|---|
| $F(x, y, \lambda) = 4xy + \lambda(x^2 + y^2 - 50)$ | Lagrange function |
| $F_x = 4y + \lambda 2x \quad = 0$ | |
| $F_y = 4x + \lambda 2y \quad = 0$ | Partials set equal to zero |
| $F_\lambda = x^2 + y^2 - 50 = 0$ | |
| $\lambda = -\dfrac{4y}{2x} = -\dfrac{2y}{x}$ | Solving $4y + \lambda 2x = 0$ for $\lambda$ (and simplifying) |
| $\lambda = -\dfrac{4x}{2y} = -\dfrac{2x}{y}$ | Solving $4x + \lambda 2y = 0$ for $\lambda$ (and simplifying) |
| $\dfrac{2y}{x} = \dfrac{2x}{y}$ | Equating the two $\lambda$'s (multiplying by $-1$) |
| $2y^2 = 2x^2$ | Multiplying both sides by $xy$ (or "cross-multiplying") |
| $y^2 = x^2$ | Dividing by 2 |
| $y = \pm x$ | Taking square roots ($+$ and $-$) |
| $x^2 + x^2 - 50 = 0$ | $x^2 + y^2 - 50 = 0$ (the third partial equation) with $y = \pm x$ |
| $2x^2 - 50 = 0$ | Simplifying |
| $2(x^2 - 25) = 0$ | Factoring |
| $\underbrace{\phantom{2(x^2-25)=0}}$ | |
| $(x + 5)(x - 5)$ | |
| $x = \pm 5$ | Solving |
| $y = \pm 5$ | From $y = \pm x$ |

Thus there are *four* critical points (all possible combinations of $x = \pm 5$ and $y = \pm 5$). We evaluate the objective function $f(x, y) = 4xy$ at each of them.

| Critical point | $f(x, y) = 4xy$ | |
|---|---|---|
| (5, 5) | 100 | 100 is the highest value of $f = 4xy$, |
| (−5, −5) | 100 | occurring at (5, 5) and at (−5, −5) |
| (5, −5) | −100 | −100 is the lowest value of $f = 4xy$, |
| (−5, 5) | −100 | occurring at (−5, 5) and at (5, −5) |

Maximum value of $f$ is 100, occurring at $\begin{cases} x = 5 \\ y = 5 \end{cases}$ and $\begin{cases} x = -5 \\ y = -5 \end{cases}$

Minimum value of $f$ is $-100$, occurring at $\begin{cases} x = -5 \\ y = 5 \end{cases}$ and $\begin{cases} x = 5 \\ y = -5 \end{cases}$

 **BE CAREFUL:** When solving an equation such as $y^2 = x^2$ in the above example, be sure to consider all possible combinations of $+$ and $-$ signs so that you find *all* of the critical points.

## Constrained Optimization of Functions of Three or More Variables

To optimize a function $f$ of any number of variables subject to a constraint $g = 0$, we proceed just as before, writing the Lagrange function $F = f + \lambda g$ and setting the partial with respect to each variable equal to zero. For example, to maximize $f(x, y, z)$ subject to $g(x, y, z) = 0$, we would write

$$F(x, y, z, \lambda) = f(x, y, z) + \lambda g(x, y, z) \qquad \text{Lagrange function}$$

and solve the partial equations

$$F_x = 0$$
$$F_y = 0$$
$$F_z = 0$$
$$F_\lambda = 0$$

### Solutions to Practice Problems

1. $2y^3 - 3x + 1 = 0$        Moving all terms to the left-hand side

2. The paint can

3. **a.** The approximate additional area for each additional foot of fence

   **b.** $|\lambda| = 100$ (using $\lambda = -x$ with $x = 100$)

   **c.** Approximately 500 more square feet of area (from $5 \cdot 100$)

## 7.5 Section Summary

To optimize (maximize or minimize) an *objective* function $f(x, y)$ subject to a *constraint* $g(x, y) = 0$, we form the *Lagrange function* $F(x, y, \lambda) = f(x, y) + \lambda g(x, y)$ and set its partials equal to zero:

$$f_x(x, y) + \lambda g_x(x, y) = 0 \qquad F_x = 0$$
$$f_y(x, y) + \lambda g_y(x, y) = 0 \qquad F_y = 0$$
$$g(x, y) = 0 \qquad F_\lambda = 0$$

We then find all solutions $(x, y)$ of these equations. The maximum or minimum (if they exist) of $f(x, y)$ subject to $g(x, y) = 0$ will occur at one of these solutions.

The absolute value of the *Lagrange multiplier* $\lambda$ found from solving these equations gives the approximate change in the *objective* function that would result from changing the *constraint* function by one unit.

## Justification of Lagrange's Method

The following is a geometric justification of Lagrange's method for maximizing an objective function $f(x, y)$ subject to a constraint $g(x, y) = 0$. In the graph below, each of the parallel curves represents a curve along which the objective function $f$ takes a constant value, with successively higher curves corresponding to higher values of the constant. Maximizing $f$ subject to $g = 0$ means finding the highest of these "objective" curves that still meets the constraint curve.

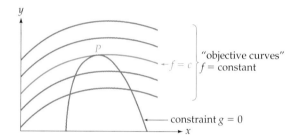

Observe that at the point where the highest objective curve meets the constraint curve (the point $P$), *the two curves have the same slope.* In Exercise 45 on page 536 we will show that the slope of the curve $f = c$ is $-f_x/f_y$, and the slope of the curve $g = 0$ is $-g_x/g_y$. Equating these slopes gives

$$-\frac{f_x}{f_y} = -\frac{g_x}{g_y} \qquad \text{or, equivalently,} \qquad -\frac{f_x}{g_x} = -\frac{f_y}{g_y} \qquad \begin{array}{l} \text{Multiplying by } f_y \\ \text{and dividing by } g_x \end{array}$$

The two sides of this last equation being equal is equivalent to equating each side to a number $\lambda$:

$$\lambda = -\frac{f_x}{g_x} \qquad \text{and} \qquad \lambda = -\frac{f_y}{g_y}$$

With a little algebra (clearing fractions, and moving everything to one side), these equations become

$$f_x + \lambda g_x = 0 \qquad \text{and} \qquad f_y + \lambda g_y = 0$$

However, these two equations, together with the constraint $g = 0$, are precisely the equations obtained from Lagrange's method by setting the partials of $F = f + \lambda g$ equal to zero. This completes the justification of Lagrange's method, and also exhibits its simplicity and power by showing all that is accomplished by the familiar steps of setting the partials equal to zero.

## 7.5 Exercises

**1–10.** Use Lagrange multipliers to maximize each function $f(x, y)$ subject to the constraint. (The maximum values *do* exist.)

**1.** $f(x, y) = 3xy, \quad x + 3y = 12$

**2.** $f(x, y) = 2xy, \quad 2x + y = 20$

**3.** $f(x, y) = 6xy, \quad 2x + 3y = 24$

**4.** $f(x, y) = 3xy, \quad 3x + 2y = 60$

**5.** $f(x, y) = xy - 2x^2 - y^2, \quad x + y = 8$

**6.** $f(x, y) = 12xy - 3y^2 - x^2, \quad x + y = 16$

**7.** $f(x, y) = x^2 - y^2 + 3, \quad 2x + y = 3$

**8.** $f(x, y) = y^2 - x^2 - 5, \quad x + 2y = 9$

**9.** $f(x, y) = \ln(xy), \quad x + y = 2e$

**10.** $f(x, y) = e^{xy}, \quad x + 2y = 8$

**11–20.** Use Lagrange multipliers to minimize each function $f(x, y)$ subject to the constraint. (The minimum values *do* exist.)

**11.** $f(x, y) = x^2 + y^2, \quad 2x + y = 15$

**12.** $f(x, y) = x^2 + y^2, \quad x + 2y = 30$

**13.** $f(x, y) = xy, \quad y = x + 8$

**14.** $f(x, y) = xy, \quad y = x + 6$

**15.** $f(x, y) = x^2 + y^2, \quad 2x + 3y = 26$

**16.** $f(x, y) = 5x^2 + 6y^2 - xy, \quad x + 2y = 24$

**17.** $f(x, y) = \ln(x^2 + y^2), \quad 2x + y = 25$

**18.** $f(x, y) = \sqrt{x^2 + y^2 + 5}, \quad 2x + y = 10$

**19.** $f(x, y) = e^{x^2 + y^2}, \quad x + 2y = 10$

**20.** $f(x, y) = 2x + y, \quad 2 \ln x + \ln y = 12$

**21–24.** Use Lagrange multipliers to maximize *and* minimize each function subject to the constraint. (The maximum and minimum values *do* exist.)

**21.** $f(x, y) = 2xy, \quad x^2 + y^2 = 8$

**22.** $f(x, y) = 2xy, \quad x^2 + y^2 = 18$

**23.** $f(x, y) = x + 2y, \quad 2x^2 + y^2 = 72$

**24.** $f(x, y) = 12x + 30y, \quad x^2 + 5y^2 = 81$

## Applied Exercises

**25–40.** Solve each using Lagrange multipliers. (The stated extreme values *do* exist.)

**25.** GENERAL: Fences A parking lot, divided into two equal parts, is to be constructed against a building, as shown in the diagram. Only 6000 feet of fence are to be used, and the side along the building needs no fence.

    **a.** What are the dimensions of the largest area that can be so enclosed?

    **b.** Evaluate and give an interpretation for $|\lambda|$.

**26.** GENERAL: Fences Three adjacent rectangular lots are to be fenced in, as shown in the diagram using 12,000 feet of fence. What is the largest total area that can be so enclosed?

**27–28.** GENERAL: Container Design A cylindrical tank without a top is to be constructed with the least amount of material (bottom plus side area). Find the dimensions if the volume is to be:

open top

$h$

**27.** 160 cubic feet.

**28.** 120 cubic feet.

**29.** GENERAL: Postal Regulations The U.S. Postal Service will accept a package if its length plus its girth is not more than 84 inches. Find the dimensions and volume of the largest package with a square end that can be mailed.

*Source:* U.S. Postal Service

length

square

girth (all the way around)

**30. GENERAL: Postal Regulations** Solve Exercise 29, but now for a package with a round end, so that the package is a cylinder rather than a rectangular solid. Compare the volume with that of Exercise 29.

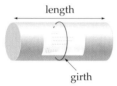
length

girth

**31. BUSINESS: Maximum Production** A company's output is given by the Cobb–Douglas production function $P = 200L^{3/4}K^{1/4}$, where $L$ and $K$ are the numbers of units of labor and capital. Each unit of labor costs $50 and each unit of capital costs $100, and $8000 is available to pay for labor and capital.

a. How many units of labor and of capital should be used to maximize production?
b. Evaluate and give an interpretation for $|\lambda|$.

**32. BUSINESS: Production Possibilities** A company manufactures two products, in quantities $x$ and $y$. Because of limited materials and capital, the quantities produced must satisfy the equation $2x^2 + 5y^2 = 32{,}500$. (This curve is called a *production possibilities curve*.) If the company's profit function is $P = 4x + 5y$ dollars, how many of each product should be made to maximize profit? Also find the maximum profit.

$2x^2 + 5y^2 = 32{,}500$
81
127

**33. GENERAL: Package Design** A metal box with a square base is to have a volume of 45 cubic inches. If the top and bottom cost 50 cents per square inch and the sides cost 30 cents per square inch, find the dimensions that minimize the cost. [*Hint:* The cost of the box is the area of each part (top, bottom, and sides) times the cost per square inch for that part. Minimize this subject to the volume constraint.]

**34. GENERAL: Building Design** A one-story building is to have 8000 square feet of floor space. The front of the building is to be made of brick, which costs $120 per linear foot, and the back and sides are to be made of cinderblock, which costs only $80 per linear foot.

a. Find the length and width that minimize the cost of the building. [*Hint:* The cost of the building is the length of the front, back, and sides, each times the cost per foot for that part. Minimize this subject to the area constraint.]
b. Evaluate and give an interpretation for $|\lambda|$.

$80 per ft

$80 per ft    8000 sq. ft    $80 per ft

$120 per ft

*Functions of Three Variables*

(The stated extreme values *do* exist.)

**35.** Minimize $f(x, y, z) = x^2 + y^2 + z^2$ subject to $2x + y - z = 12$.

**36.** Minimize $f(x, y, z) = x^2 + y^2 + z^2$ subject to $x - y + 2z = 6$.

**37.** Maximize $f(x, y, z) = x + y + z$ subject to $x^2 + y^2 + z^2 = 12$.

**38.** Maximize $f(x, y, z) = xyz$ subject to $x^2 + y^2 + z^2 = 12$.

**39. GENERAL: Building Design** A one-story storage building is to have a volume of 2000 cubic feet. The roof costs $32 per square foot, the walls $10 per square foot, and the floor $8 per square foot. Find the dimensions that minimize the cost of the building.

$z$
$y$
$x$

**40. GENERAL: Container Design** An open-top box with two parallel partitions, as in the diagram, is to have volume 64 cubic inches. Find the dimensions that require the least amount of material.

## Conceptual Exercises

**41.** True or False:  If $f(x, y)$ is the objective function and $g(x, y)$ is the constraint function, then the Lagrange function is

$$F(x, y, \lambda) = g(x, y) + \lambda f(x, y)$$

**42.** True or False:

$$|\lambda| = \left(\begin{array}{c}\text{Number of additional constraint units} \\ \text{for each additional objective unit}\end{array}\right)$$

**43.** True or False:  If the Lagrange function has no critical values, then the constrained optimization problem has no solution. (Assume for simplicity that the Lagrange function is defined for all values of its variables.)

**44.** Suppose you know that a constrained maximum problem has a solution. If the Lagrange function has one critical point, then what conclusion can you draw?

**45.** Suppose you know that a constrained maximum problem and the corresponding constrained minimum problem both have solutions. If the Lagrange function has two critical points, what conclusion can you draw?

**46.** If the maximum value of a function is a number $a$, and the maximum value of the function *subject to*

a *constraint* is a number $b$, then what can you say about the relationship between the numbers $a$ and $b$?

**47.** If the minimum value of a function is a number $a$, and the minimum value of the function *subject to a constraint* is a number $b$, then what can you say about the relationship between the numbers $a$ and $b$?

**48.** Explain why we could have defined the Lagrange function to be $F = f - \lambda g$ (instead of $F = f + \lambda g$) and still obtain the same solutions to constrained optimization problems.

## Explorations and Excursions

The following problem extends and augments the material presented in the text.

**49.** **ECONOMICS: Least Cost Rule**  A company's production is given by the Cobb–Douglas function $P = 60L^{2/3}K^{1/3}$, where $L$ and $K$ are the numbers of units of labor and capital. Each unit of labor costs $25 and each unit of capital costs $100. The company wants to produce exactly 1920 units.

    **a.** Find the numbers of units of labor and capital that meet the production requirements at the lowest cost.

    **b.** Find the marginal productivity of labor and the marginal productivity of capital. [*Hint:* This means the partials of $P$ with respect to $L$ and $K$.]

    **c.** Show that at the values found in part (a), the following relationship holds:

$$\frac{\text{Marginal productivity of labor}}{\text{Marginal productivity of capital}} = \frac{\text{Price of labor}}{\text{Price of capital}}$$

    This is called the "least cost rule."

## 7.6 TOTAL DIFFERENTIALS AND APPROXIMATE CHANGES

### Introduction

In this section we define the **total differential** of a function of several variables, and use it to approximate the change in the function resulting from changes in the independent variables. In addition to giving several applications, we will justify the interpretation of the Lagrange variable $\lambda$ stated on page 519.

## Total Differential of a Function of Two Variables

For a function $f(x)$ of one variable, we defined on page 388 the differential to be the *derivative multiplied by dx*, $dy = f'(x) \cdot dx$. For a function $f(x, y)$ of *two* variables, we define the total differential analogously as the partial derivatives multiplied by $dx$ and $dy$ and added.*

---

**Total Differential of $f(x, y)$**

For a function $f(x, y)$, the total differential $df$ is

$$df = f_x(x, y) \cdot dx + f_y(x, y) \cdot dy$$

More briefly,
$df = f_x \cdot dx + f_y \cdot dy$

---

The total differential can also be written with the partials in the $\partial$ notation:

$$df = \frac{\partial f}{\partial x} \cdot dx + \frac{\partial f}{\partial y} \cdot dy$$

Total differential in $\partial$ notation

**EXAMPLE 1**

### FINDING A TOTAL DIFFERENTIAL

Find the total differential of $f(x, y) = 5x^3 - 4xy^{-1} + 3y^4$.

*Solution*

The partials are

$$f_x = 15x^2 - 4y^{-1}$$

Partial of $5x^3 - 4xy^{-1} + 3y^4$ with respect to $x$

$$f_y = 4xy^{-2} + 12y^3$$

Partial of $5x^3 - 4xy^{-1} + 3y^4$ with respect to $y$

Then the total differential is

$$df = \underbrace{(15x^2 - 4y^{-1})}_{f_x(x, y)} \cdot dx + \underbrace{(4xy^{-2} + 12y^3)}_{f_y(x, y)} \cdot dy$$

$df = f_x\, dx + f_y\, dy$ with the above partials

---

If $dx$ and $dy$ are considered as new variables, then the total differential $df$ is a function of the *four* variables $x$, $y$, $dx$, and $dy$. The total differential of $z = f(x, y)$ can be denoted $dz$ or $df$.

---

*Technically, the partials $f_x$ and $f_y$ must be continuous. However, since we have not defined continuity for functions of two variables, we will not discuss this requirement further, except to say that it is satisfied by all functions in this section and by most functions encountered in applications.

EXAMPLE 2

**FINDING THE TOTAL DIFFERENTIAL OF A LOGARITHMIC FUNCTION**

Find the total differential of $z = \ln(x^2 + y^3)$.

*Solution*

The partials are

Partials of the denominator

$$z_x = \frac{2x}{x^2 + y^3} \qquad z_y = \frac{3y^2}{x^2 + y^3} \qquad \text{Derivative of } \ln f \text{ is } \frac{f'}{f}$$

Therefore, $dz$ is

$$dz = \frac{2x}{x^2 + y^3} \cdot dx + \frac{3y^2}{x^2 + y^3} \cdot dy \qquad \text{Partials times } dx \text{ and } dy$$

Practice Problem

Find the total differential of $g(x, y) = x^2 e^{5y}$. ➤ Solution on page 533

## Approximating Changes by Total Differentials

For a function $f(x, y)$, changing the value of $x$ by $\Delta x$ and $y$ by $\Delta y$ generally changes the value of the function. The change $\Delta f$ in the function is found by evaluating $f$ at the "changed" values and subtracting $f$ evaluated at the original values.

**Change in** $f(x, y)$

$$\Delta f = \underbrace{f(x + \Delta x, y + \Delta y)}_{\substack{f \text{ at the} \\ \text{"new" values}}} - \underbrace{f(x, y)}_{\substack{f \text{ at the} \\ \text{"old" values}}}$$

 **BE CAREFUL:** For *independent* variables (such as $x$ and $y$) we use "$\Delta$" and "$d$" interchangeably to denote changes. That is,

$$\Delta x = dx \qquad \text{and} \qquad \Delta y = dy \qquad \begin{array}{l}\text{For independent variables,} \\ \text{``$\Delta$'' = ``$d$''}\end{array}$$

However, for *dependent* variables, "$\Delta$" and "$d$" have different meanings: $\Delta$ indicates the *actual* change, and $d$ indicates the total differential.

For some functions, calculating the actual change $\Delta f$ can be complicated. The total differential provides a simple **linear approximation** for the actual change. The partials $f_x$ and $f_y$ give the rate of change of $f$ per unit change in $x$

and $y$, respectively. Changing $x$ by $\Delta x$ units changes $f$ by approximately $f_x \cdot \Delta x$ units, and changing $y$ by $\Delta y$ units changes $f$ by approximately $f_y \cdot \Delta y$ units. Therefore, changing *both* $x$ and $y$ should change $f$ by approximately the *sum* of these changes. In symbols:

---

**Differential Approximation Formula**

$$f(x + \Delta x, y + \Delta y) - f(x, y) \approx f_x \cdot \Delta x + f_y \cdot \Delta y$$

$\underbrace{\qquad\qquad\qquad\qquad\qquad}$ Change in $f$

$\underbrace{\qquad\qquad\qquad}$ Total differential of $f$
(since $\Delta x = dx$ and $\Delta y = dy$)

---

Written more compactly:

$$\Delta f \approx df$$

Since $\Delta f = f(x + \Delta x, y + \Delta y) - f(x, y)$
and $df = f_x(x, y) \cdot dx + f_y(x, y) \cdot dy$

The partials in the first formula are evaluated at $(x, y)$, and the approximation improves as $\Delta x$ and $\Delta y$ approach zero. (See pages 531–532 for a geometric explanation of the approximation $\Delta f \approx df$.)

**EXAMPLE 3**

**APPROXIMATING AN ACTUAL CHANGE BY A DIFFERENTIAL**

For $f(x, y) = x^2 + 4xy + y^3$ and values $x = 3$, $y = 2$, $\Delta x = 0.2$, and $\Delta y = -0.1$, find **a.** $\Delta f$ **b.** $df$

*Solution*

**a.** From the given values,

$$x + \Delta x = 3 + 0.2 = 3.2 \quad \text{and} \quad y + \Delta y = 2 - 0.1 = 1.9$$

The change $\Delta f$ is

$$\Delta f = f(3.2, 1.9) - f(3, 2) \qquad\qquad \Delta f = f(x + \Delta x, y + \Delta y) - f(x, y)$$

$$= \underbrace{3.2^2 + 4 \cdot (3.2) \cdot (1.9) + 1.9^3}_{f(3.2,\,1.9)} - \underbrace{(3^2 + 4 \cdot 3 \cdot 2 + 2^3)}_{f(3,\,2)} \qquad \text{Using } f(x, y) = x^2 + 4xy + y^3$$

$$= 10.24 + 24.32 + 6.859 - (9 + 24 + 8) \qquad \text{Evaluating}$$

$$= 41.419 - 41 = 0.419 \qquad\qquad \text{Change in } f \text{ is } \Delta f = 0.419$$

**b.** The total differential $df$ is

$$df = \overbrace{(2x + 4y)}^{f_x(x,\,y)} \cdot dx + \overbrace{(4x + 3y^2)}^{f_y(x,\,y)} \cdot dy$$

Partials of $x^2 + 4xy + y^3$ times $dx$ and $dy$

$$= (2 \cdot 3 + 4 \cdot 2) \cdot (0.2) + (4 \cdot 3 + 3 \cdot 2^2) \cdot (-0.1)$$

Evaluating at $x = 3$, $y = 2$, $dx = 0.2$, and $dy = -0.1$

$$= 2.8 - 2.4 = 0.4$$

Total differential is $df = 0.4$

We found $\begin{cases} \Delta f = 0.419 \\ df = 0.4 \end{cases}$

The total differential $df = 0.4$ provides a reasonably accurate approximation for the actual change $\Delta f = 0.419$. The approximation would be even more accurate for smaller values of $\Delta x$ and $\Delta y$.

Why should we bother calculating an *approximation df* when with a calculator we can easily find the *exact* change $\Delta f$? The total differential $df$ has the advantage of *linearity:* if we were to *double* the changes $\Delta x$ and $\Delta y$ in the independent variables, then the differential would also double, from 0.4 to 0.8; if we were to *halve* the changes $\Delta x$ and $\Delta y$, then the differential would also be halved, from 0.4 to 0.2. The *actual* change $\Delta f$ admits no such simple modification—whenever $\Delta x$ and $\Delta y$ change it has to be recalculated "from scratch" using the formula $\Delta f = f(x + \Delta x, y + \Delta y) - f(x, y)$. The linearity of the total differential $df$ is an advantage when you are trying to understand how various changes in $x$ and $y$ would affect the value of $f$.

**EXAMPLE 4**

**ESTIMATING ADDITIONAL PROFIT**

The American Farm Machinery Company finds that if it manufactures $x$ economy tractors and $y$ heavy-duty tractors per month, then its profit (in thousands of dollars) will be $P(x, y) = 3x^{4/3} + 0.05xy + 4y$. If the company now manufactures 125 economy tractors and 100 heavy-duty tractors per month, find an approximation for the additional profit from manufacturing 3 more economy tractors and 2 more heavy-duty tractors per month.

*Solution*

We want an estimate for the change in the profit $P(x, y)$ when production increases above the levels $x = 125$ and $y = 100$ by amounts $\Delta x = 3$ and $\Delta y = 2$. The partials are

$$P_x = 4x^{1/3} + 0.05y \qquad P_y = 0.05x + 4$$

Partials of $P = 3x^{4/3} + 0.05xy + 4y$ with respect to $x$ and $y$

The total differential is

$$dP = (4x^{1/3} + 0.05y) \cdot dx + (0.05x + 4) \cdot dy \qquad \qquad dP = P_x \cdot dx + P_y \cdot dy$$

$$= (4 \cdot 125^{1/3} + 0.05 \cdot 100) \cdot 3 + (0.05 \cdot 125 + 4) \cdot 2$$

Substituting
$x = 125, \quad y = 100,$
$\Delta x = 3, \quad \text{and} \quad \Delta y = 2$

$$\underbrace{\phantom{4 \cdot 125^{1/3}}}_{\sqrt[3]{125}\, = \, 5}$$

$$= (20 + 5) \cdot 3 + (6.25 + 4) \cdot 2 = 75 + 20.5 = 95.5 \qquad \text{In thousands of dollars}$$

Therefore, producing 3 more economy tractors and 2 more heavy-duty tractors will generate about $95,500 in additional profit.

The actual change in the profit function, found by applying the $\Delta f$ formula on page 527, is $\Delta P = P(125 + 3, 100 + 2) - P(125, 100) \approx 96.04$, so the approximation of 95.5 is indeed quite accurate.

The *linearity* of the total differential means that if sometime later the company wanted to estimate the additional profit from *doubling* these changes—making 6 more economy tractors and 4 more heavy-duty tractors—they would need only to double the estimate of $95,500 to immediately obtain $191,000.

## Estimating Errors

No physical measurement can ever be made with perfect accuracy. If you can estimate the maximum error in a measurement, then you can use differentials to estimate the resulting error in a calculation. The measurement errors may be expressed as *percentage* errors or *actual* numbers (as in Example 6, to be discussed shortly). The following example estimates the percentage error in calculating the volume of a cylinder. Such calculations have applications from predicting variations in soft-drink cans to ensuring safety margins for artificial arteries.

**EXAMPLE 5**    **ESTIMATING THE ERROR IN CALCULATING VOLUME**

A cylinder is measured to have radius $r$ and height $h$, but these measurements may be in error by up to 1%. Estimate the resulting percentage error in calculating the volume of the cylinder.

*Solution*

The height and radius being in error by 1% means that

$$\Delta r = 0.01r$$
$$\Delta h = 0.01h$$

For each, the change is 1% of the value

The volume of a cylinder is $V = \pi r^2 h$, and the total differential is

$$dV = 2\pi r h \cdot dr + \pi r^2 \cdot dh$$

Partials of $V = \pi r^2 h$ times $dr$ and $dh$

$$= 2\pi r h \cdot 0.01r + \pi r^2 \cdot 0.01h$$

Substituting $dr = \Delta r = 0.01r$ and $dh = \Delta r = 0.01h$

$$= \pi r^2 h (\underbrace{2 \cdot 0.01 + 0.01}_{0.03})$$

Factoring out $\pi r^2 h$

$$= 0.03 \underbrace{\pi r^2 h}_{} = 0.03 \cdot V$$

Change is 3% of volume

Volume $V$ of the cylinder

This result, $dV = 0.03 \cdot V$, means that the volume may be in error by as much as 3% if the radius and height are "off" by 1%.

## Geometric Visualization of *df* and $\Delta f$

A function $f(x, y)$ represents a *surface* in three-dimensional space, and the change $\Delta f$ represents the change in height when moving from one point to another *along the surface*, as shown below.

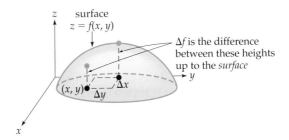

The total differential *df* represents the change in height when moving from one point to another *along the plane* that best fits the surface at the first point, as shown below. This plane is called the *tangent plane* to the surface at the point.

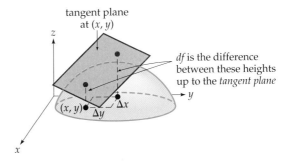

Since the tangent plane fits the surface closely near the point $(x, y)$, changes *df* in height along the *tangent plane* should be very close to changes $\Delta f$ in height along the *surface,* which is why the total differential *df* closely approximates the actual change *Δf* for small changes in *x* and *y*.

## Total Differential of a Function of Three Variables

The total differential may be generalized to apply to functions of three (or more) variables, multiplying each partial by "*d*" of that variable and adding.

---

**Total Differential of** $f(x, y, z)$

For a function $f(x, y, z)$, the total differential *df* is

$$df = f_x(x, y, z) \cdot dx + f_y(x, y, z) \cdot dy + f_z(x, y, z) \cdot dz$$

Partials times
$dx, dy,$ and $dz$

---

The total differential of a function $f(x, y, z)$ gives an estimate for the change $\Delta f$ when the variables are changed by amounts $\Delta x$, $\Delta y$, and $\Delta z$:

$$\underbrace{f(x + \Delta x, y + \Delta y, z + \Delta z) - f(x, y, z)}_{\text{Change in } f} \approx \underbrace{\frac{\partial f}{\partial x} \cdot \Delta x + \frac{\partial f}{\partial y} \cdot \Delta y + \frac{\partial f}{\partial z} \cdot \Delta z}_{\substack{\text{Total differential of } f \\ (\text{since } \Delta x = dx, \\ \Delta y = dy, \ \Delta z = dz)}}$$

or, more briefly:

$$\Delta f \approx df$$

The partials in the formula above are evaluated at $(x, y, z)$, and the approximation is increasingly accurate for smaller values of $\Delta x$, $\Delta y$, and $\Delta z$.

---

**EXAMPLE 6**

### ESTIMATING THE ERROR IN CALCULATING VOLUME

A rectangular box is measured to be 30 inches long, 24 inches wide, and 10 inches high. If the maximum errors in measuring the length, width, and height of the box are, respectively, 0.3, 0.2, and 0.1 inch, estimate the maximum error in calculating its volume.

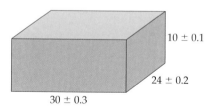

$10 \pm 0.1$

$24 \pm 0.2$

$30 \pm 0.3$

*Solution*

The volume of the box is length times width times height, $V = x \cdot y \cdot z$. We want to estimate the change in volume resulting from changing the dimensions $x = 30$, $y = 24$, and $z = 10$ by the amounts $\Delta x = 0.3$, $\Delta y = 0.2$, and $\Delta z = 0.1$. The total differential is

$$df = \overbrace{y \cdot z}^{V_x} \cdot dx + \overbrace{x \cdot z}^{V_y} \cdot dy + \overbrace{x \cdot y}^{V_z} \cdot dz \qquad \begin{array}{l} \text{Partials of} \\ V = x \cdot y \cdot z \text{ times} \\ dx, dy, \text{ and } dz \end{array}$$

$$= 24 \cdot 10 \cdot 0.3 + 30 \cdot 10 \cdot 0.2 + 30 \cdot 24 \cdot 0.1 \qquad \begin{array}{l} \text{Substituting } x = 30, \\ y = 24, \ z = 10, \ dx = 0.3, \\ dy = 0.2, \text{ and } dz = 0.1 \end{array}$$

$$= 72 + 60 + 72 = 204 \qquad \begin{array}{l} \text{Multiplying out} \\ \text{and adding} \end{array}$$

That is, the maximum error in calculating the volume is approximately 204 cubic inches.

---

While an error of 204 cubic inches may seem large, the *relative* error (the error divided by the volume, expressed as a percentage) is only

$$\frac{204}{30 \cdot 24 \cdot 10} = \frac{204}{7200} \approx 0.028 = 2.8\% \qquad \text{Relative error is 2.8\%}$$

### Solution to Practice Problem

$$g_x = 2xe^{5y} \qquad g_y = x^2 5e^{5y} \qquad\qquad \text{Partials}$$
$$dg = 2xe^{5y} \cdot dx + 5x^2 e^{5y} \cdot dy \qquad\qquad \text{Total differential}$$

## 7.6  Section Summary

For a function $f(x, y)$, the total differential $df$ is defined as the partials multiplied by $dx$ and $dy$ and added:

$$df = f_x(x, y) \cdot dx + f_y(x, y) \cdot dy \qquad\qquad df = \frac{\partial f}{\partial x} dx + \frac{\partial f}{\partial y} dy$$

The actual change in the function when $x$ and $y$ change by amounts $\Delta x$ and $\Delta y$ is

$$\Delta f = f(x + \Delta x, y + \Delta y) - f(x, y)$$

For small values of $\Delta x = dx$ and $\Delta y = dy$, the actual change $\Delta f$ can be approximated by the total differential $df$:

$$f(x + \Delta x, y + \Delta y) - f(x, y) \approx f_x(x, y) \cdot dx + f_y(x, y) \cdot dy \qquad \Delta f \approx df$$

The actual change $\Delta f$ may depend on $\Delta x$ and $\Delta y$ in very complicated ways, but the total differential $df$ is *linear* in $\Delta x = dx$ and $\Delta y = dy$, and is therefore easier to calculate.

## Justification of the Interpretation of the Lagrange Multiplier $\lambda$

In the preceding section we used Lagrange multipliers to optimize an objective function $f(x, y)$ subject to a constraint $g(x, y) = 0$. We did so by setting the partials of $F(x, y, \lambda) = f(x, y) + \lambda g(x, y)$ equal to zero, and we interpreted the absolute value of the "Lagrange multiplier" $|\lambda|$ as *the number of additional objective units per additional constraint unit*. This interpretation may be justified as follows. Setting the partials of $F$ with respect to $x$ and $y$ equal to zero gives

$$\begin{cases} f_x + \lambda g_x = 0 \\ f_y + \lambda g_y = 0 \end{cases} \quad \text{or, equivalently,} \quad \begin{cases} f_x = -\lambda g_x \\ f_y = -\lambda g_y \end{cases}$$

If we increase $x$ and $y$ by amounts $\Delta x$ and $\Delta y$, the resulting change in the objective function $f(x, y)$ can be approximated by the total differential:

$$\underbrace{f(x + \Delta x, y + \Delta y) - f(x, y)}_{\Delta f} \approx f_x \Delta x + f_y \Delta y \qquad \Delta f \approx df$$

$$= -\lambda g_x \Delta x - \lambda g_y \Delta y \qquad \begin{array}{l} \text{Substituting} \\ f_x = -\lambda g_x \text{ and} \\ f_y = -\lambda g_y \end{array}$$

$$= -\lambda \underbrace{(g_x \Delta x + g_y \Delta y)}_{dg} \qquad \begin{array}{l} \text{Factoring out } -\lambda \\ \text{leaves the total} \\ \text{differential of } g \end{array}$$

$$\approx -\lambda \cdot \Delta g \qquad \begin{array}{l} \text{Replacing } dg \text{ by } \Delta g \\ (\text{since } dg \approx \Delta g) \end{array}$$

These equations, read from beginning to end, say that $\Delta f \approx -\lambda \cdot \Delta g$, or

$$\frac{\Delta f}{\Delta g} \approx -\lambda \qquad \text{Dividing by } \Delta g$$

The left-hand side of this is the ratio of the change in the objective function $f$ to the change in the constraint function $g$. Taking absolute values and letting $\Delta x$ and $\Delta y$ approach zero (to make the approximation exact) shows that $|\lambda|$ is *the number of objective units per additional constraint unit*, as stated on page 519.

## 7.6 Exercises

**1–20.** Find the total differential of each function.

**1.** $f(x, y) = x^2 y^3$

**2.** $f(x, y) = x^4 y^{-1}$

**3.** $f(x, y) = 6x^{1/2} y^{1/3} + 8$

**4.** $f(x, y) = 100x^{0.05} y^{0.02} - 7$

**5.** $g(x, y) = \dfrac{x}{y}$

**6.** $g(x, y) = \dfrac{x}{x + y}$

**7.** $g(x, y) = (x - y)^{-1}$

**8.** $g(x, y) = \sqrt{x^2 + y^2}$

**9.** $z = \ln(x^3 - y^2)$

**10.** $z = x^2 \ln y$

**11.** $z = xe^{2y}$

**12.** $z = e^{3x - 2y}$

**13.** $w = 2x^3 + xy + y^2$

**14.** $w = 3x^2 - xy^{-1} + y^3$

**15.** $f(x, y, z) = 2x^2 y^3 z^4$

**16.** $f(x, y, z) = xy + yz + xz$

**17.** $f(x, y, z) = \ln(xyz)$

**18.** $f(x, y, z) = \ln(x^2 + y^2 + z^2)$

**19.** $f(x, y, z) = e^{xyz}$

**20.** $f(x, y, z) = e^{x^2 + y^2 + z^2}$

**21–26.** For the given function and values, find:

**a.** $\Delta f$    **b.** $df$

**21.** $f(x, y) = x^2 + xy + y^3$,  $x = 4$,  $\Delta x = dx = 0.2$,
$y = 2$,  $\Delta y = dy = -0.1$

**22.** $f(x, y) = x^3 + xy + y^3$,  $x = 5$,  $\Delta x = dx = 0.01$,
$y = 3$,  $\Delta y = dy = -0.01$

**23.** $f(x, y) = e^x + xy + \ln y$,  $x = 0$,  $\Delta x = dx = 0.05$,
$y = 1$,  $\Delta y = dy = 0.01$

**24.** $f(x, y) = \ln(x^2 + y^2)$,  $x = 6$,  $\Delta x = dx = 0.1$,
$y = 8$,  $\Delta y = dy = 0.2$

**25.** $f(x, y, z) = xy + z^2$,  $x = 3$,  $\Delta x = dx = 0.03$,
$y = 2$,  $\Delta y = dy = 0.02$,  $z = 1$,  $\Delta z = dz = 0.01$

**26.** $f(x, y, z) = x^2 + y^2 + z^2$,  $x = 3$,  $\Delta x = dx = 0.1$,
$y = 4$,  $\Delta y = dy = 0.1$,  $z = 5$,  $\Delta z = dz = 0.1$

## Applied Exercises

**27–37.** Use total differentials to solve the following exercises.

**27. GENERAL: Measurement Errors**  A rectangle is measured to be 150 feet by 100 feet, but each measurement may be "off" by half a foot. Estimate the error in calculating the area. Then estimate the error in calculating the area if each measurement is "off" by one foot.

**28. GENERAL: Telephone Calls**  For two cities with populations $x$ and $y$ (in thousands) that are 500 miles apart, the number of telephone calls per day between them can be modeled by the function $12xy$. For two cities with populations 40 thousand and 60 thousand, estimate the number of additional telephone calls if each city grows by 1 thousand people. Then estimate the number of additional calls if instead each city were to grow by only 500 people.

**29–30. BUSINESS: Profit**  An electronics company's profit in dollars from making $x$ DVD players and $y$ CD players per day is given by the following profit function $P(x, y)$. If the company currently produces 200 DVD players and 300 CD players, estimate the extra profit that would result from producing five more DVD players and four more CD players.

**29.** $P(x, y) = 2x^2 - 3xy + 3y^2$

**30.** $P(x, y) = 3x^2 - 4xy + 4y^2$

**31. GENERAL: Highway Safety**  The emergency stopping distance in feet for a truck of weight $w$ tons traveling at $v$ miles per hour on a dry road is $S = 0.027wv^2$. For a truck that weighs 4 tons and is usually driven at 60 miles per hour, estimate the extra stopping distance if it has an extra half ton of load and is traveling 5 miles per hour faster than usual.
*Source:* National Highway Traffic Safety Administration

**32. GENERAL: Scuba Diving**  The maximum duration (in minutes) of a scuba dive can be estimated by the formula  $T = \dfrac{33v}{d + 33}$,  where $v$

is the volume of air in the tank (in cubic feet at sea-level pressure) and $d$ is the depth (in feet) of the dive. For values  $v = 100$  and  $d = 67$, estimate the change in duration if an extra 20 cubic feet of air is added and the dive is 10 feet deeper.
*Source: U.S. Navy Diving Manual, Vol. 2*

**33. GENERAL: Relative Error in Calculating Area**  A rectangle is measured to have length $x$ and width $y$, but each measurement may be in error by 1%. Estimate the percentage error in calculating the area.

**34. GENERAL: Relative Error in Calculating Volume**  A rectangular solid is measured to have length $x$, width $y$, and height $z$, but each measurement may be in error by 1%. Estimate the percentage error in calculating the volume.

**35. BIOMEDICAL: Cardiac Output**  Medical researchers calculate the quantity of blood pumped through the lungs (in liters per minute) by the formula  $C = \dfrac{x}{y - z}$,  where $x$ is the amount of oxygen absorbed by the lungs (in milliliters per minute), and $y$ and $z$ are, respectively, the concentrations of oxygen in the blood just after and just before passing through the lungs (in milliliters of oxygen per liter of blood). Typical measurements are  $x = 250$, $y = 160$,  and  $z = 150$.  Estimate the error in calculating the cardiac output $C$ if each measurement may be "off" by 5 units.
*Source: T. Aherns and K. Rutherford (eds.), Essentials of Oxygenation*

**36. GENERAL: Windchill**  The windchill index announced during the winter by the weather bureau measures how cold it "feels" for a given temperature $t$ (in degrees Fahrenheit) and wind speed $w$ (in miles per hour). It is calculated by the formula  $C(t, w) = 35.74 + 0.6215t - 35.75w^{0.16} + 0.4275tw^{0.16}$.  If the temperature is 30 degrees and the wind speed is 10 miles per hour, estimate the change in the windchill temperature if the wind speed increases by 4 miles per hour and the temperature drops by 5 degrees.

*Source:* National Weather Service

**37. BIOMEDICAL: Blood Vessel Volume** A section of an artery is measured to have length 12 centimeters and diameter 0.8 centimeter. If each measurement may be off by 0.1 centimeter, find the volume with an estimate of the error.

### Conceptual Exercises

**38.** Let $f(x, y)$ be an arbitrary function. For each equation, state whether it is True or False:
   **a.** $dx = \Delta x$   **b.** $dy = \Delta y$   **c.** $df = \Delta f$

**39.** What is the total differential of a *constant* function?

**40.** What is the total differential of the *linear* function $f(x, y) = ax + by + c$ where $a$, $b$, and $c$ are constants?

**41.** What is the most general function $f(x, y)$ that satisfies $df = dx + dy$?

**42.** What is the value of the total differential of a function evaluated at a *critical point* of the function? [*Hint:* See the definition of critical point on page 488.]

**43.** If the total differential of a function $f(x, y)$ simplifies to $df = f_x(x, y) \cdot dx$, what conclusion can you draw about the function $f(x, y)$?

**44.** In Example 6 on pages 532–533 we found that the *actual* error was 204 and the *relative* error was the much smaller number $\frac{204}{7200} = 0.028$. Can the relative error ever be *greater* than the actual error?

### Explorations and Excursions

The following problem extends and augments the material presented in the text.

**45. THE SLOPE OF $f(x, y) = c$** On page 522 we used the fact that the slope in the $x$-$y$ plane of the curve defined by $f(x, y) = c$ (for constant $c$) is given by the formula $-\dfrac{f_x}{f_y}$. Verify this formula by justifying the following five steps.

   **a.** If $f(x, y) = c$ can be solved explicitly for a function $y = F(x)$, then we may write $f(x, F(x)) = c$.
   Justify: $f(x + \Delta x, F(x + \Delta x)) - f(x, F(x)) = 0$.
   **b.** Justify: $f(x + \Delta x, F(x + \Delta x) - F(x) + F(x)) - f(x, F(x)) = 0$.
   **c.** Defining $\Delta F$ by $\Delta F = F(x + \Delta x) - F(x)$, we may write the equation above as
   $$f(x + \Delta x, \Delta F + F(x)) - f(x, F(x)) = 0$$
   Then, writing $F$ for $F(x)$, this becomes
   $$f(x + \Delta x, F + \Delta F) - f(x, F) = 0$$
   Justify: $f_x \Delta x + f_y \Delta F \approx 0$.
   **d.** Justify: $\dfrac{\Delta F}{\Delta x} \approx -\dfrac{f_x}{f_y}$.
   **e.** Justify: $\dfrac{dF}{dx} = -\dfrac{f_x}{f_y}$.

This shows that the slope of $F(x)$, and therefore the slope of $f(x, y) = c$, is $-\dfrac{f_x}{f_y}$.

## 7.7 MULTIPLE INTEGRALS

### Introduction

This section discusses *integration* of functions of several variables. We define the **double integral** of a function by considering the volume under a surface $z = f(x, y)$. We then evaluate double integrals by **iterated integrals,** that is, by repeated single integrations. Finally, we use double integrals to calculate volumes, average values, and total accumulations. We restrict our attention to continuous functions (surfaces that have no holes or breaks), since most functions encountered in applications satisfy this restriction.

### Rectangular Regions, Volumes, and Double Integrals

The points $(x, y)$ in the plane for $x$ taking values between numbers $a$ and $b$ and $y$ taking values between numbers $c$ and $d$ determine a **rectangular region** $R$.

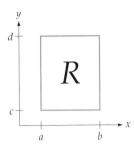

Rectangular region   $R = \{(x,y) \,|\, a \le x \le b, c \le y \le d\}$

The graph below shows a nonnegative function $f(x, y)$ defined on a rectangular region $R$. We want to find the **volume** under the surface $f$ and above the region $R$.

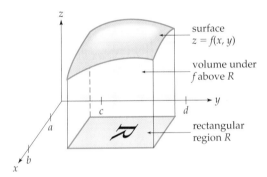

Volume under the surface   $z = f(x, y)$
lying above a rectangular region $R$

We begin by *approximating* the volume by rectangular solids ("boxes") extending from $R$ up to the surface. We divide $R$ into small rectangles by drawing lines parallel to the $x$- and $y$-axes with spacing $\Delta x$ and $\Delta y$, as shown below.* On each of these small rectangles we erect a rectangular solid with height $f(x_i, y_j)$, the height of the surface at some point $(x_i, y_j)$ in the base rectangle.

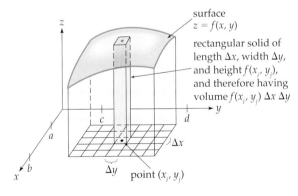

Volume under $f$ over $R$ showing one of the rectangular solids

*Technically, the parallel lines need not have equal spacing. However, we will be letting the spacings approach zero, and the final results are the same for equal or unequal spacing.

The volume of the rectangular solid is $f(x_i, y_j) \cdot \Delta x \cdot \Delta y$ (height times length times width), and the sum of the volumes of all such rectangular solids approximates the volume under $f$ above $R$:

$$\begin{pmatrix} \text{Volume under} \\ f \text{ above } R \end{pmatrix} \approx \sum f(x_i, y_j) \cdot \Delta x \cdot \Delta y \qquad \begin{array}{l} \Sigma \text{ means sum over} \\ \text{all base rectangles} \end{array}$$

The *limit* of this sum as both $\Delta x$ and $\Delta y$ approach zero (so that the base rectangles become smaller and more numerous) gives the *exact* volume, and is called the *double integral of $f$ over $R$,* denoted $\iint\limits_{R} f(x, y) \, dx \, dy$.

---

**Double Integrals**

The double integral of a continuous function $f(x, y)$ on a rectangular region $R$ is

$$\iint\limits_{R} f(x, y) \, dx \, dy = \lim_{\Delta x, \Delta y \to 0} \sum f(x_i, y_j) \cdot \Delta x \cdot \Delta y \qquad \begin{array}{l} \text{The sum is over} \\ \text{all rectangles in } R, \\ \text{each containing} \\ \text{one } (x_i, y_j) \end{array}$$

If $f(x, y)$ is nonnegative on $R$, then the double integral gives the *volume under $f$ over $R$.*

---

## Iterated Integrals

Evaluating double integrals from the definition is difficult. Fortunately, double integrals can be evaluated by two separate "single" integrations, integrating with respect to one variable at a time while holding the other variable constant. (This is analogous to partial differentiation with respect to one variable, holding the other variable constant.) Such repeated integrals are called *iterated integrals* ("iterated" means "repeated"). A proof that double integrals can be evaluated as iterated integrals can be found in a book on advanced calculus.

**EXAMPLE 1**

**EVALUATING AN ITERATED INTEGRAL**

Evaluate the iterated integral $\displaystyle\int_0^1 \int_0^2 (3x^2 + 6xy^2) \, dx \, dy$.

*Solution*

The two separate integrations will be clearer if we use parentheses:

$$\int_0^1 \left( \int_0^2 (3x^2 + 6xy^2) \, dx \right) dy \qquad \begin{array}{l} \text{An inner } x\text{-integral and} \\ \text{an outer } y\text{-integral} \end{array}$$

The inner integral gives

$$\int_0^2 (3x^2 + 6xy^2)\, dx = \left(x^3 + 6\cdot\frac{1}{2}x^2 y^2\right)\Big|_{x=0}^{x=2}$$

<u>dx means integrate</u>
with respect to $x$
holding $y$ constant

Integral
of $3x^2$

Integral
of $x$

Held
constant

$$= \underbrace{2^3 + 3\cdot 2^2 y^2}_{\substack{\text{Evaluated}\\ \text{at } x = 2}} \underbrace{-\quad 0}_{\substack{\text{And at}\\ x = 0}} \underbrace{= 8 + 12y^2}_{\text{Simplified}}$$

We now apply the outer $y$-integral to this result:

$$\int_0^1 (8 + 12y^2)\, dy = (8y + 4y^3)\Big|_{y=0}^{y=1} = 8 + 4 \quad - \quad 0 \quad = \quad 12$$

<u>Result of the</u>
inner integral

$dy$ means
integrate with
respect to $y$

$12\cdot\frac{1}{3}$

Evaluated
at $y = 1$

And at
$y = 0$

Final
answer

Therefore:

$$\int_0^1 \int_0^2 (3x^2 + 6xy^2)\, dx\, dy = 12$$

The iterated integral
equals 12

---

Always solve an iterated integral "from the inside out."

$$\int_0^1 \left(\int_0^2 (3x^2 + 6xy^2)\, dx\right) dy$$

Limits
for $y$

Limits
for $x$

First integrate
with respect to $x$

Then with
respect to $y$

In Example 1 we integrated first with respect to $x$ and then with respect to $y$. The next example shows that switching the order of integration gives the same answer, provided that we also switch the $x$ and $y$ limits of integration. That is,

$$\int_0^1 \int_0^2 (3x^2 + 6xy^2)\, dx\, dy$$

is equal to

$$\int_0^2 \int_0^1 (3x^2 + 6xy^2)\, dy\, dx$$

 **BE CAREFUL:** Be sure to switch the $x$ and $y$ limits of integration when you switch the order of integration.

**EXAMPLE 2**

**REVERSING THE ORDER OF INTEGRATION**

Evaluate $\displaystyle\int_0^2 \int_0^1 (3x^2 + 6xy^2)\, dy\, dx.$     Same as Example 1, but with the order of integration reversed

*Solution*

First we evaluate the inner $y$-integral:

$$\int_0^1 (3x^2 + 6xy^2)\, dy = (3x^2 y + 2xy^3)\Big|_{y=0}^{y=1} = 3x^2 + 2x - 0$$

Integrate with respect to $y$    $x$ held constant    $\frac{1}{3} \cdot 6$    Evaluated at $y=1$    And at $y=0$

$$= 3x^2 + 2x$$

Then we apply the outer $x$-integral to this expression:

$$\int_0^2 (3x^2 + 2x)\, dx = (x^3 + x^2)\Big|_{x=0}^{x=2} = 8 + 4 - 0 = 12$$

From inner integration    Evaluated at $x=2$    And at $x=0$    Final answer

Therefore:

$$\int_0^2 \int_0^1 (3x^2 + 6xy^2)\, dy\, dx = 12$$

Notice that Examples 1 and 2 (in which the order of integration was reversed) gave the same answer, 12. Reversing the order of integration *always* gives the same answer, provided that the function is continuous.

---

**Reversing the Order of Integration**

For a continuous $f(x, y)$

$$\int_c^d \int_a^b f(x, y)\, dx\, dy$$

is equal to

$$\int_a^b \int_c^d f(x, y)\, dy\, dx$$

---

## Double Integrals and Volumes

Earlier, we defined double integrals over rectangular regions. Double integrals can be evaluated by *either* of two iterated integrals (integrating in either order). The limits of integration are taken directly from the region $R$.

Evaluating Double Integrals

The double integral $\iint\limits_R f(x, y)\, dx\, dy$

over the region $R = \{\, (x, y)\mid a \le x \le b, c \le y \le d \,\}$

can be evaluated by finding *either* of the iterated integrals

$$\int_c^d \int_a^b f(x, y)\, dx\, dy \qquad \text{or} \qquad \int_a^b \int_c^d f(x, y)\, dy\, dx$$

The limits of integration come from the definition of the rectangle $R$.

### EXAMPLE 3

### EVALUATING A DOUBLE INTEGRAL

Evaluate $\iint\limits_R y^2 e^{-x}\, dx\, dy$ where $R = \{\, (x, y)\mid 0 \le x \le 2, -1 \le y \le 1\,\}$.

*Solution*

This double integral can be evaluated by finding either of the iterated integrals

$$\int_{-1}^1 \int_0^2 y^2 e^{-x}\, dx\, dy \qquad \text{or} \qquad \int_0^2 \int_{-1}^1 y^2 e^{-x}\, dy\, dx \qquad \text{Limits of integration come from } R$$

We find the second one, beginning with the inner integral.

$$\int_{-1}^1 y^2 e^{-x}\, dy = \left(\frac{1}{3} y^3 e^{-x}\right)\Bigg|_{y=-1}^{y=1} = \frac{1}{3} e^{-x} - \left(\frac{1}{3}(-1)e^{-x}\right)$$

Integrated — Held constant; Evaluated at $y = 1$; And at $y = -1$

$$= \frac{1}{3} e^{-x} + \frac{1}{3} e^{-x} = \frac{2}{3} e^{-x}$$

Then we integrate this with respect to $x$:

$$\int_0^2 \frac{2}{3} e^{-x}\, dx = -\frac{2}{3} e^{-x}\Bigg|_{x=0}^{x=2} = -\frac{2}{3} e^{-2} - \left(-\frac{2}{3} e^0\right) = -\frac{2}{3} e^{-2} + \frac{2}{3}$$

### Practice Problem

Show that evaluating this same double integral by the iterated integral in the *other* order gives the same answer. That is, evaluate the iterated integral

$$\int_{-1}^1 \int_0^2 y^2 e^{-x}\, dx\, dy$$

➤ Solution on page 545

The volume under a surface can be found by evaluating a double integral (since this is how double integrals were defined).

**Volume by Double Integrals**

For a nonnegative continuous function $f(x, y)$, the volume under the surface $z = f(x, y)$ and above a rectangular region $R$ in the $x$-$y$ plane is

$$\begin{pmatrix} \text{Volume under} \\ f \text{ above } R \end{pmatrix} = \iint\limits_{R} f(x, y)\, dx\, dy$$

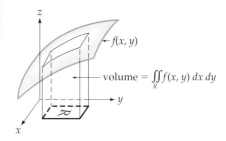

If the surface lies *below* the $x$-$y$ plane, this integral gives the *negative* of the volume.

**EXAMPLE 4**  **FINDING THE VOLUME UNDER A SURFACE**

A modernistic tent with closed sides is constructed according to the function shown below. To design ventilation and heating systems, it is necessary to know the volume under the tent. Find the volume under the tent on the indicated rectangle. All dimensions are in feet.

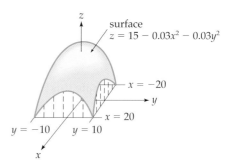

*Solution*

The volume is the integral of the function over the rectangle $R$:

$$\int_{-10}^{10} \int_{-20}^{20} (15 - 0.03x^2 - 0.03y^2)\, dx\, dy \qquad \text{Limits of integration come from } R$$

The inner integral is

$$\int_{-20}^{20} (15 - 0.03x^2 - 0.03y^2)\, dx = (15x - 0.01x^3 - 0.03y^2 x)\Big|_{x=-20}^{x=20}$$

$$= \underbrace{300 - 80 - 0.6y^2}_{\text{Evaluated at } x = 20} - \underbrace{(-300 + 80 + 0.6y^2)}_{\text{Evaluated at } x = -20} = 440 - 1.2y^2$$

Integrating this with respect to $y$:

From integrating $1.2y^2$

$$\int_{-10}^{10} (440 - 1.2y^2)\, dy = (440y - 0.4y^3)\Big|_{-10}^{10}$$

$$= 4400 - 400 - (-4400 + 400) = 8000$$

Therefore, the volume under the tent is 8000 cubic feet.

## Average Value

On page 366 the average value of a function of *one* variable over an interval was defined as the definite integral of the function divided by the length of the interval. For similar reasons, the average value of a function $f(x, y)$ of *two* variables over a region is defined as the *double* integral divided by the *area* of the region.

---

**Average Value**

$$\begin{pmatrix} \text{Average value} \\ \text{of } f \text{ over } R \end{pmatrix} = \frac{1}{\text{Area of } R} \iint\limits_R f(x, y)\, dx\, dy$$

Double integral over the region divided by the area of the region

---

For a rectangular region $R$, the area of $R$ is simply length times width.

**EXAMPLE 5**

### FINDING THE AVERAGE TEMPERATURE OVER A REGION

The temperature $x$ miles east and $y$ miles north of a weather station is $T(x, y) = 60 + 2x - 4y$ degrees. Find the average temperature over the rectangular region $R$ extending 2 miles north and south from the station and 5 miles east (as shown in the following diagram).

*Solution*

The area of the region $R$ is $4 \cdot 5 = 20$ square miles (length times width). The average temperature is the double integral divided by 20:

$$\text{Average} = \frac{1}{20} \int_{-2}^{2} \int_{0}^{5} (60 + 2x - 4y) \, dx \, dy$$

The inner integral is

$$\int_{0}^{5} (60 + 2x - 4y) \, dx = (60x + x^2 - 4yx) \Big|_{x=0}^{x=5}$$

$$= 300 + 25 - 20y - 0 = 325 - 20y$$

Evaluated at $x = 5$

The (outer) integral of this expression is

$$\int_{-2}^{2} (325 - 20y) \, dy = (325y - 10y^2) \Big|_{y=-2}^{y=2} = 650 - 40 - (-650 - 40)$$

$$= 610 + 690 = 1300$$

Finally, for the average we divide by 20 (the area of the region):

$$\frac{1300}{20} = 65$$

The average temperature over the region is 65 degrees.

## Integrating over More General Regions

We can also integrate over regions $R$ that are bounded by curves, provided that the curves as well as the function being integrated are continuous.

### Double Integrals over Regions Between Curves

Let $R$ be the region bounded by a lower curve $y = g(x)$ and an upper curve $y = h(x)$ from $x = a$ to $x = b$, as shown below. Then the double integral of $f(x, y)$ over $R$ is

$$\iint_R f(x, y) \, dx \, dy = \int_a^b \int_{g(x)}^{h(x)} f(x, y) \, dy \, dx$$

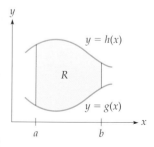

If $f$ is nonnegative, this double integral gives the volume under the surface $f(x, y)$ above $R$.

**EXAMPLE 6**

### FINDING THE VOLUME UNDER A SURFACE

Find the volume under the surface $f(x, y) = 12xy$ and above the region $R$ shown on the left.

*Solution*

The region $R$ is bounded by the upper curve $h(x) = \sqrt{x}$ and the lower curve $g(x) = x^2$ from $x = 0$ to $x = 1$. From the box on the previous page, the volume is given by:

$$\text{Volume} = \int_0^1 \int_{x^2}^{\sqrt{x}} 12xy \, dy \, dx$$

We find the inner integral first:

$$\int_{x^2}^{\sqrt{x}} 12xy \, dy = 12x \frac{1}{2} y^2 \Big|_{y=x^2}^{y=\sqrt{x}} = 6xy^2 \Big|_{y=x^2}^{y=\sqrt{x}}$$

$$= \underbrace{6x(\sqrt{x})^2}_{\substack{\text{Evaluating} \\ \text{at } y = \sqrt{x}}} - \underbrace{6x(x^2)^2}_{\substack{\text{Evaluating} \\ \text{at } y = x^2}} = \underbrace{6x^2 - 6x^5}_{\text{Simplified}}$$

Now we integrate this result with respect to $x$:

$$\int_0^1 (6x^2 - 6x^5) \, dx = \left(6 \cdot \frac{1}{3} x^3 - x^6\right) \Big|_0^1$$

$$= (2x^3 - x^6) \Big|_0^1 = 2 - 1 - (0) = 1$$

Therefore, the volume under the surface and above the region $R$ is 1 cubic unit.

---

### Solution to Practice Problem

$$\int_0^2 y^2 e^{-x} \, dx = -y^2 e^{-x} \Big|_{x=0}^{x=2} = -y^2 e^{-2} + y^2 e^0$$

$$= -y^2 e^{-2} + y^2$$

$$\int_{-1}^1 (-y^2 e^{-2} + y^2) \, dy = \left(-\frac{1}{3} y^3 e^{-2} + \frac{1}{3} y^3\right) \Big|_{-1}^1$$

$$= -\frac{1}{3} e^{-2} + \frac{1}{3} - \left(\frac{1}{3} e^{-2} - \frac{1}{3}\right) = -\frac{2}{3} e^{-2} + \frac{2}{3} \qquad \text{(same as before)}$$

## 7.7 Section Summary

The double integral of a nonnegative function $f(x, y)$ over a region $R$ gives the *volume* under the surface above $R$. (This is analogous to defining the definite integral of a function $f(x)$ of *one* variable as the *area* under the curve.)

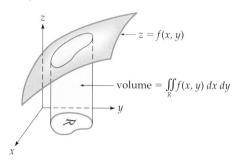

volume $= \iint\limits_R f(x, y)\, dx\, dy$

We evaluate a double integral by evaluating an iterated (repeated) integral:

$$\iint\limits_R f(x, y)\, dx\, dy \qquad \text{over} \qquad R = \{(x, y) \mid a \le x \le b, c \le y \le d\}$$

is found by evaluating either of the two iterated integrals

$$\int_c^d \int_a^b f(x, y)\, dx\, dy \qquad \text{or} \qquad \int_a^b \int_c^d f(x, y)\, dy\, dx \qquad \text{Integrating in either order}$$

Note the distinction: *double* integrals are written with $R$ (which must be specified) under the double integral, and *iterated* integrals are written with upper and lower *limits* on each integral sign.

Double integrals do more than just find volumes; they give *continuous sums* (as illustrated in Exercises 43 and 44), and when divided by the area of the region they give the *average value of the function over the region*.

Exercises 47–50 discuss "triple" integrals of functions $f(x, y, z)$ of three variables. Triple integrals are evaluated by *iterated* integrals (but *three* of them), integrating successively with respect to one variable at a time, holding all others constant.

## 7.7 Exercises

**1–6.** Evaluate each (single) integral.

**1.** $\displaystyle\int_1^{x^2} 8xy^3\, dy$

**2.** $\displaystyle\int_1^{y^2} 10x^4\, dx$

**3.** $\displaystyle\int_{-y}^{y} 9x^2y\, dx$

**4.** $\displaystyle\int_{-x}^{x} 6xy^2\, dy$

**5.** $\displaystyle\int_0^{x} (6y - x)\, dy$

**6.** $\displaystyle\int_0^{y} (4x - y)\, dx$

**7–28.** Evaluate each iterated integral.

**7.** $\displaystyle\int_0^2 \int_0^1 4xy\, dx\, dy$

**8.** $\displaystyle\int_0^2 \int_0^1 8xy\, dy\, dx$

**9.** $\displaystyle\int_0^2 \int_0^1 x\, dy\, dx$

**10.** $\displaystyle\int_0^4 \int_0^3 y\, dx\, dy$

**11.** $\displaystyle\int_0^1 \int_0^2 x^3y^7\, dx\, dy$

**12.** $\displaystyle\int_0^1 \int_0^3 x^8y^2\, dy\, dx$

**13.** $\displaystyle\int_1^3 \int_0^2 (x + y)\, dy\, dx$

**14.** $\displaystyle\int_1^2 \int_0^4 (x - y)\, dx\, dy$

**15.** $\displaystyle\int_{-1}^1 \int_0^3 (x^2 - 2y^2)\, dx\, dy$

**16.** $\displaystyle\int_{-1}^1 \int_0^3 (2x^2 + y^2)\, dy\, dx$

**17.** $\displaystyle\int_{-3}^3 \int_0^3 y^2 e^{-x}\, dy\, dx$    **18.** $\displaystyle\int_{-2}^2 \int_0^2 x e^{-y}\, dx\, dy$

**19.** $\displaystyle\int_{-2}^2 \int_{-1}^1 y e^{xy}\, dx\, dy$    **20.** $\displaystyle\int_{-1}^1 \int_{-1}^1 x e^{xy}\, dy\, dx$

**21.** $\displaystyle\int_0^2 \int_x^1 12xy\, dy\, dx$

**22.** $\displaystyle\int_0^1 \int_y^1 4xy\, dx\, dy$

**23.** $\displaystyle\int_3^5 \int_0^y (2x - y)\, dx\, dy$

**24.** $\displaystyle\int_2^4 \int_0^x (x - 2y)\, dy\, dx$

**25.** $\displaystyle\int_{-3}^3 \int_0^{4x} (y - x)\, dy\, dx$

**26.** $\displaystyle\int_{-1}^1 \int_0^{2y} (x + y)\, dx\, dy$

**27.** $\displaystyle\int_0^1 \int_{-y}^y (x + y^2)\, dx\, dy$

**28.** $\displaystyle\int_0^2 \int_{-x}^x (x^2 - y)\, dy\, dx$

**29–32.** For each double integral:
**a.** Write the *two* iterated integrals that are equal to it.
**b.** Evaluate *both* iterated integrals (the answers should agree).

**29.** $\displaystyle\iint_R 3xy^2\, dx\, dy$

with $R = \{ (x, y) \mid 0 \le x \le 2, 1 \le y \le 3 \}$

**30.** $\displaystyle\iint_R 6x^2y\, dx\, dy$

with $R = \{ (x, y) \mid 0 \le x \le 1, 1 \le y \le 2 \}$

**31.** $\displaystyle\iint_R y e^x\, dx\, dy$

with $R = \{ (x, y) \mid -1 \le x \le 1, 0 \le y \le 2 \}$

**32.** $\displaystyle\iint_R x e^y\, dx\, dy$

with $R = \{ (x, y) \mid 0 \le x \le 1, -2 \le y \le 2 \}$

**33–36.** Use integration to find the volume under each surface $f(x, y)$ above the region $R$.

**33.** $f(x, y) = x + y$
$R = \{ (x, y) \mid 0 \le x \le 2, 0 \le y \le 2 \}$

**34.** $f(x, y) = 8 - x - y$
$R = \{ (x, y) \mid 0 \le x \le 4, 0 \le y \le 4 \}$

**35.** $f(x, y) = 2 - x^2 - y^2$
$R = \{ (x, y) \mid 0 \le x \le 1, 0 \le y \le 1 \}$

**36.** $f(x, y) = x^2 + y^2$
$R = \{ (x, y) \mid 0 \le x \le 2, 0 \le y \le 2 \}$

**37–40.** Use integration to find the volume under each surface $f(x, y)$ above the region $R$.

**37.** $f(x, y) = 2xy$
over the region shown in the graph.

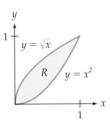

**38.** $f(x, y) = 3xy^2$
over the region shown in the graph.

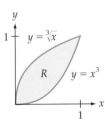

**39.** $f(x, y) = e^y$
over the region shown in the graph.

**40.** $f(x, y) = e^y$
over the region shown in the graph.

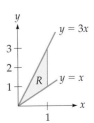

## Applied Exercises

**41. GENERAL: Average Temperature** The temperature $x$ miles east and $y$ miles north of a weather station is given by the function $f(x, y) = 48 + 4x - 2y$. Find the average temperature over the region $R$ shown below.

**42. ENVIRONMENTAL SCIENCE: Average Air Pollution** The air pollution near a chemical refinery is $f(x, y) = 20 + 6x^2y$ parts per million (ppm), where $x$ and $y$ are the numbers of miles east and north of the refinery. Find the average pollution level for the region $R$ shown below.

**43. GENERAL: Total Population** The population density (people per square mile) $x$ miles east and $y$ miles north of the center of a city is $P(x, y) = 12{,}000e^{x-y}$. Find the total population of the region $R$ shown in the following diagram. [*Hint:* Integrate the population density over the region $R$. This is an example of a double integral as a *sum*, giving a *total* population over a region.]

**44. BUSINESS: Value of Mineral Deposit** The value of an offshore mineral deposit $x$ miles east and $y$ miles north of a certain point is $f(x, y) = 4x + 6y^2$ million dollars per square mile. Find the total value of the tract shown on next column. [*Hint:* Integrate the function over the region $R$. This is an example of a double integral as a *sum*, giving a *total* value over a region.]

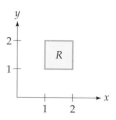

**45–46. GENERAL: Volume of a Building** To estimate heating and air conditioning costs, it is necessary to know the volume of a building.

**45.** A conference center has a curved roof whose height is $f(x, y) = 40 - 0.006x^2 + 0.003y^2$. The building sits on a rectangle extending from $x = -50$ to $x = 50$ and $y = -100$ to $y = 100$. Use integration to find the volume of the building. (All dimensions are in feet.)

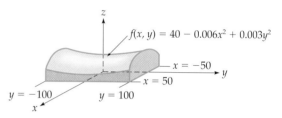

**46.** An airplane hangar has a curved roof whose height is $f(x, y) = 40 - 0.03x^2$. The building sits on a rectangle extending from $x = -20$ to $x = 20$ and $y = -100$ to $y = 100$. Use integration to find the volume of the building. (All dimensions are in feet.)

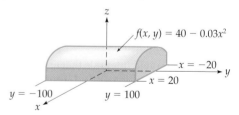

### Triple Integrals

**47–50.** Evaluate each triple iterated integral. [*Hint:* Integrate with respect to one variable at a time, treating the other variables as constants, working from the inside out.]

**47.** $\displaystyle\int_1^2 \int_0^3 \int_0^1 (2x + 4y - z^2)\, dx\, dy\, dz$

**48.** $\displaystyle\int_1^2 \int_0^3 \int_0^2 (6x - 2y + z^2)\, dx\, dy\, dz$

**49.** $\int_1^2 \int_0^2 \int_0^1 2xy^2z^3 \, dx \, dy \, dz$

**50.** $\int_1^3 \int_0^1 \int_0^2 12x^3y^2z \, dx \, dy \, dz$

### Conceptual Exercises

Assume, for simplicity, that all functions are continuous and all denominators are nonzero.

**51.** Which of the following is an *iterated* integral and which is a *double* integral?

  **a.** $\iint\limits_R f(x, y) \, dx \, dy$   **b.** $\int_a^b \int_c^d f(x, y) \, dx \, dy$

**52.** If a surface lies partly above the *x-y* plane and partly below it on a region, what does the double integral represent?

**53.** For a *double* integral there are two possible orders of integration. How many possible orders of integration are there for a *triple* integral?

**54.** True or False:
$$\int_a^b \int_c^d f(x) \, g(y) \, dx \, dy = \int_c^d f(x) \, dx \cdot \int_a^b g(y) \, dy.$$

**55.** Can $\int_a^b \int_c^d e^{f(x) + g(y)} \, dx \, dy$   be written as the product of two integrals?

**56.** Can $\int_a^b \int_c^d \dfrac{f(x)}{g(y)} \, dx \, dy$   be written as the product of two integrals?

**57.** True or False: If  $f(x, y) \le g(x, y)$   for all *x* and *y*, then $\int_a^b \int_c^d f(x, y) \, dx \, dy \le \int_a^b \int_c^d g(x, y) \, dx \, dy.$

**58.** How could you find the volume *between* two surfaces $f(x, y)$ and $g(x, y)$ over a region *R* by using one double integral? (Assume that surface *f* lies above surface *g*.)

**59.** If $f(x, y)$ is in *widgets per square blarg*, and *x* and *y* are in *blargs*, then what are the units of $\int_a^b \int_c^d f(x, y) \, dx \, dy?$

**60.** If  $\int_a^b \int_c^d f(x, y) \, dx \, dy$   is in *blargs*, and *x* and *y* are in *widgets*, then what are the units of $f(x, y)$?

## Chapter Summary with Hints and Suggestions

Reading the text and doing the exercises in this chapter have helped you to master the following concepts and skills, which are listed by section (in case you need to review them) and are keyed to particular Review Exercises. Answers for all Review Exercises are given at the back of the book, and full solutions can be found in the Student Solutions Manual.

### 7.1 Functions of Several Variables

■ Find the domain of a function of two variables. *(Review Exercises 1–4.)*

### 7.2 Partial Derivatives

■ Find the first and second partials of a function of two variables.   *(Review Exercises 5–12.)*

$$\frac{\partial}{\partial x} f(x, y) = f_x(x, y) = \lim_{h \to 0} \frac{f(x + h, y) - f(x, y)}{h}$$

$$\frac{\partial}{\partial y} f(x, y) = f_y(x, y) = \lim_{h \to 0} \frac{f(x, y + h) - f(x, y)}{h}$$

■ Evaluate the first partials of a function of two variables.   *(Review Exercises 13–16.)*

■ Solve an applied problem involving partials, and interpret the answer. *(Review Exercises 17–18.)*

$$\frac{\partial}{\partial x} f(x, y) = \left( \begin{array}{l} \text{Rate of change of } f \text{ with respect} \\ \text{to } x \text{ when } y \text{ is held constant} \end{array} \right)$$

$$\frac{\partial}{\partial y} f(x, y) = \left( \begin{array}{l} \text{Rate of change of } f \text{ with respect} \\ \text{to } y \text{ when } x \text{ is held constant} \end{array} \right)$$

### 7.3 Optimizing Functions of Several Variables

■ Find the relative extreme values of a function. *(Review Exercises 19–30.)*

$$\begin{cases} f_x = 0 \\ f_y = 0 \end{cases} \quad D = f_{xx}f_{yy} - (f_{xy})^2$$

- Solve an applied problem by optimizing a function of two variables. (*Review Exercises 31–32.*)

## 7.4 Least Squares

- Find a least squares line "by hand." (*Review Exercises 33–34.*)

- Find the least squares line for actual data, and use it to make a prediction. (*Review Exercises 35–36.*)

## 7.5 Lagrange Multipliers and Constrained Optimization

- Solve a constrained maximum or minimum problem using Lagrange multipliers. (*Review Exercises 37–42.*)

$$F(x, y, \lambda) = f(x, y) + \lambda g(x, y) \qquad \begin{cases} F_x = 0 \\ F_y = 0 \\ F_\lambda = 0 \end{cases}$$

- Find *both* extreme values in a constrained optimization problem using Lagrange multipliers. (*Review Exercises 43–44.*)

- Solve an applied problem using Lagrange multipliers. (*Review Exercises 45–48.*)

## 7.6 Total Differentials and Approximate Changes

- Find the total differential of a function. (*Review Exercises 49–54.*)

$$df = \frac{\partial f}{\partial x}\, dx + \frac{\partial f}{\partial y}\, dy$$

- Solve an applied problem by using the total differential to estimate an actual change. (*Review Exercise 55.*)

$$\Delta f = f(x + \Delta x, y + \Delta y) - f(x, y) \qquad \Delta f \approx df$$

- Estimate the relative error in an area calculation. (*Review Exercise 56.*)

## 7.7 Multiple Integrals

- Evaluate an iterated integral. (*Review Exercises 57–60.*)

$$\int_c^d \int_a^b f(x, y)\, dx\, dy = \int_a^b \int_c^d f(x, y)\, dy\, dx$$

- Find the volume under a surface above a region using a double integral. (*Review Exercises 61–64.*)

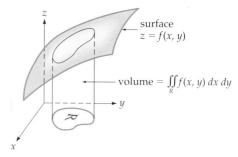

surface $z = f(x, y)$

volume $= \iint_R f(x, y)\, dx\, dy$

- Find the average population over a region using a double integral. (*Review Exercise 65.*)

- Find the total value of a region using a double integral. (*Review Exercise 66.*)

## Hints and Suggestions

- The graph of a function of two variables is represented by a *surface* above (or below) the *x-y* plane. Such three-dimensional graphs are difficult to draw "by hand" but can be shown on some graphing calculators and computer screens. Such surfaces have maximum and minimum points (just as with functions of one variable), but also a new phenomenon, *saddle points* (see page 470).

- The graph of a function of *three* or more variables would require *four* or more dimensions, and so cannot be drawn.

- When finding partial derivatives, remember which is the variable of differentiation and then treat the other variable as a constant. Other than this, the differentiation formulas are the same.

- Partials give instantaneous rates of change with respect to one variable while the other is held constant. They also give *marginals* with respect to one product while production of the other is held constant.

- The *D*-test (page 489) applies only to critical points, where the first partials are zero. These critical points must be found first, and then the *D*-test is applied to each, one at a time.

- Least squares is carried out automatically by a graphing calculator or computer when it finds the linear regression line.

- Solving a constrained optimization problem by Lagrange multipliers is often easier than eliminating one of the variables, as was done in Sections 3.3 and 3.4.

- Multiple integrals give volume under a function if the function is nonnegative.

- When finding the average value of a function of two variables, don't forget to divide by the area of the region over which you are integrating.

- **Practice for Test:** Review Exercises 3, 7, 15, 17, 19, 33, 41, 45, 49, 51, 57, 65.

## Review Exercises for Chapter 7

Practice test exercise numbers are in green.

### 7.1 Functions of Several Variables

**1–4.** For each function, state the domain.

**1.** $f(x, y) = \dfrac{\sqrt{x}}{\sqrt[3]{y}}$   **2.** $f(x, y) = \dfrac{\sqrt[3]{x}}{\sqrt{y}}$

**3.** $f(x, y) = e^{1/x} \ln y$   **4.** $f(x, y) = \dfrac{\ln y}{x}$

### 7.2 Partial Derivatives

**5–12.** For each function $f$, calculate
**a.** $f_x$,  **b.** $f_y$,  **c.** $f_{xy}$,  and  **d.** $f_{yx}$.

**5.** $f(x, y) = 2x^5 - 3x^2y^3 + y^4 - 3x + 2y + 7$

**6.** $f(x, y) = 3x^4 + 5x^3y^2 - y^6 - 6x + y - 9$

**7.** $f(x, y) = 18x^{2/3}y^{1/3}$

**8.** $f(x, y) = \ln(x^2 + y^3)$

**9.** $f(x, y) = e^{x^3 - 2y^3}$   **10.** $f(x, y) = 3x^2e^{-5y}$

**11.** $f(x, y) = ye^{-x} - x \ln y$

**12.** $f(x, y) = x^2e^y + y \ln x$

**13–16.** For each function, calculate
**a.** $f_x(1, -1)$  and  **b.** $f_y(1, -1)$.

**13.** $f(x, y) = \dfrac{x + y}{x - y}$   **14.** $f(x, y) = \dfrac{x}{x^2 + y^2}$

**15.** $f(x, y) = (x^3 + y^2)^3$   **16.** $f(x, y) = (2xy - 1)^4$

**17. BUSINESS: Marginal Productivity** A company's production is given by the Cobb–Douglas function $P(L, K) = 160L^{3/4}K^{1/4}$, where $L$ is the number of units of labor and $K$ is the number of units of capital.

**a.** Find $P_L(81, 16)$ and interpret this number.
**b.** Find $P_K(81, 16)$ and interpret this number.
**c.** From your answers to parts (a) and (b), which will increase production more, an additional unit of labor or an additional unit of capital?

**18. BUSINESS: Advertising** A clothing designer's sales $S$ depend on $x$, the amount spent on television advertising, and $y$, the amount spent on print advertising (both in thousands of dollars), according to the function

$$S(x, y) = 60x^2 + 90y^2 - 6xy + 200$$

Find $S_x(2, 3)$ and $S_y(2, 3)$, and interpret these numbers.

### 7.3 Optimizing Functions of Several Variables

**19–30.** For each function, find all relative extreme values.

**19.** $f(x, y) = 2x^2 - 2xy + y^2 - 4x + 6y - 3$

**20.** $f(x, y) = x^2 - 2xy + 2y^2 - 6x + 4y + 2$

**21.** $f(x, y) = 2xy - x^2 - 5y^2 + 2x - 10y + 3$

**22.** $f(x, y) = 2xy - 5x^2 - y^2 + 10x - 2y + 1$

**23.** $f(x, y) = 2xy + 6x - y + 1$

**24.** $f(x, y) = 4xy - 4x + 2y - 4$

**25.** $f(x, y) = e^{-(x^2+y^2)}$   **26.** $f(x, y) = e^{2(x^2+y^2)}$

**27.** $f(x, y) = \ln(5x^2 + 2y^2 + 1)$

**28.** $f(x, y) = \ln(4x^2 + 3y^2 + 10)$

**29.** $f(x, y) = x^3 - y^2 - 12x - 6y$

**30.** $f(x, y) = y^2 - x^3 + 12x - 4y$

**31. BUSINESS: Maximum Profit** A boatyard builds 18-foot and 22-foot sailboats. Each 18-foot boat costs $3000 to build, each 22-foot boat costs $5000 to build, and the company's fixed costs are $6000. The price function for the 18-foot boats is $p = 7000 - 20x$, and that for the 22-foot boats is $q = 8000 - 30y$ (both in dollars), where $x$ and $y$ are the numbers of 18-foot and 22-foot boats, respectively.

**a.** Find the company's cost function $C(x, y)$.
**b.** Find the company's revenue function $R(x, y)$.
**c.** Find the company's profit function $P(x, y)$.
**d.** Find the quantities and prices that maximize profit. Also find the maximum profit.

**32. BUSINESS: Price Discrimination** A company sells farm equipment in America and Europe, charging different prices in the two markets. The price function for harvesters sold in America is $p = 80 - 0.2x$, and the price function for harvesters sold in Europe is $q = 64 - 0.1y$ (both in thousands of dollars), where $x$ and $y$ are the numbers sold per day in America and Europe, respectively. The company's cost function is $C = 100 + 12(x + y)$ thousand dollars.

*(continues)*

a. Find the company's profit function.
b. Find how many harvesters should be sold in each market to maximize profit. Also find the price for each market.

## 7.4 Least Squares

**33–34.** For each exercise:
a. Find the least squares line "by hand."
b. Check your answer using a graphing calculator.

**33.**

| $x$ | $y$ |
|-----|-----|
| 1 | −1 |
| 3 | 6 |
| 4 | 6 |
| 5 | 10 |

**34.**

| $x$ | $y$ |
|-----|-----|
| 1 | 7 |
| 2 | 4 |
| 4 | 2 |
| 5 | −1 |

**35. GENERAL: The Aging of America** The population of Americans who are over 65 years old is growing faster than the population at large. Find the least squares line for the following data for the over-65 population, and use it to predict the size of the over-65 population in the year 2020 ($x = 6$).

*Source:* U.S. Census Bureau

|  | $x$ | $y$ = Population (millions) |
|------|-----|-----------------|
| 1970 | 1 | 20.1 |
| 1980 | 2 | 25.5 |
| 1990 | 3 | 31.2 |
| 2000 | 4 | 35.0 |
| 2010 | 5 | 40.2 |

**36. SOCIAL SCIENCES: Violent Crime** Crime, and in particular violent crime, has been decreasing in the United States. The following table gives the number of violent crimes per 1000 people in the United States for recent years. Fit a line to the data and then use it to predict the number of violent crimes per 1000 people in 2015.

*Source:* U.S. Census Bureau

| Year | $x$ | Price |
|------|-----|-------|
| 1990 | 1 | 7.3 |
| 1995 | 2 | 6.9 |
| 2000 | 3 | 5.1 |
| 2005 | 4 | 4.7 |

## 7.5 Lagrange Multipliers and Constrained Optimization

**37–42.** Use Lagrange multipliers to optimize each function subject to the given constraint. (The stated extreme values *do* exist.)

**37.** Maximize $f(x, y) = 6x^2 - y^2 + 4$
subject to $3x + y = 12$.

**38.** Maximize $f(x, y) = 4xy - x^2 - y^2$
subject to $x + 2y = 26$.

**39.** Minimize $f(x, y) = 2x^2 + 3y^2 - 2xy$
subject to $2x + y = 18$.

**40.** Minimize $f(x, y) = 12xy - 1$
subject to $y - x = 6$.

**41.** Minimize $f(x, y) = e^{x^2 + y^2}$
subject to $x + 2y = 15$.

**42.** Maximize $f(x, y) = e^{-x^2 - y^2}$
subject to $2x + y = 5$.

**43–44.** Use Lagrange multipliers to find the maximum *and* minimum values of each function subject to the given constraint. (Both extreme values *do* exist.)

**43.** $f(x, y) = 6x - 18y$ subject to $x^2 + y^2 = 40$

**44.** $f(x, y) = 4xy$ subject to $x^2 + y^2 = 32$

**45. BUSINESS: Maximum Profit** A company's profit is $P = 300x^{2/3}y^{1/3}$, where $x$ and $y$ are, respectively, the amounts spent on production and advertising. The company has a total of $60,000 to spend.

a. Use Lagrange multipliers to find the amounts for production and advertising that maximize profit.
b. Evaluate and give an interpretation for $|\lambda|$.

**46. BIOMEDICAL: Nutrition** A nursing home uses two vitamin supplements, and the nutritional value of $x$ ounces of the first together with

$y$ ounces of the second is $4x + 2xy + 8y$. The first costs $2 per ounce, the second costs $1 per ounce, and the nursing home can spend only $8 per patient per day.

**a.** Use Lagrange multipliers to find how much of each supplement should be used to maximize the nutritional value subject to the budget constraint.

**b.** Evaluate and give an interpretation for $|\lambda|$.

**47. BUSINESS: Maximum Production** A company's output is given by the Cobb–Douglas production function $P = 400L^{3/5}K^{2/5}$, where $L$ and $K$ are the number of units of labor and capital. Each unit of labor costs $50 and each unit of capital costs $60, and $1500 is available to pay for labor and capital.

**a.** How many units of labor and of capital should be used to maximize production?

**b.** Evaluate and give an interpretation for $|\lambda|$.

**48. GENERAL: Container Design** An open-top box with a square base and two perpendicular dividers, as shown in the diagram, is to have a volume of 576 cubic inches. Use Lagrange multipliers to find the dimensions that require the least amount of material.

## 7.6 Total Differentials and Approximate Changes

**49–54.** Find the total differential of each function.

**49.** $f(x, y) = 3x^2 + 2xy + y^2$

**50.** $f(x, y) = x^2 + xy - 3y^2$

**51.** $g(x, y) = \ln(xy)$

**52.** $g(x, y) = \ln(x^3 + y^3)$

**53.** $z = e^{x-y}$

**54.** $z = e^{xy}$

**55. BUSINESS: Sales** A clothing designer's sales $S$ depend on $x$, the amount spent on television advertising, and $y$, the amount spent on print advertising (all in thousands of dollars) according to the formula $S(x, y) = 60x^2 - 6xy + 90y^2 + 200$. If the company now spends 2 thousand dollars on television advertising and 3 thousand dollars on print advertising, use the total differential to estimate the change in sales if television advertising is increased by $500 and print advertising is decreased by $500. [*Hint:* $\Delta x$ and $\Delta y$ must be in thousands of dollars, just as $x$ and $y$ are.] Then estimate the change in sales if each of the changes in advertising were halved.

**56. GENERAL: Relative Error in Calculating Area** A triangular piece of real estate is measured to have length $x$ feet and altitude $y$ feet, but each measurement may be in error by 1%. Estimate the percentage error in calculating the area by using a total differential.

## 7.7 Multiple Integrals

**57–60.** Evaluate each iterated integral.

**57.** $\displaystyle\int_0^4 \int_{-1}^1 2xe^{2y}\, dy\, dx$

**58.** $\displaystyle\int_{-1}^1 \int_0^3 (x^2 - 4y^2)\, dx\, dy$

**59.** $\displaystyle\int_{-1}^1 \int_{-y}^y (x + y)\, dx\, dy$

**60.** $\displaystyle\int_{-2}^2 \int_{-x}^x (x + y)\, dy\, dx$

**61–64.** Find the volume under the surface $f(x, y)$ above the region $R$.

**61.** $f(x, y) = 8 - x - y$
$R = \{(x, y) \mid 0 \le x \le 2, 0 \le y \le 4\}$

**62.** $f(x, y) = 6 - x - y$
$R = \{(x, y) \mid 0 \le x \le 4, 0 \le y \le 2\}$

**63.** $f(x, y) = 12xy^3$ over the region shown in the graph.

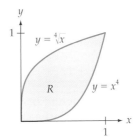

**64.** $f(x, y) = 15xy^4$ over the region shown in the graph.

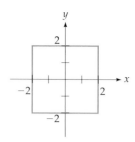

**65. GENERAL: Average Population of a Region**
The population per square mile $x$ miles east and $y$ miles north of the center of a city is $P(x, y) = 12{,}000 + 100x - 200y$. Find the *average* population over the region shown in the next column.

**66. GENERAL: Total Real Estate Value** The value of land $x$ blocks east and $y$ blocks north of the center of a town is $V(x, y) = 40 - 4x - 2y$ hundred thousand dollars per block. Find the total value of the parcel of land shown in the graph above.

## Cumulative Review for Chapters 1–7

The following exercises review some of the basic techniques that you learned in Chapters 1–7. Answers to all of these cumulative review exercises are given in the answer section at the back of the book. A graphing calculator is suggested but not required.

**1.** Draw the graph of the function
$$f(x) = \begin{cases} 2x - 1 & \text{if } x \leq 3 \\ 7 - x & \text{if } x > 3 \end{cases}$$

**2.** Simplify: $\left(\dfrac{1}{8}\right)^{-2/3}$

**3.** Use the definition of the derivative
$$f'(x) = \lim_{h \to 0} \frac{f(x + h) - f(x)}{h}$$
to find the derivative of $f(x) = \dfrac{1}{x}$.

**4.** If $f(x) = 12\sqrt[3]{x^2} - 4$, find $f'(8)$.

**5.** A camera store finds that if it sells disposable cameras at a price of $p$ dollars each, it will sell $S(p) = \dfrac{800}{p + 8}$ of them per week. Find $S'(12)$ and interpret the answer.

**6.** Find $\dfrac{d}{dx}[x^2 + (2x + 1)^4]^3$.

**7.** Make sign diagrams for the first and second derivatives and graph the function $f(x) = x^3 + 9x^2 - 48x - 148$, showing all relative extreme points and inflection points.

**8.** Make a sign diagram for the first derivative and graph the function $f(x) = \dfrac{1}{x^2 - 4x}$, showing all asymptotes and relative extreme points.

**9.** A homeowner wants to use 80 feet of fence to make a rectangular enclosure along an existing stone wall. If the side along the existing wall needs no fence, what are the dimensions of the enclosure that has the largest possible area?

**10.** An open-top box with a square base is to have a volume of 108 cubic feet. Find the dimensions of the box of this type that can be made with the least amount of material.

**11.** A spherical balloon is being inflated at the constant rate of 128 cubic feet per minute. Find how fast the radius is increasing at the moment when the radius is 4 feet.

**12.** A sum of $1000 is deposited in a bank account paying 8% interest. Find the value of the account after 3 years if the interest is compounded:
   **a.** quarterly.    **b.** continuously.

**13.** In $t$ years the population of a county is predicted to be $P(t) = 12{,}000e^{0.02t}$, and the population of a neighboring county is predicted to be $Q(t) = 9000e^{0.04t}$. In how many years will the second county overtake the first in population?

**14.** A sum is deposited in a bank account paying 6% interest compounded monthly. How many years will it take for the value to increase by 50%?

**15.** Make sign diagrams for the first and second derivatives and graph the function $f(x) = e^{-x^2/2}$, showing all relative extreme points and inflection points.

**16.** Find $\displaystyle\int (12x^2 - 4x + 1)\, dx$.

**17.** Pollution is being discharged into a lake at the rate of $18e^{0.02t}$ million gallons per year, where $t$ is the number of years from now. Find a formula for the total amount of pollution that will be discharged into the lake during the next $t$ years.

**18.** Find the area bounded by $y = 20 - x^2$ and $y = 8 - 4x$.

**19.** Find the average value of $f(x) = 12\sqrt{x}$ over the interval $[0, 4]$.

**20.** Find: **a.** $\displaystyle\int \frac{x^2\, dx}{x^3 + 1}$  **b.** $\displaystyle\int \frac{e^{\sqrt{x}}\, dx}{\sqrt{x}}$

**21.** Find by integration by parts: $\displaystyle\int xe^{4x}\, dx$.

**22.** Use the integral table on the inside back cover to find $\displaystyle\int \frac{\sqrt{4 - x^2}}{x}\, dx$. Check your answer by differentiating and then simplifying.

**23.** Evaluate $\displaystyle\int_1^\infty \frac{1}{x^3}\, dx$.

**24.** Use trapezoidal approximation with $n = 4$ trapezoids to approximate $\int_0^1 \sqrt{x^2 + 1}\, dx$. (If you use a graphing calculator, compare your answer with the value obtained by using FnInt.)

**25.** Use Simpson's Rule with $n = 4$ to approximate $\int_0^1 \sqrt{x^2 + 1}\, dx$. (If you use a graphing calculator, compare your answer with the value obtained by using FnInt.)

**26.** For the function $f(x, y) = \dfrac{3\sqrt{x} + \ln y}{x - y}$, find:

**a.** the domain
**b.** $f(4, 1)$

**27.** Find the first partial derivatives of the function $f(x, y) = x \ln y + ye^{2x}$.

**28.** Find the relative extreme values of $f(x, y) = 2x^2 - 2xy + y^2 + 4x - 6y + 12$.

**29.** Find the least squares line for the following points:

| $x$ | $y$ |
|---|---|
| 1 | $-3$ |
| 3 | 1 |
| 5 | 3 |
| 7 | 8 |

**30.** Use Lagrange multipliers to find the minimum value of $f(x, y) = 3x^2 + 2y^2 - 2xy$ subject to the constraint $x + 2y = 18$.

**31.** Find the total differential of $f(x, y) = 2x^2 + xy - 3y^2 + 4$.

**32.** Find the volume under the surface $f(x, y) = 12 - x - 2y$ over the rectangle $R = \{(x, y) \mid 0 \le x \le 2, 0 \le y \le 3\}$.

# 8

# Trigonometric Functions

### Tones, Temperature, and Trigonometry

This chapter introduces the sine and cosine functions, "wavy" curves that repeat themselves at regular intervals. Such functions have many applications. For example, musical sounds are caused by vibrations in the air, which can be picked up by a microphone and displayed on an oscilloscope. The following graph shows the sound wave from a saxophone playing a G#. This curve can be expressed as the sum of modified sine and cosine curves.*

Saxophone G#

The next graph shows the daily temperature range in New York City for the years 2006 and 2007 (the very jagged line), and the normal daily high and low temperatures (the smoother curves), which may be closely approximated by sine and cosine curves.

*Source:* National Weather Service

The following graph shows the rising level of carbon dioxide in the atmosphere as measured at the Mauna Loa Observatory in Hawaii. Carbon dioxide contributes to the "greenhouse effect," which causes global warming. This curve is a sine curve (the seasonal fluctuations) added to an exponential function (the upward trend).

*See Dorothea Bone, "Music and the Circular Functions," UMAP unit 588, COMAP, Lexington, Mass., and John R. Pierce, *The Science of Musical Sound* (Scientific American Books, W. H. Freeman & Co.).

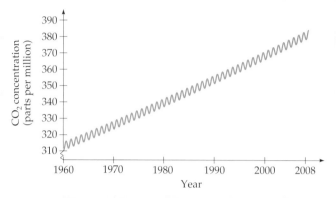

*Source:* National Oceanic and Atmospheric Administration

## 8.1 TRIANGLES, ANGLES, AND RADIAN MEASURE

### Introduction

The graph on the previous page shows the annual cycle of temperatures, varying "periodically" from winter cold to summer heat and back to winter cold. In this chapter we will study **trigonometric functions** and use them to model this and other periodic behavior, from business cycles to home heating costs. It is assumed that you have studied trigonometry at some time in the past. The first two sections review trigonometry from the beginning, but selectively rather than comprehensively, covering only the topics that will be needed for applications. Later sections discuss differentiation and integration of trigonometric functions. We begin with triangles, angles, and radian measure.

### Right Triangles

A triangle with a right (90°) angle is called a **right triangle.** The lengths of the sides of a right triangle are related by the Pythagorean Theorem.*

---

**Pythagorean Theorem**

In a right triangle, the square of the hypotenuse equals the sum of the squares of the other two sides.

$$a^2 + b^2 = c^2$$

---

*This theorem was first attributed to the school of the Greek philosopher Pythagoras (sixth century B.C.), although the Babylonian tablet "Plimpton 322" shows that it was known several centuries earlier.

A proof of the Pythagorean Theorem is outlined in Exercise 37. We first use the theorem as follows: given an equilateral triangle of side 2, cutting it in half, as shown on the left, leaves a right triangle with hypotenuse 2 and side 1. The third side, $b$, can then be found from the Pythagorean Theorem as follows.

$$1^2 + b^2 = 2^2$$    $a^2 + b^2 = c^2$  with $a = 1$  and  $c = 2$

$$1 + b^2 = 4$$    Simplifying

$$b^2 = 3$$    Subtracting 1

Therefore, the third side has length  $b = \sqrt{3}$,  leading to the following result.

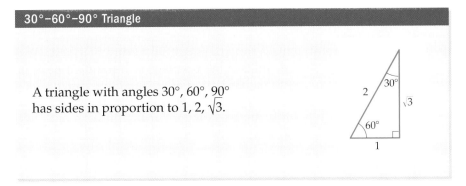

**30°–60°–90° Triangle**

A triangle with angles 30°, 60°, 90° has sides in proportion to 1, 2, $\sqrt{3}$.

We say "in proportion to" because the numbers 1, 2, $\sqrt{3}$ can be doubled, tripled, halved, or multiplied by *any* positive number, to obtain other 30°–60°–90° triangles. Note that the smallest side (1) goes opposite the smallest angle (30°), and the largest side (2) goes opposite the largest angle (90°).

We now consider a triangle with angles 45°, 45°, and 90°, as shown on the left. The two 45° angles make the triangle isosceles, so it has two equal sides, which we take to be of length 1. We find the length of the hypotenuse from the Pythagorean Theorem, as follows.

$$1^2 + 1^2 = c^2$$    $a^2 + b^2 = c^2$ with $a = b = 1$

$$2 = c^2$$    Simplifying

Therefore, the hypotenuse has length  $c = \sqrt{2}$,  giving the following result.

**45°–45°–90° Triangle**

A triangle with angles 45°, 45°, 90° has sides in proportion to 1, 1, $\sqrt{2}$.

Again, the largest side $\left(\sqrt{2}\right)$ goes opposite the largest angle (90°). These two particular right triangles will be used extensively in the next section.

## Angles and Radian Measure

Trigonometry requires a precise definition of the word **angle.** An *angle* is formed when a line segment is rotated around one of its endpoints (the **vertex**), from an **initial side** to a **terminal side.** We use an arrow to indicate the rotation.

An angle in the *x-y*-plane is in **standard position** if its vertex is at the origin and its initial side lies along the positive *x*-axis. An angle is **positive** if its rotation is *counterclockwise*, and **negative** if its rotation is *clockwise*. The first three angles in the following diagram are positive angles, and the last is a negative angle.

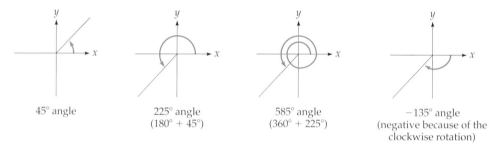

45° angle     225° angle
(180° + 45°)     585° angle
(360° + 225°)     −135° angle
(negative because of the clockwise rotation)

 **BE CAREFUL:** It is not enough to specify the initial and terminal sides of an angle—the rotation must also be specified. For example, the second, third, and fourth angles in the above diagrams have the same initial and terminal sides but are different because they have different rotations.

Angles that have the same initial and terminal sides (even if formed by different rotations) are called **coterminal angles.** Any two coterminal angles differ by a multiple of 360°.

The choice of 360 for the number of degrees in a full circle dates back to the ancient Babylonians, whose numbering system was based on 60, the "sexagesimal" system. (We still use their system for time: 60 seconds in a minute and 60 minutes in an hour.) In calculus we use another method for measuring angles, called **radian measure.** We use radian measure because it will lead to simpler formulas for differentiation and integration in Section 8.3.

The radian measure of an angle is found by placing the angle at the center of a circle of radius 1 and measuring the length of the arc of the circle cut off by the angle: the length of the arc gives the measure of the angle.

**Radian Measure**

An angle of $x$ radians is an angle that cuts off an arc of length $x$ on a circle of radius 1 centered at the vertex.

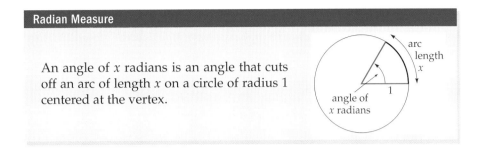

An angle whose vertex is at the center of a circle is called a **central angle.** As usual, a positive angle means counterclockwise rotation. The following diagrams show angles of $\frac{1}{2}$ radian, 1 radian, and 2 radians.

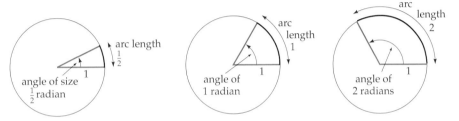

Radian measure of angle equals the arc length cut off on a unit circle.

The circumference formula $C = 2\pi r$ with $r = 1$ shows that the circumference of the unit circle is $2\pi$. Therefore, a full circle of $360°$ is equivalent to $2\pi$ radians.

$$360° = 2\pi \text{ radians}$$

Dividing each side by 2 shows that $180° = \pi$ radians, and dividing again shows that $90° = \pi/2$ radians. The degree and radian measures of the following **quadrant angles** should be memorized.

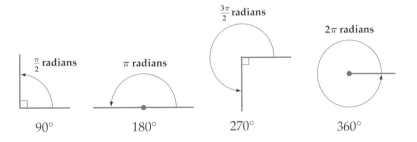

## Converting Between Degrees and Radians

Since degrees and radians are proportional, we can convert between them using the following relationship.

$$\frac{\text{Degrees}}{180} = \frac{\text{Radians}}{\pi}$$

This equation may be solved for either degrees or radians, resulting in the two formulas

$$\text{Radians} = \frac{\pi}{180} \cdot \text{degrees}$$

To change from degrees to radians, multiply by $\frac{\pi}{180}$

and

$$\text{Degrees} = \frac{180}{\pi} \cdot \text{radians}$$

To change from radians to degrees, multiply by $\frac{180}{\pi}$

**EXAMPLE 1**

**CONVERTING DEGREES TO RADIANS**

Find the radian measure of each angle:  **a.** $45°$  **b.** $150°$  **c.** $-60°$

*Solution*

To convert to radians, multiply degrees by $\frac{\pi}{180}$.

**a.** $45° = \overset{1}{45} \cdot \frac{\pi}{\underset{4}{180}} = \frac{\pi}{4}$ radians

Multiplying by $\frac{\pi}{180}$

**b.** $150° = \overset{5}{150} \cdot \frac{\pi}{\underset{6}{180}} = \frac{5\pi}{6}$ radians

**c.** $-60° = \overset{-1}{-60} \cdot \frac{\pi}{\underset{3}{180}} = -\frac{\pi}{3}$ radians

We leave $\pi$ in our answers to keep them exact. For an approximation, we could replace $\pi$ by 3.14.

**EXAMPLE 2**

**CONVERTING RADIANS TO DEGREES**

Find the degree measure of each angle:

**a.** $\frac{\pi}{6}$ radians  **b.** $\frac{3\pi}{4}$ radians  **c.** $-\frac{\pi}{2}$ radians  **d.** 1 radian

*Solution*

To convert to degrees, multiply radians by $\frac{180}{\pi}$.

**a.** $\frac{\pi}{6}$ radians $= \frac{\pi}{\underset{1}{6}} \cdot \frac{\overset{30}{180}}{\pi} = 30°$

Multiplying by $\frac{180}{\pi}$

**b.** $\dfrac{3\pi}{4}$ radians $= \dfrac{3\pi}{\underset{1}{4}} \cdot \dfrac{\overset{45}{\cancel{180}}}{\pi} = 3 \cdot 45° = 135°$

**c.** $-\dfrac{\pi}{2}$ radians $= -\dfrac{\pi}{\underset{1}{2}} \cdot \dfrac{\overset{90}{\cancel{180}}}{\pi} = -90°$

**d.** 1 radian $= 1 \cdot \dfrac{180}{\pi} \approx 57°$

1 radian ≈ 57°

## Graphing Calculator Exploration

45°
.7853981634
π/4
.7853981634

In *radian* mode

**a.** You can use a graphing calculator to convert from radians to degrees. For example, to check the answer to part (a) of Example 1, set your calculator to RADIANS (on some calculators use MODE), enter 45 degrees with the ° symbol, and press ENTER. Since your calculator gives *decimal* answers (using an approximation for $\pi$), you can check the decimal by evaluating $\pi/4$.

$(-\pi/2)^r$
−90

In *degree* mode

**b.** You can also convert from degrees to radians. For example, to check the answer to part (c) of Example 2, set your calculator for DEGREES (using MODE), enter $-\pi/2$ radians with the $^r$ symbol. The result will be in degrees. At the end, change back to RADIANS (using MODE) since that is what you will be using almost exclusively in this chapter.

**Practice Problem**

**a.** Find the radian measure of an angle of 60°.

**b.** Find the degree measure of an angle of $\dfrac{2\pi}{3}$ radians.

➤ **Solutions on page 566**

The angles 30°, 45°, and 60° occur so frequently that their radian measures should be memorized.

$\frac{\pi}{6}$ radians  $\frac{\pi}{4}$ radians  $\frac{\pi}{3}$ radians

    30°              45°             60°

These angles may be added to the quadrant angles to find corresponding angles in other quadrants. For example, the largest angle shown in the diagram on the right has measure

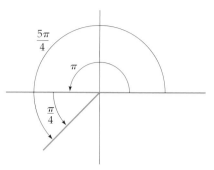

$$\pi + \frac{\pi}{4} = \frac{5\pi}{4}$$

Other angles resulting from such additions are shown in the following diagram.

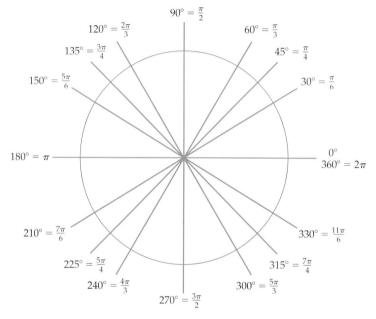

## Alternative Definition of Radian Measure

The circumference formula $C = 2\pi r$ shows that the circumference $C$ of a circle is proportional to the radius $r$. Therefore, we may define radian measure using a circle of *any* radius $r$, as long as arc length is measured in "radius" units, which means dividing the actual arc length by the radius $r$.

---

**Alternative Definition of Radian Measure**

The central angle that cuts off an arc of length $s$ on a circle of radius $r$ has radian measure $s/r$:

$$\left(\begin{matrix} \text{Radian} \\ \text{measure} \end{matrix}\right) = \frac{s}{r} \quad \begin{matrix} \leftarrow \text{arc length} \\ \leftarrow \text{radius} \end{matrix}$$

Since radian measure is a *quotient*, arc length over radius, any units (inches, feet, etc.) will cancel out. For this reason, radians are called *dimensionless* units.

## Arc Length Along a Circle

Radian measure calculates angles according to the formula

$$\text{Radians} = \frac{\text{Arc length}}{\text{Radius}}$$

Solving this formula for arc length (multiplying each side by "radius") gives the following formula.

---

### Arc Length

The arc length cut off by any central angle is

$$\begin{pmatrix} \text{Arc} \\ \text{length} \end{pmatrix} = (\text{Radians}) \cdot (\text{Radius})$$

Arc length is Radians times Radius

---

This formula gives the arc length along any circle.

## Speed of Rotation of the Earth

How fast are you moving if you "stand still" on the equator? The earth makes one full rotation ($2\pi$ radians) every 24 hours, so during 1 hour it rotates $\frac{2\pi}{24} = \frac{\pi}{12}$ radian. Since the earth's radius is approximately 4000 miles, during 1 hour a point on the equator will move a distance of

$$\begin{pmatrix} \text{Arc} \\ \text{length} \end{pmatrix} = \frac{\pi}{12} \cdot 4000 \approx 1047 \text{ miles} \qquad \text{Radians times radius}$$

That is, standing on the equator you are actually rotating around the center of the earth at a speed of over 1000 miles per hour. You don't feel it because everything around you is moving with you.

---

## Graphing Calculator Exploration

(For those familiar with the cosine function)

What is your speed (resulting from the earth's rotation) if you live somewhere other than the equator? It depends on your latitude. If you are at latitude $x$, then your speed is $1047 \cos x$.

**a.** Set your calculator for DEGREES and graph $y_1 = 1047 \cos x$ on the window [0, 90] by [−150, 1200]. Observe how the speed varies from 1047 miles per hour at the equator (0° latitude) to 0 miles per hour at the North Pole (90° latitude).

**b.** Find the latitude where you are and evaluate $y_1 = 1047 \cos x$ at that number to find your speed of rotation.

**EXAMPLE 3**  **INTERCITY DISTANCES**

The cities of Washington, D.C., and Moscow determine a central angle of 1.22 radians (at the center of the earth). Find the distance between these cities measured along the surface of the earth. (Assume that the earth is a sphere of radius 4000 miles.)

*Solution*

Since arc length is radians times radius,

$$\begin{pmatrix} \text{Distance from} \\ \text{Washington, D.C.,} \\ \text{to Moscow} \end{pmatrix} = \underbrace{1.22}_{\text{radians}} \cdot \underbrace{4000}_{\text{radius}} = 4880$$

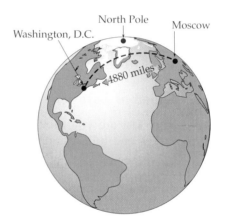

Therefore, an airplane flying between Washington, D.C., and Moscow must fly approximately 4880 miles (since the altitude of the airplane is negligible compared to the radius of the earth).

---

**Solutions to Practice Problem**

**a.** $60° = \overset{1}{60} \cdot \dfrac{\pi}{\underset{3}{180}} = \dfrac{\pi}{3}$ radians

**b.** $\dfrac{2\pi}{3}$ radians $= \dfrac{2\pi}{\underset{1}{3}} \cdot \dfrac{\overset{60}{180}}{\pi} = 120°$

## 8.1   Section Summary

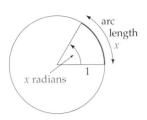

We began by finding the sides of two particular triangles: a 30°–60°–90° triangle (sides proportional to 1, 2, √3) and a 45°–45°–90° triangle (sides proportional to 1, 1, √2). We then defined the *radian measure* of an angle as the arc length cut off on a unit circle centered at the vertex of the angle.

Alternatively, using a circle of *any* radius, the radian measure is the arc length divided by the radius. To put this colloquially, if we were walking along the circumference of a circle, then walking a distance of *x* radii determines an angle of *x* radians. The diagram on page 564 shows the measures of many common angles in both degrees and radians.

To convert between degrees and radians, we use the relationship

$$\frac{\text{Degrees}}{180} = \frac{\text{Radians}}{\pi}$$

The arc length of a circle intercepted by a central angle measured in radians can be calculated from the formula

$$\left(\begin{matrix}\text{Arc} \\ \text{length}\end{matrix}\right) = (\text{Radians}) \cdot (\text{Radius})$$

## 8.1   Exercises

**1–8.** Find the degree and radian measure of each angle shown below.

**1.**

**2.**

**7.**

**8.**

**3.**

**4.**

**9–14.** For each angle:

  **i.** Find the radian measure (without using a calculator).
   **ii.** Check your answers using a graphing calculator.

  **9. a.** 30°        **b.** 225°        **c.** −180°

  **10. a.** 15°        **b.** 210°        **c.** −90°

  **11. a.** 60°        **b.** 315°        **c.** −120°

  **12. a.** 45°        **b.** 240°        **c.** −150°

  **13. a.** 540°        **b.** 135°        **c.** 1°

  **14. a.** 750°        **b.** 150°        **c.** 10°

**5.**

**6.**

**15–20.** For each angle:

  **i.** Find the degree measure (without using a calculator).
  **ii.** Check your answers using a graphing calculator.

*(See instructions on previous page.)*

**15. a.** $\dfrac{\pi}{4}$   **b.** $\dfrac{2\pi}{3}$   **c.** $-\dfrac{5\pi}{6}$

**16. a.** $\dfrac{\pi}{6}$   **b.** $\dfrac{3\pi}{2}$   **c.** $-\dfrac{5\pi}{4}$

**17. a.** $\dfrac{\pi}{3}$   **b.** $\dfrac{5\pi}{4}$   **c.** $-\dfrac{7\pi}{6}$

**18. a.** $\dfrac{3\pi}{4}$   **b.** $\dfrac{4\pi}{3}$   **c.** $-\dfrac{7\pi}{4}$

**19. a.** $\dfrac{9\pi}{4}$   **b.** $-3\pi$   **c.** 5

**20. a.** $\dfrac{7\pi}{3}$   **b.** $-5\pi$   **c.** 3

## Applied Exercises

**21. GENERAL: Arc Length** A 27-inch-long pendulum swings through an angle of $\dfrac{\pi}{9}$ radians. Find the length of arc through which it swings. (Give your answer in terms of $\pi$, and then approximate it using $\pi \approx 3.14$.)

**22. GENERAL: Building Swaying** Modern skyscrapers are designed to bend in high winds. As a general rule, a 100 mile per hour wind can cause a tall building to bend by 0.002 radian. Find the length of arc through which the 1450-foot-tall Sears Tower in Chicago would bend in such a wind. [*Note:* Because of back-and-forth motion, the swaying might be twice this great.]

**23. GENERAL: Airplanes** An airliner is flying in a circular holding pattern with a radius of 12 miles. Find the central angle (in radians) that the airliner will have flown through by flying 18 miles along this circular path.

**24. GENERAL: Distance** Find the distance along the surface of the earth between Los Angeles and Honolulu if the two cities determine a central angle of 0.64 radian. (Assume that the earth is a sphere of radius 4000 miles.)

**25. GENERAL: Distance** Find the distance along the surface of the earth between New York City and Paris if the two cities determine a central angle of 0.9 radian. (Assume that the earth is a sphere of radius 4000 miles.)

**26. GENERAL: Moon Diameter** Estimate the diameter of the moon if it is 240,000 miles from the earth and it forms an angle of 0.009 radian when viewed from the earth. [*Hint:* For very small angles, the diameter can be approximated by arc length.]

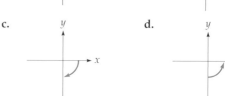

## Conceptual Exercises

**27.** True or False: All 30°–60°–90° triangles have sides of length 1, 2, $\sqrt{3}$.

**28.** True or False: Positive angles are measured *clockwise.*

**29.** True or False: 45 radians $= \dfrac{\pi}{4}$ degrees.

**30.** Which angles are in standard position?

**31.** What is the size of the smallest angle coterminal with an angle of $1000\pi$ radians?

**32.** If the shortest side of a 45°–45°–90° triangle has length 1000 kilometers, then what is the length of the longest side?

**33.** Fill in the blank: $\dfrac{180}{\pi}$ degrees = _____ radians.

**34.** Fill in the blank: $\dfrac{\pi}{180}$ radians = _____ degrees.

**35.** If a central angle of 0.123 radian cuts off an arc length of 5 on a circle, then what central angle cuts off an arc length of 10?

**36.** If you walk along the circumference of a circle a distance of 1 diameter, what is the size of the corresponding central angle (in radians)?

### *Explorations and Excursions*

The following problem extends and augments the material presented in the text.

**37. PROOF OF THE PYTHAGOREAN THEOREM**
Use the following diagram to prove the Pythagorean Theorem for the right triangle with sides $a$ and $b$ and hypotenuse $c$ as follows:

   **i.** Calculate the area of the outer square by multiplying length times width. [*Hint:* Length and width are each $a + b$.]

   **ii.** Calculate the area of the outer square again, but now as the inner square plus four right triangles, each of area $\frac{1}{2} ab$.

   **iii.** Set the two expressions for the area equal to each other and simplify, obtaining $a^2 + b^2 = c^2$.

---

### Introduction

In this section we define the **sine** and **cosine** functions, and use them to model quantities that fluctuate periodically.

### Sine and Cosine Functions

We use the Greek letter $\theta$ ("theta") to represent an angle (always measured in *radians* unless otherwise indicated). For an angle $\theta$ in standard position, we choose a point $(x, y)$ on its terminal side and define the sine and cosine functions as follows.

---

**$\sin \theta$   and   $\cos \theta$**

Let $(x, y)$ be a point on the terminal side of an angle of $\theta$ radians in standard position, and let $r$ be its distance from the origin. We define $\sin \theta$ and $\cos \theta$ as follows:

$$\sin \theta = \frac{y}{r}$$

$$\cos \theta = \frac{x}{r}$$

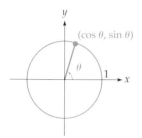

It does not matter which point on the terminal side (other than the origin) we choose; any two points $(x, y)$ and $(x', y')$ determine similar triangles, so the ratios of the sides will be the same.

If we choose the point $(x, y)$ to be a distance $r = 1$ from the origin, then the definitions of $\sin \theta$ and $\cos \theta$ simplify to

$$\sin \theta = y \qquad\qquad\qquad \sin \theta = \frac{y}{r} \text{ with } r = 1$$

$$\cos \theta = x \qquad\qquad\qquad \cos \theta = \frac{x}{r} \text{ with } r = 1$$

To put this another way, a point on the circle of radius 1 centered at the origin can be represented as $(\cos \theta, \sin \theta)$, where $\theta$ is the angle at the origin, as shown in the diagram on the left.

---

**EXAMPLE 1**

**FINDING VALUES OF SINE AND COSINE**

Evaluate $\sin \theta$ and $\cos \theta$ for the following values of $\theta$: $0$, $\dfrac{\pi}{2}$, $\pi$, and $\dfrac{3\pi}{2}$.

*Solution*

The angles are shown below with a point $(x, y)$ on the terminal side 1 unit from the origin. The sine and cosine are then just the $y$- and $x$-coordinates (since $\sin \theta = y$ and $\cos \theta = x$ when $r = 1$).

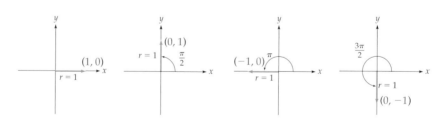

$\sin 0 = 0 \qquad\qquad \sin \dfrac{\pi}{2} = 1 \qquad\qquad \sin \pi = 0 \qquad\qquad \sin \dfrac{3\pi}{2} = -1 \qquad$ From the $y$-coordinates

$\cos 0 = 1 \qquad\qquad \cos \dfrac{\pi}{2} = 0 \qquad\qquad \cos \pi = -1 \qquad\qquad \cos \dfrac{3\pi}{2} = 0 \qquad$ From the $x$-coordinates

---

**EXAMPLE 2**

**FINDING VALUES OF SINE AND COSINE**

Find $\sin \dfrac{\pi}{6}$ and $\cos \dfrac{\pi}{6}$.

*Solution*

Draw an angle of
$\frac{\pi}{6}$ radians (30°) in
standard position.

Complete the sides
of the 30°-60°-90°
triangle with sides 1, 2, $\sqrt{3}$.

Write the coordinates
of a point based on
the triangle sides.

From these values of $x$, $y$, and $r$ we find the sine and cosine:

$$\sin \frac{\pi}{6} = \frac{1}{2} \qquad\qquad \sin \theta = \frac{y}{r}$$

$$\cos \frac{\pi}{6} = \frac{\sqrt{3}}{2} \qquad\qquad \cos \theta = \frac{x}{r}$$

---

**Practice Problem**

Find $\sin \dfrac{\pi}{4}$ and $\cos \dfrac{\pi}{4}$.

➤ Solution on page 578

Similar calculations lead to sines and cosines of the following angles. You should either memorize this table or be able to calculate these values by remembering the sides of $30°-60°-90°$ and $45°-45°-90°$ triangles.

| $\theta$ | 0 | $\dfrac{\pi}{6}$ | $\dfrac{\pi}{4}$ | $\dfrac{\pi}{3}$ | $\dfrac{\pi}{2}$ |
|---|---|---|---|---|---|
| $\sin \theta$ | 0 | $\dfrac{1}{2}$ | $\dfrac{\sqrt{2}}{2}$ | $\dfrac{\sqrt{3}}{2}$ | 1 |
| $\cos \theta$ | 1 | $\dfrac{\sqrt{3}}{2}$ | $\dfrac{\sqrt{2}}{2}$ | $\dfrac{1}{2}$ | 0 |

Notice the patterns:

$$\frac{\sqrt{0}}{2} \quad \frac{\sqrt{1}}{2} \quad \frac{\sqrt{2}}{2} \quad \frac{\sqrt{3}}{2} \quad \frac{\sqrt{4}}{2}$$

$$\frac{\sqrt{4}}{2} \quad \frac{\sqrt{3}}{2} \quad \frac{\sqrt{2}}{2} \quad \frac{\sqrt{1}}{2} \quad \frac{\sqrt{0}}{2}$$

**EXAMPLE 3**

**FINDING SINE AND COSINE IN OTHER QUADRANTS**

Find $\sin \dfrac{2\pi}{3}$ and $\cos \dfrac{2\pi}{3}$.

*Solution*

Draw an angle of
$\frac{2\pi}{3}$ radians (120°).

Note the remaining
60° angle.

Draw a perpendicular to
an axis to make a 30°-60°-90°
triangle with sides 1, 2, $\sqrt{3}$

Write the coordinates
of a point based on
the triangle sides (*with
the correct signs*).

From these values of $x$, $y$, and $r$:

$$\sin\frac{2\pi}{3} = \frac{\sqrt{3}}{2}$$

$$\cos\frac{2\pi}{3} = \frac{-1}{2} = -\frac{1}{2}$$

$\sin\theta = \dfrac{y}{r}$

$\cos\theta = \dfrac{x}{r}$

## Graphing Calculator Exploration

```
sin(2π/3)
         .8660254038
√(3)/2
         .8660254038
cos(2π/3)
               -.5
```

Your calculator should find a value of $\cos\dfrac{2\pi}{3}$ that agrees with the result found above, and a value for $\sin\dfrac{2\pi}{3}$ that agrees with $\dfrac{\sqrt{3}}{2}$ to as many decimal places as the calculator shows.

 **BE CAREFUL:** If you get incorrect answers, make sure that your calculator is (and remains) set for *radians*.

EXAMPLE 4

**FINDING SINE AND COSINE OF A NEGATIVE NUMBER**

Find $\sin\left(-\dfrac{\pi}{4}\right)$ and $\cos\left(-\dfrac{\pi}{4}\right)$.

*Solution*

Using only one graph:

From these values of $x$, $y$, and $r$:

$$\sin\left(-\frac{\pi}{4}\right) = \frac{-1}{\sqrt{2}} = -\frac{\sqrt{2}}{2} \qquad\qquad \sin\theta = \frac{y}{r}$$

Multiplying numerators and denominators by $\sqrt{2}$

$$\cos\left(-\frac{\pi}{4}\right) = \frac{1}{\sqrt{2}} = \frac{\sqrt{2}}{2} \qquad\qquad \cos\theta = \frac{x}{r}$$

## Sine and Cosine of an Acute Angle

An *acute* angle is an angle between 0 and $\pi/2$ radians. For an acute angle $\theta$, $\sin\theta$ and $\cos\theta$ can be expressed as ratios of sides of a right triangle (called hyp, opp, and adj for the *hyp*otenuse, *opp*osite side, and *adj*acent side).

**Right Triangle Formulas for  $\sin\theta$  and  $\cos\theta$**

$$\sin\theta = \frac{\text{opp}}{\text{hyp}}$$

$$\cos\theta = \frac{\text{adj}}{\text{hyp}}$$

**EXAMPLE 5**

**CALCULATING DISTANCES**

How long a ladder is needed to reach a roof 20 feet above the ground if the ladder makes a 60° angle with the ground?

*Solution*

The ladder and building form the hypotenuse and opposite side of the right triangle shown on the previous page, suggesting that we use

$$\sin \theta = \frac{\text{opp}}{\text{hyp}} \quad \text{with} \quad \theta = \frac{\pi}{3}.$$

$$\sin \frac{\pi}{3} = \frac{20}{x} \qquad\qquad \sin \theta = \frac{\text{opp}}{\text{hyp}}$$

$$\frac{\sqrt{3}}{2} = \frac{20}{x} \qquad\qquad \text{Using} \quad \sin \frac{\pi}{3} = \frac{\sqrt{3}}{2}$$

$$\sqrt{3}\,x = 40 \qquad\qquad \text{Cross-multiplying}$$

$$x = \frac{40}{\sqrt{3}} \approx 23.09 \qquad\qquad \text{Solving for } x, \text{ then using a calculator}$$

Therefore, the ladder must be about 24 feet long.

---

## Graphs of the Sine and Cosine Functions

On page 570, we saw that the coordinates of a point on a circle of radius 1 centered at the origin can be represented as $(\cos\theta, \sin\theta)$, where $\theta$ is the angle at the origin. Therefore, as $\theta$ increases, $\sin \theta$ behaves just like the $y$-coordinate of the point $(x, y)$ as it moves around the circle—first increasing from 0 to 1, then decreasing to 0 to $-1$, then increasing back to 0. Plotting $\theta$ on the horizontal axis, we obtain the graph on the left.

This pattern of values repeats itself each time the angle completes another full circle of $2\pi$ radians, giving the following graph of $\sin t$ (where we have replaced $\theta$ by $t$).

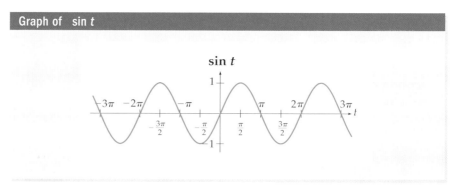

**Graph of  sin $t$**

Similarly, $\cos \theta$ behaves just like the $x$-coordinate of the point $(x, y)$ as it moves around the circle—first decreasing from 1 to 0 to $-1$, then increasing back to 1. Again, this cycle repeats itself each time the angle completes another full circle of $2\pi$ radians, giving the following graph of $\cos t$.

**Graph of  cos *t***

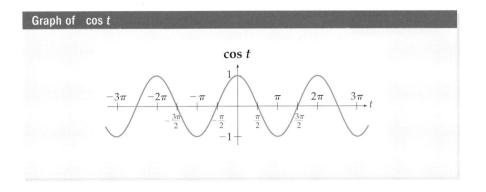

(These graphs could also have been found by the more tedious process of plotting points.)  Sin *t*  and  cos *t*  are said to be **periodic with period 2π**, meaning that they repeat their values every 2π units. We will now think of sin *t*  and  cos *t*  as *ordinary functions*, unrelated to angles, functions that oscillate between +1 and −1 in a "wavy" pattern. We will use  sin *t*  and  cos *t* to model quantities that undergo regular fluctuations, such as seasonal sales.

## Modified Sine and Cosine Curves—Changing Amplitude and Period

The **amplitude** of a sine or cosine curve is half the distance between its highest and lowest values (sometimes called its *half-height*). Therefore,  sin *t*  and cos *t*  have amplitude 1. By multiplying by constants, we may create similar "wavy" functions with any amplitude and period. For positive numbers *a* and *b*, the function  *a* sin *bt*  oscillates between the values *a* and −*a*, and so has amplitude *a*. The period of  *a* sin *bt*  is found by setting  *bt* = 2π, giving period  $t = \dfrac{2\pi}{b}$.  Similar statements hold for the function  *a* cos *bt*.

**Modified Sine and Cosine Functions**

$$\left.\begin{array}{l} y = a \sin bt \\ y = a \cos bt \end{array}\right\} \text{ have amplitude } a \text{ and period } \frac{2\pi}{b}.$$

**EXAMPLE 6**

### GRAPHING A MODIFIED SINE CURVE

Sketch the graph of $f(t) = 3 \sin 2t$.

*Solution*

The amplitude is 3 (since the sine is multiplied by 3), and the period is $\pi$ (the sine's period $2\pi$ divided by the 2). Therefore, the graph is a sine curve, but between heights 3 and $-3$, with a complete cycle occurring between 0 and $\pi$, as shown below.

## Graphing Calculator Exploration

Graph $y_1 = \cos 2x$ and $y_2 = \cos 0.5x$ on the window $[-2\pi, 2\pi]$ by $[-2, 2]$. (Be sure that your calculator is set for radians.) Notice how the number in front of the $x$ changes the period: a larger number makes the curve "wiggle" faster.

Graph $y_1 = \cos x$, $y_2 = 2 \cos x$, and $y_3 = 0.5 \cos x$ on the same window. Notice how the number in front of the *function* changes the amplitude: a larger number makes the amplitude larger.

Predict what the curve $y = 2 \cos 2x$ would look like. Then check your prediction by graphing the curve. What about $y = -2 \cos x$? What about $y = -2 \sin 2x$?

### Seasonal Sales

The sine and cosine functions with modified amplitude and period are often used to model quantities that fluctuate in a regular pattern, such as seasonal sales.

**EXAMPLE 7**  **PREDICTING SEASONAL REVENUE**

A travel agency specializing in Caribbean vacations models its daily revenue (in dollars) on day $n$ of the year by

$$R(n) = 600 + 500 \cos\left(\frac{2\pi n}{365}\right)$$

Find the company's revenue on March 1 (day $n = 60$).

*Solution*

$$R(60) = 600 + 500 \underbrace{\cos\left(\frac{2\pi \cdot 60}{365}\right)}_{0.512 \text{ (using a calculator set for radians)}} \approx 856$$

$R(n) = 600 + 500 \cos\left(\frac{2\pi n}{365}\right)$

with $n = 60$

Therefore, the travel agency's daily revenue on March 1 will be about $856.

---

Notice that the revenue in the above graph never reaches zero because of the vertical shift by 600 (see pages 59–60).

## Trigonometric Identities

A formula that holds for *all* values of the variable is called an *identity.* There are many trigonometric identities, and we will mention only a few. The first is called the *Pythagorean identity,* since it comes directly from the Pythagorean Theorem (page 558):

$$\sin^2 t + \cos^2 t = 1$$

$\sin^2 t$ means $(\sin t)^2$
$\cos^2 t$ means $(\cos t)^2$

To prove this identity, recall that $\sin t = \dfrac{y}{r}$ and $\cos t = \dfrac{x}{r}$, so

$$\sin^2 t + \cos^2 t = \left(\frac{y}{r}\right)^2 + \left(\frac{x}{r}\right)^2 = \frac{x^2 + y^2}{r^2} = \frac{r^2}{r^2} = 1 \qquad \text{Since } x^2 + y^2 = r^2$$

This and other useful identities are listed below, and the remaining proofs are developed in Exercises 52–54.

## Trigonometric Identities

$$\sin^2 t + \cos^2 t = 1$$

Pythagorean identity

$$\sin(t + 2\pi) = \sin t$$
$$\cos(t + 2\pi) = \cos t$$

Periodicity identities

$$\sin(-t) = -\sin t$$
$$\cos(-t) = \cos t$$

Negative angle identities

$$\sin t = \cos\left(\frac{\pi}{2} - t\right)$$
$$\cos t = \sin\left(\frac{\pi}{2} - t\right)$$

Complementary angle identities

$$\sin(s \pm t) = (\sin s)\cdot(\cos t) \pm (\cos s)\cdot(\sin t)$$
$$\cos(s \pm t) = (\cos s)\cdot(\cos t) \mp (\sin s)\cdot(\sin t)$$

Sum and difference identities

In the last two identities, use the upper signs throughout the equation, or the lower signs throughout.

## Solution to Practice Problem

$$\sin\frac{\pi}{4} = \frac{1}{\sqrt{2}} = \frac{\sqrt{2}}{2}$$

$$\cos\frac{\pi}{4} = \frac{1}{\sqrt{2}} = \frac{\sqrt{2}}{2}$$

## 8.2 Section Summary

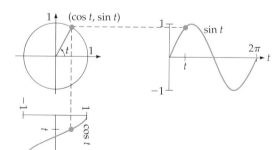

The sine and cosine functions are defined in terms of the coordinates $(x, y)$ of a point on a unit circle forming an angle of $\theta$ radians:

$$\sin\theta = y \quad \text{and} \quad \cos\theta = x$$

For acute angles, these functions can also be interpreted in terms of the sides of a right triangle:

$$\sin\theta = \frac{\text{opp}}{\text{hyp}} \quad \text{and} \quad \cos\theta = \frac{\text{adj}}{\text{hyp}}$$

For a few particular angles (such as 30°, 45°, and 60°) we can evaluate these functions *exactly*, whereas for other angles we use a calculator. We then graphed these functions and used them to model quantities that undergo regular fluctuations. The trigonometric identities will be used in the exercises and in the following section.

## 8.2 Exercises

**1–10.** Find the sine and cosine of each angle $\theta$. When $r$ is not given, calculate it from the Pythagorean Theorem.

**1.**

**2.**

**3.**

**4.**

**5.**

**6.**

**7.**

**8.**

**9.**

**10.**

**11–16.** Evaluate *without* using a calculator, leaving answers in exact form.

**11. a.** $\sin \dfrac{\pi}{3}$     **b.** $\cos \dfrac{\pi}{3}$

     **c.** $\sin \dfrac{\pi}{4}$     **d.** $\cos \dfrac{\pi}{4}$

**12. a.** $\sin \dfrac{3\pi}{4}$     **b.** $\cos \dfrac{3\pi}{4}$

     **c.** $\sin \dfrac{5\pi}{3}$     **d.** $\cos \dfrac{5\pi}{3}$

**13. a.** $\sin \dfrac{5\pi}{4}$     **b.** $\cos \dfrac{5\pi}{4}$

     **c.** $\sin \dfrac{5\pi}{6}$     **d.** $\cos \dfrac{5\pi}{6}$

**14. a.** $\sin \dfrac{11\pi}{6}$     **b.** $\cos \dfrac{11\pi}{6}$

     **c.** $\sin \dfrac{7\pi}{4}$     **d.** $\cos \dfrac{7\pi}{4}$

**15. a.** $\sin\left(-\dfrac{\pi}{3}\right)$     **b.** $\cos\left(-\dfrac{\pi}{3}\right)$

     **c.** $\sin\left(-\dfrac{\pi}{4}\right)$     **d.** $\cos\left(-\dfrac{\pi}{4}\right)$

**16. a.** $\sin \dfrac{5\pi}{2}$     **b.** $\cos \dfrac{5\pi}{2}$

     **c.** $\sin 3\pi$     **d.** $\cos 3\pi$

**17–18.** Approximate using a calculator (set for *radians*). Round answers to two decimal places.

**17. a.** $\sin \dfrac{\pi}{10}$     **b.** $\cos \dfrac{\pi}{10}$

     **c.** $\sin 1$     **d.** $\cos 1$

*(See instructions on previous page.)*

**18. a.** $\sin \dfrac{\pi}{5}$      **b.** $\cos \dfrac{\pi}{5}$

    **c.** $\sin 4$      **d.** $\cos 4$

**19–26.** Sketch the graph of each function showing the amplitude and period.

**19.** $y = 4 \sin t$      **20.** $y = 6 \cos t$

**21.** $y = \sin 4t$      **22.** $y = \cos 6t$

**23.** $y = 3 \cos 4t$      **24.** $y = 5 \sin 4t$

**25.** $y = -\cos 2t$      **26.** $y = -\sin 2t$

**27–28.** Use the sum and difference identities on page 578 to find the following sines and cosines. Do *not* use a calculator.

**27. a.** $\sin \dfrac{5\pi}{12}$      **b.** $\cos \dfrac{5\pi}{12}$

    [*Hint:* Take $s = \dfrac{\pi}{4}$ and $t = \dfrac{\pi}{6}$ with upper signs.]

**28. a.** $\sin \dfrac{7\pi}{12}$      **b.** $\cos \dfrac{7\pi}{12}$

    [*Hint:* Take $s = \dfrac{\pi}{4}$ and $t = \dfrac{\pi}{3}$ with upper signs.]

## Applied Exercises

**29. GENERAL: Ladder Height** A 30-foot ladder leaning against a wall makes a 70° angle with the ground. How high up the wall does the ladder reach?

**30. GENERAL: Kite Flying** A straight kite string makes an angle of 40° with the ground, as shown in the diagram. If 250 feet of string has been let out, how high is the kite?

**31. GENERAL: Distance** A boat sailing due west toward a point 200 feet directly north of you makes an angle of 48°, as shown in the diagram. Find the distance between you and the boat.

**32. GENERAL: Wheelchair Ramp** According to the American National Standards Institute, a wheelchair ramp should make an angle of 4.75° with the horizontal. Find the length of such a ramp that rises 2 feet.

**33. BUSINESS: Seasonal Sales** An ice cream parlor estimates that the number of ice cream cones that it will sell on day $n$ of the year is

$$S(n) = 700 - 600 \cos \dfrac{2\pi n}{365}$$

Find the number of ice cream cones that will be sold on August 1 (day 213).

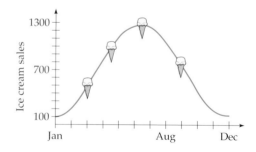

**34. BUSINESS: Seasonal Sales** A store estimates that the number of ski parkas that it will sell on day $n$ of the year is

$$S(n) = 27 + 25 \cos \dfrac{2\pi n}{365}$$

Find the number of ski parkas that will be sold on February 1 (day 32).

**35. BIOMEDICAL: Breathing** When you breathe normally, you inhale and exhale about 0.8 liter of air approximately every 4 seconds, with about 4.5 liters remaining in your lungs. Under these conditions, the total amount of air in your lungs $t$ seconds after you last inhaled is

$$L(t) = 4.9 + 0.4 \cos \frac{\pi t}{2} \quad \text{liters}$$

Find (*without* using a calculator) the amount of air in your lungs at time:

**a.** $t = 0$ (just after inhaling).
**b.** $t = 2$ (just after exhaling).

*Source: E. Batschelet, Introduction to Mathematics for Life Scientists*

 **36. COSINE AND SINE ON THE UNIT CIRCLE**

**a.** Set your calculator for PARAMETRIC equations (on some calculators, use MODE), and enter $x_1 = \cos T$, $y_1 = \sin T$. (Be sure your calculator is set for *radians*.)
**b.** GRAPH the curve on the window with $T$ in $[0, 2\pi]$, $x$ in $[-2, 2]$ and $y$ in $[-1.5, 1.5]$.
**c.** Use TRACE to move a point around the circle, with $X$ and $Y$ showing the cosine and sine of the angle. Observe how the sine and cosine change as the point moves around the circle.
**d.** Use EVALUATE to check the sine and cosine at $T$-values such as $\frac{\pi}{6}$, $\frac{\pi}{4}$, and $\frac{\pi}{3}$.
(Reset your calculator to FUNCTION mode after this exercise.)

 **37. MODIFIED SINE CURVES**

**a.** Graph the curves $y_1 = \sin x$, $y_2 = \sin 2x$, and $y_3 = \sin 0.5x$ on the window $[-2\pi, 2\pi]$ by $[-2, 2]$. (Be sure your calculator is set for *radians*.) Notice how a number in front of $x$ changes the *period*: a larger number makes the curve "wiggle faster."
**b.** Now graph $y_1 = \sin x$, $y_2 = 2 \sin x$, and $y_3 = 0.5 \sin x$ on the same window. Notice how a number in front of the function changes the *amplitude*: a larger number makes the amplitude larger.
**c.** Predict what the curve $y = 2 \sin 2x$ would look like. Then check your answer by graphing it. What about $y = -2 \sin x$? What about $y = -2 \sin 2x$?

 **38. THREE SIMILAR CURVES** Graph $y_1 = 1 - x^2$, $y_2 = \sqrt{1 - x^2}$, and $y_3 = \cos \frac{\pi x}{2}$ on the window $[-1, 1]$ by $[0, 1.3]$. Which curve is a parabola, which is a semicircle, and which is a trigonometric function? (This shows that trigonometric curves should be drawn slightly more "pointed" than circles.)

**39. GENERAL: Temperature** The mean temperature in Fairbanks, Alaska, on day $x$ of the year is approximately

$$T(x) = 36 + 22 \sin \frac{2\pi(x - 102)}{365}$$

degrees Fahrenheit.

**a.** Graph this function on the window $[0, 365]$ by $[0, 60]$. (Be sure your calculator is set for *radians*.)
**b.** Find the temperature on January 10 (day 10).
**c.** Find the temperature on July 12 (day 193).

*Source: Alaska Climate Research Center*

 **40. VERIFICATION OF** $\sin^2 x + \cos^2 x = 1$
Graph $y_1 = (\sin x)^2$, $y_2 = (\cos x)^2$, and $y_3 = y_1 + y_2$ on the window $[-2\pi, 2\pi]$ by $[-1, 2]$. Why does the fact that $y_3$ is a horizontal

line at height 1 verify the identity $\sin^2 x + \cos^2 x = 1$?

**41. GENERAL: Greenhouse Gases** The "green-house effect," which may be increasing global temperatures and, through the melting of polar ice, raising sea levels, is caused by increasing levels of carbon dioxide and other atmospheric gases. The carbon dioxide curve graphed on page 558 is given by $C(x) = 313e^{0.0041x} - 2.5 \cos(2\pi x)$ ppm (parts per million), where $x$ is the number of years since January 1, 1960.

**a.** Graph this function on the window $[50, 60]$ by $[380, 405]$.

Notice how the sine curve (seasonal variation) follows an upward trend.

**b.** Use this formula to predict the carbon dioxide levels on January 1, 2015.

*Source:* National Oceanic and Atmospheric Administration

## Conceptual Exercises

**42.** True or False: An angle of $\theta$ radians in standard position intersects a circle of radius 1 at the point $(\sin \theta, \cos \theta)$.

**43.** What is the shortest horizontal shift of a cosine curve that will turn it into a sine curve?

**44.** A modified sine curve is shifted up so that it varies between heights 4 and 10. What is the amplitude of this sine curve?

**45.** What is the maximum height of $f(x) = 8 + 3 \sin x$? What is its *minimum* height?

**46.** What is the period of $\cos(2\pi x)$?

**47.** Evaluate at sight: **a.** $\sin 4000\pi$ **b.** $\cos 4000\pi$

**48.** Evaluate at sight: **a.** $\sin 100.5\pi$ **b.** $\cos 100.5\pi$

**49.** True or False: $\sin(x - y) = -\sin(y - x)$.

**50.** True or False: $\cos(x - y) = \cos(y - x)$.

**51.** Evaluate $\sin^2(e^{3-\pi}/103^{7\pi/8}) + \cos^2(e^{3-\pi}/103^{7\pi/8})$

## Explorations and Excursions

The following problems extend and augment the material presented in the text.

**52. PROOFS OF NEGATIVE ANGLE IDENTITIES** As illustrated in the following diagram, changing the sign of an angle (from $t$ to $-t$) changes the sign of the $y$-coordinate of a point on the terminal side, but not the sign of the $x$-coordinate. Use this fact together with the definitions of $\sin t$ and $\cos t$ on page 569 to prove the negative angle identities: $\sin(-t) = -\sin t$ and $\cos(-t) = \cos t$.

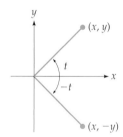

**53. PROOFS OF COMPLEMENTARY ANGLE IDENTITIES** Prove the identities

$$\sin t = \cos\left(\frac{\pi}{2} - t\right) \quad \text{and} \quad \cos t = \sin\left(\frac{\pi}{2} - t\right)$$

for an acute angle $t$ as follows.

**a.** Show that if one acute angle of a right triangle is of size $t$ radians, then the other acute angle is of size $\frac{\pi}{2} - t$ radians.

**b.** Observe that the "opposite" side for angle $t$ is the "adjacent" side for angle $\frac{\pi}{2} - t$, and vice versa. Use these facts with the expressions for $\sin t$ and $\cos t$ on page 573 to prove the identities above.

**54. PROOFS OF THE SUM AND DIFFERENCE IDENTITIES** Prove the identity

$$\sin(s + t) = (\sin s)(\cos t) + (\cos s)(\sin t)$$

for acute angles whose sum is also acute by using the diagram on the right as follows (the notation $AB$ means the length of the line segment between points $A$ and $B$, and similarly for any other points).

**a.** Show that $AB = \sin(s + t)$.
**b.** Show that $AC = \sin t$.
**c.** Show that $CD = \cos t$.

**d.** Show that $CE = \dfrac{\dfrac{CE}{DE}}{\dfrac{CD}{DE}} \cdot CD = \dfrac{\sin s}{\cos s} \cdot \cos t$.

**e.** Show that

$$AB = (\cos s)(AC + CE)$$

$$= (\cos s)\left(\sin t + \frac{\sin s}{\cos s} \cos t\right)$$

**f.** From (a) and (e), conclude that

$$\sin(s + t) = (\sin s)(\cos t) + (\cos s)(\sin t)$$

[*Note:* The identity

$$\cos(s + t) = (\cos s)(\cos t) - (\sin s)(\sin t)$$

will be proved in Exercise 91 on page 595. The identities for $\sin(s - t)$ and $\cos(s - t)$ are derived by changing $t$ to $-t$ in the identities for $\sin(s + t)$ and $\cos(s + t)$ and then using the negative angle identities.]

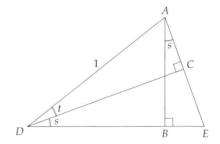

# 8.3 DERIVATIVES OF SINE AND COSINE FUNCTIONS

## Introduction

The last two sections contained no calculus, their purpose being to introduce the sine and cosine functions. In this section we will differentiate these functions, and use these derivatives to optimize functions and calculate rates of change.

## Derivatives of the Sine and Cosine Functions

The derivatives of the sine and cosine functions are given by the following formulas.

| Derivatives of $\sin t$ and $\cos t$ | |
| --- | --- |
| $\dfrac{d}{dt}\sin t = \cos t$ | The derivative of sine is cosine (for $t$ in radians) |
| $\dfrac{d}{dt}\cos t = -\sin t$ | The derivative of cosine is negative sine |

To see graphically that the derivative (the slope) of $\sin t$ is $\cos t$, observe in the following diagram that, at any particular $t$-value, the *slope* of the *sine* function is the same as the *value* (height) of the *cosine* function.

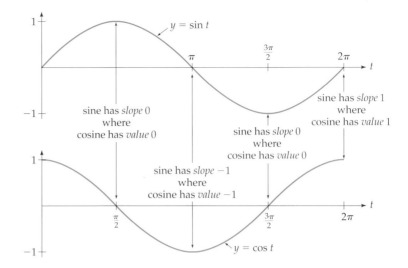

Similarly, to see that the derivative of $\cos t$ is $-\sin t$, observe in the diagram above that the *slope* of the *cosine* curve is the *negative* of the *value* of the *sine* curve. Justifications of these two formulas are given at the end of this section. The following examples show how to use these formulas.

---

**EXAMPLE 1**

**DIFFERENTIATING A QUOTIENT INVOLVING SINE**

Differentiate $f(t) = \dfrac{\sin t}{t}$.

*Solution*

$$\underset{\substack{\uparrow \\ \text{Derivative of } t}}{\overset{\substack{\text{Derivative of } \sin t}}{\ }}$$

$$\frac{d}{dt} \frac{\sin t}{t} = \frac{t \cdot \cos t - 1 \cdot \sin t}{t^2} = \frac{t \cos t - \sin t}{t^2} \qquad \text{Using the Quotient Rule}$$

---

**EXAMPLE 2**

**DIFFERENTIATING A PRODUCT INVOLVING COSINE**

Differentiate $g(t) = t^3 \cos t$.

*Solution*

$$\frac{d}{dt}(t^3 \cdot \cos t) = 3t^2 \cdot \cos t + t^3 \cdot (-\sin t) = 3t^2 \cos t - t^3 \sin t \qquad \begin{array}{l}\text{Using the}\\ \text{Product}\\ \text{Rule}\end{array}$$

$$\underset{\substack{\text{Derivative} \quad \text{Second} \qquad \text{First} \quad \text{Derivative}\\ \text{of the first} \quad \text{left alone} \quad \text{left alone} \quad \text{of } \cos t}}{\ }$$

| **Practice Problem 1** | Differentiate $f(t) = t \sin t$. | ➤ **Solution on page 590** |

These rules for differentiating sine and cosine, combined with the Chain Rule, show how to differentiate the sine and cosine of a *function*. For a differentiable function $f(t)$:

---

**Derivative of $\sin f(t)$ and $\cos f(t)$**

$$\frac{d}{dt} \sin f(t) = \cos f(t) \cdot \frac{df}{dt}$$

The derivative of the sine of a function is the cosine of the function times the derivative of the function

$$\frac{d}{dt} \cos f(t) = -\sin f(t) \cdot \frac{df}{dt}$$

The derivative of the cosine of a function is negative the sine of the function times the derivative of the function

---

**EXAMPLE 3**

**DIFFERENTIATING THE COSINE OF A FUNCTION**

Differentiate $g(t) = \cos(t^2 + 1)$.

*Solution*

$$\frac{d}{dt} \cos(\underbrace{t^2 + 1}_{f(t)}) = \underbrace{-\sin(t^2 + 1)}_{-\sin f(t)} \cdot \underbrace{2t}_{\frac{df}{dt}} \qquad \text{Using } \frac{d}{dt} \cos f = -\sin f \cdot \frac{df}{dt}$$

$$= -2t \sin(t^2 + 1)$$

---

**EXAMPLE 4**

**FINDING A SECOND DERIVATIVE**

Find the *second* derivative of $g(t) = \sin 4t$.

*Solution*

$$g'(t) = \frac{d}{dt}(\sin 4t) = \cos 4t \cdot (4) = 4 \cos 4t \qquad \text{Using } \frac{d}{dt} \sin f = \cos f \cdot \frac{df}{dt}$$

Derivative of $4t$

$$g''(t) = \frac{d}{dt}(4 \cos 4t) = 4[-\sin 4t \cdot (4)] \qquad \frac{d}{dt} \cos f = -\sin f \cdot \frac{df}{dt}$$

$$= -16 \sin 4t$$

---

| **Practice Problem 2** | Differentiate $f(t) = \sin \dfrac{t}{2}$. $\left[ \textit{Hint: } \text{Write } \dfrac{t}{2} \text{ as } \dfrac{1}{2}t. \right]$ |

➤ **Solution on page 590**

**EXAMPLE 5**

**EVALUATING A DERIVATIVE**

For $f(x) = \dfrac{\sin x}{\cos x}$, find $f'\left(\dfrac{\pi}{3}\right)$.

*Solution*

First differentiating:

Derivative of sin $x$

Derivative of cos $x$

$$f'(x) = \frac{\cos x \cdot \cos x - (-\sin x) \cdot \sin x}{\cos^2 x} \qquad \text{Using the Quotient Rule}$$

$$= \frac{\cos^2 x + \sin^2 x}{\cos^2 x} = \frac{1}{\cos^2 x} \qquad \begin{array}{l}\text{Using the identity} \\ \sin^2 x + \cos^2 x = 1\end{array}$$

Evaluating at the given $\dfrac{\pi}{3}$:

$$f'\left(\frac{\pi}{3}\right) = \frac{1}{\cos^2 \frac{\pi}{3}} = \frac{1}{\left(\cos \frac{\pi}{3}\right)^2} = \frac{1}{\left(\frac{1}{2}\right)^2} = \frac{1}{\left(\frac{1}{4}\right)} = 4 \qquad \text{Using } \cos \frac{\pi}{3} = \frac{1}{2}$$

**EXAMPLE 6**

**DIFFERENTIATING AN EXPONENTIAL TRIGONOMETRIC FUNCTION**

If $f(z) = e^{\sin z}$, find $f'(\pi)$.

*Solution*

$$f'(z) = e^{\sin z} \cos z \qquad \text{Using } \frac{d}{dt} e^f = e^f \cdot \frac{df}{dt}$$

$$f'(\pi) = \underbrace{e^{\sin \pi}}_{0} \underbrace{\cos \pi}_{-1} = e^0 \underbrace{(-1)}_{1} = -1 \qquad \text{Evaluating at } \pi$$

**Practice Problem 3**

Differentiate $f(x) = \ln(1 + \cos x)$.

➤ Solution on page 590

**EXAMPLE 7**

**DIFFERENTIATING A TRIGONOMETRIC FUNCTION**

Differentiate $f(t) = \sin^2 t^3$.

*Solution*

We write $\sin^2 t^3$ as $(\sin t^3)^2$ and then differentiate using the Generalized Power Rule.

$$\frac{d}{dt}(\sin t^3)^2 = 2(\sin t^3)^1 \cdot \underbrace{\cos t^3 \cdot 3t^2}_{\substack{\text{Derivative of} \\ \sin t^3}} = 6t^2 \sin t^3 \cos t^3$$

## Growth of Sales

In the preceding section we used sine and cosine functions to model the fluctuations in sales of seasonal goods. The *derivative* of such a sales function then gives the *rate of change of sales*, showing how rapidly sales are increasing or decreasing.

**EXAMPLE 8**

Weekly sales of
bathing suits

### FINDING GROWTH OF SEASONAL SALES

A department store predicts that the number of bathing suits that it will sell during week $t$ of the year (with $t = 1$ meaning the first week of January) is given by the formula

$$S(t) = 400 - 400 \cos \frac{2\pi t}{52}$$

Find the rate of change of the sales during the first week of May (week $t = 18$).

*Solution*

The rate of change means the derivative, and the differentiation is easier if we write $\frac{2\pi t}{52}$ as $\frac{2\pi}{52} t$, a constant times the variable.

$$S'(t) = \frac{d}{dt}\left(400 - 400 \cos \frac{2\pi}{52} t\right)$$

$$= 400 \sin \frac{2\pi}{52} t \cdot \left(\frac{2\pi}{52}\right) \qquad \text{Using } \frac{d}{dt}\cos f = -\sin f \cdot \frac{df}{dt}$$

$$= 400 \frac{2\pi}{52} \sin \frac{2\pi}{52} t \qquad \text{Rearranging}$$

Evaluating at week $t = 18$,

$$S'(18) = 400 \underbrace{\frac{2\pi}{52}}_{48.3} \underbrace{\sin\left(\frac{2\pi}{52} 18\right)}_{0.823} \approx 39.8 \qquad \begin{array}{l}\text{Using a calculator} \\ \text{set for radians}\end{array}$$

*Interpretation:* Sales will be increasing by about 40 per week.

## Graphing Calculator Exploration

dy/dx=39.776616

Verify the answer to the previous example as follows:

a. Graph the function $y_1 = 400 - 400 \cos(2\pi x/52)$ on the window $[0, 52]$ by $[-200, 1000]$.

b. Find the derivative at $x = 18$ (using $dy/dx$ or NDERIV). Your answer should agree with that found in the preceding example.

As before, we optimize a function by setting its first derivative equal to zero and solving, then using the second-derivative test to determine whether we have maximized or minimized.

EXAMPLE 9  **MAXIMIZING VOLUME**

A water trough consists of two boards, each 10 feet long and 1 foot wide, attached along one edge at an angle of $\theta$ radians, and closed at both ends, as shown below. Find the angle $\theta$ that maximizes the volume of the trough.

*Solution*

The volume will be maximized if we maximize the cross-sectional area (shown at the right).

Turning this triangle on its side gives a triangle of base 1 and height $\sin \theta$

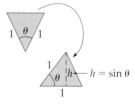

$$\left( \text{since} \quad \sin \theta = \frac{\text{opp}}{\text{hyp}} = \frac{h}{1} = h \right).$$

The area of the triangle is

$$A = \frac{1}{2} \cdot 1 \cdot \sin \theta = \frac{1}{2} \sin \theta \qquad \text{One-half base times height}$$

Differentiating,

$$A' = \frac{1}{2} \cos \theta = 0 \qquad \begin{array}{l}\text{To maximize, set the} \\ \text{derivative equal to zero}\end{array}$$

The only angle between 0 and $\pi$ whose cosine is zero (think of the graph of $\cos x$) is

$$\theta = \frac{\pi}{2} \qquad \text{Solution to} \quad \cos \theta = 0$$

The second-derivative test gives

$$A'' = -\frac{1}{2} \sin \theta \qquad \text{Differentiating} \quad A' = \frac{1}{2} \cos \theta$$

$$= -\frac{1}{2} \sin \frac{\pi}{2} = -\frac{1}{2} \qquad \text{Evaluating at } \frac{\pi}{2}$$

Negative

Since $A''$ is negative, the area is *maximized* at   $\theta = \dfrac{\pi}{2}$,   or $90°$.

Therefore, joining the boards at an angle of $90°$ will maximize the volume of the trough.

## Why We Use Radians Instead of Degrees

We can now explain why we use radians instead of degrees in calculus. On page 562 we saw that multiplying by $\dfrac{\pi}{180}$ converts degrees into radians. Therefore, the sine and cosine functions, written as functions of $t$ *degrees,* are   $\sin\!\left(\dfrac{\pi}{180}\,t\right)$   and   $\cos\!\left(\dfrac{\pi}{180}\,t\right)$.   The derivatives of these "degree" sine and cosine functions are then

$$\frac{d}{dt}\sin\!\left(\frac{\pi}{180}\,t\right) = \cos\!\left(\frac{\pi}{180}\,t\right)\cdot\frac{\pi}{180} \qquad \text{Using } \frac{d}{dt}\sin f = \cos f\cdot\frac{df}{dt}$$

and

$$\frac{d}{dt}\cos\!\left(\frac{\pi}{180}\,t\right) = -\sin\!\left(\frac{\pi}{180}\,t\right)\cdot\frac{\pi}{180} \qquad \text{Using } \frac{d}{dt}\cos f = -\sin f\cdot\frac{df}{dt}$$

Using the notations $\sin t°$ and $\cos t°$ for the "degree" sine and cosine functions, these derivative formulas become:

$$\frac{d}{dt}\sin t° = \frac{\pi}{180}\cos t° \qquad\qquad \frac{d}{dt}\sin t = \cos t$$

and

$$\frac{d}{dt}\cos t° = -\frac{\pi}{180}\sin t° \qquad\qquad \frac{d}{dt}\cos t = -\sin t$$

*radian formulas*

It is the extra factor $\pi/180$ in these formulas that makes them more complicated than the simpler *radian* formulas on page 583. This is why we use radians: they simplify the formulas for differentiation and (as we shall see on page 596) integration.

## Solutions to Practice Problems

**1.** $\dfrac{d}{dt}(t \cdot \sin t) = 1 \cdot \sin t + t \cdot \cos t = \sin t + t \cos t$

**2.** $\dfrac{d}{dt} \sin \dfrac{1}{2} t = \left( \cos \dfrac{1}{2} t \right) \left( \dfrac{1}{2} \right) = \dfrac{1}{2} \cos \dfrac{t}{2}$

**3.** $\dfrac{d}{dx} \ln(1 + \cos x) = \dfrac{-\sin x}{1 + \cos x}$

## 8.3 Section Summary

The derivatives of the sine and cosine functions are given by the following formulas.

$$\frac{d}{dt} \sin t = \cos t \qquad\qquad \frac{d}{dt} \sin f(t) = \cos f(t) \cdot \frac{df}{dt}$$

$$\frac{d}{dt} \cos t = -\sin t \qquad\qquad \frac{d}{dt} \cos f(t) = -\sin f(t) \cdot \frac{df}{dt}$$

In the formulas on the right, $f(t)$ is any differentiable function.

We used derivatives to find the rate of change of daily sales (showing how rapidly sales are changing), and to optimize functions (maximizing volume) by setting the derivative equal to zero.

## Justification of the Differentiation Formulas for $\sin t$ and $\cos t$

The differentiation formulas for the sine and cosine functions come from the definition of the derivative. Before we can show how this is done, we need to establish the following two limits.

$$\lim_{h \to 0} \frac{\sin h}{h} = 1 \qquad\qquad \frac{\sin h}{h} \text{ approaches 1 as } h \text{ approaches 0}$$

and

$$\lim_{h \to 0} \frac{\cos h - 1}{h} = 0 \qquad\qquad \frac{\cos h - 1}{h} \text{ approaches 0 as } h \text{ approaches 0}$$

Instead of presenting formal proofs of these limits, we justify them by using a graphing calculator to calculate values of the functions as the variable takes values approaching zero.

## Graphing Calculator Exploration

Find the two preceding limits as follows, using the TABLE feature of a graphing calculator (using $x$ instead of $h$ for ease of entry).

**a.** Enter $y_1 = (\sin x)/x$ and $y_2 = (\cos x - 1)/x$.

**b.** Set the TABLE feature to ASK you for the $x$-values.

**c.** Enter $x$-values approaching zero, such as 1, .1, .01, .001, .0001, ... or their negatives. You should get a table such as the one shown below. Observe that the values of $y_1 = (\sin x)/x$ approach 1 and the values of $y_2 = (\cos x - 1)/x$ approach zero.

| X | Y₁ | Y₂ | |
|---|---|---|---|
| 1 | .84147 | -.4597 | |
| .1 | .99833 | -.05 | |
| .01 | .99998 | -.005 | |
| .001 | 1 | -5E-4 | ← $-5 \times 10^{-4} = -0.0005$ |
| 1E-4 | 1 | -5E-5 | ← $-5 \times 10^{-5} = -0.00005$ |

1 E⁻4 means
1 × 10⁻⁴ = 0.0001

X =

$(\sin x)/x$    $(\cos x - 1)/x$
approaches 1   approaches 0

The derivative of $f(t) = \sin t$ is then found as follows.

$$f'(t) = \lim_{h \to 0} \frac{f(t + h) - f(t)}{h} = \lim_{h \to 0} \frac{\sin(t + h) - \sin t}{h}$$

Definition of $f'(t)$, with $f(t) = \sin t$

$$= \lim_{h \to 0} \frac{(\sin t)(\cos h) + (\cos t)(\sin h) - \sin t}{h}$$

Using the identity $\sin(s + t) = (\sin s)(\cos t) + (\cos s)(\sin t)$

$$= \lim_{h \to 0} \left( \frac{(\sin t)(\cos h) - \sin t}{h} + \frac{(\cos t)(\sin h)}{h} \right)$$

Rearranging the numerator, and then separating into two fractions

$$= \lim_{h \to 0} \left( \sin t \frac{\cos h - 1}{h} + \cos t \frac{\sin h}{h} \right)$$

Factoring sin $t$ from the first numerator, then moving sin $t$ and cos $t$ out front

$$= (\sin t) \left( \underbrace{\lim_{h \to 0} \frac{\cos h - 1}{h}}_{0} \right) + (\cos t) \left( \underbrace{\lim_{h \to 0} \frac{\sin h}{h}}_{1} \right)$$

Separating the limit into two limits, then taking sin $t$ and cos $t$ outside of the $h$ limits

$$= \underbrace{(\sin t) \cdot (0)}_{0} + \underbrace{(\cos t) \cdot (1)}_{\cos t} = \cos t$$

Using the two limits from the previous page

This is the result that we wanted, showing that the derivative of $f(t) = \sin t$ is $f'(t) = \cos t$. While the derivative formula for the cosine can be derived similarly (see Exercise 88), it can be found even more easily by using the identities $\cos t = \sin\left(\dfrac{\pi}{2} - t\right)$ and $\sin t = \cos\left(\dfrac{\pi}{2} - t\right)$ from page 578:

$$\frac{d}{dt}\cos t = \frac{d}{dt}\sin\left(\frac{\pi}{2} - t\right) = \cos\left(\frac{\pi}{2} - t\right)\cdot(-1) = -\sin t$$

$$\underbrace{\qquad\qquad}_{\substack{\text{Using}\\ \cos t = \sin\left(\frac{\pi}{2} - t\right)}} \quad \underbrace{\qquad\qquad}_{\substack{\text{Using}\\ \frac{d}{dt}\sin f = \cos f \cdot f'}} \quad \underbrace{\qquad}_{\substack{\text{Using}\\ \sin t = \cos\left(\frac{\pi}{2} - t\right)}}$$

## 8.3 Exercises

**1–60.** Differentiate each function.

**1.** $f(t) = t^3 \sin t$

**2.** $f(t) = t^2 \cos t$

**3.** $f(t) = \dfrac{\cos t}{t}$

**4.** $f(t) = (1 + \sin t)^4$

**5.** $f(t) = \sin(t^3 + 1)$

**6.** $f(t) = \cos(t^3 + t + 1)$

**7.** $f(t) = \sin 5t$

**8.** $f(t) = \cos 4t$

**9.** $f(t) = 2 \cos 3t$

**10.** $f(t) = 6 \sin 2t$

**11.** $f(t) = \cos\left(\dfrac{\pi}{180} t\right)$

**12.** $f(t) = \sin\left(\dfrac{\pi}{200} t\right)$

**13. a.** $f(t) = t^2 + \sin \pi t$
   **b.** Find $f'(0)$

**14. a.** $f(t) = t \cos \pi t$
   **b.** Find $f'(0)$

**15. a.** $f(t) = 4 \cos \dfrac{t}{2}$
   **b.** Find $f'\left(\dfrac{\pi}{3}\right)$

**16. a.** $f(t) = 48 \sin \dfrac{t}{12}$
   **b.** Find $f'(4\pi)$

**17. a.** $f(t) = \sin(\pi - t)$
   **b.** Find $f'\left(\dfrac{\pi}{2}\right)$

**18. a.** $f(t) = \cos(2\pi - t)$
   **b.** Find $f'(\pi)$

**19.** $f(t) = \cos(t + \pi)^3$

**20.** $f(t) = \sin^3(\pi + t)$

**21.** $f(t) = \cos^3(t + \pi)$

**22.** $f(t) = \sin(\pi + t)^3$

**23.** $f(t) = \sin(\pi - t)^2$

**24.** $f(t) = \sin^2(\pi - t)$

**25.** $f(t) = \sin^3 t^2$

**26.** $f(t) = \cos^4 t^2$

**27.** $f(x) = \cos \sqrt{x + 1}$

**28.** $f(x) = \sin \sqrt[3]{x}$

**29.** $f(x) = 4 \sin x + \sin 4x + \sin^4 x$

**30.** $f(x) = 4 \cos x + \cos 4x + \cos^4 x$

**31.** $f(x) = \sin e^x$

**32.** $f(x) = e^{\cos x}$

**33.** $f(z) = e^{z + \cos z}$

**34.** $f(z) = \cos(z + e^z)$

**35.** $f(z) = (1 - \cos z)^4$

**36.** $f(z) = (\sin z + \cos z)^2$

**37. a.** $f(z) = \ln(\sin z)$
   **b.** Find $f'\left(\dfrac{\pi}{2}\right)$

**38. a.** $f(z) = \cos(\ln z)$
   **b.** Find $f'(1)$

**39.** $f(z) = \sin(z + \ln z)$

**40.** $f(z) = \ln(z + \sin z)$

**41.** $f(x) = 3 \sin^2 x - 4 \cos^2 x$

**42.** $f(x) = 4 \cos^2 x + 5 \sin^2 x$

**43.** $f(x) = \dfrac{1}{x} \cos x^2$

**44.** $f(x) = x^2 \sin \dfrac{1}{x}$

**45.** $f(x) = e^{-x} \sin e^x$

**46.** $f(x) = e^x \cos e^{-x}$

**47. a.** $f(x) = \dfrac{\cos x}{\sin x}$
   **b.** Find $f'\left(\dfrac{\pi}{3}\right)$

**48. a.** $f(x) = \dfrac{\sin 2x}{\cos 2x}$
   **b.** Find $f'\left(\dfrac{\pi}{8}\right)$

**49.** $f(t) = \sin^2 t - \cos^2 t$

**50.** $f(t) = \cos^2 t + \sin^2 t$

**51. a.** $f(t) = \dfrac{1 + \sin t}{\cos t}$
   **b.** Find $f'(\pi)$

**52. a.** $f(t) = \dfrac{1 - \cos t}{\sin t}$    **b.** Find $f'\left(\dfrac{\pi}{2}\right)$

**53.** $f(x) = \sin(\cos x)$    **54.** $f(x) = \cos(\sin x)$

**55.** $f(x) = \left(\cos \dfrac{x}{2} - 1\right)^2$    **56.** $f(x) = 2x \cos \dfrac{x}{2}$

**57.** $f(x) = x^2 \sin x + 2x \cos x - 2 \sin x$

**58.** $f(x) = x^2 \cos x - 2x \sin x - 2 \cos x$

**59.** $f(t) = \sin^2(t^2 + 1)$    **60.** $f(t) = \cos^2 \sqrt{t}$

**61–66.** Find the *second* derivative of each function.

**61.** $f(t) = t \sin t$

**62.** $f(t) = \sin at + \cos bt$   (*a* and *b* are constants)

**63.** $f(x) = e^{\sin x}$

**64.** $f(x) = e^{\cos x}$

**65.** $f(z) = z \sin z + \cos z$

**66.** $f(z) = \sin z - z \cos z$

## Applied Exercises

**67–68. BUSINESS: Seasonal Sales** A manufacturer estimates that the number of air conditioners sold during week *t* of the year will be

$$S(t) = 500 - 500 \cos\left(\dfrac{\pi}{26} t\right)$$

Find the rate of change of sales at

**67.** $t = 19$  (mid-May)

**68.** $t = 37$  (mid-September)

**69. GENERAL: Maximizing Volume of a Tent**
A tent made by hanging a square sheet over a ridge pole at angle $\theta$ as shown will enclose a volume of

$$V(\theta) = 1000 \sin \theta$$

cubic feet. Find the angle $\theta$ that maximizes the volume $V(\theta)$.

**70. BUSINESS: Maximum Temperature** The temperature in a chemical refining tower after *t* hours is  $T(t) = 100 + 60(\sin t + \cos t)$  degrees Fahrenheit (for  $0 \le t \le \pi$).  Find when the temperature is the greatest.

**71. GENERAL: Honeycombs** A honeycomb consists of cells like the one shown, where $\theta$ is the angle between the upper faces and the axis of the cell. Minimizing the surface of the area is equivalent to minimizing the function

$$S(\theta) = \dfrac{\sqrt{3} - \cos \theta}{\sin \theta}$$

Minimize this function by setting its derivative equal to zero and solving, deriving the relationship  $\cos \theta = \dfrac{1}{\sqrt{3}}.$  The acute angle that solves this equation is approximately 54.7°, which is the actual angle used by honeybees.

*Source:* D'A. W. Thompson, *On Growth and Form* and E. Batschelet, *Introduction to Mathematics for Life Scientists*

**72. GENERAL: Maximizing Area**

   **a.** Find a formula for the area of a right triangle (shown on the next page) with hypotenuse 1 and one acute angle of *x* radians  $(0 < x < \pi/2)$.

**b.** Find the angle $x$ that maximizes the area. [*Hint:* Either graph the area function and use MAXIMUM, or use NDERIV to find where its derivative equals zero.]

**c.** Convert $x$ into degrees by multiplying by $\dfrac{180}{\pi}$. Did you expect to obtain this answer?

**73. ATHLETICS: Maximizing Distance of a Thrown Ball** A ball thrown with an initial velocity of 60 miles per hour at an angle of inclination with the ground of $x$ radians, as shown below, will travel a horizontal distance of

$$D(x) = 242(\sin x)(\cos x)$$

feet (neglecting air resistance).

**a.** Find the angle $x$ that maximizes this distance [*Hint:* Either graph the distance function and use MAXIMUM, or use NDERIV to find where its derivative equals zero.]

**b.** Convert $x$ into degrees by multiplying by $\dfrac{180}{\pi}$.

[*Note:* The angle that you find holds for all velocities.*]

242 sin x cos x

**74. GENERAL: Temperature** The mean temperature in Fairbanks, Alaska, on day $x$ of the year is approximately

$$T(x) = 36 + 22 \sin\left(\frac{2\pi(x - 102)}{365}\right)$$

degrees Fahrenheit.

**a.** Graph its derivative (using NDERIV) on the window [0, 365] by [−0.5, 0.5].

**b.** Find the rate of change of temperature on March 15 (day 74) and on September 12 (day 255) and interpret these answers.

(*continues*)

**c.** Find the days on which the temperature will be rising the fastest.

*Source:* Alaska Climate Research Center

**75. BIOMEDICAL: Blood Pressure** A person's blood pressure varies periodically according to the formula $p(t) = 90 + 15 \sin(2.5\pi t)$, where $t$ is the number of seconds since the beginning of a cardiac cycle.

**a.** Graph the function on the window [0, 1.6] by [0, 120].

**b.** When is blood pressure the *highest* for $0 \le t \le 1.6$, and what is the maximum blood pressure?

**c.** When is blood pressure the *lowest* for $0 \le t \le 1.6$ and what is the minimum blood pressure?

**76. ATHLETICS: Basketball** To make a basket, a basketball player must launch the ball with just the right angle and initial velocity for it to pass through the hoop. It can be shown from Newton's laws of motion that the initial velocity $v$ and angle $\theta$ must satisfy the equation

$$v = \sqrt{\dfrac{16d}{\cos^2\theta\left(\dfrac{\sin\theta}{\cos\theta} - \dfrac{h}{d}\right)}}$$

where the constants $d$ and $h$ are as shown in the following diagram. To minimize the chances of the ball bouncing off the rim, coaches recommend shooting with the smallest possible velocity, called the "soft shot." Assume that for a free throw the constants are $d = 14$ and $h = 3$, and graph the resulting function on the window [0, $\pi/2$] by [0, 100]. Then find the "soft shot" angle $\theta$ that minimizes the function.*

*Source: American Journal of Physics* **49**

---

*The result of this exercise applies to a *thrown ball*. For a ball hit by a bat, the best angle is 35°. This is because a batted ball can be given a backspin of 2000 rpm or more, giving it lift through what is called the Magnus effect. See Robert K. Adair, *The Physics of Baseball* (Harper & Row).

*In general, the angle for the soft shot is $\theta = 45° + \dfrac{1}{2}\left(\begin{array}{c}\text{Angle of elevation}\\ \text{from ball to rim}\end{array}\right)$, found by setting the derivative of $v$ equal to zero and solving for $\theta$.

## Conceptual Exercises

**77.** Find $\dfrac{d}{dx}\sin(7\pi^{23} - e^{99})$

[*Hint:* No work necessary.]

**78.** Find $\dfrac{d}{dx}\cos(e^{-44} - \ln 99\pi)$

[*Hint:* No work necessary.]

**79.** Find $\dfrac{d}{dx}(\sin^2 x + \cos^2 x)$

[*Hint:* No work necessary.]

**80.** What is the smallest positive value of $n$ such

that $\dfrac{d^n}{dx^n}\sin x = \sin x$?

**81.** What is the smallest positive value of $n$ such

that $\dfrac{d^n}{dx^n}\cos x = \cos x$?

**82.** What is the 1000th derivative of $5\sin x$?

**83.** What is the 1000th derivative of $\cos(x + 973)$?

**84.** What is the maximum angle (in degrees) that the sine or cosine curve makes with the horizontal?

**85.** Find $\displaystyle\lim_{h\to 0}\dfrac{h}{\sin h}$  [*Hint:* No work necessary if

you use a similar known limit.]

**86.** Find $\displaystyle\lim_{h\to 0}\dfrac{1 - \cos h}{h}$  [*Hint:* No work necessary

if you use a similar known limit.]

## Explorations and Excursions

The following problems extend and augment the material presented in the text.

**87. DERIVATIVE FORMULA FOR** $\cos x$  Derive the formula for the derivative of $\cos x$ (assuming that the derivative exists) from the formula for the derivative of $\sin x$ as follows.

**a.** Differentiate both sides of the identity
$\sin^2 x + \cos^2 x = 1$, obtaining

$$2\sin x \cos x + 2\cos x \dfrac{d}{dx}\cos x = 0$$

**b.** Solve this equation for $\dfrac{d}{dx}\cos x$ to obtain

$$\dfrac{d}{dx}\cos x = -\sin x.$$

**88.** Use the definition of the derivative to show that the derivative of $f(t) = \cos t$ is $f'(t) = -\sin t$. [*Hint:* The steps are similar to those on page 591, but using the identity

$\cos(t + h) = (\cos t)(\cos h) - (\sin t)(\sin h).$]

**89.** Show that the formulas on the right side of the box on page 590 reduce to the formulas on the left side in the case that the function $f$ is simply $f(t) = t$.

**90.** Why do the functions $\sin^2 x$ and $-\cos^2 x$ have the same derivative?

**91. PROOF OF** $\cos(s + t)$ **IDENTITY**  Prove the identity

$$\cos(s + t) = (\cos s)(\cos t) - (\sin s)(\sin t)$$

by differentiating the identity from page 578,

$$\sin(s + t) = (\sin s)(\cos t) + (\cos s)(\sin t)$$

with respect to $s$, holding $t$ constant.

**92. DIFFERENTIATION FORMULAS FOR** $\sin x$ **AND** $\cos x$

**a.** Verify the formula $\dfrac{d}{dx}\sin x = \cos x$ by defining $y_1 = \sin x$ and $y_2$ as its derivative (using NDERIV). Then graph $y_1$ and $y_2$ on the window $[-2\pi, 2\pi]$ by $[-2, 2]$, observing that $y_2$ appears to be the cosine curve.

**b.** Then change $y_1$ to $\cos x$ to verify that $\dfrac{d}{dx}\cos x = -\sin x$ by observing that $y_2$ appears to be the negative of the sine curve.

Note that NDERIV finds derivatives *numerically*, using the difference quotient with a small value of $h$, so this is an independent verification of the formulas.

## 8.4 INTEGRALS OF SINE AND COSINE FUNCTIONS

### Introduction

In this section we will integrate sine and cosine functions and use their integrals to find *areas under curves, total accumulations,* and *average values.*

### Integrals of Sine and Cosine

The integrals of the sine and cosine functions are given by the following formulas, which are proved by differentiating their right-hand sides.

---

**Integrals of sin *t* and cos *t***

$$\int \sin t \, dt = -\cos t + C$$
The integral of sine is negative cosine (plus a constant)

$$\int \cos t \, dt = \sin t + C$$
The integral of cosine is sine (plus a constant)

---

These formulas are easily remembered. The derivative of sine is cosine, so the integral of cosine is sine, which is just the second formula. The first formula is similar, but with a negative sign (from the differentiation formula for cosine).

---

**EXAMPLE 1**

**FINDING AREA BY INTEGRATION**

Find the area under one arch of the sine curve $y = \sin t$.

*Solution*

For the area under one arch of $\sin t$, we integrate from 0 to $\pi$:

$$\text{Area} = \int_0^\pi \sin t \, dt = -\cos t \Big|_0^\pi = -\cos \pi - (-\cos 0)$$

$$\text{Using} \quad \int \sin t \, dt = -\cos t + C \qquad \text{Evaluating at } \pi \qquad \text{And at } 0$$

$$= -(-1) - (-1) = 1 + 1 = 2$$

The area is 2 square units.

## Graphing Calculator Exploration

∫f(x)dx=2

Use a graphing calculator to verify the answer to the preceding example as follows:

**a.** Graph   $y_1 = \sin x$   on the window $[0, \pi]$ by $[-0.25, 1]$.

**b.** Integrate this function from 0 to $\pi$ (using FnInt, $\int f(x)\, dx$, or a similar command). Your answer should agree with that found in the preceding example.

The substitution method (introduced on page 390) greatly expands the class of functions that we can integrate. As before, we write the integration formulas with $x$ replaced by $u$, where $u$ stands for a differentiable function.

$$\int \sin u \, du = -\cos u + C$$

$$\int \cos u \, du = \sin u + C$$

To evaluate a particular integral, we represent part of the integrand by $u$, calculate its differential $du$, and multiply inside and outside by constants (if necessary) to complete the $du$. The right-hand side of the formula then gives the solution, after substituting back to the original variable.

**EXAMPLE 2**

**USING THE SUBSTITUTION METHOD**

Find   $\displaystyle\int x^2 \sin x^3 \, dx$.

*Solution*

The integrand involves   $\sin x^3$,   the sine of a function, suggesting the $\int \sin u \, du$   formula with   $u = x^3$.   The differential is   $du = 3x^2\, dx$   (the derivative times $dx$).

$$\int \underbrace{x^2 \sin x^3}_{\sin u} \, dx \quad = \quad \frac{1}{3}\int \underbrace{\sin x^3}_{\sin u}\ \underbrace{3x^2\, dx}_{du} \quad = \quad \frac{1}{3}\int \underbrace{\sin u \, du}_{\text{Substituting}}$$

$$u = x^3$$
$$du = 3x^2\, dx$$

Multiplying by 3 and $\frac{1}{3}$

$$= \quad \underbrace{\frac{1}{3}(-\cos u) + C}_{\substack{\text{Integrating by}\\ \int \sin u \, du = -\cos u + C}} \quad = \quad \underbrace{-\frac{1}{3}\cos x^3 + C}_{\substack{\text{Substituting back}\\ \text{to } x \text{ using } u = x^3}}$$

We can check this answer by differentiation:

$$\underbrace{\frac{d}{dx}\left(-\frac{1}{3}\cos x^3 + C\right)}_{\text{Answer}} = \underbrace{-\frac{1}{3}(-\sin x^3)(3x^2)}_{\text{Using } \frac{d}{dx}\cos f = \sin f \cdot \frac{df}{dt}} = \underbrace{x^2 \sin x^3}_{\text{Simplifying}}$$

It checks: the derivative of the answer is the integrand

**Practice Problem 1**   Find $\displaystyle\int \cos\frac{2t}{3}\, dt$.   [*Hint:* Write $\dfrac{2t}{3}$ as $\dfrac{2}{3}t$,   a constant times the variable.]

➤ Solution on page 601

## Other Substitution Formulas

An integral may require one of the earlier integration formulas (from page 390), but now with $u$ being a trigonometric function.

(A)   $\displaystyle\int u^n\, du = \frac{1}{n+1}u^{n+1} + C$        $(n \neq -1)$

(B)   $\displaystyle\int e^u\, du = e^u + C$

(C)   $\displaystyle\int u^{-1}\, du = \int \frac{du}{u} = \ln|u| + C$

---

**EXAMPLE 3**

### USING ANOTHER SUBSTITUTION FORMULA

Find $\displaystyle\int \sin^5 2t \cos 2t\, dt$.

*Solution*

The integrand includes $\sin^5 2t = (\sin 2t)^5$,   a function to a power, suggesting the $\int u^n\, du$ formula with $u = \sin 2t$.

$$\int \underbrace{(\sin 2t)^5}_{\substack{u^5 \\ u = \sin 2t \\ du = \cos 2t \cdot 2dt}} \cos 2t\, dt = \frac{1}{2}\int \underbrace{(\sin 2t)^5}_{u^5}\underbrace{2\cos 2t\, dt}_{du} = \frac{1}{2}\int \underbrace{u^5\, du}_{\text{Substituting}}$$

Multiplying by 2 and $\frac{1}{2}$

$$= \underbrace{\frac{1}{2}\frac{1}{6}u^6 + C}_{\substack{\text{Integrating} \\ \text{by formula (A)}}} = \underbrace{\frac{1}{12}\sin^6 2t + C}_{\substack{\text{Substituting back} \\ \text{using } u = \sin 2t}}$$

---

**Practice Problem 2**   Find $\displaystyle\int \cos^2 x \sin x\, dx$.

➤ Solution on page 602

**EXAMPLE 4**

**EVALUATING A DEFINITE INTEGRAL**

Evaluate the definite integral $\displaystyle\int_0^{\pi/2} e^{\sin z} \cos z \, dz$.

*Solution*

The integrand involves $e^{\sin z}$, $e$ to a function, suggesting the $\int e^u \, du$ formula with $u = \sin z$. With definite integrals, we must also change the upper and lower limits of integration to their corresponding $u$-values.

$$u = \sin \frac{\pi}{2} = 1$$

$$\int_0^{\pi/2} e^{\sin z} \cos z \, dz = \int_0^1 e^u \, du = e^u \Big|_0^1 = e^1 - e^0 = e - 1$$

$e^u$    $du$    Substituting    Integrating   Evaluating   Final

$u = \sin z$          by formula (B)        answer

$du = \cos z \, dz$

$u = \sin 0 = 0$

We leave the answer in its *exact* form, $e - 1$, although we could use $e \approx 2.718$ to obtain the approximate answer 1.718.

**EXAMPLE 5**

**INTEGRATING A QUOTIENT OF TRIGONOMETRIC FUNCTIONS**

Find $\displaystyle\int \frac{1 + \cos t}{t + \sin t} \, dt$.

*Solution*

The derivative of the denominator is the numerator: $\dfrac{d}{dt}(t + \sin t) = 1 + \cos t$, suggesting the $\displaystyle\int \frac{du}{u}$ formula with $u = t + \sin t$.

$$\int \frac{1 + \cos t}{t + \sin t} \, dt = \int \frac{du}{u} = \ln |u| + C = \ln |t + \sin t| + C$$

$u = t + \sin t$      Substituting    Integrating      Substituting back

$du = (1 + \cos t) \, dt$                by formula (C)    using $u = t + \sin t$

**Practice Problem 3**    Evaluate $\displaystyle\int \frac{\sin t}{\cos t} \, dt$.          ➤ Solution on page 602

Many applications involve the *modified* trigonometric functions sin *at* and cos *at*, with a *scale constant a* that changes the period (see page 575). These functions are easily integrated by the substitution method, and the integrals occur so often that we list them as separate formulas.

$$\int \sin at \, dt = -\frac{1}{a}\cos at + C$$

For constant $a \neq 0$

$$\int \cos at \, dt = \frac{1}{a}\sin at + C$$

For example, Practice Problem 1 on page 598 is evaluated by the second formula in one step:

$$\int \cos \underbrace{\frac{2}{3}}_{a} t \, dt = \underbrace{\frac{3}{2}}_{\frac{1}{a}}\sin \frac{2}{3}t + C \qquad \text{Using the second formula above}$$

Recall that definite integration represents *continuous summation*, summing a rate to give the *total accumulation* at that rate (see page 356). For example, integrating a sales rate (such as sales per week) from *a* to *b* gives the *total sales* from week *a* to week *b*.

---

**EXAMPLE 6**

**SUMMING TOTAL SALES OF A SEASONAL PRODUCT**

A clothing store estimates that the number of overcoats sold per week will be

$$S(t) = 30 + 30 \cos \frac{\pi t}{26} \qquad \text{Coats per week}$$

where *t* is the number of weeks into the winter. Find the total number of overcoats that will be sold during the first 8 weeks of winter.

*Solution*

To find the total for the first 8 weeks, we integrate the rate from 0 to 8:

$$\int_0^8 \left(30 + 30 \cos \overset{a}{\underset{\downarrow}{\frac{\pi}{26}}} t\right) dt = \left(30t + 30 \overset{\frac{1}{a}}{\underset{\downarrow}{\frac{26}{\pi}}}\sin \frac{\pi}{26} t\right)\Bigg|_0^8$$

$$\underbrace{\qquad}_{\text{Writing } \frac{\pi t}{26} \text{ as } \frac{\pi}{26}t} \qquad \underbrace{\qquad}_{\text{Using } \int \cos at \, dt = \frac{1}{a}\sin at + C}$$

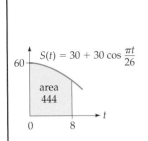

$$= \left( \underbrace{30 \cdot 8}_{240} + \overbrace{30\frac{26}{\pi} \sin \frac{8\pi}{26}}^{\approx 204} \right) - 0 \approx 444$$

$$\underbrace{\phantom{\left( 30 \cdot 8 + 30\frac{26}{\pi} \sin \frac{8\pi}{26} \right)}}_{\substack{\text{Evaluating} \\ \text{at } t = 8}} \quad \underbrace{\phantom{xxxx}}_{\substack{\text{And at} \\ t = 0}}$$

Approximately 444 overcoats will be sold during the first 8 weeks.

Integrals also give average values when multiplied by $\dfrac{1}{b - a}$ (see page 366).

**EXAMPLE 7**  **FINDING AVERAGE TEMPERATURE**

The mean daily temperature in Fairbanks, Alaska, on day $x$ of the year is

$$T(x) = 36 + 22 \sin \frac{2\pi(x - 102)}{365} \qquad \text{Degrees Fahrenheit}$$

Find the average summer temperature, June 21 to September 23 (day 172 to day 266).

*Source*: Alaska Climate Research Center

*Solution*

The average value formula (page 366) gives

$$\frac{1}{266 - 172} \int_{172}^{266} \left( 36 + 22 \sin \frac{2\pi(x - 102)}{365} \right) dx$$

The average is the integral from $a$ to $b$ divided by $(b - a)$.

We could calculate this "by hand," but instead we use a graphing calculator, graphing the function $T(x)$ on the window $[0, 365]$ by $[-20, 70]$ and using the $\int f(x)dx$ command to integrate from 172 to 266.

Dividing the result by $(266 - 172)$ gives the average summer temperature in Fairbanks:

$$\frac{5055}{266 - 172} \approx 53.8 \qquad \text{About 54 degrees!}$$

---

**Solutions to Practice Problems**

**1.** $\displaystyle\int \underbrace{\cos \tfrac{2}{3}t}_{\cos u}\, dt = \frac{3}{2} \int \underbrace{\cos \tfrac{2}{3}t}_{\cos u}\,\underbrace{\tfrac{2}{3}dt}_{du} = \frac{3}{2} \int \cos u \, du = \frac{3}{2} \sin u + C = \frac{3}{2} \sin \frac{2t}{3} + C$

$u = \tfrac{2}{3}t$

$du = \tfrac{2}{3}dt$

**2.** $\displaystyle\int \underbrace{\cos^2 x}_{u^2} \sin x \, dx = -\int \underbrace{\cos^2 x}_{u^2} \underbrace{(-\sin x) \, dx}_{du} = -\int u^2 \, du = -\tfrac{1}{3}u^3 + C = -\tfrac{1}{3}\cos^3 x + C$

$u = \cos x$
$du = -\sin x \, dx$

**3.** $\displaystyle\int \frac{\sin t}{\cos t} \, dt = -\int \frac{-\sin t}{\cos t} \, dt = -\int \frac{du}{u} = -\ln|u| + C = -\ln|\cos t| + C$

$u = \cos t$
$du = -\sin t \, dt$

## 8.4 Section Summary

The differentiation formulas for the sine and cosine functions (page 583) led immediately to the integration formulas

$$\int \sin t \, dt = -\cos t + C$$

$$\int \cos t \, dt = \sin t + C$$

These two formulas extend immediately to their "$u$ forms" by the substitution method (see page 597) and to a "scale constant" $a$ (see page 600). We use integration for *areas*, for *continuous summations*, for *total accumulations*, and for *average values*.

## 8.4 Exercises

**1–28.** Find each integral.

**1.** $\displaystyle\int (\cos t - \sin t) \, dt$

**2.** $\displaystyle\int (\sin t + \cos t) \, dt$

**11.** $\displaystyle\int \sin(\pi - t) \, dt$

**12.** $\displaystyle\int \cos(1 - t) \, dt$

**3.** $\displaystyle\int (1 + \sin x) \, dx$

**4.** $\displaystyle\int (\cos x - 1) \, dx$

**13.** $\displaystyle\int \cos \frac{2\pi(t + 20)}{365} \, dt$

**14.** $\displaystyle\int \sin \frac{\pi(t + 3)}{26} \, dt$

**5.** $\displaystyle\int (t + 1)\cos(t^2 + 2t) \, dt$

**6.** $\displaystyle\int t^2 \sin(t^3 + 1) \, dt$

**15.** $\displaystyle\int \sin^2 t \cos t \, dt$

**16.** $\displaystyle\int \cos^3 t \sin t \, dt$

**7.** $\displaystyle\int \sin \pi t \, dt$

**8.** $\displaystyle\int \cos 2t \, dt$

**17.** $\displaystyle\int \frac{\sin t}{\cos^3 t} \, dt$

**18.** $\displaystyle\int \sqrt{\sin t} \cos t \, dt$

**9.** $\displaystyle\int \cos \frac{\pi t}{2} \, dt$

**10.** $\displaystyle\int \sin \frac{2t}{5} \, dt$

**19.** $\displaystyle\int e^{1 + \sin x} \cos x \, dx$

**20.** $\displaystyle\int e^{2x} \cos e^{2x} \, dx$

**21.** $\displaystyle\int \frac{\sin t}{\cos t + 3}\, dt$

**22.** $\displaystyle\int \frac{\cos t}{1 + \sin t}\, dt$

**23.** $\displaystyle\int \frac{\sin w}{\sqrt{1 - \cos w}}\, dw$

**24.** $\displaystyle\int \sqrt[3]{1 + \sin w}\, \cos w\, dw$

**25.** $\displaystyle\int \frac{\sin y \cos y}{\sqrt{1 + \sin^2 y}}\, dy$

**26.** $\displaystyle\int \frac{\cos \sqrt{z}}{\sqrt{z}}\, dz$

**27.** $\displaystyle\int \left(x^2 - \frac{1}{x} + \cos 2x\right) dx$

**28.** $\displaystyle\int (e^{2x} + \sqrt{x} - \sin 3x)\, dx$

**29–36.** For each definite integral:

**a.** Evaluate it "by hand," leaving the answer in *exact* form.

 **b.** Check your answer to part (a) using a graphing calculator.

**29.** $\displaystyle\int_0^{1/2} \sin \pi t\, dt$

**30.** $\displaystyle\int_0^{\pi/4} \cos 2t\, dt$

**31.** $\displaystyle\int_0^{\pi} (t + \cos t)\, dt$

**32.** $\displaystyle\int_0^{\pi} (1 - \sin t)\, dt$

**33.** $\displaystyle\int_0^1 (t + \sin \pi t)\, dt$

**34.** $\displaystyle\int_0^1 (t^2 - \cos \pi t)\, dt$

**35.** $\displaystyle\int_{\pi/4}^{3\pi/4} \cos\left(t + \frac{\pi}{4}\right) dt$

**36.** $\displaystyle\int_{\pi/2}^{\pi} e^{\sin t} \cos t\, dt$

**37.** Find the area under the curve $y = \cos \dfrac{x}{2}$ and above the x-axis from $x = 0$ to $x = \pi$.

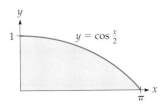

**38.** Find the area under the curve $y = x + \sin x$ and above the x-axis from $x = 0$ to $x = \pi$. Leave your answer in its *exact* form. If you are

using a graphing calculator, verify your answer numerically using FnInt or $\int f(x)dx$.

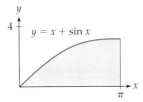

**39.** Find the area under $y = \sin x + \cos x$ and above the x-axis from $x = 0$ to $x = \pi/2$. Leave your answer in its *exact* form. If you are using a graphing calculator, verify your answer numerically using FnInt or $\int f(x)dx$.

**40.** Find the area under the curve $y = x \sin x^2$ and above the x-axis from $x = 0$ to $x = \sqrt{\pi}$. Leave your answer in its *exact* form. If you are using a graphing calculator, verify your answer numerically using FnInt or $\int f(x)dx$.

**41.** Prove the integration formulas in the box on page 600 by differentiating the right-hand sides to obtain the integrands.

**42.** Find the integral $\int \sin t \cos t\, dt$ in two ways:

**a.** Using the substitution method with $u = \sin t$.

**b.** Using the substitution method with $u = \cos t$.

**c.** Can you reconcile the two seemingly different answers? [*Hint:* Try using a trigonometric identity.]

## *Applied Exercises*

**43–44. BUSINESS: Seasonal Sales** A ski shop predicts sales of

$$S(t) = 10 + 9 \cos \frac{\pi t}{26}$$

thousand dollars during week $t$ of the year, with $t = 0$ corresponding to the beginning of January.

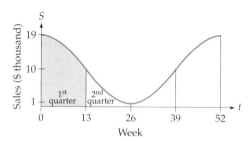

Predict the company's total sales during:

**43.** the first quarter of the year $(t = 0$ to $t = 13)$.

**44.** the second quarter of the year $(t = 13$ to $t = 26)$.

**45. GENERAL: Pollution** A chemical refinery discharges $P(t) = 5 + 3 \sin \frac{\pi t}{6}$ tons of pollution each month, where $t$ is the number of months that the plant has been in operation. For the first 18 months $(t = 0$ to $t = 18)$, find:

**a.** the *total* amount of pollutant discharged.
**b.** the *average* amount of pollutant discharged.

**46. BUSINESS: Summing an Income Stream** An investment generates a stream of income at the rate of $S(t) = 50 + 10 \sin \frac{\pi t}{6}$ thousand dollars per month, where $t$ is the number of months since the investment was made. For the first 6 months $(t = 0$ to $t = 6)$, find:

**a.** the *total* income from the income stream.
**b.** the *average* income from the income stream.

 **47. ENVIRONMENTAL: Bird Counts** The number of migratory birds flying north past a checkpoint is $300 - 300 \cos\left(\frac{\pi}{6}t\right)$ per month, where $t$ is the number of months since the beginning of the year. Find the total number for the year.

 **48. BIOMEDICAL: Blood Flow** Blood flows through an artery at the rate of $30 - 30 \cos(2.4\pi t)$ cubic centimeters per second, where $t$ is the number of seconds after the beginning of the cardiac cycle. Find the total amount of blood flowing through the artery in the first 5/6 second (one complete cardiac cycle).

 **49. PERSONAL FINANCE: Home Heating Costs** Suppose that the temperature outside of a house on day $t$ of the year is $T(t) = 60 - 30 \cos \frac{2\pi t}{365}$ degrees Fahrenheit. If it costs $1 to raise the temperature of the house by 1° for 1 day, then setting the thermostat at 75° will result in an annual heating bill equal to the area of the shaded region shown below.

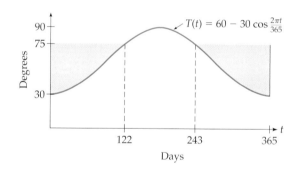

This area is given by the integral below (with the two equal areas calculated as twice the first). Use a graphing calculator to evaluate this integral to find the annual heating cost.

$$\left(\begin{array}{c}\text{Heating} \\ \text{costs at } 75°\end{array}\right) = 2 \int_0^{122} \left[75 - \left(60 - 30 \cos \frac{2\pi t}{365}\right)\right] dt$$

 **50. PERSONAL FINANCE: Home Heating Costs** (*continuation*) If the thermostat is lowered to 68°, the annual heating cost is equal to the area shown on the next page.

The graph shows $T(t) = 60 - 30 \cos \frac{2\pi t}{365}$, with the vertical axis labeled Degrees (marked at 30, 68, 90) and the horizontal axis labeled Days (marked at 107, 258, 365).

The annual heating cost is given by:

$$\left( \begin{matrix} \text{Heating} \\ \text{costs at } 68° \end{matrix} \right) = 2 \int_0^{107} \left[ 68 - \left( 60 - 30 \cos \frac{2\pi t}{365} \right) \right] dt$$

**a.** Use a graphing calculator to evaluate this integral to find the new heating cost.
**b.** Compare your answers from part (a) with your answer from Exercise 49 to find how much is saved by lowering the thermostat.

**51. BUSINESS: Hotel Occupancy** A Caribbean hotel estimates that it will have an occupancy of $200 + 100 \cos \dfrac{\pi t}{26}$ guests during week $t$ of the year. Find the *average* weekly occupancy during the first 10 weeks of the year $(t = 0$ to $t = 10)$.

**52. BUSINESS: Average Demand** A motorbike manufacturer estimates demand to be $400 + 300 \sin \dfrac{\pi t}{26}$ motorbikes per week, where $t = 0$ corresponds to the beginning of the year. Find the *average* demand over the first 20 weeks $(t = 0$ to $t = 20)$.

**53. GENERAL: Temperature in New York** The mean temperature in New York City on day $x$ of the year is

$$54 + 24 \sin \frac{2\pi(x - 101)}{365}$$

degrees Fahrenheit. Use a graphing calculator to find:

**a.** The average temperature during January and February (days 0 to 59).
**b.** The average temperature during July and August (days 181 to 243).

*Source:* National Weather Service

**54. GENERAL: Temperature in Minneapolis/ St. Paul** The mean temperature in Minneapolis/ St. Paul on day $x$ of the year is

$$45 + 32 \sin \frac{2\pi(x - 101)}{365}$$

degrees Fahrenheit. Use a graphing calculator to find:

**a.** The average temperature during January and February (days 0 to 59).
**b.** The average temperature during July and August (days 181 to 243).

*Source:* National Weather Service

## *Conceptual Exercises*

Evaluate each of the following definite integrals by thinking of the graphs of the functions, without any calculation.

**55.** $\displaystyle\int_0^{2\pi} \sin t \, dt$

**56.** $\displaystyle\int_0^1 \sin 2\pi t \, dt$

**57.** $\displaystyle\int_0^{1000} \cos \pi \, dt$

**58.** $\displaystyle\int_0^{100} \cos 2\pi \, dt$

**59.** $\displaystyle\int_{2.345}^{2.345 + 1000\pi} \cos t \, dt$

**60.** $\displaystyle\int_{-1000\pi}^{1000\pi} \sin t \, dt$

**61.** Give an approximate value for the average value of $9 + \cos x$ over a very long interval.

**62.** Give an approximate value for the average value of $\sin 4x - 2$ over a very long interval.

**63.** By just looking at the temperature function in Exercise 53, state the average annual temperature in New York City.

**64.** By just looking at the temperature function in Exercise 54, state the average annual temperature in Minneapolis/St. Paul.

## 8.5 OTHER TRIGONOMETRIC FUNCTIONS

### Introduction

In this section we define the **tangent, cotangent, secant,** and **cosecant** functions, differentiate and integrate them, and discuss some of their applications.

### The Tangent, Cotangent, Secant, and Cosecant Functions

The functions $\tan \theta$, $\cot \theta$, $\sec \theta$, and $\csc \theta$ are defined as follows:

---

**$\tan \theta$, $\cot \theta$, $\sec \theta$, and $\csc \theta$**

Let $(x, y)$ be any point on the terminal side of an angle of $\theta$ radians in standard position, and let $r$ be its distance from the origin. We define:

$$\tan \theta = \frac{\sin \theta}{\cos \theta} = \frac{y}{x}$$

$$\cot \theta = \frac{1}{\tan \theta} = \frac{x}{y}$$

$$\sec \theta = \frac{1}{\cos \theta} = \frac{r}{x}$$

$$\csc \theta = \frac{1}{\sin \theta} = \frac{r}{y}$$

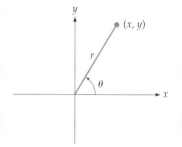

---

The domain of each function is the set of numbers such that the denominator is not zero. Unlike the sine and cosine functions, these four functions can take very large (positive) values and very small (negative) values.

---

**EXAMPLE 1**

**EVALUATING TRIGONOMETRIC FUNCTIONS AT A QUADRANT ANGLE**

Find $\tan \pi$, $\cot \pi$, $\sec \pi$, and $\csc \pi$.

*Solution*

An angle of $\pi$ radians is shown on the left, together with a point $(x, y)$ on its terminal side. From these $x$, $y$, and $r$ values, we calculate the values of the trigonometric functions:

$$\tan \pi = \frac{0}{-1} = 0 \qquad\qquad \tan \theta = \frac{y}{x}$$

$$\cot \pi = \frac{-1}{0} \quad \text{Undefined!} \qquad\qquad \cot \theta = \frac{x}{y}$$

$$\sec \pi = \frac{1}{-1} = -1 \qquad\qquad \sec \theta = \frac{r}{x}$$

$$\csc \pi = \frac{1}{0} \quad \text{Undefined!} \qquad\qquad \csc \theta = \frac{r}{y}$$

**EXAMPLE 2**

**EVALUATING TRIGONOMETRIC FUNCTIONS FROM A GRAPH**

Find $\tan \theta$, $\cot \theta$, $\sec \theta$, and $\csc \theta$ for the angle $\theta$ shown on the left.

*Solution*

The distance from the point $(-3, 4)$ to the origin is found from the Pythagorean Theorem:

$$r = \sqrt{(-3)^2 + 4^2} = \sqrt{9 + 16} = \sqrt{25} = 5$$

From these $x$, $y$, and $r$ values we calculate

$$\tan \theta = \frac{4}{-3} = -\frac{4}{3} \qquad\qquad \tan \theta = \frac{y}{x}$$

$$\cot \theta = \frac{-3}{4} = -\frac{3}{4} \qquad\qquad \cot \theta = \frac{x}{y}$$

$$\sec \theta = \frac{5}{-3} = -\frac{5}{3} \qquad\qquad \sec \theta = \frac{r}{x}$$

$$\csc \theta = \frac{5}{4} \qquad\qquad \csc \theta = \frac{r}{y}$$

**EXAMPLE 3**

**EVALUATING TRIGONOMETRIC FUNCTIONS**

Find $\tan \dfrac{\pi}{3}$, $\cot \dfrac{\pi}{3}$, $\sec \dfrac{\pi}{3}$, and $\csc \dfrac{\pi}{3}$.

*Solution*

Recall that $\sin \dfrac{\pi}{3} = \dfrac{\sqrt{3}}{2}$ and $\cos \dfrac{\pi}{3} = \dfrac{1}{2}$. See page 571

Therefore:

$$\tan \frac{\pi}{3} = \frac{\sin \dfrac{\pi}{3}}{\cos \dfrac{\pi}{3}} = \frac{\dfrac{\sqrt{3}}{2}}{\dfrac{1}{2}} = \sqrt{3} \qquad\qquad \text{Since } \tan \theta = \frac{\sin \theta}{\cos \theta}$$

$$\cot \frac{\pi}{3} = \frac{1}{\tan \dfrac{\pi}{3}} = \frac{1}{\sqrt{3}} = \frac{\sqrt{3}}{3} \qquad\qquad \text{Multiplying numerator and denominator by } \sqrt{3}$$

$$\sec \frac{\pi}{3} = \frac{1}{\cos \dfrac{\pi}{3}} = \frac{1}{\dfrac{1}{2}} = 2 \qquad\qquad \text{Since } \sec \theta = \frac{1}{\cos \theta}$$

$$\csc \frac{\pi}{3} = \frac{1}{\sin \dfrac{\pi}{3}} = \frac{1}{\dfrac{\sqrt{3}}{2}} = \frac{2}{\sqrt{3}} = \frac{2\sqrt{3}}{3} \qquad\qquad \text{Multiplying numerator and denominator by } \sqrt{3}$$

Using the [sin], [cos], and [tan] keys on your calculator, together with the [1/x] or [x⁻¹] key for reciprocals, you can evaluate trigonometric functions at other values.

**EXAMPLE 4**

**APPROXIMATING USING A CALCULATOR**

Use a calculator to approximate $\cot \dfrac{\pi}{9}$.

*Solution*

```
1/tan(π/9)
      2.747477419
```

$$\cot \frac{\pi}{9} = \frac{1}{\tan \dfrac{\pi}{9}} \approx 2.747 \qquad \text{Using a calculator (set for radians)}$$

**Practice Problem 1**

Use a calculator to find $\csc \dfrac{\pi}{7}$.

➤ **Solution on page 615**

For *acute* angles $\theta$, these trigonometric functions can be expressed as ratios of sides of a right triangle. We simply take the earlier definitions and replace $x$, $y$, and $r$ by adj, opp, and hyp.

| Right Triangle Formulas for $\tan \theta$, $\cot \theta$, $\sec \theta$, and $\csc \theta$ |
|---|

$$\tan \theta = \frac{\text{opp}}{\text{adj}}$$

$$\cot \theta = \frac{\text{adj}}{\text{opp}}$$

$$\sec \theta = \frac{\text{hyp}}{\text{adj}}$$

$$\csc \theta = \frac{\text{hyp}}{\text{opp}}$$

**EXAMPLE 5**

**ESTIMATING HEIGHT**

Estimate the height of a tree if its top makes an angle of 41° with the ground when seen from 50 feet away.

*Solution*

Since the height and the known distance (50 feet) are the opposite and adjacent sides of a right triangle, we use the tangent (opp/adj) of the angle. Letting $h$ = height of tree:

$$\tan 41° = \frac{h}{50} \qquad \tan \theta = \frac{\text{opp}}{\text{adj}}$$

Solving for $h$:

$$h = 50 \tan 41° \approx 50 \cdot 0.869 \approx 43.5 \qquad \begin{array}{l}\tan 41° \approx 0.869 \text{ (from a} \\ \text{calculator set for } degrees)\end{array}$$

The tree is about 43.5 feet high.

If an object is thrown with initial velocity $v$ (feet per second) and with an initial angle $\theta$ to the horizontal, then its height $y$ when it has gone a horizontal distance $x$ from its starting point is given by the following formula:*

$$y = x \tan \theta - \left(\frac{4x}{v} \sec \theta\right)^2$$

**EXAMPLE 6**

**HOME RUNS**

A baseball is hit at an angle of 35° with the horizontal with an initial velocity of 120 feet per second. Will it make a home run by clearing the outfield fence 400 feet away if the fence is 10 feet higher than the initial height of the ball?

*Solution*

The formula with the given data becomes

$$y = 400 \tan 35° - \left(\frac{4 \cdot 400}{120} \sec 35°\right)^2$$

| | Previous formula with distance $x = 400$, angle $\theta = 35°$, and initial velocity $v = 120$ |

0.700        1.22      Using a calculator set for *degrees*

$$\approx 15.1$$

The ball will be 15 feet high, easily clearing the 10-foot fence and making a home run.

## Graphing Calculator Exploration

What is the best angle to get the baseball as high as possible over the outfield fence (as before, neglecting air resistance and backspin)?

Set your calculator for *degrees* and enter the expression for $y$ from the preceding example into your graphing calculator as $y_1$ but with 35° replaced by $x$. Then graph $y_1$ on the window [0, 90] by [−10, 60] and use MAXIMUM to find the best angle $x$. (Afterward, reset your calculator for *radians*.)

*Neglecting air resistance and backspin. See Philippe de Mezieres, *The Mathematics of Projectiles in Sports* (Australian Academy of Sciences).

## Graphs of Tangent, Cotangent, Secant, and Cosecant Functions

The graphs of  tan $t$,  cot $t$,  sec $t$,  and  csc $t$  (for $t$ in *radians*) are shown below.

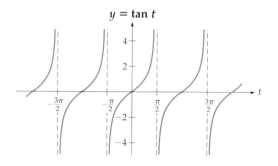

$y = \tan t$

Period: $\pi$
Undefined at:
$\pm\frac{\pi}{2}, \pm\frac{3\pi}{2}, \dots$

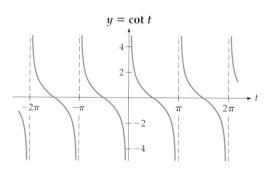

$y = \cot t$

Period: $\pi$
Undefined at:
$0, \pm\pi, \pm 2\pi, \dots$

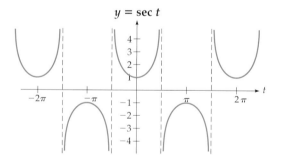

$y = \sec t$

Period: $2\pi$
Undefined at:
$\pm\frac{\pi}{2}, \pm\frac{3\pi}{2}, \dots$

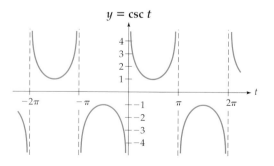

$y = \csc t$

Period: $2\pi$
Undefined at:
$0, \pm\pi, \pm 2\pi, \dots$

## Graphing Calculator Exploration

Verify the preceding graphs by setting the window of your graphing calculator to $[-2\pi, 2\pi]$ by $[-4, 4]$, and graphing successively:

$$\tan x, \quad \cot x \left(\text{as } \frac{1}{\tan x}\right), \quad \sec x \left(\text{as } \frac{1}{\cos x}\right), \quad \text{and} \quad \csc x \left(\text{as } \frac{1}{\sin x}\right)$$

 **BE CAREFUL:** Each of these functions has $x$-values where it is *undefined*. At these values, your graphing calculator screen may show "false lines," which should be ignored.

### Derivatives of Trigonometric Functions

Derivatives of the tangent, cotangent, secant, and cosecant functions (always in *radians*) are given by the formulas in the first column in the following box. Combining these formulas with the Chain Rule gives the formulas in the second column, in which $f(t)$ stands for a differentiable function.

| Derivatives of $\tan t$, $\cot t$, $\sec t$, and $\csc t$ | |
|---|---|
| $\dfrac{d}{dt} \tan t = \sec^2 t$ | $\dfrac{d}{dt} \tan f(t) = \sec^2 f(t) \cdot \dfrac{df}{dt}$ |
| $\dfrac{d}{dt} \cot t = -\csc^2 t$ | $\dfrac{d}{dt} \cot f(t) = -\csc^2 f(t) \cdot \dfrac{df}{dt}$ |
| $\dfrac{d}{dt} \sec t = \sec t \cdot \tan t$ | $\dfrac{d}{dt} \sec f(t) = \sec f(t) \cdot \tan f(t) \cdot \dfrac{df}{dt}$ |
| $\dfrac{d}{dt} \csc t = -\csc t \cdot \cot t$ | $\dfrac{d}{dt} \csc f(t) = -\csc f(t) \cdot \cot f(t) \cdot \dfrac{df}{dt}$ |

Note the pattern: adding "co" to each function and attaching a negative sign gives a new formula. For example, the derivative of the tangent is the secant squared, so the derivative of the *cotangent* is the *negative* of the *cosecant* squared.

---

**EXAMPLE 7**

**PROVING A FORMULA**

Prove the first formula: $\dfrac{d}{dt} \tan t = \sec^2 t$.

612 CHAPTER 8 TRIGONOMETRIC FUNCTIONS

*Solution*

$$\frac{d}{dt} \tan t = \frac{d}{dt}\left(\frac{\sin t}{\cos t}\right)$$

Since $\tan t = \dfrac{\sin t}{\cos t}$

Derivative of $\sin t$

Derivative of $\cos t$

$$= \frac{\cos t \cos t - (-\sin t) \sin t}{\cos^2 t}$$

Using the Quotient Rule

$$= \frac{\cos^2 t + \sin^2 t}{\cos^2 t} = \frac{1}{\cos^2 t}$$

Simplifying and using
$\sin^2 t + \cos^2 t = 1$

$$= \sec^2 t$$

Since $\dfrac{1}{\cos t} = \sec t$

This proves the first formula. The other three formulas are proved similarly (see Exercises 64–66).

<hr>

**EXAMPLE 8**

**DIFFERENTIATING THE TANGENT OF A FUNCTION**

Find $\dfrac{d}{dt} \tan(t^3 + 1)$

*Solution*

$$\frac{d}{dt} \tan(t^3 + 1) = \sec^2(t^3 + 1) \cdot (3t^2)$$

Using $\dfrac{d}{dt} \tan f = \sec^2 f \cdot \dfrac{df}{dt}$

Derivative
of $t^3 + 1$

$$= 3t^2 \sec^2(t^3 + 1)$$

<hr>

**Practice Problem 2**

Find $\dfrac{d}{dt} \sec \pi t$.

➤ Solution on page 615

## Integrals of Trigonometric Functions

The differentiation formulas in the preceding box, when read in reverse, become the following integration formulas.

**Integrals of Trigonometric Functions**

$$\int \sec^2 t \, dt = \tan t + C$$

Since $\dfrac{d}{dt} \tan t = \sec^2 t$

$$\int \csc^2 t \, dt = -\cot t + C$$

Since $\dfrac{d}{dt}(-\cot t) = \csc^2 t$

$$\int \sec t \tan t \, dt = \sec t + C$$

Since $\dfrac{d}{dt} \sec t = \sec t \tan t$

$$\int \csc t \cot t \, dt = -\csc t + C$$

Since $\dfrac{d}{dt}(-\csc t) = \csc t \cot t$

Replacing the variable $t$ by $u$ gives analogous formulas for the substitution method.

---

**EXAMPLE 9**

**INTEGRATING USING A SUBSTITUTION**

Find $\displaystyle\int \sec^2 5t \, dt$.

*Solution*

We use the first formula in the preceding box, but in its "$u$" form.

$$\int \underbrace{\sec^2 5t}_{u} \, dt = \frac{1}{5} \int \underbrace{\sec^2 5t}_{u} \underbrace{5 \, dt}_{du} = \frac{1}{5} \underbrace{\int \sec^2 u \, du}_{\text{Substituting}}$$

$$\begin{array}{c} u = 5t \\ du = 5 \, dt \end{array} \qquad \begin{array}{c} \text{Multiplying} \\ \text{and} \\ \text{dividing by 5} \end{array}$$

$$= \underbrace{\frac{1}{5} \tan u + C}_{\substack{\text{Integrating by} \\ \int \sec^2 u \, du = \tan u + C}} = \underbrace{\frac{1}{5} \tan 5t + C}_{\substack{\text{Substituting back} \\ \text{using} \;\; u = 5t}}$$

---

**EXAMPLE 10**

**INTEGRATING USING A SUBSTITUTION**

Find $\displaystyle\int \tan^3 t \, \sec^2 t \, dt$.

*Solution*

The derivative of $\tan t$ is $\sec^2 t$, so the integral is of the form $\int u^3 \, du$.

$$\int \underbrace{\tan^3 t}_{u^3} \underbrace{\sec^2 t \, dt}_{du} = \underbrace{\int u^3 \, du}_{\text{Substituting}} = \underbrace{\frac{1}{4} u^4 + C}_{\substack{\text{Integrating by} \\ \text{the Power Rule}}} = \underbrace{\frac{1}{4} \tan^4 t + C}_{\substack{\text{Substituting back} \\ \text{using} \;\; u = \tan t}}$$

$$\begin{array}{l} u = \tan t \\ du = \sec^2 t \, dt \end{array}$$

---

Exercises 69–71 show how to derive the following integration formulas.

**Integrals of  tan *t*,  cot *t*,  sec *t*  and  csc *t***

$$\int \tan t\, dt = -\ln |\cos t| + C$$

$$\int \cot t\, dt = \ln |\sin t| + C$$

$$\int \sec t\, dt = \ln |\sec t + \tan t| + C$$

$$\int \csc t\, dt = \ln |\csc t - \cot t| + C$$

## Graphing Calculator Exploration

One of the difficulties of trigonometry is that the trigonometric identities allow results to be written in many different forms. Some advanced graphing calculators can find trigonometric integrals. For example, the Texas Instruments *TI-89* graphing calculator finds the integral $\int \sec t\, dt$ as follows.

Show that this answer agrees with the third formula in the box above. [*Hint:* Begin with the answer in the screen, multiply numerator and denominator by $(\sin t + 1)$, and simplify.]

For a quick reference, all of the integration formulas used in this book are summarized on the inside back cover.

---

**EXAMPLE 11**  **FINDING AREA**

Find the area under  $y = \tan x$  from  $x = 0$  to  $x = \dfrac{\pi}{4}$.

*Solution*

Since $\tan x$ is nonnegative between $x = 0$ and $x = \dfrac{\pi}{4}$, we integrate from 0 to $\dfrac{\pi}{4}$:

$$\int_0^{\pi/4} \tan t \, dt = \underbrace{-\ln \left| \cos t \right| \Big|_0^{\pi/4}}_{\substack{\text{Integrating by the} \\ \text{first boxed formula}}} = -\ln \left| \underbrace{\cos \dfrac{\pi}{4}}_{\frac{\sqrt{2}}{2}} \right| + \ln \left| \underbrace{\cos 0}_{1} \right|$$

$$= -\ln \dfrac{\sqrt{2}}{2} + \underbrace{\ln 1}_{0} = -\ln \dfrac{\sqrt{2}}{2}$$

0.347 square units

The area is $-\ln \dfrac{\sqrt{2}}{2} \approx 0.347$ square unit.

## Solutions to Practice Problems

**1.** $\csc \dfrac{\pi}{7} \approx 2.305$          From entering $1/\sin(\pi/7)$

**2.** $\dfrac{d}{dt} \sec \pi t = (\sec \pi t)(\tan \pi t) \, \pi = \pi \sec \pi t \tan \pi t$

## 8.5 Section Summary

In this section we defined the remaining four trigonometric functions: tangent, cotangent, secant, and cosecant. If $(x, y)$ is any point on the terminal side of an angle of $\theta$ radians in standard position, and $r$ is the distance from the origin, then:

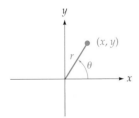

$$\tan \theta = \frac{\sin \theta}{\cos \theta} = \frac{y}{x} \qquad\qquad \sec \theta = \frac{1}{\cos \theta} = \frac{r}{x}$$

$$\cot \theta = \frac{\cos \theta}{\sin \theta} = \frac{x}{y} \qquad\qquad \csc \theta = \frac{1}{\sin \theta} = \frac{r}{y}$$

Tangent and cotangent are particularly useful in applications since they give the ratio of the sides of a right triangle with acute angle $\theta$. The graphs of these functions are shown on page 610. We often regard these as functions of a variable $t$, without regard to angles. The derivative and corresponding integral formulas are given below.

$$\frac{d}{dt}\tan t = \sec^2 t \qquad \int \sec^2 t\, dt = \tan t + C$$

$$\frac{d}{dt}\cot t = -\csc^2 t \qquad \int \csc^2 t\, dt = -\cot t + C$$

$$\frac{d}{dt}\sec t = \sec t \cdot \tan t \qquad \int \sec t \cdot \tan t\, dt = \sec t + C$$

$$\frac{d}{dt}\csc t = -\csc t \cdot \cot t \qquad \int \csc t \cdot \cot t\, dt = -\csc t + C$$

Some other integral formulas are given on page 614.

## 8.5 Exercises

**1–4.** For each angle $\theta$, find:

**a.** $\tan \theta$  **b.** $\cot \theta$  **c.** $\sec \theta$  **d.** $\csc \theta$.

**1.**

**2.**

**3.**

**4.**

**5–12.** Evaluate without using a calculator.

**5. a.** $\tan \dfrac{\pi}{6}$  **b.** $\csc \dfrac{\pi}{6}$

**6. a.** $\cot \dfrac{\pi}{6}$  **b.** $\sec \dfrac{\pi}{6}$

**7. a.** $\tan \dfrac{3\pi}{2}$  **b.** $\csc \dfrac{3\pi}{2}$

**8. a.** $\cot \dfrac{3\pi}{2}$  **b.** $\sec \dfrac{3\pi}{2}$

**9. a.** $\tan \dfrac{\pi}{4}$  **b.** $\cot \dfrac{\pi}{4}$

**10. a.** $\tan \dfrac{3\pi}{4}$  **b.** $\sec \dfrac{3\pi}{4}$

**11. a.** $\tan \dfrac{7\pi}{6}$  **b.** $\csc \dfrac{7\pi}{6}$

**12. a.** $\tan \dfrac{5\pi}{3}$  **b.** $\cot \dfrac{5\pi}{3}$

**13–16.** Use a calculator to approximate each value.

**13.** $\cot \dfrac{\pi}{5}$  **14.** $\sec \dfrac{\pi}{7}$  **15.** $\sec \dfrac{5\pi}{12}$  **16.** $\csc \dfrac{9\pi}{52}$

**17. GENERAL: Flagpole Height** Estimate the height of a flagpole if its top makes an angle of $62°$ with the ground when seen from 30 feet away.

**18. GENERAL: River Width** Estimate the width of a river if two points on opposite shores make a $35°$ angle for someone standing 40 feet down the shore.

**19. GENERAL: Slope and Angle of Inclination** Show that the slope of a line is the *tangent* of the "angle of inclination" between the line and the $x$-axis.

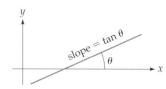

20. **ATHLETICS: Baseball** A baseball is hit so that it leaves the bat with an initial velocity of 115 feet per second and at an angle of 50° with the horizontal. Will it make a home run by clearing the outfield fence 400 feet away if the fence is 8 feet higher than the initial height of the ball? [*Hint:* Use the formula on page 609.]

21. **ATHLETICS: Football** A football is kicked from the ground with an initial velocity of 60 feet per second and at an angle of 40° with the horizontal. Will it make a field goal by clearing the 10-foot-high goal post that is 90 feet away? [*Hint:* Use the formula on page 609.]

22. **ATHLETICS: Baseball** (*continuation of 20*) At what angle should the ball be hit (with initial velocity 115 feet per second) so as to maximize its height above the outfield fence? Set your calculator for degrees and enter the height function that you used in Exercise 20 as $y_1$, but with 50° replaced by $x$. Then graph $y_1$ on an appropriate window and use MAXIMUM to maximize it and find the angle. (Afterward, reset your calculator for *radians*.)

23. **ATHLETICS: Football** (*continuation of 21*) At what angle should the football be kicked (with initial velocity 60 feet per second) so as to maximize its height over the goal post? Set your calculator for degrees and enter the height function that you used in Exercise 21 as $y_1$, but with 40° replaced by $x$. Then graph $y_1$ on an appropriate window and use MAXIMUM to maximize it and find the angle. (Afterward, reset your calculator for *radians*.)

24. Graph $y_1 = \csc^2 x$ on the window $[-2\pi, 2\pi]$ by $[-4, 4]$. What is its period? Where is it undefined? (This graph will be used in Exercise 64.)

25. Graph $y_1 = \sec x \tan x$ on the window $[-2\pi, 2\pi]$ by $[-4, 4]$. What is its period? Where is it undefined? (This graph will be used in Exercise 65.)

26. Graph $y_1 = \csc x \cot x$ on the window $[-2\pi, 2\pi]$ by $[-4, 4]$. What is its period? Where

is it undefined? (This graph will be used in Exercise 66.)

27. Derive the identity $\tan^2 t + 1 = \sec^2 t$. [*Hint:* Divide the identity $\sin^2 t + \cos^2 t = 1$ by $\cos^2 t$.]

28. Derive the identity $1 + \cot^2 t = \csc^2 t$. [*Hint:* Divide the identity $\sin^2 t + \cos^2 t = 1$ by $\sin^2 t$.]

29–36. Differentiate each function.

29. $f(t) = t \cot t$

30. $f(t) = t^2 \sec t$

31. $f(t) = \tan(t^2 - 1)$

32. $f(t) = \cot(2t - 1)$

33. $f(x) = \sec(\pi x + 1)$

34. $f(x) = \tan\left(\dfrac{x}{2} - \pi\right)$

35. $f(z) = \cot^2 z^3$

36. $f(z) = \csc^4 \sqrt{z}$

37. **GENERAL: Length of Shadow** When the sun is $x$ degrees above the horizon, a 6-foot-tall person will cast a shadow of length $S(x) = 6 \cot \dfrac{\pi x}{180}$ feet. Calculate $S'(10)$ and interpret the resulting number.

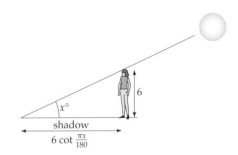

38. **GENERAL: Ladder Reach** A sufficiently long extension ladder whose base is 4 feet from a wall and that makes an angle of $x$ degrees with the ground will reach to a height of $H(x) = 4 \tan \dfrac{\pi x}{180}$ feet. Find $H'(75)$ and interpret the resulting number.

**39–48.** Find each integral.

**39.** $\displaystyle\int t^2 \sec^2(t^3 + 1)\, dt$

**40.** $\displaystyle\int \csc^2 \pi t\, dt$

**41.** $\displaystyle\int \csc \pi t \cot \pi t\, dt$

**42.** $\displaystyle\int t \sec(t^2 + 1) \tan(t^2 + 1)\, dt$

**43.** $\displaystyle\int \tan(1 - t)\, dt$

**44.** $\displaystyle\int \cot \frac{t}{2}\, dt$

**45.** $\displaystyle\int x^4 \csc x^5\, dx$

**46.** $\displaystyle\int x^{-1/2} \sec x^{1/2}\, dx$

**47.** $\displaystyle\int \cot^4 x \csc^2 x\, dx$

**48.** $\displaystyle\int \sqrt{\tan x}\, \sec^2 x\, dx$

**49–50.** Evaluate each definite integral.

**49.** $\displaystyle\int_{\pi/4}^{\pi/2} \cot t\, dt$

**50.** $\displaystyle\int_{0}^{\pi/4} \sec^2 t\, dt$

**51. a.** Find (without using a graphing calculator) the area under the curve $y = \sec x$ and above the $x$-axis from $x = 0$ to $x = \pi/4$.

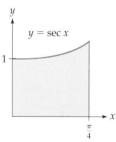

**b.** Use a graphing calculator to verify your answer (using FnInt, ∫f(x) dx, or a similar command).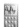

**52. a.** Find (without using a graphing calculator) the area under the curve $y = \sec x \tan x$ and above the $x$-axis between $x = 0$ and $x = \pi/4$.

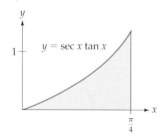

**b.** Use a graphing calculator to verify your answer (using FnInt, ∫f(x) dx, or a similar command).

## Conceptual Exercises

**53.** True or False: All six trigonometric functions have period $2\pi$.

**54.** True or False: All six trigonometric functions can be expressed in terms of the sine and cosine functions.

**55.** Which trigonometric function is the same as $\dfrac{1}{\sec \theta}$?

**56.** Which trigonometric function is the same as $\dfrac{1}{\csc \theta}$?

**57.** Find all values of $t$ that satisfy $\sec t = \frac{1}{2}$. [*Hint:* Look at the graph of the secant function.]

**58.** Find $\dfrac{d}{dt}(\sec^2 t - \tan^2 t)$.
[*Hint:* No work necessary—see Exercise 27.]

**59.** Find $\dfrac{d}{dt}(\csc^2 t - \cot^2 t)$.
[*Hint:* No work necessary—see Exercise 28.]

**60.** What does the formula $\dfrac{d}{dx}\tan t = \sec^2 t$ tell you about the sign of the slope of the tangent function? Does the graph of the tangent function verify your observation?

**61.** What does the formula $\dfrac{d}{dx}\cot t = -\csc^2 t$ tell you about the sign of the slope of the cotangent function? Does the graph of the cotangent function verify your observation?

**62.** Find $\displaystyle\int_{-\pi/4}^{\pi/4} \tan t\, dt$ [*Hint:* No work necessary—just look at the graph of the tangent function.]

## Explorations and Excursions

The following problems extend and augment the material presented in the text.

### More on Derivatives and Integrals of Trigonometric Functions

**63.** Why do $\tan^2 t$ and $\sec^2 t$ have the same derivative?

**64. a.** Derive the formula $\dfrac{d}{dt}\cot t = -\csc^2 t$ by writing $\cot t = \dfrac{\cos t}{\sin t}$ and differentiating by the Quotient Rule.

**b.** Verify this formula on a graphing calculator by entering $y_1 = \cot x$ [entered as
$(\tan x)^{-1}$], graphing its derivative (using
NDERIV), and observing that the result is
the negative of the graph of $\csc^2 x$ found in
Exercise 24.

**65. a.** Derive the formula $\dfrac{d}{dt}\sec t = \sec t \tan t$

by writing $\sec t = (\cos t)^{-1}$ and differentiating by the Generalized Power Rule.

 **b.** Verify this formula on a graphing calculator
by entering $y_1 = \sec x$ [entered as
$(\cos x)^{-1}$], graphing its derivative (using
NDERIV), and observing that the result is
the same as the graph of $\sec x \tan x$ found
in Exercise 25.

**66. a.** Derive the formula $\dfrac{d}{dt}\csc t = -\csc t \cot t$

by writing $\csc t = (\sin t)^{-1}$ and differentiating by the Generalized Power Rule.

 **b.** Verify this formula on a graphing calculator
by entering $y_1 = \csc x$ [entered as
$(\sin x)^{-1}$], graphing its derivative (using
NDERIV), and observing that the result is the
negative of the graph of $\csc x \cot x$ found
in Exercise 26.

**67.** Find the integral $\int \tan t \sec^2 t\, dt$ in two ways:

**a.** Using the substitution method with $u = \tan t$.
**b.** Using the substitution method with $u = \sec t$.
**c.** Can you reconcile the two seemingly different
answers?

**68.** Find the integral $\int \cot t \csc^2 t\, dt$ in two ways:

**a.** Using the substitution method with $u = \cot t$.
**b.** Using the substitution method with $u = \csc t$.
**c.** Can you reconcile the two seemingly different
answers?

**69.** Find $\int \tan t\, dt$ and $\int \cot t\, dt$.

[*Hint:* Write $\tan t = \dfrac{\sin t}{\cos t}$ and $\cot t = \dfrac{\cos t}{\sin t}$
and integrate by the substitution method.]

**70.** Find $\int \sec t\, dt$.

[*Hint:* Multiply the integrand by $\dfrac{\sec t + \tan t}{\sec t + \tan t}$
and integrate by the substitution method.]

**71.** Find $\int \csc t\, dt$.

[*Hint:* Multiply the integrand by $\dfrac{\csc t - \cot t}{\csc t - \cot t}$
and integrate by the substitution method.]

## Chapter Summary with Hints and Suggestions

Reading the text and doing the exercises in this chapter have helped you to master
the following concepts and skills, which are listed by section (in case you need to
review them) and are keyed to particular Review Exercises. Answers for all Review
Exercises are given at the back of the book, and full solutions can be found in the
Student Solutions Manual.

### 8.1 Triangles, Angles, and Radian Measure

■ Convert angle measures between degrees and
radians. *(Review Exercises 1–2.)*

$$\frac{\text{Degrees}}{180} = \frac{\text{Radians}}{\pi}$$

■ Calculate the length of a section of a circle.
*(Review Exercises 3–4.)*

$$\left(\begin{array}{c}\text{Arc}\\\text{length}\end{array}\right) = (\text{Radians})(\text{Radius})$$

### 8.2 Sine and Cosine Functions

■ Find the exact value of a sine or cosine function.
*(Review Exercises 5–6.)*

$$\sin t = \frac{y}{r}$$

$$\cos t = \frac{x}{r}$$

■ Sketch the graph of a modified sine or cosine function. *(Review Exercises 7–8.)*

$$y = a \sin bt \qquad y = a \cos bt$$

■ Solve an applied problem involving a trigonometric function. *(Review Exercises 9–12.)*

## 8.3 Derivatives of Sine and Cosine Functions

■ Find the derivative of a sine or cosine function. *(Review Exercises 13–20.)*

$$\frac{d}{dt} \sin t = \cos t \qquad \frac{d}{dt} \cos t = -\sin t$$

■ Solve an applied problem by differentiating a sine or cosine function. *(Review Exercises 21–26.)*

## 8.4 Integrals of Sine and Cosine Functions

■ Find the indefinite integral of a sine or cosine function. *(Review Exercises 27–36.)*

$$\int \sin t \, dt = -\cos t + C$$

$$\int \cos t \, dt = \sin t + C$$

■ Find the definite integral of a sine or cosine function. *(Review Exercises 37–38.)*

■ Find the average value of a sine or cosine function on an interval. *(Review Exercises 39–40.)*

■ Find an area involving a sine or cosine function. *(Review Exercises 41–42.)*

■ Solve an applied problem by integrating a sine or cosine function. *(Review Exercises 43–44.)*

## 8.5 Other Trigonometric Functions

■ Find the exact value of one of the other trigonometric functions. *(Review Exercises 45–46.)*

$$\tan t = \frac{\sin t}{\cos t} = \frac{y}{x}$$

$$\cot t = \frac{1}{\tan t} = \frac{x}{y}$$

$$\sec t = \frac{1}{\cos t} = \frac{r}{x}$$

$$\csc t = \frac{1}{\sin t} = \frac{r}{y}$$

■ Solve an applied problem involving one of the other trigonometric functions. *(Review Exercises 47–48.)*

■ Find the derivative of one of the other functions. *(Review Exercises 49–54.)*

$$\frac{d}{dt} \tan t = \sec^2 t$$

$$\frac{d}{dt} \cot t = -\csc^2 t$$

$$\frac{d}{dt} \sec t = \sec t \tan t$$

$$\frac{d}{dt} \csc t = -\csc t \cot t$$

■ Find the indefinite integral of one of the other trigonometric functions. *(Review Exercises 55–60.)*

■ Find the definite integral of one of the other trigonometric functions. *(Review Exercises 61–62.)*

■ Find the area under a trigonometric curve. *(Review Exercises 63–64.)*

## Sections 8.3, 8.4, and 8.5

■ Solve a more involved graphing, differentiation, integration, or numerical integration problem involving a trigonometric function. *(Review Exercises 65–84.)*

## Hints and Suggestions

■ We measure angles in *radians* because it simplifies the differentiation and integration formulas. All equivalent measures can be found by using the equation $180° = \pi$ radians.

■ Think of $\pi$ as a natural unit for radian measure, just as it occurs naturally in the formulas for the circumference and area of a circle.

■ The "triangle" definitions such as $\sin t = \dfrac{\text{opp}}{\text{hyp}}$ hold only for *acute* angles, whereas the "coordinate" definitions such as $\sin t = \dfrac{y}{r}$ hold for *all* angles.

■ We can find $\sin t$ and $\cos t$ *exactly* for some angles, such as $\dfrac{\pi}{6}, \dfrac{\pi}{4},$ and $\dfrac{\pi}{3},$ using $30°-60°-90°$ and $45°-45°-90°$ triangles. For other angles we approximate the value using a calculator.

■ Try not to think of $\sin t$ and $\cos t$ as functions of *angles*, but as ordinary "numerical" functions, so that you can use them to model any quantity that varies periodically, such as a business cycle. Each has period $2\pi$, but $\sin t$ begins at the origin and $\cos t$ begins one unit up.

■ The uses of differentiation remain the same (slopes, rates of change, marginals, and velocities) and the uses of integration remain the same (areas, total accumulations, and average values), but we can now apply these ideas to *periodic* behavior.

■ The derivative formulas for the other trigonometric functions can be derived from those for $\sin t$ and $\cos t$ by using the Quotient Rule.

■ A graphing calculator helps by graphing trigonometric functions and finding derivatives (for optimization) and definite integrals (for areas, total accumulations, and average values).

■ **Practice for Test:** Review Exercises 1, 5, 7, 9, 15, 19, 21, 25, 27, 33, 39, 41, 43, 45, 49, 55, 63.

# Review Exercises for Chapter 8

Practice test exercise numbers are in green.

## 8.1 Triangles, Angles, and Radian Measure

1. **a.** Convert $150°$ into radians.
   **b.** Convert $\dfrac{\pi}{9}$ radians into degrees.

2. **a.** Convert $120°$ into radians.
   **b.** Convert $\dfrac{\pi}{5}$ radians into degrees.

**3–4. GENERAL: Largest Clock** The largest single-faced clock in the world is the Colgate clock on the Hudson River in Jersey City, New Jersey, with a minute hand that is 26 feet long. Find the arc length through which the tip of the minute hand travels between the following times. [*Hint:* Find the angle in radians. Give an approximate answer using $\pi \approx 3.14$.]

3. Between noon and 12:40 P.M.

4. Between 3 P.M. and 3:30 P.M.

## 8.2 Sine and Cosine Functions

**5–6.** Evaluate *without* using a calculator (leaving answers in *exact* form).

5. **a.** $\sin \dfrac{2\pi}{3}$    **b.** $\cos \dfrac{2\pi}{3}$

   **c.** $\sin \dfrac{5\pi}{4}$    **d.** $\cos \dfrac{5\pi}{4}$

6. **a.** $\sin \dfrac{7\pi}{6}$    **b.** $\cos \dfrac{7\pi}{6}$

   **c.** $\sin \dfrac{3\pi}{4}$    **d.** $\cos \dfrac{3\pi}{4}$

**7–8.** Sketch the graph of each function. (If you use a graphing calculator, make a hand-drawn sketch showing the amplitude and period.)

7. $y = 7 \cos 2x$    8. $y = 6 \sin 2x$

9. **GENERAL: Building Construction** A roof is to make an angle of $30°$ with the horizontal, and rise 12 feet in the middle. Find the length of the rafters.

*(continues)*

10. **GENERAL: Water Depth** A 40-foot anchor line makes an angle of 30° with the surface. How deep is the water?

11. **GENERAL: Daylight** The number of hours of daylight in Seattle, Washington, on day $x$ of the year is approximated by the function

$$D(x) = 12.25 + 3.75 \sin \frac{2\pi(x - 81)}{365}$$

  a. Graph this function on the window [0, 365] by [0, 20]. (Be sure your calculator is set for *radians*.)
  b. Find the number of hours of daylight on May 1 (day 121).
  c. Find the number of hours of daylight on December 1 (day 335).

  *Source: Ecological Modeling 80*

12. **GENERAL: Daylight** (*continuation*) For the "daylight" function in Exercise 11, find the days with the largest and smallest amounts of daylight. [*Hint:* Use MAXIMUM and MINIMUM.]

## 8.3 Derivatives of Sine and Cosine Functions

**13–20.** Find the derivative of each function.

13. $f(t) = 2 \cos(4t - 1)$    14. $f(t) = \sin(2 - 3t)$

15. $f(x) = x^3 \sin x + 3x^2 \cos x - 6x \sin x - 6 \cos x$

16. $f(t) = \dfrac{1 + \sin t}{1 + \cos t}$    17. $f(t) = \sin^3 t^2 - \cos \dfrac{\pi}{3}$

18. $f(x) = \sin e^{x^2}$    19. $f(t) = \sqrt{\sin t}$

20. $f(t) = \sin \sqrt{t}$

21. **BUSINESS: Seasonal Sales** A manufacturer estimates that the number of overcoats sold during week $t$ of the year is

$$S(t) = 250 + 250 \cos \frac{\pi t}{26}$$

  a. Graph this function on the window [0, 52] by [−100, 600].
  b. Find the rate of change of sales at week $t = 37$ (mid September).

22. **BUSINESS: Seasonal Sales** (*continuation*) For the sales function in Exercise 21, find when the rate of increase in sales is the greatest. [*Hint:* Use NDERIV and MAXIMUM.]

23. **GENERAL: Temperature** The normal mean temperature in St. Louis, Missouri, on day $x$ of the year is approximately

$$T(x) = 50 - 26 \cos \frac{2\pi(x - 15)}{365}$$

degrees Fahrenheit.

  a. Graph this function on the window [0, 365] by [0, 90].
  b. Find the rate of change of temperature on October 15 (day 288).

*Source:* National Climatic Data Center

24. **GENERAL: Temperature** (*continuation*) For the temperature function in Exercise 23, find when the rate of increase is the greatest. [*Hint:* Use NDERIV and MAXIMUM.]

25. **GENERAL: Maximizing Volume** An enclosure made by leaning a sheet of plywood against a wall at an angle $\theta$ with the ground, will have a volume of $V(\theta) = 128 \sin \theta \, \cos \theta$ cubic feet. Find the angle $\theta$ that maximizes the volume $V(\theta)$.

26. **BIOMEDICAL: Vascular Branching** A surgeon wants to attach a new artery to an existing one at the angle $\theta$ that minimizes resistance of blood flow.

Under certain conditions, minimizing the resistance is equivalent to minimizing the function

$$f(\theta) = \frac{r^{-4} - R^{-4} \cos \theta}{\sin \theta}$$

Derive the relationship $\cos \theta = r^4/R^4$ by setting the derivative $f'$ equal to zero and solving.

*Source:* E. Batschelet, *Introduction to Mathematics for Life Scientists*

## 8.4 Integrals of Sine and Cosine Functions

**27–36.** Find each integral.

**27.** $\displaystyle\int \cos \frac{\pi t}{5} \, dt$

**28.** $\displaystyle\int t \sin(t^2 + 1) \, dt$

**29.** $\displaystyle\int \sin(2\pi - t) \, dt$

**30.** $\displaystyle\int e^{-x} \sin e^{-x} \, dx$

**31.** $\displaystyle\int e^{\cos x} \sin x \, dx$

**32.** $\displaystyle\int \frac{\cos t}{\sqrt{\sin t}} \, dt$

**33.** $\displaystyle\int \frac{\cos 3t}{1 + \sin 3t} \, dt$

**34.** $\displaystyle\int (x^3 - e^{-x} - \cos \pi x) \, dx$

**35.** $\displaystyle\int \frac{\cos(\ln y)}{y} \, dy$

**36.** $\displaystyle\int \frac{\sin \dfrac{1}{z}}{z^2} \, dz$

**37–38.** For each definite integral:
a. Evaluate it "by hand."

b. Check your answer to part (a) by evaluating using a graphing calculator.

**37.** $\displaystyle\int_{\pi/6}^{\pi/2} \sin 2t \, dt$

**38.** $\displaystyle\int_{\pi/6}^{\pi} \sin^2 t \cos t \, dt$

**39.** Find the average value of $f(t) = \sin \pi t$ on the interval $[0, 3]$. (Give an exact and also an approximate answer.)

**40.** Find the average value of $f(t) = 200 + 100 \cos \pi t$ on the interval $[0, 3]$.

**41.** Find the area under the curve $y = \sin^2 x \cos x$ and above the $x$-axis from $x = 0$ to $x = \dfrac{\pi}{2}$.

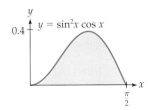

**42.** Find the area between the curves $y = \cos x$ and $y = \sin x$ from $x = 0$ to $x = \dfrac{\pi}{4}$, as shown below. (Give the answer in *exact* form and also approximate it.)

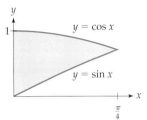

**43. BUSINESS: Total Sales** A boatyard estimates that sales during week $x$ of the year will be $S(x) = 25 - 20 \cos \dfrac{\pi x}{26}$ thousand dollars, where $x = 0$ corresponds to the beginning of the year.
a. Graph the sales function on $[0, 52]$ by $[-10, 50]$.
b. Find the total sales during the first half of the year ($x = 0$ to $x = 26$).

**44. GENERAL: Average Temperature** The temperature at time $x$ hours in a warehouse is given by the formula

$$T(x) = 60 - 10 \sin \frac{\pi x}{12}$$

degrees Fahrenheit.
a. Graph the function on $[0, 24]$ by $[0, 70]$.
b. Find the average temperature during the first 6 hours.

## 8.5 Other Trigonometric Functions

**45–46.** Evaluate *without* using a calculator (leaving answers in *exact* form).

**45. a.** $\tan \dfrac{\pi}{3}$   **b.** $\cot \dfrac{\pi}{3}$
   **c.** $\sec \dfrac{\pi}{3}$   **d.** $\csc \dfrac{\pi}{3}$

**46. a.** $\tan \dfrac{3\pi}{4}$   **b.** $\cot \dfrac{3\pi}{4}$
   **c.** $\sec \dfrac{3\pi}{4}$   **d.** $\csc \dfrac{3\pi}{4}$

**47. GENERAL: Calculating Height** Estimate the height of a monument if its top, from 600 feet away, makes an angle of 40° with the ground.

**48. GENERAL: Calculating Distance** Estimate the distance from a boat to an 80-foot-tall lighthouse if the top of the lighthouse makes an angle of 20° with the water.

**49–53.** Differentiate each function.

**49.** $f(x) = x \tan x^2$     **50.** $f(x) = x^2 \cot x$

**51.** $f(t) = \csc(\pi - t)$     **52.** $f(t) = \sec \sqrt{t}$

**53.** $f(t) = e^{\tan \pi t}$

**54.** Why do $\cot^2 t$ and $\csc^2 t$ have the same derivative?

**55–60.** Find each integral.

**55.** $\displaystyle\int \sec^2 \frac{\pi t}{2}\, dt$     **56.** $\displaystyle\int x \tan x^2\, dx$

**57.** $\displaystyle\int t^2 \sec t^3 \tan t^3\, dt$

**58.** $\displaystyle\int \csc(\pi - t) \cot(\pi - t)\, dt$

**59.** $\displaystyle\int \sqrt{\cot x}\, \csc^2 x\, dx$     **60.** $\displaystyle\int \frac{\sec^2 x}{\sqrt{\tan x}}\, dx$

**61–62.** For each definite integral:

**a.** Evaluate it "by hand."

**b.** Check your answer to part (a) by evaluating using a graphing calculator.

**61.** $\displaystyle\int_0^{\pi/4} \sec t\, dt$     **62.** $\displaystyle\int_{\pi/4}^{\pi/2} \csc^2 t\, dt$

**63. a. AREA UNDER CURVE** Find "by hand" the area under the curve $y = \cot x$ and above the $x$-axis from $x = \dfrac{\pi}{4}$ to $x = \dfrac{\pi}{2}$.

   **b.** Verify your answer to part (a) using a graphing calculator.

**64. a. AREA UNDER CURVE** Find "by hand" the area under the curve $y = \csc x$ and above the $x$-axis from $x = \dfrac{\pi}{4}$ to $x = \dfrac{\pi}{2}$.

   **b.** Verify your answer to part (a) using a graphing calculator.

## Additional Topics

**65–66. GRAPHING TRIGONOMETRIC FUNCTIONS** Graph each function for $0 \le x \le 2\pi$ (using an appropriate range of $y$-values). Make a hand-drawn sketch, showing all relative extreme points.

**65.** $f(x) = \sin^2 x$     **66.** $f(x) = \sin x + \cos x$

**67–68. IMPLICIT DIFFERENTIATION** For each equation:

**a.** Find $dy/dx$ by implicit differentiation.
**b.** Evaluate $dy/dx$ at the given point.

**67.** $\sin 2x + \cos y = 0$

Evaluate $dy/dx$ at $\left(\dfrac{\pi}{2}, \dfrac{\pi}{2}\right)$.

**68.** $\sin y - \cos 4x = 0$

Evaluate $dy/dx$ at $\left(\dfrac{\pi}{8}, 0\right)$.

**69. IMPLICIT DIFFERENTIATION: Radar Tracking** A helicopter, rising vertically at the rate of 10 feet per second, is tracked by a radar station 200 feet away, as shown in the diagram. How fast is the angle $\theta$ of the radar beam changing at the moment when the angle is $\frac{\pi}{3}$?

$$\left[ Hint: \quad \tan \theta = \frac{y}{200}. \right]$$

$\frac{dy}{dt} = 10$

$\theta$

200

**70. IMPLICIT DIFFERENTIATION: Radar Tracking**
A plane is flying at a constant altitude of 2 miles toward a radar station at the rate of 120 miles per hour, as shown in the diagram. How fast is the angle $\theta$ of the radar beam changing at the moment when the angle is $\frac{\pi}{6}$?

$$\left[ Hint: \quad \cot \theta = \frac{x}{2}. \right]$$

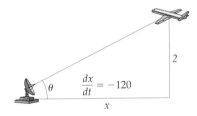

2

$\frac{dx}{dt} = -120$

$\theta$

$x$

**71–74. FUNCTIONS OF SEVERAL VARIABLES**
For each function, find the partials:
a. $f_x(x, y)$ and   b. $f_y(x, y)$.

**71.** $f(x, y) = \sin 2x - y \cos x$

**72.** $f(x, y) = \ln \cos(x - y)$

**73.** $f(x, y) = (\sin x)e^{\cos y}$

**74.** $f(x, y) = \sin(x^3 y)$

**75–80. INTEGRATION BY PARTS** Evaluate each integral by integration by parts.

**75.** $\displaystyle\int t \sin t \, dt$

**76.** $\displaystyle\int t \cos t \, dt$

**77.** $\displaystyle\int x \sec^2 x \, dx$

**78.** $\displaystyle\int x \sec x \tan x \, dx$

**79.** $\displaystyle\int e^t \sin t \, dt$

[*Hint for Exercises 79 and 80:* Integrate by parts twice and solve the resulting equation for the original integral.]

**80.** $\displaystyle\int e^t \cos t \, dt$

**81–82. NUMERICAL INTEGRATION** Approximate each integral using trapezoidal approximation with $n = 10$, $n = 100$, and $n = 500$ trapezoids. Give the most accurate approximation that your answers will support.

**81.** $\displaystyle\int_0^\pi e^{\sin x} \, dx$     **82.** $\displaystyle\int_0^{\pi/2} \sqrt{\sin x} \, dx$

**83–84. NUMERICAL INTEGRATION** Approximate each integral using Simpson's Rule with $n = 10$, $n = 100$, and $n = 500$. Give the most accurate approximation that your answers will support.

**83.** $\displaystyle\int_0^\pi e^{\sin x} \, dx$     **84.** $\displaystyle\int_0^{\pi/2} \sqrt{\sin x} \, dx$

# 9 Differential Equations

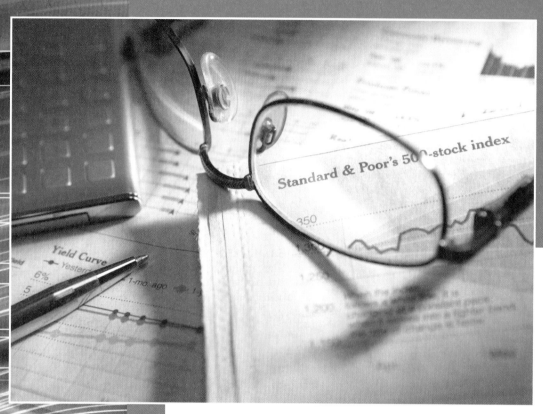

### Personal Wealth and Differential Equations

Like most people, you probably have (or soon will have) income that comes from two sources: salary and investments (such as interest on savings). From this income, you pay "necessary" expenses (housing and food), spend some of the remainder on "luxuries" (vacations and entertainment), and save the rest (increasing your investments). Given some simplifying assumptions, we can construct a mathematical model for your wealth at any time $t$. Suppose that after paying for necessities, you spend a fixed percentage (typically 75%) of the remainder on luxuries, and that your investments generate income at a fixed rate. Let

$s$ = your salary

$W(t)$ = your wealth (investments), which is a function of time $t$

$r$ = rate of interest on your wealth (savings)

$n$ = amount spent on necessities

$p$ = proportion of your income after necessities that you spend on luxuries

A model of your wealth is then given by the following "differential equation," which states that *income* minus *outflow* is what you can add to savings:

$$\underset{\substack{\text{Change} \\ \text{in wealth}}}{\frac{dW}{dt}} = \underset{\text{Salary}}{s} + \underset{\substack{\text{Interest} \\ \text{on wealth}}}{\overbrace{r \cdot W}} - (\underset{\text{Necessities}}{n} + \underset{\substack{\text{Luxuries (a fixed} \\ \text{proportion of income} \\ \text{after necessities)}}}{\underbrace{p(s + r \cdot W - n)}})$$

Income        Outflow

With a little algebra, this "differential equation" can be written

$$\frac{dW}{dt} = (1 - p)(s - n + r \cdot W)$$

The *solution* to this differential equation (assuming zero initial wealth) can be found by the methods of this chapter, giving your "wealth" function shown below and graphed on the left:

$$W(t) = \frac{s - n}{r}(e^{(1-p)rt} - 1)$$

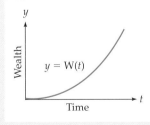

Given numbers for salary, necessities, interest rate, and proportion spent on luxuries, this solution could be used to predict when your wealth will reach a particular value (for example, enough to buy a house). It could also be used to show how your wealth would change if you lowered your "luxury proportion" from, say, 75% to 70% (see Exercise 71 on page 642).

This is but one illustration of the wide variety of applications of differential equations, which are discussed in this and the following sections.

## 9.1  SEPARATION OF VARIABLES

### Introduction

A **differential equation** is simply an equation involving derivatives. Practically any relationships involving rates of change can be expressed using differential equations, and in this chapter we will use them to model the growth of populations, the spread of epidemics, and the reduction of air pollution, to mention only a few applications. We will solve differential equations by the techniques of **separation of variables** and **integrating factors,** and for differential equations not solvable by either approach we will use **Euler's method** to obtain *approximate* solutions.

We will take $y$ to be a function of $x$, sometimes writing it as $y(x)$. We will write the derivative of $y$ as either $y'$ or $\dfrac{dy}{dx}$.

### Differential Equation   $y' = f(x)$

We have actually been solving differential equations since the beginning of Chapter 5. For example, the differential equation

$$y' = 2x$$

saying that the derivative of a function is $2x$, is solved simply by integrating:

$$y = \int 2x \, dx = x^2 + C \qquad\qquad y = x^2 + C \;\; \text{satisfies} \;\; y' = 2x$$

In general, a differential equation of the form

$$y' = f(x)$$

is solved by integrating:

$$y = \int f(x) \, dx$$

Therefore, whenever we integrate a marginal cost function to find cost, or integrate a rate to find the total accumulation, we are solving a differential equation of the form   $y' = f(x)$.

### General and Particular Solutions

The solution of the differential equation   $y' = 2x$   is   $y = x^2 + C,$   with an arbitrary constant C. We call   $y = x^2 + C$   the **general solution** because taking all possible values of the constant C gives *all* solutions of the differential equation. If we take C to be a particular number, we get a **particular solution.** Some particular solutions of the differential equation $y' = 2x$   are

$$y = x^2 + 2 \qquad \text{(taking } C = 2)$$
$$y = x^2 \qquad \text{(taking } C = 0)$$
$$y = x^2 - 2 \qquad \text{(taking } C = -2)$$

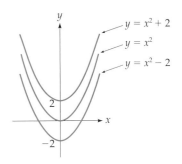

Three particular solutions from the general solution $y = x^2 + C$.

The different values of the arbitrary constant $C$ give a "family" of curves, and the general solution $y = x^2 + C$ may be thought of as the entire family.

The solution of a differential equation is a *function*. The *general* solution contains an arbitrary constant, and a *particular* solution has this constant replaced by a particular number.

### Verifying Solutions

Verifying that a function is a solution of a differential equation is simply a matter of calculating the necessary derivatives, substituting them into the differential equation, and checking that the two sides of the equation are equal.

**EXAMPLE 1**

**VERIFYING A SOLUTION OF A DIFFERENTIAL EQUATION**

Verify that $y = e^{2x} + e^{-x} - 1$ is a solution of the differential equation

$$y'' - y' - 2y = 2$$

*Solution* The differential equation involves $y$, $y'$, and $y''$, so we calculate

$$y = e^{2x} + e^{-x} - 1 \qquad \text{Given function}$$
$$y' = 2e^{2x} - e^{-x} \qquad \text{Derivative}$$
$$y'' = 4e^{2x} + e^{-x} \qquad \text{Second derivative}$$

Then we substitute these into the differential equation.

$$y'' - y' - 2y = 2$$

$$(4e^{2x} + e^{-x}) - (2e^{2x} - e^{-x}) - 2(e^{2x} + e^{-x} - 1) \stackrel{?}{=} 2 \qquad \stackrel{?}{=} \text{means the equation may not be true}$$

$$4e^{2x} + e^{-x} - 2e^{2x} + e^{-x} - 2e^{2x} - 2e^{-x} + 2 \stackrel{?}{=} 2 \qquad \text{Expanding}$$

$$4e^{2x} + e^{-x} - 2e^{2x} + e^{-x} - 2e^{2x} - 2e^{-x} + 2 \stackrel{?}{=} 2 \qquad \text{Canceling}$$

$$2 = 2 \quad \checkmark \qquad \text{It checks!}$$

Since the equation is satisfied, the given function is indeed a solution of the differential equation.

If the two sides had *not* turned out to be exactly the same, the given function $y$ would *not* have been a solution of the differential equation.

**Practice Problem 1**

Verify that $y = e^{-x} + e^{3x}$ is a solution of the differential equation

$$y'' - 2y' - 3y = 0$$

➤ **Solution on page 638**

The differential equations in Example 1 and Practice Problem 1 are called *second-order* differential equations because they involve second derivatives but no higher-order derivatives. From here on we will restrict our attention to **first-order** differential equations—that is, differential equations involving only the *first* derivative. Many first-order differential equations can be solved by a method called **separation of variables.**

## Separation of Variables

A differential equation is said to be **separable** if the variables can be "separated" by moving every $x$ and $dx$ to one side of the equation and every $y$ and $dy$ to the other side. We may then solve the differential equation by integrating both sides. Several examples will make the method clear.

**EXAMPLE 2**

**FINDING A GENERAL SOLUTION**

Find the general solution of the differential equation $\dfrac{dy}{dx} = 2xy^2$.

*Solution*

| | |
|---|---|
| $dy = 2xy^2\,dx$ | Multiplying both sides of the differential equation by $dx$ |
| $\dfrac{dy}{y^2} = 2x\,dx$ | Dividing each side by $y^2$ $(y \neq 0)$; the variables are now separated: every $y$ on one side, every $x$ on the other |
| $y^{-2}\,dy = 2x\,dx$ | In power form |
| $\displaystyle\int y^{-2}\,dy = \int 2x\,dx$ | Integrating both sides |
| $-y^{-1} = x^2 + C$ | Using the Power Rule (writing one $C$ for both integrations) |
| $\dfrac{1}{y} = -x^2 - C$ | Writing $y^{-1}$ as $\dfrac{1}{y}$ and multiplying by $-1$ |
| $y = \dfrac{1}{-x^2 - C}$ | Taking reciprocals of both sides |

This is the general solution of the differential equation. The solution may be left in this form, but if we replace the arbitrary constant $C$ by $-c$, another constant but with the opposite sign, this solution may be written

$$y = \frac{1}{-x^2 + c}$$

or, slightly shorter,

$$y = \frac{1}{c - x^2}$$

Reversing the order, giving the general solution of the differential equation

The differential equation $y' = 2xy^2$ has another solution: the function that is identically zero, $y(x) \equiv 0$. This is known as a **singular solution** and cannot be obtained from the general solution. We will not consider singular solutions further in this book except to say that $y = 0$ will generally be a solution whenever separating the variables results in $dy$ being divided by a positive power of $y$, as in the previous example.

### Separable Differential Equations

A first-order differential equation is *separable* if it can be written in the following form for some functions $f(x)$ and $g(y)$:

$$\frac{dy}{dx} = \frac{f(x)}{g(y)} \qquad \text{for } g(y) \neq 0$$

It is solved by separating variables and integrating:

$$\int g(y)\, dy = \int f(x)\, dx$$

Multiplying by $dx$ and $g(y)$ and integrating

The preceding example asked for the *general* solution of a differential equation. Sometimes we will be given a differential equation together with some additional information that selects a *particular* solution from the general solution, information that determines the value of the arbitrary constant in the general solution. This additional information is called an **initial condition.**

**EXAMPLE 3**

**FINDING A PARTICULAR SOLUTION**

Solve the differential equation $y' = \dfrac{6x}{y^2}$ with the initial condition $y(1) = 2$.

*Solution*

First we find the general solution by separating the variables.

$$\frac{dy}{dx} = \frac{6x}{y^2}$$

Replacing $y'$ by $\dfrac{dy}{dx}$

$$y^2\, dy = 6x\, dx$$

Multiplying both sides by $dx$ and $y^2$ (the variables are separated)

$$\int y^2 \, dy = \int 6x \, dx$$  Integrating both sides

$$\frac{1}{3} y^3 = 3x^2 + C$$  Using the Power Rule

$$y^3 = 9x^2 + \underbrace{3C}_{c}$$  Multiplying by 3

(3 times a constant is just another constant)

$$y^3 = 9x^2 + c$$  Replacing 3C by c

$$y = \sqrt[3]{9x^2 + c}$$  Taking cube roots

This is the general solution, with arbitrary constant $c$. The initial condition $y(1) = 2$ says that $y = 2$ when $x = 1$. We substitute these values into the general solution $y = \sqrt[3]{9x^2 + c}$ and solve for $c$.

$$2 = \sqrt[3]{9 + c}$$  $y = \sqrt[3]{9x^2 + c}$ with $x = 1$ and $y = 2$

$$8 = 9 + c$$  Cubing each side

$$-1 = c$$  Solving for c gives $c = -1$

Therefore, we replace $c$ by $-1$ in the general solution to obtain the particular solution:

$$y = \sqrt[3]{9x^2 - 1}$$  $y = \sqrt[3]{9x^2 + c}$ with $c = -1$

This solution $y = \sqrt[3]{9x^2 - 1}$ with $x = 1$ gives $y = \sqrt[3]{9 - 1} = \sqrt[3]{8} = 2$, so the initial condition $y(1) = 2$ is indeed satisfied. We could also verify that the differential equation is satisfied by substituting this solution into it.

In general, solving a differential equation with an initial condition just means finding the general solution and then using the initial condition to evaluate the constant. Several particular solutions of the differential equation $y' = 6x/y^2$ are shown below, with the solution $y = \sqrt[3]{9x^2 - 1}$ shown in color.

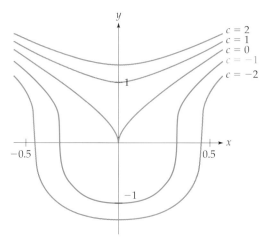

The solution $y = \sqrt[3]{9x^2 + c}$ for various values of $c$.

## Graphing Calculator Exploration

Slope field for $\dfrac{dy}{dx} = \dfrac{6x}{y^2}$ on $[-0.6, 0.6]$ by $[-1.5, 1.5]$

The same slope field with the solution $y = \sqrt[3]{9x^2 - 1}$

A graphing calculator can help you to *see* what a differential equation is saying. The differential equation in the preceding example,

$$\frac{dy}{dx} = \frac{6x}{y^2} \qquad\qquad \text{Slope equals } \frac{6x}{y^2}$$

says that the slope at any point $(x, y)$ in the plane is given by the formula $6x/y^2$. The graphing calculator program* SLOPEFLD draws a **slope field** consisting of many small line segments with the slopes specified by the differential equation. SLOPEFLD can also graph a particular solution. The slope field of the above differential equation is shown on the left top, and below it is the same slope field with the particular solution found in the preceding example.

Look back at the graph of the five solutions on the previous page and see how each of them follows the prescribed slopes in the above slope field. In fact, starting at *any* point in the plane you could draw a curve following the indicated slopes, thereby geometrically constructing a *solution curve*.

*SLOPEFLD can draw the slope field of any differential equation of the form $y' = f(x, y)$. To obtain this and other programs, go to www.cengage.com/math/berresford.

**Practice Problem 2**

Solve the differential equation and initial condition

$$\begin{cases} y' = \dfrac{6x^2}{y^4} \\ y(0) = 2 \end{cases}$$

➤ **Solution on page 639**

Recall that to solve for $y$ in the logarithmic equation

$$\ln y = f(x)$$

we exponentiate both sides and simplify

$$y = e^{f(x)} \qquad\qquad \text{Using } e^{\ln y} = y \text{ on the left side}$$

This idea will be useful in the next example.

**EXAMPLE 4**

### FINDING A PARTICULAR SOLUTION

Solve the differential equation and initial condition $\begin{cases} \dfrac{dy}{dx} = xy \\ y(0) = 2 \end{cases}$

*Solution*

$$dy = xy \, dx \qquad\qquad \text{Multiplying by } dx$$

$$\frac{dy}{y} = x \, dx$$

Dividing by $y$ ($y \neq 0$).
The variables are separated

$$\int \frac{dy}{y} = \int x \, dx$$

Integrating

$$\ln y = \frac{1}{2}x^2 + C$$

Evaluating the integrals
(assuming $y > 0$)

$$y = e^{\frac{1}{2}x^2 + C}$$

Solving for $y$ by
exponentiating

$$y = e^{\frac{1}{2}x^2} \cdot \underbrace{e^C}_{c}$$

$e$ to a sum can be
expressed as a product

$e^C$ is a positive constant $c$

$$y = ce^{x^2/2}$$

Replacing $e^C$ by $c$ and
$e^{\frac{1}{2}x^2}$ by $e^{x^2/2}$

This is the general solution. To satisfy the initial condition $y(0) = 2$, we substitute $y = 2$ and $x = 0$ and solve for $c$.

$$2 = ce^0 \qquad\qquad y = ce^{x^2/2} \text{ with } y = 2 \text{ and } x = 0$$

$$2 = c \qquad\qquad\quad \text{Since } e^0 = 1$$

Substituting $c = 2$ into the general solution gives the particular solution

$$y = 2e^{x^2/2} \qquad\qquad y = ce^{x^2/2} \text{ with } c = 2$$

We wrote the solution of the integral $\int \frac{dy}{y}$ as $\ln y$, without absolute value bars, thereby assuming that $y > 0$. *Keeping* the absolute values would have given $|y| = e^{\frac{1}{2}x^2 + C} = ce^{\frac{1}{2}x^2}$ so $y = \pm ce^{\frac{1}{2}x^2}$ where $c$ is a positive constant, showing that the general solution is $y = ce^{x^2/2}$, just as before, but now where $c$ is *any* (positive or negative) constant. We will make similar simplifying assumptions in similar situations.

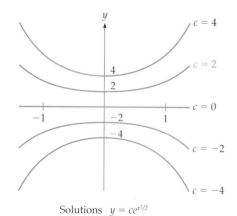

Solutions $y = ce^{x^2/2}$
for various values of $c$.

Note that the constant that was *added* in the integration step became a constant *multiplier* in the solution $y = ce^{x^2/2}$. In general, the constant may appear *anywhere* in the solution. The diagram on the left shows the solutions $y = ce^{x^2/2}$ for various values of $c$, with the particular solution $y = 2e^{x^2/2}$ shown in color.

## Graphing Calculator Exploration

The slope field for the differential equation in the preceding example, found by entering $y_1 = xy$, setting the window to $[-2, 2]$ by $[-6, 6]$, and running SLOPEFLD, is shown on the left below.

Slope field for $\dfrac{dy}{dx} = xy$

With the solution $y = 2e^{x^2/2}$

The screen on the right shows the same slope field with the particular solution found in the example. Do you see how the other solutions graphed on the previous page follow these slopes?

**Practice Problem 3**

Solve the differential equation and initial condition

$$\begin{cases} \dfrac{dy}{dx} = x^2 y \\ y(0) = 5 \end{cases}$$

➤ **Solution on page 639**

**EXAMPLE 5**

### FINDING A GENERAL SOLUTION

Find the general solution of the differential equation

$$yy' - x = 0$$

*Solution*

$$y \frac{dy}{dx} = x$$

Replacing $y'$ by $dy/dx$ and moving the $x$ to the other side

$$y \, dy = x \, dx$$

Multiplying by $dx$

$$\int y \, dy = \int x \, dx$$

Integrating

$$\frac{1}{2} y^2 = \frac{1}{2} x^2 + C$$

Using the Power Rule

$$y^2 = x^2 + \underbrace{2C}_{c}$$

Multiplying by 2

To solve for $y$ we take the square root of each side. However, there are *two* square roots, one positive and one negative.

$$y = \sqrt{x^2 + c} \quad \text{and} \quad y = -\sqrt{x^2 + c}$$

These two solutions together are the general solution:

$$y = \pm \sqrt{x^2 + c}$$

For some differential equations, the integration step requires a substitution.

**EXAMPLE 6**

**FINDING A PARTICULAR SOLUTION USING A SUBSTITUTION**

Solve the differential equation and initial condition $\begin{cases} y' = xy - x \\ y(0) = 4 \end{cases}$

*Solution*

| | |
|---|---|
| $\dfrac{dy}{dx} = xy - x$ | Replacing $y'$ by $\dfrac{dy}{dx}$ |
| $\dfrac{dy}{dx} = x(y - 1)$ | Factoring (to separate variables) |
| $\dfrac{dy}{y - 1} = x\,dx$ | Dividing by $y - 1$ and multiplying by $dx$ |
| $\displaystyle\int \dfrac{dy}{y - 1} = \int x\,dx$ | Integrating |
| $\begin{bmatrix} u = y - 1 \\ du = dy \end{bmatrix}$ | Using a substitution |
| $\displaystyle\int \dfrac{du}{u} = \int x\,dx$ | Substituting |
| $\ln u = \dfrac{1}{2}x^2 + C$ | Integrating (assuming $u > 0$) |
| $\ln(y - 1) = \dfrac{1}{2}x^2 + C$ | Substituting back to $y$ using $u = y - 1$ |
| $y - 1 = e^{\frac{1}{2}x^2 + C}$ | Solving for $y - 1$ by exponentiating |
| $y - 1 = e^{\frac{1}{2}x^2} \cdot \underbrace{e^C}_{c} = c \cdot e^{x^2/2}$ | Replacing $e^C$ by $c$ and $e^{\frac{1}{2}x^2}$ by $e^{x^2/2}$ |
| $y = ce^{x^2/2} + 1$ | Adding 1 to each side |

This is the general solution. To satisfy the initial condition $y(0) = 4$, we substitute $y = 4$ and $x = 0$.

$$4 = ce^0 + 1$$

$y = ce^{x^2/2} + 1$ with $y = 4$ and $x = 0$

$$4 = c + 1$$

Since $e^0 = 1$

This gives $c = 3$. Therefore, the particular solution is

$$y = 3e^{x^2/2} + 1$$

$y = ce^{x^2/2} + 1$ with $c = 3$

## Accumulation of Wealth

The examples so far have *given* us a differential equation to solve. In this application we will first *derive* a differential equation to represent a situation and then solve it.

 **EXAMPLE 7**

### PREDICTING WEALTH

Suppose that you have saved \$5000, and that you expect to save an additional \$3000 during each year. If you deposit these savings in a bank account paying 5% interest compounded continuously, find a formula for your bank balance after $t$ years.

*Solution*   Let $y(t)$ stand for your bank balance (in thousands of dollars) after $t$ years. Each year $y(t)$ grows by 3 (thousand dollars) plus 5% interest. This growth can be modeled by a differential equation:

$$y' = 3 + 0.05y$$

$y$ increases by 3 plus 5% of itself

Rate of change of $y$    is   3   plus    5% of $y$

Before continuing, be sure that you understand how this differential equation models the changes due to savings and interest.
    We solve it by separating variables.

$$\frac{dy}{dt} = 3 + 0.05y$$

Replacing $y'$ by $\frac{dy}{dt}$

$$\int \frac{dy}{3 + 0.05y} = \int dt$$

Dividing by $3 + 0.05y$, multiplying by $dt$, and then integrating

$$\left[ \begin{array}{l} u = 3 + 0.05y \\ du = 0.05dy \end{array} \right]$$

Using a substitution

$$20 \int \frac{du}{u} = \int dt$$

Substituting (the 20 comes from $\frac{1}{0.05} = 20$)

$$20 \ln u = t + C$$

Integrating (assuming $u > 0$)

$$\ln(3 + 0.05y) = 0.05t + \underbrace{0.05C}_{c}$$ 
Substituting $u = 3 + 0.05y$ and dividing by 20

$$\ln(3 + 0.05y) = 0.05t + c$$ 
Replacing $0.05C$ by $c$

$$3 + 0.05y = e^{0.05t + c} = e^{0.05t} \underbrace{e^c}_{k} = ke^{0.05t}$$ 
Exponentiating and then simplifying constants

$$0.05y = ke^{0.05t} - 3$$ 
Subtracting 3

$$y = \underbrace{20k}_{b}e^{0.05t} - 60$$ 
Dividing by 0.05 (Always simplify arbitrary constants)

$$y = be^{0.05t} - 60$$ 
With arbitrary constant $b$

You began at time $t = 0$ with 5 thousand dollars, which gives the initial condition $y(0) = 5$. We substitute $y = 5$ and $t = 0$.

$$5 = be^0 - 60$$ 
$y = be^{0.05t} - 60$ with $y = 5$ and $t = 0$

$$5 = b - 60$$ 
Since $e^0 = 1$

Therefore, $b = 65$, which we substitute into the general solution, obtaining a formula for your accumulated wealth after $t$ years.

$$y = 65e^{0.05t} - 60$$ 
$y = be^{0.05t} - 60$ with $b = 65$

$y = 65e^{0.05t} - 60$

For example, to find your wealth after 10 years, we evaluate the solution of the differential equation at $t = 10$.

$$y = 65e^{0.5} - 60 \approx 47.167$$ 
$y = 65e^{0.05t} - 60$ with $t = 10$

Therefore, after 10 years, you will have \$47,167 in the bank.* This wealth function is graphed on the left.

### Solutions to Practice Problems

**1.** $y = e^{-x} + e^{3x}$

$y' = -e^{-x} + 3e^{3x}$

$y'' = e^{-x} + 9e^{3x}$

Substituting these expressions into the differential equation:

$$(e^{-x} + 9e^{3x}) - 2(-e^{-x} + 3e^{3x}) - 3(e^{-x} + e^{3x}) \stackrel{?}{=} 0$$

$$e^{-x} + 9e^{3x} + 2e^{-x} - 6e^{3x} - 3e^{-x} - 3e^{3x} \stackrel{?}{=} 0$$

$$0 = 0 \checkmark \quad \text{It checks}$$

*How would this amount change if we were to replace *continuous* compounding with the more common *daily* compounding? At 5% interest with equal monthly deposits (in the middle of each month) adding up to the stated \$3000 per year, the total after 10 years would be \$47,166, differing from our figure by just \$1 after 10 years.

**2.**  $\dfrac{dy}{dx} = \dfrac{6x^2}{y^4}$

$y^4\,dy = 6x^2\,dx$

$\displaystyle\int y^4\,dy = \int 6x^2\,dx$

$\tfrac{1}{5}y^5 = 2x^3 + C$

$y^5 = 10x^3 + 5C = 10x^3 + c$

$y = \sqrt[5]{10x^3 + c}$

The initial condition gives

$2 = \sqrt[5]{0 + c}$

$2 = \sqrt[5]{c}$

$32 = c$     Raising each side to the fifth power

*Solution:* $y = \sqrt[5]{10x^3 + 32}$

**3.**  $\dfrac{dy}{dx} = x^2 y$

$\displaystyle\int \dfrac{dy}{y} = \int x^2\,dx$

$\ln y = \tfrac{1}{3}x^3 + C$

$y = e^{\frac{1}{3}x^3 + C} = e^{\frac{1}{3}x^3}\,e^C = ce^{x^3/3}$

The initial condition gives

$5 = ce^0 = c$

*Solution:* $y = 5e^{x^3/3}$

---

## 9.1 Section Summary

We solved separable differential equations by separating the variables (moving every $y$ to one side and every $x$ to the other, with $dx$ and $dy$ in the numerators) and integrating both sides. The *general* solution of a differential equation includes an arbitrary constant, while a *particular* solution results from evaluating the constant, usually by applying an initial condition. The *slope field* of a differential equation allows you to construct geometrically a solution from any point by following the slopes away from the point.

We also saw how to *derive* a differential equation from information about how a quantity changes. For example, we can "read" the differential equation as follows.

The applied exercises show how differential equations lead to important formulas in a wide variety of fields.

## 9.1 Exercises

**1–4.** Verify that the function $y$ satisfies the given differential equation.

**1.** $y = e^{2x} - 3e^x + 2$       **2.** $y = e^{5x} - 4e^x + 1$
     $y'' - 3y' + 2y = 4$        $y'' - 6y' + 5y = 5$

**3.** $y = ke^{ax} - \dfrac{b}{a}$   (for constants $a$, $b$, and $k$)
     $y' = ay + b$

**4.** $y = ax^2 + bx$   (for constants $a$ and $b$)
     $y' = \dfrac{y}{x} + ax$

**5–36.** Find the general solution of each differential equation or state that the differential equation is not separable. If the exercise says "and check," verify that your answer is a solution.

**5.** $y^2 y' = 4x$          **6.** $y^4 y' = 8x$

**7.** $y' = x + y$          **8.** $y' = xy - 1$

**9.** $y' = 6x^2 y$     and check

**10.** $y' = 12x^3 y$     and check

**11.** $y' = \dfrac{y}{x}$     and check

**12.** $y' = \dfrac{y^2}{x^2}$     and check

**13.** $yy' = 4x$          **14.** $yy' = 6x^2$

**15.** $y' = e^{xy}$          **16.** $y' = e^x + y$

**17.** $y' = 9x^2$          **18.** $y' = 6e^{-2x}$

**19.** $y' = \dfrac{x}{x^2 + 1}$          **20.** $y' = xy^2$

**21.** $y' = x^2 y$          **22.** $y' = \dfrac{x}{y}$

**23.** $y' = x^m y^n$    (for $m > 0$, $n \neq 1$)

**24.** $y' = x^m y$    (for $m > 0$)

**25.** $y' = 2\sqrt{y}$          **26.** $y' = 5 + y$

**27.** $xy' = x^2 + y^2$       **28.** $y' = \sqrt{x + y}$

**29.** $y' = xy + x$         **30.** $y' = x - 2xy$

**31.** $y' = ye^x - e^x$       **32.** $y' = ye^x - y$

**33.** $y' = ay^2$    (for constant $a > 0$)

**34.** $y' = axy$    (for constant $a$)

**35.** $y' = ay + b$    (for constants $a$ and $b$)

**36.** $y' = (ay + b)^2$    (for constants $a \neq 0$ and $b$)

**37–50.** Solve each differential equation and initial condition and verify that your answer satisfies both the differential equation and the initial condition.

**37.** $\begin{cases} y^2 y' = 2x \\ y(0) = 2 \end{cases}$    **38.** $\begin{cases} y^4 y' = 3x^2 \\ y(0) = 1 \end{cases}$

**39.** $\begin{cases} y' = xy \\ y(0) = -1 \end{cases}$    **40.** $\begin{cases} y' = y^2 \\ y(2) = -1 \end{cases}$

**41.** $\begin{cases} y' = 2xy^2 \\ y(0) = 1 \end{cases}$    **42.** $\begin{cases} y' = 2xy^4 \\ y(0) = 1 \end{cases}$

**43.** $\begin{cases} y' = \dfrac{y}{x} \\ y(1) = 3 \end{cases}$    **44.** $\begin{cases} y' = \dfrac{2y}{x} \\ y(1) = 2 \end{cases}$

**45.** $\begin{cases} y' = 2\sqrt{y} \\ y(1) = 4 \end{cases}$    **46.** $\begin{cases} y' = \sqrt{y}\, e^x - \sqrt{y} \\ y(0) = 1 \end{cases}$

**47.** $\begin{cases} y' = y^2 e^x + y^2 \\ y(0) = 1 \end{cases}$    **48.** $\begin{cases} y' = xy - 5x \\ y(0) = 4 \end{cases}$

**49.** $\begin{cases} y' = ax^2 y \\ y(0) = 2 \end{cases}$    (for constant $a > 0$)

**50.** $\begin{cases} y' = axy \\ y(0) = 4 \end{cases}$    (for constant $a > 0$)

### Applied Exercises

**51. BUSINESS: Elasticity** For a demand function $D(p)$, the elasticity of demand (see page 311) is defined as $E = \dfrac{-pD'}{D}$. Find demand functions $D(p)$ that have *constant elasticity* by solving the differential equation $\dfrac{-pD'}{D} = k$, where $k$ is a constant.

**52. BIOMEDICAL: Cell Growth** A cell receives nutrients through its surface, and its surface area is proportional to the two-thirds power of its weight. Therefore, if $w(t)$ is the cell's weight at time $t$, then $w(t)$ satisfies $w' = aw^{2/3}$, where $a$ is a positive constant. Solve this differential equation with the initial condition $w(0) = 1$ (initial weight 1 unit).
*Source: J. M. Smith, Mathematical Ideas in Biology*

**53–54. BUSINESS: Continuous Annuities** An annuity is a fund into which one makes equal payments at regular intervals. If the fund earns interest at rate $r$ compounded continuously, and deposits are made continuously at the rate of $d$ dollars per year (a "continuous annuity"), then the value $y(t)$ of the fund after $t$ years satisfies the differential equation $y' = d + ry$. (Do you see why?)

*Source:* S. Broverman, *Mathematics of Investment and Credit*

**53.** Solve the differential equation above for the continuous annuity $y(t)$ with deposit rate $d = \$1000$ and continuous interest rate $r = 0.05$, subject to the initial condition $y(0) = 0$ (zero initial value).

**54.** Solve the differential equation above for the continuous annuity $y(t)$, where $d$ and $r$ are unknown constants, subject to the initial condition $y(0) = 0$ (zero initial value).

**55. GENERAL: Time of a Murder** A medical examiner called to the scene of a murder will usually take the temperature of the body. A corpse cools at a rate proportional to the difference between its temperature and the temperature of the room. If $y(t)$ is the temperature (in degrees Fahrenheit) of the body $t$ hours after the murder, and if the room temperature is 70°, then $y$ satisfies

$$y' = -0.32(y - 70)$$

$y(0) = 98.6$ (body temperature initially 98.6°)

**a.** Solve this differential equation and initial condition.
**b.** Use your answer to part (a) to estimate how long ago the murder took place if the temperature of the body when it was discovered was 80°. [*Hint:* Find the value of $t$ that makes your solution equal 80°.]

*Source:* D. Lomen and D. Lovelock, *Exploring Differential Equations via Graphics and Data*

**56. BIOMEDICAL: Glucose Levels** Hospital patients are often given glucose (blood sugar) through a tube connected to a bottle suspended over their beds. Suppose that this "drip" supplies glucose at the rate of 25 mg per minute, and each minute 10% of the accumulated glucose is consumed by the body. Then the amount $y(t)$ of glucose (in excess of the normal level) in the body after $t$ minutes satisfies

$$y' = 25 - 0.1y \quad \text{(Do you see why?)}$$

$y(0) = 0$ (zero excess glucose at $t = 0$)

Solve this differential equation and initial condition.

**57. GENERAL: Friendships** Suppose that you meet 30 new people each year, but each year you forget 20% of all of the people that you know. If $y(t)$ is the total number of people who you remember after $t$ years, then $y$ satisfies the differential equation $y' = 30 - 0.2y$. (Do you see why?) Solve this differential equation subject to the condition $y(0) = 0$ (you knew no one at birth).

**58. ENVIRONMENTAL SCIENCE: Pollution** For more than 75 years the Flexfast Rubber Company in Massachusetts discharged toxic toluene solvents into the ground at a rate of 5 tons per year. Each year approximately 10% of the accumulated pollutants evaporated into the air. If $y(t)$ is the total accumulation of pollution in the ground after $t$ years, then $y$ satisfies

$$y' = 5 - 0.1y \quad \text{(Do you see why?)}$$

$y(0) = 0$ (initial accumulation zero)

Solve this differential equation and initial condition to find a formula for the accumulated pollutant after $t$ years.

*Source:* Greater Boston Legal Services

**59. PERSONAL FINANCE: Wealth Accumulation** Suppose that you now have \$6000, you expect to save an additional \$3000 during each year, and all of this is deposited in a bank paying 10% interest compounded continuously. Let $y(t)$ be your bank balance (in thousands of dollars) $t$ years from now.

**a.** Write a differential equation that expresses the fact that your balance will grow by 3 (thousand dollars) and also by 10% of itself. [*Hint:* See Example 7.]
**b.** Write an initial condition to say that at time zero the balance is 6 (thousand dollars).
**c.** Solve your differential equation and initial condition.
**d.** Use your solution to find your bank balance $t = 25$ years from now.

**60. PERSONAL FINANCE: Wealth Accumulation** Suppose that you now have \$2000, you expect to save an additional \$6000 during each year, and all of this is deposited in a bank paying 4% interest compounded continuously. Let $y(t)$ be your bank balance (in thousands of dollars) $t$ years from now.

**a.** Write a differential equation that expresses the fact that your balance will grow by 6 (thousand dollars) and also by 4% of itself. [*Hint:* See Example 7.]

*(continues)*

**b.** Write an initial condition to say that at time zero the balance is 2 (thousand dollars).

**c.** Solve your differential equation and initial condition.

**d.** Use your solution to find your bank balance $t = 20$ years from now.

**61. BUSINESS: Sales** Your company has developed a new product, and your marketing department has predicted how it will sell. Let $y(t)$ be the (monthly) sales of the product after $t$ months.

**a.** Write a differential equation that says that the rate of growth of the sales will be four times the one-half power of the sales.

**b.** Write an initial condition that says that at time $t = 0$ sales were 10,000.

**c.** Solve this differential equation and initial value.

**d.** Use your solution to predict the sales at time $t = 12$ months.

**62. BUSINESS: Sales** Your company has developed a new product, and your marketing department has predicted how it will sell. Let $y(t)$ be the (monthly) sales of the product after $t$ months.

**a.** Write a differential equation that says that the rate of growth of the sales will be six times the two-thirds power of the sales.

**b.** Write an initial condition that says that at time $t = 0$ sales were 1000.

**c.** Solve this differential equation and initial value.

**d.** Use your solution to predict the sales at time $t = 12$ months.

**63. BIOMEDICAL: Bacterial Colony** Let $y(t)$ be the size of a colony of bacteria after $t$ hours.

**a.** Write a differential equation that says that the rate of growth of the colony is equal to eight times the three-fourths power of its present size.

**b.** Write an initial condition that says that at time zero the colony is of size 10,000.

**c.** Solve the differential equation and initial condition.

**d.** Use your solution to find the size of the colony at time $t = 6$ hours.

**64. GENERAL: Value of a Building** Let $y(t)$ be the value of a commercial building (in millions of dollars) after $t$ years.

**a.** Write a differential equation that says that the rate of growth of the value of the building is equal to two times the one-half power of its present value.

**b.** Write an initial condition that says that at time zero the value of the building is 9 million dollars.

**c.** Solve the differential equation and initial condition.

**d.** Use your solution to find the value of the building at time $t = 5$ years.

 **65–67.** The following exercises require the use of a slope field program.

For each differential equation and initial condition:

**a.** Use SLOPEFLD or a similar program to graph the slope field for the differential equation on the window $[-5, 5]$ by $[-5, 5]$.

**b.** Sketch the slope field on a piece of paper and draw a solution curve that follows the slopes and that passes through the point $(0, 2)$.

**c.** Solve the differential equation and initial condition.

**d.** Use SLOPEFLD or a similar program to graph the slope field and the solution that you found in part (c). How good was the sketch that you made in part (b) compared with the solution graphed in part (d)?

**65.** $\begin{cases} \dfrac{dy}{dx} = \dfrac{6x^2}{y^4} \\ y(0) = 2 \end{cases}$   **66.** $\begin{cases} \dfrac{dy}{dx} = \dfrac{x^2}{y^2} \\ y(0) = 2 \end{cases}$   **67.** $\begin{cases} \dfrac{dy}{dx} = \dfrac{4x}{y^3} \\ y(0) = 2 \end{cases}$

 **68–70.** The following exercises require the use of a slope field program.

For each differential equation:

**a.** Use SLOPEFLD or a similar program to graph the slope field for the differential equation on the window $[-5, 5]$ by $[-5, 5]$.

**b.** Sketch the slope field on a piece of paper and draw a solution curve that follows the slopes and that passes through the given point.

**68.** $\dfrac{dy}{dx} = x - y^2$   **69.** $\dfrac{dy}{dx} = \dfrac{x}{y^2 + 1}$
point: $(0, 1)$     point: $(0, -1)$

**70.** $\dfrac{dy}{dx} = x \ln(y^2 + 1)$
point: $(0, -2)$

**71. PERSONAL FINANCE: Wealth Accumulation** Reread the Application Preview on page 627.

**a.** Make up some reasonable numbers of $s$ (annual salary) and $n$ (annual cost of necessities) and substitute them together with $r = 0.06$ and $p = 0.75$ into the formula for $W(t)$. Then evaluate $W(20)$ to predict your wealth in 20 years.

b. Change $p$ to 0.70 to see how your wealth $W(20)$ would change if you lived more frugally.

c. Give yourself a raise by increasing $s$ by 20% to see how that affects $W(20)$.

d. Suppose that you take out a mortgage to buy a house, thereby increasing $n$ by 15%. How does $W(20)$ change?

### Conceptual Exercises

**72–75.** The following are differential equations *stated in words*. Find the general solution of each.

**72.** The derivative of a function at each point is 6.

**73.** The derivative of a function at each point is $-2$.

**74.** The derivative of a function at each point is 0.

**75.** The derivative of a function at each point is itself.

**76–79.** Determine whether each differential equation is separable. (Do not solve it, just find whether it's separable.)

**76.** $y' = x + y$

**77.** $y' = \ln(xy)$

**78.** $y' = \ln(x + y)$

**79.** $y' = e^{x+y}$

**80.** Think of the slope field for the differential equation $\dfrac{dy}{dx} = \dfrac{6x}{y^2}$. What is the sign of the slope in quadrant I (where $x$ and $y$ are both positive)? What is the sign of the slope in each of the other three quadrants? Check your answers by looking at the slope field on page 633.

**81.** Think of the slope field for the differential equation $\dfrac{dy}{dx} = xy$. What is the sign of the slope in quadrant I (where $x$ and $y$ are both positive)? What is the sign of the slope in each of the other three quadrants? Check your answers by looking at the slope field on page 635.

### Explorations and Excursions

The following problems extend and augment the material presented in the text.

**82. BIOMEDICAL: Fick's Law** Fick's Law governs the diffusion of a solute across a cell membrane. According to Fick's Law, the concentration $y(t)$ of the solute inside the cell at time $t$ satisfies $\dfrac{dy}{dt} = \dfrac{kA}{V}(C_0 - y)$, where $k$ is the diffusion constant, $A$ is the area of the cell membrane, $V$ is the volume of the cell, and $C_0$ is the concentration outside the cell.

a. Find the general solution of this differential equation. (Your solution will involve the constants $k$, $A$, $V$, and $C_0$.)

b. Find the particular solution that satisfies the initial condition $y(0) = y_0$, where $y_0$ is the initial concentration inside the cell.

**83. BIOMEDICAL: Heart Function** In the *reservoir model*, the heart is viewed as a balloon that swells as it fills with blood (during a period called the *systole*), and then at time $t_0$ it shuts a valve and contracts to force the blood out (the *diastole*). Let $p(t)$ represent the pressure in the heart at time $t$.

a. During the diastole, which lasts from $t_0$ to time $T$, $p(t)$ satisfies the differential equation

$$\frac{dp}{dt} = -\frac{K}{R}p$$

Find the general solution $p(t)$ of this differential equation. ($K$ and $R$ are positive constants determined, respectively, by the strength of the heart and the resistance of the arteries. The differential equation states that as the heart contracts, the pressure decreases ($dp/dt$ is negative) in proportion to itself.)

b. Find the particular solution that satisfies the condition $p(t_0) = p_0$. ($p_0$ is a constant representing the pressure at the transition time $t_0$.)

c. During the systole, which lasts from time 0 to time $t_0$, the pressure $p(t)$ satisfies the differential equation

$$\frac{dp}{dt} = KI_0 - \frac{K}{R}p$$

Find the general solution of this differential equation. ($I_0$ is a positive constant representing the constant rate of blood flow into the heart while it is expanding.) [*Hint:* Use the same $u$-substitution technique that was used in Example 7.]

d. Find the particular solution that satisfies the condition $p(t_0) = p_0$.

e. In parts (b) and (d) you found the formulas for the pressure $p(t)$ during the diastole $(t_0 \le t \le T)$ and the systole $(0 \le t \le t_0)$. Since the heart behaves in a cyclic fashion, these functions must satisfy $p(T) = p(0)$. Equate the solutions at these times (use the correct formula for each time) to derive the important relationship

$$R = \frac{p_0}{I_0}\frac{1 - e^{-KT/R}}{1 - e^{-Kt_0/R}}$$

### Introduction

This section continues our study of differential equations, but with a different approach. Instead of solving individual differential equations, we will solve three important *classes* of differential equations (for **unlimited growth, limited growth,** and **logistic growth**) and remember their solutions. This will enable us to solve many problems by identifying the appropriate differential equation and then immediately writing the solution. In this section we begin *to think in terms of differential equations.*

### Proportional Quantities

The circumference of a circle is proportional to its diameter.

We say that one quantity is *proportional* to another quantity if the first quantity is a *constant multiple* of the second. That is, $y$ is proportional to $x$ if $y = ax$ for some "proportionality constant" $a$. For example, the formula $C = \pi D$ for the circumference of a circle shows that the circumference $C$ is proportional to the diameter $D$.

### Unlimited Growth

In many situations the growth of a quantity is proportional to its present size. For example, a population of cells will grow in proportion to its present size, and a bank account earns interest in proportion to its current value. If a quantity $y$ grows so that its rate of growth $y'$ is proportional to its present size $y$, then $y$ satisfies the differential equation

$$y' = ay$$

Rate of growth / is / propor-tional to / current size

We solve this differential equation by separating variables.

| | |
|---|---|
| $\dfrac{dy}{dt} = ay$ | Replacing $y'$ by $\dfrac{dy}{dt}$ |
| $\displaystyle\int \dfrac{dy}{y} = \int a\, dt$ | Dividing by $y$ ($y \neq 0$), multiplying by $dt$, and integrating |
| $\ln y = at + C$ | Integrating ($y > 0$) |
| $y = e^{at+C} = e^{at}e^C = ce^{at}$ | Solving for $y$ by exponentiating and then replacing $e^C$ by $c$ |

At time $t = 0$ this becomes

$$y(0) = ce^0 = c \qquad\qquad y(t) = ce^{at} \quad \text{with} \quad t = 0$$

Summarizing:

**Unlimited Growth**

The differential equation $\qquad y' = ay$

with initial condition $\qquad y(0) = c$

is solved by $\qquad y = ce^{at}$

Such growth, where the rate of growth is proportional to the present size, is called *unlimited growth* because the solution $y$ grows arbitrarily large. Given this result, whenever we encounter a differential equation of the form $y' = ay$, we can immediately write the solution $y = ce^{at}$, where $c$ is the initial value.

**EXAMPLE 1**  **PREDICTING ART APPRECIATION (UNLIMITED GROWTH)**

An art collection, initially worth \$20,000, continuously grows in value at the rate of 5% a year. Express this growth as a differential equation and find a formula for the value of the collection after $t$ years. Then estimate the value of the art collection after 10 years.

*Solution*

Growing continuously at the rate of 5% means that the value $y(t)$ grows by 5% of itself

$$y' \;=\; 0.05y$$

Rate of   is   5% of the
growth       current value

This differential equation is of the form $y' = ay$ (unlimited growth) with $a = 0.05$ and initial value 20,000, so we may immediately write its solution:

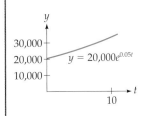

$$y(t) = 20{,}000e^{0.05t}$$

$y = ce^{at}$ with $a = 0.05$
and $c = 20{,}000$

This formula gives the value of the art collection after $t$ years. To find the value after 10 years, we evaluate at $t = 10$:

$$y(10) = 20{,}000e^{0.05 \cdot 10}$$

$$= 20{,}000e^{0.5} \approx 32{,}974$$

$y(t) = 20{,}000e^{0.05t}$ with $t = 10$

Using a calculator

In 10 years the art collection will be worth \$32,974.

The solution $ce^{at}$ is the same as the continuous compounding formula $Pe^{rt}$ (except for different letters). On page 269 we derived the formula $Pe^{rt}$ rather laboriously, using the discrete interest formula $P(1 + r/m)^{mt}$, replacing $m$ by $rn$, and taking the limit as $n \to \infty$. The present derivation, using differential equations, is much simpler and shows that $ce^{at}$ applies to *any* situation governed by the differential equation $y' = ay$.

## Limited Growth

No real population can undergo unlimited growth for very long. Restrictions of food and space would soon slow its growth. If a quantity $y(t)$ cannot grow larger than a certain fixed maximum size $M$, and if its growth rate $y'$ is proportional to how far it is from its upper limit, then $y$ satisfies

$$y' = a(M - y) \qquad\qquad a > 0$$

Rate of | is | propor- | distance below
growth | | tional to | upper bound $M$

We solve this by separating variables.

| | |
|---|---|
| $\dfrac{dy}{dt} = a(M - y)$ | Replacing $y'$ by $\dfrac{dy}{dt}$ |
| $\displaystyle\int \dfrac{dy}{M - y} = \int a\,dt$ | Dividing by $M - y$, multiplying by $dt$, and integrating |
| $\begin{bmatrix} u = M - y \\ du = -dy \end{bmatrix}$ | Using a substitution |
| $-\displaystyle\int \dfrac{du}{u} = \int a\,dt$ | Substituting |
| $-\ln u = at + C$ | Integrating $(u > 0)$ |
| $\ln(M - y) = -at - C$ | Multiplying by $-1$ and replacing $u$ by $(M - y)$ |
| $M - y = e^{-at-C} = e^{-at}e^{-C} = ce^{-at}$ | Solving for $M - y$ by exponentiating and simplifying |
| $y = M - ce^{-at}$ | Subtracting $M$ and multiplying by $-1$ |

We impose the initial condition $y(0) = 0$ (size zero at time $t = 0$).

$$0 = M - ce^0 = M - c \qquad\qquad y = M - ce^{-at} \text{ with } y = 0 \text{ and } t = 0$$

Therefore, $c = M$, which gives the solution

$$y = M - Me^{-at} = M(1 - e^{-at}) \qquad\qquad y = M - ce^{-at} \text{ with } c = M$$

Summarizing:

### Limited Growth

The differential equation

$$y' = a(M - y)$$

with initial condition

$$y(0) = 0$$

is solved by

$$y = M(1 - e^{-at})$$

This type of growth, in which the rate of growth is proportional to the distance below an upper limit $M$, is called *limited growth* because the solution $y = M(1 - e^{-at})$ asymptotically approaches the limit $M$ as $t \to \infty$. Given this result, whenever we encounter a differential equation of the form

$$y' = a(M - y)$$

with initial value zero, we may immediately write the solution

$$y = M(1 - e^{-at})$$

## Diffusion of Information by Mass Media

If a news bulletin is repeatedly broadcast over radio and television, the news spreads quickly at first, but later more slowly when most people have already heard it. Sociologists often assume that the rate at which news spreads is proportional to the number who have not yet heard the news. Let $M$ be the population of a city, and let $y(t)$ be the number of people who have heard the news within $t$ time units. Then $y$ satisfies the differential equation

$$y' \quad = \quad a(M - y)$$

Rate of growth    is    propor-tional to    number who have not heard the news

We recognize this as the differential equation for limited growth, whose solution is $y = M(1 - e^{-at})$. It remains only to determine the values of the constants $M$ and $a$.

---

**EXAMPLE 2**

### PREDICTING SPREAD OF INFORMATION (LIMITED GROWTH)

An important news bulletin is broadcast to a town of 50,000 people, and after 2 hours 30,000 people have heard the news. Find a formula for the number of people who have heard the bulletin within $t$ hours. Then find how many people will have heard the news within 6 hours.

*Solution*

If $y(t)$ is the number of people who have heard the news within $t$ hours, then $y$ satisfies

$$y' = a(50{,}000 - y) \qquad\qquad y' = a(M - y) \text{ with } M = 50{,}000$$

Number who have not heard the news

This is the differential equation for limited growth with $M = 50{,}000$, so the solution (from the preceding box) is

$$y = 50{,}000(1 - e^{-at}) \qquad\qquad y = M(1 - e^{-at}) \text{ with } M = 50{,}000$$

To find the value of the constant $a$, we use the given information that 30,000 people have heard the news within 2 hours.

$$30,000 = 50,000(1 - e^{-a \cdot 2})$$     $y = 50,000(1 - e^{-at})$ with $y = 30,000$ and $t = 2$

$$0.6 = 1 - e^{-2a}$$     Dividing each side by 50,000

$$0.4 = e^{-2a}$$     Subtracting 1 and then multiplying by $-1$

$$-0.916 \approx -2a$$     Taking natural logs (using $\ln e^x = x$ on the right)

$$a \approx 0.46$$     Dividing by $-2$

Therefore, the number of people who have heard the news within $t$ hours is

$$y(t) = 50,000(1 - e^{-0.46t})$$     $y = 50,000(1 - e^{-at})$ with $a = 0.46$

as shown in the graph. To find the number who have heard the news within 6 hours, we evaluate this solution at $t = 6$.

$$y(6) = 50,000(1 - e^{-0.46 \cdot 6}) = 50,000(1 - e^{-2.76}) \approx 46,835$$     Using a calculator

Within 6 hours about 46,800 people have heard the news.

*(Graph on left: axes labeled "People" (vertical) and "Hours" (horizontal, marked 1 2 3 4 5 6), curve labeled $y = 50,000(1 - e^{-0.46t})$, with 50,000 dashed line)*

We found the value of the constant $a$ by substituting the given data into the solution, simplifying, and taking logs. Similar steps will be required in many other problems, and in the future we shall omit the details.

## Learning Theory

Psychologists have found that there seems to be an upper limit to the number of meaningless words that a person can memorize, and that memorizing becomes increasingly difficult approaching that bound. If $M$ is this upper limit and $y(t)$ is the number of words that can be memorized in $t$ minutes, then the situation is modeled by the differential equation for limited growth.

$$y' = a(M - y)$$

Rate of increase / is / propor- tional to / upper limit $M$ minus number already memorized

This can be interpreted as saying that the rate at which new words can be memorized is proportional to the "unused memory capacity."

**EXAMPLE 3**

**PREDICTING MEMORIZATION (LIMITED GROWTH)**

Suppose that a person can memorize at most 100 meaningless words, and that after 15 minutes 10 words have been memorized. How long will it take to memorize 50 words?

*Solution*

The number $y(t)$ of words that can be memorized in $t$ minutes satisfies

$$y' = a(100 - y)$$     $y' = a(M - y)$ with $M = 100$

This is the differential equation for limited growth, so the solution is

$$y = 100(1 - e^{-at})$$

$y = M(1 - e^{-at})$
with   $M = 100$

To evaluate the constant $a$, we use the given information that 10 words have been memorized in 15 minutes.

$$10 = 100(1 - e^{-a \cdot 15})$$

$y = 100(1 - e^{-at})$  with
$y = 10$   and   $t = 15$

Solving this equation for the constant $a$ (omitting the details, which are the same as in Example 2) gives   $a = 0.007$.

$y = 100(1 - e^{-0.007t})$

$$y(t) = 100(1 - e^{-0.007t})$$

$y = 100(1 - e^{-at})$
with   $a = 0.007$

This solution (graphed on the left) gives the number of words that can be memorized in $t$ minutes. To find how long it takes to memorize 50 words, we set this solution equal to 50 and solve for $t$.

$$50 = 100(1 - e^{-0.007t})$$

$y = 100(1 - e^{-0.007t})$
with   $y = 50$

Words memorized

The slope (the rate at which additional words can be memorized) decreases near the upper limit.

Solving for $t$ (again the details are similar to those in Example 2) gives $t = 99$   minutes. Therefore, 50 words can be memorized in about 1 hour and 39 minutes.

## Logistic Growth

Some quantities grow in proportion to both their present size *and* their distance from an upper limit $M$.

$$y' = ay(M - y)$$

Rate of   is   propor-   present       upper limit $M$
growth        tional to   size       minus present size

This differential equation can be solved by separation of variables (see Exercise 65) to give the solution

$$y = \frac{M}{1 + ce^{-aMt}}$$

for constants   $a > 0$   and   $c > -1$.  This function is called the *logistic* function,* governing *logistic growth*.

*Or sometimes the *Verhulst model* or the *Beverton-Holt model*.

**Logistic Growth**

The differential equation

$$y' = ay(M - y)$$

with initial condition

$$y(0) = \frac{M}{1 + c}$$

is solved by

$$y = \frac{M}{1 + ce^{-aMt}}$$

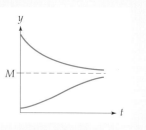

(The upper and lower curves in the box represent solutions whose initial values are, respectively, greater than or less than $M$ as determined by whether the constant $c$ is negative or positive.) As before, this result enables us to solve a differential equation "at sight," leaving only the evaluation of constants.

The lower curve in the box that rises to $M$ is called a *sigmoidal* or *S-shaped curve*, and is used to model growth that begins slowly, then becomes more rapid, and finally slows again near the upper limit. Many different quantities grow according to sigmoidal curves.

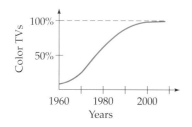

Percent of households
owning a color television
(*Source*: *Consumer USA 2008*)

Percent of households
with air conditioning
(*Source*: *Consumer USA 2008*)

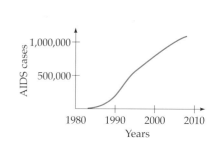

Total and Estimated U.S.
AIDS cases (*Source*:
Centers for Disease Control)

### Environmental Science

For an animal environment (such as a lake or a forest), the population that it can support will have an upper limit, called the *carrying capacity of the environment*. Ecologists often assume that an animal population grows in proportion to both its present size and its distance below the carrying capacity of the environment.

$$y' \;=\; ay(M - y)$$

Rate of    is    propor-    current    carrying capacity
growth           tional to    size    minus present size

Since this is the logistic differential equation, we know that the solution is the logistic function

$$y = \frac{M}{1 + ce^{-aMt}}$$

**EXAMPLE 4**  **PREDICTING AN ANIMAL POPULATION (LOGISTIC GROWTH)**

Ecologists estimate that an artificial lake can support a maximum of 2500 fish. The lake is initially stocked with 500 fish, and after 6 months the fish population is estimated to be 1500. Find a formula for the number of fish in the lake after $t$ months, and estimate the fish population at the end of the first year.

***Solution*** Letting $y(t)$ stand for the number of fish in the lake after $t$ months, the situation is modeled by the logistic differential equation.

$$y' = ay(2500 - y)$$

Rate of   is propor-   current       carrying capacity
growth    tional to    size        minus present size

$y' = ay(M - y)$
with $M = 2500$

The solution is the logistic function with $M = 2500$.

$$y = \frac{2500}{1 + ce^{-a2500t}} \qquad\qquad y = \frac{M}{1 + ce^{-aMt}} \text{ with } M = 2500$$

$$= \frac{2500}{1 + ce^{-bt}} \qquad\qquad \text{Replacing } a \cdot 2500 \text{ by another constant } b$$

To evaluate the constants $c$ and $b$, we use the fact that the lake was originally stocked with 500 fish.

$$500 = \frac{2500}{1 + ce^{0}} \qquad\qquad y = \frac{2500}{1 + ce^{-bt}} \text{ with } y = 500, \ t = 0$$

$$500 = \frac{2500}{1 + c} \qquad\qquad \text{Simplifying}$$

Solving this for $c$ (omitting the details—the first step is to multiply both sides by $1 + c$) gives $c = 4$, so the logistic function becomes

$$y = \frac{2500}{1 + 4e^{-bt}} \qquad\qquad y = \frac{2500}{1 + ce^{-bt}} \text{ with } c = 4$$

To evaluate $b$, we substitute the information that the population is $y = 1500$ at $t = 6$.

$$1500 = \frac{2500}{1 + 4e^{-b \cdot 6}} \qquad\qquad y = \frac{2500}{1 + 4e^{-bt}} \text{ with } y = 1500, \ t = 6$$

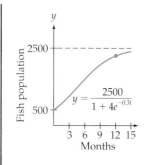

Solving for $b$ (again omitting the details—the first step is to multiply both sides by $1 + 4e^{-b \cdot 6}$) gives $b = 0.30$, so the logistic function becomes

$$y(t) = \frac{2500}{1 + 4e^{-0.3t}} \qquad\qquad y = \frac{2500}{1 + 4e^{-bt}} \text{ with } b = 0.3$$

This is the formula for the population after $t$ months. To find the population after a year, we evaluate at $t = 12$.

$$y(12) = \frac{2500}{1 + 4e^{-0.3 \cdot 12}} = \frac{2500}{1 + 4e^{-3.6}} \approx 2254 \qquad \text{Using a calculator}$$

Therefore, the population at the end of the first year is 2254, which is about 90% of the carrying capacity of the lake.

### Epidemics

Many epidemics spread at a rate proportional to both the number of people already infected (the "carriers") and also the number who have yet to catch the disease (the "susceptibles"). If $y(t)$ is the number of infected people at time $t$ from a population of size $M$, then $y(t)$ satisfies the logistic differential equation

$$y' = ay(M - y)$$

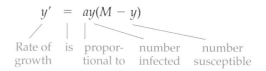

Rate of growth   is   proportional to   number infected   number susceptible

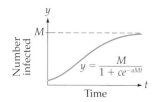

Therefore, the size of the infected population is given by the logistic function

$$y = \frac{M}{1 + ce^{-aMt}}$$

The constants are evaluated just as in Example 4, using the (initial) number of cases reported at time $t = 0$, and also the number of cases at some later time.

### Spread of Rumors

Sociologists have found that rumors spread at a rate proportional to the number who have heard the rumor (the "informed") and the number who have not heard the rumor (the "uninformed"). Therefore, in a population of size $M$, the number $y(t)$ who have heard the rumor within $t$ time units satisfies the logistic differential equation

$$y' = ay(M - y)$$

Rate of growth   is   proportional to   number informed   number uninformed

The solution $y(t)$ is then the logistic function

$$y = \frac{M}{1 + ce^{-aMt}}$$

It remains only to evaluate the constants. The spread of a rumor is analogous to the spread of a disease, with an "informed" person being one who has been "infected."

### Limited and Logistic Growth of Sales

The sales of a product whose total sales will approach an upper limit (market saturation) can be modeled by either the limited or the logistic equation. Which do you use when? For a product advertised over mass media, sales will at first grow rapidly, indicating a *limited* model. For a product becoming known only through "word of mouth," sales will at first grow slowly, indicating a *logistic* model.

**Practice Problem**

The graphs below show the total sales through day $t$ for two different products, $A$ and $B$. Which of these products was advertised and which became known only by "word of mouth"? State an appropriate differential equation for each curve.

Product A

Product B

➤ **Solution below**

**Solution to Practice Problem**

Product $A$, whose growth begins slowly, was not advertised, and the differential equation is logistic: $y' = ay(M - y)$. Product $B$, whose growth begins rapidly, was advertised, and the differential equation is limited: $y' = a(M - y)$.

## 9.2 Section Summary

The unlimited, limited, and logistic growth models are summarized in the following table. If one of these differential equations governs a particular situation, we can write the solution immediately, evaluating the constants from the given data.

**Three Models of Growth**

| Type | Differential Equation | Solution | Graph | Examples |
|---|---|---|---|---|
| **Unlimited**<br>Growth is proportional to present size | $y' = ay$ | $y = ce^{at}$ | | Investments<br>Bank accounts<br>Unlimited populations |
| **Limited**<br>Growth (starting at 0) is proportional to maximum size $M$ minus present size. | $y' = a(M - y)$ | $y = M(1 - e^{-at})$ | | Information spread by mass media<br>Memorizing random information<br>Total sales (advertised) |
| **Logistic**<br>Growth is proportional to present size and to maximum size $M$ minus present size. | $y' = ay(M - y)$ | $y = \dfrac{M}{1 + ce^{-aMt}}$ | | Confined populations<br>Epidemics<br>Rumors<br>Total sales (unadvertised) |

To decide which (if any) of the three models applies in a given situation, think of whether the growth is proportional to *size*, to *unused capacity*, or to *both* (as shown in the chart). Notice that the differential equation gives much more insight into how the growth occurs than does the solution. This is what we meant at the beginning of the section by "thinking in terms of differential equations."

# Graphing Calculator Exploration

Use the program* SLOPEFLD or a similar program to graph the slope fields for the following differential equations (one at a time) on the window [0, 5] by [0, 5]:

**a.** $y' = y$
(unlimited)

**b.** $y' = 3 - y$
(limited)

**c.** $y' = y(3 - y)$
(logistic)

Do you see how the slope field gives a "picture" of the differential equation? Each graph is on the window [0, 5] by [0, 5].

$y' = y$

$y' = 3 - y$

$y' = y(3 - y)$

*To obtain this and other programs, go to www.cengage.com/math/berresford.

## 9.2 Exercises

1. Verify that $y(t) = ce^{at}$ solves the differential equation for unlimited growth, $y' = ay$, with initial condition $y(0) = c$.

2. Verify that $y(t) = M(1 - e^{-at})$ solves the differential equation for limited growth, $y' = a(M - y)$, with initial condition $y(0) = 0$.

**3–12.** Determine the type of each differential equation: *unlimited* growth, *limited* growth, *logistic* growth, or *none* of these. (Do not solve, just identify the type.)

3. $y' = 0.02y$

4. $y' = 5(100 - y)$

5. $y' = 30(0.5 - y)$

6. $y' = 0.4y(0.01 - y)$

7. $y' = 2y^2(0.5 - y)$

8. $y' = 6y$

9. $y' = y(6 - y)$

10. $y' = 0.01(100 - y^2)$

11. $y' = 4y(0.04 - y)$

12. $y' = 4500(1 - y)$

**13–33.** Find the solution $y(t)$ by recognizing each differential equation as determining unlimited, limited, or logistic growth, and then finding the constants.

13. $y' = 6y$
    $y(0) = 1.5$

14. $y' = 0.25y$
    $y(0) = 4$

15. $y' = -y$
    $y(0) = 100$

16. $y' = \dfrac{y}{2}$
    $y(0) = 8$

17. $y' = -0.45y$
    $y(0) = -1$

18. $y' = 0$
    $y(0) = 5$

19. $y' = 2(100 - y)$
    $y(0) = 0$

20. $y' = 48(2 - y)$
    $y(0) = 0$

21. $y' = 0.05(0.25 - y)$
    $y(0) = 0$

22. $y' = \frac{2}{3}(1 - y)$
    $y(0) = 0$

23. $y' = 80 - 2y$
    $y(0) = 0$

24. $y' = 27 - 3y$
    $y(0) = 0$

[*Hint for Exercises 23–26:* Use factoring to write the differential equation in the form $y' = a(M - y)$.]

25. $y' = 2 - 0.01y$
    $y(0) = 0$

26. $y' = 6 - 8y$
    $y(0) = 0$

27. $y' = 5y(100 - y)$
    $y(0) = 10$

28. $y' = y(1 - y)$
    $y(0) = \frac{1}{2}$

29. $y' = 0.25y(0.5 - y)$
    $y(0) = 0.1$

30. $y' = \frac{1}{3}y(\frac{1}{2} - y)$
    $y(0) = \frac{1}{6}$

31. $y' = 3y(10 - y)$
    $y(0) = 20$

32. $y' = y(2 - y)$
    $y(0) = 4$

33. $y' = 6y - 2y^2$
    $y(0) = 1$

34. $y' = 3y - 6y^2$
    $y(0) = \frac{1}{6}$

[*Hint for Exercises 33–34:* Use factoring to write the differential equation in the form $y' = ay(M - y)$.]

### Applied Exercises

Write the differential equation (unlimited, limited, or logistic) that applies to the situation described. Then use its solution to solve the problem.

35. **PERSONAL FINANCE: Stamp Appreciation** The value of a stamp collection, initially worth $1500, grows continuously at the rate of 8% per year. Find a formula for its value after $t$ years.

36. **PERSONAL FINANCE: Home Appreciation** The value of a home, originally worth $25,000, grows continuously at the rate of 6% per year. Find a formula for its value after $t$ years.

37. **BUSINESS: Total Sales** A manufacturer estimates that he can sell a maximum of 100,000 digital tape recorders in a city. His total sales grow at a rate proportional to the distance below this upper limit. If after 5 months total sales are 10,000, find a formula for the total sales after $t$ months. Then use your answer to estimate the total sales at the end of the first year.

38. **BUSINESS: Product Recognition** Let $p(t)$ be the number of people in a city who have heard of a new product after $t$ weeks of advertising. The city is of size 1,000,000, and $p(t)$ grows at a rate proportional to the number of people in the city who have *not* heard of the product. If after 8 weeks 250,000 people have heard of the product, find a formula for $p(t)$. Use your formula to estimate the number of people who will have heard of the product after 20 weeks of advertising.

39. **GENERAL: Fund-Raising** In a drive to raise $5000, fund-raisers estimate that the rate of contributions is proportional to the distance from the goal. If $1000 was raised in 1 week, find a formula for the amount raised in $t$ weeks. How many weeks will it take to raise $4000?

**40. GENERAL: Learning** A person can memorize at most 40 two-digit numbers. If that person can memorize 15 numbers in the first 20 minutes, find a formula for the number that can be memorized in $t$ minutes. Use your answer to estimate how long the person will take to memorize 30 numbers.

**41. BUSINESS: Sales** A telephone company estimates the maximum market for car phones in a city to be 10,000. Total sales are proportional to both the number already sold and the size of the remaining market. If 100 phones have been sold at time $t = 0$ and after 6 months 2000 have been sold, find a formula for the total sales after $t$ months. Use your answer to estimate the total sales at the end of the first year.

**42. BIOMEDICAL: Epidemics** During a flu epidemic in a city of 1,000,000, a flu vaccine sells in proportion to both the number of people already inoculated and the number not yet inoculated. If 100 doses have been sold at time $t = 0$ and after 4 weeks 2000 doses have been sold, find a formula for the total number of doses sold within $t$ weeks. Use your formula to predict the sales after 10 weeks.

**43. SOCIOLOGY: Rumors** One person at an airport starts a rumor that a plane has been hijacked, and within 10 minutes 200 people have heard the rumor. If there are 800 people in the airport, find a formula for the number who have heard the rumor within $t$ minutes. Use your answer to estimate how many will have heard the rumor within 15 minutes.

**44. BIOMEDICAL: Epidemics** A flu epidemic on a college campus of 4000 students begins with 12 cases, and after 1 week has grown to 100 cases. Find a formula for the size of the epidemic after $t$ weeks. Use your answer to estimate the size of the epidemic after 2 weeks.

**45. ENVIRONMENTAL SCIENCE: Deer Population** A wildlife refuge is initially stocked with 100 deer, and can hold at most 800 deer. If 2 years later the deer population is 160, find a formula for the deer population after $t$ years. Use your answer to estimate when the deer population will reach 400.

**46. POLITICAL SCIENCE: Voting** Suppose that a bill in the U.S. Senate gains votes in proportion to the number of votes that it already has and to the number of votes that it does not have. If it begins with one vote (from its sponsor) and after 3 days it has 30 votes, find a formula for the number of votes that it will have after $t$ days. (*Note:* The number of votes in the Senate is 100.) When will the bill have "majority support" of 51 votes?

**47. BUSINESS: MP3 Sales** Retailers estimate the upper limit for sales of portable MP3 music players to be 22 million annually and find that sales grow in proportion to both current sales and the difference between sales and the upper limit. In 2005 sales were 16 million, and in 2008 were 19 million. Find a formula for the annual sales (in millions) $t$ years after 2005. Use your answer to predict sales in 2012.
*Source: Consumer USA 2008*

**48. BUSINESS: Digital Camera Sales** Manufacturers estimate the upper limit for sales of digital cameras to be 25 million annually and find that sales increase in proportion to both current sales and the difference between the sales and the upper limit. In 2005 sales were 6 million, and in 2008 were 20 million. Find a formula for the annual sales (in millions) $t$ years after 2005. Use your answer to predict sales in 2012.
*Source: Consumer USA 2008*

**49–52.** Solve by recognizing that the differential equation is one of the three types whose solutions we know.

**49. BIOMEDICAL: Drug Absorption** A drug injected into a vein is absorbed by the body at a rate proportional to the amount remaining in the blood. For a certain drug, the amount $y(t)$ remaining in the blood after $t$ hours satisfies $y' = -0.15y$ with $y(0) = 5$ mg. Find $y(t)$ and use your answer to estimate the amount present after 2 hours.

**50. BUSINESS: Stock Value** One model for the growth of the value of stock in a corporation assumes that the stock has a limiting "market value" $L$, and that the value $v(t)$ of the stock on day $t$ satisfies the differential equation $v' = a(L - v)$ for some constant $a$. Find a formula for the value $v(t)$ of a stock whose market value is $L = 40$ if on day $t = 10$ it was selling for $v = 30$.

**51. GENERAL: Dam Sediment** A hydroelectric dam generates electricity by forcing water through turbines. Sediment accumulating behind the dam, however, will reduce the flow and eventually require dredging. Let $y(t)$ be the amount of sediment (in thousands of tons) accumulated in $t$ years. If sediment flows in from the river at the constant rate of 20 thousand tons annually, but each year 10% of the accumulated sediment passes through the turbines, then the amount of sediment remaining satisfies the differential equation $y' = 20 - 0.1y$.

**a.** By factoring the right-hand side, write this differential equation in the form $y' = a(M - y)$. Note the value of $M$, the maximum amount of sediment that will accumulate.

**b.** Solve this (factored) differential equation together with the initial condition $y(0) = 0$ (no sediment until the dam was built).

**c.** Use your solution to find when the accumulated sediment will reach 95% of the value of $M$ found in step (a). This is when dredging is required.

**52. BIOMEDICAL: Glucose Levels** Solve Exercise 56 on page 641 by factoring the right-hand side of the differential equation to write it in the form $y' = a(M - y)$.

 **53. GENERAL: Raindrops** *(Requires Slope Field Program)* Why do larger sized raindrops fall faster than smaller ones? It depends on the resistance they encounter as they fall through the air. For large raindrops, the resistance to gravity's acceleration is proportional to the *square* of the velocity, whereas for small droplets, the resistance is proportional to the *first power* of the velocity. More precisely, their velocities obey the following differential equations, with each differential equation leading to a different *terminal velocity* for the raindrop:

**i.** $\dfrac{dv}{dt} = 32.2 - 0.1115v^2$    Downpour droplets, about 0.05 inch in diameter

**ii.** $\dfrac{dv}{dt} = 32.2 - 52.6v$    Drizzle droplets, about 0.003 inch in diameter

**iii.** $\dfrac{dv}{dt} = 32.2 - 5260v$    Fog droplets, about 0.0003 inch in diameter

(The 32.2 represents gravitational acceleration, and the other constant is determined experimentally.)

**a.** Use a slope field program to graph the slope field of differential equation (i) on the window [0, 3] by [0, 20] (using $x$ and $y$ instead of $t$ and $v$). From the slope field, must the solution curves rising from the bottom level off at a particular $y$-value? Estimate the value. This number is the terminal velocity (in feet per second) for a downpour droplet.

**b.** Do the same for differential equation (ii), but on the window [0, 0.1] by [0, 1]. What is the terminal velocity for a drizzle droplet?

**c.** Do the same for differential equation (iii), but on the window [0, 0.001] by [0, 0.01]. What is the terminal velocity for a fog droplet?

**d.** At this speed [from part (c)], how long would it take a fog droplet to fall 1 foot? This shows why fog clears so slowly.

*Source: Journal of Meteorology 6*

**54. GENERAL: Automobile Fatalities** A recent study used a logistics curve to predict the number of automobile fatalities based on the total spending on alcohol advertising in the United States in a year.

**a.** Enter the data from the following table into your graphing calculator. (Incidentally, the last line gives the situation at the time of the study: $1 billion of alcohol advertising and 20,000 alcohol-related deaths annually.)

**b.** Use the STAT then CALC then LOGISTIC operations of your calculator to fit a logistics curve to the data.

**c.** Use the logistics function to predict the number of alcohol-related deaths if alcohol advertising were eliminated ($x = 0$).

| Alcohol Advertising ($ millions) | Alcohol-related Fatalities |
|---|---|
| 325 | 12,500 |
| 850 | 18,700 |
| 1000 | 20,000 |

*Source: Review of Economics and Statistics 79*

## Conceptual Exercises

**55.** Each of the following is algebraically equivalent to the differential equations for one of our three types of growth. Which is which?

**a.** $\dfrac{y'}{M - y} = a$    **b.** $\dfrac{y'}{y} = a$    **c.** $\dfrac{y'}{y} = a(M - y)$

**56.** Which of our three types of growth would be most likely to apply to each situation?

**a.** An investment that grows by $\frac{1}{2}$% each year

**b.** A new technology, such as high-definition TV or, long ago, the automobile

**c.** A new "fad" that gets a lot of news coverage

**57.** Each of the following is an example of one of the three types of growth functions (see page 654). Which is which?

**a.** $y = (1 + e^{-t})^{-1}$    **b.** $y = \frac{1}{2}e^{t/5}$

**c.** $y = 15 - 15e^{-t}$

**58.** What is the value of each growth function at $t = 0$?

**a.** $y = \dfrac{0.03}{1 + 2e^{-3t}}$    **b.** $y = 2.5(1 - e^{-t/2})$

**c.** $y = \frac{1}{2}e^{0.04t}$

**59.** What is the limiting behavior of each growth function as $t \to \infty$?

**a.** $y = \dfrac{0.03}{1 + 2e^{-3t}}$    **b.** $y = 2.5(1 - e^{-t/2})$

**c.** $y = \frac{1}{2}e^{0.04t}$

**60.** Make up a logistic growth function that begins *higher* than $M$ and descends to $M$, as does the upper curve in the box on page 650. [*Hint:* Begin with $y = \dfrac{10}{1 + ce^{-t}}$ so that $y(0) = \dfrac{10}{1 + c}$ and choose a value for $c$ that will make $y(0) > 10$.]

### *Explorations and Excursions*

The following problems extend and augment the material presented in the text.

### *Other Growth Models*

Solve each differential equation by separation of variables.

**61. BIOMEDICAL: Individual Birthrate** If $y(t)$ is the size of a population at time $t$, then $\dfrac{y'}{y}$ is the population growth rate divided by the size of the population, and is called the *individual birthrate*. Suppose that the individual birthrate is proportional to the size of the population, $\dfrac{y'}{y} = ay$ for some constant $a$. Find a formula for the size of the population after $t$ years.

**62. BIOMEDICAL: Gompertz Curve** Another differential equation that is used to model the growth of a population $y(t)$ is $y' = bye^{-at}$, where $a$ and $b$ are constants. Solve this differential equation.

*Source: F. Brauer and C. Castillo-Chávez, Mathematical Models in Population Biology and Epidemiology*

**63. BIOMEDICAL: Allometry** Solve the differential equation of allometric growth: $y' = \dfrac{ay}{x}$ (where $a$ is a constant). This differential equation governs the relative growth rates of different parts of the same animal.

**64. GENERAL: Population Growth** Suppose that a population $y(t)$ in a certain environment grows in proportion to the square of the difference between the carrying capacity $M$ and the present population, that is, $y' = a(M - y)^2$, where $a$ is a constant. Solve this differential equation.

### *Logistic Growth Function*

**65.** Solve the logistic differential equation $y' = ay(M - y)$ as follows:

**a.** Separate variables to obtain
$$\frac{dy}{y(M - y)} = a\,dt$$

**b.** Integrate, using on the left-hand side the integration formula
$$\int \frac{dy}{y(M - y)} = \frac{1}{M}\ln\!\left(\frac{y}{M - y}\right)$$
(which may be checked by differentiation).

**c.** Exponentiate to solve for $\dfrac{y}{(M - y)}$ and then solve for $y$.

**d.** Show that the solution can be expressed as
$$y = \frac{M}{1 + ce^{-aMt}}$$

**66.** Find the inflection point of the logistic curve
$$f(x) = \frac{M}{1 + ce^{-aMx}}$$
and show that it occurs at midheight between $y = 0$ and the upper limit $y = M$. [*Hint:* Do you already know $f'(x)$?]

## 9.3 FIRST-ORDER LINEAR DIFFERENTIAL EQUATIONS

### Introduction

Not all differential equations can be solved by separation of variables. For example, the differential equation

$$y' = x + y$$

cannot be solved by separating variables (as you should check for yourself). But it can be solved by the method of *integrating factors*, which is the subject of this section. The method involves a clever use of the Product Rule, which we now review.

## Using the Product Rule Backwards

When we apply the Product Rule to an expression containing an unknown function $y$, we must write the derivative of the unknown function $y$ as $y'$. For example:

$$\frac{d}{dx}(x^2 y) = 2xy + x^2 y' \qquad \frac{d}{dx}(f \cdot g) = f' \cdot g + f \cdot g'$$

Using the Product Rule

| | | | | |
|---|---|---|---|---|
| Both are functions of $x$ | Derivative of the first | Second left alone | First left alone | Derivative of the second |

Now we will reverse this process: beginning with the two terms on the right (even if written in a different order), we will express them in the form of the derivative on the left.

---

**EXAMPLE 1**

**USING THE PRODUCT RULE BACKWARDS**

Express $x^5 y' + 5x^4 y$ as the derivative of a product.

*Solution*

$$\underbrace{x^5 y' + 5x^4 y}_{\text{Original expression}} = \underbrace{\frac{d}{dx}(x^5 y)}_{\substack{\text{Written as the} \\ \text{derivative of a product}}}$$

Recognizing $\frac{d}{dx}(x^5 y) = 5x^4 y + x^5 y'$ in the reverse order

---

**Practice Problem 1**

Express $e^{2x} y' + 2e^{2x} y$ as the derivative of a product.

➤ Solution on page 667

## First-Order Linear Differential Equations and Integrating Factors

This technique will be useful for solving a type of differential equation called **first-order linear,** which is defined as follows:

> **First-Order Linear Differential Equations**
>
> A differential equation is *first-order linear* if it can be written in the form
>
> $$y' + p(x) \cdot y = q(x)$$
>
> $y'$ and $y$ on the left, with functions of $x$ as shown
>
> where $p(x)$ and $q(x)$ are functions of $x$.

The previous equation is called the **standard form** of a first-order linear differential equation.* We solve it by multiplying both sides by an **integrating factor,** enabling us to write the left-hand side as the derivative of a product. A few examples will demonstrate the method.

**EXAMPLE 2**

**SOLVING A FIRST-ORDER LINEAR DIFFERENTIAL EQUATION**

Solve the differential equation $y' + \dfrac{2}{x}y = x.$

*Solution*

This *is* in standard form: $\quad y' + \underbrace{\dfrac{2}{x}}_{p(x)} \cdot \underbrace{y = x}_{q(x)}$

$$y' + p(x)y = q(x) \quad \text{with}$$
$$p(x) = \dfrac{2}{x} \quad \text{and} \quad q(x) = x$$

We multiply through by $x^2$. (The choice of $x^2$ will be explained shortly.)

$$x^2 y' + x^2 \underbrace{\dfrac{2}{x}}_{2x} y = \underbrace{x^2 x}_{x^3} \qquad\qquad y' + \dfrac{2}{x}y = x \quad \text{times} \quad x^2$$

$$\underbrace{x^2 y' + 2xy}_{\frac{d}{dx}(x^2 y)} = x^3 \qquad\qquad \text{Simplifying}$$

The left-hand side is now the derivative of $x^2 y$

Therefore,

$$\frac{d}{dx}(x^2 y) = x^3$$

We integrate both sides, on the left side simply dropping the $d/dx$, and on the right side using the Power Rule:

$$x^2 y = \underbrace{\frac{1}{4}x^4 + C}_{} \qquad\qquad \text{Integrating}$$

From $\displaystyle\int x^3\, dx = \frac{1}{4}x^4 + C$

Finally, we solve for the unknown function $y$ by dividing both sides by $x^2$, obtaining the general solution of the differential equation:

$$y = \frac{1}{4}x^2 + \frac{C}{x^2} \qquad\qquad x^2 y = \tfrac{1}{4}x^4 + C \quad \text{divided by } x^2$$

---

*Technically, "first-order" means that only the *first* derivative of $y$ appears, and "linear" means that the equation is linear in $y'$ and $y$. If $p(x)$ and $q(x)$ are continuous on a closed interval, then the differential equation is guaranteed to have a solution on that interval.

## How to Choose the Integrating Factor

In Example 2, the $x^2$ that we multiplied by is called an "integrating factor" because it enabled us to write the left-hand side as a derivative, which could then be integrated. How do we choose the integrating factor? There is a simple formula.

---

**Integrating Factor**

For the first-order linear differential equation

$$y' + p(x) \cdot y = q(x)$$

the integrating factor is

$$I(x) = e^{\int p(x)\, dx}$$

*e to the integral of the function multiplying $y$*

---

We omit the constant of integration in calculating the integrating factor, since it has no effect on the solution (as is shown in Exercise 55).

**EXAMPLE 3**

**FINDING AN INTEGRATING FACTOR AND SOLVING**

Solve the differential equation $y' + y = e^{3x}$.

*Solution*  The equation is already in standard form: $y' + y = e^{3x}$

with $\begin{cases} p(x) = 1 \\ q(x) = e^{3x} \end{cases}$  The integrating factor is

$p(x) = 1 \quad q(x)$

$$I(x) = e^{\int 1\, dx} = e^x$$

$I(x) = e^{\int p(x)\, dx}$

Multiplying by $I(x) = e^x$,

$$\underbrace{e^x y' + e^x y}_{\frac{d}{dx}(e^x y)} = \underbrace{e^x e^{3x}}_{e^{4x}}$$

$y' + y = e^{3x}$
multiplied by $e^x$

$$\frac{d}{dx}(e^x y) = e^{4x}$$

Writing the left-hand side as the derivative of a product

$$e^x y = \frac{1}{4} e^{4x} + C$$

Integrating both sides $\left( \text{dropping the } \dfrac{d}{dx} \right)$

From $\displaystyle\int e^{4x}\, dx = \frac{1}{4} e^{4x} + C$

$$y = \frac{1}{4} e^{3x} + C e^{-x}$$

Solving for $y$ by dividing both sides by $e^x$. This is the general solution.

Note that in the solution the constant of integration $C$ (from integrating the right-hand side) does not stand "by itself," but is multiplied by $e^{-x}$. In general, the constant $C$ may appear *anywhere* in the solution.

Finding integrating factors may involve using the properties of logarithms. For example, the integrating factor for the differential equation $y' + \dfrac{2}{x}y = x$ in Example 2 (page 660) is found as follows:

$$I(t) = e^{\int p(x)\,dx} = e^{\int \frac{2}{x}\,dx} = e^{2\int \frac{1}{x}\,dx} = e^{2\ln x}$$

$$\uparrow \qquad \underbrace{\qquad} \qquad \underbrace{\qquad}$$
$$p(x) \quad \text{Taking out} \qquad \text{Integrating}$$
$$\text{the constant} \quad 1/x \text{ gives } \ln x$$

$$= \; e^{\ln x^2} \; = \; x^2$$

$$\underbrace{\qquad} \qquad \underbrace{\qquad}$$
$$\text{Using} \qquad \text{Using}$$
$$2\ln x = \ln x^2 \quad e^{\ln z} = z$$

You should study the steps in this calculation carefully, since they occur frequently with integrating factors. The last two steps used natural logarithm properties 8 and 4 (see the inside back cover). We omitted the absolute value bars when we integrated $1/x$ to get $\ln x$, since in most applications the independent variable is positive, often representing time. Furthermore, in this particular case, including the absolute value bars leads to the same integrating factor as *without* the absolute value bars, since the $x$ is squared.

---

**Practice Problem 2**

Find the integrating factor for $y' + \dfrac{4}{x}y = x$.

➤ **Solution on page 667**

---

As before, an initial condition is used to evaluate the arbitrary constant.

---

**EXAMPLE 4**

**FINDING A PARTICULAR SOLUTION**

Solve the differential equation and initial condition $\begin{cases} xy' - 3y = x^2 \\ y(1) = 5 \end{cases}$.

*Solution*

$$y' - \frac{3}{x}y = x$$

Dividing by $x$ to write the equation in standard form with $p(x) = -3/x$

➤ Try to skip to here

$$I(x) = e^{\int \frac{-3}{x}\,dx} = e^{-3\int \frac{1}{x}\,dx} = e^{-3\ln x} = e^{\ln x^{-3}} = x^{-3}$$

Integrating factor is $x^{-3}$

$$x^{-3}y' - x^{-3}\frac{3}{x}y = x^{-3}x$$

$y' - \dfrac{3}{x}y = x$ multiplied by $x^{-3}$

$$x^{-3}y' - 3x^{-4}y = x^{-2}$$

Simplifying

$$\frac{d}{dx}(x^{-3}y) = x^{-2}$$

Expressing the left side as the derivative of a product

$$x^{-3}y = -x^{-1} + C$$

Integrating both sides

From $\int x^{-2}\,dx = -x^{-1} + C$

$$y = -x^2 + Cx^3$$

Solving for $y$ by multiplying by $x^3$

$$5 = -1 + C$$

Substituting $y = 5$ and $x = 1$ [from the given $y(1) = 5$]

Therefore $C = 6$, giving

$$y = -x^2 + 6x^3$$

$y = -x^2 + Cx^3$ with $C = 6$ (the particular solution)

The general procedure for solving first-order linear differential equations is as follows:

---

### Solving First-Order Linear Differential Equations

1. Write the differential equation in standard form $y' + p(x)\cdot y = q(x)$.

2. Find the integrating factor $I(x) = e^{\int p(x)\,dx}$ (omitting the constant of integration).

3. Multiply the equation through by the integrating factor $I(x)$.

4. Express the left-hand side as the derivative of a product, $\dfrac{d}{dx}[I(x)\cdot y]$.

5. Integrate both sides (dropping the $d/dx$ on the left side, and writing the constant of integration on the right side).

6. Solve the resulting equation for $y$. If there is an initial condition, use it to evaluate the arbitrary constant of integration.

---

This general procedure works because multiplying by the integrating factor changes the left-hand side of the differential equation into

$$e^{\int p(x)\,dx}\,y' + e^{\int p(x)\,dx}\,p(x)\cdot y \qquad \text{which equals} \qquad \frac{d}{dx}\left(e^{\int p(x)\,dx}\,y\right)$$

This last expression can then be integrated and the equation solved for $y$. The next example shows that integrating the right-hand side may require a substitution or some other integration technique.

**EXAMPLE 5**

### USING AN INTEGRATING FACTOR AND A SUBSTITUTION

Solve the differential equation and initial condition $\begin{cases} y' = 6x - 2xy \\ y(0) = 2 \end{cases}$.

*Solution*

$$y' + 2xy = 6x$$
Standard form, with $p(x) = 2x$

$$I(x) = e^{\int 2x\,dx} = e^{x^2}$$
Integrating factor

$$e^{x^2}y' + e^{x^2}2xy = e^{x^2}6x$$
$y' + 2xy = 6x$ times $e^{x^2}$

$$\frac{d}{dx}(e^{x^2}y) = e^{x^2}6x$$
Expressing the left side as the derivative of a product

$$e^{x^2}y = \int e^{x^2}6x\,dx = 3\underbrace{\int \underbrace{e^{x^2}}_{e^u} \underbrace{2x\,dx}_{du}} = \underbrace{3e^{x^2} + C}_{\text{Using } \int e^u\,du = e^u + C}$$
Integrating both sides (using the substitution method on $\int e^{x^2}6x\,dx$)

$$y = 3 + Ce^{-x^2}$$
Solving for $y$ by multiplying $e^{x^2}y = 3e^{x^2} + C$ by $e^{-x^2}$

$$2 = 3 + C$$
Substituting $y = 2$ and $x = 0$ (from the given $y(0) = 2$)

Therefore, $C = -1$, giving

$$y = 3 - e^{-x^2}$$
$y = 3 + Ce^{-x^2}$ with $C = -1$ (the particular solution)

---

## Radon Pollution

Radon is a colorless, odorless gas formed by the natural decay of uranium in the earth. When dispersed in the atmosphere it is harmless, but if allowed to concentrate in a room or basement, it becomes a serious health hazard. Indoor radon pollution is second only to smoking as a certified cause of lung cancer, killing an estimated 7000 to 30,000 Americans each year.*

**EXAMPLE 6**

### VENTING RADON POLLUTION

A 1000-cubic-foot room has a radon level of 500 pCi (picocuries) per cubic foot. A ventilation system is installed, which each hour brings in 100 cubic feet of outside air (containing 4 pCi per cubic foot), while an equal volume of air leaves the room. Assuming that the air in the room mixes thoroughly, find how soon the radon level will fall to 112 pCi per cubic foot, the safety level set by the Environmental Protection Agency (EPA).

---

*See Bernard Cohen, *Radon, a Homeowner's Guide to Detection and Control* (Consumer Reports Books). His book measures radon per *liter*, whereas we measure it per *cubic foot*. To change our numbers to "per liter," multiply them by 0.0353.

*Solution*

Let

$$y(t) = \begin{pmatrix} \text{Amount of radon in} \\ \text{room after } t \text{ hours} \end{pmatrix}$$

so

$$y(0) = 500{,}000 \qquad \text{Initial amount (500 per cubic foot times 1000 cubic feet)}$$

The concentration of radon *in the incoming air* is given as 4, and the concentration *in the room* at any time is $y/1000$ [the amount ($y$) of radon in the room divided by the volume (1000) of the room]. Multiplying these concentrations by 100, the volume of air flowing in and out, gives the rates at which radon enters and leaves the room each hour. We show these rates in an inflow-outflow diagram for a 1-hour time period.

This leads to the following differential equation:

$$y' = 400 - 0.1y$$

Rate of change of $y$   is   400 added   0.1$y$ removed

$$\begin{pmatrix} \text{Rate of} \\ \text{change} \end{pmatrix} = \begin{pmatrix} \text{Amount} \\ \text{added} \end{pmatrix} - \begin{pmatrix} \text{Amount} \\ \text{removed} \end{pmatrix}$$

$$y' + 0.1y = 400 \qquad \text{In standard form}$$

$$I(t) = e^{\int 0.1\,dt} = e^{0.1t} \qquad \text{Integrating factor}$$

$$e^{0.1t}y' + 0.1e^{0.1t}y = 400e^{0.1t} \qquad y' + 0.1y = 400 \ \text{ times } e^{0.1t}$$

$$\frac{d}{dt}(e^{0.1t}y) = 400e^{0.1t} \qquad \text{Left-hand side as the derivative of a product}$$

$$e^{0.1t}y = 400 \cdot 10e^{0.1t} + C \qquad \text{Integrating}$$

From $\displaystyle\int e^{0.1t}\,dt = \frac{1}{0.1}e^{0.1t} + C$

$$y = 4000 + Ce^{-0.1t} \qquad \text{Solving for } y \text{ by multiplying by } e^{-0.1t}$$

$$500{,}000 = 4000 + C \qquad \begin{array}{l}\text{Substituting } y = 500{,}000 \\ \text{and } t = 0 \\ \text{(from } y(0) = 500{,}000)\end{array}$$

Therefore, $C = 496,000$, giving

$$y = 4000 + 496,000e^{-0.1t}$$

$y = 4000 + Ce^{-0.1t}$
with $C = 496,000$

↖ Amount of radon
after $t$ hours

To find when the EPA limit of 112 pCi per cubic foot is reached, we solve

$$\underbrace{4000 + 496,000e^{-0.1t}}_{} \qquad = \qquad \underbrace{112,000}_{}$$

Total amount of radon
(according to formula)

Total radon at EPA safety level
(112 per cubic foot times 1000 cubic feet)

We could solve this equation for $t$ as we did in the preceding section, or we could use a graphing calculator (see the following Graphing Calculator Exploration). Either way, we would find that the radon concentration will fall to the EPA level in about 15.2 hours.

---

The amount of radon remaining in the room after $t$ hours is shown by the following graph.

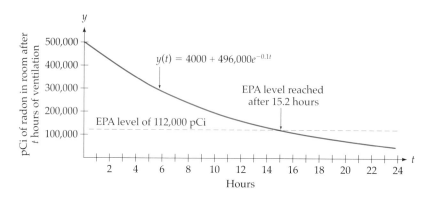

## Graphing Calculator Exploration

Graph $y_1 = 4000 + 496,000e^{-0.1x}$ and $y_2 = 112,000$ on the window $[0, 24]$ by $[0, 500,000]$. Use INTERSECT to find where they meet.

The "long-run" amount of radon in the room is found by taking the limit as $t$ approaches infinity:

$$\lim_{t \to \infty} (4000 + \underbrace{496,000e^{-0.1t}}_{}) = 4000 \qquad\qquad 4000 \text{ pCI}$$

Approaches 0
as $t \to \infty$

This same number can be found as follows: outside air (with 4 pCi per cubic foot) filling the 1000-cubic-foot room will eventually result in $4 \cdot 1000 = 4000$ pCi of radon. This agreement serves as a check of the solution.

Example 6 is a typical "mixture problem," in which two mixtures (here, outside and inside air), each having a different concentration of a particular substance (radon), are combined and you want to find the concentration at some later time. The same technique works for *any* mixture, whether it is air pollution in a room or chemicals in a container. It is generally easiest to let $y(t)$ stand for the *total amount* of the substance in the room (or container) at any time, so that the *concentration* can be expressed $\dfrac{y}{\text{Total volume}}$. Making an *inflow-outflow* diagram for the gain and loss per time unit leads to a differential equation, which you then solve. (This differential equation could also have been solved by separation of variables.)

### Solutions to Practice Problems

**1.** $e^{2x}y' + 2e^{2x}y = \dfrac{d}{dx}\left(e^{2x}y\right)$          Using $\dfrac{d}{dx}e^{2x} = 2e^{2x}$

⟶ Try to skip to here ⟶

**2.** $I(x) = e^{\int \frac{4}{x}\,dx} = e^{4\int \frac{1}{x}\,dx} = e^{4\ln x} = e^{\ln x^4} = x^4$

## 9.3 Section Summary

A first-order linear differential equation can be written in the standard form

$$y' + p(x)\cdot y = q(x) \qquad p \text{ and } q \text{ are functions of } x \text{ alone}$$

We solve first-order linear differential equations by multiplying both sides by the *integrating factor* $I(x) = e^{\int p(x)\,dx}$, expressing the left-hand side as the derivative of a product, integrating both sides, and then solving for $y$.

## 9.3 Exercises

**1–20.** Solve each first-order linear differential equation.

**1.** $y' + 2y = 8$

**2.** $y' - y = 2$

**3.** $y' - 2y = e^{-2x}$

**4.** $y' + 3y = e^{3x}$

**5.** $y' + \dfrac{5}{x}y = 24x^2$

**6.** $y' + \dfrac{4}{x}y = 12x$

**7.** $xy' - y = x^2$

**8.** $xy' - y = x$

**9.** $y' + 3x^2y = 9x^2$

**10.** $y' - 4x^3y = 8x^3$

**11.** $y' - 2xy = 0$

**12.** $y' + xy = 0$

**13.** $(x + 1)y' + y = 2x$

**14.** $y' + \dfrac{2x}{x^2 + 1}y = 3$

**15.** $y' - \dfrac{2}{x}y = 6x^3 - 9x^2$

**16.** $xy' - y = x^3 e^{x^2}$

**17.** $y' = x + y$

[*Hint for Exercises 17 and 18:* Use integration by parts or an integral table.]

**18.** $xy' + y = x^2 \ln x$

**19.** $y' + (\cos x)y = \cos x$

**20.** $y' - (\sin x)y = \sin x$

**21–28.** Solve each differential equation with the given initial condition.

**21.** $y' + 3y = 12e^x$
$y(0) = 5$

**22.** $y' + 4y = e^{-3x}$
$y(0) = 4$

**23.** $xy' + 2y = 14x^5$
$y(1) = 1$

**24.** $xy' + 4y = 10x$
$y(1) = 0$

**25.** $xy' = 2y + x^2$
$y(1) = 3$

**26.** $xy' = 3y + x^4$
$y(1) = 7$

**27.** $y' + 2xy = 4x$
$y(0) = 0$

**28.** $y' - 3x^2y = 6x^2$
$y(0) = 1$

**29–30.** Solve each differential equation in two ways: first by using an integrating factor, and then by separation of variables (as in Section 9.1).

**29.** $y' = y + 1$

**30.** $y' = xy$

**31–32.** Each exercise below has one differential equation that can be solved by an integrating factor and one that can be solved by separation of variables (as in Section 9.1). Solve each by the appropriate technique.

**31.** **a.** $y' + xy^2 = 0$
**b.** $y' = y + x^2e^x$

**32.** **a.** $yy' = x$
**b.** $y' + y = e^{-x}$

## Applied Exercises

**33. BUSINESS: Value of a Company** A company grows in value by 10% each year, and also gains 20% of a growing market estimated at $100e^{0.1t}$ million dollars, where $t$ is the number of years that the company has been in business. Therefore, the value $y(t)$ of the company (in millions of dollars) satisfies

$$y' = 0.1y + 20e^{0.1t}$$

$y(0) = 5$ (Initial value $5 million)

**a.** Solve this differential equation and initial condition to find a formula for the value of the company after $t$ years.
**b.** Use your solution to find the value of the company after 2 years.

**34. ECONOMICS: Exports** A country's cumulative exports $y(t)$ (in millions of dollars) grow in proportion to its *average* size $y/t$ over the last $t$ years, plus a fixed growth rate (10), and so satisfy

$$y' = \frac{1}{t}y + 10 \qquad \text{(for } t \geq 1)$$

$y(1) = 8$ ($8 million after 1 year)

**a.** Solve this differential equation and initial condition to find the country's cumulative exports after $t$ years.
**b.** Use your solution to find the country's cumulative exports after 5 years.

**35. BIOMEDICAL: AIDS** The total number $y(t)$ of cases of AIDS (acquired immunodeficiency syndrome) diagnosed in a certain region after $t$ years satisfies

$$y' - \frac{3}{t}y = 0 \qquad \text{(for } t \geq 1)$$

$y(1) = 125$ (125 cases after 1 year)

**a.** Solve this differential equation and initial condition. (Your solution will show that AIDS does not spread logistically, as do most epidemics, but like a power. This means that AIDS will spread more slowly, which seems to result from its being transmitted at different rates within different subpopulations.
**b.** Use your solution to predict the number of AIDS cases in the region by the year $t = 15$.

*Source:* Centers for Disease Control

**36. ENVIRONMENTAL SCIENCE: Global Warming** The burning of coal and oil is increasing the amount of carbon dioxide in the atmosphere, which is expected to trap more solar radiation and increase global temperatures (the "greenhouse effect"), possibly raising the sea level by melting polar ice and flooding coastal regions. The current 3200 billion tons of carbon dioxide in the atmosphere is growing by about 50 billion tons per year, while about 1% of the accumulated carbon dioxide is removed by natural processes. Therefore, the amount $y(t)$ of carbon dioxide (in billions of tons) $t$ years from now satisfies

$$y' = -0.01y + 50$$

$y(0) = 3200$ (Initial amount)

**a.** Solve this differential equation and initial condition.
**b.** Graph the solution on a graphing calculator and find when the amount of carbon dioxide will reach 4000 billion tons, at which time global temperatures are expected to have risen by 3 degrees Fahrenheit.
**c.** Use your solution to find the long-run amount of carbon dioxide in the atmosphere.

*Source:* Worldwatch

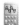

**37. ENVIRONMENTAL SCIENCE: Algae Bloom**
An algae bloom in a lake is a sudden growth of algae that consumes nutrients and blocks sunlight, killing other life in the water. Suppose that the number of tons $y(t)$ of algae after $t$ weeks satisfies

$$y' = ty + t$$
$$y(0) = 2 \qquad \text{(Initial tonnage)}$$

**a.** Solve this differential equation and initial condition.
**b.** Use your solution to find the amount of algae after 2 weeks.
**c.** Graph the solution on a graphing calculator and find when the algae bloom will reach 40 tons.

**38. BIOMEDICAL: Drug Absorption** A patient's ability to absorb a drug sometimes changes with time, and the dosage must therefore be adjusted. Suppose that the number of milligrams $y(t)$ of a drug remaining in the patient's bloodstream after $t$ hours satisfies

$$y' = -\frac{1}{t}y + t \qquad \text{(for } t \geq 1\text{)}$$
$$y(3) = 5 \qquad \text{(5 mg after 3 hours)}$$

Solve this differential equation and initial condition to find the amount remaining in the bloodstream after $t$ hours.

**39. ENVIRONMENTAL SCIENCE: Radon Pollution**
A 10,000-cubic-foot-room has an initial radon level of 800 pCi (picocuries) per cubic foot. A ventilation system is installed that each hour brings in 500 cubic feet of outside air (containing 5 pCi per cubic foot), while an equal volume of air leaves the room. Assume that the air in the room mixes thoroughly.

**a.** Find a differential equation and initial condition that govern the total amount $y(t)$ of radon in the room after $t$ hours.
**b.** Solve this differential equation and initial condition.
**c.** Graph the solution on a graphing calculator and find how soon the radon level will fall to the EPA safety level of 112 pCi per cubic foot.

**40. ENVIRONMENTAL SCIENCE: Smoke Pollution** A 5000-cubic-foot room has 400 smoke particles per cubic foot. A ventilation system is turned on that each minute brings in 50 cubic feet of outside air (containing 10 smoke particles per cubic foot), while an equal volume of air leaves the room. Assume that the air in the room mixes thoroughly.

**a.** Find a differential equation and initial condition that govern the total number $y(t)$ of smoke particles in the room after $t$ minutes.

**b.** Solve this differential equation and initial condition.
**c.** Graph the solution on a graphing calculator and find how soon the smoke level will fall to 100 smoke particles per cubic foot.

**41. ENVIRONMENTAL SCIENCE: Smoke Pollution** A 12,000-cubic-foot room has 500 smoke particles per cubic foot. A ventilation system is turned on that each minute brings in 600 cubic feet of smoke-free air, while an equal volume of air leaves the room. Also, during each minute, smokers in the room add a total of 10,000 particles of smoke to the room. Assume that the air in the room mixes thoroughly.

**a.** Find a differential equation and initial condition that govern the total number $y(t)$ of smoke particles in the room after $t$ minutes.
**b.** Solve this differential equation and initial condition.
**c.** Find how soon the smoke level will fall to 100 smoke particles per cubic foot.

**42. ENVIRONMENTAL SCIENCE: Water Purity** A 500-gallon tank is filled with water containing 0.2 ounce of impurities per gallon. Each hour, 200 gallons of water (containing 0.01 ounce of impurities per gallon) is added and mixed into the tank, while an equal volume of water is removed.

**a.** Write a differential equation and initial condition that describe the total amount $y(t)$ of impurities in the tank after $t$ hours.
**b.** Solve this differential equation and initial condition.
**c.** Use your solution to find when the impurities will reach 0.05 ounce per gallon, at which time the water may be used for drinking.
**d.** Use your solution to find the "long-run" amount of impurities in the tank.

**43. ENVIRONMENTAL SCIENCE: Water Purity**
The water in a 100,000-gallon reservoir contains 0.1 gram of pesticide per gallon. Each hour, 2000 gallons of water (containing 0.01 gram of pesticide per gallon) is added and mixed into the reservoir, and an equal volume of water is drained off.

**a.** Write a differential equation and initial condition that describe the amount $y(t)$ of pesticide in the reservoir after $t$ hours.
**b.** Solve this differential equation and initial condition.
**c.** Graph your solution on a graphing calculator and find when the amount of pesticide will reach 0.02 gram per gallon, at which time the water is safe to drink.

*(continues)*

**d.** Use your solution to find the "long-run" amount of pesticide in the reservoir.

**44. PERSONAL FINANCE: Banking** You deposit $8000 into a bank account paying 5% interest compounded continuously, and you withdraw funds continuously at the rate of $1000 per year. Therefore, the amount $y(t)$ in the account after $t$ years satisfies

$$y' = 0.05y - 1000$$
$$y(0) = 8000$$

**a.** Solve this differential equation and initial value.
**b.** Graph your solution on a graphing calculator and find how long it will take until the account is empty.

**45. BIOMEDICAL: Dieting** A person's weight $w(t)$ after $t$ days of eating $c$ calories per day can be modeled by the following differential equation

$$w' + 0.005w = \frac{c}{3500}$$

where the 0.005 represents the proportional weight loss per day when eating nothing, and 3500 is the conversion rate for calories into pounds.

**a.** If a person initially weighing 170 pounds goes on a diet of 2100 calories per day, find a formula for the person's weight after $t$ days.
**b.** Use your solution to find when the person will have lost 15 pounds.
**c.** Find the "limiting weight" that will be approached if the person continues on this diet indefinitely.

*Source: College Mathematics Journal* **18**

**46. BIOMEDICAL: Dieting** (*continuation*) Use the differential equation in Exercise 45 to find the number of calories a person weighing 170 pounds should eat each day in order to lose 6 pounds in 21 days.

## Conceptual Exercises

**47.** Is a differential equation of the form $u(x)y' + v(x)y = w(x)$ first-order linear? If so, what is its standard form?

**48.** Can we solve *any* first-order linear differential equation by the method of this section? [*Hint:* Can you always carry out every step?]

**49.** Which of the differential equations for the three types of growth summarized on page 654 are first-order linear?

**50.** Can a *separable* differential equation also be first-order linear? [*Hint:* See the definition of *separable* on page 631 and compare it to the first-order linear standard form. For what types of function $g(y)$ can these be the same?]

**51.** If you tried to solve the "unlimited growth" differential equation (see page 654) by the method of this section, what would be the integrating factor?

**52.** If you tried to solve the "limited growth" differential equation (see page 654) by the method of this section, what would be the integrating factor?

**53.** The air in a 100,000-cubic-foot room contains 483 pCi of radon per cubic foot. A ventilation system is turned on that each hour will bring in 7250 cubic feet of fresh air containing 8 pCi of radon per cubic foot, while an equal amount of air leaves the room. Assuming that the air mixes thoroughly, in the long run, what will be the total amount of radon in the room? [*Hint:* No calculation necessary.]

**54.** Explain why we cannot use an integrating factor $I(x)$ to solve the *nonlinear* differential equation $y' + p(x)y^2 = q(x)$. [*Hint:* Can we obtain the left-hand side by differentiating an expression of the form $I(x)y$ or $I(x)y^2$?]

## Explorations and Excursions

The following problems extend and augment the material presented in the text.

**55. OMITTING THE CONSTANT** Show that the constant of integration in the integrating factor can be omitted. [*Hint:* Show that multiplying by the integrating factor $e^{\int p(x)\,dx + C}$ just multiplies the differential equation by a nonzero constant, which does not change the solution.]

**56. SOLVING BY A FORMULA** Derive the formula

$$y(x) = \frac{1}{I(x)} \int I(x)q(x)\,dx \quad \text{for the solution of}$$

the first-order linear differential equation $y' + p(x)y = q(x)$, where $I(x) = e^{\int p(x)\,dx}$. (*Note:* Rather than memorizing this formula, it is better to understand the method, which generalizes to an even larger class of differential equations.)

**57–58. BIOMEDICAL: Drug Dynamics** When you swallow a pill, the medication passes through your stomach lining into your bloodstream, where some is absorbed by the cells of your body and the rest continues to circulate for future absorption. The amount $y(t)$ of medication remaining in the bloodstream after $t$ hours can be modeled by the differential equation

$$\frac{dy}{dt} = abe^{-bt} - cy$$

for constants $a$, $b$, and $c$ (respectively the dosage of the pill, the dissolution constant of the pill, and the absorption constant of the medication). For the given values of the constants:

**a.** Substitute the constants into the stated differential equation.
**b.** Solve the differential equation (with the initial condition of having no medicine in the bloodstream at time $t = 0$) to find a formula for the amount of medicine in the bloodstream at any time $t$ (hours).

**c.** Use your solution to find the amount of medicine in the bloodstream at time $t = 2$ hours.
**d.** Graph your solution on a graphing calculator and find when the amount of medication in the bloodstream is maximized.

*Source:* J. Keener and J. Sneyd, *Mathematical Physiology*

**57.** $a = 10\,\text{mg}, \quad b = 3, \quad c = 0.2$

**58.** $a = 30\,\text{mg}, \quad b = 5, \quad c = 0.3$

**59. BIOMEDICAL: Drug Dynamics** Read the explanation in the preceding exercise. Then solve the stated differential equation with the initial condition $y(0) = 0$ (initially no medication in the bloodstream). Show that your solution can be written in the form

$$y(t) = \frac{ab}{b - c}(e^{-ct} - e^{-bt}).$$

## 9.4 APPROXIMATE SOLUTIONS OF DIFFERENTIAL EQUATIONS: EULER'S METHOD

### Introduction

There are many techniques for solving differential equations besides those described in this book. Despite all of the techniques, however, there are many differential equations that cannot be solved by *any* method (as finite combinations of elementary functions). Even if a differential equation *can* be solved, its solution may be too complicated to be useful. In such cases we must settle for an **approximate solution** that is *close* to the actual solution, and one way of finding such an approximation is *Euler's method.* Leonhard Euler (1707–1783; his name is pronounced "Oiler") was a prolific Swiss mathematician in whose honor the constant $e$ was named.

### Euler's Method

A differential equation together with an initial condition is called an **initial value problem.** For an initial value problem of the form

$$y' = g(x, y)$$

$$y(a) = y_0 \qquad \text{Initial value } y = y_0 \text{ at } x = a$$

Euler's method finds an approximation to the solution on a closed interval beginning at $x = a$.* Euler observed that if the solution $y(x)$ passes through a point $(x, y)$ in the plane, then at that point the solution has slope $g(x, y)$,

---

*Technically, if $g$ and $\dfrac{\partial g}{\partial y}$ are continuous for all $x$ in an interval $[a, b]$ and all values of $y$, then the initial value problem has a solution for $x$-values in some subinterval beginning at $a$. Less stringent conditions can also be given.

since $y' = g(x, y)$. This idea is used to construct a **polygonal approximation** (consisting of connected line segments) that is tangent to the true solution at the initial point.

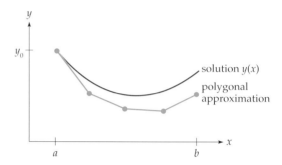

The actual (but generally unknown) solution
$y(x)$ in red and below it a polygonal approximation
with $n = 4$ segments.

To construct the Euler approximation on an interval $[a, b]$ using $n$ segments, we divide $[a, b]$ into $n$ subintervals by equally spaced points
$a = x_0 < x_1 < x_2 < \cdots < x_n = b$ with spacing (step size) $h = \dfrac{b - a}{n}$.

Beginning with $x_0 = a$, successive $x$-values are found by adding $h$ to the previous $x$-value until reaching $x_n = b$. The segments are constructed on these subintervals as follows.

The initial condition $y(a) = y_0$ gives the initial point $(x_0, y_0)$ (since $x_0 = a$), and the differential equation gives the slope at this point as $y' = g(x_0, y_0)$. The first segment is constructed from this point, $(x_0, y_0)$, with this slope, $g(x_0, y_0)$, extending a step size $h$ to the right, as illustrated below.

The first segment of the Euler approximation.

The equation of this segment comes from the point-slope form for a line, with the given point and slope.

$$y - y_0 = g(x_0, y_0) \cdot (x - x_0)$$

$y - y_1 = m(x - x_1)$ with point $(x_0, y_0)$ and slope $m = g(x_0, y_0)$.

or

$$y = y_0 + g(x_0, y_0) \cdot (x - x_0)$$

Adding $y_0$ to each side (for $x_0 \le x \le x_1$)

The $y$-coordinate $y_1$ at the end of the first segment is found by evaluating at $x = x_1$:

$$y_1 = y_0 + g(x_0, y_0) \cdot \underbrace{(x_1 - x_0)}_{h} = y_0 + g(x_0, y_0) \cdot h$$

Formula above at $x = x_1$

The second segment is constructed similarly, beginning at the previous end point $(x_1, y_1)$ and with slope $g(x_1, y_1)$ [from $y' = g(x, y)$ evaluated at this same point], as illustrated in the following diagram.

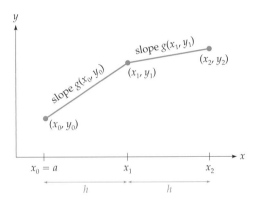

First two segments of the Euler approximation.
The slope of each segment is $g(x, y)$ evaluated
at the left-hand endpoint of the segment.

Just as before, this leads to a formula for the second segment:

$$y = y_1 + g(x_1, y_1) \cdot (x - x_1)$$

For $x_1 \le x \le x_2$

This process is then repeated, with each new segment beginning at the previous endpoint and having slope $y' = g(x, y)$ evaluated at this left endpoint. The process continues until it reaches the final $x$-value $x_n = b$. An example will illustrate the method.

**EXAMPLE 1**

**FINDING AN APPROXIMATE SOLUTION**

For the initial value problem $\begin{cases} y' = 3x - y \\ y(0) = 4 \end{cases}$, find the Euler approximation to the solution on the interval $[0, 1.5]$ using $n = 3$ segments. State the approximation to the solution at $x = 1.5$.

*Solution*

The function $g$ is

$$g(x, y) = 3x - y$$

Right-hand side of the differential equation

The step size (between successive $x$-values) is

$$h = \frac{1.5 - 0}{3} = \frac{1.5}{3} = 0.5$$

$h = \dfrac{b - a}{n}$ with $a = 0$, $b = 1.5$, and $n = 3$

The initial point is

$$x_0 = 0, \qquad y_0 = 4$$

From the given $y(0) = 4$

**Segment 1**

Slope $\qquad g(x_0, y_0) = g(0, 4) = 3 \cdot 0 - 4 = -4$

$g(x, y) = 3x - y$ with $x = 0$ and $y = 4$

Endpoint $\quad x_1 = 0.5, \quad y_1 = 4 + (-4) \cdot (0.5) = 2$

$x_1 = x_0 + h$, $y_1 = y_0 + g(x_0, y_0) \cdot h$

**Segment 2**

Slope $\qquad g(x_1, y_1) = g(0.5, 2) = 3 \cdot (0.5) - 2 = -0.5$

$g(x, y) = 3x - y$ with $x = 0.5$ and $y = 2$

Endpoint $\quad x_2 = 1, \quad y_2 = 2 + (-0.5) \cdot (0.5) = 1.75$

$x_2 = x_1 + h$, $y_2 = y_1 + g(x_1, y_1) \cdot h$

**Segment 3**

Slope $\qquad g(x_2, y_2) = g(1, 1.75) = 3 \cdot 1 - 1.75 = 1.25$

$g(x, y) = 3x - y$ with $x = 1$ and $y = 1.75$

Endpoint $\quad x_3 = 1.5, \quad y_3 = 1.75 + (1.25) \cdot (0.5) = 2.375$

$x_3 = x_2 + h$, $y_3 = y_2 + g(x_2, y_2) \cdot h$

The points (0, 4), (0.5, 2), (1, 1.75), and (1.5, 2.375) determine the polygonal approximation of the solution. The $y$-values are the approximations of the solution at the $x$-values. In particular, the approximation at the endpoint $x = 1.5$ is

$$y(1.5) \approx y_3 = 2.375$$

The polygonal approximation of the solution to this problem is graphed below.

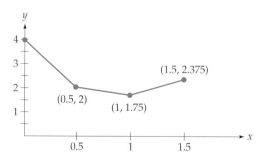

Euler approximation to the solution

of $\begin{cases} y' = 3x - y \\ y(0) = 0 \end{cases}$ with $n = 3$ segments.

## Graphing Calculator Exploration

The graphing calculator program* EULER carries out the Euler approximation for any number of segments, draws the polygonal approximation, and shows the coordinates of the points. The screen on the left shows the approximation for the initial value problem in the preceding example along with the coordinates of the final point.

*To obtain this and other programs, go to www.cengage.com/math/berresford.

### Accuracy of the Euler Approximation

In Example 1 we used only $n = 3$ segments, so we should not expect great accuracy. Small values of $n$ like 3 are only for demonstration purposes. More realistic values of $n$ are in the hundreds or thousands, with the calculations carried out on a programmable calculator or computer. The following diagram shows the Euler approximation for $n = 3$ that we found, and also the approximations for $n = 6$ and $n = 100$, together with the actual solution (found by using an integrating factor).

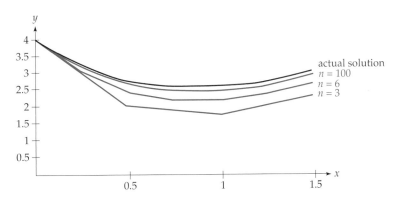

Actual solution to $\begin{cases} y' = 3x - y \\ y(0) = 4 \end{cases}$ (in red) and
the Euler approximations using 3, 6, and 100 segments.

The accuracy for $n = 100$ segments may be seen numerically from the following table.

| x-Values | Euler Approximation ($n = 100$) | Actual y-Values |
|---|---|---|
| 0 | 4 | 4 |
| 0.5 | 2.73 | 2.75 |
| 1 | 2.56 | 2.58 |
| 1.5 | 3.04 | 3.06 |

⌐— very close —⌐

Inaccuracies in Euler's approximation come from two sources. First, the *actual* solution changes slope *continuously,* whereas the approximation changes slope only at the segment endpoints. Second, the points at which the new slopes are calculated are not generally on the actual solution (except for the initial point), and so do not give the true slope of the solution curve. For greater accuracy one may increase $n$ (but this involves more calculation and therefore more roundoff errors), or use a more advanced method (such as the Runge-Kutta method).

**Practice Problem**

For the initial value problem $\begin{cases} y' = 7x - 2y \\ y(3) = 6 \end{cases}$, find:

**a.** the initial point $(x_0, y_0)$

**b.** the slope of the solution at the initial point ➤ **Solutions below**

The following differential equation cannot be solved by either of the "exact" methods of this chapter, so we use EULER. (We could also use the program listed on the next page.)

---

## Graphing Calculator Exploration

Let $y(x)$ be the value (in millions of dollars) of an electronics company $x$ years after it was founded. The company is expected to grow, but more slowly as inefficiency increases with its size. Suppose that its value $y(x)$ satisfies

$$y' = 2.6x - e^{0.3y}$$
$$y(0) = 1 \qquad \text{(Starting value is \$1 million)}$$

Use Euler's method with $n = 50$ to estimate the value of the company after 5 years.

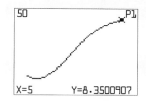

The screen on the left shows the graph from the program EULER on the interval [0, 5] beginning with $y(0) = 1$ and using 50 line segments, along with the coordinates of the final point, which gives the value of the company.

The value of the company will be $8.35 million.

---

**Solutions to Practice Problem**

**a.** $(3, 6)$

**b.** Slope: $y' = g(3, 6) = 7 \cdot 3 - 2 \cdot 6 = 21 - 12 = 9$   Since $g(x, y) = 7x - 2y$

## 9.4 Section Summary

For an initial value problem of the form $\begin{cases} y' = g(x, y) \\ y(a) = y_0 \end{cases}$, Euler's method gives an *approximation* of the solution on an interval $[a, b]$. The approximation consists of $n$ connected line segments, each on an interval of length $h = (b - a)/n$, beginning from the initial point at $x_0 = a$, with slope calculated from $y' = g(x, y)$ evaluated at the left-hand endpoint of the segment. The endpoints of these segments, after the initial point $(x_0, y_0)$ with $x_0 = a$, are calculated successively from the formulas

$$\left. \begin{array}{l} x_i = x_{i-1} + h \\ y_i = y_{i-1} + g(x_{i-1}, y_{i-1}) \cdot h \end{array} \right\} \ i = 1, 2, \ldots, n$$

The accuracy generally increases as the number of segments $n$ increases, but at the cost of more calculation. While the calculations may be carried out "by hand" (as in Example 1), they are easily done on a graphing calculator or computer.

## Calculator Program for Euler's Method

The following is a brief program for Euler's method. The small print in blue is explanation, not part of the program.

### TI-83/84 Program for Euler's Method

| | |
|---|---|
| Disp "USES FUNCTION Y₁" | Y+Y₁*H→Y |
| Prompt A, B    *beginning, ending x-values* | X+H→X |
| Prompt Y    *beginning y-value* | Disp "X=", X |
| Prompt N    *number of segments* | Disp "Y=", Y |
| (B-A)/N→H | Disp "PRESS ENTER" |
| A→X | PAUSE |
| For (I, 1, N) | END |

## 9.4 Exercises

1. If the solution to $y' = 5x - 4y$ passes through the point (4, 3), what is the slope of the solution at that point?

2. If the solution to $y' = x + 4y$ passes through the point (3, 1), what is the slope of the solution at that point?

3. For the initial value problem $\begin{cases} y' = 4xy \\ y(1) = 3 \end{cases}$, state the initial point $(x_0, y_0)$ and calculate the slope of the solution at this point.

4. For the initial value problem $\begin{cases} y' = x/y \\ y(6) = 2 \end{cases}$, state the initial point $(x_0, y_0)$ and calculate the slope of the solution at this point.

**5–10.** For each initial value problem, calculate the Euler approximation for the solution on the interval [0, 1] using $n = 4$ segments. Draw the graph of your approximation. (Carry out the calculations "by hand" with the aid of a calculator, rounding to two decimal places. Answers may differ slightly, depending on when you do the rounding.)

5. $y' = 3x - 2y$
   $y(0) = 2$

6. $y' = x + 2y$
   $y(0) = 1$

7. $y' = 4xy$
   $y(0) = 1$

8. $y' = 8x^2 - y$
   $y(0) = 2$

9. $y' + 2y = e^{4x}$
   $y(0) = 2$

10. $y' + e^y = 8x$
    $y(0) = 0$

**11–16.** For each initial value problem:

**a.** Use EULER or a similar program (see the previous page) to find the estimate for $y(2)$. Use the interval $[0, 2]$ with $n = 50$ segments.

**b.** Solve the differential equation and initial condition *exactly* by separating variables or using an integrating factor.

**c.** Evaluate the solution that you found in part (b) at $x = 2$. Compare this *actual* value of $y(2)$ with the *estimate* of $y(2)$ that you found in part (a).

**11.** $y' = \dfrac{x}{y}$
$y(0) = 1$

**12.** $y' = -xy$
$y(0) = 1$

**13.** $\dfrac{dy}{dx} = 0.2y$
$y(0) = 1$

**14.** $\dfrac{dy}{dx} = -y$
$y(0) = 1$

**15.** $y' + y = 2e^x$
$y(0) = 5$

**16.** $y' + y = e^{-x}$
$y(0) = 0$

 **17–22.** For each initial value problem, use EULER or a similar program (see the previous page) to find the approximate solution at the stated $x$-value, using 50 segments. [*Hint:* Use an interval that begins at the initial $x$-value and ends at the stated $x$-value.]

**17.** $y' = xe^{-y}$
$y(1) = 0.5$
Approximate the solution at $x = 3$

**18.** $y' = e^{x/y}$
$y(1) = 0.4$
Approximate the solution at $x = 3$

**19.** $y' = \dfrac{x + 1}{y + 1}$
$y(1) = \dfrac{1}{2}$
Approximate the solution at $x = 2.5$

**20.** $y' = 1 + \dfrac{y}{x}$
$y(2) = -1$
Approximate the solution at $x = 3.5$

**21.** $\dfrac{dy}{dx} = \sqrt{x + y}$
$y(3) = 1$
Approximate the solution at $x = 3.8$

**22.** $\dfrac{dy}{dx} = (x - y)^2$
$y(2) = 0$
Approximate the solution at $x = 2.8$

 **23–24.** Use EULER or a similar program (see the previous page) to find the approximate solution at the stated $x$-value, using the given numbers of segments.

**23.** $y' = e^{x/y}$
$y(0) = 1$
Approximate the solution at $x = 1$ using:
**a.** $n = 10$
**b.** $n = 100$
**c.** $n = 500$

**24.** $y' = x^2 + y^2$
$y(0) = 0$
Approximate the solution at $x = 2$ using:
**a.** $n = 10$
**b.** $n = 100$
**c.** $n = 500$

## Applied Exercises

**25. BUSINESS: Sales Growth** Sales of a popular new computer game are growing so that the number $y(t)$ (in thousands) sold in the first $t$ months satisfies

$$y' = 2e^{0.05y} + 2$$
$$y(0) = 0$$

Use Euler's method with $n = 3$ to estimate $y(3)$, the number sold in the first 3 months. (Carry out the calculations "by hand" with the aid of a calculator, rounding to two decimal places.)

**26. BUSINESS: Value of a Company** A biotechnology company is growing so that its value $y(t)$ (in millions of dollars) after $t$ years satisfies

$$y' = \frac{1}{5}e^{\sqrt{y}} + 1$$
$$y(0) = 1 \qquad \text{(Initially worth \$1 million)}$$

Use Euler's method with $n = 4$ to estimate $y(2)$, the company's value after 2 years. (Carry out the calculations "by hand" with the aid of a calculator, rounding to two decimal places.)

 **27. GENERAL: Dam Sediment** Sediment is flowing into a lake behind a hydroelectric dam at the rate of 1000 tons per year, while 10% of the upper layer of accumulated sediment is flushed through the turbines each year. The amount $y(t)$ of sediment after $t$ years satisfies

$$\frac{dy}{dt} = 1000 - 0.1y^{2/3}$$
$$y(0) = 0 \qquad \text{(Initially no sediment)}$$

Use EULER or a similar program with $n = 50$ to estimate $y(2)$, the amount of sediment after 2 years.

**28. ENVIRONMENTAL SCIENCE: Crop Parasite** A crop parasite is spreading so that the affected area $y(t)$ (in square miles) in $t$ years is growing according to

$$\frac{dy}{dt} = \sqrt{y^2 + 100}$$
$$y(0) = 5 \qquad \text{(Initially affects 5 square miles)}$$

Use EULER or a similar program with $n = 50$ to estimate the area affected by the parasite after 1 year.

 **29. BIOMEDICAL: Genetics** Population genetics involves initial value problems of the form

$$q' = -0.01q^2(1 - q)$$
$$q(0) = 0.9$$

in which $q(t)$ is the frequency of a gene after $t$ generations of evolution. Use EULER or a similar program with $n = 100$ to approximate $q(200)$, the gene frequency after 200 generations.

 **30. BIOMEDICAL: Epidemics** An epidemic on a campus of 5000 students begins with 5 cases. An immunization program begins immediately, with vaccine shots given to 100 students per day. The total number of cases $y(t)$ after $t$ days satisfies

$$y' = 0.0001y(5000 - y - 100t)$$
$$y(0) = 5$$

**a.** Use EULER or a similar program with $n = 50$ to estimate $y(4)$, the total number of cases after 4 days.

**b.** Find the answer to part (a) if 300 students (instead of 100) are vaccinated each day.

## Conceptual Exercises

**31.** When using Euler's method to solve $y' = g(x, y)$, suppose that the function $g(x, y)$ *increases* as $x$ increases (for every fixed value of $y$). Will the approximation *under*estimate or *over*estimate the exact solution? [*Hint:* Think of the graphs of the true solution and the approximation.]

**32.** Suppose that you use Euler's method to solve $y' = g(x, y)$ with initial condition $y(0) = 5$

on the interval [0, 2] with just *one* segment. What will be the value of $y(2)$? [*Hint:* No calculation needed—your answer will involve $g$.]

**33.** When will a differential equation of the form $y' = g(x, y)$ be *separable*? That is, what must be the form of $g(x, y)$? [*Hint:* See page 631.]

**34.** When will a differential equation of the form $y' = g(x, y)$ represent *logistic growth*? That is, what must be the form of $g(x, y)$? [*Hint:* See page 650.]

**35.** When will a differential equation of the form $y' = g(x, y)$ be *first-order linear*? That is, what must be the form of $g(x, y)$? [*Hint:* See page 659.]

**36.** What happens with Euler's method when the function $g(x, y)$ is a *constant*? To be specific, suppose that you use Euler's method to solve $y' = 7$ with initial condition $y(0) = 1$ on the interval [0, 2]. What will be your approximation for $y(2)$, and how accurate will it be? [*Hint:* What will be the slope of each segment? Since the segments are attached together, how many segments do you really need?]

**37.** In Euler's method applied to $y' = g(x, y)$, the slope of each segment is chosen to be $g$ evaluated at the *left-hand endpoint* of the segment. If $g$ is in fact a function of $x$ alone, can you think of a better way to choose the slope of the segment?

**38.** If the function $g(x, y)$ in Euler's method turns out to be a function of $x$ *alone*, then approximating the solution to $y' = g(x)$ amounts to approximating the integral of a function on an interval. Long ago (in Section 5.3) we had another way of approximating integrals (areas), using Riemann sums with left rectangles. Is there a connection?

# Chapter Summary with Hints and Suggestions

Reading the text and doing the exercises in this chapter have helped you to master the following concepts and skills, which are listed by section (in case you need to review them) and are keyed to particular Review Exercises. Answers for all Review Exercises are given at the back of the book, and full solutions can be found in the Student Solutions Manual.

## 9.1 Separation of Variables

- Find the general solution of a differential equation by separation of variables.
  *(Review Exercises 1–10.)*

- Find a particular solution of a differential equation with an initial condition.
  *(Review Exercises 11–14.)*

- Solve an applied problem involving a differential equation. *(Review Exercises 15–18.)*

■ Use a program to graph the slope field of a differential equation, and sketch the solution through a given point.
*(Review Exercises 19–20.)*

## 9.2 Further Applications of Differential Equations: Three Models of Growth

■ Choose an appropriate differential equation for an applied problem, and use it to solve the problem. *(Review Exercises 21–28.)*

## 9.3 First-Order Linear Differential Equations

■ Find the general solution of a first-order linear differential equation. *(Review Exercises 29–34.)*

$$y' + p(x) \cdot y = q(x) \qquad I(x) = e^{\int p(x)\,dx}$$

■ Find a particular solution of a first-order linear differential equation. *(Review Exercises 35–38.)*

■ Solve an applied problem involving a first-order linear differential equation.
*(Review Exercises 39–42.)*

## 9.4 Approximate Solutions of Differential Equations: Euler's Method

■ Find an Euler approximation for the solution of a differential equation. Then find the *exact* solution and compare the results. *(Review Exercises 43–44.)*

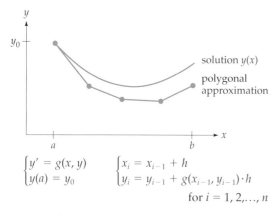

$$\begin{cases} y' = g(x, y) \\ y(a) = y_0 \end{cases} \qquad \begin{cases} x_i = x_{i-1} + h \\ y_i = y_{i-1} + g(x_{i-1}, y_{i-1}) \cdot h \end{cases}$$

$$\text{for } i = 1, 2, \dots, n$$

■ Find an Euler approximation for the solution of a differential equation.
*(Review Exercises 45–46.)*

■ Solve an applied problem by finding an Euler approximation for the solution of a differential equation. *(Review Exercises 47–48.)*

## Hints and Suggestions

■ A differential equation is an equation involving derivatives (rates of change). A solution involving an arbitrary constant is called a *general solution,* while a solution with the arbitrary constant replaced by a number is called a *particular solution.* The constant is determined by an *initial condition,* specifying the value of the solution at a particular point.

■ Solving a differential equation by separation of variables involves moving the $x$'s and $y$'s to opposite sides of the equation and integrating both sides. Many useful differential equations can be solved by this technique, but many cannot. In fact, many differential equations cannot be solved by *any* method.

■ A graphing calculator with a slope field program can show a "picture" of a differential equation of the form $\dfrac{dy}{dx} = f(x, y)$, drawing little slanted dashes with the correct slopes at many points of the plane. A solution can then be drawn through a given point following the indicated slopes.

■ With practice, you can *guess* the integrating factor that makes a first-order linear differential equation integrable. If not, use the formula on page 661.

■ Euler's method is easily carried out on a graphing calculator or a computer.

■ **Practice for Test:** Review Exercises 1, 13, 15, 19, 23, 25, 29, 35, 41, 45, 47.

# Review Exercises for Chapter 9

Practice test exercise numbers are in green.

## 9.1 Separation of Variables

**1–10.** Find the general solution of each differential equation.

**1.** $y^2 y' = x^2$    **2.** $y' = x^2 y$    **3.** $y' = \dfrac{x^3}{x^4 + 1}$

**4.** $y' = x e^{-x^2}$    **5.** $y' = y^2$    **6.** $y' = y^3$

**7.** $y' = 1 - y$    **8.** $y' = \dfrac{1}{y}$

**9.** $y' = xy - y$    **10.** $y' = x^2 + x^2 y$

**11–14.** Solve each differential equation and initial condition.

**11.** $y^2 y' = 3x^2$
$y(0) = 1$

**12.** $y' = \dfrac{y}{x^2}$
$y(1) = 1$

**13.** $y' = \dfrac{y}{x^3}$
$y(1) = 1$

**14.** $y' = \sqrt[3]{y}$
$y(1) = 0$

**15. PERSONAL FINANCE: Wealth Accumulation**
Suppose that you now have $10,000 and that you expect to save an additional $4000 during each year, and all of this is deposited in a bank paying 5% interest compounded continuously. Let $y(t)$ be your bank balance (in thousands of dollars) after $t$ years.

a. Write a differential equation and initial condition to model your bank balance.
b. Solve your differential equation and initial condition.
c. Use your solution to find your bank balance after 10 years.

**16. ENVIRONMENTAL SCIENCE: Pollution** A town discharges 4 tons of pollutant annually into a once-pristine lake, and each year bacterial action removes 25% of the accumulated pollution.

a. Write a differential equation and initial condition for the amount of pollution in the lake.
b. Solve your differential equation to find a formula for the amount of pollution in the lake after $t$ years.

**17. BIOMEDICAL: Fever Thermometers** How long should you keep a thermometer in your mouth to take your temperature? *Newton's Law of Cooling* says that the thermometer reading rises at a rate proportional to the difference between your actual temperature and the present reading. For a fever of 106 degrees Fahrenheit, the thermometer reading $y(t)$ after $t$ minutes in your mouth satisfies $y' = 2.3(106 - y)$ with $y(0) = 70$ (initially at room temperature). (The constant 2.3 is typical for household thermometers.) Solve this differential equation and initial condition.

**18. BIOMEDICAL: Fever Thermometers** (*continuation*) Use your solution $y(t)$ in Exercise 17 to calculate $y(1)$, $y(2)$, and $y(3)$, the thermometer readings after 1, 2, and 3 minutes. Do you see why 3 minutes is the usually recommended time for keeping the thermometer in your mouth?

 **19–20.** For each differential equation and initial condition:

a. Use SLOPEFLD or a similar program to graph the slope field for the differential equation on the

window $[-5, 5]$ by $[-5, 5]$.
b. Sketch the slope field on a piece of paper and draw a solution curve that follows the slopes and that passes through the point $(0, -2)$.
c. Solve the differential equation and initial condition.
d. Use SLOPEFLD or a similar program to graph the slope field and the solution that you found in part (c). How good was the sketch that you made in part (b) compared with the solution graphed in part (d)?

**19.** $\begin{cases} \dfrac{dy}{dx} = \dfrac{x^2}{y^2} \\ y(0) = -2 \end{cases}$

**20.** $\begin{cases} \dfrac{dy}{dx} = \dfrac{x}{y^2} \\ y(0) = -2 \end{cases}$

## 9.2 Further Applications of Differential Equations: Three Models of Growth

For each situation, write an appropriate differential equation (unlimited, limited, or logistic). Then find its solution and solve the problem.

**21. GENERAL: Postage Stamps** On average, the price of a first-class postage stamp grows continuously at the rate of about 5% per year. If in 2008 the price was 42 cents, estimate the price in the year 2018.

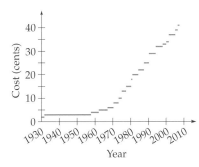

*Source:* U.S. Postal Service

**22. ECONOMICS: Computer Expenditure** The amount spent by Americans on computers and related hardware is increasing continuously at the rate of 12% per year. If in 2005 the amount was $30.6 billion, estimate the amount in the year 2015.

**23. BIOMEDICAL: Epidemics** A virus spreads through a university community of 8000 people at a rate proportional to both the number already infected and the number not yet infected. If it begins with 10 cases and grows in a week to 150 cases, estimate the size of the epidemic after 2 weeks.

**24. SOCIAL SCIENCE: Rumors** A rumor spreads through a school of 500 students at a rate proportional to both the number who have heard and the number who have not heard the rumor. If the rumor began with 2 students and within a day had spread to 75, how many students will have heard the rumor within 2 days?

**25. BUSINESS: Total Sales** A manufacturer estimates that he can sell a maximum of 10,000 DVD players in a city. His total sales grow at a rate proportional to how far they are below this upper limit. If after 7 months the total sales are 3000, find a formula for the total sales after $t$ months. Then use your answer to estimate the total sales at the end of the first year.

**26. BEHAVIORAL SCIENCE: Learning** Suppose that the maximum rate at which a mail carrier can sort letters is 60 letters per minute, and that she learns at a rate proportional to her distance from this upper limit. If after 2 weeks on the route she can sort 25 letters per minute, how many weeks will it take her to sort 50 letters per minute?

**27. BUSINESS: Advertising** A new product is advertised extensively on television to a city of 500,000 people, and the number of people who have seen the ads increases at a rate proportional to the number who have not yet seen the ads. If within 2 weeks 200,000 have seen the ads, how long must the product be advertised to reach 400,000 people?

**28. BUSINESS: Sales** A company estimates the maximum market for fax (facsimile transmission) machines in a city to be 40,000. Sales are growing in proportion to both the number already sold and the size of the remaining market. If 1000 fax machines have been sold at time $t = 0$, and after 1 year 4000 have been sold, estimate how long it will take for 20,000 to be sold.

## 9.3 First-Order Linear Differential Equations

**29–34.** Solve each first-order linear differential equation.

**29.** $xy' - 5y = 4x^7$

**30.** $xy' + \dfrac{1}{2}y = 3x$

**31.** $y' + xy = x$

**32.** $y' - 2xy = -4x$

**33.** $xy' + y = xe^x$

**34.** $xy' + y = x \ln x$

[*Hint for Exercises 33 and 34:* Use integration by parts or an integral table.]

**35–38.** Solve each first-order linear differential equation and initial condition.

**35.** $y' - xy = 0$
$y(0) = 3$

**36.** $y' + 4xy = 0$
$y(0) = 2$

**37.** $xy' + 2y = 6x$
$y(1) = 0$

**38.** $xy' - y = 4$
$y(1) = 6$

**39. BUSINESS: Investment Fund** An investment fund grows in value by 5% each year, and also attracts new capital at the rate of $10e^{0.05t}$ thousand dollars per year, where $t$ is the number of years since the fund was established. Therefore, its value $y(t)$ (in thousands of dollars) satisfies

$$y' = 0.05y + 10e^{0.05t}$$
$$y(0) = 100 \qquad \text{(Initial value \$100 million)}$$

**a.** Solve this differential equation and initial condition.
**b.** Use your answer to part (a) to find the value of the fund after 5 years.
 **c.** Graph the solution you found in part (a) on a graphing calculator, using an appropriate window. Find when the fund will reach a quarter of a million dollars. [*Hint:* Use INTERSECT with $y_2 = 250$.]

**40. GENERAL: Population Growth with Immigration** A country's population has a natural growth rate of 2%, and also accepts immigrants amounting to 1% of a neighboring population of $10e^{0.01}t$ million at time $t$ years. Therefore, the population $p(t)$ (in millions) satisfies

$$p' = 0.02p + 0.1e^{0.01t}$$
$$p(0) = 100 \qquad \text{(Initial population)}$$

**a.** Solve this differential equation and initial condition.
**b.** Use your answer to part (a) to estimate the population at time $t = 10$ years.
 **c.** Graph the solution you found in part (a) on a graphing calculator, using an approprite window. Find when the population will reach 150 million. [*Hint:* Use INTERSECT with $y_2 = 150$.]

**41. ENVIRONMENTAL SCIENCE: Air Quality** The air in a stuffy 10,000-cubic-foot lecture hall is only 10% oxygen. Each minute, 500 cubic feet of air are exchanged with the outside (containing 20% oxygen). Assume that the air in the room mixes thoroughly.

**a.** Write a differential equation and initial condition to describe the total amount $y(t)$ of oxygen in the room after $t$ minutes.

**b.** Solve the differential equation and initial condition.

**c.** Use your solution to find the amount of oxygen after 15 minutes of ventilation.

**d.** Use your solution to find the long-run oxygen content of the room.

 **e.** Graph the solution you found in part (b) on a graphing calculator, using an appropriate window. Find when the oxygen level will reach 19%. [*Hint:* Use INTERSECT with $y_2$ equal to the amount of oxygen for the 19% level.]

**42. ENVIRONMENTAL SCIENCE: Bird Population**
The natural growth rate of a population of bald eagles in a nature sanctuary is 20% per year, and two more eagles are added to the population each year. Therefore, the population $p(t)$ $t$ years from now satisfies

$$p' = 0.2p + 2$$
$$p(0) = 5 \qquad \text{(Initial population)}$$

**a.** Solve this differential equation and initial condition.

**b.** Use your solution to predict the eagle population after 2 years.

 **c.** Graph the solution you found in part (b) on a graphing calculator, using an appropriate window. Find when the eagle population will reach 24.

## 9.4 Approximate Solutions of Differential Equations: Euler's Method

**43–44.** For each initial value problem:

**a.** Calculate the Euler approximation on the interval [0, 1] using $n = 4$ segments. Carry out the calculations "by hand" with the aid of a calculator, rounding calculations to two decimal places. Draw the graph of your approximation.

**b.** Solve the differential equation and initial condition *exactly* by separating variables or using an integrating factor.

**c.** Evaluate at $x = 1$ the solution that you found in part (b). Compare this actual value of $y(1)$ (rounded to two decimal places) with the *estimate* for $y(1)$

that you found in part (a). [*Note:* Expect low accuracy using $n = 4$.]

**43.** $y' = 2 - 2y$
$y(0) = 2$

**44.** $y' = \dfrac{4x}{y}$
$y(0) = 1$

 **45–46.** For each initial value problem, use EULER or a similar program (see page 677) to find an estimate for the solution at the stated $x$-value, using 50 segments.

**45.** $y' = \dfrac{1}{y^2 + 1}$, $y(0) = 1$
Approximate the solution at $x = 2$

**46.** $y' = ye^{-x}$, $y(1) = 5$
Approximate the solution at $x = 2.5$

 **47. BUSINESS: Advertising** If a new product becomes known *both* by word of mouth *and* by advertising, then $p(t)$, the proportion of people who have heard of the new product after $t$ weeks, satisfies a differential equation and initial condition of the form

$$p' = 0.1(1 - p) + 0.1p(1 - p)$$
$$p(0) = 0.3 \quad \text{(Initially 30\% know the product)}$$

**a.** Show that the differential equation simplifies to $p' = 0.1(1 - p^2)$

**b.** Use EULER or a similar program (see page 677) with 90 segments to estimate $p(4)$, the proportion of people who have heard of the product after 4 weeks.

 **48. ENVIRONMENTAL SCIENCE: Insect Population** A colony of flying insects occupying a spherical nest grows so that the number $y(t)$ of insects after $t$ weeks satisfies

$$y' = 0.2y - 0.1\sqrt[3]{y}$$
$$y(0) = 100 \qquad \text{(Initially 100 insects)}$$

Use EULER or a similar program (see page 677) with 80 segments to estimate the population after 2 weeks.

# 10 Sequences and Series

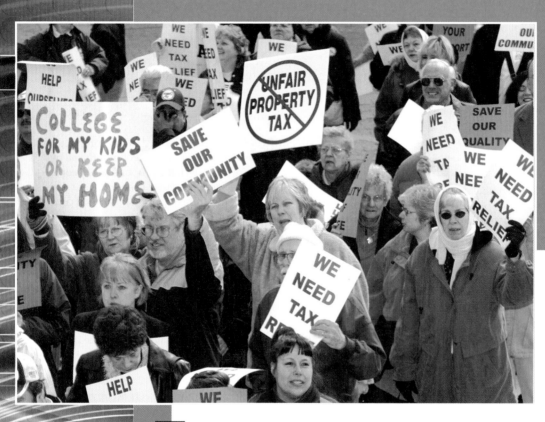

### Tax Cuts and Multiplier Effects

Tax cuts have been much in the news recently, connecting economics, politics, and social philosophy. How could we assess the effect of, say, a $30 billion tax cut on the economy? If divided equally (a highly unlikely event), it would give just $100 to each American. On a personal level, what would you do with your $100? You might spend most of it, say, 80%, amounting to $100 \cdot 0.8 = \$80$, and save the rest. (Economists call this 80% the *marginal propensity to consume,* or *MPC*.) The amount you spend would go to other people, who might similarly spend 80% of their share (in dollars, $100 \cdot 0.8 \cdot 0.8$ or $100 \cdot 0.8^2$), paying others who would in turn spend 80% (or $100 \cdot 0.8^3$), and so on in an endless chain. Therefore, the original $100 would have the effect of generating a total expenditure throughout the economy of

$$100 + 100 \cdot 0.8 + 100 \cdot 0.8^2 + 100 \cdot 0.8^3 + \cdots$$

Such "infinitely long sums" are called *infinite series* and are the subject of this chapter. This particular infinite series sums to 500, *five times* the original $100. This 5 is called the *multiplier,* since any initial amount (in place of the 100) would similarly be multiplied by 5. Scaling up to the national level, the $30 billion tax cut would also be multiplied by 5, thereby generating expenditures of $150 billion throughout the economy.

   A few caveats about the effects of cutting taxes: a tax cut deprives the government of exactly the same amount of money, and the resulting decrease in governmental spending is subject to a roughly similar multiplier, with a countervailing effect on the economy. Also, wealthy people (who can afford to save) have a lower *MPC* than poorer people (who must spend all they have), with the result that a tax cut for the wealthy will have a much smaller multiplier effect in the economy than one for the lower economic classes. We will further analyze the multiplier effect in the first section of this chapter.

## 10.1 GEOMETRIC SERIES

### Introduction

The decimal expansion of $\frac{1}{3}$ is the "infinitely long decimal" $0.333\ldots$, which represents an "infinitely long sum"

$$0.3333\ldots = \frac{3}{10} + \frac{3}{100} + \frac{3}{1000} + \frac{3}{10000} + \cdots$$

Such sums are called **infinite series** and are the subject of this chapter. This first section discusses *geometric* **series,** with applications from the multiplier effect in economics to the cumulative effect of long-term drug dosage. Later sections discuss Taylor series and Newton's method for approximating solutions to equations.

### Finite Geometric Series

A **finite series** is a sum of numbers in a particular order. The numbers are called the **terms** of the series. A **finite** *geometric* **series** is a finite series in which each successive term is obtained from the previous term by multiplying by a fixed number, denoted $r$ (for **ratio**).

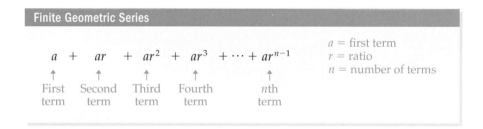

**Finite Geometric Series**

$$a \; + \; ar \; + \; ar^2 \; + \; ar^3 \; + \cdots + ar^{n-1}$$

First term, Second term, Third term, Fourth term, $n$th term

$a =$ first term
$r =$ ratio
$n =$ number of terms

 **BE CAREFUL:** Observe that the *number* of terms is always one more than the final exponent of $r$ (so, the $n$th term has exponent $n-1$).

Some examples of finite geometric series are given below.

$$2 + 4 + 8 + 16 + 32$$

First term is $a = 2$
Ratio is $r = 2$
Number of terms is $n = 5$

$$3 - 30 + 300 - 3000$$

$a = 3,\; r = -10,\; n = 4$

$$5 + \frac{5}{2} + \frac{5}{2^2} + \frac{5}{2^3} + \frac{5}{2^4} + \cdots + \frac{5}{2^{100}}$$

$a = 5,\; r = \frac{1}{2},\; n = 101$
($n$ is one more than the final power of $r$)

A series is a sum, and so it may be evaluated by adding the terms. For the first two examples above, the sums are easily found to be 62 and $-2727$ (as you may check). For the last example, adding up the 101 terms would be extremely tedious. Shortly we will develop a *formula* for summing such a series.

Not all series are geometric. The following series is *not* geometric, since successive terms are found by adding 1 to the previous denominator, *not* by multiplying by a fixed ratio.

$$\frac{1}{1} + \frac{1}{2} + \frac{1}{3} + \frac{1}{4} + \frac{1}{5} + \frac{1}{6}$$

**Practice Problem 1**

For each series below, if it is geometric, find the first term $a$, the ratio $r$, and the number of terms $n$. If the series is not geometric, say so.

**a.** $1 - \pi + \pi^2 - \pi^3 + \pi^4 - \pi^5$

**b.** $2 + \dfrac{2}{10} + \dfrac{2}{10^2} + \dfrac{2}{10^3} + \cdots + \dfrac{2}{10^8}$

**c.** $0.01 + 0.02 + 0.03 + 0.04 + 0.05$

➤ Solutions on page 694

To find a formula for the *sum* of a finite geometric series, we denote the sum by $S_n$:

$$S_n = a + ar + ar^2 + \cdots + ar^{n-1}$$

From this we subtract $r$ times $S_n$:

| | | |
|---|---|---|
| $S_n \quad = a + ar + ar^2 + \cdots + ar^{n-1}$ | | $S_n$ (sum of first $n$ terms) |
| $\underline{rS_n \quad = \quad\quad ar + ar^2 + \cdots + ar^{n-1} + ar^n}$ | | $S_n$ multiplied by $r$ |
| $S_n - rS_n = a \qquad\qquad\qquad\qquad\qquad -ar^n$ | | Subtracting (most terms cancel) |
| $S_n(1 - r) = a(1 - r^n)$ | | Factoring each side of $S_n - rS_n = a - ar^n$ |
| $S_n = a\dfrac{1 - r^n}{1 - r}$ | | Dividing each side by $1 - r$ (for $r \neq 1$) |

This formula is very useful for summing a finite geometric series.

**Sum of a Finite Geometric Series**

$$a + ar + ar^2 + \cdots + ar^{n-1} = a\frac{1 - r^n}{1 - r} \qquad (r \neq 1)$$

*Memorization hint:* The power $n$ on the right is the first power *left out* of the series on the left.

**EXAMPLE 1**

**SUMMING CUMULATIVE FINES**

In the summer of 1988, the city of Yonkers, New York, was fined for refusing to build low-income housing. The fine was $100 for the first day, $200 for the second day, $400 for the third day, and so on, with the fines doubling each day until the city council approved the housing. Find the total fine for the 14 days until the housing was approved.

*Source: New York Times*

*Solution*

$$\left(\begin{array}{c}\text{Total}\\\text{fine}\end{array}\right) = \underbrace{100}_{\text{1st day}} + \underbrace{2 \cdot 100}_{\text{2nd day}} + \underbrace{4 \cdot 100}_{\text{3rd day}} + \cdots + \underbrace{2^{13} \cdot 100}_{\text{14th day}}$$

Geometric series (final power is 1 less than the final day)

The sum formula gives

$$\left(\begin{array}{c}\text{Total}\\\text{fine}\end{array}\right) = 100\,\frac{1-2^{14}}{1-2} = 100\,\frac{1-\underbrace{16{,}384}_{16{,}383}}{-1} = 1{,}638{,}300$$

$a\,\dfrac{1-r^n}{1-r}$ with $a = 100$, $r = 2$, and $n = 14$

The total fine was $1,638,300.

**Practice Problem 2**

Sum the finite geometric series $3 + 3 \cdot 2 + 3 \cdot 2^2 + \cdots + 3 \cdot 2^5$.

➤ Solutions on page 694

## Annuities

An **annuity** is a series of equal payments at regular intervals. The total value or *amount* of an annuity, including interest, may be found from the formula for the sum of a geometric series.

**EXAMPLE 2**

**FINDING THE AMOUNT OF AN ANNUITY**

At the end of each month for 5 years you deposit $200 in a bank at 6% interest compounded monthly. Find the amount of this annuity.

*Solution*

At 6% interest compounded monthly, you receive $\frac{1}{2}$% per month, so payments are multiplied by $(1 + 0.005)$ or $(1.005)$ for each month on deposit. In 5 years there will be $5 \cdot 12 = 60$ monthly deposits, beginning with the first $200 (at the end of the first month, and so earning interest for 59 months) and ending with the final $200 (at the end of the last month, and so earning *no* interest). We sum these amounts *in reverse order*:

$$\underbrace{200}_{\substack{\text{Final}\\\text{payment}}} + 200(1.005) + 200(1.005)^2 + \cdots + \underbrace{200(1.005)^{59}}_{\substack{\text{First}\\\text{payment}}}$$

Geometric series with $a = 200$, $r = 1.005$, and $n = 60$ (final power plus 1)

The formula gives

$$\begin{pmatrix} \text{Amount} \\ \text{of annuity} \end{pmatrix} = 200 \, \frac{1 - 1.005^{60}}{1 - 1.005} \approx \underbrace{13{,}954}_{\substack{\text{Using a} \\ \text{calculator}}} \qquad a \, \frac{1 - r^n}{1 - r} \text{ with the stated} \\ \text{values of } a, r, \text{ and } n$$

The amount of the annuity is $13,954.

Note that five years of monthly payments of $200 adds up to only $60 \cdot 200 = \$12{,}000$. The remainder here is interest.

The amount of a *continuous* annuity can be found using differential equations (see Exercises 53–54 on page 641).

### Infinite Series

An **infinite series** is an expression written as a sum of infinitely many terms.*

$$a_1 + a_2 + a_3 + a_4 + \cdots$$

The following are examples of infinite series:

$$1 + 2 + 3 + 4 + 5 + \cdots$$

$$1 + \frac{1}{2} + \frac{1}{4} + \frac{1}{8} + \cdots + \frac{1}{2^n} + \cdots \qquad \frac{1}{2^n} \text{ represents a "typical" term}$$

We cannot sum an infinite series simply by adding the terms, because adding infinitely many numbers one-by-one would be impossible. Instead, we sum an infinite series by first finding the **partial sum** $S_n$ of the first $n$ terms, and then taking the *limit* as $n$ approaches infinity. The partial sums of the infinite series $a_1 + a_2 + a_3 + \cdots$ are defined as follows:

$S_1 = a_1$ — First partial sum is the first term

$S_2 = a_1 + a_2$ — Second partial sum is the sum of the first *two* terms

$S_3 = a_1 + a_2 + a_3$ — Third partial sum is the sum of the first *three* terms

$\vdots$

$S_n = a_1 + a_2 + a_3 + \cdots + a_n$ — $n$th partial sum is the sum of the first $n$ terms

$\vdots$

If the *limit* of this sequence $S_1, S_2, S_3, \dots, S_n, \dots$ as $n$ approaches infinity *exists* and equals a number $S$, then the series is said to **converge** to $S$, and $S$ is the **sum of the series.**

---

*We will sometimes begin with a term $a_0$, writing the series as $a_0 + a_1 + a_2 + \cdots$. In Section 10.3 we will consider infinite series whose terms are *functions*.

> ### Convergence of an Infinite Series
>
> For an infinite series $a_1 + a_2 + a_3 + \cdots$, if the partial sums $S_n$ approach a limit $S$,
>
> $$\lim_{n \to \infty} S_n = S$$
>
> then the series is *convergent*, and *converges* to $S$:
>
> $$S = a_1 + a_2 + a_3 + \cdots$$

An infinite series that does *not* converge is said to **diverge,** or be a **divergent series,** and therefore has no sum. The procedure of first summing a *finite* number $n$ of terms and then letting $n$ approach infinity is another example of dealing with the infinite by *dropping back to the finite and taking the limit* (see page 440).

## Infinite Geometric Series

An infinite series in which each successive term is a fixed multiple of the previous term is called an **infinite geometric series.**

> ### Infinite Geometric Series
>
> $$a + ar + ar^2 + ar^3 + \cdots$$
>
> $a$ = first term
> $r$ = ratio

The partial sums of an infinite geometric series are *finite* geometric series, for which we have a formula.

$$S_n = a \frac{1 - r^n}{1 - r}$$

Partial sum of an infinite geometric series (from page 687)

The sum of an *infinite* geometric series is then the *limit* of $S_n$ as $n$ approaches infinity. If the ratio $r$ satisfies $|r| < 1$, then $r^n$ will approach zero as $n \to \infty$.

$$\lim_{n \to \infty} r^n = 0 \qquad \text{for } |r| < 1$$

## Graphing Calculator Exploration

Verify this last statement by choosing an $r$ satisfying $|r| < 1$ and raising it to a high power. Try a few different values of $r$ and $n$.

$\left(\frac{1}{2}\right)^{25} \longrightarrow$ | (.5)^25
| 2.980232239E-8 | $\longleftarrow$ means 0.0000000298 ...
$\left(-\frac{9}{10}\right)^{200} \longrightarrow$ | (-.9)^200
| 7.05507911E-10 | $\longleftarrow$ means 0.0000000007055 ...

In each case the result is very small, and the limit is zero.

The sum $S$ of the infinite series is then found by taking the limit of the partial sums, and for $|r| < 1$,

Limit is zero

$$S = \lim_{n \to \infty} S_n = \lim_{n \to \infty} \underbrace{\left( a \frac{1 - r^n}{1 - r} \right)}_{S_n} = a \frac{1 - 0}{1 - r} = a \frac{1}{1 - r} = \frac{a}{1 - r}$$

The last expression gives a formula for the sum of a convergent infinite geometric series.

---

**Sum of an Infinite Geometric Series**

An infinite geometric series with $|r| < 1$ is convergent, and its sum is given by

$$a + ar + ar^2 + ar^3 + \cdots = \frac{a}{1 - r} \qquad \text{for } |r| < 1$$

An infinite geometric series with $|r| \geq 1$ is divergent.

---

The divergence of infinite geometric series with $|r| \geq 1$ will be shown in Exercise 82.

**EXAMPLE 3**

**SUMMING GEOMETRIC SERIES**

**a.** $\dfrac{1}{2} + \dfrac{1}{4} + \dfrac{1}{8} + \dfrac{1}{16} + \cdots = \dfrac{\frac{1}{2}}{1 - \frac{1}{2}} = \dfrac{\frac{1}{2}}{\frac{1}{2}} = 1$ $\qquad \dfrac{a}{1 - r}$ with first term $a = \frac{1}{2}$ and ratio $r = \frac{1}{2}$

**b.** $8 - \dfrac{8}{3} + \dfrac{8}{9} - \dfrac{8}{27} + \cdots = \dfrac{8}{1 - \left( -\frac{1}{3} \right)} = \dfrac{8}{\frac{4}{3}}$ $\qquad \dfrac{a}{1 - r}$ with $a = 8$ and $r = -\frac{1}{3}$

$$= 8 \cdot \dfrac{3}{4} = 6$$

**c.** $1 + \dfrac{3}{2} + \dfrac{9}{4} + \dfrac{27}{8} + \cdots \quad$ *diverges.* $\qquad r = \dfrac{3}{2}$ so $|r| > 1$

---

**Practice Problem 3**

Find the sum of the series $2 + 2\left( \dfrac{3}{5} \right) + 2\left( \dfrac{3}{5} \right)^2 + 2\left( \dfrac{3}{5} \right)^3 + \cdots$

➤ **Solution on page 694**

For medications taken over an extended period of time, the amount in the bloodstream at any time is the sum of small contributions from all past doses. This amount can be approximated by an infinite series.

EXAMPLE 4  **FINDING LONG-TERM DRUG CONCENTRATION**

A patient takes a daily dose of 150 milligrams (mg) of a drug, and each day 70% of the accumulated drug in the bloodstream is absorbed by the body. Find the long-term maximum and minimum amounts of the drug in the bloodstream.

*Solution*

One day after taking a dose, 70% of that dose will have been absorbed, so 30% of it will remain. After a second day, only 30% of that 30% will remain, with proportional reductions on subsequent days. The maximum amount in the bloodstream will occur just after a new dose, and it will be the most recent 150 mg, plus 30% of the previous 150 mg, plus $(0.30)^2$ of the dose previous to that, and so on, giving the infinite series

$$\binom{\text{Maximum}}{\text{amount}} = 150 + 150(0.3) + 150(0.3)^2 + \cdots$$

$$= \frac{150}{1 - 0.3} = \frac{150}{0.7} \approx 214 \qquad \text{Using } \frac{a}{1 - r} \text{ with } a = 150 \text{ and } r = 0.3.$$

Therefore, the long-term amount in the bloodstream just after a dose will be 214 mg. Just before the next dose, 70% of this amount will have been absorbed, leaving 30%, or $0.30 \cdot 214 \approx 64$ mg.

The long-term amount in the bloodstream will vary between 214 mg and 64 mg.

These maximum and minimum amounts are important in calculating dosages that avoid toxicity and maintain efficacy. Notice that we used an *infinite* series even though the number of past doses is certainly finite. The *approximation of the finite by the infinite* is permissible because the contribution from earlier doses rapidly becomes negligible. It also has the advantage that the infinite formula is *simpler* than the finite formula.

## Graphing Calculator Exploration

```
sum(seq(150*.3^X
,X,0,4,1))
           213.765
```

Show that the sum of the first five terms of the series in the preceding example is already very close to the sum of the *infinite* series. [*Hint:* Sum a sequence of terms of the form $150 \cdot (0.3)^x$ with $x$ going from 0 to 4 by increments of 1, as shown on the left.]

The long-term amount for medication given continuously (intravenously) may be found using differential equations (see Exercise 56 on page 641).

## Multiplier Effect in Economics

As explained more fully in the Application Preview on page 685, the **multiplier effect** shows how the effects of purchases spread throughout an economy. Suppose that you pay someone $1000. The recipient may spend a part of it (say 80%, or $800) and save the rest. The $800 spent goes to other recipients, who in turn spend a part (say 80%) for purchases of their own and save the rest. If this continues, with each recipient spending 80% of what is received, the result will be an infinite series of decreasing expenditures throughout the economy. We will see that the sum of all of these expenditures is a *multiple* of the original expenditure (hence the term "multiplier effect"). The percentage spent (here 80%) is called the "marginal propensity to consume" or *MPC*.

**EXAMPLE 5**  **FINDING THE MULTIPLIER**

If the marginal propensity to consume is 80%, find the total of all of the expenditures generated by an initial expenditure of $1000, and find the multiplier.

*Solution*

The sum of all of the expenditures, beginning with the original $1000, is

$$1000 + 1000(0.8) + 1000(0.8)^2 + \cdots = \frac{1000}{1 - 0.8}$$

$\frac{a}{1 - r}$ with $a = 1000$ and $r = 0.8$

$$= \frac{1000}{0.2} = 5000$$

The original expenditure of $1000 generated a total of $5000 of expenditures in the economy, so the multiplier is 5.

Notice that the $1000 can be replaced by *any* initial expenditure, showing that the multiplier is independent of the initial amount, depending only on the *MPC*. The multiplier is used by economists to assess the effects of tax increases and cuts.

## Graphing Calculator Exploration

How many terms in the series in the preceding example are required for the sum to reach 4500? To reach 4900? [*Hint:* Use the same operations as in the previous Graphing Calculator Exploration.]

## Summation Notation

A series may be written very compactly using the symbol $\Sigma$ (the Greek capital letter "sigma" corresponding to our S). $\Sigma$ is followed by a formula for the terms, using an "index variable" that takes the values between those indicated at the bottom and top of the $\Sigma$ (similar to the notation for definite integrals). An infinity symbol ($\infty$) at the top indicates an *infinite* series.

| **Terms written out** | **Sigma notation** | |
|---|---|---|
| $2 + 4 + 8 + \cdots + 2^{10} =$ | $\displaystyle\sum_{i=1}^{10} 2^i$ | $\leftarrow$ Last value of index $i$ is 10<br>$\leftarrow$ Formula for terms<br>$\leftarrow$ First value of index $i$ is 1 |
| $\dfrac{1}{3} + \dfrac{1}{4} + \dfrac{1}{5} + \dfrac{1}{6} + \cdots =$ | $\displaystyle\sum_{k=3}^{\infty} \dfrac{1}{k}$ | $\leftarrow \infty$ means an *infinite series*<br>$\leftarrow$ Formula for terms<br>$\leftarrow$ First value of index $k$ is 3 |

Read: the sum of $1/k$ from $k$ equals 3 to infinity

---

**Solutions to Practice Problems**

**1. a.** $a = 1, \quad r = -\pi, \quad n = 6$

   **b.** $a = 2, \quad r = \frac{1}{10}, \quad n = 9$

   **c.** Not geometric

**2.** $3 + 3 \cdot 2 + 3 \cdot 2^2 + \cdots + 3 \cdot 2^5 = 3\dfrac{1 - 2^6}{1 - 2} = 3\dfrac{1 - 64}{-1} = 3 \cdot 63 = 189$

**3.** $2 + 2\left(\dfrac{3}{5}\right) + 2\left(\dfrac{3}{5}\right)^2 + 2\left(\dfrac{3}{5}\right)^3 + \cdots = \dfrac{2}{1 - \dfrac{3}{5}} = \dfrac{2}{\dfrac{2}{5}} = 2 \cdot \dfrac{5}{2} = 5$

---

## 10.1 Section Summary

A series is *geometric* if multiplying any term by a fixed number gives the next term. The sum of a *finite* geometric series is given by the following formula:

First power omitted

$$\underbrace{\sum_{k=0}^{n-1} ar^k}_{\substack{\text{In sigma}\\\text{notation}}} = \underbrace{a + ar + ar^2 + \cdots + ar^{n-1}}_{\substack{\text{Terms written out}\\(n = \text{number of terms})}} = a\frac{1 - r^n}{1 - r} \qquad \text{for } r \neq 1$$

An *infinite* geometric series converges if the ratio $r$ satisfies $|r| < 1$, and diverges if $|r| \geq 1$. The sum of a convergent infinite geometric series is given by the following formula:

$$\underbrace{\sum_{k=0}^{\infty} ar^k}_{\substack{\text{In sigma} \\ \text{notation}}} = \underbrace{a + ar + ar^2 + ar^3 + \cdots}_{\text{Terms written out}} = \frac{a}{1-r} \quad \text{for } |r| < 1$$

Notice that the "infinite" formula is simpler than the "finite" formula. These formulas for the sums of *geometric* series are particularly important since there are no such formulas for most (nongeometric) series.

## 10.1 Exercises

**1–6.** Write out each finite series.

**1.** $\displaystyle\sum_{i=1}^{5} \frac{1}{i+1}$    **2.** $\displaystyle\sum_{i=1}^{6} \frac{i}{i+1}$

**3.** $\displaystyle\sum_{k=1}^{4} \left(-\frac{1}{3}\right)^k$    **4.** $\displaystyle\sum_{k=1}^{5} \frac{(-1)^k}{k^2}$

**5.** $\displaystyle\sum_{n=1}^{6} \frac{1-(-1)^n}{2}$    **6.** $\displaystyle\sum_{n=1}^{4} (-n)^n$

**7–12.** Write each infinite series in sigma notation, beginning with $\sum_{i=1}^{\infty}$.

**7.** $\dfrac{1}{3} + \dfrac{1}{9} + \dfrac{1}{27} + \dfrac{1}{81} + \cdots$

**8.** $\dfrac{1}{7} + \dfrac{2}{7} + \dfrac{3}{7} + \dfrac{4}{7} + \dfrac{5}{7} + \cdots$

**9.** $-2 + 4 - 8 + 16 - \cdots$

**10.** $-3 + 9 - 27 + 81 - \cdots$

**11.** $-1 + 2 - 3 + 4 - \cdots$

**12.** $-1 + 4 - 9 + 16 - \cdots$

**13–18.** Find the sum of each finite geometric series.

**13.** $1 + 2 + 2^2 + 2^3 + \cdots + 2^9$

**14.** $1 + 2 + 2^2 + 2^3 + \cdots + 2^{10}$

**15.** $3 + 3\cdot4 + 3\cdot4^2 + \cdots + 3\cdot4^5$

**16.** $2 + 2\cdot3 + 2\cdot3^2 + \cdots + 2\cdot3^9$

**17.** $3 - 3\cdot2 + 3\cdot2^2 - 3\cdot2^3 + \cdots + 3\cdot2^6$

**18.** $4 - 4\cdot3 + 4\cdot3^2 - 4\cdot3^3 + \cdots + 4\cdot3^6$

**19–40.** Determine whether each infinite geometric series converges or diverges. If it converges, find its sum.

**19.** $4 + \frac{4}{5} + \frac{4}{25} + \frac{4}{125} + \cdots$

**20.** $5 + \frac{5}{6} + \frac{5}{36} + \frac{5}{216} + \cdots$

**21.** $2 - \frac{2}{3} + \frac{2}{9} - \frac{2}{27} + \frac{2}{81} - \cdots$

**22.** $27 - \frac{27}{2} + \frac{27}{4} - \frac{27}{8} + \frac{27}{16} - \cdots$

**23.** $\frac{1}{3} + \frac{2}{3} + \frac{4}{3} + \frac{8}{3} + \frac{16}{3} + \cdots$

**24.** $\frac{1}{100} - \frac{3}{100} + \frac{9}{100} - \frac{27}{100} + \frac{81}{100} - \cdots$

**25.** $\frac{3}{2} + \frac{3^2}{2^3} + \frac{3^3}{2^5} + \frac{3^4}{2^7} + \frac{3^5}{2^9} + \cdots$

**26.** $\frac{1}{3} + \frac{1}{3^3} + \frac{1}{3^5} + \frac{1}{3^7} + \frac{1}{3^9} + \cdots$

**27.** $8 + 6 + \frac{9}{2} + \frac{27}{8} + \cdots$

**28.** $9 + 6 + 4 + \frac{8}{3} + \cdots$

**29.** $\displaystyle\sum_{i=1}^{\infty} \frac{100}{5^i}$    **30.** $\displaystyle\sum_{i=1}^{\infty} 30\cdot4^{-i}$

**31.** $\displaystyle\sum_{j=1}^{\infty} \frac{3^j}{7\cdot2^j}$    **32.** $\displaystyle\sum_{j=1}^{\infty} \frac{5\cdot2^j}{3^j}$

**33.** $\displaystyle\sum_{j=0}^{\infty} \left(-\frac{1}{2}\right)^j$    **34.** $\displaystyle\sum_{n=0}^{\infty} 3^{-n}$

**35.** $\displaystyle\sum_{k=1}^{\infty} (1.01)^k$    **36.** $\displaystyle\sum_{n=1}^{\infty} \pi^n$

**37.** $\displaystyle\sum_{k=0}^{\infty} (0.99)^k$   **38.** $\displaystyle\sum_{n=1}^{\infty} \pi^{-n}$

**39.** $\displaystyle\sum_{j=0}^{\infty} \frac{3^j}{4^{j+1}}$   **40.** $\displaystyle\sum_{i=0}^{\infty} \frac{(-1)^i}{1000}$

**41–48.** Find the value of each repeating decimal.
[*Hint:* Write each as an infinite series. For example,

$$0.\overline{25} = 0.252525\ldots = \frac{25}{100} + \frac{25}{100^2} + \frac{25}{100^3} + \cdots$$

The bar indicates the repeating part.]

**41.** $0.\overline{36} = 0.363636\ldots$   **42.** $0.\overline{63} = 0.636363\ldots$

**43.** $2.\overline{54}$ $(= 2 + 0.\overline{54})$   **44.** $1.\overline{24}$ $(= 1 + 0.\overline{24})$

**45.** $0.\overline{027}$   **46.** $0.\overline{9}$

**47.** $0.\overline{012345679}$   **48.** $0.\overline{987654320}$

(For Exercises 47 and 48, after finding the fraction, try dividing the numerator and denominator by 12345679.)

## Applied Exercises

### Finite Geometric Series

**49. a. GENERAL: Yonkers Fine** Suppose that the fine in Example 1 (beginning at $100 and doubling each day; see pages 687–688) had continued for a total of 18 days. Find the total fine.

 **b.** (Graphing calculator with series operations helpful) How many days would it have taken for the fine to reach a billion dollars?

*Source: New York Times*

**50. PERSONAL FINANCE: Annuity** A star baseball player expects his career to last 10 years, and for his retirement deposits $4000 at the end of each month in a bank account earning 12% interest compounded monthly. Find the amount of this annuity at the end of 10 years.

**51. PERSONAL FINANCE: Annuity** For your retirement "nest egg," you deposit $300 at the end of each month into a bank account paying 6% interest compounded monthly. Find the amount of this annuity at the end of your 40-year working career.

**52. PERSONAL FINANCE: Present Value of an Annuity** A retirement plan pays $1000 at retirement and every month thereafter for a total of 12 years. Find the sum of the present values of these payments (at an interest rate of 6% compounded monthly) by summing the series

$$1000 + \frac{1000}{1.005} + \frac{1000}{(1.005)^2} + \cdots + \frac{1000}{(1.005)^{143}}$$

**53. GENERAL: Million Dollar Lottery** Most state lottery jackpots are paid out over time (often 20 years), so the "real" cost to the state is the sum of the present values of the payments. Find the cost of a "million dollar" lottery by summing the present values of 240 monthly payments of $4167, beginning now, if the interest rate is 6% compounded monthly. [*Hint:* The present value of $4167 in $k$ months is $4167/(1 + 0.005)^k$.]

**54. BUSINESS: Sinking Fund** A sinking fund is an annuity designed to reach a given value at a given time in the future (often to pay off a debt or to buy new equipment). A company's $100,000 printing press is expected to last 8 years. If equal payments are to be made at the end of each quarter into an account paying 8% compounded quarterly, find the size of the quarterly payments needed to yield $100,000 at the end of 8 years. [*Hint:* Solve
$x + x(1.02) + x(1.02)^2 + \cdots + x(1.02)^{31} = 100,000.$]

**55. a. GENERAL: Saving Pennies** You make a New Year's resolution to save pennies, promising to save on successive days 1¢, 2¢, 4¢, 8¢, and so on, doubling the number each day throughout January. How much will you have saved by the end of January?

 **b.** (Graphing calculator with series operations helpful) How many days will it take for the savings to reach a million dollars?

**56. GENERAL: Chessboards** It is said that the game of chess was invented by the Grand Vizier Sissa Ben Dahir for the Indian King Shirham. When the king offered him any reward, Sissa asked for one grain of wheat to be placed on the first square of the chessboard, two on the second, four on the third, and so on, doubling each time for each of the 64 squares. Estimate the total number of grains of wheat required. (It is also said that the king, upon learning that the total was many times the world production of wheat, executed Sissa for making a fool out of him.)

 **57. GENERAL: Cycling** (Graphing calculator with series operations helpful) You plan to cycle across the United States, cycling 10 miles the first day, 10% further the second day (so 11 miles), increasing each day's distance by 10%. On which day will you reach the opposite coast 3000 miles away?

**58. PERSONAL FINANCE: Payments** (Graphing calculator with series operations helpful) A person gives you $1, waits 1 minute, gives you another dollar, waits 2 minutes, gives you another dollar, waits four minutes, etc., doubling the wait each time. How much money will you have at the end of the first year?

## Infinite Geometric Series

**59. a. BIOMEDICAL: Drug Dosage** A patient has been taking a daily dose of 12 units of insulin for an extended period of time. If 80% of the total amount in the bloodstream is absorbed by the body during each day, find the long-term maximum and minimum amounts of insulin in the bloodstream.

**b.** (Graphing calculator with series operations helpful) How many daily doses of insulin does it take for the maximum amount in the blood to reach 14.9 units?

**60. BIOMEDICAL: Drug Dosage** A patient has been taking a daily dose of $D$ milligrams of a drug for an extended period of time, and a proportion $p$ $(0 < p < 1)$ of the total amount in the bloodstream is absorbed by the body during each day. Find the long-term maximum and minimum amounts in the bloodstream.

**61. ECONOMICS: Multiplier Effect** In the United States, the marginal propensity to consume (*MPC*) is approximately 90%. Find the total economic effect of a $50 billion tax cut. Also find the multiplier.

*Source:* P. Samuelson and W. Nordhaus, *Economics*

**62. ECONOMICS: Multiplier Effect** After several years of stock market boom, in 2000 the marginal propensity to consume (*MPC*) peaked at 96%. Find the total economic effect that a $40 billion tax cut would have had at that time. Also find the multiplier.

*Source:* economics.about.com

**63. ECONOMICS: Multiplier Effect** In Japan, the marginal propensity to consume (*MPC*) is approximately 75%. Find the multiplier.

*Source:* economics.about.com

**64. ECONOMICS: Multiplier Effect** If the marginal propensity to *consume* (*MPC*) is 70%, then the marginal propensity to *save* (*MPS*) is 30%. In general, $MPS = 1 - MPC$. Show that for a given *MPC*, the multiplier is $\dfrac{1}{MPS}$.

**65–66. BUSINESS: Perpetuities** A *perpetuity* is an annuity that continues forever. The *capital value* of a perpetuity is the present value of all (future) payments. (Perpetuities are also called permanent endowments, and the continuous case was treated on page 437 using improper integrals.)

**65.** The capital value of a perpetuity that pays $300 now and every quarter from now on at an interest rate of 6% compounded quarterly is

$$300 + \frac{300}{1 + 0.015} + \frac{300}{(1 + 0.015)^2} + \cdots$$

Find this value.

**66.** The capital value of a perpetuity that pays $1000 annually, beginning now, at an interest rate of 8% compounded annually is

$$1000 + \frac{1000}{1 + 0.08} + \frac{1000}{(1 + 0.08)^2} + \frac{1000}{(1 + 0.08)^3} + \cdots$$

Find this value.

**67. a. GENERAL: Bouncing Ball** A ball dropped from a height of 6 feet bounces to two-thirds of its former height with each bounce. Find the total vertical distance that the ball travels.

**b.** (Graphing calculator with series operations helpful) At which bounce will the total vertical distance traveled by the ball first exceed 29 feet?

**68. GENERAL: Coupons** A 45¢ candy bar comes with a coupon worth 10% toward the next candy bar (which includes a coupon for 10% of the next, and so on). Therefore, the "real" value of a single 45¢ candy bar is

$$45 + 45(0.10) + 45(0.10)^2 + 45(0.10)^3 + \cdots$$

Find this value.

**69–70. BIOMEDICAL: Long-Term Population** A population (of cells or people) is such that each year a number $a$ of individuals are added (call them "immigrants"), and a proportion $p$ of the individuals who have been there die. Therefore, the proportion that survives is $(1 - p)$, so that just after an immigration the population will consist of new immigrants plus $a(1 - p)$ from the previous year's immigration plus

$a(1 - p)^2$ from the immigration before that, and so on. In the long run, the size of the population just after an immigration will then be the sum $a + a(1 - p) + a(1 - p)^2 + a(1 - p)^3 + \cdots$.

**69.** If the number of immigrants each year is 800 and the survival proportion is 0.95, find the long-run size of the population just after an immigration. Then find the long-run size just *before* an immigration.

**70.** Find the long-run size of the population just after an immigration for *any* number of immigrants $a$ and *any* survival proportion $(1 - p)$. Then find the long-run size just *before* an immigration.

## Conceptual Exercises

**71.** How many terms are in the following finite geometric series? $2 + 2 \cdot 3 + 2 \cdot 3^2 + \cdots + 2 \cdot 3^{5280}$.

**72.** What is the value of the first term $a_1$ in the following finite geometric series? $a_1 + 1 + \dfrac{1}{\pi}$.

**73.** True or False: If the partial sums $S_n$ of an infinite series all satisfy $S_n \leq 10$, then the sum $S$ also satisfies $S \leq 10$.

**74.** True or False: If the partial sums $S_n$ of an infinite series all satisfy $S_n < 1$, then the sum $S$ also satisfies $S < 1$.

**75.** What is the ratio of the following infinite geometric series? $e + \dfrac{1}{e} + \cdots$. Does it converge?

**76.** What is the 501st term in the following infinite geometric series? $\pi^{-500} + \pi^{-499} + \pi^{-498} + \cdots$. Does it converge?

**77.** Does the infinite geometric series $1{,}000{,}000 + 100{,}000 + 10{,}000 + \cdots$ converge or diverge?

**78.** Does the infinite geometric series $\frac{1}{1{,}000{,}000} + \frac{1}{100{,}000} + \frac{1}{10{,}000} + \cdots$ converge or diverge?

**79.** Applying the sum formula to the following infinite geometric series gives the result that a series of *positive* terms adds up to a *negative* number. What's wrong?

$$1 + 2 + 4 + 8 + \cdots \overset{?}{=} \frac{1}{1 - 2} = \frac{1}{-1} = -1$$

**80.** Applying the sum formula to the following infinite geometric series gives the result that a series of *integer* terms adds up to a *fraction*. What's wrong?

$$1 - 2 + 4 - 8 + \cdots \overset{?}{=} \frac{1}{1 - (-2)} = \frac{1}{3}$$

## Explorations and Excursions

The following problems extend and augment the material presented in the text.

**81. APPROXIMATING THE FINITE BY THE INFINITE** For the patient in Example 4 (page 692), the amount of the drug in the bloodstream just after the $n$th dose is

$$150 + 150(0.3) + 150(0.3)^2 + \cdots + 150(0.3)^{n-1}$$

Use a graphing calculator to sum this finite geometric series for the following values of $n$: 1, 2, 3, 4, 5, 6, and 7.

**a.** For which dose does the cumulative amount first rise above 200 milligrams?

**b.** For which dose does the cumulative amount first rise above 213 milligrams?

**82. DIVERGENCE OF GEOMETRIC SERIES** Show that the geometric series

$$a + ar + ar^2 + ar^3 + \cdots$$

diverges if $|r| \geq 1$, as follows (for $a \neq 0$):

**a.** If $r = 1$, the series becomes $a + a + a + \cdots$. Show that this series diverges by showing that the sum of the first $n$ terms is $S_n = n \cdot a$, which does not approach a limit as $n \to \infty$ (since $a \neq 0$).

**b.** If $r = -1$, then the series becomes $a - a + a - \cdots$. Show that this series diverges by showing that the partial sums $S_1, S_2, S_3, S_4, \ldots$ take values $a, 0, a, 0, \ldots$, and so do not approach any limit.

**c.** If $|r| > 1$, use the fact that $r^n$ does not approach a limit as $n \to \infty$ to show that the partial sum formula $a\dfrac{1 - r^n}{1 - r}$ does not approach a limit as $n \to \infty$.

**83.** Consider *Grandi's series* $1 - 1 + 1 - 1 + \cdots$

**a.** Show that applying the formula $\dfrac{a}{1 - r}$ gives $\dfrac{1}{2}$.

**b.** Can this formula legitimately be applied to the series?

**c.** Calculate the partial sums of this series. What would be the average of these partial sums *in the long run*? [*Note:* Such an average is called a *Cesàro sum.*]

*Source: Mathematics Magazine 56*

**84.** Derive the formula for the sum $S$ of a geometric series as follows:

$$S = a + ar + ar^2 + ar^3 + \cdots$$
$$= a + r(\underbrace{a + ar + ar^2 + \cdots}_{S}) = a + rS$$

Now solve the equation $S = a + rS$ for $S$. (Note that this derivation *assumes* that the series converges.)

## 10.2   TAYLOR POLYNOMIALS

### Introduction

Often in mathematics we approximate complicated objects by simpler objects, such as approximating a curve by its tangent line. In this section we will approximate functions such as $e^x$ and $\sin x$ by *polynomials*, called **Taylor polynomials,** named after the English mathematician Brook Taylor (1685–1731), a student of Isaac Newton.

### Factorial Notation

We begin by defining **factorial** notation. For any positive integer $n$, $n!$ (read: "$n$ factorial") is defined as the product of all integers from $n$ down to 1.

| Factorial Notation | |
|---|---|
| $n! = n \cdot (n - 1) \cdot \cdots \cdot 2 \cdot 1$ | $n!$ is the product of all integers from $n$ down to 1 |
| $0! = 1$ | Zero factorial is defined as 1 |

For example,

$$4! = 4 \cdot 3 \cdot 2 \cdot 1 = 24$$

A justification for defining $0!$ as 1 is given in Exercise 30.

### First Taylor Polynomial at   $x = 0$

We begin by finding the polynomial of degree 1 that best approximates a given function $f(x)$ near $x = 0$. A polynomial of degree 1 is a *linear* function $y = mx + b$, and it is clear from the following diagram that the best linear approximation to $f(x)$ at $x = 0$ will be the *tangent line* at $x = 0$. The tangent line has the same slope and $y$-intercept as the curve at $x = 0$, so that $m = f'(0)$ and $b = f(0)$. Substituting these into $y = mx + b$ gives $y = f'(0)x + f(0)$, or, writing it in reverse order, $y = f(0) + f'(0)x$.

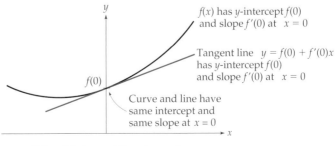

$f(x)$ and its tangent line at $x = 0$

The tangent line is called the *best linear approximation* or the *first Taylor polynomial at* $x = 0$ for the function, and is denoted $p_1(x)$.*

---

**First Taylor Polynomial at** $x = 0$

The first Taylor polynomial at $x = 0$ of $f(x)$ is

$$p_1(x) = f(0) + f'(0) \cdot x \qquad \text{Tangent line at } x = 0$$

---

**EXAMPLE 1**

**FINDING A FIRST TAYLOR POLYNOMIAL**

Find the first Taylor polynomial at $x = 0$ for $f(x) = e^x$.

*Solution*

The derivative of $f(x) = e^x$ is $e^x$, so $f(0) = f'(0) = e^0 = 1$. Therefore, the first Taylor polynomial is

$$p_1(x) = 1 + x \qquad \begin{array}{l} p_1(x) = f(0) + f'(0) \cdot x \\ \text{with } f(0) = 1 \text{ and } f'(0) = 1 \end{array}$$

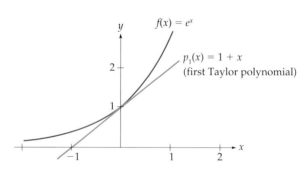

$f(x) = e^x$ and its first Taylor polynomial at $x = 0$

---

## Higher Taylor Polynomials at $x = 0$

This idea of finding the best *linear* approximation at $x = 0$ by matching $y$-intercepts and slopes (derivatives) at $x = 0$ can be extended to finding the best *quadratic* approximation by also matching the "curl" or concavity (second derivative). In fact, we may find the best *polynomial* approximation by matching higher-order derivatives of the polynomial to those of the function. For example, to approximate $f(x)$ by a *third*-degree polynomial $p(x) = a_0 + a_1 x + a_2 x^2 + a_3 x^3$ at $x = 0$, we match derivatives up to the

---

*Taylor polynomials at $x = 0$ are sometimes called Maclaurin polynomials, although there is little historical justification for this terminology, and we will not use it.

*third* order at $x = 0$, as is done in the right-hand column of the following table.

| Polynomial $p(x)$ and Its Derivatives | $p(x)$ and Its Derivatives Evaluated at $x = 0$ | $f(x)$ and Its Derivatives Evaluated at $x = 0$ | Equating the Last Two Columns |
|---|---|---|---|
| $p(x) = a_0 + a_1x + a_2x^2 + a_3x^3$ | $a_0$ | $f(0)$ | $a_0 = f(0)$ |
| $p'(x) = a_1 + 2a_2x + 3a_3x^2$ | $a_1$ | $f'(0)$ | $a_1 = f'(0)$ |
| $p''(x) = 2a_2 + 3 \cdot 2a_3x$ | $2 \cdot a_2$ | $f''(0)$ | $2!a_2 = f''(0)$ |
| $p'''(x) = 3 \cdot 2a_3$ | $3 \cdot 2a_3$ | $f'''(0)$ | $3!a_3 = f'''(0)$ |

The coefficients $a_0$, $a_1$, $a_2$, and $a_3$ of the polynomial are found by solving the equations in the extreme right column, giving

$$a_0 = f(0), \qquad a_1 = f'(0), \qquad a_2 = \frac{f''(0)}{2!}, \qquad a_3 = \frac{f'''(0)}{3!} \qquad a_n = \frac{f^{(n)}(0)}{n!}$$

Further coefficients follow the same pattern, and substituting them into the polynomial $p(x) = a_0 + a_1x + a_2x^2 + a_3x^3 + \cdots + a_nx^n$ gives the *n*th **Taylor polynomial** of an *n*-times differentiable function.

---

**Taylor Polynomials at $x = 0$**

The *n*th Taylor polynomial at $x = 0$ of $f(x)$ is

$$p_n(x) = f(0) + \frac{f'(0)}{1!}x + \frac{f''(0)}{2!}x^2 + \frac{f'''(0)}{3!}x^3 + \cdots + \frac{f^{(n)}(0)}{n!}x^n$$

---

**EXAMPLE 2**

**FINDING TAYLOR POLYNOMIALS**

Find the first three Taylor polynomials at $x = 0$ for $f(x) = e^x$.

*Solution*

All derivatives of $e^x$ are $e^x$, so $f(0) = f'(0) = f''(0) = f'''(0) = e^0 = 1$. Therefore, the first, second, and third Taylor polynomials for $e^x$ at $x = 0$ are

$$p_1(x) = 1 + x \qquad\qquad \text{Found in Example 1}$$

$$p_2(x) = 1 + x + \frac{1}{2!}x^2 \qquad\qquad f(0) + \frac{f'(0)}{1!}x + \frac{f''(0)}{2!}x^2$$
$$\text{with}\ \ f(0) = f'(0) = f''(0) = 1$$

$$p_3(x) = 1 + x + \frac{1}{2!}x^2 + \frac{1}{3!}x^3 \qquad\qquad f(0) + \frac{f'(0)}{1!}x + \frac{f''(0)}{2!}x^2 + \frac{f'''(0)}{3!}x^3$$
$$\text{with}\ \ f(0) = f'(0) = f''(0) = f'''(0) = 1$$

The following diagram shows $e^x$ and these approximating Taylor polynomials.

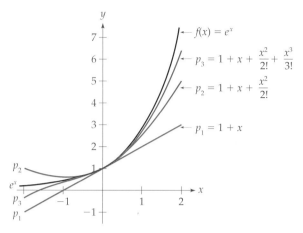

The original function $f(x) = e^x$ (red) and
its first three Taylor polynomials at $x = 0$.
Which polynomial fits $e^x$ most closely?

This graph shows two general properties of Taylor polynomials at $x = 0$:

1. Higher-order Taylor polynomials generally approximate functions more closely.

2. Each approximation is more accurate closer to $x = 0$ (and is *exact at* $x = 0$).

**Practice Problem 1**

Based on the pattern of the first three Taylor polynomials found in Example 2, find the fifth Taylor polynomial at $x = 0$ of $e^x$.

➤ **Solution on page 709**

## Graphing Calculator Exploration

Graph the fifth Taylor polynomial found in Practice Problem 1 together with $e^x$ on the window $[-3.5, 3.5]$ by $[-5, 25]$. Observe where the curves are close and where they diverge. Adjust the window to see how they diverge on a wider interval. Try adding another term, $\dfrac{x^6}{6!} = \dfrac{x^6}{720}$, to graph the *sixth* Taylor polynomial.

### Error in Taylor Approximation at $x = 0$

An approximation is useful only if we can estimate how far "off" it is. The *error* or **remainder** in approximating $f(x)$ by $p_n(x)$ is defined as

$|R_n(x)| = |f(x) - p_n(x)|$, the vertical distance between the curves at that $x$-value. If $f(x)$ is $n + 1$ times differentiable between 0 and $x$, the error can be estimated by the following formula, a proof of which may be found in a book on advanced calculus.

---

**Error in Taylor Approximation at $x = 0$**

For the $n$th Taylor polynomial $p_n(x)$ at $x = 0$ of $f(x)$, the error at $x$, $|R_n(x)| = |f(x) - p_n(x)|$, satisfies

$$|R_n(x)| \le \frac{M}{(n + 1)!} |x|^{n+1}$$

Resembles the $n + 1$st term $\dfrac{f^{(n+1)}(x)}{(n + 1)!} x^{n+1}$

where $M$ is any number such that $|f^{(n+1)}(t)| \le M$ for all $t$ between 0 and $x$.

---

**EXAMPLE 3**

**APPROXIMATING BY A TAYLOR POLYNOMIAL**

Approximate $e^{0.5}$ using the third Taylor polynomial for $e^x$, and estimate the error.

*Solution*

Using the third Taylor polynomial (found in Example 2),

$$p_3(0.5) = 1 + 0.5 + \frac{0.5^2}{2!} + \frac{0.5^3}{3!} \approx \underbrace{1.6458}$$

Estimate for $e^{0.5}$

$p_3(x) = 1 + x + \dfrac{x^2}{2!} + \dfrac{x^3}{3!}$
evaluated at $x = 0.5$

The error formula for the third Taylor polynomial requires a number $M$ such that $|f^{(4)}(t)| \le M$ for $0 \le t \le 0.5$. Since $|f^{(4)}(t)| = e^t$ is an increasing function, it is maximized at its right-hand endpoint $t = 0.5$, so

$$|f^{(4)}(t)| = e^t \le e^{0.5} < e^1 < 3 \qquad\qquad \text{Since } e \approx 2.7 < 3$$

We therefore take $M = 3$ (although smaller estimates could be found). With this value of $M$, the error formula with $n = 3$, $M = 3$, and $x = 0.5$ becomes

$$|R_3(0.5)| \le \frac{3}{4!} |0.5|^4 = \frac{3}{24}(0.0625) \approx 0.00781 \approx \underbrace{0.008}$$

Error estimate

Therefore,

$$e^{0.5} \approx 1.646 \quad \text{with an error less than } 0.008$$

We always round the error estimate *up* to the first nonzero digit (for example, an error of 0.0072 would be rounded up to 0.008) so as not to claim more accuracy than is justified. We then round the approximation to the same number of decimal places.

## Why Bother to Calculate $e^{0.5}$?

You may wonder why we approximate $e^{0.5}$ when a calculator or computer will do so at the press of a few keys. How do you think computers actually evaluate such functions? They do so by using (modified) preprogrammed Taylor polynomials.* Therefore, this section explains what happens when you press the keys. Furthermore, at some time you may need to evaluate a new or more complicated function, and Taylor polynomials enable you to do this in terms of "easier" polynomials.

Not only can we find the accuracy of a particular approximation, but we can also determine the Taylor polynomial needed to *guarantee* a given accuracy.

| EXAMPLE 4 |
| --- |

### APPROXIMATING WITHIN A GIVEN ACCURACY

Find the Taylor polynomial at $x = 0$ that approximates $e^x$ with an error less than 0.005 on the interval $-1 \le x \le 1$.

*Solution*

We want the value of $n$ that makes the error $\dfrac{M}{(n + 1)!} |x|^{n+1}$ less than 0.005. This error formula with $M = 3$ (since $|e^1| < 3$, as in Example 3) and $|x| = 1$ becomes $\dfrac{3}{(n + 1)!}$. We evaluate this formula at $n = 1, 2, 3, \ldots$ until its value first falls below 0.005.

| $n$ | $\dfrac{3}{(n+1)!}$ | |
| --- | --- | --- |
| 1 | $\dfrac{3}{2!} = \dfrac{3}{2} = 1.5$ | |
| 2 | $\dfrac{3}{3!} = \dfrac{3}{6} = 0.5$ | We could have skipped some early values of $n$ that clearly would not give small enough errors |
| 3 | $\dfrac{3}{4!} = \dfrac{3}{24} = 0.125$ | |
| 4 | $\dfrac{3}{5!} = \dfrac{3}{120} = 0.025$ | |
| 5 | $\dfrac{3}{6!} = \dfrac{3}{720} \approx 0.004$ | $< 0.005$ for the first time |

---

*Most computers use "economized" Taylor series that make the error more uniform over an interval. Most calculators, on the other hand, use a system called CORDIC, for COordinate Rotation DIgital Computer.

The first value of $n$ for which the error is small enough (less than 0.005) is $n = 5$, so the fifth Taylor polynomial (found in Practice Problem 1) provides the desired accuracy:

$$e^x \approx 1 + x + \frac{x^2}{2!} + \frac{x^3}{3!} + \frac{x^4}{4!} + \frac{x^5}{5!}$$    For $-1 \le x \le 1$, with error less than 0.005

## Graphing Calculator Exploration

(You should still have $e^x$ and its fifth Taylor polynomial at $x = 0$ in your calculator.) Use TRACE to verify that $e^x$ and the Taylor polynomial are indeed within 0.005 of each other between $x = -1$ and $x = 1$. How close are they actually? (Remember that the error formula generally *overstates* the error.)

**EXAMPLE 5**    **FINDING AN APPROXIMATION IN AN APPLICATION**

A company that produces seasonal goods produces $\sin 1.1$ thousand units on a certain day. Approximate $\sin 1.1$ by using the fifth Taylor polynomial at $x = 0$ for $\sin x$ and estimate the error.

*Solution*

The function $\sin x$ and its derivatives (from page 583) at $x = 0$ are

$$
\begin{aligned}
f(x) &= \sin x & f(0) &= 0 \\
f'(x) &= \cos x & f'(0) &= 1 \\
f''(x) &= -\sin x & f''(0) &= 0 \\
f'''(x) &= -\cos x & f'''(0) &= -1 \\
f^{(4)}(x) &= \sin x & f^{(4)}(0) &= 0 \\
f^{(5)}(x) &= \cos x & f^{(5)}(0) &= 1
\end{aligned}
$$

$\sin x$ and its derivatives evaluated at $x = 0$

Therefore, the fifth Taylor polynomial is

$$p_5(x) = 0 + \frac{1}{1!}x + 0 - \frac{1}{3!}x^3 + 0 + \frac{1}{5!}x^5$$

$f(0) + \dfrac{f'(0)}{1!}x + \cdots + \dfrac{f^{(5)}(0)}{5!}x^5$
with the preceding values for $f(0), f'(0), \ldots, f^{(5)}(0)$

$$= x - \frac{x^3}{3!} + \frac{x^5}{5!}$$

Simplifying (all even powers vanish)

Evaluating,

$$\sin 1.1 \approx 1.1 - \frac{(1.1)^3}{3!} + \frac{(1.1)^5}{5!} \approx 0.89159$$    Evaluating at $x = 1.1$

Therefore, $\sin 1.1 \approx 0.89159$. For the error formula, we need an $M$ such that $\left| f^{(6)}(t) \right| \leq M$ for all $t$ between 0 and 1.1. For the function $\sin x$, we have $\left| f^{(6)}(x) \right| = \left| -\sin x \right| \leq 1 = M$. With this $M$, the error formula becomes

$$\left| R_5(x) \right| \leq \frac{1}{6!} \left| 1.1 \right|^6 \approx 0.0024 \approx 0.003 \qquad \begin{array}{l} \dfrac{M}{(n+1)!} \left| x \right|^{n+1} \text{ with} \\[4pt] M = 1, \quad n = 5, \quad x = 1.1 \end{array}$$

Rounded *up*

Therefore,

$$\sin 1.1 \approx 0.892 \quad \text{with an error less than } 0.003$$

These are production numbers in *thousands*, so the company should produce 892 units, with a possible error of 3 units. (To be safe, it might choose to produce 895 units.)

**Practice Problem 2**

As we saw in the preceding example, the Taylor polynomial for $\sin x$ at $x = 0$ consists only of *odd* powers of $x$. Therefore, the sixth Taylor polynomial will be the same as the fifth, so we can use the error formula for the sixth Taylor polynomial to estimate the accuracy of the approximation that we found. Use the error formula for $\left| R_6(x) \right|$ to estimate the accuracy of the approximation $\sin 1.1 \approx 0.89159$ from Example 5.

➤ Solution on page 709

The function $\sin x$ and its fifth Taylor polynomial are shown in the following graph. Notice that the approximation is quite accurate for $x$-values near 0.

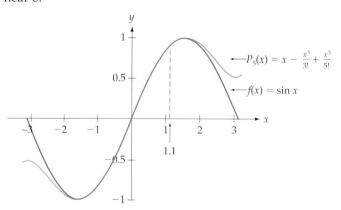

Sin $x$ and its fifth Taylor polynomial at $x = 0$

## Taylor Polynomials at $x = a$

We have been using Taylor polynomials centered *at* $x = 0$, meaning that they approximate $f(x)$ most accurately near $x = 0$. There are analogous formulas for Taylor polynomials centered at *any* number $x = a$ (provided that $f$ is $n$ times differentiable at $a$).

> **Taylor Polynomials at  $x = a$**
>
> The $n$th Taylor polynomial at  $x = a$  of  $f(x)$  is
>
> $$p_n(x) = f(a) + \frac{f'(a)}{1!}(x - a) + \frac{f''(a)}{2!}(x - a)^2 + \cdots + \frac{f^{(n)}(a)}{n!}(x - a)^n$$

Notice that for Taylor polynomials at  $x = a$:

1. The polynomial is in powers of  $(x - a)$  rather than powers of $x$.
2. The coefficients involve derivatives at  $x = a$.
3. The polynomials approximate $f(x)$ more accurately nearer  $x = a$  (and are *exact at*  $x = a$).
4. If  $a = 0$,  we obtain the Taylor polynomials at  $x = 0$.

---

**EXAMPLE 6**

**FINDING A TAYLOR POLYNOMIAL AT   $x = 4$**

Find the second Taylor polynomial for   $f(x) = \sqrt{x}$   at   $x = 4$,   and use it to approximate $\sqrt{4.2}$.

*Solution*

The function   $f(x) = \sqrt{x}$   and its derivatives at   $x = 4$   are

$$f(x) = x^{1/2} \qquad\qquad f(4) = 4^{1/2} = 2$$

$$f'(x) = \frac{1}{2}x^{-1/2} \qquad f'(4) = \frac{1}{2}4^{-1/2} = \frac{1}{2}\cdot\frac{1}{2} = \frac{1}{4}$$

$$f''(x) = -\frac{1}{4}x^{-3/2} \qquad f''(4) = -\frac{1}{4}4^{-3/2} = -\frac{1}{4}\cdot\frac{1}{8} = -\frac{1}{32}$$

The second Taylor polynomial at   $x = 4$   is

$$p_2(x) = 2 + \frac{\frac{1}{4}}{1!}(x - 4) - \frac{\frac{1}{32}}{2!}(x - 4)^2$$

$p_2(x) = f(a) + \dfrac{f'(a)}{1!}(x - a) +$ $\dfrac{f''(a)}{2!}(x - a)^2$  with  $a = 4$ and the calculated values of $f(4), f'(4),$ and $f''(4)$

$$= 2 + \frac{x - 4}{4} - \frac{(x - 4)^2}{64} \qquad\qquad \text{Simplifying}$$

Evaluating,

$$p_2(4.2) = 2 + \frac{0.2}{4} - \frac{(0.2)^2}{64} = 2.049375 \qquad \text{Evaluating at } x = 4.2$$

Approximation for $\sqrt{4.2}$

$$\sqrt{4.2} \approx 2.0494 \qquad\qquad \text{Rounding to four decimal places}$$

Actually, $\sqrt{4.2} \approx 2.04939$, so the estimate above is correct in all four decimal places. The function $f(x) = \sqrt{x}$ and its second Taylor polynomial at $x = 4$ are graphed below, showing that the approximation is very good near $x = 4$.

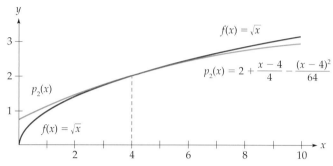

$f(x) = \sqrt{x}$ and its second Taylor polynomial at $x = 4$

## Graphing Calculator Exploration

Some advanced graphing calculators can find Taylor polynomials. For example, the Texas Instruments *TI-89* graphing calculator finds the second Taylor polynomial at $x = 4$, for the function $f(x) = \sqrt{x}$ as follows.

This answer agrees with that in the preceding example, except that it is written in reverse order.

If $f(x)$ is $n + 1$ times differentiable between $a$ and $x$, the error in approximating $f(x)$ by $p_n(x)$ can be estimated as follows:

### Error in Taylor Approximation at $x = a$

For the $n$th Taylor approximation $p_n(x)$ at $x = a$ of $f(x)$, the error at $x$, $|R_n(x)| = |f(x) - p_n(x)|$, satisfies

$$|R_n(x)| \le \frac{M}{(n + 1)!} \, |x - a|^{n+1}$$

where $M$ is any number such that $|f^{(n+1)}(t)| \le M$ for all $t$ between $a$ and $x$.

> **Solutions to Practice Problems**
>
> **1.** $1 + x + \dfrac{1}{2!}x^2 + \dfrac{1}{3!}x^3 + \dfrac{1}{4!}x^4 + \dfrac{1}{5!}x^5$, or $1 + x + \dfrac{x^2}{2!} + \dfrac{x^3}{3!} + \dfrac{x^4}{4!} + \dfrac{x^5}{5!}$
>
> **2.** $|R_6(x)| \leq \dfrac{1}{7!} |1.1|^7 = \dfrac{1.9487}{5040} \approx 0.0004$
>
> Therefore, $\sin 1.1 \approx 0.8916$ with an error of less than 0.0004.
> (The actual value of $\sin 1.1$ is 0.8912….)

## 10.2 Section Summary

Taylor polynomials approximate a curve near a single chosen point $x = a$. Any sufficiently differentiable function can be approximated by a polynomial, using the formulas

$$\sum_{k=0}^{n} \frac{f^{(k)}(0)}{k!} x^k \qquad \text{\textit{n}th Taylor polynomial at } x = 0$$
$$\text{[first term simplifies to } f(0)]$$

and

$$\sum_{k=0}^{n} \frac{f^{(k)}(a)}{k!} (x - a)^k \qquad \text{\textit{n}th Taylor polynomial at } x = a$$
$$\text{[first term simplifies to } f(a)]$$

Taylor polynomials can also be found from other Taylor polynomials by replacing the variable by a polynomial expression. The accuracy of the approximation can be found from the error formulas on pages 703 and 708.

## 10.2 Exercises

**1–8.** Find the third Taylor polynomial at $x = 0$ of each function.

**1.** $f(x) = e^{2x}$

**2.** $f(x) = \ln(1 - x)$

**3.** $f(x) = \sqrt{x + 1}$

**4.** $f(x) = \sqrt{2x + 1}$

**5.** $f(x) = 7 + x - 3x^2$

**6.** $f(x) = 2 - x + 3x^2 - x^3$

**7.** $f(x) = e^{x^2}$

**8.** $f(x) = \ln(\cos x)$

**9–12.** For each function:

**a.** Find the fourth Taylor polynomial at $x = 0$.

**b.** Graph the original function and the Taylor polynomial together on the indicated window.

**9.** $f(x) = \ln(x + 1)$
(for (b), use the window $[-2, 2]$ by $[-2, 2]$)

**10.** $f(x) = e^{-2x}$
(for (b), use the window $[-2, 2]$ by $[-2, 8]$)

**11.** $f(x) = \cos x$
(for (b), use the window $[-\pi, \pi]$ by $[-2, 2]$)

**12.** $f(x) = \cos(x + \pi)$
for (b), use the window $[-\pi, \pi]$ by $[-2, 2]$)

**13.** Find the fourth Taylor polynomial for $e^{x^2}$ by taking the second Taylor polynomial for $e^x$ (page 701) and replacing $x$ by $x^2$.

**14.** Find the third Taylor polynomial for $e^{-x}$ by taking the third Taylor polynomial for $e^x$ (page 701) and replacing $x$ by $-x$.

**15.** Find the fifth Taylor polynomial for $\sin 2x$ by taking the fifth Taylor polynomial for $\sin x$ (page 705) and replacing $x$ by $2x$.

**16.** Find the tenth Taylor polynomial for $\sin x^2$ by taking the fifth Taylor polynomial for $\sin x$ (page 705) and replacing $x$ by $x^2$.

**17. a.** For $f(x) = e^x$, find the fourth Taylor polynomial at $x = 0$.
 **b.** Use your answer in part (a) to approximate $e^{-1/2}$.
 **c.** Estimate the error by the error formula (page 703). [*Hint:* First calculate $f^{(5)}(x)$ and find an $M$ such that $|f^{(5)}(x)| \leq M$ for $-\frac{1}{2} \leq x \leq 0$, using the fact that an increasing function is maximized at its right-hand endpoint.]
 **d.** Restate the approximation and error, rounded appropriately.
 **e.** Graph $f(x) = e^x$ and its Taylor polynomial on $[-3, 3]$ by $[0, 12]$.

**18. a.** For $f(x) = \ln(x + 1)$, find the fifth Taylor polynomial at $x = 0$.
 **b.** Use your answer in part (a) to approximate $\ln \frac{3}{2}$.
 **c.** Estimate the error by the error formula (page 703). [*Hint:* Calculate $f^{(6)}(x)$ and find an $M$ such that $|f^{(6)}(x)| \leq M$ for $0 \leq x \leq \frac{1}{2}$, using the fact that a decreasing function is maximized at its left-hand endpoint.]
 **d.** Restate the approximation and error, rounded appropriately.
 **e.** Graph $f(x) = \ln(x + 1)$ and its Taylor polynomial on $[-2, 2]$ by $[-2, 2]$.

**19. a.** Find the Taylor polynomial at $x = 0$ that approximates $f(x) = \cos x$ on the interval $-1 \leq x \leq 1$ with error less than 0.0002.
 **b.** Graph $f(x) = \cos x$ and the Taylor polynomial on $[-2\pi, 2\pi]$ by $[-2, 2]$. Use TRACE to estimate the maximum difference between the two curves for $-1 \leq x \leq 1$.

**20.** For the function

$$f(x) = 1 + 2x + \frac{9}{2!}x^2 + \frac{5}{3!}x^3 - \frac{7}{4!}x^4$$

find $f'''(0)$ and $f^{(4)}(0)$. [*Hint:* No calculation is necessary.]

**21. a.** What linear function best approximates $\sqrt[3]{1 + x}$ near $x = 0$?
 **b.** Use the error formula to find an interval $[0, x]$ on which the error is less than 0.01.
 **c.** Graph $\sqrt[3]{1 + x}$ and the linear approximation on $[-2, 2]$ by $[-2, 2]$ and use TRACE to verify that the difference is less than 0.01 for the $x$-values found in part (b).

**22. a.** What quadratic function best approximates $\sqrt[3]{1 + x}$ near $x = 0$?
 **b.** Use the error formula to estimate the maximum error on $[-0.1, 0.1]$.
 **c.** Graph $\sqrt[3]{1 + x}$ and the quadratic approximation on $[-2, 2]$ by $[-2, 2]$ and use TRACE to verify that on $[-0.1, 0.1]$ the difference is less than the error estimate found in part (b).

## Taylor Polynomials at $x = a$

**23. a.** For $f(x) = \sqrt{x}$, find the third Taylor polynomial at $x = 1$.
 **b.** Graph $\sqrt{x}$ and its Taylor polynomial on $[-1, 3]$ by $[-1, 2]$.

**24. a.** For $f(x) = 3x^2 - 4x + 5$, find the second Taylor polynomial at $x = 2$.
 **b.** Multiply out the Taylor polynomial found in part (a) and show that it is equal to the original polynomial.

**25. a.** For $f(x) = \sin x$, find the third Taylor polynomial at $x = \pi$.
 **b.** Graph $\sin x$ and its Taylor polynomial on $[0, 2\pi]$ by $[-2, 2]$.

**26. a.** For $f(x) = \cos x$, find the fourth Taylor polynomial at $x = \pi$.
 **b.** Graph $\cos x$ and its Taylor polynomial on $[0, 2\pi]$ by $[-2, 2]$.

**27. a.** For $f(x) = \sqrt[3]{x}$, find the second Taylor polynomial at $x = 1$.
 **b.** Use your answer from part (a) to approximate $\sqrt[3]{1.3}$.
 **c.** Estimate the error by the error formula (page 708). [*Hint:* Find an $M$ such that $|f^{(3)}(x)| \leq M$ for $1 \leq x \leq 1.3$, using the fact that a decreasing function is maximized at its left-hand endpoint.]
 **d.** Restate the approximation and error, rounded appropriately.

**28. a.** For $f(x) = \sqrt[4]{x}$, find the second Taylor polynomial at $x = 1$.
 **b.** Use your answer from part (a) to approximate $\sqrt[4]{1.2}$.
 **c.** Estimate the error by the error formula (page 708). [*Hint:* Find an $M$ such that $|f^{(3)}(x)| \leq M$ for $1 \leq x \leq 1.2$, using the fact that a decreasing function is maximized at its left-hand endpoint.]
 **d.** Restate the approximation and error, rounded appropriately.

**29. a.** For $f(x) = \ln x$, find the sixth Taylor polynomial at $x = 1$.
 **b.** Use your answer in part (a) to approximate $\ln \frac{3}{2}$.

(*continues*)

c. Estimate the error by the error formula (page 708). [*Hint:* Find an $M$ such that $|f^{(7)}(x)| \leq M$ for $1 \leq x \leq 1.5$, using the fact that a decreasing function is maximized at its left-hand endpoint.]

d. Restate the approximation and error, rounded appropriately.

## Applied Exercises

**31. BUSINESS: Advertising** Five days after the end of an advertising campaign, a company's daily sales (as a proportion of its peak sales) will be $e^{-0.25}$. Approximate $e^{-0.25}$ by using the second Taylor polynomial at $x = 0$ for $e^x$ (rounding to two decimal places). Check your answer by using the $\boxed{e^x}$ key on your calculator.

**32. ENVIRONMENTAL SCIENCES: Deer Population** A population of deer in a wildlife refuge varies with the season, and is predicted to be $300 + 50 \cos 0.52$ on the first day of summer. Approximate this number by using the second Taylor polynomial at $x = 0$ for $\cos x$. Then check your answer using the $\boxed{\cos}$ key on your calculator (set for radians).

**33. ECONOMICS: Supply and Demand** The demand for a new toy is predicted to be $4 \cos x$, and the supply is $x + 3$ (both in thousands of units), where $x$ is the number of years $(0 \leq x \leq 2)$. Find when supply will equal demand by solving the equation

$$4 \cos x = x + 3$$

as follows. Replace $\cos x$ by its second Taylor polynomial at $x = 0$ and solve the resulting quadratic equation for the (positive) value of $x$.

**30. ZERO FACTORIAL** The relationship $(n + 1)! = (n + 1)n!$, solved for $n!$, gives

$$n! = \frac{(n + 1)!}{n + 1}$$

Evaluate this at $n = 0$ to see the reason for defining $0! = 1$.

## Conceptual Exercises

**34.** Find: **a.** $\dfrac{1000!}{999!}$ **b.** $0^{0!}$ **c.** $(0!)^0$

**35–38.** Find, if possible, the best linear approximation at $x = 0$ for each function. [*Hint:* No work necessary—just think about the graph.]

**35.** $f(x) = \sin x$      **36.** $f(x) = x$

**37.** $f(x) = \dfrac{1}{x}$      **38.** $f(x) = \ln x$

**39.** What is an upper bound for the absolute value of *any* derivative of $f(x) = \sin x$ on any interval?

**40.** Fill in the blank: If a function is *increasing* on a closed interval, then an upper bound for the function is its value at the _____-hand endpoint of the interval.

**41.** Fill in the blank: If a function is *decreasing* on a closed interval, then an upper bound for its value is its value at the _____-hand endpoint of the interval.

**42.** True or False: The $n$th Taylor approximation of a function agrees with the function for at least one value of $x$.

**43.** Suppose that $f(0) = 1$ and $f^{(n)}(0) = n!$ for $n = 1, 2, 3, \ldots$. What will be the $n$th Taylor polynomial for $f$ at $x = 0$? [*Hint:* See page 701.] Use a formula from the preceding section to express this function in a different form.

<br>

## 10.3   TAYLOR SERIES

### Introduction

In this section we return to infinite series, but with the terms of the series now being *functions*. We begin by discussing *power series*, then specializing to *Taylor series*, which may be thought of as "infinitely long" Taylor polynomials. Taylor series lead to remarkably simple infinite series for functions such as

$e^x$, $\ln x$, and $\sin x$, and enable us to find difficult integrals involving such functions.

## Power Series

In this (and the next) section we will consider infinite series whose terms involve a *variable*, such as

$$1 + x + x^2 + x^3 + \cdots$$

Replacing $x$ by $\frac{1}{2}$ gives

$$1 + \frac{1}{2} + \frac{1}{4} + \frac{1}{8} + \cdots \qquad \text{From } 1 + \frac{1}{2} + \left(\frac{1}{2}\right)^2 + \left(\frac{1}{2}\right)^3 + \cdots$$

which is a convergent geometric series with ratio $r = \frac{1}{2}$. If instead we replace $x$ by 1, we obtain

$$1 + 1 + 1 + 1 + \cdots \qquad \text{From } 1 + 1 + (1)^2 + (1)^3 + \cdots$$

which clearly diverges since the sum becomes arbitrarily large. In general, a series involving a variable may converge for some values of the variable and diverge for others. It will therefore be important to determine the values of the variable for which the series converges. The most useful series is called a **power series,** which is similar to the series $1 + x + x^2 + x^3 + \cdots$ but with the terms multiplied by coefficients. That is, a power series is of the form

$$a_0 + a_1 x + a_2 x^2 + \cdots + a_n x^n + \cdots$$

in which the coefficients $a_0$, $a_1$, $a_2$, ... , $a_n$, ... are (real) numbers. Every power series in $x$ converges for $x = 0$ (where its value is clearly $a_0$). It also converges at all points within some radius $R$ around $x = 0$, with the understanding that $R$ may be a positive number, or $R = 0$ (in which case the series converges only at $x = 0$), or $R = \infty$ (in which case the series converges for *all* values of $x$). Formally:

---

### Power Series in *x* and Radius of Convergence

For a power series in $x$, $a_0 + a_1 x + a_2 x^2 + \cdots + a_n x^n + \cdots$, exactly one of the following statements is true:

1. There is a positive number $R$ (the "radius of convergence") such that the series converges for $|x| < R$ and diverges for $|x| > R$.

2. The series converges only at $x = 0$ (radius $R = 0$).

3. The series converges for *all* $x$ (radius $R = \infty$).

---

These three cases can be shown graphically as follows.

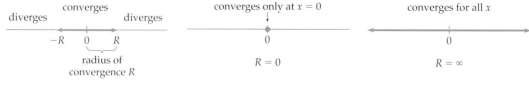

Positive radius                Zero radius                Infinite radius

For a positive radius of convergence, the series may converge at *neither* endpoint, at just *one* endpoint, or at *both* endpoints. We will generally ignore these endpoints, finding only the radius of convergence by using the following **Ratio Test.**

---

**Ratio Test**

For an infinite series $c_0 + c_1 + c_2 + \cdots$ of nonzero terms such that

the limit $r = \lim\limits_{n \to \infty} \left| \dfrac{c_{n+1}}{c_n} \right|$ exists or is infinite:

1. If $r < 1$, the series *converges.*
2. If $r > 1$ or if the limit is infinite, the series *diverges.*

---

The Ratio Test is related to the convergence of a geometric series (page 691) as follows. A geometric series converges if the ratio $r$ of successive terms satisfies $|r| < 1$, while the Ratio Test says that even if a series is *not* geometric, as long as the *limiting* ratio of successive terms is less than 1 in absolute value, then the series converges. A proof of the Ratio Test is outlined in Exercises 58–59. Finding the radius of convergence of a power series by the Ratio Test involves four steps:

1. Find expressions for successive terms $c_n$ and $c_{n+1}$ (found from $c_n$ by replacing each $n$ by $n+1$).

2. Write the quotient $\dfrac{c_{n+1}}{c_n}$ and simplify.

3. Calculate $r = \lim\limits_{n \to \infty} \left| \dfrac{c_{n+1}}{c_n} \right|$.

4. Solve the relationship $r < 1$ for $|x|$ to find the radius of convergence $R$.

**EXAMPLE 1**

**FINDING THE RADIUS OF CONVERGENCE**

Find the radius of convergence of the power series

$$\frac{1}{2}x + \frac{2}{4}x^2 + \frac{3}{8}x^3 + \cdots + \frac{n}{2^n}x^n + \cdots$$

*Solution*

Successive terms are $c_n = \dfrac{n}{2^n}x^n$ and $c_{n+1} = \dfrac{n+1}{2^{n+1}}x^{n+1}$, and their ratio is

$$\frac{c_{n+1}}{c_n} = \frac{\dfrac{n+1}{2^{n+1}}x^{n+1}}{\dfrac{n}{2^n}x^n} = \frac{n+1}{2^{n+1}} \cdot \frac{x^{n+1}}{x^n} \cdot \frac{2^n}{n} = \frac{n+1}{n} \cdot \frac{2^n}{2^{n+1}} \cdot x = \left(1 + \frac{1}{n}\right) \cdot \frac{1}{2} \cdot x \longrightarrow \frac{x}{2}$$

Inverting $\longrightarrow$ $x$

$1 + \dfrac{1}{n}$ $\qquad$ $\dfrac{1}{2}$ $\qquad$ Approaches 1 as $n \to \infty$

Therefore, $r = \lim\limits_{n \to \infty}\left|\dfrac{c_{n+1}}{c_n}\right| = \dfrac{|x|}{2}$. For convergence, this ratio must be less than 1, $\dfrac{|x|}{2} < 1$, which is equivalent to $|x| < 2$. Therefore, the radius of convergence is $R = 2$, meaning that the series converges for $|x| < 2$ and diverges for $|x| > 2$.

## Taylor Series

We now consider a special type of power series called a **Taylor series,** which may be thought of as an "infinitely long" Taylor polynomial. We begin with a function $f$ with derivatives of all orders at $x = 0$.

---

**Taylor Series at $x = 0$**

The Taylor series at $x = 0$ of $f$ is

$$f(0) + \frac{f'(0)}{1!}x + \frac{f''(0)}{2!}x^2 + \frac{f'''(0)}{3!}x^3 + \cdots + \frac{f^{(n)}(0)}{n!}x^n + \cdots$$

---

Note that the coefficients are calculated just as in Taylor polynomials.

**EXAMPLE 2**

**FINDING A TAYLOR SERIES**

Find the Taylor series at $x = 0$ for $e^x$, and find its interval of convergence.

*Solution*

The Taylor series for $e^x$ can be written down immediately from the Taylor polynomials for $e^x$ in Example 2 on page 701 simply by letting the terms continue indefinitely:

$$1 + \frac{x}{1!} + \frac{x^2}{2!} + \frac{x^3}{3!} + \frac{x^4}{4!} + \cdots \qquad \text{Taylor series for } e^x$$

For the radius of convergence, we calculate the ratio of successive terms and use the Ratio Test:

$$\frac{\dfrac{x^{n+1}}{(n+1)!}}{\dfrac{x^n}{n!}} = \frac{x^{n+1}}{(n+1)!}\cdot\frac{n!}{x^n} = \frac{x^{n+1}}{x^n}\cdot\frac{n!}{(n+1)!} \qquad c_n = \frac{x^n}{n!}$$

$$\underbrace{}_{\text{Inverting}} \qquad \underbrace{x} \qquad \underbrace{\frac{n\cdots 1}{(n+1)\cdot n \cdots 1}} = \frac{1}{n+1} \qquad c_{n+1} = \frac{x^{n+1}}{(n+1)!}$$

$$= x\cdot\frac{1}{n+1} \longrightarrow 0$$

$$\underbrace{\text{Approaches } 0}_{\text{as } n \to \infty} \quad \underbrace{r}$$

This ratio approaches 0, so $r = 0$ satisfying $|r| < 1$ for *any* fixed value of $x$. Therefore, by the Ratio Test, the series converges for *all* values of $x$ (radius of convergence $R = \infty$).

 **BE CAREFUL:** Don't confuse $r$ and $R$. *Lowercase* $r$ comes from the ratio of terms (after taking absolute values) and so may involve $x$ (as in Example 1). *Uppercase* $R$ is simply a number or the symbol $\infty$. In extreme cases they are inversely related: if $r = 0$ then $R = \infty$, while if $r = \infty$ then $R = 0$.

We have shown that the series for $e^x$ converges for every value of $x$, but we have not shown that it converges *to* $e^x$ (it might instead converge to some other value). Proving convergence to the original function requires showing that the error $|R_n(x)|$ in the $n$th Taylor polynomial approaches zero as $n \to \infty$. Such proofs are difficult, and are relegated to the Exercises (see Exercises 56 and 57). For every Taylor series in this book, and for every Taylor series you are likely to encounter in applications, if it converges, it converges to its defining function. Based on this statement, we will write the series as *equal* to its defining function wherever the series converges. For example, the series for $e^x$ *converges* for all $x$, and *equals* $e^x$:

---

**Taylor Series at $x = 0$ for $e^x$**

$$e^x = 1 + \frac{x}{1!} + \frac{x^2}{2!} + \frac{x^3}{3!} + \frac{x^4}{4!} + \cdots \qquad \text{for} \quad -\infty < x < \infty$$

---

This series is called the "Taylor series expansion" of $e^x$. Evaluating at $x = 1$ gives a remarkable series for the constant $e$:

$$e = 1 + \frac{1}{1!} + \frac{1}{2!} + \frac{1}{3!} + \frac{1}{4!} + \cdots$$

*e is the sum of the reciprocals of factorials*

## Graphing Calculator Exploration

```
sum(seq(1/N!,N,0
,7))
    2.718253968
```

**a.** Sum the first seven terms of the series above. To how many decimal places does the result agree with $e \approx 2.71828$?

**b.** How many terms are required to get agreement to *five* decimal places?

---

**EXAMPLE 3**

**FINDING THE TAYLOR SERIES FOR sin x**

Find the Taylor series at $x = 0$ for $\sin x$.

*Solution*

Differentiating $\sin x$ gives the following repeating pattern of positive and negative sines and cosines, which on the third line below are evaluated at $x = 0$.

| $f$ | $f'$ | $f''$ | $f'''$ | $f^{(4)}$ | $f^{(5)}$ | $f^{(6)}$ | $f^{(7)}$ | $\cdots$ |
|-----|------|-------|--------|-----------|-----------|-----------|-----------|----------|
| $\sin x$ | $\cos x$ | $-\sin x$ | $-\cos x$ | $\sin x$ | $\cos x$ | $-\sin x$ | $-\cos x$ | $\cdots$ |
| $0$ | $1$ | $0$ | $-1$ | $0$ | $1$ | $0$ | $-1$ | $\cdots$ |

Substituting these into the Taylor series formula (page 714) gives

$$0 + \frac{1}{1!}x + \frac{0}{2!}x^2 + \frac{-1}{3!}x^3 + \frac{0}{4!}x^4 + \frac{1}{5!}x^5 + \frac{0}{6!}x^6 + \frac{-1}{7!}x^7 + \cdots$$

Omitting the zero terms and simplifying gives the following Taylor series for $\sin x$.

$$\sin x = x - \frac{x^3}{3!} + \frac{x^5}{5!} - \frac{x^7}{7!} + \cdots \qquad \text{Alternating signs, with only odd powers}$$

---

This series converges for *all* $x$ (as is shown in Exercise 53).*

**Taylor Series at $x = 0$ for $\sin x$**

$$\sin x = x - \frac{x^3}{3!} + \frac{x^5}{5!} - \frac{x^7}{7!} + \cdots \qquad \text{for} \quad -\infty < x < \infty$$

## Operations on Taylor Series

In many ways, Taylor series behave like polynomials. For example, the Taylor series for $f(x)$ can be differentiated term by term to obtain the Taylor series for $f'(x)$ (but only for $x$ *within* the interval of convergence, avoiding the endpoints). A similar statement holds for integration. A Taylor series can also be multiplied by a constant $c$ or by a positive integer power $x^n$, or the variable may be replaced by an expression of the form $cx^n$. In each case, the resulting

---

*This series gives another reason for expressing trigonometric functions in *radians* instead of degrees. The Taylor series for $\sin x$ for *x in degrees,* written $\sin x°$, is found by multiplying each $x$ by $\frac{\pi}{180}$, obtaining

$$\sin x° = \frac{\pi}{180}x - \left(\frac{\pi}{180}\right)^3\frac{x^3}{3!} + \left(\frac{\pi}{180}\right)^5\frac{x^5}{5!} - \left(\frac{\pi}{180}\right)^7\frac{x^7}{7!} + \cdots$$

This series lacks the simplicity of the radian version.

series will be the Taylor series for the function similarly modified, and the radius of convergence will remain the same (except in the case of replacing the variable, where a finite radius of convergence may change). Two series may be added or subtracted, giving the series for the sum or difference of the original functions, and the resulting series will have a radius of convergence equal to the *smaller* of the radii of the original series.

Any power series that converges to a function must be the *Taylor* series for that function. For example, according to the formula on page 691, the geometric series $1 + x + x^2 + x^3 + \cdots$ converges to $\dfrac{1}{1-x}$ for $|x| < 1$, and so must be the *Taylor* series for that function for $|x| < 1$. Proofs of all these statements may be found in a book in advanced calculus.

---

**Taylor Series at $x = 0$ for $\dfrac{1}{1-x}$**

$$\frac{1}{1-x} = 1 + x + x^2 + x^3 + \cdots \qquad \text{for } |x| < 1$$

---

**EXAMPLE 4**

**FINDING ONE TAYLOR SERIES FROM ANOTHER**

Find the Taylor series at $x = 0$ for $\cos x$.

*Solution*

The easiest way is to differentiate the series for $\sin x$ found in Example 3:

$$\sin x = x - \frac{1}{3!}x^3 + \frac{1}{5!}x^5 - \frac{1}{7!}x^7 + \cdots \qquad \text{for } -\infty < x < \infty$$

$$\cos x = 1 - \frac{1}{3!}3x^2 + \frac{1}{5!}5x^4 - \frac{1}{7!}7x^6 + \cdots \qquad \text{Differentiating term by term by the Power Rule}$$

$$= 1 - \frac{1}{2!}x^2 + \frac{1}{4!}x^4 - \frac{1}{6!}x^6 + \cdots \qquad \text{Simplifying}$$

This gives the following Taylor series for $\cos x$.

$$\cos x = 1 - \frac{x^2}{2!} + \frac{x^4}{4!} - \frac{x^6}{6!} + \cdots \qquad \text{Alternating signs, with only even powers}$$

This series converges for *all* $x$ (since term-by-term differentiation does not change the radius of convergence).

---

**Taylor Series at $x = 0$ for $\cos x$**

$$\cos x = 1 - \frac{x^2}{2!} + \frac{x^4}{4!} - \frac{x^6}{6!} + \cdots \qquad \text{for } -\infty < x < \infty$$

---

**Practice Problem**

Find the Taylor series at $x = 0$ for $\cos x^2$. [*Hint:* Use the preceding series with a substitution.]

➤ Solution on page 721

By integrating entire series, we can construct infinite series for difficult integrals that we could not otherwise evaluate.

## Baby Booms and Echoes

An increase in births (a "baby boom") is usually followed in later generations by smaller fluctuations (called "echoes" of the boom) as the children grow up and in turn have more children. The resulting fluctuations in the number of births can be modeled by a function such as $f(t) = \dfrac{\sin t}{t}$.

**EXAMPLE 5**    **SUMMING EXCESS BIRTHS**

A country is predicted to have an excess of $\dfrac{\sin t}{t}$ million births per year (above the usual level) in year $t$. During the first $x$ years, this means an additional $\displaystyle\int_0^x \dfrac{\sin t}{t}\,dt$ million births. Find the Taylor series for this integral. Then use the series to estimate the number of extra births during the first two years.

*Solution*

There is no simple antiderivative for the function $\dfrac{\sin t}{t}$, so we express it as a Taylor series (dividing the series for $\sin t$ by $t$), which we then integrate.

$$\frac{\sin t}{t} = \frac{1}{t}\left(t - \frac{1}{3!}t^3 + \frac{1}{5!}t^5 - \frac{1}{7!}t^7 + \cdots\right) \qquad \text{Using the series for } \sin t$$

$$= 1 - \frac{1}{3!}t^2 + \frac{1}{5!}t^4 - \frac{1}{7!}t^6 + \cdots \qquad \text{Simplifying}$$

Then

$$\int_0^x \frac{\sin t}{t}\, dt = \int_0^x \left( 1 - \frac{1}{3!}t^2 + \frac{1}{5!}t^4 - \frac{1}{7!}t^6 + \cdots \right) dt$$

$$= \left( t - \frac{1}{3!}\frac{1}{3}t^3 + \frac{1}{5!}\frac{1}{5}t^5 - \frac{1}{7!}\frac{1}{7}t^7 + \cdots \right) \Bigg|_0^x \qquad \text{Integrating term by term by the Power Rule}$$

$$= x - \frac{x^3}{3 \cdot 3!} + \frac{x^5}{5 \cdot 5!} - \frac{x^7}{7 \cdot 7!} + \cdots \qquad \begin{array}{l}\text{Substituting } t = x \text{ (the evaluation at } t = 0 \text{ is zero)}\end{array}$$

This infinite series gives the *exact* value of the integral $\int_0^x \frac{\sin t}{t}\, dt$. Evaluating at $x = 2$ and approximating by the first four terms gives

$$2 - \frac{2^3}{3 \cdot 3!} + \frac{2^5}{5 \cdot 5!} - \frac{2^7}{7 \cdot 7!} \approx 1.605$$

Therefore, there will be about 1.605 million extra births during the first two years.

Although $\frac{\sin t}{t}$ is undefined at $t = 0$, its *limit* as $t \to 0$ is 1 (see pages 590–591), which is also the *value* of its Taylor series at $t = 0$.

## Graphing Calculator Exploration

Try evaluating $\int_0^2 \frac{\sin t}{t}\, dt$ on your graphing calculator. You may get a "division by zero" error, showing that some other method (such as the Taylor series) is required. Even if you do get an answer, the Taylor series has the advantage of giving a *formula* for the integral (population gain) up to *any* year $x$.

### Why Is Convergence So Important?

The radius of convergence of a Taylor series is important because only *within* that radius from the central point does the series *equal* the function. Beyond that radius the series will have no meaning, even though the *original function* may still be defined.

For example, the function $\frac{1}{1 - x}$ is defined for all $x$-values except $x = 1$, but its Taylor series $1 + x + x^2 + x^3 + \cdots$ converges only on the interval $(-1, 1)$, as we saw on page 717. Outside of this interval the *function* is still defined (its value at $x = 2$ is $\frac{1}{1-2} = \frac{1}{-1} = -1$), but its Taylor series at

$x = 2$ is $1 + 2 + 2^2 + 2^3 + \cdots$, which does not converge and is meaningless. That is, using a Taylor series outside of the radius of convergence leads to incorrect results.

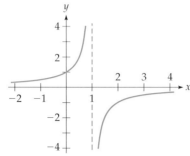

$\frac{1}{1-x}$ is defined for all $x$ except $x = 1$.

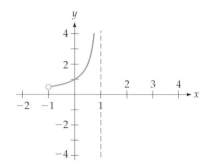

$1 + x + x^2 + x^3 + \cdots$ converges only for $-1 < x < 1$.

## Taylor Series at $x = a$

Just as with Taylor polynomials, we may define Taylor series in powers of $(x - a)$, shifting the "center point" from $x = 0$ to $x = a$. Let $f$ be a function that has derivatives of all orders at $x = a$.

---

**Taylor Series at $x = a$**

The Taylor series at $x = a$ of $f$ is

$$f(a) + \frac{f'(a)}{1!}(x - a) + \frac{f''(a)}{2!}(x - a)^2 + \cdots + \frac{f^{(n)}(a)}{n!}(x - a)^n + \cdots$$

---

Taylor *series* at $x = a$ are simply Taylor *polynomials* at $x = a$ with infinitely many terms. Taking $a = 0$ gives the Taylor series at $x = 0$, as defined earlier.

**EXAMPLE 6**

**FINDING A TAYLOR SERIES AT $x = 1$**

Find the Taylor series at $x = 1$ for $\ln x$.

*Solution*

Differentiating $f(x) = \ln x$ and evaluating at $x = 1$ gives

| $f$ | $f'$ | $f''$ | $f'''$ | $f^{(4)}$ | $\cdots$ | |
|-----|------|-------|--------|-----------|----------|---|
| $\ln x$ | $x^{-1}$ | $-x^{-2}$ | $2x^{-3}$ | $-3!x^{-4}$ | $\cdots$ | $\ln x$ and its derivatives |
| 0 | 1 | $-1$ | 2 | $-3!$ | $\cdots$ | Evaluated at $x = 1$ |

Therefore, the Taylor series at $x = 1$ for $\ln x$ is

$$(x - 1) - \frac{1}{2!}(x - 1)^2 + \frac{2}{3!}(x - 1)^3 - \frac{3!}{4!}(x - 1)^4 + \cdots$$

Taylor series with $a = 1$ and the derivatives found above

$$= (x - 1) - \frac{1}{2}(x - 1)^2 + \frac{1}{3}(x - 1)^3 - \frac{1}{4}(x - 1)^4 + \cdots$$

Simplifying

The radius of convergence is found from the Ratio Test (page 713), calculating the ratio of successive terms (and now including absolute value bars because of the alternating signs):

$$\left| \frac{\frac{1}{n+1}(x - 1)^{n+1}}{\frac{1}{n}(x - 1)^n} \right| = \left| \underbrace{\frac{n}{n+1}}_{\substack{\text{Approaches } 1 \\ \text{as } n \to \infty}} \cdot \underbrace{\frac{(x - 1)^{n+1}}{(x - 1)^n}}_{(x - 1)} \right| \to \underbrace{|x - 1|}_{r}$$

For convergence, this ratio must be less than 1: $|x - 1| < 1$, which is equivalent to $0 < x < 2$. Therefore, the Taylor series at $x = 1$ for $\ln x$ is

$$\ln x = (x - 1) - \frac{1}{2}(x - 1)^2 + \frac{1}{3}(x - 1)^3 - \frac{1}{4}(x - 1)^4 + \cdots$$

for $0 < x < 2$.

This series converges at the endpoint $x = 2$ but *not* at the endpoint $x = 0$ (see Exercises 60 and 61).

---

**Taylor Series at $x = 1$ for $\ln x$**

$$\ln x = (x - 1) - \frac{(x - 1)^2}{2} + \frac{(x - 1)^3}{3} - \frac{(x - 1)^4}{4} + \cdots \quad \text{for } 0 < x \le 2$$

---

**Solution to Practice Problem**

$$\cos x^2 = 1 - \frac{x^4}{2!} + \frac{x^8}{4!} - \frac{x^{12}}{6!} + \cdots$$

Replacing $x$ by $x^2$

## 10.3 Section Summary

Every power series in $x$, $a_0 + a_1x + a_2x^2 + \cdots + a_nx^n + \cdots$, has a *radius of convergence R* (which may be zero, a positive number, or $\infty$), and the series converges for $|x| < R$ and diverges for $|x| > R$. We find $R$ by using the Ratio Test, which may be thought of as testing whether the series is "asymptotically geometric."

A *Taylor series* is a type of power series whose coefficients are calculated from a given function $f$, using the same formulas as for Taylor polynomials:

$$f(x) = \sum_{n=0}^{\infty} \frac{f^{(n)}(0)}{n!} x^n \qquad \begin{array}{l}\text{Taylor series at } x = 0 \text{ for}\\ f(x) \text{ [the first term is } f(0)\text{]}\end{array}$$

$$f(x) = \sum_{n=0}^{\infty} \frac{f^{(n)}(a)}{n!} (x - a)^n \qquad \begin{array}{l}\text{Taylor series at } x = a \text{ for}\\ f(x) \text{ [the first term is } f(a)\text{]}\end{array}$$

A Taylor series can also be found by modifying an already existing Taylor series, as explained on pages 716–717. We derived the following Taylor series:

$$\frac{1}{1 - x} = 1 + x + x^2 + x^3 + \cdots \qquad \text{for } -1 < x < 1$$

$$e^x = 1 + \frac{x}{1!} + \frac{x^2}{2!} + \frac{x^3}{3!} + \frac{x^4}{4!} + \cdots \qquad \text{for } -\infty < x < \infty$$

$$\sin x = x - \frac{x^3}{3!} + \frac{x^5}{5!} - \frac{x^7}{7!} + \cdots \qquad \text{for } -\infty < x < \infty$$

$$\cos x = 1 - \frac{x^2}{2!} + \frac{x^4}{4!} - \frac{x^6}{6!} + \cdots \qquad \text{for } -\infty < x < \infty$$

$$\ln x = (x - 1) - \frac{(x - 1)^2}{2} + \frac{(x - 1)^3}{3} - \frac{(x - 1)^4}{4} + \cdots \quad \text{for } 0 < x \le 2$$

## 10.3 Exercises

**1–10.** Find the radius of convergence of each power series.

**1.** $1 - \dfrac{x}{2} + \dfrac{x^2}{2^2} - \dfrac{x^3}{2^3} + \cdots$

**2.** $1 + 2x + 2^2x^2 + 2^3x^3 + \cdots$

**3.** $\dfrac{x}{1 \cdot 2} + \dfrac{x^2}{2 \cdot 3} + \dfrac{x^3}{3 \cdot 4} + \dfrac{x^4}{4 \cdot 5} + \cdots$

**4.** $1 - \dfrac{x}{3} + \dfrac{x^2}{3^2} - \dfrac{x^3}{3^3} + \cdots + \dfrac{(-1)^n x^n}{3^n} + \cdots$

**5.** $\displaystyle\sum_{n=0}^{\infty} \dfrac{(-1)^n x^n}{n!}$

**6.** $\displaystyle\sum_{n=0}^{\infty} \dfrac{2^n x^n}{n!}$

**7.** $\displaystyle\sum_{n=0}^{\infty} \dfrac{n! x^n}{5^n}$

**8.** $\displaystyle\sum_{n=0}^{\infty} n! x^n$

**9.** $\displaystyle\sum_{n=0}^{\infty} \dfrac{x^{2n}}{n!}$

**10.** $\displaystyle\sum_{n=0}^{\infty} x^{n^2}$

**11–16.** Find the Taylor series at $x = 0$ for each function by calculating three or four derivatives and using the definition of Taylor series.

**11.** $\ln(1 + x)$

**12.** $e^{2x}$

**13.** $\cos 2x$

**14.** $\dfrac{1}{3 - x}$

**15.** $\sqrt{2x + 1}$

**16.** $2x^2 - 7x + 3$

**17–18.** Find the Taylor series at $x = 0$ for each function in *two* ways:

**a.** By calculating derivatives and using the definition of Taylor series

**b.** By modifying a known Taylor series

**17.** $e^{x/5}$                   **18.** $\sin 2x$

**19.** Find the Taylor series at $x = \dfrac{\pi}{2}$ for $\sin x$.

**20.** Find the Taylor series at $x = 1$ for $\sqrt{x}$.

**21–28.** Find the Taylor series at $x = 0$ for each function by modifying one of the Taylor series from this section.

**21.** $\dfrac{x^2}{1 - x}$              **22.** $\dfrac{1}{1 + 5x}$

**23.** $\sin x^2$                **24.** $e^{-x^2}$

**25.** $\dfrac{e^x - 1}{x}$             **26.** $\sin x + \cos x$

[*Hint:* For Exercise 26, use two series from this section, writing the terms in order of increasing powers of $x$.]

**27.** $\dfrac{x - \sin x}{x^3}$          **28.** $\dfrac{1 - \cos x}{x^2}$

**29.** Find the Taylor series at $x = 0$ for $\ln(x + 1)$ by integrating both sides of

$$\frac{1}{1 + x} = 1 - x + x^2 - x^3 + \cdots \quad \text{for } |x| < 1$$

**30.** Find the Taylor series at $x = 0$ for $\dfrac{1}{(1 - x)^2}$ by differentiating both sides of

$$\frac{1}{1 - x} = 1 + x + x^2 + x^3 + \cdots \quad \text{for } |x| < 1$$

**31. a.** Find the Taylor series at $x = 0$ for $\dfrac{1}{1 + x^2}$ by modifying one of the series derived in this section.

    **b.** Use the Ratio Test to find the radius of convergence of the series from part (a). Notice that although the original function $\dfrac{1}{1 + x^2}$ is defined for *all* values of $x$, the series is defined only for a much narrower set of $x$-values.

**32. HYPERBOLIC SINE FUNCTION** Use the Taylor series at $x = 0$ for $e^x$ to find the series for $\dfrac{e^x - e^{-x}}{2}$. (This function is called the *hyperbolic*

*sine,* written $\sinh x$. Notice how similar its Taylor series is to that of $\sin x$.)

**33. HYPERBOLIC COSINE FUNCTION** Use the Taylor series at $x = 0$ for $e^x$ to find the series for $\dfrac{e^x + e^{-x}}{2}$. (This function is called the *hyperbolic*

*cosine,* written $\cosh x$. Note how similar its Taylor series is to that of $\cos x$. The graph of $\cosh x$ is called a "catenary." The St. Louis Gateway Arch was built in this shape: see Exercise 48 on page 273.)

**34. SERIES CALCULATION OF sin 1**

    **a.** Use a calculator set for *radians* to find $\sin 1$ (rounded to five decimal places).

    **b.** Estimate $\sin 1$ by using the first three terms of the Taylor series

$$\sin x = x - \frac{x^3}{3!} + \frac{x^5}{5!} - \cdots$$

evaluated at $x = 1$ (rounded to five decimal places). [*Note:* Computers use series similar to this to calculate values of trigonometric functions.]

**35. SERIES CALCULATION OF $e^{0.5}$**

    **a.** Use a calculator to find $e^{0.5}$ (rounded to five decimal places).

    **b.** Estimate $e^{0.5}$ by using the first four terms of the Taylor series

$$e^x = 1 + x + \frac{x^2}{2!} + \frac{x^3}{3!} + \frac{x^4}{4!} + \cdots$$

evaluated at $x = 0.5$ (rounded to five decimal places).

     **c.** (Requires sequence and series operations) Set your calculator to find the sum of the series in part (b) up to any number of terms. How many terms are required for the sum (rounded to five decimal places) to agree with the value found in part (a)?

 **36. SERIES CALCULATION OF $e^2$** (Requires sequence and series operations)

    **a.** Use your calculator to find $e^2$ rounded to six decimal places.

    **b.** The Taylor series for $e^x$ evaluated at $x = 2$ gives $e^2 = \displaystyle\sum_{n=0}^{\infty} \frac{2^n}{n!}$. Set your calculator to find the sum of this series up to any number of terms. How many terms are required for the sum (rounded to six decimal places) to agree with your answer to part (a)?

**37. TAYLOR SERIES FOR sin $x$ AND cos $x$**
Differentiate the series

$$\cos x = 1 - \frac{x^2}{2!} + \frac{x^4}{4!} - \frac{x^6}{6!} + \cdots$$

and check that the resulting series is the negative of the series for sin $x$, showing (again) that

$$\frac{d}{dx} \cos x = -\sin x.$$

**38. TAYLOR SERIES FOR $e^x$** Differentiate the series

$$e^x = 1 + \frac{x}{1!} + \frac{x^2}{2!} + \frac{x^3}{3!} + \frac{x^4}{4!} + \cdots$$

showing (again) that $e^x$ is its own derivative.

## Applied Exercises

**39. GENERAL: Bell-Shaped Curve** The "normal distribution" or "bell-shaped curve" is used to predict many things, including IQs, incomes, and manufacturing errors. The area under this curve from 0 up to any number $x$ is given by the integral

$$\frac{1}{\sqrt{2\pi}} \int_0^x e^{-t^2/2}\, dt.$$

**a.** Find the Taylor series at 0 for $e^{-t^2/2}$.
    [*Hint:* Modify a known series.]

**b.** Integrate this series from 0 to $x$ and multiply by 0.4 $\left(\text{which approximates } \dfrac{1}{\sqrt{2\pi}}\right)$, obtaining a Taylor series for $\dfrac{1}{\sqrt{2\pi}} \displaystyle\int_0^x e^{-t^2/2}\, dt$.

**c.** Estimate $\dfrac{1}{\sqrt{2\pi}} \displaystyle\int_0^1 e^{-t^2/2}\, dt$ by using the first three terms of the series found in part (b) evaluated at $x = 1$. Your answer finds, for example, the proportion of people with IQs between 100 and 115 (see page 450).

**40. SOCIAL SCIENCE: Baby Booms** If the birthrate (thousands of births per year) rises by an amount $\dfrac{1 - \cos t}{t}$ (above the usual level) in year $t$, the number of excess births during the first $x$ years will be $\displaystyle\int_0^x \frac{1 - \cos t}{t}\, dt$.

**a.** Find the Taylor series at 0 for $\dfrac{1 - \cos t}{t}$.
    [*Hint:* Modify a known series.]

**b.** Integrate this series from 0 to $x$, obtaining a Taylor series for the integral $\displaystyle\int_0^x \frac{1 - \cos t}{t}\, dt$.

**c.** Estimate $\displaystyle\int_0^1 \frac{1 - \cos t}{t}\, dt$ by using the first three terms of the series found in part (b) evaluated at $x = 1$.

**41. BUSINESS: Sales Fluctuations** As the pace of change in modern society quickens, popular fashions may fluctuate increasingly rapidly. Suppose that sales (above a certain minimum level) for a fashion item are $\cos t^2$ in year $t$, so that extra sales during the first $x$ years are $\int_0^x \cos t^2\, dt$ (in thousands).

**a.** Find the Taylor series at 0 for $\cos t^2$.
    [*Hint:* Modify a known series.]

**b.** Integrate this series from 0 to $x$, obtaining a Taylor series for the integral $\int_0^x \cos t^2\, dt$.

**c.** Estimate $\int_0^1 \cos t^2\, dt$ by using the first three terms of the series found in part (b) evaluated at $x = 1$.

**42. GENERAL: Reactor Temperature** Under emergency conditions, the temperature in a nuclear containment vessel is expected to rise at the rate of $200e^{t^2}$ degrees per hour, so that the temperature change in the first $x$ hours will be $200 \int_0^x e^{t^2}\, dt$ degrees. Estimate the temperature rise as follows:

**a.** Find the Taylor series at 0 for $e^{t^2}$.
    [*Hint:* Modify a known series.]

**b.** Integrate this series from 0 to $x$ and multiply by 200, obtaining a Taylor series for $200 \int_0^x e^{t^2}\, dt$.

**c.** Estimate the temperature change in the first half hour by using the first three terms of the series found in part (b) evaluated at $x = \frac{1}{2}$.

(*continues*)

## Conceptual Exercises

**43.** Can a power series in $x$ diverge at 3 but converge at 4?

**44.** What is the maximum number of terms in the Taylor series for a polynomial of degree 7?

**45.** Fill in the blank: If in the Ratio Test you find $r = \dfrac{|x|}{17}$, then the radius of convergence is _____.

**46.** Suppose that $f(0) = 1$ and $f^{(n)}(0) = n!$ for $n = 1, 2, 3, \ldots$. What will be the Taylor series for $f$ at $x = 0$? [*Hint:* See page 714.] Do you know a formula from earlier in the chapter to express this function in a different form?

**47–50.** Find, by inspection, the sum of each series.

**47.** $1 - \dfrac{1}{1!} + \dfrac{1}{2!} - \dfrac{1}{3!} + \cdots$ [*Hint:* See page 715.]

**48.** $\pi - \dfrac{\pi^3}{3!} + \dfrac{\pi^5}{5!} - \dfrac{\pi^7}{7!} + \cdots$ [*Hint:* See page 716.]

**49.** $1 - \dfrac{\pi^2}{2!} + \dfrac{\pi^4}{4!} - \dfrac{\pi^6}{6!} + \cdots$ [*Hint:* See page 717.]

**50.** $1 - \dfrac{1}{2} + \dfrac{1}{3} - \dfrac{1}{4} + \cdots$ [*Hint:* See page 721.]

**51.** If a Taylor series at $x = a$ converges on the interval [3, 17] but on no larger interval, what is the value of $a$?

**52.** Fill in the blank: If a Taylor series at $x = 2$ converges at $x = 5$, then the radius of convergence is at least _____.

## Explorations and Excursions

The following problems extend and augment the material presented in the text.

### More on Convergence of Series

**53. TAYLOR SERIES FOR sin $x$** Use the Ratio Test to show that the Taylor series
$$\sin x = \sum_{n=0}^{\infty} \frac{(-1)^n x^{2n+1}}{(2n+1)!}$$ converges for all $x$.

**54. TAYLOR SERIES AND POWER SERIES** For a power series that converges to a function $f$,
$$f(x) = a_0 + a_1 x + a_2 x^2 + a_3 x^3 + \cdots$$
show that the coefficients must be $a_0 = f(0)$, $a_1 = f'(0), a_2 = \dfrac{f''(0)}{2!}, \ldots$. Since these are exactly the Taylor series coefficients, this shows that a power series that converges to a function *is* the Taylor series for that function. [*Hint:* For $a_0$, just evaluate the given equation at 0. For $a_1, a_2, \ldots$,

differentiate one or more times and then evaluate at $x = 0$. We are assuming the fact that a power series can be differentiated term by term inside its radius of convergence.]

**55. nth TERM TEST** For a convergent infinite series $S = c_1 + c_2 + c_3 + \cdots$, let $S_n$ be the sum of the first $n$ terms: $S_n = c_1 + c_2 + c_3 + \cdots + c_n$.

**a.** Show that $S_n - S_{n-1} = c_n$.
**b.** Take the limit of the equation in part (a) as $n \to \infty$ to show that $\lim_{n \to \infty} c_n = 0$. This proves the $n$th term test for infinite series:

> In a convergent infinite series, the $n$th term must approach zero as $n \to \infty$.

Or equivalently:

> If the $n$th term of a series does *not* approach zero as $n \to \infty$, then the series diverges.

**56. TAYLOR SERIES FOR $e^x$** Show that the Taylor series for $e^x$, $\sum_{n=0}^{\infty} \dfrac{x^n}{n!}$, converges to $e^x$ by showing that the error for the $n$th Taylor approximation approaches zero as follows:

**a.** Use the Ratio Test to show that the series $\sum_{n=0}^{\infty} \dfrac{|x|^n}{n!}$ converges for all $x$.
**b.** Use the $n$th term test (Exercise 55) to conclude that $\dfrac{|x|^n}{n!} \to 0$ as $n \to \infty$.
**c.** Use part (b) to show that
$$|R_n(x)| \le \frac{e^{|x|}}{(n+1)!}|x|^{n+1} \to 0$$
as $n \to \infty$ for any fixed value of $x$.

**57. TAYLOR SERIES FOR sin $x$** Show that the Taylor series for $\sin x$, $\sum_{n=0}^{\infty} \dfrac{(-1)^n x^{2n+1}}{(2n+1)!}$, converges to $\sin x$ by showing that the error for the $n$th Taylor approximation approaches zero as follows:

**a.** Use the Ratio Test to show that the series $\sum_{n=0}^{\infty} \dfrac{|x|^n}{n!}$ converges for all $x$.
**b.** Use the $n$th term test (Exercise 55) to conclude that $\dfrac{|x|^{2n+1}}{(2n+1)!} \to 0$ as $n \to \infty$.
[*Hint:* If the $n$th term approaches zero, then so does the $2n + 1$st.]
**c.** Use part (b) to show that
$$|R_{2n}(x)| \le \frac{1}{(2n+1)!}|x|^{2n+1} \to 0$$
as $n \to \infty$ for any fixed value of $x$.

**58. RATIO TEST** Prove part 1 of the Ratio Test (page 713) for positive terms as follows:

a. If $\lim\limits_{n \to \infty} \dfrac{c_{n+1}}{c_n} = r < 1,$ and if $s$ is a number between $r$ and 1, so that $r < s < 1,$ explain why there is an integer $N$ such that

$$\dfrac{c_{n+1}}{c_n} \le s \quad \text{for all} \quad n \ge N.$$

b. Show that this last inequality implies that

$$c_{N+1} \le s \cdot c_N, \quad c_{N+2} \le s^2 \cdot c_N, \quad c_{N+3} \le s^3 \cdot c_N, \ldots$$

c. Show that $\sum_{n=0}^{\infty} c_{N+n}$ converges by showing that

$$\sum_{n=0}^{\infty} c_{N+n} \le \sum_{n=0}^{\infty} s^n \cdot c_N = c_N \sum_{n=0}^{\infty} s^n = c_N \dfrac{1}{1-s}$$

It can be shown that if a series of positive terms remains below a (finite) number, then the series converges.

d. Lastly, show that $\sum_{n=0}^{\infty} c_n$ converges.

It can be shown that if a series of positive terms converges, then the series obtained by changing the signs of some or all of the terms will also converge.

**59.** Prove part 2 of the Ratio Test (page 713) as follows:

a. If $\lim\limits_{n \to \infty} \left| \dfrac{c_{n+1}}{c_n} \right| = r > 1,$ explain why there is an integer $N$ such that $\left| \dfrac{c_{n+1}}{c_n} \right| \ge 1$ for all $n \ge N.$

b. Show that this last inequality implies that

$$|c_{N+1}| \ge |c_N|, \quad |c_{N+2}| \ge |c_N|, \quad |c_{N+3}| \ge |c_N|, \ldots$$

c. Then show that these inequalities imply that $\lim\limits_{n \to \infty} c_n \ne 0,$ so $\sum_{n=0}^{\infty} c_n$ diverges by the $n$th term test (Exercise 55).

**60. TAYLOR SERIES FOR ln $x$** Show that the Taylor series for ln $x$,

$$(x-1) - \dfrac{(x-1)^2}{2} + \dfrac{(x-1)^3}{3} - \dfrac{(x-1)^4}{4} + \cdots$$

diverges at $x = 0$ by evaluating the series at 0 and explaining why it diverges.

**61. TAYLOR SERIES FOR ln $x$** Show that the Taylor series for ln $x$,

$$(x-1) - \dfrac{(x-1)^2}{2} + \dfrac{(x-1)^3}{3} - \dfrac{(x-1)^4}{4} + \cdots$$

converges at $x = 2$ by carrying out the following steps.

a. Evaluate the series at $x = 2$ to obtain

$$1 - \dfrac{1}{2} + \dfrac{1}{3} - \dfrac{1}{4} + \dfrac{1}{5} - \cdots$$

b. Group the terms in pairs to obtain

$$\left(1 - \dfrac{1}{2}\right) + \left(\dfrac{1}{3} - \dfrac{1}{4}\right) + \left(\dfrac{1}{5} - \dfrac{1}{6}\right) + \cdots \text{ and}$$

explain why this expresses the series as a sum of positive terms.

c. Then group the terms as

$$1 - \left(\dfrac{1}{2} - \dfrac{1}{3}\right) - \left(\dfrac{1}{4} - \dfrac{1}{5}\right) - \left(\dfrac{1}{6} - \dfrac{1}{7}\right) - \cdots \text{ and}$$

explain why this shows that the sum must be less than 1.

It can be shown that if a series of positive terms remains below a (finite) number, then the series converges.

**62. USING SERIES TO FIND LIMITS\*** Carry out the following steps to prove that if $a > 0,$ then $\lim\limits_{x \to \infty} x e^{-ax} = 0.$

a. Show that $e^x > \dfrac{x^2}{2}$ for $x > 0.$

[*Hint:* Begin with the Taylor series

$$e^x = 1 + x + \dfrac{x^2}{2!} + \dfrac{x^3}{3!} + \cdots \text{ and drop all but}$$

the third term on the right.]

b. Replace $x$ by $ax$ to obtain $e^{ax} > \dfrac{a^2 x^2}{2}.$

c. Take the reciprocal of each side to obtain

$$e^{-ax} < \dfrac{2}{a^2 x^2}.$$

d. Multiply each side by $x$ to obtain $x e^{-ax} < \dfrac{2}{a^2 x}.$

e. Let $x \to \infty$ to obtain the claimed result.

**63. USING SERIES TO FIND LIMITS\*** Prove that if $a > 0,$ then $\lim\limits_{x \to \infty} x^2 e^{-ax} = 0.$

[*Hint:* Use steps similar to those in Exercise 62, but in step (a) keep only the fourth term, and in step (d) multiply by $x^2.$]

**64. RADIUS OF CONVERGENCE** Show that the power series $\sum_{n=0}^{\infty} a_n x^n$ has radius of convergence $R = \lim\limits_{n \to \infty} \left| \dfrac{a_n}{a_{n+1}} \right|,$ provided that this limit exists. [*Hint:* Use the Ratio Test.]

\*The results of Exercises 62 and 63 will be used to find the mean and variance of an exponential random variable on pages 773 and 780.

## 10.4 NEWTON'S METHOD

### Introduction

Many times we have set a function equal to zero and solved for $x$. For example, setting the first or second derivative equal to zero is the first step in finding relative extreme points and inflection points. Newton's method is a procedure for *approximating* the solutions to such equations in case the *exact* solutions are difficult or impossible to find. The method was invented by Isaac Newton, one of the originators of calculus, and has many applications, such as finding the internal rate of return (interest rate) on a loan.*

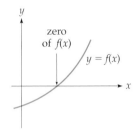

### Zeros of a Function

Recall from page 40 that an $x$-value at which a function $f(x)$ equals zero is called a **zero** of the function, or equivalently, a **root** of the equation $f(x) = 0$. A zero of a function is not simply the number 0, but is a value that, when substituted into the function, gives the *result* zero. Graphically, the zeros of a function are the $x$-values at which the graph meets the $x$-axis.

---

**EXAMPLE 1**

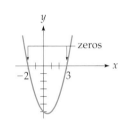

Zeros of $f(x) = x^2 - x - 6$ are the $x$-values where the curve crosses the $x$-axis.

### FINDING ZEROS BY FACTORING

Find the zeros of $f(x) = x^2 - x - 6$.

*Solution*

$$f(x) = x^2 - x - 6 = (x - 3) \cdot (x + 2) = 0$$

Factoring and setting equal to zero

The zeros are

$$\begin{cases} x = 3 \\ x = -2 \end{cases}$$

From setting $(x - 3)$ and $(x + 2)$ equal to zero

These zeros are shown on the graph on the left.

---

### Newton's Method

Only relatively simple functions can be factored. For more complicated functions, the zeros can only be *approximated,* and one of the most useful techniques is Newton's method. The method begins with an **initial approximation** $x = x_0$, which may be merely a "guess" (more on this later). The function is then replaced by its first Taylor polynomial at $x_0$ (the tangent line

---

*Newton's method is also called the Newton–Raphson method, after Joseph Raphson, a contemporary and friend of Issac Newton. Raphson first published the method, giving credit to Newton, who had described it in an unpublished manuscript many years earlier.

at $x_0$), since it is the best linear approximation for $f(x)$ near $x_0$. That is, instead of solving $f(x) = 0$, we set the *Taylor polynomial* equal to zero:

$$f(x_0) + f'(x_0) \cdot (x - x_0) = 0$$
First Taylor polynomial at $x = x_0$ set equal to zero

$$f'(x_0) \cdot (x - x_0) = -f(x_0)$$
Subtracting $f(x_0)$ from each side

$$x - x_0 = -\frac{f(x_0)}{f'(x_0)}$$
Dividing each side by $f'(x_0)$ [assuming $f'(x_0) \neq 0$)]

$$x = x_0 - \frac{f(x_0)}{f'(x_0)}$$
Solving for $x$ by adding $x_0$ to each side

This last formula gives the $x$-value (which we call $x_1$) where the first Taylor polynomial (the tangent line) crosses the $x$-axis. This $x_1$ should be a better approximation than the initial appoximation $x_0$ for the zero of the function $f(x)$, as shown below.

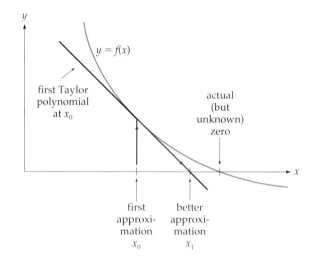

Initial approximation $x_0$ and the better approximation $x_1$,
which is closer to the actual zero.

The same formula can then be applied to the "better" approximation $x_1$ to obtain an even better approximation $x_2$, and then again to $x_2$ for a still better $x_3$, and so on:

$$x_1 = x_0 - \frac{f(x_0)}{f'(x_0)}$$

$$x_2 = x_1 - \frac{f(x_1)}{f'(x_1)}$$
Same formula applied to successive approximations

$$x_3 = x_2 - \frac{f(x_2)}{f'(x_2)}$$

$$\vdots$$

Each use of the formula is called an "iteration" ("to iterate" means "to repeat"). If the results of the iterations $x_1, x_2, x_3, \ldots$ approach a limit, we say that the iterations "converge" to this limit. In summary:

---

**Newton's Method**

To approximate a solution to $f(x) = 0$, choose an initial approximation $x_0$, and calculate $x_1, x_2, x_3, \ldots$ using

$$x_{n+1} = x_n - \frac{f(x_n)}{f'(x_n)} \qquad \text{for } n = 0, 1, 2, \ldots$$

If the numbers $x_0, x_1, x_2, \ldots$ converge, they converge to a solution of $f(x) = 0$.

---

**EXAMPLE 2**

**APPROXIMATING BY NEWTON'S METHOD**

Approximate $\sqrt{2}$ by using three iterations of Newton's method.

*Solution*

We approximate $\sqrt{2}$ by seeking a zero $f(x) = x^2 - 2$, which equals zero at $x = \sqrt{2}$. Since $\sqrt{2}$ lies between 1 and 2, we choose our initial approximation to be $x_0 = 1$ (we could just as well have chosen $x_0 = 2$ or $x_0 = 1.5$). Calculating $f(x)$ and $f'(x)$:

$$f(x) = x^2 - 2 \qquad \text{The function whose zero we want}$$

$$f'(x) = 2x \qquad \text{Its derivative}$$

The iteration formula is

$$x_{n+1} = x_n - \frac{x_n{}^2 - 2}{2x_n} \qquad x_{n+1} = x_n - \frac{f(x_n)}{f'(x_n)} \text{ with the } f \text{ and } f' \text{ above}$$

The approximations are

$$x_0 = 1 \qquad \text{Initial approximation}$$

$$x_1 = 1 - \frac{1^2 - 2}{2 \cdot 1} = 1.5 \qquad \begin{array}{l} x_1 = x_0 - \dfrac{f(x_0)}{f'(x_0)} \\ \text{applied to } x_0 = 1 \end{array}$$

$$x_2 = 1.5 - \frac{1.5^2 - 2}{2 \cdot 1.5} \approx 1.4167 \qquad \begin{array}{l} x_2 = x_1 - \dfrac{f(x_1)}{f'(x_1)} \\ \text{applied to } x_1 = 1.5 \end{array}$$

$$x_3 = 1.4167 - \frac{1.4167^2 - 2}{2 \cdot 1.4167} \approx 1.4142 \qquad \begin{array}{l} \text{Iteration formula} \\ \text{applied to } x_2 = 1.4167 \end{array}$$

The $x_3$ is our approximation for $\sqrt{2}$, the zero of $f(x) = x^2 - 2$:

$$\sqrt{2} \approx 1.4142$$

This estimate is correct to four decimal places (after only three iterations).

The approximations $x_0$, $x_1$, $x_2$, and $x_3$ to $\sqrt{2}$ are shown in the graph below. The convergence is rapid enough that the second and third iterations require a magnified view.

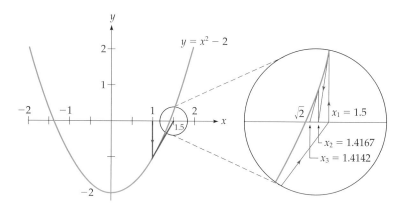

$f(x) = x^2 - 2$ and the iterations $x_0$, $x_1$, $x_2$, and $x_3$ approaching $\sqrt{2}$.

**Practice Problem 1**

Carry out one more iteration "by hand" with the formula above, using $x_3 = 1.4142$, keeping (if possible) nine decimal places of accuracy.

➤ Solution on page 734

Observe that the number of correct digits in successive approximations roughly *doubled* with each iteration:

$x_0 = 1$        Initial "guess"
$x_1 = 1.5$        1 digit correct
$x_2 = 1.4167$        3 digits correct
$x_3 = 1.4142$        5 digits correct
$x_4 = 1.414213562$        10 digits correct (from Practice Problem 1)

How do you decide when to stop the iterations? Generally, you first choose a degree of accuracy, say 0.001, and stop when two successive approximations differ by less than your chosen accuracy.

How do you choose the function to use in the iterations? The number in Example 1, $\sqrt{2}$, satisfies $x^2 = 2$, or equivalently $x^2 - 2 = 0$, so we sought the zeros of the function $f(x) = x^2 - 2$. Other roots require different equations. For example, $\sqrt[3]{11}$ satisfies $x^3 = 11$, so we want to find a zero of $f(x) = x^3 - 11$, beginning with, perhaps, $x_0 = 2$.

How do you choose the initial estimate $x_0$? There is no general rule—you just try to find a number close to the number that you are seeking. The better the initial estimate, the faster the given accuracy will be achieved. If a function has *several* zeros, the one that the iterations converge to will depend upon the initial choice $x_0$. In Example 2, the equation $x^2 - 2 = 0$ actually has *two* solutions, $\sqrt{2}$ and $-\sqrt{2}$, corresponding to the two places where the curve crosses the $x$ axis (see the graph above). The initial choice of $x_0 = 1$

led to an approximation for $\sqrt{2}$, but choosing $x_0 = -1$ would have led to an approximation for $-\sqrt{2}$.

**Practice Problem 2**

To approximate $\sqrt[5]{30}$, what function $f(x)$ would you use, and what would be a good initial estimate $x_0$? ➤ **Solution on page 734**

### Calculator Program for Newton's Method

The program NEWTON for the *TI-83/84* graphing calculator carries out Newton's method for any number of iterations (entered as the number $N$) and is listed below.*

> #### *TI-83/84 Program NEWTON*
>
> Disp "USES FUNCTION Y₁"
> Input "NUM OF ITERNS?", N
> Input "INIT GUESS?", X
> For (I, 1, N)
> X − Y₁/NDERIV (Y₁, X, X) → X
> Disp X
> End

The ZERO and SOLVE operations in a graphing calculator use Newton's method or one similar to it. The "initial guess" that these operations request is simply the initial $x_0$ of Newton's method.

**EXAMPLE 3**

**APPROXIMATING WITHIN A GIVEN TOLERANCE**

Approximate the solution to $e^x = 2 - 2x$, continuing until two successive iterations agree to nine decimal places.

*Solution*

We find the initial estimate $x_0$ from a quick sketch of the graphs of $e^x$ and $2 - 2x$, as shown on the left. They intersect at an $x$-value closer to 0 than to 1, so we choose $x_0 = 0$ as our initial estimate (other choices are also possible). We write the equation $e^x = 2 - 2x$ with zero on the right:

$$e^x - 2 + 2x = 0$$

$\underbrace{\phantom{e^x - 2 + 2x}}_{f(x)}$

Definition of $f(x)$

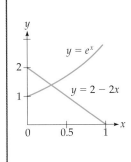

---

*Although this program is short enough to be typed directly into a calculator, it may be obtained in the same way as the others by going to www.cengage.com/math/berresford.

Newton's method may also be carried out as follows. Enter the function in Y₁, then (if 0 is the initial guess) enter the two statements in the home screen 0 → X and X − Y₁/NDERIV(Y₁, X, X) → X and then repeatedly press ENTER.

```
INIT GUESS? 0
        .3333333148
        .3149922861
        .3149230588
agree { .3149230578
        .3149230578
               Done
```

We could carry out the calculations "by hand," using the iteration formula

$$x_{n+1} = x_n - \frac{e^{x_n} - 2 + 2x_n}{e^{x_n} + 2}$$

$$x_{n+1} = x_n - \frac{f(x_n)}{f'(x_n)}$$

beginning with $x_0 = 0$, but instead we use the graphing calculator program NEWTON (see the previous page). The results, using 5 iterations, are shown on the left.

The last two answers clearly agree to nine decimal places, giving the approximation

$$x = 0.3149230578 \qquad \text{Approximation to the solution of } e^x = 2 - 2x$$

## Internal Rate of Return on a Loan

By federal law, any loan agreement must disclose the rate of interest. If the amount of the loan and the repayments are specified, finding the implied interest rate (called the **internal rate of return**) can involve an equation whose roots can only be approximated.*

**EXAMPLE 4**  **FINDING THE INTERNAL RATE OF RETURN**

A bank loan of $2500 is to be repaid in three payments of $1000 at the end of the first, second, and third years. Find the internal rate of return.

*Solution*

The internal rate of return is defined as the interest rate $r$ for which the present values of the payments add up to the amount of the loan. To find the present value at the time of the loan, we divide the payment after $n$ years by $(1 + r)^n$:

$$2500 = \frac{1000}{(1 + r)^1} + \frac{1000}{(1 + r)^2} + \frac{1000}{(1 + r)^3}$$

Amount of loan    Present value of first payment    Present value of second payment    Present value of third payment

Dividing by 1000 and replacing $1 + r$ by $x$:

$$2.5 = \frac{1}{x} + \frac{1}{x^2} + \frac{1}{x^3} \qquad x = 1 + r$$

$$\underbrace{2.5x^3 - x^2 - x - 1}_{f(x)} = 0$$

Multiplying by $x^3$ and moving all terms to the left

Definition of $f(x)$

*See H. Paley, P. Colwell, and R. Cannaday, "Internal Rates of Return," UMAP Module 640, COMAP.

Again, we could calculate the iterations "by hand" using the formula on page 729 with the $f(x)$ above and its derivative beginning with $x = 1$, but instead we use the graphing calculator program NEWTON.

## Graphing Calculator Exploration

```
NUM OF ITRNS? 4
INIT GUESS? 1
      1.111111049
      1.097250629
      1.097010329
      1.097010257
            Done
```

Use the graphing calculator program NEWTON with $y_1 = 2.5x^3 - x^2 - x - 1$, beginning with $x_0 = 1$ to obtain the results shown on the left.

The iterations give $x \approx 1.097$. To find $r$, we subtract 1 (since we defined $x = 1 + r$), giving $r \approx 0.097$.

The internal rate of return is 9.7%.

Converting 0.097 to a percent

To state this result in terms of bank interest, an account with $2500 at 9.7% would yield exactly the same amount at the end of three years as another account (also at 9.7%) with three $1000 payments made at the end of the first, second, and third years.

### The Iterations May Not Converge

Although Newton's method often converges rapidly, there are functions for which it converges very slowly, or not at all. For example, if the derivative is *zero* at any approximation, then the iteration formula will involve division by zero, and so the process will stop. (This can often be remedied by choosing a different starting point.) The following two diagrams illustrate other problems: the first shows what can happen if there is an inflection point between two successive approximations, and the second shows the difficulties caused by a nearby critical point.

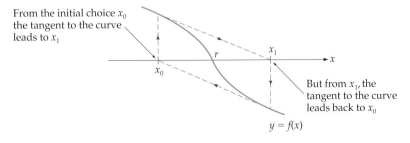

For this curve, the approximations will alternate between $x_0$ and $x_1$, never approaching the true root $r$ in the center of the diagram.

Starting at $x_0$, the tangent line to the curve leads to $x_1$ far to the *right*, instead of to the actual root $r$, which is to the left.

Although these diagrams show difficulties that can occur, Newton's method is generally very successful in applications, usually doubling the number of correct digits with each successive iteration.

---

**Solutions to Practice Problems**

**1.** $x_4 = 1.4142 - \dfrac{1.4142^2 - 2}{2 \cdot 1.4142} \approx 1.414213562$

**2.** Use $f(x) = x^5 - 30$ [or $f(x) = 30 - x^5$] and $x_0 = 2$ (although other values are possible).

---

## 10.4 Section Summary

Newton's method is a procedure for approximating the *zeros* of a function, or equivalently, the *roots* or *solutions* of $f(x) = 0$. An initial estimate $x_0$ is successively "improved" using the formula

$$x_{\text{new}} = x_{\text{old}} - \frac{f(x_{\text{old}})}{f'(x_{\text{old}})}$$

The convergence is often quite rapid, but is affected by such considerations as the accuracy of the initial estimate, roundoff errors, and how closely the function is approximated by its first Taylor polynomial (the tangent line).

Besides finding internal rates of return, the method is useful for solving many other equations, such as finding break-even points (where revenue = cost) or market equilibrium points (where supply = demand), as we will see in the exercises.

## 10.4 Exercises

**1–4.** Use Newton's method beginning with the given $x_0$ to find the first two approximations $x_1$ and $x_2$. Carry out the calculation "by hand" with the aid of a calculator, rounding to two decimal places.

**1.** $x^3 + x - 4 = 0$
   $x_0 = 1$

**2.** $x^3 + 2x - 4 = 0$
   $x_0 = 1$

**3.** $e^x - 3x = 0$
   $x_0 = 0$

**4.** $x - \cos x = 0$
   $x_0 = 0$

**5–6.** Use Newton's method to approximate each root, continuing until two successive iterations agree to three decimal places. Carry out the calculation "by hand" with the aid of a calculator, rounding to three decimal places.

**5.** $\sqrt{5}$   **6.** $\sqrt[3]{22}$

**7–10.** Use Newton's method to approximate the root of each equation, beginning with the given $x_0$ and continuing until two successive approximations agree to three decimal places. Carry out the calculation "by hand" with the aid of a calculator, rounding to three decimal places.

**7.** $x^3 + 3x - 8 = 0$   **8.** $x^3 - 3x + 3 = 0$
$x_0 = 2$   $x_0 = -2$

**9.** $e^x + 3x + 2 = 0$   **10.** $e^x + 4x - 4 = 0$
$x_0 = -1$   $x_0 = 0$

**11–12.** Use NEWTON or a similar program to approximate each root, continuing until two successive iterations agree to nine decimal places.

**11.** $\sqrt[3]{130}$   **12.** $\sqrt{130}$

**13–20.** Use NEWTON or a similar program to approximate the root of each equation beginning with the given $x_0$ and continuing until two successive approximations agree to nine decimal places.

**13.** $x^5 + 2x - 1 = 0$   **14.** $x^5 + 5x - 4 = 0$
$x_0 = 0$   $x_0 = 1$

**15.** $x + \ln x - 5 = 0$   **16.** $x^2 + \ln x - 3 = 0$
$x_0 = 2$   $x_0 = 2$

**17.** $e^{x^2} + x - 3 = 0$   **18.** $e^{2x} + 3x = 0$
$x_0 = 1$   $x_0 = 0$

**19.** $x + \sin x - 2 = 0$
$x_0 = 1$

**20.** $x^2 + \cos x - 4 = 0$
$x_0 = 1$

**21–24.** Use a graphing calculator to graph $f(x)$ and $g(x)$ together on a reasonable window and estimate the $x$-value where the curves meet. Then use Newton's method to approximate the solution of $f(x) = g(x)$, beginning with your estimate and continuing until two successive iterations agree to nine decimal places. (You may check your answer using INTERSECT, which uses a method similar to Newton's.)

**21.** $f(x) = x$,   $g(x) = 2 \cos x$

**22.** $f(x) = \sin x$,   $g(x) = 1 - x$

**23.** $f(x) = \ln x$,   $g(x) = -x$

**24.** $f(x) = e^x$,   $g(x) = 2 - x$

**25.** Use Newton's method (either "by hand" or using a graphing calculator) to solve $x^2 + 3 = 0$, beginning with $x_0 = 1$. Explain why it doesn't work.

**26.** Apply Newton's method (either "by hand" or using a graphing calculator) to the function $f(x) = x^{1/3}$ beginning with $x_0 = 1$. The graph below may help explain what goes wrong. [*Hint:* See the diagram on the page 733.] Does choosing another starting point help?

**27.** What happens when Newton's method is applied to $f(x) = 2x - 4$?  (Begin at any number other than 2.)

**28.** What happens when Newton's method is applied to a linear function $f(x) = mx + b$  with  $m \neq 0$?

**29.** Use NEWTON or a similar program to calculate the first 20 or so iterations for the zero of $f(x) = x^{10}$, beginning with $x_0 = 1$. Notice how slowly the values converge to the actual zero, $x = 0$. Can you see why from the following graph? How many iterations does it take to reach 0.001 (rounded)? Try using ZERO on your graphing calculator to find the root, starting at $x = 1$.

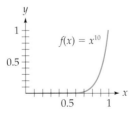

**30. SQUARE ROOTS BY ITERATION** Show that finding $\sqrt{N}$ by applying Newton's method to the equation $x^2 - N = 0$ leads to the iteration formula

$$x_{n+1} = \frac{1}{2}\left(x_n + \frac{N}{x_n}\right).$$

## Applied Exercises

**31. ECONOMICS: Supply and Demand** The supply and demand functions for a product are $S(p) = p$ and $D(p) = 10e^{-0.1p}$, where $p$ is its price (in dollars). Find the "market equilibrium price" $p$ at which $S(p) = D(p)$. Use Newton's method with two iterations, beginning with $x_0 = 4$. Carry out the calculation "by hand" with the aid of a calculator, rounding to two decimal places.

**32. BUSINESS: Cost and Revenue** The cost and revenue functions for a company are $C(x) = 50 + 2x$ and $R(x) = 2x^{3/2}$, where $x$ is the quantity (in thousands). Find the "break-even" quantity $x$ at which $C(x) = R(x)$. Use Newton's method with two iterations, beginning with $x_0 = 10$. Carry out the calculation "by hand" with the aid of a calculator, rounding to one decimal place.

 **33. ECONOMICS: Supply and Demand** The supply and demand functions for a product are $S(p) = 2e^{0.5p}$ and $D(p) = 10 - p$, where $p$ is the price of the product (in dollars) and $0 < p < 10$. Find the "market equilibrium price" $p$ at which $S(p) = D(p)$. Use NEWTON or a similar program, beginning with some initial guess between 0 and 10, and continuing until the iterations agree to two decimal places.

 **34. BUSINESS: Cost and Revenue** The cost and revenue functions for a company are $C(x) = 50 + 40\sqrt{x}$ and $R(x) = 3x$, where $x$ is the quantity $(x \geq 100)$. Find the "break-even" quantity $x$ at which $C(x) = R(x)$. Use NEWTON or a similar program, beginning with an initial guess of at least 100, and continuing until the iterations agree when rounded to the nearest unit.

 **35. ECONOMICS: Internal Rate of Return** A $15,000 loan is to be repaid in four installments of $5000 each at the end of the first, second, third, and fourth years. Use NEWTON or a similar program to find the internal rate of return (to the nearest tenth of a percent).

 **36. ECONOMICS: Internal Rate of Return** An $8000 loan is to be repaid in installments of $1000, $2000, $3000, and $4000 at the end of the first, second, third, and fourth years, respectively. Use NEWTON or a similar program to find the internal rate of return (to the nearest tenth of a percent).

 **37. BIOMEDICAL: Epidemics** An epidemic is subject to two effects: linear growth, as infected people move into the area, and exponential decline, as existing cases are cured. Find when the two effects will be in equilibrium by solving $5t = 30e^{-0.2t}$, where $t$ is the number of weeks since the epidemic began. Use NEWTON or a similar program, beginning with some appropriate initial value and continuing until the iterations agree to two decimal places.

 **38. SOCIAL SCIENCE: Immigration** Immigration into a country is occurring at the rate of $2e^{0.1t}$ thousand people in year $t$, and the government sets an upper limit of $20 - t$ thousand people in year $t$ (for $0 \leq t \leq 20$). Find when this limit will be reached by solving $2e^{0.1t} = 20 - t$. Use NEWTON or a similar program, beginning with some appropriate initial estimate and continuing until the iterations agree to two decimal places.

## Conceptual Exercises

**39.** What is the zero of the linear function $f(x) = mx + b$ with $m \neq 0$?

**40.** Find all zeros of: **a.** $\ln x$ **b.** $e^x$ **c.** $\sin x$

**41.** If 13 is a zero of $f(x)$, then name a zero of $g(x) = f(x + 17)$.

**42.** What happens in Newton's method if your initial estimate is actually a *solution* to $f(x) = 0$?

**43.** What happens in Newton's method if the function is *linear* with nonzero slope? Assume that you do not begin *at* the solution.

**44.** How many iterations will Newton's method take if applied to the absolute value function? (See page 56.) Assume that you do not begin at 0.

**45.** Let $f$ be a function that has one zero, has positive slope, and is concave up. Will Newton's method make estimates to the *right* or the *left* of the zero? Assume that you do not begin *at* the solution. [*Hint:* Make a sketch.]

**46.** Let $f$ be a function that has one zero, has negative slope, and is concave down. Will Newton's method make estimates to the *right* or the *left* of the zero? Assume that you do not begin *at* the solution. [*Hint:* Make a sketch.]

**47.** To estimate the solution to $f(x) = g(x)$, one person might apply Newton's method to $f(x) - g(x)$ and another might apply it to $g(x) - f(x)$. If both use the same initial point, how will their iterations differ?

**48.** What is the internal rate of return if you borrow $100 and pay back $110 a year later?

## 10.1 Geometric Series

- Determine whether an infinite geometric series converges, and if so, find its sum.
  *(Review Exercises 1–8.)*

$$a + ar + ar^2 + ar^3 + \cdots = \frac{a}{1-r} \quad \text{for} \quad |r| < 1$$

- Find the value of a repeating decimal.
  *(Review Exercises 9–10.)*

- Find the value of an annuity.
  *(Review Exercises 11–12.)*

$$a + ar + ar^2 + \cdots + ar^{n-1} = a\frac{1-r^n}{1-r}$$

- Solve an applied problem using an infinite series.
  *(Review Exercises 13–14.)*

## 10.2 Taylor Polynomials

- Find a Taylor polynomial for a function.
  *(Review Exercises 15–18.)*

$$p_n(x) = f(0) + \frac{f'(0)}{1!}x + \frac{f''(0)}{2!}x^2 + \cdots + \frac{f^{(n)}(0)}{n!}x^n$$

- Find a Taylor polynomial for a function, use it to approximate a value, and estimate the error.
  *(Review Exercises 19–20.)*

$$|R_n(x)| \le \frac{M}{(n+1)!}|x|^{n+1} \quad \text{for} \quad |f^{(n+1)}| \le M$$

- Find the best linear approximation for a function.
  *(Review Exercise 21.)*

- Find a Taylor polynomial that approximates a function within a given accuracy.
  *(Review Exercise 22.)*

- Find a Taylor polynomial of $x = a$ for function and graph both.
  *(Review Exercises 23–24.)*

- Approximate a value using a Taylor polynomial and estimate the error.
  *(Review Exercise 25.)*

- Solve an applied problem using a Taylor polynomial and check using a calculator.
  *(Review Exercises 26–28.)*

## 10.3 Taylor Series

- Find the radius of convergence of a power series.
  *(Review Exercises 29–32.)*

$$r = \lim_{n \to \infty} \left| \frac{c_{n+1}}{c_n} \right|$$

- Use one Taylor series to find another.
  *(Review Exercises 33–36.)*

$$\frac{1}{1-x} = 1 + x + x^2 + x^3 + \cdots \quad \text{on } (-1, 1)$$

$$e^x = 1 + \frac{x}{1!} + \frac{x^2}{2!} + \frac{x^3}{3!} + \frac{x^4}{4!} + \cdots \quad \text{on } (-\infty, \infty)$$

$$\sin x = x - \frac{x^3}{3!} + \frac{x^5}{5!} - \frac{x^7}{7!} + \cdots \quad \text{on } (-\infty, \infty)$$

$$\cos x = 1 - \frac{x^2}{2!} + \frac{x^4}{4!} - \frac{x^6}{6!} + \cdots \quad \text{on } (-\infty, \infty)$$

$$\ln x = (x - 1) - \frac{(x-1)^2}{2} + \frac{(x-1)^3}{3} - \cdots$$
$$\text{on } (0, 2]$$

- Find a Taylor series in two ways: from the definition using derivatives and by modifying a known series. *(Review Exercises 37–38.)*

$$f(0) + \frac{f'(0)}{1!}x + \frac{f''(0)}{2!}x^2 + \frac{f'''(0)}{3!}x^3 + \cdots$$

- Find a Taylor series at $x = a$. Compare it to the original function. *(Review Exercise 39.)*

$$f(a) + \frac{f'(a)}{1!}(x - a) + \frac{f''(a)}{2!}(x - a)^2 + \cdots$$

- Verify two trigonometric identities using Taylor series. *(Review Exercise 40.)*

- Solve an applied problem by using a Taylor series. *(Review Exercises 41–42.)*

## 10.4 Newton's Method

- Use Newton's method "by hand" to approximate a root within a given accuracy. *(Review Exercises 43–46.)*

$$x_{\text{new}} = x_{\text{old}} - \frac{f(x_{\text{old}})}{f'(x_{\text{old}})}$$

- Use Newton's method with a calculator or computer to approximate a root within a given accuracy. *(Review Exercises 47–52.)*

- Solve an applied problem by using Newton's method. *(Review Exercises 53–55.)*

### Hints and Suggestions

- In a finite geometric series, the number of terms is always one more than the final power of $x$.

- In mathematics, the terms *series* and *sequence* mean different things. Terms in a *series* are separated by + or − signs, and those in a *sequence* by *commas* (as in a sequence of approximations from Newton's method).

- Compare the formulas for the sum of finite and infinite geometric series: the *infinite* formula is simpler than the *finite* formula.

- For an infinite geometric series, be sure to check convergence ($|r| < 1$) before using the formula, or you may get an incorrect answer.

- The *first* Taylor polynomial is the best *linear* approximation, the *second* Taylor polynomial is the best *quadratic* approximation (at a point), and so on. Taylor polynomials are found by matching values and derivatives at a point.

- A Taylor polynomial at $x = a$ is *exact at* $x = a$, and less accurate further from $x = a$ (as judged by the error formula). The accuracy generally improves as the degree increases.

- The error formula for Taylor polynomials is easy to remember, since it closely resembles the first term left off.

- Taylor series are most easily found by modifying existing Taylor series.

- In practice, Newton's method for approximating the root of an equation generally converges very quickly, typically doubling the number of digits of accuracy with each iteration.

- A graphing calculator helps by summing many terms of a series, but it cannot sum an *infinite* series (unless you give it a formula). It can also *graph* a function together with several Taylor polynomials to show how close they are. It is also useful for calculating iterations of Newton's method.

- **Practice for Test:** Review Exercises 1, 3, 7, 13, 19, 23, 29, 35, 41, 47, 53.

# Review Exercises for Chapter 10

Practice test exercise numbers are in green.

## 10.1 Geometric Series

**1–8.** Determine whether each infinite geometric series converges or diverges. If it converges, find its sum.

1. $\frac{3}{10} + \frac{9}{100} + \frac{27}{1000} + \cdots$

2. $\frac{2}{5} - \frac{4}{25} + \frac{8}{125} - \cdots$

3. $\frac{3}{100} - \frac{9}{100} + \frac{27}{100} - \cdots$

4. $\frac{2}{3} - \frac{4}{9} + \frac{8}{27} - \cdots$

5. $10 + 8 + \frac{32}{5} + \frac{128}{25} + \cdots$

6. $\sum_{i=0}^{\infty} \left(-\frac{3}{4}\right)^i$

7. $\sum_{k=0}^{\infty} 6(-0.2)^k$

8. $\sum_{n=0}^{\infty} \frac{2^n}{3}$

**9–10.** Find the value of each repeating decimal.

9. $0.\overline{54} = 0.545454\ldots$

10. $0.\overline{72} = 0.727272\ldots$

11. **PERSONAL FINANCE: Annuity** A star basketball player preparing for his parents' retirement plans to deposit $6000 at the end of each month for 8 years into an account paying 6% compounded monthly. Find the amount of this annuity.

12. **BUSINESS: Annuities** Derive the formula

$$\text{Amount} = D\left[\frac{(1 + i)^n - 1}{i}\right]$$

for the amount of an annuity consisting of $n$ payments of $D$ dollars each, with payments made at the end of each period, at compound interest rate $i$ per period. (The formula in the brackets is denoted $S_{\overline{n}i}$ (read "s angle $n$ at $i$") and is calculated in tables.)

**13. BIOMEDICAL: Drug Dosage** A patient has been taking daily doses of 0.26 milligram (mg) of digoxin for an extended period of time, and 65% of the drug in the bloodstream is absorbed by the body during each day.

  **a.** Find the maximum and minimum long-term amounts in the bloodstream.

 **b.** How many doses does it take for the cumulative amount in the bloodstream to reach 0.39 mg? To reach 0.399 mg?

**14. ECONOMICS: Banking Reserves** When a bank receives a deposit, it keeps only a small part (say 10%) as its "reserves," lending out the rest. These loans result in other purchases, with the sums eventually deposited in other banks, which in turn keep 10% reserves and lend out the other 90%. This process continues, with an original deposit of $1000 generating deposits totaling

$$1000 + 1000(0.90) + 1000(0.90)^2 + \cdots$$

  **a.** Sum this series and find the "multiplier" by which the original deposit is multiplied. This illustrates how the banking system multiplies money.

 **b.** (Graphing calculator with series operations helpful) How many terms of the preceding series are required for the sum to surpass $9000? To surpass $9900?

## 10.2 Taylor Polynomials

**15–18.** For each function:

**a.** Find the third Taylor polynomial at $x = 0$.

 **b.** Graph the original function and the Taylor polynomial together on the indicated window.

**15.** $f(x) = \sqrt{x + 9}$
(for (b), use window $[-15, 15]$ by $[-6, 6]$)

**16.** $f(x) = e^{3x}$
(for (b), use window $[-1, 1]$ by $[-1, 8]$)

**17.** $f(x) = \ln(2x + 1)$
(for (b), use window $[-1, 1]$ by $[-5, 5]$)

**18.** $f(x) = \sin^2 x$
(for (b), use window $[-3, 3]$ by $[-2, 2]$)

**19. a.** For $f(x) = \sqrt{x + 1}$, find the third Taylor polynomial at $x = 0$.
  **b.** Use your answer to part (a) to approximate $\sqrt{2}$.

*(continues)*

**c.** Estimate the error by the error formula. [*Hint:* Calculate $f^{(4)}(x)$ and find an $M$ such that $|f^{(4)}(x)| \leq M$ for $0 \leq x \leq 1$.]

**d.** Restate the approximation and error, rounded appropriately.

**20. a.** For $f(x) = \cos x$, find the fourth Taylor polynomial at $x = 0$.
  **b.** Use your answer to part (a) to approximate $\cos 1$.
  **c.** Estimate the error by the error formula. [*Hint:* Calculate $f^{(5)}(x)$ and find an $M$ such that $|f^{(5)}(x)| \leq M$ for $0 \leq x \leq 1$.]
  **d.** Restate the approximation and error, rounded appropriately.

**21.** What linear function best approximates $\sqrt[n]{1 + x}$ near $x = 0$?

**22.** Find the Taylor polynomial at $x = 0$ that approximates $e^x$ on the interval $-1 \leq x \leq 1$ with error less than 0.0001.

**23. a.** For $f(x) = \sin x$, find the fourth Taylor polynomial at $x = \pi/2$.
 **b.** Graph $\sin x$ and its Taylor polynomial together on the window $[-2, 5]$ by $[-2, 2]$.

**24. a.** For $f(x) = x^{3/2}$, find the third Taylor polynomial at $x = 4$.
  **b.** Graph $x^{3/2}$ and its Taylor polynomial together on the window $[-2, 15]$ by $[-5, 60]$.

**25. a.** For $f(x) = x^{2/3}$, find the second Taylor polynomial at $x = 1$.
  **b.** Use your answer to part (a) to approximate $(1.2)^{2/3}$.
  **c.** Calculate $f^{(3)}(x)$ and find an $M$ such that $|f^{(3)}(x)| \leq M$ for $1 \leq x \leq 1.2$.
  **d.** Estimate the error by the error formula.
  **e.** Restate the approximation and error, rounded appropriately.

**26. BIOMEDICAL: Drug Dosage** Two hours after a patient takes a 1-gram dose of penicillin, the amount of penicillin in the patient's bloodstream will be $e^{-0.22}$ gram. Approximate $e^{-0.22}$ by using the second Taylor polynomial for $e^x$ (rounded to two decimal places). Check your answer by using $\boxed{e^x}$ on a calculator.

**27. GENERAL: Temperature** The mean temperature in New York City on December 25 is predicted to be $55 - 24 \cos 0.45$. Approximate this number by using the second Taylor polynomial at $x = 0$ for $\cos x$. Then check your answer using the $\boxed{\cos}$ key on a calculator.

**28. BUSINESS: Sales** Just after introducing a new product, a company's weekly sales are predicted to be $200e^{0.1t^2}$ units.

a. Find the fourth Taylor polynomial at $t = 0$ for this function. [*Hint:* Begin with the second Taylor polynomial for $e^x$ and use substitution, then multiply by 200.]

b. Use your Taylor polynomial to approximate the sales in week $t = 2$.

c. Use a calculator to evaluate the original function at $t = 2$ and compare answers.

## 10.3 Taylor Series

**29–32.** Find the radius of convergence of each power series.

**29.** $\displaystyle\sum_{n=0}^{\infty} \frac{nx^n}{3^n}$

**30.** $\dfrac{x}{\sqrt{1}} + \dfrac{x^2}{\sqrt{2}} + \dfrac{x^3}{\sqrt{3}} + \dfrac{x^4}{\sqrt{4}} + \cdots$

**31.** $\displaystyle\sum_{n=0}^{\infty} \frac{n!x^n}{6^n}$

**32.** $\displaystyle\sum_{n=1}^{\infty} \frac{(-1)^n x^n}{(n-1)!}$

**33–36.** For each function, find the Taylor series at $x = 0$ by modifying a known Taylor series.

**33.** $\dfrac{x^3}{1 - x^2}$

**34.** $x \cos 2x$

**35.** $xe^{x^2}$

**36.** $xe^{-x}$

**37–38.** For each function, find the Taylor series at $x = 0$ in two ways:

a. By calculating derivatives and using the definition of Taylor series.

b. By modifying a known Taylor series

**37.** $\dfrac{1}{1 - 2x}$

**38.** $e^{-x/2}$

**39. a.** Find the Taylor series at $x = \pi/2$ for $\cos x$.

 **b.** Graph $\cos x$ and the sum of the first three nonzero terms of the Taylor series from part (a) on the window $[-3, 6]$ by $[-2, 2]$.

**40.** Verify the negative angle identities (page 578) $\sin(-x) = -\sin x$ and $\cos(-x) = \cos x$ by using the Taylor series for the sine and cosine functions.

**41. GENERAL: Commuter Traffic** The number of commuter cars passing through a toll plaza per hour is $\dfrac{1}{1 + t^2}$ thousand, where $t$ is the number of hours after 9 a.m. Therefore, the total number of cars (in thousands) within $x$ hours of 9 a.m. is given by the integral $\displaystyle\int_0^x \frac{1}{1 + t^2}\, dt$.

a. Find the Taylor series at $t = 0$ for $\dfrac{1}{1 + t^2}$ by modifying an existing series.

b. Integrate this series from 0 to $x$, thereby finding a Taylor series for $\displaystyle\int_0^x \frac{1}{1 + t^2}\, dt$.

c. Estimate the number of cars passing through the toll plaza during the first half hour after 9 a.m. by evaluating the first three terms of the series at $x = \frac{1}{2}$.

d. Compare your answer to that found by using FnInt or $\int$ f(x) dx on your graphing calculator to integrate the original function from 0 to 0.5.

**42. BIOMEDICAL: Gene Frequency** Under certain circumstances, the number of generations necessary to increase the frequency of a gene from 0.2 to 0.5 is

$$2.5 \int_{0.2}^{0.5} \frac{1}{x^2(1 - x)}\, dx$$

rounded to the nearest integer.

a. Use the Taylor series at $x = 0$ for $1/(1 - x)$ to find an infinite series for $1/(x^2(1 - x))$.

b. Integrate this series term by term to find a series expression for $\displaystyle\int \frac{1}{x^2(1 - x)}\, dx$.

c. Use the first five terms of this series to estimate the definite integral

$$\int_{0.2}^{0.5} \frac{1}{x^2(1 - x)}\, dx$$

d. How many generations are necessary to raise the gene frequency from 0.2 to 0.5? Compare this solution to that found for Example 3 on pages 424–425.

## 10.4 Newton's Method

**43–44.** Use Newton's method to approximate each root, continuing until two successive iterations agree to three decimal places. Carry out the calculations "by hand" with the aid of a calculator, rounding to three decimal places.

**43.** $\sqrt{26}$

**44.** $\sqrt[3]{30}$

**45–46.** Use Newton's method to approximate the root of each equation, beginning with the given $x_0$ and continuing until two successive iterations agree to three decimal places. Carry out the calculations "by hand" with the aid of a calculator, rounding to three decimal places.

**45.** $x^3 + 3x - 3 = 0$
$x_0 = 1$

**46.** $e^x + 10x - 4 = 0$
$x_0 = 0$

 **47–48.** Use NEWTON or a similar program to approximate each root, continuing until two successive iterations agree to nine decimal places.

47. $\sqrt[3]{100}$

**48.** $\sqrt{10}$

 **49–52.** Use NEWTON or a similar program to approximate the root of each equation, beginning with the given $x_0$ and continuing until two successive approximations agree to nine decimal places.

**49.** $e^x - 5x - 3 = 0$
$x_0 = 1$

**50.** $e^x + 4x - 100 = 0$
$x_0 = 3$

**51.** $x^3 = 5 \ln x + 10$
$x_0 = 2$

**52.** $x^4 = 2 + \sin x$
$x_0 = 0$

 **53. ECONOMICS: Internal Rate of Return** For an investment of $20,000, you receive three payments of $9000 at the end of the first, second, and third years. Find the internal rate of return (to the nearest tenth of a percent).

 **54. ECONOMICS: Supply and Demand** The supply and demand functions for a product are $S(p) = 20 + 3p$ and $D(p) = 50 - 8\sqrt{p}$ (for $0 \le p \le 25$), where $p$ is the price (in dollars) of the product. Find the "market equilibrium price" $p$ at which $S(p) = D(p)$. Use Newton's method, beginning with some reasonable initial guess and continuing until the iterations agree to two decimal places.

 **55. BUSINESS: Cost and Revenue** The cost and revenue functions for a company are $C(x) = 3x + 8000$ and $R(x) = 4x\sqrt{x}$, where $x$ is the quantity. Find the "break-even" quantity, the $x$ at which $C(x) = R(x)$. Use Newton's method, beginning with some reasonable initial value and continuing until the iterations agree to the nearest whole unit.

# Probability

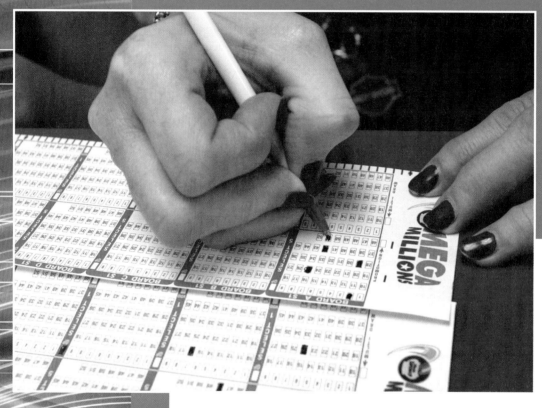

### Coincidences

We often hear of "amazing" coincidences. For example, a woman named Evelyn Marie Adams won the New Jersey Lottery twice, in 1985 and in 1986, an event that was widely reported to have a probability of 1 in 17 trillion.

Actually, the figure 1 in 17 trillion is misleading; such events are not all that unlikely. In fact, given enough tries, the most outrageous things are virtually certain to happen. For example, if a coincidence is defined as an event with a one-in-a-million chance of happening to you today, then in the United States, with about 300 million people, we should expect more than 300 coincidences each day, and more than 100,000 in a year.

Returning to the supposed 1-in-17-trillion double lottery winning, that figure is the right answer to the wrong question: What is the probability that a *preselected* person who buys just *two* tickets for separate lotteries will win on both? The more relevant question is: What is the probability that *some* person, among the many millions who buy lottery tickets (most buying multiple tickets), will win twice in a lifetime? It has been calculated* that such a double winning is likely to occur once in seven years, with the likelihood approaching certainty for longer periods.

Some knowledge of probability is necessary for an understanding of the world, if only to cast doubt on the misleading statements that one often hears. In this first section we lay the foundations for the probability discussed in later sections. For further information on probability, see the two books listed below.**

*See Persi Diaconis and Frederick Mosteller, "Methods for Studying Coincidences," *Journal of the American Statistical Association* **84**.

**For a readable introduction to probability, see Warren Weaver, *Lady Luck,* Dover Publications. For a more complete exposition, see William Feller, *An Introduction to Probability Theory and Its Applications,* vol. 1, 3rd ed., John Wiley.

## 11.1 DISCRETE PROBABILITY

### Introduction

Probability measures the *likelihood* of an event: a 90% probability of rain means that rain is very likely, and should occur on about nine of ten similar days. Probability affects our lives in many ways: insurance rates are calculated from probabilities of disasters, election results are predicted from random samples, and airplanes are overbooked based on estimated likelihoods of "no-shows." Probability is a vast and fascinating subject, and in this chapter we will discuss only a few aspects of it that are related to calculus. We begin with **discrete probability,** in which probabilities are *added,* with later sections discussing **continuous probability,** in which probabilities are summed by *integration.*

In this section we will also define the important concept of a (discrete) *random variable,* along with its mean, variance, and standard deviation. Then we will introduce the *Poisson random variable* and use it to model numbers of rare events, such as frequencies of hurricanes or the number of people living to 100.

### Elementary Events and Their Probabilities

Probability begins with an experiment, real or imagined, whose outcome depends on chance. In the experiment of rolling one die (the singular of dice), the possible outcomes are the numbers of dots 1, 2, 3, 4, 5, and 6, which are called **elementary events.** An **event** is a set consisting of elementary events. For example, the event of *rolling at least a 5* on a die would be represented

$$E_1 = \{5, 6\} \qquad \text{Rolling at least 5 means rolling 5 or 6}$$

The event of rolling an *even* number would be represented

$$E_2 = \{2, 4, 6\} \qquad \text{Rolling evens means rolling 2, 4, or 6}$$

In general:

> **Elementary Events, Events, and Sample Spaces**
>
> Given an experiment, the possible outcomes, exactly one of which must occur, are called *elementary events*, denoted $E_1, E_2, E_3, \dots, E_n$. An *event* is a set of elementary events. The set of *all* elementary events is called the *sample space.*

If there are infinitely many elementary events, we denote them $E_1, E_2, E_3, \dots$. Further examples of experiments and their sample spaces are given below.

| Experiment | Sample Space | |
|---|---|---|
| | Elementary events ↓ ↓ ↓ ↓ | |
| Tossing a coin twice | {HH, HT, TH, TT} | "HT" means "heads then tails" |
| Counting defects in a manufacturing process | {0, 1, 2, 3, ... } | An infinite sample space |

To each elementary event we assign a number that represents the **probability** of that event. In practice, we choose the number to be the long-run proportion of times that we expect the event to occur, but mathematically there are only two requirements: each probability must be between 0 and 1 (inclusive), and they must add to 1. Formally:

---

**Probabilities of Elementary Events**

The elementary events $E_1, E_2, \ldots, E_n$ are assigned probabilities $P(E_i) = p_i$ (for $i = 1, 2, \ldots, n$) such that for each $i$,

$$0 \leq p_i \leq 1 \qquad \text{Probabilities between 0 and 1}$$

and

$$p_1 + p_2 + \cdots + p_n = 1 \qquad \text{Probabilities sum to 1}$$

---

If there are infinitely many elementary events $E_1, E_2, \ldots$, then the sum of the probabilities will be an *infinite series* that must sum to 1. The condition that the probabilities sum to 1 can be interpreted as meaning that *something* must happen.

---

**EXAMPLE 1**

**ASSIGNING PROBABILITIES TO ELEMENTARY EVENTS**

For the experiment of rolling a die, the six possible faces are equally likely, each expected to occur about one-sixth of the time, so we assign probability $\frac{1}{6}$ to each.

$$P(1) = \frac{1}{6}, \qquad P(2) = \frac{1}{6}, \qquad P(3) = \frac{1}{6},$$

$$P(4) = \frac{1}{6}, \qquad P(5) = \frac{1}{6}, \qquad P(6) = \frac{1}{6}$$

*P* is read "the probability of"

---

**EXAMPLE 2**

**ASSIGNING PROBABILITIES TO ELEMENTARY EVENTS**

For the experiment of tossing a (fair) coin twice, the four elementary events HH, HT, TH, TT should each occur about one-quarter of the time, leading to the assignment

$$P(\text{HH}) = \frac{1}{4}, \qquad P(\text{HT}) = \frac{1}{4}, \qquad P(\text{TH}) = \frac{1}{4}, \qquad P(\text{TT}) = \frac{1}{4}$$

---

**EXAMPLE 3**

**ASSIGNING UNEQUAL PROBABILITIES**

In America, newborn babies are slightly more often male than female, the ratio being about 52 to 48. Therefore, for the experiment of observing the sex of a newborn baby, with elementary events M and F, we would assign probabilities

$$P(\text{M}) = 0.52 \quad \text{and} \quad P(\text{F}) = 0.48 \qquad \text{Probabilities need not be equal}$$

An event consists of elementary events, and the *probability* of the event is found by adding the probabilities of the constituent elementary events.

---

**Probability of an Event**

The probability $P(E)$ of an event $E$ is the sum of the probabilities of the elementary events in $E$.

---

**EXAMPLE 4**

**FINDING THE PROBABILITY OF AN EVENT**

For one roll of a die, find the probability of rolling an even number.

*Solution*

The event "rolling an even number" is made up of the elementary events 2, 4, and 6, which have probability $\frac{1}{6}$ each (from Example 1). Therefore, the probability of rolling an even number is the *sum* of the probabilities of these elementary events:

$$P\left(\begin{array}{c}\text{Rolling an}\\\text{even number}\end{array}\right) = P(2) + P(4) + P(6) = \frac{1}{6} + \frac{1}{6} + \frac{1}{6} = \frac{3}{6} = \frac{1}{2}$$

Elementary events

---

**EXAMPLE 5**

**FINDING THE PROBABILITY OF AN EVENT**

For two tosses of a coin, find the probability of at least one head.

*Solution*

The event "at least one head" consists of the elementary events HH, HT, and TH, with probability $\frac{1}{4}$ each (from Example 2). Adding these probabilities gives

$$P\left(\begin{array}{c}\text{At least}\\\text{one head}\end{array}\right) = P(\text{HH}) + P(\text{HT}) + P(\text{TH}) = \frac{1}{4} + \frac{1}{4} + \frac{1}{4} = \frac{3}{4}$$

---

## Random Variables

Often in probability we are interested in some numerical quantity, such as the number of heads in two tosses of a coin, or the sum of the faces when rolling two dice. Such quantities are called **random variables.** (Technically, a random variable is a *function defined on the sample space*, since for each elementary event it takes a numerical value.) If we calculate the probability that a random variable takes each of its possible values, the collection of these probabilities is called the *probability distribution* of the random variable.

> **Random Variables and Probability Distributions**
>
> A *random variable* is a numerical quantity whose value is determined by the outcome that occurs. Its *probability distribution* gives the probability that it takes each of its possible values.

An example should make matters clear.

**EXAMPLE 6**

**FINDING THE PROBABILITY DISTRIBUTION OF A RANDOM VARIABLE**

For two tosses of a (fair) coin, find the probability distribution of

$$X = \left( \begin{array}{c} \text{Number} \\ \text{of heads} \end{array} \right).$$

*Solution*

The possible values of $X$ are 0, 1, and 2 (the possible numbers of heads in two tosses). The probability that $X$ takes any one of these values is calculated by finding the elementary events for which $X$ takes that value and adding their probabilities.

$$P(X = 0) = \frac{1}{4}$$
$X = 0$ (no heads) only for TT, which occurs with probability $\frac{1}{4}$

$$P(X = 1) = \frac{1}{2}$$
$X = 1$ for HT and TH, which occur with probabilities $\frac{1}{4} + \frac{1}{4} = \frac{1}{2}$

$$P(X = 2) = \frac{1}{4}$$
$X = 2$ only for HH, which occurs with probability $\frac{1}{4}$

These three probabilities make up the probability distribution for $X$.

A probability distribution may be shown as a **bar graph,** with bars whose heights (and also areas) show the probability of each possible value. The probabilities adding to 1 means that the total area under the graph must be 1.

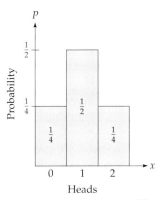

Probability distribution of $X = \left( \begin{array}{c} \text{Number} \\ \text{of heads} \end{array} \right)$
in two tosses of a coin (from Example 6)

EXAMPLE 7

**FINDING A PROBABILITY FOR A RANDOM VARIABLE**

For two tosses of a coin and $X = \left(\begin{array}{c} \text{Number} \\ \text{of heads} \end{array}\right)$, find $P(0 \le X \le 1)$.

*Solution*

$0 \le X \le 1$ means that $X = 0$ or $X = 1$, so we *add* these probabilities:

$$P(0 \le X \le 1) = P(X = 0) + P(X = 1) = \frac{1}{4} + \frac{1}{2} = \frac{3}{4} \qquad \text{The } \frac{1}{4} \text{ and } \frac{1}{2} \text{ are from Example 6}$$

We interpret this probability as meaning that if you were to toss a pair of coins many times, in about three-quarters of the tosses you would get either no heads or one head.

The probability $P(0 \le X \le 1)$ is given by the first two bars in the graph of the probability distribution.

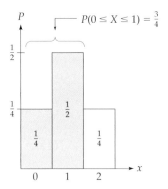

Calculating probabilities as areas under a graph will be very important in later sections.

## Expected Value

Suppose that a game of chance pays $8 with probability $\frac{1}{4}$ and $4 with probability $\frac{3}{4}$. If you were to play this game many times, what would be your *long-term average winnings per play?* We multiply each prize by its probability (after all, winning $8 with probability $\frac{1}{4}$ means winning $8 about every fourth time, for an *average* of $\frac{1}{4} \cdot 8 = \$2$ each time).

$$\left(\begin{array}{c} \text{Average} \\ \text{winnings} \end{array}\right) = 8 \cdot \frac{1}{4} + 4 \cdot \frac{3}{4} = 2 + 3 = 5 \qquad \text{Each prize times its probability}$$

This answer, $5, is called the *expected value* of the game, meaning that *in the long run* your winnings should average about $5 per game. Therefore, $5 would be a *fair price* to charge for playing this game. For *any* random variable $X$, its **mean** or **expected value** is written $E(X)$ or $\mu$ ("mu," the Greek letter m), and is found by multiplying the possible values of $X$ by their probabilities and adding.

> **Expected Value of a Random Variable**
>
> A random variable $X$ taking values $x_1, x_2, \ldots, x_n$ with probabilities $p_1, p_2, \ldots, p_n$ has an *expected value* or *mean*
>
> $$\mu = E(X) = \sum_{i=1}^{n} x_i \cdot p_i \qquad\qquad x_1 p_1 + \cdots + x_n p_n$$

The expected value is a **weighted average,** with the values of $X$ weighted by their probabilities. If a random variable has infinitely many possible values, then the sum will be an infinite series.*

**EXAMPLE 8**

**FINDING THE EXPECTED VALUE OF A RANDOM VARIABLE**

For two tosses of a coin and $\quad X = \begin{pmatrix} \text{Number} \\ \text{of heads} \end{pmatrix}, \quad$ find $\quad E(X)$.

*Solution*

Using the probabilities calculated in Example 6 on page 747:

$$E(X) = 2 \cdot \frac{1}{4} \quad + \quad 1 \cdot \frac{1}{2} \quad + \quad 0 \cdot \frac{1}{4} \qquad \begin{array}{l}\text{Values times their} \\ \text{probabilities, added}\end{array}$$

$$\underset{P(X=2)}{\uparrow} \qquad \underset{P(X=1)}{\uparrow} \qquad \underset{P(X=0)}{\uparrow}$$

$$= \frac{1}{2} + \frac{1}{2} = 1$$

That is, in two tosses of a fair coin, the expected number of heads is $E(X) = 1$.

We interpret this expected value as meaning that repeating this experiment (tossing two coins) many times and keeping track of the numbers of heads, in the long run you should average about 1 head per repetition (just as you might have expected).

For this random variable, the expected value 1 is the "middle number" of its possible values. In general, this will happen whenever the probability distribution is symmetric around a central value.

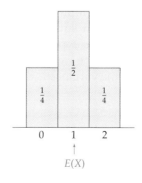

*In this case there is an added technicality that the series $\sum_{i=1}^{\infty} |x_i| \cdot p_i$ must converge.

**EXAMPLE 9**  **FINDING THE EXPECTED VALUE OF REAL ESTATE**

You are considering buying a piece of property whose value $V$ depends on whether oil is found on it. You estimate that with probability 20% it will be worth $300,000 and with probability 80% it will be worth only $10,000. Find the expected value $E(V)$ of the property.

*Solution*

$$E(V) = 300{,}000 \cdot 0.2 + 10{,}000 \cdot 0.8$$

Values times their probabilities, added

$$= 60{,}000 + 8000 = 68{,}000$$

$68,000

This expected value of $68,000 could then be compared with the selling price to decide whether it is a "good buy."

## Variance

For a random variable $X$, the possible values may be clustered closely around the mean, or they may be distributed widely on either side of the mean. The *spread* or *dispersion* of a random variable around its mean is measured by the **variance.**

---

**Variance of a Random Variable**

A random variable $X$ taking values $x_1, x_2, \ldots, x_n$ with probabilities $p_1, p_2, \ldots, p_n$, and having mean $\mu$, has *variance*

$$\text{Var}(X) = \sum_{i=1}^{n} (x_i - \mu)^2 p_i \qquad (x_1 - \mu)^2 p_1 + \cdots + (x_n - \mu)^2 p_n$$

---

The variance is the differences between the possible values and the mean, squared, multiplied by their probabilities, and added. It is sometimes called the **mean square deviation.** The squaring of the differences $x_i - \mu$ ensures that a positive difference does not cancel a negative one. Obviously, finding the variance requires first finding the mean $\mu$. (An alternative formula for the variance is developed in Exercise 47.)

**EXAMPLE 10** **FINDING THE VARIANCE OF A RANDOM VARIABLE**

For two tosses of a fair coin and $X = \begin{pmatrix} \text{Number} \\ \text{of heads} \end{pmatrix}$, find $\text{Var}(X)$.

*Solution*

In Example 6 we found the possible values of $X$ and their probabilities, and in Example 8 we found the mean $\mu = 1$. From these,

$$\text{Var}(X) = (0-1)^2 \cdot \frac{1}{4} + (1-1)^2 \cdot \frac{1}{2} + (2-1)^2 \cdot \frac{1}{4}$$

Values minus $\mu = 1$, squared, and multiplied by their probabilities

$$= \frac{1}{4} + 0 + \frac{1}{4} = \frac{1}{2}$$

Simplified

A particular variance such as $\text{Var}(X) = \frac{1}{2}$ has little intuitive meaning by itself. Variance may be understood *comparatively:* a variance of 2 would indicate a greater spread away from the mean than a variance of $\frac{1}{2}$.

If the units of the random variable are "dollars," then the variance has units "dollars squared," which is difficult to interpret. The **standard deviation** of $X$, written $\sigma(X)$, is the square root of the variance, and so has the same units as the random variable. [The symbol $\sigma$ ("sigma") is the Greek letter s.]

**Standard Deviation of a Random Variable**

The standard deviation of a random variable is the square root of the variance:

$$\sigma(X) = \sqrt{\text{Var}(X)}$$

The standard deviation is sometimes called the **root mean square deviation** of the random variable. Standard deviations will be used extensively in Section 11.4.

**EXAMPLE 11**

**FINDING THE STANDARD DEVIATION OF A RANDOM VARIABLE**

For two tosses of a fair coin and $X = \left(\begin{array}{c}\text{Number}\\ \text{of heads}\end{array}\right)$, find $\sigma(X)$.

*Solution*

$$\sigma(X) = \sqrt{\text{Var}(X)} = \sqrt{0.5} \approx 0.707$$

Using $\text{Var}(X) = \frac{1}{2}$ from Example 10

## Poisson Distribution

Until now we have defined random variables by imagining an experiment (rolling a die, tossing a coin) and finding the probability distribution. We now turn our attention to certain classic random variables that are useful in applications. The first of these is the **Poisson random variable,** named for the French mathematician Siméon Poisson (1781–1840). A Poisson random variable has an infinite number of possible values, 0, 1, 2, 3, ... , and its probability distribution depends on a constant or *parameter a* > 0, which is also its mean.

**Poisson Random Variable**

For a Poisson random variable $X$ with mean  $a > 0$,

$$P(X = k) = e^{-a}\frac{a^k}{k!} \qquad\qquad \text{for}\quad k = 0, 1, 2, \ldots$$

---

**EXAMPLE 12**

**FINDING PROBABILITIES FOR A POISSON RANDOM VARIABLE**

For the Poisson random variable $X$ with mean  $a = 3$,  find  $P(X = 0)$, $P(X = 1)$, and  $P(X = 2)$.

*Solution*

$$P(X = 0) = e^{-3}\frac{3^0}{0!} = e^{-3}\frac{1}{1} \approx 0.05 \qquad\qquad e^{-a}\frac{a^k}{k!} \;\;\text{with}\;\; a = 3 \;\;\text{and}\;\; k = 0$$

$$P(X = 1) = e^{-3}\frac{3^1}{1!} = e^{-3}\frac{3}{1} \approx 0.15 \qquad\qquad e^{-a}\frac{a^k}{k!} \;\;\text{with}\;\; a = 3 \;\;\text{and}\;\; k = 1$$

$$P(X = 2) = e^{-3}\frac{3^2}{2!} = e^{-3}\frac{9}{2} \approx 0.22 \qquad\qquad e^{-a}\frac{a^k}{k!} \;\;\text{with}\;\; a = 3 \;\;\text{and}\;\; k = 2$$

---

The probability distribution of a Poisson random variable with mean  $a = 3$  is graphed below.

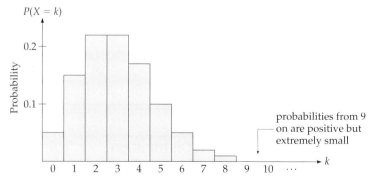

Probability distribution for the
Poisson random variable with mean  $a = 3$

To show that the probabilities for a Poisson random variable sum to 1, we factor $e^{-a}$ out of the sum and recognize the remaining sum as the Taylor series for $e^a$ (from page 715):

$$\underbrace{\sum_{k=0}^{\infty} e^{-a}\frac{a^k}{k!}}_{\substack{\text{Sum of the} \\ \text{probabilities}}} = \underbrace{e^{-a}}_{\substack{\text{Factored} \\ \text{out}}}\underbrace{\left(1 + \frac{a}{1} + \frac{a^2}{2!} + \frac{a^3}{3!} + \cdots\right)}_{\substack{\text{Taylor series for } e^a, \\ \text{which equals } e^a}} = \underbrace{e^{-a}(e^a)}_{\substack{\text{Replacing} \\ \text{the series} \\ \text{by } e^a}} = \underbrace{e^0 = 1}_{\substack{\text{Adding} \\ \text{exponents}}}$$

Therefore, the probabilities *do* sum to 1. Exercise 13 shows that a Poisson random variable $X$ has mean  $E(X) = a$.  (The variance is also $a$, so the standard deviation is $\sqrt{a}$, although we will not use these facts.)

Poisson random variables are often used to model independent events that occur rarely, such as the number of defective items in a manufacturing process or people living to age 100.* The Poisson distribution is sometimes (although inaccurately) called the **law of rare events**.

**EXAMPLE 13**    **HURRICANE PREDICTION**

During the years 1900 through 2007, 184 hurricanes reached the United States, for an average of 1.7 per year during those 108 years. Use the Poisson distribution with mean  $a = 1.7$  to find the probability that three or fewer hurricanes will reach the United States during a particular year.

*Source:* National Oceanic and Atmospheric Administration

*Solution*

If  $X = \begin{pmatrix}\text{Number of} \\ \text{hurricanes}\end{pmatrix}$,  then we want to find  $P(X \leq 3)$,  where $X$ is a Poisson random variable with  $a = 1.7$.

$P(X \leq 3)$

$= P(X = 0) + P(X = 1) + P(X = 2) + P(X = 3)$      $\begin{array}{l}X \leq 3 \text{ means} \\ X = 0, 1, 2, \text{ or } 3\end{array}$

$= e^{-1.7}\left(\dfrac{1.7^0}{0!} + \dfrac{1.7^1}{1!} + \dfrac{1.7^2}{2!} + \dfrac{1.7^3}{3!}\right)$      $\begin{array}{l}e^{-a}\left(\dfrac{a^0}{0!} + \dfrac{a^1}{1!} + \dfrac{a^2}{2!} + \dfrac{a^3}{3!}\right) \\ \text{with } a = 1.7\end{array}$

$\approx 0.907$      Using a calculator

Therefore, with probability about 91% at most three hurricanes will reach the United States in any given year.

---

*Two events are *independent* if the occurrence of one does not make the occurrence of the other more or less likely. For further examples that fit the Poisson distribution, see F. Thorndike, "Applications of Poisson's Probability Summation," *Bell System Technical Journal* **5**.

## Graphing Calculator Exploration

```
poissoncdf(1.7,3
)
        .9068105662
```

A graphing calculator with a command like POISSONCDF (in the probability DISTRibution menu) finds the probability that a Poisson random variable takes values *less than or equal to* a given number. To find the probability in the preceding example, we would use the POISSONCDF command with the given mean (1.7) and the maximum number of occurrences (3). The screen on the left verifies the answer we found.

What is the number such that with probability 99% at most that many hurricanes will reach the United States in a given year? [*Hint:* Experiment with POISSONCDF, replacing 3 by larger numbers.]

Calculations such as this are used to determine insurance premiums and land values in hurricane-prone areas. The Poisson distribution with $a = 1.7$ is graphed below. The entire area is 1, and the shaded portion shows the probability $P(X \leq 3)$.

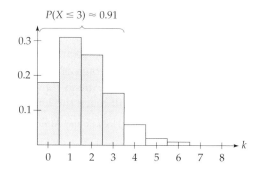

$P(X \leq 3) \approx 0.91$

**Practice Problem**

In a shipment of 100 light bulbs, an average of 2 bulbs will be defective. Suppose that the number of defective bulbs in each shipment is modeled by a Poisson random variable.

**a.** What should be the value of the parameter $a$?

**b.** Find the probability that a given shipment has no defective bulbs.

➤ **Solutions below**

**Solutions to Practice Problem**

**a.** $a = 2$    **b.** $P(X = 0) = e^{-2}\dfrac{2^0}{0!} = e^{-2} \approx 0.135$

## 11.1 Section Summary

For an experiment, the possible outcomes (exactly one of which must occur) are called *elementary events,* and collectively they make up the *sample space.* The elementary events are assigned nonnegative probabilities that sum to 1, usually based on long-run proportions. An *event* consists of one or more elementary events, and its probability is the sum of the probabilities of its constituent elementary events.

A *random variable* (represented by a capital letter, such as $X$, $Y$, or $Z$) takes values that depend on the elementary events. Its *probability distribution* consists of the probabilities that it takes these various values. The *mean* or *expected value* $E(X)$ gives an average value for $X$, and the *variance* $\text{Var}(X)$ and *standard deviation* $\sigma(X)$ measure the *deviation* of the values away from the mean.

$$E(X) = \mu = \sum_{i=1}^{n} x_i p_i$$

For possible values $x_1, x_2, \dots, x_n$ with probabilities $p_1, p_2, \dots, p_n$

$$\text{Var}(X) = \sum_{i=1}^{n} (x_i - \mu)^2 p_i$$

$$\sigma(X) = \sqrt{\text{Var}(X)}$$

A *Poisson* random variable with mean $a > 0$ has probability distribution

$$P(X = k) = e^{-a}\frac{a^k}{k!} \qquad \text{for } k = 0, 1, 2, \dots$$

Poisson random variables are used to model the frequencies of independent rare events, with the parameter $a$ set equal to the average number of occurrences.

## 11.1  Exercises

**1.** For the event of rolling one die, find

**a.** $P\left(\begin{matrix}\text{Rolling at}\\\text{least a 3}\end{matrix}\right)$  **b.** $P\left(\begin{matrix}\text{Rolling an}\\\text{odd number}\end{matrix}\right)$

**2.** For the event of tossing a coin twice, find

**a.** $P\left(\begin{matrix}\text{Tossing exactly}\\\text{one head}\end{matrix}\right)$  **b.** $P\left(\begin{matrix}\text{Tossing at most}\\\text{one head}\end{matrix}\right)$

**3–4.** For the event of tossing a coin *three times*, find

**3. a.** $P\left(\begin{matrix}\text{Tossing exactly}\\\text{one head}\end{matrix}\right)$  **b.** $P\left(\begin{matrix}\text{Tossing exactly}\\\text{two heads}\end{matrix}\right)$

**4. a.** $P\left(\begin{matrix}\text{Tossing exactly}\\\text{three heads}\end{matrix}\right)$  **b.** $P\left(\begin{matrix}\text{Tossing more}\\\text{heads than tails}\end{matrix}\right)$

**5–6.** Find the expected value, variance, and standard deviation of each random variable. Notice how the variance and standard deviation are affected by the "spread" of the values.

**5. a.** $X$ takes values 9 and 11, each with probability $\frac{1}{2}$.

    **b.** $Y$ takes values 0 and 20, each with probability $\frac{1}{2}$.

**6. a.** $X$ takes values 19 and 21, each with probability $\frac{1}{2}$.

    **b.** $Y$ takes values 0 and 40, each with probability $\frac{1}{2}$.

**7–10.** For each spinner, the arrow is spun and then stops, pointing in one of the numbered sectors, with probability proportional to the area of the sector. Let $X$ stand for the number so chosen. Find

**a.** the probability distribution of $X$.
**b.** $E(X)$.
**c.** $\text{Var}(X)$.

**7.**

**8.**

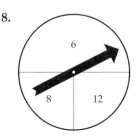

*(See instructions on previous page.)*

**9.**          **10.**

**11.** For a spinner such as that described in Exercises 7–10, what is the probability that the arrow ends up pointing

 **a.** in the top 30°?          **b.** in the top 10°?

 **c.** exactly upward?

**12.** *(continuation)* If lines are drawn from the center of the spinner to 360 points evenly spaced around the circumference, what is the probability that the arrow ends up falling exactly on one of these lines?

**13.** Show that the mean of a Poisson random variable $X$ with parameter $a$ is $E(X) = a$.

**14.** A die is rolled, and $X$ is defined as the number of dots on the face that lands upward. Find

 **a.** $E(X)$ (Your answer shows that the mean need not be a possible value.)
 **b.** $\text{Var}(X)$

**15.** For a Poisson random variable $X$ with mean 2, find (to three decimal places)

 **a.** $P(X = 0)$          **b.** $P(X = 1)$
 **c.** $P(X = 2)$          **d.** $P(X = 3)$

**16.** For a Poisson random variable $X$ with mean 1.5, find (to three decimal places)

 **a.** $P(X = 0)$          **b.** $P(X = 1)$
 **c.** $P(X = 2)$          **d.** $P(X = 3)$

 **17.** For a Poisson random variable $X$ with mean 2, find (to three decimal places)

 **a.** $P(X \le 1)$          **b.** $P(X \le 2)$
 **c.** $P(X \le 3)$          **d.** $P(X > 3)$

 *[Hint for (d):* Use your answer to (c).*]*

 **18.** For a Poisson random variable $X$ with mean 1.5, find (to three decimal places)

 **a.** $P(X \le 1)$          **b.** $P(X \le 2)$
 **c.** $P(X \le 3)$          **d.** $P(X > 3)$

 *[Hint for (d):* Use your answer to (c).*]*

## *Applied Exercises*

**19.** SOCIAL SCIENCE: **Family Composition** A young couple hopes to have four children.

 **a.** State the sample space for the sexes of the children listed in birth order, using B for boy and G for girl. (For example, GGBB means two girls then two boys.)
 **b.** If each possible outcome is equally likely, what is the probability of each?
 **c.** Find the probability distribution of the random variable $X = \begin{pmatrix} \text{Number} \\ \text{of girls} \end{pmatrix}$. Make a bar graph for the probability distribution.
 **d.** Find the probability $P(X \ge 1)$.
 **e.** Find $E(X)$ and $\text{Var}(X)$.

**20.** BUSINESS: **Expected Value of a Company** A venture capitalist buys a biotechnology company, estimating that with probability 0.2 it will be worth $2,000,000, with probability 0.7 it will be worth $300,000, and with probability 0.1 it will be worthless. Find the expected value of the company.

**21.** BUSINESS: **Price of an Insurance Policy** An insurance company predicts that a policy it plans to offer will cost the company the following amounts in claims: $10,000 with probability $\frac{1}{10}$, $2000 with probability $\frac{1}{2}$, $1000 with probability $\frac{3}{10}$, and nothing at all with probability $\frac{1}{10}$. If it wants to sell the policy for $500 more than the expected cost of claims, what price should it set?

**22.** GENERAL: **Lawsuits** A lawyer in a court case estimates that with probability 0.25 she will win $100,000, with probability 0.4 she will win $20,000, and with probability 0.35 she will win nothing. Her client is offered a settlement of $45,000 to drop the suit. Based on expected value, should she advise her client to accept the settlement?

## Applied Exercises Using Poisson Random Variables

**23. BUSINESS: Quality Control** An automobile assembly line produces 200 cars per day, and an average of 4 per day fail inspection. Use the Poisson distribution to find the probability that on a given day, 2 or fewer cars will fail inspection.

**24. GENERAL: Misprints** An 11-chapter book has 35 misprints, for an average of 3.2 per chapter. Use the Poisson distribution to find the probability that a given chapter has 2 or fewer misprints.

**25. BIOMEDICAL: Bacilli** In an examination of 1000 phagocytes (white blood cells), an average of 1.93 bacilli were found per phagocyte. Use the Poisson distribution to find the probability that a given phagocyte has 2 or fewer bacilli.

**26. GENERAL: Spacial Poisson Distribution** A paper company produces rolls of paper with an average of 0.2 flaw per square yard. Use the Poisson distribution to find the probability that a given square yard has no flaws.

**27. GENERAL: Hurricanes** An average of 1.7 hurricanes reach the United States each year. Use the Poisson distribution to find the probability that no hurricane reaches the United States in a given year.
*Source:* National Oceanic and Atmospheric Administration

**28. SOCIAL SCIENCE: Centenarians** In a group of 100 twenty-year-old Americans, an average of 2.15 will live to be 100 years old ("centenarians"). Use the Poisson distribution to find the probability that in such a group of 100, at least one will be a centenarian. [*Hint:* Find the probability that *none* will be a centenarian, and subtract it from 1.]
*Source: National Vital Statistics Report,* 2007

**29–36.** A graphing calculator with series operations is useful but not necessary.

**29. BUSINESS: Quality Control** In a box of 100 screws, an average of 2 are defective. The manufacturer offers a refund if 5 or more are defective. Use the Poisson distribution to find the probability of a refund.
[*Hint:* First find $P(X < 5)$.]

**30. GENERAL: Parking** A college parking lot has an average of six empty parking places at any time. If eight cars arrive at the same time, use the Poisson distribution to find the probability that they can all be accommodated.
[*Hint:* First find the probability that the cars *cannot* be accommodated.]

**31. GENERAL: Traffic** A certain mathematics professor driving to work encounters, on average, 2 traffic jams per month. Use the Poisson distribution to find the probability that in a given month he will encounter:
**a.** no traffic jams.
**b.** at least 1 traffic jam. [*Hint:* Use your answer to part (a).]
**c.** 3 or fewer traffic jams.

**32. BIOMEDICAL: Blood Type** People with blood type AB+ can receive blood from people with *any* blood type and so are called *universal recipients.* On the average, 3 people in 100 have this blood type. Use the Poisson distribution to find the probability that in a group of 100 people:
**a.** none will be universal recipients.
**b.** 3 or fewer will be universal recipients.
*Source:* U.S. Census Bureau

**33. BIOMEDICAL: Blood Type** People with blood type O− can give blood to people with *any* blood type and so are called *universal donors.* On average, 7 people in 100 have this blood type. Use the Poisson distribution to find the probability that in a group of 100 people:
**a.** no more than 3 will be universal donors.
**b.** no more than 5 will be universal donors.
*Source:* U.S. Census Bureau

**34. SOCIAL SCIENCE: Crime** On the average in the United States, there will be 1.4 robberies per 1000 people in a year. Use the Poisson distribution to find the probability that in a neighborhood of 1000 residents there will be:
**a.** no robberies.
**b.** no more than 2 robberies.
*Source: Statistical Abstract of the United States,* 2008

**35. SOCIAL SCIENCE: Crime** On average in the United States, there will be 4.2 motor vehicle thefts per 1000 people in a year. Use the Poisson distribution to find the probability that in a neighborhood of 1000 residents there will be:
**a.** no automobile thefts.
**b.** no more than 6 automobile thefts.
*Source: Statistical Abstract of the United States,* 2008

**36. GENERAL: Airplane Accidents** During the years 1990 to 2007 there were, on average, 2.7 serious commercial airplane accidents per year. Use the Poisson distribution to find the probability that in any particular year the number of accidents will not exceed 4.
*Source:* National Transportation Safety Board

## Conceptual Exercises

**37.** If a sample space has 275 equally likely elementary events, what is the probability of each?

**38.** If a sample space contains just two elementary events, and one occurs twice as often as the other, what are their probabilities?

**39.** Find a simple expression for $P(X \le 5) - P(X < 5)$.

**40.** If $P(X \le 10) = \frac{2}{3}$, then which of the following can you find and what is its value: $P(X \ge 10)$ or $P(X > 10)$?

**41.** True or False: If a random variable has mean $\mu$, then one of its values must be $\mu$.

**42.** Can the mean of a random variable be greater than each of its values?

**43.** If a random variable $X$ has mean 5, then what is the mean of $X + 100$? Of $2X$?

**44.** If a random variable $X$ has variance 5, then what is the variance of $2X$? If $X$ has *standard deviation* 5, then what is the standard deviation of $2X$?

**45.** What can you say about a random variable whose variance is *zero*? To be specific, what can you say about $X$ if $E(X) = 5$ and $Var(X) = 0$?

**46.** If $X$ is a Poisson random variable and $P(X = 0) = e^{-7}$, what is its mean?

## Explorations and Excursions

The following problem extends and augments the material presented in the text.

*Alternative Formula for Variance*

**47.** For a random variable $X$ with mean $\mu$ that takes values $x_1, x_2, \ldots, x_n$ with probabilities $p_1, p_2, \ldots, p_n$, prove the following "alternative" formula for the variance

$$\text{Var}(X) = \sum_{i=1}^{n} x_i^2 \cdot p_i - \mu^2$$

by giving a justification for each numbered equality in the following derivation.

$$\text{Var}(X) \overset{1}{=} \sum_{i=1}^{n} (x_i - \mu)^2 p_i \overset{2}{=} \sum_{i=1}^{n} (x_i^2 - 2x_i\mu + \mu^2) p_i$$

$$\overset{3}{=} \sum_{i=1}^{n} x_i^2 p_i - \sum_{i=1}^{n} 2x_i\mu p_i + \sum_{i=1}^{n} \mu^2 p_i$$

$$\overset{4}{=} \sum_{i=1}^{n} x_i^2 p_i - 2\mu \sum_{i=1}^{n} x_i p_i + \mu^2 \sum_{i=1}^{n} p_i$$

$$\overset{5}{=} \sum_{i=1}^{n} x_i^2 p_i - 2\mu\mu + \mu^2 1 \overset{6}{=} \sum_{i=1}^{n} x_i^2 p_i - \mu^2$$

## 11.2 CONTINUOUS PROBABILITY

### Discrete Random Variables

The random variables that we have encountered so far have taken values that are separated from each other, and can be written out in a list, such as $0, 1, 2, 3, \ldots$. Such random variables are called *discrete* random variables.* For example, the Poisson random variable (see page 752) is a discrete random variable. The probability distribution of a discrete random variable may be shown as a bar graph over the possible values, with total area 1.

### Continuous Random Variables

Other random variables take values that make up a *continuous interval* of the real line. For example, the proportion of time that you study today can be any number in the interval $[0, 1]$, and the duration of a telephone call can be any number in the interval $(0, \infty)$. For a **continuous random variable,** the probability

---

*Do not confuse the word "discrete," which means *separated* or *distinct,* with the word "discreet," which means *prudent* or *showing self-restraint.*

distribution is given by a nonnegative function over the interval of possible values, with total area 1. Such a function is called a **probability density function.**

---

**Probability Density Function**

A probability density function on an interval $[A, B]$ is a function $f$ such that

1. $f(x) \geq 0$   for $A \leq x \leq B$

2. $\displaystyle\int_A^B f(x)\, dx = 1$

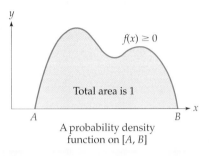

A probability density function on $[A, B]$

---

The interval may be infinite, in which case  $A = -\infty$,  or  $B = \infty$,  or both, with the integral above being an *improper* integral. For simplicity, we define $f(x)$ to be zero for all $x$-values outside of the interval $[A, B]$.

**EXAMPLE 1**

### CONSTRUCTING A PROBABILITY DENSITY FUNCTION

Find the value of the constant $a$ that makes   $f(x) = ax(1 - x)$   a probability density function on the interval $[0, 1]$.

*Solution*

To find the area under the curve, we integrate over the given interval:

$$a \underbrace{\int_0^1 x(1 - x)\, dx}_{\substack{\text{Moved} \\ \text{outside}}} = a \underbrace{\int_0^1 (x - x^2)\, dx}_{\substack{\text{Multiplied} \\ \text{out}}} = a \underbrace{\left(\frac{1}{2} x^2 - \frac{1}{3} x^3\right)\Big|_0^1}_{\substack{\text{Integrating by} \\ \text{the Power Rule}}}$$

$$= a \Big(\underbrace{\frac{1}{2} - \frac{1}{3}}_{\substack{\text{Evaluating} \\ \text{at 1}}} \underbrace{-\ \ 0}_{\substack{\text{and} \\ \text{at 0}}}\Big) = \underbrace{\frac{a}{6}}_{\substack{\text{Area under} \\ \text{the curve}}} \qquad \substack{\text{Must} \\ \text{equal 1}}$$

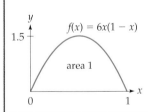

For this area $a/6$ to equal 1, we must have   $a = 6$.   The probability density function with $a$ replaced by 6 is then   $f(x) = 6x(1 - x)$   on $[0, 1]$, which is graphed on the left and is clearly nonnegative.*

---

*This probability density function is an example of a *beta density*, defined more generally as $f(x) = ax^{\alpha}(1 - x)^{\beta}$  for positive constants $\alpha$ and $\beta$ and  $0 \leq x \leq 1$,  where $a$ is a constant that makes the area under the curve equal to 1. Therefore, the probability density function in Example 1 is a beta density with constants  $\alpha = 1$,  $\beta = 1$,  and  $a = 6$.  Beta distributions are of secondary importance compared to the other distributions discussed in this chapter.

For *discrete* random variables, probabilities were given by *areas of bars*. For *continuous* random variables, probabilities are given by *areas under curves*. More precisely, the probability that $X$ takes values between two given numbers is found by integrating the probability density function between those two numbers.

---

**Probabilities from Probability Density Functions**

For a continuous random variable $X$ with probability density function $f$,

$$P(a \le X \le b) = \int_a^b f(x)\, dx$$

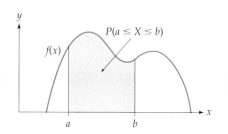

$P(a \le X \le b)$ is the area under the probability density function from $a$ to $b$.

---

**EXAMPLE 2**  **FINDING A PROBABILITY FOR A CONTINUOUS RANDOM VARIABLE**

Let $X$ be defined as the number of inches of rain in a town during the first week of April in each year. Suppose that a meteorologist has used past weather records to determine that the probability density function for $X$ is $f(x) = 6x(1 - x)$ on $[0, 1]$. Find the probability that the rainfall is between 0 and 0.4 inch.

*Solution*

To find $P(0 \le X \le 0.4)$ we integrate the probability density function from 0 to 0.4:

$$P(0 \le X \le 0.4) = \int_0^{0.4} 6x(1 - x)\, dx \qquad \text{Integrating the probability density function}$$

$$= \int_0^{0.4} (6x - 6x^2)\, dx \qquad \text{Multiplying out}$$

$$= (3x^2 - 2x^3)\Big|_0^{0.4} \qquad \text{Using the Power Rule}$$

$$= (0.48 - 0.128) - 0 = 0.352 \qquad \text{Evaluating}$$

Therefore, the probability that the rainfall will be between 0 and 0.4 inch is about 35%.

On pages 348–351 we defined definite integrals by summing areas of rectangles under curves. Therefore, finding continuous probabilities as areas under curves is a natural extension of finding discrete probabilities as sums of areas of rectangles.

**a.** Find the value of $a$ that makes $f(x) = ax^2$ a probability density function on $[0, 1]$.

**b.** For $X$ with this probability density function, find $P(0 \leq X \leq 0.5)$.

➤ **Solutions on page 767**

## One-Point Probabilities Are Zero

For a continuous random variable $X$, the probability that $X$ takes any particular single value is zero: $P(X = a) = 0$. This is because $P(X = a)$ is found by integrating the probability density function "from $a$ to $a$," which gives zero:

$$P(X = a) = \underbrace{P(a \leq X \leq a)}_{\substack{\text{Equivalent} \\ \text{to } X = a}} = \int_a^a f(x)\, dx = 0$$

Same upper and lower limits

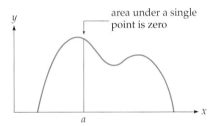

area under a single point is zero

Therefore, adding one or both endpoints to the inequality in $P(a < X < b)$ does not change the probability:

$$P(a < X < b) = P(a \leq X \leq b) = \int_a^b f(x)\, dx \qquad \text{Endpoints do not change the probability}$$

In continuous probability, what does zero probability mean? Zero probability does *not* mean that the event is impossible, only that it is "overwhelmingly unlikely." Think of choosing a number from the interval $[0, 1]$ by some means such as throwing a dart at a line segment. The probability of choosing (hitting) exactly one single preselected number $a$ is $P(X = a)$, which is zero, as we saw above. The probability being zero means only that predicting the exact point in advance, while theoretically possible, is overwhelmingly unlikely. (In *discrete* probability, however, zero probability *does* mean that the event is impossible. Intuitively, the difference is that in continuous probability a random variable has so many more possible values that the probability must be spread over them much more "thinly.")

## Mean, Variance, and Standard Deviation

The mean and variance of a continuous random variable are defined by *integrals* that are analogous to the sums on pages 749 and 750.

> ### Mean, Variance, and Standard Deviation
>
> For a continuous random variable $X$ with probability density function $f$ on $[A, B]$,
>
> $$\mu = E(X) = \int_A^B x f(x)\, dx \qquad \begin{array}{l}\text{Mean or expected value is the} \\ \text{integral of } x \text{ times } f(x)\end{array}$$
>
> $$\text{Var}(X) = \int_A^B (x - \mu)^2 f(x)\, dx \qquad \begin{array}{l}\text{Variance is } x \text{ minus the mean,} \\ \text{squared, multiplied by } f(x), \\ \text{and integrated}\end{array}$$
>
> $$\sigma(X) = \sqrt{\text{Var}(X)} \qquad \begin{array}{l}\text{Standard deviation is the} \\ \text{square root of the variance}\end{array}$$

If the interval is infinite, with $A = -\infty$, or $B = \infty$, or both, then the integrals are *improper* integrals.* As before, the mean $\mu = E(X)$ gives an *average* value, and the variance $\text{Var}(X)$ and standard deviation $\sigma(X)$ measure the *spread* or *dispersion* of the values away from this average. Also as before, finding the variance involves first finding the mean.

---

**EXAMPLE 3**  **FINDING THE EXPECTED VALUE OF A CONTINUOUS RANDOM VARIABLE**

For the random variable in Example 2, $X = \left(\begin{array}{c}\text{Inches of} \\ \text{rainfall}\end{array}\right)$ with probability density function $f(x) = 6x(1 - x)$ on $[0, 1]$, find $E(X)$.

*Solution*

$$E(X) = \underbrace{\int_0^1 x \cdot 6x(1 - x)\, dx}_{\substack{\text{Integrating } x \text{ times the} \\ \text{probability density function}}} = \underbrace{\int_0^1 (6x^2 - 6x^3)\, dx}_{\substack{\text{Multiplying} \\ \text{out}}}$$

$$= \underbrace{\left(2x^3 - \frac{6}{4}x^4\right)\Big|_0^1}_{\substack{\text{Integrating by} \\ \text{the Power Rule}}} = \underbrace{2 - \frac{3}{2}}_{\text{Evaluating}} = \underbrace{\frac{1}{2}}_{E(X)}$$

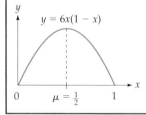

The expected value $\mu = E(X) = \frac{1}{2}$ means that the week's rainfall should average half an inch. It is reasonable that the mean is the "middle value" of the interval $[0, 1]$, since this probability density function is symmetric about this number.

---

*In this case there is an added technicality that the improper integral $\int |x| f(x)\, dx$ over the infinite interval must converge.

We could find the *variance* of the rainfall by evaluating the integral

$$\text{Var}(X) = \int_0^1 \left(x - \frac{1}{2}\right)^2 6x(1-x)\,dx \qquad \int_A^B (x-\mu)^2 f(x)\,dx \quad \text{with} \quad \mu = \tfrac{1}{2}$$

There is, however, an easier formula, which is proved at the end of this section. A discrete version of it was proved in Exercise 47 on page 758.

---

**Alternative Formula for Variance**

For a continuous random variable $X$ with mean $\mu$ and probability density function $f$ on $[A, B]$,

$$\text{Var}(X) = \int_A^B x^2 f(x)\,dx - \mu^2 \qquad \begin{array}{l}\text{Integrate } x^2 \text{ times } f(x), \text{ then} \\ \text{subtract the square of the mean}\end{array}$$

---

**EXAMPLE 4**

## FINDING THE VARIANCE AND STANDARD DEVIATION

Find the variance and standard deviation of the random variable with probability density function $f(x) = 6x(1-x)$ on $[0, 1]$.

*Solution*

In Example 3 we found that the mean of this random variable is $\mu = \tfrac{1}{2}$. To use the alternative formula for variance, we first integrate $x^2$ times the probability density function:

$$\int_0^1 x^2 \underbrace{6x(1-x)}_{f(x)}\,dx = \int_0^1 (6x^3 - 6x^4)\,dx = \left(\frac{6}{4}x^4 - \frac{6}{5}x^5\right)\Bigg|_0^1$$

$$= \frac{3}{2} - \frac{6}{5} = \frac{3}{10}$$

Then

$$\text{Var}(X) = \underbrace{\frac{3}{10}}_{\substack{\text{Preceding} \\ \text{integral}}} - \underbrace{\left(\frac{1}{2}\right)^2}_{\mu^2} = \frac{3}{10} - \frac{1}{4} = \frac{1}{20} = 0.05 \qquad \begin{array}{l}\int_A^B x^2 f(x)\,dx - \mu^2 \\ \text{with} \quad \mu = \tfrac{1}{2}\end{array}$$

$$\sigma(X) = \sqrt{0.05} \approx 0.224 \qquad \begin{array}{l}\text{Square root of the} \\ \text{variance}\end{array}$$

Therefore, $\text{Var}(X) = 0.05$ and $\sigma(X) \approx 0.224$

## Graphing Calculator Exploration

```
fnInt((X-0.5)²6X
(1-X),X,0,1)
            .05
```

Verify that the earlier formula (on page 762) for variance gives the same answer as in Example 4 by using a graphing calculator to evaluate $\int_0^1 (x - 0.5)^2 6x(1 - x)\, dx$.

The variance and standard deviation will be used extensively in Section 11.4.

### Cumulative Distribution Functions

Because we so often integrate a probability density function, it is useful to have a name for the integral up to a number $x$: it is called the **cumulative distribution function.**

---

**Cumulative Distribution Function**

For a continuous random variable $X$ with probability density function $f$ on $[A, B]$, the *cumulative distribution function* is

$$F(x) = P(X \le x) = \int_A^x f(t)\, dt$$

$F(x)$ is the probability that $X$ is less than or equal to $x$, and is found by integrating the probability density function from $A$ to $x$

---

 **BE CAREFUL:** We use uppercase letters (such as $F$, $G$, and $H$) for cumulative distribution functions, and lowercase letters (such as $f$, $g$, and $h$) for probability density functions.

Graphically, the cumulative distribution function $F(x)$ is the *area* under the probability density function $f$ up to $x$:

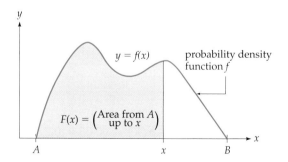

The curve $y = f(x)$ is the probability density function. The area under it up to $x$ is the cumulative distribution function $F(x)$.

Since integrating $f$ gives $F$, differentiating $F$ gives $f$. (To be completely precise, this relationship holds at $x$-values where $f(x)$ is continuous.)

$$\frac{d}{dx} F(x) = f(x)$$

The derivative of the cumulative distribution function $F(x)$ is the probability density function $f(x)$.

Probabilities $P(a \le X \le b)$, which we found by integrating the probability density function, can also be found by *subtracting* values of the cumulative distribution function.

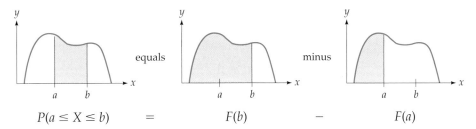

$P(a \le X \le b) \quad = \quad F(b) \quad - \quad F(a)$

In words:

---

**Probabilities from the Cumulative Distribution Function**

For a continuous random variable $X$ with cumulative distribution function $F$,

$$P(a \le X \le b) = F(b) - F(a)$$

Either or both $\le$ may be replaced by $<$

---

**EXAMPLE 5**

**FINDING AND USING A CUMULATIVE DISTRIBUTION FUNCTION**

For a random variable $W$ with probability density function
$f(x) = \frac{1}{9}x^2$ on $[0, 3]$:

**a.** Find the cumulative distribution function $F(x)$.

**b.** Use $F(x)$ to find $P(1 \le W \le 2)$.

*Solution*

**a.** The cumulative distribution function is the integral of the probability density function up to $x$. We use $t$ to avoid confusion with the upper limit $x$.

$$F(x) = \int_0^x \frac{1}{9} t^2 \, dt = \left( \frac{1}{9} \frac{1}{3} t^3 \right) \Big|_0^x = \frac{1}{27} x^3 \qquad F(x) = \int_A^x f(t) \, dt$$

Therefore, the cumulative distribution function is $F(x) = \frac{1}{27} x^3$ on $[0, 3]$.

**b.** To find $P(1 \le W \le 2)$, we evaluate $F(x) = \frac{1}{27}x^3$ at $x = 1$ and $x = 2$ and subtract:

$$P(1 \le W \le 2) = F(2) - F(1) = \underbrace{\frac{1}{27}2^3 - \frac{1}{27}1^3}_{\substack{\text{Values of } F \\ \text{subtracted}}} = \underbrace{\frac{8}{27} - \frac{1}{27}}_{\substack{\text{Evaluating } F \\ \text{at 2 and 1}}} = \frac{7}{27}$$

Therefore, $P(1 \le W \le 2) = \frac{7}{27}$.

---

## Graphing Calculator Exploration

```
Y1(2)-Y1(1)
     .2592592593
Ans▶Frac
          7/27
```

To check the answer to the preceding example:

**a.** Define $y_1$ as the definite integral of $\frac{1}{9}x^2$ from 0 to $x$ by entering $y_1 = \text{FnInt}(x^2/9, x, 0, x)$.

**b.** Then calculate $y_1(2) - y_1(1)$.

---

In some applications it is easier to *first* find the cumulative distribution function and *then* find the probability density function by differentiation.

**EXAMPLE 6**   **FINDING AND USING A CUMULATIVE DISTRIBUTION FUNCTION**

A dart is thrown at a circular target of radius 12 inches, hitting a point at random. Let $X$ be defined as

$$X = \left( \begin{array}{c} \text{Distance of dart} \\ \text{from center of circle} \end{array} \right)$$

Find:  **a.** the cumulative distribution function $F(x)$
        **b.** the probability density function $f(x)$
        **c.** the mean $E(X)$

*Solution*

A point being chosen "at random" from a region means that the probability of the point being in any subregion is proportional to the *area* of that subregion. (After all, the dart should have twice the probability of landing in an area twice as large.) The dart landing within $x$ inches of the center means that it lands inside a circle of radius $x$, and the probability of this, $P(X \le x)$, is the area of the inner circle ($\pi x^2$) divided by the area of the entire target ($\pi 12^2$):

**a.** $P(X \le x) = \dfrac{\pi x^2}{\pi 12^2} = \dfrac{x^2}{144}$

Therefore, $F(x) = \dfrac{x^2}{144}$ for $0 \le x \le 12$.

**b.** $f(x) = \dfrac{d}{dx}\left(\dfrac{1}{144}\,x^2\right) = \dfrac{1}{72}\,x$

Differentiating $F(x)$ to find $f(x)$

Therefore, $f(x) = \dfrac{1}{72}\,x$ for $0 \le x \le 12$.

**c.** $E(X) = \displaystyle\int_0^{12} x \cdot \dfrac{1}{72}\,x\,dx = \dfrac{1}{72}\displaystyle\int_0^{12} x^2\,dx = \dfrac{1}{72}\cdot\dfrac{1}{3}x^3\,\Big|_0^{12} = \dfrac{12^3}{216} = 8$

The expected value $E(X) = 8$ means that if many darts are thrown, their average distance from the center will be about 8 inches.

---

### Solutions to Practice Problem

**a.** $a\displaystyle\int_0^1 x^2\,dx = a\dfrac{1}{3}x^3\,\Big|_0^1 = \dfrac{a}{3}$

For the integral to equal 1, we must have $a = 3$.

**b.** $P(0 \le X \le 0.5) = \displaystyle\int_0^{0.5} 3x^2\,dx = x^3\,\Big|_0^{0.5} = (0.5)^3 = 0.125$

---

## 11.2 Section Summary

A random variable is *continuous* if its possible values form an *interval* $[A, B]$, which may be infinite, and if its probabilities are found by integrating its *probability density function* (a nonnegative function whose integral over $[A, B]$ is 1):

$$P(a \le X \le b) = \int_a^b f(x)\,dx$$

Either or both $\le$ may be replaced by $<$

The *cumulative distribution function* for a random variable is

$$F(x) = P(X \le x) = \int_A^x f(t)\,dt$$

$A$ is the lowest possible value of the random variable

Differentiating the cumulative distribution function $F(x)$ gives the probability density function $f(x)$. Probabilities $P(a \le X \le b)$ may also be found by subtracting values of the cumulative distribution function:

$$P(a \le X \le b) = F(b) - F(a)$$

Either or both $\le$ may be replaced by $<$

The mean gives the *average* value for the random variable, and the *variance* and *standard deviation* indicate how widely the values are spread around the mean:

$$\mu = E(X) = \int_A^B x f(x)\,dx$$

$$\text{Var}(X) = \int_A^B (x - \mu)^2 f(x)\,dx = \int_A^B x^2 f(x)\,dx - \mu^2$$

Both formulas give the same answer

$$\sigma(X) = \sqrt{\text{Var}(X)}$$

The height of the probability density function at a point does not, by itself, give a probability, but it can be interpreted in a *relative* sense: if the probability density function is twice as high at one $x$-value as at another, the random variable is about twice as likely to be near the first $x$-value as near the other (where by "near" we mean "within a small fixed distance of").

## Proof of the Alternative Formula for Variance

To prove the alternative formula for variance stated on page 763, we begin with the *definition* of variance given on page 762:

$$\text{Var}(X) = \int_A^B (x - \mu)^2 f(x)\,dx = \int_A^B (x^2 - 2x\mu + \mu^2) f(x)\,dx$$

Expanding $(x - \mu)^2$

$$= \int_A^B x^2 f(x)\,dx - 2\mu \underbrace{\int_A^B x f(x)\,dx}_{\substack{\mu \\ \text{(by definition} \\ \text{of mean)}}} + \mu^2 \underbrace{\int_A^B f(x)\,dx}_{\substack{1 \\ \text{(by property 2} \\ \text{on page 759)}}}$$

Separating into 3 integrals, and taking out constants

$$= \int_A^B x^2 f(x)\,dx \underbrace{- 2\mu^2 + \mu^2}_{-\mu^2} = \int_A^B x^2 f(x)\,dx - \mu^2$$

This last expression is the alternative formula for the variance

## 11.2 Exercises

**1–10.** Find the value of the constant $a$ that makes each function a probability density function on the stated interval.

**1.** $ax^2(1 - x)$ on $[0, 1]$      **2.** $a\sqrt{x}$ on $[0, 1]$

**3.** $ax^2$ on $[0, 3]$           **4.** $ax(2 - x)$ on $[0, 2]$

**5.** $\dfrac{a}{x}$ on $[1, e]$         **6.** $a \sin x$ on $[0, \pi]$

**7.** $axe^x$ on $[0, 1]$   [*Hint:* Integrate by parts.]

**8.** $a \ln x$ on $[1, e]$   [*Hint:* Integrate by parts using $u = \ln x$.]

 **9.** $ae^{-x^2}$ on $[-1, 1]$

 **10.** $\dfrac{a}{1 + x^2}$ on $[-1, 1]$

**11–16.** For each probability density function $f(x)$, find
**a.** $E(X)$    **b.** $\text{Var}(X)$    **c.** $\sigma(X)$

**11.** $f(x) = 3x^2$ on $[0, 1]$

**12.** $f(x) = \frac{1}{2}x$ on $[0, 2]$

**13.** $f(x) = 12x^2(1 - x)$ on $[0, 1]$

**14.** $f(x) = 20x^3(1 - x)$ on $[0, 1]$

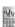 **15.** $\dfrac{4}{\pi}\dfrac{1}{1+x^2}$ on $[0, 1]$    **16.** $1.34e^{-x^2}$ on $[0, 1]$

**17.** If $X$ has probability density function
$f(x) = \frac{1}{12}x$ on $[1, 5]$, find
   **a.** the cumulative distribution function $F(x)$
   **b.** $P(3 \le X \le 5)$

**18.** If $X$ has probability density function
$f(x) = \frac{1}{10}$ on $[0, 10]$, find
   **a.** the cumulative distribution function $F(x)$
   **b.** $P(3 \le X \le 7)$

**19.** If $X$ has probability density function
$f(x) = \dfrac{2}{(1+x)^2}$ on $[0, 1]$, find $P\left(X \le \dfrac{1}{3}\right)$.

**20.** If $X$ has probability density function
$f(x) = \frac{1}{6}x^{-1/2}$ on $[1, 16]$, find $P(4 \le X \le 9)$.

**21.** If $X$ has cumulative distribution function
$F(x) = \frac{1}{3}x - \frac{1}{3}$ on $[1, 4]$, find $P(2 \le X \le 3)$.

**22.** If $X$ has cumulative distribution function
$F(x) = \frac{1}{8}x^{3/2}$ on $[0, 4]$, find $P(1 \le X \le 4)$.

**23.** If $X$ has cumulative distribution function
$F(x) = x^2$ on $[0, 1]$, find
   **a.** $P(0.5 \le X \le 1)$
   **b.** the probability density function $f(x)$

**24.** If $X$ has cumulative distribution function
$F(x) = \frac{1}{2}\sqrt{x} - \frac{1}{2}$ on $[1, 9]$, find
   **a.** $P(4 \le X \le 9)$
   **b.** the probability density function $f(x)$

**25.** For any cumulative distribution function $F(x)$ on
the interval $[A, B]$, show that
   **a.** $F(A) = 0$
   **b.** $F(B) = 1$

**26.** For any cumulative distribution function $F(x)$,
show that if $a \le b$, then $F(a) \le F(b)$.

 **27.** If $X$ has probability density function $f(x) = \dfrac{2}{\pi}\dfrac{1}{1+x^2}$ on $[-1, 1]$, find $P\left(-\dfrac{1}{2} \le X \le \dfrac{1}{2}\right)$.

 **28.** If $X$ has probability density function $f(x) = 0.67e^{-x^2}$ on $[-1, 1]$, find $P\left(-\frac{1}{2} \le X \le \frac{1}{2}\right)$.

## Applied Exercises

**29. SOCIAL SCIENCE: Voter Turnout** If the proportion of registered voters who actually vote in an election is a random variable $X$ with probability density function $f(x) = 6x(1 - x)$, find $P(0.4 \le X \le 0.6)$.

**30. BEHAVIORAL SCIENCE: Learning** If the number of days required for a worker to learn a new skill is a random variable $X$ with probability density function $f(x) = \frac{4}{27}x^2(3 - x)$ on $[0, 3]$, find the expected time $E(X)$.

**31. BIOMEDICAL: Life Expectancy** In a 4-year study, the number of years that a patient survives after an experimental medical procedure is a random variable $X$ with probability density function $f(x) = \frac{3}{64}x^2$ on $[0, 4]$. Find
   **a.** the expected survival time $E(X)$
   **b.** $P(2 \le X \le 4)$

**32. GENERAL: Milk Freshness** If the shelf life of a carton of milk (in days) is a random variable $X$ with probability density function $f(x) = \frac{1}{32}x$ on $[0, 8]$, find
   **a.** the expected shelf life $E(X)$
   **b.** the time at which the probability of spoilage is only 25% (that is, find $b$ such that $P(X < b) = 0.25$)

**33. ATHLETICS: Darts** A dart hits a point at random on a circular dart board of radius 9 inches. For
$$X = \left(\begin{array}{c}\text{Distance of dart} \\ \text{from center}\end{array}\right), \quad \text{find}$$
   **a.** the cumulative distribution function $F(x)$
   **b.** the probability density function $f(x)$
   **c.** $E(X)$
   **d.** $P(3 \le X \le 6)$ using the *probability density function*
   **e.** $P(3 \le X \le 6)$ using the *cumulative distribution function*

**34. BUSINESS: Product Failure** A spherical ball bearing of radius 1 inch has a flaw located randomly somewhere within it. If the flaw is within half an inch of the center, the ball bearing will fracture when it is used. Find
   **a.** the probability that the ball bearing will fracture. [*Hint:* Use the volumes of the "inner" and entire spheres.]
   **b.** the expected distance from the flaw to the center of the ball bearing.

 **35. BUSINESS: Profitability** If the number of years until a new store makes a profit (or goes out of business) is a random variable

$X$ with probability density function $f(x) = 0.75x(2 - x)$ on $[0, 2]$, find

**a.** the expected number of years $E(X)$
**b.** the variance and standard deviation
**c.** $P(X \geq 1.5)$

 **36. BUSINESS: Newspaper Advertising** If the proportion of a newspaper devoted to advertising is a random variable $X$ with probability density function $f(x) = 20x^3(1 - x)$ on $[0, 1]$, find

**a.** the expected proportion $E(X)$
**b.** the variance and standard deviation
**c.** the probability that more than 80% is devoted to advertising

**37. BIOMEDICAL: Weibull Aging Model** The probability that the fruit fly *Drosophila melanogaster* lives longer than $t$ days is given by the *Weibull survivorship function*

$$S(t) = e^{-(0.02t)^{3.3}}$$

**a.** Graph this function on the window $[-5, 105]$ by $[-0.5, 1.5]$.
**b.** What is the probability that a fly has a life span of more than 30 days?

*Source: H. McCallum, Population Parameters: Estimation for Ecological Models*

 **38. ATHLETICS: Baseball** Based on an analysis of five years of 9-inning major league baseball games, the number of hits per team per game can be approximated by a random variable $X$ with probability density function $f(x) = 3(.05x)^2(1 - .05x)^3$ on $[0, 20]$. Find $P(10 \leq X \leq 20)$.

*Source: J. Albert, J. Bennett, and J. Cochran, Anthology of Statistics in Sports*

**39. ATHLETICS: Baseball** Based on an analysis of 1782 major league baseball games, the winning margin in a game can be approximated by a random variable $X$ with probability density function $f(x) = 0.55e^{-0.37x}$ on $[1, 11]$. Find $P(1 \leq X \leq 3)$.

*Source: J. Albert, J. Bennett, and J. Cochran, Anthology of Statistics in Sports*

**40–41. GENERAL: Richest Americans** A study of the richest Americans (those with a net worth of at least $600 million) found that their net worth $X$ (in millions of dollars) has probability density function $f(x) = 1250x^{-2.1}$ on $[600, \infty]$.

*Source: Contemporary Physics 46*

**40.** Find $P(600 \leq X \leq 1000)$ (that is, that the net worth of such a person is between $600 million and $1 billion).

**41.** Find $P(1000 \leq X \leq 5000)$ (that is, that the net worth of such a person is between $1 billion and $5 billion).

## Conceptual Exercises

**42.** Can a cumulative distribution function $F$ satisfy $F(2) > F(3)$?

**43.** Explain what it means for an event in *discrete* probability to have probability 0 as opposed to an event in *continuous* probability to have probability zero.

**44.** If $X$ has probability density function $f(x)$, then what is the probability density function for $X + 5$? [*Hint:* Think of how the graph is shifted.]

**45.** If $X$ has cumulative distribution function $F(x)$, then what is the cumulative distribution function for $X - 7$? [*Hint:* Think of how the graph is shifted.]

**46.** Fill in the blanks: If $f$ is a probability density function on $[A, B]$ and $F$ is the cumulative distribution function, then $F(A) = \_\_\_\_$ and $F(B) = \_\_\_\_$.

**47.** In Example 6 (pages 766–767), the dart could land between 0 and 12 inches from the center, but the mean was 8. Explain in intuitive terms why the mean was *more* than 6, halfway to the edge.

**48.** Suppose that a probability density function on the interval $[A, B]$ has the shape shown below. Would you expect the mean to be closer to $A$ or to $B$?

**49.** True or False: If a random variable $X$ has probability density function $f$ on $[A, B]$, then $A \leq E(X) \leq B$.

**50.** If a continuous random variable $X$ has mean 5, then what is the mean of $X + 100$? Of $2X$?

**51.** If a continuous random variable $X$ has variance 5, then what is the variance of $2X$? If $X$ has *standard deviation* 5, then what is the standard deviation of $2X$?

## 11.3   UNIFORM AND EXPONENTIAL RANDOM VARIABLES

### Introduction

The preceding section introduced the essentials of continuous probability: continuous random variables, probability density functions, cumulative distribution functions, mean, variance, and standard deviation. However, little was said about how to *use* random variables to model particular situations. Although a complete answer to this important question would require further courses in probability, statistics, and mathematical modeling, many situations can be modeled by just a few standard types of random variables. In this section we will study two of the most useful, *uniform* and *exponential* random variables, and in the next section we will study *normal* random variables.

### Uniform Random Variables

Suppose that you want to choose a point at random from an interval $[0, B]$. If all points are to be "equally likely," the probability density function must be constant over the interval $[0, B]$, and if the area under it is to be 1, its uniform height must be $1/B$, as shown on the right.

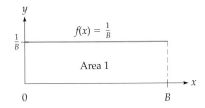

Uniform probability density function on $[0, B]$

The resulting probability density function, $f(x) = 1/B$ for $0 \le x \le B$, is called the **uniform probability density function** on $[0, B]$. A random variable with this probability density function is called a **uniform random variable** on $[0, B]$ and is said to be **uniformly distributed** between 0 and $B$. The mean of a uniform random variable on $[0, B]$ is easily calculated from the formula $E(X) = \int_A^B x f(x)\, dx$.

$$E(X) = \int_0^B x \frac{1}{B}\, dx = \frac{1}{B} \int_0^B x\, dx = \left( \frac{1}{B} \frac{1}{2} x^2 \right) \Big|_0^B = \frac{1}{B} \frac{1}{2} B^2 - 0 = \frac{B}{2}$$

$\underbrace{\phantom{xx}}_{f(x)}$  $\underbrace{\phantom{xxx}}_{\substack{\text{Moved} \\ \text{outside}}}$  $\overbrace{\phantom{xxxx}}^{\substack{\text{Integrating by} \\ \text{the Power} \\ \text{Rule}}}$  $\overbrace{\phantom{xxxx}}^{\substack{\text{Evaluating} \\ \text{at } B \text{ and at } 0}}$  $\underbrace{\phantom{xx}}_{E(X)}$

The mean $\mu = E(X) = B/2$ is the midpoint of the interval $[0, B]$, as should be expected. For the variance, we use the alternative formula from page 763, first integrating $x^2$ times the probability density function:

$$\int_0^B x^2 \frac{1}{B}\, dx = \frac{1}{B} \int_0^B x^2\, dx = \frac{1}{B} \frac{1}{3} x^3 \Big|_0^B = \frac{1}{B} \frac{1}{3} B^3 = \frac{B^2}{3}$$

Then

$$\text{Var}(X) = \underbrace{\frac{B^2}{3}}_{} - \underbrace{\left( \frac{B}{2} \right)^2}_{} = \frac{B^2}{3} - \frac{B^2}{4} = \underbrace{\frac{B^2}{12}}_{\text{Var}(X)} \qquad \text{Var}(X) = \int_A^B x^2 f(x)\, dx - \mu^2$$

$\underbrace{\phantom{xxxxxxxxx}}_{\substack{\text{Calculated } \mu \\ \text{above}}}$

Summarizing:

---

**Uniform Random Variable**

A uniform random variable $X$ on the interval $[0, B]$ has probability density function

$$f(x) = \frac{1}{B} \qquad \text{for } 0 \leq x \leq B$$

Its mean, variance, and standard deviation are

$$E(X) = \frac{B}{2} \qquad \text{Var}(X) = \frac{B^2}{12} \qquad \sigma(X) = \sqrt{\text{Var}(X)} = \frac{B}{2\sqrt{3}}$$

---

**EXAMPLE 1**

### FINDING THE MEAN AND VARIANCE OF A UNIFORM RANDOM VARIABLE

A campus shuttle bus departs every 18 minutes. For a person arriving at a random time, find the mean and variance of the wait for the bus.

*Solution*

Arriving at a random time means that all waiting times between 0 and 18 minutes are equally likely, so $X = \left(\begin{array}{c}\text{Waiting} \\ \text{time}\end{array}\right)$ is uniformly distributed on the interval $[0, 18]$. Therefore, the preceding formulas for the uniform distribution give

$$E(X) = \frac{18}{2} = 9 \qquad\qquad E(X) = \frac{B}{2} \text{ with } B = 18$$

$$\text{Var}(X) = \frac{18^2}{12} = 27 \qquad\qquad \text{Var}(X) = \frac{B^2}{12} \text{ with } B = 18$$

$$\sigma(X) = \sqrt{27} \approx 5.2 \qquad\qquad \text{Standard deviation}$$

---

The fact that the average wait is 9 minutes for buses that come every 18 minutes should be intuitively reasonable. The standard deviation of 5.2 minutes gives a rough indication of the expected deviation from this mean, although an exact interpretation of this number is difficult.

## Exponential Random Variables

We next define **exponential random variables,** which are used to model waiting times that are potentially unlimited in duration, such as the time

Exponential probability
density function with
parameter $a$.

until an electronic component fails or the time between the arrival of telephone calls. An exponential random variable takes values in the interval $[0, \infty)$ and has probability density function $f(x) = ae^{-ax}$ for $x \geq 0$ with parameter $a > 0$, as shown on the left.

To verify that $f(x)$ is a probability density function, we must show that the area under it is 1. The integral is improper, so we first integrate from 0 to B:

$$\int_0^B f(x)\,dx = a \underbrace{\int_0^B e^{-ax}\,dx}_{\substack{\text{Moved}\\ \text{outside}}} = a \underbrace{\left(-\frac{1}{a}\right)e^{-ax}\Big|_0^B}_{\substack{\text{Using formula 4}\\ \text{(inside back cover)}}} = \underbrace{-e^{-aB} + e^0}_{\text{Evaluating}} = \underbrace{1 - e^{-aB}}_{\substack{\text{Reversing}\\ \text{the order}}}$$

Then we let $B \to \infty$:

$$\int_0^\infty f(x)\,dx = \lim_{B \to \infty}\underbrace{(1 - e^{-aB})}_{\to 0} = 1 \qquad\qquad \text{Area is 1}$$

To find the mean $E(X) = \displaystyle\int_0^\infty xf(x)\,dx$, we use integration by parts:

$$\int \underbrace{x}_{u}\ \underbrace{ae^{-ax}dx}_{dv} = \underbrace{x}_{u}\underbrace{(-e^{-ax})}_{v} - \int \underbrace{-e^{-ax}}_{v}\ \underbrace{dx}_{du} = \underbrace{-xe^{-ax} - \frac{1}{a}e^{-ax}}_{\substack{\text{Using formula 4}\\ \text{(inside back cover)}}} \qquad \begin{array}{l}\text{Indefinite}\\ \text{integral}\end{array}$$

$$\left[\begin{array}{ll} u = x & dv = ae^{-ax}\,dx \\ du = dx & v = \int ae^{-ax}\,dx = -e^{-ax} \end{array}\right]$$

Evaluating from 0 to B:

$$\left(-xe^{-ax} - \frac{1}{a}e^{-ax}\right)\Big|_0^B = -Be^{-aB} - \frac{1}{a}e^{-aB} + \frac{1}{a}$$

Finally, taking the limit as $B \to \infty$ (using $\displaystyle\lim_{B \to \infty} Be^{-aB} = 0$, from Exercise 62 on page 726),

$$E(X) = \lim_{B \to \infty}\left(\underbrace{-Be^{-aB}}_{\substack{\to 0\\ \text{as } B \to \infty}} - \underbrace{\frac{1}{a}e^{-aB}}_{\substack{\to 0\\ \text{as } B \to \infty}} + \frac{1}{a}\right) = \underbrace{\frac{1}{a}}_{E(X)}$$

A similar calculation (see Exercise 29) shows that $\text{Var}(X) = 1/a^2$, and therefore $\sigma(X) = 1/a$. These results are easy to remember: both the mean and standard deviation are $1/a$, the reciprocal of the parameter. Summarizing:

---

**Exponential Random Variable**

An exponential random variable $X$ with parameter $a > 0$ takes values in $[0, \infty)$ and has probability density function

$$f(x) = ae^{-ax} \quad \text{for } x \geq 0$$

Its mean, variance, and standard deviation are

$$E(X) = \frac{1}{a}$$

$$\text{Var}(X) = \frac{1}{a^2}$$

$$\sigma(X) = \frac{1}{a}$$

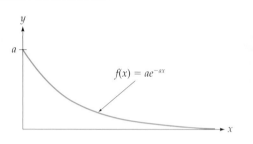

---

**Practice Problem 1**

Find the mean, variance, and standard deviation of the random variable with probability density function $\quad f(x) = 2e^{-2x} \quad$ for $x \geq 0$.
[*Hint:* Solve by inspection.]                          ➤ **Solution on page 777**

Exponential random variables are used to model waiting times for events whose *non*occurrence up to any time $t$ has no effect on the probability of the occurrence in any later time interval. (For example, your *not* winning the lottery last week has no effect on your chances this week, since each week's game is independent, so your time between winnings would be an exponential random variable.) In practice, the value of the parameter $a$ is found by setting the theoretical mean $1/a$ equal to the observed average based on past experience.

**EXAMPLE 2**    **FINDING PROBABILITIES OF COMPUTER FAILURE**

A company's computer occasionally breaks down and based on past records, the time $X$ between failures averages 50 days. Model $X$ by an exponential random variable and find:

**a.** $P(X \leq 20)$     **b.** $P(X > 20)$     **c.** $P(30 \leq X \leq 60)$

*Solution*

The value of the parameter $a$ is found by setting the theoretical mean $1/a$ equal to the observed average, here 50:

$$\frac{1}{a} = 50 \quad \text{so} \quad a = \frac{1}{50} = 0.02 \qquad \text{Parameter } a = 0.02$$

Therefore, the random variable $X$ has probability density function

$$f(x) = 0.02e^{-0.02x} \qquad \text{on } [0, \infty) \qquad\qquad f(x) = ae^{-ax} \text{ with } a = 0.02$$

For part (a), we integrate this function from 0 to 20 to find $P(X \le 20)$:

$$0.02 \underbrace{\int_0^{20} e^{-0.02x} \, dx}_{\substack{\text{Moved} \\ \text{outside}}} = 0.02 \underbrace{\left(\frac{1}{-0.02}\right) e^{-0.02x} \Big|_0^{20}}_{\substack{\text{Using formula 4} \\ \text{(inside back cover)}}} = \underbrace{-e^{-0.4} + e^0}_{\text{Evaluating}} \underbrace{\approx 0.33}_{\substack{\text{Using a} \\ \text{calculator}}}$$

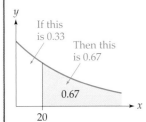

If this is 0.33

Then this is 0.67

0.67

20

Therefore, with probability about $\frac{1}{3}$, the next failure will occur within 20 days.

For part (b), $P(X > 20)$, we could integrate from 20 to $\infty$, but it is easier to find the probability of all *other* values, $P(X \le 20)$, which we found in part (a) to be 0.33, and subtract it from 1:

$$P(X > 20) = 1 - P(X \le 20) \approx 1 - 0.33 = 0.67$$

Therefore, with probability about $\frac{2}{3}$, the next failure will occur some time after 20 days.

For part (c), $P(30 \le X \le 60)$, we could evaluate the integral $\int_{30}^{60} 0.02e^{-0.02x} \, dx$ "by hand." Instead, we use a graphing calculator as shown on the left. The value 0.248 means that with probability about $\frac{1}{4}$ the next failure of the computer will occur some time between 30 and 60 days (1 to 2 months) after the preceding failure. Such calculations are used to determine the need for a backup computer system.

∫f(x)dx=.24761742

on [0, 100] by [−0.005, 0.02]

---

**EXAMPLE 3**

### FINDING AND USING A CUMULATIVE DISTRIBUTION FUNCTION

Find the cumulative distribution function for the probability density function in Example 2, and use it to find $P(30 \le X \le 60)$.

*Solution*

For the cumulative distribution function $F(x) = P(X \le x)$, we integrate the probability density function up to $x$:

$$F(x) = \int_0^x 0.02e^{-0.02t} \, dt = -e^{-0.02t} \Big|_0^x$$

$$= -e^{-0.02x} + 1 = 1 - e^{-0.02x}$$

$y$

1

0.5

$F(x) = 1 - e^{-0.02x}$

100    200    $x$

Therefore, the cumulative distribution function is $F(x) = 1 - e^{-0.02x}$ for $0 \le x < \infty$, as shown on the left.

We then find $P(30 \le X \le 60)$ by evaluating $F(x)$ at 60 and at 30 and subtracting:

$$P(30 \le X \le 60) = F(60) - F(30) = \underbrace{1 - e^{-0.02(60)}}_{F(60)} - \underbrace{(1 - e^{-0.02(30)})}_{F(30)}$$

$$= -e^{-1.2} + e^{-0.6} \approx 0.248 \qquad \text{Same answer as in Example 2(c)}$$

## Relationship Between Exponential and Poisson Random Variables

There is a useful relationship between the *Poisson* random variable discussed on pages 751–753 and the *exponential* random variables discussed here. If the time *intervals* between events are independent *exponential* random variables with parameter *a*, then the total *number* of events to occur up to time *t* will be a *Poisson* random variable with parameter *at*. For instance, in the computer failure situation described in Examples 2 and 3, the times between failures are independent exponential random variables with parameter 0.02, so the total number of failures within *t* days will be Poisson with parameter (mean) 0.02*t*. Such ideas are used in designing service queues, such as waiting lines for customers in a bank or docking facilities for ships in a port.

## Median Value of a Random Variable

The mean or expected value of a random variable gives a "representative" or "middle" value for the random variable. There is another possible choice for a "representative" value—the **median,** which is a number *m* such that the random variable is equally likely to be above *m* as below *m*. More precisely:

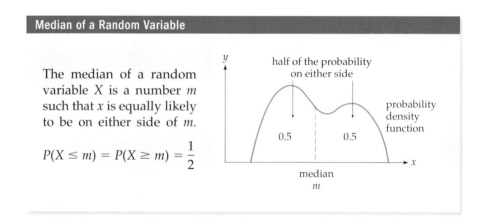

**Median of a Random Variable**

The median of a random variable $X$ is a number $m$ such that $x$ is equally likely to be on either side of $m$.

$$P(X \leq m) = P(X \geq m) = \frac{1}{2}$$

half of the probability on either side

0.5        0.5

probability density function

median
$m$

**EXAMPLE 4**        **FINDING THE MEDIAN OF A CONTINUOUS RANDOM VARIABLE**

Find the median time between failures for the computer in Example 2.

*Solution*

The median is most easily found by setting the cumulative distribution function $F(x)$ equal to $\frac{1}{2}$ and solving for $x$. We could do this "by hand," solving $1 - e^{-0.02x} = 0.5$ (using the cumulative distribution function found in Example 3), but instead we use a graphing calculator.

## Graphing Calculator Exploration

Intersection
X=34.657359 Y=.5

Define $y_1 = 1 - e^{-0.02x}$ and $y_2 = 0.5$, graph them on the window [0, 100] by [0, 1], and use INTERSECT to find where they meet.

The median waiting time is about 35 days, meaning that about half of the time the failure will occur *before* 35 days, and half of the time *after* 35 days.

### Median Versus the Mean

In the computer failure example, the median $m \approx 35$ was smaller than the mean $\mu = 50$ (calculated from the formula $\mu = \frac{1}{a} = \frac{1}{0.02} = 50$). Intuitively, the mean is larger because such waiting times can occasionally be very long, and these very large values increase the mean but not the median.

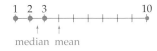

median   mean

To clarify the difference between mean and median, think of the numbers 1, 2, 3, and 10. The mean (the average) is $\frac{1 + 2 + 3 + 10}{4} = \frac{16}{4} = 4,$ but three of the four numbers are *below* this mean. For the median value, we could choose any number between 2 and 3 (such as 2.5), since two values are below and two are above this number. If we were to replace the 10 by an even larger number, this would increase the mean but not the median. In general, the mean is influenced by the values of the highest and lowest numbers, but the median is not.

**Practice Problem 2**

**a.** A news commentator recently read the statistic that the mean family income in the United States is about $40,000, and that 65% of families are below this level. He then said that this must be wrong, because only 50% of the families can be below the mean. Is he right?

**b.** In the 1995 major league baseball strike, it was revealed that the mean salary for major league baseball players was $1.2 million, while the median salary was only $500,000. What does this tell you about the distribution of salaries among the players?

➤ **Solutions below**

**Solutions to Practice Problems**

**1.** The probability density function is exponential with parameter $a = 2,$ so

$$E(X) = \frac{1}{a} = \frac{1}{2} \qquad \text{Var}(X) = \frac{1}{a^2} = \frac{1}{4} \qquad \sigma(X) = \frac{1}{2}$$

**2. a.** He is wrong. It is the *median*, not the mean, that has 50% of the people below it and 50% above it.

   **b.** Baseball salaries are distributed very unequally, with a few players having extraordinarily high salaries that "skew" the mean, but most having much lower salaries, half of them under $500,000.

A *uniform* random variable takes values in a finite interval, with all values equally likely. An *exponential* random variable takes values in the infinite interval $[0, \infty)$, with decreasing probabilities for higher values.

| Random Variable | Mean $E(X)$ | Variance $Var(X)$ | Probability Density Function | |
|---|---|---|---|---|
| Uniform on $[0, B]$ | $\dfrac{B}{2}$ | $\dfrac{B^2}{12}$ | $f(x) = \dfrac{1}{B}$ for $0 \leq x \leq B$ | |
| Exponential (parameter $a > 0$) | $\dfrac{1}{a}$ | $\dfrac{1}{a^2}$ | $f(x) = ae^{-ax}$ for $0 \leq x < \infty$ | |

How do we choose the constants for these random variables in a given application? For the uniform distribution, $B$ is chosen as the largest possible value of the random variable. For the exponential distribution, the parameter $a$ is determined by setting the theoretical mean $1/a$ equal to the *observed average value* based on past experience.

The *median* of a random variable $X$ is a number $m$ such that the random variable is equally likely to be above it as below it:  $P(X \geq m) = P(X \leq m)$. The median is most easily found by setting the cumulative distribution function equal to $\frac{1}{2}$ and solving.

**11.3** Exercises

### Exercises on Uniform Random Variables

**1.** If $X$ is a uniform random variable on the interval $[0, 10]$, find
  **a.** the probability density function $f(x)$
  **b.** $E(X)$        **c.** $Var(X)$
  **d.** $\sigma(X)$        **e.** $P(8 \leq X \leq 10)$

**2.** If $X$ is a uniform random variable on the interval $[0, 0.01]$, find

  **a.** the probability density function $f(x)$
  **b.** $E(X)$        **c.** $Var(X)$
  **d.** $\sigma(X)$        **e.** $P(X \geq 0.005)$

**3.** For a uniform random variable on the interval $[0, B]$, find
  **a.** the cumulative distribution function $F(x)$
  **b.** the median

## Applied Exercises for Uniform Random Variables

**4. GENERAL: Waiting Time** Buses arrive at a station every 12 minutes, so if you arrive at a random time, your wait is uniformly distributed on the interval [0, 12]. Find

**a.** $E(X)$    **b.** $\text{Var}(X)$
**c.** $\sigma(X)$    **d.** $P(X \le 3)$

**5. GENERAL: Accidents** A highway patrol station responds to emergencies within 30 miles of its location on a highway. Suppose that an accident occurs on this part of the highway and that the distance $X$ between the accident and the station is uniformly distributed over the interval [0, 30]. Find

**a.** $E(X)$    **b.** $\text{Var}(X)$
**c.** $\sigma(X)$    **d.** $P(X > 24)$

**6. GENERAL: Waiting Time** Your calculus teacher is often late to class, and the number of minutes you wait is uniformly distributed on the interval [0, 10] (since you leave after 10 minutes). Find

**a.** $E(X)$    **b.** $\text{Var}(X)$
**c.** $\sigma(X)$    **d.** $P(X \ge 9)$

**7. Uniform Random Variable on [A, B]** For any two numbers $A$ and $B$ with $A < B$, a *uniform random variable on the interval* [A, B] has probability density function

$$f(x) = \frac{1}{B - A} \quad \text{for} \quad A \le x \le B$$

Find

**a.** the cumulative distribution function $F(x)$
**b.** the mean $E(X)$
**c.** the variance $\text{Var}(X)$

**8. GENERAL: Roundoff Errors** When rounding numbers to the nearest integer, the error $X$ in the rounding is uniformly distributed on the interval [−0.5, 0.5]. Use Exercise 7 to find

**a.** $E(X)$    **b.** $\text{Var}(X)$
**c.** $\sigma(X)$    **d.** $P(-0.25 \le X \le 0.25)$

## Exercises on Exponential Random Variables

**9–12.** For the exponential random variables with the following densities, find

**a.** $E(X)$    **b.** $\text{Var}(X)$    **c.** $\sigma(X)$
[*Hint:* Solve by inspection.]

**9.** $f(x) = 5e^{-5x}$ on $[0, \infty)$

**10.** $f(x) = e^{-x}$ on $[0, \infty)$

**11.** $f(x) = \frac{1}{2}e^{-x/2}$ on $[0, \infty)$

**12.** $f(x) = 0.001e^{-0.001x}$ on $[0, \infty)$

**13–16.** State the probability density function for an exponential random variable with

**13.** Mean 3    **14.** Mean 0.01
**15.** Variance $\frac{1}{4}$    **16.** Standard deviation $\frac{1}{4}$

**17.** For an exponential random variable with parameter $a$, find

**a.** the cumulative distribution function $F(x)$
**b.** the median

**18.** Find the probability that an exponential random variable is less than its mean.

## Applied Exercises for Exponential Random Variables

**19. BUSINESS: Employee Safety** Records indicate that the average time between accidents on a factory floor is 20 days. If the time between accidents is an exponential random variable, find the probability that the time between accidents is less than a month (30 days).

**20. GENERAL: Emergency Calls** The number of minutes between calls to an emergency 911 center is exponentially distributed with mean 8. There is one emergency operator who takes 2 minutes to deal with each call. What is the probability that the next caller will get a busy signal?

21. **BUSINESS: Quality Control** The time until failure of a TV picture tube is an exponential random variable with mean 10 years. What is the probability that the tube will fail during the 2-year guarantee period?

22. **POLITICAL SCIENCE: Supreme Court Vacancies** The length of time between vacancies on the Supreme Court is exponentially distributed with mean 2 years. Find the probability that at least 4 years will elapse between vacancies.
    *Source: New York Times*

23. **GENERAL: Powerball** Powerball is a multistate lottery with drawings twice a week, in which there may or may not be jackpot winners. On average, 4 weeks elapse between jackpot wins. Find the probability that the time between jackpot wins is:

    **a.** no more than one week.
    **b.** no more than six weeks.

    (Note that here you are approximating a discrete-time event by a continuous-time model, as is often done in practice. Incidentally, the largest powerball jackpot was $315 million, won on Christmas day of 2002.)
    *Source:* power-ball.com

24. **BUSINESS: Product Reliability** The lifetime of a light bulb is exponentially distributed. What must the mean lifetime be if the manufacturer wants the bulb to last at least 750 hours with probability 0.90?

**MEDIAN:** Find the median of the random variable with the given probability density function.

25. $f(x) = 2x$ on $[0, 1]$

26. $f(x) = 3x^2$ on $[0, 1]$

27. $f(x) = \frac{1}{2}x$ on $[0, 2]$

28. $f(x) = \frac{1}{6}x^{-1/2}$ on $[0, 16]$

29. Find the variance of an exponential random variable with parameter $a$. [*Hint:* You will need to integrate by parts twice, and to use the results of Exercises 62 and 63 on page 726.]

30. **GENERAL: Telephone Calls** You follow the telephone company's recommendation to let a number ring for ten rings (50 seconds) before hanging up. If the time it takes a government bureaucrat to answer his phone is exponentially distributed with mean 45 seconds, what is the probability that you will reach him?

31. **GENERAL: Cash Machines** The number of minutes between arrivals of customers at an automated teller machine is exponentially distributed with mean 7 minutes. If a customer's transactions take 3 minutes, what is the probability that the next arrival will have to wait?

32. **BIOMEDICAL: ICU Admissions** The number of hours between the arrivals of new patients in an intensive care unit is exponentially distributed with mean 8 hours. Find the median time between arrivals.

33. **GENERAL: Airplane Maintenance** The amount of flight time between failures of an airplane engine is exponentially distributed with mean 700 hours. If the engine is inspected every 100 hours of flight time, what is the probability that the engine will fail before it is inspected?

34. **BUSINESS: Computer Failure** The number of days between failures of a company's computer system is exponentially distributed with mean 10 days. What is the probability that the next failure will occur between 7 and 14 days after the last failure?

35. **GENERAL: Light Bulbs** The life of a light bulb is exponentially distributed with mean 1000 hours. What is the probability that the light bulb will burn out sometime between 800 and 1200 hours?

**36–37. BIOMEDICAL: Survival Time** For a person who has received treatment for a life-threatening disease, such as cancer, the number of years of life after the treatment (the *survival time*) can be modeled by an exponential random variable.

36. Suppose that the average survival time for a group of patients is 5 years. Find the probability that a randomly selected patient survives for no more than 7 years.

37. The *survival function* $S(x)$ gives the probability that a randomly selected patient will survive for at least $x$ years after treatment has ended. Show that $S(x) = 1 - F(x)$, where $F(x)$ is the cumulative distribution function for the exponential random variable, and find a formula for $S(x)$ in terms of the exponential parameter $a$.

## Conceptual Exercises

38. An angle is uniformly distributed between 0 and 180 degrees. What is its probability density function expressed in *radians*?

**39.** If a uniform random variable (as described on page 772) has mean 10, what is its probability density function?

**40.** If a uniform random variable (as described on page 772) has variance 3, what is its probability density function?

**41.** If $X$ is uniformly distributed on $[0, 2]$, what is the *probability* density function for $3X$?

**42.** If $X$ is an exponential random variable with mean $\frac{1}{5}$, what is its probability density function?

**43.** Consider a uniform random variable (as described on page 772) on the interval $[0, 1]$. What is the probability density function of an *exponential* random variable that has the same mean?

**44.** Fill in the blank: If $X$ is an exponential random variable with parameter 4, then $2X$ is an

exponential random variable with parameter ___. [*Hint:* What is the mean of $X$? What is the mean of $2X$? Then what is the parameter for $2X$?]

**45.** In a daily lottery that may or may not award a jackpot, the intervals between jackpots are independent exponential random variables with mean 10 days. Then the total number of jackpots in the next 30 days is a random variable. What kind of a random variable is it? [*Hint:* See the top paragraph on page 776.]

**46.** Consider the incomes of three people: you, your math teacher, and Bill Gates (the richest man in the world). Which will be greater: the median or the mean?

**47.** The number of children in an American family is a random variable with mean 2. For this random variable, which will be greater: the median or the mean?

## 11.4 NORMAL RANDOM VARIABLES

### Introduction

Studies show that heights of adults cluster around a mean of 65 inches, with larger and smaller heights occurring with lower probabilities. Similarly, IQs cluster around a mean of 100, with values farther away from the mean occurring with lower probabilities. Measurements that cluster around a mean $\mu$, with values farther from the mean occurring with lower probabilities, are often modeled by a **normal random variable.** The probability density function of a normal random variable is often referred to as the *bell-shaped curve,* but more formally called the **Gaussian probability density function,** after the German mathematician Carl Friedrich Gauss (1777–1855). Although normal random variables are continuous, they are often used to model discrete quantities.

### Normal Random Variables

The probability density function of a normal random variable depends on two parameters, denoted $\mu$ and $\sigma > 0$, which are, respectively, the mean and standard deviation of the random variable.

**Normal Random Variable**

A normal random variable with mean $\mu$ and standard deviation $\sigma$ has probability density function

$$f(x) = \frac{1}{\sigma\sqrt{2\pi}} e^{-\frac{1}{2}\left(\frac{x-\mu}{\sigma}\right)^2} \qquad \text{for} \quad -\infty < x < \infty$$

A normal random variable is a **standard** normal random variable if it has mean $\mu = 0$ and standard deviation $\sigma = 1$. A standard normal random variable will be denoted by the letter $Z$. Any normal random variable $X$ can be **standardized** by subtracting the mean and dividing by the standard deviation:

$$Z = \frac{X - \mu}{\sigma} \qquad \begin{array}{l} \mu = \text{mean} \\ \sigma = \text{standard deviation} \end{array}$$

The probability density function for a *standard* normal random variable $Z$ takes the following simpler form:

**Standard Normal Random Variable**

A standard normal random variable $Z$ has probability density function

$$f(x) = \frac{1}{\sqrt{2\pi}}\, e^{-\frac{1}{2}x^2} \qquad \text{for} \quad -\infty < x < \infty$$

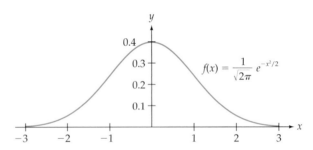

Normal probability density function with mean
$\mu = 0$ and standard deviation $\sigma = 1$

The area under the normal probability density function is 1, as is shown in Exercise 41 on pages 789–790. The mean $\mu$ gives the location of the central peak of this probability density function. The following graph shows three normal probability density functions with different means.

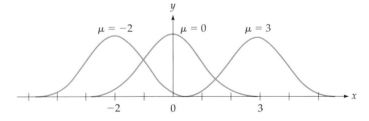

Three normal probability density
functions with different means $\mu$

The standard deviation $\sigma$ measures how the values *spread* away from the mean, so the peak of the probability density function will be higher and more pointed if $\sigma$ is small, and lower and more rounded if $\sigma$ is large.

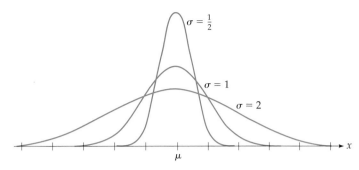

Three normal probability density
functions with different standard deviations $\sigma$

**IMPORTANT NOTE:** For those *not* using graphing calculators, turn now to the Appendix at the back of the book for normal probabilities calculated by *tables*.

## Probabilities for Normal Random Variables

The probability that a normal random variable takes values between two numbers is the area under its probability density function between the two numbers.

The integral of the normal probability density function cannot be expressed in terms of elementary functions, so we cannot find the area by integration. However, we can *approximate* the area by *numerical* integration, as in Section 6.4. The results can be listed in a table (as in the Appendix), or they can be found directly on a graphing calculator, using the probability DISTRibution menu and the NORMALCDF command. The NORMALCDF command finds the probability that a normal random variable takes values between any two given numbers.

**EXAMPLE 1**

**FINDING A STANDARD NORMAL PROBABILITY**

For a standard normal random variable $Z$, find   $P(0 \leq Z \leq 1.24)$.

*Solution*

We use the NORMALCDF command with the given upper and lower limits 0 and 1.24.

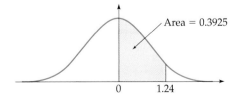

Therefore,   $P(0 \leq Z \leq 1.24) \approx 0.3925$.

## Graphing Calculator Exploration

Use the NORMALCDF command to find $P(0 \leq Z \leq 0.52)$ for a standard normal random variable $Z$. (Answer: 0.1985)

On some calculators, you can show such results graphically using the SHADENORM command:

Window: $[-3, 3]$ by $[-0.2, 0.8]$ Use CLRDRAW afterward to clear the screen

leads to

The values of a normal random variable are said to be *normally distributed*.

 **EXAMPLE 2**

### FINDING THE PROBABILITY OF MEN'S HEIGHTS

The heights of American men are approximately normally distributed, with mean $\mu = 70$ inches and standard deviation $\sigma = 3.1$ inches. Find the proportion of men who are less than 6 feet tall.

*Source:* G. Salvendy, *Handbook of Human Factors and Ergonomics,* 3rd ed.

*Solution*

If $X =$ height, we want $P(X < 72)$.          72 inches = 6 feet

For the lower limit, we enter a sufficiently small number to include essentially *all* of the area to the left of the upper limit (here, 0 will do). We use the NORMALCDF command, entering the given mean and standard deviation after the upper and lower limits.

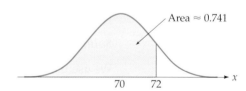

Therefore, approximately 74% of American men are under 6 feet tall.

Such calculations are used for designing automobiles and other consumer products.

Why did we enter the mean and variance in Example 2 but not in Example 1? Because the calculator will *assume* that $\mu = 0$ and $\sigma = 1$ (standard normal, which was the case in Example 1) unless told otherwise. How do we pick a lower limit to include essentially *all* of the area below the upper limit? As a "rule of thumb," choose a convenient number at least five standard deviations below the mean. In this example, five standard deviations would be $5 \cdot 2.6 = 13$, and our lower limit of 0 was far more than this distance below the mean of 70. A similar rule of thumb applies when there is no upper limit: go at least five standard deviations *above* the mean.

## Graphing Calculator Exploration

The weights of American women are approximately normally distributed, with mean $\mu = 151.8$ pounds and standard deviation $\sigma = 38.8$ pounds. Find the proportion of women who weigh more than 140 pounds. (Answer: 62%)

*Source:* G. Salvendy, *Handbook of Human Factors and Ergonomics,* 3rd ed.

**EXAMPLE 3**  **FINDING THE PROBABILITY OF ACCEPTABLE CHOLESTEROL LEVELS**

Cholesterol is a fat-like substance that, in the bloodstream, can increase the risk of heart disease. Cholesterol levels below 200 (milligrams per deciliter of blood) are generally considered acceptable. A survey by the U.S. Department of Health found that cholesterol levels for one age group were approximately normally distributed, with mean $\mu = 222$ and standard deviation $\sigma = 28$. What proportion of the people in the study had acceptable cholesterol levels?

*Source:* Rogosin Institute

### Solution

Letting $X = \left(\begin{array}{c}\text{Cholesterol} \\ \text{level}\end{array}\right)$, we want $P(X < 200)$. We again use 0 for the lower limit. (Do you see why it satisfies the "at least five standard deviations below the mean" rule?) The NORMALCDF command with these numbers gives the following result.

Therefore, only about 22% of the people in the study had acceptable cholesterol levels.

## Inverse Probabilities for Normal Random Variables

We may *reverse* the process of finding probabilities, finding instead the value that results in a *given* probability. A graphing calculator command like INVNORM finds the number such that, with a given probability, a normal random variable is *below* that number.

EXAMPLE 4  **FINDING THE HEIGHT FOR A GIVEN PROBABILITY**

Women's heights are approximately normally distributed, with mean $\mu = 64.6$ inches and standard deviation $\sigma = 2.9$ inches. Find the height that marks the tallest 10%.

*Source:* G. Salvendy, *Handbook of Human Factors and Ergonomics,* 3rd ed.

*Solution*

The height that marks the tallest 10% means that 90% are below that height. The INVNORM command with the given numbers provides the following answer.

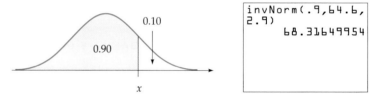

Therefore, a height of about 68.3 inches, or 5 feet 8 inches, marks the tallest 10% of women.

Measuring distance from the mean in *standard deviation units* leads to some useful probabilities for a normal random variable X.

With probability 68%, X is within 1 standard deviation of its mean.
With probability 95%, X is within 2 standard deviations of its mean.
With probability 99.7%, X is within 3 standard deviations of its mean.

These results are shown graphically as follows.

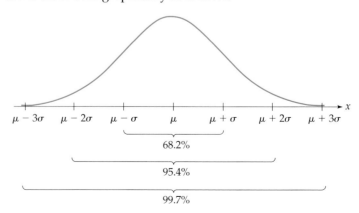

For example, from the fact that men's heights are normally distributed with $\mu = 70$ and $\sigma = 3.1$, we can conclude

About 68% of men have heights
$70 \pm 3.1$ inches (that is, heights
66.9 inches to 73.1 inches).

Mean $\pm 1$ standard deviation

About 95% of men have heights
$70 \pm 2(3.1)$ inches (that is, heights
63.8 inches to 76.2 inches).

Mean $\pm 2$ standard deviations

About 99.7% of men have heights
$70 \pm 3(3.1)$ inches (that is, heights
60.7 inches to 79.3 inches).

Mean $\pm 3$ standard deviations

## 11.4 Section Summary

Normal random variables are used to model values that cluster symmetrically around a mean, $\mu$, with the probability decreasing as the distance from $\mu$ increases. The probability density function of a normal random variable with mean $\mu$ and standard deviation $\sigma$ is

$$f(x) = \frac{1}{\sigma\sqrt{2\pi}} e^{-\frac{1}{2}\left(\frac{x-\mu}{\sigma}\right)^2} \qquad \text{for } -\infty < x < \infty$$

The mean $\mu$ gives the location of the central peak, and the standard deviation $\sigma$ measures the *spread* of the values around this mean (larger $\sigma$ meaning greater spread).

The probability density of a *standard* normal random variable ($\mu = 0$ and $\sigma = 1$) is

$$f(x) = \frac{1}{\sqrt{2\pi}} e^{-\frac{1}{2}x^2} \qquad \text{for } -\infty < x < \infty$$

Probabilities for a normal random variable can be found either in a *table* (by first *standardizing*, defining $Z = \dfrac{X - \mu}{\sigma}$ and then using the standard normal table, as explained in the Appendix) or by using a *graphing calculator* with an operation like NORMALCDF. Similarly, we can find the $x$-value such that with given probability the random variable is below that value either by using tables or a graphing calculator operation like INVNORM.

## 11.4 Exercises

**1–10.** Find each probability for a standard normal random variable Z.

**1.** $P(0 \leq Z \leq 1.95)$

**2.** $P(-1.23 \leq Z \leq 0)$

**3.** $P(-2.55 \leq Z \leq 0.48)$

**4.** $P(-1.5 \leq Z \leq 1.05)$

**5.** $P(1.4 \leq Z \leq 2.8)$

**6.** $P(1 \leq Z \leq 2)$

**7.** $P(Z < -3.1)$

**8.** $P(Z > 3)$

**9.** $P(Z \leq 3.45)$

**10.** $P(Z \geq -3.5)$

11. If $X$ is normal with mean 9 and standard deviation 2, find $P(10 < X < 11)$.

12. If $X$ is normal with mean 3 and standard deviation $\frac{1}{2}$, find $P(2 \le X \le 3.5)$.

13. If $X$ is normal with mean $-5$ and standard deviation 2, find $P(X \le -6)$.

14. If $X$ is normal with mean 0.05 and standard deviation 0.01, find $P(X > 0.04)$.

## Applied Exercises

15. **SOCIAL SCIENCE: Workload** A motor vehicle office can process only 250 license applications per day. If the number of applications on a given day is normally distributed with mean 210 and standard deviation 30, what is the probability that the office can handle the day's applications? [*Hint:* Find $P(0 \le X \le 250)$.]

16. **BIOMEDICAL: Hospital Stays** A patient's medical insurance will pay for at most 20 days in the hospital. Past experience with similar cases shows that the length of hospital stay is normally distributed with mean 17 days and standard deviation 2 days. What is the probability that the insurance coverage will be enough? [*Hint*: Find $P(0 \le X \le 20)$.]

17. **BUSINESS: Advertising** The percentage of television viewers who watch a certain show is a normal random variable with mean 22% and standard deviation 3%. The producer guarantees to advertisers that the viewer percentage will be at least 20%, or else the advertisers will receive free air time. What is the probability that the producers will have to give free air time? [*Hint:* Find $P(X < 20)$.]

18. **PSYCHOLOGY: Learning** The time that a rat takes to "solve" a maze is normally distributed with mean 3 minutes and standard deviation 30 seconds. Find the probability that the rat will take less than 2 minutes.

19. **BUSINESS: Orders** The number of orders received by a company each day is normally distributed with mean 2000 and variance 160,000. If the orders exceed 2500, the company will have to hire extra help. Find the probability that extra help will be needed. [*Hint:* First find $\sigma$.]

20. **BUSINESS: Airline Overbooking** In order to avoid empty seats, airlines generally sell more tickets than there are seats. Suppose that the number of ticketed passengers who show up for a flight with 100 seats is a normal random variable with mean 97 and standard deviation 6. What is the probability that at least one person will have to be "bumped" (denied space on the plane)?

21. **BIOMEDICAL: Weights** Weights of men are approximately normally distributed with mean 190 pounds and standard deviation 40. The minimum and maximum weights in order to be an FBI Police Officer (providing security at FBI offices) are 117 and 238 pounds. What proportion of men fall in this range?

*Source:* Federal Bureau of Investigation; G. Salvendy, *Handbook of Human Factors and Ergonomics*, 3rd ed.

22. **GENERAL: SAT Scores** Test scores on the 2007 SAT I: Reasoning Test in Mathematics were approximately normally distributed with mean 515 and standard deviation 114. What score would put you in the top 20%?

*Source:* College Board

23. **SOCIAL SCIENCE: IQs** Although intelligence cannot be measured on a single scale, IQ (Intelligence Quotient) scores are still used in some situations. IQ scores are normally distributed with mean 100 and standard deviation 15. Find the IQ score that places you in the top 2%.

24. **BUSINESS: Product Reliability** The number of years that an electrical generator will last is a normal random variable. Company A makes generators with a mean lifetime of 8 years and a standard deviation of 1 year. Company B makes generators with a mean lifetime of 7 years and a standard deviation of 2 years. Which company's generators have a higher probability of lasting at least 10 years?

25. **BUSINESS: Quality Control** A soft drink bottler fills 32-ounce bottles with a high-speed bottling machine. Assume that the amount of the soft drink put into the bottle is a normal random variable. The mean amount that the machine puts into the bottles is adjustable, but the standard deviation is 0.25 ounce. What should be the mean amount such that with probability 98% a bottle contains at least 32 ounces?

26. **ECONOMICS: Portfolio Management** Modern portfolio theory generally assumes that the yield on a diversified portfolio of investments will be

normally distributed around its historic mean. What is the probability that a diversified portfolio will yield at least 5% if the historic mean is 8% and standard deviation 1.5%?

## Conceptual Exercises

In the following exercises, assume that all random variables are normal and that $Z$ is *standard* normal. No calculation is necessary.

**27.** What are the mean and standard deviation of $2Z$?

**28.** What are the mean and standard deviation of $Z + 5$?

**29.** What are the mean and standard deviation of $2Z + 5$?

**30.** What are the mean and standard deviation of $2(Z + 5)$?

**31.** If $X$ and $Y$ both have mean 0 but $\sigma(X) = 2$ and $\sigma(Y) = 3$, then which is greater: $P(X > 4)$ or $P(Y > 4)$?

**32.** If $X$ and $Y$ both have standard deviation 1 but $E(X) = 2$ and $E(Y) = 3$, then which is greater: $P(X > 4)$ or $P(Y > 4)$?

**33.** If $X$ and $Y$ both represent the heights of people, but $X$ is in feet and $Y$ is in inches, which has the greater mean? Which has the greater standard deviation?

**34.** If $X$ and $Y$ both represent the temperature at noon tomorrow but $X$ is in degrees Fahrenheit and $Y$ is in degrees Celsius (or Centigrade), which has the greater mean? Which has the greater standard deviation? [*Hint:* The relationship is $X = \frac{9}{5}Y + 32$.]

**35.** If $P(X > 47.8) = \frac{1}{2}$, then what do you know about $X$?

**36.** If $\frac{1}{2}X + 2$ is *standard* normal, then what do you know about $X$?

## Explorations and Excursions

The following problems extend and augment the material presented in the text.

### Calculus of the Normal Distribution

**37.** Find the $x$-value at which the standard normal probability density function $f(x) = \dfrac{1}{\sqrt{2\pi}} e^{-x^2/2}$ is maximized. Also find the value of the function at this $x$-value.

**38.** Find the $x$-value at which the (nonstandard) normal probability density function $f(x) = \dfrac{1}{\sigma\sqrt{2\pi}} e^{-\frac{1}{2}\left(\frac{x-\mu}{\sigma}\right)^2}$ (for constants $\mu$ and $\sigma$) is maximized. Also find the maximum value of the function at this $x$-value.

**39.** Find the inflection points of the standard normal probability density function $f(x) = \dfrac{1}{\sqrt{2\pi}} e^{-x^2/2}$.

**40.** Find the $x$-coordinates of the inflection points of the (nonstandard) normal probability density function $f(x) = \dfrac{1}{\sigma\sqrt{2\pi}} e^{-\frac{1}{2}\left(\frac{x-\mu}{\sigma}\right)^2}$ (for constants $\mu$ and $\sigma$).

**41.** Verify that the area under the standard normal curve is 1 by carrying out the following steps. (*Requires sections 6.3, 7.7, and 8.5.*)

a. Define $I = \displaystyle\int_{-\infty}^{\infty} e^{-\frac{1}{2}x^2}\,dx$ and show that
$$I = 2 \int_0^{\infty} e^{-\frac{1}{2}x^2}\,dx. \quad [\text{\textit{Hint:} Use symmetry.}]$$

b. Show that $I^2 = 4 \displaystyle\int_0^{\infty} e^{-\frac{1}{2}x^2}\,dx \cdot \int_0^{\infty} e^{-\frac{1}{2}y^2}\,dy.$

[*Hint:* $I^2$ consists of two copies of $I$ expressed with different variables and multiplied together.]

c. Show that $I^2 = 4 \displaystyle\int_0^{\infty} \left( \int_0^{\infty} e^{-\frac{1}{2}(x^2 + y^2)}\,dy \right) dx.$

[*Hint:* Move one integral (which is a constant) inside the other and use some algebra.]

d. Show that $I^2 = 4 \displaystyle\int_0^{\infty} \left( \int_0^{\infty} e^{-\frac{1}{2}(x^2 + x^2 s^2)} x\,ds \right) dx.$

[*Hint:* In the inner integral in step (c), use the substitution $y = x \cdot s$ so that $dy = x \cdot ds$]

e. Show that $I^2 = 4 \displaystyle\int_0^{\infty} \left( \int_0^{\infty} e^{-\frac{1}{2}x^2(1 + s^2)} x\,ds \right) dx$
$$= 4 \int_0^{\infty} \left( \int_0^{\infty} e^{-\frac{1}{2}x^2(1 + s^2)} x\,dx \right) ds.$$

[*Hint:* Use some algebra and then reverse the order of integration.]

f. Show that $I^2 = 4 \displaystyle\int_0^{\infty} \dfrac{1}{1 + s^2}\,ds.$ [*Hint:* In the inner integral of the last part of step (e), use the $\int e^u\,du$ formula with $u = -\frac{1}{2}x^2(1 + s^2)$ and $du = -x(1 + s^2)\,dx$ and evaluate (by limits) at 0 and $\infty$ to show that the inner integral equals $\dfrac{1}{1 + s^2}$.]

(continues)

**g.** Define $s = \tan\theta$ so that $s = 0$ corresponds to $\theta = 0$, $s = \infty$ corresponds to $\theta = \pi/2$, $ds = \sec^2\theta\, d\theta$, and $1 + s^2 = 1 + \tan^2\theta = \sec^2\theta$. [*Hint:* The last equation comes from dividing the identity $\sin^2\theta + \cos^2\theta = 1$ by $\cos^2\theta$.]

**h.** Show that

$$I^2 = 4\int_0^{\pi/2} \frac{1}{1 + \tan^2\theta}\sec^2\theta\, d\theta$$

$$= 4\int_0^{\pi/2} \frac{1}{\sec^2\theta}\sec^2\theta\, d\theta = 4\int_0^{\pi/2} d\theta = 4\cdot\frac{\pi}{2}$$

$$= 2\pi$$

[*Hint:* In the integral from step (f), use the substitution from step (g).]

**i.** Show that $I = \sqrt{2\pi}$. [*Hint:* Since $I^2 = 2\pi$ from step (h).]

**j.** Show that $\dfrac{1}{\sqrt{2\pi}}\displaystyle\int_{-\infty}^{\infty} e^{-\frac{1}{2}x^2}\, dx = 1$, as was to be shown. [*Hint:* Use step (i) with the definition of $I$ (from step (a)) and divide by $\sqrt{2\pi}$.]

---

# Chapter Summary with Hints and Suggestions

Reading the text and doing the exercises in this chapter have helped you to master the following concepts and skills, which are listed by section (in case you need to review them) and are keyed to particular Review Exercises. Answers for all Review Exercises are given at the back of the book, and full solutions can be found in the Student Solutions Manual.

## 11.1 Discrete Probability

■ Find $E(X)$, $\text{Var}(X)$, and $\sigma(X)$ for a discrete random variable. *(Review Exercises 1–2.)*

$$\mu = E(X) = \sum_{i=1}^{n} x_i p_i$$

$$\text{Var}(X) = \sum_{i=1}^{n} (x_i - \mu)^2 p_i = \sum_{i=1}^{n} x_i^2 p_i - \mu^2$$

$$\sigma(X) = \sqrt{\text{Var}(X)}$$

■ Solve an applied problem using expected value. *(Review Exercises 3–4.)*

■ Solve an applied problem using a Poisson random variable. *(Review Exercises 5–10.)*

$$P(X = k) = e^{-a}\frac{a^k}{k!} \qquad k = 0, 1, 2, \ldots$$

$$E(X) = a$$

## 11.2 Continuous Probability

■ Make a function into a probability density function. *(Review Exercises 11–16.)*

$$\int_A^B f(x)\, dx = 1$$

Area = 1

■ Find a probability and the cumulative distribution function. *(Review Exercise 17.)*

$$P(a \le X \le b) = \int_a^b f(x)\, dx$$

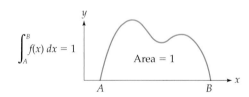

$P(a \le X \le b)$

$$F(x) = \int_A^x f(t)\, dt$$

■ Find a probability and the probability density function. *(Review Exercise 18.)*

$$P(a \leq X \leq b) = F(b) - F(a) \qquad f(x) = \frac{d}{dx} F(x)$$

■ Find $E(X)$, Var$(X)$, and $\sigma(X)$ for a continuous random variable. *(Review Exercises 19–20.)*

$$\mu = E(X) = \int_A^B x f(x)\, dx$$

$$\text{Var}(X) = \int_A^B (x - \mu)^2 f(x)\, dx$$

$$= \int_A^B x^2 f(x)\, dx - \mu^2$$

$$\sigma(X) = \sqrt{\text{Var}(X)}$$

■ Solve an applied problem involving a continuous random variable.
*(Review Exercises 21–22.)*

■ Find $E(X)$, Var$(X)$, and $\sigma(X)$ for a continuous random variable.
*(Review Exercises 23–24.)*

■ Solve an applied problem involving a continuous random variable.
*(Review Exercises 25–26.)*

## 11.3 Uniform and Exponential Random Variables

■ Find the probability density function, $E(X)$, Var$(X)$, $\sigma(X)$, and a probability for a uniform random variable. *(Review Exercises 27–30.)*

$$f(x) = \frac{1}{B}$$

$$E(X) = \frac{B}{2} \qquad \text{Var}(X) = \frac{B^2}{12} \qquad \sigma(X) = \frac{B}{2\sqrt{3}}$$

■ Solve an applied problem using an exponential random variable. *(Review Exercises 31–32.)*

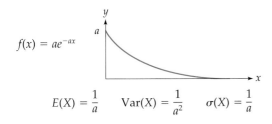

$$f(x) = ae^{-ax}$$

$$E(X) = \frac{1}{a} \qquad \text{Var}(X) = \frac{1}{a^2} \qquad \sigma(X) = \frac{1}{a}$$

■ Find the median of a continuous random variable. *(Review Exercises 33–34.)*

$$P(X \leq m) = P(X \geq m) = \frac{1}{2}$$

■ Solve an applied problem involving an exponential random variable.
*(Review Exercises 35–36.)*

## 11.4 Normal Random Variables

■ Find a probability for a standard normal random variable. *(Review Exercises 37–40.)*

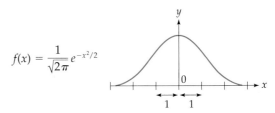

$$f(x) = \frac{1}{\sqrt{2\pi}} e^{-x^2/2}$$

■ Find a probability for a normal random variable with a given mean and variance.
*(Review Exercises 41–42.)*

$$f(x) = \frac{1}{\sigma\sqrt{2\pi}} e^{-\frac{1}{2}\left(\frac{x-\mu}{\sigma}\right)^2}$$

$$Z = \frac{X - \mu}{\sigma}$$

■ Solve an applied problem using a normal random variable. *(Review Exercises 43–46.)*

### Hints and Suggestions

■ In discrete probability, probabilities are found by *adding*. In continuous probability, they are found by *integrating* a probability density function (or by subtracting values of a cumulative distribution function).

■ The *expected value* or *mean* of a random variable gives a "representative" value, while the *variance* and *standard deviation* measure the "spread" or "dispersion" of the values away from the mean.

- Poisson random variables are often used to model the number of occurrences of "rare events," with the parameter a set equal to the *average number of occurrences.*

- The alternative formula for the variance (page 763) is generally easier to use than the definition (page 762). The corresponding alternative variance formula for a *discrete* random variable is given in Exercise 47 on page 758.

- The mean and the median are both used as "representative" numbers for a random variable. The mean is influenced by the extreme values, but the median is not. For example, if $X$ takes values {1, 2, 99} with equal probability, the median would be 2, but the mean would be 34 (showing the influence of the 99).

- Uniform random variables are used to model waiting times for which all numbers in a *fixed interval* are *equally likely*. The parameter $B$ is set equal to the largest possible value.

- Exponential random variables are used to model waiting times that can be *arbitrarily long*. The parameter $a$ is found by setting the mean $1/a$ equal to the observed average value based on past experience.

- Normal random variables are used to model measurements that cluster symmetrically around the mean. About 68% of the probability is within one standard deviation of the mean, and about 95% is within two standard deviations of the mean.

- A graphing calculator helps by finding medians and integrating probability density functions to find probabilities, especially for the normal distribution.

- **Practice for Test:** Review Exercises 1, 5, 13, 17, 19, 21, 23, 25, 27, 29, 31, 33, 35, 41, 45.

# Review Exercises for Chapter 11

Practice test exercise numbers are in green.

## 11.1 Discrete Probability

**1–2.** For each spinner, the arrow is spun and then stops, pointing in one of the numbered sectors. Let $X$ stand for the number so chosen. If the probability of each number is proportional to the area of its sector, find
**a.** $E(X)$   **b.** $\text{Var}(X)$   **c.** $\sigma(X)$

1.

2.

3. **BUSINESS: Land Value** A builder estimates a parcel of land to be worth $2,000,000 if he can get a zoning variance, and only $50,000 if he cannot. He believes that the probability of getting the variance is 40%. What is the expected value of the land?

4. **BUSINESS: Land Value** A land speculator must choose between two pieces of property, whose values depend on whether they contain oil. Property A will be worth $1,200,000 with probability 10% (that is, if oil is found), and only $2000 with probability 90%. Property B will be worth $1,000,000 with probability 15%, and only $1000 with probability 85%. On the basis of expected value, which property should the speculator buy?

5. **BUSINESS: Quality Control** Computer disks are manufactured in batches of 1000, with an average of two defective disks per batch. Use the Poisson distribution to find the probability that a batch has two or fewer defective disks.

6. **BUSINESS: Insurance** According to a recent insurance company study, the number of major natural "catastrophes" (such as earthquakes and

floods) resulting in at least $5 million of claims has recently averaged 2.63 per month. Use the Poisson distribution to estimate the probability that the number $X$ in a given month satisfies $2 \leq X \leq 4$.

*Source:* Insurance Information Institute

7. **GENERAL: Mail Delivery**  On an average day you receive 6 letters. Use a Poisson random variable to find the probability that on a given day you receive no mail.

8. **GENERAL: Cookies**  Suppose that 100 cookies are to be made from dough containing 500 chocolate chips. Use a Poisson random variable to find the probability that a given cookie has no chips.

9. **GENERAL: Earthquakes**  From 1900 through 2007 there were 547 major (Richter 6 or more) earthquakes worldwide, for an average of 5.1 per year. Use a Poisson random variable to estimate the probability that in a given year the number of earthquakes will be:

   a. 3 or fewer.
   b. 5 or fewer.
   c. 6 or fewer.
   d. 7 or more.

   *Source:* U.S. Geological Survey

10. **BUSINESS: Quality Control**  In a shipment of 500 television sets, an average of two sets will be defective. The dealer will return the shipment if more than four sets are defective. Use the Poisson distribution to find the probability of a return. [*Hint:* First find $P(X \leq 4)$ and then subtract it from 1.]

## 11.2 Continuous Probability

**11–16.**  Find the value of $a$ that makes each function a probability density function on the given interval.

11. $ax^2(2 - x)$ on $[0, 2]$          12. $a\sqrt{9 - x}$ on $[0, 9]$

13. $\dfrac{a}{(1 + x)^2}$ on $[0, 1]$          14. $a \sin x$ on $[0, \pi]$

 15. $a\sqrt{1 - x^2}$ on $[-1, 1]$    16. $ae^{-|x|}$ on $[-2, 2]$

17. For the probability density function $f(x) = \frac{1}{9}x^2$ on $[0, 3]$, find
   a. $P(1 \leq X \leq 2)$ using the probability density function
   b. the cumulative distribution function $F(x)$
   c. $P(1 \leq X \leq 2)$ using the cumulative distribution function

18. For the cumulative distribution function $F(x) = \frac{1}{4}\sqrt{x} - \frac{1}{4}$ on $[1, 25]$, find
   a. $P(4 \leq X \leq 9)$ using the cumulative distribution function
   b. the probability density function $f(x)$
   c. $P(4 \leq X \leq 9)$ using the probability density function

19. For the probability density function $f(x) = \frac{3}{2}(1 - x^2)$ on $[0, 1]$, find
   a. $E(X)$          b. $\text{Var}(X)$          c. $\sigma(X)$

20. For the probability density function $f(x) = \frac{2}{3}x^{-3/2}$ on $[1, 16]$, find
   a. $E(X)$          b. $\text{Var}(X)$          c. $\sigma(X)$

21. **GENERAL: Electrical Demand**  The amount of electricity (in millions of kilowatt-hours) used by a city is a random variable $X$ with probability density function $f(x) = \frac{1}{32}x$ on $[0, 8]$. Find
   a. the expected demand $E(X)$
   b. the power level that, with probability 0.95, will be sufficient [that is, find $b$ such that $P(X \leq b) = 0.95$]

22. **BEHAVIORAL SCIENCE: Learning**  In a test of memory, the proportion $X$ of items recalled is a random variable with probability density function $f(x) = 12x(1 - x)^2$ on $[0, 1]$. Find
   a. $P\left(X \leq \frac{1}{2}\right)$          b. $E(X)$

 23. For the probability density function
   $$f(x) = \frac{2}{\pi}\sqrt{1 - x^2} \text{ on } [-1, 1], \text{ find}$$
   a. $E(X)$          b. $\text{Var}(X)$          c. $\sigma(X)$

 24. For the probability density function $f(x) = 504x^3(1 - x)^5$ on $[0, 1]$, find
   a. $E(X)$          b. $\text{Var}(X)$          c. $\sigma(X)$

 25. **BUSINESS: Manufacturing Flaws**  A flaw is randomly located in a computer disk. The $x$-coordinate of the flaw has probability density function $f(x) = \frac{1}{2\pi}\sqrt{4 - x^2}$ on $[-2, 2]$. Find the probability that the $x$-coordinate of the flaw lies between $x = -1$ and $x = 1$.

26. **BEHAVIORAL SCIENCE:  Time to Complete a Task**  The fraction of a minute that a person takes to complete a task has probability density function $f(x) = 105x^2(1 - x)^4$ for $x$ in $[0, 1]$. Find the probability that the time lies between $x = 0.25$ and $x = 0.5$ minute.

## 11.3 Uniform and Exponential Random Variables

**27.** If $X$ is a uniform random variable on the interval $[0, 50]$, find

   **a.** the probability density function $f(x)$
   **b.** $E(X)$    **c.** $\text{Var}(X)$    **d.** $\sigma(X)$
   **e.** $P(15 \le X \le 45)$

**28.** GENERAL: Rainfall Rain is forecast, and you estimate that the actual number of inches $X$ of rain is uniformly distributed on the interval $[0, 2]$. Find

   **a.** $E(X)$    **b.** $\text{Var}(X)$
   **c.** $\sigma(X)$    **d.** $P(X < 0.5)$

**29.** For the exponential random variable with probability density function $f(x) = 0.01e^{-0.01x}$ on $[0, \infty)$, find

   **a.** $E(X)$    **b.** $\text{Var}(X)$

**30.** State the probability density function for an exponential random variable with mean 4.

**31.** GENERAL: Airplane Flight Delays For flights from U.S. airports, the "taxi-out" time (the number of minutes between leaving the gate and takeoff) can be approximated by an exponential random variable with mean 20. Find the probability that for a given flight the taxi-out time is more than:

   **a.** 1 hour (60 minutes)
   **b.** 3 hours (180 minutes)

   *Source:* U.S. Bureau of Transportation Statistics

**32.** BUSINESS: Quality Control The "life" of a typical incandescent light bulb is exponentially distributed with mean 1000 hours. Find the probability that the bulb will burn out within the first 500 hours.

   *Source:* General Electric

**33–34.** Find the median of the random variable with the given probability density function.

**33.** $f(x) = \frac{3}{8}x^2$ on $[0, 2]$

**34.** $f(x) = 3e^{-3x}$ on $[0, \infty)$

 **35.** GENERAL: Fire Alarms The time between calls to a fire department is an exponential random variable with mean 4 hours. If the fire company is out of the firehouse for 1 hour on each call, what is the probability that the company is out when the next call comes in?

 **36.** GENERAL: Ship Arrivals The time between the arrival of ships in a port is an exponential random variable with mean 5 hours. If each arrival requires 1 hour of docking time with the harbor's tugboat, what is the probability that the next arriving ship will have to wait?

## 11.4 Normal Random Variables

**37–40.** Find the indicated probability for a standard normal random variable $Z$.

**37.** $P(0.95 \le Z \le 2.54)$

**38.** $P(-1.11 \le Z \le 1.44)$

**39.** $P(Z < 2.5)$       **40.** $P(Z > 1)$

**41.** If $X$ is normal with mean 12 and standard deviation 2, find $P(10 < X \le 15)$.

**42.** If $X$ is normal with mean 1 and standard deviation 0.4, find $P(0 < X \le 1.8)$.

**43.** BUSINESS: Refunds The "life" of an automobile tire is normally distributed with mean 50,000 miles and standard deviation 5000. If the company guarantees its tires for 40,000 miles, what is the probability that a tire will outlast its guarantee?

**44.** SOCIAL SCIENCE: Smoking In a large-scale study carried out in the 1960s among male smokers 35 to 45 years of age, the number of cigarettes smoked daily was approximately normal with mean 28 and standard deviation 10. What proportion had a two-pack-a-day habit (40 or more cigarettes)?

   *Source: National Cancer Institute Monograph No. 19*

**45.** BIOMEDICAL: Weights Heights of women are approximately normally distributed with mean 64.6 inches and standard deviation 2.9. The minimum and maximum heights in order to be an FBI Police Officer (providing security at FBI offices) are 60 and 72 inches. What proportion of women fall in this range?

   *Sources: Federal Bureau of Investigation; G. Salvendy, Handbook of Human Factors and Ergonomics, 3rd ed.*

**46.** GENERAL: SAT Scores Test scores on the 2007 SAT Critical Reasoning Test (Verbal) were approximately normally distributed with mean 502 and standard deviation 113. What score will put you in the top 30%?

   *Source: College Board*

## Cumulative Review for Chapters 1–11

The following exercises review some of the basic concepts and techniques that you learned in Chapters 1–11. Answers to all of these Cumulative Review exercises are given in the answer section at the back of the book. A graphing calculator is suggested but not required.

**1. a.** Find an equation for the horizontal line through the point $(-2, 5)$.
   **b.** Find an equation for the vertical line through the point $(-2, 5)$.

**2.** Simplify: $\left(\frac{1}{4}\right)^{-3/2}$

**3.** Use the definition of the derivative,
$$f'(x) = \lim_{h \to 0} \frac{f(x + h) - f(x)}{h}, \quad \text{to find the deriv-}$$
ative of $f(x) = 3x^2 - 7x + 1$.

**4.** The temperature in an industrial refining tank after $t$ hours is $T(t) = 12\sqrt{t^3} + 225$ degrees Fahrenheit. Find $T'(4)$ and $T''(4)$ and interpret these values.

**5.** Find $\dfrac{d}{dx}\sqrt[3]{x^3 + 8}$.

**6.** Find the derivative of $f(x) = (2x + 3)^3(3x + 2)^4$.

**7.** Make sign diagrams for the first and second derivatives and then graph the function $f(x) = x^3 - 12x^2 - 60x + 400$, showing all relative extreme points and inflection points.

**8.** Make sign diagrams for the first and second derivatives and graph the function $f(x) = \sqrt[3]{x} + 1$ showing all relative extreme points and inflection points.

**9.** Make a sign diagram for the first derivative and graph the function $f(x) = \dfrac{1}{x^2 - 4}$ showing all asymptotes and relative extreme points.

**10.** A gardener wants to enclose two identical rectangular pens along an existing wall. If the side along the wall needs no fence, what is the largest total area that can be enclosed using only 300 feet of fence?

**11.** An open-top box is to be made from a square sheet of metal 12 inches on each side by cutting a square from each corner and folding up the sides. Find the volume of the largest box that can be made in this way.

**12.** For $xy - 3y = 5$, use implicit differentiation to find $\dfrac{dy}{dx}$ at the point $(4, 5)$.

**13.** A circular oil slick is expanding such that its area is growing at the rate of 3 square miles per day. Find how fast the radius is growing at the moment when the radius is 2 miles.

**14.** A sum of $5000 is deposited in a bank account earning 6% interest. Find the value after 4 years if the interest is compounded:
   **a.** semiannually.      **b.** continuously.

**15.** An investment grows by 9% compounded continuously. How long will it take to increase by 75%?

**16.** How much must you deposit now in a bank account paying 8% compounded quarterly to have $100,000 in 25 years?

**17.** Find
   **a.** $\dfrac{d}{dx}\left(\dfrac{e^{2x}}{x}\right)$      **b.** $\dfrac{d}{dx}\ln(x^3 + 1)$

**18.** Consumer demand for a product is $D(p) = 12{,}000e^{-0.04p}$ units, where $p$ is the selling price in dollars. Find the price that maximizes consumer expenditure.
   [*Hint*: Expenditure is $E(p) = p \cdot D(p)$.]

**19.** The gross domestic product of a country after $t$ years is $G(t) = 3 + 5e^{0.06t}$. Find the relative rate of change after 2 years.

**20.** Find
   **a.** $\displaystyle\int (12x^3 - 4x + 5)\,dx$

   **b.** $\displaystyle\int 6e^{-2x}\,dx$

**21.** Evaluate $\displaystyle\int_{1}^{e} \frac{2}{x}\,dx$.

**22.** A company's marginal cost function is $MC(x) = 3x^{-1/2}$ thousand dollars, where $x$ is the number of units. Find the total cost of producing units 25 through 100.

**23.** Radioactive waste is being deposited into a new underground storage vault at the rate of $120e^{0.15t}$ tons per year, where $t$ is the number of years that the depository has been open. Find a formula for the total amount of radioactive waste deposited during the first $t$ years.

**24.** Find the area bounded by $y = x^2 + 4$ and $y = 6x + 4$.

**25.** The population of a city is predicted to be $P(t) = 60e^{0.04t}$ thousand people, where $t$ is the number of years from now. Find the average population over the next 10 years (year 0 to year 10).

**26.** Find

**a.** $\displaystyle\int \frac{x\,dx}{\sqrt{x^2 + 1}}$    **b.** $\displaystyle\int_{0}^{1} xe^{-x^2}\,dx$.

**27.** Find by integration by parts: $\displaystyle\int x^2 \ln x\,dx$.

**28.** Use the integral table on the inside back cover to find $\displaystyle\int \frac{x}{\sqrt{x-1}}\,dx$.

**29.** Evaluate $\displaystyle\int_{0}^{\infty} e^{-2x}\,dx$.

**30.** Use trapezoidal approximation with $n = 4$ trapezoids to approximate $\displaystyle\int_{0}^{1} e^{\sqrt{x}}\,dx$.
(If you are using a graphing calculator, compare your answer with that using FnInt.)

**31.** Use Simpson's Rule with $n = 4$ to approximate $\displaystyle\int_{0}^{1} e^{\sqrt{x}}\,dx$.
(If you are using a graphing calculator, compare your answer with that using FnInt.)

**32.** Find the first partial derivatives of $f(x, y) = xe^{y} - y \ln(x^2 + 1)$.

**33.** Find the extreme values of $f(x, y) = 3x^2 + 2y^2 - 2xy - 8x - 4y + 15$.

**34.** Find the least squares line for the following points:

| $x$ | $y$ |
|---|---|
| 2 | 4 |
| 4 | 3 |
| 6 | 0 |
| 8 | $-1$ |

**35.** Use Lagrange multipliers to find the maximum value of $f(x, y) = 100 - 2x^2 - 3y^2 + 2xy$ subject to the constraint $2x + y = 18$.

**36.** Find the total differential of $f(x, y) = 5x^2 - 2xy + 2y^3 - 12$.

**37.** Find the volume under the surface $f(x, y) = x^2 + y^2$ over the rectangle $R = \{(x, y) \mid 0 \le x \le 3, 0 \le y \le 3\}$.

**38. a.** Convert $135°$ into radians.

**b.** Convert $\dfrac{5\pi}{6}$ into degrees.

**39.** Evaluate *without* using a calculator.

**a.** $\sin \dfrac{3\pi}{4}$    **b.** $\cos \dfrac{3\pi}{4}$

**c.** $\sin \dfrac{5\pi}{6}$    **d.** $\cos \dfrac{5\pi}{6}$.

**40.** Sketch the graph of $f(x) = 2 \cos 4x$.

**41.** Find $\dfrac{d}{dx}(\sin x \cos x)$.

**42.** Find $\displaystyle\int e^{\cos x} \sin x\,dx$

**43.** Find the area under $y = \cos x$ from $x = 0$ to $x = \pi/2$.

**44.** Evaluate *without* using a calculator.

**a.** $\tan \dfrac{5\pi}{6}$    **b.** $\cot \dfrac{5\pi}{6}$

**c.** $\sec \dfrac{3\pi}{4}$    **d.** $\csc \dfrac{3\pi}{4}$.

**45.** Find

**a.** $\dfrac{d}{dx}\tan(x^2 + 1)$

**b.** $\displaystyle\int \sec(2t + 1) \tan(2t + 1)\,dt$

**46. a.** Find the general solution to the differential equation $y' = x^3 y$.

   **b.** Then find the particular solution that satisfies initial $y(0) = 2$.

**47.** Solve the first-order linear differential equation and initial condition $\begin{cases} y' + 6xy = 0 \\ y(0) = 4 \end{cases}$

**48.** Find the Euler approximation on the interval $[0, 1]$ using $n = 4$ segments for the following initial value problem, and draw a graph of your approximation. $\begin{cases} y' = \dfrac{12x}{y} \\ y(0) = 2 \end{cases}$

**49.** Determine whether the geometric series converges, and if so, find its sum:

$$\tfrac{4}{5} + \tfrac{16}{25} + \tfrac{64}{125} + \cdots$$

**50.** Find the third Taylor polynomial at $x = 0$ for $f(x) = e^{2x}$.

**51.** Find the Taylor series at $x = 0$ for $f(x) = \sin x^2$ by modifying a known Taylor series. What is its radius of convergence?

**52.** Use Newton's method to approximate $\sqrt{105}$, continuing until two successive iterations agree to five decimal places.

**53.** A department store receives umbrellas in batches of 120, and, on average, four are defective. Use a Poisson random variable to find the probability that three or fewer are defective.

**54.** If $X$ is a uniform random variable on the interval $[0, 18]$, find:

   **a.** $E(X)$          **b.** $\text{Var}(X)$

**55.** Suppose that the waiting time for an elevator in a tall building is exponentially distributed with mean 1 minute. Find the probability that your wait will be $1\tfrac{1}{2}$ minutes or less.

**56.** If a random variable $X$ is normally distributed with mean 12 and standard deviation 2, find $P(11 \leq X \leq 13)$.

# Appendix

## NORMAL PROBABILITIES FROM TABLES

### Introduction

This Appendix is for readers of Section 11.4 who do not use graphing calculators. You should read Section 11.4 through "Normal Random Variables," which ends on page 783, then read this Appendix, and then return to page 786, just after Example 4.

### Standard Normal Random Variables

A normal random variable with mean $\mu = 0$ and standard deviation $\sigma = 1$ is called a *standard* normal random variable. Its probability density function

$$f(x) = \frac{1}{\sqrt{2\pi}} e^{-\frac{1}{2}x^2} \qquad \text{for } -\infty < x < \infty$$

Probability density function for a standard normal random variable

was graphed on page 782. Throughout this section, the letter $Z$ will always stand for a *standard normal random variable.* The probability that a standard normal random variable $Z$ lies between two numbers is the area under the probability density function between the two numbers.

---

**Probabilities for Standard Normal Random Variables**

For a standard normal random variable $Z$,

$$P(a \le Z \le b) = \int_a^b \frac{1}{\sqrt{2\pi}} e^{-\frac{1}{2}x^2}\, dx$$

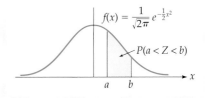

---

The integral of the normal probability density function cannot be expressed in terms of elementary functions, so we cannot find the area by integration. However, we can *approximate* the area by *numerical* integration as in Section 6.4, and then list the results in a table. One such table, on page A7 at the end of this Appendix, gives the areas under this curve from 0 to any positive number $x$. The first two examples show how to use the table.

**EXAMPLE 1**

**FINDING A PROBABILITY FOR A STANDARD NORMAL RANDOM VARIABLE**

Find $P(0 \leq Z \leq 1.24)$.

*Solution*

The table on page A7 gives the probability that a standard normal random variable $Z$ is between 0 and any higher number. For the given number 1.24, we find the *row* headed 1.2 and the *column* headed .04 (the second decimal place), which intersect at the table value of .3925:

From the table at the end of the Appendix: the row headed 1.2 and the column headed .04 lead to the table value .3925

Therefore:

$$P(0 \leq Z \leq 1.24) = .3925 \qquad \text{Also written } 0.3925$$

row ⌐ column ⌐ table

**Practice Problem 1**

Find $P(0 \leq Z \leq 0.52)$                                     ➤ **Solution on page A6**

We will use the notation

$$A(x) = \begin{pmatrix} \text{Area under the standard} \\ \text{normal probability density} \\ \text{function from 0 to } x \end{pmatrix}$$

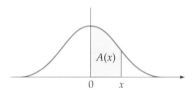

That is, $A(x)$ is the number found in the standard normal table in the row and column determined by $x$. For instance, in Example 1 we found $A(1.24) = 0.3925$, and in Practice Problem 1 you found $A(0.52) = 0.1985$.

The probabilitiy that $Z$ is between two numbers can be found by adding and subtracting values from the table, using the symmetry of the curve about $x = 0$ and the fact that the area under each half of the curve is 0.5 (since the total area is 1). The following example illustrates the technique, with the calculation shown graphically as well.

**EXAMPLE 2**

**FINDING PROBABILITIES FROM A TABLE**

Find:  **a.** $P(-0.75 \leq Z \leq 0)$     **b.** $P(Z \leq 0.65)$

 **c.** $P(-1.2 \leq Z \leq 0.68)$     **d.** $P(0.65 \leq Z \leq 1.45)$

 **e.** $P(Z > 0.95)$

*Solution*

**a.** $P(-0.75 \leq Z \leq 0)$ $= A(0.75) = 0.2734$

From the table on page A7

 =

Using the symmetry of the curve about $x = 0$

**b.** $P(Z \leq 0.65)$ $= \frac{1}{2} + A(0.65) = 0.5 + 0.2422 = 0.7422$

= 

Using the fact that the area to the left of $x = 0$ is $\frac{1}{2}$

**c.** $P(-1.2 \leq Z \leq 0.68)$ $= A(1.2) + A(0.68) = 0.3849 + 0.2518 = 0.6367$

 =

Using symmetry

**d.** $P(0.65 \leq Z \leq 1.45)$ $= A(1.45) - A(0.65) = 0.4265 - 0.2422 = 0.1843$

 found from

Area up to 1.45 minus area up to 0.65

**e.** $P(Z > 0.95)$ $= \frac{1}{2} - A(0.95) = 0.5 - 0.3289 = 0.1711$

 found from

Area beyond 0.95 is $\frac{1}{2}$ minus area from 0 to 0.95

**Practice Problem 2**

Find:  **a.** $P(0.5 \leq Z \leq 1.25)$

 **b.** $P(-2.1 \leq Z \leq 0.75)$     ➤ Solutions on page A6

### Probabilities for Nonstandard Random Variables

How do we find probabilities for a *non*standard normal random variable with mean $\mu$ and standard deviation $\sigma$? Instead of developing tables for all possible means and standard deviations, we use a process called *standardization* that changes a nonstandard random variable into one with mean 0 and standard deviation 1, so we can use the standard normal table. Standardizing is achieved by defining a new random variable by subtracting the mean $\mu$ and dividing by the standard deviation $\sigma$.

---

**Standardizing Normal Random Variables**

A normal random variable $X$ with mean $\mu$ and standard deviation $\sigma$ is *standardized* by defining

$$Z = \frac{X - \mu}{\sigma}$$

Subtracting the mean and dividing by the standard deviation

$Z$ is then a normal random variable with mean 0 and standard deviation 1.

---

The values of a normal random variable are said to be *normally distributed.*

**EXAMPLE 3**

### STANDARDIZING TO FIND A PROBABILITY

The heights of American men are approximately normally distributed, with mean $\mu = 70$ inches and standard deviation $\sigma = 3.1$ inches. Find the proportion of American men who are less than 6 feet tall.

*Source:* G. Salvendy, *Handbook of Human Factors and Ergonomics,* 3rd ed.

*Solution*

If $X =$ height, we want $P(X < 72)$
Standardizing:

72 inches = 6 feet

$$X < 72 \quad \text{is equivalent to} \quad \underbrace{\frac{X - \mu}{\sigma}}_{Z} < \underbrace{\frac{72 - 70}{3.1}}_{0.65}$$

Subtracting mean $\mu = 70$ and dividing by $\sigma = 3.1$

Therefore:

$$P(\underbrace{X < 72}_{\text{Normal}}) = P(\underbrace{Z < 0.65}_{\substack{\text{Standard} \\ \text{normal}}}) = 0.7422$$

$P(Z < 0.65) = 0.7422$ from Example 2b

That is, approximately 74% of American men are under 6 feet tall.

Such calculations are used for designing automobiles and other consumer products.

**Practice Problem 3**

The weights of American women are approximately normally distributed, with mean 151.8 pounds and standard deviation 38.8 pounds. Find the proportion of women who weigh more than 140 pounds.

*Source:* G. Salvendy, *Handbook of Human Factors and Ergonomics,* 3rd ed.

➤ **Solution on page A6**

**EXAMPLE 4**

### FINDING THE PROBABILITY OF ACCEPTABLE CHOLESTEROL LEVELS

Cholesterol is a fat-like substance that, in the bloodstream, can increase the risk of heart disease. Cholesterol levels below 200 (milligrams per deciliter of blood) are generally considered acceptable. A survey by the U.S. Department of Health found that cholesterol levels for one age group were approximately normally distributed with mean $\mu = 222$ and standard deviation $\sigma = 28$. What proportion of people in the study had acceptable cholesterol levels?

*Source:* Rogosin Institute

*Solution*

Letting $X = \left(\begin{array}{c}\text{Cholesterol}\\\text{level}\end{array}\right),$ we want $P(X \le 200).$ The inequality

$$X \le 200$$

with $X$ standardized becomes

$$\underbrace{\frac{X - \mu}{\sigma}}_{Z} \le \underbrace{\frac{200 - 222}{28}}_{-0.79} \qquad \text{Using } \mu = 222 \text{ and } \sigma = 28$$

Therefore:

$$P(X \le 200) = P(Z \le -0.79) = \tfrac{1}{2} - A(0.79) = 0.5000 - 0.2852 = 0.2148$$

$\underbrace{\qquad}_{\substack{\text{Normal}}}$ $\underbrace{\qquad}_{\substack{\text{Standard}\\\text{normal}}}$ $\underbrace{\qquad}_{\substack{\text{Using the}\\\text{graph below,}\\\text{as in}\\\text{Example 2}}}$ $\underbrace{\qquad}_{\substack{\text{From the}\\\text{standard}\\\text{normal}\\\text{table}}}$ $\underbrace{\qquad}_{\text{Answer}}$

Therefore, only about 21% of the people in the study had acceptable cholesterol levels.

### Inverse Probabilities for Normal Random Variables

The standard normal table can also be used "in reverse" to find the values that result in a given probability.

**EXAMPLE 5**

**FINDING THE HEIGHT FOR A GIVEN PROBABILITY**

Women's heights are approximately normally distributed, with mean $\mu = 64.6$ inches and standard deviation $\sigma = 2.9$ inches. Find the height that marks the tallest 10%.

*Source:* G. Salvendy, *Handbook of Human Factors and Ergonomics,* 3rd ed.

*Solution*

We begin by finding the value of the *standard* normal random variable $Z$ that marks the top $10\% = 0.10$ (or, equivalently, the number $z$ with area 0.40 between 0 and $z$, as shown on the right).

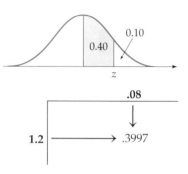

To find this $z$-value, we look for .40 *inside* the standard normal table. The value that comes closest to .40 is .3997, and the "outside" value that gives it is 1.28. Therefore:

$$z = 1.28$$

*z*-value of the top 10% (using *z* to represent a value of the random variable *Z*)

$$\frac{x - 64.6}{2.9} = 1.28$$

Replacing *z* by $\dfrac{x - \mu}{\sigma}$ with the given $\mu$ and $\sigma$

$$x - 64.6 = 3.712$$

Multiplying by 2.9

$$x = 68.312$$

Adding 64.6

Therefore, a height of about 68.3 inches, or 5 feet 8 inches, marks the tallest 10% of women.

You should now return to Section 11.4, just after Example 4 on page 786.

**Solutions to Practice Problems**

**1.** $P(0 \leq Z \leq 0.52) = 0.1985$

**2. a.** $P(0.5 \leq Z \leq 1.25) = A(1.25) - A(0.5) = 0.3944 - 0.1915 = 0.2029$

    **b.** $P(-2.1 \leq Z \leq 0.75) = A(2.1) + A(0.75) = 0.4821 + 0.2734 = 0.7555$

**3.** $P(X > 140) = P\left(\dfrac{X - \mu}{\sigma} > \dfrac{140 - 151.8}{38.8}\right) = P(Z > -0.30) = \frac{1}{2} + A(0.30)$

    $= 0.5 + 0.1179 = 0.6179$

About 62% of women weigh more than 140 pounds.

Area Under the Standard
Normal Density from 0 to $x$

| x | .00 | .01 | .02 | .03 | .04 | .05 | .06 | .07 | .08 | .09 |
|---|---|---|---|---|---|---|---|---|---|---|
| **0.0** | .0000 | .0040 | .0080 | .0120 | .0160 | .0199 | .0239 | .0279 | .0319 | .0359 |
| **0.1** | .0398 | .0438 | .0478 | .0517 | .0557 | .0596 | .0636 | .0675 | .0714 | .0754 |
| **0.2** | .0793 | .0832 | .0871 | .0910 | .0948 | .0987 | .1026 | .1064 | .1103 | .1141 |
| **0.3** | .1179 | .1217 | .1255 | .1293 | .1331 | .1368 | .1406 | .1443 | .1480 | .1517 |
| **0.4** | .1554 | .1591 | .1628 | .1664 | .1700 | .1736 | .1772 | .1808 | .1844 | .1879 |
| **0.5** | .1915 | .1950 | .1985 | .2019 | .2054 | .2088 | .2123 | .2157 | .2190 | .2224 |
| **0.6** | .2258 | .2291 | .2324 | .2357 | .2389 | .2422 | .2454 | .2486 | .2518 | .2549 |
| **0.7** | .2580 | .2612 | .2642 | .2673 | .2704 | .2734 | .2764 | .2794 | .2823 | .2852 |
| **0.8** | .2881 | .2910 | .2939 | .2967 | .2996 | .3023 | .3051 | .3078 | .3106 | .3133 |
| **0.9** | .3159 | .3186 | .3212 | .3238 | .3264 | .3289 | .3315 | .3340 | .3365 | .3389 |
| **1.0** | .3413 | .3438 | .3461 | .3485 | .3508 | .3531 | .3554 | .3577 | .3599 | .3621 |
| **1.1** | .3643 | .3665 | .3686 | .3708 | .3729 | .3749 | .3770 | .3790 | .3810 | .3820 |
| **1.2** | .3849 | .3869 | .3888 | .3907 | .3925 | .3944 | .3962 | .3980 | .3997 | .4015 |
| **1.3** | .4032 | .4049 | .4066 | .4082 | .4099 | .4115 | .4131 | .4147 | .4162 | .4177 |
| **1.4** | .4192 | .4207 | .4222 | .4236 | .4251 | .4265 | .4279 | .4292 | .4306 | .4319 |
| **1.5** | .4332 | .4345 | .4357 | .4370 | .4382 | .4394 | .4406 | .4418 | .4429 | .4441 |
| **1.6** | .4452 | .4463 | .4474 | .4484 | .4495 | .4505 | .4515 | .4525 | .4535 | .4545 |
| **1.7** | .4554 | .4564 | .4573 | .4582 | .4591 | .4599 | .4608 | .4616 | .4625 | .4633 |
| **1.8** | .4641 | .4649 | .4656 | .4664 | .4671 | .4678 | .4686 | .4693 | .4699 | .4706 |
| **1.9** | .4713 | .4719 | .4726 | .4732 | .4738 | .4744 | .4750 | .4756 | .4761 | .4767 |
| **2.0** | .4772 | .4778 | .4783 | .4788 | .4793 | .4798 | .4803 | .4808 | .4812 | .4817 |
| **2.1** | .4821 | .4826 | .4830 | .4834 | .4838 | .4842 | .4846 | .4850 | .4854 | .4857 |
| **2.2** | .4861 | .4864 | .4868 | .4871 | .4875 | .4878 | .4881 | .4884 | .4887 | .4890 |
| **2.3** | .4893 | .4896 | .4898 | .4901 | .4904 | .4906 | .4909 | .4911 | .4913 | .4916 |
| **2.4** | .4918 | .4920 | .4922 | .4925 | .4927 | .4929 | .4931 | .4932 | .4934 | .4936 |
| **2.5** | .4938 | .4940 | .4941 | .4943 | .4945 | .4946 | .4948 | .4949 | .4951 | .4952 |
| **2.6** | .4953 | .4955 | .4956 | .4957 | .4959 | .4960 | .4961 | .4962 | .4963 | .4964 |
| **2.7** | .4965 | .4966 | .4967 | .4968 | .4969 | .4970 | .4971 | .4972 | .4973 | .4974 |
| **2.8** | .4974 | .4975 | .4976 | .4977 | .4977 | .4978 | .4979 | .4979 | .4980 | .4981 |
| **2.9** | .4981 | .4982 | .4982 | .4983 | .4984 | .4984 | .4985 | .4985 | .4986 | .4986 |
| **3.0** | .4987 | .4987 | .4987 | .4988 | .4988 | .4989 | .4989 | .4989 | .4990 | .4990 |
| **3.1** | .4990 | .4991 | .4991 | .4991 | .4992 | .4992 | .4992 | .4992 | .4993 | .4993 |
| **3.2** | .4993 | .4993 | .4994 | .4994 | .4994 | .4994 | .4994 | .4995 | .4995 | .4995 |
| **3.3** | .4995 | .4995 | .4995 | .4996 | .4996 | .4996 | .4996 | .4996 | .4996 | .4997 |
| **3.4** | .4997 | .4997 | .4997 | .4997 | .4997 | .4997 | .4997 | .4997 | .4997 | .4998 |
| **3.5** | .4998 | .4998 | .4998 | .4998 | .4998 | .4998 | .4998 | .4998 | .4998 | .4998 |

# Answers to Selected Exercises

## Exercises 1.1 page 16

**1.** $\{x \mid 0 \leq x < 6\}$    **3.** $\{x \mid x \leq 2\}$   **5. a.** Increase by 15 units

**b.** Decrease by 10 units   **7.** $m = -2$   **9.** $m = \frac{1}{3}$   **11.** $m = 0$   **13.** Slope is undefined.

**15.** $m = 3$, $(0, -4)$   **17.** $m = -\frac{1}{2}$, $(0, 0)$   **19.** $m = 0$, $(0, 4)$

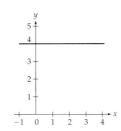

**21.** Slope and $y$-intercept do not exist.   **23.** $m = \frac{2}{3}$, $(0, -4)$   **25.** $m = -1$, $(0, 0)$

**27.** $m = 1$, $(0, 0)$   **29.** $m = \frac{1}{3}$, $\left(0, \frac{2}{3}\right)$   **31.** $m = \frac{2}{3}$, $(0, -1)$

**33.** $y = -2.25x + 3$   **35.** $y = 5x + 3$   **37.** $y = -4$   **39.** $x = 1.5$   **41.** $y = -2x + 13$

**43.** $y = -1$   **45.** $y = -2x + 1$   **47.** $y = \frac{3}{2}x - 2$

**49.** $y = -x + 5$, $y = -x - 5$, $y = x + 5$, $y = x - 5$

**51.** Substituting $(0, b)$ into $y - y_1 = m(x - x_1)$ gives $y - b = m(x - 0)$, or $y = mx + b$.

**53. a.**  **b.**

on $[-5, 5]$ by $[-5, 5]$    on $[-5, 5]$ by $[-5, 5]$

**55.** Low: $[0, 8)$; average: $[8, 20)$; high: $[20, 40)$; critical $[40, \infty)$

**57. a.** 3 minutes 38.28 seconds    **b.** In about 2033

**59. a.** $y = 4.1x + 51.1$    **b.** Sales are increasing by 4.1 million units per year.    **c.** 112.6 million units

**61. a.** $y = \frac{9}{5}x + 32$    **b.** 68°

**63. a.** $V = 50{,}000 - 2200t$    **65. a.**     **67. a.**

**b.** \$39,000

**c.**

on $[0, 20]$ by $[0, 50{,}000]$

**b.** Men: 28.04 years; women: 26.88 years    **b.** 28%    **c.** 18%

**c.** Men: 28.62 years; women: 27.84 years

**69. a.** $y = 45.5x + 661.9$    **b.** Revenues are increasing by \$45.5 million per year.    **c.** \$1162 million

**71. a.** $y = 2.03x + 63.9$    **b.** The male life expectancy is increasing by 2.03 years per decade, which is 0.203 year (or about 2.4 months) per year.    **c.** 77.1 years

**73. a.** $y = -0.848x + 73.65$    **b.** Future longevity decreases by 0.848 year (or about 10.2 months) per year.    **c.** 52.5 years    **d.** It would not make sense to use the regression line to predict future longevity at age 90 because the line predicts $-2.7$ years of life remaining.    **75.** False. Infinity is not a number.

**77.** $m = \dfrac{y_2 - y_1}{x_2 - x_1}$ for any two points $(x_1, y_1)$ and $(x_2, y_2)$ on the line   *or*   $m = \left(\begin{array}{c}\text{Amount that the line rises} \\ \text{when } x \text{ increases by 1}\end{array}\right)$

**79.** False. For a vertical line the slope is *undefined.*    **81.** True    **83.** $\frac{4}{3}$, or $-\frac{4}{3}$ if the ladder slopes downward

**85.** $(-b/m, 0)$, $m \neq 0$    **87.** Smaller populations increase toward the carrying capacity

## Exercises 1.2 page 29

**1.** 64    **3.** $\frac{1}{16}$    **5.** 8    **7.** $\frac{8}{5}$    **9.** $\frac{1}{32}$    **11.** $\frac{8}{27}$    **13.** 1    **15.** $\frac{4}{9}$    **17.** 5    **19.** 125    **21.** 8

**23.** 4    **25.** $-32$    **27.** $\frac{125}{216}$    **29.** $\frac{9}{25}$    **31.** $\frac{1}{4}$    **33.** $\frac{1}{2}$    **35.** $\frac{1}{8}$    **37.** $\frac{1}{4}$    **39.** $-\frac{1}{2}$    **41.** $\frac{1}{4}$    **43.** $\frac{4}{5}$

**45.** $\frac{64}{125}$    **47.** $-243$    **49.** 2.14    **51.** 274.37    **53.** 0.977 (rounded)    **55.** 2.720 (rounded)

**57.** $4x^{-5}$    **59.** $2x^{-4/3}$    **61.** $3x^{-3/2}$    **63.** $3x^{-2}$    **65.** $x^{10}$    **67.** $z^{27}$    **69.** $x^8$    **71.** $w^5$    **73.** $y^5/x$

**75.** $27y^4$    **77.** $u^2v^2w^2$    **79.** 25.6 feet    **81.** Costs will be multiplied by 2.3.    **83.** 125 beats per minute

**85.** About 42.6 thousand work-hours, or 42,600 work-hours, rounded to the nearest hundred hours.

**87. a.** About 32 times more ground motion    **b.** About 12.6 times more ground motion

**89.** About 312 mph

**91.**     $x \approx 18.2$.    Therefore, the land area must be increased by a factor of more than 18 to double the number of species.

on $[0, 100]$ by $[0, 4]$

**93.** 60 mph    **95. a.** $y = 11.43x^{2.369}$    **b.** 4120    **97. a.** $y = 632.5x^{0.5040}$    **b.** 2118

**99.** 3, since $\sqrt{9}$ means the *principal* square root. (To get $\pm 3$ you would have to write $\pm \sqrt{9}$.)

**101.** False.  $\dfrac{2^6}{2^2} = \dfrac{64}{4} = 16,$ while $2^{6/2} = 2^3 = 8.$ (The correct statement is $\dfrac{x^m}{x^n} = x^{m-n}.$)

**103.** All nonnegative values of $x$.     **105.** All values of $x$ except 0.

## Exercises 1.3 page 45

**1.** Yes     **3.** No     **5.** No     **7.** No     **9.** Domain: $\{ x \mid x \le 0 \text{ or } x \ge 1 \}$; Range: $\{ y \mid y \ge -1 \}$

**11. a.** $f(10) = 3$     **b.** $\{ x \mid x \ge 1 \}$     **c.** $\{ y \mid y \ge 0 \}$     **13. a.** $h(-5) = -1$     **b.** $\{ z \mid z \ne -4 \}$     **c.** $\{ y \mid y \ne 0 \}$

**15. a.** $h(81) = 3$     **b.** $\{ x \mid x \ge 0 \}$     **c.** $\{ y \mid y \ge 0 \}$     **17. a.** $f(-8) = 4$     **b.** $\mathbb{R}$     **c.** $\{ y \mid y \ge 0 \}$

**19. a.** $f(0) = 2$     **b.** $\{ x \mid -2 \le x \le 2 \}$     **c.** $\{ y \mid 0 \le y \le 2 \}$

**21. a.** $f(-25) = 5$     **b.** $\{ x \mid x \le 0 \}$     **c.** $\{ y \mid y \ge 0 \}$

**23.**      **25.**      **27.**      **29.**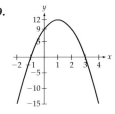

**31. a.** (20, 100)     **33. a.** $(-40, -200)$     **35.** $x = 7,\ x = -1$     **37.** $x = 3,\ x = -5$

**b.**      **b.**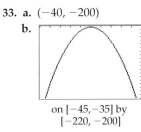

on [15, 25] by [100, 120]

on $[-45, -35]$ by $[-220, -200]$

**39.** $x = 4,\ x = 5$     **41.** $x = 0,\ x = 10$

**43.** $x = 5,\ x = -5$     **45.** $x = -3$

**47.** $x = 1,\ x = 2$     **49.** No solutions

**51.** No solutions     **53.** $x = -4,\ x = 5$

**55.** $x = 4,\ x = 5$     **57.** $x = -3$

**59.** No (real) solutions

**61.** $x = 1.14,\ x = -2.64$     **63. a.** The slopes are all 2, but the $y$ intercepts differ     **b.** $y = 2x - 8$

**65.** $C(x) = 4x + 20$     **67.** $P(x) = 15x + 500$     **69. a.** 17.7 lbs/in.$^2$     **b.** 15,765 lbs/in.$^2$     **71.** 132 feet

**73. a.** 400 cells     **b.** 5200 cells     **75.** About 208 mph     **77.** 2.92 seconds

**79. a.** Break even at 40 units and at 200 units.     **b.** Profit maximized at 120 units. Max profit is $12,800 per week.

**81. a.** Break even at 20 units and at 80 units.     **b.** Profit maximized at 50 units. Max profit is $1800 per day.

**83.** $v = \dfrac{c}{w + a} - b$

**85. a.**

**87. a.** $f(x) = 100x - x^2$ or $-x^2 + 100x$     **b.** $50

**89. a.** $y = 0.514x^2 - 4.5x + 23.2$     **b.** 43.3

**91. a.** $y = 0.51x^2 - 3.70x + 11.96$

**b.** 13.7. Since "1 in 8" is 12.5%, this prediction is fairly close.

**b.** About 40% (from 39.8)

**c.** About 60% (from 60.4)

**93.** Yes—many parabolas cross the $x$-axis twice. No—that would violate the vertical line test.

**95.** $f(5) = 9$ (halfway between 7 and 11)     **97.** $x$ is in prendles and $y$ is in blargs.

**99.** No—that would violate the vertical line test. [*Note:* A *parabola* is a geometric shape and so *may* open sideways, but a quadratic function, being a *function*, must pass the vertical line test.]

## Exercises 1.4 page 65

**1.** Domain: $\{\,x \mid x > 0 \text{ or } x < -4\,\}$; Range: $\{\,y \mid y > 0 \text{ or } y < -2\,\}$  **3. a.** $f(-3) = 1$  **b.** $\{\,x \mid x \neq -4\,\}$
**c.** $\{\,y \mid y \neq 0\,\}$  **5. a.** $f(-1) = -\frac{1}{2}$  **b.** $\{\,x \mid x \neq 1\,\}$  **c.** $\{\,y \mid y \leq 0 \text{ or } y \geq 4\,\}$  **7. a.** $f(2) = 1$
**b.** $\{\,x \mid x \neq 0, x \neq -4\,\}$  **c.** $\{\,y \mid y > 0 \text{ or } y \leq -3\,\}$  **9. a.** $g\left(-\frac{1}{2}\right) = \frac{1}{2}$  **b.** $\mathbb{R}$  **c.** $\{\,y \mid y > 0\,\}$
**11.** $x = 0$, $x = -3$, $x = 1$  **13.** $x = 0$, $x = 2$, $x = -2$  **15.** $x = 0$, $x = 3$  **17.** $x = 0$, $x = 5$
**19.** $x = 0$, $x = 3$  **21.** $x = 0$, $x = 1$  **23.** $x \approx -1.79$, $x = 0$, $x \approx 2.79$

**25.**   **27.**   **29.**   **31.**
Includes $(6, -5)$
Excludes $(6, 0)$

**33.** Polynomial  **35.** Exponential  **37.** Polynomial  **39.** Rational  **41.** Piecewise linear
**43.** Polynomial  **45.** None (not a polynomial because of the fractional exponent)
**47. a.** $y_4$  **b.** $y_1$  **49. a.** $(7x - 1)^5$  **b.** $7x^5 - 1$  **c.** $x^{25}$  **51. a.** $\dfrac{1}{x^2 + 1}$  **b.** $\left(\dfrac{1}{x}\right)^2 + 1$  **c.** $x$
**c.**
on $[-3, 3]$ by $[0, 5]$

**53. a.** $\sqrt{x^3 - x^2} - 1$  **b.** $(\sqrt{x} - 1)^3 - (\sqrt{x} - 1)^2$  **c.** $\sqrt{\sqrt{x} - 1} - 1$

**55. a.** $\left(\dfrac{x^3 - 1}{x^3 + 1}\right)^2 - \dfrac{x^3 - 1}{x^3 + 1}$  **b.** $\dfrac{(x^2 - x)^3 - 1}{(x^2 - x)^3 + 1}$  **c.** $(x^2 - x)^2 - (x^2 - x)$

**d.** $(0, 1)$, because $a^0 = 1$ for any constant $a \neq 0$.

**57. a.** $f(g(x)) = x$  **b.** $g(f(x)) = x$  **59.** $5x^2 + 10xh + 5h^2$ or $5(x^2 + 2xh + h^2)$
**61.** $2x^2 + 4xh + 2h^2 - 5x - 5h + 1$  **63.** $10x + 5h$ or $5(2x + h)$  **65.** $4x + 2h - 5$  **67.** $14x + 7h - 3$
**69.** $3x^2 + 3xh + h^2$  **71.** $\dfrac{-2}{(x + h)x}$  **73.** $\dfrac{-2x - h}{x^2(x + h)^2}$ or $\dfrac{-2x - h}{x^2(x^2 + 2xh + h^2)}$ or $\dfrac{-2x - h}{x^4 + 2x^3h + x^2h^2}$
**75. a.** 2.70481  **b.** 2.71815  **c.** 2.71828  **d.** Yes, 2.71828
**77.** Shifted left 3 units and up 6 units  **79.** About 680 million people
**81. a.** $300  **b.** $500  **c.** $2000

on $[-10, 10]$ by $[-10, 10]$

**d.**

**83. a.** $f\left(\frac{2}{3}\right) = 7$,  $f\left(1\frac{1}{3}\right) = 14$,  $f(4) = 29$,  and  $f(10) = 53$

**b.**

**85.** $R(v(t)) = 2(60 + 3t)^{0.3}$,  $R(v(10)) \approx 7.714$ million dollars
**87. a.** About 1 million cells  **b.** No
**89. a.** $y = 374.6 \cdot 1.477^x$  **b.** 18547
**91. a.** $y = 1753.7 \cdot 3.1948^x$  **b.** 583,696
**93.** One will have "missing points" at the excluded $x$-values.

**95.** A slope of 1 corresponds to a tax rate of 100%, which means that every dollar earned would be paid as the tax.
**97.** $f(f(x)) = x + 2a$  **99.** $f(x + 10)$ is shifted 10 units to the *left*.

**101.** False. $f(x + h) = (x + h)^2 = x^2 + 2xh + h^2$

**103. a.**  Note that each line segment in this graph includes its left endpoint but excludes its right endpoint, and so should be drawn like •——∘ .

$y = \text{INT}(x)$ on
$[-5, 5]$ by $[-5, 5]$

**b.** Domain: $\mathbb{R}$; range: $\{ \dots, -3, -2, -1, 0, 1, 2, 3, \dots \}$—that is, the set of all integers

**105. a.** $f(g(x)) = acx + ad + b$    **b.** Yes

## Chapter 1 Review Exercises page 71

**1.** $\{ x \mid 2 < x \le 5 \}$ ∘———• 2   5

**2.** $\{ x \mid -2 \le x < 0 \}$ •———∘ -2   0

**3.** $\{ x \mid x \ge 100 \}$ ———• 100

**4.** $\{ x \mid x \le 6 \}$ •——— 6

**5.** Hurricane: $[74, \infty)$; storm: $[55, 74)$; gale: $[38, 55)$; small craft warning: $[21, 38)$

**6. a.** $(0, \infty)$    **b.** $(-\infty, 0)$    **c.** $[0, \infty)$    **d.** $(-\infty, 0]$    **7.** $y = 2x - 5$    **8.** $y = -3x + 3$    **9.** $x = 2$

**10.** $y = 3$    **11.** $y = -2x + 1$    **12.** $y = 3x - 5$    **13.** $y = 2x - 1$    **14.** $y = -\frac{1}{2}x + 1$

**15. a.** $V = 25{,}000 - 3000t$    **b.** \$13,000    **16. a.** $V = 78{,}000 - 5000t$    **b.** \$38,000

**17. a.** $y = 237x - 101$    **b.** The cost is increasing by \$237 every 2 years, so the yearly increase is \$118.50.
    **c.** \$491.50    **d.** \$1795    **e.** \$73,725 in 2001 and \$269,250 in 2012

**18.** 36    **19.** $\frac{3}{4}$    **20.** 8    **21.** 10    **22.** $\frac{1}{27}$    **23.** $\frac{1}{1000}$    **24.** $\frac{9}{4}$    **25.** $\frac{64}{27}$    **26.** 13.97    **27.** 112.32

**28. a.** About 42 lbs    **b.** About 1630 lbs    **29. a.** About 127 lbs    **b.** About 7185 lbs

**30. a.** $y = 11.64x^{2.797}$    **b.** 7297.3

**31. a.** $f(11) = 2$    **b.** $\{ x \mid x \ge 7 \}$    **c.** $\{ y \mid y \ge 0 \}$    **32. a.** $g(-1) = \frac{1}{2}$    **b.** $\{ t \mid t \ne -3 \}$    **c.** $\{ y \mid y \ne 0 \}$

**33. a.** $h(16) = \frac{1}{8}$    **b.** $\{ w \mid w > 0 \}$    **c.** $\{ y \mid y > 0 \}$    **34. a.** $w(8) = \frac{1}{16}$    **b.** $\{ z \mid z \ne 0 \}$    **c.** $\{ y \mid y > 0 \}$

**35.** Yes    **36.** No

**37.**     **38.**     **39.**     **40.**

**41.** $x = 0$, $x = -3$    **42.** $x = 5$, $x = -1$    **43.** $x = -2$, $x = 1$    **44.** $x = 1$, $x = -1$

**45. a.** Vertex $(5, -50)$    **46. a.** Vertex: $(-7, -64)$    **47.** $C(x) = 45 + 0.25x$

**b.**     **b.** 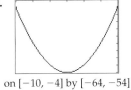    **48.** $I(t) = 800t$

on $[2, 8]$ by $[-50, -40]$    on $[-10, -4]$ by $[-64, -54]$

**49.** $T(x) = 70 - \dfrac{x}{300}$

**50.** $C(t) = 27 + 0.58t$; in about 5 years from 2000, so in 2005.

**51. a.** Break even at 15 and 65 units    **b.** Profit maximized at 40 units. Max profit is \$1250 per week.

**52. a.** Break even at 150 and 450 units.    **b.** Profit maximized at 300 units. Max profit is \$67,500 per month.

**53. a.** $y = 1.39x^2 - 9.11x + 66.1$    **b.** 156.3

**54. a.** $f(-1) = 1$  **b.** $\{ x \mid x \neq 0 \text{ and } x \neq 2 \}$  **c.** $\{ y \mid y > 0 \text{ or } y \leq -3 \}$

**55. a.** $f(-8) = \frac{1}{2}$  **b.** $\{ x \mid x \neq 0 \text{ and } x \neq -4 \}$  **c.** $\{ y \mid y > 0 \text{ or } y \leq -4 \}$

**56. a.** $g\left(\frac{3}{2}\right) = 27$  **b.** $\mathbb{R}$  **c.** $\{ y \mid y > 0 \}$  **57. a.** $g\left(\frac{5}{3}\right) = 32$  **b.** $\mathbb{R}$  **c.** $\{ y \mid y > 0 \}$

**58.** $x = 0, \quad x = 1, \quad x = -3$  **59.** $x = 0, \quad x = 2, \quad x = -4$  **60.** $x = 0, \quad x = 5$  **61.** $x = 0, \quad x = 2$

**62.**   **63.**   **64.**   **65.**

**66. a.** $f(g(x)) = \left(\frac{1}{x}\right)^2 + 1 = \frac{1}{x^2} + 1$  **b.** $g(f(x)) = \frac{1}{x^2 + 1}$

**67. a.** $f(g(x)) = \sqrt{5x - 4}$  **b.** $g(f(x)) = 5\sqrt{x} - 4$  **68. a.** $f(g(x)) = \frac{x^3 + 1}{x^3 - 1}$  **b.** $g(f(x)) = \left(\frac{x + 1}{x - 1}\right)^3$

**69. a.** $f(g(x)) = 2^{x^2}$  **b.** $g(f(x)) = (2^x)^2 = 2^{2x}$  **70.** $4x + 2h - 3$  **71.** $\dfrac{-5}{(x + h)x}$

**72.** $A(p(t)) = 2(18 + 2t)^{0.15}, \quad A(p(4)) \approx \$3.26 \text{ million}$

**73. a.** $x = -1, \quad x = 0, \quad x = 3$  **74. a.** $x = -3, \quad x = 0, \quad x = 1$  **75. a.** $y = 2625 \cdot 1.329^x$
**b.**   **b.** **b.** \$10,874

on $[-5, 5]$ by $[-5, 5]$    on $[-5, 5]$ by $[-5, 5]$

## Exercises 2.1 page 90

**1.**

| $x$ | $5x - 7$ | $x$ | $5x - 7$ |
|---|---|---|---|
| 1.9 | 2.5 | 2.1 | 3.5 |
| 1.99 | 2.95 | 2.01 | 3.05 |
| 1.999 | 2.995 | 2.001 | 3.005 |

**a.** $\lim\limits_{x \to 2^-} (5x - 7) = 3$  **b.** $\lim\limits_{x \to 2^+} (5x - 7) = 3$  **c.** $\lim\limits_{x \to 2} (5x - 7) = 3$

**3.**

| $x$ | $\dfrac{x^3 - 1}{x - 1}$ | $x$ | $\dfrac{x^3 - 1}{x - 1}$ |
|---|---|---|---|
| 0.9 | 2.71 | 1.1 | 3.31 |
| 0.99 | 2.97 | 1.01 | 3.03 |
| 0.999 | 2.997 | 1.001 | 3.003 |

**a.** $\lim\limits_{x \to 1^-} \left(\dfrac{x^3 - 1}{x - 1}\right) = 3$  **b.** $\lim\limits_{x \to 1^+} \left(\dfrac{x^3 - 1}{x - 1}\right) = 3$  **c.** $\lim\limits_{x \to 1} \left(\dfrac{x^3 - 1}{x - 1}\right) = 3$

**5.** 7.389 (rounded)  **7.** $-0.25$  **9.** 1  **11.** 2  **13.** 8  **15.** 2  **17.** $\sqrt{2}$  **19.** 6  **21.** $5x^3$

**23.** 4    **25.** $\frac{1}{2}$    **27.** $-9$    **29.** $2x$    **31.** $4x^2$    **33. a.** 1    **b.** 3    **c.** Does not exist
**35. a.** $-1$    **b.** $-1$    **c.** $-1$    **37. a.** $-1$    **b.** 2    **c.** Does not exist
**39. a.** $-2$    **b.** $-2$    **c.** $-2$    **41. a.** 0    **b.** 0    **c.** 0
**43. a.** $-1$    **b.** 1    **c.** Does not exist
**45.** $\lim\limits_{x \to -\infty} f(x) = 0$; $\lim\limits_{x \to 3^-} f(x) = \infty$ and $\lim\limits_{x \to 3^+} f(x) = -\infty$, so that $\lim\limits_{x \to 3} f(x)$ does not exist; and $\lim\limits_{x \to \infty} f(x) = 0$.
**47.** $\lim\limits_{x \to -\infty} f(x) = 0$; $\lim\limits_{x \to 0^-} f(x) = \infty$ and $\lim\limits_{x \to 0^+} f(x) = \infty$, so that $\lim\limits_{x \to 0} f(x) = \infty$; and $\lim\limits_{x \to \infty} f(x) = 0$.
**49.** $\lim\limits_{x \to -\infty} f(x) = 2$; $\lim\limits_{x \to 1^-} f(x) = -\infty$ and $\lim\limits_{x \to 1^+} f(x) = \infty$, so that $\lim\limits_{x \to 1} f(x)$ does not exist; and $\lim\limits_{x \to \infty} f(x) = 2$.
**51.** $\lim\limits_{x \to -\infty} f(x) = 1$; $\lim\limits_{x \to -2^-} f(x) = \infty$ and $\lim\limits_{x \to -2^+} f(x) = \infty$, so that $\lim\limits_{x \to 2} f(x) = \infty$; and $\lim\limits_{x \to \infty} f(x) = 1$.
**53.** Continuous    **55.** Discontinuous, (3) is violated    **57.** Discontinuous, (1) is violated
**59.** Discontinuous, (2) is violated
**61. a.**

**65.** Continuous
**67.** Discontinuous at $x = 1$

**63. a.**

**b.** $\lim\limits_{x \to 3^-} f(x) = 3$, $\lim\limits_{x \to 3^+} f(x) = 3$
**c.** Yes

**b.** $\lim\limits_{x \to 3^-} f(x) = 3$, $\lim\limits_{x \to 3^+} f(x) = 4$
**c.** No, (2) is violated

**69.** Discontinuous at $x = -1$, $x = 0$, and $x = 1$    **71.** Discontinuous at $x = 4$
**73.** Continuous    **75.** Continuous    **77.** Discontinuous at $x = 6$
**79.** The two functions are *not* equal to each other, since at $x = 1$ one is defined and the other is not (see pages 52–53).
**81.** 1.11 (dollars)    **83.** 100
**85.** As $x$ approaches $c$, the function is approaching $\lim\limits_{x \to c} f(x)$ even if the value of the function *at c* is different, so the limit is where it's "going."
**87.** False. The value of the function *at* 2 has nothing to do with the limit as $x$ *approaches* 2.
**89.** False. *Both* one-sided limits would have to exist and agree to guarantee that the limit exists.
**91.** False. On the left side the limit exists and equals 2 (as we saw in Example 4 on page 82), but on the right side the denominator of the fraction is zero. Therefore, one side of the equation is defined and the other is not.
**93.** True. The third requirement for continuity at $x = 2$ is that the *limit* and the *value* at 2 must agree, so if one is 7 the other must be 7.    **95.** None of the limits exist.

## Exercises 2.2 page 105

**1.** At $P_1$: positive slope    **3.** At $P_1$: positive slope    **5.** At $P_1$: slope is 3
At $P_2$: negative slope        At $P_2$: negative slope        At $P_2$: slope is $-\frac{1}{2}$
At $P_3$: zero slope            At $P_3$: zero slope
**7.** Your graph should look roughly like the following:

**9. a.** 5    **b.** 4    **c.** 3.5    **d.** 3.1    **e.** 3.01    **f.** 3
**11. a.** 13    **b.** 11    **c.** 10    **d.** 9.2    **e.** 9.02    **f.** 9
**13. a.** 5    **b.** 5    **c.** 5    **d.** 5    **e.** 5    **f.** 5
**15. a.** 0.2247    **b.** 0.2361    **c.** 0.2426    **d.** 0.2485    **e.** 0.2498    **f.** 0.25
**17.** 3    **19.** 5    **21.** 9    **23.** $\frac{1}{4}$    **25.** $f'(x) = 2x - 3$
**27.** $f'(x) = -2x$    **29.** $f'(x) = 9$

**31.** $f'(x) = \frac{1}{2}$    **33.** $f'(x) = 0$    **35.** $f'(x) = 2ax + b$    **37.** $f'(x) = 5x^4$    **39.** $f'(x) = \dfrac{-2}{x^2}$

**41.** $f'(x) = \dfrac{1}{2\sqrt{x}}$    **43.** $f'(x) = 3x^2 + 2x$

**45. a.** $y = x + 1$    **b.**     **47. a.**

on window
$[-10, 10]$ by $[-10, 10]$

**49. a.** $f'(x) = 3$    **b.** The graph of $f(x) = 3x - 4$ is a straight line with slope 3.
**51. a.** $f'(x) = 0$    **b.** The graph of $f(x) = 5$ is a horizontal straight line with slope 0.
**53. a.** $f'(x) = m$    **b.** The graph of $f(x) = mx + b$ is a straight line with slope $m$.
**55. a.** $f'(x) = 2x - 8$    **b.** Decreasing at the rate of 4 degrees per minute (since $f'(2) = -4$)
   **c.** Increasing at the rate of 2 degrees per minute (since $f'(5) = 2$)
**57. a.** $f'(x) = 4x - 1$
   **b.** $f'(5) = 19$. *Interpretation:* When 5 words have been memorized, the memorization time is increasing at the rate of 19 seconds per word.
**59. a.** $f'(x) = x - 3.7$
   **b.** $-2.7$. *Interpretation:* In 1940 the percentage of immigrants was decreasing by 2.7 percentage points per decade (so about 0.27 of a percentage point per year).
   **c.** Increasing by 3.3 percentage points per decade (so about a third of a percentage point per year).
**61.** The *average* rate of change requires *two* $x$-values and is the change in the function-values divided by the change in the $x$-values. The *instantaneous* rate of change is at a *single* $x$-value, and is found from the formula

$$f'(x) = \lim_{h \to 0} \frac{f(x + h) - f(x)}{h}.$$

**63.** Substituting $h = 0$ would make the denominator zero, and we can't divide by zero. That's why we need to do some algebra on the difference quotient—to cancel the terms that are zero so that afterward we can evaluate by direct substitution.
**65.** The units of $x$ are *blargs,* and the units of $f$ are *prendles.*
**67.** The patient's health is deteriorating during the first day (temperature is rising above normal). The patient's health is improving during the second day (temperature is falling).

## Exercises 2.3 page 120

**1.** $4x^3$    **3.** $500x^{499}$    **5.** $\frac{1}{2}x^{-1/2}$    **7.** $2x^3$    **9.** $2w^{-2/3}$    **11.** $-6x^{-3}$    **13.** $8x - 3$

**15.** $-\frac{1}{2}x^{-3/2} = -\dfrac{1}{2\sqrt{x^3}}$    **17.** $-2x^{-4/3} = -\dfrac{2}{\sqrt[3]{x^4}}$    **19.** $2\pi r$    **21.** $\frac{1}{2}x^2 + x + 1$    **23.** $\frac{1}{2}x^{-1/2} + x^{-2}$

**25.** $4x^{-1/3} + 4x^{-4/3}$    **27.** $-5x^{-3/2} - 3x^{2/3}$    **29.** $1 + 2x$    **31.** 80    **33.** 3    **35.** 27    **37.** 1

**39. a.** $y = 4x - 7$      **41. a.** $y = 2x - 6$
   **b.**

**43.** For $y_1 = 5$ and window $[-10, 10]$ by $[-10, 10]$, your calculator screen should look like the following:

**45. a.** 8; when purchasing 8 items, the cost of the last item is about $8.
   **b.** 4; when purchasing 64 items, the cost of the last item is about $4.
**47.** 4.01, which is very close to 4
**49. a.** 300; in 2030 this population group will be increasing by 300 thousand per year.
   **b.** $-300$; in 2010 this population group will be decreasing by 300 thousand per year.
**51.** Increasing by about 8000 people per additional day
**53.** 10; in 2010 the percentage of households will be increasing by 10% per year.
**55.** Increasing by about 6 phrases per hour
**57. a.** $MU(x) = 50x^{-1/2}$   **b.** 50   **c.** 0.05
**59. a.** $f(12) \approx 40$. *Interpretation:* The probability of a high school graduate quitting smoking is about 40%.
   $f'(12) \approx 1.8$. *Interpretation:* The probability of quitting increases by about 1.8% for each additional year of education.
   **b.** $f(16) \approx 60$. *Interpretation:* The probability of a college graduate quitting smoking is about 60%.
   $f'(16) \approx 8.5$. *Interpretation:* The probability of quitting increases by about 8.5% for each additional year of education.
**61. a.** $36,600
   **b.** 7300; in 2017–18 private college tuition will be increasing by $7300 every five years.
   **c.** $1460
**63.** $f(x) = 2$ will have a graph that is a horizontal line at height 2, and the slope (the derivative) of a horizontal line is zero. A function that stays constant will have a rate of change of zero, so its derivative (instantaneous rate of change) will be zero.
**65.** If $f$ has a particular rate of change, then the rate of change of $2 \cdot f(x)$ will be *twice* as large, and the rate of change of $c \cdot f(x)$ will be $c$ times as large, so the derivative of $c \cdot f(x)$ will be $c \cdot f'(x)$, which is just the Constant Multiple Rule.
**67.** Since $-f$ slopes down by the same amount that $f$ slopes up, the slope of $-f$ should be the *negative* of the slope of $f$. The constant multiple rule with $c = -1$ also says that the slope of $-f$ will be the negative of the slope of $f$.
**69.** Evaluating first would give a constant, and the derivative of a constant is zero, so evaluating and *then* differentiating would always give zero, regardless of the function and number. This supports the idea that we should always differentiate and *then* evaluate to obtain anything meaningful.
**71.** Each additional year of education increases life expectancy by 1.7 years.

## Exercises 2.4 page 136

**1.** $10x^9$   **3.** $9x^8 + 4x^3$   **5.** $5x^4 + 2x$   **7.** $15x^2 - 1$   **9.** $9\sqrt{x} + \dfrac{1}{\sqrt{x}}$   **11.** $4x^3$   **13.** $9x^2 + 8x + 1$

**15.** $4x^3 + 9x^2 - 2x$   **17.** $-6x^2 + 4x - 1$   **19.** 1   **21.** $36t + 8t^{1/3}$ (after simplification)

**23.** $7z^6 - 1$ (after simplification)   **25.** $6\sqrt{z} + \dfrac{1}{\sqrt{z}} + 10$   **27.** $6x^5$   **29.** $-\dfrac{3}{x^4}$ or $-3x^{-4}$   **31.** $\dfrac{x^4 - 3}{x^4}$

**33.** $-\dfrac{2}{(x-1)^2}$   **35.** $\dfrac{5}{(2+x)^2}$ (after simplification)   **37.** $\dfrac{4t}{(t^2+1)^2}$   **39.** $\dfrac{2s^3 + 3s^2 + 1}{(s+1)^2}$ (after simplification)

**41.** $\dfrac{x^2 + 2x - 5}{(x + 1)^2}$ (after simplification)     **43.** $\dfrac{2x^5 + 4x^3}{(x^2 + 1)^2}$ (after simplification)     **45.** $-\dfrac{t^2 + 4t + 5}{(t^2 + t - 3)^2}$

**47.** $y = 3x^{-1}, \dfrac{dy}{dx} = -3x^{-2}, \dfrac{dy}{dx} = -\dfrac{3}{x^2}$     **49.** $y = \dfrac{3}{8}x^4, \dfrac{dy}{dx} = \dfrac{3}{2}x^3, \dfrac{dy}{dx} = \dfrac{3x^3}{2}$

**51.** $y = \dfrac{1}{3}x^2 - \dfrac{5}{3}x, \dfrac{dy}{dx} = \dfrac{2}{3}x - \dfrac{5}{3}, \dfrac{dy}{dx} = \dfrac{2x - 5}{3}$     **53.** $3x^2\dfrac{x^2 + 1}{x + 1} + (x^3 + 2)\dfrac{x^2 + 2x - 1}{(x + 1)^2}$

**55.** $\dfrac{3x^6 + 13x^4 + 18x^2 - 2x}{(x^2 + 2)^2}$     **57.** $\dfrac{x^{-1/2}}{(x^{1/2} + 1)^2} = \dfrac{1}{\sqrt{x}(\sqrt{x} + 1)^2}$     **59.** $MAR(x) = \dfrac{xR'(x) - R(x)}{x^2}$

**61. a.** $C'(x) = \dfrac{100}{(100 - x)^2}$     **b.** Increasing by 4¢ per additional percentage of purity

 **c.** Increasing by 25¢ per additional percentage of purity

**63. b.** Rates of change are 4 and 25

**65. a.** $6 + \dfrac{45}{x}$     **b.** $-\dfrac{45}{x^2}$     **c.** $-\dfrac{1}{20}$, so the average cost is decreasing by 5¢ per additional clock.

**67.** Increasing at the rate of 7 degrees per hour     **69. b.** 7     **c.** About 104.5 degrees

**71. a.** \$33,928, \$46,301 (from 33.928, 46.301)

 **b.** \$1342, so in 2010 the per capita national debt will be growing by \$1342 per year;
 \$1141, so in 2020 the per capita national debt will be growing by \$1141 per year.

**73. a.** $y' = -2x^{-3} = -\dfrac{2}{x^3}$   (which is undefined at $x = 0$)

 **b.** Your calculator should give "Error" but may, incorrectly, give "0."

**75.** False. The product rule gives the correct right-hand side.     **77.** True

**79.** The right-hand side multiplies out to   $g \cdot f' + f \cdot g'$,   which agrees with the product rule.

**81.** False. This would be equivalent to saying that the derivative (instantaneous rate of change) of a product is the product of the derivatives. The *Product Rule* gives the correct way of finding the derivative of a product.

**83.** $\dfrac{d}{dx}(fgh) = f'(gh) + f(gh)' = f'gh + fg'h + fgh'$     **85.** $2f(x)f'(x)$

**87.** $\dfrac{dy}{dx} = \dfrac{R}{\left(1 + \dfrac{R - 1}{K}x\right)^2} > 0.$   As the population increases, the number of offspring increases.

## Exercises 2.5 page 149

**1. a.** $4x^3 - 6x^2 - 6x + 5$     **b.** $12x^2 - 12x - 6$     **c.** $24x - 12$     **d.** $24$

**3. a.** $1 + x + \dfrac{1}{2}x^2 + \dfrac{1}{6}x^3 + \dfrac{1}{24}x^4$     **b.** $1 + x + \dfrac{1}{2}x^2 + \dfrac{1}{6}x^3$     **c.** $1 + x + \dfrac{1}{2}x^2$     **d.** $1 + x$

**5. a.** $\dfrac{5}{2}x^{3/2}$     **b.** $\dfrac{15}{4}x^{1/2}$     **c.** $\dfrac{15}{8}x^{-1/2}$     **d.** $-\dfrac{15}{16}x^{-3/2}$     **7. a.** $-\dfrac{2}{x^3}$ or $-2x^{-3}$     **b.** $-\dfrac{2}{27}$

**9. a.** $\dfrac{1}{x^3} = x^{-3}$     **b.** $\dfrac{1}{27}$     **11. a.** $x^{-4}$     **b.** $\dfrac{1}{81}$     **13.** $12x^2 + 2$     **15.** $12x^{-7/3}$

**17.** $\dfrac{2x - 2}{(x^2 - 2x + 1)^2} = \dfrac{2}{(x - 1)^3}$     **19.** $2\pi$     **21.** $90$     **23.** $-720$     **25.** $3$

**27.** $20x^3 - 12x^2 + 6x - 2$     **29.** $\dfrac{2x^5 - 4x^3 - 6x}{(x^2 + 1)^4} = \dfrac{2x^3 - 6x}{(x^2 + 1)^3}$     **31.** $\dfrac{-32x - 16}{(4x^2 + 4x + 1)^2} = \dfrac{-16}{(2x + 1)^3}$

**33. a.** 54 mph     **b.** $-42$ mph or 42 mph south     **c.** 24 mi/hr$^2$     **35.** 310 ft/sec, 61 ft/sec$^2$

**37. a.** About 10.2 seconds   **b.** At about 326 feet per second. [*Note:* This is about 222 miles per hour. More realistically, the ball would be slowed by air resistance, going no faster than its *terminal velocity*. For a steel ball the size of a marble, this would be about 200 mph.]   **c.** $-32$ feet per second per second

**39. a.** $-32t + 1280$   **b.** 40 seconds   **c.** 25,600 feet

**41.** $D'(8) = 24$: After 8 years the debt is growing by \$24 billion per year.
$D''(8) = 1$: After 8 years the debt will be growing increasingly rapidly, with the rate of growth growing by about \$1 billion per year.

**43.** $L(10) = 93$;   *Interpretation:* By 2100 sea levels may have risen by 93 centimeters (about 3 feet).   $L'(10) = 12.6$; *Interpretation:* In 2100 sea levels will be rising by about 12.1 centimeters per decade, or about 1.26 cm (about half an inch) per year.   $L''(10) = 1.06$.   *Interpretation:* The rise in the sea level will be speeding up (by about 1 cm per decade per decade).

**45. a.**  Approximately 22° and 18°

on [0, 50] by [0, 40]

**b.** Each successive 1-mph increase in wind speed lowers the windchill index, but less so as wind speed rises.
**c.** $y_2(15) \approx -0.4°$  and  $y_2(30) \approx -0.2°$.   *Interpretation:* At a wind speed of 15 mph, each additional mph decreases the windchill index by about 0.4°, whereas at a wind speed of 30 mph, each additional mph decreases the windchill index by only about 0.2°.

**47. a.** Negative   **b.** Negative   **49. a.** Negative   **b.** Positive   **51.** True   **53. a.** iii   **b.** i   **c.** ii

**55.** 100!   **57.** $\dfrac{d^2}{dx^2}(fg) = \dfrac{d}{dx}(f'g + fg') = f''g + f'g' + f'g' + fg'' = f''g + 2f'g' + fg''$

## Exercises 2.6 page 160

*Note:* For Exercises 1 through 9, there are other possible correct answers.

**1.** $f(x) = \sqrt{x}, g(x) = x^2 - 3x + 1$   **3.** $f(x) = x^{-3}, g(x) = x^2 - x$   **5.** $f(x) = \dfrac{x+1}{x-1}, g(x) = x^3$

**7.** $f(x) = x^4, g(x) = \dfrac{x+1}{x-1}$   **9.** $f(x) = \sqrt{x} + 5, g(x) = x^2 - 9$   **11.** $6x(x^2+1)^2$

**13.** $4(2x^2 - 7x + 3)^3(4x - 7)$   **15.** $4(3z^2 - 5z + 2)^3(6z - 5)$   **17.** $\frac{1}{2}(x^4 - 5x + 1)^{-1/2}(4x^3 - 5)$

**19.** $\frac{1}{3}(9z - 1)^{-2/3}(9) = 3(9z - 1)^{-2/3}$   **21.** $-8x(4 - x^2)^3$   **23.** $-12w^2(w^3 - 1)^{-5}$   **25.** $4x^3 - 4(1 - x)^3$

**27.** $-\frac{1}{2}(3x^2 - 5x + 1)^{-3/2}(6x - 5)$   **29.** $-\frac{2}{3}(9x + 1)^{-5/3}(9) = -6(9x + 1)^{-5/3}$

**31.** $-\frac{2}{3}(2x^2 - 3x + 1)^{-5/3}(4x - 3)$   **33.** $3[(x^2 + 1)^3 + x]^2[6x(x^2 + 1)^2 + 1]$

**35.** $6x(2x + 1)^5 + 30x^2(2x + 1)^4 = 6x(2x + 1)^4(7x + 1)$

**37.** $6(2x + 1)^2(2x - 1)^4 + 8(2x + 1)^3(2x - 1)^3 = 2(2x + 1)^2(2x - 1)^3(14x + 1)$

**39.** $2(3z^2 - z + 1)^4 + 8z(3z^2 - z + 1)^3(6z - 1) = 2(3z^2 - z + 1)^3(27z^2 - 5z + 1)$

**41.** $-6\dfrac{(x+1)^2}{(x-1)^4}$   **43.** $2x(1 + x^2)^{1/2} + x^3(1 + x^2)^{-1/2}$   **45.** $\dfrac{1}{4}x^{-1/2}(1 + x^{1/2})^{-1/2}$

**47. a.** $4x(x^2 + 1)$   **b.** $4x^3 + 4x$   **49. a.** $-\dfrac{3}{(3x+1)^2}$   **b.** $-3(3x + 1)^{-2}$

**51.** $20(x^2 + 1)^9 + 360x^2(x^2 + 1)^8$   **53.** $MC(x) = 4x(4x^2 + 900)^{-1/2}$   $MC(20) = \frac{8}{5} = 1.60$   **55.** $x = 27$

**57.** $S'(25) \approx 2.1$. *Interpretation:* At income level \$25,000, status increases by about 2.1 units for each additional \$1000 of income.

**59.** $R'(50) = 32\frac{1}{3}$   **61.** 26 mg   **63.** $P'(2) = 0.24$; pollution is increasing by about 0.24 ppm per year.

**65. b.** 0.0456   **c.** 39.47, so about 40 years

**67.** False. There should not be a prime of the first $g$ on the right-hand side.

**69.** The Generalized Power Rule is a special case of the Chain Rule, when the *outer* function is just a *power* of the variable.

**71.** True.

**73.** False. The *outer* function, $\sqrt{x}$, was not differentiated. The correct right-hand side is $\frac{1}{2}[g(x)]^{-1/2} \cdot g'(x)$.

**75.** No. Since instantaneous rates of change are derivatives, this would be saying that $\frac{d}{dx}[f(x)]^2 = [f'(x)]^2$ where $f(x)$ is the length of a side. The Chain Rule gives the correct derivative of $[f(x)]^2$.

**77.** $\frac{d}{dx}L(g(x)) = L'(g(x))\, g'(x) = \frac{1}{g(x)}g'(x) = \frac{g'(x)}{g(x)}$

**79.** Solve for $F(h)$ to find $F(h) = \frac{f(x+h) - f(x)}{h}$ and use the continuity of $F$ at 0 to obtain $F(0) = \lim_{h \to 0} F(h) = \lim_{h \to 0}\frac{f(x+h) - f(x)}{h}$, which is the definition of the derivative on page 100.

## Exercises 2.7 page 168

**1.** $-2, 0$, and $2$   **3.** $-3$ and $3$

**5.** $\lim_{h \to 0}\frac{f(x+h) - f(x)}{h}$ simplifies to $\lim_{h \to 0}\frac{|2h|}{h}$, which gives $\begin{cases} 2 & \text{for } h > 0 \\ -2 & \text{for } h < 0 \end{cases}$ so the limit (and therefore the derivative at $x = 0$) does not exist.

**7.** $\lim_{h \to 0}\frac{f(x+h) - f(x)}{h}$ simplifies to $\lim_{h \to 0}\frac{h^{2/5}}{h} = \lim_{h \to 0}\frac{1}{h^{3/5}}$ which does not exist. Therefore, the derivative at $x = 0$ does not exist.

**9.** If you got a numerical answer, it is wrong, since the function is undefined at $x = 0$, so the derivative at $x = 0$ does not exist. (For an explanation, see the Graphing Calculator Exploration on page 165.)

**11. a.** The formula comes from substituting the given function and $x$-value into the definition of the derivative and simplifying.

   **b.** 3.16, 31.6, 316 (rounded)   **c.** No, no   **d.**

on [0, 1] by [0, 1]

**13.** One explanation: At a corner point, a proposed "tangent line" can tip back and forth (see page 104), and so there is no well-defined slope.

**15.** At a discontinuity, the values of the function take a sudden *jump*, and so a (steady) rate of change cannot be defined.

**17.** True.   **19.** One possibility is:

## Chapter 2 Review Exercises page 171

**1.**

| x | 4x + 2 | x | 4x + 2 |
|---|---|---|---|
| 1.9 | 9.6 | 2.1 | 10.4 |
| 1.99 | 9.96 | 2.01 | 10.04 |
| 1.999 | 9.996 | 2.001 | 10.004 |

**a.** $\lim\limits_{x \to 2^-} (4x + 2) = 10$    **b.** $\lim\limits_{x \to 2^+} (4x + 2) = 10$    **c.** $\lim\limits_{x \to 2} (4x + 2) = 10$

**2.**

| x | $\dfrac{\sqrt{x+1}-1}{x}$ | x | $\dfrac{\sqrt{x+1}-1}{x}$ |
|---|---|---|---|
| $-0.1$ | 0.513 | 0.1 | 0.488 |
| $-0.01$ | 0.501 | 0.01 | 0.499 |
| $-0.001$ | 0.500 | 0.001 | 0.500 |

**a.** $\lim\limits_{x \to 0^-} \dfrac{\sqrt{x+1}-1}{x} = 0.5$    **b.** $\lim\limits_{x \to 0^+} \dfrac{\sqrt{x+1}-1}{x} = 0.5$    **c.** $\lim\limits_{x \to 0} \dfrac{\sqrt{x+1}-1}{x} = 0.5$

**3. a.** 3    **b.** $-2$    **c.** Does not exist    **4. a.** $-1$    **b.** $-1$    **c.** $-1$    **5.** 5    **6.** $\pi$    **7.** 4

**8.** 2    **9.** $\frac{1}{2}$    **10.** $-3$    **11.** $2x^2$    **12.** $-x^2$

**13.** $\lim\limits_{x \to -\infty} f(x) = 0$; $\lim\limits_{x \to -2^-} f(x) = \infty$ and $\lim\limits_{x \to -2^+} f(x) = \infty$, so that $\lim\limits_{x \to -2} f(x) = \infty$; and $\lim\limits_{x \to \infty} f(x) = 0$.

**14.** $\lim\limits_{x \to -\infty} f(x) = 3$; $\lim\limits_{x \to 2^-} f(x) = -\infty$ and $\lim\limits_{x \to 2^+} f(x) = \infty$, so that $\lim\limits_{x \to 2} f(x)$ does not exist;

and $\lim\limits_{x \to \infty} f(x) = 3$.

**15.** Continuous    **16.** Continuous    **17.** Discontinuous at $x = -1$    **18.** Continuous

**19.** Discontinuous at $x = 0$ and at $x = -1$    **20.** Discontinuous at $x = 3$ and at $x = -3$

**21.** Discontinuous at $x = 5$    **22.** Continuous    **23.** $4x + 3$    **24.** $6x + 2$    **25.** $-\dfrac{3}{x^2} = -3x^{-2}$

**26.** $\dfrac{2}{\sqrt{x}}$    **27.** $10x^{2/3} + 2x^{-3/2}$    **28.** $10x^{3/2} + 2x^{-4/3}$    **29.** $-16$    **30.** $-9$    **31.** 1    **32.** $\frac{1}{2}$

**33. a.** 20.  *Interpretation:* Costs are increasing by about $20 per additional license.

    **b.** 10.  *Interpretation:* Costs are increasing by about $10 per additional license.

**34.** $f'(10) \approx -2.3$  (thousand hours).  *Interpretation:* After 10 planes, the construction time is decreasing by about 2300 hours for each additional plane built.

**35. a.** $\dfrac{dA}{dr} = 2\pi r$    **b.** As the radius increases, the area "grows by a circumference."

**36. a.** $V' = \frac{4}{3}\pi r^2 \cdot 3 = 4\pi r^2$    **b.** As radius increases, volume "grows by a surface area."

**37.** $40x^3 + 6$ (after simplification)    **38.** $15x^4 - 2x$    **39.** $2x(x^2 - 5) + (x^2 + 5)2x = 4x^3$    **40.** $4x^3$

**41.** $(4x^3 + 2x)(x^5 - x^3 + x) + (x^4 + x^2 + 1)(5x^4 - 3x^2 + 1) = 9x^8 + 5x^4 + 1$

**42.** $(5x^4 + 3x^2 + 1)(x^4 - x^2 + 1) + (x^5 + x^3 + x)(4x^3 - 2x) = 9x^8 + 5x^4 + 1$

**43.** $\dfrac{2}{(x+1)^2}$    **44.** $\dfrac{-2}{(x-1)^2}$    **45.** $\dfrac{-10x^4}{(x^5-1)^2}$    **46.** $\dfrac{12x^5}{(x^6+1)^2}$

**47. a.** $-\dfrac{1}{x^2}$    **b.** $-x^{-2}$ (after simplification)    **c.** By simplifying to $f(x) = 2 + x^{-1}$, the derivative is $-x^{-2}$.

**48.** $S'(6) = -10,$ so at a price of \$6 each, sales will decrease by about 10 for each dollar increase in price.

**49. a.** $AP(x) = \dfrac{6x - 200}{x}$ **b.** $MAP(x) = \dfrac{200}{x^2}$ **c.** $MAP(10) = 2$, so average profit is increasing by about \$2 per additional unit.

**50. a.** $7.5 + \dfrac{50}{x}$ **b.** $-\dfrac{50}{x^2}$ **c.** $-\dfrac{1}{50}$, so the average cost is decreasing by 2¢ per additional mouse.

**51.** $9x^{-1/2} + 2x^{-5/3}$ **52.** $-4x^{-4/3} - 3x^{-1/2}$ **53.** $2x^{-4}$ **54.** $6x^{-5}$ **55.** $-24$ **56.** $60$

**57.** $480$ **58.** $\frac{3}{8}$ **59.** $15$ **60.** $70$

**61.** $P(10) = 200, P'(10) = 20, P''(10) = 9.$ *Interpretation:* 10 years from now the population will be 200 thousand, growing at the rate of 20 thousand per year, and the growth will be accelerating.

**62.** Velocity 2500 ft/sec; acceleration 150 ft/sec$^2$ **63. a.** 347.25 feet

**64.** $3(4z^2 - 3z + 1)^2(8z - 3)$ **65.** $4(3z^2 - 5z - 1)^3(6z - 5)$ **66.** $-5(100 - x)^4$ **67.** $-4(1000 - x)^3$

**68.** $\frac{1}{2}(x^2 - x + 2)^{-1/2}(2x - 1)$ **69.** $\frac{1}{2}(x^2 - 5x - 1)^{-1/2}(2x - 5)$ **70.** $2(6z - 1)^{-2/3}$ **71.** $(3z + 1)^{-2/3}$

**72.** $-2(5x + 1)^{-7/5}$ **73.** $-6(10x + 1)^{-8/5}$ **74.** $2x(2x - 1)^4 + 8x^2(2x - 1)^3 = 2x(2x - 1)^3(6x - 1)$

**75.** $5(x^3 - 2)^4 + 60x^3(x^3 - 2)^3 = 5(x^3 - 2)^3(13x^3 - 2)$ **76.** $3x^2(x^3 + 1)^{1/3} + x^5(x^3 + 1)^{-2/3}$

**77.** $4x^3(x^2 + 1)^{1/2} + x^5(x^2 + 1)^{-1/2}$ **78.** $3[(2x^2 + 1)^4 + x^4]^2[16x(2x^2 + 1)^3 + 4x^3]$

**79.** $2[(3x^2 - 1)^3 + x^3][18x(3x^2 - 1)^2 + 3x^2]$ **80.** $\frac{1}{2}[(x^2 + 1)^4 - x^4]^{-1/2}[8x(x^2 + 1)^3 - 4x^3]$

**81.** $[(x^3 + 1)^2 + x^2]^{-1/2}[3x^2(x^3 + 1) + x]$ **82.** $12(3x + 1)^3(4x + 1)^3 + 12(3x + 1)^4(4x + 1)^2$

**83.** $6x(x^2 + 1)^2(x^2 - 1)^4 + 8x(x^2 + 1)^3(x^2 - 1)^3$ **84.** $\dfrac{-20}{x^2}\left(\dfrac{x + 5}{x}\right)^3 = \dfrac{-20(x + 5)^3}{x^5}$

**85.** $-20\left(\dfrac{x + 4}{x}\right)^4 \dfrac{1}{x^2} = -20\dfrac{(x + 4)^4}{x^6}$ **86.** $20(2w^2 - 4)^4 + 320w^2(2w^2 - 4)^3$

**87.** $24(3w^2 + 1)^3 + 432w^2(3w^2 + 1)^2$ **88.** $6z(z + 1)^3 + 18z^2(z + 1)^2 + 6z^3(z + 1)$

**89.** $12z^2(z + 1)^4 + 32z^3(z + 1)^3 + 12z^4(z + 1)^2$

**90. a.** $6x^2(x^3 - 1)$ **b.** $6x^5 - 6x^2$ **91. a.** $\dfrac{-3x^2}{(x^3 + 1)^2}$ **b.** $-(x^3 + 1)^{-2}(3x^2)$

**92.** $P'(5) = 3.$ *Interpretation:* When producing 5 tons, profit increases by about 3 thousand dollars for each additional ton.

**93.** $V'(8) = 17.496.$ *Interpretation:* Value increased by about \$17.50 for each additional percentage of interest.

**94. a.** $P(5) - P(4) \approx 2.73,$ $P(6) - P(5) \approx 3.23,$ both of which are near 3 **b.** At about $x = 7.6$

**95.** $x \approx 16$ **96.** $0.08$

**97.** $N'(96) = -250.$ *Interpretation:* At age 96, the number of survivors is decreasing by about 250 people per year.

**98.** $x = -3, x = 1, x = 3$ **99.** $x = 2, x = -2$ **100.** $x = 0, x = 3.5$ **101.** $x = 0, x = 3$

**102.** $\lim\limits_{h \to 0} \dfrac{f(x + h) - f(x)}{h}$ simplifies to $\lim\limits_{h \to 0} \dfrac{|5h|}{h}$, which gives $\begin{cases} 5 & \text{for } h > 0 \\ -5 & \text{for } h < 0 \end{cases}$ and so the limit (and therefore the derivative at $x = 0$) does not exist.

**103.** $\lim\limits_{h \to 0} \dfrac{f(x + h) - f(x)}{h}$ simplifies to $\lim\limits_{h \to 0} \dfrac{h^{3/5}}{h} = \lim\limits_{h \to 0} \dfrac{1}{h^{2/5}}$ which does not exist. Therefore, the derivative at $x = 0$ does not exist.

## Exercises 3.1 page 188

**1. a.** $(-\infty, -2)$ and $(0, \infty)$    **b.** $(-2, 0)$    **3.** All but 3 (where the function is undefined)    **5.** 4 and $-4$

**7.** $-1$ and 5    **9.** $0, -4,$ and 1    **11.** 3    **13.** No CNs    **15.** $\frac{1}{3}$ and $-1$

**17.**

| $f' > 0$ | $f' = 0$ | $f' < 0$ | $f' = 0$ | $f' > 0$ |
|---|---|---|---|---|
| | $x = -1$ | | $x = 3$ | |
| ↗ | → | ↘ | → | ↗ |
| | rel max $(-1, 15)$ | | rel min $(3, -17)$ | |

Open intervals of increase: $(-\infty, -1)$ and $(3, \infty)$; open intervals of decrease: $(-1, 3)$

**19.**

| $f' < 0$ | $f' = 0$ | $f' > 0$ | $f' = 0$ | $f' < 0$ | $f' = 0$ | $f' > 0$ |
|---|---|---|---|---|---|---|
| | $x = -4$ | | $x = 0$ | | $x = 1$ | |
| ↘ | → | ↗ | → | ↘ | → | ↗ |
| | rel min $(-4, -64)$ | | rel max $(0, 64)$ | | rel min $(1, 61)$ | |

Open intervals of increase: $(-4, 0)$ and $(1, \infty)$; open intervals of decrease: $(-\infty, -4)$ and $(0, 1)$

**21.**

| $f' > 0$ | $f' = 0$ | $f' < 0$ | $f' = 0$ | $f' > 0$ | $f' = 0$ | $f' < 0$ |
|---|---|---|---|---|---|---|
| | $x = 0$ | | $x = 1$ | | $x = 2$ | |
| ↗ | → | ↘ | → | ↗ | → | ↘ |
| | rel max $(0, 1)$ | | rel min $(1, 0)$ | | rel max $(2, 1)$ | |

Increase: $(-\infty, 0)$ and $(1, 2)$; decrease: $(0, 1)$ and $(2, \infty)$

**23.**

| $f' < 0$ | $f' = 0$ | $f' > 0$ | $f' = 0$ | $f' > 0$ |
|---|---|---|---|---|
| | $x = 0$ | | $x = 1$ | |
| ↘ | → | ↗ | → | ↗ |
| | rel min $(0, 0)$ | | neither $(1, 1)$ | |

Increase: $(0, 1)$ and $(1, \infty)$; decrease: $(-\infty, 0)$

**25.**

| $f' < 0$ | $f' = 0$ | $f' > 0$ |
|---|---|---|
| | $x = 1$ | |
| ↘ | → | ↗ |
| | rel min $(1, 0)$ | |

Increase: $(1, \infty)$; decrease: $(-\infty, 1)$

**27.**

| $f' < 0$ | $f' = 0$ | $f' > 0$ | $f' = 0$ | $f' < 0$ | $f' = 0$ | $f' > 0$ |
|---|---|---|---|---|---|---|
| | $x = -2$ | | $x = 0$ | | $x = 2$ | |
| ↘ | → | ↗ | → | ↘ | → | ↗ |
| | rel min $(-2, 0)$ | | rel max $(0, 16)$ | | rel min $(2, 0)$ | |

Increase: $(-2, 0)$ and $(2, \infty)$; decrease: $(-\infty, -2)$ and $(0, 2)$

**29.**

| $f' < 0$ | $f' = 0$ | $f' > 0$ | $f' = 0$ | $f' < 0$ | $f' = 0$ | $f' > 0$ |
|---|---|---|---|---|---|---|
| | $x = 0$ | | $x = 2$ | | $x = 4$ | |
| ↘ | → | ↗ | → | ↘ | → | ↗ |
| | rel min $(0, 0)$ | | rel max $(2, 16)$ | | rel min $(4, 0)$ | |

Increase: $(0, 2)$ and $(4, \infty)$; decrease: $(-\infty, 0)$ and $(2, 4)$

**31.**

| $f' > 0$ | $f' = 0$ | $f' < 0$ | $f' = 0$ | $f' > 0$ | $f' = 0$ | $f' > 0$ |
|---|---|---|---|---|---|---|
| | $x = 0$ | | $x = 2$ | | $x = 5$ | |
| ↗ | → | ↘ | → | ↗ | → | ↗ |
| | rel max $(0, 0)$ | | rel min $(2, -108)$ | | neither $(5, 0)$ | |

Increase: $(-\infty, 0)$, $(2, 5)$, and $(5, \infty)$; decrease: $(0, 2)$

**33.**

| $f' < 0$ | $f$ und | $f' < 0$ |
|---|---|---|
| | $x = -3$ | |
| ↘ | | ↘ |

**35.**

| $f' < 0$ | $f$ und | $f' < 0$ |
|---|---|---|
| | $x = 2$ | |
| ↘ | | ↘ |

**37.**

| $f' > 0$ | $f$ und | $f' > 0$ |
|---|---|---|
| | $x = -2$ | |
| ↗ | | ↗ |

**39.**

| $f' > 0$ | $f$ und | $f' < 0$ |
|---|---|---|
| | $x = 2$ | |
| ↗ | | ↘ |

**41.**

| $f' > 0$ | $f$ und | $f' > 0$ | $f' = 0$ | $f' < 0$ | $f$ und | $f' < 0$ |
|---|---|---|---|---|---|---|
| | $x = -1$ | | $x = 1$ | | $x = 3$ | |
| ↗ | | ↗ | → | ↘ | | ↘ |
| | | | rel max $(1, -3)$ | | | |

**43.**

| $f' > 0$ | $f' = 0$ | $f' < 0$ |
|---|---|---|
| | $x = 0$ | |
| ↗ | → | ↘ |
| | rel max | |
| | $(0, 2)$ | |

**45.**

| $f' < 0$ | $f' = 0$ | $f' > 0$ |
|---|---|---|
| | $0$ | |
| ↘ | → | ↗ |
| | rel min | |
| | $(0, 0)$ | |

**47.**

| $f' < 0$ | $f' = 0$ | $f' > 0$ | $f' = 0$ | $f' < 0$ |
|---|---|---|---|---|
| | $x = -1$ | | $x = 1$ | |
| ↘ | → | ↗ | → | ↘ |
| | rel min | | rel max | |
| | $(-1, -1)$ | | $(1, 1)$ | |

**49.**

| $f' > 0$ | $f$ und | $f' > 0$ | $f' = 0$ | $f' < 0$ | $f$ und | $f' < 0$ |
|---|---|---|---|---|---|---|
| | $x = -1$ | | $x = 0$ | | $x = 1$ | |
| ↗ | | ↗ | → | ↘ | | ↘ |
| | | | rel max | | | |
| | | | $(0, -3)$ | | | |

**51.**

| $f' < 0$ | $f$ und | $f' < 0$ | $f' = 0$ | $f' < 0$ | $f$ und | $f' < 0$ |
|---|---|---|---|---|---|---|
| | $x = -3$ | | $x = 0$ | | $x = 3$ | |
| ↘ | | ↘ | → | ↘ | | ↘ |

**53.**

| $f' > 0$ | $f$ und | $f' > 0$ | $f' = 0$ | $f' < 0$ | $f$ und | $f' < 0$ |
|---|---|---|---|---|---|---|
| | $x = -1$ | | $x = 1$ | | $x = 3$ | |
| ↗ | | ↗ | → | ↘ | | ↘ |
| | | | rel max | | | |
| | | | $(1, -1)$ | | | |

**55.**

| $f' > 0$ | $f$ und | $f' > 0$ | $f' = 0$ | $f' < 0$ | $f$ und | $f' < 0$ |
|---|---|---|---|---|---|---|
| | $x = -1$ | | $x = 0$ | | $x = 1$ | |
| ↗ | | ↗ | → | ↘ | | ↘ |
| | | | rel max | | | |
| | | | $(0, 0)$ | | | |

**57.**

| $f' < 0$ | $f$ und | $f' < 0$ |
|---|---|---|

$x = -2$

↘           ↘

**59.**

| $f' < 0$ | $f$ und | $f' > 0$ | $f' = 0$ | $f' < 0$ | $f$ und | $f' < 0$ |
|---|---|---|---|---|---|---|

$x = 0$      $x = 2$      $x = 3$

↘      ↗      →      ↘      ↘

rel max
$(2, -1)$

**61.**

| $f' > 0$ | $f' = 0$ | $f' < 0$ | $f' = 0$ | $f' > 0$ | $f' = 0$ | $f' < 0$ |
|---|---|---|---|---|---|---|

$x = -1$      $x = 0$      $x = 1$

↗    →    ↘    →    ↗    →    ↘

rel max        rel min        rel max
$(-1, 1)$       $(0, 0)$       $(1, 1)$

**63.** $f'(x) = 2ax + b = 0$ at $x = \dfrac{-b}{2a}$

**65.**

**67. i.** $f(0) = 0$, giving $d = 0$    **ii.** $f'(x) = 3ax^2 + 2bx + c$ so $f'(0) = c$, giving $c = 0$
    **iii.** $f(100) = 5$ means $a \cdot 100^3 + b \cdot 100^2 = 5$, and $f'(100) = 0$ means $3a \cdot 100^2 + 2b \cdot 100 = 0$, giving
    $a = -0.00001$ and $b = 0.0015$. Therefore, $f(x) = -0.00001x^3 + 0.0015x^2$.
    **iv.**

**69. a.**

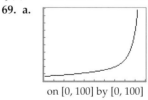

on $[0, 100]$ by $[0, 100]$

**71. a.** $\displaystyle\lim_{x \to \infty} AC(x) = \lim_{x \to \infty} 3 + \dfrac{50}{x} = 3$ and $MC(x) = (3x + 50)' = 3$

**b.**

**73.** False. The original function must also be *defined* at the critical number.
**75.** False. For example, see the graph on page 178.
**77.** Decreasing on $(-\infty, 1)$ and increasing on $(1, \infty)$. [*Remember:* Wherever the derivative is positive (above the $x$-axis) the function will be increasing.]

**79.** In a rational function, the denominator of the derivative is the *square* of the denominator of the original function (from the quotient rule). Therefore, both denominators will be zero at the same $x$-values, making both the function and its derivative undefined at the same $x$-values.

**81.** True. See the diagrams on pages 182 and 183.

**83.** If $p(c) = 0$ and $q(c) = 0$, they may have a common factor that can be canceled out so that $\lim\limits_{x \to c} \dfrac{p(x)}{q(x)}$ may exist: see Example 4 and the following Graphing Calculator Exploration on page 82.

**85. d.**

on [0, 10] by [0, 10]

## Exercises 3.2 page 203

**1.** Point 2    **3.** Points 3 and 5    **5.** Points 4 and 6

**7. a.**

| $f' > 0$ | $f' = 0$ | $f' < 0$ | $f' = 0$ | $f' > 0$ |
|---|---|---|---|---|

$x = -3$ $\qquad$ $x = 1$

↗ $\quad$ → $\quad$ ↘ $\quad$ → $\quad$ ↗

rel max $\qquad$ rel min
$(-3, 32)$ $\qquad$ $(1, 0)$

**b.**

$f'' < 0$ $\quad$ $f'' = 0$ $\quad$ $f'' > 0$

$x = -1$

con dn $\qquad$ con up
IP $(-1, 16)$

**c.**

**9. a.**

$f' > 0$ $\quad$ $f' = 0$ $\quad$ $f' > 0$

$x = 1$

↗ $\quad$ → $\quad$ ↗

neither
$(1, 5)$

**b.**

$f'' < 0$ $\quad$ $f'' = 0$ $\quad$ $f'' > 0$

$x = 1$

con dn $\qquad$ con up
IP $(1, 5)$

**c.**

**11. a.**

$f' < 0$ $\quad$ $f' = 0$ $\quad$ $f' > 0$ $\quad$ $f' = 0$ $\quad$ $f' > 0$

$x = 0$ $\qquad$ $x = 3$

↘ $\quad$ → $\quad$ ↗ $\quad$ → $\quad$ ↗

rel min $\qquad$ neither
$(0, 2)$ $\qquad$ $(3, 29)$

**b.**

$f'' > 0$ $\quad$ $f'' = 0$ $\quad$ $f'' < 0$ $\quad$ $f'' = 0$ $\quad$ $f'' > 0$

$x = 1$ $\qquad$ $x = 3$

con up $\qquad$ con dn $\qquad$ con up
IP $(1, 13)$ $\qquad$ IP $(3, 29)$

**c.**

**13. a.**

| $f' < 0$ | $f' = 0$ | $f' > 0$ | $f' = 0$ | $f' > 0$ |
|---|---|---|---|---|
| | $x = -3$ | | $x = 0$ | |

↘  →  ↗  →  ↗

rel min　　　　neither
$(-3, -12)$　　$(0, 15)$

**b.**

| $f'' > 0$ | $f'' = 0$ | $f'' < 0$ | $f'' = 0$ | $f'' > 0$ |
|---|---|---|---|---|
| | $x = -2$ | | $x = 0$ | |

con up　　　　con dn　　　　con up
IP $(-2, -1)$　　IP $(0, 15)$

**c.**

**15. a.**

| $f' < 0$ | $f' = 0$ | $f' > 0$ | $f' = 0$ | $f' < 0$ |
|---|---|---|---|---|
| | $x = 0$ | | $x = 4$ | |

↘  →  ↗  →  ↘

rel min　　　　rel max
$(0, 0)$　　　$(4, 256)$

**b.**

| $f'' > 0$ | $f'' = 0$ | $f'' > 0$ | $f'' = 0$ | $f'' < 0$ |
|---|---|---|---|---|
| | $x = 0$ | | $x = 3$ | |

con up　　　　con up　　　　con dn
IP $(3, 162)$

**c.**

**17. a.**

| $f' > 0$ | $f' = 0$ | $f' > 0$ |
|---|---|---|
| | $x = -2$ | |

↗  →  ↗

neither
$(-2, 0)$

**b.**

| $f'' < 0$ | $f'' = 0$ | $f'' > 0$ |
|---|---|---|
| | $x = -2$ | |

con dn　　　　con up
IP $(-2, 0)$

**c.**

**19. a.**

| $f' > 0$ | $f' = 0$ | $f' < 0$ | $f' = 0$ | $f' > 0$ |
|---|---|---|---|---|
| | $x = 1$ | | $x = 3$ | |

↗  →  ↘  →  ↗

rel max　　　　rel min
$(1, 4)$　　　$(3, 0)$

**b.**

| $f'' < 0$ | $f'' = 0$ | $f'' > 0$ |
|---|---|---|
| | $x = 2$ | |

con dn　　　　con up
IP $(2, 2)$

**c.**

**21. a.**

| $f' > 0$ | $f'$ und | $f' > 0$ |
|---|---|---|
| | $x = 0$ | |

↗  ↗

neither
$(0, 0)$

**b.**

| $f'' > 0$ | $f''$ und | $f'' < 0$ |
|---|---|---|
| | $x = 0$ | |

con up　　　　con dn
IP $(0, 0)$

**c.**

**23. a.**

| $f' < 0$ | $f'$ und | $f' > 0$ |
|---|---|---|
| | $x = 0$ | |

$\searrow$   $\nearrow$

rel min
$(0, 2)$

**b.**

| $f'' < 0$ | $f''$ und | $f'' < 0$ |
|---|---|---|
| | $x = 0$ | |

con dn      con dn

**c.**

**25. a.**

| $f'$ und | $f' > 0$ |
|---|---|
| $x = 0$ | |

$\nearrow$

**b.**

| $f''$ und | $f'' < 0$ |
|---|---|
| $x = 0$ | |

con dn

**c.**

**27. a.**

| $f' < 0$ | $f'$ und | $f' > 0$ |
|---|---|---|
| | $x = 1$ | |

$\searrow$   $\nearrow$

rel min
$(1, 0)$

**b.**

| $f'' < 0$ | $f''$ und | $f'' < 0$ |
|---|---|---|
| | $x = 1$ | |

con dn      con dn

**c.**

**29. a.**

| $f' < 0$ | $f$ und | $f' < 0$ | $f$ und | $f' < 0$ |
|---|---|---|---|---|
| | $x = -1$ | | $x = 1$ | |

$\searrow$   $\searrow$   $\searrow$

**b.**

| $f'' < 0$ | $f$ und | $f'' > 0$ | $f'' = 0$ | $f'' < 0$ | $f$ und | $f'' > 0$ |
|---|---|---|---|---|---|---|
| | $x = -1$ | | $x = 0$ | | $x = 1$ | |

con dn        con up    IP $(0, 0)$    con dn        con up

**c.**

**31.**

| $f'$ und | $f' > 0$ |
|---|---|
| $x = 0$ | $\nearrow$ |

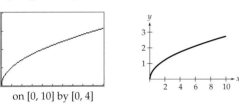

on $[0, 10]$ by $[0, 4]$

**33.**

| $f'$ und | $f' < 0$ |
|---|---|
| $x = 0$ | $\searrow$ |

on $[0, 10]$ by $[0, 4]$

**35.**

| $f' < 0$ | $f'$ und | $f' > 0$ | $f' = 0$ | $f' < 0$ |
|---|---|---|---|---|
| | $x = 0$ | | $x = 1$ | |

$\searrow$   $\nearrow$   $\rightarrow$   $\searrow$

rel min      rel max
$(0, 0)$      $(1, 3)$

on $[-2, 10]$ by $[-10, 10]$

**37.**

| $f' > 0$ | $f'$ und | $f' < 0$ | $f' = 0$ | $f' > 0$ |
|---|---|---|---|---|
| | $x = 0$ | | $x = 1$ | |

$\nearrow$   $\searrow$   $\rightarrow$   $\nearrow$

rel max      rel min
$(0, 0)$      $(1, -2)$

on $[-2, 10]$ by $[-10, 10]$

**39.**

| $f' > 0$ | $f' = 0$ | $f' < 0$ | $f'$ und | $f' > 0$ | $f' = 0$ | $f' < 0$ |
|---|---|---|---|---|---|---|
| | $x = -1$ | | $x = 0$ | | $x = 1$ | |
| ↗ | → | ↘ | | ↗ | → | ↘ |
| | rel max $(-1, 2)$ | | rel min $(0, 0)$ | | rel max $(1, 2)$ | |

on $[-5, 5]$ by $[-10, 5]$

**41.** Relative maximum at $x = 1$; relative minimum at $x = 3$

**43.** Relative minimum at $x = 0$ and at $x = 2$; relative maximum at $x = 1$

**45.** Relative maximum at $x = -3$; relative minimum at $x = 3$

**47.** $-0.77, 0,$ and $0.77$

**49.**

**51.**

**53.**

**55.**

**57. a.**

| $f' > 0$ | $f' = 0$ | $f' < 0$ | $f' = 0$ | $f' > 0$ |
|---|---|---|---|---|
| | $x = 1$ | | $x = 5$ | |
| ↗ | → | ↘ | → | ↗ |
| | rel max $(1, 32)$ | | rel min $(5, 0)$ | |

| $f'' < 0$ | $f'' = 0$ | $f'' > 0$ |
|---|---|---|
| | $x = 3$ | |
| con dn | | con up |
| | IP $(3, 16)$ | |

**b.**

**c.** The decline in revenue first begins to slow at $x = 3$ years.

**59. a.**

| $f' = 0$ | $f' < 0$ | $f' = 0$ | $f' > 0$ |
|---|---|---|---|
| $0$ | | $x = 3$ | |
| | ↘ | → | ↗ |
| | | rel min $(3, 85)$ | |

| $f'' = 0$ | $f'' < 0$ | $f'' = 0$ | $f'' > 0$ |
|---|---|---|---|
| $0$ | | $x = 2$ | |
| | con dn | | con up |
| | | IP $(2, 96)$ | |

**b.**

**c.** The decline in temperature first begins to slow at $x = 2$ hours.

**61.**

| $f'$ und | $f' > 0$ |
|---|---|
| $x = 0$ | ↗ |

rel min
(0, 0)

| $f''$ und | $f'' < 0$ |
|---|---|
| $x = 0$ | |

con dn

**63. a.**

| $S'$ und | $S' > 0$ |
|---|---|
| $i = 0$ | ↗ |

| $S''$ und | $S'' < 0$ |
|---|---|
| $i = 0$ | |

con dn

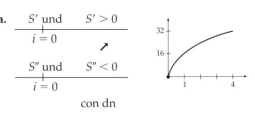

**b.** Concave down. Status increases more slowly at higher income levels.

**65. a.**

| $f' > 0$ | $f' = 0$ | $f' < 0$ | $f' = 0$ | $f' > 0$ |
|---|---|---|---|---|
| ↗ | $x = 3$ → | ↘ | $x = 9$ → | ↗ |

rel max      rel min
(3, 78)      (9, 42)

| $f'' < 0$ | $f'' = 0$ | $f'' > 0$ |
|---|---|---|
| | $x = 6$ | |

con dn     con up
    IP (6, 60)

**b.**

**c.** The inflection point in 2003 is the first sign that the decline in AT&T's net income is beginning to slow.

**67.** (50, 2.5). The curve is concave up (slope increasing) before $x = 50$ and concave down (slope decreasing) after $x = 50$. Therefore, the slope is maximized at $x = 50$.

**69.** Concave up for $a > 0$ and concave down for $a < 0$.

**71.** False. The sign of the second derivative must actually *change* (*positive* on one side and *negative* on the other) for the point to be an inflection point.

**73.** Yes. The second derivative is positive (above the *x*-axis) on one side and negative (below the *x*-axis) on the other, so the concavity *does* change.

**75.** The inflection point was the first sign of the coming downturn. Alternative answer: The sales were growing fastest at the inflection point.

**77.** True

**79.** $f''(x) = 2a$, therefore: For $a > 0, f'' > 0$, so $f$ is concave up.
                         For $a < 0, f'' < 0$, so $f$ is concave down.

**81.** Where the curve is concave *up,* it lies *above* its tangent line, and where it is concave *down,* it lies *below* its tangent line, so *at* an inflection point it must cross its tangent line.

## Exercises 3.3 page 217

**1.** Max $f$ is 12 (at $x = 1$), min $f$ is $-8$ (at $x = -1$).     **3.** Max $f$ is 16 (at $x = -2$), min $f$ is $-16$ (at $x = 2$).

**5.** Max $f$ is 9 (at $x = 1$), min $f$ is 0 (at $x = 0$ and at $x = -2$).

**7.** Max $f$ is 20 (at $x = 1$), min $f$ is $-12$ (at $x = -3$).

**9.** Max $f$ is 81 (at $x = 3$), min $f$ is $-16$ (at $x = 2$).

**11.** Max $f$ is 4 (at $x = 2$), min $f$ is $-50$ (at $x = 5$).     **13.** Max $f$ is 5 (at $x = 0$), min $f$ is 0 (at $x = 5$).

**15.** Max $f$ is 4 (at $x = -1$ and at $x = 2$), min $f$ is 0 (at $x = 0$ and at $x = 3$).

**17.** Max $f$ is 1 (at $x = 0$), min $f$ is 0 (at $x = -1$ and at $x = 1$).

**19.** Max $f$ is $\frac{1}{2}$ (at $x = 1$), min $f$ is $-\frac{1}{2}$ (at $x = -1$).     **21. a.** The number is $\frac{1}{2}$.     **b.** The number is 3.

**23. a.** Both at endpoints     **b.** One at a critical number (the maximum) and one at an endpoint (the minimum)
    **c.** Both at critical numbers     **d.** Yes; for example, [2, 10]

**25.** On the 20th day     **27.** 31 mph     **29.** 52     **31.** 2/3     **33.** 36 years     **35.** 12 miles from A toward B

**37.** Produce 40 per day, price = $400, max profit = $6500

**39.** Rent 40 per day, price = $120, max revenue = $4800

**41.** 400 feet along the building and 200 feet perpendicular to the building

**43.** Each is 200 yards parallel to the river and 150 yards perpendicular to the river

**45.** 3 inches high with a base 12 inches by 12 inches; volume: 432 cubic inches

**47.** The numbers are 25 and 25.     **49.** $r = 110/\pi \approx 35$ yards, $x = 110$ yards

**51.** **a.** At time 10 hours; 1,500,000 bacteria (since $N(t)$ is in thousands)

     **b.** At time 5 hours; growing by 75,000 bacteria per hour (inflection point)

**53.** Remove a square of side $x \approx 0.96$ inch; volume $\approx 15$ square inches

**55.** A function can have at most *one* absolute maximum value (the highest value it attains) but it can have several *relative* maximum values (values that are high *relative* to their neighboring values).

**57.** False. Think of a function like the first graph on page 207.

**59.** For example

There are many others.

**61.** $f$ has a relative minimum at $x = 5$.     **63.** True     **65.** $r = 2$ cm

## Exercises 3.4 page 227

**1.** Price: $14,400; sell 16 cars per day (from $x = 2$ price reductions)

**3.** Ticket price: $150; number sold: 450 (from $x = 5$ price reductions)

**5.** Rent the cars for $90, and expect to rent 54 cars (from $x = 2$ price increases)

**7.** 25 trees per acre (from $x = 5$ extra trees per acre)     **9.** Base: 2 feet by 2 feet; height: 1 foot

**11.** Base: 14 inches by 14 inches; height: 28 inches; volume: 5488 cubic inches

**13.** 50 feet along the driveway and 100 feet perpendicular to the driveway; cost: $800

**15.** 6.4%     **17.** 16 years

**19.** [*Hint:* If area is $A$ (a constant) and one side is $x$, then show that the perimeter is $P = 2x + 2\dfrac{A}{x}$, which is minimized at $x = \sqrt{A}$. Then show that this means the rectangle is a square.]

**21.** The page should be 8 inches wide and 12 inches tall.

**23.** **e.**

      **f.** Price: $325; quantity: 35 bicycles     **g.** Price: $350; quantity: 40 bicycles

**25.** Revenue is price times quantity.

**27.** $65. (*Negative* three price reductions means three price *increases* of $5, so $50 goes up to $65.)

**29.** A linear cost function will have a positive slope (the unit cost), and so will be minimized at its left-hand endpoint. At this point, where $x = 0$, nothing is produced, which is not a realistic solution for a business situation.

**31.** Producing more will generally drive the price down, which will eventually reduce profits.

**33.** No. Revenue and profit will generally be maximized at different points, since one has price subtracted and the other does not. A careful look at the graphs on page 212 will show that revenue and profit are maximized at different $x$-values.

**35.** $R' = 2cpx - 3cx^2 = xc(2p - 3x)$, which is zero when $x = \tfrac{2}{3}p$. The second-derivative test will show that $R$ is maximized.

## Exercises 3.5 page 236

1. Lot size: 400 boxes; 10 orders during the year     3. Lot size: 500 bottles; 20 orders during the year
5. Lot size: 40 cars per order; 20 orders during the year
7. Produce 1000 games per run; 2 production runs during the year
9. Produce 40,000 tapes per run; 25 runs for the year
11. Population: 20,000; yield: 40,000 (from $p = 200$)     13. Population: 75,000; yield: 2250 (from $p = 75$)
15. Population: 26,000 metric tons; yield: 84,500 metric tons (from $p = 26$)
17. Population: 625,000; yield: 625,000 (from $p = 625$)
19. Population: 100,000 metric tons; yield: 50,000 metric tons (from $p = 1000$)
21. Because of the larger number of items on hand, storage costs will increase and reorder costs will decrease.
23. Lot size should decrease.
25. If $f(p) < p$ for all values of $p$, next year's population will always be smaller than this year's population, so the animal will eventually become extinct.
27. Animal populations, such as fish in the sea, can be very difficult to estimate. Also, one year's population depends on factors other than last year's population, such as weather, predators, disease, and changing environmental conditions.

## Exercises 3.6 page 247

1. $\dfrac{dy}{dx} = \dfrac{2x}{3y^2}$     3. $\dfrac{dy}{dx} = \dfrac{3x^2}{2y}$     5. $\dfrac{dy}{dx} = \dfrac{3x^2 + 2}{4y^3}$     7. $\dfrac{dy}{dx} = -\dfrac{x + 1}{y + 1}$     9. $\dfrac{dy}{dx} = -\dfrac{2y}{x}$ (after simplification)

11. $\dfrac{dy}{dx} = \dfrac{-y + 1}{x}$     13. $\dfrac{dy}{dx} = -\dfrac{y - 1}{2x}$ (after simplification)     15. $\dfrac{dy}{dx} = \dfrac{1}{3y^2 - 2y + 1}$

17. $\dfrac{dy}{dx} = -\dfrac{y^2}{x^2}$ (after simplification)     19. $\dfrac{dy}{dx} = \dfrac{3x^2}{2(y - 2)}$     21. $\dfrac{dy}{dx} = 2$     23. $\dfrac{dy}{dx} = -3$     25. $\dfrac{dy}{dx} = -1$

27. $\dfrac{dy}{dx} = -4$     29. $\dfrac{dp}{dx} = -\dfrac{2}{2p + 1}$     31. $\dfrac{dp}{dx} = \dfrac{1}{24p + 4}$     33. $\dfrac{dp}{dx} = -\dfrac{p}{3x}$ (after simplification)

35. $\dfrac{dp}{dx} = -\dfrac{p + 5}{x + 2}$

37. $\dfrac{dp}{dx} = -\dfrac{1}{2}$. *Interpretation:* The rate of change of price with respect to quantity is $-1/2$, so price decreases by about \$0.50 or 50¢ when quantity increases by 1.

39. $\dfrac{dx}{dp} = -1$, so sales will decrease by 1 for each \$1 increase in price.

41. $\dfrac{dh}{dp} = \dfrac{1}{25} = 0.04$, so a person will work for an extra $\frac{1}{25}$ hour (2.4 minutes) for an extra \$1 in pay (at the given pay rate).

43. **a.** $\dfrac{ds}{dr} = \dfrac{3r^2}{2s} = 8$     **b.** $\dfrac{dr}{ds} = \dfrac{2s}{3r^2} = \dfrac{1}{8}$     **c.** $\dfrac{ds}{dr} = 8$ means that the rate of change of sales with respect to research expenditures is 8, so that increasing research by \$1 million will increase sales by about \$8 million (at these levels of $r$ and $s$).
$\dfrac{dr}{ds} = \dfrac{1}{8}$ means that the rate of change of research expenditures with respect to sales is $\dfrac{1}{8}$, so that increasing sales by \$1 million will increase research by about $\dfrac{1}{8}$ million dollars (at these levels of $r$ and $s$).

45. $3x^2 \dfrac{dx}{dt} + 2y \dfrac{dy}{dt} = 0$     47. $2x \dfrac{dx}{dt} y + x^2 \dfrac{dy}{dt} = 0$     49. $6x \dfrac{dx}{dt} - 7 \dfrac{dx}{dt} y - 7x \dfrac{dy}{dt} = 0$

51. $2x \dfrac{dx}{dt} + \dfrac{dx}{dt} y + x \dfrac{dy}{dt} = 2y \dfrac{dy}{dt}$     53. Decreasing by $72\pi \approx 226$ in$^3$/hr

55. Increasing by $32\pi \approx 101$ cm$^3$/week     57. Growing by \$16,000 per day

**59.** Increasing by 400 cases per year     **61.** Slowing by $\frac{1}{2}$ mm/sec per year     **63.** Yes (65.8 mph)

**65.** $\dfrac{dx}{dt} = 2,$ so sales are increasing at the rate of 2 per week.     **67.** $y^3 - x^2 + 1 = 0$

**69.** $\dfrac{dy}{dx}$ would be the distance driven per gallon of gas, or miles per gallon. $\dfrac{dx}{dy}$ would be the amount of gas used per mile of driving, or gallons per mile (which, you would hope, is a small fraction).

**71.** No. The point $(10, 10)$ does not satisfy the original equation and so is not on the curve. Substituting $(10, 10)$ gives a meaningless result.

**73.** $\dfrac{dP}{dt} = -1500,$ where $P$ is the population and $t$ is time in years.

**75.** $\dfrac{dR}{dt} = 2\,\dfrac{dP}{dt},$ where $R$ stands for revenue, $P$ for profit, and $t$ for time in years.

## Chapter 3 Review Exercises page 252

**1.**

**2.**

**3.**

**4.**

**5.**

**6.**

**7.**

**8.**

**9.**

**10.**

**11.**

**12.**

**13.**

**14.**

**15.**

**16.**

**17.**

**18.**

**19.**

**20.**

**21.** Max $f$ is 220 (at $x = 5$), min $f$ is $-4$ (at $x = 1$).     **22.** Max $f$ is 130 (at $x = 5$), min $f$ is $-32$ (at $x = 2$).

**23.** Max $f$ is 64 (at $x = 0$), min $f$ is $-64$ (at $x = 4$).

**24.** Max $f$ is 6401 (at $x = 10$), min $f$ is 1 (at $x = 0$ and $x = 2$).

**25.** Max $h$ is 4 (at $x = 9$), min $h$ is 0 (at $x = 1$).

**26.** Max $f$ is 10 (at $x = 0$), min $f$ is 0 (at $x = 10$ and $x = -10$).

**27.** Max $g$ is 25 (at $w = 3$ and $w = -3$), min $g$ is 0 (at $w = 2$ and $w = -2$).

**28.** Max $g$ is 16 (at $x = 4$), min $g$ is 0 (at $x = 0$ and $x = 8$).

**29.** Max $f$ is $\frac{1}{2}$ (at $x = 1$), min $f$ is $-\frac{1}{2}$ (at $x = -1$).     **30.** Max $f$ is $\frac{1}{4}$ (at $x = 2$), min $f$ is $-\frac{1}{4}$ (at $x = -2$).

**31.** $v = 2c$, which means that the tugboat should travel through the water at twice the speed of the current.

**32.** $v = \sqrt[4]{\dfrac{aw^2}{3b}}$     **33.** 3600 square feet      **34.** 1800 square feet

**35.** 15 cubits (gilded side) by 135 cubits     **36.** Base: 10 inches by 10 inches; height: 5 inches

**37.** Radius: 2 inches; height: 4 inches     **38.** Radius: 2 inches; height: 2 inches

**39.** Price: $2400 each; quantity: 9 per week     **40.** 5 weeks     **41.** $x = \frac{3}{4}$ mile

**42. a.** $t \approx 0.5 = 50\%$    **b.** $2    **43.** Radius $\approx 1.2$ inches; height $\approx 4.8$ inches

**44.** $x \approx 1.59$ inches; volume $\approx 33.07$ cubic inches     **45.** $x \approx 1.13$ inches; volume $\approx 12.13$ cubic inches

**46.** 600 per run; $1\frac{1}{2}$ runs per year (or 3 runs in 2 years)     **47.** Lot size: 50; 10 orders during the year

**48.** Population: 150,000 ($p = 150$); yield: 450,000

**49.** Population: $p = 900$ (thousand); yield: 900 (thousand)

**50.** $\dfrac{dy}{dx} = \dfrac{-6x - 4y}{4x + y}$     **51.** $\dfrac{dy}{dx} = \dfrac{-y^2}{2xy - 1}$ (after simplification)     **52.** $\dfrac{dy}{dx} = \dfrac{-2y^2 + 6xy}{4xy - 3x^2}$

**53.** $\dfrac{dy}{dx} = \dfrac{y^{1/2}}{x^{1/2}}$ (after simplification)     **54.** $\dfrac{dy}{dx} = -1$     **55.** $\dfrac{dy}{dx} = \dfrac{1}{7}$     **56.** $\dfrac{dy}{dx} = -\dfrac{1}{6}$     **57.** $\dfrac{dy}{dx} = 1$

**58.** 600 in$^3$/hr     **59.** Increasing by $4200 per day     **60.** Increasing by $45,000 per month

**61. a.** Decreasing by about 0.31 cm$^3$/min    **b.** Decreasing by about 0.05 cm$^3$/min

## Cumulative Review for Chapters 1–3 page 255

**1.** $y = -\frac{1}{2}x + 1$     **2.** $\frac{5}{2}$     **3.** 20.086

**4. a.**     **b.** 4     **5.** $f'(x) = 4x - 5$    **6.** $f'(x) = 12x^{1/2} + 6x^{-3}$

    **c.** 1

    **d.** Does not exist     **7.** $f'(x) = 9x^8 + 10x^4 - 8x^3$    **8.** $f'(x) = \dfrac{11}{(3x - 2)^2}$

    **e.** Discontinuous at $x = 3$

**9.** $P'(8) = 1200$, so in 8 years the population will be increasing by 1200 people per year.

    $P''(8) = -50$, so in 8 years the rate of growth will be slowing by 50 people per year per year.

**10.** $\dfrac{2x}{\sqrt{2x^2 - 5}}$

**11.** $12(3x + 1)^3(4x + 1)^3 + 12(3x + 1)^4(4x + 1)^2 = 12(3x + 1)^3(4x + 1)^2(7x + 2)$

**12.** $\dfrac{12(x - 2)^2}{(x + 2)^4}$

**13.**

| $f' > 0$ | $f' = 0$ | $f' < 0$ | $f' = 0$ | $f' > 0$ |

$x = -2 \qquad\qquad x = 10$

↗   →   ↘   →   ↗

rel max      rel min
$(-2, 464)$     $(10, -400)$

| $f'' < 0$ | $f'' = 0$ | $f'' > 0$ |

$x = 4$

con dn      con up

IP $(4, 32)$

**14.**

| $f' < 0$ | $f'$ und | $f' > 0$ |

$x = 0$

↘      ↗

rel min
$(0, -1)$

| $f'' < 0$ | $f''$ und | $f'' < 0$ |

$x = 0$

con dn      con dn

**15.** 15,000 square feet     **16.** Price: $170; quantity: 18 per day

**17.** $\dfrac{dy}{dx} = \dfrac{-x^2 - 3y^2}{6xy + 1}$ (after simplification). Evaluating at $(1, 2)$ gives $\dfrac{dy}{dx} = -1$.     **18.** $\dfrac{2}{\pi} \approx 0.64$ ft/min

## Exercises 4.1 page 269

**1. a.** 7.389    **b.** 0.135    **c.** 1.649
**3. a.** $e^3$    **b.** $e^2$    **c.** $e^5$
**5.**                 **7.**               **9.** 5.697 (rounded)

**11. a.** $e^x$    **b.** $e^x$    **c.** $e^x$    **d.** $e^x$    **e.** $e^x$ will exceed any power of $x$ for large enough values of $x$.
**13. a.** $2144    **b.** $2204    **c.** $2226    **15. a.** $2196    **b.** $2096
**17.** The annual yield should be 9.69% (based on the nominal rate of 9.25%).
**19.** $8629    **21.** $72.65    **23.** $751.31    **25.** 10% compounded quarterly (yielding 10.38%, better than 10.30%)
**27. a.** $2758    **b.** $12,456    **29.** 7.28 billion
**31. a.** 0.53 (the chances are better than 50–50)    **b.** 0.70 (quite likely)
**33. a.** 0.267 or 26.7%    **b.** 0.012 or 1.2%    **35. a.** 1.3 milligrams    **b.** 0.84 milligram
**37.** 208    **39. a.** About 153 degrees    **b.** About 123 degrees    **41.** 38    **43.** 6.5%
**45.** By about 25%    **47. b.** In about 2087 (from $x \approx 82$)    **49.** $59,079
**51.** The $x$-axis $(y = 0)$ is a horizontal asymptote.    **53.** $e^x$    **55.** That its growth is proportional to its size.

**57.** With 5% *monthly* you must wait until the end of the month to receive interest, while for 5% compounded *continuously* you begin receiving interest right away, so the interest begins earning interest without any delay.

**59.** Depreciation *by a fixed percentage* gives the bigger decrease in the first year, and *straight line* gives the bigger decrease in the last year.

## Exercises 4.2 page 286

**1. a.** 2    **b.** 4    **c.** $-1$    **d.** $-2$    **e.** $\frac{1}{2}$    **f.** $-\frac{1}{2}$    **3. a.** 10    **b.** $\frac{1}{2}$    **c.** $\frac{4}{3}$    **d.** 0    **e.** 1    **f.** $-3$

**5.** $\ln x$    **7.** $2\ln x$ or $\ln x^2$    **9.** $\ln x$    **11.** $3x$    **13.** $7x$    **15.** Domain: $\{\, x \mid x > 1 \text{ or } x < -1 \,\}$; Range: $\mathbb{R}$

**17. a.** 2.9 years (from 35 months)    **b.** 1.7 years (from 20.5 months)    **19. a.** 15.7 years    **b.** 3.2 years

**21.** 1.9 years    **23.** About 17.1 years    **25.** 77 days    **27.** 0.58 or 58%    **29.** About 4 weeks

**31.** About 31,400 years    **33.** 1.7 million years    **35.** About 79 days    **37.** About 5 years (from 4.67)

**39. a.** 35 years    **b.** 55.5 years

**41. a.** About 11.6 years (from 46.6 quarters)    **b.** About 6.8 years (from 27.2 quarters)

**43. a.** About 9 days    **b.** About 11 days

**45.** About 5300 years old. (For reference, this means that Iceman lived 2000 years before King Tutankhamen.)

**47.** About 6.5 years    **49.** In 7 years and 10 months    **51.** $\ln 1 = 0.$    **53.** $\ln \dfrac{x}{y} = \ln x - \ln y.$    **55.** $x = \log y$

**57.** $\ln(-2) = x$ would mean that $e^x = -2,$ which is impossible because $e^x$ is always positive.

**59.** *Continuous* compounding would give the shortest, and *annual* would give the longest.

**61. b.** About 39 hours    **63.** Solving $0.94 = (1 - \frac{1}{1000})^x$ gives $x = \dfrac{\ln 0.94}{\ln 0.999} \approx 61.84,$ so about 62 generations.

**65. a.** 12 years    **b.** 11.9 years    **67.** $\dfrac{\ln k}{\ln(1 + r)}$

## Exercises 4.3 page 301

**1.** $2x \ln x + x$    **3.** $\dfrac{2}{x}$    **5.** $\dfrac{1}{2}x^{-1}$    **7.** $\dfrac{6x}{x^2 + 1}$    **9.** $\dfrac{1}{x}$    **11.** $\dfrac{xe^x - 2e^x}{x^3}$ or $\dfrac{e^x(x - 2)}{x^3}$

**13.** $(3x^2 + 2)e^{x^3 + 2x}$    **15.** $x^2 e^{x^3/3}$    **17.** $1 + e^{-x}$    **19.** 2    **21.** $e^{1 + e^x}e^x$ or $e^{1 + x + e^x}$    **23.** $ex^{e-1}$    **25.** 0

**27.** $\dfrac{4x^3}{x^4 + 1} - 2e^{x/2} - 1$    **29.** $2x \ln x + 2xe^{x^2}$    **31.** $e^x\left(\ln x^2 + \dfrac{2}{x}\right)$    **33.** $\dfrac{-2}{x}$    **35.** $6e^{2t}(e^{2t} + 1)^2$

**37.** $\dfrac{t + \frac{1}{t}}{\sqrt{t^2 + 2\ln t}}$    **39.** $\dfrac{e^z(z^2 - 2z - 1)}{(z^2 - 1)^2}$    **41.** $\dfrac{20e^{-2z}}{(1 + e^{-2z})^2}$    **43.** $\dfrac{-4}{(e^x - e^{-x})^2}$    **45. a.** $\dfrac{1 - 5\ln x}{x^6}$    **b.** 1

**47. a.** $\dfrac{4x^3}{x^4 + 48}$    **b.** $\dfrac{1}{2}$    **49. a.** $\dfrac{e^x - 3}{e^x - 3x}$    **b.** $-2$    **51. a.** $5\ln x + 5$    **b.** $5\ln 2 + 5 \approx 8.466$

**53. a.** $\dfrac{xe^x - e^x}{x^2}$    **b.** $\dfrac{2e^3}{9} \approx 4.463$    **55.** $-4x^3 e^{-x^5/5} + x^8 e^{-x^5/5}$ or $x^3 e^{-x^5/5}(x^5 - 4)$    **57.** $f^{(n)}(x) = k^n e^{kx}$

**59.**

on $[-2, 2]$ by $[-1, 2]$

rel max: $(0, 1)$
IP: $\left(\frac{1}{2}, 0.61\right)$
$\left(-\frac{1}{2}, 0.61\right)$

**61.**
on $[-5, 5]$ by $[-1, 4]$

rel min: $(0, 0)$
IP: $(1, 0.69)$
$(-1, 0.69)$

**63.**

on $[-1, 8]$ by $[-1, 3]$

rel min: $(0, 0)$
rel max: $(2, 0.54)$
IP: $(0.59, 0.19)$
$(3.41, 0.38)$

**65.**
on $[-2, 2]$ by $[-2, 2]$

rel max: $(-0.37, 0.37)$
rel min: $(0.37, -0.37)$

**67.** $\dfrac{dy}{dx} = \dfrac{ye^x}{2y - e^x}$     **69.** 9479, so annual salary increases by about $9480 per additional year of calculus (after the first year).

**71. a.** Increasing by $50 per year     **b.** Increasing by about $82.44 per year

**73.** Increasing by about 0.134 billion $(= 134$ million) people per year

**75. a.** Decreasing by 0.06 mg/hr     **b.** Decreasing by 0.054 mg/hr

**77. a.** Increasing by about 81.4 (thousand) sales per week     **b.** Increasing by about 33 (thousand) sales per week

**79.** $p = \$100$     **81. a.** $R(x) = 400xe^{-0.20x}$     **b.** Quantity: $x = 5$ (thousand); price: $p = \$147.15$

**83. a.** After 15 minutes the temperature of the beer is 57.5 degrees and increasing at the rate of 43.8 degrees per hour.
**b.** After 1 hour the temperature of the beer is 69.1 degrees and increasing at the rate of 3.2 degrees per hour.

**85.** 2.3 seconds

**87. a.**     **b.** $f(35) = 132.4$, $f'(x) = 0.834$. At age 35, the fastest man's time is 2 hours 12.4 minutes and increasing at about 0.834 minute per year.
**c.** $f(80) = 234.9$, $f'(80) = 7.76$. At age 80, the fastest man's time is 3 hours 54.9 minutes and increasing at about 7.76 minutes per year.

**89.** $e^x$     **91.** 0 (since $e^5$ is a constant)     **93.** $\dfrac{1}{x}$     **95.** 0 (since $\ln 5$ is a constant)

**97.** $e^x$ and $\ln x$ are inverse functions, so $e^{\ln x} = x$, and the derivative of $x$ is 1.

**99.** When $N < K$, $\ln(K/N)$ is positive and the population is increasing, while when $N > K$, $\ln(K/N)$ is negative and the population is decreasing.

**101.** $r = a/b$     **105.** $e^{0.802x}$ (rounded)

**107. a.** $(\ln 10)10^x$     **b.** $(\ln 3)(2x)3^{x^2+1}$     **c.** $(\ln 2)3 \cdot 2^{3x}$     **d.** $(\ln 5)6x \cdot 5^{3x^2}$     **e.** $-(\ln 2)2^{4-x}$

**109. a.** $\dfrac{1}{(\ln 2)x}$     **b.** $\dfrac{2x}{(\ln 10)(x^2 - 1)}$     **c.** $\dfrac{4x^3 - 2}{(\ln 3)(x^4 - 2x)}$

## Exercises 4.4 page 315

**1. a.** $2/t$     **b.** 2 and 0.2     **3. a.** 0.2     **b.** 0.2     **5. a.** $2t$     **b.** 20     **7. a.** $-2t$     **b.** $-20$

**9. a.** $\dfrac{1}{2(t - 1)}$     **b.** $\frac{1}{10}$     **11.** 0.50, so the stock was appreciating by about 50% per year.

**13.** 0.0071 or 0.71%     **15. a.** 0.012 or 1.2%     **b.** Yes, in about 15.3 years

**17. a.** $E(p) = \dfrac{5p}{200 - 5p}$     **b.** Inelastic $\left(E = \frac{1}{3}\right)$     **19. a.** $E(p) = \dfrac{2p^2}{300 - p^2}$     **b.** Unit-elastic $(E = 1)$

**21. a.** $E(p) = 1$     **b.** Unit-elastic $(E = 1)$     **23. a.** $E(p) = \dfrac{3p}{2(175 - 3p)}$     **b.** Elastic $(E = 3)$

**25. a.** $E(p) = 2$     **b.** Elastic $(E = 2)$     **27. a.** $E(p) = 0.01p$     **b.** Elastic $(E = 2)$

**29.** Lower the price $(E = 8)$     **31.** No $(E = 1.25)$     **33.** Yes $\left(E = \frac{3}{8} = 0.375\right)$     **35.** Lower its price $(E = 7.2)$

**37.** $E = 0.112$     **39. a.** $E \approx 0.35$     **b.** Raise the price     **c.** $p \approx \$20,400$     **41.** $E \approx 0.52$ (from 8.65%/16.7%)

**43.** 1.2     **45.** $\dfrac{n}{x}$     **47.** $E(p) = 2p$     **49.** Increase strongly

**51.** Inelastic for cigarettes (they are habit forming), and elastic for jewelry

**53.** $E(p) = \dfrac{-pa(-c)e^{-cp}}{ae^{-cp}} = cp$     **55.** $E_s(p) = n$

## Chapter 4 Review Exercises page 319

**1. a.** $18,845.41     **b.** $18,964.81     **2.** The second bank $(1.015^4 \approx 1.0614 < e^{0.0598} \approx 1.0616)$

**3. a.** $V(t) = 800,000(0.8)^t$     **b.** $327,680     **4.** Drug B     **5.** In about 2060 (from $x \approx 60.2$ years after 2000)

**6.** $4^8 = 65,536$ megabits, which is enough to hold two-hundred fifty-six 16-volume encyclopedias on one chip.

**7. a.** 7.1 years (from 14.2 half-years)     **b.** 4.2 years (from 8.3 half-years)

**8. a.** 9.9 years    **b.** 5.8 years    **9.** 50.7 million years    **10.** 1.85 million years    **11.** 2.3 hours

**12.** 13.7 years    **13. a.** In about 6.25 years (from $x \approx 25$ quarters)    **b.** In about 6.2 years

**14. a.** About 11 days    **b.** About 16 days    **15.** $\dfrac{1}{x}$    **16.** $\dfrac{4x}{x^2 - 1}$    **17.** $\dfrac{-1}{1 - x}$ or $\dfrac{1}{x - 1}$

**18.** $\dfrac{x}{x^2 + 1}$    **19.** $\dfrac{1}{3x}$    **20.** 1    **21.** $\dfrac{2}{x}$    **22.** $\ln x$    **23.** $-2xe^{-x^2}$    **24.** $-e^{1-x}$    **25.** $2x$

**26.** $2x(\ln x)e^{x^2 \ln x - x^2/2}$    **27.** $10x + 2\ln x + 2$    **28.** $6x^2 + 3\ln x + 3$    **29.** $6x^2 - 3e^{2x} - 6xe^{2x}$

**30.** $4 - 4xe^{2x} - 4x^2e^{2x}$

**31.**    rel min: $(0, \ln 4) \approx (0, 1.4)$

IP: $(2, \ln 8) \approx (2, 2.1)$

$(-2, \ln 8) \approx (-2, 2.1)$

**32.**    rel max: $(0, 16)$

IP: $(2, 16e^{-1/2}) \approx (2, 9.7)$

$(-2, 16e^{-1/2}) \approx (-2, 9.7)$

**33. a.** Increasing by 136 thousand per week    **b.** Increasing by 55 thousand per week

**34. a.** Decreasing by 0.12 mg per hour    **b.** Decreasing by 0.08 mg per hour

**35.** Decreasing by 33.3% per second

**36. a.** Increasing by 3.5 degrees per hour    **b.** Increasing by 2.1 degrees per hour

**37. a.** Increasing by 6667 per hour    **b.** Increasing by 816 per hour    **38.** 25 years

**39. a.** $R(x) = 200xe^{-0.25x}$    **b.** Quantity $x = 4$ (thousand); price $p = \$73.58$

**40. a.** $R(x) = 5x - x \ln x$    **b.** Quantity $x = e^4 \approx 54.60$; price $= \$1$    **41.** Price $= \$50$

**42.**    rel min: $(0, 0)$

rel max: $(4, 4.69)$

IP: $(2, 2.17)$

$(6, 3.21)$

on $[-2, 10]$ by $[-1, 10]$

**43.**    rel max: $(-0.72, 0.12)$

rel min: $(0.72, -0.12)$

IP: $(-0.43, 0.07)$

$(0.43, -0.07)$

on $[-2, 2]$ by $[-2, 2]$

**44.** $p \approx \$769$    **45.** $x \approx 130$ (in thousands)    **46.** 0.0033 or 0.33%    **47.** 0.0031 or 0.31%

**48.** Raise prices  $(E = 0.8)$    **49.** Raise prices  $(E = 0.7)$    **50.** $E = 0.44$    **51.** 0.104 or 10.4%

**52. a.** $E \approx 1.29$    **b.** Lower the price    **c.** About \$8700 (from $p \approx 8.7$)

## Exercises 5.1 page 333

**1.** $\frac{1}{5}x^5 + C$    **3.** $\frac{3}{5}x^{5/3} + C$    **5.** $\frac{2}{3}u^{3/2} + C$    **7.** $-\frac{1}{3}w^{-3} + C$    **9.** $2\sqrt{z} + C$    **11.** $x^6 + C$

**13.** $4x^2 - 5x + C$    **15.** $2x^4 - x^3 + 2x + C$    **17.** $4x^{3/2} + \frac{3}{2}x^{2/3} + C$    **19.** $6x^{8/3} + 24x^{-2/3} + C$

**21.** $6t^{5/3} + 3t^{1/3} + C$    **23.** $-2z^{-2} + 2z^{1/2} + C = -\dfrac{2}{z^2} + 2\sqrt{z} + C$    **25.** $\frac{1}{3}x^3 - x^2 + x + C$

**27.** $3x^4 - 4x^3 + C$    **29.** $\frac{2}{3}w^{3/2} + 4w^{5/2} + C$    **31.** $2x^3 - 3x^2 + x + C$    **33.** $\frac{1}{3}x^3 + x^2 - 8x + C$

**35.** $\frac{1}{3}r^3 - r + C$    **37.** $\frac{1}{2}x^2 - x + C$    **39.** $\frac{1}{4}t^4 + t^3 + \frac{3}{2}t^2 + t + C$

**41. b.**

on $[-3, 3]$ by $[-5, 5]$

**43.** $C(x) = 8x^{5/2} - 9x^{5/3} + x + 4000$    **45.** $R(x) = 9x^{4/3} + 2x^{3/2}$

**47. a.** $D(t) = -0.08t^3 + 9t^2$    **b.** 820 feet    **49. a.** $6t^{1/2}$    **b.** 30 words

**51. a.** $P(t) = 16t^{5/2}$    **b.** 512 tons    **c.** No

**53.** $P(x) = -\frac{1}{3}x^3 + 25x^2 - 300x + 31{,}000$; 35,500 thousand (or 35.5 million)

**55.** $I(x) = \frac{800}{3}x^3 - 2400x^2 + 5400x + 4200$; \$7800 million (or \$7.8 billion)

**57.** $\dfrac{1}{x}$    **59.** $C$ (a constant)    **61.** They differ by a constant (since they are integrals of the same function).

**63. a.** $\int x \cdot x \, dx = \int x^2 \, dx = \frac{1}{3}x^3 + C$

   **b.** $x \cdot \int x \, dx = x\left(\frac{1}{2}x^2 + C\right) = \frac{1}{2}x^3 + Cx$   [*Note:* The $C$'s in parts (a) and (b) may differ.]

   **c.** No, so you *cannot* move a variable across the integral sign.

**65. a.** $\int \frac{x}{x} \, dx = \int 1 \, dx = x + C$

   **b.** $\dfrac{\int x \, dx}{\int x \, dx} = \dfrac{\frac{1}{2}x^2 + C}{\frac{1}{2}x^2 + C'}$   [*Note:* The prime is to indicate that constant $C'$ in the denominator may be different from the constant $C$ in the numerator.]

   **c.** No, so the integral of a quotient is *not* the quotient of the integrals.

**67.** Cost   **69. a.** Differentiate   **b.** Integrate   **c.** Differentiate   **d.** Integrate

## Exercises 5.2 page 345

**1.** $\frac{1}{3}e^{3x} + C$   **3.** $4e^{x/4} + C$   **5.** $20e^{0.05x} + C$   **7.** $-\frac{1}{2}e^{-2y} + C$   **9.** $-2e^{-0.5x} + C$   **11.** $9e^{2x/3} + C$

**13.** $-5 \ln|x| + C$   **15.** $3 \ln|x| + C$   **17.** $\frac{3}{2}\ln|v| + C$   **19.** $2e^{2x} - 3x^2 + C$   **21.** $\frac{1}{3}e^{3x} - 3 \ln|x| + C$

**23.** $6e^{0.5t} - 2 \ln|t| + C$   **25.** $\frac{1}{3}x^3 + \frac{1}{2}x^2 + x + \ln|x| - x^{-1} + C$   **27.** $250e^{0.02t} - 200e^{0.01t} + C$

**29.** $\frac{1}{2}e^w - \frac{1}{4}w^2 + C$   **31.** $\frac{1}{2}z^2 + \ln|z| + C$   **33.** $e^x + \ln|x| + C$   **35.** $\frac{1}{2}x^2 + 2x + \ln|x| + C$

**37.** $t + 2 \ln|t| + 3t^{-1} + C$   **39.** $\frac{1}{3}x^3 - 3x^2 + 12x - 8 \ln|x| + C$   **41. a.** $360e^{0.05t} - 355$   **b.** About 624 cases

**43. a.** $50 \ln t$ (since $t > 1$, absolute value bars are not needed)   **b.** No (about 170 sold)

**45. a.** $1000e^{0.02t} - 1000$   **b.** In about 2031 (23 years from 2008)   **47. a.** $500e^{0.4x} - 500$   **b.** About $3195

**49. a.** $60e^{-0.2t} + 10$   **b.** In about 5 hours   **51. a.** $-4000e^{-0.2t} + 4000$   **b.** About $3\frac{1}{2}$ years

**53. a.** $8000e^{0.05t} - 3000$   **b.** About $10,190   **55. a.** $610e^{0.01t} - 610$   **b.** In about 2029 (21 years from 2008)

**57. a.** $3.2e^{0.5t} - 3.2$   **b.** $61.70   **59.** $e^x + C$   **61.** $\frac{1}{e+1}x^{e+1} + C$  (using the Power Rule for $x$ to a constant power)

**63.** $e^{-1}x + C$   [$e^{-1}$ is a constant (it has no variable) and the integral of a constant is the constant times $x$ plus $C$.]

**65.** $\displaystyle\int e^{ax+b} \, dx = \int e^{ax}e^b \, dx = e^b \int e^{ax} \, dx = e^b\left(\frac{1}{a}e^{ax} + C\right) = \frac{1}{a}e^{ax}e^b + Ce^b = \frac{1}{a}e^{ax+b} + C'$ (where $C'$ is another constant)

**67.** No. The box applies only to $\displaystyle\int \frac{1}{x} \, dx$ and not to $\displaystyle\int \frac{1}{x^5} \, dx$. For the latter, use the Power Rule.

## Exercises 5.3 page 360

**1.** 2.75 square units   **3.** 4.15 square units   **5.** 0.760 square unit

**7. i.** 2.8 square units   **ii.** 3 square units

**9. i.** 4.411 square units (or 4.412, depending on rounding)   **ii.** $\frac{14}{3} \approx 4.667$ square units

**11. i.** 0.719 square units   **ii.** $\ln 2 \approx 0.693$ square units

**13. i.** for left rectangles: 2.9, 2.99, 2.999
   for midpoint rectangles: 3, 3, 3
   for right rectangles: 3.1, 3.01, 3.001
   **ii.** 3 square units

**15. i.** for left rectangles: 4.515, 4.652, 4.665
   for midpoint rectangles: 4.668, 4.667, 4.667
   for right rectangles: 4.815, 4.682, 4.668
   **ii.** $\frac{14}{3} \approx 4.667$ square units

**17. i.** for left rectangles: 0.719, 0.696, 0.693
   for midpoint rectangles: 0.693, 0.693, 0.693
   for right rectangles: 0.669, 0.691, 0.693
   **ii.** $\ln 2 \approx 0.693$ square unit

**19.** 9 square units

**21.**

8 square units

**23.**

ln 2  square unit

**25.** 160 square units    **27.** 19 square units
**29.** 2 square units    **31.** 16 square units
**33.** ln 5 square units    **35.** ln 2 $+ \frac{7}{3}$ square units
**37.** 4 square units (from $2e^{\ln 3} - 2$)
**39.** $2e - 2$ square units    **41.** 5 square units
**43.** $\frac{15}{32}$ square unit

**45.** $4 + \ln 2$ square units    **47.** $\frac{111}{100}$    **49.** 13    **51.** $\frac{3}{4}$    **53.** 1 (from $\ln e - \ln 1$)    **55.** $78 + \ln 3$
**57.** $-3 \ln 2$    **59.** $4e^3 - 4$    **61.** $5e - 5e^{-1}$    **63.** 1 (from $e^{\ln 3} - e^{\ln 2}$)    **65.** $\frac{7}{2} + \ln 2$    **67.** 1.107
**69.** 2.925    **71.** 92.744
**73. a.** 9    **b.** Completing the calculation: $= \left(\frac{1}{3}3^3 + C\right) - \left(\frac{1}{3}0^3 + C\right) = \frac{1}{3}3^3 + C - \frac{1}{3}0^3 - C = \frac{1}{3}\cdot 27 = 9$
   **c.** The $C$ always cancels because it is both added and subtracted in the evaluation step.

**75.** $3\frac{1}{2} + \frac{\pi}{4}$    **77.** 132 units    **79.** 411 checks    **81.** $-300e^{-2} + 300 \approx \$259.40$
**83.** $225e^{0.8} - 225 \approx 276$, so about \$2.76    **85.** $26e^{0.1} - 26 \approx 2.7$ million metric tons
**87.** $-30e^{-2} + 30 \approx 26$ words    **89. a.** $\frac{1}{2}, \frac{1}{3}, \frac{1}{4}, \frac{1}{5}$    **b.** Area $= \dfrac{1}{n+1}$    **91.** 87 milligrams

**93.** 586 cars    **95.** 4023 people    **97.** Underestimate    **99.** Overestimate

**101.** The second friend is right: indefinite integrals are antiderivatives plus constants, and definite integrals are limits of Riemann sums. The Fundamental Theorem of Integral Calculus shows how to evaluate one using the other, but the two ideas are not the same.
**103.** 0   (since the integral is "from 1 to 1")
**105.** The Fundamental Theorem of Integral Calculus requires that the function be continuous on the interval, and this function is not even *defined* at $x = 0$, so the theorem cannot be applied. This particular integral cannot be found.
**107.** $\dfrac{a}{-b+1}B^{-b+1} - \dfrac{a}{-b+1}A^{-b+1} = \dfrac{a}{1-b}\left(B^{1-b} - A^{1-b}\right)$    **109.** $-4{,}000{,}000e^{-0.6} + 4{,}000{,}000 \approx \$1{,}804{,}753$

## Exercises 5.4 page 375

**1.** 3    **3.** 4    **5.** $\frac{1}{5}$    **7.** 5    **9.** $\dfrac{104}{3}$ or $34\frac{2}{3}$    **11.** 2    **13.** 3    **15.** $e - 1$    **17.** $10(e^{0.1} - 1) \approx 1.05$
**19.** $\ln 2$    **21.** $\dfrac{1}{n+1}$    **23.** $a + b$    **25.** About 0.845    **27.** 318    **29.** 70°    **31.** About 25.6 tons
**33.** \$3194.53    **35.** \$9.8 billion (from 98 hundred million)    **37.** 6 square units    **39.** $\frac{1}{2}e^4 - e^2 + \frac{1}{2}$ square units

**41. a.**

   **b.** 9 square units

**43. a.**

   **b.** 18 square units

**45.** 4 square units    **47.** 32 square units
**49.** 32 square units    **51.** 32 square units
**53.** $\frac{1}{12}$ square unit    **55.** 2 square units
**57.** 1250 square units
**59.** 5.694 square units (rounded)

**61.** About 123 million    **63. a.** 10    **b.** \$6000    **65.** About \$529 billion
**67.** About \$139 thousand    **69.** 400,000 lives    **71.** $(2x + 5)e^{x^2+5x}$    **73.** $\dfrac{2x + 5}{x^2 + 5x}$
**75.** The curve rises more steeply in the beginning and so spends "more of the time" higher up, and so its average should be higher than the "middle" $y$-value.
**77.** $\dfrac{y_1 + y_2}{2}$   (since the average height of a line is at its midpoint)    **79.** The average value will be greater than 7.
**81.** $-c$    **83.** The additional sales during the month resulting from the advertising.

## Exercises 5.5 page 386

**1.** $60,000    **3.** $10,000    **5.** $7500    **7.** $5285 (rounded)    **9.** $100    **11.** $160,000

**13. a.** $x = 500$    **b.** $50,000    **c.** $25,000    **15. a.** $x = 50$    **b.** $2500    **c.** $7500

**17. a.** $x \approx 119.48$    **b.** $10,065    **c.** $3446 (all rounded)    **19.** 0.52    **21.** 0.35    **23.** 0.2

**25.** $1 - \dfrac{2}{n+1} = \dfrac{n-1}{n+1}$    **27.** 0.46    **29.** 0.28    **31.** $L(x) = x^{2.13}$, Gini index $\approx 0.36$

**33.** $4(x^5 - 3x^3 + x - 1)^3(5x^4 - 9x^2 + 1)$    **35.** $\dfrac{4x^3}{x^4 + 1}$    **37.** $3x^2 e^{x^3}$

**39.** When demand is 0, the price is 1000, or equivalently, at a price of 1000 the demand is 0, so 1000 is such a high price that *no one* will buy the product.

**41.** When demand is 0, the price is 0, or equivalently, at a price of 0 the demand is 0, so whatever they are, you can't even *give* them away.

**43.** Upward, since at a higher price producers should be willing to make more of them.

**45.** No. Such a curve would mean, for example, that the lowest paid 10% of the population is making *more* than 10% of the income, but then they can't be the lowest paid 10%.

**47.** Good thing: everyone has the same income, so no one is poorer than anyone else. Bad thing: everyone is paid the same, no matter whether they are lazy or hard working, so ambition is not rewarded. (Other answers are possible.)

## Exercises 5.6 page 398

**1.** $\frac{1}{10}(x^2 + 1)^{10} + C$    **3.** $\frac{1}{20}(x^2 + 1)^{10} + C$    **5.** $\frac{1}{5}e^{x^5} + C$    **7.** $\frac{1}{6}\ln(x^6 + 1) + C$

**9.** $u = x^3 + 1, du = 3x^2\,dx$: the powers in the integrand and the $du$ do not match.

**11.** $u = x^4, du = 4x^3\,dx$: the powers in the integrand and the $du$ do not match.    **13.** $\frac{1}{24}(x^4 - 16)^6 + C$

**15.** $-\frac{1}{2}e^{-x^2} + C$    **17.** $\frac{1}{3}e^{3x} + C$    **19.** Cannot be found by our substitution formulas

**21.** $\frac{1}{5}\ln|1 + 5x| + C$    **23.** $\frac{1}{4}(x^2 + 1)^{10} + C$    **25.** $\frac{1}{5}(z^4 + 16)^{5/4} + C$

**27.** Cannot be found by our substitution formulas    **29.** $\frac{1}{24}(2y^2 + 4y)^6 + C$    **31.** $\frac{1}{2}e^{x^2 + 2x + 5} + C$

**33.** $\frac{1}{12}\ln|3x^4 + 4x^3| + C$    **35.** $-\frac{1}{12}(3x^4 + 4x^3)^{-1} + C$    **37.** $-\frac{1}{2}\ln|1 - x^2| + C$    **39.** $\frac{1}{16}(2x - 3)^8 + C$

**41.** $\frac{1}{2}\ln(e^{2x} + 1) + C$    **43.** $\frac{1}{2}(\ln x)^2 + C$    **45.** $2e^{x^{1/2}} + C$    **47.** $\frac{1}{4}x^4 + \frac{1}{3}x^3 + C$

**49.** $\frac{1}{6}x^6 + \frac{2}{5}x^5 + \frac{1}{4}x^4 + C$    **51.** $\frac{1}{2}e^9 - \frac{1}{2}$    **53.** $\frac{1}{2}\ln 2$    **55.** $32\frac{2}{3}$    **57.** $-\ln 2$    **59.** $3e^2 - 3e$

**61. a.** $u^n u'$    **b.** $\int u^n u'\,dx$    **63. a.** $\dfrac{u'}{u}$    **b.** $\int \dfrac{u'}{u}\,dx$    **65.** $\dfrac{1}{2}\ln(2x + 1) + 50$    **67.** $\dfrac{1}{2}$ million

$\qquad\qquad$ └── agree ──┘ $\qquad\qquad\qquad$ └── agree ──┘

**69.** $\frac{1}{3}\ln 5 - \frac{1}{3}\ln 2 \approx 0.305$ million    **71.** $20\frac{1}{3}$ units    **73.** About 346    **75.** $\frac{1}{2}\ln 10 \approx 1.15$ tons

**77.** 56 pounds    **79.** Yes—one is as easy (or as hard) as the other.

**81.** It simply changes the "name" of the variable, changing, for example, $\int e^{x^2}\,dx$ into $\int e^{u^2}\,du$, which is just as difficult.

**83.** It ignores the fact that $dx$ and $du = 2x\,dx$ are different, so one cannot be substituted for the other.

**85.** $\frac{1}{6}(x - 5)^6 + \frac{6}{5}(x - 5)^5 + C$    **87.** $\frac{1}{8}(x - 2)^8 + \frac{2}{7}(x - 2)^7 + C$

**89.** $x - 2 + \ln|x - 2| + C = x + \ln|x - 2| + C'$    **91.** $\frac{3}{7}(x - 4)^{7/3} + 3(x - 4)^{4/3} + C$

**93.** $\frac{2}{3}(x + 2)^{3/2} - 4(x + 2)^{1/2} + C$

## Chapter 5 Review Exercises page 403

**1.** $8x^3 - 4x^2 + x + C$    **2.** $3x^4 + 3x^2 - 3x + C$    **3.** $4x^{3/2} - 5x + C$    **4.** $6x^{4/3} - 2x + C$

**5.** $6x^{5/3} - 2x^2 + C$    **6.** $2x^{5/2} - 3x^2 + C$    **7.** $\frac{1}{3}x^3 - 16x + C$    **8.** $x^3 + x^2 + 4x + C$

**9.** $C(x) = 2x^{1/2} + 4x + 20{,}000$     **10. a.** $P(t) = 200t^{3/2} + 40{,}000$     **b.** 52,800 people

**11.** $2e^{x/2} + C$     **12.** $-\frac{1}{2}e^{-2x} + C$     **13.** $4\ln|x| + C$     **14.** $2\ln|x| + C$     **15.** $2e^{3x} - 6\ln|x| + C$

**16.** $\frac{1}{2}x^2 - \ln|x| + C$     **17.** $3x^3 + 2\ln|x| + 2e^{3x} + C$     **18.** $-x^{-1} + \ln|x| - e^{-x} + C$

**19. a.** $1500e^{0.05t} - 1500$     **b.** In about 2072 (64 years after 2008)

**20. a.** $2000e^{0.1t} - 2000$     **b.** About 5.6 years     **21. a.** $550e^{0.02t} - 550$     **b.** About 2038 (30 years after 2008)

**22. a.** $200\ln x$     **b.** In about 20 months     **23.** 36     **24.** 90     **25.** 4     **26.** $\ln 5$

**27.** $1 - e^{-2}$     **28.** $2e - 2$     **29.** $20e^5 - 100e + 80$     **30.** $25e^{0.4} - 50e^{0.2} + 25$     **31.** 13 square units

**32.** 36 square units     **33.** $6e^6 - 6$ square units     **34.** $2e^2 - 2$ square units     **35.** $\ln 100$ square units

**36.** $\ln 1000$ square units

**37.** 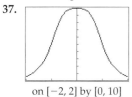     Area $\approx 21.4$ square units     **38.** 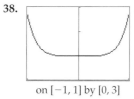     Area $\approx 2.54$ square units

on $[-2, 2]$ by $[0, 10]$     on $[-1, 1]$ by $[0, 3]$

**39.** 64 pounds     **40.** 9 words     **41.** About 6.3 degrees     **42.** $1640     **43.** $653.39

**44.** About 143 pages     **45. a.** 2.28 square units     **b.** $\frac{8}{3} \approx 2.667$ square units

**46. a.** 4.884 square units     **b.** $\frac{16}{3} \approx 5.333$ square units

**47. a.** for left rectangles: 5.899, 7.110, 7.239     **b.** $e^2 - e^{-2} \approx 7.254$ square units
       for midpoint rectangles: 7.206, 7.253, 7.254
       for right rectangles: 8.801, 7.400, 7.268

**48. a.** for left rectangles: 1.506, 1.398, 1.387     **b.** $\ln 4 \approx 1.386$ square units
       for midpoint rectangles: 1.383, 1.386, 1.386
       for right rectangles: 1.281, 1.375, 1.385

**49.** $\frac{4}{3}$ square units     **50.** 108 square units     **51.** $\frac{1}{6}$ square unit     **52.** $\frac{3}{10}$ square unit

**53.** 2 square units     **54.** $\frac{1}{2}$ square unit     **55.** About 17.13 square units     **56.** About 0.496 square unit

**57.** $\frac{1}{3}\ln 4$     **58.** $\frac{28}{3}$ or $9\frac{1}{3}$     **59.** About 4.72     **60.** About 2.77     **61.** About 493 million

**62.** About $5800     **63.** About $629,000 (from 6.29 hundred thousand dollars)     **64.** About $49.95

**65.** 27 square meters     **66.** About $934 billion     **67.** $480,000

**68.** $160,000     **69.** About $5623     **70.** About $611     **71.** About 0.56     **72.** About 0.43

**73.** About 0.23     **74.** About 0.68     **75.** $\frac{1}{4}(x^3 - 1)^{4/3} + C$     **76.** $\frac{1}{6}(x^4 - 1)^{3/2} + C$

**77.** Cannot be integrated by our substitution formulas

**78.** Cannot be integrated by our substitution formulas

**79.** $-\frac{1}{3}\ln|9 - 3x| + C$     **80.** $-\frac{1}{2}\ln|1 - 2x| + C$     **81.** $\frac{1}{3}(9 - 3x)^{-1} + C$     **82.** $\frac{1}{2}(1 - 2x)^{-1} + C$

**83.** $\frac{1}{2}(8 + x^3)^{2/3} + C$     **84.** $(9 + x^2)^{1/2} + C$     **85.** $-\frac{1}{2}(w^2 + 6w - 1)^{-1} + C$     **86.** $-\frac{1}{2}(t^2 - 4t + 1)^{-1} + C$

**87.** $\frac{2}{3}(1 + \sqrt{x})^3 + C$     **88.** $(1 + x^{1/3})^3 + C$     **89.** $\ln|e^x - 1| + C$     **90.** $\ln|\ln x| + C$     **91.** $\frac{61}{3}$ or $20\frac{1}{3}$

**92.** 2     **93.** 2     **94.** $\frac{5}{12}$     **95.** $\ln 7$     **96.** $-\ln 2$     **97.** $\frac{1}{4}e - \frac{1}{4}$     **98.** $\frac{1}{5}e - \frac{1}{5}$     **99.** 8 square units

**100.** 5 square units     **101.** $\frac{1}{4} - \frac{1}{4}e^{-4} = \frac{1}{4}(1 - e^{-4}) \approx 0.25$     **102.** $\frac{1}{4}\ln 13 \approx 0.64$

**103.** $C(x) = (2x + 9)^{1/2} + 97$     **104.** $\ln 28 \approx 3.33$ degrees


## Exercises 6.1 page 418

**1.** $\frac{1}{2}e^{2x} + C$     **3.** $\frac{1}{2}x^2 + 2x + C$     **5.** $\frac{2}{3}x^{3/2} + C$     **7.** $\frac{1}{5}(x + 3)^5 + C$     **9.** $\frac{1}{2}xe^{2x} - \frac{1}{4}e^{2x} + C$

**11.** $\frac{1}{6}x^6 \ln x - \frac{1}{36}x^6 + C$     **13.** $(x + 2)e^x - e^x + C$     **15.** $\frac{2}{3}x^{3/2} \ln x - \frac{4}{9}x^{3/2} + C$

**17.** $\frac{1}{6}(x - 3)(x + 4)^6 - \frac{1}{42}(x + 4)^7 + C$     **19.** $-2te^{-0.5t} - 4e^{-0.5t} + C$     **21.** $-t^{-1}\ln t - t^{-1} + C$

**23.** $\frac{1}{10}s(2s + 1)^5 - \frac{1}{120}(2s + 1)^6 + C$     **25.** $-\frac{1}{2}xe^{-2x} - \frac{1}{4}e^{-2x} + C$     **27.** $2x(x + 1)^{1/2} - \frac{4}{3}(x + 1)^{3/2} + C$

**29.** $\frac{1}{a}xe^{ax} - \frac{1}{a^2}e^{ax} + C$     **31.** $\frac{1}{n + 1}x^{n+1}\ln ax - \frac{1}{(n + 1)^2}x^{n+1} + C$     **33.** $x\ln x - x + C$

**35.** $\dfrac{1}{2}x^2e^{x^2} - \dfrac{1}{2}e^{x^2} + C$    **37. a.** $\dfrac{1}{2}e^{x^2} + C$  (by substitution)    **b.** $\dfrac{1}{4}(\ln x)^4 + C$  (by substitution)

    **c.** $\dfrac{1}{3}x^3 \ln 2x - \dfrac{1}{9}x^3 + C$  (by parts)    **d.** $\ln(e^x + 4) + C$  (by substitution)    **39.** $e^2 + 1$    **41.** $9 \ln 3 - 3 + \dfrac{1}{9}$

**43.** $\dfrac{2^6}{30} = \dfrac{32}{15}$    **45.** $4 \ln 4 - 3$    **47. a.** $\dfrac{1}{6}x(x-2)^6 - \dfrac{1}{42}(x-2)^7 + C$    **b.** $\dfrac{1}{7}(x-2)^7 + \dfrac{1}{3}(x-2)^6 + C$

**49.** Using $u = x^n$ and $dv = e^x dx$, the result follows immediately.    **51.** $x^2e^x - 2xe^x + 2e^x + C$

**53.** $R(x) = 4xe^{x/4} - 16e^{x/4} + 16$    **55.** \$105.7 million    **57.** $-14e^{-2.5} + 4 \approx 2.85$ milligrams

**59.** $2 \ln 2 - 1 + \dfrac{1}{4} \approx 0.64$ square unit    **61.** $72 \ln 6 - 24 + \dfrac{1}{9} \approx 105$ thousand customers

**63.** About 5.7 liters    **65. (b)** $\int u \, dv = uv - \int v \, du$    **67.** $v = e^x$    **69.** $v = x$    **71.** $u = \ln x$

**73. a.** The result follows immediately.    **b.** [Hint: Think of the C.]    **75.** $-x^2e^{-x} - 2xe^{-x} - 2e^{-x} + C$

**77.** $(x+1)^2e^x - 2(x+1)e^x + 2e^x + C = (x^2+1)e^x + C$  (after simplification)    **79.** $\dfrac{1}{3}x^3(\ln x)^2 - \dfrac{2}{9}x^3 \ln x + \dfrac{2}{27}x^3 + C$

**81.** $2e^2 - 2 \approx 12.78$    **83.** $-x^2e^{-x} - 2xe^{-x} - 2e^{-x} + C = -e^{-x}(x^2 + 2x + 2) + C$

**85.** $\dfrac{1}{2}x^3e^{2x} - \dfrac{3}{4}x^2e^{2x} + \dfrac{3}{4}xe^{2x} - \dfrac{3}{8}e^{2x} + C$    **87.** $\dfrac{1}{3}(x-1)^3e^{3x} - \dfrac{1}{3}(x-1)^2e^{3x} + \dfrac{2}{9}(x-1)e^{3x} - \dfrac{2}{27}e^{3x} + C$

## Exercises 6.2 page 429

**1.** Formula 12, $a = 5$, $b = -1$    **3.** Formula 14, $a = -1$, $b = 7$    **5.** Formula 9, $a = -1$, $b = 1$

**7.** $\dfrac{1}{6} \ln \left| \dfrac{3+x}{3-x} \right| + C$    **9.** $-\dfrac{1}{x} - 2 \ln \left| \dfrac{x}{2x+1} \right| + C$    **11.** $-x - \ln |1 - x| + C$    **13.** $\ln \left| \dfrac{2x+1}{x+1} \right| + C$

**15.** $\dfrac{x}{2}\sqrt{x^2 - 4} - 2 \ln \left| x + \sqrt{x^2 - 4} \right| + C$    **17.** $-\ln \left| \dfrac{1 + \sqrt{1 - z^2}}{z} \right| + C$

**19.** $\dfrac{1}{2}x^3e^{2x} - \dfrac{3}{4}x^2e^{2x} + \dfrac{3}{4}xe^{2x} - \dfrac{3}{8}e^{2x} + C$    **21.** $-\dfrac{1}{100}x^{-100} \ln x - \dfrac{1}{10{,}000}x^{-100} + C$

**23.** $\dfrac{1}{3} \ln \left| \dfrac{x}{x+3} \right| + C$    **25.** $\dfrac{1}{8} \ln \left| \dfrac{z^2 - 2}{z^2 + 2} \right| + C$    **27.** $\dfrac{x}{2}\sqrt{9x^2 + 16} + \dfrac{8}{3} \ln \left| 3x + \sqrt{9x^2 + 16} \right| + C$

**29.** $-\dfrac{1}{2} \ln \left| \dfrac{2 + \sqrt{4 - e^{2t}}}{e^t} \right| + C$    **31.** $\dfrac{1}{2} \ln \left| \dfrac{e^t - 1}{e^t + 1} \right| + C$    **33.** $\dfrac{1}{4} \ln \left| x^4 + \sqrt{x^8 - 1} \right| + C$

**35.** $\dfrac{1}{3} \ln \left| \dfrac{\sqrt{x^3 + 1} - 1}{\sqrt{x^3 + 1} + 1} \right| + C$    **37.** $\dfrac{1}{2} \ln \left| \dfrac{e^t - 1}{e^t + 1} \right| + C$    **39.** $2xe^{x/2} - 4e^{x/2} + C$

**41.** $\dfrac{1}{4} \ln \left| \dfrac{e^{-x} + 4}{e^{-x}} \right| + C = \dfrac{1}{4} \ln(1 + 4e^x) + C$    **43.** $\dfrac{15}{2} - 8 \ln 8 + 8 \ln 4 \approx 1.95$    **45.** $\dfrac{1}{2} \ln \dfrac{1}{2} - \dfrac{1}{2} \ln \dfrac{1}{3} \approx 0.203$

**47.** $-4 + 5 \ln 3 \approx 1.49$    **49.** $\dfrac{1}{2} \ln |2x + 6| + C$    **51.** $\dfrac{x}{2} - \dfrac{3}{2} \ln |2x + 6| + C$    **53.** $-\dfrac{1}{3}(1 - x^2)^{3/2} + C$

**55.** $\sqrt{1 - x^2} - \ln \left| \dfrac{1 + \sqrt{1 - x^2}}{x} \right| + C$    **57.** $\dfrac{1}{2}\left( \ln |x + 1| - \dfrac{1}{3} \ln |3x + 1| \right) - \dfrac{1}{2} \ln \left| \dfrac{3x+1}{x+1} \right| + C$

**59.** $\ln \left| x + \sqrt{x^2 + 1} \right| - \ln \left| \dfrac{1 + \sqrt{x^2 + 1}}{x} \right| + C$    **61.** $x + 2 \ln |x - 1| + C$

**63.** $-x^2e^{-x} - 2xe^{-x} - 2e^{-x} + 2$ million sales    **65.** 24 generations    **67.** $C(x) = \ln\left(x + \sqrt{x^2 + 1}\right) + 2000$

**69.** Formula 9 with $x = e^{50t}$    **71.** Formula 18 with $x = t^{50}$    **73.** 476 uses

**75.** By a formula (either 22 or 23) and by integration by parts (with $u = \ln t$ and $dv = dt$)

**77.** Formula 10 [with denominator $(x + 1)(x - 1) = x^2 - 1$] and formula 15

## Exercises 6.3 page 440

**1.** 0    **3.** 1    **5.** Does not exist    **7.** Does not exist    **9.** 0    **11.** 0    **13.** Does not exist    **15.** 0

**17.** $\dfrac{1}{2}$    **19.** $\dfrac{1}{8}$    **21.** Divergent    **23.** 100    **25.** 20    **27.** $\dfrac{1}{2}$    **29.** Divergent    **31.** $\dfrac{1}{3}$    **33.** $\dfrac{1}{3}$

**35.** Divergent    **37.** 1    **39.** Divergent    **41.** $\int_0^\infty e^{\sqrt{x}}\,dx$ diverges and $\int_0^\infty e^{-x^2}\,dx$ converges to 0.88623

**43.** $200,000    **45. a.** $10,000    **b.** $9999.55    **47.** About $3,963,000 (from 3963 thousand)

**49.** 1,000,000 barrels (from 1000 thousand)    **51.** 2 square units    **53.** $\dfrac{1}{a}$ square units    **55.** 0.61 or 61%

**57.** 0.30 or 30%    **59.** 20,000    **61. (b)** $\displaystyle\lim_{x\to\infty}\frac{1}{x^3}$ exists (and equals zero)

**63. (a)** $\displaystyle\lim_{x\to-\infty} e^{3x}$ exists (and equals zero)    **65.** False (see Example 4 on page 435)

**67.** Diverge [*Hint:* Think of the graph.]    **69.** True [*Hint:* Think of the graph.]    **71.** $D/r$    **73.** $40,992

## Exercises 6.4 page 453

Some answers may vary depending on rounding.
  **1. a.** 8.75    **b.** 8.667    **c.** 0.083    **d.** 1%    **3. a.** 0.697    **b.** 0.693    **c.** 0.004    **d.** 0.6%
  **5.** 1.154    **7.** 0.743    **9.** 0.593    **11.** 8.6968    **13.** 2.925    **15.** 0.4772 or about 48%
  **17.** About 70 billion tons (from 69.7)    **19.** 8.667    **21.** 0.693    **23.** 1.148    **25.** 0.747
  **27.** 0.593    **29.** 8.69678496    **31.** 2.92530    **33. a.** $-\displaystyle\int_1^0 \frac{t}{1+t^3}\,dt = \int_0^1 \frac{t}{1+t^3}\,dt$    **b.** 0.374
  **35.** 821 feet    **37.** About 1062 trillion dollars
  **39.** The second derivative of a linear function is zero, so the error formula gives zero error, making the approximation exact. (Alternatively, the area under the graph of a linear function *is* a trapezoid, so approximating by a trapezoid will give the exact answer.)
  **41.** 9    **43.** $\frac{1}{2}\cdot 0 + 4 + 5 + 3 + \frac{1}{2}\cdot 0 = 12$ square units
  **45.** $n = 2$ (Since the method is exact for *any n*, use the smallest number possible.)
  **47.** The justification follows from carrying out the indicated steps.

## Chapter 6 Review Exercises page 457

  **1.** $\frac{1}{2}xe^{2x} - \frac{1}{4}e^{2x} + C$    **2.** $-xe^{-x} - e^{-x} + C$    **3.** $\frac{1}{9}x^9 \ln x - \frac{1}{81}x^9 + C$    **4.** $\frac{4}{5}x^{5/4}\ln x - \frac{16}{25}x^{5/4} + C$
  **5.** $\frac{1}{6}(x-2)(x+1)^6 - \frac{1}{42}(x+1)^7 + C$    **6.** $\frac{1}{5}(x+3)(x-1)^5 - \frac{1}{30}(x-1)^6 + C$    **7.** $2t^{1/2}\ln t - 4t^{1/2} + C$
  **8.** $\frac{1}{4}x^4 e^{x^4} - \frac{1}{4}e^{x^4} + C$    **9.** $x^2 e^x - 2xe^x + 2e^x + C$    **10.** $x(\ln x)^2 - 2x\ln x + 2x + C$
  **11.** $\dfrac{1}{n+1}x(x+a)^{n+1} - \dfrac{1}{(n+1)(n+2)}(x+a)^{n+2} + C$    **12.** $-\dfrac{1}{n+1}x(1-x)^{n+1} - \dfrac{1}{(n+1)(n+2)}(1-x)^{n+2} + C$
  **13.** $4e^5 + 1$    **14.** $\frac{1}{4}e^2 + \frac{1}{4}$    **15.** $-\ln|1-x| + C$    **16.** $-\frac{1}{2}e^{-x^2} + C$    **17.** $\frac{1}{4}x^4 \ln 2x - \frac{1}{16}x^4 + C$
  **18.** $(1-x)^{-1} + C$    **19.** $\frac{1}{2}(\ln x)^2 + C$    **20.** $\frac{1}{2}\ln(e^{2x} + 1) + C$    **21.** $2e^{\sqrt{x}} + C$    **22.** $\frac{1}{8}(e^{2x}+1)^4 + C$
  **23.** $-15{,}000e^{-0.5} + 10{,}000 \approx 902$ million dollars    **24.** 6.78 hundred gallons (from $25 - 10e^{0.6}$)
  **25.** $\dfrac{1}{10}\ln\left|\dfrac{5+x}{5-x}\right| + C$    **26.** $\dfrac{1}{4}\ln\left|\dfrac{x-2}{x+2}\right| + C$    **27.** $2\ln|x-2| - \ln|x-1| + C$
  **28.** $-\ln\left|\dfrac{x-1}{x-2}\right| + C$ or $\ln\left|\dfrac{x-2}{x-1}\right| + C$    **29.** $\ln\left|\dfrac{\sqrt{x+1}-1}{\sqrt{x+1}+1}\right| + C$    **30.** $\dfrac{2x-4}{3}\sqrt{x+1} + C$
  **31.** $\ln\left|x + \sqrt{x^2+9}\right| + C$    **32.** $\ln\left|x + \sqrt{x^2+16}\right| + C$    **33.** $\dfrac{z^2-2}{3}\sqrt{z^2+1} + C$ (from formula 13)
  **34.** $e^t - 2\ln(e^t + 2) + C$    **35.** $\frac{1}{2}x^2 e^{2x} - \frac{1}{2}xe^{2x} + \frac{1}{4}e^{2x} + C$
  **36.** $x(\ln x)^4 - 4x(\ln x)^3 + 12x(\ln x)^2 - 24x\ln x + 24x + C$    **37.** $\ln\left|\dfrac{2x+1}{x+1}\right| + 1000$
  **38.** 1305 (from $750 + 800\ln 80 - 800\ln 40$)    **39.** $\dfrac{1}{4}$    **40.** $\dfrac{1}{5}$    **41.** Divergent    **42.** Divergent
  **43.** $\frac{1}{2}$    **44.** $2e^{-2}$    **45.** Divergent    **46.** Divergent    **47.** 5    **48.** $10e^{-10}$    **49.** $\frac{1}{4}$    **50.** $\frac{1}{5}$
  **51.** $\frac{1}{2}$    **52.** $\frac{1}{4}$    **53.** 1    **54.** 1    **55.** $\frac{1}{3}$    **56.** $\frac{1}{2}$    **57.** $60,000    **58.** 0.35 or 35%    **59.** 240 thousand

**60.** 14,000 metric tons    **61.** $\int_1^\infty \frac{1}{x^3}\,dx$  converges to 0.5    **62.** $\int_1^\infty \frac{1}{\sqrt[3]{x}}\,dx$  diverges    **63.** 1.102    **64.** 1.09

**65.** 1.204    **66.** 0.852    **67.** 0.570    **68.** 1.313    **69.** 1.089    **70.** 1.075    **71.** 1.195    **72.** 0.856

**73.** 0.528    **74.** 1.348    **75.** 1.0894    **76.** 1.0747    **77.** 1.1951    **78.** 0.8556    **79.** 0.5285    **80.** 1.3357

**81.** 1.089429    **82.** 1.074669    **83.** 1.194958    **84.** 0.855624    **85.** 0.527887    **86.** 1.347855

**87. a.** $-\int_1^0 \frac{1}{1+t^2}\,dt = \int_0^1 \frac{1}{1+t^2}\,dt$    **b.** 0.783    **88. a.** $-\int_1^0 \frac{1}{1+t^4}\,dt = \int_0^1 \frac{1}{1+t^4}\,dt$    **b.** 0.862

## Exercises 7.1 page 471

**1.** $\{(x, y) \mid x \neq 0, y \neq 0\}$    **3.** $\{(x, y) \mid x \neq y\}$    **5.** $\{(x, y) \mid x > 0, y \neq 0\}$
**7.** $\{(x, y, z) \mid x \neq 0, y \neq 0, z > 0\}$    **9.** 3    **11.** 4    **13.** $-2$    **15.** 1    **17.** $e^{-1} + e$    **19.** $e^{-1}$
**21.** 0    **23.** 0.0157    **25.** 45 minutes    **27.** 472.7
**29.** $P(2L, 2K) = a(2L)^b(2K)^{1-b} = a2^bL^b2^{1-b}K^{1-b} = 2aL^bK^{1-b} = 2P(L, K)$    **31.** 20,000 calls

**33.** $C(x, y) = 210x + 180y + 4000$    **35. a.** $V = xyz$    **b.** $M = xy + 2xz + 2yz$

**37. a.**
  **b.** A given wind speed will lower the windchill further on a colder day than on a warmer day.
  **c.** For the lowest curve: $dy/dx \approx -0.63$, meaning that at 20 degrees and 10 mph of wind, windchill drops by about 0.63 degrees for each additional 1 mph of wind. For the highest curve: $dy/dx \approx -0.33$, meaning that at 50 degrees and 10 mph of wind, windchill drops by only about 0.33 degrees for each additional 1 mph of wind.
  **d.** Yes—the effect of wind on the windchill index is greater on a colder day.
**39.** 0.64 or 64%    **41.** Coordinate system (b). (The other is a *left*-handed coordinate system.)
**43.** (a) has a relative minimum point and (b) has a relative maximum point.
**45.** False—the order is backwards. See page 462 for the correct definition.    **47.** Satisfies (a)    **49.** Satisfies (b)

## Exercises 7.2 page 485

**1. a.** $3x^2 + 6xy^2 - 1$    **b.** $6x^2y - 6y^2 + 1$    **3. a.** $6x^{-1/2}y^{1/3}$    **b.** $4x^{1/2}y^{-2/3}$

**5. a.** $5x^{-0.95}y^{0.02}$    **b.** $2x^{0.05}y^{-0.98}$    **7. a.** $-(x + y)^{-2}$    **b.** $-(x + y)^{-2}$    **9. a.** $\dfrac{3x^2}{x^3 + y^3}$    **b.** $\dfrac{3y^2}{x^3 + y^3}$
**11. a.** $6x^2e^{-5y}$    **b.** $-10x^3e^{-5y}$    **13. a.** $ye^{xy}$    **b.** $xe^{xy}$

**15. a.** $\dfrac{x}{x^2 + y^2}$  or  $x(x^2 + y^2)^{-1}$    **b.** $\dfrac{y}{x^2 + y^2}$  or  $y(x^2 + y^2)^{-1}$    **17. a.** $3v(uv - 1)^2$    **b.** $3u(uv - 1)^2$

**19. a.** $ue^{(u^2 - v^2)/2}$    **b.** $-ve^{(u^2 - v^2)/2}$    **21.** $18, -10$    **23.** $0, 2e$    **25.** $1\frac{1}{2}$
**27. a.** $30x - 4y^3$    **b. and c.** $-12xy^2$    **d.** $-12x^2y + 36y^2$
**29. a.** $-2x^{-5/3}y^{2/3}$    **b. and c.** $2x^{-2/3}y^{-1/3} - 12y^2$    **d.** $-2x^{1/3}y^{-4/3} - 24xy$

**31. a.** $ye^x$    **b. and c.** $e^x - \dfrac{1}{y}$    **d.** $xy^{-2}$    **33.** All three are $36x^2y^2$.    **35. a.** $y^2z^3$    **b.** $2xyz^3$    **c.** $3xy^2z^2$
**37. a.** $8x(x^2 + y^2 + z^2)^3$    **b.** $8y(x^2 + y^2 + z^2)^3$    **c.** $8z(x^2 + y^2 + z^2)^3$
**39. a.** $2xe^{x^2+y^2+z^2}$    **b.** $2ye^{x^2+y^2+z^2}$    **c.** $2ze^{x^2+y^2+z^2}$    **41.** $-14$    **43.** $4e^6$
**45. a.** $P_x = 4x - 3y + 150$    **b.** \$50 (profit per additional DVD player)    **c.** $P_y = -3x + 6y + 75$
  **d.** \$75 (profit per additional CD player)
**47. a.** 250 (the marginal productivity of labor is 250, so production increases by about 250 for each additional unit of labor)
  **b.** 108 (the marginal productivity of capital is 108, so production increases by about 108 for each additional unit of capital)    **c.** Labor

**49.** $S_x = -0.1$   (sales fall by 0.1 for each dollar price increase)

$S_y = 0.4y$   (sales rise by $0.4y$ for each additional advertising dollar above the level $y$)

**51. a.** 0.52 (status increases by about 0.52 unit for each additional $1000 of income)

   **b.** 5.25 (status increases by 5.25 units for each additional year of education)

**53. a.** 97.2 (skid distance increases by about 97 feet for each additional ton)

   **b.** 12.96 (skid distance increases by about 13 feet for each additional mph)

**55. a.** 0.001152      **b.** 3.69, so about 4 more games      **57.** 0      **59.** $\dfrac{\partial f}{\partial x} = 3, \quad \dfrac{\partial f}{\partial y} = -2$

**61.** $\dfrac{\partial f}{\partial x}$ is negative; $\dfrac{\partial f}{\partial y}$ is positive.      **63.** $f(x, y) = -x + y$      **65.** $f(x, y) = \pi x + ey + \ln 2$

**67. a.** $\dfrac{\partial D}{\partial E} > 0, \ \dfrac{\partial D}{\partial T} < 0$;   interpretation: Your desire to complete the task should increase with the expectation of

success and decrease with the time needed to complete the task.

**69. a.** Rate at which butter sales change as butter prices rise

   **b.** Negative: as prices rise, sales will fall.

   **c.** Rate at which butter sales change as margarine prices rise

   **d.** Positive: as margarine prices rise, people will switch to butter, so butter sales will rise.

**71.** Harmonic      **73.** Not harmonic

**75.** From taking the limit as $h \to 0$:   $\dfrac{\partial f}{\partial x} = 4x - 3$  and  $\dfrac{\partial f}{\partial y} = 6y + 7$

**77.** From taking the limit as $h \to 0$:   $\dfrac{\partial f}{\partial x} = 10x - 3$  and  $\dfrac{\partial f}{\partial y} = 7$

## Exercises 7.3 page 497

**1.** Rel min value:  $f = 5$  at  $x = 0, \ y = -1$      **3.** Rel min value:  $f = -12$  at  $x = -2, \ y = 2$

**5.** Rel max value:  $f = 23$  at  $x = 5, \ y = 2$      **7.** No rel extreme values [saddle point at $(2, -4)$]

**9.** No rel extreme values      **11.** Rel min value:  $f = 1$  at  $x = 0, \ y = 0$

**13.** Rel min value:  $f = 0$  at  $x = 0, \ y = 0$

**15.** Rel max value:  $f = 3$  at  $x = 1, \ y = -1$  [saddle point at $(-1, -1)$]

**17.** Rel max value:  $f = 17$  at  $x = -1, \ y = -2$  [saddle point at $(-1, 2)$]

**19.** No rel extreme values [saddle point at $(2, 6)$]

**21.** 10 units of product A, sell for $7000 each; 7 units of product B, sell for $13,000 each.

Maximum profit: $22,000

**23. a.** $P = -0.2x^2 + 16x - 0.1y^2 + 12y - 20$

   **b.** 40 cars in America, sell for $12,000; 60 cars in Europe, sell for $10,000

**25.** 6 hours of practice and 1 hour of rest

**27. a.** $x = 1200, \ p = \$6, \ R = \$7200$      **b.** $x = 800, \ y = 800, \ p = \$4,$   revenue $= \$3200$ for each

   **c.** Duopoly (1600 versus 1200)      **d.** Duopoly

**29.** Sell the sedans for $19,200, selling 12 per day, and sell the SUVs for $23,200, selling 7 per day.

**31. a.** $P = -0.2x^2 + 16x - 0.1y^2 + 12y - 0.1z^2 + 8z - 22$      **b.** 40 in America, 60 in Europe, 40 in Asia

**33.** Rel min value:  $f = -1$  at  $x = 1, \ y = 1$  [saddle point at $(0, 0)$]

**35.** Rel max value:  $f = 32$  at  $x = 4, \ y = 4$  [saddle point at $(0, 0)$]

**37.** Rel min value:  $f = -162$  at  $x = 3, \ y = 18$  and at  $x = -3, \ y = -18$  [saddle point at $(0, 0)$]

**39.** Saddle point      **41.** True

**43.** The function must have a relative maximum or minimum at the critical point.

**45.** Very little. Depending on the sign of $D$, there may be a relative minimum point or a saddle point.

**47.** False. No conclusion can be drawn. For example, at $(0, 0)$,  $x^4 + y^4$  has a relative minimum,  $-x^4 - y^4$

has a relative maximum, and  $x^4 - y^4$  has a saddle point, and in each case  $D = 0$.

## Exercises 7.4 page 508

*Note:* Your answers may differ slightly depending on the stage at which you do the rounding.
   **1.** $y = 3.5x - 1.67$     **3.** $y = -0.79x + 6.6$     **5.** $y = 2.4x + 6.9$     **7.** $y = -2.1x + 7.6$
   **9.** $y = 2.2x + 5$; prediction: 16 million     **11.** $y = -8x + 125$; prediction: 85 arrests
**13.** $y = -0.009x + 0.434$; prediction: 0.380     **15.** $y = -2.1x + 33.3$; prediction: 18.6%
**17.** $y = -0.16x + 71.6$     **19.** $y = 1.09e^{0.63x}$     **21.** $y = 17.45e^{-0.47x}$     **23.** $y = 0.98e^{0.78x}$
**25.** $y = 16.95e^{-0.52x}$     **27.** $y = 649e^{0.145x}$; prediction: $1,340,000 (from 1340)
**29.** $y = 110e^{0.453x}$; prediction: about $1059
**31.** Elderly: $y = 0.232e^{0.548x}$; children: $y = 0.15x + 1.47$; the elderly population is larger at $x = 4$, which
   corresponds to 2045.
**33.** False (see, for example, the graph on page 504)     **35.** True
**37.** $y = ax + b + 5$ (as can be proved from the formulas on page 502)
**39.** The formulas on page 502 give $a = 0/0$, which is undefined.
**41.** The third column requires the logarithm of that number, and logs of 0 or negative numbers are *undefined*.
   Therefore, the $y$'s must be positive.

## Exercises 7.5 page 523

   **1.** Max $f = 36$ at $x = 6$, $y = 2$     **3.** Max $f = 144$ at $x = 6$, $y = 4$
   **5.** Max $f = -28$ at $x = 3$, $y = 5$     **7.** Max $f = 6$ at $x = 2$, $y = -1$
   **9.** Max $f = 2$ (from $\ln e^2$) at $x = e$, $y = e$     **11.** Min $f = 45$ at $x = 6$, $y = 3$
**13.** Min $f = -16$ at $x = -4$, $y = 4$     **15.** Min $f = 52$ at $x = 4$, $y = 6$
**17.** Min $f = \ln 125$ at $x = 10$, $y = 5$     **19.** Min $f = e^{20}$ at $x = 2$, $y = 4$
**21.** Max $f = 8$ at $x = 2$, $y = 2$ and at $x = -2$, $y = -2$;
   Min $f = -8$ at $x = 2$, $y = -2$ and at $x = -2$, $y = 2$
**23.** Max $f = 18$ at $x = 2$, $y = 8$; Min $f = -18$ at $x = -2$, $y = -8$
**25. a.** 1000 feet perpendicular to building, 3000 feet parallel to building
   **b.** $|\lambda| = 1000$; each additional foot of fence adds about 1000 square feet of area
**27.** $r \approx 3.7$ feet, $h \approx 3.7$ feet
**29.** End: 14 inches by 14 inches; length $= 28$ inches; volume $= 5488$ cubic inches
**31. a.** $L = 120$, $K = 20$     **b.** $|\lambda| \approx 1.9$; output increases by about 1.9 for each additional dollar
**33.** Base: 3 inches by 3 inches; height: 5 inches     **35.** Min $f = 24$ at $x = 4$, $y = 2$, $z = -2$
**37.** Max $f = 6$ at $x = 2$, $y = 2$, $z = 2$     **39.** Base: 10 feet by 10 feet; height: 20 feet
**41.** False. It should be $F(x, y, \lambda) = f(x, y) + \lambda g(x, y)$.
**43.** True. A solution can occur only at a critical point.
**45.** The maximum occurs at one critical point, and the minimum occurs at the other.     **47.** $a \leq b$
**49. a.** $L = 64$, $K = 8$     **b.** $40L^{-1/3}K^{1/3}, 20L^{2/3}K^{-2/3}$
   **c.** $\dfrac{40L^{-1/3}K^{1/3}}{20L^{2/3}K^{-2/3}} \stackrel{?}{=} \dfrac{25}{100}$ (and now simplify and substitute $L = 64$, $K = 8$)

## Exercises 7.6 page 534

   **1.** $df = 2xy^3 \cdot dx + 3x^2y^2 \cdot dy$     **3.** $df = 3x^{-1/2}y^{1/3} \cdot dx + 2x^{1/2}y^{-2/3} \cdot dy$     **5.** $dg = \dfrac{1}{y} \cdot dx - \dfrac{x}{y^2} \cdot dy$

   **7.** $dg = -(x - y)^{-2} \cdot dx + (x - y)^{-2} \cdot dy$     **9.** $dz = \dfrac{3x^2}{x^3 - y^2} \cdot dx - \dfrac{2y}{x^3 - y^2} \cdot dy$

**11.** $dz = e^{2y} \cdot dx + 2xe^{2y} \cdot dy$     **13.** $dw = (6x^2 + y) \cdot dx + (x + 2y) \cdot dy$

**15.** $df = 4xy^3z^4 \cdot dx + 6x^2y^2z^4 \cdot dy + 8x^2y^3z^3 \cdot dz$     **17.** $df = \dfrac{1}{x} dx + \dfrac{1}{y} dy + \dfrac{1}{z} dz$

**19.** $df = yze^{xyz} \cdot dx + xze^{xyz} \cdot dy + xye^{xyz} \cdot dz = e^{xyz}(yz \cdot dx + xz \cdot dy + xy \cdot dz)$

**21. a.** $\Delta f = 0.479$     **b.** $df = 0.4$
**23. a.** $\Delta f \approx 0.112$     **b.** $df = 0.11$     **25. a.** $\Delta f = 0.1407$     **b.** $df = 0.14$
**27.** 125 square feet; 250 square feet     **29.** \$4300     **31.** About 113 feet     **33.** 2%
**35.** 0.5 liter per minute
**37.** Volume is $1.92\pi \approx 6.03 \text{ cm}^3$ with a maximum error of $0.496\pi \approx 1.56 \text{ cm}^3$.     **39.** 0
**41.** $f(x, y) = x + y + c$ where $c$ is a constant.     **43.** It is independent of $y$ (since $f_y = 0$).
**45. a.** $f$ is being evaluated at two points along the curve; each of these points gives $f = c$, and $c - c = 0$.
   **b.** Subtracting and adding $F(x)$.
   **c.** Approximating the change $\Delta f = f(x + \Delta x, F + \Delta F) - f(x, F)$ by the total differential $df = f_x \Delta x + f_y \Delta F$.
   **d.** Subtracting $f_y \Delta F$ and dividing by $f_y$ and $\Delta x$.
   **e.** Taking the limit as $\Delta x \to 0$ causes $\dfrac{\Delta F}{\Delta x}$ to approach $\dfrac{dF}{dx}$ and the approximation to become exact.

## Exercises 7.7 page 546

**1.** $2x^9 - 2x$     **3.** $6y^4$     **5.** $2x^2$     **7.** 4     **9.** 2     **11.** $\frac{1}{2}$     **13.** 12
**15.** 14     **17.** $-9e^{-3} + 9e^3$     **19.** 0     **21.** $-12$     **23.** 0     **25.** 72     **27.** $\frac{1}{2}$
**29. a.** $\displaystyle\int_1^3 \int_0^2 3xy^2 \, dx \, dy$ and $\displaystyle\int_0^2 \int_1^3 3xy^2 \, dy \, dx$     **b.** Both equal 52
**31. a.** $\displaystyle\int_0^2 \int_{-1}^1 ye^x \, dx \, dy$ and $\displaystyle\int_{-1}^1 \int_0^2 ye^x \, dy \, dx$     **b.** Both equal $2e - 2e^{-1}$     **33.** 8 cubic units
**35.** $\frac{4}{3}$ cubic units     **37.** $\frac{1}{6}$ cubic unit     **39.** $\frac{1}{2}e^2 - e + \frac{1}{2}$ cubic units     **41.** 45 degrees $\left(\text{from } \frac{540}{12}\right)$
**43.** About 180,200 people     **45.** 900,000 cubic feet     **47.** 14     **49.** 10
**51.** (a) is a double integral and (b) is an iterated integral.
**53.** 6     **55.** Yes: $\displaystyle\int_c^d e^{f(x)} \, dx \cdot \int_a^b e^{g(y)} \, dy$     **57.** True     **59.** Widgets

## Chapter 7 Review Exercises page 551

**1.** $\{(x, y) \mid x \geq 0, y \neq 0\}$     **2.** $\{(x, y) \mid y > 0\}$     **3.** $\{(x, y) \mid x \neq 0, y > 0\}$     **4.** $\{(x, y) \mid x \neq 0, y > 0\}$
**5. a.** $10x^4 - 6xy^3 - 3$     **b.** $-9x^2y^2 + 4y^3 + 2$     **c. and d.** $-18xy^2$
**6. a.** $12x^3 + 15x^2y^2 - 6$     **b.** $10x^3y - 6y^5 + 1$     **c. and d.** $30x^2y$
**7. a.** $12x^{-1/3}y^{1/3}$     **b.** $6x^{2/3}y^{-2/3}$     **c. and d.** $4x^{-1/3}y^{-2/3}$
**8. a.** $\dfrac{2x}{x^2 + y^3}$     **b.** $\dfrac{3y^2}{x^2 + y^3}$     **c. and d.** $\dfrac{-6xy^2}{(x^2 + y^3)^2}$
**9. a.** $3x^2e^{x^3 - 2y^3}$     **b.** $-6y^2e^{x^3 - 2y^3}$     **c. and d.** $-18x^2y^2e^{x^3 - 2y^3}$
**10. a.** $6xe^{-5y}$     **b.** $-15x^2e^{-5y}$     **c. and d.** $-30xe^{-5y}$
**11. a.** $-ye^{-x} - \ln y$     **b.** $e^{-x} - \dfrac{x}{y}$     **c. and d.** $-e^{-x} - \dfrac{1}{y}$
**12. a.** $2xe^y + yx^{-1}$     **b.** $x^2e^y + \ln x$     **c. and d.** $2xe^y + x^{-1}$     **13. a.** $\frac{1}{2}$     **b.** $\frac{1}{2}$
**14. a.** 0     **b.** $\frac{1}{2}$     **15. a.** 36     **b.** $-24$     **16. a.** 216     **b.** $-216$
**17. a.** 80: rate at which production increases for each additional unit of labor
   **b.** 135: rate at which production increases for each additional unit of capital     **c.** Capital
**18.** $S_x = 222$: rate at which sales increase for each additional \$1000 in TV ads.
   $S_y = 528$: rate at which sales increase for each additional \$1000 in print ads.
**19.** Min $f = -13$ at $x = -1$, $y = -4$     **20.** Min $f = -8$ at $x = 4$, $y = 1$
**21.** Max $f = 8$ at $x = 0$, $y = -1$     **22.** Max $f = 6$ at $x = 1$, $y = 0$
**23.** No rel extreme values $\left(\text{saddle point at } x = \frac{1}{2}, y = -3\right)$
**24.** No rel extreme values $\left(\text{saddle point at } x = -\frac{1}{2}, y = 1\right)$     **25.** Max $f = 1$ at $x = 0$, $y = 0$
**26.** Min $f = 1$ at $x = 0$, $y = 0$     **27.** Min $f = 0$ at $x = 0$, $y = 0$

**28.** Min $f = \ln 10$ at $x = 0$, $y = 0$

**29.** Max $f = 25$ at $x = -2$, $y = -3$ (saddle point at $x = 2$, $y = -3$)

**30.** Min $f = -20$ at $x = -2$, $y = 2$ (saddle point at $x = 2$, $y = 2$)

**31. a.** $C(x, y) = 3000x + 5000y + 6000$    **b.** $R(x, y) = 7000x - 20x^2 + 8000y - 30y^2$

   **c.** $P(x, y) = -20x^2 + 4000x - 30y^2 + 3000y - 6000$

   **d.** Make 100 18-foot boats, sell for $5000 each, and 50 22-foot boats, sell for $6500 each;

     max profit: $269,000.

**32. a.** $P(x, y) = -0.2x^2 + 68x - 0.1y^2 + 52y - 100$

   **b.** America: sell 170 for $46,000 each; Europe: sell 260 for $38,000 each (since prices are in thousands)

**33.** $y = 2.6x - 3.2$    **34.** $y = -1.8x + 8.4$    **35.** $y = 4.97x + 15.5$; prediction: 45.3 million

**36.** $y = -0.96x + 8.4$; prediction: 2.64    **37.** Max $f = 292$ at $x = 12$, $y = -24$

**38.** Max $f = 156$ at $x = 10$, $y = 8$    **39.** Min $f = 90$ at $x = 7$, $y = 4$

**40.** Min $f = -109$ at $x = -3$, $y = 3$    **41.** Min $f = e^{45}$ at $x = 3$, $y = 6$

**42.** Max $f = e^{-5}$ at $x = 2$, $y = 1$

**43.** Max $f = 120$ at $x = 2$, $y = -6$; Min $f = -120$ at $x = -2$, $y = 6$

**44.** Max $f = 64$ at $x = 4$, $y = 4$ and at $x = -4$, $y = -4$;

   Min $f = -64$ at $x = 4$, $y = -4$ and at $x = -4$, $y = 4$

**45. a.** $40,000 for production, $20,000 for advertising

   **b.** $|\lambda| \approx 159$: production increases by about 159 units for each additional dollar

**46. a.** $\frac{1}{2}$ ounce of the first and 7 ounces of the second

   **b.** $|\lambda| = 9$: each additional dollar results in about 9 additional nutritional units

**47. a.** $L = 18, K = 10$    **b.** $|\lambda| \approx 3.79$; output increases by about 3.79 units for each additional dollar

**48.** Base: 12 inches by 12 inches; height: 4 inches    **49.** $df = (6x + 2y) \cdot dx + (2x + 2y) \cdot dy$

**50.** $df = (2x + y) \cdot dx + (x - 6y) \cdot dy$    **51.** $dg = \dfrac{1}{x}\,dx + \dfrac{1}{y}\,dy = \dfrac{dx}{x} + \dfrac{dy}{y}$

**52.** $dg = \dfrac{3x^2}{x^3 + y^3}\,dx + \dfrac{3y^2}{x^3 + y^3}\,dy$    **53.** $dz = e^{x-y}\,dx - e^{x-y}\,dy$    **54.** $dz = ye^{xy}\,dx + xe^{xy}\,dy$

**55.** Sales would decrease by about $153,000 (from $dS = -153$); sales would decrease by about $76,500.

**56.** $2\%$    **57.** $8e^2 - 8e^{-2}$    **58.** $10$    **59.** $\frac{4}{3}$    **60.** $\frac{32}{3}$ or $10\frac{2}{3}$    **61.** 40 cubic units

**62.** 24 cubic units    **63.** $\frac{5}{6}$ cubic unit    **64.** $\frac{8}{9}$ cubic unit    **65.** $12,000\left(\text{from }\dfrac{192,000}{16}\right)$

**66.** $640 hundred thousand, or $64,000,000

## Cumulative Review for Chapters 1–7 page 554

**1.**

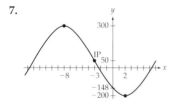

**2.** 4    **3.** $\dfrac{-1}{x^2}$ (but found using the *definition*)    **4.** 4

**5.** $S'(12) = -2$: each $1 price increase (above $12) decreases sales by 2 per week

**6.** $3[x^2 + (2x + 1)^4]^2[2x + 8(2x + 1)^3]$

**7.**

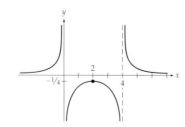

**8.**

**9.** 40 feet parallel to wall, 20 feet perpendicular to wall

**10.** Base: 6 feet by 6 feet; height: 3 feet    **11.** $\dfrac{2}{\pi} \approx 0.64$ foot per minute    **12. a.** \$1268.24    **b.** \$1271.25

**13.** In about 14.4 years    **14.** About 6.8 years

**15.**

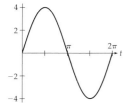

**16.** $4x^3 - 2x^2 + x + C$    **17.** $900e^{0.02t} - 900$ million gallons

**18.** $85\frac{1}{3}$ square units    **19.** 16

**20. a.** $\frac{1}{3}\ln|x^3 + 1| + C$    **b.** $2e^{x^{1/2}} + C$    **21.** $\frac{1}{4}xe^{4x} - \frac{1}{16}e^{4x} + C$

**22.** $\sqrt{4 - x^2} - 2\ln\left|\dfrac{2 + \sqrt{4 - x^2}}{x}\right| + C$    **23.** $\dfrac{1}{2}$

**24.** 1.15148 [compared with actual (rounded) value of 1.14779]

**25.** 1.14778 [compared with actual (rounded) value of 1.14779]

**26. a.** $\{(x, y)\,|\, x \geq 0, y > 0, \ x \neq y\}$    **b.** 2    **27.** $f_x = \ln y + 2ye^{2x}, \ f_y = \dfrac{x}{y} + e^{2x}$

**28.** Min $f = 2$ at $x = 1, \ y = 4$; no relative max    **29.** $y = 1.75x - 4.75$

**30.** Min $f = 90$ at $x = 4, \ y = 7$    **31.** $df = (4x + y)\,dx + (x - 6y)\,dy$    **32.** 48 cubic units

## Exercises 8.1 page 567

**1.** $90°, \dfrac{\pi}{2}$    **3.** $45°, \dfrac{\pi}{4}$    **5.** $540°, 3\pi$    **7.** $-45°, -\dfrac{\pi}{4}$    **9. a.** $\dfrac{\pi}{6}$   **b.** $\dfrac{5\pi}{4}$   **c.** $-\pi$

**11. a.** $\dfrac{\pi}{3}$   **b.** $\dfrac{7\pi}{4}$   **c.** $-\dfrac{2\pi}{3}$    **13. a.** $3\pi$   **b.** $\dfrac{3\pi}{4}$   **c.** $\dfrac{\pi}{180}$    **15. a.** $45°$   **b.** $120°$   **c.** $-150°$

**17. a.** $60°$   **b.** $225°$   **c.** $-210°$    **19. a.** $405°$   **b.** $-540°$   **c.** $\dfrac{900°}{\pi}$    **21.** $3\pi \approx 9.4$ inches

**23.** $\dfrac{3}{2}$ radians    **25.** 3600 miles    **27.** False. The sides are in *proportion* to $1, 2, \sqrt{3}$.

**29.** False. The units are reversed.    **31.** 0    **33.** 1    **35.** 0.246 radian

**37. i.** $(a + b)^2 = a^2 + 2ab + b^2$    **ii.** $c^2 + 4 \cdot \frac{1}{2}ab = c^2 + 2ab$

   **iii.** $a^2 + 2ab + b^2 = c^2 + 2ab$   so   $a^2 + b^2 = c^2$

## Exercises 8.2 page 579

**1.** $\sin\theta = \frac{24}{25}$    **3.** $\sin\theta = \frac{5}{13}$    **5.** $\sin\theta = \frac{21}{29}$    **7.** $\sin\theta = -\frac{3}{5}$ $\left(\substack{\text{after} \\ \text{simplification}}\right)$

   $\cos\theta = \frac{7}{25}$       $\cos\theta = \frac{12}{13}$       $\cos\theta = -\frac{20}{29}$      $\cos\theta = -\frac{4}{5}$

**9.** $\sin\theta = -\frac{7}{25}$ $\left(\substack{\text{after} \\ \text{simplification}}\right)$      **11. a.** $\dfrac{\sqrt{3}}{2}$   **b.** $\dfrac{1}{2}$   **c.** $\dfrac{\sqrt{2}}{2}$   **d.** $\dfrac{\sqrt{2}}{2}$

   $\cos\theta = \frac{24}{25}$

**13. a.** $-\dfrac{\sqrt{2}}{2}$   **b.** $-\dfrac{\sqrt{2}}{2}$   **c.** $\dfrac{1}{2}$   **d.** $-\dfrac{\sqrt{3}}{2}$

**15. a.** $-\dfrac{\sqrt{3}}{2}$   **b.** $\dfrac{1}{2}$   **c.** $-\dfrac{\sqrt{2}}{2}$   **d.** $\dfrac{\sqrt{2}}{2}$    **17. a.** 0.31    **b.** 0.95    **c.** 0.84    **d.** 0.54

**19.**     **21.**     **23.**

**25.**

**27. a.** $\dfrac{\sqrt{6}}{4} + \dfrac{\sqrt{2}}{4}$ **b.** $\dfrac{\sqrt{6}}{4} - \dfrac{\sqrt{2}}{4}$ $\left(\begin{smallmatrix}\text{after}\\\text{simplification}\end{smallmatrix}\right)$

**29.** About 28.2 feet **31.** About 299 feet **33.** 1219 cones

**35. a.** 5.3 liters **b.** 4.5 liters

**37. a.**
on $[-2\pi, 2\pi]$ by $[-2, 2]$

**b.**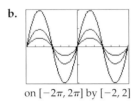
on $[-2\pi, 2\pi]$ by $[-2, 2]$

**39. a.**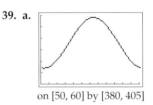
on $[50, 60]$ by $[380, 405]$

**b.** $14°$
**c.** $58°$

**41. b.** About 390 ppm (from 389.7) **43.** Shifting the cosine curve to the right by $\pi/2$ will turn it into a sine curve.
**45.** Maximum height is 11, and minimum height is 5. **47. a.** 0 **b.** 1 **49.** True
**51.** 1 (since $\sin^2 x + \cos^2 x = 1$) **53.** The identities follow from the indicated arguments.

## Exercises 8.3 page 592

**1.** $3t^2 \sin t + t^3 \cos t$ **3.** $\dfrac{-t \sin t - \cos t}{t^2}$ **5.** $3t^2 \cos(t^3 + 1)$ **7.** $5 \cos 5t$ **9.** $-6 \sin 3t$

**11.** $-\dfrac{\pi}{180} \sin\left(\dfrac{\pi}{180} t\right)$ **13. a.** $2t + \pi \cos \pi t$ **b.** $\pi$ **15. a.** $-2 \sin \dfrac{t}{2}$ **b.** $-1$

**17. a.** $-\cos(\pi - t)$ **b.** 0 **19.** $-3(t + \pi)^2 \sin(t + \pi)^3$ **21.** $-3[\cos^2(t + \pi)][\sin(t + \pi)]$

**23.** $-2(\pi - t) \cos(\pi - t)^2$ **25.** $6t \sin^2 t^2 \cos t^2$ **27.** $-\frac{1}{2}(x + 1)^{-1/2} \sin(x + 1)^{1/2}$

**29.** $4 \cos x + 4 \cos 4x + 4(\sin^3 x)(\cos x)$ **31.** $e^x \cos e^x$ **33.** $(1 - \sin z) e^{z + \cos z}$ **35.** $4(1 - \cos z)^3 \sin z$

**37. a.** $\dfrac{\cos z}{\sin z}$ **b.** 0 **39.** $\left(1 + \dfrac{1}{z}\right) \cos(z + \ln z)$ **41.** $14 \sin x \cos x \left(\begin{smallmatrix}\text{after}\\\text{simplification}\end{smallmatrix}\right)$

**43.** $-x^{-2} \cos x^2 - 2 \sin x^2$ **45.** $-e^{-x} \sin e^x + \cos e^x$ **47. a.** $\dfrac{-1}{\sin^2 x} \left(\begin{smallmatrix}\text{after}\\\text{simplification}\end{smallmatrix}\right)$ **b.** $-\dfrac{4}{3}$

**49.** $4 \sin t \cos t \left(\begin{smallmatrix}\text{after}\\\text{simplification}\end{smallmatrix}\right)$ **51. a.** $\dfrac{1 + \sin t}{\cos^2 t} \left(\begin{smallmatrix}\text{after}\\\text{simplification}\end{smallmatrix}\right)$ **b.** 1 **53.** $-\cos(\cos x) \sin x$

**55.** $-\left(\cos \dfrac{x}{2} - 1\right) \sin \dfrac{x}{2}$ **57.** $x^2 \cos x$ **59.** $4t \sin(t^2 + 1) \cos(t^2 + 1)$ **61.** $2 \cos t - t \sin t \left(\begin{smallmatrix}\text{after}\\\text{simplification}\end{smallmatrix}\right)$

**63.** $-(\sin x) e^{\sin x} + (\cos^2 x) e^{\sin x}$ **65.** $\cos z - z \sin z$ **67.** Increasing by about 45 sales per week

**69.** $\dfrac{\pi}{2}$ or $90°$ **71.** The result follows from the indicated steps. **73. a.** $x \approx 0.7854$ (radian) **b.** $45°$

**75. a.**

**b.** At $t = 0.2$ and $t = 1$ second, blood pressure is 105.
**c.** At $t = 0.6$ and $t = 1.4$ second, blood pressure is 75.
**77.** 0 (since you are differentiating a constant)
**79.** 0 (since $\sin^2 x + \cos^2 x = 1$) **81.** 4 **83.** $\cos(x + 973)$ **85.** 1
**87.** The formula follows from the indicated steps. **89.** Use $f = t$ and $f' = 1$

**91.** The identity follows by differentiation, as indicated.

## Exercises 8.4 page 602

**1.** $\sin t + \cos t + C$ **3.** $x - \cos x + C$ **5.** $\dfrac{1}{2} \sin(t^2 + 2t) + C$ **7.** $-\dfrac{1}{\pi} \cos \pi t + C$

**9.** $\dfrac{2}{\pi} \sin \dfrac{\pi t}{2} + C$ **11.** $\cos(\pi - t) + C$ **13.** $\dfrac{365}{2\pi} \sin \dfrac{2\pi(t + 20)}{365} + C$ **15.** $\dfrac{1}{3} \sin^3 t + C$

**17.** $\frac{1}{2}\cos^{-2}t + C$    **19.** $e^{1+\sin x} + C$    **21.** $-\ln|\cos t + 3| + C$    **23.** $2(1 - \cos w)^{1/2} + C$

**25.** $(1 + \sin^2 y)^{1/2} + C$    **27.** $\frac{1}{3}x^3 - \ln|x| + \frac{1}{2}\sin 2x + C$    **29.** $\frac{1}{\pi} \approx 0.318$    **31.** $\frac{1}{2}\pi^2 \approx 4.935$

**33.** $\frac{1}{2} + \frac{2}{\pi} \approx 1.137$    **35.** $-1$    **37.** 2 square units    **39.** 2 square units

**41.** $\frac{d}{dt}\left(-\frac{1}{a}\cos at + C\right) = -\frac{1}{a}(-\sin at)\cdot a = \sin at$  and similarly for the other formula

**43.** \$204,000 (from  $130 + \frac{234}{\pi}$,  rounded to the nearest thousand)

**45. a.** About 101 tons $\left(\text{from } 90 + \frac{36}{\pi}\right)$    **b.** About 5.6 tons per month    **47.** 3600 birds

**49.** About \$6668    **51.** About 277 guests per week    **53. a.** About 32.3°    **b.** About 75.6°
**55.** 0 (Integrating over a whole period includes an area above the $x$-axis and an equal (negative) area *below* the $x$-axis.)
**57.** $-1000$ (since  $\cos \pi$  is a constant equal to $-1$)    **59.** 0 (since you are integrating over many whole periods)
**61.** 9 (since, in the long run,  $\cos x$  will average out to zero)    **63.** 54° Fahrenheit

## Exercises 8.5 page 616

**1. a.** $\frac{5}{12}$    **b.** $\frac{12}{5}$    **c.** $\frac{13}{12}$    **d.** $\frac{13}{5}$    **3. a.** $\frac{8}{15}$    **b.** $\frac{15}{8}$    **c.** $-\frac{17}{15}$    **d.** $-\frac{17}{8}$

**5. a.** $\frac{1}{\sqrt{3}} = \frac{\sqrt{3}}{3}$    **b.** 2    **7. a.** Undefined    **b.** $-1$    **9. a.** 1    **b.** 1    **11. a.** $\frac{1}{\sqrt{3}} = \frac{\sqrt{3}}{3}$    **b.** $-2$

**13.** 1.376 (approx)    **15.** 3.864 (approx)    **17.** About 56 feet

**19.** Slope $= \frac{\Delta y}{\Delta x} = \frac{\text{opp}}{\text{adj}} = \tan\theta$  from diagram

**21.** Yes, it will cross the goal post at a height of 14.2 feet.
**23.** 51.3° (for a height of 20.25 feet)

**25.**

Period $2\pi$; undefined at $\pm\frac{\pi}{2}, \pm\frac{3\pi}{2}, \dots$

on $[-2\pi, 2\pi]$ by $[-4, 4]$

**27.** Result follows from the hint and the definitions of  $\tan t$  and  $\sec t$.    **29.** $\cot t - t\csc^2 t$
**31.** $2t\sec^2(t^2 - 1)$    **33.** $\pi\sec(\pi x + 1)\tan(\pi x + 1)$    **35.** $-6z^2\cot z^3\csc^2 z^3$
**37.** $S'(10) \approx -3.5$  (be sure your calculator is set for radians). Shadow shortens by about 3.5 feet for each additional degree of sun elevation (near 10° elevation).

**39.** $\frac{1}{3}\tan(t^3 + 1) + C$    **41.** $-\frac{1}{\pi}\csc\pi t + C$    **43.** $\ln|\cos(1 - t)| + C$    **45.** $\frac{1}{5}\ln|\csc x^5 - \cot x^5| + C$

**47.** $-\frac{1}{5}\cot^5 x + C$    **49.** $-\ln\frac{\sqrt{2}}{2} \approx 0.347$    **51.** $\ln(\sqrt{2} + 1) \approx 0.881$ square unit

**53.** False. The tangent and cotangent functions have period $\pi$.    **55.** $\cos\theta$    **57.** There are none.
**59.** 0. Exercise 28 shows that  $\csc^2 t - \cot^2 t = 1$,  and differentiating 1 gives 0.
**61.** The slope is always negative and, in fact, always less than or equal to $-1$ (agreeing with the graph of  $\cot t$).
**63.** Because of the identity  $1 + \tan^2 t = \sec^2 t$  (differentiating each side)
**65.** Follows from the Generalized Power Rule (and the definitions of $\sec t$ and $\tan t$)
**67. a.** $\frac{1}{2}\tan^2 t + C$    **b.** $\frac{1}{2}\sec^2 t + C$    **c.** The equality  $\tan^2 t + 1 = \sec^2 t$  shows that $\frac{1}{2}\tan^2 t$  and  $\frac{1}{2}\sec^2 t$ differ by only a constant, so the two answers are the same (because of the $+C$).
**69.** $-\ln|\cos t| + C$, $\ln|\sin t| + C$    **71.** $\ln|\csc t - \cot t| + C$

## Chapter 8 Review Exercises page 621

**1. a.** $\dfrac{5\pi}{6}$  **b.** $20°$  **2. a.** $\dfrac{2\pi}{3}$  **b.** $36°$  **3.** $\dfrac{104\pi}{3} \approx 109$ feet  **4.** $26\pi \approx 82$ feet

**5. a.** $\dfrac{\sqrt{3}}{2}$  **b.** $-\dfrac{1}{2}$  **c.** $-\dfrac{\sqrt{2}}{2}$  **d.** $-\dfrac{\sqrt{2}}{2}$  **6. a.** $-\dfrac{1}{2}$  **b.** $-\dfrac{\sqrt{3}}{2}$  **c.** $\dfrac{\sqrt{2}}{2}$  **d.** $-\dfrac{\sqrt{2}}{2}$

**7.**   **8.**   **9.** 24 feet  **10.** 20 feet

**11. a.** 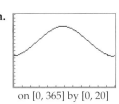  **b.** About 14.6 hours  **c.** About 8.7 hours

on [0, 365] by [0, 20]

**12.** Daylight is maximized on day 172 (June 21) with 16 hours of daylight, and minimized on day 355 (December 21) with 8.5 hours of daylight.

**13.** $-8\sin(4t - 1)$  **14.** $-3\cos(2 - 3t)$  **15.** $x^3\cos x$

**16.** $\dfrac{\cos t + \sin t + 1}{(1 + \cos t)^2}$ $\left(\begin{array}{c}\text{after}\\ \text{simplification}\end{array}\right)$  **17.** $6t\sin^2 t^2 \cdot \cos t^2$

**18.** $2xe^{x^2}\cos e^{x^2}$  **19.** $\tfrac{1}{2}(\sin t)^{-1/2}\cos t$  **20.** $\tfrac{1}{2}t^{-1/2}\cos t^{1/2}$

**21. a.** 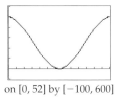  **b.** Increasing by about 29 sales per week

on [0, 52] by [−100, 600]

**22.** Week 39, sales increasing by about 30 sales per week

**23. a.**   **b.** Decreasing by about 0.45 degree per day

on [0, 365] by [0, 90]

**24.** On day 106 (April 16), when temperature is rising by about 0.45 degree per day

**25.** $\dfrac{\pi}{4}$ or $45°$

**26.** Follows from setting $R' = 0$ and solving  **27.** $\dfrac{5}{\pi}\sin\dfrac{\pi t}{5} + C$  **28.** $-\dfrac{1}{2}\cos(t^2 + 1) + C$

**29.** $\cos(2\pi - t) + C$  **30.** $\cos e^{-x} + C$  **31.** $-e^{\cos x} + C$  **32.** $2(\sin t)^{1/2} + C$  **33.** $\tfrac{1}{3}\ln(1 + \sin 3t) + C$

**34.** $\dfrac{1}{4}x^4 + e^{-x} - \dfrac{1}{\pi}\sin \pi x + C$  **35.** $\sin(\ln y) + C$  **36.** $\cos z^{-1} + C$  **37.** $\dfrac{3}{4}$  **38.** $-\dfrac{1}{24}$  **39.** $\dfrac{2}{3\pi} \approx 0.212$

**40.** 200  **41.** $\tfrac{1}{3}$ square unit  **42.** $\sqrt{2} - 1 \approx 0.414$ square unit

**43. a.**   **44. a.** 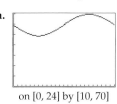  **45. a.** $\sqrt{3}$  **b.** $\dfrac{1}{\sqrt{3}} = \dfrac{\sqrt{3}}{3}$  **c.** 2  **d.** $\dfrac{2}{\sqrt{3}} = \dfrac{2\sqrt{3}}{3}$

on [0, 52] by [−10, 50]   on [0, 24] by [10, 70]

**46. a.** $-1$  **b.** $-1$  **c.** $\dfrac{-2}{\sqrt{2}} = -\sqrt{2}$  **d.** $\dfrac{2}{\sqrt{2}} = \sqrt{2}$

**47.** About 503 feet  **48.** About 220 feet

**b.** $650,000 (from 650 thousand)  **b.** About $53.6°$

**49.** $\tan x^2 + 2x^2\sec^2 x^2$  **50.** $2x\cot x - x^2\csc^2 x$  **51.** $\csc(\pi - t)\cot(\pi - t)$

**52.** $\tfrac{1}{2}t^{-1/2}\sec t^{1/2}\tan t^{1/2}$  **53.** $(\pi\sec^2 \pi t)e^{\tan \pi t}$

**54.** The identity $1 + \cot^2 t = \csc^2 t$ shows that the functions differ by only a constant, so the derivatives must agree (the derivative of each is $-2\cot t\csc^2 t$).

**55.** $\dfrac{2}{\pi}\tan\dfrac{\pi t}{2} + C$  **56.** $-\dfrac{1}{2}\ln|\cos x^2| + C$  **57.** $\dfrac{1}{3}\sec t^3 + C$  **58.** $\csc(\pi - t) + C$

**59.** $-\dfrac{2}{3}(\cot x)^{3/2} + C$     **60.** $2(\tan x)^{1/2} + C$     **61.** $\ln(\sqrt{2}+1) \approx 0.881$

**62.** 1     **63.** $-\ln\dfrac{\sqrt{2}}{2} \approx 0.347$ square unit     **64.** $-\ln(\sqrt{2}-1) \approx 0.881$ square unit

**65.**

**66.**

**67. a.** $y' = \dfrac{2\cos 2x}{\sin y}$     **b.** $-2$

**68. a.** $y' = -\dfrac{4\sin 4x}{\cos y}$     **b.** $-4$

**69.** $\dfrac{1}{80}$ radian per second

**70.** 15 radians per hour     **71. a.** $2\cos 2x + y\sin x$     **b.** $-\cos x$     **72. a.** $-\dfrac{\sin(x-y)}{\cos(x-y)}$     **b.** $\dfrac{\sin(x-y)}{\cos(x-y)}$

**73. a.** $(\cos x)e^{\cos y}$     **b.** $-(\sin x)(\sin y)e^{\cos y}$     **74. a.** $3x^2y\cos x^3y$     **b.** $x^3\cos x^3y$

**75.** $-t\cos t + \sin t + C$     **76.** $t\sin t + \cos t + C$     **77.** $x\tan x + \ln|\cos x| + C$

**78.** $x\sec x - \ln|\sec x + \tan x| + C$     **79.** $\frac{1}{2}e^t(\sin t - \cos t)$     **80.** $\frac{1}{2}e^t(\sin t + \cos t)$

**81.** 6.209     **82.** 1.198     **83.** 6.208758036     **84.** 1.198

## Exercises 9.1 page 640

**1.** Check that $(4e^{2x} - 3e^x) - 3(2e^{2x} - 3e^x) + 2(e^{2x} - 3e^x + 2) \overset{?}{=} 4$     **3.** Check that $kae^{ax} \overset{?}{=} a\left(ke^{ax} - \dfrac{b}{a}\right) + b$

**5.** $y = \sqrt[3]{6x^2 + c}$     **7.** Not separable     **9.** $y = ce^{2x^3}$   Check that $c6x^2e^{2x^3} \overset{?}{=} 6x^2(ce^{2x^3})$

**11.** $y = cx$  (since $e^{\ln x} = x$)  Check that $c \overset{?}{=} \dfrac{cx}{x}$     **13.** $y = \sqrt{4x^2 + c}$  and  $y = -\sqrt{4x^2 + c}$

**15.** Not separable     **17.** $y = 3x^3 + C$     **19.** $y = \dfrac{1}{2}\ln(x^2 + 1) + C$     **21.** $y = ce^{x^3/3}$

**23.** $y = \left(\dfrac{1-n}{m+1}x^{m+1} + c\right)^{1/(1-n)}$     **25.** $y = (x + c)^2$     **27.** Not separable

**29.** $y = ce^{x^2/2} - 1$     **31.** $y = ce^{e^x} + 1$     **33.** $y = \dfrac{1}{c - ax}$     **35.** $y = ce^{ax} - \dfrac{b}{a}$

**37.** $y = \sqrt[3]{3x^2 + 8}$  Check that  $(3x^2 + 8)^{\frac{2}{3}} \cdot \dfrac{1}{3}(3x^2 + 8)^{-\frac{2}{3}} \cdot 6x \overset{?}{=} 2x$  and  $y(0) = \sqrt[3]{8} = 2$

**39.** $y = -e^{x^2/2}$  Check that  $-xe^{x^2/2} \overset{?}{=} x(-e^{x^2/2})$  and  $y(0) = -e^0 = -1$

**41.** $y = (1 - x^2)^{-1}$   Check that  $-(1 - x^2)^{-2}(-2x) \overset{?}{=} 2x[(1 - x^2)^{-1}]^2$  and  $y(0) = (1 - 0)^{-1} = 1$

**43.** $y = 3x$  (using $e^{\ln x} = x$)  Check that  $3 \overset{?}{=} \dfrac{3x}{x}$  and  $y(1) = 3 \cdot 1 = 3$

**45.** $y = (x + 1)^2$  or  $y = (x - 3)^2$  Check that  $2(x + 1) \overset{?}{=} 2\sqrt{(x + 1)^2}$  and  $y(1) = 2^2 = 4$  and that
  $2(x - 3) \overset{?}{=} 2\sqrt{(x - 3)^2}$  and  $y(1) = (-2)^2 = 4$

**47.** $y = \dfrac{1}{2 - e^x - x}$   Check that  $\dfrac{e^x + 1}{(2 - e^x - x)^2} \overset{?}{=} \left(\dfrac{1}{2 - e^x - x}\right)^2 e^x + \left(\dfrac{1}{2 - e^x - x}\right)^2$  and  $y(0) = \dfrac{1}{2 - 1} = 1$

**49.** $y = 2e^{ax^3/3}$   Check that  $2ax^2e^{ax^3/3} \overset{?}{=} ax^22e^{ax^3/3}$  and  $y(0) = 2e^0 = 2$

**51.** $D(p) = cp^{-k}$  (for any constant $c$)     **53.** $y = 20{,}000e^{0.05t} - 20{,}000$

**55. a.** $y = 28.6e^{-0.32t} + 70$     **b.** About 3.28 hours     **57.** $y = 150 - 150e^{-0.2t}$

**59. a.** $y' = 3 + 0.10y$     **b.** $y(0) = 6$     **c.** $y(t) = 36e^{0.1t} - 30$
  **d.** $y(25) = 408.570$  thousand dollars, or \$408,570 (rounded)

**61. a.** $y' = 4y^{1/2}$     **b.** $y(0) = 10{,}000$     **c.** $y(t) = (2t + 100)^2$     **d.** 15,376

**63. a.** $y' = 8y^{3/4}$     **b.** $y(0) = 10{,}000$     **c.** $y(t) = (2t + 10)^4$     **d.** 234,256

**65. c.** $y = \sqrt[5]{10x^3 + 32}$     **67. c.** $y = \sqrt[4]{8x^2 + 16}$     **69. a.**      **b.**

**d.**      **d.**

**73.** $y = -2x + C$     **75.** $y = ke^x$     **77.** Not separable     **79.** Separable

**81.** Quadrant I: positive; II: negative; III: positive; IV: negative

**83. a.** $p(t) = Ce^{-Kt/R}$ for $t_0 \le t \le T$     **b.** $p(t) = p_0 e^{-K(t-t_0)/R}$ for $t_0 \le t \le T$

**c.** $p(t) = I_0 R - Ce^{-Kt/R}$ for $0 \le t \le t_0$     **d.** $p(t) = I_0 R - (I_0 R - p_0)e^{K(t_0-t)/R}$ for $0 \le t \le t_0$

## Exercises 9.2 page 655

**1.** $y' = cae^{at} = a(ce^{at}) = ay$     **3.** Unlimited     **5.** Limited     **7.** None     **9.** Logistic     **11.** Logistic
$y(0) = ce^0 = c$

**13.** $y = 1.5e^{6t}$     **15.** $y = 100e^{-t}$     **17.** $y = -e^{-0.45t}$     **19.** $y = 100(1 - e^{-2t})$     **21.** $y = 0.25(1 - e^{-0.05t})$

**23.** $y = 40(1 - e^{-2t})$     **25.** $y = 200(1 - e^{-0.01t})$     **27.** $y = \dfrac{100}{1 + 9e^{-500t}}$     **29.** $y = \dfrac{0.5}{1 + 4e^{-0.125t}}$

**31.** $y = \dfrac{10}{1 - \frac{1}{2}e^{-30t}} = \dfrac{20}{2 - e^{-30t}}$     **33.** $y = \dfrac{3}{1 + 2e^{-6t}}$

**35.** $y' = 0.08y$     **37.** $y' = a(100,000 - y)$     **39.** $y' = a(5000 - y)$     **41.** $y' = ay(10,000 - y)$

$y = 1500e^{0.08t}$     $y = 100,000(1 - e^{-0.021t})$     $y = 5000(1 - e^{-0.223t})$     $y = \dfrac{10,000}{1 + 99e^{-0.535t}}$

About 22,276     About 7.2 weeks     About 8612 sales

**43.** $y' = ay(800 - y)$     **45.** $y' = ay(800 - y)$     **47.** $y' = ay(22 - y)$     **49.** $y = 5e^{-0.15t}$

$y = \dfrac{800}{1 + 799e^{-0.558t}}$     $y = \dfrac{800}{1 + 7e^{-0.280t}}$     $y = \dfrac{22}{1 + 0.375e^{-0.288t}}$     About 3.7 mg

About 675 people     About 6.9 years     About 21 million sales

**51. a.** $y' = 0.1(200 - y)$ with $M = 200$     **b.** $y = 200(1 - e^{-0.1t})$     **c.** 30 years  [from solving
$200(1 - e^{-0.1t}) = 0.95 \cdot 200$]

**53. a.** About 17 feet per second     **b.** About 0.6 foot per second     **c.** About 0.006 foot per second

**d.** About $\dfrac{1}{0.006} \approx 167$ seconds, or about 2.8 minutes

**55. a.** Limited     **b.** Unlimited     **c.** Logistic     **57. a.** Logistic     **b.** Unlimited     **c.** Limited

**59. a.** Approaches 0.03     **b.** Approaches 2.5     **c.** Approaches infinity

**61.** $y = \dfrac{1}{c - at}$     **63.** $y = ce^{a \ln x} = cx^a$     **65.** The solution follows from the indicated steps.

## Exercises 9.3 page 667

**1.** $y = 4 + Ce^{-2x}$     **3.** $y = -\frac{1}{4}e^{-2x} + Ce^{2x}$     **5.** $y = 3x^3 + Cx^{-5}$     **7.** $y = x^2 + Cx$     **9.** $y = 3 + Ce^{-x^3}$

**11.** $y = Ce^{x^2}$     **13.** $y = \dfrac{x^2 + C}{x + 1}$     **15.** $y = 3x^4 - 9x^3 + Cx^2$     **17.** $y = -x - 1 + Ce^x$     **19.** $y = 1 + Ce^{-\sin x}$

**21.** $y = 3e^x + 2e^{-3x}$     **23.** $y = 2x^5 - x^{-2}$     **25.** $y = x^2 \ln x + 3x^2$     **27.** $y = 2 - 2e^{-x^2}$     **29.** $y = Ce^x - 1$

**31. a.** $y = \dfrac{1}{x^2/2 + C}$ (by separation of variables)     **b.** $y = \dfrac{1}{3}x^3 e^x + Ce^x$ (by using an integrating factor)

**33. a.** $y = 20te^{0.1t} + 5e^{0.1t}$     **b.** About \$55 million     **35. a.** $y = 125t^3$     **b.** 421,875

**37. a.** $y = 3e^{t^2/2} - 1$     **b.** About 21 tons     **c.** In about 2.3 weeks (2 weeks and 2 days)

**39. a.** $y' = 2500 - 0.05y, y(0) = 8,000,000$     **b.** $y = 50,000 + 7,950,000e^{-0.05t}$     **c.** About 40.1 hours

**41. a.** $y' = 10,000 - 0.05y, y(0) = 6,000,000$     **b.** $y = 200,000 + 5,800,000e^{-0.05t}$     **c.** About 35 minutes

**43. a.** $y' = 20 - 0.02y$, $y(0) = 10{,}000$    **b.** $y = 1000 + 9000e^{-0.02t}$    **c.** About 110 hours    **d.** 1000 grams
**45. a.** $w(t) = 120 + 50e^{-0.005t}$    **b.** About 71 days    **c.** 120 pounds

**47.** Yes. Just divide through by $u(x)$ to put it in standard form    $y' + \dfrac{v(x)}{u(x)}y = \dfrac{w(x)}{u(x)}$.

**49.** Unlimited (since it can be written in the form    $y' - ay = 0$)    and limited (since it can be written in the form    $y' + ay = aM$).    **51.** $I(t) = e^{-at}$    **53.** 800,000 pCi (the volume times the concentration in the incoming air)

**55.** Using    $e^{\int p(x)\,dx+C} = e^{\int p(x)\,dx}e^{C}$    multiplies the differential equation by a positive constant $e^{C}$, which, when divided out, leaves the same equation as when the $C$ is omitted.

**57. a.** $y' = 30e^{-3t} - 0.2y$    **b.** $y = 10.7(e^{-0.2t} - e^{-3t})$    **c.** 7.1 milligrams    **d.** $0.97 \approx 1$ hour (8.2 mg)

**59.** *Hint:* An intermediate step is    $ye^{ct} = \dfrac{ab}{c-b}e^{(c-b)t} + K$

## Exercises 9.4 page 677

**1.** 8    **3.** $(1, 3)$, slope $= 12$
**5.**     **7.**     **9.**     **11. a.** 2.226
     **b.** $y = \sqrt{x^2 + 1}$
     **c.** $y(2) = \sqrt{5} \approx 2.236$
**13. a.** 1.489
     **b.** $y = e^{0.2x}$
     **c.** $y(2) = e^{0.4} \approx 1.492$

**15. a.** 7.836    **b.** $y = e^x + 4e^{-x}$    **c.** $y(2) \approx 7.930$    **17.** 1.73    **19.** 2.24    **21.** 2.83
**23. a.** 2.3096    **b.** 2.3309    **c.** 2.3327    **25.** $y(3) \approx 13{,}490$ (from 13.49 thousand)
**27.** $y(2) \approx 1981$ tons    **29.** $q(200) \approx 0.65$
**31.** Underestimate, since the slopes of the segments will underestimate the true slope on the interval, as in the diagram on page 672.
**33.** $g(x, y) = \dfrac{f(x)}{g(y)}$    (using $g(x, y)$ and $g(y)$ for two different functions).    **35.** $g(x, y) = q(x) - p(x) \cdot y$
**37.** Take the slope to be $g$ evaluated at the *midpoint* of the segment. Alternatively, take the slope to be the *average* of $g$ calculated at the two endpoints of the segment. (Other answers are possible.)

## Chapter 9 Review Exercises page 680

**1.** $y = \sqrt[3]{x^3 + c}$    **2.** $y = ce^{x^3/3}$    **3.** $y = \dfrac{1}{4}\ln(x^4 + 1) + C$    **4.** $y = -\dfrac{1}{2}e^{-x^2} + C$    **5.** $y = \dfrac{1}{c-x}$
**6.** $y = \dfrac{1}{\sqrt{c-2x}}$ and $y = -\dfrac{1}{\sqrt{c-2x}}$    **7.** $y = 1 + ce^{-x}$ or $y = 1 - e^{c-x}$
**8.** $y = \sqrt{2x + c}$ and $y = -\sqrt{2x + c}$    **9.** $y = ce^{\frac{1}{2}x^2 - x}$    **10.** $y = ce^{x^3/3} - 1$    **11.** $y = \sqrt[3]{3x^3 + 1}$
**12.** $y = e^{1-x^{-1}}$ (from $ee^{-x^{-1}}$)    **13.** $y = e^{\frac{1}{2} - \frac{1}{2}x^{-2}}$ (from $e^{\frac{1}{2}}e^{-\frac{1}{2}x^{-2}}$)    **14.** $y = \left(\frac{2}{3}x - \frac{2}{3}\right)^{3/2}$ or $y = -\left(\frac{2}{3}x - \frac{2}{3}\right)^{3/2}$
**15. a.** $y' = 4 + 0.05y$, $y(0) = 10$    **b.** $y = 90e^{0.05t} - 80$    **c.** 68.385 thousand or \$68,385
**16. a.** $y' = 4 - 0.25y$, $y(0) = 0$    **b.** $y = 16 - 16e^{-0.25t}$    **17.** $y = 106 - 36e^{-2.3t}$
**18.** $y(1) \approx 102.4$    **19. c.** $y = \sqrt[3]{x^3 - 8}$    **20.** $y = \sqrt[3]{\frac{3}{2}x^2 - 8}$
     $y(2) \approx 105.6$    **d.**    **d.**
     $y(3) \approx 105.96$

**21.** $y' = 0.05y$       **22.** $y' = 0.12y$        **23.** $y' = ay(8000 - y)$        **24.** $y' = ay(500 - y)$
 $y = 42e^{0.05t}$       $y = 30.6e^{0.12t}$        $y = \dfrac{8000}{1 + 799e^{-2.73t}}$ ($t$ in weeks)        $y = \dfrac{500}{1 + 249e^{-3.78t}}$
 About 69¢       About $102 billion        About 1819 cases        About 443

**25.** $y' = a(10,000 - y)$       **26.** $y' = a(60 - y)$       **27.** $y' = a(500,000 - y)$
 $y = 10,000(1 - e^{-0.051t})$       $y = 60(1 - e^{-0.269t})$       $y = 500,000(1 - e^{-0.255t})$ ($t$ in weeks)
 About 4577       About 6.7 weeks       About 6.3 weeks

**28.** $y' = ay(40,000 - y)$       **29.** $y = 2x^7 + Cx^5$       **30.** $y = 2x + Cx^{-1/2}$       **31.** $y = 1 + Ce^{-x^2/2}$
 $y = \dfrac{40,000}{1 + 39e^{-1.47t}}$       **32.** $y = 2 + Ce^{x^2}$       **33.** $y = e^x - x^{-1}e^x + Cx^{-1}$       **34.** $y = \tfrac{1}{2}x \ln x - \tfrac{1}{4}x + Cx^{-1}$
 About $2\tfrac{1}{2}$ years       **35.** $y = 3e^{x^2/2}$       **36.** $y = 2e^{-2x^2}$       **37.** $y = 2x - 2x^{-2}$       **38.** $y = -4 + 10x$

**39. a.** $y = 10te^{0.05t} + 100e^{0.05t}$       **b.** $192,604 [from $y(5) \approx 192.604$]       **c.** About 7.33 years
**40. a.** $p = -10e^{0.01t} + 110e^{0.02t}$       **b.** About 123 million people       **c.** About 19.4 years
**41. a.** $y' = 100 - 0.05y,\ y(0) = 1000$       **b.** $y(t) = 2000 - 1000e^{-0.05t}$
   **c.** About 1528 cubic feet (so the oxygen percentage has risen from 10% to 15.28%)
   **d.** 2000 cubic feet       **e.** About 46 minutes
**42. a.** $p = 15e^{0.2t} - 10$       **b.** 12 eagles       **c.** About 4 years

**43. a.**       **b.** $y = 1 + e^{-2x}$       **44. a.**       **b.** $y = \sqrt{4x^2 + 1}$
      **c.** $y(1) = 1 + e^{-2} \approx 1.14$       **c.** $y(1) = \sqrt{5} \approx 2.24$

**45.** $y(2) \approx 1.70$       **46.** $y(2.5) \approx 6.68$
**47. a.** $0.1(1 - p) + 0.1p(1 - p) = 0.1(1 - p)(1 + p) = 0.1(1 - p^2)$       **b.** $p(4) \approx 0.61$, or 61%       **48.** $y(2) \approx 148$

## Exercises 10.1 page 695

**1.** $\dfrac{1}{2} + \dfrac{1}{3} + \dfrac{1}{4} + \dfrac{1}{5} + \dfrac{1}{6}$       **3.** $-\dfrac{1}{3} + \dfrac{1}{9} - \dfrac{1}{27} + \dfrac{1}{81}$       **5.** $1 + 0 + 1 + 0 + 1 + 0$       **7.** $\displaystyle\sum_{i=1}^{\infty} \dfrac{1}{3^i}$ or $\displaystyle\sum_{i=1}^{\infty} \left(\dfrac{1}{3}\right)^i$

**9.** $\displaystyle\sum_{i=1}^{\infty} (-2)^i$       **11.** $\displaystyle\sum_{i=1}^{\infty} (-1)^i \cdot i$       **13.** $2^{10} - 1 = 1023$       **15.** $4^6 - 1 = 4095 \left(\text{from } 3 \cdot \dfrac{1 - 4^6}{1 - 4}\right)$

**17.** $1 + 2^7 = 129 \left(\text{from } 3 \cdot \dfrac{1 - (-2)^7}{1 - (-2)}\right)$       **19.** 5       **21.** $\tfrac{3}{2}$       **23.** Diverges       **25.** 6       **27.** 32       **29.** 25

**31.** Diverges       **33.** $\tfrac{2}{3}$       **35.** Diverges       **37.** 100       **39.** 1       **41.** $\tfrac{4}{11}$       **43.** $2\tfrac{6}{11}$ or $\tfrac{28}{11}$       **45.** $\tfrac{1}{37}$
**47.** $\tfrac{1}{81}$       **49. a.** $26,214,300       **b.** 24 days       **51.** $597,447       **53.** $584,541
**55. a.** $21,474,836.47       **b.** 27 days       **57.** On the 37th day       **59. a.** 15 units, 3 units       **b.** 4 doses
**61.** $500 billion; multiplier = 10       **63.** 4       **65.** $20,300       **67. a.** 30 feet       **b.** At the 9th bounce
**69.** 16,000; 15,200       **71.** 5281       **73.** True       **75.** $1/e^2$; yes
**77.** Converges   $(r = 1/10)$       **79.** $r = 2$,   so the sum formula does not apply
**81. a.** Dose 3       **83. a.** $\dfrac{1}{1 - (-1)} = \dfrac{1}{2}$
   **b.** Dose 5       **b.** No: $r = -1$  violates  $|r| < 1$.
         **c.** Partial sums are $1, 0, 1, 0, \ldots$, which in the long run average to $1/2$.

## Exercises 10.2 page 709

**1.** $1 + 2x + 2x^2 + \tfrac{4}{3}x^3$       **3.** $1 + \tfrac{1}{2}x - \tfrac{1}{8}x^2 + \tfrac{1}{16}x^3$       **5.** $7 + x - 3x^2$       **7.** $1 + x^2$

**9. a.** $x - \frac{1}{2}x^2 + \frac{1}{3}x^3 - \frac{1}{4}x^4$    **11. a.** $1 - \frac{1}{2}x^2 + \frac{1}{24}x^4$    **13.** $1 + x^2 + \frac{1}{2}x^4$    **15.** $2x - \frac{2^3}{3!}x^3 + \frac{2^5}{5!}x^5$

**b.**

on $[-2, 2]$ by $[-2, 2]$

**b.**

on $[-\pi, \pi]$ by $[-2, 2]$

**17. a.** $1 + x + \frac{1}{2!}x^2 + \frac{1}{3!}x^3 + \frac{1}{4!}x^4$

**b.** 0.60677

**c.** $\left| R_4\left(-\frac{1}{2}\right) \right| \le \frac{1}{5!} \left| -\frac{1}{2} \right|^5 < 0.0003$

(using $M = 1$, from $f^{(5)}(x) = e^x$ evaluated at the right-hand endpoint 0)

**d.** $e^{-1/2} \approx 0.6068$   with error less than 0.0003 (actual value: $e^{-1/2} \approx 0.6065$)

**e.**

on $[-3, 3]$ by $[0, 12]$

**19. a.** $p_6(x) = 1 - \frac{1}{2!}x^2 + \frac{1}{4!}x^4 - \frac{1}{6!}x^6$ has error $|R_6(x)| \le \frac{1}{7!}|1|^8 < 0.00022$ on $[-1, 1]$.

**b.**   (Actual error for $-1 \le x \le 1$ is less than 0.00003.)

on $[-2\pi, 2\pi]$ by $[-2, 2]$

**21. a.** $1 + \frac{1}{3}x$    **b.** $[0, 0.3]$    **23. a.** $1 + \frac{1}{2}(x - 1) - \frac{1}{8}(x - 1)^2 + \frac{1}{16}(x - 1)^3$    **25. a.** $-(x - \pi) + \frac{1}{6}(x - \pi)^3$

**c.**

on $[-2, 2]$ by $[-2, 2]$

**b.**

on $[-1, 3]$ by $[-1, 2]$

**b.**

on $[0, 2\pi]$ by $[-2, 2]$

**27. a.** $1 + \frac{1}{3}(x - 1) - \frac{1}{9}(x - 1)^2$    **b.** 1.09

**c.** $|R_2(1.3)| \le \frac{10/27}{3!} |1.3 - 1|^3 < 0.002$ $\left( \text{using } M = \frac{10}{27}, \text{ from } f^{(3)}(x) = \frac{10}{27}x^{-8/3} \text{ evaluated at } x = 1 \right)$

**d.** $\sqrt[3]{1.3} \approx 1.09$   with error less than 0.002 (actually, $\sqrt[3]{1.3} \approx 1.091$)

**29. a.** $(x - 1) - \frac{1}{2}(x - 1)^2 + \frac{1}{3}(x - 1)^3 - \frac{1}{4}(x - 1)^4 + \frac{1}{5}(x - 1)^5 - \frac{1}{6}(x - 1)^6$

**b.** 0.40469

**c.** $|R_6(1.5)| \le \frac{720}{7!} |1.5 - 1|^7 < 0.002$

**d.** $\ln 1.5 \approx 0.405$   with error less than 0.002 (actually, $\ln \frac{3}{2} \approx 0.4055$)

**31.** 0.78    **33.** $x = \frac{1}{2}$ year    **35.** $x$

**37.** $f(x) = \frac{1}{x}$   is not defined at $x = 0$   and so has no linear approximation there.

**39.** 1   (or any number greater than 1, but the best answer is 1)

**41.** Left    **43.** $1 + x + x^2 + \cdots + x^n = \frac{1 - x^{n+1}}{1 - x}$

## Exercises 10.3 page 722

**1.** $R = 2$    **3.** $R = 1$    **5.** $R = \infty$    **7.** $R = 0$    **9.** $R = \infty$    **11.** $x - \frac{x^2}{2} + \frac{x^3}{3} - \frac{x^4}{4} + \cdots$

**13.** $1 - \frac{4x^2}{2!} + \frac{16x^4}{4!} - \cdots$    **15.** $1 + x - \frac{x^2}{2!} + \frac{3x^3}{3!} - \frac{5 \cdot 3x^4}{4!} + \cdots$    **17.** $1 + \frac{x}{5} + \frac{x^2}{5^2 \cdot 2!} + \frac{x^3}{5^3 \cdot 3!} + \cdots$

**19.** $1 - \dfrac{1}{2!}\left(x - \dfrac{\pi}{2}\right)^2 + \dfrac{1}{4!}\left(x - \dfrac{\pi}{2}\right)^4 - \cdots$     **21.** $x^2 + x^3 + x^4 + \cdots$     **23.** $x^2 - \dfrac{x^6}{3!} + \dfrac{x^{10}}{5!} - \dfrac{x^{14}}{7!} + \cdots$

**25.** $1 + \dfrac{x}{2!} + \dfrac{x^2}{3!} + \dfrac{x^3}{4!} + \cdots$     **27.** $\dfrac{1}{3!} - \dfrac{x^2}{5!} + \dfrac{x^4}{7!} - \cdots$     **29.** $x - \dfrac{1}{2}x^2 + \dfrac{1}{3}x^3 - \dfrac{1}{4}x^4 + \cdots$

**31. a.** $1 - x^2 + x^4 - x^6 + \cdots$     **b.** $R = 1$     **33.** $1 + \dfrac{x^2}{2!} + \dfrac{x^4}{4!} + \dfrac{x^6}{6!} + \cdots$

**35. a.** $1.64872$     **b.** $1.64583$     **c.** 7 terms (the last having exponent 6)

**37.** $\dfrac{d}{dx}\left(1 - \dfrac{x^2}{2!} + \dfrac{x^4}{4!} - \dfrac{x^6}{6!} + \cdots\right) = -2\dfrac{x}{2!} + 4\dfrac{x^3}{4!} - 6\dfrac{x^3}{6!} + \cdots = -x + \dfrac{x^3}{3!} - \dfrac{x^5}{5!} + \cdots$

$$= -\left(x - \dfrac{x^3}{3!} + \dfrac{x^5}{5!} - \cdots\right) = -\sin x$$

**39. a.** $1 - \dfrac{1}{2}t^2 + \dfrac{1}{2^2 2!}t^4 - \dfrac{1}{2^3 3!}t^6 + \cdots$     **b.** $0.4x - \dfrac{0.4}{3 \cdot 2}x^3 + \dfrac{0.4}{5 \cdot 2^2 \cdot 2!}x^5 - \dfrac{0.4}{7 \cdot 2^3 \cdot 3!}x^7 + \cdots$

   **c.** Approximately 0.34, so about 34% have IQs in the specified range

**41. a.** $1 - \dfrac{t^4}{2!} + \dfrac{t^8}{4!} - \cdots$     **b.** $x - \dfrac{x^5}{5 \cdot 2!} + \dfrac{x^9}{9 \cdot 4!} - \cdots$

   **c.** Approximately 0.905 (thousand), so about 905 sales

**43.** No. (If the radius of convergence from 0 does not extend beyond 3, it cannot reach 4.)

**45.** 17     **47.** $e^{-1} = 1/e$     **49.** $-1$   (from $\cos \pi$)     **51.** 10 (the midpoint of the interval)

**53.** The absolute ratio of successive terms is $\left|\dfrac{x^2}{(2n + 3)(2n + 2)}\right| \to 0$  as  $n \to \infty$.

**55. a.** $S_n - S_{n-1} = (c_1 + c_2 + c_3 + \cdots + c_{n-1} + c_n) - (c_1 + c_2 + c_3 + \cdots + c_{n-1}) = c_n$

   **b.** Taking limits: $\lim\limits_{n \to \infty}(S_n - S_{n-1}) = \lim\limits_{n \to \infty} c_n$.  Since the series converges, the limit on the left-hand side is

   $S - S = 0$,  so the limit on the right-hand side must also be zero: $\lim\limits_{n \to \infty} c_n = 0$.

**57.** The result follows from the indicated steps.

**59. a.** If $\left|\dfrac{c_{n+1}}{c_n}\right|$ approaches  $r > 1$,  then $\left|\dfrac{c_{n+1}}{c_n}\right|$  must eventually remain above 1.

   **b.** $\left|\dfrac{c_{n+1}}{c_n}\right| \geq 1$  for  $n = N$  implies that   $|c_{N+1}| \geq |c_N|$.  Then,  $\left|\dfrac{c_{n+1}}{c_n}\right| \geq 1$  for  $n = N + 1$   implies that

   $|c_{N+2}| \geq |c_{N+1}|$,  which together with  $|c_{N+1}| \geq |c_N|$  implies that  $|c_{N+2}| \geq |c_N|$.  Similar reasoning
   leads to the other inequalities.

   **c.** Terms beyond the $N$th being (in absolute value) at least of size  $|c_N| \neq 0$  means that the $n$th term cannot
   approach zero, so the series diverges.

**61. b.** Each parenthesis (and also the next term) is positive, so it is a sum of positive terms.
   **c.** The series begins with 1, with positive quantities subtracted, and so it must be less than 1.

**63.** The results of the steps are as follows:

   **a.** $e^x > \dfrac{x^3}{6}$     **b.** $e^{ax} > \dfrac{a^3 x^3}{6}$     **c.** $e^{-ax} < \dfrac{6}{a^3 x^3}$     **d.** $x^2 e^{-ax} < \dfrac{6}{a^3 x}$     **e.** Letting $x \to \infty$ gives the claimed result.

## Exercises 10.4 page 734

**1.** $x_1 = 1.5, x_2 = 1.39$     **3.** $x_1 = 0.5, x_2 = 0.61$     **5.** 2.236     **7.** 1.513     **9.** $-0.814$

**11.** 5.065797019     **13.** 0.486389036     **15.** 3.693441359     **17.** 0.869647810     **19.** 1.106060158

**21.** 1.029866529     **23.** 0.567143290

**25.** The iterations give $-1, 1, -1, 1, \ldots$. Reason:  $x^2 + 3 = 0$   has no solutions (the left-hand side is always
   positive).

**27.** It converges in one step to 2.     **29.** 102 iterations     **31.** \$5.67

**33.** $2.61 **35.** 12.6% **37.** 3.18 weeks **39.** $-b/m$ **41.** $-4$
**43.** It will converge to the zero in one step. **45.** To the right of the zero
**47.** The iterations will be exactly the same (because changing the sign of the function will not change the $f/f'$ in the iteration formula).

## Chapter 10 Review Exercises page 738

**1.** $\frac{3}{7}$ **2.** $\frac{2}{7}$ **3.** Diverges **4.** $\frac{2}{5}$ **5.** 50 **6.** $\frac{4}{7}$ **7.** 5 **8.** Diverges **9.** $\frac{6}{11}$ **10.** $\frac{8}{11}$
**11.** $736,971 **12.** Follows from summing $D + D(1 + i) + D(1 + i)^2 + \cdots + D(1 + i)^{n-1}$
**13. a.** 0.4 milligram and 0.14 milligram **b.** Just after the 4th dose; just after the 6th dose
**14. a.** Sum = 10,000, multiplier = 10 **b.** 22 terms; 44 terms
**15. a.** $3 + \frac{1}{6}x - \frac{1}{216}x^2 + \frac{1}{3888}x^3$ **16. a.** $1 + 3x + \frac{9}{2}x^2 + \frac{9}{2}x^3$ **17. a.** $2x - 2x^2 + \frac{8}{3}x^3$ **18. a.** $x^2$
**b.**  **b.**  **b.**  **b.**
on $[-15, 15]$ by $[-6, 6]$     on $[-1, 1]$ by $[-1, 8]$     on $[-1, 1]$ by $[-5, 5]$     on $[-3, 3]$ by $[-2, 2]$

**19. a.** $1 + \frac{1}{2}x - \frac{1}{8}x^2 + \frac{1}{16}x^3$
**b.** 1.4375
**c.** $|R_3(1)| \leq \dfrac{15/16}{4!} |1|^4 < 0.04$ (using $|f^{(4)}(x)| = \dfrac{15}{16}(x + 1)^{-7/2} \leq \dfrac{15}{16} = M$ for $0 \leq x \leq 1$)
**d.** $\sqrt{2} \approx 1.44$ with error less than 0.04

**20. a.** $1 - \dfrac{1}{2}x^2 + \dfrac{1}{24}x^4$ **b.** 0.54167 **c.** $|R_4(1)| \leq \dfrac{1}{5!} |1|^5 < 0.009$ (using $|f^{(5)}(x)| = |-\sin x| \leq 1 = M$)
**d.** $\cos 1 \approx 0.542$ with error less than 0.009 (actual value: $\cos 1 \approx 0.540$)

**21.** $1 + \dfrac{1}{n}x$ **22.** $p_7(x) = 1 + x + \dfrac{1}{2!}x^2 + \dfrac{1}{3!}x^3 + \cdots + \dfrac{1}{7!}x^7$ has error $|R_7(x)| \leq \dfrac{3}{8!} |1|^8 < 0.00008$ on $[-1, 1]$

**23. a.** $1 - \dfrac{1}{2!}\left(x - \dfrac{\pi}{2}\right)^2 + \dfrac{1}{4!}\left(x - \dfrac{\pi}{2}\right)^4$ **24. a.** $8 + 3(x - 4) + \frac{3}{16}(x - 4)^2 - \frac{1}{128}(x - 4)^3$
**b.**  **b.**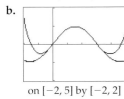
on $[-2, 5]$ by $[-2, 2]$     on $[-2, 15]$ by $[-5, 60]$

**25. a.** $1 + \dfrac{2}{3}(x - 1) - \dfrac{1}{9}(x - 1)^2$ **b.** 1.12889 **c.** $f^{(3)}(x) = \dfrac{8}{27}x^{-7/3}$, so $M = \dfrac{8}{27}$
**d.** $|R_2(1.2)| \leq \dfrac{8/27}{3!} |0.2|^3 < 0.0004$
**e.** $(1.2)^{2/3} \approx 1.1289$ with error less than 0.0004 (actually, $(1.2)^{2/3} \approx 1.1292$)
**26.** 0.80 gram **27.** 33 degrees **28. a.** $200 + 20t^2 + t^4$ **b.** 296 units **c.** $200e^{0.4} \approx 298$
**29.** $R = 3$ **30.** $R = 1$ **31.** $R = 0$ **32.** $R = \infty$ **33.** $x^3 + x^5 + x^7 + x^9 + \cdots$
**34.** $x - \dfrac{2^2x^3}{2!} + \dfrac{2^4x^5}{4!} - \cdots$ **35.** $x + x^3 + \dfrac{x^5}{2!} + \dfrac{x^7}{3!} + \cdots$ **36.** $x - x^2 + \dfrac{x^3}{2!} - \dfrac{x^4}{3!} + \cdots$
**37.** $1 + 2x + 2^2x^2 + 2^3x^3 + \cdots$ **38.** $1 - \dfrac{x}{2} + \dfrac{x^2}{2^2 2!} - \dfrac{x^3}{2^3 3!} + \cdots$

**39. a.** $-\left(x - \dfrac{\pi}{2}\right) + \dfrac{1}{3!}\left(x - \dfrac{\pi}{2}\right)^3 - \dfrac{1}{5!}\left(x - \dfrac{\pi}{2}\right)^5 + \cdots$  **b.**

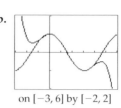

on $[-3, 6]$ by $[-2, 2]$

**40.** $\sin(-x) = -x - \dfrac{(-x)^3}{3!} + \dfrac{(-x)^5}{5!} - \dfrac{(-x)^7}{7!} + \cdots = -x + \dfrac{x^3}{3!} - \dfrac{x^5}{5!} + \dfrac{x^7}{7!} - \cdots = -\sin x,$  and similarly for cosine

**41. a.** $1 - t^2 + t^4 - t^6 + \cdots$  **b.** $x - \frac{1}{3}x^3 + \frac{1}{5}x^5 - \frac{1}{7}x^7 + \cdots$
  **c.** Approximately 0.465 (thousand), so about 465 cars  **d.** 0.464

**42. a.** $\dfrac{1}{x^2}(1 + x + x^2 + x^3 + x^4 + \cdots) = \dfrac{1}{x^2} + \dfrac{1}{x} + 1 + x + x^2 + \cdots$

  **b.** $\dfrac{-1}{x} + \ln x + x + \dfrac{1}{2}x^2 + \dfrac{1}{3}x^3 + \cdots$

  **c.** $\left.\left(\dfrac{-1}{x} + \ln x + x + \dfrac{1}{2}x^2 + \dfrac{1}{3}x^3\right)\right|_{0.2}^{0.5} \approx 4.36$

  **d.** $2.5 \cdot 4.36 = 10.9$, so 11 generations are necessary, which agrees with the solution of Example 3 in Section 2 of Chapter 6.

**43.** 5.099  **44.** 3.107  **45.** 0.818  **46.** 0.269  **47.** 4.641588834  **48.** 3.162277660
**49.** $-0.475711445$  **50.** 4.411048937  **51.** 2.435766883  **52.** $-1.033503908$  **53.** 16.6%
**54.** $4.40  **55.** 165

## Exercises 11.1 page 755

**1. a.** $\frac{2}{3}$  **b.** $\frac{1}{2}$  **3. a.** $\frac{3}{8}$  **b.** $\frac{3}{8}$
**5. a.** $E(X) = 10$, $\text{Var}(X) = 1$  $\sigma(X) = 1$  **b.** $E(Y) = 10$, $\text{Var}(Y) = 100$, $\sigma(Y) = 10$
**7. a.** $P(X = 4) = \frac{1}{3}$, $P(X = 7) = \frac{1}{3}$, $P(X = 10) = \frac{1}{3}$  **b.** 7  **c.** 6
**9. a.** $P(X = 2) = \frac{1}{2}$, $P(X = 4) = \frac{1}{4}$, $P(X = 8) = \frac{1}{4}$  **b.** 4  **c.** 6

**11. a.** $\dfrac{1}{12}\left(\text{from }\dfrac{30°}{360°}\right)$  **b.** $\dfrac{1}{36}$  **c.** 0

**13.** $E(X) = e^{-a}\left(0 \cdot 1 + 1 \cdot \dfrac{a}{1} + 2 \cdot \dfrac{a^2}{2!} + 3 \cdot \dfrac{a^3}{3!} + \cdots\right) = e^{-a}\left(a + \dfrac{a^2}{1!} + \dfrac{a^3}{2!} + \cdots\right)$

  $= e^{-a}a\left(1 + \dfrac{a}{1!} + \dfrac{a^2}{2!} + \cdots\right) = e^{-a}ae^a = a$

**15. a.** 0.135  **b.** 0.271  **c.** 0.271  **d.** 0.180  **17. a.** 0.406  **b.** 0.677  **c.** 0.857  **d.** 0.143
**19. a.** {BBBB, BBBG, BBGB, BBGG, BGBB, BGBG, BGGB, BGGG, GBBB, GBBG, GBGB, GBGG, GGBB, GGBG, GGGB, GGGG}

  **b.** $\frac{1}{16}$

  **c.** $P(X = 0) = \frac{1}{16}$
  $P(X = 1) = \frac{1}{4}$
  $P(X = 2) = \frac{3}{8}$
  $P(X = 3) = \frac{1}{4}$
  $P(X = 4) = \frac{1}{16}$

  **d.** $\frac{15}{16}$

  **e.** $E(X) = 2$,  $\text{Var}(X) = 1$

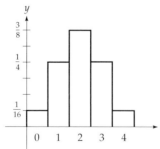

**21.** $2800  **23.** 0.238  **25.** 0.70
**27.** $e^{-1.7} \approx 0.18$, or about 18%.
**29.** 0.053, or less than 6%
**31. a.** 0.135, or about 14%
  **b.** About 86% (from 100–14%)
  **c.** 0.857, or about 86%

**33. a.** 0.082, or about 8%   **b.** 0.301, or about 30%   **35. a.** 0.015, or about 1.5%   **b.** 0.867, or about 87%

**37.** $\frac{1}{275}$   **39.** $P(X = 5)$   **41.** False. (See Example 9.)   **43.** 105; 10

**45.** $X = 5$. (In general, if the variance is zero, then the random variable is a *constant*.)

**47. 1:** Definition of variance   **2:** Expanding $(x_i - \mu)^2$   **3:** Separating into three sums
**4:** Taking out constants   **5:** Using $\sum_{i=1}^{n} x_i p_i = \mu$   and   $\sum_{i=1}^{n} p_i = 1$
**6:** Simplifying and combining the last two terms

## Exercises 11.2 page 768

**1.** $a = 12$   **3.** $a = \frac{1}{9}$   **5.** $a = 1$   **7.** $a = 1$   **9.** $a \approx 0.670$   **11. a.** $\frac{3}{4}$   **b.** $\frac{3}{80} = 0.0375$   **c.** 0.194

**13. a.** $\frac{3}{5} = 0.6$   **b.** $\frac{1}{25} = 0.04$   **c.** $\frac{1}{5} = 0.2$   **15. a.** 0.441   **b.** 0.079   **c.** 0.280

**17. a.** $\frac{1}{24}x^2 - \frac{1}{24}$   **b.** $\frac{2}{3}$   **19.** $\frac{1}{2}$   **21.** $\frac{1}{3}$   **23. a.** 0.75   **b.** $f(x) = 2x$ on $[0, 1]$

**25. a.** $F(A) = P(X \le A) = P(A \le X \le A) = 0$
   **b.** $F(B) = P(X \le B) = P(A \le X \le B) = 1$ (because $A$ and $B$ are the smallest and largest possible values)

**27.** 0.590   **29.** 0.296   **31. a.** 3   **b.** $\frac{7}{8} = 0.875$

**33. a.** $F(x) = \frac{1}{81}x^2$ on $[0, 9]$   **b.** $f(x) = \frac{2}{81}x$ on $[0, 9]$   **c.** 6   **d. and e.** $\frac{1}{3}$

**35. a.** 1   **b.** $\frac{1}{5} = 0.2$; 0.447   **c.** 0.156, or about 16%

**37. a.**   **39.** 0.537, or about 54%   **41.** 0.473, or about 47%

**43.** In *discrete* probability it means that the event *cannot* occur, while in *continuous* probability it means that the event is *overwhelmingly unlikely* (like guessing one number from infinitely many).

**45.** $F(x + 7)$   (because the graph is shifted 7 units to the left)

   **b.** 0.83 (rounded)

**47.** There is more area in the *outer* 6 inches of the target than in the *inner* 6 inches, so it is more likely that the dart will land *beyond* 6 inches, making the mean *greater* than 6.

**49.** True   **51.** 20; 10

## Exercises 11.3 page 778

**1. a.** $f(x) = \frac{1}{10}$ on $[0, 10]$   **b.** 5   **c.** $\frac{25}{3}$   **d.** $\sqrt{25/3} \approx 2.89$   **e.** $\frac{1}{5} = 0.2$

**3. a.** $F(x) = \frac{x}{B}$ on $[0, B]$   **b.** $\frac{B}{2}$   **5. a.** 15   **b.** 75   **c.** $5\sqrt{3} \approx 8.7$   **d.** $\frac{1}{5} = 0.2$

**7. a.** $F(x) = \frac{x - A}{B - A}$ on $[A, B]$   **b.** $\frac{A + B}{2}$   **c.** $\frac{(B - A)^2}{12}$ (after simplification)

**9. a.** $\frac{1}{5}$   **b.** $\frac{1}{25}$   **c.** $\frac{1}{5}$   **11. a.** 2   **b.** 4   **c.** 2   **13.** $f(x) = \frac{1}{3}e^{-x/3}$ on $[0, \infty)$

**15.** $f(x) = 2e^{-2x}$ on $[0, \infty)$   **17. a.** $F(x) = 1 - e^{-ax}$ on $[0, \infty)$   **b.** $\frac{\ln 2}{a} \approx \frac{0.693}{a}$ $\left(\text{from } \frac{\ln 0.5}{-a}\right)$   **29.** $\frac{1}{a^2}$

**19.** 0.777   **21.** 0.181   **23. a.** 0.22 or 22%   **b.** 0.78 or 78%   **25.** 0.707   **27.** 1.41

**31.** 0.35 or 35%   **33.** 0.13 or 13%   **35.** 0.15 or 15%

**37.** If   $X = \left(\begin{array}{c}\text{Survival}\\\text{time}\end{array}\right)$,   then $S(x) = P(X \ge x) = 1 - P(X < x) = 1 - F(x) = 1 - (1 - e^{-ax}) = e^{-ax}$

**39.** $f(x) = \frac{1}{20}$ on $[0, 20]$   **41.** $f(x) = \frac{1}{6}$ on $[0, 6]$   **43.** $f(x) = 2e^{-2x}$ for $0 \le x < \infty$

**45.** A Poisson random variable with parameter 3   **47.** The mean (because there are some very large families)

## Exercises 11.4 page 787

**1.** 0.4744   **3.** 0.679   **5.** 0.0782   **7.** 0.001   **9.** 0.9997   **11.** 0.15   **13.** 0.3085

**15.** 0.909, or about 91%   **17.** 0.2525, or about 25%   **19.** 0.1056, or about 11%

**21.** 0.851, or about 85%   **23.** 131   **25.** 32.51 oz.

**27.** Mean 0 and standard deviation 2   **29.** Mean 5 and standard deviation 2   **31.** $P(Y > 4)$

**33.** $Y$ has the greater mean and the greater standard deviation.   **35.** It has mean 47.8.

**37.** $x = 0$, $f(0) = \dfrac{1}{\sqrt{2\pi}}$   **39.** $\left(1, \dfrac{1}{\sqrt{2\pi}}e^{-1/2}\right)$ and $\left(-1, \dfrac{1}{\sqrt{2\pi}}e^{-1/2}\right)$ $\left(\text{Note that the } y\text{-coordinates equal } \dfrac{1}{\sqrt{2\pi e}}.\right)$

**41.** The steps follow from the given hints.

## Chapter 11 Review Exercises page 792

**1. a.** 6   **b.** 18   **c.** 4.24   **2. a.** 3   **b.** 5   **c.** 2.24   **3.** $830,000

**4.** Property B ($150,850 versus $121,800)   **5.** 0.677   **6.** 0.612   **7.** 0.0025 or about one quarter of 1%

**8.** $e^{-5} \approx 0.0067$   or about seven tenths of 1 percent

**9. a.** 0.251, or about 25%   **b.** 0.598, or about 60%   **c.** 0.747, or about 75%   **d.** About 25% (using part (c))

**10.** About 0.05 or 5%   **11.** $a = \frac{3}{4}$   **12.** $a = \frac{1}{18}$   **13.** $a = 2$   **14.** $a = \frac{1}{2}$   **15.** $a \approx 0.637$   **16.** $a \approx 0.578$

**17. a.** $\frac{7}{27}$   **b.** $\frac{1}{27}x^3$ on [0, 3]   **c.** $\frac{7}{27}$   **18. a.** $\frac{1}{4}$   **b.** $\frac{1}{8}x^{-1/2}$ on [1, 25]   **c.** $\frac{1}{4}$

**19. a.** $\frac{3}{8} = 0.375$   **b.** $\frac{19}{320} \approx 0.0594$   **c.** 0.244

**20. a.** 4   **b.** 12   **c.** 3.46   **21. a.** 5.33   **b.** 7.8 (both in million kilowatt-hours)

**22. a.** $\frac{11}{16} = 0.6875$   **b.** $\frac{2}{5} = 0.4$   **23. a.** 0   **b.** 0.25   **c.** 0.5   **24. a.** 0.4   **b.** 0.0218   **c.** 0.1477

**25.** About 0.609   **26.** About 0.530   **27. a.** $\frac{1}{50}$   **b.** 25   **c.** $\frac{625}{3}$ $\left(\text{from } \frac{2500}{12}\right)$   **d.** 14.4   **e.** $\frac{3}{5} = 0.6$

**28. a.** 1.   **b.** $\frac{1}{3}$   **c.** 0.577   **d.** $\frac{1}{4}$   **29. a.** 100   **b.** 10,000   **30.** $f(x) = 0.25e^{-0.25x}$ on $[0, \infty)$

**31. a.** 0.0498, or about 5%

    **b.** 0.00012, or about .012% [*Note:* Although this is a very small proportion, it predicts that of the 7.5 million flights annually, more than 900 will have such runway delays, which agrees with flight records.]

**32.** 0.393, or about 39%   **33.** 1.59   **34.** 0.231   **35.** About 0.22 or 22%   **36.** About 0.18 or 18%

**37.** 0.1655   **38.** 0.7916   **39.** 0.9938   **40.** 0.1587   **41.** 0.7745   **42.** 0.971

**43.** 0.977 or about 98%   **44.** 0.115 or about 12%   **45.** 0.938, or about 94%   **46.** 561.25, so about 561

## Cumulative Review for Chapters 1–11 page 795

**1. a.** $y = 5$   **b.** $x = -2$   **2.** 8   **3.** $f'(x) = 6x - 7$   (but found from the *definition* of the derivative)

**4.** $T'(4) = 36$,   so after 4 hours the temperature is rising at the rate of 36 degrees per hour.   $T''(4) = \frac{9}{2} = 4.5$, so after 4 hours the temperature is rising increasingly fast, with the rate of increase rising by 4.5 degrees per hour per hour.

**5.** $f'(x) = x^2(x^3 + 8)^{-2/3}$   **6.** $f'(x) = 6(2x + 3)^2(3x + 2)^4 + 12(2x + 3)^3(3x + 2)^3 = 6(2x + 3)^2(3x + 2)^3(7x + 8)$

**7.**

$f' > 0$    $f' = 0$    $f' < 0$    $f' = 0$    $f' > 0$

$x = -2$      $x = 10$

rel max      rel min
$(-2, 464)$     $(10, -400)$

$f'' < 0$    $f'' = 0$    $f'' > 0$

$x = 4$

con dn      con up

IP (4, 32)

**8.**

$f' > 0$    $f'$ und    $f' > 0$

$x = 0$

neither
(0, 1)

$f'' > 0$    $f''$ und    $f'' < 0$

$x = 0$

con up      con dn

IP (0, 1)

**9.**

| $f' > 0$ | $f'$ und | $f' > 0$ | $f'$ und | $f' < 0$ | $f'$ und | $f' < 0$ |
|---|---|---|---|---|---|---|
| | $x = -2$ | | $x = 0$ | | $x = 2$ | |

↗     ↗     →     ↘     ↘

rel max
$(0, -1/4)$

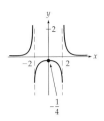

**10.** 7500 square feet    **11.** 128 cubic inches    **12.** $\dfrac{dy}{dx} = -5$    **13.** $\dfrac{3}{4\pi} \approx 0.24$ mile per day

**14. a.** $6333.85    **b.** $6356.25    **15.** About 6.22 years    **16.** $13,803.30

**17. a.** $\dfrac{2xe^{2x} - e^{2x}}{x^2}$    **b.** $\dfrac{3x^2}{x^3 + 1}$    **18.** $p = \$25$    **19.** 0.039 or 3.9%

**20. a.** $3x^4 - 2x^2 + 5x + C$    **b.** $-3e^{-2x} + C$    **21.** 2    **22.** 30 thousand dollars or $30,000

**23.** $800e^{0.15t} - 800$    **24.** 36 square units    **25.** 73.8 thousand people

**26. a.** $(x^2 + 1)^{1/2} + C$    **b.** $-\dfrac{1}{2}e^{-1} + \dfrac{1}{2} \approx 0.316$    **27.** $\dfrac{1}{3}x^3 \ln x - \dfrac{1}{9}x^3 + C$    **28.** $\dfrac{2x + 4}{3}\sqrt{x - 1} + C$

**29.** $\dfrac{1}{2}$    **30.** 1.98 (rounded) (compare with 2.000 from FnInt)

**31.** 1.99 (rounded) (compare with 2.000 from FnInt)    **32.** $\dfrac{\partial f}{\partial x} = e^y - y\dfrac{2x}{x^2 + 1}; \dfrac{\partial f}{\partial y} = xe^y - \ln(x^2 + 1)$

**33.** Min $f = 3$ at $x = 2$, $y = 2$. No max value.

**34.** $y = -0.9x + 6$    **35.** Max $f = 10$ at $x = 7$, $y = 4$    **36.** $df = (10x - 2y)\,dx + (-2x + 6y^2)\,dy$

**37.** 54 cubic units    **38. a.** $\dfrac{3\pi}{4}$    **b.** $150°$    **39. a.** $\dfrac{\sqrt{2}}{2}$    **b.** $-\dfrac{\sqrt{2}}{2}$    **c.** $\dfrac{1}{2}$    **d.** $-\dfrac{\sqrt{3}}{2}$

**40.**

**41.** $\cos^2 x - \sin^2 x$    **42.** $-e^{\cos x} + C$    **43.** 1 square unit

**44. a.** $-\dfrac{\sqrt{3}}{3}$    **b.** $-\sqrt{3}$    **c.** $-\sqrt{2}$    **d.** $\sqrt{2}$

**45. a.** $2x \sec^2(x^2 + 1)$    **b.** $\frac{1}{2}\sec(2t + 1) + C$

**46. a.** $y = Ce^{x^4/4}$    **b.** $y = 2e^{x^4/4}$    **47.** $y = 4e^{-3x^2}$

**48.**

**49.** 4    **50.** $1 + 2x + 2x^2 + \dfrac{4}{3}x^3$

**51.** $x^2 - \dfrac{1}{3!}x^6 + \dfrac{1}{5!}x^{10} - \dfrac{1}{7!}x^{14} + \cdots$ with $R = \infty$    **52.** 10.24695

**53.** 0.433    **54. a.** $E(X) = 9$    **b.** $\text{Var}(X) = 27$    **55.** 0.777    **56.** 0.383

# Index

# DEFINITION OF THE DERIVATIVE

$$f'(x) = \lim_{h \to 0} \frac{f(x+h) - f(x)}{h}$$

# DIFFERENTIATION FORMULAS

Power Rule:
$$\frac{d}{dx} x^n = nx^{n-1}$$

Constant Multiple Rule:
$$\frac{d}{dx} [cf(x)] = cf'(x)$$

Sum-Difference Rule:
$$\frac{d}{dx} [f(x) \pm g(x)] = f'(x) \pm g'(x)$$

Product Rule:
$$\frac{d}{dx} [f(x)g(x)] = f'(x)g(x) + f(x)g'(x)$$

Quotient Rule:
$$\frac{d}{dx} \left[ \frac{f(x)}{g(x)} \right] = \frac{g(x)f'(x) - g'(x)f(x)}{[g(x)]^2}$$

Generalized Power Rule:
$$\frac{d}{dx} [f(x)]^n = n[f(x)]^{n-1} f'(x)$$

Chain Rule:
$$\frac{d}{dx} [f(g(x))] = f'(g(x))g'(x)$$

$$\frac{dy}{dx} = \frac{dy}{du} \frac{du}{dx} \qquad \text{with } y = f(u) \text{ and } u = g(x)$$

Logarithmic Formulas:
$$\frac{d}{dx} \ln x = \frac{1}{x} \qquad\qquad \frac{d}{dx} \ln f(x) = \frac{f'(x)}{f(x)}$$

Exponential Formulas:
$$\frac{d}{dx} e^x = e^x \qquad\qquad \frac{d}{dx} e^{f(x)} = e^{f(x)} f'(x)$$

# AREA AND VOLUME FORMULAS

| Rectangle | Circle | Rectangular solid | Cylinder | Sphere |
|---|---|---|---|---|
| Area $= l \cdot w$ | Area $= \pi r^2$ | Volume $= l \cdot w \cdot h$ | Volume $= \pi r^2 l$ | Volume $= \frac{4}{3} \pi r^3$ |
| Perimeter $= 2l + 2w$ | Circumference $= 2\pi r$ | | | Surface area $= 4\pi r^2$ |

# PROPERTIES OF NATURAL LOGARITHMS

**1.** $\ln 1 = 0$

**2.** $\ln e = 1$

**3.** $\ln e^x = x$

**4.** $e^{\ln x} = x$

**5.** $\ln (MN) = \ln M + \ln N$

**6.** $\ln \left( \frac{1}{N} \right) = -\ln N$

**7.** $\ln \left( \frac{M}{N} \right) = \ln M - \ln N$

**8.** $\ln (M^P) = P \cdot \ln M$